T0177675

École de Physique des Houches
Session XCV, 2–27 August 2010

Quantum Theory from Small to Large Scales

Edited by

Jürg Fröhlich, Manfred Salmhofer, Vieri Mastropietro,
Wojciech De Roeck, Leticia F. Cugliandolo

OXFORD
UNIVERSITY PRESS

OXFORD
UNIVERSITY PRESS

Great Clarendon Street, Oxford, OX2 6DP,
United Kingdom

Oxford University Press is a department of the University of Oxford.
If furthers the University's objective of excellence in research, scholarship,
and education by publishing worldwide. Oxford is a registered trade mark of
Oxford University Press in the UK and in certain other countries

First Edition published in 2012

Impression: 1

British Library Cataloguing in Publication Data

Data available

Library of Congress Cataloging in Publication Data

Data available

ISBN 978–0–19–965249–5

Printed and bound by
CPI Group (UK) Ltd, Croydon, CR0 4YY

École de Physique des Houches

Service inter-universitaire commun à l'Université Joseph Fourier de Grenoble et à
l'Institut National Polytechnique de Grenoble

Subventionné par l'Université Joseph Fourier de Grenoble,
le Centre National de la Recherche Scientifique, le
Commissariat à l'Énergie Atomique

Directeur:
Leticia F. Cugliandolo, Université Pierre at Marie Curie – Paris VI, France

Directeurs scientifiques de la session XCV:
Jürg Fröhlich, ETH Zürich, Switzerland
Manfred Salmhofer, Universität Heidelberg, Heidelberg, Germany
Vieri Mastropietro, Universita di Roma "Tor Vergata", Rome, Italy
Wojciech De Roeck, Universität Heidelberg, Heidelberg, Germany
Leticia F. Cugliandolo, LPTHE, Université Paris VI, Paris, France

Previous sessions

LXXVII	2002	Slow relaxations and nonequilibrium dynamics in condensed matter
LXXVIII	2002	Accretion discs, jets and high energy phenomena in astrophysics
LXXIX	2003	Quantum entanglement and information processing
LXXX	2003	Methods and models in neurophysics
LXXXI	2004	Nanophysics: Coherence and transport
LXXXII	2004	Multiple aspects of DNA and RNA
LXXXIII	2005	Mathematical statistical physics
LXXXIV	2005	Particle physics beyond the Standard Model
LXXXV	2006	Complex systems
LXXXVI	2006	Particle physics and cosmology: the fabric of spacetime
LXXXVII	2007	String theory and the real world: from particle physics to astrophysics
LXXXVIII	2007	Dynamos
LXXXIX	2008	Exact methods in low-dimensional statistical physics and quantum computing
XC	2008	Long-range interacting systems
XCI	2009	Ultracold gases and quantum information
XCII	2009	New trends in the physics and mechanics of biological systems
XCIII	2009	Modern perspectives in lattice QCD: quantum field theory and high performance computing
XCIV	2010	Many-body physics with ultra-cold gases
XCV	2010	Quantum theory from small to large scales

Publishers

- Session VIII: Dunod, Wiley, Methuen
- Sessions IX and X: Herman, Wiley
- Session XI: Gordon and Breach, Presses Universitaires
- Sessions XII–XXV: Gordon and Breach
- Sessions XXVI–LXVIII: North Holland
- Session LXIX–LXXVIII: EDP Sciences, Springer
- Session LXXIX–LXXXVIII: Elsevier
- Session LXXXIX– : Oxford University Press

Preface

Introduction

The purpose of this school has been to introduce a mixed audience of researchers—young and not so young, women and men, mathematically minded and more oriented towards practical calculations and heuristic arguments—to a number of problems, mathematical techniques, computational methods and recent results in the following areas of fundamental physical theory, all connected to non-relativistic quantum theory:

1. Foundations of quantum statistical mechanics, open systems, quantum mechanical transport theory (including an analysis of entropy production, entropy fluctuation, Onsager relations, Kubo formulae, return to equilibrium, non-equilibrium steady states, fundamental laws of thermodynamics, equipartition, quantum Brownian motion).
2. Mathematical methods useful in coping with some problems in area 1) (such as operator algebras, random matrix theory, supersymmetry methods, expansion- and spectral methods, multi-scale analysis).
3. Mesoscopic physics; such as quantum pumps, quantum dots, counting statistics, dissipation-free transport.
4. Equilibrium statistical mechanics of quantum systems with many degrees of freedom; (in particular, quantum lattice systems, cold Bose gases, Bose–Einstein condensation, quantum phase transitions, theory of graphene; (spin) quantum Hall effect, topological insulators).
5. Mathematical methods to study problems in area 4), such as functional integration, renormalization group methods, operator methods, analysis of global and local symmetries.
6. Quantum electrodynamics with non-relativistic matter (Pauli–Fierz model), atomic spectroscopy, polaron-type models—along with mathematical methods, such as operator-theoretic renormalization group methods, iterative perturbation theory, methods of spectral and scattering theory.
7. Foundations of quantum theory and quantum information theory.

Apart from confronting the audience with new results from quantum physics, one idea behind our choice of topics has been to illustrate how a comparatively small number of mathematical structures and techniques—such as multi-scale analysis passing 'from small to large scales'—underlie the solution of a large diversity of physical problems from many different areas in quantum physics. This is, we believe, one of the fascinating aspects of mathematical physics.

Indeed, our school has been a *School of Mathematical Physics*, understood broadly, mostly addressing problems of fundamental physical theory of the twentieth century that, mathematically speaking, have remained open until recently or now.

What do we mean by 'fundamental physical theory of the twentieth century'? After the discovery of the correct formula for the spectral energy density of black-body radiation in 1900, *Planck* identified four fundamental constants of nature, which, he was convinced, have the same fundamental significance and numerical value throughout the entire universe; Boltzmann's constant k_B related to Avogadro's number N_A (and to Faraday's constant F), Planck's constant $\hbar$, the speed of light c, and the Planck length ℓ_P (related to Newton's gravitational constant G_N). These fundamental constants symbolize—as Planck foresaw—revolutionary changes in theoretical physics that were to take place in the twentieth century and appear to continue into the twentyfirst century, with vast technological applications. It might be appropriate to recall that theorists who were mathematical physicists, in the best sense of this denomination, played a leading rôle in triggering and developing these changes, among them *Einstein, Born, Heisenberg, Jordan, Dirac, Schrödinger*, and *Pauli*.

What revolutionary changes do $N_A, \hbar, c$, and ℓ_P stand for?

- N_A stands for the atomistic constitution of matter (as opposed to a description of matter as a continuous medium) and for statistical mechanics;
- $\hbar$ stands for quantum mechanics;
- c stands for the special theory of relativity and for Minkowski space–time;
- (c, ℓ_P) stand for general relativity and a description of gravitational forces in terms of space–time curvature.

In mathematical terms, we can interpret the constants $N_A^{-1}, \hbar, c^{-1}$ and ℓ_P as playing the rôle of deformation parameters.

- N_A^{-1} is connected to the deformation of Hamiltonian continuum theories of matter (such as Vlasov theory and Euler's theory of incompressible fluids) to atomistic theories of matter (such as the Newtonian mechanics of systems of identical point-particles);
- $\hbar$ is the parameter associated with the deformation of classical Hamiltonian theories to quantum theories.

Mathematically, N_A^{-1} and $\hbar$ are parameters appearing in the deformation of Abelian, associative algebras of functions on phase space to non-Abelian, associative 'algebras of observables'. These two deformations, from *continuum* to *atomistic* and from *classical* to *quantum*, give rise to a departure from a realistic, deterministic description of nature to a fundamentally probabilistic description; (a development exploited, e.g., in quantum information theory and quantum computing).

- The parameter c^{-1} appears in the deformation of Galilean to Poincaré symmetry and of the geometry of Newtonian space–time to that of Minkowski space–time.
- The pair (c^{-1}, ℓ_P) appears in the deformation of flat Minkowski space–time to the curved space-time of general relativity, providing a geometrical description of gravitational fields.

Planck understood that the main goal of theoretical physics in the twentieth century and beyond is to find a unified description of nature in which all of these four constants have their natural non-zero values. This goal has not been reached yet! However, there are theories describing limiting regimes of nature where only one or two of these constants can be regarded as negligibly small, while the remaining ones are positive. (In the nineteenth century, only one of the three constants ℓ_P, c^{-1}, N_A^{-1}, at a time, was assigned a non-zero value.)

- $N_A^{-1}, \hbar > 0; c^{-1}, \ell_P \to 0$: Atomistic, non-relativistic quantum mechanics. (There are quantum theories describing matter as a continuous medium, i.e., with $N_A^{-1} = 0$, such as the Gross-Pitaevskii or Hartree equation of a Bose–Einstein condensate. These theories are realistic and deterministic. The probabilistic nature of quantum mechanics only appears when $\hbar$ *and* N_A^{-1} are taken to be positive.—The Gross–Pitaevskii and Hartree theories appear in limiting regimes of the quantum mechanics of Bose gases, as is reviewed in the lectures of Erdös and Seiringer.)
- $\hbar, c^{-1} > 0; \ell_P \to 0$: Relativistic quantum field theory (without gravity), one of the key concerns of many theorists, ever since the late 1920s, that led to the discovery of numerous new concepts and techniques including ones directly relevant for topics of our school, such as the renormalization group.
- $N_A^{-1}, \hbar, \ell_P > 0; c^{-1} \to 0$: Non-relativistic theories of quantum gravity (some of which of marginal relevance in astrophysics).

A mathematically consistent theory encompassing regimes where $\hbar, c^{-1}$ and ℓ_P must all be taken to be positive, sometimes called *M-theory* or 'relativistic quantum gravity', remains to be discovered. Thus, the reductionist programme to discover a unified description of nature on all scales, from very small to very large, has not reached its ultimate goal yet. Fundamental physical theory must continue!

So, why a Les Houches school on non-relativistic quantum (many-body) theory, rather than one on M-theory or other blueprints of fundamental theories? The answer to this question is obvious: Reductionism is but one side of the coin. Indeed, in order to be able to derive predictions relevant for our experience from the 'equations' of some fundamental theory, we must understand the main properties of their solutions on all scales, from microscopic ones at which such equations are formulated to macroscopic ones where phenomena appear that we are interested in describing quantitatively. This assignment remains pertinent in the context of well established theoretical frameworks, such as non-relativistic atomistic quantum theory. There are many fascinating physical phenomena—some of potentially huge technological importance—that have been or are awaiting a mathematically coherent description in that particular framework. The study of such phenomena is fascinating and important in itself. But, apart from that, it may lead to the discovery of interesting new concepts and powerful new technical tools. Often it forces one to borrow concepts and methods from other areas of fundamental physics. Who would doubt that the study of electron gases in solids at low temperatures is an exercise in non-relativistic atomistic quantum theory? Yet, a theoretical analysis of the quantum Hall effect leads one to make use of topological field theories, current algebras and the chiral anomaly, that is of concepts and objects that were first introduced in the study of relativistic quantum field theory. Theoretical studies of graphene reveal a low-energy spectrum of quasi-particles with all the properties of $(2+1)$-dimensional relativistic Dirac fermions. Majorana fermions appear in the analysis of surface states of certain insulating materials dubbed three-dimensional topological insulators. In other materials, excitations may be encountered that resemble the hypothetical 'axions' of particle theory. The Goldstone modes in an interacting Bose gas exhibiting Bose-Einstein condensation or in a Néel-ordered anti-ferromagnet obey a dispersion law that is linear in the size of their wave-vectors at long wave-lengths—as if they were relativistic, massless scalar particles (but with another interpretation of the meaning of c). Concepts borrowed from the theory of non-Abelian gauge fields, originally introduced in particle physics, play an interesting and fruitful rôle in some of the studies alluded to above.

Thus, the study of a large diversity of phenomena, all belonging to the realm of non-relativistic atomistic quantum theory $(N_A^{-1}, \hbar > 0, c^{-1}, \ell_P \to 0)$, may lead one to discover a wealth of fascinating *emergent properties* first encountered in other areas of fundamental physical theory. Along the way, one employs tools and methods originally developed to explore other areas of physics, such as relativistic quantum field theory. Conversely, the emergence of, for example, relativistic symmetries in certain limiting regimes of condensed matter physics (such as superfluid Helium) may open up a new perspective in the search for certain types of fundamental theories (e.g. in cosmology).

Large, strongly correlated quantum systems are notoriously hard to treat in a mathematically reliable way. In many of the cases where rigorous mathematical understanding has been reached, it has provided us with useful new intuition into important mechanisms and structural properties that reveal essential features of such systems in a much clearer way than excruciatingly detailed approximate calculations. Concise mathematical understanding of physical phenomena, such as various forms of magnetism, superconductivity, the quantum Hall effect, and so on, is therefore not merely of technical importance but tends to lead to a better understanding of the basic concepts and physical mechanisms underlying such phenomena. Thus, much fascinating work for mathematical physicists (and many interesting summer schools like ours) lie ahead of us.

Perhaps, it is appropriate to say a few words about some of the purposes of the school, other than purely scientific ones, the organizers had in mind when they decided to put together this particular session. Here are some examples of such purposes:

- To get participants excited and enthusiastic about the beauty and depth of theoretical and mathematical physics and to open views at fascinating scientific landscapes from different angles that have proven to be fruitful and productive in the past.
- To confront the audience with interesting and topical problems in some active areas of theoretical physics and to encourage younger colleagues to attempt to choose some of these problems for further study.
- To teach the audience some mathematical techniques that have proven to be useful in coping with physical problems of the sort discussed at our school.

- To forge interactions and collaboration among participants and promote friendship among people from many diverse backgrounds and countries.
- Last but not least, to demonstrate with concrete examples that the approach and the methods of mathematical physics are not a superfluous luxury, but can help to understand and solve physical problems of great interest and considerable importance.

It is up to the participants of our school to decide whether we or they have achieved some of these goals.

As indicated repeatedly, our school has been a school of mathematical physics, understood in a broad sense. Why would one want to become a mathematical physicist? Here are some answers (among many others):

- Interest in the way nature works and in exploring the structure and contents of physical theories.
- Interest in applying mathematics to the theoretical description of physical phenomena. Here is a piece of advice: adapt the mathematical methods to the physical problems you have chosen to study, rather than mutilating these problems until some mathematical methods that you happen to master can be applied! (The latter is a danger often threatening endeavours of mathematical physicists.)
- Interest in showing you are brighter than anybody else. (Examples of some famous theorists show that this is not the worst motivation. But it is obviously no guarantee of success.)

As mathematical physicists, we would, however, most often hesitate to work on problems of applied science or engineering using the methods and meeting the standards of mathematical rigour common in mathematical physics. If, at some point in our careers, we feel like contributing to the solution of problems in the applied sciences or in engineering we will trade our hat of a mathematical physicist for the hat of a computational scientist or an applied mathematician, or whatever. In the applied sciences and in engineering, what counts is finding reliable, concrete, but most often only approximate solutions to complex problems—mathematical rigour is usually not a concern.

For the organizers, this school of mathematical physics has not only brought about a broad survey of and initiation in new directions of research in quantum theory—from small to large scales—it has also been an intense and largely positive human experience in the wonderful scenery beneath the Mont Blanc. We hope and dare expect that most participants, too, enjoyed the school as much as we did and found it useful.

Acknowledgements

First of all, we wish to thank all the speakers at the school for their heroic efforts in preparing their lectures. We especially thank all those among the speakers who have gone through the pain of writing the lecture notes published in this book.

Finally, we thank Philippe Blanchard for having agreed to present an evening lecture to a wider public on matters related to the subject of the school, and Jean Zinn-Justin for having contributed a seminar on quantum field- and particle theory.

This session of the Les Houches Summer School and the publication of this book would not have been possible without the generous support of several organizations. We gratefully acknowledge support by

- *Deutsche Forschungsgemeinschaft* via the focused research groups
  FOR718: Analysis and Stochastics in Complex Physical Systems and
  FOR723: Functional Renormalization Group for Correlated Fermion Systems.
- The *Center for Theoretical Studies* at ETH Zürich.
- *Ruprecht-Karls-Universität Heidelberg.*
- The *INTELBIOMAT* programme of the *European Science Foundation.*
- *Annales Henri Poincaré.*
- The *Daniel Iagolnitzer Foundation.*
- The *European Research Council* via the ERC starting grant *CoMBoS—Collective Phenomena in Classical and Quantum Many Body Systems.*

Last but not least, we would like to thank the entire staff of *L'Ecole de Physique des Houches* for their efficient and friendly support before, during, and after the school, which was essential for the success of our school and the good atmosphere in which it took place.

J. Fröhlich
M. Salmhofer
V. Mastropietro
W. De Roeck
L. F. Cugliandolo

Contents

List of participants

Organizers

FRÖHLICH Jürg
Institute of Theoretical Physics, HIT K23.7, Wolfgang-Pauli-Strasse ETH-Hönggerberg CH-8093 Zurich, Switzerland.

SALMHOFER Manfred
Universität Heidelberg, Theoretische Physik, Philosophenweg 19, D-69120 Heidelberg, Germany.

MASTROPIETRO Vieri
Universita di Roma 'Tor Vergata', Dipartimento di Matematica, Via della Ricerca Scientifica, 00133 Roma, Italy.

DE ROECK Wojciech
Universität Heidelberg, Theoretische Physik, Philosophenweg 16, D-69120 Heidelberg, Germany.

CUGLIANDOLO Leticia F.
LPTHE, Tour 24-5ème étage, 4 place Jussieu, F-75232 Paris Cedex 05, France.

Lecturers

BACH Volker
Institut für Analysis und Algebra, Carl-Friedrich-Gauss-Fakultaet, Technische Universitaet Braunschweig, Pockelsstr. 14, 38106 Braunschweig, Germany.

BLANCHARD Philippe
Fakultät für Physik, Universitt Bielefeld, Universitätsstr. 25, 33615 Bielefeld, Germany.

ERDŐS László
University of Munich, Dept. of Mathematics, Theresienstr. 39, D-80333, Germany.

FELDMAN Joel
University of British Columbia, Department of Mathematics, 1984 Mathematics Road, Vancouver, BC, Canada V6T 1Z2.

GIULIANI Alessandro
University of Roma Tre, Department of Mathematics, L.go S. Leonardo Murialdo 1, 00146 Rome, Italy.

GRAF Gian Michele
Institut für Theoretische Physik, ETH Zürich, HIT K 42.1, Wolfgang-Pauli-Str. 27, 8093 Zürich, Switzerland.

HASTINGS MATTHEW B.
Microsoft Research Station Q, CNSI Building Room 2243, University of California, Santa Barbara, CA, 93106, USA.

PILLET CLAUDE-ALAIN
Département de mathématiques et mécanique, Centre de Physique Théorique (UMR 6207), UFR Sciences et Techniques, Université du Sud Toulon-Var, 83957 La Garde Cedex, France.

PIZZO ALESSANDRO
University of California, Mathematics Department at UC Davis, Mathematic Sciences Building, One Shields Avenue, Davis, C A, 95616 USA.

ROSCH ACHIM
Institute of Theoretical Physics, University of Cologne, Zülpicher Str. 77, 50937 Cologne, Germany.

SEIRINGER ROBERT
Department of Mathematics & Statistics, McGill University, 805 Sherbrooke Street West, Montreal Quebec H3A 2K6, Canada.

SIGAL ISRAEL MICHAEL
University of Toronto, Department of Mathematics, 40 St. George Street, Bahen Centre, Toronto, ON, Canada M5S 2E4.

SPENCER THOMAS
Institute for Advanced Study, Princeton, NJ 08540, USA.

WOLF MICHAEL MARC
Zentrum Mathematik, M5, Technische Universität München, Boltzmannstrasse 3, 85748 Garching, Germany.

YAU HORNG-TZER
Harvard University, Mathematics department, 1 Oxford Street, Cambridge, MA 02138, USA.

ZHANG SHOUCHENG
Physics Department, Stanford University, LAM, Room 303, McCullough Bldg, 476 Lomita Mall, Stamford, CA 94305-4045, USA.

PARTICIPANTS

AJANKI OSKARI
University of Helsinki, Department of Mathematics and Statistics, PO Box 68, 00014 Helsinki, Finland.

ANSARI MOHAMMAD
Institute for Quantum Computing, University of Waterloo, 200 University Avenue West, Waterloo, N2L 3GI, ON, Canada.

BACHMANN SVEN
University of California Davis, Dept. of Mathematics, 1 Shields Avenue, Davis, CA 95616 USA.

BACSI ADAM
BME TTK Fizika Tanszék Budapest University of Technology and Economics, Physics department, Budafoki UT 8, H-1111 Budapest, Hungary.

BAND RAM
Department of Mathematics, University of Bristol, Howard House, Queen's Avenue, Bristol, BS8 1QT UK.

BEZVERSHENKO IULIIA
Bogolyubov Institute for Theoretical Physics, Metrolohichna str., 14b, Kyiv, 03680 Ukraine.

BIERI SAMUEL
Institute of Theoretical Physics, HIT, Wolfgang-Pauli-Strasse 27, ETH-Hönggerberg CH-8093 Zürich, Switzerland.

BLOIS CINDY
University of British Columbia, Dept. Of Mathematics, Room 121, 1984 Mathematics Road, Vancouver, BC, V6T 1Z2 Canada.

CENATIEMPO SERENA
University of Rome 'La Sapienza' Dpt. of Physics Edificio Fermi, Room 314, Piazzale Aldo Moro 5, 00185 Rome Italy.

DE CAUSMAECKER KAREN
Department of Physics and Astronomy, Krijgslaan 281 S9, 9000 Gent, Belgium.

DE MELO FERNANDO
Instituut voor Theoretische Fysica, Katholieke Universiteit Leuven Celestijnenlaan 200D B-3001 Heverlee, Belgium.

DYBALSKI WOJCIECH
Zentrum Mathematik, Bereich M5, Technische Universität München, D-85747 Garching bei München, Germany.

EGLI DANIEL
ETII Hönggerberg, ITP, HIT K22.3, Wolfgang-Pauli-Str. 27 8093 Zürich, Switzerland.

FAUSER MICHAEL
Technische Universität München Zentrum Mathematik-M7, Boltzmannstrasse 3 85747 Garching Bei München, Germany.

GAMAYUN OLEKSANDR
Bogolyubov Institute for Theoretical Physics, 14-b, Metrologichna Str., Kiev, 03680 Ukraine.

GEIGER TOBIAS
Physikalisches Institut/Abteilung Buchleitner, Hermann-Herder Str. 3, 79104 Freiburg, Germany.

GREENBLATT RAFAEL
Dipartamento di Matematica, Università degli Studi Roma Tre, Largo San Leonardo Murialdo 1 Roma 00146 Italy.

HANSON JACK
Princeton University, Dept. of Physics, Jadwin Hall, Washington Road, Princeton, NJ 08544 USA.

HASLER DAVID
Ludwig Maximilians University, Theresienstrasse 37, 80333 München, Germany.

HUTCHINSON JOANNA
University of Bristol, Dpt. of Mathematics, 5th floor, Howard House, Queens Avenue, Bristol BS8 1QT, UK.

IMBRIE JOHN
Dept. of Mathematics, Kerchof Hall, University of Virginia, Charlottesville VA 22904-4137, USA.

JÄKEL CHRISTIAN
Cardiff University, School of Mathematics, Senghennydd Road CF24 4AG Cardiff, UK.

JÖRG DAVID
Universität Heidelberg, Institut für Theoretische Physik, Philosophenweg 19, 69120 Heidelberg, Germany.

KNÖRR HANS KONRAD
Johannes Gutenberg-Universität Mainz, FB08 - Institut für Mathematik, Staudingerweg 9, 55099 Mainz, Germany.

KNOWLES ANTTI
Harvard University, Department of Mathematics, 1 Oxford Street, Cambridge MA, 02138, USA.

KRÖNKE SVEN
Institut für Theoretische Physik, Technische Universität Dresden, Zellescher Weg 17, 01069 Dresden, Germany.

LAJKO MIKLOS
MTA-SZFKI 29-33 Konkoly-Thege M. u. 1121-Budapest, Hungary.

LEIN MAX
Zentrum Mathematik Lehrstuhl M5 Boltzmannstrasse 3, 85747 Garching, Germany.

LEMM MARIUS
Ludwig-Maximilians-Universität, Steinickeweg 4, 80798 München, Germany.

LIGABO MARILENA
University of Bari, Math Dept. (Campus), via Orabona 70125 Bari, Italy.

LIPOVSKY JIRI
Academy of Sciences of the Czech Republic, Nuclear Physics Institute, Dept. of Theoretical Physics, Rez 250 68, Czech Republic.

LOHMANN MARTIN
ETH Zurich Dpt. of Mathematics, Rämistr. 101 CH-8092 Zurich Switzerland.

LU LONG
Universität Heidelberg, Theoretische Physik, Philosophenweg 19, D-69120 Heidelberg, Germany.

LUHRMANN Jonas
Baurat-Gerber-Str. 8, 37073 Göttingen, Germany.

NAPIORKOWSKI Marcin
University of Warsaw, Faculty of Physics, Dpt. of Mathematical Methods in Physics, Hoza 74, 00-682, Warsaw Poland.

OGATA Yoshiko
Department of Mathematical Sciences, University of Tokyo, 3-8-1 Komaba, Tokyo, 153-8914, Japan.

PHAN THANH Nam
University of Copenhagen, Department of Mathematical Science, Universitetsparken 5, DK-2100 Copenhagen, Denmark.

PORTA Marcello
Dipartimento di Fisica, Università di Roma 'Sapienza', Piazzale Aldo Moro 5, 00185 Roma, Italy.

SCHERER Daniel
Theoretisch-Physikalisches Institut, FSU Jena Max-Wien-Platz 1, 07743 Jena, Germany.

SCHNELLI Kevin
ETH Zürich, ITP, HIT K 22.3, Wolfgang Pauli Str. 27, CH 8093 Zürich, Switzerland.

SERI Marcello
Dept. Mathematik, Bismarckstr. 1 1/2, 91054 Erlangen, Germany.

SIMONELLA Sergio
Università degli Studi di Roma 'La Sapienza', Piazzale Aldo Moro 5, 00185 Rome, Italy.

SLOBODENIUK Artur
Bogolyubov Institute for Theoretical Physics, 14-B Metrologichna Str., Kyiv, Ukraine.

SNIZHKO Kyrylo
Taras Shevchenko National University of Kyiv, Department of Physics, Academician Glushkov avenue, 2, Building 1, Kyiv, Ukraine.

TAJ David
Département de Physique, Université de Fribourg, Ch. du Musée 3, CH-1700, Fribourg, Switzerland.

TRENDELKAMP-SCHROER Benjamin
TU-Dresden, Institut fuer Theoretische Physik, Zellescher Weg 17, 01069 Dresden, Germany.

VERSHYNINA Anna
University of California Davis, Dept. of Mathematics, 1 Shields Avenue, Davis, CA 95616 USA.

VIGNES-TOURNERET Fabien
Université Lyon 1, Institut Camille Jordan, Bâtiment Braconnier, 43 Boulevard du 11 Novembre 1918, 69622 Villeurbanne cedex, France.

WALTER Michael
Georg-August-Universität Göttingen, Mathematisches Institut, Bunsenstrae 3–5, D-37073 Göttingen, Germany.

WANG Zhituo
Laboratoire de Physique Théorique d'Orsay, Ch 128 G, Université Paris-Sud 11, 91400 Orsay-ville, France.

WILHELM Lukas
Weierstrass Institute, Mohrenstrasse 39, 10117 Berlin, Germany.

ZHOU Gang
ETH Zürich, Institute for Theoretical Physics, Wolf-Gang-Pauli-Strasse 21, HIT K 12.1, 8092 Zürich, Switzerland.

Part I

Long lectures

1

Lecture notes on quantum Brownian motion

László ERDŐS[1]

Institute of Mathematics, University of Munich
Theresienstr. 39, D-80333 Munich, Germany

[1] Partially supported by SFB-TR 12 Grant of the German Research Council.

Chapter Contents

Preface

Einstein's kinetic theory of the Brownian motion, based upon light water molecules continuously bombarding a heavy pollen, provided an explanation of diffusion from Newtonian mechanics. Since the discovery of quantum mechanics it has been a challenge to verify the emergence of diffusion from the Schrödinger equation.

The first step in this program is to verify the linear Boltzmann equation as a certain scaling limit of a Schrödinger equation with random potential. In the second step, one considers a longer time scale that corresponds to infinitely many Boltzmann collisions. The intuition is that the Boltzmann equation then converges to a diffusive equation similarly to the central limit theorem for Markov processes with sufficient mixing. In this chapter we present the mathematical tools to rigorously justify this intuition. The new material relies on joint papers with H.-T. Yau and M. Salmhofer.

1.1 Overview of the rigorous derivations of diffusions

The fundamental equations governing the basic laws of physics, the Newton and the Schrödinger equations, are time reversible and have no dissipation. It is remarkable that dissipation is nevertheless ubiquitous in nature, so that almost all macroscopic phenomenological equations are dissipative. The oldest such example is perhaps the equation of heat conductance found by Fourier. This investigation has led to the heat equation, the simplest type of diffusion equation:

$$\partial_t u = \Delta_x u, \tag{1.1}$$

where $u(x,t)$ denotes the temperature at position $x \in \mathbb{R}^d$ and time t. One key feature of the diffusion equations is their inhomogeneous scaling; the time scales as the square of the spatial distance:

$$t \sim x^2; \qquad \text{time} \sim (\text{distance})^2.$$

In this chapter we will explore how diffusion equations emerge from first principle physical theories such as the classical Newtonian dynamics and the quantum Schrödinger dynamics. In Section 1.1 we give an overview of existing mathematical results on the derivation of diffusion (and related) equations. In Sections 1.2–1.4 we discuss the basic formalism and present a few well-known preliminary facts on stochastic, classical, and quantum dynamics. An experienced reader can skip these sections. In Section 1.5 we introduce our main model, the random Schrödinger equation or the quantum Lorentz gas and its lattice version, the Anderson model. In Section 1.6 we formulate our main theorems that state that the random Schrödinger equation exhibits diffusion and after a certain scaling and limiting procedure it can be described by a heat equation.

The remaining sections contain the sketch of the proofs. Since these proofs are quite long and complicated, we will not only have to omit many technical details, but also only briefly mention several essential ideas. We will primarily focus on the most important aspect: the classification and the estimates of the so-called non-repetition

Feynman graphs. Estimates of other Feynman graphs, such as recollision graphs and graphs with higher order pairings, will only be discussed very superficially.

Our new results presented in this review were obtained in collaboration with H.-T. Yau and M. Salmhofer.

1.1.1   The founding fathers: Brown, Einstein, Boltzmann

The story of diffusion starts with R. Brown in 1827 who observed almost two centuries ago that the motion of a wildflower pollen suspended in water was erratic (Brown 1828). He saw a picture similar to Fig. 1.1 under his microscope (the abrupt velocity changes in this picture are exaggerated to highlight the effect, see later for more precise explanation). He 'observed many of them very evidently in motion' and argued that this motion was not due to water currents but 'belonged to the particle itself'. First he thought that this activity characterizes only 'living' objects like pollens, but later he found that small rock or glass particles follow the same pattern. He also noted that this apparently chaotic motion never seems to stop.

This picture led to the kinetic explanation by A. Einstein in 1905 (Einstein 1905) that such a motion was created by the constant kicks on the relatively heavy pollen by the light water molecules. Independently, M. Smoluchowski (von Smoluchowski 1906) arrived at a similar theory. It should be noted that at that time even the atomic-molecular structure of matter was not universally accepted. Einstein's theory was strongly supported by Boltzmann's kinetic theory, which, however, was phenomenological and seriously debated at the time. Finally in 1908 Perrin (Perrin 1909) (awarded the Nobel prize in 1926) experimentally verified Einstein's theory and used it, among others, to give a precise estimate on the Avogadro number. These experiments gave the strongest evidence for atoms and molecules at that time.

We should make a remark on Boltzmann's theory although this will not be the subject of this chapter. Boltzmann's kinetic theory postulates that in a gas of interacting particles at relatively low density the collisions between the particles are statistically independent (Ansatz of *molecular chaos*). This enabled him to write up

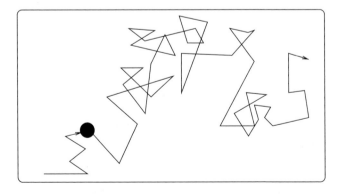

Fig. 1.1 Brown's picture under the microscope.

the celebrated (nonlinear) Boltzmann equation for the time evolution of the single particle phase space density $f_t(x, v)$. The main assumption is that the collision rate is simply the product of the particle densities with the appropriate velocities, thus he arrived at the following equation:

$$\partial_t f_t(x, v) + v \cdot \nabla_x f_t(x, v) = \int \sigma(v, v_1) \big[f_t(x, v') f_t(x, v_1') - f_t(x, v) f_t(x, v_1) \big]. \quad (1.2)$$

Here v, v_1 is the pair of incoming velocities of the colliding particles and v', v_1' are the outgoing velocities. In the case of elastic hard balls, the pair (v', v_1') is uniquely determined by (v, v_1) (plus a randomly chosen contact vector) due to energy and momentum conservation, for other type of collisions the kernel $\sigma(v, v_1)$ describes microscopic details of the collision mechanism. The integration is for the random contact vector and all other variables apart from x, v, subject to constraints (momentum and energy conservation) encoded in $\sigma(v, v_1)$.

In addition to the highly nontrivial Ansatz of independence, this theory also assumes the molecular nature of gases and fluids in contrast to their more intuitive continuous description. Nowadays we can easily accept that gases and fluids on large scales (starting from micrometre scales and above—these are called *macroscopic scales*) can be characterized by continuous density functions, while on much smaller scales (nanometres and below, these are called *microscopic scales*) the particle nature dominates. But in Boltzmann's time the particle picture was strongly debated. The ingenuity of Boltzmann's theory is that he could combine these two pictures. The Boltzmann equation is an equation of continuous nature (as it operates with density functions) but its justification (especially the determination of the collision kernel) follows an argument using particles. No wonder it gave rise to so much controversy especially before experiments were available. Moreover, Boltzmann theory implies irreversibility (entropy production) that was for long thought to be more a philosophical than a physical question.

After this detour on Boltzmann we return to the diffusion models. Before we discuss individual models, let me mention the *key conceptual difficulty* behind all rigorous derivations of diffusive equations. Note that the Hamiltonian dynamics (either classical or quantum) is reversible and deterministic. The diffusion equation (1.1), and also the Boltzmann equation, is irreversible: there is permanent loss of information along their evolution (recall that the heat equation is usually not solvable backward in time). Moreover, these equations exhibit a certain chaotic nature (Brown's picture). How can these two contradicting facts be reconciled?

The short answer is that the loss of information is due to a *scale separation* and the *integration of microscopic degrees of freedom*. On the microscopic (particle) level the dynamics remains reversible. The continuous (fluid) equations live on the macroscopic (fluid) scale: they are obtained by neglecting (more precisely, integrating out) many degrees of freedom on short scales. Once we switch to the fluid description, the information in these degrees of freedom is lost forever.

This two-level explanation foreshadows that for a rigorous mathematical description one would need a scale separation parameter, usually called ε that describes the ratio between the typical microscopic and macroscopic scales. In practice this ratio is of order

$$\varepsilon = \frac{1 \ \text{Angstrom}}{1 \ \text{cm}} = 10^{-8},$$

but mathematically we will consider the $\varepsilon \to 0$ so-called *scaling limit*.

Once the scale separation is put on a solid mathematical ground, we should note another key property of the macroscopic evolution equations we will derive, namely their Markovian character. Both the heat equation (1.1) and the nonlinear Boltzmann equation (1.2) give the time derivative of the macroscopic density at any fixed time t in terms of the density at the *same time* only. That is these evolution equations express a process where the future state depends only on the present state; the dependence of the future on the past is only indirect through the present state. We will call this feature the *Markovian property* or in short *Markovity*. Note that colliding particles do build up a memory along their evolution: the state of the recolliding particles will remember their previous collisions. This effect will have to be supressed to obtain a Markovian evolution equation, as was already recognized by Boltzmann in his Ansatz. There are essentially two ways to reduce the rate of recollisions: either by going to a low density situation or by introducing a small coupling constant.

Apart from the rigorous formulation of the scaling limit, we thus will have to cope with the main technical difficulty: **controlling memory (recollision) effects**. Furthermore, specifically in quantum models, we will have to **control interference effects** as well.

Finally, we should remark that Einstein's model is simpler than Boltzmann's, as light particles do not interact (roughly speaking, Boltzmann's model is similar to Fig. 1.2 but the light particles can also collide with each other and not only with the heavy particle). Thus the verification of the Ansatz of molecular chaos from first principle Hamiltonian dynamics is technically easier in Einstein's model. Still, Einstein's model already addresses the key issue: how does diffusion emerge from Hamiltonian mechanics?

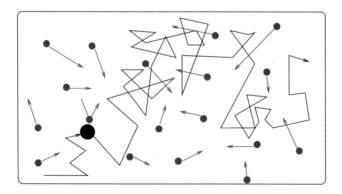

Fig. 1.2 Einstein's explanation of Brown's picture.

1.1.2 Mathematical models of diffusion

The first step to set up a correct mathematical model is to recognize that some stochasticity has to be added. Although we expect that for a **'typical'** initial configuration the diffusion equations are correct, this certainly will **not hold for all** initial configuration. We expect that after opening the door between a warm and a cold room, the temperature will equilibrate (thermalize) but certainly there exists a 'bad' initial configuration of the participating $N \sim 10^{23}$ particles such that all 'warm' particles will, maybe after some time, head towards the cold room and vice versa; that is the two room temperatures will be exchanged instead of thermalization. Such configurations are extremely rare; their measure in all reasonable sense goes to zero very fast as $N \to \infty$, but nevertheless they are there and prevent us from deriving diffusion equations for all initial configurations. It is, however, practically hopeless to describe all 'bad' initial configurations; they may not be recognizable by glancing at the initial state. The stochastic approach circumvents this problem by precisely formulating what we mean by saying that the 'bad events are rare' without identifying them.

The stochasticity can be added in several different ways and this largely determines the technical complications involved in the derivation of the diffusion. In general, the more stochasticity is added, the easier the derivation is.

The easiest is if the **dynamics itself is stochastic**; the typical example being the classical random walk and its scaling limit, the Wiener process (theorized by Wiener (Wiener 1923)). In the typical examples, the random process governing the dynamics of the system has no correlation. The whole microscopic dynamics is Markovian (in some generalizations the dynamics of the one particle densities is not fully Markovian, although it typically has strong statistical mixing properties). Of course this dynamics is not Hamiltonian (no energy conservation and no reversibility), and thus it is not particularly surprising that after some scaling limit one obtains a diffusion equation. The quantum counterpart of the classical random walk is a Schrödinger equation with a Markovian time dependent random potential, see (Pillet 1985) and the recent proof of quantum diffusion in this model by Kang, and Schenker (Kang *et al.* 2009).

The next level of difficulty is a Hamiltonian system with a **random Hamiltonian**. Some physical data in the Hamiltonian (e.g. potential or magnetic field) is randomly selected (describing a disordered system), but they are frozen forever and after that the Hamiltonian evolution starts (this is in sharp contrast to the stochastic dynamics, where the system is subject to fresh random inputs along its evolution). A typical model is the *Lorentz gas*, where a single Hamiltonian particle is moving among fixed random scatterers. Recollisions with the same scatterer are possible and this effect potentially ruins the Markovity. Thus this model is usually considered in the weak coupling limit, that is the coupling parameter λ between the test-particle and the random scatterers is set to converge to zero, $\lambda \to 0$, and simultaneously a long time limit is considered. For a weak coupling, the recollision becomes a higher order effect and may be neglected. It turns out that if

$$t \sim \lambda^{-2} \tag{1.3}$$

then one can prove a nontrivial Markovian diffusive limit dynamics (see Kesten and Papanicolaou (Kesten, and Papanicolaou 1980/81), extended more recently to somewhat longer times by Komorowski and Ryzhik (Komorowski and Ryzhik 2006)). The relation (1.3) between the coupling and the time scale is called the *van Hove limit*. Similar supressing of the recollisions can also be achieved by the *low density limit* (see later).

The corresponding quantum model (*Anderson model*) is the main focus of the current chapter. We will show that in the van Hove limit the linear Boltzmann equation arises after an appropriate rescaling, and for longer time scales the heat equation emerges. In particular, we describe the behaviour of the quantum evolution in the presumed *extended states regime* of the Anderson model up to a certain time scale. We will use a mathematically rigorous perturbative approach. We mention that supersymmetric methods (Efetov 1999) offer a very attractive alternative approach to studying quantum diffusion, although the mathematical control of the resulting functional integrals is difficult. Effective models that are reminiscent of the saddle point approximations of the supersymmetric approach are more accessible to rigorous mathematics. Recently Disertori, Spencer, and Zirnbauer have proved a diffusive estimate for the two-point correlation functions in a three-dimensional supersymmetric hyperbolic sigma model (Disertori *et al.* 2010) at low temperatures and localization was also established in the same model at high temperatures (Disertori, and Spencer 2010).

The following level of difficulty is a **deterministic Hamiltonian with random initial data of many degrees of freedom**. The typical example is Einstein's model, where the test-particle is immersed in a heat bath. The heat bath is called *ideal gas*, if it is characterized by a Hamiltonian H_{bath} of noninteracting particles. The initial state is typically a temperature equilibrium state, $\exp\left(-\beta H_{bath}\right)$ that is the initial data of the heat-bath particles are given by this equilibrium measure at inverse temperature β. Again, some scaling limit is necessary to reduce the chance of recollisions, one can for example consider the limit $m/M \to \infty$, where M and m are the mass of the test particle and the heat-bath particles, respectively. In this case, a collision with a single light particle does not have sizable effect on the motion of the heavy particle (this is why the abrupt changes in velocity in Fig. 1.1 are an exaggeration). Therefore, the rate of collisions has to be increased in parallel with $m/M \to 0$ to have a sizeable total collision effect. Similarly to the van Hove limit, there is a natural scaling limit, where nontrivial limiting dynamics was proven by Dürr, Goldstein, and Lebowitz (Dürr, Goldstein, and Lebowitz 1980/81). On the quantum side, this model corresponds to a quantum particle immersed in a phonon (or possibly photon) bath. On the kinetic scale the Boltzmann equation was proven in (Erdős 2002). Recently De Roeck and Fröhlich (De Roeck and Fröhlich 2011) (see also (De Roeck, Fröhlich, and Pizzo 2010) for an earlier toy model) have proved quantum diffusion for a related model in $d \geq 4$ dimensions, where the mass of the quantum particle was large and an additional internal degree of freedom ('spin') was coupled to the heat bath to enhance the decay of time correlations.

We remark that the problem becomes enormously more complicated if the **heat-bath particles can interact among each other**, that is if we are facing a truly interacting many-body dynamics. In this case there is even no need to distinguish between the tracer particle and the heat-bath particles, in the simplest model one just considers identical particles interacting via a two-body interaction and one investigates the single-particle density function $f_t(x, v)$. In a certain scaling limit regime, the model should lead to the nonlinear Boltzmann equation (1.2). This has only been proven in the classical model for short time by Lanford (Lanford 1976). His work appeared in 1975 and since then nobody has extended the proof to include longer time scales. We mention that the complications are partly due to the fact that the nonlinear Boltzmann equation itself does not have a satisfactory existence theory for long times. The corresponding quantum model is an unsolved problem; although via an order by order expansion (Benedetto, Castella, Esposito, and Pulvirenti 2008) there is no doubt on the validity of the nonlinear Boltzmann equation starting from a weakly coupled interacting many-body model, the expansion cannot be controlled up to date. Lukkarinen and Spohn (Lukkarinen and Spohn 2011) have studied a weakly nonlinear cubic Schrödinger equation with a random initial data (drawn from a phonon bath) near equilibrium. They proved that the space-time covariance of the evolved random wave function satisfies a nonlinear Boltzmann-type equation in the kinetic limit, for short times. While this is still a one-particle model and only fluctuations around the equilibrium are studied, its nonlinear character could be interpreted as a first step towards understanding the truly many-body Boltzmann equation.

Table 1.1 Microscopic models for diffusion

	CLASSICAL MECHANICS	QUANTUM MECHANICS
Stochastic dynamics (no memory)	Random walk (Wiener)	Random kick model with zero time corr. potential (Pillet, Schenker–Kang)
Hamiltonian particle in a random environment (one body)	Lorentz gas: particle in random scatterers (Kesten–Papanicolaou) (Komorowski–Ryzhik)	Anderson model or quantum Lorentz gas (Spohn, Erdős–Yau, Erdős–Salmhofer–Yau Disertori–Spencer–Zirnbauer)
Hamiltonian particle in a heat bath (randomness in the many-body data)	Einstein's kinetic model (Dürr–Goldstein–Lebowitz)	Electron in phonon or photon bath (Erdős, Erdős–Adami, Spohn–Lukkarinen De Roeck–Fröhlich)
Periodic Lorentz gas (randomness in the one-body initial data)	Sinai billiard (Bunimovich–Sinai)	Ballistic (Bloch waves, easy)
Many-body interacting Hamiltonian	Nonlinear Boltzmann eq (short time: Lanford)	Quantum NL Boltzmann (unsolved)

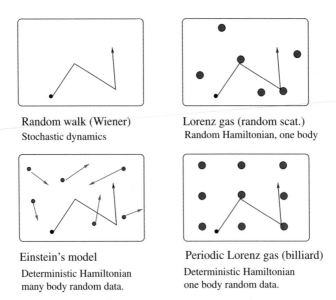

| Random walk (Wiener) | Lorenz gas (random scat.) |
| Stochastic dynamics | Random Hamiltonian, one body |

Einstein's model	Periodic Lorenz gas (billiard)
Deterministic Hamiltonian	Deterministic Hamiltonian
many body random data.	one body random data.

Fig. 1.3 Microscopic models for diffusion.

Finally, the most challenging (classical) model is the **deterministic Hamiltonian with a random initial data of a few degrees of freedom**, the typical example being the various mathematical billiards. The simplest billiard is the hard-core periodic Lorentz gas (also called Sinai's billiard (Bunimovich, and Sinai 1980/81)), where the scatterers are arranged in a periodic lattice and a single point particle ('billiard ball') moves among these scatterers according to the usual rules of specular reflections. The methods of proofs here originate in a more dynamical system than in statistical mechanics and they have a strong geometric nature (e g. convexity of the scatterers is heavily used),

All these models have natural quantum mechanical analogues; for the Hamiltonian systems they simply follow from standard quantization of the classical Hamiltonian. These models are summarized in Table 2.1 below and they are also illustrated in Fig. 1.3. We note that the quantum analogue of the periodic Lorentz gas is ballistic due to the Bloch waves (a certain diagonalization procedure similar to the Fourier transform), thus in this case the classical and quantum models behave differently; the quantum case being relatively trivial.

1.2 Some facts on stochastic processes

Our goal is to understand the stochastic features of quantum dynamics. A natural way to do so is to compare quantum dynamics with a much simpler stochastic dynamics whose properties are well-known or much easier to establish. In this section the most important ingredients of elementary stochastic processes are collected and we will introduce the Wiener process. Basic knowledge of probability theory (random variables, expectation, variance, independence, Gaussian normal random variable, characteristic function) is assumed.

1.2.1 The central limit theorem

Let v_i, $i = 1, 2, \ldots$, be a sequence of independent, identically distributed (denoted by i.i.d. for brevity) random variables with zero expectation, $\mathbf{E}\, v_i = 0$ and finite variance, $\mathbf{E}\, v_i^2 = \sigma^2$. The values of v_i are either in $\mathbb{R}^d$ or $\mathbb{Z}^d$. In case of $d > 1$, the variance σ^2 is actually a matrix (called *covariance matrix*), defined as $\sigma_{ab}^2 = \mathbf{E}\, v^{(a)} v^{(b)}$ where $v = (v^{(1)}, v^{(2)}, \ldots, v^{(d)})$. The v_i's should be thought of as velocities or steps of a walking particle at discrete time i. For simplicity we will work in $d = 1$ dimensions, the extension to the general case is easy. A natural question is: where is the particle after n steps if n is big?

Let

$$S_n := \sum_{i=1}^{n} v_i$$

be the location of the particle after n steps. It is clear that the expectation is zero, $\mathbf{E}\, S_n = 0$, and the variance is

$$\mathbf{E}\, S_n^2 = \mathbf{E}\left(\sum_{i=1}^{n} v_i \right)^2 = \sum_{i=1}^{n} \mathbf{E}\, v_i^2 = n\sigma^2 \qquad (1.4)$$

(note that independence via $\mathbf{E}\, v_i v_j = 0$ for $i \neq j$ has been used).

This suggests we should rescale S_n and define

$$X_n := \frac{S_n}{\sqrt{n}} = \frac{\sum_{i=1}^{n} v_i}{\sqrt{n}}$$

then $\mathbf{E}\, X_n = 0$ and $\mathbf{E}\, X_n^2 = \sigma^2$. The surprising fact is that the distribution of X_n tends to a universal one (namely to the Gaussian normal distribution) as $n \to \infty$, no matter what the initial distribution of v_i was! This is the central limit theorem, a cornerstone of probability theory and one of the fundamental theorems of nature.

Theorem 1.1 *Under the conditions $\mathbf{E}\, v_i = 0$ and $\mathbf{E}\, v_i^2 = \sigma^2 < \infty$ the distribution of the rescaled sum X_n of the i.i.d random variables v_i converges to the normal distribution, that is*

$$X_n \Longrightarrow X$$

in the sense of probability distributions where X is a random variable distributed according to $N(0, \sigma^2)$.

We recall that the density of the normal distribution in $\mathbb{R}^1$ is given by

$$f(x) = \frac{1}{\sqrt{2\pi}\sigma} e^{-\frac{x^2}{2\sigma^2}},$$

and the convergence is the standard weak convergence of probability measures (or convergence in distribution or convergence in law):

Definition 1.2 *A sequence of random variables X_n converges to the random variable X in distribution, $X_n \Longrightarrow X$ if*

$$\boldsymbol{E}G(X_n) \to \boldsymbol{E}G(X)$$

for any continuous bounded function $G : \mathbb{R} \to \mathbb{R}$

In particular, X_n converges in distribution to $N(0, \sigma^2)$ if

$$\boldsymbol{E}\,G(X_n) \to \int G(x) f(x) \mathrm{d}x.$$

Analogous definitions hold in higher dimensions.

Recall that we will have two scales: a microscopic and a macroscopic one. In this example, the microscopic scale corresponds to one step of the walking particle and we run the process up to a total of n units of the microscopic time. The macroscopic time will be the natural time scale of the limit process and it will be kept order one even as $n \to \infty$. Recalling that we introduced $\varepsilon \ll 1$ as a scaling parameter, we now reformulate the central limit theorem in this language.

Let $T > 0$ be a macroscopic time (think of it as an unscaled, order one quantity) and let

$$n := [T\varepsilon^{-1}],$$

where $[\cdot]$ denotes the integer part. Let again $\boldsymbol{E}\,v_i = 0$, $\boldsymbol{E}\,v_i^2 = \sigma^2$. From the central limit theorem it directly follows that

$$\widetilde{X}_T^\varepsilon := \frac{1}{\varepsilon^{-1/2}} \sum_{i=1}^{[T\varepsilon^{-1}]} v_i \Longrightarrow X_T, \qquad \text{as } \varepsilon \to 0, \tag{1.5}$$

where X_T is a normal Gaussian variable with variance $T\sigma^2$ and with density function

$$f_T(X) = \frac{1}{\sqrt{2\pi T}\sigma} \exp\left(-\frac{X^2}{2\sigma^2 T}\right). \tag{1.6}$$

Note that the normalization in (1.5) is only by $\sqrt{\varepsilon^{-1}}$ and not by $\sqrt{n} \approx \sqrt{T\varepsilon^{-1}}$, therefore the limit depends on T (but not on ε, of course).

Since the Gaussian function $f_T(X)$ gives the probability of finding X_T at location x, we have

$$f_T(X)\mathrm{d}X \approx \mathrm{Prob}\{\widetilde{X}_T^\varepsilon \text{ is located at } X + \mathrm{d}X \text{ at time } T\}.$$

Note that we used macroscopic space and time coordinates, X and T. Translating this statement into the original process in microscopic coordinates we see that

$$f_T(X)\mathrm{d}X \approx \mathrm{Prob}\{\text{finding } S_n \text{ at } \varepsilon^{-1/2}(X + \mathrm{d}X) \text{ at time } n \approx \varepsilon^{-1}T\}.$$

Note that the space and time are rescaled differently; if (X, T) denote the macroscopic space/time coordinates and (x, t) denote the microscopic coordinates, then

$$x = \varepsilon^{-1/2} X, \qquad t = \varepsilon^{-1} T.$$

This is the typical *diffusive scaling*, where

$$\text{time} = (\text{distance})^2. \tag{1.7}$$

Finally, we point out that the limiting density function $f_T(X)$ satisfies the heat equation with *diffusion coefficient* σ^2:

$$\partial_T f_T(X) = \sigma^2 \partial_X^2 f_T(X). \tag{1.8}$$

In fact, the Gaussian function (1.6) is the fundamental solution to the heat equation also in higher dimensions.

The emergence of the normal distribution and the heat equation is universal; note that apart from the variance (and the zero expectation) no other information on the distribution of the microscopic step v was used. The details of the microscopic dynamics are wiped out, and only the essential feature, the heat equation (or the normal distribution) with a certain diffusion coefficient, remains visible after the $\varepsilon \to 0$ scaling limit.

We remark that the central limit theorem (with essentially the same proof) is valid if v_i's have some short time correlation. Assume that v_i is a stationary sequence of random variables with zero mean, $\mathbf{E} \, v_i = 0$, and let

$$R(i, j) := \mathbf{E} \, v_i v_j$$

the correlation function. By stationarity, $R(i, j)$ depends only on the difference, $i - j$, so let us use the notation

$$R(i - j) := \mathbf{E} \, v_i v_j$$

This function is also called the *velocity-velocity autocorrelation function.*

Theorem 1.3 *Assume that the velocity-velocity autocorrelation function is summable, that is*

$$\sum_{k=-\infty}^{\infty} R(k) < \infty.$$

Then the rescaled sums

$$\widetilde{X}_T^\varepsilon := \varepsilon^{1/2} \sum_{i=1}^{[T\varepsilon^{-1}]} v_i$$

converge in distribution to a normal random variable,

$$\widetilde{X}_T^\varepsilon \Longrightarrow X_T, \qquad \varepsilon \to 0,$$

whose density function $f_T(X)$ satisfies the heat equation

$$\partial_T f = D \Delta f_T$$

with a diffusion coefficient

$$D := \sum_{k=-\infty}^{\infty} R(k),$$

assuming that D is finite. This relation between the diffusion coefficient and the sum (or integral) of the velocity-velocity autocorrelation function is called the Green–Kubo formula.

Finally we mention that all results remain valid in higher dimensions, $d > 1$. The role of the diffusion coefficient is taken over by a diffusion matrix D_{ij} defined by the correlation matrix of the single step distribution:

$$D_{ij} = \mathbf{E} \, v^{(i)} v^{(j)}$$

where $v = (v^{(1)}, v^{(2)}, \ldots, v^{(d)})$. The limiting heat equation (1.8) in the uncorrelated case is modified to

$$\partial_T f_T(X) = \sum_{i,j=1}^{d} D_{ij} \partial_i \partial_j f_T(X).$$

In particular, for random vectors v with i.i.d. components the covariance matrix is constant D times the identity, and we obtain the usual heat equation

$$\partial_T f_T(X) = D \Delta_X f_T(X)$$

where D is called the *diffusion coefficient*.

1.2.2 Markov processes and their generators

Let X_t, $t \geq 0$, be a continuous-time stochastic process, that is X_t is a one-parameter family of random variables, the parameter is usually called time. The *state space* of the process is the space from where X_t takes its values, in our case $X_t \in \mathbb{R}^d$ or $\mathbb{Z}^d$. As usual, $\mathbf{E}$ will denote the expectation with respect to the distribution of the process.

Definition 1.4 *X_t is a Markov process if for any $t \geq \tau$*

$$Dist(X_t \mid \{X_s\}_{s \in [0,\tau]}) = Dist(X_t \mid X_\tau)$$

that is if the conditional distribution of X_t conditioned on the family of events X_s in times $s \in [0, \tau]$ is the same as conditioned only on the event at time τ.

In simple terms it means that the state, X_t, of the system at time t depends on the past between times $[0, \tau]$ only via the state of the system at the last time of the past interval, τ. All necessary information about the past is condensed in the last moment.

The process depends on its initial value, X_0. Let $\mathbf{E}_x$ denote the expectation value assuming that the process started from the point $x \in \mathbb{R}^d$ or $\mathbb{Z}^d$, that is

$$\mathbf{E}_x \varphi(X_t) = \mathbf{E}\{\varphi(X_t) \mid X_0 = x\}.$$

Here and in the sequel φ will denote functions on the state space, these are also called *observables*. We will not specify exactly the space of observables.

Markov processes are best described by their generators:

Definition 1.5 *The generator of the Markov process X_t is an operator acting on the observables φ and it is given by*

$$(\mathcal{L}\varphi)(x) := \frac{d}{d\varepsilon}\Big|_{\varepsilon=0+0} \mathbf{E}_x \, \varphi(X_\varepsilon).$$

This definition is a bit lousy since the function space on which $\mathcal{L}$ act is not defined. Typically it is either the space of continuous functions or the space of L^2-functions on the state space, but for both concepts an extra stucture—topology or measure—is needed on the state space. Moreover, the generator is typically an unbounded operator, not defined on the whole function space but only on a dense subset. For this chapter we will leave these technicalities aside, but when necessary, we will think of the state space $\mathbb{R}^d, \mathbb{Z}^d$ equipped with their natural measures and for the space of functions we will consider the L^2-space.

The key fact is that the generator tells us everything about the Markov process itself. Let us demonstrate it by answering two natural questions in terms of the generator.

Question 1: Starting from $X_0 = x$, what is the distribution of X_t?

Answer: Let φ be an observable (function on the state space) and define

$$f_t(x) := \mathbf{E}_x \, \varphi(X_t). \tag{1.9}$$

We wish to derive an evolution equation for f_t:

$$\partial_t f_t(x) = \lim_{\varepsilon \to 0} \frac{1}{\varepsilon} \mathbf{E}_x \Big[\mathbf{E}_{X_\varepsilon} \varphi(\widetilde{X}_t) - \mathbf{E}_x \, \varphi(X_t) \Big],$$

since by the Markov property

$$\mathbf{E}_x \, \varphi(X_{t+\varepsilon}) = \mathbf{E}_x \mathbf{E}_{X_\varepsilon} (\varphi(\widetilde{X}_t)),$$

where $\widetilde{X}_t$ is a new copy of the Markov process started at the (random) point X_ε. Thus

$$\partial_t f_t(x) = \lim_{\varepsilon \to 0} \frac{1}{\varepsilon} \mathbf{E}_x \Big[f_t(X_\varepsilon) - f_t(x) \Big] = \frac{d}{d\varepsilon}\Big|_{\varepsilon=0+0} \mathbf{E}_x \, f_t(X_\varepsilon) = (\mathcal{L}f_t)(x).$$

Therefore f_t, defined in (1.9), solves the initial value problem

$$\partial_t f_t = \mathcal{L}f_t \qquad \text{with} \quad f_0(x) = \varphi(x).$$

Formally one can write the solution as

$$f_t = e^{t\mathcal{L}}\varphi.$$

If the observable is a Dirac delta function at a fixed point y in the state space,

$$\varphi(x) = \delta_y(x),$$

then the solution to the above initial value problem is called the *transition kernel* of the Markov process and it is denoted by $p_t(x, y)$:

$$\partial_t p_t(\cdot, y) = \mathcal{L}p_t(\cdot, y) \qquad \text{with} \quad p_0(\cdot, y) = \delta_y.$$

The intuitive meaning of the transition kernel is

$$p_t(x, y)\mathrm{d}y := \mathrm{Prob}\{\text{After time } t \text{ the process is at } y + \mathrm{d}y \text{ if it started at } x \text{ at time } 0\}.$$

Question 2: Suppose that the initial value of the process, $X_0 = x$, is distributed according to a density function $\psi(x)$ on the state space. What is the probability that after time t the process is at y?
Answer:

$$g_t(y) := \mathrm{Prob}(X_t \text{ at } y \text{ after time } t) = \int \psi(x) p_t(x, y)\mathrm{d}x.$$

It is an exercise to check (at least formally), that g_t solves the following initial value problem

$$\partial_t g_t = \mathcal{L}^* g_t \qquad \text{with} \quad g_0 = \psi,$$

where $\mathcal{L}^*$ denotes the adjoint of $\mathcal{L}$ (with respect to the standard scalar product of the L^2 of the state space).

1.2.3   Wiener process and its generator

The Wiener process is the rigorous mathematical construction of Brownian motion. There are various ways to introduce it, we choose the shortest (but not the most intuitive) definition.

First we need the concept of the *Gaussian process*. We recall that the centred Gaussian random variables have the remarkable property that all their moments are determined by the covariance matrix (a random variable is called *centred* if its expectation is zero, $\mathbf{E}\,X = 0$). If $(X_1, X_2, \ldots, X_{2k})$ is a centred Gaussian vector-valued random variable, then the higher moments can be computed by **Wick's formula** from the second moments (or *two-point correlations*)

$$\mathbf{E}\,X_1 X_2 \ldots X_{2k} = \sum_{\pi \text{ pairing}} \prod_{(i,j)\in\pi} \mathbf{E}\,X_i X_j, \tag{1.10}$$

where the summation is over all possible (unordered) pairings of the indices $\{1, 2, \ldots 2k\}$.

A stochastic process $X_t \in \mathbb{R}^d$ is called *Gaussian*, if any finite dimensional projection is a Gaussian vector-valued random variable, that is for any $t_1 < t_2 < \ldots < t_k$, the vector $(X_{t_1}, X_{t_2}, \ldots, X_{t_n}) \in \mathbb{R}^{dn}$ is a Gaussian random variable.

Definition 1.6 *A continuous Gaussian stochastic process* $W_t = (W_t^{(1)}, \ldots, W_t^{(d)}) \in \mathbb{R}^d$, $t \geq 0$, *is called the d-dimensional (standard) Wiener process if it satisfies the following*

    i) $W_0 = 0$
    ii) $\boldsymbol{E} W_t = 0$
    iii) $\boldsymbol{E} W_s^{(a)} W_t^{(b)} = \min\{s, t\} \delta_{ab}$.

From Wick's formula it is clear that all higher moments of W_t are determined. It is a fact that the above list of requirements determines the Wiener process uniquely. We can now extend the central limit theorem to processes.

Theorem 1.7 *Recall the conditions of Theorem 1.1. The stochastic process*

$$\widetilde{X}_T^\varepsilon := \varepsilon^{1/2} \sum_{i=1}^{[T\varepsilon^{-1}]} v_i$$

(defined already in (1.5)) converges in distribution (as a stochastic process) to the Wiener process

$$\widetilde{X}_T^\varepsilon \Longrightarrow W_T$$

as $\varepsilon \to 0$.

The proof of this theorem is not trivial. It is fairly easy to check that the moments converge, that is for any $k \in \mathbb{N}^d$ multi-index

$$\mathbf{E}\,[\widetilde{X}_T^\varepsilon]^k \to \mathbf{E}\,W_T^k$$

and the same holds even for different times:

$$\mathbf{E}\,\big[\widetilde{X}_{T_1}^\varepsilon\big]^{k_1} \big[\widetilde{X}_{T_2}^\varepsilon\big]^{k_2} \ldots \big[\widetilde{X}_{T_m}^\varepsilon\big]^{k_m} \to \mathbf{E}\,W_{T_1}^{k_1} W_{T_2}^{k_2} \ldots W_{T_m}^{k_m},$$

but this is not quite enough. The reason is that a continuous time process is a collection of uncountable many random variables and the joint probability measure on such a big space is not necessarily determined by finite dimensional projections. If the process has some additional regularity property, then yes. The correct notion is the *stochastic equicontinuity* (or Kolmogorov condition) which provides the necessary compactness (also called tightness in this context). We will not go into more details here, see any standard book on stochastic processes.

Theorem 1.8 *The Wiener process on* $\mathbb{R}^d$ *is Markovian with generator*

$$\mathcal{L} = \frac{1}{2}\Delta.$$

Idea of the proof. We work in $d = 1$ dimension for simplicity. The Markovity can be easily checked from the explicit formula for the correlation function. The most important ingredient is that the Wiener process has *independent increments:*

$$\mathbf{E}\,(W_t - W_s)W_u = 0 \qquad (1.11)$$

if $u \le s \le t$; that is the increment in the time interval $[s, t]$ is independent of the past. The formula (1.11) is trivial to check from property (iii).

Now we compute the generator using Definition 1.5

$$\frac{\mathrm{d}}{\mathrm{d}\varepsilon}\Big|_{\varepsilon=0+0}\mathbf{E}_0\varphi(W_\varepsilon) = \frac{\mathrm{d}}{\mathrm{d}\varepsilon}\Big|_{\varepsilon=0+0}\mathbf{E}_0\left(\varphi(0) + \varphi'(0)W_\varepsilon + \frac{1}{2}\varphi''(0)W_\varepsilon^2 + \dots\right) = \frac{1}{2}\varphi''(0).$$

Here we used that $\mathbf{E}_0 W_\varepsilon = 0$, $\mathbf{E}_0 W_\varepsilon^2 = \varepsilon$ and the dots refer to irrelevant terms that are higher order in ε. $\square$

Finally we remark that one can easily define a Wiener process with any nontrivial (non-identity) diffusion coefficient matrix D as long as it is positive definite (just write $D = A^*A$ and apply the linear transformation $W \to AW$ to the standard Wiener process). The generator is then

$$\mathcal{L} = \frac{1}{2}\sum_{i,j=1}^{d} D_{ij}\partial_i\partial_j.$$

1.2.4   Random jump process on the sphere S^{d-1} and its generator

The state space of this process is the unit sphere $S = S^{d-1}$ in $\mathbb{R}^d$. We are given a function

$$\sigma(v, u) : S \times S \to \mathbb{R}_+$$

which is interpreted as jump rate from the point v to u.

The process will be parametrized by a continuous time t, but, unlike the Wiener process, its trajectories will not be continuous, rather piecewise constant with some jumps at a discrete set of times. A good intuitive picture is that the particle jumping on the unit sphere has a random clock (so-called *exponential clock*) in its pocket. The clock emits a signal with the standard exponential distribution, that is

$$\text{Prob}\{\text{there is a signal at } t + \mathrm{d}t\} = e^{-t}\mathrm{d}t,$$

and when it ticks, the particle, being at location v, jumps to any other location $u \in S$ according to the distribution $u \to \sigma(v, u)$ (with the appropriate normalization):

$$\text{Prob}\{\text{the particle from } v \text{ jumps to } u + \mathrm{d}u\} = \frac{\sigma(v, u)\mathrm{d}u}{\int \sigma(v, u)\mathrm{d}u}.$$

The transition of the process at an infinitesimal time increment is given by

$$v_{t+\varepsilon} = \begin{cases} u + du & \text{with probability} \quad \varepsilon\sigma(v_t, u)du \\ v_t & \text{with probability} \quad 1 - \varepsilon \int \sigma(v_t, u)du \end{cases} \qquad (1.12)$$

up to terms of order $O(\varepsilon^2)$ as $\varepsilon \to 0$.

Thus the generator of the process can be computed from Definition 1.5. Let

$$f_t(v) := \mathbf{E}_v \varphi(v_t),$$

assuming that the process starts from some given point $v_0 = v$ and φ is an arbitrary observable. Then

$$\partial_t f_t(v) = \lim_{\varepsilon \to 0+0} \frac{1}{\varepsilon} \mathbf{E}_v \left(\mathbf{E}_{v_\varepsilon} \varphi(\tilde{v}_t) - \mathbf{E}_v \varphi(\tilde{v}_t) \right)$$

$$= \int du \, \sigma(v, u) \left[\mathbf{E}_u \varphi(\tilde{v}_t) - \mathbf{E}_v \varphi(\tilde{v}_t) \right]$$

$$= \int du \, \sigma(v, u) \left[f_t(u) - f_t(v) \right],$$

where we used the Markov property in the first line and the infinitesimal jump rate from (1.12) in the second line. Note that with probability $1 - \varepsilon\sigma(v, u)$ we have $v_\varepsilon = v$. Thus the generator of the random jump process is

$$(\mathcal{L}f)(v) := \int du \, \sigma(v, u) \left[f(u) - f(v) \right]. \qquad (1.13)$$

Note that it has two terms; the first term is called the *gain term* the second one is the *loss term*. The corresponding evolution equation for the time dependent probability density of the jump process,

$$\partial_t f_t(v) = \int du \, \sigma(v, u) \left[f_t(u) - f_t(v) \right] \qquad (1.14)$$

is called the *linear Boltzmann equation in velocity space*.

The elements of the state space S will be interpreted as velocities of a moving particle undergoing fictitious random collisions. Only the velocities are recorded. A velocity distribution $f(v)$ is called *equilibrium distribution*, if $\mathcal{L}f \equiv 0$; it is fairly easy to see that, under some nondegeneracy condition on σ, the equilibrium exists uniquely.

1.3   Classical mechanics of a single particle

Classical mechanics of a single particle in d-dimensions is given by a Hamiltonian (energy function) defined on the phase space $\mathbb{R}^d \times \mathbb{R}^d$:

$$H(v, x) := \frac{1}{2} v^2 + U(x). \qquad (1.15)$$

Here x is the position, v is the momentum coordinate. For the majority of this chapter we will not distinguish between momentum and velocity, since we almost always consider the standard kinetic energy $\frac{1}{2}v^2$.

The Hamiltonian equation of motions is the following set of $2d$ coupled first order differential equations:

$$\dot{x}(t) = \partial_v H = v \qquad \dot{v}(t) = -\partial_x H = -\nabla U(x). \tag{1.16}$$

For example, in case of the free evolution, when the potential is zero, $U \equiv 0$, we have

$$x(t) = x_0 + v_0 t \tag{1.17}$$

that is linear motion with a constant speed.

Instead of describing each point particle separately, we can think of a continuum of (noninteracting) single particles, described by a phase space density $f(x, v)$. This function is interpreted to give the number of particles with given velocity at a given position, more precisely

$$\int_\Delta f(x, v)\mathrm{d}x\mathrm{d}v = \text{Number of particles at } x \text{ with velocity } v \text{ such that } (x, v) \in \Delta.$$

Another interpretation of the phase space density picture is that a single particle with velocity v_0 at x_0 can be described by the measure

$$f(x, v) = \delta(x - x_0)\delta(v - v_0),$$

and it is then natural to consider more general measures as well.

The system evolves with time, so does its density, that is we consider the time dependent phase space densities, $f_t(x, v)$. For example in the case of a single particle governed by (1.16) with initial condition $x(0) = x_0$ and $v(0) = v_0$, the evolution of the phase space density is given by

$$f_t(x, v) = \delta(x - x(t))\delta(v - v(t)),$$

where $(x(t), v(t))$ is the Hamiltonian trajectory computed from (1.16).

It is easy to check that if the point particle trajectories satisfy (1.16), then f_t satisfies the following evolution equation

$$(\partial_t + v \cdot \nabla_x)f_t(x, v) = \nabla U(x) \cdot \nabla_v f_t(x, v), \tag{1.18}$$

which is called the *Liouville equation*. The left-hand side is called the *free streaming term*. The solution to

$$(\partial_t + v \cdot \nabla_x)f_t(x, v) = 0 \tag{1.19}$$

is given by the linear transport solution

$$f_t(x, v) = f_0(x - vt, v),$$

where f_0 is the phase space density at time zero. This corresponds to the free evolution (1.17) in the Hamiltonian particle representation.

1.3.1 The linear Boltzmann equation

The linear Boltzmann equation is a phenomenological combination of the free flight equation (1.19) and the jump process on the sphere of the velocity space (1.14)

$$(\partial_t + v \cdot \nabla_x) f_t(x, v) = \int \sigma(u, v) [f_t(x, u) - f_t(x, v)] du. \qquad (1.20)$$

Note that we actually have the adjoint of the jump process (observe that u and v are interchanged compared with (1.14)), so the solution determines:

$$f_t(x, v) = \text{Prob}\{\text{the process is at } (x, v) \text{ at time } t\},$$

given the initial probability density $f_0(x, v)$ (Question 2 in Section 1.2.2).

The requirement that the velocities are constrained to the sphere corresponds to energy conservation. Of course there is no real Hamiltonian dynamics behind this equation: the jumps are stochastic.

Notice that the free flight plus collision process standing behind the Boltzmann equation is actually a random walk in continuous time. In the standard random walk the particle 'jumps' in a new random direction after every unit time. In the Boltzmann process the jumps occur at random times ('exponential clock'), but the effect on a long time scale is the same. In particular, the long time evolution of the linear Boltzmann equation is diffusion (Wiener process) in position space.

The following theorem formulates this fact more precisely. We recall that to translate the velocity process into position space, one has to integrate the velocity, that is will consider

$$x_t = \int_0^t v_s ds.$$

Theorem 1.9 *Let v_t be a random velocity jump process given by the generator (1.14). Then*

$$X_\varepsilon(T) =: \varepsilon^{1/2} \int_0^{T/\varepsilon} v_t \, dt \to W_T \qquad (\text{in distribution,})$$

where W_T is a Wiener process with diffusion coefficient being the velocity autocorrelation

$$D = \int_0^\infty R(t) dt, \qquad R(t) := \boldsymbol{E} v_0 v_t.$$

Here $\boldsymbol{E}$ is with respect to the equilibrium measure of the jump process on S^{d-1}.

1.4 Quantum mechanics of a single particle

1.4.1 Wavefunction, Wigner transform

The state space of a quantum particle in d-dimensions is $L^2(\mathbb{R}^d)$ or $L^2(\mathbb{Z}^d)$. Its elements are called L^2-wavefunctions and they are usually denoted by $\psi = \psi(x) \in L^2(\mathbb{R}^d)$ or $L^2(\mathbb{Z}^d)$. We assume the normalization, $\|\psi\|_2 = 1$. We recall the interpretation of the wave function: the position space density $|\psi(x)|^2 \mathrm{d}x$ gives the probability of finding an electron at a spatial domain:

$$\int_\Delta |\psi(x)|^2 \mathrm{d}x = \mathrm{Prob}\{\text{the particle's position is in } \Delta\}.$$

Similarly, the Fourier transform of ψ,

$$\widehat{\psi}(v) := \int_{\mathbb{R}^d} e^{-iv \cdot x} \psi(x) \mathrm{d}x,$$

determines the momentum space density:

$$\int_\Delta |\widehat{\psi}(v)|^2 \mathrm{d}v = \mathrm{Prob}\{\text{the particle's momentum is in } \Delta\}.$$

In the lattice case, that is when $\mathbb{Z}^d$ replaces $\mathbb{R}^d$ as the physical space, the Fourier transform is replaced with Fourier series. In these notes we neglect all 2π's that otherwise enter the definition of the Fourier transform.

By the Heisenberg uncertainty principle, one cannot simultaneously determine the position and the momentum of a quantum particle, thus the concept of classical phase space density does not generalize directly to quantum mechanics. Nevertheless one can define a substitute for it, namely the *Wigner transform*. For any L^2-wavefunction ψ we define the Wigner transform of ψ as

$$W_\psi(x, v) := \int \overline{\psi}\left(x + \frac{z}{2}\right) \psi\left(x - \frac{z}{2}\right) e^{ivz} \mathrm{d}z,$$

and we still interpret it as 'quantum phase space density'.

It is easy to check that W_ψ is always real but in general is not positive (thus it cannot be the density of a positive measure—in coincidence with the Heisenberg principle). However, its marginals reconstruct the position and momentum space densities, as the following formulas can be easily checked:

$$\int W_\psi(x, v) \mathrm{d}v = |\psi(x)|^2, \qquad \int W_\psi(x, v) \mathrm{d}x = |\widehat{\psi}(v)|^2.$$

In particular, for normalized wave functions $\|\psi\|_2 = 1$, we have

$$\iint W_\psi(x, v) \mathrm{d}v \mathrm{d}x = 1. \tag{1.21}$$

We remark that for the lattice case some care is needed for the proper definition of the Wigner transform, since $x \pm \frac{z}{2}$ may not be a lattice site. The precise definition in this case is

$$W_\psi(x,v) := \sum_{\substack{y,z\in\mathbb{Z}^d \\ y+z=2x}} e^{iv(y-z)}\bar\psi(y)\psi(z), \tag{1.22}$$

where $\psi \in \ell^2(\mathbb{Z}^d)$, that is $y, z \in \mathbb{Z}^d$, but $x \in (\mathbb{Z}/2)^d$. The formulas for marginal of the Wigner transform modify as follows:

$$\int W_\psi(x,v)\mathrm{d}v = 2^d|\psi(x)|^d$$

if $x \in \mathbb{Z}^d$ and it is zero if $x \in (\mathbb{Z}/2)^d \setminus \mathbb{Z}^d$. We still have

$$\int_{(\mathbb{Z}/2)^d} W_\psi(x,v)\mathrm{d}x = 2^{-d}\sum_{x\in(\mathbb{Z}/2)^d} W_\psi(x,v) = |\hat\psi(v)|^2$$

and

$$\int_{(\mathbb{Z}/2)^d}\mathrm{d}x\int\mathrm{d}v W_\psi(x,v) = \|\psi\|^2.$$

Often we will use the Wigner transform in the Fourier representation, by which we will always mean Fourier transform in the first (x) variable only, that is with the convention

$$\hat f(\xi) = \int e^{-ix\xi}f(x)\mathrm{d}x,$$

we define

$$\widehat{W}_\psi(\xi,v) := \int e^{-ix\xi}W_\psi(x,v)\mathrm{d}x.$$

After a simple calculation, we have

$$\widehat{W}_\psi(\xi,v) = \overline{\hat\psi\left(v-\frac{\xi}{2}\right)}\hat\psi\left(v+\frac{\xi}{2}\right). \tag{1.23}$$

In the discrete case we have $\xi \in R^d/(2\cdot 2\pi\mathbb{Z}^d)$ and

$$\widehat{W}(\xi,v) = \overline{\hat\psi\left(v-\frac{\xi}{2}\right)}\hat\psi\left(v+\frac{\xi}{2}\right).$$

More generally, if $J(x,v)$ is a classical phase space observable, the scalar product

$$\langle J, W_\psi\rangle = \int J(x,v)W_\psi(x,v)\mathrm{d}x\mathrm{d}v$$

can be interpreted as the expected value of J in state ψ. Recall that 'honest' quantum mechanical observables are self-adjoint operators $\mathcal{O}$ on $L^2(\mathbb{R}^d)$ and their expected value is given by

$$\langle \psi, \mathcal{O}\psi \rangle = \int \overline{\psi}(x)\,(\mathcal{O}\psi)(x)\mathrm{d}x.$$

For a large class of observables there is a natural relation between observables $\mathcal{O}$ and their phase space representations (called *symbols*) that are functions on the phase space like $J(x, v)$. For example, if J depends only on x or only on v, then the corresponding operator is just the standard quantization of J, that is

$$\int J(x)W_\psi(x, v)\mathrm{d}x\mathrm{d}v = \langle \psi, J\psi \rangle$$

where J is a multiplication operator on the right-hand side, or

$$\int J(v)W_\psi(x, v)\mathrm{d}x\mathrm{d}v = \langle \psi, J(-i\nabla)\psi \rangle,$$

and similar relations hold for the Weyl quantization of any symbol $J(x, v)$.

We also remark that the map $\psi \to W_\psi$ is invertible, that is one can fully reconstruct the wave function from its Wigner transform. On the other hand, not every real function of two variables (x, v) is the Wigner transform of some wavefunction.

1.4.2 Hamilton operator and the Schrödinger equation

The quantum dynamics is generated by the *Hamilton operator*

$$H = -\frac{1}{2}\Delta_x + U(x)$$

acting on $\psi \in L^2(\mathbb{R}^d)$. The first term is interpreted as the kinetic energy and it is the quantization of the classical kinetic energy $\frac{1}{2}v^2$ (compare with (1.15)). The momentum operator is $p = -i\nabla$ and we set the mass equal one, so momentum and velocity coincide. (To be pedantic, we should use the word momentum everywhere instead of velocity; since the momentum is the canonical quantity, then the dispersion relation, $e(p)$, defines the kinetic energy as a function of the momentum operator, and its gradient, $v = \nabla e(p)$ is the physical velocity. Since we will mostly use the dispersion relation $e(p) = \frac{1}{2}p^2$, the momentum and velocity coincide. However, for the lattice model the velocity and the momentum will differ.)

The evolution of a state is given by the Schrödinger equation

$$i\partial_t\psi_t = H\psi_t = \left(-\frac{1}{2}\Delta + U\right)\psi_t$$

with a given initial data $\psi_{t=0} = \psi_0$. Formally, the solution is

$$\psi_t = e^{-itH}\psi_0.$$

If H is self-adjoint in L^2, then the unitary group e^{-itH} can be defined by the spectral theorem. Almost all Hamilton operators in mathematical physics are self-adjoint, however the self-adjointness usually requires some proof. We will neglect this issue here, but we only mention that self-adjointness is more than the symmetry of

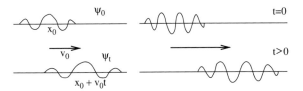

Fig. 1.4 Evolution of slower (left) and a faster (right) wave packet.

the operator, because H is typically unbounded when issues about the proper domain of the operator become relevant.

Note the complex i in the Schrödinger equation, it plays an absolutely crucial role. It is responsible for the wave-like character of quantum mechanics. The quantum evolution are governed by *phase* and *dispersion*. Instead of giving precise explanations, look at Fig. 1.4: the faster the wave oscillates, the faster it moves.

We can also justify this picture by some sketchy calculation (all these can be made rigorous in the so-called semiclassical regime). We demonstrate that the free ($U \equiv 0$) evolution of

$$\psi_0(x) := e^{iv_0 x} A(x - x_0)$$

(with some fixed x_0, v_0) after time t is supported around $x_0 + v_0 t$. Here we tacitly assume that A is not an oscillatory function (e.g. positive) and the only oscillation is given explicitly in the phase $e^{iv_0 x}$, mimicking a plane wave.

We compute the evolution of ψ_0:

$$\psi_t(x) = \int e^{iv(x-y)} e^{-itv^2/2} \underbrace{e^{iv_0 y} A(y - x_0)}_{=:\psi_0(y)} \, dy dv$$

$$= e^{iv_0 x_0} \int e^{iv(x-x_0)} e^{-itv^2/2} \widehat{A}(v - v_0) \, dv \sim \int e^{i\Phi(v)} \widehat{A}(v - v_0) \, dv$$

with a phase factor

$$\Phi(v) := v(x - x_0) - \frac{1}{2} tv^2.$$

We apply the stationary phase argument: the integral is concentrated around the stationary points of the phase. Thus

$$0 = \nabla_v \Phi = x - x_0 - tv$$

gives $x = x_0 + vt \approx x_0 + v_0 t$ if $\widehat{A}$ is sufficiently localized.

This argument can be made rigorous if a scale separation Ansatz is used. We introduce a small parameter ε and assume that

$$\psi_0(x) = \varepsilon^{d/2} e^{iv_0 x} A(\, \varepsilon(x - x_0)\,),$$

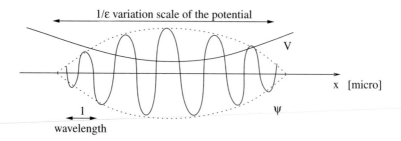

Fig. 1.5 Semiclassical setup: short wavelength, large scale envelope and potential.

that is the wave has a slowly varying envelope (amplitude) and fast phase. The prefactor is chosen to keep the normalization $\|\psi_0\| = 1$ independently of ε. Such states and their generalizations of the form

$$\varepsilon^{d/2} e^{iS(\varepsilon x)/\varepsilon} A(\ \varepsilon(x - x_0)\ ) \tag{1.24}$$

are called *WKB-states*. These states are characterized by a slowly varying amplitude and wavelength profile and by a fast oscillation. A similar picture holds if the potential is nonzero, but slowly varying, that is $U(x)$ is replaced with $U(\varepsilon x)$.

1.4.3   Semiclassics

The WKB states and the rescaled potential belong to the *semiclassical theory* of quantum mechanics. A quantum mechanical system is in the *semiclassical regime* whenever all the data (potential, magnetic field, initial profile etc.) live on a much bigger scale than the quantum wavelength.

The scale separation parameter is usually explicitly included in the Schrödinger equation, for example:

$$i\partial_t \psi_t(x) = \left[-\frac{1}{2}\Delta_x + U(\varepsilon x)\right]\psi_t(x). \tag{1.25}$$

This equation is written in microscopic coordinates. It becomes more familiar if we rewrite it in macroscopic coordinates

$$(X, T) = (x\varepsilon, t\varepsilon),$$

where (1.25) takes the form

$$i\varepsilon\partial_T \Psi_T(X) = \left[-\frac{\varepsilon^2}{2}\Delta_X + U(X)\right]\Psi_T(X).$$

Traditionally ε is denoted h in this equation and is referred to the (effective) Planck constant, whose smallness in standard units justifies taking the limit $h \to 0$ (the so-called *semiclassical limit*). No matter in what coordinates and units, the essential signature of the semiclassical regime is the scale separation between the wavelength and any other lengthscales.

The Wigner transform $W_\psi(x, v)$, written originally in microscopic coordinates, can also be rescaled:

$$W_\psi^\varepsilon(X, V) := \varepsilon^{-d} W_\psi\left(\frac{X}{\varepsilon}, V\right).$$

Note that apart from the rescaling of the X variable (V variable is unscaled!), a normalizing prefactor ε^{-d} is added to keep the integral of W^ε normalized (see (1.21))

$$\iint W_\psi^\varepsilon(X, V)\mathrm{d}X\mathrm{d}V = 1.$$

The following theorem is a simple prototype of the theorems we wish to derive in more complicated situations. It shows how the quantum mechanical Schrödinger evolution can be approximated by a classical equation in a certain limiting regime:

Theorem 1.10 *Consider a sequence of initial states ψ_0^ε whose rescaled Wigner transform has a weak limit:*

$$\lim_{\varepsilon \to 0} W_{\psi_0^\varepsilon}^\varepsilon(X, V) \rightharpoonup W_0(X, V).$$

For example, the WKB states given in (1.24) have a weak limit

$$W_0(X, V) = |A(X)|^2 \delta(V - \nabla S(X)).$$

Let ψ_t^ε denote the solution to the semiclassical Schrödinger equation (1.25) with initial state ψ_0^ε. Then the weak limit of the rescaled Wigner transform of the solution at time $t = T/\varepsilon$,

$$W_T(X, V) := \lim_{\varepsilon \to 0} W_{\psi_{T/\varepsilon}^\varepsilon}^\varepsilon(X, V)$$

satisfies the Liouville equation:

$$(\partial_T + V \cdot \nabla_X)W_T(X, V) = \nabla U(X) \cdot \nabla_V W_T(X, V). \tag{1.26}$$

Recall that the Liouville equation is equivalent to the classical Hamilton dynamics (see the derivation of (1.18)).

In this way, the Liouville equation (1.26) mimics the semiclassical quantum evolution on large scales (for a proof, see e.g. (Markowich and Ringhofer 1989)). Notice that the weak limit is taken; this means that the small scale structure of the Wigner transform $W_{\psi_t^\varepsilon}$ is lost. Before the weak limit, the Wigner transform of ψ_t^ε carries all information about ψ_t^ε, but the weak limit means testing it against functions that live on the macroscopic scale. This phenomenon will later be essential to understand why the irreversibility of the Boltzmann equation does not contradict the reversibility of Schrödinger dynamics.

1.5   Random Schrödinger operators

1.5.1   Quantum Lorentz gas or the Anderson model

Now we introduce our basic model, the quantum Lorentz gas. Physically, we wish to model electron transport in a disordered medium. We assume that electrons are non-interacting, thus we can focus on the evolution of a single electron. The disordered medium is modelled by a potential describing random impurities (other models are also possible, for example one could include random perturbations in the kinetic energy term). This can be viewed as the quantum analogue of the classical Lorentz gas, where a classical particle is moving among random obstacles.

Definition 1.11   *The model of a single particle described by the Schrödinger equation on $\mathbb{R}^d$ or on $\mathbb{Z}^d$,*

$$i\partial_t \psi_t(x) = H\psi_t(x), \qquad H = -\Delta_x + \lambda V_\omega(x),$$

is called the **quantum Lorentz gas***, if $V_\omega(x)$ is a random potential at location x.*

Notation: The subscript ω indicates that $V(x)$ is a random variable that also depends on an element ω in the probability space.

More generally, we can consider

$$H = H_0 + \lambda V, \tag{1.27}$$

where H_0 is a deterministic Hamiltonian, typically $H_0 = -\Delta$. We remark that a periodic background potential can be added, that is

$$H = -\Delta_x + U_{per}(x) + \lambda V_\omega(x)$$

can be investigated with similar methods. We recall that the operator with a periodic potential and no randomness,

$$H_0 = -\Delta_x + U_{per}(x),$$

can be more or less explicitly diagonalized by using the theory of Bloch waves. The transport properties are similar to those of the free Laplacian. This is called the *periodic quantum Lorentz gas* and it is the quantum analogue of the Sinai billiard. However, while the dynamics of the Sinai billiard has a very complicated structure, its quantum version is quite straightforward (the theory of Bloch waves is fairly simple).

We also remark that for the physical content of the model, it is practically unimportant whether we work on $\mathbb{R}^d$ or on its lattice approximation $\mathbb{Z}^d$. The reason is that we are investigating long time, large distance phenomenon; the short scale structure of the space does not matter. However, technically $\mathbb{Z}^d$ is much harder (a bit unusual, since typically $\mathbb{R}^d$ is harder as one has to deal with the ultraviolet regime). If one works on $\mathbb{Z}^d$, then the Laplace operator is interpreted as the discrete Laplace operator on $\mathbb{Z}^d$, that is

$$(\Delta f)(x) := 2d\, f(x) - \sum_{|e|=1} f(x+e). \tag{1.28}$$

In Fourier space this corresponds to the dispersion relation

$$e(p) = \sum_{j=1}^{d}(1 - \cos p^{(j)}) \tag{1.29}$$

on the torus $p \in [-\pi, \pi]^d$.

The random potential can be fairly arbitrary, the only important requirement is that it cannot have a long-distance correlation in space. For convenience, we will assume i.i.d. random potential for the lattice case with a standard normalization:

$$\{V(x) \ : \ x \in \mathbb{Z}^d\} \quad \text{i.i.d} \qquad \mathbf{E}\,V(x) = 0, \quad \mathbf{E}\,V^2(x) = 1.$$

The coupling constant λ can then be used to adjust the strength of the randomness. We can also write this random potential as

$$V(x) = \sum_{\alpha \in \mathbb{Z}^d} v_\alpha \delta(x - \alpha), \tag{1.30}$$

where $\{v_\alpha \ : \ \alpha \in \mathbb{Z}^d\}$ is a collection of i.i.d. random variables and δ is the usual lattice delta function.

For continuous models, the random potential can, for example, be given as follows;

$$V(x) = \sum_{\alpha \in \mathbb{Z}^d} v_\alpha B(x - \alpha),$$

where B is a nice (smooth, compactly supported) single site potential profile and $\{v_\alpha \ : \ \alpha \in \mathbb{Z}^d\}$ is a collection of i.i.d. random variables. It is also possible to let the randomness perturb the location of the obstacles instead of their strength, that is

$$V(x) = \sum_{\alpha \in \mathbb{Z}^d} B(x - y_\alpha(\omega)),$$

where, for example, $y_\alpha(\omega)$ is a random point in the unit cell around $\alpha \in \mathbb{Z}^d$, or even more canonically, the collection $\{y_\alpha(\omega)\}_\alpha$ is just a Poisson point process in $\mathbb{R}^d$. The combination of these two models is also possible and meaningful, actually in our work (Erdős, Salmhofer, and Yau 2007a) we consider

$$V(x) = \int B(x - y)\mathrm{d}\mu_\omega(y) \tag{1.31}$$

where μ_ω is a Poisson point process with homogeneous unit density and with i.i.d. coupling constants (weights), that is

$$\mu_\omega = \sum_\alpha v_\alpha(\omega)\delta_{y_\alpha(\omega)}, \tag{1.32}$$

where $\{y_\alpha(\omega) : \alpha = 1, 2, \ldots\}$ is a realization of the Poission point process and $\{v_\alpha\}$ are independent (from each other and from y_α as well) real valued random variables.

The lattice model $-\Delta + \lambda V$ with i.i.d. random potentials (1.30) is called the *Anderson model*. It was invented by Anderson (Anderson 1958) who was the first to realize that electrons move quite differently in disordered media than in free space or in a periodic background. The main phenomenon Anderson discovered was the *Anderson localization*, asserting that at sufficiently strong disorder (or in $d = 1, 2$ at any nonzero disorder) the electron transport stops. We will explain this in more detail later, here we just remark that Anderson was awarded the Nobel Prize in 1977 for this work.

However, in $d \geq 3$ dimensions electron transport is expected despite the disorder if the disorder is weak. This is actually what we experience in real life; the electric wire is conducting although it does not have a perfect periodic lattice structure. However, the nature of the electric transport changes: we expect that on large scales it can be described by a diffusive equation (like the heat equation) instead of a wave-type equation (like the Schrödinger equation).

Note that the free Schrödinger equation is ballistic (due to wave coherence). This means that if we define the *mean square displacement* by the expectation value of the observable x^2 at state ψ_t,

$$\langle x^2 \rangle_t := \int \mathrm{d}x |\psi_t(x)|^2 x^2 \quad \left(= \int \mathrm{d}x \big| e^{-it\Delta} \psi_0(x) \big|^2 x^2 \right), \tag{1.33}$$

then it behaves as

$$\langle x^2 \rangle_t \sim t^2$$

for large $t \gg 1$ as it can be easily computed. In contrast, the mean square displacement for the heat equation scales as t and not as t^2 (see equation (1.7)).

Thus the long time transport of the free Schrödinger equation is very different from the heat equation. We nevertheless claim that, in a weakly disordered medium, the long time Schrödinger evolution can be described by a diffusion (heat) equation. Our results answer the intriguing question of how the diffusive character emerges from a wave-type equation.

1.5.2   Known results about the Anderson model

We will not give a full account of all known mathematical results on the Anderson model, we will just mention the main points. The main issue is to study the dichotomic nature of the Anderson model, namely that at low dimension ($d = 1, 2$) or at high disorder ($\lambda \geq \lambda_0(d)$) or near the spectral edges of the unperturbed operator H_0 the time evolved state remains localized, while at high dimension ($d \geq 3$), at low disorder and away from the spectral edges it is delocalized.

There are several signatures of (de)localization and each of them can be used for rigorous definition. We list three approaches:

i) *Spectral approach.* If H has pure point (PP) spectrum then the system is in the localized regime, if H has absolutely continuous (AC) spectrum then it is in the delocalized regime. (The singular continuous spectrum, if appears at all, gives rise to anomalous diffusions.) It is to be noted that even in the pure point regime the spectrum is dense.

ii) *Dynamical approach.* One considers the mean square displacement (1.33) and the system is in the localized regime if

$$\sup_{t \geq 0} \langle x^2 \rangle_t < \infty$$

(other observables can also be considered).

iii) *Conductor or Insulator?* This is the approach closest to physics and mathematically it has not been sufficiently elaborated (for a physical review, see (Lee and Ramakrishnan 1985)). In this approach one imposes an external voltage to the system and computes the currect and the ohmic resistance.

These criteria to distinguish between localized and delocalized regimes are not fully equivalent (especially in $d = 1$ dimensional systems there are many anomalous phenomena due to the singular continuous spectrum), but we do not go into more detail.

The mathematically most studied approach is the spectral method. The main results are the following:

i) In $d = 1$ dimensions all eigenfunctions are localized for all $\lambda \neq 0$ (Goldsheid, Molchanov and Pastur 1997).

ii) In $d \geq 1$ localization occurs for large λ or for energies near the edges of the spectrum of H_0 (see (1.27)). This was first proven by the groundbreaking work of Fröhlich and Spencer (Fröhlich and Spencer 1983) on the exponential decay of the resolvent via the *multiscale method* (the absence of AC spectrum was then proved by Fröhlich, Martinelli, Spencer and Scoppola (Fröhlich, Martinelli, Scoppola, and Spencer 1985) and the exponential localization by Simon and Wolff (Simon and Wolff 1986)). Later a different method was found by Aizenman and Molchanov (Aizenman and Molchanov 1993) (*fractional power method*). The spectrum is not only pure point, but the eigenfunctions decay exponentially (*strong localization*).

iii) Many more localization results were obtained by variants of these methods for various models, including magnetic field, acoustic waves, localized states along sample edges and so on. We will not attempt to give a list of references here.

Common in all localization results is that the random potential dominates, that is in one way or another the free evolution H_0 is treated as a perturbation.

It is important to understand that the transport in the free Schrödinger equation is due to the coherence of the travelling wave. Random potential destroys this coherence; the stronger the randomness is, the surer is the destruction. This picture is essential to understand why random potentials may lead to localization even at energies that belong to the regime of classical transport. For example in $d = 1$ any small randomness stops transport, although the classical dynamics is still ballistic at energies that are higher than the maximal peak of the potential. In other words, Anderson localization is a truly quantum phenomenon, it cannot be explained by a heuristics based upon classically trapped particles.

1.5.3 Major open conjectures about the Anderson model

All the previous results were for the localization regime, where one of the two main methods (multiscale analysis or fractional power method) was applicable. The complementary regimes remain unproven. Most notable is the following list of open questions and conjectures:

i) **Extended states conjecture** For small $\lambda \leq \lambda_0(d)$ and in dimension $d \geq 3$ the spectrum is absolutely continuous away from the edges of H_0. In particular, there exists a threshold value (called *mobility edge*) near the edges of the unperturbed spectrum that separates the two spectral types.
   This conjecture has been proven only in the following three cases:

   a) Bethe lattice (infinite binary tree) that roughly corresponds to $d = \infty$ (Klein (Klein 1994), recently different proofs were obtained in (Aizenman, Sims, and Warzel 2006) and (Froese, Hasler, and Spitzer 2007)).
   b) Sufficiently decaying random potential, $\mathbf{E}\, V_x^2 = |x|^{-\alpha}$, for $\alpha > 1$ (Bourgain (Bourgain 2003)). The decay is sufficiently strong such that the typical number of collisions with obstacles is finite. Note that the randomness is not stationary.
   c) In $d = 2$ with a constant magnetic field the spectrum cannot be everywhere pure point (Germinet, Klein and Schenker (Germinet, Klein, and Schenker 2007)).

ii) **Two dimensional localization** In $d = 2$ dimensions all eigenfunctions are localized for all λ. (that is the model in $d = 2$ dimensions behaves as in $d = 1$).

iii) **Quantum Brownian motion conjecture** For small λ and $d \geq 3$, the location of the electron is governed by a heat equation in a vague sense:

$$\partial_t |\psi_t(x)|^2 \sim \Delta_x |\psi_t(x)|^2 \quad \Longrightarrow \quad \langle\, x^2\,\rangle_t \sim t \qquad t \gg 1. \tag{1.34}$$

The precise formulation of the first statement requires a scaling limit. The second statement about the diffusive mean square displacement is mathematically precise, but what really stands behind it is a diffusive equation that on a large scales mimics the Schrödinger evolution. Moreover, the dynamics of the quantum particle converges to the Brownian motion as a process as well; this means that the joint distribution of the quantum densities $|\psi_t(x)|^2$ at different times $t_1 < t_2 < \ldots < t_n$ converges to the corresponding finite dimensional marginals of the Wiener process.

   Note that the 'quantum Brownian motion conjecture' is much stronger than the 'Extended states conjecture', since the former more precisely describes how the states extend. All these three open conjectures have been outstanding for many years and we seem to be far from their complete solutions.

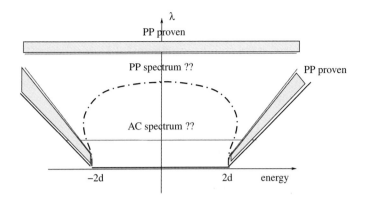

Expected spectrum at small, nonzero disorder $0 < \lambda \ll 1$

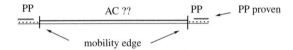

Fig. 1.6 Phase diagram of the Anderson model in $d = 3$.

Fig. 1.6 depicts the expected phase diagram of the Anderson model in dimensions $d \geq 3$. The picture shows the different spectral types (on the horizontal energy axis) at various disorder. The grey regimes indicate what has actually been proven.

Our main result, explained in the next sections, is that the 'quantum Brownian motion conjecture' in the sense of (1.34) holds *in the scaling limit* up to times $t \sim \lambda^{-2-\kappa}$. More precisely, we will fix a positive number κ and we will consider the family of random Hamiltonians

$$H = H_\lambda = -\frac{1}{2}\Delta + \lambda V$$

parametrized by λ. We consider their long time evolution up to times $t \sim \lambda^{-2-\kappa}$ and we take $\lambda \to 0$ limit. We show that, after appropriately rescaling the space and time, the heat equation emerges. Note that the time scale depends on the coupling parameter. This result is of course far from either of the conjectures i) and iii) above, since those conjectures require fixing λ (maybe small but fixed) and letting $t \to \infty$.

We emphasize that the typical number of collisions is $\lambda^2 t$. The reason is that the quantum rate of collision with a potential that scales as $O(\lambda)$ is of order λ^2. This follows from simple scattering theory: if

$$\widetilde{H} = -\Delta + \lambda V_0$$

is a Hamiltonian with a single bump potential (that is V_0 is smooth, compactly supported) and ψ_{in} denotes the incoming wave, then after scattering the wave splits into two parts;

$$e^{-it\widetilde{H}}\psi_{in} = \beta e^{it\Delta}\psi_{in} + \psi_{sc}(t), \qquad t \gg 1. \tag{1.35}$$

Here $\psi_{sc}(t)$ is the scattered wave (typically spherical if V_0 is spherical) while the first term describes the transmitted wave that did not notice the obstacle (apart from its amplitude being scaled down by a factor β). Elementary calculation then shows that the scattered and transmitted waves are (almost) orthogonal and their amplitudes satisfy

$$\|\psi_{sc}(t)\|^2 = O(\lambda^2), \qquad \beta^2 = 1 - O(\lambda^2). \tag{1.36}$$

Therefore the incoming wave scatters with a probability $O(\lambda^2)$.

Thus up to time t, the particle encounters $\lambda^2 t$ collisions. In our scaling

$$n := \text{Number of collisions} \sim \lambda^2 t \sim \lambda^{-\kappa} \to \infty.$$

The asymptotically infinite number of collisions is necessary to detect the diffusion (Brownian motion, heat equation), similarly as the Brownian motion arises from an increasing number of independent 'kicks' (steps in random walk, see Theorem 1.7).

1.6   Main result

In this section we formulate our main result precisely, that is what we mean by the 'Quantum Brownian motion conjecture' holds in the scaling limit up to times $t \sim \lambda^{-2-\kappa}$. Before going into the precise formulation or the sketch of any proof, we will explain why it is a difficult problem.

1.6.1   Why is this problem difficult?

The dynamics of a single quantum particle among random scatterers (obstacles) is a multiple scattering process. The quantum wave function bumps into an obstacle along its evolution, and according to standard scattering theory, it decomposes into two parts: a wave that goes through the obstacle without noticing it and a scattering wave (1.35). The scattered wave later bumps into another obstacle etc, giving rise to a complicated multiple scattering picture, similar to the one on Fig. 1.7. Obviously, the picture becomes more and more complicated as the number of collisions, n, increases. Finally, when we perform a measurement, we select a domain in space and we compute the wave function at the point. By the superposition principle of quantum mechanics, at that point *all* elementary scattering waves have to be added up; and eventually there are exponentially many of them (in the parameter λ^{-1}). They are quantum waves, that is complex valued functions, so they must be added together with their phases. The cancellations in this sum due to the phases are essential even to get the right order of magnitude. Thus one way or another one must trace all elementary scattering waves with an enormous precision (the wavelength is order one in microscopic units but the waves travel at distance λ^{-2} between two collisions!).

It is important to emphasize that this system is **not semiclassical**. Despite the appearance of a small parameter, what matters is that the typical wavelength remains comparable with the spatial variation lengthscale of the potential. Thus the process remains quantum mechanical and cannot be described by any semiclassical method.

1.6.2   Scales

Since the collision rate is $O(\lambda^2)$, and the typical velocity of the particle is unscaled (order 1), the typical distance between two collisions (the so-called *mean free path*) is $L = O(\lambda^{-2})$ and the time elapsed between consecutive collisions is also $O(\lambda^{-2})$.

Kinetic scale

In order to see some nontrivial effects of the collisions, we have to run the dynamics at least up to times of order λ^{-2}. Within that time, the typical distance covered is also of order λ^{-2} (always understood in microscopic coordinates). Thus the simplest scale that yields a nontrivial result is the so-called **kinetic scaling**,

$$t = \lambda^{-2}T, \qquad x = \lambda^{-2}X, \qquad n = \lambda^2 t = O(1),$$

that is the space and time are scaled by λ^2 and the typical number of collisions $\lambda^2 t$ is of order 1.

One convenient way to quantify Fig. 1.7 is that one works in a large quadratic box whose size is at least comparable with the mean free path (otherwise the boundary would have a too big an effect). Thus one can consider the box

$$\Lambda = [0, L]^d$$

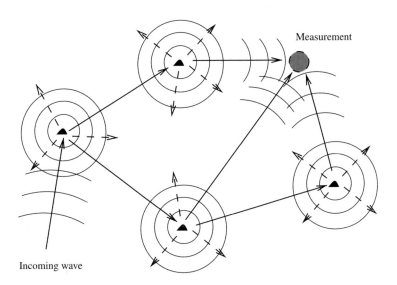

Fig. 1.7   Schematic picture of multiple scattering.

and put $|\Lambda| = L^d$ obstacles in it, for example one at every lattice site. For definiteness, we work with the Anderson model, that is

$$H = -\Delta_x + \lambda \sum_{\alpha \in \mathbb{Z}^d} v_\alpha \delta(x - \alpha).$$

Each elementary scattering wave corresponds to a particular (ordered) sequence of random obstacles. Since the total number of obstacles is L^d, after n collisions we have $\sim |\Lambda|^n$ elementary waves to sum up in Fig. 1.7, thus

$$\psi_t = \sum_A \psi_A, \qquad (1.37)$$

where $A = (\alpha_1, \alpha_2, \ldots, \alpha_n)$ is a sequence of obstacle locations ($\alpha_i \in \mathbb{Z}^d$) and ψ_A describes the elementary scattering wave that has undergone the collisions with obstacles at $\alpha_1, \alpha_2, \ldots \alpha_n$ in this order. The precise definition of ψ_A will come later.

These waves must be summed up together with their phase. It is easy to see from standard scattering theory that a spherical wave decays as $\lambda|\text{distance}|^{-(d-1)/2}$ (the prefactor λ comes from the fact that the total amplitude of the scattering is scaled down by a factor of λ due to the coupling constant in front of the potential, see (1.36)). Since the amplitudes multiply, and since the typical mean free path is λ^{-2}, after n collisions the typical amplitude of ψ_A is

$$|\psi_A| \sim \left[\lambda \left(\lambda^2 \right)^{\frac{d-1}{2}} \right]^n = \lambda^{dn}.$$

Thus if we try to sum up these waves in (1.37) neglecting their phases, then

$$\sum_A |\psi_A| \sim |\Lambda|^n \lambda^{dn} = \left(\lambda^{-2d} \right)^n \lambda^{dn} = \lambda^{-dn} \to \infty.$$

However, if we can sum up these waves *assuming* that their phases are independent, that is they could be summed up as independent random variables, then it is the variance that is additive (exactly as in the central limit theorem in (1.4)):

$$|\psi_t|^2 = \left| \sum_A \psi_A \right|^2 \approx \sum_A |\psi_A|^2 \sim |\Lambda|^n \lambda^{2dn} = O(1). \qquad (1.38)$$

Thus it is essential to extract a strong independence property from the phases. Since phases are determined by the random obstacles, there is some hope that at least most of the elementary waves are roughly independent. This will be our main technical achievement, although it will be formulated in a different language.

We remark that in the physics literature, the independence of phases is often postulated as an Ansatz under the name of 'random phase approximation'. Our result will mathematically justify this procedure.

Diffusive scale

Now we wish to go beyond the kinetic scale and describe a system with potentially infinitely many collisions. Thus we choose a longer time scale and we rescale the space appropriately. We obtain the following **diffusive scaling**:

$$t = \lambda^{-\kappa}\lambda^{-2}T, \qquad x = \lambda^{-\kappa/2}\lambda^{-2}T, \qquad n = \lambda^2 t = \lambda^{-\kappa}.$$

Notice that the time is rescaled by an additional factor $\lambda^{-\kappa}$ compared with the kinetic scaling, and space is rescaled by the square root of this additional factor. This represents the idea that the model scales diffusively with respect to the units of the kinetic scale. The total number of collisions is still $n = \lambda^2 t$, and now it tends to infinity.

If we try to translate this scaling into the elementary wave picture, then first we have to choose a box that is larger than the largest space scale, that is $L \geq \lambda^{-2-\kappa/2}$. The total number of elementary waves to sum up is

$$|\Lambda|^n = (\lambda^{-2-\frac{\kappa}{2}})^{dn} \sim \lambda^{-\lambda^{-\kappa}}$$

that is superexponentially large. Even with the assumption that only the variances have to be summable (see (1.38)), we still get a superexponentially divergent sum:

$$|\psi_t|^2 \approx \sum_A |\psi_A|^2 \sim |\Lambda|^n \lambda^{2nd} = \left[\lambda^{-2-\kappa/2}\right]^{nd} \lambda^{2nd} = \lambda^{-nd\kappa/2} = \lambda^{-(const)\lambda^{-\kappa}}$$

We will have to perform a **renormalization** to prevent this blow-up.

1.6.3 Kinetic scale: (linear) Boltzmann equation

The main result on the kinetic scale is the following theorem:

Theorem 1.12 Boltzmann equation in the kinetic scaling limit *Let the dimension be at least $d \geq 2$. Consider the random Schrödinger evolution on $\mathbb{R}^d$ or $\mathbb{Z}^d$*

$$i\partial_t \psi_t = H\psi_t, \qquad H = H_\lambda = -\frac{1}{2}\Delta + \lambda V(x),$$

where the potential is spatially uncorrelated (see more precise conditions below). Consider the kinetic rescaling

$$t = \lambda^{-2}\mathcal{T}, \qquad x = \lambda^{-2}\mathcal{X},$$

with setting $\varepsilon := \lambda^2$ to be the scale separation parameter in space.

Then the weak limit of the expectation of the rescaled Wigner transform of the solution ψ_t exists,

$$\boldsymbol{E}W^\varepsilon_{\psi_{\mathcal{T}/\varepsilon}}(\mathcal{X},\mathcal{V}) \rightharpoonup F_{\mathcal{T}}(\mathcal{X},\mathcal{V})$$

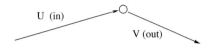

U (in)

V (out)

Fig. 1.8 Incoming and outgoing momentum in a collision.

and F_T satisfies the linear Boltzmann equation,

$$\left(\partial_T + \nabla e(V) \cdot \nabla_X\right) F_T(X, V) = \int dU \sigma(U, V) \left[F_T(X, U) - F_T(X, V)\right].$$

Here $e(V)$ is the dispersion relation of the free kinetic energy operator given by $e(V) = \frac{1}{2}V^2$ for the continuous model and by (1.29) for the discrete model. The collision kernel $\sigma(U, V)$ is explicitly computable and is model dependent (see below).

The velocities U, V are interpreted as incoming and outgoing velocities, respectively, see Fig. 1.8. The collision kernel always contains an *onshell condition*, that is a term $\delta(e(U) - e(V))$ guaranteeing energy conservation (see the examples below). Thus the right-hand side of the linear Boltzmann equation is exactly the generator of the jump process on a fixed energy shell (in case of $e(V) = \frac{1}{2}V^2$ it is on the sphere) as defined in Section 1.2.4.

This theorem has been proven in various related models.

- **Continuous model on $\mathbb{R}^d$** The random potential is chosen to be a homogeneous Gaussian field, this means that the random variable $V(x)$ is Gaussian and it is stationary with respect to the space translations. We assume that $\mathbf{E}\, V(x) = 0$, then the distribution of $V(x)$ is uniquely determined by the two-point correlation function

$$R(x - y) := \mathbf{E}V(x)V(y),$$

  and we assume that R is a Schwartz function. The dispersion relation of the free Laplacian on $\mathbb{R}^d$ is

$$e(p) := \frac{1}{2}p^2,$$

  and the Boltzmann collision kernel is computed from the microscopic data as

$$\sigma(U, V) = \delta(e(U) - e(V))|\widehat{R}(U - V)|. \tag{1.39}$$

  This result was first obtained by Spohn (Spohn 1977) for short macroscopic time, $T \leq T_0$, and later extended to arbitrary times in a joint work with Yau (Erdős and Yau 2000) with a different method. We assume that the dimension $d \geq 2$, with a special nondegeneracy condition on the initial data in case of $d = 2$ forbidding concentration of mass at zero momentum. The $d = 1$ case is special; due to strong interference effects no Markovian limiting dynamics is expected.
- **Discrete model + Non Gaussian** The method of (Erdős and Yau 2000) was extended by Chen in two different directions (Chen 2005). He considered the discrete model, that is the original Anderson model and he allowed non-Gaussian randomness

as well. In the case of Laplacian on $\mathbb{Z}^d$ the dispersion relation is more complicated (1.29), but the collision kernel is simpler;

$$\sigma(U,V) = \delta(e(U) - e(V)), \tag{1.40}$$

that is it is simply the uniform measure on the energy surface given by the energy of the incoming velocity.

- **Phonon model** In this model the random potential is replaced by a heat bath of non-interacting bosonic particles that interact only with the single electron. *Formally*, this model leads to an electron in a time dependent random potential, but the system is still Hamiltonian. This model is the quantum analogue of Einstein's picture. The precise formulation of the result is a bit more complicated, since the phonon bath has to be mathematically introduced, so we will not do it here. The interested reader can consult with (Erdős 2002). More recently, De Roeck and Fröhlich (De Roeck and Fröhlich 2010) have proved diffusion for any times in $d \geq 4$ dimensions with an additional strong spin coupling that enhances the loss of memory.

- **Cubic Schrödinger equation with random initial data** A nonlinear version of the Boltzmann equation was proven in (Lukkarinen and Spohn 2011) where the Schrödinger equation contained a weak cubic nonlinearity and the initial data was drawn from near thermal equilibrium.

- **Wave propagation in random medium** It is well known that in a system of harmonically coupled oscillators the wave propagates ballistically. If the masses of the oscillators deviate randomly from a constant value, the wave propagation can change. The effect is similar to the disturbances in the free Schrödinger propagation; the ballistic transport is a coherence effect that can be destroyed by small perturbations. This model, following the technique of (Erdős and Yau 2000), has been worked out in (Lukkarinen and Spohn 2007).

- **Low density models** There is another way to reduce the effects of possible recollisions apart from introducing a small coupling constant: one can consider a low density of obstacles. Let V_0 be a radially symmetric Schwartz function with a sufficiently small weighted Sobolev norm:

$$\|\langle x \rangle^{N(d)} \langle \nabla \rangle^{N(d)} V_0\|_\infty \text{ is sufficiently small for some } N(d) \text{ sufficiently large.}$$

The Hamiltonian of the model is given by

$$H = -\frac{1}{2}\Delta_x + \sum_{\alpha=1}^{M} V_0(x - x_\alpha) \qquad \text{in a box } [-L, L]^d, \quad L \gg \varepsilon^{-1}$$

acting on $\mathbb{R}^d$, $d \geq 3$, where $\{x_\alpha\}_{\alpha=1,\dots M}$ denotes a collection of random i.i.d. points with density

$$\varrho := \frac{M}{L^d} \to 0.$$

The kinetic scaling corresponds to choosing

$$x \sim \varepsilon^{-1}, \quad t \sim \varepsilon^{-1}, \qquad \varepsilon = \varrho,$$

and then letting $\varepsilon \to 0$. In this case convergence to the linear Boltzmann equation in the spirit of Theorem 1.12 was proven in (Erdős and Yau 1998), (Eng and Erdős 2005) (although these papers consider the $d = 3$ case, the extension to higher dimensions is straightforward). The dispersion relation is $e(p) = \frac{1}{2}p^2$ and the collision kernel is

$$\sigma(U, V) = |T(U, V)|^2 \delta(U^2 - V^2)$$

where $T(U, V)$ is the quantum scattering cross section for $-\frac{1}{2}\Delta + V_0$.

It is amusing the compare the weak coupling and the low density scenarios; they indeed describe different physical models (Fig. 1.9). Although in both cases the result is the linear Boltzmann equation, the microscopic properties of the model influence the collision kernel. In particular, in the low density model the full single-bump scattering information is needed (while the function $\hat{R}$ (1.39) in the weak coupling model can be viewed as the Born approximation of the full scattering cross section). The scaling is chosen such that the typical number of collisions remains finite in the scaling limit. Note that in both models the wavelength is comparable with the spatial variation of the potential; both live on the microscopic scale. Thus neither model is semiclassical.

We make one more interesting remark. When compared with the corresponding classical Hamiltonian, the low density model yields the (linear) Boltzmann equation both in classical and quantum mechanics (although the collision kernels are different). The weak coupling model yields the Boltzmann equation when starting from quantum mechanics, however it yields a random walk on the energy shell (for the limiting velocity process v_t) when starting from classical mechanics (Kesten, and Papanicolaou

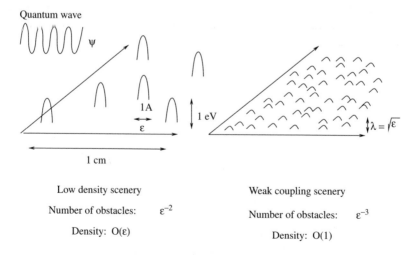

Fig. 1.9 Low density and weak coupling scenarios.

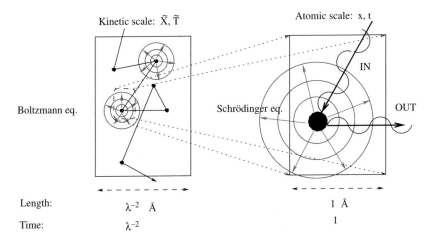

Fig. 1.10 Macroscopic and microscopic scales.

1980/81) and (Komorowski and Ryzhik 2006). For further references, see (Erdős and Yau 2000).

Figure 1.10 schematically depicts the microscopic and macroscopic scales in Theorem 1.12. The right side of the picture is under a microscope which 'sees' scales comparable with the wavelength (Angstrom scale). Only one obstacle and only one outgoing wave are pictured. In reality the outgoing wave goes in all directions. On this scale the Schrödinger equation holds:

$$i\partial_t \psi_t(x) = \left[-\Delta_x + \lambda V(x)\right]\psi_t(x).$$

On the left side we zoomed our 'microscope' out. Only a few obstacles are pictured that are touched by a single elementary wave function. All other trajectories are also possible. The final claim is that on this scale the dynamics can be described by the Boltzmann equation

$$\left(\partial_{\mathcal{T}} + \nabla e(V) \cdot \nabla_{\mathcal{X}}\right) F_{\mathcal{T}}(\mathcal{X}, V) = \int \mathrm{d}U \sigma(U, V)\left[F_{\mathcal{T}}(\mathcal{X}, U) - F_{\mathcal{T}}(\mathcal{X}, V)\right].$$

Actually the obstacles (black dots) on the left picture are only fictitious: recall that there are no physical obstacles behind the Boltzmann equation. It represents a process, where physical obstacles are replaced by a stochastic dynamics: the particle has an exponential clock and it randomly decides to change its trajectory that is to behave as if there were an obstacle (recall Section 1.2.4). It is crucial to make this small distinction because it is exactly the fictitious obstacles that are the signatures that all recollisions have been eliminated, hence this guarantees the Markov property for the limit dynamics.

The main message of the result is that the long time ($t = T\varepsilon^{-1}$) Schrödinger evolution can modelled by a finite time (T) Boltzmann evolution on the macroscopic scale. Of course the detailed short scale information is lost, which explains irreversibility.

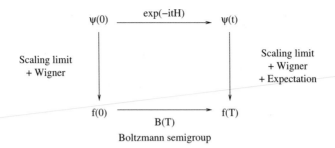

Fig. 1.11 Boltzmann semigroup models, Schrödinger evolution.

The effective limiting equation is classical, but quantum features are retained in the collision kernel.

Note that this approximation is not just conceptually interesting but it also gives rise to an enormous computational simplification. Solving the Schrödinger equation directly for a long time in an complicated environment is a computationally impossible task. In contrast, the Boltzmann equation needs to be solved on $O(1)$ times and there is no complicated environment. This is expressed schematically in the 'commutative diagram' of Fig. 1.11. The top horizontal line is the true (Schrödinger) evolution. The theorem says that one can 'translate' the quantum initial data by Wigner transform and scaling limit into a phase space density $f(0) = f_0(X, V)$ and one can use the much simpler (and shorter time) Boltzmann evolution to f_0 to obtain the Wigner transform (after expectation and scaling limit) of the true evolution ψ_t.

1.6.4   Diffusive scale: heat equation

We consider the same model with a weakly coupled random potential as before. For definiteness, we work on the lattice $\mathbb{Z}^d$, $d \geq 3$, the Hamiltonian is given by

$$H = -\Delta + \lambda V,$$

where the discrete Laplacian is given by (1.28) with dispersion relation (1.29) and where the random potential is

$$V(x) = \sum_{\alpha \in \mathbb{Z}^d} v_\alpha \delta(x - \alpha).$$

The random variables v_α are i.i.d. with moments $m_k = \mathbf{E}\, v_\alpha^k$ satisfying

$$m_1 = m_3 = m_5 = 0, \quad m_{2d} < \infty.$$

Let

$$\psi_t := e^{-itH}\psi_0,$$

and we rescale the Wigner transform as before:

$$W_{\psi_t} \to W_{\psi_t}^\varepsilon = \varepsilon^{-d} W_{\psi_t}(X/\varepsilon, v).$$

We remark that in case of the discrete lattice $\mathbb{Z}^d$ the definition of the Wigner function $W_\psi(x,v)$ is given by (1.22). Our main result is the following

Theorem 1.13 (Quantum diffusion in the discrete case) *For any dimension $d \geq 3$ there exists $\kappa_0(d) > 0$ (in $d = 3$ one can choose $\kappa_0(3) = \frac{1}{10000}$), such that for any $\kappa \leq \kappa_0(d)$ and any $\psi_0 \in L^2(\mathbb{Z}^d)$ the following holds. In the diffusive scaling*

$$t = \lambda^{-\kappa}\lambda^{-2}T, \qquad x = \lambda^{-\kappa/2}\lambda^{-2}X, \qquad \varepsilon = \lambda^{-\kappa/2-2},$$

we have that

$$\int_{\{e(v)=e\}} \boldsymbol{E}\, W_{\psi_t}^\varepsilon(X,v)\,\mathrm{d}v \rightharpoonup f_T(X,e) \qquad \text{weakly as } \lambda \to 0, \tag{1.41}$$

and for almost all $e > 0$ the limiting function satisfies the heat equation

$$\partial_T f_T(X,e) = \nabla_X \cdot D(e)\nabla_X f_T(X,e) \tag{1.42}$$

with a diffusion matrix given by

$$D_{ij}(e) = \Big\langle \nabla e(v) \otimes \nabla e(v) \Big\rangle_e, \qquad \langle f(v)\rangle_e = \text{Average of } f \text{ on } \{v : e(v) = e\} \tag{1.43}$$

and with initial state

$$f_0(X,e) := \delta(X)\int \delta(e(v) - e)|\widehat{\psi_0}(v)|^2\mathrm{d}v.$$

The weak convergence in (1.46) means that for any Schwartz function $J(x,v)$ on $\mathbb{R}^d \times \mathbb{R}^d$ we have

$$\lim_{\varepsilon \to 0} \int_{\left(\frac{\varepsilon}{2}\mathbb{Z}\right)^d} \mathrm{d}X \int \mathrm{d}v J(X,v)\boldsymbol{E}\, W_{\psi(\lambda^{-\kappa-2}T)}^\varepsilon(X,v) = \int_{\mathbb{R}^d} \mathrm{d}X \int \mathrm{d}v J(X,v) f_T(X,e(v))$$

uniformly in T on $[0,T_0]$, where T_0 is an arbitrary fixed number.

This result is proven in (Erdős, Salmhofer, and Yau 2007a). The related continuous model is treated in (Erdős, Salmhofer, and Yau 2007b, 2008). A concise summary of the ideas can be found in the expository paper (Erdős, Salmhofer, and Yau 2006). Note again that weak limit means that only macroscopic observables can be controlled. Moreover, the theorem does not keep track of the velocity any more, since on the largest diffusive scale the velocity of the particle is not well defined (similarly as the Brownian motion has no derivative). Thus we have to integrate out the velocity variable on a fixed energy shell (the energy remains macroscopic observable). This justifies the additional velocity integration in (1.41).

The diffusion matrix is the velocity autocorrelation matrix (computed in equilibrium) obtained from the Boltzmann equation

$$D_{Boltz}(e) := \int_0^\infty \mathcal{E}_e\left[\nabla e(v(0)) \otimes \nabla e(v(t))\right]\mathrm{d}t, \tag{1.44}$$

similarly to Theorem 1.3. Here $\mathcal{E}_e$ denotes the expectation value of the Markov process $v(t)$ described by the linear Boltzmann equation as its generator if the initial velocity $v(0)$ is distributed according to the equilibrium measure of the jump process with generator (1.13) on the energy surface $\{e(v) = e\}$ with the Boltzmann collision kernel (1.40). Since this kernel is the uniform measure on the energy surface, the Boltzmann velocity process has no memory and thus it is possible to compute the velocity autocorrelation just by averaging with respect to the uniform measure on $\{e(v) = e\}$:

$$D_{Boltz}(e) = \int_0^\infty \mathcal{E}_e \left[\nabla e(v(0)) \otimes \nabla e(v(t))\right] dt = \left\langle \nabla e(v) \otimes \nabla e(v) \right\rangle_e. \qquad (1.45)$$

In particular, Theorem 1.13 states that the diffusion matrix $D(e)$ obtained from the quantum evolution coincides with the velocity autocorrelation matrix of the linear Boltzmann equation.

For completeness, we state the result also for the continuum model.

Theorem 1.14 (Quantum diffusion in the continuum case) *Let the dimension be at least $d \geq 3$ and $\psi_0 \in L^2(\mathbb{R}^d)$. Consider the diffusive scaling*

$$t = \lambda^{-\kappa} \lambda^{-2} T, \qquad x = \lambda^{-\kappa/2} \lambda^{-2} X, \qquad \varepsilon = \lambda^{-\kappa/2-2}.$$

Let

$$\psi_t = e^{itH} \psi_0, \qquad H = -\frac{1}{2}\Delta + \lambda V,$$

where the random potential is given by (1.31) with a single site profile B that is a spherically symmetric Schwartz function with 0 in the support of its Fourier transform, $0 \in \text{supp } \hat{B}$. For $d \geq 3$ there exists $\kappa_0(d) > 0$ (in $d = 3$ one can choose $\kappa_0(3) = 1/370$) such that for any $\kappa < \kappa_0(d)$ we have

$$\int_{\{e(v)=e\}} \mathbf{E}\, W_{\psi_t}^\varepsilon(X, v)\, dv \rightharpoonup f_T(X, e) \qquad (weakly\ as\ \lambda \to 0), \qquad (1.46)$$

and for almost all energies $e > 0$ the limiting function satisfies the heat equation

$$\partial_T f_T(X, e) = D_e \Delta_X f_T(X, e)$$

with initial state

$$f_0(X, e) := \delta(X) \int \delta(e(v) - e) |\hat{\psi}_0(v)|^2 dv.$$

The diffusion coefficient is given by

$$D_e := \frac{1}{d} \int_0^\infty \mathcal{E}_e \left[v(0) \cdot v(t)\right] dt, \qquad (1.47)$$

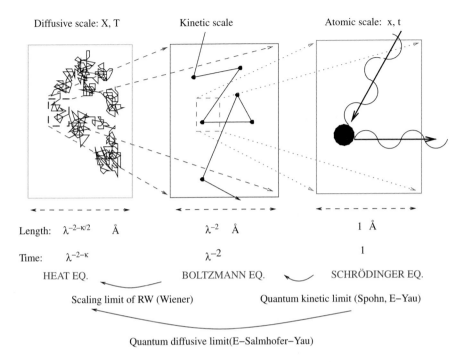

Diffusive scale: X, T   Kinetic scale   Atomic scale: x, t

Length: $\lambda^{-2-\kappa/2}$ Å λ^{-2} Å   1 Å

Time: $\lambda^{-2-\kappa}$ λ^{-2} 1

HEAT EQ.   BOLTZMANN EQ.   SCHRÖDINGER EQ.

Scaling limit of RW (Wiener)   Quantum kinetic limit (Spohn, E–Yau)

Quantum diffusive limit(E–Salmhofer–Yau)

Fig. 1.12 Three scales: diffusive, kinetic, and atomic.

where $\mathcal{E}_e$ is the expectation value for the random jump process on the energy surface $\{e(v) = e\}$ with generator

$$\sigma(u, v) = |\widehat{B}(u - v)|^2 \delta(e(u) - e(v)).$$

The condition $0 \in \operatorname{supp} \widehat{B}$ is not essential for the proof, but the statement of the theorem needs to be modified if the support is separated away from zero. In this case, the low momentum component of the initial wave function moves ballistically since the diameter of the energy surface is smaller than the minimal range of $\widehat{B}$. The rest of the wave function still moves diffusively.

Figure 1.12 shows schematically the three scales we discussed. Notice that going from the Boltzmann scale to the diffusive scale is essentially the same as going from the random walk to Brownian motion in Wiener's construction (Theorems 1.3 and 1.7). However it is misleading to try to prove Theorem 1.13 by a two-step limiting argument, first using the kinetic limit (Theorem 1.12) then combining it with Wiener's argument. There is only one limiting parameter in the problem, one cannot take first the kinetic limit, then the additional $\lambda^{-\kappa} \to \infty$ limit modelling the long kinetic time scale. The correct procedure is to consider the Schrödinger evolution and run it up to the diffusive time scale. In this way, the intermediate Boltzmann scale can give only a hint of what to expect but it cannot be used for proof.

In fact, this is not only a mathematical subtlety. Correlations that are small on the kinetic scale, and were neglected in the first limit, can become relevant on longer time scales. Theorem 1.13 shows that it is not the case; at least up to times $t \sim \lambda^{-2-\kappa}$ the correlations are not strong enough to destroy the diffusive picture coming from the long time limit of the Boltzmann equation (in particular the diffusion coefficient can be correctly computed from the Boltzmann equation). However, this should not be taken for granted, and in fact it is not expected to hold in $d = 2$ dimensions for exponentially long times, $t \sim \exp(\lambda^{-1})$. Theorem 1.12 on the kinetic limit is valid in $d = 2$ as well and Wiener's argument is dimension independent. On the other hand if the diffusive evolution were true up to any time scales then in particular the state would be delocalized, in contrast to the conjecture that the random Schrödinger operator in $d = 2$ dimensions is always localized.

1.7   Feynman graphs (without repetition)

1.7.1   Derivation of Feynman graphs

Feynman graphs are extremely powerful graphical representations of perturbation expansions. They have been primarily used in many-body theories (they were invented for QED), but they are very convenient for our problem as well to organize perturbation expansion. They have been used by physicist to study random Schrödinger operators, a good overview from the physics point of view is (Vollhardt, and Wölfle 1980).

We will derive the Feynman graphs we need for this presentation. Recall that

$$H = -\Delta + \lambda V, \qquad V = \sum_{\alpha \in \mathbb{Z}^d} V_\alpha, \qquad \mathbb{E}V_\alpha = 0, \tag{1.48}$$

where V_α is a single-bump potential around $\alpha \in \mathbb{Z}^d$, for example

$$V_\alpha(x) = v_\alpha \delta(x - \alpha).$$

We consider the potential as a small perturbation of $-\Delta$ and we expand the unitary evolution by the identity (Duhamel formula)

$$\psi_t = e^{-itH}\psi_0 = e^{it\Delta}\psi_0 - i\lambda \int_0^t e^{-i(t-s)H}Ve^{is\Delta}\psi_0 \, ds. \tag{1.49}$$

This identity can be seen by differentiating $U(t) := e^{-itH}e^{itH_0}$ where $H = H_0 + \lambda V$, $H_0 = -\Delta$:

$$\frac{dU}{dt} = e^{-itH}(-iH + iH_0)e^{itH_0} = e^{-itH}(-i\lambda V)e^{itH_0},$$

thus we can integrate back from 0 to t

$$U(t) = I + \int_0^t ds \, e^{-i(t-s)H}(-i\lambda V)e^{i(t-s)H_0}$$

and multiply by e^{-itH_0} from the right.

We can iterate the expansion (1.49) by repeating the same identity for $e^{-i(t-s)H}$. For example, the next step is

$$e^{-itH}\psi_0 = e^{it\Delta}\psi_0 - i\lambda \int_0^t e^{i(t-s)\Delta}Ve^{is\Delta}\psi_0\,ds$$

$$+ \lambda^2 \int_0^t ds_1 \int_0^{s_1} ds_2\, e^{-i(t-s_1-s_2)H}Ve^{is_2\Delta}Ve^{is_1\Delta}\psi_0$$

and so on. We see that at each step we obtain a new *fully expanded* term (second term), characterized by the absence of the unitary evolution of H; only free evolutions $e^{-is\Delta}$ appear in these terms. There is always one last term that contains the full evolution and that will be estimated trivially after a sufficient number of expansions. More precisely, we can expand

$$\psi_t = e^{-itH}\psi_0 = \sum_{n=0}^{N-1} \psi^{(n)}(t) + \Psi_N(t), \qquad (1.50)$$

where

$$\psi^{(n)}(t) := (-i\lambda)^n \int_{\mathbb{R}_+^{n+1}} ds_0 ds_1 \ldots ds_n\, \delta\Big(t - \sum_{j=0}^n s_j\Big) e^{is_0\Delta}Ve^{is_1\Delta}V \ldots Ve^{is_n\Delta}\psi_0$$

and

$$\Psi_N(t) := (-i\lambda) \int_0^t ds\, e^{-i(t-s)H}V\psi^{(N-1)}(s). \qquad (1.51)$$

Recalling that each V is a big summation (1.48), we arrive at the following *Duhamel formula*

$$\psi_t = \sum_{n=0}^{N-1} \sum_A \psi_A + \text{full evolution term}, \qquad (1.52)$$

where the summation is over collision histories:

$$A := (\alpha_1, \alpha_2, \cdots, \alpha_n), \qquad \alpha_j \in \mathbb{Z}^d,$$

and each elementary wave function is given by

$$\psi_A := \psi_A(t) = (-i\lambda)^n \int d\mu_{n,t}(\underline{s})\, e^{is_0\Delta}V_{\alpha_1} \cdots V_{\alpha_n}e^{is_n\Delta}\psi_0. \qquad (1.53)$$

Here, for simplicity, we introduced the notation

$$\int d\mu_{n,t}(\underline{s}) := \int_{\mathbb{R}_+^{n+1}} \delta\Big(t - \sum_{j=0}^n s_j\Big) ds_0 ds_1 \ldots ds_n \qquad (1.54)$$

for the integration over the simplex $s_0 + s_1 + \ldots + s_n = t$, $s_j \geq 0$.

If the cardinality of A is $n = |A|$, then we say that the elementary wave function is of order n and their sum for a fixed n is denoted by

$$\psi^{(n)}(t) := \sum_{A \,:\, |A|=n} \psi_A(t).$$

Note that the full evolution term Ψ_N has a similar structure, but the leftmost unitary evolution $e^{is_0\Delta}$ is replaced with $e^{-is_0 H}$.

We remark that not every elementary wave function has to be expanded up to the same order. The expansion can be adjusted at every step, that is depending on the past collision history, encoded in the sequence A, we can decide if an elementary wave function ψ_A should be expanded further or it we should stop the expansion. This will be essential to organize our expansion without overexpanding: once the collision history A indicates that ψ_A is already small (e.g. it contains recollisions), we will not expand it further.

Each ψ_A can be symbolically represented as shown on Fig. 1.13, where lines represent the free propagators and bullets represent the potential terms (collisions). Note the times between collisions is not recorded in the graphical picture since they are integrated out.

We need to compute expectation values of quadratic functionals of ψ like the Wigner transform. For simplicity we will show how the L^2-norm can be treated, the Wigner transform is very similar (see next section).

Suppose for the moment that there is no repetition in A, that is all α_j's are distinct. To compute $\mathbf{E}\|\psi\|^2 = \mathbf{E}\,\overline{\psi}\,\psi$, we need to compute

$$\sum_{n,n'} \sum_{A \,:\, |A|=n} \sum_{B \,:\, |B|=n'} \mathbf{E}\,\langle \psi_B \,, \psi_A \rangle.$$

This is zero unless B is a permutation of A, because individual potentials have zero expectation. Higher order moments of the same potential V_α are excluded since we assumed that A and B have no repetition. Thus the only nonzero contributions come from second moments, that is from pairings of each element of A with an element of B, in particular, $|A| = |B|$ is forced, that is $n = n'$.

In conclusion, with the tacit assumption that there are no repetitions, we can express $\mathbf{E}\|\psi\|^2$ as a summation over the order n and over permutations between the sets A and B with $|A| = |B| = n$:

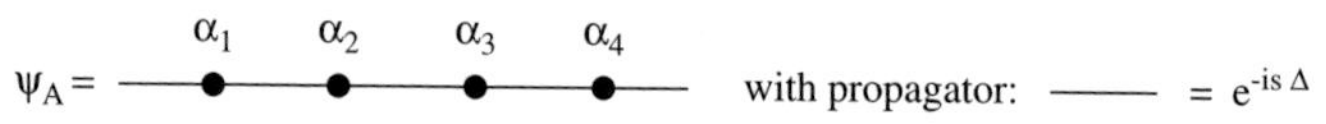

$$\psi_A = \text{—}\bullet\text{—}\bullet\text{—}\bullet\text{—}\bullet\text{—} \qquad \text{with propagator:} \quad \text{———} = e^{-is\,\Delta}$$

with labels α_1, α_2, α_3, α_4.

Fig. 1.13   Representation of a wave function with collision history A.

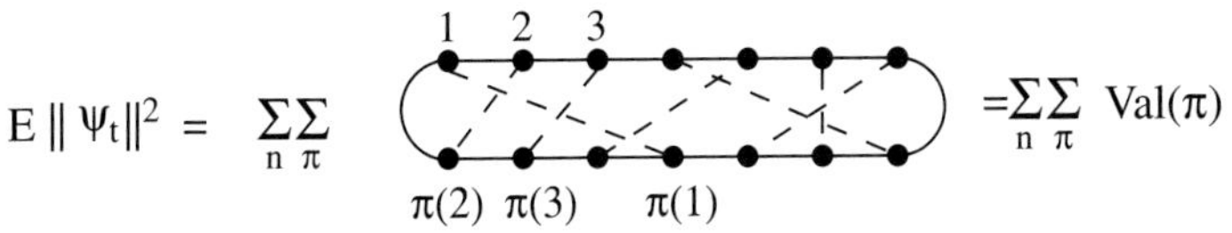

$$\mathbf{E}\,\|\,\Psi_t\|^2 \;=\; \sum_n \sum_\pi \;\; \begin{array}{c} \scriptstyle 1 \;\; 2 \;\; 3 \\ \text{(graph)} \\ \scriptstyle \pi(2)\;\;\pi(3)\qquad \pi(1) \end{array} \;=\; \sum_n \sum_\pi \;\mathrm{Val}(\pi)$$

Fig. 1.14 Representation of the L^2-norm as the sum of the Feynman graphs.

$$\sum_{n,n'} \sum_{A\,:|A|=n}^* \sum_{B\,:|B|=n'}^* \mathbf{E}\,\langle \psi_B,\psi_A\rangle = \sum_n \sum_{\pi\in S_n} \sum_{A\,:|A|=n}^* \mathbf{E}\,\langle \psi_{\pi(A)},\psi_A\rangle =: \sum_n \sum_{\pi\in S_n} \mathrm{Val}(\pi),$$

$$(1.55)$$

where S_n is set of permutations on n elements and upper star denotes summation over non-repetitive sequences. $\mathrm{Val}(\pi)$ is defined by the last formula by summing up all sequences A; this summation will be computed explicitly and it will give rise to delta functions among momenta.

Drawing the two horizontal lines that represent ψ_A and ψ_B parallel and connecting the paired bullets, we obtain Fig. 1.14, where the first summation is over the order n, the second summation is over all permutations $\pi\in S_n$, a graph called the *Feynman graph of π*. We will give an explicit formula for the value of each Feynman graph, here denoted by $\mathrm{Val}(\pi)$.

1.7.2 L^2-norm vs. Wigner transform

Eventually we need to express

$$\mathbf{E}\langle J, W_\psi\rangle := \int J(x,v)\,\mathbf{E}\,W_\psi(x,v)\,\mathrm{d}x\mathrm{d}v = \int \widehat{J}(\xi,v)\,\mathbf{E}\,\widehat{W}_\psi(\xi,v)\,\mathrm{d}\xi\mathrm{d}v = \mathbf{E}\langle \widehat{J}, \widehat{W}_\psi\rangle$$

(recall that a hat means Fourier transform only in the x variable). However, we have

Lemma 1.15 *For any Schwartz function J, the quadratic functional $\psi \mapsto \mathbf{E}\langle \widehat{J}, \widehat{W}_\psi\rangle$ is continuous in the L^2-norm.*

Proof. Let $\psi, \phi \in L^2$ and set

$$\Omega := \mathbf{E}\Big(\langle \widehat{J}, \widehat{W}_\psi\rangle - \langle \widehat{J}, \widehat{W}_\phi\rangle\Big) = \mathbf{E}\int \mathrm{d}\xi\mathrm{d}v \widehat{J}(\xi,v)\Big[\widehat{W}_\psi(\xi,v) - \widehat{W}_\phi(\xi,v)\Big].$$

Let $v_\pm = v \pm \frac{\xi}{2}$ and write

$$\big[\,\cdot\,\big] = \overline{\widehat{\psi}(v_-)}\big(\widehat{\psi}(v_+) - \widehat{\phi}(v_+)\big) + \big(\overline{\widehat{\psi}(v_-)} - \overline{\widehat{\phi}(v_-)}\big)\widehat{\phi}(v_+).$$

Thus

$$|\Omega| \le \int \mathrm{d}\xi\mathrm{d}v |\widehat{J}(\xi,v)|\,\mathbf{E}\big|[\,\cdot\,]\big| \le \int \mathrm{d}\xi \sup_v |\widehat{J}(\xi,v)| \int \mathrm{d}v\, \mathbf{E}\big|[\,\cdot\,]\big|$$

$$\le \int \mathrm{d}\xi \sup_v |\widehat{J}(\xi,v)| \int \mathrm{d}v\, \mathbf{E}\Big[|\widehat{\psi}(v_-)|\cdot|\widehat{\psi}(v_+) - \widehat{\phi}(v_+)| + |\widehat{\psi}(v_-) - \overline{\widehat{\phi}(v_-)}|\cdot|\widehat{\phi}(v_+)|\Big].$$

By Schwarz inequality,

$$\int \mathrm{d}v \mathbf{E}\Big(|\widehat{\psi}(v_-)| \cdot |\widehat{\psi}(v_+) - \widehat{\phi}(v_+)|\Big) \leq \Big(\int \mathrm{d}v \mathbf{E}|\widehat{\psi}(v_-)|^2\Big)^{1/2} \Big(\int \mathrm{d}v \mathbf{E}|\widehat{\psi}(v_+) - \widehat{\phi}(v_+)|^2\Big)^{1/2}$$

$$= \Big(\int \mathrm{d}v \mathbf{E}|\widehat{\psi}(v)|^2\Big)^{1/2} \Big(\int \mathrm{d}v \mathbf{E}|\widehat{\psi}(v) - \widehat{\phi}(v)|^2\Big)^{1/2}$$

$$= \Big(\mathbf{E}\|\widehat{\psi}\|^2\Big)^{1/2} \Big(\mathbf{E}\|\widehat{\psi} - \widehat{\phi}\|^2\Big)^{1/2}. \tag{1.56}$$

Thus

$$|\Omega| \leq \Big(\int \mathrm{d}\xi \sup_v |\widehat{J}(\xi, v)|\Big) \Big(\sqrt{\mathbf{E}\|\psi\|^2} + \sqrt{\mathbf{E}\|\phi\|^2}\Big) \Big(\mathbf{E}\|\widehat{\psi} - \widehat{\phi}\|^2\Big)^{1/2},$$

which completes the proof of the lemma. □

   This lemma, in particular, guarantees that it is sufficient to consider nice initial data, that is we can assume that ψ_0 is a Schwartz function. More importantly, this lemma guarantees that it is sufficient to control the Feynman diagrammatic expansion of the L^2-norm of ψ; the same expansion for the Wigner transform will be automatically controlled. Thus all estimates of the error terms can be done at the level of the L^2-norm; the more complicated arguments appearing in the Wigner transform are necessary only for the explicit computation of the main term.

1.7.3   Stopping the expansion

We will never work with infinite expansions, see (1.50), but we will have to control the last, fully expanded term. This will be done by using the unitarity of the full evolution $e^{-i(t-s)H}$ in the definition of Ψ_N, see (1.51). More precisely, we have:

Lemma 1.16. (Unitarity bound) *Assume that* $\|\psi_0\| = 1$. *According to* (1.50), *we write* $\psi(t) = \Phi_N(t) + \Psi_N(t)$ *with* $\Phi_N = \sum_{n=0}^{N-1} \psi^{(n)}$ *containing the fully expanded terms. Then*

$$\Big|\mathbf{E}\Big[\langle J, W_{\psi(t)} \rangle - \langle J, W_{\Phi_N(t)} \rangle\Big]\Big| \leq \Big(1 + \sqrt{\mathbf{E}\|\Phi_N\|^2}\Big) t \Big[\sup_{0 \leq s \leq t} \mathbf{E}\|\lambda V \psi^{(N-1)}(s)\|^2\Big]^{1/2}.$$

In other words, the difference between the true wave function $\psi(t)$ and its N-th order approximation Φ_N *can be expressed in terms of fully expanded quantities*, but we have to pay an additional price t. This additional t factor will be compensated by a sufficiently long expansion, that is by choosing N large enough so that $\psi^{(N-1)}(s)$ is small. In practice we will combine this procedure by stopping the expansion for each elementary wave function ψ_A separately, depending on the collision history. Once ψ_A is sufficiently small to compensate for the t factor lost in the unitarity estimate, we can stop its expansion.

Proof. We apply Lemma 1.15 with $\psi = \psi(t)$ and $\phi = \Phi_N$, so that $\psi - \phi = \Psi_N$. Then we have, for any N, that

$$\left| \mathbf{E}\big[\langle J, W_{\psi(t)} \rangle - \langle J, W_{\Phi_N(t)} \rangle \big] \right| \leq \left(1 + \sqrt{\mathbf{E} \|\Phi_N\|^2} \right) \left(\mathbf{E} \|\Psi_N\|^2 \right)^{1/2},$$

since $\|\psi_t\| = \|\psi_0\| = 1$. Then using the unitarity, that is that

$$\left\| e^{-i(t-s)H} \lambda V \psi^{(N-1)}(s) \right\| = \left\| \lambda V \psi^{(N-1)}(s) \right\|,$$

we can estimate

$$\mathbf{E} \|\Psi_N\|^2 = \mathbf{E} \left\| \int_0^t ds\, e^{-i(t-s)H} \lambda V \psi^{(N-1)}(s) \right\|^2$$

$$\leq \mathbf{E} \left(\int_0^t ds \left\| \lambda V \psi^{(N-1)}(s) \right\| \right)^2$$

$$\leq t \int_0^t ds\, \mathbf{E} \left\| \lambda V \psi^{(N-1)}(s) \right\|^2, \qquad (1.57)$$

which proves Lemma 1.16. □

1.7.4 Outline of the proof of the Boltzmann equation with Feynman graphs

The main observation is that among the $n!$ possible permutations of order n, only one, the identity permutation, contributes to the limiting Boltzmann equation. Actually it has to be like that, since this is the only graph whose collision history can be interpreted classically, since here $A = B$ as sets, so the two wave-functions ψ_A and ψ_B visit the same obstacles in the same order. If there were a discrepancy between the order in which the obstacles in A and B are visited, then no classical collision history could be assigned to that pairing.

The Feynman graph of the identity permutation is called *the ladder graph*, see Fig. 1.15. Its value can be computed explicitly and it resembles to the n-th term in the Taylor series of the exponential function.

The ladder with n pairings corresponds to a classical process with n collisions. If the collision rate is $\sigma \lambda^2$, then the classical probability of n collisions is given by the (unnormalized) Poisson distribution formula

$$\text{Prob } (n \text{ collisions}) = \frac{(\sigma \lambda^2 t)^n}{n!}.$$

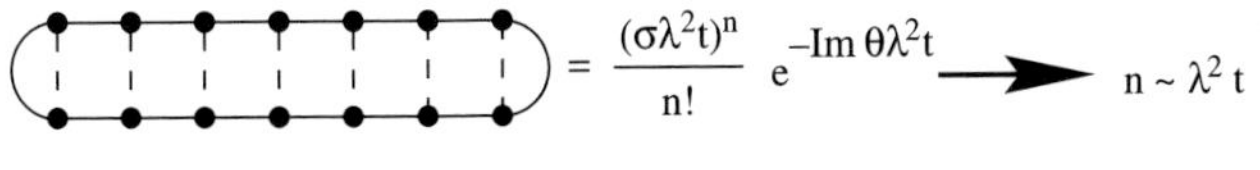

Fig. 1.15 Contribution of the ladder graph.

This is clearly summable for the kinetic scale $t \sim \lambda^{-2}$, and the typical number of collisions is the value n giving the largest term, that is $n \sim \lambda^2 t$. The exponential damping factor $e^{-\mathrm{Im}(\theta)\lambda^2 t}$ in Fig. 1.15 comes from renormalization (see later in Section 1.8.4 where the constants σ and θ will also be defined).

We note that from the Duhamel formula leading to the Feynman graphs, it is not at all clear that the value of the ladder graphs is of order one (in the limiting parameter λ). According to the Duhamel representation, they are highly oscillatory integrals of the form (written in momentum space)

$$\mathrm{Val}(id_n) \sim \lambda^{2n} \left(\int_0^t e^{is_1 p_1^2} \int_0^{s_1} e^{is_2 p_2^2} \cdots \int_0^{s_n} \right) \left(\int_0^t e^{-is_1' p_1^2} \int_0^{s_1'} e^{-is_2' p_2^2} \cdots \int_0^{s_n'} \right).$$
$$(1.58)$$

Here id_n is the identity permutation in S_n which generates the ladder. If one estimates these integrals naively, neglecting all oscillations, one obtains

$$\mathrm{Val}(id_n) \leq \frac{\lambda^{2n} t^{2n}}{(n!)^2},$$

which is completely wrong, since it does not respect the power counting that $\lambda^2 t$ should be the relevant parameter.

Assuming that the ladder can be computed precisely, with all oscillations taken into account, one still has to deal with the remaining $n! - 1$ nonladder graphs. Their explicit formulas will reveal that their values are small, but there are many of them, so it is not clear at all if they are still summable to something negligible.

It is instructive to compare the value of the non-ladder graphs with the ladder graph because that is of order one (see Section 1.8.6 for a detailed computation). Figure 1.16 contains schematic estimates of simple crossing graphs that can be checked by direct computation (once the explicit formula for $\mathrm{Val}(\pi)$ is established in the next section). It indicates that the value of a nonladder graph is a positive power of λ that is related with the combinatorial complexity of the permutation (e.g. the number of 'crosses').

Accepting for the moment that all non-ladder graphs are at least by an order λ^2 smaller than the ladder graphs, we obtain the following majorizing estimate for the summation of the terms in the Duhamel expansion:

$$\sum_n \sum_\pi |\mathrm{Val}(\pi)| \leq \sum_n \frac{(\sigma \lambda^2 t)^n}{n!} \left(\underbrace{1}_{A=B} + \underbrace{(n!-1)(\mathrm{small})}_{A \neq B} \right).$$

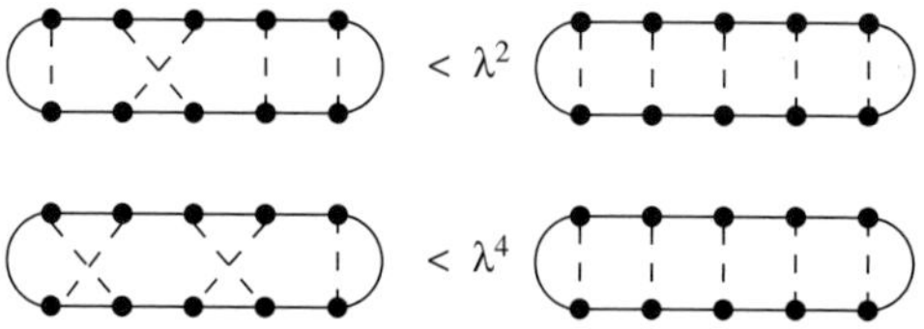

Fig. 1.16   Comparing the value of various crossing graphs.

This series clearly converges for short kinetic time ($\lambda^2 t = T \leq T_0$) but not for longer times, since the two $n!$'s cancel each other and the remaining series is geometric. This was the basic idea of Spohn's proof (Spohn 1977) and it clearly shows the limitation of the method to short macroscopic times.

The source of the $1/n!$ is actually the time-ordered multiple time integration present in (1.54); the total volume of the simplex

$$\{(s_1, s_2, \ldots, s_n) \ : \ 0 \leq s_1 \leq \ldots \leq s_n \leq t\} \tag{1.59}$$

is $t^n/n!$. However, (1.58) contains two such multiple time integrations, but it turns out that only one of them is effective. The source of the competing $n!$ is the pairing that comes from taking the expectation, for example in case of Gaussian randomness this appears explicitly in the Wick theorem. The reason for this mechanism is the fact that in quantum mechanics the physical quantities are always quadratic in ψ, thus we have to pair two collision histories in all possible ways even if only one of them has a classical meaning. This additional $n!$ is very typical in all perturbative quantum mechanical arguments.

This was only the treatment of the fully expanded terms, the error term in (1.52) needs a separate estimate. As explained in detail in Lemma 1.16, here we simply use the unitarity of the full evolution

$$\left\| \int_0^t e^{i(t-s)H} \underbrace{\int V e^{is_1 \Delta} V e^{is_2 \Delta} \ldots \mathrm{d}\mu_{n,s}(\underline{s})}_{\psi_s^\#} \right\| \leq \underbrace{t}_{price} \sup_s \|\psi_s^\#\|. \tag{1.60}$$

In this estimate we lose a factor of t, but we end up with a fully expanded term $\psi_s^\#$ which can also be expanded into Feynman graphs.

How to go beyond Spohn's argument that is valid only up to short macroscopic time? Notice that 'most' graphs have many 'crosses', hence they are expected to be much smaller, due to phase decoherence than the naive 'one-cross' estimate indicates. One can thus distinguish graphs with one, two and more crossings. From the graphs with more crossings, additional λ^2-factors can be obtained. On the other hand, the combinatorics of the graphs with a few crossings is much smaller than $n!$.

On kinetic scale the following estimate suffices (here N denotes the threshold so that the Duhamel expansion is stopped after N collisions):

$$\mathbf{E}\|\psi_t\|^2 \leq \sum_{n=0}^{N-1} \frac{(\sigma\lambda^2 t)^n}{n!} \Big(\underbrace{1}_{\text{ladder}} + \underbrace{n\lambda^2}_{\text{one cross}} + \underbrace{n!\lambda^4}_{\text{rest}} \Big)$$

$$+ \underbrace{t}_{unit.price} \frac{(\sigma\lambda^2 t)^N}{N!} \Big(\underbrace{1}_{\text{ladder}} + \underbrace{N\lambda^2}_{\text{one cross}} + \underbrace{N^2\lambda^4}_{\text{two cross}} + \underbrace{N!\lambda^6}_{\text{rest}} \Big).$$

Finally, we can optimize

$$N = N(\lambda) \sim (\log \lambda)/(\log \log \lambda)$$

to get the convergence. This estimate gives the kinetic (Boltzmann) limit for all fixed T ($t = T\lambda^{-2}$). This concludes the outline of the proof of Theorem 1.12.

For the proof of Theorem 1.13, that is going to diffusive time scales, one needs to classify essentially **all** diagrams; it will not be enough to separate a few crosses and treat all other graphs identically.

1.8 Key ideas of the proof of the diffusion (Theorems 1.13 and 1.14)

We start with an apologetic remark. For pedagogical reasons, we will present the proof of a mixed model that does not actually exists. The reason is that the dispersion relation is simpler in the continuum case, but the random potential is simpler in the discrete case. The simplicity of the dispersion relation is related to the fact that in continuum the level sets of the dispersion relation, $\{v : e(v) = e\}$, are convex (spheres), thus several estimates are more effective. The simplicity of the random potential in the discrete case is obvious: no additional form-factor B and no additional Poisson point process needs to be treated. Moreover, the ultraviolet problem is automatically absent, while in the continuous case the additional large momentum decay needs to be gained from the decay of $\hat{B}$. Therefore, we will work in the discrete model (i.e. no form factors B, no Poisson points, and no large momentum problem), and we will denote the dispersion relation in general by $e(p)$. All formulas hold for both dispersion relations, except some estimates that will be commented on. The necessary modifications from this 'fake' model to the two real models are technically sometimes laborous, but not fundamental. We will comment on them at the end of the proof.

1.8.1 Stopping rules

The Duhamel formula has the advantage that it can be stopped at a different number of terms depending on the collision history of each elementary wave-function. Thus the threshold N does not have to be chosen uniformly for each term in advance; looking at each term

$$(-i\lambda)^n \int e^{-is_0 H} V_{\alpha_1} \cdots V_{\alpha_n} e^{is_n \Delta} \psi_0 \, \mathrm{d}\mu_{n,t}(\underline{s})$$

separately, we can decide if we wish to expand $e^{-is_0 H}$ further, or we rather use the unitary estimate (1.60). The decision may depend on the sequence $(\alpha_1, \ldots, \alpha_n)$.

Set $K := (\lambda^2 t)\lambda^{-\delta}$ (with some $\delta > 0$) as an absolute upper threshold for the number of collisions in an expanded term (notice that K is a much larger than the typical number of collisions). We stop the expansion if we *either* have reached K expanded potential terms *or* we see a repeated α-label. This procedure has the advantage that repetition terms do not pile up. This stopping rule leads us to the following version of the Duhamel formula:

$$\psi_t = \sum_{n=0}^{K-1} \psi_{n,t}^{nr} + \int_0^t e^{-i(t-s)H} \underbrace{\left(\psi_{*K,s}^{nr} + \sum_{n=0}^{K} \psi_{*n,s}^{rep} \right)}_{=:\psi_s^{err}} ds \qquad (1.61)$$

$$\text{with} \qquad \psi_{*n,t}^{nr} := \sum_{A \,:\, \text{nonrep.}} \psi_{*A,t}, \qquad \psi_{*n,s}^{rep} := \sum_{\substack{|A|=n \\ \text{first rep. at } \alpha_n}} \psi_{*A,s} \qquad (1.62)$$

Here ψ^{nr} contains collision histories with no repetition, while the repetition terms ψ^{rep} contain exactly one repetition at the last obstacle, since once such a repetition is found, the expansion is stopped. In particular, the second sum above runs over $A = (\alpha_1, \ldots \alpha_n)$ where $\alpha_n = \alpha_j$ for some $j < n$ and this is the first repetition. Actually the precise definition of the elementary wave-functions in (1.62) is somewhat different from (1.53) (this fact is indicated by the stars), because the expansion starts with a potential term while ψ_A in (1.53) starts with a free evolution.

The backbone of our argument is the following three theorems:

Theorem 1.17 (Error terms are negligible) *We have the following estimate*

$$\sup_{s \le t} E \| \psi_s^{err} \| = o(t^{-2}).$$

In particular, by the unitarity estimate this trivially implies that

$$E \left\| \int_0^t e^{-i(t-s)H} \psi_s^{err} ds \right\|^2 = o(1). \qquad (1.63)$$

Theorem 1.18 (Only the ladder contributes) *For* $n \le K = (\lambda^2 t)\lambda^{-\delta} = O(\lambda^{-\kappa-\delta})$ *we have*

$$E \| \psi_{n,t}^{nr} \|^2 = \text{Val}(id_n) + o(1) \qquad (1.64)$$

$$E W_{\psi_{n,t}^{nr}} = \text{Val}_{\text{Wig}}(id_n) + o(1). \qquad (1.65)$$

Here Val_{Wig} denotes the value of the Feynman graphs that obtained by expanding the Wigner transform instead of the L^2-norm of ψ_t. Note that (1.65) follows from (1.64) by (the proof of) Lemma 1.15.

Theorem 1.19 (Wigner transform of the main term) *The total contribution of the ladder terms up to K collisions,*

$$\sum_{n=0}^{K} \text{Val}_{\text{Wig}}(id_n),$$

satisfies the heat equation.

In the following sections we will focus on the non-repetition terms with $n \le K$, that is on Theorem 1.18. We will only discuss the proof for the L^2-norm (1.64), the proof of (1.65) is a trivial modification. Theorem 1.17 is a fairly long case by case study, but it essentially relies on the same ideas that are behind Theorem 1.18. Some of these cases are sketched in Section 1.10. Finally, Theorem 1.19 is a nontrivial but fairly explicit calculation that we sketch in Section 1.11.

1.8.2 Feynman diagrams in the momentum-space formalism

Now we express the value of a Feynman diagram explicitly in momentum representation. We recall the formula (1.53) and Fig. 1.13 for the representation of the collision history of an elementary wavefunction. We will express it in momentum space. This means that we assign a running momentum, $p_0, p_1, \ldots, p_n$ to each edge of the graph in Fig. 1.13. Note that we work on the lattice, so all momenta are on the torus, $p \in \mathbf{T}^d = [-\pi, \pi]^d$; in the continuum model, the momenta would run over all $\mathbb{R}^d$. The propagators thus become multiplications with $e^{-is_j e(p_j)}$ and the potentials become convolutions:

$$
\widehat{\psi}_{A,t}(p) = \int_{(\mathbf{T}^d)^n} \prod_{j=1}^n dp_j \int d\mu_{n,t}(\underline{s}) \, e^{-is_0 e(p)} \widehat{V}_{\alpha_1}(p - p_1) e^{-is_1 e(p_1)} \widehat{V}_{\alpha_2}(p_1 - p_2) \ldots \widehat{\psi}_0(p_n)
$$

$$
= \underbrace{e^{\eta t}}_{\eta := 1/t} \int_{(\mathbf{T}^d)^n} \prod_{j=1}^n dp_j \int_{-\infty}^{\infty} d\alpha \, e^{-i\alpha t} \prod_{j=1}^n \frac{1}{\alpha - e(p_j) + i\eta} \widehat{V}_{\alpha_j}(p_{j-1} - p_j) \widehat{\psi}_0(p_n).
$$

(1.66)

In the second step we used the identity

$$
\int_{\mathbb{R}_+^{n+1}} \delta\left(t - \sum_{j=0}^n s_j\right) ds_0 ds_1 \ldots ds_n \, e^{-is_0 e(p_0)} e^{-is_1 e(p_1)} \ldots e^{-is_n e(p_n)}
$$

$$
= e^{\eta t} \int_{\mathbb{R}} d\alpha \, e^{-i\alpha t} \prod_{j=1}^n \frac{1}{\alpha - e(p_j) + i\eta}
$$

(1.67)

that holds for any $\eta > 0$ (we will choose $\eta = 1/t$ in the applications to neutralize the exponential prefactor). This identity can be obtained from writing

$$
\delta\left(t - \sum_{j=0}^n s_j\right) = \int e^{-i\alpha(t - \sum_{j=0}^n s_j)} d\alpha
$$

and integrating out the times. (Note that the letter α is used here and in the sequel as the conjugate variable to time, while previously α_j denoted obstacles. We hope this does not lead to confusion, unfortunately this convention was adopted in our papers as well.)

The L^2-norm is computed by doubling the formula (1.66). We will use the letters $p_1, \ldots p_n$ for the momenta in ψ_A and primed momenta for ψ_B:

$$\mathbf{E}\|\psi_t^{nr}\|^2 = \mathbf{E}\left\|\sum_{A\,:\,\text{nonrep}}\psi_A\right\|^2 = \sum_{A,B\,\text{nonrep}}\mathbf{E}\langle\psi_A,\psi_B\rangle = \sum_n\sum_{\pi\in\Pi_n}\text{Val}(\pi),$$

$$\text{Val}(\pi) := e^{2\eta t}\int_{-\infty}^{\infty}\mathrm{d}\alpha\mathrm{d}\beta\,e^{i(\alpha-\beta)t}\prod_{j=0}^{n}\frac{1}{\alpha-e(p_j)-i\eta}\frac{1}{\beta-e(p_j')+i\eta} \qquad (1.68)$$

$$\times\lambda^{2n}\prod_{j=1}^{n}\delta\Big((p_{j-1}-p_j)-(p_{\pi(j)-1}'-p_{\pi(j)}')\Big)|\widehat{\psi}_0(p_n)|^2\delta(p_n-p_n')\prod_{j=0}^{n}\mathrm{d}p_j\mathrm{d}p_j'.$$

After the pairing, here we computed explicitly the pair expectations:

$$\mathbf{E}\sum_{\alpha\in\mathbb{Z}^d}\overline{\widehat{V}_\alpha(p)}\widehat{V}_\alpha(q) = \sum_{\alpha\in\mathbb{Z}^d}e^{i\alpha(p-q)} = \delta(p-q), \qquad (1.69)$$

and we also used the fact that only terms with $B = \pi(A)$ are nonzero. This formula holds for the simplest random potential (1.30) in the discrete model; for the continuum model (1.31) we have

$$\mathbf{E}\sum_\alpha\overline{\widehat{V}_\alpha(p)}\widehat{V}_\alpha(q) = |\widehat{B}(p)|^2\delta(p-q). \qquad (1.70)$$

According to the definition of V_ω in the continuum model (1.32), α labels the realizations of the Poisson point process in the last formula (1.70).

The presence of the additional delta function, $\delta(p_n - p_n')$, in (1.68) is due to the fact that we compute the L^2-norm. We remark that the analogous formula in the expansion for Wigner transform differs only in this factor; the arguments of the two wavefunctions in the momentum representation of the Wigner transform are shifted by a fixed value ξ (1.23), thus the corresponding delta function will be $\delta(p_n - p_n' - \xi)$.

A careful reader may notice that the non-repetition condition on A imposes a restriction on the summation in (1.69); in fact the summation for α_j is not over the whole $\mathbb{Z}^d$, but only for those elements that are distinct from the other α_i's. These terms we will add to complete the sum (1.69) and then we will estimate their effect separately. They correspond to higher order cumulants and their contribution is negligible, but technically they cause serious complications; essentially a complete cumulant expansion needs to be organized. We will shortly comment on them in Section 1.10.1; we will neglect this issue for the moment.

This calculation gives that

$$\mathbf{E}\|\psi_t^{nr}\|^2 = \sum_n\sum_{\pi\in S_n}\text{Val}(\pi)$$

as it is pictured in Fig. 1.14. More concisely,

$$\text{Val}(\pi) = \lambda^{2n}e^{2\eta t}\int\mathrm{d}\mathbf{p}\mathrm{d}\mathbf{p}'\mathrm{d}\alpha\mathrm{d}\beta\,e^{it(\alpha-\beta)}\prod_{j=0}^{n}\frac{1}{\alpha-e(p_j)-i\eta}\frac{1}{\beta-e(p_j')+i\eta}\,\Delta_\pi(\mathbf{p},\mathbf{p}')\,|\widehat{\psi}_0(p_n)|^2,$$

$$(1.71)$$

where $\mathbf{p}$ and $\mathbf{p}'$ stand for the collection of integration momenta and

$$\Delta_\pi(\mathbf{p}, \mathbf{p}') := \delta(p_n - p'_n) \prod_{j=1}^{n} \delta\Big((p_{j-1} - p_j) - (p'_{\pi(j)-1} - p'_{\pi(j)})\Big)$$

contains the product of all delta functions. These delta functions can be obtained from the graph: they express the Kirchoff law at each pair of vertices, that is the signed sum of the four momenta attached to any paired vertices must be zero. It is one of the main advantages of the Feynman graph representation that the complicated structure of momentum delta functions can be easily read off from the graph.

We remark that in (1.71) we omitted the integration domains; the momentum integration runs through the momentum space, that is each p_j and p'_j is integrated over $\mathbb{R}^d$ or $\mathbf{T}^d$, depending whether we consider the continuum or the lattice model. The $d\alpha$ and $d\beta$ integrations always run through the reals. We will adopt this short notation in the future as well:

$$\int d\mathbf{p}d\mathbf{p}'d\alpha d\beta = \int_{(\mathbf{T}^d)^{n+1}} \prod_{j=0}^{n} dp_j \int_{(\mathbf{T}^d)^{n+1}} \prod_{j=0}^{n} dp'_j \int_{\mathbb{R}} d\alpha \int_{\mathbb{R}} d\beta.$$

We will also often use momentum integrations without indicating their domains, which is always $\mathbb{R}^d$ or $\mathbf{T}^d$.

1.8.3   Lower order examples

In this section we compute explicitly a few low order diagrams for the Wigner transform. Recalling that $\Phi_N = \sum_{n=0}^{N-1} \psi^{(n)}$ represents the fully expanded terms and recalling the Wigner transform in momentum space (1.23), we can write the (rescaled) Wigner transform of Φ_N as follows:

$$\widehat{W}^\varepsilon_{\Phi_N}(\xi, v) = \sum_{n'=0}^{N-1} \sum_{n=0}^{N-1} \overline{\widehat{\psi}_t^{(n')}\Big(v - \frac{\varepsilon}{2}\xi\Big)} \widehat{\psi}_t^{(n)}\Big(v + \frac{\varepsilon}{2}\xi\Big) =: \sum_{n,n'=0}^{N-1} \widehat{W}^\varepsilon_{n,n',t}(\xi, v).$$

Set $k_n = v + \frac{\varepsilon}{2}\xi$ and $k'_{n'} := v - \frac{\varepsilon}{2}\xi$, then we can write

$$\widehat{\psi}_t(k_n) = (-i\lambda)^n \int d\mu_{n,t}(\underline{s}) e^{-is_n e(k_n)} \int \prod_{j=0}^{n-1} \Big(dk_j \, e^{-is_j e(k_j)}\Big) \prod_{j=1}^{n} \widehat{V}(k_j - k_{j-1})\widehat{\psi}_0(k),$$

which can be doubled to obtain $\widehat{W}^\varepsilon_{n,n',t}$. Notice that we shifted all integration variables, that is we use $k_j = v_j + \frac{\varepsilon}{2}\xi$.

Case 1: $n = n' = 0$. In this case there is no need for taking expectation and we have

$$\widehat{W}^\varepsilon_{n,n',t}(\xi, v) = e^{it(e(k'_0) - e(k_0))} \widehat{\psi}_0(k_0) \overline{\widehat{\psi}_0(k'_0)} = e^{it(e(k'_0) - e(k_0))} \widehat{W}^\varepsilon_0(\xi, v)$$

where W^ε_0 is the Wigner transform of the initial wave function ψ_0. Suppose, for simplicity, that the initial wavefunction is unscaled, that is it is supported near the origin in position space, uniformly in λ. Then it is easy to show that its rescaled Wigner transform converges to the Dirac delta measure in position space:

$$\lim_{\varepsilon \to 0} W^\varepsilon_{\psi_0}(X, v) \mathrm{d}X \mathrm{d}v = \delta(X)|\widehat{\psi}_0(v)|^2 \mathrm{d}X \mathrm{d}v$$

since for any fixed ξ,

$$\widehat{\psi}_0(k_0)\overline{\widehat{\psi}_0(k_0')} \to |\widehat{\psi}_0(v)|^2 \tag{1.72}$$

as $\varepsilon \to 0$. In the sequel we will use the continuous dispersion relation $e(k) = \frac{1}{2}k^2$ for simplicity because it produces explicit formulas. In the general case, similar formulas are obtained by Taylor expanding $e(k)$ in ε up to the first order term and higher order terms are negligible but need to be estimated.

Since

$$e(k_0') - e(k_0) = \frac{1}{2}\left(v - \frac{\varepsilon}{2}\xi\right)^2 - \frac{1}{2}\left(v + \frac{\varepsilon}{2}\xi\right)^2 = -\varepsilon v \cdot \xi,$$

we get that in the kinetic limit, when $\lambda^2 t = T$ is fixed and $\varepsilon = \lambda^2$

$$\lim_{\varepsilon \to 0} \int \widehat{J}(\xi, v)\widehat{W}^\varepsilon_{n,n',t}(\xi, v)\mathrm{d}\xi \mathrm{d}v = \int \widehat{J}(\xi, v)|\widehat{\psi}_0(v)|^2 e^{-iTv \cdot \xi} \mathrm{d}\xi \mathrm{d}v = \int J(Tv, v)|\widehat{\psi}_0(v)|^2 \mathrm{d}v,$$

by using dominated convergence and (1.72). The last formula is the weak formulation of the evolution of initial phase space measure $\delta(X)|\widehat{\psi}_0(v)|^2 \mathrm{d}X \mathrm{d}v$ under the free motion along straight lines.

Case 2: $n = 0$, $n' = 1$ or $n = 1$, $n' = 0$. Since there is one single potential in the expansion, these terms are zero after taking the expectation:

$$\mathbf{E}\,\widehat{W}^\varepsilon_{1,0,t}(\xi, v) = \mathbf{E}\,\widehat{W}^\varepsilon_{0,1,t}(\xi, v) = 0.$$

Case 3: $n = n' = 1$. We have

$$\widehat{W}^\varepsilon_{1,1,t}(\xi, v) = (-i\lambda)\int_0^t \mathrm{d}s\, e^{-i(t-s)e(k_1)} \int \mathrm{d}k_0 \widehat{V}(k_1 - k_0)e^{-ise(k_0)}\widehat{\psi}_0(k_0)$$

$$\times (i\lambda)\int_0^t \mathrm{d}s'\, e^{i(t-s')e(k_1')} \int \mathrm{d}k_0' \overline{\widehat{V}(k_1' - k_0')}e^{is'e(k_0')}\overline{\widehat{\psi}_0(k_0')}. \tag{1.73}$$

The expectation value acts only on the potentials, and we have (using now the correct continuum potential)

$$\mathbf{E}\widehat{V}(k_1 - k_0)\overline{\widehat{V}(k_1' - k_0')} = |\widehat{B}(k_1 - k_0)|^2 \delta(k_1' - k_0' - (k_1 - k_0)).$$

After collecting the exponents, we obtain

$$\mathbf{E}\widehat{W}^\varepsilon_{1,1,t}(\xi, v) = \lambda^2 \int \mathrm{d}k_0 \mathrm{d}k_0' |\widehat{B}(k_1 - k_0)|^2 \delta(k_1' - k_0' - (k_1 - k_0))$$

$$\times e^{it(e(k_1') - e(k_1))} Q_t(e(k_1) - e(k_0))\overline{Q_t}(e(k_1') - e(k_0'))\widehat{\psi}_0(k_0)\overline{\widehat{\psi}_0(k_0')},$$

where we introduced the function

$$Q_t(a) := \frac{e^{ita} - 1}{ia}.$$

Consider first the case $\xi = 0$, that is when the Wigner transform becomes just the momentum space density. Then $v = k_1 = k_1'$ and thus $k_0 = k_0'$ by the delta function, so $e(k_1) - e(k_0) = e(k_1') - e(k_0')$ and we get

$$\mathbf{E}\widehat{W}^{\varepsilon}_{1,1,t}(0, v) = \lambda^2 \int \mathrm{d}k_0 |\widehat{B}(k_1 - k_0)|^2 |\widehat{\psi}_0(k_0)|^2 |Q_t(e(k_1) - e(k_0))|^2.$$

Clearly

$$t^{-1}|Q_t(A)|^2 \to 2\pi\delta(A)$$

as $t \to \infty$ (since $\int \left(\frac{\sin x}{x}\right)^2 \mathrm{d}x = \pi$), so we obtain that in the limit $t \to \infty$, $\lambda^2 t = T$ fixed,

$$\mathbf{E}\widehat{W}^{\varepsilon}_{1,1,t}(0, v) \to T \int \mathrm{d}k_0 |\widehat{B}(k_1 - k_0)|^2 2\pi\delta(e(k_1) - e(k_0)) |\widehat{\psi}_0(k_0)|^2$$

$$= T \int \mathrm{d}k_0 \sigma(k_1, k_0) |\widehat{\psi}_0(k_0)|^2 \qquad (1.74)$$

(recall $k_1 = v$ in this case), where the collision kernel is given by

$$\sigma(k_1, k_0) := |\widehat{B}(k_1 - k_0)|^2 2\pi\delta(e(k_1) - e(k_0)).$$

A similar but somewhat more involved calculation gives the result for a general ξ, after testing it against a smooth function J. Then it is easier to use (1.73) directly, and we get, after taking expectation and changing variables,

$$\mathbf{E}\widehat{W}^{\varepsilon}_{1,1,t}(\xi, v) = \lambda^2 \int_0^t \mathrm{d}s \int_0^t \mathrm{d}s' \int \mathrm{d}v_0 \, e^{i\Phi/2} |\widehat{B}(v - v_0)|^2 \widehat{W}^{\varepsilon}_0(\xi, v_0)$$

with a total phase factor

$$\Phi := -(t - s)\left(v + \frac{\varepsilon}{2}\xi\right)^2 - s\left(v_0 + \frac{\varepsilon}{2}\xi\right)^2 + (t - s')\left(v - \frac{\varepsilon}{2}\xi\right)^2 + s'\left(v_0 - \frac{\varepsilon}{2}\xi\right)^2$$

$$= (s - s')(v^2 - v_0^2) - 2\varepsilon\left[\left(t - \frac{s + s'}{2}\right)v - \frac{s + s'}{2}v_0\right] \cdot \xi.$$

After changing variables: $b = s - s'$ and $T_0 = \varepsilon\frac{s+s'}{2}$, we notice that the first summand gives

$$\int \mathrm{d}b \, e^{ib(v^2 - v_0^2)} = 2\pi\delta(v^2 - v_0^2), \qquad (1.75)$$

and from the second we have

$$\mathbf{E}\widehat{W}^{\varepsilon}_{1,1,t}(\xi,v) = \int_0^T \mathrm{d}T_0 \int \mathrm{d}v_0\, e^{-i\left[(T-T_0)v - T_0 v_0\right]\cdot\xi} \sigma(v,v_0)\widehat{W}^{\varepsilon}_0(\xi,v_0).$$

These steps, especially (1.75), can be made rigorous only if tested against a smooth function J, so we have

$$\mathbf{E}\langle J, W_{1,1,t}\rangle \to \int_0^T \mathrm{d}T_0 \int \mathrm{d}v_0 J\Big((T-T_0)v - T_0 v_0, v\Big)\sigma(v,v_0)|\widehat{\psi}_0(v_0)|^2.$$

The dynamics described in the first variable of J is a free motion with velocity v_0 up to time T_0, then a fictitious collision happens that changes the velocity from v_0 to v (and this process is given by the rate $\sigma(v,v_0)$) and then free evolution continues during the remaining time $T - T_0$ with velocity v. This is exactly the one 'gain' collision term in the Boltzmann equation.

Case 4: $n = 0$, $n' = 2$ or $n = 2$, $n' = 0$.

The calculation is similar to the previous case, we just record the result:

$$\mathbf{E}\widehat{W}_{2,0,t}(\xi,v) = -\lambda^2 t\widehat{\psi}_0(k_1)\overline{\widehat{\psi}_0(k_1')}\, e^{-it\left(e(k_1) - e(k_1')\right)} \int \mathrm{d}q |\widehat{B}(k_1 - q)|^2 tR\big(t(e(k_1) - e(q))\big),$$

where

$$R(u) = \frac{e^{iu} - iu - 1}{u^2}.$$

Simple calculation shows that $tR(tu) \to \pi\delta(u)$ as $t \to \infty$. Thus we have

$$\mathbf{E}\widehat{W}_{2,0,t}(\xi,v) \to -\frac{T}{2}|\widehat{\psi}_0(v)|^2 e^{-iTv\cdot\xi}\sigma(v),$$

where we defined

$$\sigma(v) := \int \sigma(v,q)\mathrm{d}q.$$

Similar result holds for $\widehat{W}_{0,2,t}$ and combining it with (1.74), we obtain the conservation of the L^2-norm up to second order in λ, since with $\xi = 0$ we have

$$\int \mathrm{d}v \left[\mathbf{E}\widehat{W}_{1,1,t}(0,v) + \mathbf{E}\widehat{W}_{2,0,t}(0,v) + \mathbf{E}\widehat{W}_{0,2,t}(0,v)\right] = 0.$$

For general ξ and after testing against a smooth function, we obtain

$$\mathbf{E}\langle J, W_{2,0,t}\rangle \to -\frac{T}{2}\int \mathrm{d}v J(Tv, v)\sigma(v)|\widehat{\psi}_0(v)|^2$$

which corresponds to the loss term in the Boltzmann equation.

Apart from revealing how the Boltzmann equation emerges from the quantum expansion, the above calculation carries another important observation. Notice that terms with $n \neq n'$ did give rise to non-negligible contributions despite the earlier statement (1.55) that $n = n'$ is forced by the nonrepetition rule (which is correct) and that repetitive collision sequences are negligible (which is, apparently, not correct). The next section will explain this.

1.8.4   Self-energy renormalization

The previous simple explicit calculations showed that not all repetition terms are neglible, in fact immediate recollisions are of order one and they—but only they—have to be treated differently. We say that a collision sequence $(\alpha_1, \alpha_2, \ldots)$ has an *immediate recollision* if $\alpha_i = \alpha_{i+1}$ for some i. This modification must also be implemented in the stopping rule, such that immediate recollisions do not qualify recollisions for the purpose of collecting reasons to stop the expansion (see Section 1.8.1).

Fortunately immediate repetitions appear very locally in the Feynman graph and they can be resummed according to the schematic picture of Fig. 1.17. The net effect is that the free propagator $e(p) = \frac{1}{2}p^2$ needs to be changed to another propagator $w(p)$ that differs from $e(p)$ by an $O(\lambda^2)$ amount. More precisely, using the time independent formalism, second line of (1.66), we notice that the first term in Fig. 1.17 is represented by

$$\frac{1}{\alpha - e(p) - i\eta} \tag{1.76}$$

(which is also often called propagator). The second term carries a loop integration (momentum q), and its contribution is

$$\frac{1}{(\alpha - e(p) - i\eta)^2} \int \frac{\lambda^2 |\widehat{B}(p-q)|^2}{\alpha - e(q) - i\eta} \, \mathrm{d}q.$$

This formula is written for the continuum model and the integration domain is the whole momentum space $\mathbb{R}^d$. For the lattice model $\widehat{B}$ is absent and the integration is over $\mathbf{T}^d$.

The third term carries two independent loop integration (momenta q and q'), and its contribution is

$$\frac{1}{(\alpha - e(p) - i\eta)^3} \int \frac{\lambda^2 |\widehat{B}(p-q)|^2 \mathrm{d}q}{\alpha - e(q) - i\eta} \int \frac{\lambda^2 |\widehat{B}(p-q')|^2 \mathrm{d}q'}{\alpha - e(q') - i\eta}.$$

Setting

$$\Theta_\eta(p, \alpha) := \frac{|\widehat{B}(p-q)|^2 \mathrm{d}q}{\alpha - e(q) - i\eta}$$

and noticing that due to the almost singularities of the $(\alpha - e(p) - i\eta)^{-k}$ prefactors the main contribution comes from $\alpha \sim e(p)$, we can set

$$\theta(p) := \lim_{\eta \to 0+} \Theta(p, e(p)). \tag{1.77}$$

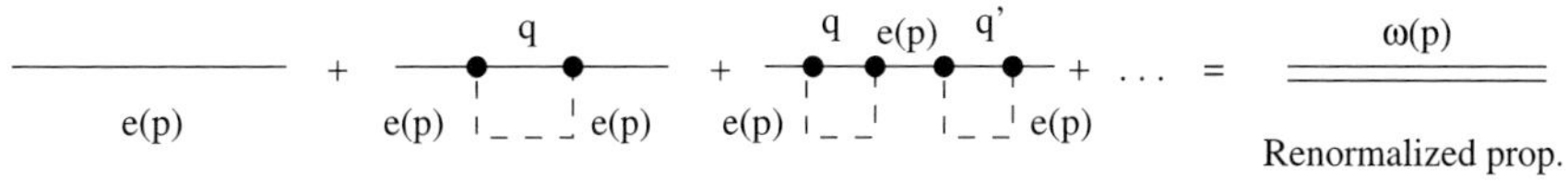

Fig. 1.17 Schematic picture of the renormalization.

(similar, in fact easier, formulas hold for the lattice model). Therefore, modulo negligible errors, the sum of the graphs in Fig. 1.17 give rise to the following geometric series:

$$\frac{1}{\alpha - e(p) - i\eta} + \frac{\lambda^2\theta(p)}{(\alpha - e(p) - i\eta)^2} + \frac{\left(\lambda^2\theta(p)\right)^2}{(\alpha - e(p) - i\eta)^3} + \ldots = \frac{1}{\alpha - \left(e(p) + \lambda^2\theta(p)\right) - i\eta}$$

This justifies defining the *renormalized propagator*

$$w(p) := e(p) + \lambda^2\theta(p)$$

and the above calculation indicates that all immediate recollisions can be taken into account by simply replacing $e(p)$ with $w(p)$. This is in fact can be rigorously proved up to the leading order we are interested.

An alternative way to see the renormalization is to reorganize how the original Hamiltonian is split into main and perturbation terms:

$$H = \underbrace{e(p) + \lambda^2\theta(p)}_{w(p)} + \lambda V - \lambda^2\theta(p).$$

The precise definition of the correction term $\theta(p)$ is determined by the following self-consistent equation:

$$\theta(p) := \int \frac{dq}{w(p) - w(q) + i0}.$$

The formula (1.77) is in fact only the solution to this equation up to order λ^2, but for our purposes such precision is sufficient. The imaginary part of θ can also be computed as

$$\sigma(p) := \eta \int \frac{dq}{|w(p) - w(q) + i\eta|^2} \to \mathrm{Im}\theta(p) \qquad \eta \to 0$$

and notice that it is not zero. In particular, the renormalization *regularizes the propagator*: while the trivial supremum bound on the original propagator is

$$\sup_p \left| \frac{1}{\alpha - e(p) - i\eta} \right| \le \eta^{-1}$$

the similar bound on the renormalized propagator is much better:

$$\left| \frac{1}{\alpha - w(p) - i\eta} \right| \le \frac{1}{\lambda^2 + \eta}. \tag{1.78}$$

Strictly speaking, this bound does not hold if $p \approx 0$ since $\operatorname{Im}\theta(p)$ vanishes at the origin, but such regime in the momentum space has a small volume, since it is given by an approximate point singularity, while the (almost) singularity manifold of (1.76) is large, it has codimension one.

The bound (1.78) will play a crucial role in our estimates. Recall that due to the exponential prefactor $e^{2\eta t}$ in (1.71), eventually we will have to choose $\eta \sim 1/t$. Thus in the diffusive scaling, when $t \gg \lambda^{-2}$, the bound (1.78) is a substantial improvement.

The precise formulas are not particularly interesting; the main point is that after renormalization only the ladder has classical contribution and gives the limiting equation. If one uses the renormalized propagators, then one can assume that no immediate repetitions occur in the expansion (in practice they do occur, but they are algebraically cancelled out by the renormalization of the propagator). Only after this renormalization will the value of the ladder graph given in Fig. 1.15 be correct with the exponential damping factor. From now on we will assume that the renormalization is performed, the propagator is $w(p)$ and there is no immediate recollision. In particular, the statements of the key Theorems 1.17 and 1.18 are understood *after this renormalization*.

1.8.5   Control of the crossing terms

The key to the proof of Theorem 1.18 is a good classification of all Feynman diagrams based upon the complexity of the permutation $\pi : A \to B$. This complexity is expressed by a degree $d(\pi)$ that is defined, temporarily, as follows. We point out that this is a slightly simplified definition, the final definition is a bit more involved and is given later in Definition 1.23.

Definition 1.20 *Given a permutation $\pi \in S_n$ on $\{1, 2, \ldots, n\}$, an index i is called* **ladder index** *if $|\pi(i) - \pi(i-1)| = 1$ or $|\pi(i) - \pi(i+1)| = 1$. The* **degree of the permutation** *π is defined as*

$$d(\pi) = \#\{non\text{-}ladder\ indices\}. \tag{1.79}$$

Two examples are shown in Fig. 1.18. The top pictures show the pairing in the Feynman diagrams, the bottom pictures show the graph of the permutation as a function

$$\pi : \{1, 2, \ldots, n\} \to \{1, 2, \ldots, n\},$$

where the set $\{1, 2, \ldots, n\}$ is naturally embedded into the reals and we conveniently joined the discrete points of the graph of π. The dotted region encircles the ladder indices: notice that a long antiladder also has many ladder indices. This indeed shows that it is not really the number of total crosses in the Feynman graph that is responsible for the smallness of the value of the Feynman graph, rather the disordered structure and that is more precisely expressed by the nonladder indices.

The main philosophy is that most permutations have a high degree (combinatorial complexity). The key estimates are the following two lemmas. The first lemma

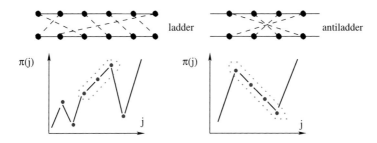

Fig. 1.18 Ladder and antiladder.

estimates the number of permutations with a given degree; the proof is an elementary combinatorial counting and will not be presented here.

Lemma 1.21 *The number of permutations with a given degree is estimated as*

$$\#\{\pi \ : \ d(\pi) = d\} \leq (Cn)^d.$$

The second lemma is indeed the hard part of the proof; it claims that the value of the Feynman graph decreases polynomially (in λ) as the degree increases. Thus we can *gain a λ factor per each nonladder vertex* of the Feynman graph.

Lemma 1.22 *There exists some positive κ, depending on the dimension d, such that*

$$\mathrm{Val}(\pi) \leq (C\lambda)^{\kappa d(\pi)}. \tag{1.80}$$

Combining these two lemmas, we can easily control the series $\sum_\pi \mathrm{Val}(\pi)$:

$$\sum_{\pi \in S_n} \mathrm{Val}(\pi) = \sum_{d=0}^{\infty} \sum_{\pi:d(\pi)=d} \mathrm{Val}(\pi) = \sum_d C^d n^d \lambda^{\kappa d} < \infty$$

if $n \leq K \sim \lambda^{-\kappa}$ (assume $\delta = 0$ for simplicity). Since $n \sim \lambda^2 t$, we get convergence for $t \leq c\lambda^{-2-\kappa}$, that is the κ from Lemma 1.22 determines the time scale for which our proof is valid.

We remark that, although for technical reasons we can prove (1.80) only for very small κ, it should be valid up to $\kappa = 2$ but not beyond. To see this, recall that the best possible estimate for a single Feynman graph is $O(\lambda^{2n})$ and their total number is $n!$. Since

$$\lambda^{2n} n! \approx (\lambda^2 n)^n \approx (\lambda^4 t)^n,$$

the summation over all Feynman graphs will diverge if $t \geq \lambda^{-4}$. This means that this method is limited up to times $t \ll \lambda^{-4}$. Going beyond $t \sim \lambda^{-4}$ requires a second resummation procedure, namely the resummation of the so-called *four-legged* subdiagrams. In other words, one cannot afford to estimate each Feynman graph individually, a certain cancellation mechanism among them has to be found. Similar resummations have been done in many-body problems in euclidean (imaginary time) theories but not in real time. In the current problem it is not clear even on the intuitive level which diagrams cancel each other.

1.8.6   An example

In this very concrete example we indicate why the cross is smaller than the ladder by a factor λ^2, that is we justify the first estimate in Fig. 1.16. We recall that the value of a Feynman graph of order n is

$$\text{Val}(\pi) = \lambda^{2n} e^{2\eta t} \int \mathbf{dp} \mathbf{dp'} d\alpha d\beta \, e^{it(\alpha-\beta)} \prod_{j=0}^{n} \frac{1}{\alpha - \overline{\omega}(p_j) - i\eta} \frac{1}{\beta - \omega(p'_j) + i\eta} \Delta_\pi(\mathbf{p}, \mathbf{p'}),$$

(1.81)

where the delta function comes from momentum conservation. Notice that the integrand is a function that are almost singular on the level sets of the dispersion relation, that is on the manifolds

$$\alpha = \text{Re}\,\omega(p_j) = e(p_j) + \lambda^2 \text{Re}\,\theta(p_j), \qquad \beta = \text{Re}\,\omega(p'_j) = e(p'_j) + \lambda^2 \text{Re}\,\theta(p'_j).$$

For the typical values of α and β, these are manifolds of codimension d in the $2(n+1)d$ dimensional space of momenta. The singularities are regularized by the imaginary parts of the denominators, $\text{Im}\,\omega(p_j) + \eta$, which are typically positive quantities of order $O(\lambda^2 + \eta)$. The main contribution to the integral comes from regimes of integration near these 'almost singularity' manifolds. We remark that in the continuum model the momentum space extends to infinity, so in principle one has to control the integrals in the large momentum (ultraviolet) regime as well, but this is only a technicality.

The main complication in evaluating and estimating the integral (1.81) comes from the delta functions in $\Delta_\pi(\mathbf{p}, \mathbf{p'})$ since they may enhance singularity overlaps and may increase the value of the integral. Without these delta functions, each momentum integration dp_j and dp'_j could be performed independently and the total value of the Feynman graph would be very small, of order $\lambda^{2n} |\log \lambda|^{2(n+1)}$. It is exactly the overlap of these (almost) singularities, forced by the delta functions, that is responsible for the correct size of the integral.

To illustrate this effect, we consider the simplest example for a cross and compare it with the direct pairing (ladder). The notations are found in Fig. 1.19. For simplicity, we assume that $\lambda^2 = \eta$, that is we are in the kinetic regime and the extra regularization coming from the renormalization, $\text{Im}\,\theta$, is not important.

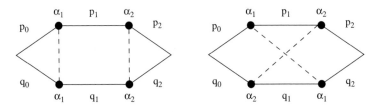

Fig. 1.19 Ladder and cross.

In case of the ladder, the delta functions force all paired momenta to be exactly the same (assuming that $p_0 = p'_0$ since we compute the L^2-norm of ψ_t), that is

$$\text{Ladder} \quad \Longrightarrow \quad p_j = p'_j, \quad \forall j = 0, 1, 2.$$

In contrast, the crossing delta functions yield the following relations:

$$p'_0 = p_0, \quad p'_1 = p_0 - p_1 + p_2, \quad p'_2 = p_2.$$

Now we can compute explicitly:

$$\text{Val(ladder)} = \lambda^4 \, e^{2\eta t} \int d\alpha d\beta \, e^{i(\alpha-\beta)t} \int \frac{1}{\alpha - \omega(p_0) - i\eta} \frac{1}{\alpha - \omega(p_1) - i\eta} \frac{1}{\alpha - \omega(p_2) - i\eta}$$

$$\times \frac{1}{\beta - \omega(p_0) + i\eta} \frac{1}{\beta - \omega(p_1) + i\eta} \frac{1}{\beta - \omega(p_2) + i\eta} dp_0 dp_1 dp_2.$$

$$(1.82)$$

Choosing $\eta = 1/t$, with a simple analysis one can verify that the main contribution comes from the regime where $|\alpha - \beta| \lesssim t^{-1}$, thus effectively all singularities overlap (with a precision η). We have

$$\int \frac{dp}{|\alpha - \omega(p) + i\eta|^2} \sim \frac{1}{|\text{Im}\,\omega|} \sim \lambda^{-2}. \tag{1.83}$$

The intermediate relations are not completely correct as they stand because ω depends on the momentum and its imaginary part actually vanishes at $p = 0$. However, the volume of this region is small. Moreover, in the continuum case the ultraviolet regime needs attention as well, but in that case the decaying form factor $|\widehat{B}|^2$ is also present. A more careful calculation shows that the final relation in (1.83) is nevertheless correct.

Thus, effectively, we have

$$\text{Val(Ladder)} \sim \lambda^4 (\lambda^{-2})^2 \sim O(1) \tag{1.84}$$

modulo logarithmic corrections. Here (1.83) has been used twice and the last pair of denominators integrated out by $d\alpha d\beta$ collecting a logarithmic term:

$$\int \frac{d\alpha}{|\alpha - \omega(p_0) - i\eta|} = O(|\log \eta|) \tag{1.85}$$

(recall that $\eta = 1/t \sim \lambda^2$). This estimate again is not completely correct as it stands, because the integral in (1.85) is logarithmically divergent at infinity, but the ultraviolet regime is always trivial in this problem. Technically, for example, here one can save a little more α decay from the other denominators in (1.82). Both inequalities (1.83) and (1.85) hold both in the discrete and continuum case. In Appendix 1.13 we list them and some more complicated related estimates that will also be used more precisely.

As a rule of thumb, we should keep in mind that the main contribution in all our integrals come from the regimes where:

(i) The two dual variables to the time are close with a precision $1/t$, that is

$$|\alpha - \beta| \le 1/t;$$

(ii) The momenta are of order one and away from zero (that is there is no ultraviolet or infrared issue in this problem);

(iii) The variables α, β are also of order one, that is there is no divergence at infinity for their integrations.

By an *alternative argument* one can simply estimate all (but one) β-denominators in (1.82) by an L^∞-bound (1.78), that is by $O(\lambda^{-2})$, then integrate out $d\beta$ and then all α-denominators, using (1.85) and the similar L^1-bound

$$\int \frac{dp}{|\alpha - w(p) - i\eta|} = O(|\log \eta|) \tag{1.86}$$

for the momentum integrals. The result is

$$\mathrm{Val(Ladder)} \lesssim \lambda^4 (\lambda^{-2})^2 (\log \lambda)^4 = (\log \lambda)^4$$

which coincides with the previous calculation (1.84) modulo logarithmic terms. In fact, one can verify that these are not only upper estimates for the ladder, but in fact the value of the n-th order ladder is

$$\mathrm{Val(Ladder)} \sim O_n(1) \tag{1.87}$$

as $\lambda \to 0$, modulo $|\log \lambda|$ corrections.

The similar calculation for the crossed diagram yields:

$$\mathrm{Val(cross)} = \lambda^4 \int d\alpha d\beta \; e^{i(\alpha - \beta)t} \int \frac{1}{\alpha - w(p_0) - i\eta} \frac{1}{\alpha - w(p_1) - i\eta} \frac{1}{\alpha - w(p_2) - i\eta}$$

$$\times \frac{1}{\beta - w(p_0) + i\eta} \frac{1}{\beta - w(p_0 - p_1 + p_2) + i\eta} \frac{1}{\beta - w(p_2) + i\eta} dp_0 dp_1 dp_2. \tag{1.88}$$

Assuming again that the main contribution is from the regime where $\alpha \sim \beta$ (with precision $1/t$), we notice that spherical singularities of the two middle denominators overlap only at a point singularity

$$\int dp_1 \frac{1}{|\alpha - w(p_1) - i\eta|} \frac{1}{|\alpha - w(p_0 - p_1 + p_2) + i\eta|} \sim \frac{1}{|p_0 + p_2| + \lambda^2} \tag{1.89}$$

(see also Fig. 1.20 and Lemma 1.28 for a more precise estimate). In three dimensions the point singularity is harmless (will disappear by the next integration using (1.112) from Lemma 1.28), and we thus obtain

$$\mathrm{Val(cross)} \le \lambda^2 \mathrm{Val(ladder)}$$

(modulo logarithms).

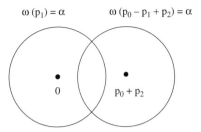

Fig. 1.20 Overlap of the shifted level sets in the p_1-space.

We emphasize that inequality (1.89) in this form holds only for the continuous dispersion relation (see (1.111) in Appendix 1.13), that is *if the level sets of the dispersion relation are convex*, since it relies on the fact that a convex set and its shifted copy overlap transversally, thus a small neighbourhood of these sets (where $|\alpha - e(p)|$ is small) have a small intersection. This is wrong for the level sets of the discrete dispersion relation which, for a certain range of α, is not a convex set (see Fig. 1.21). However, (1.89) holds with an additional factor $\eta^{-3/4}$ on the left hand side (see (1.114)), which is a weaker estimate than in the continuous case but it is still useful because it is stronger than the trivial estimate η^{-1} obtained by by taking the L^∞-norm of one propagator on the left-hand side of (1.89).

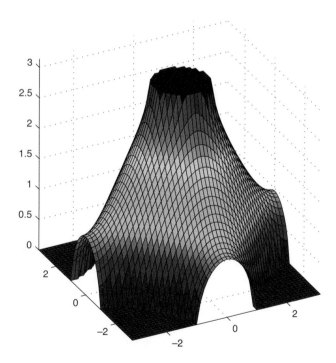

Fig. 1.21 A level set of the discrete dispersion relation $e(p)$ from (1.29).

1.9   Integration of general Feynman graphs

1.9.1   Formulas and the definition of the degree

We introduce a different encoding for the permutation π by using permutation matrices. To follow the standard convention of labelling the rows and columns of a matrix by positive natural numbers, we will shift all indices in (1.71) by one, that is from now on we will work with the formula

$$\mathrm{Val}(\pi) = \lambda^{2n} e^{2\eta t} \int \mathrm{dp}\,\mathrm{dp'}\,\mathrm{d}\alpha\,\mathrm{d}\beta \; e^{it(\alpha-\beta)} \prod_{j=1}^{n+1} \frac{1}{\alpha - \overline{\omega}(p_j) - i\eta} \, \frac{1}{\beta - \omega(p_j') + i\eta} \, \Delta_\pi(\mathbf{p}, \mathbf{p}') ,$$

(1.90)

where

$$\Delta_\pi(\mathbf{p}, \mathbf{p}') := \delta(p_{n+1} - p_{n+1}') \prod_{j=1}^{n} \delta\Big((p_{j+1} - p_j) - (p_{\pi(j)+1}' - p_{\pi(j)}') \Big).$$

We introduce a convenient notation. For any $(n+1) \times (n+1)$ matrix M and for any vector of momenta $\mathbf{p} = (p_1, \ldots p_{n+1})$, we let $M\mathbf{p}$ denote the following $(n+1)$-vector of momenta

$$M\mathbf{p} := \Big(\sum_{j=1}^{n+1} M_{1j} p_j, \; \sum_{j=1}^{n+1} M_{2j} p_j, \ldots \Big).$$

(1.91)

The permutation $\pi \in S_n$ acting on the indices $\{1, 2, \ldots, n\}$ in (1.90) will be encoded by an $(n+1) \times (n+1)$ matrix $M(\pi)$ defined as follows

$$M_{ij}(\pi) := \begin{cases} 1 & \text{if} & \tilde{\pi}(j-1) < i \leq \tilde{\pi}(j) \\ -1 & \text{if} & \tilde{\pi}(j) < i \leq \tilde{\pi}(j-1) \\ 0 & \text{otherwise,} \end{cases}$$

(1.92)

where, by definition, $\tilde{\pi}$ is the **extension** of π to a permutation of $\{0, 1, \ldots, n+1\}$ by $\tilde{\sigma}(0) := 0$ and $\tilde{\sigma}(n+1) := n+1$. In particular $[M\mathbf{p}]_1 = p_1$, $[M\mathbf{p}]_{n+1} = p_{n+1}$. It is easy to check that

$$\Delta_\pi(\mathbf{p}, \mathbf{p}') = \prod_{j=1}^{n+1} \delta\Big(p_j' - [M\mathbf{p}]_j \Big),$$

(1.93)

in other words, each p'-momentum can be expressed as a linear combination of p-momenta, the matrix M encodes the corresponding coefficients, and these are all the relations among the p and p' momenta that are enforced by Δ_π. In particular, all p-momenta are independent.

The rule to express p'-momenta in terms of p-momenta is transparent in the graphical representation of the Feynman graph: the momentum p_j appears in those p_i'-momenta which fall into its 'domain of dependence', that is the section between the image of the two endpoints of p_j, and the sign depends on the ordering of these images (Fig. 1.22). Notice that the roles of $\mathbf{p}$ and $\mathbf{p}'$ are symmetric, we could have expressed

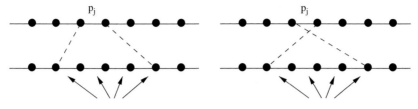

These primed momenta are equal $+p_j$ $+ \dots$ These primed momenta are equal $-p_j$ $+ \dots$

Fig. 1.22 Domain of dependencies of the momenta.

the p-momenta in terms of p'-momenta as well. It follows from this symmetry that $M(\pi^{-1}) = [M(\pi)]^{-1}$ and

$$\Delta_\pi(\mathbf{p}, \mathbf{p}') = \prod_{j=1}^{n+1} \delta\left(p_j - [M^{-1}\mathbf{p}']_j \right)$$

also holds.

A little linear algebra and combinatorics reveals that M is actually a **totally unimodular** matrix, which means that all its subdeterminants are 0 or ± 1. This will mean that the Jacobians of the necessary changes of variables are always controlled.

The following definition is crucial. It establishes the necessary concepts to measure the complexity of a permutation.

Definition 1.23. (Valley, peak, slope, and ladder) *Given a permutation $\pi \in \mathcal{S}_n$ let $\tilde{\sigma}$ be its extension. A point $(j, \pi(j))$, $j \in I_n = \{1, 2, \dots, n\}$, on the graph of π is called* **peak** *if $\pi(j) < \min\{\tilde{\pi}(j-1), \tilde{\pi}(j+1)\}$, it is called* **valley** *if $\pi(j) > \max \{\tilde{\pi}(j-1), \tilde{\pi}(j+1)\}$. Furthermore, if $\pi(j) - 1 \in \{\tilde{\pi}(j-1), \tilde{\pi}(j+1)\}$ and $(j, \pi(j))$ is not a valley, then the point $(j, \pi(j))$, $j \in I_n$, is called* **ladder**. *Finally, a point $(j, \pi(j))$, $j \in I_n$, on the graph of π is called* **slope** *if it is not a peak, valley, or ladder.*

Let $I = \{1, 2, \dots, n+1\}$ denote the set of row indices of M. This set is partitioned into five disjoint subsets, $I = I_p \cup I_v \cup I_\ell \cup I_s \cup I_{last}$, such that $I_{last} := \{n+1\}$ is the last index, and $i \in I_p, I_v, I_\ell$ or I_s depending on whether $(\pi^{-1}(i), i)$ is a peak, valley, ladder or slope, respectively. The cardinalities of these sets are denoted by $p := |I_p|$, $v := |I_v|$, $\ell := |I_\ell|$ and $s := |I_s|$. The dependence on π is indicated as $p = p(\pi)$ etc. if necessary. We define the **degree** *of the permutation π as*

$$\deg(\pi) := d(\pi) := n - \ell(\pi) . \tag{1.94}$$

Remarks: (i) The terminology of peak, valley, slope, ladder comes from the graph of the permutation $\tilde{\pi}$ drawn in a coordinate system where the axis of the dependent variable, $\pi(j)$, is oriented downward (see Fig. 1.23). It immediately follows from the definition of the extension $\tilde{\pi}$ that the number of peaks and valleys is the same:

$$p = v.$$

By the partitioning of I, we also have

$$p + v + \ell + s + 1 = n + 1.$$

(ii) The nonzero entries in the matrix $M(\sigma)$ follow the same geometric pattern as the graph: each downward segment of the graph corresponds to a column with a few consecutive 1's, upward segments correspond to columns with (-1)'s. These blocks of nonzero entries in each column will be called the **tower** of that column. In Fig. 1.23 we also pictured the towers of $M(\pi)$ as rectangles.

(iii) Because our choice of orientation of the vertical axis follows the convention of labelling rows of a matrix, a peak is a local minimum of $j \to \pi(j)$. We fix the convention that the notions 'higher' or 'lower' for objects related to the vertical axis (e.g. row indices) always refer to the graphical picture. In particular the 'bottom' or the 'lowest element' of a tower is located in the row with the highest index.

Also, a point on the graph of the function $j \to \pi(j)$ is traditionally denoted by $(j, \pi(j))$, where the first coordinate j runs on the horizontal axis, while in the labelling of the (i, j)–matrix element M_{ij} of a matrix M the first coordinate i labels rows, that is it runs vertically. To avoid confusion, we will always specify whether a double index (i, j) refers to a point on the graph of π or a matrix element.

(iv) We note that for the special case of the identity permutation $\pi = id = id_n$ we have $I_p = I_s = I_v = \emptyset$, and $I_\ell = \{1, 2, \ldots, n\}$. In particular, $\deg(id) = 0$ and $\deg(\pi) \geq 2$ for any other permutation $\pi \neq id$.

(v) Note that the definition in (1.94) slightly differs from the preliminary definition of the degree given in (1.79). Not every index participating in a ladder ia defined as a ladder index; the top index of the ladder is excluded (to avoid overcounting) and the bottom index of a ladder is also excluded if it is a valley, since the valleys will play a special role. The precise definition of the degree given in (1.94) is not canonical, other variants are also possible; the essential point is that long ladders should reduce the degree.

An example is shown on Fig. 1.23 with $n = 8$. The matrix corresponding to the permutation on this figure is the following (zero entries are left empty)

$$M(\pi) := \begin{pmatrix} 1 & & & & & \\ & 1 & & & & \\ & & 1 & & & \\ & & 1 & & -1 & 1 \\ & & 1 & & -1 & 1 \\ & & 1 & -1 & & 1 \\ & & 1 & -1 & & 1 \\ & & & & & 1 \end{pmatrix} \begin{matrix} 1\ \ell \\ 2\ \ell \\ 3\ p \\ 4\ \ell \\ 5\ s \\ 6\ \ell \\ 7\ v \\ 8\ s \\ 9\ (last) \end{matrix} \qquad (1.95)$$

The numbers on the right indicate the column indices and the letters show whether it is peak/slope/valley/ladder or last. In this case $I_p = \{3\}$, $I_v = \{7\}$, $I_s = \{5, 8\}$. $I_\ell = \{1, 2, 4, 6\}$, $I_{last} = \{9\}$ and $\deg(\sigma) = 4$.

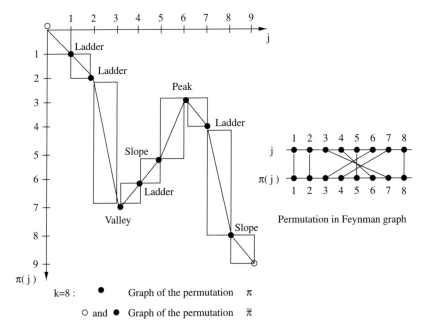

Fig. 1.23 Graph of a permutation with the towers.

1.9.2 Failed attempts to integrate out Feynman diagrams

To estimate the value of a Feynman graph, based upon (1.90) and (1.93), our task is to successively integrate out all p_j's in the multiple integral of the form

$$Q(M) := \lambda^{2n} \int d\alpha d\beta \int d\mathbf{p} \prod_{i=1}^{n+1} \frac{1}{|\alpha - \omega(p_i) - i\eta|} \frac{1}{|\beta - \omega\left(\sum_{j=1}^{n+1} M_{ij} p_j\right) + i\eta|}$$

As usual, the unspecified domains of integrations for the α and β variables is $\mathbb{R}$, and for the $d\mathbf{p}$ momentum variables is $(\mathbf{T}^d)^{n+1}$. As one p_j is integrated out, the matrix M needs to be updated and effective algorithm is needed to keep track of these changes.

Why is it hard to compute or estimate quite such an explicit integral? The problem is that there is nowhere to start: each integration variable p_j may appear in many denominators: apparently there is no 'easy' integration variable at the beginning.

As a **first attempt**, we can try to decouple the interdependent denominators by Schwarz or Hölder inequalities. It turns out that they cannot be used effectively, since by their application we will lose the nonoverlapping effects imposed by the crossing. If one tries to decouple these integrals trivially, one obtains the ladder and gains nothing. In contrast to a complicated crossing diagram, the ladder is easily computable because $p_i = p_i'$ means that the integrals decouple:

$$Q(M) \le \lambda^{2n} \int d\alpha d\beta \int d\mathbf{p} \frac{1}{|\alpha - \omega(p_{n+1}) - i\eta|} \frac{1}{|\beta - \omega(p_{n+1}) + i\eta|} \tag{1.96}$$

$$\times \left[\prod_{i=1}^{n} \frac{1}{|\alpha - \omega(p_i) - i\eta|^2} + \frac{1}{|\beta - \omega\left(\sum_{j=1}^{n} \widetilde{M}_{ij} p_j\right) + i\eta|^2} \right],$$

where $\widetilde{M}$ denotes the $n \times n$ upper minor of M and we used that M and $\widetilde{M}$ differ only with an entry 1 in the diagonal, so in particular $\sum_{j=1}^{n+1} M_{n+1,j} p_j = p_{n+1}$ and $\det(M) = \det(\widetilde{M})$. Changing variables in the second term $(p'_i = \sum_{j=1}^{n} \widetilde{M}_{ij} p_j)$ and using that the Jacobian is one $(\det(M) = \pm 1)$, we see that the first and second terms in the parenthesis are exactly the same. Thus

$$Q(M) \le 2\lambda^{2n} \int d\alpha d\beta \int d\mathbf{p} \frac{1}{|\alpha - \omega(p_{n+1}) - i\eta|} \frac{1}{|\beta - \omega(p_{n+1}) + i\eta|} \prod_{i=1}^{n} \frac{1}{|\alpha - \omega(p_i) - i\eta|^2},$$

and we can succesively integrate out the momenta $p_1, p_2, \ldots p_n$ and then finally α, β. Using the inequalities (1.83) and (1.85), we obtain (with the choice $\eta = 1/t$)

$$|Val(\pi)| \le e^{2\eta t} Q(M) \le C\lambda^{2n} \lambda^{-2n} (|\log \eta|^2) = C(|\log \eta|^2)$$

that is essentially the same estimate as the value of the ladder (1.87).

As a **second attempt**, we can trivially estimate all (but one) β-denominators, by their L^∞ norm

$$\left| \frac{1}{\beta - \omega(\ldots) + i\eta} \right| \le \frac{1}{\lambda^2 |\mathrm{Im}\omega|} \sim \lambda^{-2} \tag{1.97}$$

(see (1.78) and the remark afterward on its limitations), and then integrate out all p_j one by one by using the L^1-bound (1.86)

$$\int \frac{d p}{|\alpha - \omega(p) + i\eta|} = O(|\log \eta|).$$

This gives

$$|Val(\pi)| \le O(|\log \eta|^{n+2}),$$

that is it is essentially order 1 with logarithmic corrections. Note that this second method gives a worse exponent for the logarithm than the first one, nevertheless this method will be the good starting point.

The second attempt is a standard integration procedure in diagrammatic perturbation theory. The basic situation is that given a large graph with momenta assigned to the edges, each edge carries a propagator, that is a function depending in this momenta, and momenta are subject to momentum conservation (Kirchoff law) at each vertex. The task is to integrate out all momenta. The idea is to use L^∞-bounds for the propagator on certain edges in the graph to decouple the rest and use independent L^1-bounds for the remaining integrations. Typically one constructs a spanning tree in the graph, then each additional edge creates a loop. The Kirchoff delta functions amount to expressing all tree momenta in terms of loop momenta in a trivial way. Then one uses L^∞-bound on 'tree'-propagators to free up all Kirchoff delta functions, integrate out the 'tree'-variables freely and then use L^1-bound on the 'loop'-propagators that are now independent.

This procedure can be used to obtain a rough bound on values of very general Feynman diagrams subject to Kirchoff laws. We will formalize this procedure in Appendix 1.14. Since typically L^1 and L^∞ bounds on the propagators scale with a different power of the key parameter of the problem (in our case λ), we will call this method *the power counting estimate*.

In our case simple graph theoretical counting shows that

$$\text{Number of tree momenta} = \text{number of loop momenta} = n + 1.$$

In fact, after identifying the paired vertices in the Feynman diagram (after closing the two long horizontal lines at the two ends), we obtain a new graph where one can easily see that the edges in the lower horizontal line, that is the edges corresponding p'_j-momenta, form a spanning tree and the edges of all p_j-variables form loops. Thus the delta functions in (1.93) represent the way how the tree momenta are expressed in terms of the loop momenta.

Since each L^∞-bound on 'tree'-propagators costs λ^{-2} by (1.97), the total estimate would be of order

$$\lambda^{2n}\lambda^{-2(n+1)}$$

with actually logarithmic factors. But due to the additional α, β integrations, one L^∞-bound can be saved (modulo logarithm) by using (1.85). So the total estimate is, in fact,

$$\lambda^{2n}\lambda^{-2n} = O(1)$$

modulo logarithmic factors, that is the same size as for the ladder diagrams. The conclusion is that even in the second attempt, with the systematic power counting estimate, we did not gain from the crossing structure of the permutation either.

1.9.3 The new algorithm to integrate out Feynman diagrams

In our new algorithm, we use the L^∞-bound on the propagators for a carefully selected subset of the 'tree'-variables: the ones that lie 'above the peak'. The selection is shown in Fig. 1.24 for a concrete example. Notice that the segments in the horizontal axis correspond to $p_1, p_2, \ldots, p_{n+1}$, that is the loop momenta, and the segments in the vertical axis correspond to the tree momenta.

We will first explain the main idea, then we will work out a concrete example to illustrate the algorithm. After drawing the graph of the permutation, a structure of valleys and peaks emerges. By carefully examining the relation between this graph and the matrix M, one notices that if one estimates only those 'tree'-propagators that lie above a peak by the trivial L^∞-bound, then all the remaining propagators can be successively integrated out by selecting the integration variables p_j in an appropriate

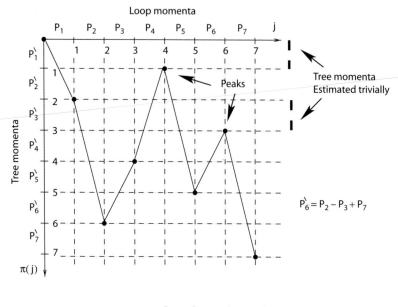

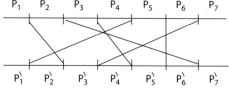

Fig. 1.24 Valleys and peaks determine the order of integration.

order (dictated by the graph). Each integration involves no more than two propagators at one time, hence it can be estimated by elementary calculus (using estimates collected in Appendix 1.13).

In fact, this is correct if there are no ladder indices; but momenta corresponding to ladder indices can be integrated out locally (see later). As it turns out, ladder indices are neutral; their integration yields an $O(1)$ factor. The gain comes from nonladder indices, and this will justify the definition of the degree (1.94).

More precisely, after an easy bookkeeping we notice that in this way we gain roughly as many λ^2 factors as many slopes and valleys we have (no gain from peaks or ladders). Since the number of peaks and valleys are the same,

$$p = v,$$

and the peaks, valleys, slopes, and ladders altogether sum up to n,

$$p + v + \ell + s = n$$

(see Remark (i) after Definition 1.23), we see that

$$v + s = n - p - \ell \geq \frac{1}{2}(n - \ell) = \frac{1}{2}d(\pi),$$

since $n - 2p - \ell = n - p - v - \ell = s \geq 0$. Thus we gain at least $\lambda^{d(\pi)}$. This would prove (1.80) with $\kappa = 1$. After various technical estimates that were neglected here, the actual value of κ is reduced, but it still remains positive and this completes the proof of Lemma 1.22.

An example without ladder

As an example, we will show the integration procedure for the graph on Fig. 1.24. This is a permutation which has no ladder index for simplicity, we will comment on ladder indices afterwards. The total integral is (for simplicity, we neglected the $\pm i\eta$ regularization in the formulas below to highlight the structure better)

$$\lambda^{12} \int d\alpha d\beta d\mathbf{p} \frac{1}{|\alpha - \omega(p_1)|} \frac{1}{|\alpha - \omega(p_2)|} \frac{1}{|\alpha - \omega(p_3)|} \frac{1}{|\alpha - \omega(p_4)|} \frac{1}{|\alpha - \omega(p_5)|} \frac{1}{|\alpha - \omega(p_6)|} \frac{1}{|\alpha - \omega(p_7)|}$$

$$\times \frac{1}{|\beta - \omega(p_1)|} \frac{1}{|\beta - \omega(p_1 - p_4 + p_5)|} \frac{1}{|\beta - \omega(p_2 - p_4 + p_5)|} \frac{1}{|\beta - \omega(p_2 - p_4 + p_5 - p_6 + p_7)|}$$

$$\times \frac{1}{|\beta - \omega(p_2 - p_3 + p_5 - p_6 + p_7)|} \frac{1}{|\beta - \omega(p_2 - p_3 + p_7)|} \frac{1}{|\beta - \omega(p_7)|}. \tag{1.98}$$

Notice that every variable appears in at least three denominators, for a graph of order n, typically every momentum appears in $(const)n$ different denominators, so there is no way to start the integration procedure (an integral with many almost singular denominators is impossible to estimate directly with a sufficient precision).

Now we estimate those β-factors by L^∞-norm (that is by λ^{-2} according to (1.97)) whose corresponding primed momenta lie right above a peak; in our case these are the factors

$$\frac{1}{|\beta - \omega(p_1')|} \frac{1}{|\beta - \omega(p_3')|}$$

that is the first and the third β-factor in (1.98). After this estimate, they will be removed from the integrand. We will lose a factor $(\lambda^{-2})^p = \lambda^{-4}$ recalling that $p = 2$ is the number of peaks. We are left with the integral

$$\lambda^{12}(\lambda^{-2})^p \int d\alpha d\beta d\mathbf{p} \frac{1}{|\alpha - \omega(p_1)|} \frac{1}{|\alpha - \omega(p_2)|} \frac{1}{|\alpha - \omega(p_3)|} \frac{1}{|\alpha - \omega(p_4)|} \frac{1}{|\alpha - \omega(p_5)|} \frac{1}{|\alpha - \omega(p_6)|}$$

$$\times \frac{1}{|\alpha - \omega(p_7)|} \frac{1}{|\beta - \omega(p_1 - p_4 + p_5)|} \frac{1}{|\beta - \omega(p_2 - p_4 + p_5 - p_6 + p_7)|}$$

$$\times \frac{1}{|\beta - \omega(p_2 - p_3 + p_5 - p_6 + p_7)|} \frac{1}{|\beta - \omega(p_2 - p_3 + p_7)|} \frac{1}{|\beta - \omega(p_7)|}. \tag{1.99}$$

Now the remaining factors can be integrated out by using the following generalized version of (1.89) (see Appendix 1.13)

$$\sup_{\alpha, \beta} \int \frac{dp}{|\alpha - \omega(p) + i\eta|} \frac{1}{|\beta - \omega(p + u) - i\eta|} \leq \frac{|\log \eta|}{|u| + \lambda^2}. \tag{1.100}$$

Suppose we can forget about the possible point singularity, that is neglect the case when $|u| \ll 1$. Then the integral (1.100) is $O(1)$ modulo an irrelevant log factor. Then the successive integration is done according to graph: we will get rid of the factors $|\beta - w(p_1')|^{-1}$, $|\beta - w(p_2')|^{-1}$, $|\beta - w(p_3')|^{-1}$, and so on in this order. The factors with p_1' and p_3' have already been removed, so the first nontrivial integration will eliminate

$$\frac{1}{|\beta - w(p_2')|} = \frac{1}{|\beta - w(p_1 - p_4 + p_5)|}. \tag{1.101}$$

Which integration variable to use? Notice that the point (1,2) was not a peak, that means that there is a momentum (in this case p_1) such that p_2' is the primed momenta with the largest index that still depends on p_1 (the 'tower' of p_1 ends at p_2'). This means that among all the remaining β-factors no other β-factor, except for (1.101), involves p_1! Thus only two factors involve p_1 and not more, so we can integrate out the p_1 variable by using (1.100)

$$\int dp_1 \frac{1}{|\alpha - w(p_1)|} \frac{1}{|\beta - w(p_1 - p_4 + p_5)|} = O(1)$$

(modulo logs and modulo the point singularity problem).

In the next step we are then left with

$$\lambda^{12}(\lambda^{-2})^p \int d\alpha d\beta d\mathbf{p} \frac{1}{|\alpha - w(p_2)|} \frac{1}{|\alpha - w(p_3)|} \frac{1}{|\alpha - w(p_4)|} \frac{1}{|\alpha - w(p_5)|} \frac{1}{|\alpha - w(p_6)|}$$

$$\times \frac{1}{|\alpha - w(p_7)|} \frac{1}{|\beta - w(p_2 - p_4 + p_5 - p_6 + p_7)|}$$

$$\times \frac{1}{|\beta - w(p_2 - p_3 + p_5 - p_6 + p_7)|} \frac{1}{|\beta - w(p_2 - p_3 + p_7)|} \frac{1}{|\beta - w(p_7)|}.$$

Since we have already taken care of the β-denominators with p_1', p_2', the next one would be the β-denominator with p_3', but this was estimated trivially (and removed) at the beginning. So the next one to consider is

$$\frac{1}{|\beta - w(p_4')|} = \frac{1}{|\beta - w(p_2 - p_4 + p_5 - p_6 + p_7)|}.$$

Since p_4' is not above a peak, there is a p-momentum whose tower has the lowest point at the level 4, namely p_4. From the graph we thus conclude that p_4 appears only in this β-factor (and in one α-factor), so again it can be integrated out:

$$\int dp_4 \frac{1}{|\alpha - w(p_4)|} \frac{1}{|\beta - w(p_2 - p_4 + p_5 - p_6 + p_7)|} \leq O(1)$$

(modulo logs and point singularity).

We have then

$$\lambda^{12}(\lambda^{-2})^p \int d\alpha d\beta d\mathbf{p} \frac{1}{|\alpha - \omega(p_2)|} \frac{1}{|\alpha - \omega(p_3)|} \frac{1}{|\alpha - \omega(p_5)|} \frac{1}{|\alpha - \omega(p_6)|}$$

$$\times \frac{1}{|\alpha - \omega(p_7)|} \frac{1}{|\beta - \omega(p_2 - p_3 + p_5 - p_6 + p_7)|} \frac{1}{|\beta - \omega(p_2 - p_3 + p_7)|} \frac{1}{|\beta - \omega(p_7)|}.$$

Next,

$$\frac{1}{|\beta - \omega(p_5')|} = \frac{1}{|\beta - \omega(p_2 - p_3 + p_5 - p_6 + p_7)|}$$

includes even two variables (namely p_5, p_6) that do not appear in any other β-denominators (because the towers of p_5 and p_6 end at the level 5, in other words because right below the row of p_5' there is a valley). We can freely choose which one to integrate out, say we choose p_5, and perform

$$\int dp_5 \frac{1}{|\alpha - \omega(p_5)|} \frac{1}{|\beta - \omega(p_2 - p_3 + p_5 - p_6 + p_7)|} \leq O(1).$$

We are left with

$$\lambda^{12}(\lambda^{-2})^p \int d\alpha d\beta d\mathbf{p} \frac{1}{|\alpha - \omega(p_2)|} \frac{1}{|\alpha - \omega(p_3)|} \frac{1}{|\alpha - \omega(p_6)|}$$

$$\times \frac{1}{|\alpha - \omega(p_7)|} \frac{1}{|\beta - \omega(p_2 - p_3 + p_7)|} \frac{1}{|\beta - \omega(p_7)|}. \tag{1.102}$$

Finally

$$\frac{1}{|\beta - \omega(p_6')|} = \frac{1}{|\beta - \omega(p_2 - p_3 + p_7)|}$$

can be integrated out either by p_2 or p_3 (both towers end at the level of p_6'), for example

$$\int dp_2 \frac{1}{|\alpha - \omega(p_2)|} \frac{1}{|\beta - \omega(p_2 - p_3 + p_7)|} \leq O(1).$$

The last β-factor is eliminated by the β integration by (1.85) and then in the remaining integral,

$$\lambda^{12}(\lambda^{-2})^p \int d\alpha d\mathbf{b} \frac{1}{|\alpha - \omega(p_3)|} \frac{1}{|\alpha - \omega(p_6)|} \frac{1}{|\alpha - \omega(p_7)|},$$

one can integrate out each remaining momenta one by one. We have thus shown that the value of this permutation is

$$\mathrm{Val}(\pi) \leq \lambda^{12}(\lambda^{-2})^p = \lambda^8$$

modulo logs and point singularities.

In general, the above procedure gives

$$\lambda^{2n-2p} \leq \lambda^n = \lambda^{d(\pi)}$$

if there are no ladders, $\ell = 0$.

General algorithm including ladder indices and other fine points

Finally, we show how to deal with ladder indices. The idea is that first one has to integrate out the consecutive ladder indices after taking a trivial Schwarz inequality to decouple the α and β-factors of the consecutive ladder indices and then proceed similarly to (1.96). It is important that only propagators with ladder indices will be Schwarzed, for the rest of the integrand we will use the successive integration procedure explained in the previous section. In this way the ladder indices remain neutral for the final bookkeeping, in fact, the integral

$$\int \frac{dp}{|\alpha - \omega(p) - i\eta|^2} \sim \lambda^{-2}$$

exactly compensates the λ^2 prefactor carried by the ladder index. Actually, one needs to take care that not only the λ-powers but even the constants cancel each other as well, since an error C^n would not be affordable when n, being the typical number of the collisions, is $\lambda^{-\kappa}$. Therefore the above bound will be improved to

$$\int \frac{\lambda^2}{|\alpha - \omega(p) - i\eta|^2} dp = 1 + O(\lambda^{1-12\kappa}), \tag{1.103}$$

(see (1.116)), and it is this point where the careful choice of the remormalized propagator $\omega(p)$ plays a crucial role. This is also one error estimate which further restricts the value of κ. After integrating out the ladder indices, we perform the successive estimates made in Section 1.9.3.

There were two main issues have been swept under the rug (among several other less important ones ...). First, there are several additional log factors floating around. In fact, they can be treated generously, since with this procedure we gain a $\lambda^{(const.)d(\pi)}$ factor and the number of log factors is also comparable with $d(\pi)$. However, here the key point again is that by the ladder integration in (1.103) we do not lose any log factor; even not a constant factor!

The second complication, the issue of the point singularities, is much more serious and this accounts for the serious reduction of the value of κ in the final theorem. In higher dimensions, $d \geq 3$, one point singularity of the form $(|p| + \eta)^{-1}$ is integrable, but it may happen that the *same* point singularity arises from different integrations of the form (1.100) along our algorithm. This would yield a high multiplicity point singularity whose integral is large.

First, we indicate with an example that overlapping point singularities indeed do occur; they certainly would occur in the ladders, had we not integrated out the ladders separately.

The structure of the integral for a set of consecutive ladder indices, $\{k, k + 1, k + 2, \ldots k + m\}$ is as follows:

$$\Omega = \int dp_k dp_{k+1} \ldots dp_{k+m} \frac{1}{|\alpha - \omega(p_k)|} \frac{1}{|\alpha - \omega(p_{k+1})|} \cdots \frac{1}{|\alpha - \omega(p_{k+m})|}$$

$$\times \frac{1}{|\beta - \omega(p_k + u)|} \frac{1}{|\beta - \omega(p_{k+1} + u)|} \cdots \frac{1}{|\beta - \omega(p_{k+m} + u)|}. \tag{1.104}$$

Here we used the fact that if the consecutive ladders are the points $(k, s), (k + 1, s + 1), (k + 2, s + 2), \ldots$, then the corresponding momenta are related as

$$p_k - p'_s = p_{k+1} - p'_{s+1} = p_{k+2} - p'_{s+2} = \cdots$$

that is one can write $p'_{s+i} = p_{k+i} + u$ with a fixed vector u (that depends on all other momenta but not on $p_k, p_{k+1}, \ldots p_{k+m}$, e.g. $u = p_2 - p_4 + p_9$).

Using (1.100) successively, we obtain

$$\Omega \leq \left(\frac{1}{|u| + \lambda^2} \right)^{m+1},$$

that is the **same** point singularity arises from each integration. We call this phenomenon *accumulation of point singularities*. Since u is a linear combination of other momenta, that need to be integrated out, at some point we would be faced with

$$\int \left(\frac{1}{|u| + \lambda^2} \right)^{m+1} du \sim \left(\lambda^{-2} \right)^{m-2}. \tag{1.105}$$

Since $m + 1$ consecutive ladder indices carry a factor $\lambda^{2(m+1)}$, we see that from a consecutive ladder sequence we might gain only λ^6, irrespective of the length of the sequence. It turns out that even this calculation is a bit too optimistic; the formula (1.105) is only rough caricature, also propagators depending on u are present in the u integration. In reality, eventually, the ladder indices are integrated out to $O(1)$.

Based upon this example, one may think that the accumulation of point singularities occurs only in the presence of the consecutive ladders; once they are taken care of separately (and not gaining but also not losing from them), the remaining graph can be integated out without any accumulation of point singularities. We conjecture that this is indeed the case, but we could not quite prove this. Instead, we estimate several more β-factors (propagators with 'tree momenta') by the trivial L^∞-bound to surely exclude this scenario. They are carefully chosen (see the concept of 'uncovered slope indices' in Definition 10.3 of (Erdős, Salmhofer, and Yau 2008)) so that after their trivial removal, along the successive integration algorithm for the remaining propagators, indeed no accumulation of point singularity occurs.

This completes the sketch of the proof of estimating the nonrepetition Feynman diagrams, that is the proof of Theorem 1.18.

1.10   Feynman graphs with repetitions

In this short section we indicate how to prove Theorem 1.17. Recall that ψ_s^{err} contains terms with many $(n \gg \lambda^{-\kappa})$ potentials and terms with (not immediate) recollisions.

The estimate of the terms with a large number of collisions is relatively easy; these are still nonrepetition graphs, so the integration procedure from Section 1.9 applies. Permutations with degree $d(\pi) \geq C/\kappa$ have a much smaller contribution than the required $o(t^{-2})$ error. Permutations with low complexity contain macroscopically long ladders (that is ladders with length cn with some positive constant c) and these can be computed quite precisely. The precise calculation reveals a factor $1/(cn)!$ due to the time-ordered integration (see discussion around (1.59)) and the value of such graphs can be estimated by

$$\frac{(\lambda^2 t)^n}{(cn)!} e^{-c'\lambda^2 t}.$$

The combinatorics of such low complexity graphs is at most $n^{C/\kappa}$, so their total contribution is negligible if $n \gg \lambda^2 t \sim \lambda^{-\kappa}$.

The repetition terms from ψ_s^{err} require a much more laborous treatment. The different repetition patterns are shown on Fig. 1.25. When one of the repetitions shows up (apart from immediate repetition that were renormalized), we stop the expansion to reduce complications (see (1.61)). Actually the stopping rule is a bit more involved, because the repetition pattern has to collect sufficient 'repetitions' to render that term sufficiently small even after paying the price for the unitary estimate, but we will not go into these details. Finally, each term we compute by 'bare hand', after applying a certain graph surgery to reduce the number of different cases. One example of this reduction is shown on Fig. 1.26 while Fig. 1.27 shows the result of an explicit estimate. The explicit estimates rely on the fact that the repetition imposes a certain momentum restriction that reduces the volume of the maximal overlap of singularities, reminiscent of the mechanism behind the crossing estimate, Section 1.8.6. Unfortunately, even after the graph surgery reduction, still a considerable number of similar but not identical cases have to be estimated on a case by case basis.

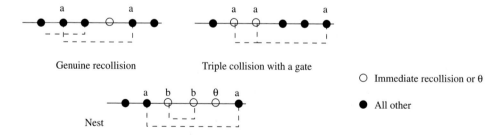

Fig. 1.25   Various repetition patterns.

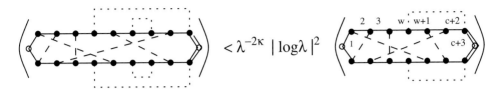

$$< \lambda^{-2\kappa} \, |\log\lambda\,|^2$$

Fig. 1.26 Removal of a gate.

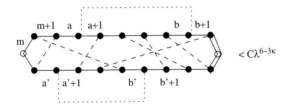

$$< C\lambda^{6-3\kappa}$$

Fig. 1.27 Two sided recollision.

1.10.1 Higher order pairings: the lumps

We also mention how to deal with the higher order pairings that we postponed from Section 1.8.2. Recall that the stopping rule monitored whether we have a (non-immediate) repetition in the collision sequence $A = (\alpha_1, \alpha_2, \ldots, \alpha_n)$. This procedure has a side effect when computing the expectation:

$$\mathbf{E}\prod \overline{\widehat{V}_{\alpha_j}(p_{j+1} - p_j)} \widehat{V}_{\alpha_j}(q_{j+1} - q_j) = \sum_{\alpha_\ell \neq \alpha_k} \prod_j e^{i\alpha_j [p_{j+1} - p_j - (q_{j+1} - q_j)]}$$

The non-repetition restriction $\alpha_\ell \neq \alpha_k$ destroys the precise delta function $\sum_\alpha e^{i\alpha p} = \delta(p)$. This requires us to employ the following connected graph formula that expresses the restricted sum as a linear combination of momentum delta functions:

Lemma 1.24. (connected graph formula) *Let $\mathcal{A}_n$ be the set of partitions of $\{1, \ldots, n\}$. There exist explicit coefficients $c(k)$ such that*

$$\sum_{\alpha_\ell \neq \alpha_k} e^{iq_j\alpha_j} = \sum_{\mathbf{A}\in\mathcal{A}_n} \prod_\nu c(|A_\nu|)\delta\Big(\sum_{\ell\in A_\nu} q_\ell\Big) \qquad \mathbf{A} = (A_1, A_2, \ldots),$$

where the summation is over all partitions $\mathbf{A}$ of the set $\{1, \ldots, n\}$, and $A_1, A_2, \ldots$, denote the elements of a fixed partition, in particular they are disjoint sets and $A_1 \cup A_2 \cup \ldots = \{1, \ldots, n\}$.

The appearence of nontrivial partitioning sets means that instead of pairs we really have to consider 'hyperpairs', that is subsets (called *lumps*) of size more than 2.

Lumps have large combinatorics, but the value of the corresponding Feynman graph is small since in a large lump many more obstacle indices must coincide than in a pairing. Nevertheless, their treatment would require setting up a separate notation and run the algorithm of Section 1.9 for general lumps instead of pairs.

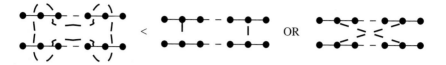

Fig. 1.28   Breaking up lumps.

To avoid these complications and repeating a similar proof, we invented a graph surgery that reduces lumps to pairs. The idea is that lumps are artificially broken up into pairs (that is permutations) and the corresponding Feynman graphs can be compared. The break-up is not unique, for example. Fig. 1.28 shows two possible break-ups of a lump of four elements into two pairs.

There are many break-ups possible, but the key idea is that we choose the break-up that gives the biggest $d(\pi)$. The following combinatorial statement shows a lower estimate on the maximal degree, demonstrating that every lump can be broken up into an appropriate pairing with relatively high degree.

Proposition 1.25 *Let* $\mathbf{B} = \{B_1, B_2, \ldots\}$ *be a partition of vertices.*

$$s(\mathbf{B}) := \frac{1}{2} \sum \{ |B_j| \; : \; |B_j| \geq 4 \}$$

Then there exists a permutation π, *compatible with* $\mathbf{B}$ *such that* $d(\pi) \geq \frac{1}{2} s(\mathbf{B})$.

Using this lemma we reduce the estimate of hyperpairs to the previous case involving only pairs. This completes the very sketchy ideas of the proof of Theorem 1.17.

1.11   Computation of the main term

In this section we sketch the computation of the main term, that is the proof of Theorem 1.19. It is an explicit but nontrivial computation. The calculation below is valid only in the discrete case as it uses the fact that for any incoming velocity u, the collision kernel $\sigma(u, v)$ is constant in the outgoing velocity, that is the new velocity is independent of the old one (this was the reason why the diffusion coefficient could be directly computed (see (1.45)) without computing the velocity autocorrelation function via the Green–Kubo formula (1.47) as in the continuous case.

The computation of the main term for the continuous case relates the diffusion coefficient to the underlying Boltzmann process. This procedure is conceptually more general but technically somewhat more involved (see Section 6 of (Erdős, Salmhofer, and Yau 2007a)), so here we present only the simpler method for the discrete model.

The Fourier transform of the rescaled Wigner transform $W(X/\varepsilon, V)$ is given by

$$\widehat{W_t}(\varepsilon\xi, v) = \overline{\widehat{\psi_t}\left(v + \frac{\varepsilon\xi}{2} \right)} \widehat{\psi_t}\left(v - \frac{\varepsilon\xi}{2} \right),$$

where recall that $\varepsilon = \lambda^{2+\kappa/2}$ is the space rescaling and $t = \lambda^{-2-\kappa}T$.

We want to test the Wigner transform against a macroscopic observable, that is compute

$$\langle \mathcal{O}, \mathbf{E}\widehat{W}_t \rangle = \langle \mathcal{O}(\xi, v), \mathbf{E}\widehat{W}_t(\varepsilon\xi, v) \rangle = \int dv d\xi \, \mathcal{O}(\xi, v) \, \mathbf{E}\widehat{W}_t(\varepsilon\xi, v).$$

Recall from Lemma 1.15 that the Wigner transform enjoys the following continuity property:

$$\langle \mathcal{O}, \mathbf{E}\widehat{W}_\psi \rangle - \langle \mathcal{O}, \mathbf{E}\widehat{W}_\phi \rangle \leq C\sqrt{\mathbf{E}\|\psi\|^2 + \mathbf{E}\|\phi\|^2}\sqrt{\mathbf{E}\|\psi - \phi\|^2},$$

in particular, by using (1.63), it is sufficient to compute the Wigner transform of

$$\sum_{n=0}^{K} \psi_{n,t}^{nr}.$$

We can express this Wigner transform in terms of Feynman diagrams, exactly as for the L^2-calculation. The estimate (1.65) in our key Theorem 1.18 implies that only the ladder diagrams matter (after renormalization). Thus we have

$$\langle \mathcal{O}, \mathbf{E}\widehat{W}_t \rangle \approx \sum_{k \leq K} \lambda^{2k} \int_{\mathbb{R}} d\alpha d\beta \, e^{it(\alpha-\beta)+2t\eta} \tag{1.106}$$

$$\times \int d\xi dv \mathcal{O}(\xi, v) R_\eta\left(\alpha, v + \frac{\varepsilon\xi}{2}\right) \overline{R_\eta\left(\beta, v - \frac{\varepsilon\xi}{2}\right)}$$

$$\times \prod_{j=2}^{k}\left[\int dv_j R_\eta\left(\alpha, v_j + \frac{\varepsilon\xi}{2}\right)\overline{R_\eta\left(\beta, v_j - \frac{\varepsilon\xi}{2}\right)}\right]$$

$$\times \int dv_1 R_\eta\left(\alpha, v_1 + \frac{\varepsilon\xi}{2}\right)\overline{R_\eta\left(\beta, v_1 - \frac{\varepsilon\xi}{2}\right)}\overline{\widehat{W}_0(\varepsilon\xi, v_1)},$$

with the renormalized propagator

$$R_\eta(\alpha, v) := \frac{1}{\alpha - e(v) - \lambda^2\theta(v) + i\eta}.$$

We perform each dv_j integral. The key technical lemma is the following:

Lemma 1.26 *Let* $f(p) \in C^1(\mathbb{R}^d)$ *,* $a := (\alpha + \beta)/2$*,* $\lambda^{2+4\kappa} \leq \eta \leq \lambda^{2+\kappa}$ *and fix* $r \in \mathbb{R}^d$ *with* $|r| \leq \lambda^{2+\kappa}/4$*. Then we have*

$$\int \frac{\lambda^2 f(v)}{\left(\alpha - e(v-r) - \lambda^2\bar{\theta}(v-r) - i\eta\right)\left(\beta - e(v+r) - \lambda^2\theta(v+r) + i\eta\right)} \, dv$$

$$= -2\pi i \int \frac{\lambda^2 f(v)\,\delta(e(v) - a)}{(\alpha - \beta) + 2(\nabla e)(v)\cdot r - 2i\lambda^2\mathcal{I}(a)} \, dv + o(\lambda^{1/4})$$

where

$$\mathcal{I}(a) := Im \int \frac{dv}{a - e(v) - i0} = \int \delta(e(v) - a)\, dv,$$

in particular, $\mathcal{I}(e(p)) = Im\, \theta(p)$.

The proof of Lemma 1.26 relies on the following (approximate) identity

$$\frac{1}{(\alpha - \bar{g}(v - r) - i0)(\beta - g(v + r) + i0)} \approx \frac{1}{\alpha - \beta + g(v + r) - \bar{g}(v - r)}$$

$$\times \left[\frac{1}{\beta - g(v) + i0} - \frac{1}{\alpha - \bar{g}(v) - i0} \right]$$

and on careful Taylor expansions. □

Accepting this lemma, we change variables $a = (\alpha + \beta)/2$, $b = (\alpha - \beta)/\lambda^2$ and choose $\eta \ll t^{-1}$ in (1.106). Then we get

$$\langle \mathcal{O}, \mathbf{E}\widehat{W_t} \rangle \approx \sum_{k \leq K} \int d\xi da db\, e^{it\lambda^2 b} \left(\prod_{j=1}^{k+1} \int \frac{-2\pi i F^{(j)}(\xi, v_j)\, \delta(e(v_j) - a)}{b + \lambda^{-2}\varepsilon(\nabla e)(v_j) \cdot \xi - 2i\mathcal{I}(a)} dv_j \right)$$

with $F^{(1)} := \widehat{W_0}$, $F^{(k+1)} := \mathcal{O}$, and $F^{(j)} \equiv 1$, $j \neq 1, k+1$. Here W_0 is the rescaled Wigner transform of the initial state.

Let $d\mu_a(v)$ be the normalized surface measure of the level surface $\Sigma_a := \{e(v) = a\}$ defined via the co-area formula, that is the integral of any function h w.r.t. $d\mu_a(v)$ is given by

$$\langle h \rangle_a := \int h(v) d\mu_a(v) := \frac{\pi}{\mathcal{I}(a)} \int_{\Sigma_a} h(q) \frac{dm(q)}{|\nabla e(q)|}$$

where $dm(q)$ is the Lebesgue measure on the surface Σ_a. Often this measure is denoted by $\delta(e(v) - a)dv$, that is

$$\int h(v) d\mu_a(v) = \frac{\pi}{\mathcal{I}(a)} \int h(v)\delta(e(v) - a)dv.$$

Let $H(v) := \frac{\nabla e(v)}{2\mathcal{I}(a)}$. Then we have

$$\langle \mathcal{O}, \mathbf{E}\widehat{W_t} \rangle \approx 2\mathcal{I}(a) \sum_{k \leq K} \int d\xi \int_{\mathbb{R}} da db\, e^{i2t\lambda^2 \mathcal{I}(a)b} \left(\prod_{j=1}^{k+1} \int \frac{-i F^{(j)}(\xi, v_j)}{b + \lambda^{-2}\varepsilon H(v_j) \cdot \xi - i} d\mu_a(v_j) \right).$$

We expand the denominator up to second order

$$\int \frac{-i}{b + \varepsilon \lambda^{-2} H(v) \cdot \xi - i} \, d\mu_a(v) \qquad (1.107)$$

$$= \frac{-i}{b-i} \int \left[1 - \frac{\varepsilon \lambda^{-2} H(v) \cdot \xi}{b-i} + \frac{\varepsilon^2 \lambda^{-4} [H(v) \cdot \xi]^2}{(b-i)^2} + O\left((\varepsilon \lambda^{-2} |\xi|)^3 \right) \right] d\mu_a(v).$$

After summation over k, and recalling that $\xi = O(1)$ due the decay in the observable $\mathcal{O}$, the effect of the last (error) term is $K(\varepsilon \lambda^{-2})^3 = \lambda^{-\kappa} (\lambda^{\kappa/2})^3 = o(1)$, thus we can keep only the first three terms on the right-hand side of (1.107).

The linear term cancels out by symmetry: $H(v) = -H(-v)$. To compute the quadratic term, we define

$$D(a) := 4\mathcal{I}(a) \int d\mu_a(v) \, H(v) \otimes H(v),$$

which is exactly the diffusion matrix (1.43). Thus

$$\langle \mathcal{O}, \mathbf{E}\widehat{W}_t \rangle \approx \sum_{k \leq K} \int d\xi \int_{\mathbb{R}} da \, 2\mathcal{I}(a) \int \widehat{W}_0(\varepsilon \xi, v_1) d\mu_a(v_1) \int \mathcal{O}(\xi, v) d\mu_a(v)$$

$$\times \int_{\mathbb{R}} db \, e^{2i\lambda^2 \mathcal{I}(a)tb} \left(\frac{-i}{b-i} \right)^{k+1} \left[1 + \frac{\varepsilon^2 \lambda^{-4} \langle \xi, D(a)\xi \rangle}{4\mathcal{I}(a)} \frac{1}{(b-i)^2} \right]^{k-1}.$$

Setting

$$B^2 := \frac{\varepsilon^2 \lambda^{-4} \langle \xi, D(a)\xi \rangle}{4\mathcal{I}(a)},$$

the arising geometric series can be summed up:

$$\sum_{k=0}^{\infty} \left(\frac{-i}{b-i} \right)^{k+1} \left[1 + \frac{B^2}{(b-i)^2} \right]^{k+1} = (-i) \frac{(b-i)^2 + B^2}{(b-i)^3 + i(b-i)^2 + iB^2}.$$

We now perform the db integration analytically by using the formula (with $A := 2\lambda^2 \mathcal{I}(a)$):

$$(-i) \int_{\mathbb{R}} db \, e^{itAb} \frac{(b-i)^2 + B^2}{(b-i)^3 + i(b-i)^2 + iB^2} = 2\pi e^{-tAB^2} + o(1)$$

from the dominant residue $b = iB^2$. We can then compute

$$tAB^2 = \varepsilon^2 \lambda^{-4-\kappa} \frac{T}{2} \langle \xi, D(a)\xi \rangle = \frac{T}{2} \langle \xi, D(a)\xi \rangle.$$

In particular, this formula shows that to get a nontrivial limit, ε has to be chosen as $\varepsilon = O(\lambda^{-2-\kappa/2})$, that is the diffusive scaling emerges from the calculation. Finally, we have

$$\langle \mathcal{O}, \mathbf{E}\widehat{W} \rangle \approx \int \mathrm{d}\xi \int_{\mathbb{R}} \mathrm{d}a \, \mathcal{I}(a) \Big(\int \mathcal{O}(\xi, v) \mathrm{d}\mu_a(v) \Big) \langle \widehat{W_0} \rangle_a \exp \Big(-\frac{T}{2} \langle \xi, D(a)\xi \rangle \Big)$$

where

$$f(T, \xi, a) := \langle \widehat{W_0} \rangle_a \exp \Big(-\frac{T}{2} \langle \xi, D(a)\xi \rangle \Big)$$

is the solution of the heat equation (1.42) in Fourier space. This completes the sketch of the calculation of the main term and hence the proof of Theorem 1.19. ☐

1.12 Conclusions

As a conclusion, we informally summarize the main achievements.

(1) We rigorously proved diffusion from a Hamiltonian quantum dynamics in an environment of fixed time independent random scatterers in the weak coupling regime. The quantum dynamics is described up to a time scale $t \sim \lambda^{-2-\kappa}$ with some $\kappa > 0$, that is well beyond the kinetic time scale. The typical number of collisions converges to infinity.

(2) We identified the quantum diffusion constant (or matrix) and we showed that it coincides with the diffusion constant (matrix) obtained from the long term evolution of the Boltzmann equation. This shows that the two-step limit (kinetic limit of quantum dynamics followed by the scaling limit of the random walk) yields the same result, up to the leading order, as the one-step limit (diffusive limit of quantum dynamics).

(3) We controlled the interferences of random waves in a multiple scattering process with infinitely many collisions with random obstacles in the extended states regime.

(4) In agreement with (2), we showed that quantum interferences and memory effects do not become relevant up to our scale. We remark that this is expected to hold for any κ in $d = 3$, but *not expected* to hold for $d = 2$ (localization).

(5) As our main technical achievement, we classified and estimated Feynman graphs up to *all orders*. We gained an extra λ-power *per each nonladder vertex* compared with the usual power counting estimate relying on the L^{∞} and L^1-bounds for the propagators and on the tree and loop momenta.

1.13 Appendix: Inequalities

In this appendix we collect the precise bounds on integrals of propagators that are used in the actual proofs. Their proofs are more or less elementary calculations, however the calculations are much easier for the continuous dispersion relation. For details, see the Appendix B of (Erdős, Salmhofer, and Yau 2007a) for the continuous model and in Appendix A of (Erdős, Salmhofer, and Yau 2007b) for the discrete case.

We define the weighted Sobolev norm

$$\|f\|_{m,n} := \sum_{|\alpha| \le n} \|\langle x \rangle^m \partial^\alpha f(x)\|_\infty$$

with $\langle x \rangle := (2 + x^2)^{1/2}$ (here α is a multi-index).

Lemma 1.27. (Continuous dispersion relation (Erdős, Salmhofer, and Yau 2007a)) *Let $e(p) = \frac{1}{2}p^2$. Suppose that $\lambda^2 \ge \eta \ge \lambda^{2+4\kappa}$ with $\kappa \le 1/12$. Then we have,*

$$\int \frac{|h(p-q)|\mathrm{d}p}{|\alpha - \omega(p) + i\eta|} \le \frac{C\|h\|_{2d,0} |\log \lambda| \log\langle \alpha \rangle}{\langle \alpha \rangle^{1/2} \langle |q| - \sqrt{2|\alpha|}\rangle}, \tag{1.108}$$

and for $0 \le a < 1$

$$\int \frac{|h(p-q)|\mathrm{d}p}{|\alpha - e(p) + i\eta|^{2-a}} \le \frac{C_a \|h\|_{2d,0} \, \eta^{-2(1-a)}}{\langle \alpha \rangle^{a/2} \langle |q| - \sqrt{2|\alpha|}\rangle}. \tag{1.109}$$

For $a = 0$ and with $h := \widehat{B}$, the following more precise estimate holds. There exists a constant C_0, depending only on finitely many $\|B\|_{k,k}$ norms, such that

$$\int \frac{\lambda^2 |\widehat{B}(p-q)|^2 \, \mathrm{d}p}{|\alpha - \overline{\omega}(p) - i\eta|^2} \le 1 + C_0 \lambda^{-12\kappa} \left[\lambda + |\alpha - \omega(q)|^{1/2}\right]. \tag{1.110}$$

One can notice that an additional decaying factor $h(p-q)$ must be present in the estimates due to the possible (but irrelevant) ultraviolet divergence. In the applications $h = \widehat{B}$. Moreover, an additional decaying factor in α was also saved, this will help to eliminate the logarithmic divergence of the integral of the type

$$\int_{\mathbb{R}} \frac{\mathrm{d}\alpha}{|\alpha - c + i\eta|} \sim |\log \eta|$$

modulo the logarithmic divergence at infinity.

The following statement estimates singularity overlaps:

Lemma 1.28 *Let $\mathrm{d}\mu(p) = \mathbf{1}(|p| \le \zeta)\mathrm{d}p$, in applications $\zeta \sim \lambda^{-\kappa}$. For any $|q| \le C\lambda^{-1}$*

$$I_1 := \int \frac{\mathrm{d}\mu(p)}{|\alpha - e(p) + i\eta| \, |\beta - e(p+q) + i\eta|} \le \frac{C\zeta^{d-3}|\log \eta|^2}{\|q\|} \tag{1.111}$$

$$I_2 := \int \frac{\mathrm{d}\mu(p)}{|\alpha - e(p) + i\eta| \, |\beta - e(p+q) + i\eta|} \frac{1}{\|p - r\|} \le \frac{C\eta^{-1/2}\zeta^{d-3}|\log \eta|^2}{\|q\|} \tag{1.112}$$

uniformly in r, α, β. Here $\|q\| := |q| + \eta$.

To formulate the statement in the discrete case, we redefine

$$\|q\| := \eta + \min\{|q - \gamma| \; : \; \gamma \text{ is a critical point of } e(p)\}$$

for any momentum q on the d-dimensional torus. It is easy to see that there are 2^d critical points of the discrete dispersion relation $e(p) = \sum_{j=1}^{d}(1 - \cos p^{(j)})$.

Lemma 1.29. (Discrete dispersion relation (Erdős, Salmhofer, and Yau 2007b)) *The following bounds hold for the dispersion relation $e(p) = \sum_{j=1}^{d}(1 - \cos p^{(j)})$ in $d \geq 3$ dimensions.*

$$\int \frac{\mathrm{d}p}{|\alpha - e(p) + i\eta|} \frac{1}{\|p - r\|} \leq c|\log \eta|^3 , \tag{1.113}$$

$$I := \int \frac{\mathrm{d}p}{|\alpha - e(p) + i\eta|} \frac{1}{|\beta - e(p + q) + i\eta|} \leq \frac{c\eta^{-3/4}|\log \eta|^3}{\|q\|} \tag{1.114}$$

$$I(r) := \int \frac{\mathrm{d}p}{|\alpha - e(p) + i\eta|} \frac{1}{|\beta - e(p + q) + i\eta|} \frac{1}{\|p - r\|} \leq \frac{c\eta^{-7/8}|\log \eta|^3}{\|q\|} \tag{1.115}$$

uniformly in r, α, β. With the (carefully) renormalized dispersion relation it also holds that

$$\sup_{\alpha} \int \frac{\lambda^2 \, \mathrm{d}p}{|\alpha - \overline{\omega}(p) - i\eta|^2} \leq 1 + C_0 \lambda^{1-12\kappa} \tag{1.116}$$

(compare with (1.110)).

Because the gain in the integrals (1.114), (1.115) compared with the trivial estimate $\eta^{-1}|\log \eta|$, is quite weak (compare with the much stronger bounds in Lemma 1.28 for the continuous case), we need one more inequality that contains four denominators. This special combination occurs in the estimates of the recollision terms. The proof of this inequality is considerably more involved than the above ones, see (Erdős and Salmhofer 2007). The complication is due to the lack of convexity of the level sets of the dispersion relation (see Fig. 1.21). We restrict our attention to the $d = 3$ dimensional case, as it turns out, this complication is less serious in higher dimensions.

For any real number α we define

$$\|\alpha\| := \min\{|\alpha|, |\alpha - 2|, |\alpha - 3|, |\alpha - 4|, |\alpha - 6|\} \tag{1.117}$$

in the $d = 3$ dimensional model. The values $0, 2, 4, 6$ are the critical values of $e(p)$. The value $\alpha = 3$ is special, for which the level surface $\{e(p) = 3\}$ has a flat point. In general, in $d \geq 3$ dimensions, $\|\alpha\|$ is the minimum of $|\alpha - d|$ and of all $|\alpha - 2m|$, $0 \leq m \leq d$.

Lemma 1.30 *(Four Denominator Lemma (Erdős and Salmhofer 2007))* *For any $\Lambda > \eta$ there exists C_Λ such that for any $\alpha \in [0, 6]$ with $\|\alpha\| \geq \Lambda$,*

$$\mathcal{I} = \int \frac{\mathrm{d}p\mathrm{d}q\mathrm{d}r}{|\alpha - e(p) + i\eta||\alpha - e(q) + i\eta||\alpha - e(r) + i\eta||\alpha - e(p - q + r + u) + i\eta|} \leq C_\Lambda |\log \eta|^{14} \tag{1.118}$$

uniformly in u.

The key idea behind the proof of Lemma 1.30 is to study the decay property of the Fourier transform of the level sets $\Sigma_a := \{p : e(p) = a\}$, that is the quantity

$$\widehat{\mu}_a(\xi) = \int_{\Sigma_a} e^{ip\xi}\mathrm{d}p := \int_{\Sigma_a} \frac{e^{ip\xi}}{|\nabla e(p)|}\mathrm{d}m_a(p),$$

where m_a is the uniform (euclidean) surface measure on Σ_a. Defining

$$I(\xi) = \int \frac{e^{ip\xi}\mathrm{d}p}{|\alpha - e(p) + i\eta|},$$

we get by the co-area formula that

$$I(\xi) = \int_0^6 \frac{\mathrm{d}a}{|\alpha - a + i\eta|}\widehat{\mu}_a(\xi).$$

From the convolution structure of $\mathcal{I}$, we have

$$\mathcal{I} = \int I(\xi)^4 e^{-iu\xi}\mathrm{d}\xi \le \int |I(\xi)|^4 \mathrm{d}\xi \le \left(\int_0^6 \frac{\mathrm{d}a}{|\alpha - a + i\eta|}\right)^4 \sup_a \int |\widehat{\mu}_a(\xi)|^4 \mathrm{d}\xi.$$

The first factor on the right-hand side is of order $|\log\eta|^4$, so the key question is the decay of the Fourier transform, $\widehat{\mu}_a(\xi)$, of the level set, for large ξ. It is well known (and follows easily from stationary phase calculation) that for surfaces $\Sigma \subset \mathbb{R}^3$ whose Gauss curvature is strictly separated away from zero (e.g. strictly convex surfaces), the Fourier transform

$$\widehat{\mu}_\Sigma(\xi) := \int_\Sigma e^{ip\xi}\,\mathrm{d}p$$

decays as

$$|\widehat{\mu}_\Sigma(\xi)| \le \frac{C}{|\xi|}$$

for large ξ. Thus for such surfaces the L^4-norm of the Fourier transform is finite. For surfaces, where only one principal curvature is nonvanishing, the decay is weaker, $|\xi|^{-1/2}$. Our surface Σ_a even has a flat point for $a = 3$. A detailed analysis shows that although the decay is not sufficient uniformly in all direction ξ, the exceptional directions and their neighbourhoods are relatively small so that the L^4-norm is still finite. The precise statement is more involved, here we flash up the main result. Let $K = K(p)$ denote the Gauss curvature of the level set Σ_a and let $\nu(p)$ denote the outward normal direction at $p \in \Sigma_a$.

Lemma 1.31. (Decay of the Fourier transform of Σ_a, (Erdős and Salmhofer 2007)) *For $\nu \in S^2$ and $r > 0$ we have*

$$\widehat{\mu}_a(r\nu) \lesssim \frac{1}{r} + \frac{1}{r^{3/4}|D(\nu)|^{1/2} + 1},$$

where $D(\nu) = \min_j |\nu(p_j) - \nu|$ where p_j's are finitely many points on the curve $\Gamma = \{K = 0\} \cap \Sigma_a$, at which the neutral direction of the Gauss map is parallel with Γ.

This lemma provides sufficient decay for the L^4-norm to be only logarithmically divergent, which can be combined with a truncation argument to obtain (1.118).

We stated all above estimates (with the exception of the precise (1.110) and (1.116)) for the bare dispersion relation $e(p)$. The modification of these estimates for the renormalized dispersion relation $\omega(p)$ is straightforward if we are prepared to lose an additional $\eta^{-\kappa}$ factor by using

$$\frac{1}{|\alpha - \omega(p) + i\eta|} \leq \frac{1}{|\alpha - e(p) + i\eta|} + \frac{c\lambda^2}{|\alpha - e(p) + i\eta||\alpha - \omega(p) + i\eta|} \leq \frac{1 + c\lambda^2\eta^{-1}}{|\alpha - e(p) + i\eta|}.$$

1.14   Appendix: Power counting for integration of Feynman diagrams

Lemma 1.32 *Let Γ be a connected oriented graph with set of vertices $V(\Gamma)$ and set of edges $E(\Gamma)$. Let $N = |V(\Gamma)|$ be the number of vertices and $K = |E(\Gamma)|$ be the number of edges. Let $p_e \in \mathbb{R}^d$ denote a (momentum) variable associated with the edge $e \in E(\Gamma)$ and $\boldsymbol{p}$ denotes the collection $\{p_e : e \in E(\Gamma)\}$. Since the graph is connected, $K \geq N - 1$. Let $R : \mathbb{R}^d \to \mathbb{C}$ denote a function with finite L^1 and L^∞ norms, $\|R\|_1$ and $\|R\|_\infty$. For any vertex $w \in V(\Gamma)$, let*

$$\Omega_w := \int \Delta_w(\boldsymbol{p}) \prod_{e \in E(\Gamma)} R(p_e) dp_e,$$

where

$$\Delta_w(\boldsymbol{p}) = \prod_{\substack{v \in V(\Gamma) \\ v \neq w}} \delta\Big(\sum_{e\,:\,e \sim v} \pm p_e \Big) \tag{1.119}$$

is a product of delta functions expressing Kirchoff laws at all vertices but w. Here $e \sim v$ denotes the fact that the edge e is adjacent to the vertex v and $\pm$ indicates that the momenta have to be summed up according to the orientation of the edges with respect to the vertex. More precisely, if the edge e is outgoing with respect to the vertex v, then the sign is minus, otherwise it is plus.

Then the integral Ω_w is independent of the choice of the special vertex w and the following bound holds

$$|\Omega_w| \leq \|R\|_1^{K-N+1} \|R\|_\infty^{N-1}.$$

Proof. First notice that the arguments of the delta functions for *all* $v \in V(\Gamma)$,

$$\sum_{e\,:\,e \sim v} \pm p_e,$$

sum up to zero (every p_e, $e \in E(\Gamma)$, appears exactly twice, once with plus once with minus). This is the reason why one delta function had to be excluded from the product (1.119), otherwise they would not be independent. It is trivial linear algebra to see that, independently of the choice of w, all Δ_w determine the same linear subspace in the space of momenta, $(\mathbb{R}^d)^{E(\Gamma)} = \mathbb{R}^{dK}$.

Let T be a spanning tree in Γ and let $\mathcal{T} = E(T)$ denote the edges in Γ that belong to T. Let $\mathcal{L} = E(\Gamma) \setminus \mathcal{T}$ denote the remaning edges, called 'loops'. Momenta associated with T, $\{p_e : e \in \mathcal{T}\}$, are called tree-momenta, the rest are called loop-momenta. Under Kirchoff's law at the vertices, all tree momenta can be expressed in terms of linear combinations (with coefficients ± 1) of the loop momenta:

$$p_e = \sum_{a \in \mathcal{L}} \sigma_{e,a} p_a, \qquad \forall e \in \mathcal{T}, \tag{1.120}$$

where the coefficients $\sigma_{e,a}$ can be determined as follows. Consider the graph $T \cup \{a\}$, that is adjoin the edge a to the spanning tree. This creates a single loop L_a in the graph that involves some elements of $\mathcal{T}$. We set $\sigma_{e,a} = 1$ if $e \in L_a$ and the orientation of e and a within this loop coincide. We set $\sigma_{e,a} = -1$ if the orientation of e and a are opposite, and finally $\sigma_{e,a} = 0$ if $e \notin L_a$.

It is easy to check that the $N - 1$ linear relations (1.120) are equivalent to the ones determined by the product (1.119) of delta functions. For, if all tree momenta are defined as linear combinations of the loop momenta given by (1.120), then the Kirchoff law is trivially satisfied. To check this, notice that when summing up the edge momenta at each vertex, only those loop momenta p_a appear whose loop L_a contains v. If $v \in L_a$, then p_a appears exactly twice in the sum, once with plus and once with minus, since the momenta p_a flowing along the loop L_a once enters and once exits at the vertex v. Thus the relations (1.120) imply the relations in (1.119), and both sets of relations have the same cardinality, $N - 1$. Since the relations (1.120) are obviously independent (the tree momenta are all different), we obtain that both sets of relations determine the same subspace (of codimension $d(N - 1)$) of $\mathbb{R}^{dK}$.

Thus we can rewrite

$$\Omega = \Omega_w = \int \left(\prod_{e \in E(\Gamma)} R(p_e) \mathrm{d}p_e \right) \prod_{e \in \mathcal{T}} \delta\left(p_e - \sum_{a \in \mathcal{L}} \sigma_{e,a} p_a \right);$$

in particular, Ω_w is independent of w.

Now we estimate all factors $R(p_e)$ with $e \in \mathcal{T}$ by L^∞-norm, then we integrate out all tree momenta and we are left with the L^1-norms of the loop momenta:

$$|\Omega| \le \|R\|_\infty^{N-1} \int \left(\prod_{e \in \mathcal{L}} |R(p_e)| \mathrm{d}p_e \right) \left(\prod_{e \in \mathcal{T}} \delta\left(p_e - \sum_{a \in \mathcal{L}} \sigma_{e,a} p_a \right) \mathrm{d}p_e \right) = \|R\|_\infty^{N-1} \|R\|_1^{K-N+1}$$

completing the proof of Lemma 1.32. $\square$

Acknowledgement

The author is grateful to M. Salmhofer for many suggestions to improve this presentation and for making his own lecture notes available from which, in particular, the material of Section 1.7.2 and 1.8.3 was borrowed.

References

Adami, R. L. Erdős, (2008): Rate of decoherence for an electron weakly coupled to a phonon gas. *J. Statis. Physics* **132**, (2): 301–328.

Aizenman, M. and S. Molchanov (1993): Localization at large disorder and at extreme energies: an elementary derivation, *Comm. Math. Phys.* **157**: 245–278.

Aizenman, M. R. Sims, S. Warzel, (2006): Absolutely continuous spectra of quantum tree graphs with weak disorder. *Comm. Math. Phys.* **264**: 371–389.

Anderson, P. (1958): Absences of diffusion in certain random lattices, *Phys. Rev.* **109**: 1492–1505.

Benedetto, D. F. Castella, R. Esposito and M. Pulvirenti (2008): From the N-body Schrodinger equation to the quantum Boltzmann equation: a term-by-term convergence result in the weak coupling regime. *Comm. Math. Phys.* **277**: 1–44.

Brown, R. Philosophical Magazine N. S. **4** 161–173 (1828), and **6**: 161–166 (1829).

Bourgain, J. (2003): Random lattice Schrödinger operators with decaying potential: some higher dimensional phenomena. *Lecture Notes in Mathematics*, Vol. 1807: 70–99.

Bunimovich, L., Y. Sinai (1980/81): Statistical properties of Lorentz gas with periodic configuration of scatterers. *Comm. Math. Phys.* **78** (4): 479–497.

Chen, T. (2005): Localization Lengths and Boltzmann Limit for the Anderson Model at Small Disorders in Dimension 3. *J. Stat. Phys.* **120**: 279–337.

De Roeck, W., J. Fröhlich: Diffusion of a massive quantum particle coupled to a quasi-free thermal medium in dimension $d \geq 4$. *Comm. Math. Phys.* **303** (3), 613–707 (2011). *Preprint.* http://arxiv.org/abs/0906.5178

De Roeck, W., J. Fröhlich, A. Pizzo (2010): Quantum Brownian motion in a simple model system. *Comm. Math. Phys.* **293**, (2): 361–398.

Disertori, M., Spencer, T.: Anderson localization for a supersymmetric sigma model. *Preprint Comm. Math. Phys.* **300** (3), 659–671 (2010). arXiv:0910.3325.

Disertori, M., Spencer, T., Zirnbauer, M.: Quasi-diffusion in a 3D Supersymmetric Hyperbolic Sigma Model. *Comm. Math. Phys.* **300** (2), 435–486 (2010). *Preprint* arXiv:0901.1652.

Dürr, D., S. Goldstein, J. Lebowitz (1980/81): A mechanical model of Brownian motion. *Comm. Math. Phys.* **78** (4): 507–530.

Einstein, A. (1905): Über die von der molekularkinetischen Theorie der Wärme geforderte Bewegung von in ruhenden Flüssigkeiten suspendierten Teilchen, **17**, 549–560, and: Zur Theorie der Brownschen Bewegung, (1906) *Ann. der Physik*, **19**: 180.

Efetov, K., 1999: *Supersymmetry in Disorder and Chaos*, Cambridge University Press.

Eng D. and L. Erdős, (2005) The Linear Boltzmann Equation as the Low Density Limit of a Random Schrödinger Equation. *Rev. Math. Phys*, **17**, (6): 669–743.

Erdős, L. (2002): Linear Boltzmann equation as the long time dynamics of an electron weakly coupled to a phonon field. *J. Stat. Phys.*, **107**: 1043–1128.

Erdős, L., Salmhofer, M. (2007): Decay of the Fourier transform of surfaces with vanishing curvature. *Math. Z.* **257**, (2): 261–294.

Erdős, L. and H.-T. Yau, (1998): Linear Boltzmann equation as scaling limit of quantum Lorentz gas. Advances in Differential Equations and Mathematical Physics. *Contemporary Mathematics* **217**: 137–155.

Erdős, L. and H.-T. Yau, (2000): Linear Boltzmann equation as the weak coupling limit of the random Schrödinger equation, *Comm. Pure Appl. Math.* **LIII**: 667–735.

Erdős, L., Salmhofer, M. and H.-T. Yau, (2006): Towards the quantum Brownian motion. Lecture Notes in Physics, **690**, Mathematical Physics of Quantum Mechanics, Selected and Refereed Lectures from QMath9. Eds. *Joachim Asch and Alain Joye.* pp. 233–258.

Erdős, L., Salmhofer, M. and H.-T. Yau, (2007a): Quantum diffusion of the random Schrodinger evolution in the scaling limit II. The recollision diagrams. *Comm. Math. Phys.* **271**: 1–53.

Erdős, L., Salmhofer, M. and H.-T. Yau, (2007b): Quantum diffusion for the Anderson model in scaling limit. *Ann. Inst. H. Poincare* **8**: 621–685.

Erdős, L., Salmhofer, M. and H.-T. Yau, (2008): Quantum diffusion of the random Schrodinger evolution in the scaling limit. *Acta Math.* **200**, (2): 211–277.

Froese, R., Hasler, D. W. and Spitzer, (2007): Absolutely continuous spectrum for the Anderson model on a tree: a geometric proof of Klein's theorem. *Comm. Math. Phys.* **269**: 239–257.

Fröhlich, J., Martinelli, F. Scoppola E. and Spencer, T. (1985): Constructive proof of localization in the Anderson tight binding model. *Comm. Math. Phys.* **101**, (1): 21–46.

Fröhlich, J. and T. Spencer, (1983): Absence of diffusion in the Anderson tight binding model for large disorder or low energy, *Comm. Math. Phys.* **88**: 151–184.

Germinet, F., Klein, A. Schenker J. (2007): Dynamical delocalization in random Landau Hamiltonians. *Ann. of Math.* **166**: 215–244.

Ho, T. G., Landau L. J. and Wilkins A. J. (1993): On the weak coupling limit for a Fermi gas in a random potential. *Rev. Math. Phys.* **5**, (2): 209–298.

Ya, I., Goldsheid, Molchanov S. A. and Pastur, L. A. (1997): A pure point spectrum of the one dimensional Schrödinger operator. *Funct. Anal. Appl.* **11**: 1–10.

Kang, Y., Schenker J. (2009): Diffusion of wave packets in a Markov random potential. *J. Stat. Phys.* **134**: 1005–1022.

Kesten, H., Papanicolaou G. (1980/81): A limit theorem for stochastic acceleration. *Comm. Math. Phys.* **78**: 19–63.

Klein, A. (1994): Absolutely continuous spectrum in the Anderson model on the Bethe lattice, *Math. Res. Lett.* **1**: 399–407.

Komorowski, T., Ryzhik L. (2006): Diffusion in a weakly random Hamiltonian flow. *Comm. Math. Phys.* **263**: 277–323.

Lanford, O. (1976): On the derivation of the Boltzmann equation. *Astérisque* **40**: 117–137.

Lee, P. A., Ramakrishnan, T. V. (1985): Disordered electronic systems. *Rev. Mod. Phys.* **57**: 287–337.

Lukkarinen, J., and Spohn, H. (2007): Kinetic limit for wave propagation in a random medium. *Arch. Ration. Mech. Anal.* **183**: 93–162.

Lukkarinen, J. and H. Spohn, Weakly nonlinear Schrödinger equation with random initial data. *Invent. Math.* **183**, 79–188 (2011). *Preprint* arXiv:0901.3283.

Markowich, P. and Ringhofer, C. (1989): Analysis of the quantum Liouville equation. *Z. Angew. Math. Mech*, **69**: 121–129.

Perrin, J. B. (1909): Mouvement brownien et réalité moléculaire. *Annales de chimie et de physiqe*, VIII 18: 5–114.

Pillet, C.–A. (1985): Some results on the quantum dynamics of a particle in a Markovian potential. *Comm. Math. Phys.* **102** (2): 237–254.

Simon, B. and Wolff, T. (1986): Singular continuous spectrum under rank one perturbations and localization for random Hamiltonians. Comm. Pure Appl. Math. **39**(1): 75–90.

von Smoluchowski, M. (1906): Zur kinetischen Theorie der Bwownschen Molekularbewegung und der Suspensionen. *Ann. der Physik*, **21**: 756–780.

Spohn, H., (1977): Derivation of the transport equation for electrons moving through random impurities. *J. Statist. Phys.* **17** (6): 385–412.

Vollhardt, D. Wölfle, P. (1980): Diagrammatic, self-consistent treatment of the Anderson localization problem in $d \leq 2$ dimensions. *Phys. Rev. B* **22**: 4666–4679.

Wiener, N. (1923): Differential space. *J. Math and Phys.* **58**: 131–174.

2

The temporal ultraviolet limit

Tadeusz BALABAN[1], Joel FELDMAN[2],*,
Horst KNÖRRER and Eugene TRUBOWITZ[3]

[1]Department of Mathematics, Rutgers University
110 Frelinghuysen Rd
Piscataway, NJ 08854-8019, USA

[2]Department of Mathematics, University of British Columbia
Vancouver, BC, Canada V6T1Z2

[3]Mathematik, ETH
8092 Zürich, Switzerland

*Lecture given by Joel Feldman.

Chapter Contents

2.1 Introduction

2.1.1 The physical setting

These lectures[1] concern the first, relatively small, step in a programme whose long-term goal is the mathematically rigorous construction of a standard model of a gas of bosons. Even this first step is too long and complicated to present completely here. But I will outline it and highlight a couple of the tools employed that tend to crop up quite commonly in constructions of quantum field theories and many-body models. The model of our gas of bosons is based on the following assumptions.

- Each particle in the gas has a kinetic energy. The corresponding quantum mechanical observable is an operator h. The most commonly used h is $-\frac{1}{2m}\Delta$, which corresponds to the classical kinetic energy $\frac{\mathbf{p}^2}{2m}$. (Balaban *et al.* 2010c) allows a more general class of operators like this.
- The particles in the gas interact with each other through a translationally invariant, exponentially decaying, strictly positive definite two-body potential, $2v(\mathbf{x}, \mathbf{y})$.
- The system is in the thermodynamic equilibrium given by the grand canonical ensemble with temperature $T > 0$ and chemical potential $\mu \in \mathbb{R}$. We shall not place any further restrictions on T and μ. But the most interesting temperatures are small and the most interesting chemical potentials are small and positive.

2.1.2 The Physics of interest

I'll formulate the model mathematically, carefully, later. But to get a first hint both of the expected physical behaviour and of the formalism that we shall use, consider the following, formal, functional integral representation of the partition function for this system. This representation is commonly used in the physics literature. See, for example, (Negele and Orland 1988, (2.66)).

$$\operatorname{Tr} e^{-\frac{1}{kT}(H-\mu N)} = \int \prod_{\substack{\mathbf{x}\in\mathbb{R}^3 \\ 0<\tau\le\frac{1}{kT}}} \frac{d\alpha_\tau(\mathbf{x})^* \wedge d\alpha_\tau(\mathbf{x})}{2\pi i} \; e^{\mathcal{A}(\alpha^*,\alpha)} \tag{2.1}$$

where H is the Hamiltonian, N is the number operator and the 'action'

$$\mathcal{A}(\alpha^*, \alpha) = \int_0^{\frac{1}{kT}} d\tau \int_{\mathbb{R}^3} d^3\mathbf{x} \; \alpha_\tau(\mathbf{x})^* \frac{\partial}{\partial\tau}\alpha_\tau(\mathbf{x}) - \int_0^{\frac{1}{kT}} d\tau \; K\left(\alpha_\tau^*, \alpha_\tau\right) \tag{2.2}$$

with

$$K(\alpha^*, \alpha) = \iint d\mathbf{x} d\mathbf{y} \; \alpha(\mathbf{x})^* h(\mathbf{x}, \mathbf{y})\alpha(\mathbf{y}) - \mu \int d\mathbf{x} \; \alpha(\mathbf{x})^*\alpha(\mathbf{x})$$

$$+ \iint d\mathbf{x} d\mathbf{y} \; \alpha(\mathbf{x})^*\alpha(\mathbf{x}) \, v(\mathbf{x}, \mathbf{y}) \, \alpha(\mathbf{y})^*\alpha(\mathbf{y}) \tag{2.3}$$

---

[1] These notes expand upon lectures given by Joel Feldman.

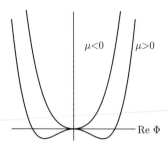

Fig. 2.1   Graph of the effective potential.

and $h(\mathbf{x}, \mathbf{y})$ being the kernel of the operator h. In the integral on the right-hand side of (2.1), there is a two-parameter family of integration variables. The first parameter, τ, runs over the 'time' interval $\left(0, \frac{1}{kT}\right]$ (the reason for the half open, half closed time interval is that there is a periodicity condition $\alpha_0(\mathbf{x}) = \alpha_{\frac{1}{kT}}(\mathbf{x})$) and the second parameter, $\mathbf{x}$, runs over 'space', $\mathbb{R}^d$. For each τ and $\mathbf{x}$, there is an integration variable, $\alpha_\tau(\mathbf{x})$, that runs over the complex plane, $\mathbb{C}$. For a complex variable $z = x + iy$, $\frac{dz \wedge dz^*}{2\pi i}$ is the usual Euclidean measure $\frac{1}{\pi} dx dy$.

Thus the 'measure' for the integral on the right hand side of (2.1) is a Lebesgue measure in uncountably many variables. It clearly has no mathematical meaning. But it is still a useful source of intuition. If $\alpha_\tau(\mathbf{x}) = \Phi \in \mathbb{C}$ is a constant, independent of τ and $\mathbf{x}$, the action $\mathcal{A}(\alpha^*, \alpha)$ simplifies to minus the integral over τ and $\mathbf{x}$ of the 'naive effective potential' $\hat{v}(0)|\Phi|^4 - \mu|\Phi|^2$ where $\hat{v}(0) = \int d\mathbf{y}\, v(\mathbf{x}, \mathbf{y})$ (recall that $v(\mathbf{x}, \mathbf{y})$ is translation invariant) and we have assumed and that h annihilates constants and that $\hat{v}(0) > 0$. This effective potential is graphed above in Fig. 3.1. Its minimum is

- nondegenerate at the point $\Phi = 0$ when $\mu < 0$ and
- degenerate along the circle $|\Phi| = \sqrt{\frac{\mu}{2\hat{v}(0)}}$ when $\mu > 0$.

This suggests that, if the temperature is low so that fluctuations about the minimum are small, each integration variable $\alpha_\tau(\mathbf{x})$ tends to be localized about 0 when $\mu < 0$ and tends to be localized about $|\alpha_\tau(\mathbf{x})| = \sqrt{\frac{\mu}{2\hat{v}(0)}}$ when $\mu > 0$. To help us glean some more detailed intuition from the formal functional integral, we introduce 'Euclidean time evolving' annihilation and creation operators

$$a(\tau, \mathbf{x}) = e^{(H - \mu N)\tau} a(\mathbf{x}) e^{-(H - \mu N)\tau} \qquad a^\dagger(\tau, \mathbf{x}) = e^{(H - \mu N)\tau} a^\dagger(\mathbf{x}) e^{-(H - \mu N)\tau}$$

and notation for 'expectation values' both in the physical Hilbert space and with respect to the functional integral

$$\langle f(a^\dagger, a) \rangle = \frac{\mathrm{Tr}\left(e^{-\frac{1}{kT}(H - \mu N)} f(a^\dagger, a)\right)}{\mathrm{Tr}\, e^{-\frac{1}{kT}(H - \mu N)}}$$

$$\langle\!\langle f(\alpha^*, \alpha) \rangle\!\rangle = \frac{\int \prod_{\mathbf{x}, \tau} \frac{d\alpha_\tau(\mathbf{x})^* \wedge d\alpha_\tau(\mathbf{x})}{2\pi i}\; e^{\mathcal{A}(\alpha^*, \alpha)} f(\alpha^*, \alpha)}{\int \prod_{\mathbf{x}, \tau} \frac{d\alpha_\tau(\mathbf{x})^* \wedge d\alpha_\tau(\mathbf{x})}{2\pi i}\; e^{\mathcal{A}(\alpha^*, \alpha)}}$$

We will use two more functional integral representations similar to the representation (2.1) for the partition function. They are for the one and two point correlation functions

$$\langle a^{(\dagger)}(\tau, \mathbf{x}) \rangle = \langle\!\langle \alpha_\tau(\mathbf{x})^{(*)} \rangle\!\rangle \tag{2.4}$$

$$\langle a^\dagger(\tau, \mathbf{x})\, a(\tau', \mathbf{x}') \rangle = \langle\!\langle \alpha_\tau(\mathbf{x})^* \alpha_{\tau'}(\mathbf{x}') \rangle\!\rangle \tag{2.5}$$

The first is valid for $\frac{1}{kT} \geq \tau \geq 0$ and the second is valid for $\frac{1}{kT} \geq \tau > \tau' \geq 0$. Actually, (2.4) is two formulae at once—one when the bracketed exponents are included and one when the bracketed exponents are omitted. Let us try to compute these expectation values, at least approximately.

(1) **The one point function for** $\mu < 0$**:** First consider $\mu < 0$. The one point function (2.4) is zero by symmetry considerations. This can be seen by using either side of (2.4). On the right-hand side, make the change of variables which rotates each integration variable by a fixed angle θ. That is

$$\alpha_\tau(\mathbf{x}) \to e^{i\theta} \alpha_\tau(\mathbf{x}) \qquad \alpha_\tau(\mathbf{x})^* \to e^{-i\theta} \alpha_\tau(\mathbf{x})^*$$

As both the measure $\frac{d\alpha_\tau(\mathbf{x})^* \wedge d\alpha_\tau(\mathbf{x})}{2\pi i}$ and the action $\mathcal{A}(\alpha^*, \alpha)$ are invariant under this change of variables, we have

$$\langle\!\langle \alpha_\tau(\mathbf{x})^{(*)} \rangle\!\rangle = e^{(-)i\theta} \langle\!\langle \alpha_\tau(\mathbf{x})^{(*)} \rangle\!\rangle \qquad \Longrightarrow \qquad \langle\!\langle \alpha_\tau(\mathbf{x})^{(*)} \rangle\!\rangle = 0$$

For the corresponding argument on the left-hand side, we unitarily transform the Hilbert space using the operator $e^{iN\theta}$. By cyclicity of the trace

$$\mathrm{Tr}\left(e^{-\frac{1}{kT}(H - \mu N)} a^{(\dagger)}(\tau, \mathbf{x}) \right) = \mathrm{Tr}\left(e^{-iN\theta} e^{-\frac{1}{kT}(H - \mu N)} a^{(\dagger)}(\tau, \mathbf{x}) e^{iN\theta} \right)$$

$$= \mathrm{Tr}\left(e^{-\frac{1}{kT}(H - \mu N)} e^{-iN\theta} a^{(\dagger)}(\tau, \mathbf{x}) e^{iN\theta} \right)$$

$$= e^{(-)i\theta} \mathrm{Tr}\left(e^{-\frac{1}{kT}(H - \mu N)} a^{(\dagger)}(\tau, \mathbf{x}) \right)$$

The critical step was the second equality, where we used that $H - \mu N$ commutes with the number operator N. That is, the Hamiltonian conserves particle number. For the third equality, we used that

$$e^{-iN\theta} a^{(\dagger)}(\tau, \mathbf{x}) e^{iN\theta} = e^{(-)i\theta} a^{(\dagger)}(\tau, \mathbf{x})$$

Once again, we have

$$\langle a^{(\dagger)}(\tau, \mathbf{x}) \rangle = e^{(-)i\theta} \langle a^{(\dagger)}(\tau, \mathbf{x}) \rangle \qquad \Longrightarrow \qquad \langle a^{(\dagger)}(\tau, \mathbf{x}) \rangle = 0$$

It would appear that this argument also implies $\langle a^{(\dagger)}(\tau, \mathbf{x}) \rangle = 0$ when $\mu > 0$. But there is a subtlety when $\mu > 0$ that we will discuss shortly.

(2) **The two point function for** $\mu < 0$**:** Now let's move on to the two point function (2.5) when $\mu < 0$. We are expecting the most important contributions to the functional integral to come from $\alpha_\tau(\mathbf{x}) \approx 0$. So approximate the action $\mathcal{A}$ by dropping all terms of degree strictly bigger than two in the integration variables. That is, drop the quartic, $v(\mathbf{x}, \mathbf{y})$ part of (2.3). This turns the action into

a quadratic function of the integration variables. Using (the natural formal analogue of) part (a) of Lemma 2.33 with $D = -\frac{\partial}{\partial \tau} + h - \mu$, we have

$$\langle\!\langle \alpha_\tau(\mathbf{x})^* \alpha_{\tau'}(\mathbf{x}') \rangle\!\rangle = \left(-\frac{\partial}{\partial \tau} + h - \mu \right)^{-1} \left((\tau, \mathbf{x}), (\tau', \mathbf{x}') \right)$$

The right-hand side is the kernel of the operator inverse of $-\frac{\partial}{\partial \tau} + h - \mu$. Because h is translation invariant we can use the Fourier transform to compute it.

$$\left(-\frac{\partial}{\partial \tau} + h - \mu \right)^{-1} \left((\tau, \mathbf{x}), (\tau', \mathbf{x}') \right) =$$

$$kT \sum_{k_0 \in 2\pi kT\mathbb{Z}} \int_{\mathbb{R}^3} \frac{d^3\mathbf{k}}{(2\pi)^3} e^{ik_0(\tau - \tau') - i\mathbf{k}\cdot(\mathbf{x} - \mathbf{x}')} \frac{1}{-ik_0 + \hat{h}(\mathbf{k}) - \mu}$$

(If you were expecting minus this answer, it is probably because you forgot that the usual two–point function is defined to be $-\langle\!\langle \alpha_\tau(\mathbf{x})^* \alpha_{\tau'}(\mathbf{x}') \rangle\!\rangle$.) The sum over k_0 can be evaluated exactly using a contour integral trick (see, for example, (Fetter and Walecka, 1971, (25.32)–(25.35))) giving

$$\langle\!\langle \alpha_\tau(\mathbf{x})^* \alpha_{\tau'}(\mathbf{x}') \rangle\!\rangle = \int_{\mathbb{R}^3} \frac{d^3\mathbf{k}}{(2\pi)^3} e^{-i\mathbf{k}\cdot(\mathbf{x} - \mathbf{x}')} e^{(\hat{h}(\mathbf{k}) - \mu)(\tau - \tau')} \left(e^{\frac{1}{kT}(\hat{h}(\mathbf{k}) - \mu)} - 1 \right)^{-1}$$

For large $\mathbf{k}$ the integrand is bounded in absolute value by the exponential of minus a constant times $|\mathbf{k}|^2$, since $\tau - \tau' < \frac{1}{kT}$. Furthermore the denominator never vanishes, because $\mu < 0$. Both the last two sentences remain true even if, in $e^{(\hat{h}(\mathbf{k}) - \mu)(\tau - \tau')}$ and $\left(e^{\frac{1}{kT}(\hat{h}(\mathbf{k}) - \mu)} - 1 \right)^{-1}$, $\mathbf{k}$ is given a fixed, not too big, imaginary part. Consequently, $\langle\!\langle \alpha_\tau(\mathbf{x})^* \alpha_{\tau'}(\mathbf{x}') \rangle\!\rangle$ decays exponentially to zero as $|\mathbf{x} - \mathbf{x}'| \to \infty$.

(3) **The one point function for $\mu > 0$:** We have already seen that when $\mu > 0$ the naive effective potential takes its minimum value on the circle $|\Phi| = \sqrt{\frac{\mu}{2\hat{v}(0)}}$ in the complex plane. This suggests that the integration variables $\alpha_\tau(\mathbf{x})$ would like to stay near that circle. But nothing in the integral favours any phase of Φ over any other phase. Something very similar happens in magnetic materials. Indeed it can be useful to pretend that each $\alpha_\tau(\mathbf{x})$ represents the needle of a magnetic compass. As $\mu > 0$ and the temperature is very low, the length of each needle is essentially fixed at $\sqrt{\frac{\mu}{2\hat{v}(0)}}$. But its orientation, the argument of $\alpha_\tau(\mathbf{x})$, is free. If we now subject the system to an external magnetic field that favours one particular direction, all of $\alpha_\tau(\mathbf{x})$'s will take values near a single Φ on the circle. If the temperature is low enough, this will remain the case even if the strength of the magnetic field is then reduced to zero. The same thing happens if, instead of applying a weak bulk magnetic field, we impose boundary conditions near infinity that favour one particular phase of Φ. The moral is that the behaviour of the system, and in particular the one and two point functions, can be expected to depend not only on the action, but also on the limiting process used to carefully define the system. This is a very common phenomenon in symmetry breaking scenarios.

So let's assume that our limiting process favours one particular Φ. Make a change of variables

$$\alpha_\tau(\mathbf{x}) = \Phi + \beta_\tau(\mathbf{x}) \qquad \alpha_\tau(\mathbf{x})^* = \Phi^* + \beta_\tau(\mathbf{x})^* \tag{2.6}$$

We are expecting $\beta_\tau(\mathbf{x})$ to be small. Under this change of variables, the $K(\alpha^*, \alpha)$ of (2.3) becomes, supressing the τ subscripts and recalling that the kinetic energy operator h annihilates constants,

$$K(\alpha^*, \alpha) = \iint d\mathbf{x}d\mathbf{y} \; \beta(\mathbf{x})^* h(\mathbf{x}, \mathbf{y})\beta(\mathbf{y})$$

$$+ \int d\mathbf{x} \left[-\mu|\Phi|^2 + \hat{v}(0)|\Phi|^4 \right]$$

$$+ \int d\mathbf{x} \; \beta(\mathbf{x})^* \left[-\mu + 2\hat{v}(0)|\Phi|^2 \right]\Phi + \int d\mathbf{x} \; \Phi^* \left[-\mu + 2\hat{v}(0)|\Phi|^2 \right]\beta(\mathbf{x})$$

$$+ \int d\mathbf{x} \; \beta(\mathbf{x})^* \left[-\mu + 2\hat{v}(0)|\Phi|^2 \right]\beta(\mathbf{x}) + 2|\Phi|^2 \iint d\mathbf{x}d\mathbf{y} \; \beta(\mathbf{x})^* v(\mathbf{x}, \mathbf{y})\beta(\mathbf{y})$$

$$+ (\Phi^*)^2 \iint d\mathbf{x}d\mathbf{y} \; \beta(\mathbf{x})v(\mathbf{x}, \mathbf{y})\beta(\mathbf{y}) + \Phi^2 \iint d\mathbf{x}d\mathbf{y} \; \beta(\mathbf{x})^* v(\mathbf{x}, \mathbf{y})\beta(\mathbf{y})^*$$

$$+ O(|\beta|^3) + O(|\beta|^4).$$

In computing the one and two point functions, the constant (i.e. independent of β) term in the second row will appear both in the numerator and in the denominator and so will cancel out. So we may as well drop it. The two degree one terms in the third row and the first degree two term in the fourth row are zero because $|\Phi^2| = \frac{\mu}{2\hat{v}(0)}$. We drop all terms of degree three and four in β, β^*, by way of approximation. So we end up with the action

$$\tilde{A}(\beta^*, \beta) = \int_0^{\frac{1}{kT}} d\tau \int d^3\mathbf{x} \; \beta_\tau(\mathbf{x})^* \frac{\partial}{\partial \tau}\beta_\tau(\mathbf{x}) - \int_0^{\frac{1}{kT}} d\tau \; \tilde{K}(\beta_\tau^*, \beta_\tau) \tag{2.7}$$

where

$$\tilde{K}(\beta^*, \beta) = \iint d\mathbf{x}d\mathbf{y} \; \beta(\mathbf{x})^* \left[h(\mathbf{x}, \mathbf{y}) + 2|\Phi|^2 v(\mathbf{x}, \mathbf{y}) \right]\beta(\mathbf{y})$$

$$+ (\Phi^*)^2 \iint d\mathbf{x}d\mathbf{y} \; \beta(\mathbf{x})v(\mathbf{x}, \mathbf{y})\beta(\mathbf{y}) + \Phi^2 \iint d\mathbf{x}d\mathbf{y} \; \beta(\mathbf{x})^* v(\mathbf{x}, \mathbf{y})\beta(\mathbf{y})^*.$$

This action is, of course, no longer invariant under $\beta \to e^{i\theta}\beta$, $\beta^* \to e^{-i\theta}\beta^*$. But it is still invariant under $\beta \to -\beta$, $\beta^* \to -\beta^*$. Hence

$$\langle\!\langle \alpha_\tau(\mathbf{x})^{(*)} \rangle\!\rangle = \Phi^{(*)} + \langle\!\langle \beta_\tau(\mathbf{x})^{(*)} \rangle\!\rangle = \Phi^{(*)}.$$

This is nonzero and shows us that conservation of particle number has been broken.

(4) **The two point function for $\mu > 0$:** By making a change of variables $\alpha_\tau(\mathbf{x}) \rightarrow \alpha_\tau(\mathbf{x})e^{i\theta}$ we may always arrange that the favoured Φ has phase zero, so that it is positive. So for simplicity, we now set $\Phi = \sqrt{\frac{\mu}{2\hat{v}(0)}}$, which we denote $\sqrt{n_0}$. To compute the two point functions, using the approximate action (2.7) we apply (the natural formal analogue) of part (b) of Lemma 2.33 with

$$D = -\frac{\partial}{\partial\tau} + h + 2n_0 v \qquad V = W = n_0 v.$$

Note that, because v and h are translationally invariant, D and $V = W$ commute with each other and we may also compute with these operators using Fourier transforms. In particular, in momentum space, the operators D, $D^t = \frac{\partial}{\partial\tau} + h + 2n_0 v$ and V are multiplication by $-ik_0 + \hat{h}(\mathbf{k}) + 2n_0\hat{v}(\mathbf{k})$, $ik_0 + \hat{h}(\mathbf{k}) + 2n_0\hat{v}(\mathbf{k})$ and $n_0\hat{v}(\mathbf{k})$, respectively. Hence the kernel of $(DD^t - 4V^2)^{-1}$ is

$$(DD^t - 4V^2)^{-1}((\tau, \mathbf{x}), (\tau', \mathbf{x}'))$$

$$= kT \sum_{k_0 \in 2\pi kT\mathbb{Z}} \int_{\mathbb{R}^3} \frac{d^3k}{(2\pi)^3} e^{ik_0(\tau-\tau')-ik\cdot(\mathbf{x}-\mathbf{x}')} \frac{1}{k_0^2 + [\hat{h}(\mathbf{k}) + 2n_0\hat{v}(\mathbf{k})]^2 - 4n_0^2\hat{v}(\mathbf{k})^2}$$

$$= kT \sum_{k_0 \in 2\pi kT\mathbb{Z}} \int_{\mathbb{R}^3} \frac{d^3k}{(2\pi)^3} e^{ik_0(\tau-\tau')-ik\cdot(\mathbf{x}-\mathbf{x}')} \frac{1}{k_0^2 + \hat{h}(\mathbf{k})[\hat{h}(\mathbf{k}) + 4n_0\hat{v}(\mathbf{k})]}$$

Combining (2.6), (three variants of) (2.65) and (2.67),

$$\langle\!\langle \alpha_\tau(\mathbf{x})^*\alpha_{\tau'}(\mathbf{x}') \rangle\!\rangle = n_0 + \langle\!\langle \beta_\tau(\mathbf{x})^*\beta_{\tau'}(\mathbf{x}') \rangle\!\rangle$$

$$= n_0 + kT \sum_{k_0 \in 2\pi kT\mathbb{Z}} \int_{\mathbb{R}^3} \frac{d^3k}{(2\pi)^3} e^{ik_0(\tau-\tau')+ik\cdot(\mathbf{x}-\mathbf{x}')} \frac{ik_0 + \hat{h}(\mathbf{k}) + 2n_0\hat{v}(\mathbf{k})}{k_0^2 + \hat{h}(\mathbf{k})[\hat{h}(\mathbf{k}) + 4n_0\hat{v}(\mathbf{k})]}$$

$$\langle\!\langle \alpha_\tau(\mathbf{x})\alpha_{\tau'}(\mathbf{x}') \rangle\!\rangle = n_0 + \langle\!\langle \beta_\tau(\mathbf{x})\beta_{\tau'}(\mathbf{x}') \rangle\!\rangle$$

$$= n_0 - kT \sum_{k_0 \in 2\pi kT\mathbb{Z}} \int_{\mathbb{R}^3} \frac{d^3k}{(2\pi)^3} e^{ik_0(\tau-\tau')+ik\cdot(\mathbf{x}-\mathbf{x}')} \frac{2n_0\hat{v}(\mathbf{k})}{k_0^2 + \hat{h}(\mathbf{k})[\hat{h}(\mathbf{k}) + 4n_0\hat{v}(\mathbf{k})]}$$

In contrast to the case $\mu < 0$, these expectation values converge to n_0, rather than zero, as $|\mathbf{x} - \mathbf{x}'| \rightarrow \infty$. This is called 'long range order'. Note also that the integrands have poles at

$$k_0 = \pm iE(\mathbf{k}) \qquad \text{where } E(\mathbf{k}) = \sqrt{\hat{h}(\mathbf{k})[\hat{h}(\mathbf{k}) + 4n_0\hat{v}(\mathbf{k})]}$$

This $E(\mathbf{k})$ is the (approximate) 'single–particle excitation energy'. When $\hat{h}(\mathbf{k}) = \frac{k^2}{2m}$,

$$E(\mathbf{k}) \approx c|\mathbf{k}| \quad \text{with} \quad c = \sqrt{\frac{2n_0\hat{v}(0)}{m}} \quad \text{when} \quad \mathbf{k} \approx 0$$

This 'linear dispersion relation' is used, because of the Landau theory of superfluidity, as a signal that the interacting Bose gas is superfluid. The ideal Bose gas has a quadratic dispersion relation and is not superfluid.

The above discussion suggests that there will be a phase transition. For μ below some critical value (which will probably not be exactly zero, because of renormalization effects) the expected value of a single annihilation or creation operator will be zero, just as you would expect from conservation of particle number. But, when the temperature is low enough, for μ above the critical point, it will be Φ for some complex number of modulus $|\Phi| \approx \sqrt{\frac{\mu}{2\hat{v}(0)}} \neq 0$ (despite an action which conserves particle number) and its precise value (i.e. which allowed Φ it is) will depend on the limiting process used to define the model. So we have to be very careful about how we define the model.

2.1.3 A rigorous starting point

To carefully define the model, for example to carefully define the partition function on the left-hand side of (2.1), you take a limit of obviously well-defined approximations. One way to get a (pretty) obviously well-defined approximate partition function is to replace space, $\mathbb{R}^3$, by a finite number of points, say $X = \mathbb{Z}^3/L\mathbb{Z}^3$. However, even for an approximate model with space having only a finite number of points, the functional integral on the right-hand side of the corresponding (2.1) still has uncountably many integration variables, because time is still $(0, \frac{1}{kT}]$, and so is still not really defined.

At this point, I am just going to quote a theorem (I'll give the important parts of the proof in Section 2.4) which says that, when X is finite, you can get a rigorous representation of the partition function by taking a limit of a sequence of integrals, with each integral in the sequence having only finitely many integration variables. To get finitely many integration variables, you replace 'time', $(0, \frac{1}{kT}]$, by a finite number of points too. The theorem, proven in (Balaban *et al.* 2008b, Theorem 2.2) is the following.

Theorem 2.1 *Suppose that* $\mathrm{R}(\varepsilon), \mathrm{r}(\varepsilon) \to \infty$ *as* $\varepsilon \to 0$ *at suitable rates*[1]. *For each fixed finite* X,

$$\mathrm{Tr}\, e^{-\frac{1}{kT}(H-\mu N)} = \lim_{\varepsilon \to 0} \int \prod_{\tau \in \varepsilon\mathbb{Z} \cap (0, \frac{1}{kT}]} \left[d\mu_{\mathrm{R}(\varepsilon)}(\alpha_\tau^*, \alpha_\tau)\, I_0(\varepsilon; \alpha_{\tau-\varepsilon}^*, \alpha_\tau) \right] \qquad (2.8)$$

with the convention that $\alpha_0 = \alpha_{\frac{1}{kT}}$. *Here,*

$$d\mu_{\mathrm{R}(\varepsilon)}(\alpha^*, \alpha) = \prod_{\mathbf{x} \in X} \frac{d\alpha^*(\mathbf{x}) \wedge d\alpha(\mathbf{x})}{2\pi\imath}\, e^{-\alpha^*(\mathbf{x})\alpha(\mathbf{x})}\, \chi\big(|\alpha(\mathbf{x})| < \mathrm{R}(\varepsilon)\big)$$

denotes an unnormalized Gaussian measure, cut off at radius $\mathrm{R}(\varepsilon)$, *and*

$$I_0(\varepsilon; \alpha^*, \beta) = \zeta_\varepsilon(\alpha, \beta)\, e^{\langle \alpha^*, j(\varepsilon)\beta \rangle - \varepsilon\, \langle \alpha^*\beta\,,\, v\, \alpha^*\beta \rangle}$$

---

[1] One can think of $\mathrm{R}(\varepsilon) \sim \frac{1}{\sqrt[4]{\varepsilon}}$ and of $\mathrm{r}(\varepsilon)$ as a power of $\ln \frac{1}{\varepsilon}$ or as a small power of $\frac{1}{\varepsilon}$.

with

$$j(\varepsilon) = e^{-\varepsilon(h-\mu)}$$

and $\zeta_\varepsilon(\alpha, \beta)$ being the characteristic function of

$$\{\alpha, \beta : \mathbb{C}^X \to \mathbb{C} \mid \|\alpha - \beta\|_\infty < r(\varepsilon)\}.$$

We write the ($\mathbb{R}$–style) scalar product[2], $\langle f, g \rangle = \sum_{\mathbf{x} \in X} f(\mathbf{x}) g(\mathbf{x})$ for any two fields $f, g : X \to \mathbb{C}$.

Now the integrals in this theorem do not look very much like the functional integral on the right-hand side of (2.1). In fact, one has a lot of freedom in choosing the integrand in (2.8) and I have deliberately chosen the integrand to make the next steps easy, rather than to make it look like the integrand of (2.1). Here is how to see that the integral of (2.8) is actually not so different from the integral of (2.1).

- First observe that (2.8) has one complex integration variable for each 'space–time' point $(\mathbf{x}, \tau)$ with $\mathbf{x} \in X$ and $\tau \in \varepsilon\mathbb{Z} \cap (0, \frac{1}{kT}]$, a finite approximation to the 'time set' $(0, \frac{1}{kT}]$.
- In contrast to the integration variables of (2.1), each complex integration variable of (2.8) does not run over all $\mathbb{C}$, because of the cutoff functions $\chi(|\alpha(\mathbf{x})| < R(\varepsilon))$, which restrict each integration variable to a finite disk in $\mathbb{C}$, and $\zeta_\varepsilon(\alpha_{\tau-\varepsilon}, \alpha_\tau)$, which restricts the time-derivative of $\alpha_\tau(\mathbf{x})$. But in the limit $\varepsilon \to 0$, these cutoffs disappear.
- Consider the total exponent

$$- \sum_{\substack{\mathbf{x} \in X \\ \tau \in \varepsilon\mathbb{Z} \cap (0, \frac{1}{kT}]}} \alpha_\tau(\mathbf{x})^* \alpha_\tau(\mathbf{x}) + \sum_{\tau \in \varepsilon\mathbb{Z} \cap (0, \frac{1}{kT}]} \left[\langle \alpha_{\tau-\varepsilon}^*, e^{-\varepsilon(h-\mu)} \alpha_\tau \rangle - \varepsilon \left\langle \alpha_{\tau-\varepsilon}^* \alpha_\tau, v \alpha_{\tau-\varepsilon}^* \alpha_\tau \right\rangle \right]$$

of (2.8) (including the part of the exponent hidden inside the measure $d\mu_{R(\varepsilon)}$). Expand the exponential in powers of ε, keeping only $1\!\!\!/ - \varepsilon(h-\mu)$ and throwing away all contributions of order at least ε^2. This gives exactly

$$\varepsilon \sum_{\tau \in \varepsilon\mathbb{Z} \cap (0, \frac{1}{kT}]} \left[\left\langle \alpha_{\tau-\varepsilon}^*, \frac{\alpha_\tau - \alpha_{\tau-\varepsilon}}{\varepsilon} \right\rangle - \left\langle \alpha_{\tau-\varepsilon}^*, (h-\mu) \alpha_\tau \right\rangle - \left\langle \alpha_{\tau-\varepsilon}^* \alpha_\tau, v \alpha_{\tau-\varepsilon}^* \alpha_\tau \right\rangle \right]$$

In the limit $\varepsilon \to 0$, $\varepsilon \sum_{\tau \in \varepsilon\mathbb{Z} \cap (0, \frac{1}{kT}]}$ becomes $\int_0^{\frac{1}{kT}} d\tau$ and $\frac{\alpha_\tau - \alpha_{\tau-\varepsilon}}{\varepsilon}$ becomes $\frac{\partial}{\partial \tau} \alpha_\tau$ and we get $\mathcal{A}(\alpha^*, \alpha)$.

To get from the integral

$$\int \prod_{\tau \in \varepsilon\mathbb{Z} \cap (0, \frac{1}{kT}]} \left[d\mu_{R(\varepsilon)}(\alpha_\tau^*, \alpha_\tau) \, I_0(\varepsilon; \alpha_{\tau-\varepsilon}^*, \alpha_\tau) \right] \qquad (2.9)$$

---

[2] Thus the usual scalar product over $\mathbb{C}^{|X|}$ is $\langle f^*, g \rangle$.

of the rigorous starting point, (2.8), to the full construction and analysis of the model of interest, we still need to execute several steps.

- *Step 1:* Take the temporal ultraviolet limit, $\varepsilon \to 0$. Of course Theorem 2.1 tells us that the limit exists and tells us that it is the approximate partition function. But that information by itself is virtually useless. We need to develop a picture of the limiting value we can work with in later steps.
- *Step 2:* Take the spatial infrared limit (i.e. the thermodynamic limit) $X \to \mathbb{Z}^3$ and possibly the temporal infrared limit $\frac{1}{kT} \to \infty$ (i.e. $T \to 0$).
- *Step 3:* Get properties of the limit, like symmetry breaking.

In these notes, we shall just discuss Step 1, the temporal ultraviolet limit. That is only an extremely small part of full construction. In fact, steps 2 and 3 can be expected to be exceptionally long and arduous and research on them has barely begun. Nonetheless, step 1 is not only a necessary step, but its treatment provides a useful glimpse, in a relatively simple setting, at techniques that are suitable for the later steps, and other models, as well. For a different, earlier, treatment of the ultraviolet limit in some related models see (Ginibre 1965; Ginibre 1971; Brydges and Federbush 1976; Brydges and Federbush 1977).

In the initial integral, (2.9), there is one complex integration variable, $\alpha_\tau(\mathbf{x})$, for each 'space–time' point $(\mathbf{x}, \tau)$ with $\mathbf{x} \in X$ and $\tau \in \varepsilon\mathbb{Z} \cap (0, \frac{1}{kT}]$. Recall that X is the finite discrete torus $\mathbb{Z}^3/(L\mathbb{Z})^3$, for some large $L \in \mathbb{N}$. Figure 2.2 contains one dot for each of the integration variable labels, $(\mathbf{x}, \tau)$. (Ignore the the difference between light and dark dots for a minute.) You will notice an asymmetry in that figure—the distance, ε, between dots in the τ direction is miniscule compared to the distance, 1, between dots in the X direction. In Step 1, we eliminate that asymmetry. We shall 'integrate out' all integration variables $\alpha_\tau(\mathbf{x})$ for which $(\mathbf{x}, \tau)$ is located at one of the lighter dots in Fig. 2.2, leaving the integration variables $\alpha_\tau(\mathbf{x})$ for which $(\mathbf{x}, \tau)$ is located at one of the darker dots. That is, the final result for Step 1 is a representation of the partition function as an integral having α_τ as an integration variable only if $\tau \in \theta\mathbb{Z}$ where θ is some fixed constant, independent of ε. Thus the set of integration variables for the final result of Step 1 looks like the set of integration variables for a classical spin system (in four dimensions). In fact, the final result of Step 1 looks somewhat like the classical

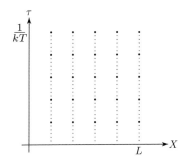

Fig. 2.2 The integration variables.

N-vector spin system for which Balaban proved the existence of the infrared limit and of symmetry breaking in (Balaban 1995a, 1995b, 1996a, 1996b, 1996c, 1998a, 1998b, 1998c). However there are substantial technical differences between the output of Step 1 and the class of models that Balaban considered. So one cannot execute Steps 2 and 3 simply by saying 'Balaban already did it'.

To execute step 1, we repeatedly apply a simple version of a renormalization group procedure, called 'decimation'. In each decimation step we integrate out all α_τ's having every second remaining value of τ. In the first decimation step, we integrate out $\alpha_{\tau'}$ with $\tau' = \varepsilon$, 3ε, 5ε, $\cdots$. The integral with respect to these variables factorizes into the product, over $\tau = 2\varepsilon$, 4ε, 6ε, $\cdots$, of the independent integrals

$$\int d\mu_{R(\varepsilon)}(\alpha^*_{\tau-\varepsilon}, \alpha_{\tau-\varepsilon})\, I_0(\varepsilon; \alpha^*_{\tau-2\varepsilon}, \alpha_{\tau-\varepsilon})\, I_0(\varepsilon; \alpha^*_{\tau-\varepsilon}, \alpha_\tau)$$

That is, assuming that $\frac{1}{kT} \in 2\varepsilon\mathbb{N}$,

$$\int \prod_{\tau \in \varepsilon\mathbb{Z} \cap (0, \frac{1}{kT}]} \left[d\mu_{R(\varepsilon)}(\alpha^*_\tau, \alpha_\tau)\, I_0(\varepsilon; \alpha^*_{\tau-\varepsilon}, \alpha_\tau) \right]$$

$$= \int \prod_{\tau \in 2\varepsilon\mathbb{Z} \cap (0, \frac{1}{kT}]} \left[d\mu_{R(\varepsilon)}(\alpha^*_\tau, \alpha_\tau)\, I_1(\varepsilon; \alpha^*_{\tau-2\varepsilon}, \alpha_\tau) \right] \qquad (2.10)$$

where

$$I_1(\varepsilon; \alpha^*_{\tau-2\varepsilon}, \alpha_\tau) = \int d\mu_{R(\varepsilon)}(\alpha^*_{\tau-\varepsilon}, \alpha_{\tau-\varepsilon})\, I_0(\varepsilon; \alpha^*_{\tau-2\varepsilon}, \alpha_{\tau-\varepsilon})\, I_0(\varepsilon; \alpha^*_{\tau-\varepsilon}, \alpha_\tau). \qquad (2.11)$$

In the second decimation step, we integrate out $\alpha_{\tau'}$ with $\tau' = 2\varepsilon$, 6ε, 10ε, $\cdots$ in the integral on the right hand side of (2.10). The integral with respect to these variables factorizes into the product, over $\tau = 4\varepsilon$, 8ε, 12ε, $\cdots$, of the independent integrals

$$\int d\mu_{R(\varepsilon)}(\alpha^*_{\tau-2\varepsilon}, \alpha_{\tau-2\varepsilon})\, I_1(\varepsilon; \alpha^*_{\tau-4\varepsilon}, \alpha_{\tau-2\varepsilon})\, I_1(\varepsilon; \alpha^*_{\tau-2\varepsilon}, \alpha_\tau).$$

That is, assuming that $\frac{1}{kT} \in 4\varepsilon\mathbb{N}$,

$$\int \prod_{\tau \in \varepsilon\mathbb{Z} \cap (0, \frac{1}{kT}]} \left[d\mu_{R(\varepsilon)}(\alpha^*_\tau, \alpha_\tau)\, I_0(\varepsilon; \alpha^*_{\tau-\varepsilon}, \alpha_\tau) \right]$$

$$= \int \prod_{\tau \in 2\varepsilon\mathbb{Z} \cap (0, \frac{1}{kT}]} \left[d\mu_{R(\varepsilon)}(\alpha^*_\tau, \alpha_\tau)\, I_1(\varepsilon; \alpha^*_{\tau-2\varepsilon}, \alpha_\tau) \right]$$

$$= \int \prod_{\tau \in 4\varepsilon\mathbb{Z} \cap (0, \frac{1}{kT}]} \left[d\mu_{R(\varepsilon)}(\alpha^*_\tau, \alpha_\tau)\, I_2(\varepsilon; \alpha^*_{\tau-4\varepsilon}, \alpha_\tau) \right]$$

Fig. 2.3 The integration variables, again.

where

$$I_2(\varepsilon; \alpha^*_{\tau-4\varepsilon}, \alpha_\tau) = \int d\mu_{R(\varepsilon)}(\alpha^*_{\tau-2\varepsilon}, \alpha_{\tau-2\varepsilon})\, I_1(\varepsilon; \alpha^*_{\tau-4\varepsilon}, \alpha_{\tau-2\varepsilon})\, I_1(\varepsilon; \alpha^*_{\tau-2\varepsilon}, \alpha_\tau)$$

$$= \int \prod_{\tau' \in \varepsilon\mathbb{Z} \cap (\tau-4\varepsilon, \tau)} d\mu_{R(\varepsilon)}(\alpha^*_{\tau'}, \alpha_{\tau'}) \prod_{\tau \in \varepsilon\mathbb{Z} \cap (\tau-4\varepsilon, \tau]} I_0(\varepsilon; \alpha^*_{\tau'-\varepsilon}, \alpha_{\tau'})$$

In general, for $n \geq 1$, $\varepsilon > 0$, set

$$I_n(\varepsilon; \alpha^*, \beta) = \int \prod_{\tau \in \varepsilon\mathbb{Z} \cap (0, 2^n\varepsilon)} d\mu_{R(\varepsilon)}(\alpha^*_\tau, \alpha_\tau) \prod_{\tau \in \varepsilon\mathbb{Z} \cap (0, 2^n\varepsilon]} I_0(\varepsilon; \alpha^*_{\tau-\varepsilon}, \alpha_\tau) \qquad (2.12)$$

with $\alpha_0 = \alpha$ and $\alpha_{2^n\varepsilon} = \beta$. If, as in Fig. 2.3, below, $\frac{1}{kT} = p\theta$ and $\varepsilon = 2^{-m}\theta$, then

$$\int \prod_{\tau \in \varepsilon\mathbb{Z} \cap (0, \frac{1}{kT}]} \left[d\mu_{R(\varepsilon)}(\alpha^*_\tau, \alpha_\tau)\, I_0(\varepsilon; \alpha^*_{\tau-\varepsilon}, \alpha_\tau) \right] = \int \prod_{\ell=1}^{p} \left[d\mu_{R(\varepsilon)}(\phi^*_\ell, \phi_\ell)\, I_m(\varepsilon; \phi^*_{\ell-1}, \phi_\ell) \right]$$

$$(2.13)$$

with the convention $\phi_0 = \phi_p$. I have renamed $\alpha_{\ell\theta} = \phi_\ell$.

Combining (2.8) and (2.13) we get

$$\mathrm{Tr}\, e^{-\frac{1}{kT}(H-\mu N)} = \lim_{m \to \infty} \int \prod_{\ell=1}^{p} \left[d\mu_{R(2^{-m}\theta)}(\phi^*_\ell, \phi_\ell)\, I_m(2^{-m}\theta; \phi^*_{\ell-1}, \phi_\ell) \right]$$

So far we have just made a trivial rearrangement of the order of integration. But (Balaban *et al.* 2010c) have shown that

- $I_\theta(\alpha^*, \beta) = \lim_{m \to \infty} I_m(2^{-m}\theta; \alpha^*, \beta)$ exists
- and that the partition function can be written as

$$\mathrm{Tr}\, e^{-\frac{1}{kT}(H-\mu N)} = \int \prod_{\ell=1}^{p} \left[\Pi_{\mathbf{x} \in X} \frac{d\phi_\ell(\mathbf{x})^* \phi_\ell(\mathbf{x})}{2\pi\imath} e^{-\phi_\ell(\mathbf{x})^* \phi_\ell(\mathbf{x})} \right] I_\theta(\phi^*_{\ell-1}, \phi_\ell)$$

- and that, if θ was chosen sufficiently small, I_θ may be written as the sum of a dominant part (which is shown to have a logarithm, which I will describe in more detail below) plus (ugly) terms indexed by proper subsets of X and which are nonperturbatively small, exponentially in the size of the subsets.

We call the dominant term the 'stationary phase approximation' (SP), because it is obtained by restricting all domains of integration in our functional integrals, simply by fiat, to appropriate neighbourhoods of stationary points. I'll describe this process in more detail in Section 2.2. The dominant contribution looks just like a perturbation of the original $e^{\langle \alpha^*, j(\varepsilon)\beta \rangle - \varepsilon \langle \alpha^*\beta, v\,\alpha^*\beta \rangle}$ in our starting point (2.8). Here is the precise form of the dominant contribution to $I_n(\varepsilon; \alpha^*, \beta)$.

$$I_n^{(SP)}(\varepsilon; \alpha^*, \beta) = Z_{2^n\varepsilon}(\varepsilon)^{|X|}\, e^{\langle \alpha^*, j(2^n\varepsilon)\beta \rangle + V_{2^n\varepsilon}(\varepsilon; \alpha^*, \beta) + \mathcal{E}_{2^n\varepsilon}(\varepsilon; \alpha^*, \beta)} \tag{2.14}$$

where, for every δ that is an integer multiple of ε,

$$V_\delta(\varepsilon; \alpha^*, \beta) = -\varepsilon \sum_{\tau \in \varepsilon \mathbb{Z} \cap [0,\delta)} \left\langle \left[j(\tau)\alpha^*\right]\left[j(\delta - \tau - \varepsilon)\beta\right],\, v\left[j(\tau)\alpha^*\right]\left[j(\delta - \tau - \varepsilon)\beta\right]\right\rangle. \tag{2.15}$$

The normalization constant $Z_\delta(\varepsilon)$ is chosen so that $\mathcal{E}_\delta(\varepsilon; 0,0) = 0$. It is extremely close to 1. (See (Balaban *et al.* 2010b, Appendix C).) The function $\mathcal{E}_\delta(\varepsilon; \alpha^*, \beta)$ is defined for real numbers $0 < \varepsilon \le \delta \le \Theta$ such that $\delta = 2^n\varepsilon$ for some integer $n \ge 0$. It is determined by the recursion relation

$$\mathcal{E}_\varepsilon(\varepsilon; \alpha^*, \beta) = 0$$

$$\mathcal{E}_{2\delta}(\varepsilon; \alpha^*, \beta) = \mathcal{E}_\delta(\varepsilon; \alpha^*, j(\delta)\beta) + \mathcal{E}_\delta(\varepsilon; j(\delta)\alpha^*, \beta)$$

$$+ \log \frac{\int d\mu_{\mathrm{r}(\delta)}(z^*, z)\, e^{\partial \mathcal{A}_\delta(\varepsilon; \alpha^*, \beta; z^*, z)}}{\int d\mu_{\mathrm{r}(\delta)}(z^*, z)} \tag{2.16}$$

where

$$\begin{aligned} \partial \mathcal{A}_\delta(\varepsilon; \alpha^*, \beta; z_*, z) = &\left[V_\delta(\varepsilon; \alpha^*, j(\delta)\beta + z) - V_\delta(\varepsilon; \alpha^*, j(\delta)\beta)\right] \\ &+ \left[V_\delta(\varepsilon; j(\delta)\alpha^* + z_*, \beta) - V_\delta(\varepsilon; j(\delta)\alpha^*, \beta)\right] \\ &+ \left[\mathcal{E}_\delta(\varepsilon; \alpha^*, j(\delta)\beta + z) - \mathcal{E}_\delta(\varepsilon; \alpha^*, j(\delta)\beta)\right] \\ &+ \left[\mathcal{E}_\delta(\varepsilon; j(\delta)\alpha^* + z_*, \beta) - \mathcal{E}_\delta(\varepsilon; j(\delta)\alpha^*, \beta)\right]. \end{aligned} \tag{2.17}$$

The motivation for this recursion relation comes from the stationary phase construction and is given in Section 2.2. In Section 2.3, I will outline the argument that the functions $\mathcal{E}_\delta(\varepsilon; \alpha^*, \beta)$ are

- analytic function of the fields,
- of degree at least two in each of α^* and β,
- perturbatively small corrections.

2.2   Motivation for the stationary phase approximation

The functions $I_n(\varepsilon; \alpha^*, \beta)$ of (2.12) can also be defined recursively by

$$I_{n+1}(\varepsilon; \alpha^*, \beta) = \int d\mu_{\mathrm{R}(\varepsilon)}(\phi^*, \phi)\, I_n(\varepsilon; \alpha^*, \phi) I_n(\varepsilon; \phi^*, \beta). \tag{2.18}$$

One of the morals of (Balaban *et al.* 2010c) is that the integrand is highly oscillatory and that the dominant contributions may be extracted using the stationary phase by discarding contributions far away from the critical point of the ('free part') of the exponent.

By way of motivation for the stationary phase approximation, and in particular for the recursive definition (2.16) of $\mathcal{E}_\delta(\varepsilon;\alpha^*,\beta)$, replace I_n by

$$I_n^{(SP)}(\varepsilon;\alpha^*,\beta) = \mathcal{Z}_{\varepsilon_n}(\varepsilon)^{|X|}\; e^{\langle\alpha^*,j(\varepsilon_n)\beta\rangle + \mathcal{V}_{\varepsilon_n}(\varepsilon;\alpha^*,\beta) + \mathcal{E}_{\varepsilon_n}(\varepsilon;\alpha^*,\beta)}$$

in the recursion relation (2.18). Here, $\varepsilon_n = 2^n\varepsilon$. (Start with $n = 0$, $\mathcal{Z}_\varepsilon(\varepsilon) = 1$ and $\mathcal{E}_\varepsilon(\varepsilon;\alpha^*,\beta) = 0$. Then, aside from the cutoff function $\zeta_\varepsilon(\alpha,\beta)$, which is going to incorporated by our choice of domain of integration, $I_0^{(SP)}(\varepsilon;\alpha^*,\beta)$ is the same as $I_0(\varepsilon;\alpha^*,\beta)$.) The resulting integral

$$\int d\mu_{R(\varepsilon)}(\phi^*,\phi)\; I_n^{(SP)}(\varepsilon;\alpha^*,\phi)\, I_n^{(SP)}(\varepsilon;\phi^*,\beta)$$

$$= \mathcal{Z}_{\varepsilon_n}(\varepsilon)^{2|X|}\int d\mu_{R(\varepsilon)}(\phi^*,\phi)\; e^{\langle\alpha^*,j(\varepsilon_n)\phi\rangle + \langle\phi^*,j(\varepsilon_n)\beta\rangle}\; e^{\mathcal{V}_{\varepsilon_n}(\varepsilon;\alpha^*,\phi) + \mathcal{V}_{\varepsilon_n}(\varepsilon;\phi^*,\beta)}$$

$$e^{\mathcal{E}_{\varepsilon_n}(\varepsilon;\alpha^*,\phi) + \mathcal{E}_{\varepsilon_n}(\varepsilon;\phi^*,\beta)}$$

$$= \mathcal{Z}_{\varepsilon_n}(\varepsilon)^{2|X|}\left[\prod_{\mathbf{x}\in X}\int_{|\phi(\mathbf{x})|<R(\varepsilon)}\frac{d\phi^*(\mathbf{x})\wedge d\phi(\mathbf{x})}{2\pi\imath}\right]e^{\mathcal{A}(\alpha^*,\beta\,;\,\phi^*,\phi)}$$

$$= \mathcal{Z}_{\varepsilon_n}(\varepsilon)^{2|X|}\left[\prod_{\mathbf{x}\in X}\int_{\substack{|\phi(\mathbf{x})|<R(\varepsilon)\\ \phi_*(\mathbf{x})=\phi(\mathbf{x})^*}}\frac{d\phi_*(\mathbf{x})\wedge d\phi(\mathbf{x})}{2\pi\imath}\right]e^{\mathcal{A}(\alpha^*,\beta\,;\,\phi_*,\phi)}$$

$$(2.19)$$

with

$$\mathcal{A}(\alpha^*,\beta\,;\,\phi_*,\phi) = -\langle\phi_*\,,\,\phi\rangle + \langle\alpha^*,j(\varepsilon_n)\phi\rangle + \langle\phi_*,j(\varepsilon_n)\beta\rangle$$
$$+ \mathcal{V}_{\varepsilon_n}(\varepsilon;\alpha^*,\phi) + \mathcal{V}_{\varepsilon_n}(\varepsilon;\phi_*,\beta)$$
$$+ \mathcal{E}_{\varepsilon_n}(\varepsilon;\alpha^*,\phi) + \mathcal{E}_{\varepsilon_n}(\varepsilon;\phi_*,\beta).$$

Here we have written $\mathcal{A}$ as a function of four independent complex fields α^*,β,ϕ_*, and ϕ. The activity in the penultimate line of (2.19) is obtained simply by evaluating $\mathcal{A}(\alpha^*,\beta;\phi_*,\phi)$ with $\phi_* = \phi^*$, the complex conjugate of ϕ. But in the last line, we introduce, for each $\mathbf{x}\in X$, a new, complex integration variable $\phi_*(\mathbf{x})$. That is, $(\phi(\mathbf{x}),\phi_*(\mathbf{x}))\in\mathbb{C}^2$. To get equality between the second last line and the last line of (2.19), we build the condition $\phi_*(\mathbf{x}) = \phi(\mathbf{x})^*$ into the domain of integration. The reason for introducing independent complex fields ϕ_* and ϕ lies in the fact that the critical point (where the first order derivatives with respect to ϕ_* and ϕ vanish) of the quadratic part

$$-\langle\phi_*\,,\,\phi\rangle + \langle j(\varepsilon_n)\alpha^*,\phi\rangle + \langle\phi_*,j(\varepsilon_n)\beta\rangle$$
$$= -\langle\phi_* - j(\varepsilon_n)\alpha^*\,,\,\phi - j(\varepsilon_n)\beta\rangle + \underbrace{\langle j(\varepsilon_n)\alpha^*,j(\varepsilon_n)\beta\rangle}_{\langle\alpha^*,j(\varepsilon_{n+1})\beta\rangle} \qquad (2.20)$$

of $\mathcal{A}$ is 'not real'. Precisely, the critical point is

$$\phi_*^{\mathrm{crit}} = j(\varepsilon_n)\,\alpha^*, \qquad \phi^{\mathrm{crit}} = j(\varepsilon_n)\,\beta$$

and in general $\left(\phi_*^{\mathrm{crit}}\right)^* \neq \phi^{\mathrm{crit}}$. To do stationary phase, we introduce the 'fluctuation variables' $z_*(\mathbf{x})$, $z(\mathbf{x})$ and make the change of variables

$$\phi_*(\mathbf{x}) = \phi_*^{\mathrm{crit}}(\mathbf{x}) + z_*(\mathbf{x}), \qquad \phi(\mathbf{x}) = \phi^{\mathrm{crit}}(\mathbf{x}) + z(\mathbf{x}). \tag{2.21}$$

Under this change of variables the domain of integration

$$\left\{ \left(\phi_*(\mathbf{x}), \phi(\mathbf{x}) \right) \mid \phi_*(\mathbf{x}) = \phi(\mathbf{x})^*, \; |\phi(\mathbf{x})| < \mathrm{R}(\varepsilon) \right\}$$

is transformed into

$$M(\mathbf{x}) = \left\{ (z_*(\mathbf{x}), z(\mathbf{x})) \mid \left(\phi_*^{\mathrm{crit}}(\mathbf{x}) + z_*(\mathbf{x}) \right)^* = \phi^{\mathrm{crit}}(\mathbf{x}) + z(\mathbf{x}) \right.$$

$$\left. \text{and } \left| \phi^{\mathrm{crit}}(\mathbf{x}) + z(\mathbf{x}) \right| < \mathrm{R}(\varepsilon) \right\}.$$

After the change of variables, the integral (2.19) is over a real $2|X|$ dimensional subset in the complex $2|X|$ dimensional space of fields z_*, z.

The first step in the stationary phase approximation is to replace, for each $\mathbf{x} \in X$, the domain of integration $M(\mathbf{x})$ by the neighbourhood

$$D(\mathbf{x}) = \left\{ (z_*(\mathbf{x}), z(\mathbf{x})) \in \mathbb{C}^2 \;\middle|\; |z_*(\mathbf{x})| \leq \mathrm{r}(\varepsilon_n), \; |z(\mathbf{x})| \leq \mathrm{r}(\varepsilon_n), \right.$$

$$\left. \left(z_*(\mathbf{x}) + \phi_*^{\mathrm{crit}}(\mathbf{x}) \right)^* = z(\mathbf{x}) + \phi^{\mathrm{crit}}(\mathbf{x}) \right\} \tag{2.22}$$

of the critical point. In (Balaban *et al.* 2010c) we justify this approximation by the observation that, whenever $(z_*(\mathbf{x}), z(\mathbf{x})) \notin D(\mathbf{x})$ for some $\mathbf{x} \in X$, the integrand is extremely small. I will sketch the reasons for this in Section 2.2.2, below. Observe that, in general, first, the critical point $z(\mathbf{x}) = z_*(\mathbf{x}) = 0$ is not in $D(\mathbf{x})$, and, second, $z_*(\mathbf{x}) \neq z(\mathbf{x})^*$ on $D(\mathbf{x})$.

The quadratic part (2.20) of the effective action $\mathcal{A}\left(\alpha^*, \beta; \phi_*^{\mathrm{crit}} + z_*, \phi^{\mathrm{crit}} + z \right)$ in the new variables is

$$- \langle j(\varepsilon_n)\alpha^* + z_*, \, j(\varepsilon_n)\beta + z \rangle + \langle \alpha^*, \, j(\varepsilon_n)\left(j(\varepsilon_n)\beta + z \right) \rangle$$

$$+ \langle j(\varepsilon_n)\left(j(\varepsilon_n)\alpha^* + z_* \right), \, \beta \rangle$$

$$= - \langle z_*, z \rangle + \langle \alpha^*, \, j(\varepsilon_{n+1})\beta \rangle.$$

(This is why we introduced the $j(\varepsilon)$ in Theorem 2.1.) Inserting the change of variables (2.21), we see that the part of (2.19) near the critical point is,

$$\mathcal{Z}_{\varepsilon_n}(\varepsilon)^{2|X|} \left[\prod_{\mathbf{x} \in X} \int_{D(\mathbf{x})} \frac{dz_*(\mathbf{x}) \wedge dz(\mathbf{x})}{2\pi\imath} \right] e^{\tilde{\mathcal{A}}(\alpha^*, \beta; \, z_*, z)} \tag{2.23}$$

where

$$\tilde{A}(\alpha^*, \beta; z_*, z) = -\langle z_*, z \rangle + \langle \alpha^*, j(\varepsilon_{n+1})\beta \rangle$$

$$+ \mathcal{V}_{\varepsilon_n}(\varepsilon; \alpha^*, \phi^{\mathrm{crit}} + z) + \mathcal{V}_{\varepsilon_n}(\varepsilon; \phi_*^{\mathrm{crit}} + z_*, \beta)$$

$$+ \mathcal{E}_{\varepsilon_n}(\varepsilon; \alpha^*, \phi^{\mathrm{crit}} + z) + \mathcal{E}_{\varepsilon_n}(\varepsilon; \phi_*^{\mathrm{crit}} + z_*, \beta)$$

$$= -\langle z_*, z \rangle + \langle \alpha^*, j(\varepsilon_{n+1})\beta \rangle + \mathcal{V}_{\varepsilon_{n+1}}(\varepsilon; \alpha^*, \beta)$$

$$+ \mathcal{E}_{\varepsilon_n}(\varepsilon; \alpha^*, \phi^{\mathrm{crit}}) + \mathcal{E}_{\varepsilon_n}(\varepsilon; \phi_*^{\mathrm{crit}}, \beta) + \partial \mathcal{A}_{\varepsilon_n}(\varepsilon; \alpha^*, \beta; z_*, z)$$

with the part of $\tilde{A}(\alpha^*, \beta; z_*, z)$ that is of degree at least one in (z_*, z) being (except for the explicit $-\langle z_*, z \rangle$)

$$\partial \mathcal{A}_\delta(\varepsilon; \alpha^*, \beta; z_*, z) = \left[\mathcal{V}_\delta(\varepsilon; \alpha^*, j(\delta)\beta + z) - \mathcal{V}_\delta(\varepsilon; \alpha^*, j(\delta)\beta) \right]$$

$$+ \left[\mathcal{V}_\delta(\varepsilon; j(\delta)\alpha^* + z_*, \beta) - \mathcal{V}_\delta(\varepsilon; j(\delta)\alpha^*, \beta) \right]$$

$$+ \left[\mathcal{E}_\delta(\varepsilon; \alpha^*, j(\delta)\beta + z) - \mathcal{E}_\delta(\varepsilon; \alpha^*, j(\delta)\beta) \right]$$

$$+ \left[\mathcal{E}_\delta(\varepsilon; j(\delta)\alpha^* + z_*, \beta) - \mathcal{E}_\delta(\varepsilon; j(\delta)\alpha^*, \beta) \right].$$

We have used that

$$\mathcal{V}_{\varepsilon_n}(\varepsilon; \alpha^*, \phi^{\mathrm{crit}}) + \mathcal{V}_{\varepsilon_n}(\varepsilon; \phi_*^{\mathrm{crit}}, \beta) = \mathcal{V}_{\varepsilon_n}(\varepsilon; \alpha^*, j(\varepsilon_n)\beta) + \mathcal{V}_{\varepsilon_n}(\varepsilon; j(\varepsilon_n)\alpha^*, \beta)$$

$$= \mathcal{V}_{\varepsilon_{n+1}}(\varepsilon; \alpha^*, \beta).$$

(The definition (2.15) of $\mathcal{V}_\delta(\varepsilon; \alpha^*, \beta)$ has been rigged to give this.) Apply Stokes' Theorem, once for each $\mathbf{x} \in X$, to replace the domain $D(\mathbf{x})$ with the union of

$$\left\{ (z_*(\mathbf{x}), z(\mathbf{x})) \mid z_*(\mathbf{x}) = z(\mathbf{x})^*, \ |z(\mathbf{x})| \le \mathrm{r}(\varepsilon_n) \right\}$$

(which contains the critical point) and a 'side boundary'. This is done in Lemma 2.2 below. (Choose $r = \mathrm{r}(\varepsilon_n)$ and $\rho(\mathbf{x}) = \phi_*^{\mathrm{crit}}(\mathbf{x})^* - \phi^{\mathrm{crit}}(\mathbf{x}) = (j(\varepsilon_n)(\alpha - \beta))(\mathbf{x}).$) This gives that (2.23) is the sum of

$$\mathcal{Z}_{\varepsilon_n}(\varepsilon)^{2|X|} \left[\prod_{\mathbf{x} \in X} \int_{|z(\mathbf{x})| \le \mathrm{r}(\varepsilon_n)} \frac{dz^*(\mathbf{x}) \wedge dz(\mathbf{x})}{2\pi \imath} \right] e^{\tilde{A}(\alpha^*, \beta; z^*, z)} \tag{2.24}$$

and $\mathcal{Z}_{\varepsilon_n}(\varepsilon)^{2|X|}$ times

$$\sum_{\substack{R \subset X \\ R \ne \emptyset}} \left[\prod_{\mathbf{x} \in R} \int_{C(\mathbf{x})} \frac{dz_*(\mathbf{x}) \wedge dz(\mathbf{x})}{2\pi i} \right] \left[\prod_{\mathbf{x} \in X \setminus R} \int_{|z(\mathbf{x})| \le \mathrm{r}(\varepsilon_n)} \frac{dz(\mathbf{x})^* \wedge dz(\mathbf{x})}{2\pi i} \right] e^{\tilde{A}(\alpha^*, \beta; z^*, z)} \Bigg|_{\substack{z_*(\mathbf{x}) = z(\mathbf{x})^* \\ \text{for } \mathbf{x} \in X \setminus R}}$$

where, for each $\mathbf{x} \in X$, $C(\mathbf{x})$ is a two real dimensional submanifold of $\mathbb{C}^2$ whose boundary is the union of 'circles' $\partial D(\mathbf{x})$ and

$$\left\{ (z_*(\mathbf{x}), z(\mathbf{x})) \in \mathbb{C}^2 \ \middle| \ z_*^*(\mathbf{x}) = z(\mathbf{x}), \ |z(\mathbf{x})| = \mathrm{r}(\varepsilon_n) \right\}.$$

The second step in the stationary phase approximation is to ignore all but the first term. That is, to replace (2.23) with (2.24). In (Balaban *et al.* 2010c) we argue that $-z_*(\mathbf{x})z(\mathbf{x})$ has an extremely large negative real part whenever $(z_*(\mathbf{x}), z(\mathbf{x})) \in C(\mathbf{x})$ (see part (b) of Lemma 2.2, below) and that this replacement introduces a nonperturbatively small error.

Thus, the stationary phase approximation for

$$\int d\mu_{\mathrm{R}(\varepsilon)}(\phi^*, \phi)\, I_n^{(\mathrm{SP})}(\varepsilon;\, \alpha^*, \phi)\, I_n^{(\mathrm{SP})}(\varepsilon;\, \phi^*, \beta)$$

is (2.24), which can also be written as

$$\mathcal{Z}_{\varepsilon_n}(\varepsilon)^{2|X|} e^{\langle \alpha^*, j(\varepsilon_{n+1})\beta \rangle + \mathcal{V}_{\varepsilon_{n+1}}(\varepsilon;\, \alpha^*, \beta)}$$

$$e^{\mathcal{E}_{\varepsilon_n}(\varepsilon;\, \alpha^*, j(\varepsilon_n)\beta)\, +\mathcal{E}_{\varepsilon_n}(\varepsilon;\, j(\varepsilon_n)\alpha^*, \beta)} \int d\mu_{\mathrm{r}(\varepsilon_n)}(z^*, z)\, e^{\partial \mathcal{A}_{\varepsilon_n}(\varepsilon;\, \alpha^*, \beta; z_*, z)}$$

This is indeed of the desired form, namely (2.14) with n replaced by $n + 1$, if

$$\mathcal{Z}_{\varepsilon_{n+1}}(\varepsilon) = \mathcal{Z}_{\varepsilon_n}(\varepsilon)^2 \int_{|z| < \mathrm{r}(\varepsilon_n)} \frac{dz^* \wedge dz}{2\pi i}\, e^{-|z|^2}$$

and $\mathcal{E}_{\varepsilon_{n+1}}(\varepsilon;\, \alpha^*, \beta)$ is given by the recursion relation (2.16).

2.2.1   Stokes' Theorem

We next give a short discussion and proof of the version of Stokes' Theorem that we used above. The setting is that we are given a radius $r > 0$ and a complex vector $\rho \in \mathbb{C}^X$ that obeys $|\rho(\mathbf{x})| < 2r$ for all $\mathbf{x} \in X$ and we wish to 'move the domain of integration' from the initial domain $D_{\mathbb{C}} = \bigtimes_{\mathbf{x} \in X} D_{\mathbb{C}}(\mathbf{x})$, where

$$D_{\mathbb{C}}(\mathbf{x}) = \left\{ (z_*(\mathbf{x}), z(\mathbf{x})) \in \mathbb{C}^2 \;\middle|\; |z_*(\mathbf{x})| \le r,\; |z(\mathbf{x})| \le r,\; z(\mathbf{x}) - z_*(\mathbf{x})^* = \rho(\mathbf{x}) \right\}$$

(see (2.22) above) to the final domain $D_{\mathbb{R}} = \bigtimes_{\mathbf{x} \in X} D_{\mathbb{R}}(\mathbf{x})$, where

$$D_{\mathbb{R}}(\mathbf{x}) = \left\{ (z_*(\mathbf{x}), z(\mathbf{x})) \in \mathbb{C}^2 \;\middle|\; z_*^*(\mathbf{x}) = z(\mathbf{x}),\; |z(\mathbf{x})| \le r \right\}$$

We start by taking a closer look at $D_{\mathbb{C}}(\mathbf{x})$. At each point of $D_{\mathbb{C}}(\mathbf{x})$, the value of the variable $z_*(\mathbf{x})$ is completely determined by the value of the variable $z(\mathbf{x})$ through $z_*(\mathbf{x}) = z(\mathbf{x})^* - \rho(\mathbf{x})^*$. The set of allowed values of the variable $z(\mathbf{x})$ is precisely the intersection of the two discs $|z(\mathbf{x})| \le r$ and $|z(\mathbf{x}) - \rho(\mathbf{x})| \le r$. The two discs overlap because of the hypothesis $|\rho(\mathbf{x})| < 2r$. At each point of the corresponding final domain $D_{\mathbb{R}}(\mathbf{x})$, the value of the variable $z_*(\mathbf{x})$ is again completely determined by the value of the variable $z(\mathbf{x})$, through $z_*(\mathbf{x}) = z(\mathbf{x})^*$, and the set of allowed values of the variable $z(\mathbf{x})$ can be thought of as being precisely the intersection of the two discs $|z(\mathbf{x})| \le r$ and $|z(\mathbf{x}) - 0| \le r$, which happen to coincide.

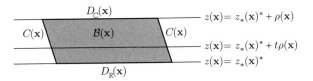

Fig. 2.4 The domain of integration for Stokes' theorem.

It is a simple matter to interpolate between $D_{\mathbb{C}}(\mathbf{x})$ and $D_{\mathbb{R}}(\mathbf{x})$. Define, for each $0 \leq t \leq 1$,

$$D_t(\mathbf{x}) = \left\{ (z_*(\mathbf{x}), z(\mathbf{x})) \in \mathbb{C}^2 \,\Big|\, |z_*(\mathbf{x})| \leq r, \ |z(\mathbf{x})| \leq r, \ z(\mathbf{x}) - z_*(\mathbf{x})^* = t\rho(\mathbf{x}) \right\}$$

Once again, at each point of $D_t(\mathbf{x})$, the value of the variable $z_*(\mathbf{x})$ is completely determined by the value of the variable $z(\mathbf{x})$, this time through $z_*(\mathbf{x}) = z(\mathbf{x})^* - t\rho(\mathbf{x})^*$, and the set of allowed values of the variable $z(\mathbf{x})$ is precisely the intersection of the two discs $|z(\mathbf{x})| \leq r$ and $|z(\mathbf{x}) - t\rho(\mathbf{x})| \leq r$. When $t = 1$, $D_t(\mathbf{x}) = D_{\mathbb{C}}(\mathbf{x})$ and when $t = 0$, $D_t(\mathbf{x}) = D_{\mathbb{R}}(\mathbf{x})$. Hence $\mathcal{B}(\mathbf{x}) = \bigcup_{0 \leq t \leq 1} D_t(\mathbf{x})$ is a the three (real) dimensional set whose boundary is the union of $D_{\mathbb{C}}(\mathbf{x})$ (that's the part of the boundary with $t = 1$) and $D_{\mathbb{R}}(\mathbf{x})$ (that's the part of the boundary with $t = 0$) and the two (real) dimensional surface $C(\mathbf{x}) = \bigcup_{0 < t < 1} \partial D_t(\mathbf{x})$ (that's the part of the boundary with $0 < t < 1$) where

$$\partial D_t(\mathbf{x}) = \left\{ (z_*(\mathbf{x}), z(\mathbf{x})) \in \mathbb{C}^2 \,\Big|\, \max\left\{ |z_*(\mathbf{x})|, |z(\mathbf{x})| \right\} = r, \ z(\mathbf{x}) - z_*(\mathbf{x})^* = t\rho(\mathbf{x}) \right\}$$

The surface $C(\mathbf{x})$ then joins the curves bounding $D_{\mathbb{R}}(\mathbf{x})$ and $D_{\mathbb{C}}(\mathbf{x})$.

Lemma 2.2

(a) *Let $f(\alpha_*, \beta; z_*, z)$ be a function that is analytic in the variables α_*, β in a neighbourhood of the origin in $\mathbb{C}^{2X}$ and in the variables $(z_*, z) \in \bigtimes_{\mathbf{x} \in X} \mathcal{P}(\mathbf{x})$, with, for each $\mathbf{x} \in X$, $\mathcal{P}(\mathbf{x})$ an open neighbourhood of $\mathcal{B}(\mathbf{x})$. Then*

$$\int_{D_{\mathbb{C}}} \prod_{\mathbf{x} \in X} \left[\frac{dz_*(\mathbf{x}) \wedge dz(\mathbf{x})}{2\pi i} e^{-z_*(\mathbf{x})z(\mathbf{x})} \right] e^{f(\alpha_*, \beta; z_*, z)}$$

$$= \sum_{R \subset X} \prod_{\mathbf{x} \in R} \left(\int_{C(\mathbf{x})} \frac{dz_*(\mathbf{x}) \wedge dz(\mathbf{x})}{2\pi i} e^{-z_*(\mathbf{x})z(\mathbf{x})} \right)$$

$$\prod_{\mathbf{x} \in X \setminus R} \left(\int_{|z(\mathbf{x})| \leq r} \frac{dz(\mathbf{x})^* \wedge dz(\mathbf{x})}{2\pi i} e^{-z_*(\mathbf{x})z(\mathbf{x})} \right) e^{f(\alpha_*, \beta; z_*, z)} \Bigg|_{\substack{z_*(\mathbf{x}) = z(\mathbf{x})^* \cdot \\ \text{for } \mathbf{x} \in X \setminus R}}$$

(b) *We have*

$$\mathrm{Re}\left(z_*(\mathbf{x}) z(\mathbf{x}) \right) \geq \frac{1}{2}\left(r^2 - |\rho(\mathbf{x})|^2 \right)$$

for all $(z_*(\mathbf{x}), z(\mathbf{x})) \in C(\mathbf{x})$. *Furthermore the area of* $C(\mathbf{x})$ *is bounded by* $4\pi r|\rho|$. *That is,*

$$\left| \int_{C(\mathbf{x})} \frac{dz_*(\mathbf{x}) \wedge dz(\mathbf{x})}{2\pi i} \, f\big(z_*(\mathbf{x}), z(\mathbf{x})\big) \right| \le 2r|\rho| \sup_{C(\mathbf{x})} \big| f\big(z_*(\mathbf{x}), z(\mathbf{x})\big) \big|.$$

Proof:

(a) We apply Stokes' theorem once for each point $\mathbf{x} \in X$ to the differential form

$$\omega = \bigwedge_{\mathbf{x} \in X} \frac{dz_*(\mathbf{x}) \wedge dz(\mathbf{x})}{2\pi i} \, \exp\Big\{ - \langle z_*, z \rangle + f(\alpha_*, \beta; z_*, z) \Big\}.$$

Since ω is a holomorphic $2|X|$ form in $\mathbb{C}^{2|X|}$, $d\omega = 0$ and

$$\int_D \omega = \sum_{R \subset X} \int_{M_R} \omega \qquad \text{where} \qquad M_R = \prod_{\mathbf{x} \notin R} D_{\mathbb{R}}(\mathbf{x}) \times \prod_{\mathbf{x} \in R} C(\mathbf{x}).$$

(b) Let $(z_*, z) \in C(\mathbf{x})$. We suppress the dependence on $\mathbf{x}$. There is a $0 \le t \le 1$ such that $\max\{|z_*|, |z|\} = r$ and $z_* = z^* - t\rho^*$. So

$$z_* z = |z|^2 - t\rho^* z$$
$$z_* z = |z_*|^2 + t\rho z^* - |t\rho|^2.$$

Adding and taking the real part,

$$2\mathrm{Re}\,(z_* z) = |z|^2 + |z_*|^2 - t^2 |\rho|^2 \ge r^2 - |\rho|^2.$$

By construction, $C(\mathbf{x})$ is contained in the union of the two cylinders

$$U = \big\{ \, (re^{-i\theta} - t\rho^*, re^{i\theta}) \mid \theta \in [0, 2\pi], \ t \in [0, 1] \, \big\}$$
$$L = \big\{ \, (re^{i\theta}, re^{-i\theta} + t\rho) \mid\mid \theta \in [0, 2\pi], \ t \in [0, 1] \, \big\}.$$

The upper cylinder contains the part of $C(\mathbf{x})$ with $|z(\mathbf{x})| = r$ and the lower cylinder contains the part with $z_*(\mathbf{x}) = r$. We'll bound the integral over the upper cylinder. On U, we have $dz = ire^{i\theta} \, d\theta$ and $dz_* = -ire^{-i\theta} \, d\theta - \rho^* dt$, which gives

$$dz_* \wedge dz = -i\rho^* re^{i\theta} \, dt \wedge d\theta$$

since $d\theta \wedge d\theta = 0$. Hence

$$\left| \int_U \frac{dz_* \wedge dz}{2\pi i} \, f(z_*, z) \right|$$

$$\le \frac{r|\rho|}{2\pi} \int_0^{2\pi} d\theta \int_0^1 dt \, \big| f(re^{-i\theta} - t\rho^*, re^{i\theta}) \big| \le r|\rho| \sup_U \big| f(z_*, z) \big|. \qquad \Box$$

2.2.2 The Error

We finish off this section by hinting at why the error introduced by the stationary phase approximation is extremely small. We consider the case $n = 0$. The initial functional integral representation (2.8) may be written

$$\text{Tr } e^{-\frac{1}{kT}(H-\mu N)} = \lim_{\varepsilon \to 0} \int \prod_{\tau \in \varepsilon \mathbb{Z} \cap (0, \frac{1}{kT}]} \left\{ \left[\prod_{\mathbf{x} \in X} \frac{d\alpha_\tau^*(\mathbf{x}) \wedge d\alpha_\tau(\mathbf{x})}{2\pi \imath} \chi(|\alpha_\tau(\mathbf{x})| < R(\varepsilon)) \right] \right.$$

$$\left. e^{-\frac{1}{2}\langle \alpha_{\tau-\varepsilon}^*, \alpha_{\tau-\varepsilon} \rangle} I_0(\varepsilon; \alpha_{\tau-\varepsilon}^*, \alpha_\tau) e^{-\frac{1}{2}\langle \alpha_\tau^*, \alpha_\tau \rangle} \right\}$$

where

$$I_0(\varepsilon; \alpha^*, \beta) = e^{\langle \alpha^*, j(\varepsilon)\beta \rangle} e^{-\varepsilon \langle \alpha^* \beta, v \alpha^* \beta \rangle} \zeta_\varepsilon(\alpha, \beta). \tag{2.25}$$

(a) We first discuss why inserting the 'time derivative small field characteristic functions' $\zeta_\varepsilon(\alpha, \beta)$, with $\alpha = \alpha_{\tau-\varepsilon}$ and $\beta = \alpha_\tau$ for the various different values of τ, (which are not present in the formal functional integral (2.1)) introduced only a very small error, which tends to zero quickly in the limit $\varepsilon \to 0$. The critical observation is that the quadratic part of the exponent of $e^{-\frac{1}{2}\langle \alpha^*, \alpha \rangle} I_0(\varepsilon; \alpha^*, \beta) e^{-\frac{1}{2}\langle \beta^*, \beta \rangle}$ obeys

$$\text{Re} \left\{ -\frac{1}{2}\langle \alpha^*, \alpha \rangle + \langle \alpha^*, j(\varepsilon)\beta \rangle - \frac{1}{2}\langle \beta^*, \beta \rangle \right\} \approx \text{Re} \left\{ -\frac{1}{2}\langle \alpha^*, \alpha \rangle + \langle \alpha^*, \beta \rangle - \frac{1}{2}\langle \beta^*, \beta \rangle \right\}$$

$$= -\frac{1}{2}\|\alpha - \beta\|_{L^2}^2$$

which generates a factor of order $e^{-\frac{1}{2}r(\varepsilon)^2}$ when (α, β) is not in the support of $\zeta_\varepsilon(\alpha, \beta)$. This factor will be miniscule, because we shall choose $r(\varepsilon) = \frac{1}{(\varepsilon v)^{1/20}}$ where v is a small positive constant (and $\frac{1}{20}$ is a randomly chosen small positive number).

(b) A similar mechanism generates small factors whenever the difference $\beta - \alpha$ (now think of this as $\alpha_\tau - \alpha_{\tau-2\varepsilon}$) between the two arguments of

$$I_1(\varepsilon; \alpha^*, \beta) = \int d\mu_{R(\varepsilon)}(\phi^*, \phi) \, I_0(\varepsilon; \alpha^*, \phi) I_0(\varepsilon; \phi^*, \beta)$$

is larger than roughly $r(2\varepsilon)$. Consequently, we use the stationary phase approximation for this integral only when the 'time derivative small field condition' $\|\alpha - \beta\|_\infty \leq r(2\varepsilon)$ is satisfied. The change of variables (2.21) expresses I_1 as

$$I_1(\varepsilon; \alpha^*, \beta) = e^{\langle \alpha^*, j(2\varepsilon)\beta \rangle} \left[\prod_{\mathbf{x} \in X} \int_{M(\mathbf{x})} \frac{dz_*(\mathbf{x}) \wedge dz(\mathbf{x})}{2\pi \imath} e^{-z_*(\mathbf{x})z(\mathbf{x})} \right] e^{\tilde{A}(\alpha^*, \beta; z_*, z)}$$

$$\zeta_\varepsilon(\alpha, j(\varepsilon)\beta + z) \, \zeta_\varepsilon((j(\varepsilon)\alpha^* + z_*)^*, \beta).$$

The characteristic function $\zeta_\varepsilon\big(\alpha, j(\varepsilon)\beta + z\big)$ limits the domain of integration to z's obeying

$$\|z + j(\varepsilon)\beta - \alpha\|_\infty < r(\varepsilon).$$

Since $\|\alpha - \beta\|_\infty \le r(2\varepsilon) = \frac{1}{2^{1/20}}r(\varepsilon)$ and $\|j(\varepsilon)\beta - \beta\|_\infty \le \text{const}\,\varepsilon\,R(\varepsilon) \ll r(\varepsilon)$ (by our choice of $R(\varepsilon)$ – see part (d), below), this condition is more or less equivalent to $\|z\|_\infty < r(\varepsilon)$. Indeed, on the difference between the domain $\|z + j(\varepsilon)\beta - \alpha\|_\infty < r(\varepsilon)$ and the domain $\|z\|_\infty < r(\varepsilon)$, the integrand is extremely small, for reasons like those given in part (a), above. Similarly, the condition imposed by the second ζ_ε is roughly equivalent to $\|z_*\|_\infty < r(\varepsilon)$. The two conditions $\|z\|_\infty \le r(\varepsilon)$ and $\|z_*\|_\infty \le r(\varepsilon)$ are built into the domains of integration $D(\mathbf{x})$ in (2.22).

(c) The 'time derivative small field condition' $\|\alpha - \beta\|_\infty \le r(2\varepsilon) = \frac{1}{2^{1/20}}r(\varepsilon)$ is also used to ensure that $-z_*(\mathbf{x})z(\mathbf{x})$ has an extremely large negative real part whenever $(z_*(\mathbf{x}), z(\mathbf{x}))$ lies on $C(\mathbf{x})$, the side of the Stokes' 'cylinder'. This may be seen from part (b) of Lemma 2.2, with $r = r(\varepsilon)$ and $\rho = (\phi_*^{\text{crit}})^* - \phi^{\text{crit}} = j(\varepsilon)[\alpha - \beta]$.

(d) Another mechanism, which is similar in spirit to, but different from, the supression of large time derivatives, arises from the $e^{-\varepsilon\langle\alpha^*\beta,\,v\,\alpha^*\beta\rangle}$ in (2.25). When $\alpha \approx \beta$ (i.e. when the time derivative is small), the exponent is roughly

$$-\varepsilon\,\langle\alpha^*\alpha,\,v\,\alpha^*\alpha\rangle \le -\varepsilon\mathfrak{v}_1\,\langle\alpha^*\alpha,\,\alpha^*\alpha\rangle = -\varepsilon\mathfrak{v}_1\sum_{\mathbf{x}\in X}|\alpha(\mathbf{x})|^4$$

where $\mathfrak{v}_1$ is the smallest eigenvalue of the integral operator with kernel $v(\mathbf{x}, \mathbf{y})$. Recall that we have assumed that the integral operator with kernel $v(\mathbf{x}, \mathbf{y})$ is strictly positive. So if for some $\mathbf{x} \in X$, we have $|\alpha(\mathbf{x})| \ge R(\varepsilon)$, then

$$e^{-\varepsilon\langle\alpha^*\alpha,\,v\,\alpha^*\alpha\rangle} \le e^{-\mathfrak{v}_1\varepsilon R(\varepsilon)^4}.$$

The large field cutoff $R(\varepsilon)$ is chosen so that this is, again, minuscule when ε is small. For example, $R(\varepsilon) = \frac{1}{(\varepsilon v)^{3/10}}$, (with $\frac{3}{10}$ a randomly chosen number that is strictly bigger than, but close to $\frac{1}{4}$) does the job.

2.3   Bounds on the stationary phase approximation

In this section, we outline the proof of some bounds on the $\mathcal{E}_\delta(\varepsilon; \alpha^*, \beta)$'s of (2.16). The bounds are expressed in terms of a family of norms on analytic functions of $\{\alpha^*(\mathbf{x}), \beta(\mathbf{x}) \mid \mathbf{x} \in X\}$.

An analytic function $f(\alpha^*, \beta)$ of α^* and β may be expanded in a power series

$$f(\alpha^*, \beta) = \sum_{k,\ell\ge 0}\ \sum_{\substack{\mathbf{x}_1,\cdots,\mathbf{x}_k\in X \\ \mathbf{y}_1,\cdots,\mathbf{y}_\ell\in X}} a(\mathbf{x}_1,\cdots,\mathbf{x}_k\,;\,\mathbf{y}_1,\cdots,\mathbf{y}_\ell)\,\alpha(\mathbf{x}_1)^*\cdots\alpha(\mathbf{x}_k)^*\,\beta(\mathbf{y}_1)\cdots\beta(\mathbf{y}_\ell)$$

(with the coefficients $a(\mathbf{x}_1, \cdots, \mathbf{x}_k ; \mathbf{y}_1, \cdots, \mathbf{y}_\ell)$ invariant under permutations of $\mathbf{x}_1$, $\cdots$, $\mathbf{x}_k$ and of $\mathbf{y}_1$, $\cdots$, $\mathbf{y}_\ell$). For the functions of interest, the 'symmetric coefficient system' $a(\mathbf{x}_1, \cdots, \mathbf{x}_k ; \mathbf{y}_1, \cdots, \mathbf{y}_\ell)$ will be translation invariant (recall that X is the finite discrete torus $\mathbb{Z}^3/(L\mathbb{Z})^3$, for some large $L \in \mathbb{N}$) but otherwise exponentially decaying (uniformly in L). We tailor our norms to these two characteristics by defining the norm

$$\|f(\alpha^*, \beta)\|_\delta = \sum_{k,\ell \geq 0} \max_{\mathbf{x} \in X} \max_{1 \leq i \leq k+\ell} \sum_{\substack{(\vec{\mathbf{x}}, \vec{\mathbf{y}}) \in X^k \times X^\ell \\ (\vec{\mathbf{x}}, \vec{\mathbf{y}})_i = \mathbf{x}}} w_\delta(\vec{\mathbf{x}} ; \vec{\mathbf{y}}) \, |a(\vec{\mathbf{x}} ; \vec{\mathbf{y}})| \tag{2.26}$$

with the 'weight system'

$$w_\delta(\vec{\mathbf{x}} ; \vec{\mathbf{y}}) = \kappa(\delta)^{k+\ell} \, e^{md(\vec{\mathbf{x}}, \vec{\mathbf{y}})} \qquad \text{for } (\vec{\mathbf{x}}, \vec{\mathbf{y}}) \in X^k \times X^\ell \tag{2.27}$$

where $\tau(\vec{\mathbf{x}}, \vec{\mathbf{y}})$ is the minimal length of a tree whose set of vertices contains the set $\{\mathbf{x}_1, \cdots, \mathbf{x}_k, \mathbf{y}_1, \cdots, \mathbf{y}_\ell\}$. We refer to (2.27) as the weight system with mass m that associates the constant weight factor $\kappa(\delta)$ to the fields α^* and β. During the course of the proof, we will use other similar norms, with different weights. Roughly speaking, for $\|f(\alpha^*, \beta)\|_\delta$ to be finite, each coefficient $a(\mathbf{x}_1, \cdots, \mathbf{x}_k ; \mathbf{y}_1, \cdots, \mathbf{y}_\ell)$

- must decay a bit better than exponentially with rate m when one argument is held fixed and at least one other argument is moved far away and
- must be of size smaller than $\frac{1}{\kappa(\delta)^{k+\ell}}$. (The weight $\kappa(\delta)$ will be chosen shortly.)

If $\|f(\alpha^*, \beta)\|_\delta$ is finite, then $f(\alpha^*, \beta)$ is analytic, and bounded by $|X| \|f(\alpha^*, \beta)\|_\delta$, on the domain $\{\, (\alpha^*, \beta) \in \mathbb{C}^{2|X|} \mid |\alpha(\mathbf{x})|, |\beta(\mathbf{x})| < \kappa(\delta) \text{ for all } \mathbf{x} \in X \,\}$.

The decay properties of $\mathcal{E}_n$'s arise from the decay properties of the operators $j(\tau) = e^{-\tau(\mathbf{h} - \mu)}$ and v in the initial I_0 of Theorem 2.1. In general, we capture the decay properties of any operator $\mathcal{A}$ on $L^2(X) = \mathbb{C}^X$, with kernel $\mathcal{A}(\mathbf{x}, \mathbf{y})$ (i.e. that maps $\varphi(\mathbf{x}) \in L^2(X)$ to $(\mathcal{A}\varphi)(\mathbf{x}) = \sum_{\mathbf{y} \in X} \mathcal{A}(\mathbf{x}, \mathbf{y})\varphi(\mathbf{y}) \in L^2(X))$, by using the weighted L^1–L^∞ operator norm

$$\|\|\mathcal{A}\|\| = \max \left\{ \sup_{\mathbf{x} \in X} \sum_{\mathbf{y} \in X} e^{md(\mathbf{x}, \mathbf{y})} |\mathcal{A}(\mathbf{x}, \mathbf{y})| \,,\, \sup_{\mathbf{y} \in X} \sum_{\mathbf{x} \in X} e^{md(\mathbf{x}, \mathbf{y})} |\mathcal{A}(\mathbf{x}, \mathbf{y})| \right\} \tag{2.28}$$

where $d(\mathbf{x}, \mathbf{y})$ is the metric on $X = \mathbb{Z}^3/L\mathbb{Z}^3$. Some useful properties of this norm are given in

Lemma 2.3

(a) *For any two operators* $\mathcal{A}, \mathcal{B} : L^2(X) \to L^2(X)$

$$\|\|\mathcal{A}\mathcal{B}\|\| \leq \|\|\mathcal{A}\|\| \, \|\|\mathcal{B}\|\|$$

(b) *For any operator* $\mathcal{A} : L^2(X) \to L^2(X)$ *and any complex number* α

$$\|\|e^{\alpha \mathcal{A}}\|\| \leq e^{|\alpha| \, \|\|\mathcal{A}\|\|} \qquad \|\|e^{\alpha \mathcal{A}} - \mathbb{1}\|\| \leq |\alpha| \, \|\|\mathcal{A}\|\| e^{|\alpha| \, \|\|\mathcal{A}\|\|}.$$

Proof:

(a) By the triangle inequality, for each $\mathbf{x} \in X$,

$$\sum_{\mathbf{y} \in X} e^{md(\mathbf{x},\mathbf{y})} |(\mathcal{AB})(\mathbf{x},\mathbf{y})| \leq \sum_{\mathbf{y},\mathbf{z} \in X} e^{md(\mathbf{x},\mathbf{z})} |\mathcal{A}(\mathbf{x},\mathbf{z})| e^{md(\mathbf{z},\mathbf{y})} |\mathcal{B}(\mathbf{z},\mathbf{y})|$$

$$\leq \sum_{\mathbf{z} \in X} e^{md(\mathbf{x},\mathbf{z})} |\mathcal{A}(\mathbf{x},\mathbf{z})| \ \|\!|\!|\mathcal{B}|\!|\!|$$

$$\leq \|\!|\!|\mathcal{A}|\!|\!| \ \|\!|\!|\mathcal{B}|\!|\!|.$$

The other bound, in which one sums over $\mathbf{x}$ rather than $\mathbf{y}$, is similar.

(b) By part (a),

$$\|\!|\!|e^{\alpha \mathcal{A}}|\!|\!| \leq \sum_{n=0}^{\infty} \frac{1}{n!} \|\!|\!|\alpha^n \mathcal{A}^n|\!|\!| \leq \sum_{n=0}^{\infty} \frac{1}{n!} |\alpha|^n \|\!|\!|\mathcal{A}|\!|\!|^n = e^{|\alpha| \ \|\!|\!|\mathcal{A}|\!|\!|}$$

and

$$\|\!|\!|e^{\alpha \mathcal{A}} - \mathbb{1}|\!|\!| \leq \sum_{n=1}^{\infty} \frac{1}{n!} \|\!|\!|\alpha^n \mathcal{A}^n|\!|\!| \leq \sum_{n=1}^{\infty} \frac{1}{n!} |\alpha|^n \|\!|\!|\mathcal{A}|\!|\!|^n \leq |\alpha| \ \|\!|\!|\mathcal{A}|\!|\!| e^{|\alpha| \ \|\!|\!|\mathcal{A}|\!|\!|}.$$
$$\square$$

Corollary 2.4 *Let* $\tau \geq 0$.

$$\|\!|\!|j(\tau)|\!|\!| \leq e^{\tau(\|\!|\!|\mathbf{h}|\!|\!| + \mu)} \qquad \|\!|\!|j(\tau) - \mathbb{1}|\!|\!| \leq \tau(\|\!|\!|\mathbf{h}|\!|\!| + |\mu|) e^{\tau(\|\!|\!|\mathbf{h}|\!|\!| + |\mu|)}.$$

Proof: Write $j(\tau) = e^{\tau\mu} e^{-\tau\mathbf{h}}$ and $j(\tau) - \mathbb{1} = e^{\tau\mu}(e^{-\tau\mathbf{h}} - \mathbb{1}) + e^{\tau\mu} - \mathbb{1}$. By the previous Lemma

$$\|\!|\!|j(\tau)|\!|\!| = e^{\tau\mu} \|\!|\!|e^{-\tau\mathbf{h}}|\!|\!| \leq e^{\tau\mu} e^{\tau\|\!|\!|\mathbf{h}|\!|\!|}$$

and

$$\|\!|\!|j(\tau) - \mathbb{1}|\!|\!| \leq e^{\tau\mu} \|\!|\!|e^{-\tau\mathbf{h}} - \mathbb{1}|\!|\!| + \|\!|\!|e^{\tau\mu} - \mathbb{1}|\!|\!| \leq \tau \|\!|\!|\mathbf{h}|\!|\!| e^{\tau\mu} e^{\tau\|\!|\!|\mathbf{h}|\!|\!|} + |e^{\tau\mu} - \mathbb{1}|. \qquad \square$$

The quantities relevant for the estimates of $\mathcal{E}_\delta(\varepsilon; \alpha^*, \beta)$, in addition to the radii $r(\delta)$, of the domain of integration, and $\kappa(\delta)$, of the domain of analyticity, are the norm $\|\!|\!|v|\!|\!|$ of the interaction, the decay rate m, a constant K_j such that

$$\|\!|\!|j(\tau)|\!|\!| \leq e^{K_j\tau} \quad \text{and} \quad \|\!|\!|j(\tau) - \mathbb{1}|\!|\!| \leq K_j \tau \, e^{K_j\tau} \quad \text{for } \tau \geq 0 \qquad (2.29)$$

(see Corollary 2.4) and a constant $0 < \Theta \leq 1$ that bounds the range of θ's (see eqn. (2.13)) for which the constructions work. In (Balaban *et al.* 2010b, Hypothesis 1.1) we give a set of hypotheses on these constants. (For the full temporal ultraviolet limit, not just the stationary phase approximation, see (Balaban *et al.* 2010c, Appendix F).) For the purpose of this chapter, I'll just make one reasonably specific choice. I'll allow any $K_j, \mathrm{m} > 0$ and view them just as fixed constants. Then I'll pick sufficiently small (depending on K_j and m) $0 < \mathfrak{v}, \Theta \le 1$ and allow any interaction v with $\|\|v\|\| < \mathfrak{v}$. Then I'll set

$$r(\delta) = \frac{1}{(\delta\mathfrak{v})^{\frac{1}{20}}} \qquad \kappa(\delta) = \frac{1}{(\delta\mathfrak{v})^{\frac{3}{10}}} \tag{2.30}$$

Think of the exponents $\frac{1}{20}$ and $\frac{3}{10}$ as being just a tiny bit bigger than 0 and $\frac{1}{4}$, respectively.

I will outline the proof of

Theorem 2.5 *Under the above hypotheses, there is a constant K_E such that*

$$\left\| \mathcal{E}_\delta(\varepsilon; \alpha^*, \beta) \right\|_\delta \le K_E \delta^2 \|\|v\|\|^2 r(\delta)^2 \kappa(\delta)^6$$

for all $0 \le \varepsilon \le \delta \le \Theta$ for which $\frac{\delta}{\varepsilon}$ is a power of 2. The function $\mathcal{E}_\delta(\varepsilon; \alpha^*, \beta)$ has degree at least two both[3] in α^* and β.

In (Balaban *et al.* 2010b, Theorem 1.4), we also prove

Theorem 2.6 *The limit*

$$\mathcal{E}_\theta(\alpha^*, \beta) = \lim_{m \to \infty} \mathcal{E}_\theta(2^{-m}\theta; \alpha^*, \beta)$$

exists uniformly in $0 \le \theta \le \Theta$. It fulfils the estimate

$$\left\| \mathcal{E}_\theta(\alpha^*, \beta) \right\|_\theta \le K_E \, \theta^2 \|\|v\|\|^2 r(\theta)^2 \kappa(\theta)^6$$

and has degree at least two in both α^* and β.

The proof of Theorem 2.6 uses the same techniques as the proof of Theorem 2.5. So I won't discuss the former at all.

Remark 2.7 *Theorem 2.5 implies that $\left\| \mathcal{E}_\delta(\varepsilon; \alpha^*, \beta) \right\|_\delta \le K_E \left(\frac{\|\|v\|\|}{\mathfrak{v}} \right)^2$ for all $0 \le \varepsilon \le \delta \le \Theta$. In particular $\mathcal{E}_\delta(\varepsilon; \alpha^*, \beta)$ is analytic and bounded pointwise by $K_E |X| \left(\frac{\|\|v\|\|}{\mathfrak{v}} \right)^2$ on $\left\{ (\alpha^*, \beta) \in \mathbb{C}^{2|X|} \mid |\alpha(\mathbf{x})|, |\beta(\mathbf{x})| < (\delta\mathfrak{v})^{-\frac{3}{10}} \text{ for all } \mathbf{x} \in X \right\}$. The coefficients in its power series expansion decay exponentially at rate at least m.*

We formulate the recursion relation (2.16) that defines $\mathcal{E}_{\varepsilon_n}(\varepsilon; \alpha^*, \beta)$ more abstractly.

Definition 2.8 *Let $0 \le \varepsilon \le \delta$. For an action $\mathcal{E}(\alpha^*, \beta)$ we set*

$$\mathfrak{R}_{\delta,\varepsilon}[\mathcal{E}](\alpha^*, \beta) = \mathcal{E}(\alpha^*, j(\delta)\beta) + \mathcal{E}(j(\delta)\alpha^*, \beta) + \log \frac{\int d\mu_{r(\delta)}(z^*, z) \, e^{\partial A_{\delta,\varepsilon}(\mathcal{E}; \alpha^*, \beta; z^*, z)}}{\int d\mu_{r(\delta)}(z^*, z)}$$

---

[3] By this we mean that every monomial appearing in its power series expansion contains a factor of the form $\alpha^*(\mathbf{x}_1) \, \alpha^*(\mathbf{x}_2) \, \beta(\mathbf{x}_3) \, \beta(\mathbf{x}_4)$.

whenever the logarithm is defined. Here

$$\partial \mathcal{A}_{\delta,\varepsilon}(\mathcal{E}; \alpha^*, \beta; z_*, z) = \big[\mathcal{V}_\delta(\varepsilon; \alpha^*, j(\delta)\beta + z) - \mathcal{V}_\delta(\varepsilon; \alpha^*, j(\delta)\beta)\big]$$
$$+ \big[\mathcal{V}_\delta(\varepsilon; j(\delta)\alpha^* + z_*, \beta) - \mathcal{V}_\delta(\varepsilon; j(\delta)\alpha^*, \beta)\big]$$
$$+ \big[\mathcal{E}(\alpha^*, j(\delta)\beta + z) - \mathcal{E}(\alpha^*, j(\delta)\beta)\big]$$
$$+ \big[\mathcal{E}(j(\delta)\alpha^* + z_*, \beta) - \mathcal{E}(j(\delta)\alpha^*, \beta)\big].$$

The recursion relation (2.16) is equivalent to

$$\mathcal{E}_\varepsilon(\varepsilon; \alpha^*, \beta) = 0$$
$$\mathcal{E}_{\varepsilon_{n+1}}(\varepsilon; \alpha^*, \beta) = \mathfrak{R}_{\varepsilon_n,\varepsilon}\big[\mathcal{E}_{\varepsilon_n}(\varepsilon; \alpha^*, \beta)\big] \tag{2.31}$$

To prove Theorem 2.5, we perform induction on n to successively bound $\mathcal{E}_{\varepsilon_n}(\varepsilon; \cdot)$ for $n = 0, \cdots, \log_2 \frac{\Theta}{\varepsilon}$. The heart of the induction step is given in Proposition 2.11. Proposition 2.11, in turn, is an application of a corollary to (Balaban *et al.* 2010a, Theorem 3.4), which, specialized to the current setting, says

Theorem 2.9 *Let $\kappa > 0$ and denote by $\| \cdot \|_{\mathrm{fl}}$ the norm[4] with weight system of mass m that assigns the weight $\kappa > 0$ to the fields α^* and β and the weight $4\mathrm{r}(\delta)$ to the fields z_* and z. If $f(\alpha^*, \beta; z_*, z)$ is an analytic function on a neighbourhood of the origin in $\mathbb{C}^{4|X|}$ that obeys $\|f\|_{\mathrm{fl}} < \frac{1}{16}$, then there is an analytic function $g(\alpha^*, \beta)$ such that*

$$\frac{\int e^{f(\alpha^*,\beta;z^*,z)} \, d\mu_{\mathrm{r}(\delta)}(z^*, z)}{\int e^{f(0,0;z^*,z)} \, d\mu_{\mathrm{r}(\delta)}(z^*, z)} = e^{g(\alpha^*,\beta)} \tag{2.32}$$

and

$$\|g\|_{\mathrm{fl}} \leq \frac{\|f\|_{\mathrm{fl}}}{1 - 16\|f\|_{\mathrm{fl}}}.$$

I'll give an outline of the proof of this theorem in Section 2.5. See Theorem 2.29. The corollary that we shall use is (Balaban *et al.* 2010a, Corollary 3.5), which, again specialized to the current setting, says

Corollary 2.10 *Let $f(\alpha^*, \beta; z_*, z)$ be an analytic function on a neighbourhood of the origin in $\mathbb{C}^{4|X|}$ that obeys $\|f\|_{\mathrm{fl}} < \frac{1}{32}$. Define, for each complex number ζ with $|\zeta| \|f\|_{\mathrm{fl}} < \frac{1}{16}$, the function $G(\zeta) = G(\zeta; \alpha^*, \beta)$ by the condition*

$$\frac{\int e^{\zeta f(\alpha^*,\beta;z^*,z)} \, d\mu_{\mathrm{r}(\delta)}(z^*, z)}{\int e^{\zeta f(0,0;z^*,z)} \, d\mu_{\mathrm{r}(\delta)}(z^*, z)} = e^{G(\zeta;\alpha^*,\beta)} \tag{2.33}$$

---

[4] The 'fl' in $\| \cdot \|_{\mathrm{fl}}$ stands for fluctutation. This norm is defined just as in (2.26), except that there are four fields, α^*, β, z_* and z, instead of two, and the $\kappa(\delta)^{k+\ell}$ of (2.27) is replaced by $\kappa^{k+\ell}\big(4\mathrm{r}(\delta)\big)^{n_*+n}$, where k is the number of α^* fields, ℓ is the number of β fields, n_* is the number of z_* fields and n is the number of z fields.

as in Theorem 2.9. Then $G(\zeta)$ is a (Banach space valued) analytic function of ζ and, for each $n \in \mathbb{N}$, the $g(\alpha^*, \beta) = G(1; \alpha^*, \beta)$ of Theorem 2.9 obeys

$$\left\| g - \frac{dG}{d\zeta}(0) - \cdots - \frac{1}{n!}\frac{d^nG}{d\zeta^n}(0) \right\|_\text{fl} \le \left(\frac{\|f\|_\text{fl}}{\frac{1}{20} - \|f\|_\text{fl}} \right)^{n+1}.$$

We have $G(0) = 0$ and

$$\frac{dG}{d\zeta}(0) = \int \left[f(\alpha^*, \beta; z^*, z) - f(0, 0; z^*, z) \right] d\mu_{\mathrm{r}(\delta)}(z^*, z).$$

If the symmetric coefficient system $a(\vec{\mathbf{x}}_*, \vec{\mathbf{x}}; \vec{\mathbf{y}}_*, \vec{\mathbf{y}})$ of f obeys $a(\vec{\mathbf{x}}_*, \vec{\mathbf{x}}; \vec{\mathbf{y}}_*, \vec{\mathbf{y}}) = 0$ whenever $\vec{\mathbf{y}} = \vec{\mathbf{y}}_*$, then $\frac{dG}{d\zeta}(0) = 0$.

Proof: The proof of the bound in this corollary is a short, straight–forward application of the Cauchy integral formula. For the details, see (Balaban *et al.* 2010a, Corollary 3.5).

The left-hand side is 1 when $\alpha^* = \beta = 0$, so $G(0) = 0$. To show that $\frac{dG}{d\zeta}(0) = 0$, under the specified conditions on the coefficient system, expand $f(\alpha^*, \beta; z^*, z)$ in powers of the fields α^*, β, z^* and z. This expresses $\int f(\alpha^*, \beta; z^*, z)\, d\mu_{\mathrm{r}(\delta)}(z^*, z)$ as a sum of terms, with each term being some coefficient (depending on α^* and β) times

$$\int \prod_{\mathbf{x} \in X} \left\{ \left[z(\mathbf{x})^* \right]^{n_\mathbf{x}} z(\mathbf{x})^{m_\mathbf{x}} \right\} d\mu_{\mathrm{r}(\delta)}(z^*, z)$$

Switching to polar coordinates, $z(\mathbf{x}) = \rho(\mathbf{x})e^{i\theta(\mathbf{x})}$,

$$\int \prod_{\mathbf{x} \in X} \left\{ \left[z(\mathbf{x})^* \right]^{n_\mathbf{x}} z(\mathbf{x})^{m_\mathbf{x}} \right\} d\mu_{\mathrm{r}(\delta)}(z^*, z)$$

$$= \prod_{\mathbf{x} \in X} \frac{1}{\pi} \int_0^{r(\delta)} d\rho(\mathbf{x}) \int_0^{2\pi} d\theta(\mathbf{x})\, e^{-\rho(\mathbf{x})^2} \rho(\mathbf{x})^{n_\mathbf{x}+m_\mathbf{x}+1} e^{i(m_\mathbf{x}-n_\mathbf{x})\theta(\mathbf{x})}.$$

(2.34)

Unless $m_\mathbf{x} = n_\mathbf{x}$ for every $\mathbf{x} \in X$, the right-hand side is zero because of the $\theta(\mathbf{x})$ integrals. When $m_\mathbf{x} = n_\mathbf{x}$ for every $\mathbf{x} \in X$, the coefficient multiplying this integral is zero because of the hypothesis on the symmetric coefficient system. Hence

$$\int f(\alpha^*, \beta; z^*, z)\, d\mu_{\mathrm{r}(\delta)}(z^*, z) = \int f(0, 0; z^*, z)\, d\mu_{\mathrm{r}(\delta)}(z^*, z) = 0.$$

$\square$

For the induction step, we use

Proposition 2.11 *For all $0 \le \varepsilon \le \delta \le \Theta/2$, with δ an integer multiple of ε, the following holds:*

Let $\mathcal{E}(\alpha^*, \beta)$ be an analytic function which has degree at least two both in α^* and in β and which obeys $\|\mathcal{E}(\alpha^*, \beta)\|_\delta \le 4\, e^{5\delta K_j} \delta \|v\|\, \mathrm{r}(\delta)\, \kappa(2\delta)^3$. Then $\mathfrak{R}_{\delta, \varepsilon}\big[\mathcal{E}\big](\alpha^*, \beta)$ is well defined, has degree at least two both in α^* and in β, and satisfies the estimate

$$\left\|\Re_{\delta,\varepsilon}[\mathcal{E}]\right\|_{2\delta} \leq 2^{20}\, e^{10\delta K_j}\delta^2\, \|v\|^2\, \mathrm{r}(\delta)^2\, \kappa(2\delta)^6 + 2\, e^{2\delta K_j}\left(\frac{\kappa(2\delta)}{\kappa(\delta)}\right)^4 \|\mathcal{E}\|_\delta.$$

Proof: Observe that the functions

$$\mathcal{V}_\delta(\varepsilon;\,\alpha^*, j(\delta)\beta + z) - \mathcal{V}_\delta(\varepsilon;\,\alpha^*, j(\delta)\beta) \quad \text{and} \quad \mathcal{E}(\alpha^*, j(\delta)\beta + z) - \mathcal{E}(\alpha^*, j(\delta)\beta)$$

both have degree at least two in α^*, degree at least one in z and do not depend on z_*. Similarly, $\mathcal{V}_\delta(\varepsilon;\, j(\delta)\alpha^* + z_*, \beta) - \mathcal{V}_\delta(\varepsilon;\, j(\delta)\alpha^*, \beta)$ and $\mathcal{E}(j(\delta)\alpha^* + z_*, \beta) - \mathcal{E}(j(\delta)\alpha^*, \beta)$ have degree at least two in β, degree at least one in z_* and do not depend on z. Since the integral of any monomial against $d\mu_{\mathrm{r}(\delta)}(z^*, z)$ is zero unless there are the same number of z's and z^*'s (see (2.34)),

$$\int d\mu_{\mathrm{r}(\delta)}(z^*, z)\ \partial\mathcal{A}_{\delta,\varepsilon}(\mathcal{E};\,\alpha^*, \beta;\, z^*, z) = 0 \tag{2.35}$$

and $\log \dfrac{\int d\mu_{\mathrm{r}(\delta)}(z^*,z)\, e^{\partial\mathcal{A}_{\delta,\varepsilon}(\mathcal{E};\,\alpha^*,\beta;z^*,z)}}{\int d\mu_{\mathrm{r}(\delta)}(z^*,z)}$ has degree at least two both in α^* and in β. This implies that $\Re_{\delta,\varepsilon}[\mathcal{E}](\alpha^*, \beta)$ has degree at least two both in α^* and in β.

We apply Corollary 2.10, with $\kappa = \kappa(2\delta)$. Clearly, $\|f(\alpha^*,\beta)\|_{2\delta} = \|f(\alpha^*,\beta)\|_{\mathrm{fl}}$ for functions that are independent of the fluctuation fields z_*, z. To apply the corollary, we need to bound $\|\partial\mathcal{A}_{\delta,\varepsilon}(\mathcal{E};\,\alpha^*, \beta;\, z_*, z)\|_{\mathrm{fl}}$.

We'll first bound

$$\mathcal{V}_\delta(\varepsilon;\,\alpha^*, j(\delta)\beta + z) - \mathcal{V}_\delta(\varepsilon;\,\alpha^*, j(\delta)\beta)$$

$$= \varepsilon \sum_{\tau\in\varepsilon\mathbb{Z}\cap[0,\delta)} \left[\langle \gamma_{*\tau}\, g_{\tau+\varepsilon},\, v\, \gamma_{*\tau}\, g_{\tau+\varepsilon}\rangle - \langle \gamma_{*\tau}\, \hat{g}_{\tau+\varepsilon},\, v\, \gamma_{*\tau}\, \hat{g}_{\tau+\varepsilon}\rangle\right]$$

with

$$\gamma_{*\tau} = j(\tau)\alpha^* \quad g_\tau = j(2\delta - \tau)\beta \quad \hat{g}_\tau = j(\delta - \tau)\big(j(\delta)\beta + z\big) = j(2\delta - \tau)\beta + j(\delta - \tau)z.$$

Expand out $\hat{g}_\tau$ as a sum of two terms, as in the last equation, expressing the summand $\langle \gamma_{*\tau}\, g_{\tau+\varepsilon},\, v\, \gamma_{*\tau}\, g_{\tau+\varepsilon}\rangle - \langle \gamma_{*\tau}\, \hat{g}_{\tau+\varepsilon},\, v\, \gamma_{*\tau}\, \hat{g}_{\tau+\varepsilon}\rangle$ itself as a sum of three terms, each of which is (except for a minus sign) of the form

$$\langle (\Gamma_1\gamma_1)(\Gamma_2\gamma_2),\, v\, (\Gamma_3\gamma_3)(\Gamma_4\gamma_4)\rangle$$

$$= \sum_{\substack{x_1,x_2,x_3,x_4\in X \\ y_1,y_2\in X}} \left[\prod_{\ell=1,2}\Gamma_\ell(y_1, x_\ell)\gamma_\ell(x_\ell)\right] v(y_1, y_2)\left[\prod_{\ell=3,4}\Gamma_\ell(y_2, x_\ell)\gamma_\ell(x_\ell)\right]$$

with

$$\Gamma_1 = \Gamma_2 = j(\tau - \varepsilon) \quad \gamma_1 = \gamma_3 = \alpha^* \quad (\Gamma_3,\gamma_3), (\Gamma_4,\gamma_4) \in \big\{\big(j(2\delta - \tau),\beta\big),\, \big(j(\delta - \tau), z\big)\big\}$$

and with at least one of (Γ_3, γ_3), (Γ_4, γ_4) being $(j(\delta - \tau), z)$. In general

$$\sum_{\substack{\mathbf{x}_2,\mathbf{x}_3,\mathbf{x}_4\in X \\ \mathbf{y}_1,\mathbf{y}_2\in X}} e^{m\tau(\mathbf{x}_1,\mathbf{x}_2,\mathbf{x}_3,\mathbf{x}_4)} \left| \left[\prod_{\ell=1,2} \Gamma_\ell(\mathbf{y}_1,\mathbf{x}_\ell)\kappa_\ell \right] v(\mathbf{y}_1,\mathbf{y}_2) \left[\prod_{\ell=3,4} \Gamma_\ell(\mathbf{y}_2,\mathbf{x}_\ell)\kappa_\ell \right] \right|$$

$$\leq \sum_{\substack{\mathbf{x}_2,\mathbf{x}_3,\mathbf{x}_4 \\ \mathbf{y}_1,\mathbf{y}_2}} \left| \left[\prod_{\ell=1,2} e^{md(\mathbf{x}_\ell,\mathbf{y}_1)}\Gamma_\ell(\mathbf{y}_1,\mathbf{x}_\ell)\kappa_\ell \right] v(\mathbf{y}_1,\mathbf{y}_2) e^{md(\mathbf{y}_1,\mathbf{y}_2)} \left[\prod_{\ell=3,4} e^{md(\mathbf{x}_\ell,\mathbf{y}_2)}\Gamma_\ell(\mathbf{y}_2,\mathbf{x}_\ell)\kappa_\ell \right] \right|$$

$$\leq \prod_{\ell=1}^{4} \kappa_\ell \sum_{\substack{\mathbf{x}_2\in X \\ \mathbf{y}_1,\mathbf{y}_2\in X}} \left[\prod_{\ell=1,2} e^{md(\mathbf{x}_\ell,\mathbf{y}_1)}|\Gamma_\ell(\mathbf{y}_1,\mathbf{x}_\ell)| \right] |v(\mathbf{y}_1,\mathbf{y}_2)| e^{md(\mathbf{y}_1,\mathbf{y}_2)} \||\Gamma_3|\| \, \||\Gamma_4|\|$$

$$\leq \prod_{\ell=1}^{4} \kappa_\ell \sum_{\mathbf{x}_2,\mathbf{y}_1\in X} \left[\prod_{\ell=1,2} e^{md(\mathbf{x}_\ell,\mathbf{y}_1)}|\Gamma_\ell(\mathbf{y}_1,\mathbf{x}_\ell)| \right] \||v|\| \, \||\Gamma_3|\| \, \||\Gamma_4|\|$$

$$\leq \||v|\| \prod_{\ell=1}^{4} \kappa_\ell \||\Gamma_\ell|\|.$$

To get from the first line to the second line, we used that the set of vertices of the tree in Fig. 3.5 below contains $\mathbf{x}_1$, $\mathbf{x}_2$, $\mathbf{x}_3$, and $\mathbf{x}_4$ so that

$$\tau(\mathbf{x}_1,\mathbf{x}_2,\mathbf{x}_3,\mathbf{x}_4) \leq d(\mathbf{x}_1,\mathbf{y}_1) + d(\mathbf{x}_2,\mathbf{y}_1) + d(\mathbf{y}_1,\mathbf{y}_2) + d(\mathbf{y}_2,\mathbf{x}_3) + d(\mathbf{y}_2,\mathbf{x}_4)$$

The bounds when $\mathbf{x}_2$ or $\mathbf{x}_3$ or $\mathbf{x}_4$ is fixed instead of $\mathbf{x}_1$ are the same.

As

$$\||j(\tau)|\| \leq e^{K_j\delta} \qquad \||j(2\delta - \tau - \varepsilon)|\| \leq e^{2K_j\delta} \qquad \||j(\delta - \tau - \varepsilon)|\| \leq e^{K_j\delta}$$

and α^*, β and z have weights $\kappa(2\delta)$, $\kappa(2\delta)$ and $4r(\delta)$, respectively, we have, for each $\tau \in \varepsilon\mathbb{Z} \cap (0, \delta]$,

$$\left\| \langle \gamma_{*\tau} g_{\tau+\varepsilon}, \, v \gamma_{*\tau} g_{\tau+\varepsilon} \rangle - \langle \gamma_{*\tau} \hat{g}_{\tau+\varepsilon}, \, v \gamma_{*\tau} \hat{g}_{\tau+\varepsilon} \rangle \right\|_{\mathrm{fl}} \leq 12e^{5K_j\delta} \||v|\| \, \mathrm{r}(\delta) \, \kappa(2\delta)^3.$$

Here, we have assumed that $\Theta v \leq \frac{1}{2^{10}}$ so that $4\mathrm{r}(\delta) \leq \kappa(2\delta)$. Summing over τ and multiplying by ε gives

$$\left\| \mathcal{V}_\delta(\varepsilon; \alpha^*, j(\delta)\beta + z) - \mathcal{V}_\delta(\varepsilon; \alpha^*, j(\delta)\beta) \right\|_{\mathrm{fl}} \leq 12e^{5K_j\delta} \, \delta \||v|\| \, \mathrm{r}(\delta) \, \kappa(2\delta)^3.$$

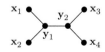

Fig. 2.5 A longer tree.

Similarly

$$\left\| \mathcal{V}_\delta(\varepsilon; \, j(\delta)\alpha^* + z_*, \, \beta) - \mathcal{V}_\delta(\varepsilon; \, j(\delta)\alpha^* + z_*, \, \beta) \right\|_{\mathrm{fl}} \le 12 e^{5K_j\delta}\,\delta\,\|\!|\!|v|\!|\!|\,\mathrm{r}(\delta)\,\kappa(2\delta)^3.$$

Next, we bound $\mathcal{E}(\alpha^*, \, j(\delta)\beta + z) - \mathcal{E}(\alpha^*, \, j(\delta)\beta)$. For any analytic function $f(\alpha^*, \beta)$,

$$\left\| f(\alpha^*, \, j(\delta)\beta + z) - f(\alpha^*, \, j(\delta)\beta) \right\|_{\mathrm{fl}} \le \left\| f(\alpha^*, \, j(\delta)\beta + z) \right\|_{\mathrm{fl}}$$

since the symmetric coefficient system for $f(\alpha^*, \, j(\delta)\beta + z) - f(\alpha^*, \, j(\delta)\beta)$ is precisely the symmetric coefficient system for $f(\alpha^*, \, j(\delta)\beta + z)$, but with the coefficients for terms having no z's replaced by 0. So, by Proposition 2.32,[5]

$$\left\| f(\alpha^*, \, j(\delta)\beta + z) - f(\alpha^*, \, j(\delta)\beta) \right\|_{\mathrm{fl}} \le \left\| f(\alpha^*, \, j(\delta)\beta + z) \right\|_{\mathrm{fl}} \le \left\| f(\alpha^*, \beta) \right\|_\delta$$

since

$$\|\!|\!|j(\delta)|\!|\!|\,\kappa(2\delta) + 4\mathrm{r}(\delta) \le e^{\delta K_j}\kappa(2\delta) + 4\mathrm{r}(\delta) = \left[e^{\delta K_j}\frac{1}{2^{\frac{3}{10}}} + 4(\delta\mathfrak{v})^{\frac{1}{4}} \right]\kappa(\delta) \le \kappa(\delta)$$

if Θ and $\mathfrak{v}$ are small enough. In particular $\left\| \mathcal{E}(\alpha^*, \, j(\delta)\beta + z) - \mathcal{E}(\alpha^*, \, j(\delta)\beta) \right\|_{\mathrm{fl}} \le \|\mathcal{E}\|_\delta$. Similarly $\left\| \mathcal{E}(j(\delta)\alpha^* + z_*, \, \beta) - \mathcal{E}(j(\delta)\alpha^*, \, \beta) \right\|_{\mathrm{fl}} \le \|\mathcal{E}\|_\delta$.

Combining the bounds of the previous two paragraphs and then using the hypothesis that $\|\mathcal{E}\|_\delta \le 4 e^{5\delta K_j}\delta\|\!|\!|v|\!|\!|\,\mathrm{r}(\delta)\,\kappa(2\delta)^3$, we get

$$\left\| \partial A_{\delta,\varepsilon}(\mathcal{E}; \, \cdot) \right\|_{\mathrm{fl}} \le 24\, e^{5\delta K_j}\delta\|\!|\!|v|\!|\!|\,\mathrm{r}(\delta)\,\kappa(2\delta)^3 + 2\|\mathcal{E}\|_\delta \le 2^5\, e^{5\delta K_j}\delta\|\!|\!|v|\!|\!|\,\mathrm{r}(\delta)\,\kappa(2\delta)^3$$

$$\le 2^5\, e^{5\delta K_j}\frac{1}{2^{\frac{9}{10}}}(\delta\mathfrak{v})^{\frac{1}{20}} \le \frac{1}{64} \tag{2.36}$$

if $\Theta \le 1$ and $\Theta\mathfrak{v}$ is small enough. Finally, by (2.35) and Corollary 2.10

$$\left\| \log \frac{\int d\mu_{\mathrm{r}(\delta)}(z^*, z)\, e^{\partial A_{\delta,\varepsilon}(\mathcal{E}; \, \alpha^*,\beta;z_*,z)}}{\int d\mu_{\mathrm{r}(\delta)}(z^*, z)} \right\|_{2\delta} \le \frac{\left\| \partial A_{\delta,\varepsilon}(\mathcal{E}; \, \cdot) \right\|_{\mathrm{fl}}^2}{\left(\frac{1}{20} - \left\| \partial A_{\delta,\varepsilon}(\mathcal{E}; \, \cdot) \right\|_{\mathrm{fl}} \right)^2}$$

$$\le 2^{20}\, e^{10\delta K_j}\delta^2\|\!|\!|v|\!|\!|^2\,\mathrm{r}(\delta)^2\,\kappa(2\delta)^6.$$

Combining this estimate and the estimate of Lemma 2.12, below, with $f = \mathcal{E}$, we get the desired bound on $\left\| \mathfrak{R}_{\delta,\varepsilon}[\mathcal{E}] \right\|_{2\delta}$. $\qquad\square$

Lemma 2.12 *Let $f(\alpha^*, \beta)$ be an analytic function that has degree at least two both in α^* and β. Then*

$$\left\| f(\alpha^*, j(\delta)\beta) \right\|_{2\delta}, \, \left\| f(j(\delta)\alpha^*, \beta) \right\|_{2\delta} \le e^{2\delta K_j} \left(\frac{\kappa(2\delta)}{\kappa(\delta)} \right)^4 \|f\|_\delta.$$

---

[5] Actually, by an obvious generalization of Proposition 2.32, since the g of Proposition 2.32 is a function of a single field. See (Balaban *et al.* 2010a, Corollary A.2) for such a generalization.

Proof: Introduce the auxiliary norm $\|\cdot\|_{\text{aux}}$ that uses the weight system of mass m that associates the constant weight factor $\kappa(\delta)$ to the field α_* and the constant weight factor $e^{-\delta K_j}\kappa(\delta)$ to the field β. Since, by (2.29), $\||j(\delta)\|| e^{-\delta K_j}\kappa(\delta) \le \kappa(\delta)$., part (i) of Proposition 2.32 gives

$$\left\|f\left(\alpha^*, j(\delta)\beta\right)\right\|_{w_{\text{aux}}} \le \|f\|_\delta.$$

As $f\left(\alpha^*, j(\delta)\beta\right)$ has degree at least two both in α^* and β and $e^{-\delta K_j}\kappa(\delta) \ge \kappa(2\delta)$, if Θ is small enough,

$$\left\|f\left(\alpha^*, j(\delta)\beta\right)\right\|_{2\delta} \le \left(\frac{\kappa(2\delta)}{\kappa(\delta)}\right)^2 \left(\frac{\kappa(2\delta)}{e^{-\delta K_j}\kappa(\delta)}\right)^2 \left\|f\left(\alpha^*, j(\delta)\beta\right)\right\|_{w_{\text{aux}}} \le e^{2\delta K_j}\left(\frac{\kappa(2\delta)}{\kappa(\delta)}\right)^4 \|f\|_\delta.$$

The estimate on $\left\|f\left(j(\delta)\alpha^*, \beta\right)\right\|_{2\delta}$ is similar. $\square$

Proof of Theorem 2.5 Choose $K_E = 2^{21}e^{10K_j}$. We write $\delta = \varepsilon_n = 2^n\varepsilon$ and prove the statement by induction on n. In the case $n = 0$ there is nothing to prove. For the induction step from n to $n+1$, set $\delta = \varepsilon_n$. The hypothesis of Proposition 2.11, with $\mathcal{E} = \mathcal{E}_\delta$, is satisfied since, by the inductive hypothesis,

$$\left\|\mathcal{E}_\delta\right\|_\delta \le K_E\,\delta^2 \||v\||^2\, \mathrm{r}(\delta)^2 \kappa(\delta)^6 = K_E\,(\delta\mathfrak{v})^{\frac{1}{20}} 2^{\frac{9}{10}}\,\delta\||v\||\,\mathrm{r}(\delta)\kappa(2\delta)^3 \le \delta\||v\||\,\mathrm{r}(\delta)\kappa(2\delta)^3 \tag{2.37}$$

if $\Theta\mathfrak{v}$ has been chosen small enough. Using (2.31) and Proposition 2.11, we see that

$$\left\|\mathcal{E}_{\varepsilon_{n+1}}\right\|_{\varepsilon_{n+1}} \le 2^{20}\,e^{10\delta K_j}\delta^2\||v\||^2\,\mathrm{r}(\delta)^2\,\kappa(2\delta)^6 + 2\,e^{2\delta K_j}\left(\frac{\kappa(2\delta)}{\kappa(\delta)}\right)^4 K_E\,\delta^2\||v\||^2\,\mathrm{r}(\delta)^2\,\kappa(\delta)^6$$

$$= \left[2^{20}\,e^{10\delta K_j} + 2\,e^{2\delta K_j}\left(\frac{\kappa(\delta)}{\kappa(2\delta)}\right)^2 K_E\right]\delta^2\||v\||^2\,\mathrm{r}(\delta)^2\,\kappa(2\delta)^6$$

$$= \frac{1}{2}\left[\frac{2^{19}\,e^{10\delta K_j}}{K_E} + e^{2\delta K_j}\left(\frac{\kappa(\delta)}{\kappa(2\delta)}\right)^2\right]\left(\frac{\mathrm{r}(\delta)}{\mathrm{r}(2\delta)}\right)^2 K_E(2\delta)^2\||v\||^2\,\mathrm{r}(2\delta)^2\,\kappa(2\delta)^6$$

$$= \frac{1}{2}\left[\frac{1}{4} + e^{2\delta K_j}2^{\frac{6}{10}}\right]2^{\frac{1}{10}} K_E(2\delta)^2\||v\||^2\,\mathrm{r}(2\delta)^2\,\kappa(2\delta)^6$$

$$\le K_E(2\delta)^2\||v\||^2\,\mathrm{r}(2\delta)^2\,\kappa(2\delta)^6 \qquad \text{(if Θ has been chosen small enough)}$$

$$= K_E\,\varepsilon_{n+1}^2\||v\||^2\,\mathrm{r}(\varepsilon_{n+1})^2\,\kappa(\varepsilon_{n+1})^6 \qquad\qquad\qquad \square$$

2.4   Functional integrals

In this section, I will outline the proof of a functional integral representation of the partition function like that of Theorem 2.1. It is an example of the class of rigorous functional integral representations in which the object of interest is expressed as a limit of finite dimensional integrals. At the end of this section, I will mention, and provide references to, a couple of other classes of rigorous functional integral representations that are used in mathematical physics.

I remind you that we have decided to approximate the left-hand side of (2.1) by replacing space $\mathbb{R}^3$ by a finite number of points, say $X = \mathbb{Z}^3/L\mathbb{Z}^3$ and that the Hamiltonian is

$$H = \int d\mathbf{x}d\mathbf{y}\ \psi^\dagger(\mathbf{x})\,\mathrm{h}(\mathbf{x},\mathbf{y})\,\psi(\mathbf{y}) + \int d\mathbf{x}_1 d\mathbf{x}_2\ \psi^\dagger(\mathbf{x}_1)\psi^\dagger(\mathbf{x}_2)\,v(\mathbf{x}_1,\mathbf{x}_2)\,\psi(\mathbf{x}_1)\psi(\mathbf{x}_2)$$

$$(2.38)$$

with $\int d\mathbf{x}$ just meaning $\sum_{\mathbf{x}\in X}$. We are still assuming that the kinetic energy operator $\mathrm{h} \geq 0$ and that the two-body potential $2v(\mathbf{x},\mathbf{y})$ is strictly positive when viewed as the kernel of an integral operator. I have claimed that then the partition function is (pretty obviously) well-defined. Let's check that this is indeed the case. The Hilbert space of all states of this system is

$$\mathcal{F} = \bigoplus_{n=0}^{\infty} \mathcal{F}_n \text{ with } \mathcal{F}_n = L_s^2(X^n) = \mathbb{C}^{|X|^n}/S_n.$$

Here

- a vector in the n-particle subspace $\mathcal{F}_n$ is a function $f(\mathbf{x}_1,\cdots,\mathbf{x}_n)$, with each argument $\mathbf{x}_j$ running over X, that is invariant under permutation of its arguments,
- the inner product between two n-particle vectors $f, g \in \mathcal{F}_n$ is

$$\langle f,g\rangle_{\mathcal{F}_n} = \int_{X^n} d\mathbf{x}_1 \cdots d\mathbf{x}_n\ \overline{f(\mathbf{x}_1,\cdots,\mathbf{x}_n)}\,g(\mathbf{x}_1,\cdots,\mathbf{x}_n)$$

  where $\int_X d\mathbf{x}\ f(\mathbf{x})$ just means $\sum_{\mathbf{x}\in X} f(\mathbf{x})$,
- $\mathcal{F}_0 = \mathbb{C}$,
- the inner product between two vectors $\mathbf{f} = (f_n)_{n\geq 0}$ and $\mathbf{g} = (g_n)_{n\geq 0}$ in $\mathcal{F}$ is

$$\langle \mathbf{f},\mathbf{g}\rangle_{\mathcal{F}} = \sum_{n\geq 0} \langle f_n, g_n\rangle_{\mathcal{F}_n}.$$

Now both H and N map the n-particle space $\mathcal{F}_n$ (which is finite dimensional) into itself. We'll show that the, positive, operator $e^{-\frac{1}{kT}(H-\mu N)}$ is trace class by bounding, for each non-negative integer n, the trace of the restriction of $e^{-\frac{1}{kT}(H-\mu N)}$ to $\mathcal{F}_n$ and then observe that the bound is easily summable over n. Now the restriction of N to $\mathcal{F}_n$ is just $n\mathbb{1}$ and the following lemma provides a lower bound on $H \restriction \mathcal{F}_n$.

Lemma 2.13 *There are constants $C, D > 0$ such that the restriction of H to $\mathcal{F}_n$ is bounded below by $(Cn - D)n\mathbb{1}$.*

Proof: Use $\psi^\dagger(\mathbf{x})$, $\psi(\mathbf{x})$ to denote the annihilation and creation operators at $\mathbf{x} \in X$. By the commutation relations $[\psi(\mathbf{x}), \psi^\dagger(\mathbf{x}')] = \delta_{\mathbf{x},\mathbf{x}'}$, the interaction

$$V = \int d\mathbf{x}_1 d\mathbf{x}_2 \; \psi^\dagger(\mathbf{x}_1)\psi^\dagger(\mathbf{x}_2)\, v(\mathbf{x}_1,\mathbf{x}_2)\, \psi(\mathbf{x}_1)\psi(\mathbf{x}_2)$$

$$= \int d\mathbf{x}_1 d\mathbf{x}_2 \; \psi^\dagger(\mathbf{x}_1)\psi(\mathbf{x}_1)\, v(\mathbf{x}_1,\mathbf{x}_2)\, \psi^\dagger(\mathbf{x}_2)\psi(\mathbf{x}_2) - \int d\mathbf{x} \; \psi^\dagger(\mathbf{x})\, v(\mathbf{x},\mathbf{x})\, \psi(\mathbf{x})$$

$$= \int d\mathbf{x}_1 d\mathbf{x}_2 \; n(\mathbf{x}_1)\, v(\mathbf{x}_1,\mathbf{x}_2)\, n(\mathbf{x}_2) - \int d\mathbf{x} \; v(\mathbf{x},\mathbf{x})\, n(\mathbf{x}) \tag{2.39}$$

where $n(\mathbf{x}) = \psi^\dagger(\mathbf{x})\psi(\mathbf{x})$ is the local number operator at $\mathbf{x}$. Now, restricted to $\mathcal{F}_n$, $\{\, n(\mathbf{x}) \mid \mathbf{x} \in X \,\}$ is a family of commuting, bounded self-adjoint operators on the finite dimensional Hilbert space $\mathcal{F}_n$ (that is, they are self-adjoint matrices). So there is an orthonormal basis $\{\delta_Y\}$ for $\mathcal{F}_n$ consisting of simultaneous eigenvectors for all of the $n(\mathbf{x})$'s. We denote the eigenvalues $\mu_Y(\mathbf{x})$. (All this is easy to find and is given in (Balaban *et al.* 2008a), but we don't need the explicit formulae.) So, for any $\varphi = \sum_Y \varphi_Y \delta_Y$,

$$\left\langle \varphi \,,\, \int_{X^2} d\mathbf{x}_1 d\mathbf{x}_2 \; n(\mathbf{x}_1) v(\mathbf{x}_1,\mathbf{x}_2) n(\mathbf{x}_2)\, \varphi \right\rangle$$

$$= \sum_{Y_1,Y_2} \overline{\varphi_{Y_1}}\, \varphi_{Y_2} \left\langle \delta_{Y_1} \,,\, \int_{X^2} d\mathbf{x}_1 d\mathbf{x}_2 \; n(\mathbf{x}_1) v(\mathbf{x}_1,\mathbf{x}_2) n(\mathbf{x}_2)\, \delta_{Y_2} \right\rangle$$

$$= \int_{X^2} d\mathbf{x}_1 d\mathbf{x}_2 \; v(\mathbf{x}_1,\mathbf{x}_2) \sum_{Y_1,Y_2} \overline{\varphi_{Y_1}}\, \varphi_{Y_2} \left\langle n(\mathbf{x}_1)\delta_{Y_1} \,,\, n(\mathbf{x}_2)\, \delta_{Y_2} \right\rangle$$

$$= \int_{X^2} d\mathbf{x}_1 d\mathbf{x}_2 \; v(\mathbf{x}_1,\mathbf{x}_2) \sum_{Y_1,Y_2} \overline{\varphi_{Y_1}}\, \varphi_{Y_2}\, \mu_{Y_1}(\mathbf{x}_1)\mu_{Y_2}(\mathbf{x}_2) \langle \delta_{Y_1} \,,\, \delta_{Y_2} \rangle$$

$$= \sum_Y |\varphi_Y|^2 \int_{X^2} d\mathbf{x}_1 d\mathbf{x}_2 \; \mu_Y(\mathbf{x}_1) v(\mathbf{x}_1,\mathbf{x}_2)\mu_Y(\mathbf{x}_2).$$

By hypothesis, v is a strictly positive operator on $L^2(X)$. Denote by $\lambda_0 > 0$ its smallest eigenvalue. Then

$$\int_{X^2} d\mathbf{x}_1 d\mathbf{x}_2 \; \mu_Y(\mathbf{x}_1) v(\mathbf{x}_1,\mathbf{x}_2)\mu_Y(\mathbf{x}_2) \geq \lambda_0 \int_X d\mathbf{x}\; \mu_Y^2(\mathbf{x}) \geq \frac{\lambda_0}{|X|} \left(\int_X d\mathbf{x}\; \mu_Y(\mathbf{x}) \right)^2$$

$$= \frac{\lambda_0}{|X|}\, n^2$$

by Cauchy-Schwarz and the fact that, on $\mathcal{F}_n$, $\int_X d\mathbf{x}\; n(\mathbf{x}) = n$. Hence

$$\left\langle \varphi \,,\, \int_{X^2} d\mathbf{x}_1 d\mathbf{x}_2 \; n(\mathbf{x}_1) v(\mathbf{x}_1,\mathbf{x}_2) n(\mathbf{x}_2)\, \varphi \right\rangle \geq \frac{\lambda_0}{|X|}\, n^2 \sum_Y |\varphi_Y|^2 = \frac{\lambda_0}{|X|}\, n^2 \|\varphi\|^2.$$

Since, on $\mathcal{F}_n$ the $n(\mathbf{x})$'s are positive operators adding up to n, every $0 \leq \mu_Y(\mathbf{x}) \leq n$ and

$$\|\psi(\mathbf{x})\|^2_{\mathcal{F}_n \to \mathcal{F}_{n-1}} = \|\psi(\mathbf{x})^\dagger \psi(\mathbf{x})\|_{\mathcal{F}_n} = \|n(\mathbf{x})\|_{\mathcal{F}_n} \leq n$$

$$\|\psi(\mathbf{x})^\dagger\|_{\mathcal{F}_{n-1} \to \mathcal{F}_n} = \|\psi(\mathbf{x})\|_{\mathcal{F}_n \to \mathcal{F}_{n-1}} \leq \sqrt{n}.$$

Consequently

$$\|H_0\| = \left\| \int d\mathbf{x} d\mathbf{y} \ \psi^\dagger(\mathbf{x}) \, \mathrm{h}(\mathbf{x}, \mathbf{y}) \, \psi(\mathbf{y}) \right\| \leq n \int d\mathbf{x} d\mathbf{y} \ |\mathrm{h}(\mathbf{x}, \mathbf{y})|. \tag{2.40}$$

We can easily do better than this, but we don't need to. The lemma follows with $C = \frac{\lambda_0}{|X|}$. □

Now back to the trace. Since the dimension of $\mathcal{F}_n$ is less than $|X|^n$ and every eigenvalue of $(H - \mu N) \restriction \mathcal{F}_n$ is at least $Cn^2 - Dn - \mu n$, we have

$$\mathrm{Tr}_{\mathcal{F}_n} \ e^{-\frac{1}{kT}(H - \mu N)} \leq e^{-\frac{1}{kT}(Cn^2 - Dn - \mu n)} |X|^n.$$

This is obviously summable over n.

2.4.1   A rigorous version of the functional integral

So we now know that, when X is finite, the partition function $\mathrm{Tr}_{\mathcal{F}} \ e^{-\frac{1}{kT}(H - \mu N)}$ is well-defined. I'll now outline the proof of a functional integral representation for $\mathrm{Tr}_{\mathcal{F}} \ e^{-\frac{1}{kT}(H - \mu N)}$ that is similar to that of Theorem 2.1, but whose integrand looks a lot more like the $e^{\mathcal{A}(\alpha^*, \alpha)}$ with the $\mathcal{A}(\alpha^*, \alpha)$ of (2.2).

We use the notation $\beta = \frac{1}{kT}$ and, for any $p \in \mathbb{N}$,

$$\mathcal{T}_p = \left\{ \tau = q\frac{\beta}{p} \ \middle| \ q = 1, \cdots, p \right\}$$

$$\varepsilon_p = \frac{\beta}{p}$$

$$d\mu_{p,r}(\alpha^*, \alpha) = \prod_{\tau \in \mathcal{T}_p} \prod_{\mathbf{x} \in X} \left[\frac{d\alpha^*_\tau(\mathbf{x}) \wedge d\alpha_\tau(\mathbf{x})}{2\pi\imath} \chi(|\alpha_\tau(\mathbf{x})| < r) \right]$$

$$K(\alpha^*, \alpha) = \iint d\mathbf{x} d\mathbf{y} \ \alpha(\mathbf{x})^* h(\mathbf{x}, \mathbf{y}) \alpha(\mathbf{y}) - \mu \int d\mathbf{x} \ \alpha(\mathbf{x})^* \alpha(\mathbf{x})$$
$$+ \iint d\mathbf{x} d\mathbf{y} \ \alpha(\mathbf{x})^* \alpha(\mathbf{x}) \, v(\mathbf{x}, \mathbf{y}) \, \alpha(\mathbf{y})^* \alpha(\mathbf{y}).$$

Theorem 2.14 *Suppose that the sequence* $\mathrm{R}(p) \to \infty$ *as* $p \to \infty$ *at a suitable rate. Precisely*

$$\lim_{p \to \infty} p \, e^{-\frac{1}{2}\mathrm{R}(p)^2} = 0 \quad and \quad \mathrm{R}(p) < p^{\frac{1}{24|X|}}.$$

Then

$$\mathrm{Tr}\, e^{-\beta\,(H-\mu N)} = \lim_{p\to\infty} \int d\mu_{p,\mathrm{R}(p)}(\alpha^*,\alpha) \prod_{\tau\in\mathcal{T}_p} e^{-\int dy\,[\alpha_\tau^*(\mathrm{y})-\alpha_{\tau-\varepsilon_p}^*(\mathrm{y})]\alpha_\tau(\mathrm{y})} e^{-\varepsilon_p K(\alpha_{\tau-\varepsilon_p}^*,\alpha_\tau)}$$

with the convention that $\alpha_0 = \alpha_\beta$.

Almost all of the rest of this section is used to outline the proof of Theorem 2.14.

2.4.2   The main ingredients—coherent states

The first main ingredient in the proof is the use of coherent states. I'll give the formulae only for the case $|X| = 1$, because then they are short and clean. The general case is very similar. If $|X| = 1$, then

$$\mathcal{F} = \bigoplus_{n=0}^{\infty} \mathcal{F}_n \text{ with } \mathcal{F}_n = \mathbb{C}.$$

Let $e_n = 1 \in \mathbb{C} = \mathcal{F}_n$. We can think of each vector in $\mathcal{F}$ as a sequence $(v_0, v_1, v_2, \cdots)$ of complex numbers. Then e_n is the sequence all of whose components are zero, except for that with index n, which is 1. For each $\alpha \in \mathbb{C}$ the coherent state

$$|\,\alpha\,\rangle = \sum_{n=0}^{\infty} \frac{1}{\sqrt{n!}} \alpha^n e_n \in \mathcal{F} \tag{2.41}$$

is an eigenvector for the field (or annihilation) operator

$$\psi e_n = \sqrt{n}\, e_{n-1}.$$

To check this, we compute

$$\psi\,|\,\alpha\,\rangle = \sum_{n=1}^{\infty} \frac{1}{\sqrt{(n-1)!}} \alpha^n e_{n-1} = \alpha\,|\,\alpha\,\rangle. \tag{2.42}$$

The action of the creation operator

$$\psi^\dagger e_n = \sqrt{n+1}\, e_{n+1}$$

on the αth coherent state vector is

$$\psi^\dagger\,|\,\alpha\,\rangle = \sum_{n=0}^{\infty} \frac{\sqrt{n+1}}{\sqrt{n!}} \alpha^n e_{n+1} = \frac{\partial}{\partial\alpha} \sum_{n=0}^{\infty} \frac{1}{\sqrt{(n+1)!}} \alpha^{n+1} e_{n+1} = \frac{\partial}{\partial\alpha}\,|\,\alpha\,\rangle. \tag{2.43}$$

Because the e_n's form an orthonormal basis, the inner product between two coherent states is

$$\langle\,\alpha\,|\,\gamma\,\rangle = \sum_{m,n=0}^{\infty} \frac{\bar{\alpha}^m}{\sqrt{m!}} \frac{\gamma^n}{\sqrt{n!}} \delta_{m,n} = \sum_{n=0}^{\infty} \frac{1}{n!}(\bar{\alpha}\gamma)^n = e^{\bar{\alpha}\gamma}$$

For general X, there is a similar coherent state $|\alpha\rangle$ for each $|X|$-component complex vector $\alpha \in \mathbb{C}^{|X|}$. The inner product between two such coherent states is

$$\langle \alpha | \gamma \rangle = e^{\int dy \, \overline{\alpha(y)} \, \gamma(y)}.$$

2.4.3   The main ingredients—approximate resolution of the identity

One of our main tools is going to be the analogue for coherent states of the identity that for any orthonormal basis

$$\mathbb{1}v = \sum_n (e_n, v) \, e_n.$$

Formally, the corresponding statement for coherent states is

$$\mathbb{1} = \int \prod_{\mathbf{x} \in X} \left[\frac{d\alpha^*(\mathbf{x}) \wedge d\alpha(\mathbf{x})}{2\pi\imath} \right] e^{-\int dy \, |\alpha(y)|^2} \, |\alpha\rangle\langle\alpha|$$

where $|\alpha\rangle\langle\alpha|$ is the linear operator that maps $v \in \mathcal{F}$ to the inner product of v and $|\alpha\rangle$ times the vector $|\alpha\rangle$. The integral 'sums' over all possible coherent states and the exponential $e^{-\int dy \, |\alpha(y)|^2} = \frac{1}{\||\alpha\rangle\|^2}$ turns the coherent states into unit vectors. Here is a rigorous version of the resolution of the identity for coherent states.

Theorem 2.15 *For each $r > 0$, let*

$$I_r = \prod_{\mathbf{x} \in X} \left[\int_{|\alpha(\mathbf{x})| < r} \frac{d\alpha^*(\mathbf{x}) \wedge d\alpha(\mathbf{x})}{2\pi\imath} \right] e^{-\int dy \, |\alpha(y)|^2} \, |\alpha\rangle\langle\alpha|. \tag{2.44}$$

Then

(a) $0 < I_r < \mathbb{1}$.
(b) I_r *commutes with* N.
(c) *The operator norm of* I_r *is bounded by one for all* r *and, for each vector* $v \in \mathcal{F}$, $I_r v$ *converges to* v *as* $r \to \infty$. *That is,* I_r *convergences strongly to* $\mathbb{1}$.
(d) *For all* n *and* r, *the operator norms*

$$\left\| (\mathbb{1} - I_r) \upharpoonright \mathcal{F}_n \right\| \leq |X| \, 2^{n+1} \, e^{-r^2/2} \qquad \left\| I_r \upharpoonright \mathcal{F}_n \right\| \leq \frac{1}{n!} \left(|X| r^2 \right)^n.$$

Remark 2.16 *Observe, from part (d), that if $\frac{n}{r^2} \ll 1$ then $I_r \upharpoonright \mathcal{F}_n \approx \mathbb{1}$ while if $\frac{n}{r^2} \gg 1$, then $I_r \upharpoonright \mathcal{F}_n \approx 0$ (use that $n! \approx n^n$). So we can think of I_r as being, very roughly, projection onto $\bigoplus_{n=0}^{r^2} \mathcal{F}_n$.*

Proof: We first observe that when you apply the operator $e^{-\int dy \, |\alpha(y)|^2} \, |\alpha\rangle\langle\alpha|$ to some vector v, the resulting vector is of norm at most $\frac{1}{\||\alpha\rangle\|^2} \, \||\alpha\rangle\| \, \||\alpha\rangle\| \, \|v\| \leq \|v\|$. So the integrand of the right-hand side of (2.44) is of operator norm at most one. As it is also continuous in α and the domain of integration is of finite volume, the right-hand side of (2.44) is obviously well-defined.

The proof is easy once one has an orthonormal basis of eigenvectors for I_r—and it is easy to just guess such a basis. Again, to simplify the notation, I'll just give the proof for $|X| = 1$. Then

$$\mathcal{F} = \bigoplus_{n=0}^{\infty} \mathcal{F}_n \quad \text{with } \mathcal{F}_n = \mathbb{C}$$

and $\{e_m = 1 \in \mathbb{C} = \mathcal{F}_m \mid m = 0, 1, 2, 3, \cdots\}$ is an orthonormal basis for $\mathcal{F}$. If part (b) of the Theorem is true, then each of the $\mathcal{F}_m$'s, which has basis $\{e_m\}$, will be left invariant by I_r. So each e_m will be an eigenvector. To verify that this is indeed the case, and to find the corresponding eigenvalues, we compute $I_r e_m$.

Recall that

$$|\alpha\rangle = \sum_{n=0}^{\infty} \frac{1}{\sqrt{n!}} \alpha^n e_n.$$

So

$$I_r e_m = \int_{|\alpha|<r} \frac{d\bar{\alpha} \wedge d\alpha}{2\pi i} e^{-|\alpha|^2} |\alpha\rangle\langle\alpha | e_m\rangle = \int_{|\alpha|<r} \frac{d\bar{\alpha} \wedge d\alpha}{2\pi i} e^{-|\alpha|^2} |\alpha\rangle \frac{1}{\sqrt{m!}} \bar{\alpha}^m$$

$$= \sum_{n=0}^{\infty} \frac{1}{\sqrt{n!}\sqrt{m!}} e_n \int_{|\alpha|<r} \frac{d\bar{\alpha} \wedge d\alpha}{2\pi i} e^{-|\alpha|^2} \bar{\alpha}^m \alpha^n.$$

Now switch to polar coordinates. That is, make the change of variables $\alpha = \rho e^{i\theta}$. Recalling that if $z = x + iy$, then $\frac{d\bar{z} \wedge dz}{2\pi i} = \frac{dx \wedge dy}{\pi}$,

$$I_r e_m = \sum_{n=0}^{\infty} \frac{1}{\sqrt{n!}\sqrt{m!}} e_n \int_0^{2\pi} d\theta \int_0^r d\rho \frac{1}{\pi} e^{-\rho^2} \rho^{m+n+1} e^{i\theta(n-m)}$$

$$= \frac{1}{m!} e_m \int_0^r d\rho \, 2 e^{-\rho^2} \rho^{2m+1} = \frac{1}{m!} e_m \int_0^{r^2} dt \, e^{-t} t^m \qquad \text{where } t = \rho^2$$

$$= \left\{ 1 - \frac{1}{m!} \int_{r^2}^{\infty} e^{-t} t^m \, dt \right\} e_m.$$

This tells us that each e_m is an eigenvector of I_r with eigenvalue $1 - \frac{1}{m!} \int_{r^2}^{\infty} e^{-t} t^m \, dt$, which is always between 0 and 1 and which tends to zero as $r \to \infty$. Parts (a), (b), and (c) follow. For part (d) , just bound

$$\frac{1}{n!} \int_{r^2}^{\infty} e^{-t} t^n \, dt = 2^n \int_{r^2}^{\infty} e^{-t} \frac{1}{n!} \left(\frac{t}{2}\right)^n \, dt \leq 2^n \int_{r^2}^{\infty} e^{-t} e^{t/2} \, dt = 2^{n+1} e^{-r^2/2}$$

and

$$\frac{1}{n!} \int_0^{r^2} e^{-t} t^n \, dt \leq \frac{1}{n!} r^{2n} \int_0^{\infty} e^{-t} \, dt = \frac{1}{n!} r^{2n}. \qquad \square$$

2.4.4   The Main ingredients—Trace

Formally, the analogue of

$$\operatorname{Tr} B = \sum_n (e_n, Be_n)$$

for coherent states is

$$\operatorname{Tr} B = \int \prod_{\mathbf{x} \in X} \left[\frac{d\alpha^*(\mathbf{x})\, d\alpha(\mathbf{x})}{2\pi\imath} \right] e^{-\int d\mathbf{y}\, |\alpha(\mathbf{y})|^2} \langle\, \alpha \mid B \mid \alpha \,\rangle.$$

Our next main tool for the proof is the following rigorous version of that formula.

Proposition 2.17 *Let B be a bounded operator on $\mathcal{F}$. For all $r > 0$, BI_r is trace class and*

$$\operatorname{Tr} BI_r = \prod_{\mathbf{x} \in X} \left[\int_{|\alpha(\mathbf{x})| < r} \frac{d\alpha^*(\mathbf{x})\, d\alpha(\mathbf{x})}{2\pi\imath} \right] e^{-\int d\mathbf{y}\, |\alpha(\mathbf{y})|^2} \langle\, \alpha \mid B \mid \alpha \,\rangle.$$

Proof: As usual, I'll just give the proof for $|X| = 1$.

Recall that, by definition, BI_r is trace class when the eigenvalues of the operator square root of $I_r B^* BI_r$ (all of which are non-negative) are summable. There is a theorem which says that a product of a trace class operator (in our case I_r) and a bounded operator (in our case B) is trace class. In our case, we can also easily check directly that BI_r is trace class. Here is the argument. By the min–max principle (Reed and Simon 1978, Theorem XIII.1, with $H = -I_r B^* BI_r$) the $(n+1)^{\text{st}}$ eigenvalue of $I_r B^* BI_r$, counting from largest to smallest, is

$$\inf_{\varphi_1, \cdots \varphi_n \in \mathcal{F}} \ \ \sup_{\substack{\psi \in \mathcal{F},\, \|\psi\|=1 \\ \psi \perp \varphi_1, \cdots, \varphi_n}} \ \langle\, \psi \mid I_r B^* BI_r \mid \psi \,\rangle \leq \sup_{\substack{\psi \in \oplus_{m \geq n} \mathcal{F}_m \\ \|\psi\|=1}} \ \langle\, \psi \mid I_r B^* BI_r \mid \psi \,\rangle$$

$$\leq \|B\|^2 \sup_{m \geq n} \left(\frac{1}{m!} r^{2m} \right)^2$$

by part (d) of Theorem 2.15. Hence the $(n+1)^{\text{st}}$ eigenvalue of the operator square root, again counting from largest to smallest, is at most $\|B\| \sup_{m \geq n} \frac{1}{m!} r^{2m}$ and this is clearly summable over n.

So BI_r is trace class and the trace itself is

$$\operatorname{Tr} BI_r = \sum_m \langle\, e_m \mid BI_r \mid e_m \,\rangle$$

$$= \sum_m \int_{|\alpha| < r} \frac{d\bar{\alpha} \wedge d\alpha}{2\pi\imath} e^{-|\alpha|^2} \langle\, e_m \mid B \mid \alpha \,\rangle\langle\, \alpha \mid e_m \,\rangle$$

$$= \int_{|\alpha|<r} \frac{d\bar{\alpha} \wedge d\alpha}{2\pi\imath} e^{-|\alpha|^2} \sum_m \langle \alpha | e_m \rangle \langle e_m | B | \alpha \rangle$$

$$= \int_{|\alpha|<r} \frac{d\bar{\alpha} \wedge d\alpha}{2\pi\imath} e^{-|\alpha|^2} \langle \alpha | B | \alpha \rangle .$$

Moving the sum over m inside the integral is justified by the Lebesgue dominated convergence theorem, since

$$\sum_m |\langle \alpha | e_m \rangle \langle e_m | B | \alpha \rangle| \leq \| |\alpha \rangle \| \, \| B | \alpha \rangle \| \leq \|B\| e^{|\alpha|^2}. \qquad \square$$

2.4.5 Consolidation—where we are now

Combining Theorem 2.15 and Proposition 2.17, we now have:

Lemma 2.18 *Assume that* $\lim_{p\to\infty} p \, e^{-\frac{1}{2}\mathrm{R}(p)^2} = 0$. *Then*

$$\mathrm{Tr} \, e^{-\beta(H-\mu N)} =$$

$$\lim_{p\to\infty} \prod_{\substack{x\in X \\ \tau\in\mathcal{T}_p}} \left[\int_{|\alpha_\tau(x)|<\mathrm{R}(p)} \frac{d\alpha_\tau^*(x)\, d\alpha_\tau(x)}{2\pi\imath} \, e^{-|\alpha_\tau(x)|^2} \right] \prod_{\tau\in\mathcal{T}_p} \left\langle \alpha_{\tau-\frac{\beta}{p}} \left| e^{-\frac{\beta}{p}(H-\mu N)} \right| \alpha_\tau \right\rangle .$$

Proof: By Theorem 2.15,

$$\mathrm{Tr} \, e^{-\beta(H-\mu N)} = \mathrm{Tr} \, \prod_{\tau\in\mathcal{T}_p} e^{-\frac{\beta}{p}(H-\mu N)} \mathbb{1} = \lim_{p\to\infty} \mathrm{Tr} \, \prod_{\tau\in\mathcal{T}_p} e^{-\frac{\beta}{p}(H-\mu N)} \mathrm{I}_{\mathrm{R}(p)}.$$

To justify the last step

- Let $\varepsilon > 0$.
- Denote by P_n the orthogonal projector onto $\oplus_{m=0}^n \mathcal{F}_n$. Use Lemma 2.13 to select an $n \in \mathbb{N}$, independent of p, such that

$$\left| \mathrm{Tr} \, (\mathbb{1} - P_n) \prod_{\tau\in\mathcal{T}_p} e^{-\frac{\beta}{p}(H-\mu N)} \mathbb{1} \right| < \tfrac{\varepsilon}{4} \qquad \left| \mathrm{Tr} \, (\mathbb{1} - P_n) \prod_{\tau\in\mathcal{T}_p} e^{-\frac{\beta}{p}(H-\mu N)} \mathrm{I}_{\mathrm{R}(p)} \right| < \tfrac{\varepsilon}{4} .$$

- Express

$$P_n \prod_{\tau\in\mathcal{T}_p} e^{-\frac{\beta}{p}(H-\mu N)} \mathbb{1} - P_n \prod_{\tau\in\mathcal{T}_p} e^{-\frac{\beta}{p}(H-\mu N)} \mathrm{I}_{\mathrm{R}(p)}$$

as the telescoping sum over $1 \leq \ell \leq p$ of

$$P_n \prod_{\tau\in\mathcal{T}_p} e^{-\frac{\beta}{p}(H-\mu N)} I_\tau^\ell \qquad \text{with } I_\tau^\ell = \begin{cases} \mathbb{1} & \text{if } \tau < \ell\frac{\beta}{p} \\ \mathbb{1} - \mathrm{I}_{\mathrm{R}(p)} & \text{if } \tau = \ell\frac{\beta}{p} \\ \mathrm{I}_{\mathrm{R}(p)} & \text{if } \tau > \ell\frac{\beta}{p}. \end{cases}$$

- By Lemma 2.13 and part (d) of Theorem 2.15, the trace of each of the p terms in the telescoping sum is bounded by a constant C, which depends on $|X|$ and n, times $e^{-R(p)^2/2}$.
- Finally, by hypothesis, if p is sufficiently large then $Cpe^{-R(p)^2/2} < \frac{\varepsilon}{2}$.

It now suffices to substitute in the definition (2.44) of I_r and apply Proposition 2.17.

$\square$

2.4.6   The main ingredients—perturbation theory

Lemma 2.18 has given us a functional integral representation for the partition function, but has not told us very much about what the integrand looks like. The next step is to exploit the fact that $\frac{\beta}{p}$ is very small when p is large to help us understand what

$$\left\langle \alpha_{\tau-\frac{\beta}{p}} \,\Big|\, e^{-\frac{\beta}{p}(H-\mu N)} \,\Big|\, \alpha_\tau \right\rangle$$

looks like.

Proposition 2.19 *There are constants C, c such that the following holds. For each $\varepsilon > 0$, there is an analytic function $F(\varepsilon, \alpha^*, \beta)$ such that*

$$\left\langle \alpha \,\Big|\, e^{-\varepsilon(H-\mu N)} \,\Big|\, \beta \right\rangle = e^{F(\varepsilon, \alpha^*, \beta)}$$

on the domain $\|\alpha\|_\infty$, $\|\beta\|_\infty < C\frac{1}{\sqrt{\varepsilon}}$, where, as usual $\|\alpha\|_\infty = \max_{\mathbf{x} \in X} |\alpha(\mathbf{x})|$. Write

$$F(\varepsilon, \alpha^*, \beta) = \int_X d\mathbf{x} \, \alpha(\mathbf{x})^* \beta(\mathbf{x}) - \varepsilon K(\alpha^*, \beta) + \mathcal{F}_0(\varepsilon, \alpha^*, \beta)$$

where $K(\alpha^*, \beta)$ was defined in (2.3). Then

$$|\mathcal{F}_0(\varepsilon, \alpha^*, \beta)| \leq c\, \varepsilon^2 (\Phi^2 + \|v\|_{1,\infty}^2 \Phi^6)$$

for all $0 < \varepsilon \leq 1$ and $\|\alpha\|_\infty, \|\beta\|_\infty \leq \Phi \leq \frac{1}{2} C \frac{1}{\sqrt{\varepsilon}}$.

Idea of Proof $\left\langle \alpha \,\big|\, e^{-\varepsilon(H-\mu N)} \,\big|\, \beta \right\rangle$ is an entire function of α^* and β and a C^∞ function of ε for $\varepsilon \geq 0$. (Just plug in the definitions (2.41) of $|\alpha\rangle$ and $|\beta\rangle$ in terms of the standard basis and use that the operator norm of $e^{-\varepsilon(H-\mu N)}(H - \mu N)^m$ restricted to $\mathcal{F}_n$ is bounded by a constant times $(n+1)^m$ for all integers $m, n \geq 0$ and $\varepsilon \geq 0$.) But $\left\langle \alpha \,\big|\, e^{-\varepsilon(H-\mu N)} \,\big|\, \beta \right\rangle$ can take the value zero (see (Balaban *et al.* 2008a, Example 3.12)), so its logarithm need not be everywhere defined. Since $\left\langle \alpha | \beta \right\rangle = e^{\int \alpha^*(\mathbf{x})\beta(\mathbf{x})\,d\mathbf{x}} \neq 0$, continuity implies that the matrix element has the representation

$$\left\langle \alpha \,\Big|\, e^{-\varepsilon(H-\mu N)} \,\Big|\, \beta \right\rangle = e^{F(\varepsilon, \alpha^*, \beta)} \tag{2.45}$$

in some neighbourhood of 0, with $F(\varepsilon, \alpha^*, \beta)$ is analytic in α^*, β. But is the neighbourhood big enough and what can we say about F? To go further, differentiate (2.45) with respect to ε to give

$$e^{F(\varepsilon, \alpha^*, \beta)} \frac{\partial F}{\partial \varepsilon}(\varepsilon, \alpha^*, \beta) = -\left\langle \alpha \,\Big|\, (H - \mu N)e^{-\varepsilon(H-\mu N)} \,\Big|\, \beta \right\rangle.$$

Now, downstairs on the right-hand side, substitute in the definitions of H and N in terms of the annihilation and creation operators $\psi(\mathbf{x})$ and $\psi^\dagger(\mathbf{x})$ (see eqn (2.38)) and use

$$\psi(\mathbf{x})\,|\,\alpha\,\rangle = \alpha(\mathbf{x})\,|\,\alpha\,\rangle \qquad \psi^\dagger(\mathbf{x})\,|\,\alpha\,\rangle = \frac{\partial}{\partial\alpha(\mathbf{x})}\,|\,\alpha\,\rangle$$

This gives us the differential equation

$$\frac{\partial}{\partial\varepsilon}F = -\mathcal{K}(\alpha^*,\frac{\partial}{\partial\alpha^*})F - \iint_X d\mathbf{x}d\mathbf{y}\; \alpha(\mathbf{x})^*\alpha(\mathbf{y})^*\,v(\mathbf{x},\mathbf{y})\frac{\partial\;F}{\partial\alpha(\mathbf{x})^*}\frac{\partial\;F}{\partial\alpha(\mathbf{y})^*}$$

where

$$\mathcal{K}(\alpha^*,\frac{\partial}{\partial\alpha^*}) = \iint_X d\mathbf{x}d\mathbf{y}\; \alpha(\mathbf{x})^*\,h(\mathbf{x},\mathbf{y})\frac{\partial}{\partial\alpha(\mathbf{y})^*} - \mu\int_X d\mathbf{x}\; \alpha(\mathbf{x})^*\frac{\partial}{\partial\alpha(\mathbf{x})^*}$$

$$+ \iint_X d\mathbf{x}d\mathbf{y}\; \alpha(\mathbf{x})^*\alpha(\mathbf{y})^*\,v(\mathbf{x},\mathbf{y})\frac{\partial}{\partial\alpha(\mathbf{x})^*}\frac{\partial}{\partial\alpha(\mathbf{y})^*}.$$

The details are in (Balaban *et al.* 2008a, Lemma 3.8)). As F also satisfies the initial condition

$$F(0,\alpha^*,\beta) = \ln\langle\alpha\,|\,\beta\rangle = \int_X d\mathbf{x}\,\alpha(\mathbf{x})^*\beta(\mathbf{x})$$

we now have a first order initial value problem for F, viewed as a function of ε. It is tedious but straightforward to convert this into a system of integral equations for coefficients in the Taylor expansion of $F(\varepsilon,\alpha^*,\beta)$ in powers of α^* and β. (See (Balaban *et al.* 2008a, Lemma 3.9).) The system can be solved and bounded by iteration. The details are in (Balaban *et al.* 2008a, Lemmas 3.8 and 3.9 and Proposition 3.6). $\qquad\square$

2.4.7 Finishing off the proof of Theorem 2.14

So we now have

$$\mathrm{Tr}\; e^{-\beta\,(H-\mu N)} = \lim_{p\to\infty}\int d\mu_{p,\mathrm{R}(p)}(\alpha^*,\alpha)\prod_{\tau\in\mathcal{T}_p}e^{-\int dy\,[\alpha_\tau^*(\mathbf{y})-\alpha_{\tau-\varepsilon_p}^*(\mathbf{y})]\alpha_\tau(\mathbf{y})}e^{-\varepsilon_p K(\alpha_{\tau-\varepsilon_p}^*,\alpha_\tau)}$$

$$\prod_{\tau\in\mathcal{T}_p}e^{-\mathcal{F}_0(\varepsilon_p,\alpha_{\tau-\varepsilon_p}^*,\alpha_\tau)}$$

and we just have to show that discarding the $\mathcal{F}_0$'s does not affect the value of the $p\to\infty$ limit. Before I outline the argument that this is the case, I'll make two remarks:

- If the functional integral representation is going to be used as a starting point for a renormalization group construction, it may not be necessary to show that discarding the $\mathcal{F}_0$'s does not affect the value of the $p\to\infty$ limit. The bound on $\mathcal{F}_0$ provided by Proposition 2.19 may be adequate in itself.

- Observe that the sum

$$\sum_{\tau \in \mathcal{T}_p} \mathcal{F}_0(\varepsilon_p, \alpha^*_{\tau - \varepsilon_p}, \alpha_\tau)$$

has p terms and that each term is of order $\varepsilon^2 = \frac{1}{p^2}$. So the sum is of order $\frac{1}{p}$. This is a first hint that these terms disappear in the limit $p \to \infty$. But it does not prove anything, since the volume of the domain of integration is growing like $R(p)^{2|X|p}$. (The order ε^2 in the bound of Proposition 2.19 is also multiplied by a $\Phi^6 \leq R(p)^6$, but it grows relatively slowly with p.)

Now here is an outline of the argument that we may discard the $\mathcal{F}_0$'s. The details are in (Balaban *et al.* 2008a). Let $r > 0$. Define, for $\mathcal{I} : \mathbb{C}^{2|X|} \to \mathbb{C}$, the seminorm

$$\|\mathcal{I}\|_r = \sup_{\substack{\alpha, \phi \in \mathbb{C}^X \\ |\alpha|_X, |\phi|_X \leq r}} |\mathcal{I}(\alpha, \phi)|$$

and, for $\mathcal{I}, \mathcal{J} : \mathbb{C}^{2|X|} \to \mathbb{C}$, with $\|\mathcal{I}\|_r, \|\mathcal{J}\|_r < \infty$, the '$r$–product' of $\mathcal{I}, \mathcal{J}$

$$(\mathcal{I} *_r \mathcal{J})(\alpha, \gamma) = \int \mathcal{I}(\alpha, \phi) \mathcal{J}(\phi, \gamma) \, d\mu_r(\phi^*, \phi)$$

where

$$d\mu_r(\phi^*, \phi) = \prod_{\mathbf{x} \in X} \left[\frac{d\phi^*(\mathbf{x}) \wedge d\phi(\mathbf{x})}{2\pi \imath} \chi(|\phi(\mathbf{x})| < r) \right].$$

The qth power with respect to this product is denoted

$$\mathcal{I}^{*_r q} = \overbrace{\mathcal{I} *_r \mathcal{I} *_r \cdots *_r \mathcal{I}}^{q \text{ factors}}.$$

For each $\varepsilon > 0$, set

$$\mathcal{I}_\varepsilon(\alpha, \phi) = e^{-\frac{1}{2}\|\alpha\|^2 - \frac{1}{2}\|\phi\|^2} e^{F(\varepsilon, \alpha^*, \phi)} = e^{-\frac{1}{2}\|\alpha\|^2 - \frac{1}{2}\|\phi\|^2} \left\langle \alpha \left| e^{-\varepsilon(H - \mu N)} \right| \phi \right\rangle$$

$$\tilde{\mathcal{I}}_\varepsilon(\alpha, \phi) = e^{-\frac{1}{2}\|\alpha\|^2 - \frac{1}{2}\|\phi\|^2} e^{F(\varepsilon, \alpha^*, \phi) - \mathcal{F}_0(\varepsilon, \alpha^*, \phi)}$$

$$= \exp \left\{ -\frac{1}{2}\|\alpha\|^2 - \frac{1}{2}\|\phi\|^2 + \int d\mathbf{x} \, \alpha^*(\mathbf{x})\phi(\mathbf{x}) - \varepsilon K(\alpha^*, \phi) \right\}.$$

Lemma 2.18 and Proposition 2.19 state that, for $R(p)$ obeying $\lim\limits_{p \to \infty} pe^{-\frac{1}{2}R(p)^2} = 0$,

$$\operatorname{Tr} e^{-\beta K} = \lim_{p \to \infty} \int d\mu_r(\phi^*, \phi) \, \mathcal{I}_\varepsilon^{*_r p}(\phi, \phi) \Big|_{\substack{r = R(p) \\ \varepsilon = \beta/p}}$$

and we would like to have

$$\text{Tr}\, e^{-\beta K} = \lim_{p \to \infty} \int d\mu_r(\phi^*, \phi)\, \tilde{\mathcal{I}}_\varepsilon^{*,p}(\phi, \phi)\Big|_{\substack{r=\text{R}(p) \\ \varepsilon=\beta/p}}$$

instead. By Lemma 2.13, the operator $H - \mu N$ is bounded below. Say $H - \mu N \geq -K_0 \mathbb{1}$. Then, for any $q \in \mathbb{N}$,

$$\mathcal{I}_\varepsilon^{*,q}(\alpha, \phi) = e^{-\frac{1}{2}\|\alpha\|^2 - \frac{1}{2}\|\phi\|^2} \left\langle \alpha \left| \left(e^{-\varepsilon K} I_r\right)^{q-1} e^{-\varepsilon K} \right| \phi \right\rangle$$

implies that

$$\|\mathcal{I}_\varepsilon^{*,q}\|_r \leq e^{-\frac{1}{2}\|\alpha\|^2 - \frac{1}{2}\|\phi\|^2} \|\alpha\| \left(e^{\varepsilon K_0} \|I_r\|\right)^{q-1} e^{\varepsilon K_0} \|\phi\| = e^{q\varepsilon K_0} \|I_r\|^{q-1} \leq e^{q\varepsilon K_0}$$

for all $r > 0$, by part (c) of Theorem 2.15. The difference

$$\|\mathcal{I}_\varepsilon - \tilde{\mathcal{I}}_\varepsilon\|_r = \sup_{|\alpha|_x, |\phi|_x \leq r} \left| e^{-\frac{1}{2}\|\alpha\|^2 - \frac{1}{2}\|\phi\|^2} \left\langle \alpha \left| e^{-\varepsilon K} \right| \phi \right\rangle \left[1 - e^{-\mathcal{F}_0(\varepsilon, \alpha^*, \phi)}\right] \right|$$

$$\leq e^{\varepsilon K_0} \sup_{|\alpha|_x, |\phi|_x \leq r} \left| \left[1 - e^{-\mathcal{F}_0(\varepsilon, \alpha^*, \phi)}\right] \right|$$

$$\leq e^{\varepsilon K_0} \,\text{const}\varepsilon^2 r^6 |X| \, e^{\text{const}\varepsilon^2 r^6 |X|}$$

by Proposition 2.19 (assuming that $1 \leq r \leq \frac{\text{const}}{\sqrt{\varepsilon}}$). The following proposition is, naturally, proven by induction in (Balaban *et al.* 2008b, Proposition 3.16).

Proposition 2.20 *Let* $K_0, \varepsilon, \zeta > 0$ *and* $0 < \kappa < 1$ *and* $r, C_\beta \geq 1$ *obey*

$$C_\beta \left(\pi r^2\right)^{3|X|} \zeta^{1-\kappa} \leq \varepsilon$$

Let $\mathcal{I}, \tilde{\mathcal{I}} : \mathbb{C}^{2|X|} \to \mathbb{C}$ *obey*

$$\|\mathcal{I} - \tilde{\mathcal{I}}\|_r \leq \zeta \qquad \|\mathcal{I}^{*,q}\|_r \leq e^{q\varepsilon K_0} \text{ for all } q \in \mathbb{N}.$$

Then, for all $q \in \mathbb{N}$ *with* $q \leq \frac{C_\beta}{\varepsilon}$,

$$\|\tilde{\mathcal{I}}^{*,q}\|_r \leq e^{q\varepsilon(K_0 + \zeta^\kappa)}$$

$$\|\tilde{\mathcal{I}}^{*,q} - \mathcal{I}^{*,q}\|_r \leq \zeta^\kappa e^{q\varepsilon(K_0 + \zeta^\kappa)}$$

$$\int d\mu_r(\phi^*, \phi) \left| \tilde{\mathcal{I}}^{*,q}(\phi, \phi) - \mathcal{I}^{*,q}(\phi, \phi) \right| \leq \zeta^\kappa e^{q\varepsilon(K_0 + \zeta^\kappa)}.$$

It now suffices to apply Proposition 2.20 with $\zeta = \varepsilon^{3/2}$, $r = R(p)$, $p = \frac{\beta}{\varepsilon}$, $\kappa = \frac{1}{12}$ and $C_\beta = \beta$. Since

$$C_\beta \left(\pi r^2\right)^{3|X|} \zeta^{1-\kappa} = \beta\left(\pi R(\frac{\beta}{\varepsilon})^2\right)^{3|X|} \varepsilon^{\frac{33}{24}} \le \varepsilon \qquad \text{if } R(p) < p^{\frac{1}{24|X|}} \text{ and } \varepsilon \text{ is small enough}$$

$$e^{\varepsilon K_0} \operatorname{const} \varepsilon^2 R(\frac{\beta}{\varepsilon})^6 |X| \, e^{\operatorname{const} \varepsilon^2 R(\frac{\beta}{\varepsilon})^6 |X|} \le \varepsilon^{\frac{3}{2}} \quad \text{if } R(p) < p^{\frac{1}{24}} \text{ and } \varepsilon \text{ is small enough}$$

the hypotheses of Proposition 2.20 are satisfied.

2.4.8 Cylinder set measures

There are other rigorous functional integral representations used in quantum mechanics, quantum field theory, and condensed matter physics. Probably the most elegant and powerful class of such representations use cylinder set measures. Cylinder set measures refer to measures on infinite dimensional vector spaces that are built by taking a limit of a collection of probability measures defined on finite dimensional subspaces of the vector space. The measures on the different subspaces have to be consistent with each other in a natural sense that I will now explain. Let $\mathcal{I}$ be a countable set and take as our vectors space

$$\mathbb{R}^{\mathcal{I}} = \left\{ \vec{x} = (x_i)_{i \in \mathcal{I}} \mid x_i \in \mathbb{R} \text{ for all } i \in \mathcal{I} \right\}.$$

For each finite subset $I \subset \mathcal{I}$ define the subspace $\mathbb{R}^I$ of $\mathbb{R}^{\mathcal{I}}$ by

$$\mathbb{R}^I = \left\{ (x_i)_{i \in \mathcal{I}} \in \mathbb{R}^{\mathcal{I}} \mid x_i = 0 \text{ for all } i \in \mathcal{I} \setminus I \right\}$$

and define the natural projection $P_I : \mathbb{R}^{\mathcal{I}} \to \mathbb{R}^I$ by

$$(P_I \vec{x})_j = \begin{cases} x_j & \text{if } j \in I \\ 0 & \text{if } j \in \mathcal{I} \setminus I \end{cases}$$

Suppose that we are given, for each finite $I \subset \mathcal{I}$, a probability measure μ_I on $\mathbb{R}^I$. This family of measures is said to be consistent if for each pair of finite subsets $I, I' \subset \mathcal{I}$ obeying $I \subset I'$ and for each measurable $A \subset \mathbb{R}^I$, we have

$$\mu_{I'}\left(\left\{ \vec{x} \in \mathbb{R}^{I'} \mid P_I \vec{x} \in A \right\}\right) = \mu_I(A).$$

The theorem that 'takes the limit' is:

Theorem 2.21 (Kolmogorov's theorem) *Let $\mathcal{I}$ be a countable set and let a probability measure μ_I on $\mathbb{R}^I$ be given for each finite set $I \subset \mathcal{I}$ so that the family of μ_I's are consistent. Then there are a probability measure space $(X, \mathcal{F}, \mu)$ and random variables $\{f_\alpha\}_{\alpha \in \mathcal{I}}$ so that μ_I is the joint probability distribution of $\{f_\alpha\}_{\alpha \in I}$. That is*

$$\mu_I(A) = \mu\left(\left\{ x \in X \mid P_I\big(f_\alpha(x)\big)_{\alpha \in \mathcal{I}} \in A \right\}\right)$$

for all measurable $A \subset \mathbb{R}^I$. Moreover this space is unique in the sense that if $(X', \mathcal{F}', \mu')$ and $\{f'_\alpha\}_{\alpha \in I}$ also have these properties and if $\mathcal{F}$ (and respectively, $\mathcal{F}'$) is the smallest σ–field which respect to which the f_α (respectively f'_α) are measurable, then there is an isomorphism of the probability measure spaces under which each f_α corresponds to f'_α.

A very convenient tool for constructing cylinder measures is (Minlos 1959):

Theorem 2.22 (Minlos' theorem) *A necessary and sufficient condition for a function $\Phi : \mathcal{S}(\mathbb{R}^\nu) \to \mathbb{C}$ to be the Fourier transform*

$$\Phi(\varphi) = \int e^{iT(\varphi)} \, d\mu(T)$$

of a cylinder set probability measure μ on $\mathcal{S}'(\mathbb{R}^\nu)$ is that $\Phi(0) = 1$, Φ be positive definite and Φ be continuous in the Fréchet topology on $\mathcal{S}(\mathbb{R}^\nu)$.

Here

- $\mathcal{S}(\mathbb{R}^\nu)$ is Schwartz space, the space of all C^∞ functions on $\mathbb{R}^\nu$ all of whose derivatives decay faster than any polynomial at infinity,
- $\mathcal{S}'(\mathbb{R}^\nu)$ is the space of tempered distributions, the space of all continuous linear functions on $\mathcal{S}(\mathbb{R}^\nu)$,
- a cylinder set measure on $\mathcal{S}'(\mathbb{R}^\nu)$ is a measure on the σ–field generated by the functions $\{ T \mapsto T(\varphi) \mid \varphi \in \mathcal{S}(\mathbb{R}^\nu) \}$,
- $\Phi : \mathcal{S}(\mathbb{R}^\nu) \to \mathbb{C}$ is positive definite if $\sum_{i,j=1}^n \overline{z_i} z_j \Phi(\varphi_i - \varphi_j) \geq 0$ for all $n \in \mathbb{N}$, $z_1, \cdots, z_n \in \mathbb{C}$ and $\varphi_1, \cdots, \varphi_n \in \mathcal{S}(\mathbb{R}^n)$.

For an exposition on cylinder set measures, see (Gel'fand and Vilenkin 1968; Simon 2005). For applications of cylinder set measures to Brownian motion and Wiener processes, see (Nelson 1964, Appendix A) and (Durrett 2010). For applications of cylinder set measures to Schrödinger operators, see (Simon 2005). For applications of cylinder set measures to field theory and statistical mechanics, see (Ginibre 1971; Fröhlich 1974; Feldman and Osterwalder 1976; Glimm and Jaffe 1987).

2.4.9 A warning about complex measures

It is critical that cylinder measures are (real-valued) probability measures. The point is that complex measures must have finite total mass. (When you compute the measure of a complicated set by cutting it up into countable many disjoint sets and adding up the measures of the pieces, it is important that it not matter what order you do the sum in. And that is the case only if the sum is absolutely convergent.) That dramatically limits the class of complex measures on infinite dimensional vector spaces. In particular, in our case, the exponent $\mathcal{A}(\alpha^*, \alpha)$ is complex and as a result, $e^{\mathcal{A}(\alpha^*, \alpha)}$ oscillates wildly. In contrast to the Wiener measure, $\frac{1}{\text{const}} \prod \frac{d\alpha^*_\tau(\mathbf{x}) \wedge d\alpha_\tau(\mathbf{x})}{2\pi i} e^{\text{part of} \mathcal{A}(\alpha^*, \alpha)}$ cannot be turned into an ordinary well-defined complex measure on some space of paths.

Here is a well-known example, due to Cameron (Cameron 1960, 1963), that illustrates the phenomenon. Let $\left[C_{ij}\right]_{i,j \in \mathbb{N}}$ be a 'matrix' with infinitely many rows and columns. Assume that C has real entries, is symmetric and is strictly positive definite in the sense that $\sum_{i,j} \alpha_i C_{i,j} \alpha_j > 0$ for all nonzero, real 'vectors' $[\alpha_i]_{i \in \mathbb{N}}$ having only finitely many nonzero components. A simple example of such a matrix is the identity matrix

$$\delta_{i,j} = \begin{cases} 1 & \text{if } i = j \\ 0 & \text{if } i \neq j \end{cases}$$

Another example is $-\Delta_{i,j} + m^2 \delta_{i,j}$ where $\Delta_{i,j}$ is the discrete Laplacian on $\mathbb{Z}^3$ and i, j refers to some arbitrary ordering of the points in $\mathbb{Z}^3$. Fix some $\sigma \in \mathbb{C}$ with $\operatorname{Re} \sigma > 0$. Consider, for each $n \in \mathbb{N}$, the measure

$$d\mu_n(\vec{\alpha}) = \frac{e^{-\frac{1}{2}\sigma \vec{\alpha} \cdot C \vec{\alpha}} \, d^n \vec{\alpha}}{\int_{\mathbb{R}^n} e^{-\frac{1}{2}\sigma \vec{\alpha} \cdot C \vec{\alpha}} \, d^n \vec{\alpha}}$$

on $\mathbb{R}^n$. Here, in computing $\vec{\alpha} \cdot C \vec{\alpha}$, set $\alpha_j = 0$ for all $j > n$. If σ is real, this is a legitimate probability measure. If, in addition, C is diagonal it is trivial to apply Kolmogorov's theorem and create a cylinder set measure. (For many other C's you can also create a cylinder set measures, with more work.) If $\operatorname{Im} \sigma \neq 0$, μ_n is still a legitimate (complex) measure on $\mathbb{R}^n$ and is still normalized so that $\int_{\mathbb{R}^n} d\mu_n(\vec{\alpha}) = 1$. In particular, if we write $C_n = [C_{ij}]_{1 \le i,j \le n}$, then, by making an orthogonal change of variables so as to diagonalize C_n, it is easy to see that

$$\int_{\mathbb{R}^n} e^{-\frac{1}{2}\sigma \vec{\alpha} \cdot C \vec{\alpha}} \, d^n \vec{\alpha} = \left[\left(\frac{\pi}{\sigma}\right)^n \frac{1}{\det C_n} \right]^{1/2} \neq 0.$$

(We aren't going to care which square root is used.) The total mass of μ_n is

$$\int_{\mathbb{R}^n} |d\mu_n(\vec{\alpha})| = \frac{\int_{\mathbb{R}^n} \left| e^{-\frac{1}{2}(\sigma \vec{\alpha} \cdot C \vec{\alpha})} \right| d^n \vec{\alpha}}{\left| \int_{\mathbb{R}^n} e^{-\frac{1}{2}\sigma \vec{\alpha} \cdot C \vec{\alpha}} \, d^n \vec{\alpha} \right|} = \frac{\int_{\mathbb{R}^n} e^{-\frac{1}{2}(\operatorname{Re}\sigma) \vec{\alpha} \cdot C \vec{\alpha}} \, d^n \vec{\alpha}}{\left| \int_{\mathbb{R}^n} e^{-\frac{1}{2}\sigma \vec{\alpha} \cdot C \vec{\alpha}} \, d^n \vec{\alpha} \right|} = \left\{ \frac{|\sigma|}{\operatorname{Re}\sigma} \right\}^{n/2}.$$

Since $\operatorname{Im} \sigma \neq 0$, this tends to infinity as $n \to \infty$ and we can't get a legitimate complex measure in the limit $n \to \infty$. Another model computation of this type which is closer to the integral of Theorem 2.14 is given in (Balaban *et al.* 2008a, Appendix A).

2.4.10 Grassmann integrals

Another class of functional integrals are Grassmann integrals. They are often used in fermionic models. Grassmann integrals are certain linear functionals defined on Grassmann algebras, which are a particularly simple class of algebras. The reason that the linear functionals are called 'integrals' is that they have, up to signs, all of the usual algebraic properties of integrals, including, for example, integration by parts. For discussions of Grassmann integrals, see (Berezin 1966; Feldman *et al.* 2002) and (Salmhofer 1999, Appendix B).

2.5 A simple high temperature expansion

High temperature expansions are extremely widely used tools in rigorous treatments of quantum field theories and statistical mechanical systems (and not just at high temperatures). This section is concerned with a very simple example of such an expansion. There are many other high temperature expansions. At the end of this section, I'll mention some others and give some references.

2.5.1 Motivation—a renormalization group construction protocol

Here is a cartoon description of a commonly used procedure for constructing and analysing quantum field theories and models in condensed matter physics and statistical mechanics.

- Express the quantities of interest as functional integrals like

$$\mathcal{G}(\Psi) = \ln \frac{\int e^{\mathcal{A}(\Psi,\Phi)} \, d\mu(\Phi)}{\int e^{\mathcal{A}(0,\Phi)} \, d\mu(\Phi)}.$$

- Factor the measure $d\mu(\Phi) = \prod_{\ell=1}^{\infty} d\mu_\ell(\varphi_\ell)$ to express

$$\mathcal{G}(\Psi) = \ln \frac{\int e^{\mathcal{A}(\Psi,\varphi_1,\varphi_2,\cdots)} \prod_{\ell=1}^{\infty} d\mu_\ell(\varphi_\ell)}{\int e^{\mathcal{A}(0,\varphi_1,\varphi_2,\cdots)} \prod_{\ell=1}^{\infty} d\mu_\ell(\varphi_\ell)}.$$

- Do the integrals one at a time. Define the 'effective action at scale n' by

$$\mathcal{A}_n(\Psi,\varphi_{n+1},\varphi_{n+2},\cdots) = \ln \frac{\int e^{\mathcal{A}(\Psi,\varphi_1,\varphi_2,\cdots)} \prod_{\ell=1}^{n} d\mu_\ell(\varphi_\ell)}{\int e^{\mathcal{A}(0,\varphi_1,\cdots,\varphi_n,0,\cdots)} \prod_{\ell=1}^{n} d\mu_\ell(\varphi_\ell)}$$

Then

$$\mathcal{A}_n(\psi) = \ln \frac{\int e^{\mathcal{A}_{n-1}(\psi,\varphi)} \, d\mu_n(\varphi)}{\int e^{\mathcal{A}_{n-1}(0,\varphi)} \, d\mu_n(\varphi)} \tag{2.46}$$

where $\varphi = \varphi_n$ and $\psi = (\Psi,\varphi_{n+1},\varphi_{n+2},\cdots)$.

The decimation procedure of Sections 2.2 and 2.3 was like this. To be able to implement such a procedure, you have to be able to prove bounds on integrals like in equation (2.46). In this section, we'll derive such bounds.

2.5.2 The main theorem

Let $X(=$ space$)$ be a finite set. Let $d\mu_0(z)$ be a normalized measure on $\mathbb{C}$ that is supported in $|z| \leq r$ for some constant r. We endow $\mathbb{C}^X$ with the ultralocal product measure

$$d\mu(\varphi) = \prod_{\mathbf{x} \in X} d\mu_0(\varphi(\mathbf{x}))$$

Theorem *Let w and W be weight systems for 1 and 2 fields, respectively, that obey*

$$W(\vec{\mathbf{x}},\vec{\mathbf{y}}) \geq (4r)^{n(\vec{y})} w(\vec{\mathbf{x}}).$$

Let $F : \mathbb{C}^{|X|} \times \mathbb{C}^{|X|} \to \mathbb{C}$ be analytic on a neighbourhood of the origin. If $F(\psi, \varphi)$ obeys $\|F\|_W < \frac{1}{16}$, then there is an analytic function $f(\psi)$ such that

$$\frac{\int e^{F(\psi,\varphi)} \, d\mu(\varphi)}{\int e^{F(0,\varphi)} \, d\mu(\varphi)} = e^{f(\psi)} \quad \text{and} \quad \|f\|_w \leq \frac{\|F\|_W}{1 - 16\|F\|_W}.$$

(I'll fill in the missing definitions later and then restate the Theorem and call it Theorem 2.30.)

2.5.3   Outline of the Proof—Algebra

We'll first do some algebra and end up with an explicit (but messy) formula for $f(\psi)$ in terms of $F(\psi, \varphi)$. After that we'll introduce the norms and do the bounds which show that the formula makes sense and that the Theorem is true. We use the notation

$$\mathbf{x} \in X = \text{space, a finite set}$$

$$\vec{\mathbf{x}} \in \mathcal{X} = \text{multispace} = \bigcup_{n \geq 0} X^n = \{ (\mathbf{x}_1, \cdots, \mathbf{x}_n) \in X^n \mid n \geq 0 \}$$

and, for $\vec{\mathbf{x}} = (\mathbf{x}_1, \cdots, \mathbf{x}_n) \in X^n$, $\vec{\mathbf{y}} = (\mathbf{y}_1, \cdots, \mathbf{y}_m) \in X^m$ and $\varphi : X \to \mathbb{C}$,

$$n(\vec{\mathbf{x}}) = n$$

$$\vec{\mathbf{x}} \circ \vec{\mathbf{y}} = (\mathbf{x}_1, \cdots, \mathbf{x}_n, \mathbf{y}_1, \cdots, \mathbf{y}_m) \in X^{n+m}$$

$$\varphi(\vec{\mathbf{x}}) = \varphi(\mathbf{x}_1)\varphi(\mathbf{x}_2) \cdots \varphi(\mathbf{x}_n)$$

$$\text{supp}(\vec{\mathbf{x}}) = \{\mathbf{x}_1, \cdots, \mathbf{x}_n\} \subset X.$$

By hypothesis, $F : \mathbb{C}^{|X|} \times \mathbb{C}^{|X|} \to \mathbb{C}$ is analytic on a neighbourhood of the origin. We shall end up showing that $f : \mathbb{C}^{|X|} \to \mathbb{C}$ is analytic too. So there are unique expansions

$$F(\psi, \varphi) = \sum_{\vec{\mathbf{x}}, \vec{\mathbf{y}} \in \mathcal{X}} A(\vec{\mathbf{x}}, \vec{\mathbf{y}}) \, \psi(\vec{\mathbf{x}})\varphi(\vec{\mathbf{y}}) \qquad f(\psi) = \sum_{\vec{\mathbf{x}} \in \mathcal{X}} a(\vec{\mathbf{x}}) \, \psi(\vec{\mathbf{x}}) \qquad (2.47)$$

with $A(\vec{\mathbf{x}}, \vec{\mathbf{y}})$, $a(\vec{\mathbf{x}})$ invariant under permutations of the components of $\vec{\mathbf{x}}$ and under permutations of the components of $\vec{\mathbf{y}}$. We are about to do an integral over φ. Hide the ψ dependence of F by setting

$$\alpha(\vec{\mathbf{y}}) = \sum_{\vec{\mathbf{x}} \in \mathcal{X}} A(\vec{\mathbf{x}}, \vec{\mathbf{y}}) \, \psi(\vec{\mathbf{x}}).$$

With this notation

$$F(\psi, \varphi) = \sum_{\vec{\mathbf{y}} \in \mathcal{X}} \alpha(\vec{\mathbf{y}}) \, \varphi(\vec{\mathbf{y}}).$$

By factoring $e^{F(\psi,0)}$ out of the integral in the numerator of (2.32) and $e^{F(0,0)}$ out of the integral in the denominator of (2.32) and moving $F(\psi,0) - F(0,0)$ into $f(\psi)$, we may assume that $F(\psi,0) = 0$. (Check yourself that the bound is preserved by this operation.) Expand the exponential to give

$$e^{F(\psi,\varphi)} = \sum_{\ell=0}^{\infty} \frac{1}{\ell!} F(\psi,\varphi)^{\ell} = 1 + \sum_{\ell=1}^{\infty} \frac{1}{\ell!} \sum_{\vec{y}_1,\cdots,\vec{y}_\ell \in \mathcal{X}} \alpha(\vec{y}_1)\cdots\alpha(\vec{y}_\ell)\,\varphi(\vec{y}_1)\cdots\varphi(\vec{y}_\ell). \quad (2.48)$$

The decay properties of the coefficients $a(\vec{x})$ in (2.47) are extremely important. Those coefficients are going to built out of products like $A(\vec{x}_1,\vec{y}_1)\cdots A(\vec{x}_\ell,\vec{y}_\ell)$ with $\vec{x} = (\vec{x}_1,\cdots,\vec{x}_\ell)$. We are told (it is built into the norm $\|\cdot\|_W$) that $A(\vec{x},\vec{y})$ decays as the components of $(\vec{x},\vec{y})$ are separated. But that does not in general imply that $A(\vec{x}_1,\vec{y}_1)\cdots A(\vec{x}_\ell,\vec{y}_\ell)$ decays as the components of $\vec{x} = (\vec{x}_1,\cdots,\vec{x}_\ell)$ are separated. But if we know in addition, for example, that, for each $1 \le j \le \ell - 1$, $\vec{y}_j$ has a component that is equal to some component of $\vec{y}_{j+1}$, then $A(\vec{x}_1,\vec{y}_1)\cdots A(\vec{x}_\ell,\vec{y}_\ell)$ does decay as the components of $\vec{x} = (\vec{x}_1,\cdots,\vec{x}_\ell)$ are separated.

We now build some machinery to keep track of such component overlaps. Define the incidence graph $G(\vec{y}_1,\cdots,\vec{y}_\ell)$ to be the labelled graph with

- vertices $\{1,\cdots,\ell\}$ and
- an edge between $i \ne j$ when $\operatorname{supp}\vec{y}_i \cap \operatorname{supp}\vec{y}_j \ne \emptyset$.

For a subset of $Z \subset X$, denote by $\mathcal{C}(Z)$ the set of all ordered tuples $(\vec{y}_1,\cdots,\vec{y}_n)$ such that

- $Z = \operatorname{supp}\vec{y}_1 \cup \cdots \cup \operatorname{supp}\vec{y}_n$.
- $G(\vec{y}_1,\cdots,\vec{y}_n)$ is connected.

We call such a tuple a connected cover of Z. Now reorganize the ℓth term of (2.48) according to the supports of the connected components of $G(\vec{y}_1,\cdots,\vec{y}_\ell)$.

$$\sum_{\vec{y}_1,\cdots,\vec{y}_\ell \in \mathcal{X}} \alpha(\vec{y}_1)\cdots\alpha(\vec{y}_\ell)\,\varphi(\vec{y}_1)\cdots\varphi(\vec{y}_\ell)$$

$$(2.49)$$

$$= \sum_{n=1}^{\ell} \frac{1}{n!} \sum_{\substack{Z_1,\cdots,Z_n \subset X \\ \text{pairwise disjoint} \\ \text{nonempty}}} \sum_{\substack{I_1 \cup \cdots \cup I_n = \{1,\cdots,\ell\} \\ I_1,\cdots,I_n \text{ pairwise disjoint}}} \sum_{\substack{\vec{y}_1,\cdots,\vec{y}_\ell \\ (\vec{y}_i)_{i \in I_j} \in \mathcal{C}(Z_j)}} \alpha(\vec{y}_1)\cdots\alpha(\vec{y}_\ell)\,\varphi(\vec{y}_1)\cdots\varphi(\vec{y}_\ell).$$

Fix, for the moment, pairwise disjoint nonempty subsets $Z_1,\cdots,Z_n$ of X and $\ell \ge n$. Then

$$\sum_{\substack{I_1 \cup \cdots \cup I_n = \{1,\cdots,\ell\} \\ I_1,\cdots,I_n \text{ disjoint}}} \sum_{\substack{\vec{y}_1,\cdots,\vec{y}_\ell \\ (\vec{y}_i, i \in I_j) \in \mathcal{C}(Z_j)}} \alpha(\vec{y}_1)\cdots\alpha(\vec{y}_\ell)\,\varphi(\vec{y}_1)\cdots\varphi(\vec{y}_\ell)$$

$$= \sum_{\substack{k_1,\cdots,k_n \ge 1 \\ k_1+\cdots+k_n=\ell}} \sum_{\substack{I_1,\cdots,I_n \subset \{1,\cdots,\ell\} \\ I_1,\cdots,I_n \text{ disjoint} \\ |I_j|=k_j}} \sum_{\substack{\vec{y}_1,\cdots,\vec{y}_\ell \\ (\vec{y}_i, i \in I_j) \in \mathcal{C}(Z_j)}} \alpha(\vec{y}_1)\cdots\alpha(\vec{y}_\ell)\,\varphi(\vec{y}_1)\cdots\varphi(\vec{y}_\ell) \quad (2.50)$$

$$= \sum_{\substack{k_1,\cdots,k_n \geq 1 \\ k_1+\cdots+k_n=\ell}} \frac{\ell!}{k_1!\cdots k_n!} \sum_{\substack{(\vec{y}_1,\cdots,\vec{y}_{k_1})\in\mathcal{C}(Z_1) \\ \vdots \\ (\vec{y}_{\ell-k_n+1},\cdots,\vec{y}_\ell)\in\mathcal{C}(Z_n)}} \alpha(\vec{y}_1)\cdots\alpha(\vec{y}_\ell)\,\varphi(\vec{y}_1)\cdots\varphi(\vec{y}_\ell).$$

As the measure μ factorizes with each factor normalized, and the different Z_j's are disjoint,

$$\int \varphi(\vec{y}_1)\cdots\varphi(\vec{y}_\ell)\,\mathrm{d}\,\mu(\varphi) = \prod_{j=1}^{n} \int \varphi(\vec{y}_{p_{j-1}+1})\cdots\varphi(\vec{y}_{p_j})\,\mathrm{d}\,\mu(\varphi) \qquad (2.51)$$

(where $p_0 = 0$ and, for $1 \leq j \leq n$, $p_j = k_1 + \cdots + k_j$).

Substituting (2.50) into (2.49) and then (2.49) into (2.48) and then integrating and applying (2.51) gives

$$\int e^{F(\psi,\varphi)}\,\mathrm{d}\,\mu(\varphi) = 1 + \sum_{\ell=1}^{\infty} \frac{1}{\ell!} \sum_{n=1}^{\ell} \frac{1}{n!} \sum_{\substack{Z_1,\cdots,Z_n \subset X \\ \text{pairwise disjoint} \\ \text{nonempty}}} \sum_{\substack{k_1,\cdots,k_n \geq 1 \\ k_1+\cdots+k_n=\ell}} \frac{\ell!}{k_1!\cdots k_n!} \cdots$$

$$= 1 + \sum_{n=1}^{\infty} \sum_{\ell=n}^{\infty} \frac{1}{n!} \sum_{\substack{Z_1,\cdots,Z_n \subset X \\ \text{pairwise disjoint} \\ \text{nonempty}}} \sum_{\substack{k_1,\cdots,k_n \geq 1 \\ k_1+\cdots+k_n=\ell}} \frac{1}{k_1!\cdots k_n!} \cdots$$

$$= 1 + \sum_{n=1}^{\infty} \frac{1}{n!} \sum_{\substack{Z_1,\cdots,Z_n \subset X \\ \text{pairwise disjoint} \\ \text{nonempty}}} \sum_{k_1,\cdots,k_n \geq 1} \frac{1}{k_1!\cdots k_n!} \cdots$$

so that

$$\int e^{F(\psi,\varphi)}\,\mathrm{d}\,\mu(\varphi) = 1 + \sum_{n=1}^{\infty} \frac{1}{n!} \sum_{\substack{Z_1,\cdots,Z_n \subset X \\ \text{pairwise disjoint}}} \prod_{j=1}^{n} \Phi(Z_j) \qquad (2.52)$$

where, for $\emptyset \neq Z \subset X$,

$$\Phi(Z) = \sum_{k=1}^{\infty} \frac{1}{k!} \sum_{(\vec{y}_1,\cdots,\vec{y}_k)\in\mathcal{C}(Z)} \alpha(\vec{y}_1)\cdots\alpha(\vec{y}_k) \int \varphi(\vec{y}_1)\cdots\varphi(\vec{y}_k)\,\mathrm{d}\,\mu(\varphi) \qquad (2.53)$$

and $\Phi(\emptyset) = 0$.

We next rewrite (2.52) so that it looks like 'the sum of the values of all Feynman diagrams'. We do this so that we can use the standard fact that 'the logarithm of the sum of the values of all Feynman diagrams is the sum of the values of all connected Feynman diagrams'. If we define

$$\zeta(Z, Z') = \begin{cases} 0 & \text{if } Z \cap Z' \neq \emptyset \\ 1 & \text{if } Z \text{ and } Z' \text{ are disjoint} \end{cases}$$

and $G_n = \{ \{i,j\} \subset \mathbb{N}^2 \mid 1 \le i < j \le n \}$ is the complete graph on $\{1, \cdots, n\}$, then

$$\int e^{F(\psi,\varphi)} \, d\mu(\varphi) = 1 + \sum_{n=1}^{\infty} \frac{1}{n!} \sum_{Z_1, \cdots, Z_n \subset X} \prod_{\{i,j\} \in G_n} \zeta(Z_i, Z_j) \prod_{j=1}^{n} \Phi(Z_j)$$

$$= 1 + \sum_{n=1}^{\infty} \frac{1}{n!} \sum_{Z_1, \cdots, Z_n} \left(\sum_{g \subset G_n} \prod_{\{i,j\} \in g} \big(\zeta(Z_i, Z_j) - 1\big) \right) \prod_{j=1}^{n} \Phi(Z_j)$$

$$= 1 + \sum_{n=1}^{\infty} \frac{1}{n!} \sum_{Z_1, \cdots, Z_n \subset X} \rho(Z_1, \cdots, Z_n) \prod_{j=1}^{n} \Phi(Z_j)$$

where

$$\rho(Z_1, \cdots, Z_n) = \begin{cases} 1 & \text{if } n = 1 \\ \sum_{g \subset G_n} \prod_{\{i,j\} \in g} \big(\zeta(Z_i, Z_j) - 1\big) & \text{if } n \ge 2. \end{cases}$$

Define

$$\rho^T(Z_1, \cdots, Z_n) = \begin{cases} 1 & \text{if } n = 1 \\ \sum_{g \in \mathcal{C}_n} \prod_{\{i,j\} \in g} \big(\zeta(Z_i, Z_j) - 1\big) & \text{if } n \ge 2 \end{cases}$$

where $\mathcal{C}_n$ is the set of connected subgraphs of G_n. By a standard argument, outlined in the motivation below,

$$\ln \int e^{F(\psi,\varphi)} d\mu = \sum_{n=1}^{\infty} \frac{1}{n!} \sum_{Z_1, \cdots, Z_n \subset X} \rho^T(Z_1, \cdots, Z_n) \prod_{j=1}^{n} \Phi(Z_j). \tag{2.54}$$

(By 'ln' we just mean that the exponential of the right-hand side is $\int e^F(\psi, \varphi) \, d\mu$.)

Motivation Define the value of the graph $g \subset G_n$ to be

$$\text{Val}(g) = \begin{cases} \sum_{Z \subset X} \Phi(Z) & \text{if } n = 1 \\ \sum_{Z_1, \cdots, Z_n} \prod_{\{i,j\} \in g} C(Z_i, Z_j) \prod_{j=1}^{n} \Phi(Z_j) & \text{if } n > 1 \end{cases}$$

where $C(Z_i, Z_j) = \zeta(Z_i, Z_j) - 1$. If the connected components of $g \in \mathcal{G}_n$ are $g_1, \cdots, g_m$, then

$$\text{Val}(g) = \prod_{j=1}^{m} \text{Val}(g_m).$$

Directly from the definitions,

$$\int e^{F(\psi,\varphi)} d\mu = 1 + \sum_{n=1}^{\infty} \frac{1}{n!} \sum_{g \subset G_n} \text{Val}(g). \tag{2.55}$$

On the other hand, the exponential of the right-hand side of (2.54) is

$$\exp\left\{ \sum_{n=1}^{\infty} \frac{1}{n!} \sum_{g \in C_n} \text{Val}(g) \right\} = \prod_{n=1}^{\infty} \prod_{g \in C_n} e^{\frac{1}{n!} \text{Val}(g)} \tag{2.56}$$

If you expand out the exponential and the two products, you will get the sum of the values of all graphs, with the value of each graph given as the product of the values of its connected components. To complete the proof that the right-hand sides of (2.55) and (2.56) are equal, you just have to check carefully that the combinatorial coefficients match up. See, for example, (Salmhofer 1999, Section 2.4). ☐

Equation (2.54) provides a formula for $f(\psi) = \ln \int e^{F(\psi,\varphi)} \, d\mu(\varphi)$. We now just unravel all of the definitions to extract the coefficient system $\{a(\vec{x})\}_{\vec{x} \in \mathcal{X}}$, of (2.47), for $f(\psi)$. Recall from (2.53) that

$$\Phi(Z) = \sum_{k=1}^{\infty} \frac{1}{k!} \sum_{(\vec{y}_1,\cdots,\vec{y}_k) \in \mathcal{C}(Z)} \alpha(\vec{y}_1) \cdots \alpha(\vec{y}_k) \int \varphi(\vec{y}_1) \cdots \varphi(\vec{y}_k) \, d\mu(\varphi)$$

and substitute in

$$\alpha(\vec{y}) = \sum_{\vec{x} \in \mathcal{X}} A(\vec{x}, \vec{y}) \, \psi(\vec{x})$$

to give

$$\Phi(Z) = \sum_{k=1}^{\infty} \frac{1}{k!} \sum_{\substack{(\vec{y}_1,\cdots,\vec{y}_k) \in \mathcal{C}(Z). \\ \vec{x}_1,\cdots,\vec{x}_k \in \mathcal{X}}} A(\vec{x}_1, \vec{y}_1) \cdots A(\vec{x}_k, \vec{y}_k) \psi(\vec{x}_1) \cdots \psi(\vec{x}_k) \int \varphi(\vec{y}_1) \cdots \varphi(\vec{y}_k) \, d\mu(\varphi)$$

So, if we set, for each $(\vec{x}, \vec{y}) \in \mathcal{X}^2$,

$$\tilde{A}(\vec{x}, \vec{y}) = \sum_{k=1}^{\infty} \frac{1}{k!} \sum_{\substack{(\vec{y}_1,\cdots,\vec{y}_k) \in \mathcal{C}(\text{supp } \vec{y}) \\ \vec{y}_1 \circ \cdots \circ \vec{y}_k = \vec{y}}} \sum_{\substack{\vec{x}_1,\cdots,\vec{x}_k \\ \vec{x}_1 \circ \cdots \circ \vec{x}_k = \vec{x}}} A(\vec{x}_1, \vec{y}_1) \cdots A(\vec{x}_k, \vec{y}_k) \int \varphi(\vec{y}) \, d\mu(\varphi) \tag{2.57}$$

we have

$$\Phi(Z)(\psi) = \sum_{\substack{(\vec{x},\vec{y}) \in \mathcal{X}^2 \\ \text{supp } \vec{y} = Z}} \tilde{A}(\vec{x}, \vec{y}) \, \psi(\vec{x}).$$

Recall, from (2.54), that

$$\ln \int e^{F(\psi,\varphi)} d\mu = \sum_{n=1}^{\infty} \frac{1}{n!} \sum_{Z_1,\cdots,Z_n \subset X} \rho^T(Z_1,\cdots,Z_n) \prod_{j=1}^{n} \Phi(Z_j).$$

Therefore,

$$\ln \int e^{F(\psi,\varphi)} d\mu(\varphi) = \sum_{\vec{\mathbf{x}} \in \mathcal{X}} a(\vec{\mathbf{x}}) \ \psi(\vec{\mathbf{x}})$$

where, for $\vec{\mathbf{x}} \in \mathcal{X}$,

$$a(\vec{\mathbf{x}}) = \sum_{n=1}^{\infty} \frac{1}{n!} \sum_{\substack{\vec{\mathbf{x}}_1,\cdots,\vec{\mathbf{x}}_n \in \mathcal{X} \\ \vec{\mathbf{x}}_1 \circ \cdots \circ \vec{\mathbf{x}}_n = \vec{\mathbf{x}}}} \sum_{\vec{\mathbf{y}}_1,\cdots,\vec{\mathbf{y}}_n \in \mathcal{X}} \rho^T(\mathrm{supp}\ \vec{\mathbf{y}}_1,\cdots,\mathrm{supp}\ \vec{\mathbf{y}}_n) \prod_{j=1}^{n} \tilde{A}(\vec{\mathbf{x}}_j,\vec{\mathbf{y}}_j). \quad (2.58)$$

Also

$$f(\psi) = \ln \frac{\int e^{F(\psi,\varphi)} d\mu(\varphi)}{\int e^{F(0,\varphi)} d\mu(\varphi)} = \sum_{\substack{\vec{\mathbf{x}} \in \mathcal{X} \\ n(\vec{\mathbf{x}}) > 0}} a(\vec{\mathbf{x}}) \ \psi(\vec{\mathbf{x}})$$

Now the $a(\vec{\mathbf{x}})$ of (2.58) might not be invariant under permutations of the components of $\vec{\mathbf{x}}$. We can of course symmetrize, but that will not be necessary for doing the estimates.

This brings us to the end of the algebraic part of the proof. We next specify the class of norms that are used in Theorem 2.29. This class generalizes the norms of (2.26) and (2.27).

2.5.4 Norms

Definition 2.23 (Weight system for one field) *A weight system for one field is a function $w : \mathcal{X} \to (0,\infty)$ that satisfies:*

(a) $w(\vec{\mathbf{x}})$ *is invariant under permutations of the components of $\vec{\mathbf{x}}$.*
(b) $\qquad w(\vec{\mathbf{x}} \circ \vec{\mathbf{x}}') \le w(\vec{\mathbf{x}}) w(\vec{\mathbf{x}}')$
$\qquad$ *for all $\vec{\mathbf{x}}, \vec{\mathbf{x}}' \in \mathcal{X}$ with $\mathrm{supp}(\vec{\mathbf{x}}) \cap \mathrm{supp}(\vec{\mathbf{x}}') \ne \emptyset$.*

Example 2.24 (Weight systems)

(a) If $\kappa : X \to (0,\infty)$ (called a *weight factor*) then

$$w(\vec{\mathbf{x}}) = \kappa(\vec{\mathbf{x}}) = \prod_{\ell=1}^{n(\vec{\mathbf{x}})} \kappa(\mathbf{x}_\ell)$$

is a weight system for one field.

(b) Let $d : X \times X \to \mathbb{R}_{\ge 0}$ be a metric. The length of a tree T with vertices in X is the sum of the lengths of all edges of T (where the length of an edge is the distance between its vertices). For a subset $S \subset X$, denote by $\tau(S)$ the length of the shortest tree in X whose set of vertices contains S. Then

$$w(\vec{\mathbf{x}}) = e^{\tau(\operatorname{supp}(\vec{\mathbf{x}}))}$$

is a weight system for one field.

(c) If $w_1(\vec{\mathbf{x}})$ and $w_2(\vec{\mathbf{x}})$ are weight systems for one field, then so is

$$w_3(\vec{\mathbf{x}}) = w_1(\vec{\mathbf{x}})w_2(\vec{\mathbf{x}}).$$

Definition 2.25 (Norms for functions of one field) *Let $f(\psi)$ be a function which is defined and analytic on a neighbourhood of the origin in $\mathbb{C}^{|X|}$. Then f has a unique expansion of the form $f(\psi) = \sum_{\vec{\mathbf{x}} \in \mathcal{X}} a(\vec{\mathbf{x}}) \, \psi(\vec{\mathbf{x}})$ with $a(\vec{\mathbf{x}})$ invariant under permutations of the components of $\vec{\mathbf{x}}$. (We call $a = \{ \, a(\vec{\mathbf{x}}) \mid \vec{\mathbf{x}} \in \mathcal{X} \, \}$ the symmetric coefficient system for f.) If $w(\vec{\mathbf{x}})$ is a weight system for one field, we define*

$$\|f\|_w = \|a\|_w \equiv \sum_{n \geq 0} \max_{\substack{1 \leq i \leq n \\ \mathbf{z} \in X}} \sum_{\substack{\vec{\mathbf{x}} \in X^n \\ \mathbf{x}_i = \mathbf{z}}} w(\vec{\mathbf{x}}) \, |a(\vec{\mathbf{x}})|.$$

Here $\mathbf{x}_i$ is the ith component of the n–tuple $\vec{\mathbf{x}}$. The term in the above sum with $n = 0$ is simply $w(-) \, |a(-)|$ where $-$ denotes the 0–tuple.

Remark 2.26 *If*

$$f(\psi) = \sum_{\vec{\mathbf{x}} \in \mathcal{X}} a(\vec{\mathbf{x}}) \, \psi(\vec{\mathbf{x}})$$

with $a(\vec{\mathbf{x}})$ not necessarily invariant under permutations of the components of $\vec{\mathbf{x}}$, then

$$\|f\|_w \leq \|a\|_w \equiv \sum_{n \geq 0} \max_{\substack{1 \leq i \leq n \\ \mathbf{z} \in X}} \sum_{\substack{\vec{\mathbf{x}} \in X^n \\ \mathbf{x}_i = \mathbf{z}}} w(\vec{\mathbf{x}}) \, |a(\vec{\mathbf{x}})|.$$

Definition 2.27 (Weight system for two fields) *A weight system for two fields is a function $W : \mathcal{X}^2 \rightarrow (0, \infty)$ that satisfies:*

(a) $W(\vec{\mathbf{x}}, \vec{\mathbf{y}})$ *is invariant under permutations of the components of $\vec{\mathbf{x}}$ and is invariant under permutations of the components of $\vec{\mathbf{y}}$.*

(b) $\quad W(\vec{\mathbf{x}} \circ \vec{\mathbf{x}}', \vec{\mathbf{y}} \circ \vec{\mathbf{y}}') \leq W(\vec{\mathbf{x}}, \vec{\mathbf{y}}) W(\vec{\mathbf{x}}', \vec{\mathbf{y}}')$
    whenever $\operatorname{supp}(\vec{\mathbf{x}}, \vec{\mathbf{y}}) \cap \operatorname{supp}(\vec{\mathbf{x}}', \vec{\mathbf{y}}') \neq \emptyset$.

Definition 2.28 (Norms for functions of two fields)
    Let

$$F(\psi, \varphi) = \sum_{(\vec{\mathbf{x}}, \vec{\mathbf{y}}) \in \mathcal{X}^2} A(\vec{\mathbf{x}}, \vec{\mathbf{y}}) \, \psi(\vec{\mathbf{x}}) \varphi(\vec{\mathbf{y}})$$

with $A(\vec{\mathbf{x}}, \vec{\mathbf{y}})$ invariant under permutations of the components of $\vec{\mathbf{x}}$ and under permutations of the components of $\vec{\mathbf{y}}$. If $W(\vec{\mathbf{x}}, \vec{\mathbf{y}})$ is a weight system for two fields, we define

$$\|F\|_W = \|A\|_W \equiv \sum_{n,m \geq 0} \max_{\substack{1 \leq i \leq n+m \\ \mathbf{z} \in X}} \sum_{\substack{(\vec{\mathbf{x}},\vec{\mathbf{y}}) \in X^n \times X^m \\ (\vec{\mathbf{x}},\vec{\mathbf{y}})_i = \mathbf{z}}} W(\vec{\mathbf{x}},\vec{\mathbf{y}}) \, |A(\vec{\mathbf{x}},\vec{\mathbf{y}})|$$

Here $(\vec{\mathbf{x}},\vec{\mathbf{y}})_i$ is the ith component of the $(n+m)$–tuple $(\vec{\mathbf{x}},\vec{\mathbf{y}})$. The term in the above sum with $n = m = 0$ is simply $W(-,-) \, |A(-,-)|$.

2.5.5 Review of the main theorem

Recall that

- $X(= \text{space})$ is a finite set and,
- $d\mu_0(z)$ is a normalized measure on $\mathbb{C}$ that is supported in $|z| \leq r$ for some constant r and,
- we endow $\mathbb{C}^X$ with the ultralocal product measure

$$d\mu(\varphi) = \prod_{\mathbf{x} \in X} d\mu_0\big(\varphi(\mathbf{x})\big).$$

We now have all of the definitions required to state

Theorem 2.29 *Let w and W be weight systems for 1 and 2 fields, respectively, that obey*

$$W(\vec{\mathbf{x}},\vec{\mathbf{y}}) \geq (4r)^{n(\vec{\mathbf{y}})} w(\vec{\mathbf{x}})$$

Let $F : \mathbb{C}^{|X|} \times \mathbb{C}^{|X|} \to \mathbb{C}$ be analytic on a neighbourhood of the origin. If $F(\psi,\varphi)$ obeys $\|F\|_W < \frac{1}{16}$, then there is an analytic function $f(\psi)$ such that

$$\frac{\int e^{F(\psi,\varphi)} \, d\mu(\varphi)}{\int e^{F(0,\varphi)} \, d\mu(\varphi)} = e^{f(\psi)}$$

and

$$\|f\|_w \leq \frac{\|F\|_W}{1 - 16\|F\|_W}.$$

2.5.6 Outline of the proof of Theorem 2.29—Bounds

Step 1—organizing the sums. Recall, from (2.58), that

$$a(\vec{\mathbf{x}}) = \sum_{n=1}^{\infty} \frac{1}{n!} \sum_{\substack{\vec{\mathbf{x}}_1,\cdots,\vec{\mathbf{x}}_n \in \mathcal{X} \\ \vec{\mathbf{x}}_1 \circ \cdots \circ \vec{\mathbf{x}}_n = \vec{\mathbf{x}}}} \sum_{\vec{\mathbf{y}}_1,\cdots,\vec{\mathbf{y}}_n \in \mathcal{X}} \rho^T(\text{supp } \vec{\mathbf{y}}_1, \cdots, \text{supp } \vec{\mathbf{y}}_n) \prod_{j=1}^{n} \tilde{A}(\vec{\mathbf{x}}_j, \vec{\mathbf{y}}_j). \quad (2.59)$$

The bound

$$\left| \rho^T(\text{supp } \vec{\mathbf{y}}_1, \cdots, \text{supp } \vec{\mathbf{y}}_n) \right| \leq \#\{ \text{ spanning trees in } G(\vec{\mathbf{y}}_1, \cdots, \vec{\mathbf{y}}_n) \}.$$

is due to Rota (Rota 1964). For a simple proof see (Simon 1993, Theorem V.7.A.6). A spanning tree for a graph is just a tree with the same set of vertices as the graph. Hence

$$
\begin{aligned}
|a(\vec{\mathbf{x}})| &\le \sum_{n=1}^{\infty} \frac{1}{n!} \sum_{\substack{\vec{\mathbf{x}}_1,\cdots,\vec{\mathbf{x}}_n \in \mathcal{X} \\ \vec{\mathbf{x}}_1 \circ \cdots \circ \vec{\mathbf{x}}_n = \vec{\mathbf{x}}}} \sum_{\vec{\mathbf{y}}_1,\cdots,\vec{\mathbf{y}}_n \in \mathcal{X}} \sum_{\substack{T \text{ spanning tree} \\ \text{for } G(\vec{\mathbf{y}}_1,\cdots,\vec{\mathbf{y}}_n)}} \prod_{j=1}^{n} |\tilde{A}(\vec{\mathbf{x}}_j, \vec{\mathbf{y}}_j)| \\[2mm]
&\le \sum_{n=1}^{\infty} \frac{1}{n!} \sum_{\substack{T \text{ labelled tree with} \\ \text{vertices } 1,\cdots,n}} \sum_{\vec{\mathbf{y}} \in \mathcal{X}} |\tilde{A}|_T(\vec{\mathbf{x}}, \vec{\mathbf{y}})
\end{aligned}
\tag{2.60}
$$

where

$$
|\tilde{A}|_T(\vec{\mathbf{x}}, \vec{\mathbf{y}}) = \sum_{\substack{\vec{\mathbf{y}}_1,\cdots,\vec{\mathbf{y}}_n \in \mathcal{X} \\ \vec{\mathbf{y}} = \vec{\mathbf{y}}_1 \circ \cdots \circ \vec{\mathbf{y}}_n \\ T \subset G(\vec{\mathbf{y}}_1,\cdots,\vec{\mathbf{y}}_n)}} \sum_{\substack{\vec{\mathbf{x}}_1,\cdots,\vec{\mathbf{x}}_n \in \mathcal{X} \\ \vec{\mathbf{x}} = \vec{\mathbf{x}}_1 \circ \cdots \circ \vec{\mathbf{x}}_n}} \prod_{\ell=1}^{n} |\tilde{A}(\vec{\mathbf{x}}_\ell, \vec{\mathbf{y}}_\ell)|.
$$

Recall, from (2.57), that

$$
\tilde{A}(\vec{\mathbf{x}}, \vec{\mathbf{y}}) = \sum_{k=1}^{\infty} \frac{1}{k!} \sum_{\substack{(\vec{\mathbf{y}}_1,\cdots,\vec{\mathbf{y}}_k) \in \mathcal{C}(\text{supp } \vec{\mathbf{y}}) \\ \vec{\mathbf{y}}_1 \circ \cdots \circ \vec{\mathbf{y}}_k = \vec{\mathbf{y}}}} \sum_{\substack{\vec{\mathbf{x}}_1,\cdots,\vec{\mathbf{x}}_k \\ \vec{\mathbf{x}}_1 \circ \cdots \circ \vec{\mathbf{x}}_k = \vec{\mathbf{x}}}} A(\vec{\mathbf{x}}_1, \vec{\mathbf{y}}_1) \cdots A(\vec{\mathbf{x}}_k, \vec{\mathbf{y}}_k) \int \varphi(\vec{\mathbf{y}}) \, d\mu(\varphi)
$$

For each $(\vec{\mathbf{y}}_1, \cdots, \vec{\mathbf{y}}_k)$, $G(\vec{\mathbf{y}}_1, \cdots, \vec{\mathbf{y}}_k)$ is connected and hence contains at least one tree. So

$$
\begin{aligned}
|\tilde{A}(\vec{\mathbf{x}}, \vec{\mathbf{y}})| &\le \sum_{k=1}^{\infty} \frac{1}{k!} \sum_{\substack{T \text{ labelled tree} \\ \text{with vertices} \\ 1,\cdots,k}} \sum_{\substack{\vec{\mathbf{y}}_1,\cdots,\vec{\mathbf{y}}_k \in \mathcal{X} \\ \vec{\mathbf{y}} = \vec{\mathbf{y}}_1 \circ \cdots \circ \vec{\mathbf{y}}_k \\ T \subset G(\vec{\mathbf{y}}_1,\cdots,\vec{\mathbf{y}}_k)}} \sum_{\substack{\vec{\mathbf{x}}_1,\cdots,\vec{\mathbf{x}}_k \in \mathcal{X} \\ \vec{\mathbf{x}} = \vec{\mathbf{x}}_1 \circ \cdots \circ \vec{\mathbf{x}}_k}} r^{n(\vec{\mathbf{y}})} \prod_{\ell=1}^{n} |A(\vec{\mathbf{x}}_\ell, \vec{\mathbf{y}}_\ell)| \\[2mm]
&= \sum_{k=1}^{\infty} \frac{1}{k!} \sum_{\substack{T \text{ labelled tree} \\ \text{with vertices} \\ 1,\cdots,k}} r^{n(\vec{\mathbf{y}})} |A|_T(\vec{\mathbf{x}}, \vec{\mathbf{y}})
\end{aligned}
\tag{2.60'}
$$

where

$$
|A|_T(\vec{\mathbf{x}}, \vec{\mathbf{y}}) = \sum_{\substack{\vec{\mathbf{y}}_1,\cdots,\vec{\mathbf{y}}_k \in \mathcal{X} \\ \vec{\mathbf{y}} = \vec{\mathbf{y}}_1 \circ \cdots \circ \vec{\mathbf{y}}_k \\ T \subset G(\vec{\mathbf{y}}_1,\cdots,\vec{\mathbf{y}}_k)}} \sum_{\substack{\vec{\mathbf{x}}_1,\cdots,\vec{\mathbf{x}}_n \in \mathcal{X} \\ \vec{\mathbf{x}} = \vec{\mathbf{x}}_1 \circ \cdots \circ \vec{\mathbf{x}}_k}} \prod_{\ell=1}^{k} |A(\vec{\mathbf{x}}_\ell, \vec{\mathbf{y}}_\ell)|.
$$

Step 2—bound on B_T

Lemma 2.30 *Let ω be an arbitrary weight system for two fields and define the weight system ω' by*

$$\omega'(\vec{\mathbf{x}}, \vec{\mathbf{y}}) = 2^{n(\vec{\mathbf{y}})}\omega(\vec{\mathbf{x}}, \vec{\mathbf{y}}).$$

Let T be a labelled tree with vertices $1, \cdots, n$ and coordination numbers $d_1, \cdots, d_n$ (meaning that vertex j has d_j lines attached to it). Let B be any (not necessarily symmetric) coefficient system for two fields with $B(-,-) = 0$. We define a new coefficient system B_T by

$$B_T(\vec{\mathbf{x}}, \vec{\mathbf{y}}) = \sum_{\substack{\vec{\mathbf{y}}_1, \cdots, \vec{\mathbf{y}}_n \in \mathcal{X} \\ \vec{\mathbf{y}} = \vec{\mathbf{y}}_1 \circ \cdots \circ \vec{\mathbf{y}}_n \\ T \subset G(\vec{\mathbf{y}}_1, \cdots, \vec{\mathbf{y}}_n)}} \sum_{\substack{\vec{\mathbf{x}}_1, \cdots, \vec{\mathbf{x}}_n \in \mathcal{X} \\ \vec{\mathbf{x}} = \vec{\mathbf{x}}_1 \circ \cdots \circ \vec{\mathbf{x}}_n}} \prod_{\ell=1}^{n} B(\vec{\mathbf{x}}_\ell, \vec{\mathbf{y}}_\ell)$$

Then

$$\left\| B_T \right\|_\omega \leq d_1! \cdots d_n! \, \|B\|_{\omega'}^n$$

Outline of Proof. (The details are in (Balaban *et al.* 2009, Section III).

Ingredient 1:

- For each $1 \leq \ell \leq n$, think of $(\vec{\mathbf{x}}_\ell, \vec{\mathbf{y}}_\ell)$ as the locations of (two species of) stars in a galaxy.
- In computing $\left\| B_T \right\|_\omega = \sum_{N,M \geq 0} \max_{\substack{1 \leq i \leq N+M \\ \mathbf{z} \in X}} \sum_{\substack{(\vec{\mathbf{x}}, \vec{\mathbf{y}}) \in X^N \times X^M \\ (\vec{\mathbf{x}}, \vec{\mathbf{y}})_i = \mathbf{z}}} \omega(\vec{\mathbf{x}}, \vec{\mathbf{y}}) \left| B_T(\vec{\mathbf{x}}, \vec{\mathbf{y}}) \right|$, we must

  hold fixed the location of one star (the ith) and sum over the locations of all other stars. Suppose, for example, that we have chosen $i = 1$ so that the fixed star is in galaxy $\ell = 1$.
- View 1 as the root of the tree T.
- Then the set of vertices of T is endowed with a natural partial ordering under which 1 is the smallest vertex.
- For each vertex $2 \leq \ell \leq n$, denote by $\pi(\ell)$ the predecessor vertex of ℓ under this partial ordering, as illustrated in Fig. 2.6.

$$\pi(7) = \pi(3) = \pi(4) = 2$$
$$\pi(2) = \pi(5) = 6$$
$$\pi(6) = 1$$

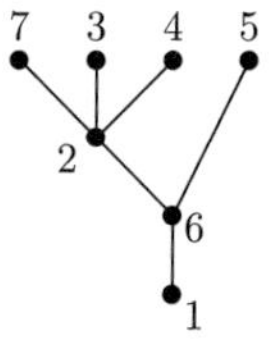

Fig. 2.6 A Sample tree partial ordering.

- The condition that $T \subset G(\vec{\mathbf{y}}_1, \cdots, \vec{\mathbf{y}}_n)$ ensures that, for each $2 \leq \ell \leq n$, the support of $\vec{\mathbf{y}}_\ell$ intersects the support of $\vec{\mathbf{y}}_{\pi(\ell)}$, so that at least one of the $n(\vec{\mathbf{y}}_\ell)$ components of $\vec{\mathbf{y}}_\ell$ takes the same value (in X) as some component of $\vec{\mathbf{y}}_{\pi(\ell)}$.
- Write $n(\vec{\mathbf{y}}_\ell) = n_\ell$.
- The product over $2 \leq \ell \leq n$ of the number of choices of which $\vec{\mathbf{y}}$–star in galaxy ℓ is at the same location of which $\vec{\mathbf{y}}$–star in galaxy $\pi(\ell)$ is

$$\prod_{\ell=2}^{n} [n_\ell n_{\pi(\ell)}] = \prod_{\ell=1}^{n} n_\ell^{d_\ell} = \prod_{\ell=1}^{n} \frac{n_\ell^{d_\ell}}{d_\ell!} d_\ell! \leq d_1! \cdots d_n! \prod_{\ell=1}^{n} 2^{n_\ell}$$

by using first year calculus and Stirling. (Alternatively, just use $\frac{n^d}{d!} \leq e^n$. This gives a slightly weaker theorem, but the change is insignificant.)

   Ingredient 2:

- Since T is connected,

$$\omega(\vec{\mathbf{x}}, \vec{\mathbf{y}}) \leq \prod_{\ell=1}^{n} \omega(\vec{\mathbf{x}}_\ell, \vec{\mathbf{y}}_\ell)$$

for all $\vec{\mathbf{x}}_1, \cdots, \vec{\mathbf{x}}_n \in \mathcal{X}$ and $\vec{\mathbf{y}}_1, \cdots, \vec{\mathbf{y}}_n \in \mathcal{X}$ under consideration, by the second condition of Definition 2.27. So we may absorb each factor $\omega(\vec{\mathbf{x}}_\ell, \vec{\mathbf{y}}_\ell)$ into $B(\vec{\mathbf{x}}_\ell, \vec{\mathbf{y}}_\ell)$ and it suffices to consider $\omega = 1$.

   Ingredient 3:

- Iteratively apply

$$\sum_{\substack{\vec{\mathbf{x}}_\ell, \vec{\mathbf{y}}_\ell \in \mathcal{X} \\ \vec{\mathbf{y}}_{\ell, m_\ell} = \vec{\mathbf{y}}_{\pi(\ell), p_\ell}}} \sum_{\vec{\mathbf{x}}_\ell \in \mathcal{X}} 2^{n(\vec{\mathbf{y}}_\ell)} |B(\vec{\mathbf{x}}_\ell, \vec{\mathbf{y}}_\ell)| \leq \|B\|_{\omega'}$$

starting with the largest ℓ's, in the partial ordering of T, and ending with $\ell = 1$. (For $\ell = 1$, substitute $\vec{\mathbf{x}}_{1,1} = \mathbf{x}$ for $\vec{\mathbf{y}}_{\ell, m_\ell} = \vec{\mathbf{y}}_{\pi(\ell), p_\ell}$.)

$\square$

Step 3—sum over n (or k) and T

Lemma 2.31 *Let* $0 < \varepsilon < \frac{1}{8}$. *Then*

$$\sum_{n=1}^{\infty} \frac{1}{n!} \sum_{\substack{d_1, \cdots, d_n \\ d_1 + \cdots + d_n = 2(n-1)}} \sum_{\substack{T \text{ labelled tree} \\ \text{with coordination} \\ \text{numbers } d_1, \cdots, d_n}} d_1! \cdots d_n! \, \varepsilon^n \leq \frac{\varepsilon}{1 - 8\varepsilon}.$$

Proof: Each line of a tree is connected to exactly two vertices. So the sum, $d_1 + \cdots + d_n$, of all coordination numbers is exactly twice the number of lines in the tree. The number of lines in a tree of n vertices is exactly $n = -1$. So we must have $d_1 + \cdots + d_n = 2(n-1)$. That accounts for the condition on the second sum.

By the Cayley formula, the number of labelled trees on $n \geq 2$ vertices with specified coordination numbers $(d_1, d_2, \cdots, d_n)$ is

$$\frac{(n-2)!}{\prod_{j=1}^{n}(d_j - 1)!}$$

Therefore

$$\sum_{n=2}^{\infty} \frac{1}{n!} \sum_{\substack{d_1,\cdots,d_n \\ d_1+\cdots+d_n=2(n-1)}} \sum_{\substack{T \text{ labelled tree} \\ \text{with coordination} \\ \text{numbers } d_1,\cdots,d_n}} d_1! \cdots d_n!\, \varepsilon^n \leq \sum_{n=2}^{\infty} \sum_{\substack{d_1,\cdots,d_n \\ d_1+\cdots+d_n=2(n-1)}} d_1 \cdots d_n\, \varepsilon^n.$$

The number of possible choices of coordination numbers $(d_1, d_2, \cdots, d_n)$ subject to the constraint $d_1 + d_2 + \cdots + d_n = 2(n-1)$ is

$$\binom{2(n-1)-1}{n-1} = \binom{2n-3}{n-1} \leq 2^{2n-3}$$

and $d_1 \cdots d_n \leq 2^n$. (Any maximizer must have $d_j \leq 2$ for every $1 \leq j \leq n$.) Therefore

$$\sum_{n=2}^{\infty} \frac{1}{n!} \sum_{\substack{d_1,\cdots,d_n \\ d_1+\cdots+d_n=2(n-1)}} \sum_{\substack{T \text{ labelled tree} \\ \text{with coordination} \\ \text{numbers } d_1,\cdots,d_n}} d_1! \cdots d_n!\, \varepsilon^n \leq \sum_{n=2}^{\infty} 2^{2n-3}\, 2^n \varepsilon^n = \frac{8\varepsilon^2}{1 - 8\varepsilon}.$$

For $n = 1$, $d_1 = 0$ and the number of trees is 1, so the $n = 1$ term is ε. So the full sum is bounded by $\varepsilon + \frac{8\varepsilon^2}{1-8\varepsilon} = \frac{\varepsilon}{1-8\varepsilon}$. □

Step 4–bound on $\|a\|$ *in terms of* $\|\tilde{A}\|$. We introduce, for each $\sigma > 0$, the auxiliary weight system

$$W_\sigma(\vec{\mathbf{x}}, \vec{\mathbf{y}}) = W(\vec{\mathbf{x}}, \vec{\mathbf{y}}) \left(\frac{\sigma}{4r}\right)^{n(\vec{\mathbf{y}})}.$$

Clearly

$$W_{4r}(\vec{\mathbf{x}}, \vec{\mathbf{y}}) = W(\vec{\mathbf{x}}, \vec{\mathbf{y}}) \quad \text{and} \quad w(\vec{\mathbf{x}}) \leq W_1(\vec{\mathbf{x}}, \vec{\mathbf{y}}) \tag{2.61}$$

for all $(\vec{\mathbf{x}}, \vec{\mathbf{y}}) \in \mathcal{X}^2$.
    We now prove

$$\|a\|_w \leq \frac{\|\tilde{A}\|_{W_2}}{1 - 8\|\tilde{A}\|_{W_2}}. \tag{2.62}$$

Recall from (2.60) that

$$|a(\vec{\mathbf{x}})| \le \sum_{n=1}^{\infty} \frac{1}{n!} \sum_{\substack{T \text{ labelled tree with} \\ \text{vertices } 1,\cdots,n}} \sum_{\vec{\mathbf{y}} \in \mathcal{X}} |\tilde{A}|_T(\vec{\mathbf{x}}, \vec{\mathbf{y}}).$$

Therefore, by (2.61) and Lemma 2.30, with $\omega = W_1$ and $\omega' = W_2$,

$$\|a\|_w \le \sum_{n=1}^{\infty} \frac{1}{n!} \sum_{\substack{T \text{ labelled tree with} \\ \text{vertices } 1,\cdots,n}} \left\| \,|\tilde{A}|_T \right\|_{W_1}$$

$$\le \sum_{n=1}^{\infty} \frac{1}{n!} \sum_{\substack{d_1,\cdots,d_n \\ d_1+\cdots+d_n=2(n-1)}} \sum_{\substack{T \text{ labelled tree} \\ \text{with coordination} \\ \text{numbers } d_1,\cdots,d_n}} \left\| \,|\tilde{A}|_T \right\|_{W_1}$$

$$\le \sum_{n=1}^{\infty} \frac{1}{n!} \sum_{\substack{d_1,\cdots,d_n \\ d_1+\cdots+d_n=2(n-1)}} \sum_{\substack{T \text{ labelled tree} \\ \text{with coordination} \\ \text{numbers } d_1,\cdots,d_n}} d_1! \cdots d_n! \, \left\| \,|\tilde{A}| \right\|_{W_2}^n.$$

Now apply Lemma 2.31 with $\varepsilon = \left\| \,|\tilde{A}| \right\|_{W_2} = \|\tilde{A}\|_{W_2}$ to get

$$\|a\|_w \le \frac{\|\tilde{A}\|_{W_2}}{1 - 8\|\tilde{A}\|_{W_2}}.$$

Step 5—bound on $\|\tilde{A}\|$ in terms of $\|A\|$. We now prove

$$\|\tilde{A}\|_{W_2} \le \frac{\|A\|_W}{1 - 8\|A\|_W} = \frac{\|F\|_W}{1 - 8\|F\|_W}. \qquad (2.63)$$

Note that combining (2.62) and (2.63) yields the final bound

$$\|f\|_w \le \|a\|_w \le \frac{\|\tilde{A}\|_{W_2}}{1 - 8\|\tilde{A}\|_{W_2}} \le \frac{\frac{\|F\|_W}{1-8\|F\|_W}}{1 - 8\frac{\|F\|_W}{1-8\|F\|_W}} = \frac{\|F\|_W}{1 - 16\|F\|_W}.$$

Recall from (2.60') that

$$|\tilde{A}(\vec{\mathbf{x}}, \vec{\mathbf{y}})| \le \sum_{k=1}^{\infty} \frac{1}{k!} \sum_{\substack{T \text{ labelled tree with} \\ \text{vertices } 1,\cdots,k}} r^{n(\vec{\mathbf{y}})} |A|_T(\vec{\mathbf{x}}, \vec{\mathbf{y}})$$

By construction, $\left\| \, r^{n(\vec{\mathbf{y}})} |A|_T(\vec{\mathbf{x}}, \vec{\mathbf{y}}) \, \right\|_{W_2} = \left\| \,|A|_T \right\|_{W_{2r}}$. Hence, by Lemma 2.30, with $\omega = W_{2r}$ followed by Lemma 2.31,

$$\|\tilde{A}\|_{W_2} \le \sum_{k=1}^{\infty} \frac{1}{k!} \sum_{\substack{T \text{ labelled tree with} \\ \text{vertices } 1,\cdots,k}} \left\| \,|A|_T \right\|_{W_{2r}}$$

$$\leq \sum_{k=1}^{\infty} \frac{1}{k!} \sum_{\substack{d_1,\cdots,d_k \\ d_1+\cdots+d_k=2(k-1)}} \sum_{\substack{T \text{ labelled tree} \\ \text{with coordination} \\ \text{numbers } d_1,\cdots,d_k}} d_1! \cdots d_k! \, \|A\|_{W_{4r}}^k$$

$$\leq \frac{\|A\|_W}{1 - 8\|A\|_W}$$

since $W_{4r} = W$. This gives (2.63). $\qquad\square$

2.5.7 Changes of variables

In this subsection we provide a couple of tools that may used to prove bounds on 'complicated' functions that are constructed from 'simple' functions using changes of variables.

For $\kappa > 0$, we denote by w_κ the weight system, for functions of one field, ψ, with mass m that associates the constant weight factor κ to the field ψ. That is

$$w_\kappa(\mathbf{x}_1, \cdots, \mathbf{x}_n) = e^{\mathrm{m}\,\tau(\{\mathbf{x}_1,\cdots,\mathbf{x}_n\})}\kappa^n$$

Similarly, for $\kappa, \lambda > 0$, we denote by $w_{\kappa,\lambda}$ the weight system, for functions of two fields, ψ and ϕ, with mass m that associates the constant weight factor κ to the field ψ and the constant weight factor λ to the field ϕ. To simplify notation, we write $\|g(\psi)\|_\kappa$ and $\|f(\psi, \phi)\|_{\kappa,\lambda}$ for $\|g(\psi)\|_{w_\kappa}$ and $\|f(\psi, \phi)\|_{w_{\kappa,\lambda}}$, respectively.

Proposition 2.32 *Let g be an analytic function on a neighbourhood of the origin in $\mathbb{C}^X$.*

1. *Let J be an operator on $\mathbb{C}^X$ with kernel $J(\mathbf{x}, \mathbf{y})$. Define $\tilde{g}$ by*

$$\tilde{g}(\psi) = g(J\psi).$$

   Let $\kappa > 0$ and set $\kappa' = \kappa \|\!|J|\!\|$. ($\|\!|J|\!\|$ was defined in (2.28).) Then $\|\tilde{g}\|_\kappa \leq \|g\|_{\kappa'}$.
2. *Define f by*

$$f(\psi; \phi) = g(\psi + \phi).$$

   Then $\|f\|_{\kappa,\lambda} = \|g\|_{\kappa+\lambda}$.

Proof:

(i) Let $a(\vec{\mathbf{x}})$ be a symmetric coefficient system for g. Define, for each $n \geq 0$,

$$\tilde{a}(\mathbf{x}_1, \cdots, \mathbf{x}_n) = \sum_{\mathbf{y}_1,\cdots,\mathbf{y}_n \in X} a(\mathbf{y}_1, \ldots, \mathbf{y}_n) \prod_{\ell=1}^{n} J(\mathbf{y}_\ell, \mathbf{x}_\ell).$$

Then $\tilde{a}(\vec{\mathbf{x}})$ is a symmetric coefficient system for $\tilde{g}$. Since

$$\tau(\{\mathbf{x}_1, \cdots, \mathbf{x}_n\}) \leq \tau(\{\mathbf{y}_1, \cdots, \mathbf{y}_n\}) + \sum_{\ell=1}^{n} d(\mathbf{y}_\ell, \mathbf{x}_\ell)$$

we have

$$e^{m\,\tau(\{\mathbf{x}_1,\cdots,\mathbf{x}_n\})} \le e^{m\,\tau(\{\mathbf{y}_1,\cdots,\mathbf{y}_n\})} \prod_{\ell=1}^{n} e^{md(\mathbf{y}_\ell,\mathbf{x}_\ell)}$$

and hence

$$w_\kappa(\mathbf{x}_1,\cdots,\mathbf{x}_n)\big|\tilde{a}(\mathbf{x}_1,\cdots,\mathbf{x}_n)\big|$$

$$\le \sum_{\mathbf{y}_1,\cdots,\mathbf{y}_n \in X} w_{\kappa'}(\mathbf{y}_1,\cdots,\mathbf{y}_n)\big|a(\mathbf{y}_1,\cdots,\mathbf{y}_n)\big| \prod_{\ell=1}^{n} \Big[\frac{\kappa}{\kappa'}e^{md(\mathbf{y}_\ell,\mathbf{x}_\ell)}\big|J(\mathbf{y}_\ell,\mathbf{x}_\ell)\big|\Big].$$

We are to bound

$$\|\tilde{g}\|_\kappa = \sum_{n\ge 0} \max_{\mathbf{x}\in X} \max_{1\le j\le n} \sum_{\substack{\mathbf{x}_1,\cdots,\mathbf{x}_n \in X^n \\ \mathbf{x}_j=\mathbf{x}}} w_\kappa(\mathbf{x}_1,\cdots,\mathbf{x}_n)\,\big|\tilde{a}(\mathbf{x}_1,\cdots,\mathbf{x}_n)\big|$$

$$\le \sum_{n\ge 0} \max_{\mathbf{x}\in X} \max_{1\le j\le n} \sum_{\substack{\mathbf{x}_1,\cdots,\mathbf{x}_n\in X \\ \mathbf{x}_j=\mathbf{x}}} \sum_{\mathbf{y}_1,\cdots,\mathbf{y}_n\in X} \tag{2.64}$$

$$w_{\kappa'}(\mathbf{y}_1,\cdots,\mathbf{y}_n)\big|a(\mathbf{y}_1,\cdots,\mathbf{y}_n)\big| \prod_{\ell=1}^{n}\Big[\frac{\kappa}{\kappa'}e^{md(\mathbf{y}_\ell,\mathbf{x}_\ell)}\big|J(\mathbf{y}_\ell,\mathbf{x}_\ell)\big|\Big].$$

Fix any $n\ge 0$, $\mathbf{x}\in X$ and $1\le j\le n$. By the definitions of κ' and $|||J|||$, for each $\ell\ne j$ and $\mathbf{y}_\ell\in X$,

$$\sum_{\mathbf{x}_\ell\in X} \frac{\kappa}{\kappa'}e^{m\,d(\mathbf{y}_\ell,\mathbf{x}_\ell)}J(\mathbf{y}_\ell,\mathbf{x}_\ell) = \sum_{\mathbf{x}_\ell\in X} \frac{1}{|||J|||}e^{m\,d(\mathbf{y}_\ell,\mathbf{x}_\ell)}J(\mathbf{y}_\ell,\mathbf{x}_\ell) \le 1.$$

Therefore

$$\sum_{\substack{\mathbf{x}_1,\cdots,\mathbf{x}_n\in X \\ \mathbf{x}_j=\mathbf{x}}} \sum_{\mathbf{y}_1,\cdots,\mathbf{y}_n\in X} w_{\kappa'}(\mathbf{y}_1,\cdots,\mathbf{y}_n)\big|a(\mathbf{y}_1,\cdots,\mathbf{y}_n)\big| \prod_{\ell=1}^{n}\Big[\frac{\kappa}{\kappa'}e^{md(\mathbf{y}_\ell,\mathbf{x}_\ell)}\big|J(\mathbf{y}_\ell,\mathbf{x}_\ell)\big|\Big]$$

$$\le \sum_{\mathbf{y}\in X} \frac{\kappa}{\kappa'}e^{m\,d(\mathbf{y},\mathbf{x})}J(\mathbf{y},\mathbf{x}) \sum_{\substack{\mathbf{y}_1,\cdots,\mathbf{y}_n\in X \\ \mathbf{y}_j=\mathbf{y}}} w_{\kappa'}(\mathbf{y}_1,\cdots,\mathbf{y}_n)\big|a(\mathbf{y}_1,\cdots,\mathbf{y}_n)\big|$$

$$\le \sum_{\mathbf{y}\in X} \frac{\kappa}{\kappa'}e^{m\,d(\mathbf{y},\mathbf{x})}J(\mathbf{y},\mathbf{x}) \max_{\mathbf{y}\in X} \sum_{\substack{\mathbf{y}_1,\cdots,\mathbf{y}_n\in X \\ \mathbf{y}_j=\mathbf{y}}} w_{\kappa'}(\mathbf{y}_1,\cdots,\mathbf{y}_n)\big|a(\mathbf{y}_1,\cdots,\mathbf{y}_n)\big|$$

$$\le \max_{\mathbf{y}\in X} \sum_{\substack{\mathbf{y}_1,\cdots,\mathbf{y}_n\in X \\ \mathbf{y}_j=\mathbf{y}}} w_{\kappa'}(\mathbf{y}_1,\cdots,\mathbf{y}_n)\big|a(\mathbf{y}_1,\cdots,\mathbf{y}_n)\big|$$

since, once again, $\sum_{\mathbf{y} \in X} \frac{\kappa}{\kappa'} e^{\mathrm{m}\, d(\mathbf{y}, \mathbf{x})} J(\mathbf{y}, \mathbf{x}) = \sum_{\mathbf{y} \in X} \frac{1}{\|J\|} e^{\mathrm{m}\, d(\mathbf{y}, \mathbf{x})} J(\mathbf{y}, \mathbf{x}) \leq 1$. Consequently, (2.64) is bounded by

$$\sum_{n \geq 0} \max_{1 \leq j \leq n} \max_{\mathbf{y} \in X} \sum_{\substack{\mathbf{y}_1, \cdots, \mathbf{y}_n \in X \\ \mathbf{y}_j = \mathbf{y}}} w_{\kappa'}(\mathbf{y}_1, \cdots, \mathbf{y}_n) \big| a(\mathbf{y}_1, \cdots, \mathbf{y}_n) \big|. = \|g\|_{\kappa'}.$$

This proves part (i) of the Proposition.

(ii) Let $a(\vec{\mathbf{u}})$ be a symmetric coefficient system for g. Since a is invariant under permutation of its $\vec{\mathbf{u}}$ components,

$$g(\psi + \phi) = \sum_{\vec{\mathbf{u}} \in \mathcal{X}} a(\vec{\mathbf{u}})\, (\psi + \phi)(\vec{\mathbf{u}}) = \sum_{\vec{\mathbf{x}}, \vec{\mathbf{y}} \in \mathcal{X}} a(\vec{\mathbf{x}} \circ \vec{\mathbf{y}}) \binom{n(\vec{\mathbf{x}}) + n(\vec{\mathbf{y}})}{n(\vec{\mathbf{y}})} \psi(\vec{\mathbf{x}}) \phi(\vec{\mathbf{y}})$$

so that

$$a_+(\vec{\mathbf{x}}; \vec{\mathbf{y}}) = a(\vec{\mathbf{x}} \circ \vec{\mathbf{y}}) \binom{n(\vec{\mathbf{x}}) + n(\vec{\mathbf{y}})}{n(\vec{\mathbf{y}})}$$

is a symmetric coefficient system for f. We have

$$\|f\|_{\kappa, \lambda} = \sum_{k, \ell \geq 0} \max_{\mathbf{p} \in X} \max_{1 \leq i \leq k + \ell} \sum_{\substack{\vec{\mathbf{x}} \in X^k,\ \vec{\mathbf{y}} \in X^\ell \\ (\vec{\mathbf{x}}, \vec{\mathbf{y}})_i = \mathbf{p}}} w_{\kappa, \lambda}(\vec{\mathbf{x}}; \vec{\mathbf{y}}) \big| a_+(\vec{\mathbf{x}}; \vec{\mathbf{y}}) \big|$$

$$= \sum_{k, \ell \geq 0} \max_{\mathbf{p} \in X} \max_{1 \leq i \leq k + \ell} \sum_{\substack{\vec{\mathbf{x}} \in X^k,\ \vec{\mathbf{y}} \in X^\ell \\ (\vec{\mathbf{x}}, \vec{\mathbf{y}})_i = \mathbf{p}}} e^{\mathrm{m}\, \tau(\mathrm{supp}(\vec{\mathbf{x}}, \vec{\mathbf{y}}))} \kappa^k \lambda^\ell \binom{k + \ell}{\ell} \big| a(\vec{\mathbf{x}} \circ \vec{\mathbf{y}}) \big|$$

$$= \sum_{k, \ell \geq 0} \binom{k + \ell}{\ell} \kappa^k \lambda^\ell \max_{\mathbf{p} \in X} \max_{1 \leq i \leq k + \ell} \sum_{\substack{\vec{\mathbf{x}} \in X^k,\ \vec{\mathbf{y}} \in X^\ell \\ (\vec{\mathbf{x}}, \vec{\mathbf{y}})_i = \mathbf{p}}} e^{\mathrm{m}\, \tau(\mathrm{supp}(\vec{\mathbf{x}}, \vec{\mathbf{y}}))} \big| a(\vec{\mathbf{x}} \circ \vec{\mathbf{y}}) \big|$$

$$= \sum_{k, \ell \geq 0} \binom{k + \ell}{\ell} \kappa^k \lambda^\ell \max_{\mathbf{p} \in X} \max_{1 \leq i \leq k + \ell} \sum_{\substack{\vec{\mathbf{u}} \in X^{k + \ell} \\ \vec{\mathbf{u}}_i = \mathbf{p}}} e^{\mathrm{m}\, \tau(\mathrm{supp}(\vec{\mathbf{u}}))} \big| a(\vec{\mathbf{u}}) \big|$$

$$= \sum_{n \geq 0} (\kappa + \lambda)^n \max_{\mathbf{p} \in X} \max_{1 \leq i \leq n} \sum_{\substack{\vec{\mathbf{u}} \in X^n \\ \vec{\mathbf{u}}_i = \mathbf{p}}} e^{\mathrm{m}\, \tau(\mathrm{supp}(\vec{\mathbf{u}}))} \big| a(\vec{\mathbf{u}}) \big|$$

$$= \|g\|_{\kappa + \lambda}. \qquad \qquad \square$$

2.5.8   Other Related High Temperature Expansions

The expansion treated in the main body of this section is just one of many similar expansions that are widely used in the construction and analysis of quantum field theories and many-body theories. Here are a few classes of such expansions. Don't take too seriously the names that I have assigned them — they do not have universally accepted meanings.

- *Cluster expansions* are expansions for unnormalized Schwinger functions, like for example $\int \varphi(z_1) \cdots \varphi(z_n) e^{-\mathcal{V}} \, d\mu(\varphi)$, that are used for proving the convergence of the infinite volume limit and of decay properties of the normalized Schwinger functions. See (Abdesselam and Rivasseau 1995; Brydges 1986; Glimm *et al.* 1973; Glimm and Jaffe 1987; Rivasseau 1991).
- *Mayer expansions* are used to implement cancellations between the numerator and denominator in expressions like

$$\frac{\int \varphi(z_1) \cdots \varphi(z_n) e^{-\mathcal{V}} \, d\mu(\varphi)}{\int e^{-\mathcal{V}} \, d\mu(\varphi)}$$

(in the limit as the volume tends to infinity, such numerators and denominators tend to behave like the exponential of a constant times the volume) and are used for proving the convergence of the infinite volume limit and of decay properties of the normalized Schwinger functions. See (Abdesselam and Rivasseau 1995; Brydges 1986; Glimm *et al.* 1973; Glimm and Jaffe 1987; Rivasseau 1991).
- *Polymer expansions.* In its simplest form, a polymer expansion looks like

$$1 + \sum_{n=1}^{\infty} \frac{1}{n!} \sum_{\substack{X_1, \cdots X_n \subset X \\ X_j \neq \emptyset \text{ for all } 1 \leq j \leq n \\ X_i \cap X_j = \emptyset \text{ for all } i \neq j}} \prod_{\ell=1}^{n} A(X_\ell).$$

Here each nonempty subset X_i of the world X is called a polymer and the function $A(X_i)$ is called a polymer activity. Two polymers X_i and X_j are said to be compatible if they are disjoint. Our expansion (2.52) was of this form. See (Brydges 1986; Cammarota 1982; Pordt 1998; Kotecký and Preiss 1986; Salmhofer 1999; Simon 1993) for lots of others.

2.6   Appendix: Complex Gaussian integrals

Integrals of polynomials times exponentials of quadratic functions are called Gaussian integrals and can be evaluated exactly. Lemma 2.33, below, does so in our setting, where the integration variables are complex. To have integrals that actually exist, we consider an arbitrary, but finite, number, L, of complex integation variables. To compactify notation, we write

$$\vec{\alpha} = (\alpha_1, \alpha_2, \cdots, \alpha_L) \qquad \langle \vec{\beta}, \vec{\alpha} \rangle = \sum_{\ell=1}^{L} \beta_\ell \alpha_\ell.$$

Note that $\langle \vec{\beta}, \vec{\alpha} \rangle$ is not the usual complex inner product. We deliberately do not include a complex conjugate on the right-hand side, so that all complex conjugates in our formulae appear explicitly.

To further compactify notation, we evaluate a generating functional. By repeatedly differentiating the conclusion of the lemma with respect to components of $\vec{j}_*$ and $\vec{j}$ and then setting $\vec{j}_* = \vec{j} = 0$, you can create any polynomial you like downstairs on the left hand side. To be precise, suppose that $\mathcal{A}(\alpha^*, \alpha)$ is some action and set

$$\mathcal{S}(\vec{j}_*, \vec{j}, \vec{\alpha}^*, \vec{\alpha}) = \langle \vec{j}_*, \alpha \rangle + \langle \vec{\alpha}^*, \vec{j} \rangle.$$

Define the expectation of $f(\vec{\alpha}^*, \vec{\alpha})$ to be

$$\langle f(\vec{\alpha}^*, \vec{\alpha}) \rangle = \frac{\int \prod_{\ell=1}^{L} \frac{d\alpha_\ell^* \wedge d\alpha_\ell}{2\pi i} \; e^{\mathcal{A}(\alpha^*, \alpha)} f(\vec{\alpha}^*, \vec{\alpha})}{\int \prod_{\ell=1}^{L} \frac{d\alpha_{\ell}* \wedge d\alpha_\ell}{2\pi i} \; e^{\mathcal{A}(\alpha^*, \alpha)}}$$

Then, for each $1 \le m \le L$,

$$\langle \alpha_m \rangle = \frac{\partial}{\partial j_{*m}} \left\langle e^{\mathcal{S}(\vec{j}_*, \vec{j}, \alpha^*, \alpha)} \right\rangle \Big|_{\vec{j}_* = \vec{j} = 0} \qquad \langle \alpha_m^* \rangle = \frac{\partial}{\partial j_m} \left\langle e^{\mathcal{S}(\vec{j}_*, \vec{j}, \alpha^*, \alpha)} \right\rangle \Big|_{\vec{j}_* = \vec{j} = 0}$$

and, for each $1 \le m, n \le L$,

$$\langle \alpha_m^* \alpha_n \rangle = \frac{\partial^2}{\partial j_m \partial j_{*n}} \left\langle e^{\mathcal{S}(\vec{j}_*, \vec{j}, \alpha^*, \alpha)} \right\rangle \Big|_{\vec{j}_* = \vec{j} = 0}. \tag{2.65}$$

Lemma 2.33 *Let $L \in \mathbb{N}$.*

(a) *Let D be an $L \times L$ matrix whose real part, $D + D^*$, is strictly positive. That is, all of the eigenvalues of $\left[D_{\ell, \ell'} + \overline{D_{\ell', \ell}} \right]_{1 \le \ell, \ell' \le L}$ are strictly positive. Let $\vec{j}_*, \vec{j} \in \mathbb{C}^L$ and set*

$$\mathcal{A}(\alpha^*, \alpha) = -\langle \vec{\alpha}^*, D\vec{\alpha} \rangle \qquad and \qquad \mathcal{J}(\vec{j}_*, \vec{j}) = \left\langle e^{\mathcal{S}(\vec{j}_*, \vec{j}, \alpha^*, \alpha)} \right\rangle.$$

Then

$$\mathcal{J}(\vec{j}_*, \vec{j}) = e^{\mathcal{C}(\vec{j}_*, \vec{j})} \qquad where \qquad \mathcal{C}(\vec{j}_*, \vec{j}) = \langle \vec{j}_*, D^{-1}\vec{j} \rangle.$$

(b) *Let D, V and W be $L \times L$ matrices with V and W self–transpose. That is, $V_{\ell, \ell'} = V_{\ell', \ell}$ and $W_{\ell, \ell'} = W_{\ell', \ell}$. Assume that the matrix*

$$\begin{bmatrix} \frac{1}{2}(D + D^*) & V + \overline{W} \\ W + \overline{V} & \frac{1}{2}(D + D^*)^t \end{bmatrix}$$

is strictly positive. Let $\vec{j}_*, \vec{j} \in \mathbb{C}^L$ and set

$$\mathcal{A}(\alpha^*, \alpha) = -\langle \vec{\alpha}^*, D\vec{\alpha} \rangle - \langle \vec{\alpha}^*, V\vec{\alpha}^* \rangle - \langle \vec{\alpha}, W\vec{\alpha} \rangle \qquad and$$

$$\mathcal{J}(\vec{j}_*, \vec{j}) = \left\langle e^{\mathcal{S}(\vec{j}_*, \vec{j}, \alpha^*, \alpha)} \right\rangle.$$

Then

$$\mathcal{J}(\vec{j}_*, \vec{j}) = e^{\mathcal{D}(\vec{j}_*, \vec{j})}$$

where

$$\mathcal{D}(\vec{j}_*,\vec{j}) = \left\langle \vec{j}_*, \left(D - 4V(D^t)^{-1}W\right)^{-1}\vec{j} \right\rangle - \left\langle WD^{-1}\vec{j}, \left(D - 4V(D^t)^{-1}W\right)^{-1}\vec{j} \right\rangle$$

$$- \left\langle \vec{j}_*, \left(D - 4V(D^t)^{-1}W\right)^{-1} V(D^t)^{-1}\vec{j}_* \right\rangle. \qquad (2.66)$$

In the special case that $V = W$ and V commutes with D, $\mathcal{D}$ simplifies to

$$\mathcal{D}(\vec{j}_*,\vec{j}) = \left\langle D\vec{j}_*, \left(DD^t - 4V^2\right)^{-1}\vec{j} \right\rangle - \left\langle V\vec{j}, \left(DD^t - 4V^2\right)^{-1}\vec{j} \right\rangle$$

$$- \left\langle V\vec{j}_*, \left(D^t D - 4V^2\right)^{-1}\vec{j}_* \right\rangle. \qquad (2.67)$$

Proof:

(a) The positivity condition on $D + D^*$ ensures that D is invertible. (Otherwise, there would be a nonzero vector $\vec{v}$ with $D\vec{v} = 0$ and hence $\left\langle \vec{v}^*, (D + D^*)\vec{v} \right\rangle = 0$, which would contradict the strict positivity of $D + D^*$.) We start by completing the square of the exponent in the numerator.

$$\mathcal{A}(\alpha^*,\alpha) + \mathcal{S}(\vec{j}_*,\vec{j},\alpha^*,\alpha) = -\left\langle \vec{\alpha}^*, D\vec{\alpha} \right\rangle + \left\langle \vec{j}_*, \alpha \right\rangle + \left\langle \vec{\alpha}^*, \vec{j} \right\rangle$$

$$= -\left\langle \left(\vec{\alpha}^* - (D^t)^{-1}\vec{j}_*\right), D\left(\vec{\alpha} - D^{-1}\vec{j}\right) \right\rangle + \left\langle \vec{j}_*, D^{-1}\vec{j} \right\rangle.$$

At this stage, we have that

$$\mathcal{J}(\vec{j}_*,\vec{j}) = e^{\mathcal{C}(\vec{j}_*,\vec{j})} \frac{\int \prod_{\ell=1}^{L} \frac{d\alpha_\ell^* \wedge d\alpha_\ell}{2\pi i}\, e^{-\left\langle (\vec{\alpha}^* - (D^t)^{-1}\vec{j}_*), D(\vec{\alpha} - D^{-1}\vec{j}) \right\rangle}}{\int \prod_{\ell=1}^{L} \frac{d\alpha_\ell^* \wedge d\alpha_\ell}{2\pi i}\, e^{-\left\langle \vec{\alpha}^*, D\vec{\alpha} \right\rangle}}. \qquad (2.68)$$

So it remains only to prove that the ratio of two integrals is exactly one. If $\vec{\alpha}$ and $\vec{\alpha}^*$ were independent integration variables, the change of variables $\vec{\alpha} \to \vec{\alpha} + D^{-1}\vec{j}$, $\vec{\alpha}^* \to \vec{\alpha}^* + (D^t)^{-1}\vec{j}_*$ would convert the numerator into exactly the integral that is in the denominator and we would be done. Unfortunately, $\vec{\alpha}$ and $\vec{\alpha}^*$ are not independent and, usually, $(D^t)^{-1}\vec{j}_*$ is not the complex conjugate of $D^{-1}\vec{j}$. Fortunately, with a little trickery, we can legitimately make the desired change of variables. First introduce a new, independent, vector of complex variables $\vec{\alpha}_*$. There is, in general, no requirement that $\vec{\alpha}_*$ be the complex conjugate of $\vec{\alpha}$. Replace all $\vec{\alpha}^*$'s in the integral of the numerator by $\vec{\alpha}_*$ and choose as the domain of integration

$$D_\alpha = \left\{ (\vec{\alpha}_*, \vec{\alpha}) \in \mathbb{C}^{2L} \mid \vec{\alpha} = \vec{\alpha}_*^* \right\}.$$

This recovers the original integral. That is,

$$\int_{\mathbb{C}^L} \prod_{\ell=1}^{L} \frac{d\alpha_\ell^* \wedge d\alpha_\ell}{2\pi i}\, e^{-\left\langle (\vec{\alpha}^* - (D^t)^{-1}\vec{j}_*), D(\vec{\alpha} - D^{-1}\vec{j}) \right\rangle}$$

$$= \int_{D_\alpha} \prod_{\ell=1}^{L} \frac{d\alpha_{*\ell} \wedge d\alpha_\ell}{2\pi i}\, e^{-\left\langle (\vec{\alpha}_* - (D^t)^{-1}\vec{j}_*), D(\vec{\alpha} - D^{-1}\vec{j}) \right\rangle}.$$

Now make the change of variables

$$\vec{\alpha}_* = \vec{z}_* + (D^t)^{-1}\vec{j}_* \qquad \vec{\alpha} = \vec{z} + D^{-1}\vec{j}.$$

This gives

$$\int \prod_{\ell=1}^{L} \frac{d\alpha_\ell^* \wedge d\alpha_\ell}{2\pi i} \, e^{-<(\vec{\alpha}^* - (D^t)^{-1}\vec{j}_*)\,,\,D\,(\vec{\alpha} - D^{-1}\vec{j})>} = \int_{D_1} \prod_{\ell=1}^{L} \frac{dz_{*\ell} \wedge dz_\ell}{2\pi i} \, e^{-<\vec{z}_*\,,\,D\,\vec{z}>}$$

with the domain

$$D_1 = \{ \, (\vec{z}_*, \vec{z}) \in \mathbb{C}^{2L} \mid \vec{z} = \vec{z}_*^* + \vec{\rho} \, \} \qquad \text{with } \vec{\rho} = (D^*)^{-1}\vec{j}_*^* - D^{-1}\vec{j}.$$

In the next paragraph, we will use Stokes' theorem to show that we may replace the domain D_1 with the domain

$$D_0 = \{ \, (\vec{z}_*, \vec{z}) \in \mathbb{C}^{2L} \mid \vec{z} = \vec{z}_*^* \, \}.$$

Once that is done, we will have shown, this time legitimately, that the integral of the numerator in (2.68) is the same as the integral of the denominator, completing the proof.

Here are the details of the application of Stokes' theorem. Let R be a large cutoff radius and define, for each $0 \leq t \leq 1$,

$$D_{t,R} = \{ \, (\vec{z}_*, \vec{z}) \in \mathbb{C}^{2L} \mid \vec{z} = \vec{z}_*^* + t\vec{\rho}, \ \max_{1 \leq \ell \leq L} |z_\ell| \leq R \, \}.$$

Think of $\mathcal{B} = \bigcup_{t=0}^{1} D_t$ as a solid 'cylinder'. The boundary of $\mathcal{B}$ is the union of the top $D_{1,R}$ (which approachs D_1 as $R \to \infty$) and the bottom $D_{0,R}$ (which approachs D_0 as $R \to \infty$) and the side

$$C_R = \bigcup_{\ell=1}^{L} C_{R,\ell} \quad \text{with} \quad C_{R,\ell} = \bigcup_{0 \leq t \leq 1} \{ \, (\vec{z}_*, \vec{z}) \in \mathbb{C}^{2L} \mid \vec{z} = \vec{z}_*^* + t\vec{\rho},$$

$$\max_{1 \leq \ell' \leq L} |z_{\ell'}| \leq R, \ |z_\ell| = R \, \}.$$

By Stokes' theorem, for any $2L$–form ω,

$$\int_{\mathcal{B}} d\omega = \int_{D_{1,R}} \omega - \int_{D_{0,R}} \omega + \int_{C_R} \omega$$

if $D_{1,R}$ and $D_{0,R}$ are oriented in the usual way and C_R is oriented suitably. In our case, the form $\omega = \bigwedge_{\ell=1}^{L} \frac{dz_{*\ell} \wedge dz_\ell}{2\pi i} \, e^{-<\vec{z}_*\,,\,D\,\vec{z}>}$ obeys $d\omega = 0$ (i.e. is closed) because $e^{-<\vec{z}_*\,,\,D\,\vec{z}>}$ is an analytic function of $\vec{z}_*$ and $\vec{z}$. Hence

$$\int_{D_{1,R}} \omega = \int_{D_{0,R}} \omega - \int_{C_R} \omega.$$

So we just have to show that $\int_{C_R} \omega$ converges to zero as $R \to \infty$. We start by bounding the integrand, or rather the real part of the exponent of the integrand. At any point on the side, C_R, $R \le |\vec{z}| \le \sqrt{L}R$ and there is a $0 \le t \le 1$ such that $\vec{z}_* = \vec{z}^* - t\vec{\rho}^*$ so that

$$\text{Re}\,\langle \vec{z}_*, D\vec{z} \rangle = \text{Re}\,\langle \vec{z}^*, D\vec{z} \rangle - t\text{Re}\,\langle \vec{\rho}^*, D\vec{z} \rangle \tag{2.69}$$

$$= \frac{1}{2}\langle \vec{z}^*, (D + D^*)\vec{z} \rangle - t\text{Re}\,\langle \vec{\rho}^*, D\vec{z} \rangle$$

$$\ge \frac{1}{2}\lambda_0 R^2 - \sqrt{L}\,|\vec{\rho}|\,\|D\|\,R$$

where λ_0 is the smallest eigenvalue of $D + D^*$, assumed strictly positive, and $\|D\|$ is the operator norm of the matrix D.

We next bound the volume of the domain of integration. It suffices to do so for $C_{R,1}$. The other $C_{R,\ell}$'s can be treated in the same way. On $C_{R,1}$, we have $|z_1| = R$. We may parametrize $z_1 = Re^{i\theta}$, with θ running over $[0, 2\pi]$. Then $z_{*,1} = Re^{-i\theta} - t\rho_1^*$ and

$$dz_1 = iRe^{i\theta}\,d\theta \qquad dz_{*,1} = -iRe^{-i\theta}\,d\theta - \rho_1^*\,dt \qquad \frac{dz_{*,1} \wedge dz_1}{2\pi i} = -\frac{\rho_1^* R}{2\pi}e^{i\theta}\,dt \wedge d\theta.$$

For all the other ℓ's, we may parametrize $z_\ell = x_\ell + iy_\ell$ with (x_ℓ, y_ℓ) running over $x_\ell^2 + y_\ell^2 \le R^2$. Then $z_{*,\ell} = x_\ell - iy_\ell - t\rho_\ell^*$ and

$$dz_\ell = dx_\ell + idy_\ell \qquad dz_{*,\ell} = dx_\ell - idy_\ell - \rho_\ell^*\,dt.$$

Since there is already a dt in $\frac{dz_{*,1} \wedge dz_1}{2\pi i}$ and $dt \wedge dt = 0$,

$$\bigwedge_{\ell=1}^{L} \frac{dz_{*\ell} \wedge dz_\ell}{2\pi i} = -\frac{\rho_1^* R}{2\pi}e^{i\theta}\,dt \wedge d\theta \bigwedge_{\ell=2}^{L} \frac{dx_\ell \wedge dy_\ell}{\pi}$$

and

$$\left| \int_{C_{R,1}} \omega \right| \le |\rho_1|R \int_0^1 dt \int_0^{2\pi} \frac{d\theta}{2\pi} \iint_{x_2^2 + y_2^2 \le R^2} \frac{dx_2\,dy_2}{2\pi} \cdots$$

$$\iint_{x_L^2 + y_L^2 \le R^2} \frac{dx_L\,dy_L}{2\pi}\,\sup e^{-\text{Re}\,\langle \vec{z}_*, D\vec{z} \rangle}$$

$$\le |\rho_1|R\left(\frac{1}{2}R^2\right)^{L-1} e^{-\frac{1}{2}\lambda_0 R^2 + \sqrt{L}\,|\vec{\rho}|\,\|D\|\,R}$$

This easily converges to zero as $R \to \infty$.

(b) Once again, the main step is completing the square for the exponent of the numerator. We start my multiplying out

$$-\langle(\vec{\alpha}^* - \vec{J}_*), D(\vec{\alpha} - \vec{J})\rangle - \langle(\vec{\alpha}^* - \vec{J}_*), V(\vec{\alpha}^* - \vec{J}_*)\rangle - \langle(\vec{\alpha} - \vec{J}), W(\vec{\alpha} - \vec{J})\rangle$$

$$= -\langle\vec{\alpha}^*, D\vec{\alpha}\rangle - \langle\vec{\alpha}^*, V\vec{\alpha}^*\rangle - \langle\vec{\alpha}, W\vec{\alpha}\rangle + \langle\vec{\alpha}^*, (D\vec{J} + 2V\vec{J}_*)\rangle$$

$$+ \langle(D^t\vec{J}_* + 2W\vec{J}), \vec{\alpha}\rangle - \langle\vec{J}_*, D\vec{J}\rangle - \langle\vec{J}_*, V\vec{J}_*\rangle - \langle\vec{J}, W\vec{J}\rangle$$

The first three terms are exactly $\mathcal{A}(\vec{\alpha}^*, \vec{\alpha})$. The next two terms will form $\mathcal{S}(\vec{J}_*, \vec{J}, \vec{\alpha}^*, \vec{\alpha})$ exactly provided

$$D\vec{J} + 2V\vec{J}_* = \vec{J}$$

$$D^t\vec{J}_* + 2W\vec{J} = \vec{J}_*.$$

Solving this pair of linear equations gives

$$\vec{J}_* = (D^t - 4WD^{-1}V)^{-1}(\vec{J}_* - 2WD^{-1}\vec{J})$$

$$\vec{J} = (D - 4V(D^t)^{-1}W)^{-1}(\vec{J} - 2V(D^t)^{-1}\vec{J}_*).$$

Substituting this in,

$$\langle\vec{J}_*, D\vec{J}\rangle + \langle\vec{J}_*, V\vec{J}_*\rangle + \langle\vec{J}, W\vec{J}\rangle = \frac{1}{2}\langle\vec{J}_*, \vec{J}\rangle + \frac{1}{2}\langle\vec{J}_*, \vec{J}\rangle$$

$$= \frac{1}{2}\langle(\vec{J}_* - 2WD^{-1}\vec{J}), (D - 4V(D^t)^{-1}W)^{-1}\vec{J}\rangle$$

$$+ \frac{1}{2}\langle\vec{J}_*, (D - 4V(D^t)^{-1}W)^{-1}(\vec{J} - 2V(D^t)^{-1}\vec{J}_*)\rangle$$

$$= \langle\vec{J}_*, (D - 4V(D^t)^{-1}W)^{-1}\vec{J}\rangle - \langle WD^{-1}\vec{J}, (D - 4V(D^t)^{-1}W)^{-1}\vec{J}\rangle$$

$$- \langle\vec{J}_*, (D - 4V(D^t)^{-1}W)^{-1}V(D^t)^{-1}\vec{J}_*\rangle$$

$$= \mathcal{D}(\vec{J}_*, \vec{J}).$$

Thus

$$\mathcal{A}(\vec{\alpha}^*, \vec{\alpha}) + \mathcal{S}(\vec{J}_*, \vec{J}, \vec{\alpha}^*, \vec{\alpha}) = \mathcal{A}(\vec{\alpha}^* - \vec{J}_*, \vec{\alpha} - \vec{J}) + \mathcal{D}(\vec{J}_*, \vec{J}).$$

The rest of the proof is very much like that of part (a), using in place of $\mathrm{Re}\langle\vec{z}^*, D\vec{z}\rangle = \frac{1}{2}\langle\vec{z}^*, (D + D^*)\vec{z}\rangle$ in the bound (2.69),

$$\mathrm{Re}\left\{\langle\vec{z}^*, D\vec{z}\rangle + \langle\vec{z}^*, V\vec{z}^*\rangle + \langle\vec{z}, W\vec{z}\rangle\right\} = \frac{1}{2}[(\vec{z}^*)^* \vec{z}^t]$$

$$\begin{bmatrix} \frac{1}{2}(D + D^*) & V + \overline{W} \\ W + \overline{V} & \frac{1}{2}(D + D^*)^t \end{bmatrix} \begin{bmatrix} \vec{z} \\ \vec{z}^* \end{bmatrix}.$$

□

References

Abdesselam, A. and Rivasseau, V. (1995). Trees, forests and jungles: a botanical garden for cluster expansions. In V. Rivasseau (ed.) *Constructive Physics (Palaiseau, 1994), Lecture Notes in Physics*, 446, pp. 7–36. Springer.

Balaban, T. (1995a). A low temperature expansion and 'spin wave picture' for classical *N*-vector models. In *Constructive Physics (Palaiseau, 1994), Lecture Notes in Physics*, 446, pp. 201–218. Springer.

Balaban, T. (1995b). A low temperature expansion for classical *N*-vector models. I. A renormalization group flow. *Comm. Math. Phys.*, **167**: 103–154.

Balaban, T. (1996a). Localization expansions. I. Functions of the 'background' configurations. *Comm. Math. Phys.*, **182**: 33–82.

Balaban, T. (1996b). A low temperature expansion for classical *N*-vector models. II. Renormalization group equations. *Comm. Math. Phys.*, **182**: 675–721.

Balaban, T. (1996c). The variational problems for classical *N*-vector models. *Comm. Math. Phys.*, **175**: 607–642.

Balaban, T. (1998a). The large field renormalization operation for classical *N*-vector models. *Comm. Math. Phys.*, **198**: 493–534.

Balaban, T. (1998b). A low temperature expansion for classical *n*-vector models. III. A complete inductive description, fluctuation integrals. *Comm. Math. Phys.*, **196**: 485–521.

Balaban, T. (1998c). Renormalization and localization expansions. II. Expectation values of the 'fluctuation' measures. *Comm. Math. Phys.*, **198**: 1–45.

Balaban, T., Feldman, J., Knörrer, H., and Trubowitz, E. (2008a). A functional integral representation for many boson systems. I: The partition function. *Annales Henri Poincaré*, **9**: 1229–1273.

Balaban, T., Feldman, J., Knörrer, H., and Trubowitz, E. (2008b). A functional integral representation for many boson systems. II: Correlation functions. *Annales Henri Poincaré*, **9**: 1275–1307.

Balaban, T., Feldman, J., Knörrer, H., and Trubowitz, E. (2009). Power series representations for bosonic effective actions. *Journal of Statistical Physics*, **134**: 839–857.

Balaban, T., Feldman, J., Knörrer, H., and Trubowitz, E. (2010a). Power series representations for complex bosonic effective actions. I. a small field renormalization group step. *Journal of Mathematical Physics*, **51**: 053305.

Balaban, T., Feldman, J., Knörrer, H., and Trubowitz, E. (2010b). Power series representations for complex bosonic effective actions. II. a small field renormalization group flow. *Journal of Mathematical Physics*, **51**: 053306.

Balaban, T., Feldman, J., Knörrer, H., and Trubowitz, E. (2010c). The temporal ultraviolet limit for complex bosonic many-body models. *Annales Henri Poincaré*, **11**: 151–350.

Berezin, F. A. (1966). The method of second quantization. In *Pure Appl. Phys., vol 24*. Academic Press.

Brydges, D. C. (1986). A short course on cluster expansions. In K. Osterwalder and R. Stora (ed.) *Phénomènes critiques, systèmes aléatoires, théories de jauge (Les Houches, 1984)*, pp. 129–183. North-Holland.

Brydges, D. C. and Federbush, P. (1976). The cluster expansion in statistical physics. *Commun. Math. Phys.*, **49**: 233–246.

Brydges, D. C. and Federbush, P. (1977). The cluster expansion for potentials with exponential fall-off. *Commun. Math. Phys.*, **53**: 19–30.

Cameron, R. H. (1960). A family of integrals serving to connect the Wiener and Feynman integrals. *J. Math. and Phys.*, **39**: 126–140.

Cameron, R. H. (1962/1963). The Ilstow and Feynman integrals. *J. Analyse Math.*, **10**: 287–361.

Cammarota, C. (1982). Decay of correlations for infinite range interactions in unbounded spin systems. *Comm. Math. Phys.*, **85**: 517–528.

Durrett, R. (2010). *Probability: Theory and examples.* Cambridge University Press.

Feldman, J., Knörrer, H., and Trubowitz, E. (2002). Fermionic functional integrals and the renormalization group. In *CRM Monograph Series of the American Mathematical Society, 16.* AMS.

Feldman, J. and Osterwalder, K. (1976). The Wightman axioms and the mass gap for weakly coupled $\left(\varphi^4\right)_3$ quantum field theories. *Annals of Physics*, **97**: 80–135.

Fetter, A.L. and Walecka, J.D. (1971). *Quantum Theory of Many-Particle Systems.* McGraw-Hill.

Fröhlich, J. (1974). Verification of axioms for euclidean and relativistic fields and Haag's theorem in a class of $P(\phi)_2$ models. *Annales Henri Poincaré*, **21**: 271–317.

Gel'fand, I. M. and Vilenkin, N.Ya. (1968). *Generalized Functions. Applications of Harmonic Analysis, Vol 4.* Acad. Press.

Ginibre, J. (1965). Reduced density matrices of quantum gases. I. Limit of infinite volume. *J. Math. Phys.*, **6**: 238–251.

Ginibre, J. (1971). Some applications of functional integration in statistical mechanics. In C. DeWitt and R. Stora (ed.) *Statistical Mechanics and Quantum Field Theory (Les Houches, 1970)*, pp. 327–427. Gordon and Breach.

Glimm, J. and Jaffe, A. (1987). *Quantum Physics: A Functional Integral Point of View.* Springer-Verlag.

Glimm, J., Jaffe, A., and Spencer, T. (1973). The particle structure of the weakly coupled $P(\phi)_2$ model and other applications of high temperature expansions, Part II: The cluster expansion. In G. Velo and A. S. Wightman (ed.) *Constructive Quantum Field Theory (Erice 1973), Lecture Notes in Physics, Vol 25*, pp. 199–242. Springer.

Kotecký, R. and Preiss, D. (1986). Cluster expansion for abstract polymer models. *Commun. Math. Phys.*, **103**: 491.

Minlos, R. A. (1959). Generalized random processes and their extension to a measure. *Trudy Moskov. Mat. Obšč*, **8**: 497–518.

Negele, J. W. and Orland, H. (1988). *Quantum Many-Particle Systems.* Addison–Wesley.

Nelson, E. (1964). Feynman integrals and the Schrödinger equation. *Journal of Mathematical Physics*, **5**: 332–343.

Pordt, A. (1998). Polymer expansions in particle physics. In H. Meyer-Ortmanns and A. Klümper *Field Theoretical Tools for Polymer and Particle Physics (Lecture Notes in Physics, vol. 508)*, pp. 45–67. Springer.

Reed, M. and Simon, B. (1978). *Methods of Modern Mathematical Physics, IV: Analysis of Operators*. Academic Press.

Rivasseau, V. (1991). *From Perturbative to Constructive Renormalization*. Princeton University Press.

Rota, G.-C. (1964). On the foundations of combinatorial theory, I. Theory of Möbius functions. *Z. Wahrscheinlichkeitstheorie Verw. Gebiete*, **2**: 340.

Salmhofer, M. (1999). Renormalization, an introduction. In *Texts and Monographs in Physics*. Springer.

Simon, B. (1993). *The Statistical Mechanics of Lattice Gases, Volume 1*. Princeton University Press.

Simon, B. (2005). *Functional Integration and Quantum Physics*. AMS Chelsea Publishing.

3
Locality in quantum systems

Matthew B. HASTINGS

Microsoft Research Station Q,
Elings Hall, University of California
Santa Barbara, CA 93106, USA

Chapter Contents

Preface

These lecture notes focus on the application of ideas of locality, in particular Lieb–Robinson bounds, to quantum many-body systems. We consider applications including correlation decay, topological order, a higher dimensional Lieb–Schultz–Mattis theorem, and a nonrelativistic Goldstone theorem. The emphasis is on trying to show the ideas behind the calculations. As a result, the proofs are only sketched with an emphasis on the intuitive ideas behind them, and in some cases we use techniques that give very slightly weaker bounds for simplicity.

3.1 Introduction and notation

The basic problem studied in quantum many-body theory is to find the properties of the ground state of a given Hamiltonian. Expressed mathematically, the Hamiltonian H is a Hermitian matrix, and the ground state, which we write ψ_0, is the eigenvector of this matrix with the lowest eigenvalue. The properties we are interested in studying are expectation values of various observables: given a Hermitian matrix O, we would like to compute the expectation value $\langle \psi_0, O\psi_0 \rangle$. A note on notation: we write an inner product of two vectors as $\langle v, w \rangle$, rather than the notation $\langle v|w \rangle$, as the first notation is more common in math and also will be more clear given the large number of absolute value signs we will also use later). Sometimes we will write the expectation value in the ground state simply as $\langle O \rangle$.

The above paragraph sets out some of the mathematics of quantum mechanics, but it must seem quite dry. Someone reading that who is not familiar with quantum theory could be excused for thinking that the problem is essentially one of linear algebra, and that the basic tool employed by quantum physicists is a linear algebra package to diagonalize large matrices. In fact, while the method of 'exact diagonalization' on a computer is an important technique in studying quantum systems, it is only a small part of how physical problems are studied. The feature that distinguishes the study of quantum many-body systems is *locality* of interactions. Consider a typical Hamiltonian, such as the one-dimensional transverse field Ising model:

$$H = -J \sum_{i=1}^{N-1} S_i^z S_{i+1}^z + B \sum_{i=1}^{N} S_i^x. \tag{3.1}$$

The very notation we employ to describe this Hamiltonian implicitly assumes local interactions. To be more precise, throughout these notes, we have in mind quantum systems on a finite size lattice. We associate a D dimensional Hilbert space with each lattice site, and the Hilbert space of the whole system is the tensor product of these spaces. We use N to represent the number of lattice sites, so that the Hilbert space on which a Hamiltonian such as (3.1) is defined is D^N dimensional. A term such as $S_i^z S_{i+1}^z$ is a short-hand notation for the term $I_1 \otimes I_2 \otimes ... \otimes I_{i-1} \otimes S_i^z \otimes S_{i+1}^z \otimes I_{i+2} \otimes ... \otimes I_N$, where I_j is the identity operator on site j. This local structure of the interaction terms will greatly constrain the properties of ψ_0, in particular if there is a spectral gap.

We will usually use symbols $i, j, k, \dots$ to denote lattice sites and we use letters $X, Y, Z, \dots$ to denote sets of lattice sites. We use Λ to denote the set of all lattice sites. We use $|X|$ to denote the cardinality of a set X.

We say that an operator O is 'supported on set A' if we can write O as a tensor product of two operators:

$$O = I_{\Lambda \setminus A} \otimes P, \tag{3.2}$$

where $I_{\Lambda \setminus A}$ is the identity operator on the sites not in set A (note that $\Lambda \setminus A$ denotes the set of sites not in A) and P is some operator defined on the $D^{|A|}$ dimensional Hilbert space on set A. For example, the operator $S_i^z S_{i+1}^z$ is supported on the set of sites $\{i, i+1\}$. Colloquiually, instead of saying that O is supported on set A, one often hears that O 'acts on set A', although we will avoid that terminology.

Finally, we define $\|O\|$ to represent the 'operator norm' of an operator O. If O is Hermitian, the operator norm is equal to the absolute value of the largest eigenvalue of O. For arbitrary operators O, we define

$$\|O\| = \max_{\psi, |\psi|=1} |O\psi|, \tag{3.3}$$

where the maximum is taken over vectors ψ with norm 1, and $|O\psi|$ denotes the norm of the vector $O\psi$.

Using two simple assumptions, that the Hamiltonian has local interactions and that the Hamiltonian has a spectral gap, we will be able to prove a wide variety of results about the ground state of the Hamiltonian. We will consider two different cases of a spectral gap. In one case, we consider a Hamiltonian with a unique ground state, ψ_0, which has energy E_0, with the next lowest energy state having energy E_1 with $E_1 - E_0 \geq \Delta E$. Here, ΔE is called the 'spectral gap'. In the other case, we consider a Hamiltonian with several degenerate or approximately degenerate low energy states, ψ_0^a, for $a = 1, \dots, k$. Here, k is the number of different low energy states. Let these states have energies E_0^a. Then, assume that there is a gap ΔE separating these ground states from higher energy states. That is, the $k + 1$-st smallest eigenvalue of the Hamiltonian is greater than or equal to $E_0^a + \Delta E$ for all $a = 1, \dots, k$.

We will begin by considering properties of correlations in these systems, and prove an exponential decay of correlation functions. We then consider how the ground state of the system changes under a change in the Hamiltonian, and use this to prove a non-relativistic variant of Goldstone's theorem. Finally, we apply these techniques to more interesting systems with topological order, beginning with the Lieb–Schultz–Mattis theorem. However, before we can do any of this, we need to more precisely define the locality properties of the Hamiltonian and to prove a set of bounds on the propagation of information through the system called 'Lieb–Robinson bounds'.

3.2   Locality and Lieb–Robinson bounds

To define locality, we need a notion of distance. We introduce a metric on the lattice, $\text{dist}(i, j)$. For example, on a square lattice one may consider a Manhattan metric or a Euclidean metric. In one dimension on a system with open boundary conditions, it is natural to consider the metric $\text{dist}(i,j) = |i - j|$, while for a one dimensional system with periodic boundary conditions and N sites it is natural to consider the metric $\text{dist}(i,j) = \min_n |i - j + nN|$, where the minimum ranges over integers n. We define the distance between two sets A, B to be

$$\text{dist}(A, B) = \min_{i \in A, j \in B} \text{dist}(i,j). \tag{3.4}$$

We define the diameter of a set A by

$$\text{diam}(A) = \max_{i,j \in A} \text{dist}(i,j). \tag{3.5}$$

We consider Hamiltonians

$$H = \sum_Z H_Z, \tag{3.6}$$

where H_Z is supported on a set Z. We will be interested in studying problems in which $\|H_Z\|$ decays rapidly with the diameter of the set Z. For example, in the one-dimensional transverse field Ising model Hamiltonian (3.1), with metric $\text{dist}(i,j) = |i - j|$, the only terms that appears have diameter 0 or 1 (the magnetic field term has diameter zero and the Ising interaction has diameter one), so $\|H_Z\| = 0$ for $\text{diam}(Z) > 1$. These interactions are 'finite range'; we will also be able to consider cases in which the interactions have a slower decay, such as an exponential decay (one can prove various results about power law decay also, which we will not consider in these notes).

Note that the reason one chooses one particular metric over another is to make the terms in the Hamiltonian local with respect to that metric. Note that Hamiltonian (3.1) has open boundary conditions, in that site N does not interact with site 1. Instead, if we considered a one-dimensional transverse field Ising

$$H = -J \sum_{i=1}^{N-1} S_i^z S_{i+1}^z - J S_N^z S_1^z + B \sum_{i=1}^{N} S_i^x, \tag{3.7}$$

with periodic boundary conditions, the term $S_N^z S_1^z$ would have diameter $N - 1$ with respect to the metric $\text{dist}(i,j) = |i - j|$. However, if we instead consider the metric $\text{dist}(i,j) = \min_n |i - j + nN|$ suggested above, then all of the interactions have diameter 0 or 1 again.

As an aside, one might wonder why we do not just pick the metric $\text{dist}(i,j) = 0$ for all i,j. In this case, all of the interactions in the Hamiltonian have diameter zero and are therefore 'local'. However, the kinds of results we will prove below have to do with exponential decay of various quantities with distance (for example, decay of correlations as a function of spacing between operators), and all of the results would become trivial in the case of the metric $\text{dist}(i,j) = 0$. Thus, we want to pick a metric such that our Hamiltonian is local, but such that the set of sites Λ has a large diameter.

It is not necessary that the system be defined on a regular lattice. For example, we could imagine a graphical structure to the interactions. Suppose each site is associated with the vertex of a graph, and the Hamiltonian is a sum of terms H_Z where each Z contains only two sites and H_Z is non-vanishing only if there is an edge of the graph connecting the sites in Z. Then, the shortest path metric on the graph gives us a metric for which H_Z is non-vanishing only if Z has diameter 0 or 1.

We now consider the Lieb–Robinson bounds. These are bounds describing time evolution of operators in a local Hamiltonian. While most of our interest in these notes is in static properties of systems (such as correlation functions at a given instant in time), it turns out that an understanding of the dynamical properties as a function of time is very useful to prove results about the statics. The bounds were first proven in (Lieb and Robinson 1972). Then, a proof which does not involve the dimension of the Hilbert space dimension on a given site was given in (Hastings 2004). More general lattices were considered in (Nachtergaele and Sims 2006), but only using a proof which depended on the Hilbert space dimension again. The dimension independent proof for arbitrary lattices was given in (Hastings and Koma 2006) and this is what we follow.

Various sorts of Lieb–Robinson bounds can be proven. We consider first the case of exponentially decaying interactions in some detail, then we consider either kinds of decay in the interaction (either finite range interactions or other types of decay). We define the time-dependence of operators by the Heisenberg evolution:

$$O(t) \equiv \exp(iHt)O\exp(-iHt). \tag{3.8}$$

Then,

Theorem 3.1 *Suppose for all sites i, the following holds:*

$$\sum_{X \ni i} \|H_X\| |X| \exp[\mu \operatorname{diam}(X)] \leq s < \infty, \tag{3.9}$$

for some positive constants μ, s. Let A_X, B_Y be operators supported on sets X, Y, respectively. Then, if $\operatorname{dist}(X, Y) > 0$,

$$\|[A_X(t), B_Y]\| \leq 2\|A_X\| \|B_Y\| \sum_{i \in X} \exp[-\mu \operatorname{dist}(i, Y)] \left[e^{2s|t|} - 1 \right] \tag{3.10}$$

$$\leq 2\|A_X\| \|B_Y\| |X| \exp[-\mu \operatorname{dist}(X, Y)] \left[e^{2s|t|} - 1 \right].$$

Before proving the theorem, let us describe the physical meaning of this theorem. Suppose A_X is some operator. Let $B_l(X)$ denote the ball of radius l about set X. That is, $B_l(X)$ is the set of sites i, such that $\operatorname{dist}(i, X) \leq l$. Following (Bravyi, Hastings, and Verstraete 2006), define

$$A_X^l(t) = \int dU \, U A_X(t) U^\dagger, \tag{3.11}$$

where the integral is over unitaries supported on the set of sites $\Lambda \setminus B_l(X)$ with the Haar measure. Then, A_X^l is supported on $B_l(X)$. Since $U A_X(t) U^\dagger = A_X(t) + U[A_X(t), U^\dagger]$, we have

$$\|A_X^l(t) - A_X(t) \le \int dU \|[A_X, U]\|. \tag{3.12}$$

Using the Lieb–Robinson bound (3.10) to bound the right-hand side of the above equation, we see that $A_X^l(t)$ is exponentially close to A_X if l is sufficiently large compared to $2st/\mu$. Thus, to exponential accuracy, we can approximate a time-evolved operator such as $A_X(t)$ by an operator supported on the set $B_l(X)$. That is, the 'leakage' of the operator outside the light-cone is small.

Another comment regarding the assumptions on the theorem: because the term $\exp[\mu\,\mathrm{diam}(Z)]$ grows exponentially in Z, we need the norm of the terms H_Z to decay exponentially in Z. On finite dimensional lattices (such as a hypercubic lattice), the cardinality of Z, $|Z|$, is bounded by a power in the diameter of Z. So, on such lattices, if the terms H_Z in the Hamiltonian decay exponentially in $\mathrm{diam}(Z)$, we can find a μ such that the assumptions of the theorem are satisfied. By expressing the theorem as we have, however, the theorem can be applied to models defined on, for example, arbitrary graphs, where the cardinality of a set might not be bounded by a power of its diameter.

One final remark: we have implicitly assumed that all of the operators are bosonic, in that operators supported on disjoint sets commute with each other. One can straightforwardly generalize all of this to the case of fermionic operators also, so that two fermionic operators which are supported on disjoint sets anti-commute but that two bosonic operators, or one bosonic and one fermionic operator, commute on disjoint sets. In this case, one can instead prove a bound on the anti-commutator of two fermionic operators at different times.

3.2.1 Proof of Lieb–Robinson bound

The proof of the Lieb–Robinson bound we now give can be straightforwardly adapted to time-dependent Hamiltonians, though we only present the proof in the time-independent case for simplicity of notation.

Recall that we assume that $|\Lambda|$ is finite. If it is necessary to consider the infinite volume limit, we take the limit after deriving the desired Lieb–Robinson bounds which hold uniformly in the size of the lattice. Our essential tool is the series expansion (3.27) below for the commutator $[A(t), B]$. Let A be supported on X and B be supported on Y. We assume $t > 0$ because negative t can be treated in the same way. Let $\epsilon = t/N$ with a large positive integer N, and let

$$t_n = \frac{t}{N}n \quad \text{for } n = 0, 1, \ldots, N. \tag{3.13}$$

Then we have

$$\|[A(t), B]\| - \|[A(0), B]\| = \sum_{i=0}^{N-1} \epsilon \times \frac{\|[A(t_{n+1}), B]\| - \|[A(t_n), B]\|}{\epsilon}. \tag{3.14}$$

In order to obtain the bound (3.22) below, we want to estimate the summand in the right-hand side. To begin with, we note that the identity, $\|U^*OU\| = \|O\|$, holds for any observable O and for any unitary operator U. Using this fact, we have

$$\|[A(t_{n+1}), B]\| - \|[A(t_n), B]\| = \|[A(\epsilon), B(-t_n)]\| - \|[A, B(-t_n)]\|$$

$$\leq \|[A + i\epsilon[H_\Lambda, A], B(-t_n)]\| - \|[A, B(-t_n)]\| + \mathcal{O}(\epsilon^2)$$

$$= \|[A + i\epsilon[I_X, A], B(-t_n)]\| - \|[A, B(-t_n)]\| + \mathcal{O}(\epsilon^2)$$

$$\tag{3.15}$$

with

$$I_X = \sum_{Z: Z \cap X \neq \emptyset} H_Z, \tag{3.16}$$

where we have used

$$A(\epsilon) = A + i\epsilon[H_\Lambda, A] + \mathcal{O}(\epsilon^2) \tag{3.17}$$

and a triangle inequality. Further, by using

$$A + i\epsilon[I_X, A] = e^{i\epsilon I_X} A e^{-i\epsilon I_X} + \mathcal{O}(\epsilon^2), \tag{3.18}$$

we have

$$\|[A + i\epsilon[I_X, A], B(-t_n)]\| \leq \|[e^{i\epsilon I_X} A e^{-i\epsilon I_X}, B(-t_n)]\| + \mathcal{O}(\epsilon^2)$$

$$= \|[A, e^{-i\epsilon I_X} B(-t_n) e^{i\epsilon I_X}]\| + \mathcal{O}(\epsilon^2)$$

$$\leq \|[A, B(-t_i) - i\epsilon[I_X, B(-t_n)]]\| + \mathcal{O}(\epsilon^2)$$

$$\leq \|[A, B(-t_n)]\| + \epsilon \|[A, [I_X, B(-t_n)]]\| + \mathcal{O}(\epsilon^2). \quad (3.19)$$

Substituting this into the right-hand side in the last line of (3.15), we obtain

$$\|[A(t_{n+1}), B]\| - \|[A(t_n), B]\| \leq \epsilon \|[A, [I_X, B(-t_n)]]\| + \mathcal{O}(\epsilon^2)$$

$$\leq 2\epsilon \|A\| \|[I_X(t_n), B]\| + \mathcal{O}(\epsilon^2). \quad (3.20)$$

Further, substituting this into the right-hand side of (3.14) and using (3.16), we have

$$\|[A(t), B]\| - \|[A(0), B]\| \leq 2\|A\| \sum_{n=0}^{N-1} \epsilon \times \|[I_X(t_n), B]\| + \mathcal{O}(\epsilon)$$

$$\leq 2\|A\| \sum_{Z: Z \cap X \neq \emptyset} \sum_{n=0}^{N-1} \epsilon \times \|[H_Z(t_n), B]\| + \mathcal{O}(\epsilon). \quad (3.21)$$

Since $H_Z(t)$ is a continuous function of the time t for a finite volume, the sum in the right-hand side converges to the integral in the limit $\epsilon \downarrow 0$ $(N \uparrow \infty)$ for any fixed finite lattice Λ. In consequence, we obtain

$$\|[A(t), B]\| - \|[A(0), B]\| \leq 2\|A\| \sum_{Z:Z\cap X\neq\emptyset} \int_0^{|t|} ds\, \|[H_Z(s), B]\|. \tag{3.22}$$

We define

$$C_B(X, t) := \sup_{A\in\mathcal{A}_X} \frac{\|[A(t), B]\|}{\|A\|}, \tag{3.23}$$

where $\mathcal{A}_X$ is the set of observables supported on the set X. Then we have

$$C_B(X, t) \leq C_B(X, 0) + 2 \sum_{Z:Z\cap X\neq\emptyset} \|H_Z\| \int_0^{|t|} ds\, C_B(Z, s) \tag{3.24}$$

from the above bound (3.22). Assume $\mathrm{dist}(X, Y) > 0$. Then we have $C_B(X, 0) = 0$ from the definition of $C_B(X, t)$, and note that

$$C_B(Z, 0) \leq 2\|B\|, \tag{3.25}$$

for $Z \cap Y \neq \emptyset$ and

$$C_B(Z, 0)) = 0 \tag{3.26}$$

otherwise. Using these facts and the above bound (3.24) iteratively, we obtain

$$
\begin{aligned}
C_B(X, t) &\leq 2 \sum_{Z_1:Z_1\cap X\neq\emptyset} \|H_{Z_1}\| \int_0^{|t|} ds_1\, C_B(Z_1, s_1) \\
&\leq 2 \sum_{Z_1:Z_1\cap X\neq\emptyset} \|H_{Z_1}\| \int_0^{|t|} ds_1\, C_B(Z_1, 0) \\
&\quad + 2^2 \sum_{Z_1:Z_1\cap X\neq\emptyset} \|H_{Z_1}\| \sum_{Z_2:Z_2\cap Z_1\neq\emptyset} \|H_{Z_2}\| \int_0^{|t|} ds_1 \int_0^{|s_1|} ds_2\, C_B(Z_2, s_2) \\
&\leq 2\|B\|(2|t|) \sum_{Z_1:Z_1\cap X\neq\emptyset, Z_1\cap Y\neq\emptyset} \|H_{Z_1}\| \\
&\quad + 2\|B\| \frac{(2|t|)^2}{2!} \sum_{Z_1:Z_1\cap X\neq\emptyset} \|H_{Z_1}\| \sum_{Z_2:Z_2\cap Z_1\neq\emptyset, Z_2\cap Y\neq\emptyset} \|H_{Z_2}\| \\
&\quad + 2\|B\| \frac{(2|t|)^3}{3!} \sum_{Z_1:Z_1\cap X\neq\emptyset} \|H_{Z_1}\| \sum_{Z_2:Z_2\cap Z_1\neq\emptyset} \|H_{Z_2}\| \sum_{Z_3:Z_3\cap Z_2\neq\emptyset, Z_3\cap Y\neq\emptyset} \|H_{Z_3}\| + \cdots
\end{aligned}
\tag{3.27}
$$

We now bound each term in equation (3.27), using the assumption on the decay of terms in the Hamiltonian (3.9). The first term is bounded by $2(2|t|) \sum_{i\in X} \exp(-\mu\,\mathrm{dist}(i, Y))$. The second term is bounded by

$$2\frac{(2|t|)^2}{2!}\sum_{i\in X}\sum_{Z_1\ni i}\sum_{j\in Z_1}\sum_{Z_2\ni j,\,Z_2\cap Y\neq\emptyset}1.\tag{3.28}$$

Recall that $\mathrm{dist}(i,Y)\le\mathrm{dist}(i,j)+\mathrm{dist}(j,Y)$. Thus, $\exp[-\mu\mathrm{dist}(i,Y)]\exp[\mu\mathrm{dist}(i,j)]\exp[\mu\mathrm{dist}(j,Y)]\ge1$. Thus, the second term is bounded by

$$2\frac{(2|t|)^2}{2!}\exp[-\mathrm{dist}(i,Y)]\sum_{i\in X}\sum_{Z_1\ni i}\sum_{j\in Z_1}\exp[\mu\mathrm{dist}(i,j)]\sum_{Z_2\ni j,\,Z_2\cap Y\neq\emptyset}\exp[\mu\mathrm{dist}(j,Y)]$$

$$\le2\frac{(2|t|)^2}{2!}\exp[-\mathrm{dist}(i,Y)]\sum_{i\in X}\sum_{Z_1\ni i}\sum_{j\in Z_1}\exp(\mu\mathrm{dist}(i,j))\sum_{Z_2\ni j,\,Z_2\cap Y\neq\emptyset}\exp[\mu\mathrm{diam}(Z_2)]$$

$$\le2\frac{(2|t|)^2}{2!}\exp[-\mathrm{dist}(i,Y)]\sum_{i\in X}\sum_{Z_1\ni i}\sum_{j\in Z_1}\exp[\mu\mathrm{dist}(i,j)]s$$

$$\le2\frac{(2|t|)^2}{2!}\exp[-\mathrm{dist}(i,Y)]\sum_{i\in X}\sum_{Z_1\ni i}|Z_1|\exp[\mu\mathrm{diam}(Z_1)]s$$

$$\le\exp[-\mu\mathrm{dist}(i,Y)]2\frac{(2|t|)^2}{2!}s^2.\tag{3.29}$$

Proceeding in this fashion, we bound the n-th term in the series (3.27) by $2\exp[-\mu\mathrm{dist}(i,Y)](2s|t|)^n/n!$. Adding these together, we arrive at the bound (3.10).

3.2.2   The Lieb–Robinson velocity and finite-range interactions

The bound (3.10) is not in quite the most convenient form for later use. The most convenient form of the theorem is that, supposing the bound (3.9) holds, then there is a constant v_{LR} depending only on s,μ such that for $t<\mathrm{dist}(X,Y)/v_{LR}$, we have

$$\|[A_X(t),B_Y]\|\le\frac{v_{LR}|t|}{l}g(l)|X|\|A_X\|\|B_Y\|,\tag{3.30}$$

where $l=\mathrm{dist}(X,Y)$ and $g(l)$ decays exponentially in l. One may choose, for example, $v_{LR}=4s/\mu$ and equation (3.30) will follow from equation (3.10). Similarly, the operator $A_X(t)$ can be approximated by an operator $A_X^l(t)$ supported on the set of sites within distance $l=v_{LR}t$ of the set X up to an error bounded by $\frac{v_{LR}|t|}{l}g(l)|X|\|A_X\|$.

The reader should note that in almost all applications of the Lieb–Robinson bound, the error term on the right-hand side of equation (3.30) associated with the 'leakage' outside the light-cone is negligible. For example, in the next section, we will prove exponential decay of correlation functions. The proof will involve summing up various error terms. The most important error terms will not arise from the Lieb–Robinson bound. In fact, for clarity, the reader may wish to pretend that such error terms are zero on the first reading of any of the proofs in the rest of the paper, and only later worry about how large these errors are.

Other types of interaction decay can be consider similarly. Suppose the interaction is finite range. For example, suppose $H_Z = 0$ for $\text{diam}(Z) > R$ for some interaction range R. Then, we can find constants μ, s such that the assumption (3.9) is satisfied. However, we can do even better in the case of finite range interaction. Suppose each term in the Hamiltonian only has support on a set of two sites, and suppose we have $R = 1$ for simplicity and suppose $\|H_Z\| \le J$ for some constant J. We again use the series (3.27). However, we note that we have now bounded $C_B(X,t)$ by a sum over paths on the lattice starting at sites in X and ending at sites in Y, with each path weighted by $(2J|t|)^l/l!$, where l is the length of the path. Such a weighted sum over paths is a well-studied problem in statistical physics. In this case, one finds that $g(l)$ decays faster than exponentially in l (roughly, $g(l)$ is $\exp(-\text{const} * l^2)$ as one may verify). Other cases of finite range interaction can be handled similarly.

3.2.3 Other types of decaying interactions

One may also consider other types of decaying interaction, slower than exponential. This subsection will be useful in considering quasi-adiabatic continuation later, but may be skipped on first reading.

The important things to remember from the calculations in the rest of the section are that one can treat decay other than exponential, and that decays which are slower than exponential decay give rise to Lieb–Robinson bounds with error terms that also decay slower than exponential. That is, we find error terms on the right-hand side of the Lieb–Robinson bounds consisting of a function which is exponentially growing in time, multiplied by a function which decays slower than exponentially in space. As a result, these types of decay do *not* give rise to a Lieb–Robinson velocity; for example, a function $\exp(t)/l^3$ is only small for t which is logarithmically large in l, while a function $\exp(t)\exp(-\sqrt{l})$ is only small for t of order $\sqrt{l}$.

A further important thing to remember is that the sum over sites in equation (3.27) can worsen the bounds beyond what one might expect. That is, the decay in the error bound may be not as rapid as the decay in the interactions. For example, if we have a two-dimensional lattice, with an interaction H that is a sum of H_Z with each H_Z having $|Z| = 2$ (i.e. each term in the interaction acts on only two sites), with $\|H_Z\| \sim 1/\text{diam}(Z)$, then we find that if the distance between a site i and a set Y is equal to l, then the first term in equation (3.27) is of order $1/l$. The second term is of order $\sum_j (1/\text{dist}(i,j))(1/\text{dist}(j,Y))$ where the sum is over sites j. However, one may see (by replacing the sum by an integral) that the second term diverges logarithmically as the lattice size tends to infinity. So, in fact in this case no Lieb–Robinson bound can be proven. One needs a faster power-law decay than this to prove the Lieb–Robinson bound for a two-dimensional lattice. General consideration of different power-law decays was given in (Hastings and Koma 2006).

One way of treating these was considered in (Hastings 2004). We define:

Definition 3.2 *A function $K(l)$ is* **reproducing** *for a given lattice Λ if, for any pair of sites i, j we have*

$$\sum_m K(\text{dist}(i,m))K(\text{dist}(m,j) \le \lambda K(\text{dist}(i,j)), \tag{3.31}$$

for some constant λ.

For a square lattice in D dimensions and a shortest-path metric, a powerlaw $K(l) \sim l^{-\alpha}$ is reproducing for sufficiently large α. An exponential decay is *not* reproducing. However an exponential multiplying a sufficiently fast decaying power is. Using this definition and equation (3.27), suppose that $\|H_Z\| \le K(\text{diam}(Z))$ for some reproducing K. Then, we find that equation (3.27) is bounded by

$$2K(\text{dist}(i,Y))(2|t| + \frac{(2|t|)^2}{2!}\lambda + \frac{(2|t|)^3}{3!}\lambda^2 + \ldots \tag{3.32}$$

$$\le 2K(\text{dist}(i,Y))\frac{\exp(2\lambda|t|) - 1}{\lambda}.$$

Note, as mentioned above, that if K decays slower than exponentially, then this function does not lead to a Lieb–Robinson velocity; the time t at which we have a meaningful bound will grower slower than exponentially in $\text{dist}(i,Y)$.

A useful trick is that if we start with a function which is not reproducing (such as an exponential) in many cases we can bound it by a slightly slower decay exponential times a sufficiently fast power law to arrive at a reproducing decay function.

3.3 Correlation decay

We now consider the decay of correlations in the ground state of a Hamiltonian with a spectral gap and with a Lieb–Robinson bound. Such a decay was first proven in (Hastings 2004). In addition to the interest in this result for itself, we consider it because it introduces a combination of two techniques which will be particularly useful. We combine Lieb-Robinson bounds with a set of tools using the Fourier transform. Physically, one may imagine that a spectral gap ΔE sets a time scale, ΔE^{-1}, and then the Lieb–Robinson bound allows us to define a length scale $v_{LR}\Delta E^{-1}$. The techniques we introduce combining the Lieb–Robinson bound with the Fourier transform make this statement precise.

The statement of the theorem with a unique ground state (Hastings 2004) is

Theorem 3.3 *For a quantum lattice system with a unique ground state and a spectral gap ΔE, and any operators A_X, B_Y supported on sets X, Y, we have*

$$\left|\langle \psi_0, A_X B_Y \psi_0 \rangle - \langle \psi_0, A_X \psi_0 \rangle \langle \psi_0, B_Y \psi_0 \rangle\right|$$

$$\le C\left\{\exp(-l\Delta E/2v_{LR}) + \min(|X|, |Y|)g(l)\right\}\|O_A\|\|O_B\|, \tag{3.33}$$

for some constant C, where $l = \text{dist}(X,Y)$. We can also consider systems with multiple ground states ψ_0^a as above, with a spectral gap ΔE separating those states from the rest of the spectrum. In this case (Hastings 2004), we can instead obtain bound the quantity $\langle \psi_0^a, A_X B_Y \psi_0^a \rangle - \langle \psi_a^a, A_X P_0 B_Y \psi_0^a \rangle$, where

$$P_0 = \sum_a |\psi_0^a\rangle\langle\psi_0^a| \tag{3.34}$$

is the projector onto the ground state subspace. The bound on this quantity $\langle \psi_0^a, A_X B_Y \psi_0^a \rangle - \langle \psi_a^a, A_X P_0 B_Y \psi_0^a \rangle$ is equal to the right-hand side of equation (3.33) plus an additional term which is proportional to the energy difference between the ground states, and in fact this result reduces to equation (3.33) in the case of a single ground state.

Before sketching the proof of the theorem, we discuss the application of the theorem in various settings. Consider the transverse field Ising model Hamiltonian in the paramagnetic phase, $B \gg J$, with spin-1/2 on each site. In this case, there is a unique ground state and a spectral gap, so that all correlations decay exponentially in space. On the other hand, we may consider the Hamiltonian in the ferromagnetic phase, $J \gg B$. For $B = 0$, the model has two exactly degenerate ground states, corresponding to all spins pointing up or all spins pointing down. In fact, the Hamiltonian has a particular symmetry. It commutes with the unitary operator $\prod_i(2S_i^x)$. This operator flips all the spins. So, the eigenstates of the Hamiltonian can be chosen to be eigenstates of this particular unitary. To do this, at $B = 0$ we may choose the ground states to be symmetric and anti-symmetric combinations of the states with all spins up or down:

$$\psi_\pm = \frac{1}{\sqrt{2}}\Big(|\uparrow\uparrow\uparrow\,...\rangle \pm |\downarrow\downarrow\downarrow\,...\rangle\Big). \tag{3.35}$$

For $B > 0$ but B still much smaller than J, there starts to be a small splitting (exponentially small in system size) between the two lowest energy states, and the gap to the rest of the spectrum remains open. Now, consider a local operator such as S_i^z. We will see what the action of this operator is in the ground state sector. That is, we will construct a 2-by-2 matrix M whose matrix elements are the matrix elements of S_i^z in the ground state. For $B = 0$, we have

$$\langle \psi_+, S_i^z \psi_+ \rangle = \langle \psi_-, S_i^z \psi_- \rangle = 0, \tag{3.36}$$

and

$$\langle \psi_+, S_i^z \psi_- \rangle = \langle \psi_-, S_i^z \psi_+ \rangle = 1/2, \tag{3.37}$$

so

$$M = \begin{pmatrix} 0 & 1/2 \\ 1/2 & 0 \end{pmatrix}. \tag{3.38}$$

For $B > 0$, we instead find that

$$M = \begin{pmatrix} 0 & m \\ m & 0 \end{pmatrix}, \tag{3.39}$$

where m is the 'order parameter' (m is greater than zero in the ferromagnetic phase and $m < 1/2$ for $B > 0$).

The nontrivial matrix elements of S_i^z in the ground state sector are what give rise to the long range correlations: the theorem above shows that $\langle \psi_+, S_i^z S_j^z \psi_+ \rangle$ approaches m^2 when $\mathrm{dist}(i,j)$ gets large. This behavior will contrast strongly with the topologically ordered case below.

We only sketch the proof in the case of a unique ground state (the reader is invited to consider the case of multiple ground states). Define B_Y^+ to be the positive energy part of B_Y. That is, let $\{\psi_i\}$ be a basis of eigenstates of H, with ψ_i having energy E_i. In this basis, we define B_Y^+ to have matrix elements

$$(B_Y^+)_{ij} = (B_Y)_{ij}\theta(E_i - E_j), \tag{3.40}$$

where $\theta(x)$ is the step function: $\theta(x) = 1$ for $x > 0$, $\theta(0) = 1/2$, and $\theta(x) = 0$ for $x < 0$. Without loss of generality, assume that $\langle \psi_0, A_X \psi_0 \rangle = \langle \psi_0, B_Y \psi_0 \rangle = 0$. Then, we find that

$$\langle \psi_0, A_X B_Y \psi_0 \rangle = \langle \psi_0, A_X B_Y^+ \psi_0 \rangle = \langle \psi_0, [A_X, B_Y^+] \psi_0 \rangle, \tag{3.41}$$

since $B_Y^+ \psi_0 = B_Y \psi_0$ as there are no states with energy less than ψ_0. Our strategy is to construct an approximation to B_Y^+, which we call $\tilde{B}_Y^+$, which has two properties. First, $\tilde{B}_Y^+$ has small commutator with A_X. Second,

$$B_Y^+|\psi_0\rangle \approx \tilde{B}_Y^+|\psi_0\rangle, \tag{3.42}$$

and

$$\langle \psi_0|\tilde{B}_Y^+ \approx \langle \psi_0|B_Y^+ = 0, \tag{3.43}$$

where the error in the approximation is discussed below. Then, by making the error in the approximations small, and by making the commutator $[A_X, \tilde{B}_Y^+]$ small in operator norm, we will be able to bound $\langle \psi_0, [A_X, B_Y^+] \psi_0 \rangle$ as follows: we first show (using (3.42,3.43)) that $\langle \psi_0, [A_X, B_Y^+]\psi_0 \rangle$ is close to $\langle \psi_0, [A_X, \tilde{B}_Y^+]\psi_0 \rangle$. Then, we show that that quantity is small using a bound on the operator norm of the commutator.

To do this, we define $\tilde{B}_Y^+$ by

$$\tilde{B}_Y^+ = \frac{1}{2\pi} \lim_{\epsilon \to 0^+} \int dt B_Y(t) \frac{1}{it + \epsilon} \exp[-(t\Delta E)^2/(2q)], \tag{3.44}$$

where q is a constant that we choose later. Note that as $q \to \infty$, we find that the Gaussian on the right-hand side of equation (3.44) gets broader, and $\tilde{B}_Y^+$ converges to B_Y^+. So, taking q large will make it easier to satisfy (3.42,3.43). Conversely, for q large, we will be able to show that $\|[A_X, \tilde{B}_Y^+]\|$ is small, since for large times t the integral

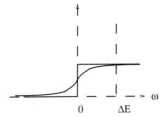

Fig. 3.1 Plot of a step function (solid line) and a sketch of a step function convolved with a Gaussian (solid line). Dashed line at ΔE shows that the difference between the functions is small at sufficiently large frequency.

over t in (3.44) is cut off by the Gaussian, and for short times t the commutator of A_X with $B_Y(t)$ is bounded by the Lieb–Robinson bound.

Using the energy gap, one may show that

$$\left| B_Y^+ |\psi_0\rangle - \tilde{B}_Y^+ |\psi_0\rangle \right| \le C \exp(-q/2) \|B_Y\|. \tag{3.45}$$

To show this, we bound the absolute value of the matrix element $\langle \psi_i, (B_Y^+ - \tilde{B}_Y^+)\psi_0\rangle$ for some $i > 0$. This is equal to

$$|\langle \psi_i, (B_Y^+ - \tilde{B}_Y^+)\psi_0\rangle| = |(B_Y)_{i0}| \left| 1 - \int dt \exp[i(E_i - E_0)t] \frac{1}{it + \epsilon} \exp[-(t\Delta E)^2/(2q)] \right|. \tag{3.46}$$

The integral in the above equation is equal to the Fourier transform of the function $\frac{1}{it+\epsilon} \exp[-(t\Delta E)^2/(2q)]$. This Fourier transform is the convolution of the Fourier transform of the step function with a Gaussian. In Fig. 4.1, we sketch the Fourier transform of this function, as well as the step function. For a narrow width of the Gaussian in frequency (which occurs if q is large), the Fourier transform of the two functions are very close for energy above ΔE. One can bound the difference in the Fourier transform by $C \exp(-q/2)$. So, $|(B_Y^+ - \tilde{B}_Y+)\psi_0|^2 \le \sum_{i>0} |(B_Y)_{i0}|^2 \exp(-q)$, so equation (3.45) follows. Similarly, one can bound the quantity in (3.43) in the same way.

One may also bound $\|[A_X, \tilde{B}_Y^+]\|$ by a triangle inequality as

$$\|[A_X, \tilde{B}_Y^+]\| \le \frac{1}{2\pi} \lim_{\epsilon \to 0^+} \int dt \|[A_X, B_Y(t)]\| \left| \frac{1}{it + \epsilon} \right| \exp[-(t\Delta E)^2/(2q)]. \tag{3.47}$$

To bound this integral, we split the integral over t into times t less than l/v_{LR} and $t > l/v_{LR}$. For $t < l/v_{LR}$, we use the Lieb–Robinson bound to bound the integral by a constant times $g(l)|X|\|A_X\|\|B_Y\|$. For $t > l/v_{LR}$, we use the Gaussian to bound it by a constant times

$$\|A_X\|\|B_Y\| \exp[-(l\Delta E/v_{LR})^2/2q]. \tag{3.48}$$

We now pick q to minimize the sum of terms in (3.45,3.48). Picking $q = l\Delta E/v_{LR}$ is the best possible choice, and gives the error bound in the theorem.

The most important thing to remember from this sketch is: we combine Lieb–Robinson bounds and Fourier transforms. We have some function which is not smooth (like the step function). We approximate it by some smooth function (the step function convolved with a Gaussian) to obtain a function whose Fourier transform decays rapidly in time (in this case as $1/t$ times a Gaussian). This then allows us to apply the Lieb–Robinson bounds. The same ideas are behind the technique of quasi-adiabatic continuation later.

3.4   Topological order

There are many different properties that characterize 'topological order'. Consider a Hamiltonian such as Kitaev's toric code model (Kitaev 2003). We will not review this model here, since it is well-explained elsewhere. However, this model has several unique properties. On a torus, the model has four exactly degenerate ground states, with an energy gap to the rest of the spectrum. Surprisingly, however, on other topologies, the ground state degeneracy is different. For example, the model has a unique ground state on a sphere. This contrasts strongly with the case of the transverse field Ising model mentioned above, which has either one ground state (in the paramagnetic phase) or two ground states (in the ferromagnetic phase) independent of the topology of the lattice. This property of the ground state degeneracy depending upon topology is one characteristic of a Hamiltonian with topological order. Another characteristic is certain corrections to the entanglement entropy (Levin and Wen 2006; Kitaev and Preskill 2006). We will not consider these corrections here.

In fact, we would like to regard both of these properties (the ground state degeneracy and the entropy) as being secondary to one particular defining feature of topological order which we now explain. Suppose a model has k different ground states, ψ_0^a for $a = 1, \ldots, k$. Then, given any operator O, we can project this operator O into the ground state sector, defining a k-by-k matrix M whose matrix elements are

$$M_{ab} = \langle \psi_0^a, O\psi_0^b \rangle. \tag{3.49}$$

In the case of the toric code on a torus, the matrix M is a 4-by-4 matrix. Surprisingly, if the operator O is local in the sense that the diameter of the support of O is sufficiently small compared to L, then the matrix M is equal to a constant times the identity matrix. In the case of the toric code, in fact the matrix M_{ab} is equal to a multiple of the identity matrix so long as the diameter of the support of O is less than $L/2$, so one can consider operators O which act on a very large number of sites, and yet still the operator acts just as the identity operator when projected into the ground state sector. This behaviour contrasts strongly with the case of the transverse field Ising model in the ferromagnetic phase, for which the operator S_i^z acts nontrivially on the ground state sector as discussed above.

If we slightly perturb the toric code, the model remains in the same phase (Bravyi, Hastings, and Michalakis 2010; Bravyi and Hastings 2010). The four ground states will no longer be exactly degenerate, but the difference in energy between them will be exponentially small as a function of system size (see (Bravyi, Hastings, and Michalakis 2010; Bravyi and Hastings 2010) for general upper bounds on the energy splitting). Further, this topological order property will be slightly weakened: we will instead have a property that if operator O is supported on a set of diameter sufficiently small compared to L (for example, a diameter of at most $L/2$ will suffice for the toric code), then the corresponding matrix M is exponentially close to a multiple the identity matrix (again, see (Bravyi, Hastings, and Michalakis 2010; Bravyi and Hastings 2010) for general upper bounds).

This property, that local operators are close to a multiple of the identity matrix when projected into the ground state sector was identified in the Hall effect (Wen and Niu 1990) and in other topologically ordered states (Freedman, Kitaev, Larsen, and Wang 2001). Note that this property in fact ensures that the splitting between the ground states is small: the Hamiltonian itself is a sum of local operators, so by this property of topological order, the Hamiltonian has almost the same expectation value in each of the different ground states.

We can quantify this topological order as in (Bravyi, Hastings, and Verstraete 2006). We say that a system has (l, ϵ) topological order if, for any operator O supported on a set of diameter as most l, the corresponding matrix M is within ϵ of a multiple of the identity. That is, for some complex number z, we have $\|M - zI\| \leq \epsilon$.

One interesting property of this viewpoint about topological order is that it is a property of a set of states, rather than a property of a Hamiltonian. Given any set of orthogonal states, ψ_0^a, we can ask whether these states are topologically ordered, independently of whether or not the states happen to be the ground states of some local Hamiltonian.

Note that it is *not* the case that every state ψ_0^1 has some partner state ψ_0^2 which gives us a pair of topologically order states. For example, consider a system of spin-1/2 spins and consider the state with all spins up, $|\psi_0^1\rangle = |\uparrow\uparrow\uparrow \ldots\rangle$. There is no state $|\psi_0^2\rangle$ that is orthogonal to $|\psi_0^1\rangle$ such that the two states ψ_0^a for $a = 1, 2$ are topologically ordered.

3.4.1 Topological order under time evolution

Note that we can use the Lieb–Robinson bound to describe the behaviour of topological order under time evolution (Bravyi, Hastings, and Verstraete 2006). Suppose a state ϕ has (l, ϵ) topological order for some l, ϵ. Thus, there exists another state ϕ' which is the partner of state ϕ. Let us evolve ϕ for time t under some Hamiltonian H to obtan $\psi = \exp(-iHt)\phi$ (one can consider also time-dependent Hamiltonians). Define $\psi' = \exp(-iHt)\phi'$. We now wish to show thhat ψ, ψ' retain some memory of the topological order in ϕ, ϕ': the length scale l will be smaller and the error ϵ will be larger in a way that we can bound quantitatively.

Let O be any local operator supported on a set of diameter $l - m$, for some $m \leq v_{LR}t$. We project this operator O into the two dimensional space of states spanned by ψ' and ψ. The result is equal to the projection of the operator $\exp(-iHt)O\exp(iHt)$ into the space of states spanned by ϕ' and ϕ. However, the operator $\exp(-iHt)O\exp(iHt)$ can be approximated, by the Lieb–Robinson bound, by an operator supported on a set of diameter $\mathrm{diam}(O) + m \leq l$ up to an error $g(m)|X|\|A_X\|$ as given by equation (3.30). We bound $|X|$ by some constant times l^d for a d-dimensional lattice. Thus, we find that ψ, ψ' have $(l - m, \epsilon + g(m)\mathrm{const}.l^d)$ topological order for any $m \leq v_{LR}t$. Thus, topological order cannot be completely destroyed in a short-time. For example, suppose that the interactions in the Hamiltonian are such that $g(l)$ decays exponentially in l, and suppose that the initial states ϕ, ϕ' have (l, ϵ) topological order with $l = L/2$ and ϵ exponentially small in L. Then, choosing $m = L/4$, we find that for times up to $L/(4v_{LR})$, the states ϕ, ϕ' have $(L/4, \epsilon')$ topological order, with ϵ' still being exponentially small in L.

Conversely, topological order also cannot be produced in a short time. If we start with a state such as $|\psi_0^1\rangle = |\uparrow\uparrow\uparrow\ldots\rangle$ and evolve for time t under some Hamiltonian H, we cannot produce (l, ϵ) topological order for l large compared to $v_{LR}t$ and ϵ small.

3.5 Fourier transforms

As seen by our analysis of correlation functions, Fourier transforms play a key role in the application of techniques of Lieb–Robinson bounds to quantum many-body systems. In this section, we collect a few useful facts about Fourier transforms.

First, the Fourier transform of a Gaussian is a Gaussian. There is an 'uncertainty principle' at work here: the narrower the Gaussian in time, the wider the Gaussian in frequency, and vice-versa.

Second, the Fourier transform of the product of two functions is the convolutions of their Fourier transform. This was used, for example, to estimate the term $\exp(-q/2)$ in the calculation of correlation function decay.

While we have noticed the general principle that there is a tradeoff between the spread of a function in time and in frequency, it will be useful in some applications to have functions which have compact support in frequency. That is, we would like to find a function $\tilde{g}(\omega)$ which vanishes for $|\omega| > 1$, with $\tilde{g}(0) = 1$, and such that the Fourier transform $g(t)$ decays as rapidly in time as possibly. In the classic paper (Ingham 1934), it is shown how to construct such functions $g(t)$ such that

$$|g(t)| \leq \mathcal{O}(\exp(-|t|\epsilon(|t|))), \tag{3.50}$$

for *any* monotonically decreasing positive function $\epsilon(y)$ such that

$$\int_1^\infty \frac{\epsilon(y)}{y}\,dy \tag{3.51}$$

is convergent. Further, it was shown that this is the optimal possible decay. For example, the function $\epsilon(y)$ may be chosen to be

$$\epsilon(y) = 1/\log(2+y)^2. \tag{3.52}$$

Thus, this function $g(t)$ has so-called 'subexponential decay' (Dziubanski and Hernández 1998). A function $f(t)$ is defined to have subexponential decay if, for any $\alpha < 1$, $|f(t)| \leq C_\alpha \exp(-t^\alpha)$, for some C_α which depends on α.

Thus, while we cannot quite obtain exponential decay in time and compact support in frequency, we can come very close to it, obtaining functions which 'almost decay exponentially'.

We will find it useful also in the next section to have a function $\tilde{F}(\omega)$ such that $\tilde{F}(\omega) = 1/\omega$ for $|\omega| \geq 1$ and such that $\tilde{F}(\omega)$ is odd and the Fourier transform $F(t)$ decays rapidly in time. We can do this using the functions $g(t)$ above as follows. First, we assume without loss of generality that the function $g(t)$ above is an even function of t (if not, simply take the even part of the function). Then, define the even function $f(t)$ by

$$f(t) = \delta(t) - g(t), \tag{3.53}$$

where $\delta(t)$ is the Dirac δ-function (note that $f(t)$ is thus a distribution rather than a function). Thus that the Fourier transform $\tilde{f}(\omega)$ has the property that $\tilde{f}(\omega) = 0$ for $\omega = 0$ and $\tilde{f}(\omega) = 1$ for $|\omega| \geq 1$. Then, define $F(t)$ by

$$F(t) = \frac{i}{2} \int du f(u)\mathrm{sgn}(t - u), \tag{3.54}$$

where $\mathrm{sgn}(t - u)$ is the sign function: $\mathrm{sgn}(t - u) = 1$ for $t > u$, $\mathrm{sgn}(t - u) = -1$ for $t < u$, and $\mathrm{sgn}(0) = 0$ (since we convolve $f(u)$ against $\mathrm{sgn}(t - u)$, the resulting $F(t)$ is a function, rather than a distribution). We now show the time decay of $F(t)$ and we show that the Fourier transform $\tilde{F}(\omega)$ is equal to $-1/\omega$ for $|\omega| \geq 1$, as desired (this calculation is directly from (Hastings 2004)).

Lemma 3.4 *Let $F(t)$ be as defined in 3.54. Let $\tilde{F}(\omega)$ be the Fourier transform of $F(t)$. Then,*

$$|F(t)| \leq |\int_{|t|}^{\infty} f(u)du|, \tag{3.55}$$

and

$$\tilde{F}(\omega) = \frac{-1}{\omega}\tilde{f}(\omega). \tag{3.56}$$

Proof: *Assume, without loss of generality, that $t \geq 0$. Then, we have $|F(t)| \leq |\int_t^{\infty} f(u)du|/2 + |\int_{-\infty}^t f(u)du|/2$. Since $\tilde{f}(0) = 0$, we have $|\int_{-\infty}^t f(u)du| = |\int_t^{\infty} f(u)du|$. Thus, $|F(t)| \leq |\int_t^{\infty} f(u)du|$.*
We have

$$\tilde{F}(\omega) = \frac{i}{2}\int dt \exp(i\omega t)\int du f(u)\mathrm{sgn}(t - u). \tag{3.57}$$

Integrating by parts in t, we have

$$\tilde{F}(\omega) = \frac{-1}{\omega}\int dt \exp(i\omega t)\int du f(u)\delta(t - u) \tag{3.58}$$

$$= \frac{-1}{\omega}\tilde{f}(\omega).$$

Note that $\lim_{t\to\pm\infty}\left(\int du f(u)\text{sgn}(t-u)\right)=0$, *so the contributions to the integration by parts from the upper and lower limits of integration vanish.*   □

Finally, given the decay of $f(t)$, it follows that $F(t)$ decays subexponentially also.

A side note: we can also use this idea of compactly supported functions in the correlation decay calcuation done previously. Using the functions $\tilde{g}(\omega)$ described above, we can construct, for example, a family of functions $\tilde{f}(\omega,\epsilon)$ such that $\lim_{\epsilon\to 0}\tilde{f}(\omega,\epsilon)=1$ for $\omega\geq 1$ and $\lim_{\epsilon\to 0}\tilde{f}(\omega,\epsilon)=0$ for $\omega\leq -1$, and with the Fourier transform $g(t,\epsilon)$ decaying subexponentially at large times. We construct this family by taking the Fourier transform of the function $g(t)/(it+\epsilon)$; the Fourier transform of this function converges, as $\epsilon\to 0^+$, to the convolution of the Fourier transform of $f(t)$ with a step function. This family of functions still is singular in the limit $t,\epsilon\to 0$ (a singularity like $1/(it+\epsilon)$, just as we encountered in the Gaussian function $\exp[-(t\Delta E)^2/2q]/(it+\epsilon)$ in the correlation decay calculation). Such a singularity is in fact unavoidable given the large ω behaviour of the function. This approach will not give quite as tight bounds on the correlation decay (the bounds will be subexponential rather than exponential), but the calculation is a little simpler since we will have only error terms involving the bound on the commutator $\|[A_X,\tilde{B}_Y^+]\|$, while with these compactly supported functions the difference $\tilde{B}_Y^+\psi_0 - B_Y^+\psi_0$ will vanish.

3.6   Quasi-adiabatic continuation

We now consider the problem of how the ground state of a local Hamiltonian changes as a parameter in the Hamiltonian is changed. Suppose we have a parameter dependent Hamiltonian, H_s, where s is some real number. Suppose that $H_s=\sum_Z H_Z(s)$, with $H_Z(s)$ being differentiable. Suppose further that we have uniform bounds on the locality properties of the Hamiltonian (for example, for all s, the exponential decay (3.9) holds or some other similar assumption holds uniformly in s). The main idea of this section can be summarized in a single sentence as follows: if such a Hamiltonian H_s has a lower bound on the spectral gap which is uniform in s and has a unique ground state, $\psi_0(s)$, then we can define a Hermitian operator, called the quasi-adiabatic continuation operator $\mathcal{D}_s$, which is local (in some slightly weaker sense), such that either $\partial_s\psi_0(s)=i\mathcal{D}_s\psi_0(s)$ or $\partial_s\psi_0(s)\approx i\mathcal{D}_s\psi_0(s)$ (whether we want exact or approximate equality depends upon the application). We will also present a generalization to the case of multiple ground states, and use this result to prove a Goldstone theorem. Finally, we will use these ideas to discuss what we mean by a 'phase' of a quantum system, and to explore the stability of topological order under perturbations.

In the next section, we will use the quasi-adiabatic continuation operator defined in this section to evolve states along paths in parameter space such that *the Hamiltonian does not necessarily have a spectral gap for $s>0$*. This continuation along these paths will allow us to prove a higher dimensional Lieb–Schultz–Mattis theorem. The fact that we continue along paths which might not have a gap is in fact essential to the proof of that theorem: the results in this section show that continuing along a gapped path implies that one remains in the ground state, while our goal in the next section is to construct a state which is *different* from the ground state but still has low energy to prove a variational result.

We now define the quasi-adiabatic continuation operator:

Definition 3.5 *Given a parameter-dependent Hamiltonian, H_s, an operator O, and function $F(t)$, we define the* **quasi-adiabatic continuation operator** *to be the operator $\mathcal{D}(H_s, O)$ defined by*

$$i\mathcal{D}(H_0, O) = \int F(\Delta Et) \exp(iH_s t) O \exp(-iH_s t) dt, \tag{3.59}$$

where F is an odd function of time so that $\mathcal{D}$ is Hermitian.

Note that since F is odd, its Fourier transform $\tilde{F}(\omega)$ obeys

$$\tilde{F}(0) = 0, \tag{3.60}$$

which will be useful in discussions of Berry phase later.

Given a parameter dependent Hamiltonian $H_s = \sum_Z H_Z(s)$, we define

$$\mathcal{D}_s = \mathcal{D}(H_s, \partial_s H_s). \tag{3.61}$$

We also sometimes write $\mathcal{D}_s^Z = \mathcal{D}(H_s, \partial_s H_Z(s))$, so that

$$\mathcal{D}_s = \sum_Z \mathcal{D}_s^Z. \tag{3.62}$$

We will use two different types of functions $F(t)$ in the definition of the quasi-adiabatic continuation operator. The first type of function is the function $F(t)$ constructed in the previous section, such that the Fourier transform of $F(t)$ obeys $\tilde{F}(\omega) = -1/\omega$ for $|\omega| \geq 1$. This will give, as we now show,

$$\partial_s \psi_0(s) = i\mathcal{D}_s \psi_0(s). \tag{3.63}$$

We call the quasi-adiabatic continuation operator arising from such a function $F(t)$ an 'exact quasi-adiabatic continuation operator'. The second type of function $F(t)$ will lead to only approximate equality

$$\partial_s \psi_0(s) \approx i\mathcal{D}_s \psi_0(s), \tag{3.64}$$

and we call this the 'Gaussian quasi-adiabatic continuation operator'.

Let us first show

$$\partial_s \psi_0(s) = i\mathcal{D}_s \psi_0(s) \tag{3.65}$$

in the case of an exact quasi-adiabatic continuation operator. We have

$$iD_s\psi_0(s) = \int F(\Delta Et)\exp(iH_st)\Big(\partial_s H_s\Big)\exp(-iH_st)\mathrm{d}t\psi_0(s) \tag{3.66}$$

$$= \sum_{i\neq 0}|\psi_i(s)\rangle\langle\psi_i(s)|\int F(\Delta Et)\exp(iH_st)\Big(\partial_s H_s\Big)\exp(-iH_st)\mathrm{d}t\psi_0(s)$$

$$= \sum_{i\neq 0}|\psi_i(s)\rangle\langle\psi_i(s),\Big(\partial_s H_s\Big)\psi_0(s)\rangle\int F(\Delta Et)\exp[i(E_i(s)-E_0(s))t]\mathrm{d}t$$

$$= \sum_{i\neq 0}\frac{1}{E_0(s)-E_i(s)}|\psi_i(s)\rangle\langle\psi_i(s),\Big(\partial_s H_s\Big)\psi_0(s)\rangle$$

$$= \partial_s\psi_0(s),$$

where $\psi_i(s)$ for $i > 0$ denote excited states of the Hamiltonian with energy $E_i(s)$. The second line follows by inserting the identity as $\sum_i |\psi_i(s)\rangle\langle\psi_i(s)|$, and noting that property 3.60 implies that the term with $i = 0$ is absent. The third line follows by using the fact that the $\psi_s(s)$ are eigenstates. The fourth line follows from the properties of $F(t)$. The fifth line is ordinary perturbation theory.

In the case of multiple ground states, one can generalize this result as follows. We instead find that if we have ground states $\psi_0^a(s)$ which are not exactly degenerate, then

$$\partial_s\psi_0^a(s) = iD_s\psi_0(s) + \sum_b Q_{ab}\psi_0^b(s), \tag{3.67}$$

where Q_{ab} are the matrix elements of some anti-Hermitian matrix Q (if the ground states are exactly degenerate, then $\partial_s\psi_0^a(s)$ may be ill-defined). There is an important Berry phase property: if the ground states are exactly degenerate, then the Berry phase arising from the quasi-adiabatic evolution is the same as the usual non-Abelian Berry phase (Hastings 2007), while small corrections to this result occur if there is ground state splitting. Similarly, if $P_0(s)$ is the projector onto the ground state sector of H_s, then

$$\partial_s P_0(s) = i[D_s, P_0(s)]. \tag{3.68}$$

The other important property that the quasi-adiabatic continuation operator, D_s has, in addition to (3.66), is that it is local. Let us assume that $\|\partial_s H_Z(s)\|$ is bounded by a constant times $\|H_Z(s)\|$ to fix a normalization on how rapidly the Hamiltonian changes. Then if the original Hamiltonian H has a superpolynomial decay in its interactions (so that $\|H_Z\|$ decays superpolynomially in $\mathrm{diam}(Z)$), and the lattice is finite dimensional, then D_s also has a superpolynomial decay: we can write $D_s = \sum_Z D_Z(s)$, where $D_Z(s)$ is supported on Z, with $\|D_Z\|$ decaying superpolynomially in $\mathrm{diam}(Z)$. Note that $D_Z(s)$ is *not* the same thing as $\mathcal{D}_s^Z$. Indeed, $\mathcal{D}_s^Z$ is not supported on Z.

To prove the locality of D_Z, one uses the Lieb–Robinson bounds and the super-polynomial decay of the function $F(t)$. Before sketching the proof, let us give the basic idea. Consider a given $\mathcal{D}_s^Z$. in integral in equation (3.59) is small, while at short time we can approximate $\exp(iH_s t)(\partial_s H_Z(s))\exp(-iH_s t)$ by an operator supported near Z. More precisely, we will decompose $\mathcal{D}_s^Z = \sum_{l=0}^{\infty} O_l(Z)$, where $O_l(Z)$ is supported on the set of sites within distance l of Z as follows. We define

$$O_0(Z) = \int\int F(\Delta Et)\partial_s H_Z(s))^0(t)\mathrm{d}t, \tag{3.69}$$

where, following equation (3.11), $(\partial_s H_Z(s))^0(t)$ denotes an approximation to $(\partial_s H_Z(s))(t) \equiv \exp(iH_s t)(\partial_s H_Z(s))\exp(-iH_s t)$ which is localized on set Z. We define, for $l > 0$,

$$O_l(Z) = \int F(\Delta Et)\Big((\partial_s H_Z(s))^l(t) - (\partial_s H_Z(s))^{l-1}(t)\Big)\mathrm{d}t. \tag{3.70}$$

Summing over l recovers the desired result. Now, we define $D_Z(s)$ to be the sum over Y of the $O_l(Y)$ which are supported on Y.

As mentioned, one can also consider Gaussian quasi-adiabatic continuation operators. These were the first type of quasi-adiabatic continuation operators considered (Hastings 2004), while the exact operators will be considered later (Osborne 2007). The Gaussian quasi-adiabatic continuation operators in some cases lead to tighter bounds. For example, in the Lieb–Schultz–Mattis theorem, they lead to tighter bounds than the exact operators due to the faster time decay. On the other hand, in many cases the exact operators are much more convenient. *At this point, we make a deliberate choice due to the nature of these lecture notes. Rather than consider the Gaussian operators in detail (which leads to an enormous number of triangle inequalities in the actual calculations, potentially obscuring the physics), we will only use the exact quasi-adiabatic continuation operators.* This will lead to slightly less tight results in many cases, but the improvement in clarity (and generality in considering topological phases later), seems well worth it.

3.6.1 Lieb–Robinson bounds for quasi-adiabatic continuation

One particular advantage of considering the exact quasi-adiabatic continuation operators is that we have a Lieb–Robinson bound for them. We have shown that the norm of the terms $D_Z(s)$ decays superpolynomially in $\mathrm{diam}(Z)$ (indeed, it decays subexponentially if the original Hamiltonian is a finite dimensional lattice with exponentially decaying interactions). This implies (see the previous discussion on reproducing functions) that we have a Lieb–Robinson bound for quasi-adiabatic evolution; that is, if we evolve an operator O under the equation of motion

$$\partial_s O(s) = i[\mathcal{D}_s, O], \tag{3.71}$$

we can prove a Lieb–Robinson bound for $O(s)$. This is a particular advantage compared to the Gaussian case, where the proof of Lieb–Robinson bounds for quasi-adiabatic evolution is much more difficult and the bounds are weaker (Hastings and Michalakis 2009).

Note that, as discussed previously in subsection (3.2.3), we do not actually obtain a finite Lieb–Robinson velocity. That is, if we want to bound the commutator $\|[A_X(s), B_Y]\|$, where X is supported on X and B is supported on Y, the largest value of s for which we obtain a meaningful bound grows slower than linearly in $\operatorname{dist}(X, Y)$. This is not a problem in most applications, since in general we will be considering path lengths of order unity, while we will often consider distances between sets X, Y which are of order system size.

3.6.2   Goldstone's Theorem

We now present an application to a nonrelativistic Goldstone theorem. This theorem is perhaps not that surprising, but the results here (originally in (Hastings 2007)) are more general and simpler than previous nonrelativistic results (Wreszinski 1981/87).

Above, we have described the clustering of correlation functions, proving that the connected correlated function, $\langle AB \rangle - \langle AP_0 B \rangle$, of two operators A, B with support on sets X, Y is exponentially small in the distance $\operatorname{dist}(X, Y)$. Here, for a system with k different ground states. $\langle O \rangle \equiv k^{-1} \sum_a \langle \psi_0^a, O\psi_0^a \rangle$. For a parameter dependent Hamiltonian, we define $\langle O \rangle_s \equiv k^{-1} \sum_a \langle \psi_0^a(s), O\psi_0^a(s) \rangle$.

The goal now is to prove (or at least sketch the proof of) a stronger statement about the decay of correlation functions in gapped systems with a continuous symmetry showing that the expectation value $\langle AP_0 B \rangle$ is small also.

Goldstone's theorem is a statement that a system with a spontaneously broken continuous symmetry has gapless excitations. We first need to define a continuous symmetry, or equivalently, a conserved charge. This means that

Definition 3.6 *We say that a lattice Hamiltonian H has a conserved charge if the following holds. For every site i, there is an operator q_i supported on site i, with q_i having integer eigenvalues. Let $Q = \sum_i q_i$. Then, we require that*

$$[Q, H] = 0. \tag{3.72}$$

Further, we assume that $\|q_i\| \leq q_{max}$ for some q_{max} (this is a technical point, needed in the later bounds). Our nonrelativistic Goldstone's theorem will be the contrapositive of the usual statement of Goldstone's theorem: we will show that the presence of a gap (between a degenerate ground state sector and the rest of the spectrum) bounds the correlation functions.

For any set X, we define $R(\theta, X) = \prod_{i \in X} \exp[iq_i \theta]$. We consider operators $\phi_X, \bar{\phi}_Y$ with support on sets X, Y which transform as vectors as follows under this $U(1)$ symmetry: $R(-\theta, X)\phi_X R(\theta, X) = \exp[i\theta]\phi_X$ and $R(-\theta, Y)\bar{\phi}_Y R(\theta, Y) = \exp[-i\theta]\bar{\phi}_Y$.

For example, in a Bose system with conserved particle, the q_i can represent the particle number on a given site and the operators $\phi_X, \bar{\phi}_Y$ can represent creation and annihilation operators for the bosons. For a spin system, the q_i can represent the z component of the spin on a site and the $\phi_X, \bar{\phi}_Y$ can represent raising and lowering spin operators on sites.

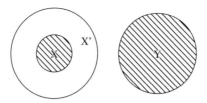

Fig. 3.2 Illustration of the geometry we consider. X, Y are shown as shaded regions, while X' includes everything within the outer circle around X.

We do not require the states in the ground state sector to be degenerate with each other, simply the existence of a gap between that sector and the rest of the spectrum. This result is stronger than that in (Hastings and Koma 2006) as it is valid in arbitrary dimension; it is also stronger than other previous results (Wreszinski 1981/87) which either required a unique ground state or else assumed an ergodic property which is equivalent to requiring the vanishing of the matrix elements in the ground state sector in which case the decay or correlations becomes equivalent to clustering.

We will show that, for a local Hamiltonian on a finite dimensional lattice with a gap ΔE between the ground state sector and the rest of the spectrum that $\langle \phi_X \overline{\phi}_Y \rangle$ is superpolynomially small in $\mathrm{dist}(X, Y)$ (in (Hastings 2007), stronger exponential results are obtained). To show this, we define a set of parameter dependent Hamiltonians $\mathcal{H}_\theta$ as follows. Let X' denote the set of sites i such that $\mathrm{dist}(X, i) \leq \mathrm{dist}(X, Y)/2$, as shown in Fig. 4.2. Then define $H_\theta = R(X', \theta) H R(X', -\theta)$. Clearly, then, as $R(X', -\theta)$ is a unitary transformation, $\mathcal{H}_\theta$ has the same spectrum of $\mathcal{H}$ and the ground states of $\mathcal{H}_\theta$ are given by $\psi_0^a(\theta)\rangle = R(X', \theta)\psi_0(\theta)\rangle$.

Thus,

$$\partial_\theta \langle \phi_X \overline{\phi}_Y \rangle_\theta = \partial_\theta \langle R(X', -\theta)\phi_X \overline{\phi}_Y R(X', \theta)\rangle \tag{3.73}$$

$$= \partial_\theta \exp[i\theta]\langle \phi_X \overline{\phi}_Y \rangle$$

$$= i\langle \phi_X \overline{\phi}_Y \rangle,$$

where we used the fact that $X \subset X'$ while $Y \cap X' = 0$ so that $[\overline{\phi}_Y, R(X', \theta)] = 0$ and where we evaluate the derivatives at $\theta = 0$.

Using equation (3.68), however,

$$\partial_\theta \langle \phi_X \overline{\phi}_Y \rangle_\theta = \frac{1}{q} \partial_\theta \mathrm{Tr}(P_0(\theta)\phi_X \overline{\phi}_Y) \tag{3.74}$$

$$= \frac{1}{q} \mathrm{Tr}(P_0(\theta)[\phi_X \overline{\phi}_Y, \mathcal{D}_\theta)]).$$

However, recall that $\mathcal{D}_\theta$ is a sum of terms $\mathcal{D}_Z(\theta)$, arising from the different terms $\partial_\theta H_Z(\theta)$. We have $H_Z(\theta) = R(X', \theta) H_Z R(X', -\theta)$. However, if Z is a subset of X' or if Z is a subset of the complement of X', then $H_Z(\theta) = H_Z(0)$. To see this, note that if H_Z is a subset of the complement of X', then H_Z commutes with $R(X', \theta)$. If H_Z

is a subset of X', then $H_Z(\theta) = R(\Lambda, \theta)H_Z R(\Lambda, -\theta)$, where Λ is the set of all sites. Since $H_Z(\theta)$ commutes with Q, $R(\Lambda, \theta)H_Z R(\Lambda, -\theta) = H_Z$. Thus, the only terms that contribute to $\mathcal{D}_\theta$ are indeed those where Z intersects both X' and the complement of X'. However, the corresponding terms $\mathcal{D}_Z(\theta)$ have small commutator with $\phi_X \bar{\phi}_Y$ by the locality of the quasi-adiabatic evolution operator: we can approximate $\mathcal{D}_\theta$ by an operator localized near the boundary of X', a distance l from sets X and Y and so the commutator $[\phi_X \bar{\phi}_Y, \mathcal{D}_\theta)]$ can be shown to be superpolynomially small after summing over Z.

It is interesting to note that the assumption of a finite dimensional lattice is necessary in this derivation (it comes in when we sum over Z, and is needed to bound the sum of terms by the number of terms times a bound on the norm of each term). We sketch a system which is not finite dimensional, and show how a Goldstone theorem may fail in this case. Consider a random graph with V nodes each having coordination number 3. Consider a set of V spin-1/2 spins, with Hamiltonian

$$H = -\sum_{i,j} J_{ij} \vec{S}_i \cdot \vec{S}_j, \tag{3.75}$$

where the interaction matrix J_{ij} equals 1 if the nodes i, j are connected by an edge on the graph, and zero otherwise. The interaction is ferromagnetic, so pointing all spins up (or in any other direction) gives a ground state. Further, the Hamiltonian is local, using a shortest path metric on the graph to define $\mathrm{dist}(i, j)$. However, a random graph of this form is typically an expander graph (Alon 1986/2003) with a gap in the spectrum of the graph Laplacian, so a spin-wave theory calculation (Ashcroft and Mermin 1976) gives a gap in the magnon spectrum. Thus, this system has a set of degenerate ground states and a gap. However, the spin correlations do not decay, as $\langle \vec{S}_i \cdot \vec{S}_j \rangle = 1/4$ for all i, j.

3.7   Lieb–Schultz–Mattis in higher dimensions

The Lieb–Schultz–Mattis theorem, proven in 1961 (Lieb *et al.* 1961), is a theorem about the spectrum on one-dimensional quantum spin systems with symmetries. We present the theorem in slightly more general form for theories with a conserved $U(1)$ charge, as considered later by Affleck and Lieb (Affleck and Lieb 1986).

Consider a one-dimensional Hamiltonian H, with finite-range interactions (one can consider also sufficiently rapidly decaying interactions; we do not consider this case in order to make the discussion as simple as possible but we encourage the reader to work out what kinds of decay would still allow the theorem to be proven). Assume that the Hamiltonian is translationally invariant, with periodic boundary conditions. Let T be the translation operator, so $[T, H] = 0$.

Then,

Theorem 3.7 *Consider a one-dimensional, periodic, translationally invariant Hamiltonian with finite-range interactions, with N sites, conserved charge Q and ground state ψ_0. Define the ground state filling factor ρ by*

$$\rho = \langle \psi_0, Q\psi_0 \rangle / N. \tag{3.76}$$

Assume that ρ is not an integer. Then, either the ground state is degenerate or the gap between the ground state and the first excited state is bounded by

$$\Delta E \leq \text{const.}/N, \tag{3.77}$$

where the constant depends only on the strength J of the interactions in H and the range of the interactions.

Let us give some examples of the application of this theorem. Consider the one-dimensional Heisenberg model:

$$H = \sum_i \vec{S}_i \cdot \vec{S}_{i+1}, \tag{3.78}$$

with spin-1/2 on each site. We identify the charge q_i by

$$q_i = S_i^z + 1/2, \tag{3.79}$$

so that q_i has integer eigenvalues. The Heisenberg model has not just the $U(1)$ invariance, but instead has a full $SU(2)$ invariance (invariance under rotation). So, if the ground state is nondegenerate, then the ground state has spin 0. Let us indeed assume that it is true that the ground state has spin 0 (one can also prove this by other means), so also the ground state has $S^z = 0$, which corresponds to $\rho = 1/2$, hence ρ is non-integer. Thus, this model meets the conditions of the theorem and so must obey the conclusions: there must be a state within energy of order $1/N$ of the ground state. In fact, the lowest energy state has energy of order $1/N$ above the ground state, and corresponds to a 'spinon' excitation. The model has a continuous energy spectrum in the infinite N limit.

Another model is the Majumdar–Ghosh model (Majumdar and Ghosh 1969):

$$H = \sum_i \vec{S}_i \cdot \vec{S}_{i+1} + (1/2)H = \sum_i \vec{S}_i \cdot \vec{S}_{i+2}. \tag{3.80}$$

This model also meets the conditions of the theorem (it again has $SU(2)$ invariance and the ground state turns out to have total spin 0). So, it must meet the conclusions of the theorem. However, this model meets the conclusions of the theorem in a different way. It has two exactly degenerate ground states as sketched in Figure 3.3, and then a gap to the rest of the spectrum. One ground state has spins 1 and 2 in a singlet, spins 3 and 4 in a singlet, and so on. The other ground state is translated by one, so it has spins 2 and 3 in a singlet, and so on, and finally spins N and 1 in a singlet. So, both ground states are products of singlets. If the constant 1/2 is changed to some number near 1/2, then there appears an exponentially small splitting between the two ground states, and a gap to the rest of the spectrum, which still meets the conclusions of the theorem.

Finally, one can also consider spin systems with spin 1 per site. These systems do *not* meet the conditions of the theorem, since if they have total $S^z = 0$, then they have integer ρ. Hence, the theorem does not imply the existence of a gap for these systems. In fact, such systems may have a unique ground state and a spectral gap, the so-called 'Haldane gap' (Haldane 1983).

We now prove the one-dimensional Lieb–Schultz–Mattis theorem. The proof is variational. First, note that given that $[H, Q] = 0$, without loss of generality we can assume that each term in H commutes with Q. To prove this, let $H = \sum_Z H_Z$. Then,

$$H = \sum_Z H'_Z, \tag{3.81}$$

where

$$H'_Z \equiv \frac{1}{2\pi} \int_0^{2\pi} d\theta \exp(i\theta Q) H_Z \exp(-i\theta Q). \tag{3.82}$$

However, each term H'_Z commutes with Q. So, from now on, without loss of generality, we assume that every term in H commutes with Q.

Define a state ψ_{LSM} by

$$\psi_{LSM} = \left(\prod_{j=1}^{N} \exp(2\pi i \frac{j}{N} q_j) \right) \psi_0. \tag{3.83}$$

One may show that

$$\langle \psi_{LSM}, H\psi_{LSM} \rangle - \langle \psi_0, H\psi_0 \rangle \leq \text{const.}/N. \tag{3.84}$$

To show this, we first show for every Z that

$$\langle \psi_{LSM}, H_Z \psi_{LSM} \rangle - \langle \psi_0, H_Z \psi_0 \rangle \leq \text{const.}/N^2, \tag{3.85}$$

as may be proven using the fact that Z has bounded diameter and that H_Z commutes with Q (we encourage the reader to work through the detailed proof; the bound on the right-hand side will depend on the diameter of Z). We then sum equation (3.85) over Z to arrive at equation (3.84).

So, we have shown that ψ_{LSM} is close in energy to the ground state. We now show that ψ_{LSM} is orthogonal to the ground state, which will complete the variational proof

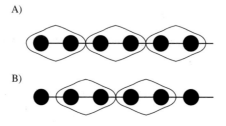

A)

B)

Fig. 3.3   A) One ground state of Majumdar–Ghosh model. Circles indicate lattice sites. Light line around circles indicate that they are in a singlet. B) Another ground state.

of this one-dimensional Lieb–Schultz–Mattis theorem. Note that we can assume that the ground state is an eigenvector of T (otherwise, since T commutes with H, the ground state is degenerate), so that

$$T\psi_0 = z\psi_0, \tag{3.86}$$

for some complex number z with $|z| = 1$. Now consider $T\psi_{LSM}$. We will show that ψ_{LSM} is also an eigenvector of T but with an eigenvalue different from z, which will imply that ψ_{LSM} is orthogonal to ψ_0, completing the proof. We have:

$$T\psi_{LSM} = T\left(\prod_{j=1}^{N} \exp(2\pi i \frac{j}{N} q_j)\right)\psi_0 \tag{3.87}$$

$$= \left\{T\left(\prod_{j=1}^{N} \exp(2\pi i \frac{j}{N} q_j)\right)T^{-1}\right\}T\psi_0$$

$$= z\left\{T\left(\prod_{j=1}^{N} \exp(2\pi i \frac{j}{N} q_j)\right)T^{-1}\right\}\psi_0$$

$$= z\left(\prod_{j=1}^{N} \exp(2\pi i \frac{j}{N} q_{j-1})\right)\psi_0,$$

where we define $q_0 = q_N$ (recall that we have periodic boundary conditions). Thus, shifting the summation variable j by 1 and recalling that q_j has integer eigenvalues so $\exp(2\pi i q_N) = 1$, we have

$$T\psi_{LSM} = z\left(\prod_{j=1}^{N} \exp(2\pi i \frac{j+1}{N} q_j)\right)\psi_0 \tag{3.88}$$

$$= z\left(\prod_{j=1}^{N} \exp(2\pi i \frac{j}{N} q_j)\right)\exp(2\pi i Q/N)\psi_0$$

$$= z\left(\prod_{j=1}^{N} \exp(2\pi i \frac{j}{N} q_j)\right)\exp(2\pi i \rho)\psi_0$$

$$= z\exp(2\pi i \rho)\psi_{LSM}.$$

In the above equation, we have assumed, without loss of generality, that ψ_0 is an eigenvector of Q (otherwise the ground state degeneracy follows automatically since $[Q, H] = 0$), so $\exp(2\pi i Q/N)\psi_0 = \exp(2\pi i \rho)\psi_0$. However, since we assume that ρ is not an integer, we find that ψ_{LSM} is an eigenvector T with eigenvalue $z\exp(2\pi i \rho)\psi_{LSM}$ which differs from z, completing the proof.

A useful exercise for the reader is the following: consider the state which is the symmetric combination of the two ground states of the Majumdar–Ghosh model mentioned above, and call this state ψ_0. It is an eigenvector of T with eigenvalue $+1$. Now, construct the state ψ_{LSM} and verify that it is close to (within distance $1/N$) the state which is the anti-symmetric combination of the two ground states and that it is an eigenvector of T with eigenvalue -1. This helps give an intuitive explanation of what is going on in the above proof: for states which are dimerized, meaning that the ground state is a product of singlets as in the Majumdar–Ghosh model, the operator $\prod_{j=1}^{N} \exp(2\pi i \frac{j}{N} q_j)$, called the Lieb–Schultz–Mattis twist operator, counts the number of dimers crossing a given line. That is, this operator multiplies the state by minus one if an odd number cross and by plus one otherwise, up to $1/N$ corrections. The higher dimensional proof we discuss below re-uses some ideas of twist operators, and can be intuitively viewed along the same lines of 'counting dimers' when applied to systems with dimerized ground states.

The above proof technique simply does not work in two dimensions. Suppose we have a two-dimensional system, with size L in each direction which is periodic in one direction. The energy of the state ψ_{LSM} is a *constant* amount above the ground state, and does not go to zero as L goes to infinity. The reason is that the energy per site is of order $1/L^2$, while there are a total of L^2 sites. In contrast, in the one-dimensional case, there are only L sites. This different scaling between one and two dimensions should be very familiar from statistical mechanics: two dimensions is the 'lower critical dimension' to break a continuous symmetry.

There is another physical reason why the one dimensional proof fails in two dimensions. In one dimension, there are only two possibilities (Affleck and Lieb 1986). Either, the system has a continuous spectrum, or, if it has a degenerate ground state and a spectral gap (as in the case of the Majumdar–Ghosh model), there is a discrete symmetry breaking, with a local order parameter. For example, in the Majumdar–Ghosh model, for any i, the operator $\vec{S}_i \cdot \vec{S}_{i+1}$ has non trivial action in the ground state subspace. In contrast, in two-dimensional system, there might also be topological order. We might have a system with degenerate ground states but for which no local operator has non trivial action in the ground state subspace (i.e. every local operator is close to a multiple of the identity when projected into the ground state subspace).

Topological order can arise in spin systems; one set of proposals involves the idea that a so-called 'short-range resonating valence bond' state describes the ground state (Sutherland 1988/89). Such states are a liquid-like superposition of various singlet configurations, that are in many ways very physically similar to dimer models or to the toric code. So, already on physical grounds we expect that we will need some new technique to prove a Lieb–Schultz–Mattis theorem in higher dimensions. Conversely, we get a nice payoff from such a theorem in more than one dimension: it will rule out the possibility (for certain systems which obey the conditions of the theorem) of having a unique ground state and a spectral gap. If we show then that some system does have a spectral gap then either there is ordinary order (some local operator has nontrivial action in the ground state subspace) or there is topological order. So, such a theorem can be a route to proving topological order.

We now sketch the higher dimensional proof (Hastings 2004, 2005). This is intended only to be a sketch. The statement is that:

Theorem 3.8 *Consider a Hamiltonian H, defined on a finite-dimensional lattice with finite interaction range R and a bound on interaction strength J. Let H have translation invariance in one direction with periodic boundary conditions and have a length L in that direction. Let the lattice have a total of N sites, with N bounded by a constant times a polynomial in L. Let H have conserved charge Q and ground state ψ_0. Define the ground state filling factor ρ by*

$$\rho = \langle \psi_0, Q\psi_0 \rangle / L. \tag{3.89}$$

Assume that ρ is not an integer. Then, either the ground state is degenerate or the gap between the ground state and the first excited state is bounded by

$$\Delta E \leq \mathrm{const.}\log(L)/L, \tag{3.90}$$

where the constant depends only on R,J, q_{max}, and the lattice geometry.

The theorem can be extended to sufficiently rapidly decaying interactions also. The bound that N is at most a polynomial times L implies that the theorem works for aspect ratios of order unity (for example, an L-by-L square lattice in two dimensions) or even aspect ratios which are quite far from unity (an L-by-L^3 square lattice, for example). Note that the bound is slightly weaker than in one dimension (we have $\log(L)/L$ instead of $1/L$).

Finally, the fact that $\rho = \langle \psi_0, Q\psi_0 \rangle / L$ is a minor annoyance in the statement of the theorem. In a spin-1/2 system on a square latttice with odd width (i.e., an L-by-M lattice with translational invariance in the first direction and with M even) we do indeed find that ρ is noninteger. However, for such a system on a lattice of even width, the theorem does not work. In fact, there are counterexamples to a conjectured theorem with even width (a spin ladder, consisting of an L-by-2 system of spin-1/2 spins with Heisenberg $\vec{S}_i \cdot \vec{S}_j$ interactions between nearest neighbour spins has a unique ground state and a gap). However, we expect that if a two-dimensional system on an L-by-M lattice has translation invariance and periodic boundary conditions in both directions, then the goes to zero as both L and M get large. This has not been proven yet. Still, the theorem above covers a wide variety of cases with a minimal number of assumptions (only one direction of translation invariance required).

Further, the translation invariance in at least one direction is a necessary condition. The reader is invited to work out a counter-example if no translation invariance is assumed.

The sketch of the proof is as follows. It again is variational. It is also a proof by contradiction. That is, we assume that the Hamiltonian has a spectral gap ΔE and use this assumed spectral gap to construct a variational state which has low energy. If the initial gap is large enough (larger than a constant times $\log(L)/L$) the variational state will have energy less than $\log(L)/L$ proving the theorem by contradiction.

We begin by defining a parameter-dependent family of Hamiltonians, $H(\theta_1, \theta_2)$. These Hamiltonians are defined by 'twisting the boundary conditions' in one particular direction, the direction in which the lattice is translation invariant. Let us label the coordinate of a site i in this direction by $x(i)$, with $0 \leq x(i) < L$, as shown in Figure 3.4.

To define the flux insertion operator, we need to define the Hamiltonian with twisted boundary conditions. Let Q_X be defined by

$$Q_X = \sum_i^{1 \le x(i) \le L/2} q_i, \qquad (3.91)$$

where $x(i)$ is the $\hat{x}$-coordinate of site i. That is, Q_X is the total charge in the half of the system to the left of the vertical line with $x = L/2 + 1$ and to the right of $x = 0$. Let

$$H(\theta_1, \theta_2) = \sum_Z H_Z(\theta_1, \theta_2), \qquad (3.92)$$

where $H_Z(\theta_1, \theta_2)$ is defined as follows. If the set Z is within distance R of the vertical line $x = 0$, then $H_Z(\theta_1, \theta_2) = \exp(i\theta_1 Q_X) H_Z \exp(-i\theta_1 Q_X)$; if the set Z is within distance R of the vertical line $x = L/2$, then $H_Z(\theta_1, \theta_2) = \exp(-i\theta_2 Q_X) H_Z \exp(i\theta_2 Q_X)$; otherwise, $H_Z(\theta_1, \theta_2) = H_Z$. Note that,

$$H(\theta, -\theta) = \exp(i\theta Q_X) H \exp(-i\theta Q_X). \qquad (3.93)$$

This unitary equivalences implies that $H(\theta, -\theta)$ has the same spectrum as H which will be useful below.

The introduction of the two different vertical lines is an important technical trick. We now define an operator W_1 which generates the quasi-adiabatic evolution along the path where θ_1 evolves from 0 to 2π and $\theta_2 = 0$. That is, define $\mathcal{D}_\theta^1$ to generate the quasi-adiabatic evolution for $H_s = H(s, 0)$ and let

$$W_1 = \exp\left(\int_0^{2\pi} d\theta \mathcal{D}_\theta^1\right), \qquad (3.94)$$

where the exponential is θ ordered. Similarly, let W_2 generate quasi-adiabatic evolution along the path where θ_2 evolves from 0 to 2π and $\theta_1 = 0$ and let $\mathcal{D}_\theta^2$ generate the quasi-adiabatic evolution for $H_s = H(0, -s)$. Finally, let W generate quasi-adiabatic evolution along the path $\theta_1 = -\theta_2 = \theta$ as θ evolves from 0 to 2π.

An important point: we do *not* assume that the gap remains open along the paths above used to define W_1, W_2. We simply use the assumed initial gap at $\theta_1 = \theta_2 = 0$ and then evolve quasi-adiabatically as if the gap remained open along the path.

Using locality of the quasi-adiabatic evolution operators, if the gap is sufficiently large, one may show that

$$W_1 W_2 \approx W_2 W_1 \approx W, \qquad (3.95)$$

where the approximation means that $W_1 W_2 - W$ is small in operator norm. The size that the gap needs to be and the magnitude of the error $\|W_1 W_2 - W\|$ both depend on the quasi-adiabatic evolution operator we use. Roughly, the quasi-adiabatic evolution operator $\mathcal{D}^1$ is supported near the line $x(i) = 0$, up to a length scale which is inversely proportional to the gap. We need this length scale to be small compared to $L/4$ so that we can approximate the operator $\mathcal{D}^1$ by an operator supported within distance $L/4$ of the line $x(i) = 0$; we make the same approximation for $\mathcal{D}^2$ so that in this case, the operators $\mathcal{D}^1$ and $\mathcal{D}^2$ can be approximated by operators supported on disjoint sets. Using exact quasi-adiabatic evolution operators, one finds that the gap needs to be at least $f(l)/L$ for some function f growing slower than any polynomial, while for Gaussian operators one can choose $f(L)$ to be a constant times $\log(L)$.

We consider the variational state:

$$W_1\psi_0. \tag{3.96}$$

We wish to show that this state has low energy. Note that W_1 is unitary. The Hamiltonian H is a sum of terms H_Z. We will show that each term H_Z has roughly the same expectation value in the state $W_1\psi_0$ as it does in the ground state. If Z is far from the line $x(i) = 0$ (for example, far can mean that the distance is at least $L/4$), then this follows directly from the locality of the quasi-adiabatic evolution operator: H_Z almost commutes with W_1:

$$\langle W_1\psi_0, H_Z W_1\psi_0 \rangle \approx \langle \psi_0, H_Z \psi_0 \rangle. \tag{3.97}$$

Conversely, if H_Z is near the line $x(i) = 0$ (say, the distance is less than $L/4$), H_Z almost commutes with W_2. So:

$$\langle W_1\psi_0, H_Z W_1\psi_0 \rangle \approx \langle \psi_0, W_1^\dagger W_2^\dagger H_Z W_2 W_1 \psi_0 \rangle \tag{3.98}$$

$$\approx \langle \psi_0, W^\dagger H_Z W \psi_0 \rangle.$$

Note, however, that W describes evolution along a path of Hamiltonians which are all unitarily equivalent to $H(0,0)$ so all Hamiltonians along this path have the same spectral gap. Further, $H(2\pi, -2\pi) = H(0,0)$ So, $W\psi_0 = w\psi_0$ for some complex

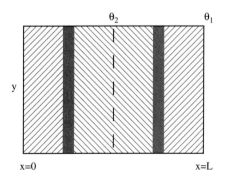

Fig. 3.4 Twists in boundary conditions are applied at two places, at $x(i) = 0 = L$ (along the boundary of the systems) and at $x(i) = L/2$ (along the dashed line). We want to approximate W_1 by an operator supported on the upward slanting grey lines (near $x(i) = 0$) and to approximate the operator W_2 by an operator supported on the downward slanting grey lines, so that W_1, W_2 will approximately commute. This requires a gap sufficiently large compared to $1/L$.

number w with $|w| = 1$ (we will worry about this phase w in the next paragraph, for now it doesn't matter). So, $\langle W_1\psi_0, H_Z W_1\psi_0\rangle \approx \langle\psi_0, H_Z\psi_0\rangle$ for all Z.

This proves that $W_1\psi_0$ is a low energy state. We now need to show that $W_1\psi_0$ is orthogonal to the ground state. We do this by showing that it has a different expectation value for the translation operator, T, than the ground state does. Suppose $\langle\psi_0, T\psi_0\rangle = z$ for some complex number z. Then, consider $\langle\psi_0, W_1^\dagger T W_1\psi_0\rangle$. This equals $\langle\psi_0, W_1^\dagger (T W_1 T^{-1}) T\psi_0\rangle = z\langle\psi_0, W_1^\dagger (T W_1 T^{-1})\psi_0\rangle$. This is approximately equal to $z\langle\psi_0, W_2^\dagger W_1^\dagger (T W_1 T^{-1}) W_2\psi_0\rangle$. Note that $W_2^\dagger W_1^\dagger$ is close to $W^\dagger$, so $W_1 W_2\psi_0$ is close to $w\psi_0$ where w is some phase as mentioned above. Similarly, $W_2^\dagger W_1^\dagger (T W_1 T^{-1}) W_2\psi_0$ is close to $w'\psi_0$ for some other complex number w' with $|w'| = 1$. To see this, note that $(T W_1 T^{-1})$ describes quasi-adiabatic evolution where we twist the boundary conditions along the line $x(i) = 1$ rather than along $x(i) = 0$, so $(T W_1 T^{-1}) W_2$ is close to some operator W' which describes quasi-adiabatic evolution of a Hamiltonian with twisted boundary condition by θ along line $x(i) = 1$ and by $-\theta$ along $x(i) = L/2$. So, $\langle W_1\psi_0, T W_1\psi_0\rangle$ is close to $\overline{w}w'z$. We now just need to work out the phases w, w'.

However, using the property (3.60), one can show that $w = \exp(2\pi i\langle\psi_0, Q_X\psi_0\rangle)$. Similarly, the phase w' depends on the expectation value of q_i summed over $1 < x(i) \leq L/2$. So,

$$\overline{w}'w = \exp(2\pi i\langle\psi_0, \sum_{i,x(i)=1} q_i\psi_0\rangle). \tag{3.99}$$

This expectation value is noninteger by assumption. So, $W_1\psi_0$ has a different expectation value for T than ψ_0, so it is orthogonal to ψ_0.

Some heuristic comments on why physically our proof of the theorem is necessarily a proof by contradiction. That is, why we assumed a gap at the beginning of the proof. We use the idea of twisting the boundary conditions. A state such as an anti-ferromagnet will strongly resist this twist in boundary conditions; that is, the ground state energy will change by an amount of order unity when we impose this boundary twist in a two-dimensional system (and by an even larger amount in higher dimensions). Thus, for a state such as an anti-ferromagnetic (which is gapless), there is no reason to expect that the procedure we described of twisting boundary conditions and following the quasi-adiabatic evolution of the state along the path will give any useful results. However, suppose we have a system which has a gap. By the theorem we have just sketched, such a system cannot have a unique ground state and then a gap to the next lowest energy state. So, we instead want to consider a state with a degenerate ground state and a gap. Such a system could be a valence bond solid or a resonating valence bond system, among other possiblities. In such a system, twisting the boundary conditions does not lead to a large energy cost. Instead, it typically costs only an exponentially small amount of energy to twist the boundary conditions. Then, when we start at one ground state and quasi-adiabatically evolve it, twisting the boundary angle θ from 0 to 2π, we transform it a state close to an orthogonal ground state. This is analogous to the discussion in one dimension, where starting with one of the ground states of the Majumdar–Ghosh model and constructing the state ψ_{LSM}, gave us something close to the other ground state. So, the physical idea of the higher-dimensional proof is that if there is a sufficiently large gap from the ground state sector to the rest of the spectrum, then we can use the quasi-adiabatic continuation to construct a unitary that transforms one ground state into another.

3.8 What is a phase?

What is a phase of a quantum many-body system? We are used to the idea that physical systems have distinct phases, with phase transitions between them. For example, water can appear as ice, water, or steam (and further, there are many distinct phase of ice). This discussion of the properties of water is a discussion of systems at nonzero temperature, while our focus in this notes is on quantum systems at zero temperature, but many of the same phenomena occur in both cases. For example, we do not actually consider steam and water to be distinct phases of matter. While usually water turns into steam by being boiled (a phase transition, where the energy is nonanalytic in the thermodynamic limit), we can also move from water to steam without any phase transition, by following a path in the two-dimensional plane of temperature and pressure. A similar phenomenon occurs in the transverse field Ising model. Suppose we consider the model with an additional parallel magnetic field so that the Hamiltonian is

$$H = -J \sum_{i=1}^{N-1} S_i^z S_{i+1}^z + B \sum_{i=1}^{N} S_i^x + H \sum_{i=1}^{N} S_i^z. \qquad (3.100)$$

Suppose $J \gg B$ and $H > 0$. Then, the system has a unique ground state; at $B = 0$, this ground state is the state with all spins up, while for $B > 0$, there are quantum fluctuations about this state.

When H changes sign, the ground state changes from all spins up to all spins down in the case $B = 0$, crossing a phase transition. This is a zero temperature phase transition. In the case of $B = 0$, this phase transition is a level crossing: at $H = 0$, there are two exactly degenerate ground states. For $B \neq 0$, this level crossing becomes

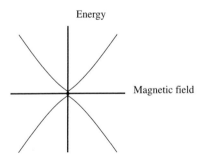

Fig. 3.5 Energy of lowest two states as a function of parallel magnetic field H. This is a sketch. The crossing is an avoided crossing but the splitting between states is exponentially small at $H = 0$. At H of order $1/N$, there is another avoided crossing as the energy gap becomes of order unity.

an avoided crossing as in Figure 3.5. Recall that we said that for $B \neq 0$ but $H = 0$, the system has two ground states with an exponentially small splitting between them. The behaviour of the two lowest energy states as a function of H can be roughly understood in the following toy model. Suppose that H is very small. Then we can focus on just the two lowest energy states and use equation (3.39) to arrive at the following two-by-two Hamiltonian (we obtain this Hamiltonian by projecting the term $\sum_{i=1}^{N} S_i^z$ into the ground state subspace):

$$H = \begin{pmatrix} t & HmN \\ HmN & -t, \end{pmatrix}, \tag{3.101}$$

where t is some exponentially small splitting between the two lowest states and $N = |\Lambda|$ is the size of the system.

By a change of basis (going to symmetric and anti-symmetric combinations of the two ground states), we arrive instead at

$$H = \begin{pmatrix} HmN & t \\ t & -HmN \end{pmatrix}. \tag{3.102}$$

The two different basis vectors here correspond to the spin up and spin down ground states.

While this Hamiltonian is valid for small H, for larger H we need to worry about the excited states. However, this Hamiltonian already reveals the essential point, namely that since t is exponentially small as a function of N, in the limit of $N \to \infty$, the ground states energy per site is a nonanalytic function of H. However, this behaviour is a lot like changing from water to steam by boiling: we can also move from $H < 0$ to $H > 0$ without crossing a phase transition. Instead, one should follow the path of first making H large and negative, then decreasing J until $J << B$, then changing the sign of H (which does not involve a phase transition since $J << B$) and then increasing J, as shown in Fig. (3.6).

We are motivated by this analysis to adopt the following definition of a quantum phase: two Hamiltonians, H_0 and H_1 describe systems in the same quantum phase if both H_0 and H_1 have a spectral gap, and one can find a smooth path H_s connecting

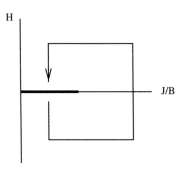

Fig. 3.6 Path to follow. Thickened line on axis denotes B/J less than the critical value.

H_0 and H_1 which keeps the gaps open and keeps the Hamiltonian local. (As a technical point for those interested, in some cases it may be more appropriate to consider a 'stable limit' when describing equality of quantum phases, as in the case of topologically ordered phases of free fermion systems (Kitaev 2009)).

Now, one may choose to have a more refined notion of a quantum phase, which takes into account symmetries; for example, one might wish to insist that there is a path connecting H_0 to H_1 which respects symmetries such as an Ising symmetry. We do not consider this kind of restriction on the definition here. Our interest is instead the case of systems which are in distinct phases even without any assumptions on the symmetry. That is, Hamiltonians which *cannot* be connected by such a smooth path.

We can use the ideas developed in these notes to prove that certain Hamiltonians cannot be connected by a smooth path of local Hamiltonians without closing the gap or without the path length being long (where 'long' means that the path length diverges as the system size gets large). Consider, for example, a toric code on a torus. This has four ground states and a spectral gap. Call this Hamiltonian H_0. Consider instead a Hamiltonian consisting of two copies of the transverse field Ising model on a torus with $J \gg B$. Call this Hamiltonian H_1 (if one wants a better statement of H_1, break the square lattice into two different sublattices, and have interactions only between spins on a given sublattice, so that way we have two copies of the transverse field Ising model with the same number of degrees of freedom as in the toric code system). Note that H_0, H_1 both have four ground states and a gap. So, can we find a path connecting H_0 to H_1? The answer is no. If Such a path existed, then we could use quasi-adiabatic continuation to evolve the four ground states of H_0 to produce some linear combination of the four ground states of H_1. However, this would imply that (recall the discussion in subsection 3.4.1 and the fact that we have a Lieb–Robinson bound for the quasi-adiabatic continuation operators) the ground states of H_1 are also topologically ordered. Since this is not true, no such path can exist. The reason we need to assume that the path is not 'long' is that the length of the path plays the role of time in the Lieb–Robinson bound, and recall that in the discussion of the behaviour of topological order under time evolution we only showed that topological order could not appear after evolution for a short time, but not for arbitrary time.

We can use similar arguments to show that Hamiltonians cannot be connected by such a smooth path even when the ground state is unique (again, without closing the gap and without the path being long). One of the key properties of systems like the toric code is the particular set of expectation values they have for certain operators they have called string operators; these are operators which are products of single-site operators around a loop which are analogous to Wilson loops in gauge theories. One can define 'dressed operators' by quasi-adiabatically continuing these string operators along a path. Thus, given a string operator O for a Hamiltonian H_0, we can define an operator

$$\tilde{O} \equiv UOU^\dagger, \tag{3.103}$$

where

$$U = \mathcal{S} \exp(i \int_0^1 ds \mathcal{D}_s),$$ (3.104)

where the calligraphic s in front of the exponential denotes that it is an s-ordered exponential. Then, the operator $\tilde{O}$ has the same expectation value in the ground state of H_1 as O does in the ground state of H_0. This dressed operator O is precisely equal to the operator $O(s = 1)$ as described by the evolution of equation (3.71).

Further, if two operators O, O' anti-commute with each other, then the operators $\tilde{O}, \tilde{O}'$ also anti-commute with each other. Now, consider the transverse field Ising model in the phase $B \gg J$. This has a unique ground state. Similarly, the toric code on a sphere has a unique ground state. However, these two models cannot be connected by a continuous path. To show this, note that if they were connected, then we could also connect the toris code to the transverse field Ising model in the phase $B \neq 0, J = 0$. However, the ground state of this model with $B \neq 0, J = 0$ can be shown to be inconsistent with the properties of such dressed string operators. Here is a sketch (this sketch depends on properties of the toric code which are not discussed in these notes and need to be read elsewhere): Consider an electric loop operator, indicated as the solid line in Fig. 3.7(A). Call this operator E. Let the dashed line represent a magnetic loop operator, which we call M. These operators commute with each and both have expectation value 1 in the toric code ground state. In Fig. 3.7(B), we show that the operator E can be written as a product of two different operators, which we call E_1 and E_2 which act on part of the loop. Note that $\{E_1, M\} = \{E_2, M\} = 0$. Hence, the expectation value of $E_1 M E_2$ is equal to minus 1 in the toric code ground state. Now consider the dressed operators $\tilde{E}_1, \tilde{E}_2, \tilde{M}$ assuming the existence of a path connecting the toric code to the transverse field Ising model. Since $\tilde{E}_1 \tilde{E}_2$ would have expectation value 1 in the transverse field Ising model ground state, ψ_1, the state $\tilde{E}_1 \tilde{E}_2 \psi_1$ must be simply the product state of all spins pointing along the transverse magnetic field. However, using locality of the dressed operators, $\tilde{E}_1$ can be approximated by an operator supported near the support of E_1. Hence, the

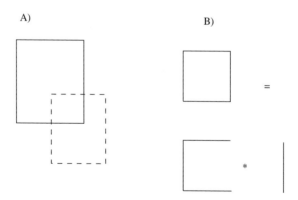

A)

B)

=

*

Fig. 3.7

state $\tilde{E}_2 \psi_1$ must have all of the spins which are far from the support of E_1 pointing approximately along the transverse magnetic field. Thus, the state $\tilde{E}_2 \psi_1$ has all of its spins, except those near the upper and lower ends of the support of E_2 (those close to the support of both E_1 and E_2) pointing approximately along the magnetic field. Similarly, the operator $\tilde{M}$ can be approximated by an operator supported near the support of M; however, this means that $\tilde{M}$ can be approximated by an operator, which we call $\tilde{M}'$, which is not supported near the upper and lower ends of E_2. As noted, the spins away from the upper and lower ends of the support of E_2 are aligned approximately along the magnetic field, and those are the only spins in the support of $\tilde{M}'$. Since $\tilde{M}\psi_1 = \psi_1$, we have $\tilde{M}'\psi_1$ close to ψ_1 and so, using the fact that the spins in the support of $\tilde{M}'$ are almost aligned with the field in the state $\tilde{E}_2 \psi_1$ and in the state ψ_1, we find that $\tilde{M}'\tilde{E}_2 \psi_1$ is close to $\tilde{E}_2 \psi_1$. Thus, $\tilde{M}\tilde{E}_2 \psi_1$ is close to $\tilde{E}_1 \psi_1$. Hence, we find that the expectation value of $\tilde{E}_1 \tilde{M} \tilde{E}_2$ is close to unity in the transverse field Ising model, while it was close to minus 1 in the toric code, giving a contradiction.

One may also choose to take into account other (anti-unitary) symmetries such as time-reversal symmetry. These symmetries are well-understood in the non-interacting case (Kitaev 2009), but only a limited understanding has been obtained in the interacting case (Fidkowski and Kitaev 2009). In general, even without symmetries the classification of phases of matter of interacting systems under the definition above is only in the earliest stages. One interesting case is that lattice Hamiltonians are known, the so-called 'Levin–Wen models' (Levin and Wen 2005), which realize certain two-dimensional unitary topological quantum field theories (TQFTs), in particular those theories which are quantum doubles. Thus, a classification of interacting phases of matter requires a classification of these TQFTs. Some results on classification of TQFTs were obtained in (Rowell *et al.* 2009). However, it is likely that the full classification of lattice models includes many other phases in addition to those described by TQFTs, so this classification is a problem for the future.

3.9 Stability of topologically ordered phase

In this section, we briefly mention certain recent results on the stability of quantum phases. The analysis throughout these notes has always dealt with systems with a gap. In some cases, we were able to prove either the absence of a gap or the degeneracy of a ground states (such as in the Lieb–Schultz–Mattis theorem). However, the reader may be wondering: how do we know a system has a gap? Similarly, suppose we have proven that a certain model, such as a toric code or Levin–Wen model, is in a topologically nontrivial phase and cannot be connected to a topologically trivial phase without closing the gap. However, what happens if we slightly perturb the Hamiltonian? Suppose we consider a Hamiltonian

$$H = H_0 + sV, \tag{3.105}$$

where H_0 is some unperturbed Hamiltonian describing a topologically nontrivial phase, s is some real number, and V is a perturbation. Does the model remain in the same phase for sufficiently small s? Does the gap remain open?

The interesting question here is to consider the case in which V is a sum of local terms. Thus, we want

$$V = \sum_Z V_Z, \tag{3.106}$$

where the operator norms $\|V_Z\|$ decay rapidly as a function of the diameter of the set Z (just as we required a similar decay on the norms of the terms $\|H_Z\|$ in the Hamiltonian H_0). A very elementary result is that for any given system size, there is an s_0 such that for $|s| < s_0$ the gap remains open: simply use the fact that for any given system size, the norm of the operator V is finite, and the gap for $s > 0$ is lower bounded by

$$\Delta E(s) \geq \Delta E(0) - 2s\|V\|, \tag{3.107}$$

where $\Delta E(s)$ denotes the gap as a function of s and the factor of 2 in front of the second term occurs because the ground state energy increases by at most $s\|V\|$ while the first excited state energy decreases by at most the same amount.

So, we may take $s_0 = \Delta E(0)/4\|V\|$. However, such a bound, while elementary, is also fairly useless, since it leads to an s_0 which tends to zero as the system size tends to infinity. Instead, we want a bound which is a uniform function of system size. Such bounds were provided in (Bravyi, Hastings, and Michalakis 2010; Bravyi and Hastings 2010). We will not review them here, except to note that using such bounds one can then prove (using quasi-adiabatic continuation) that many of the properties of the topologically ordered system (such as ground state splitting, braiding, fusion rules, etc.) remain the same in this phase. An interesting open question is to understand the behaviour of topological entanglement entropy as a function of perturbation. Perhaps some smoothed definition of topological entanglement entropy exists (smoothing over different boundaries?) such that it can also be proven to be invariant under perturbations? These problems, and problems like the classification of different phases of lattice quantum systems, are problems for the future.

Acknowledgements

I thank S. Bravyi, T. Koma, T. Loring, S. Michalakis, F. Verstraete, and X.-G. Wen for useful collaboration on these and related ideas.

References

Affleck, I. and Lieb, E. H. (1986). *Lett.Math.Phys.* 12, 57.

Alon, N. Combinatorica (1986). J. Friedman, Conf. Proc. of the Annual ACM Symposium on Theory of Computing, 720 (2003); J. Friedman, **6(2)**, 83, preprint cs/0405020.

Ashcroft, N. W. and Mermin, N. D. Solid State Physics, Chapter 33, (Harcourt Brace College Publishers, New York 1976).

Bravyi, S. Hastings, M. B. and Verstraete, F. (2006). 'Lieb-Robinson bounds and the generation of correlations and topological quantum order', *Phys. Rev. Lett.* **97**, 050401.

Bravyi, S. and Hastings, M. B. (2010). 'A short proof of stability of topological order under local perturbations', e-print arXiv:1001:4363.

Bravyi, S. Hastings, M. B. and Michalakis, S. (2010). 'Topological quantum order: stability under local perturbations', e-print arXiv:1001:0344.

Dziubanski, J. and Hernández, E. (1998). 'Band-limited wavelets with subexponential decay', **41**, 398.

Fidkowski, L. and Kitaev, A. arXiv:0904.2197.

Freedman, M. H. Kitaev, A. Larsen, M. J. and Wang, Z. (2001). 'Topological Quantum computation', e-print quant-ph/0101025.

Haldane, F. D. M. (1983). *Phys. Rev. Lett.* **50**, 1153.

Hastings, M. B. Preprint (2010) arXiv:1001.5280.

Hastings M. B. (2004). *Phys. Rev. B* **69**, 104431.

Hastings, M. B. (2004). 'Locality in quantum and Markov dynamics on lattices and networks', *Phys. Rev. Lett.* **93**, 140402.

Hastings, M. B. (2005). *Europhys. Lett.* **70**, 824.

Hastings M. B. and Koma, T. (2006). *Comm. Math. Phys.* **265**, 781.

Hastings, M. B. (2007). 'Quasi-adiabatic continuation in gapped spin and fermion systems: Goldstone's theorem and flux periodicity', preprint cond-mat/0612538, JSTAT, P05010.

Hastings, M. B. and Michalakis, S. (2009) 'Quantization of hall conductance for interacting electrons without averaging assumptions', *Comm. Math. Phys.* Preprint arXiv:0911.4706.

Ingham, A. E. (1934). 'A note on Fourier transforms', *J. London Math. Soc.* **9**, 29.

Kitaev, A. (2003). 'Fault-tolerant quantum computation by anyons', *Ann. Phys.* **303**, 2.

Kitaev, A. and Preskill, J. (2006). hep-th/0510092, *Phys. Rev. Lett.* **96**, 110404.

Kitaev, A. Preprint (2009). arXiv:0901.2686.

Lieb, E. H. Schultz, T. D. and Mattis, D. C. (1961). *Ann. Phys.* (N. Y.) **16**, 407.

Lieb E. H. and Robinson, D. W. (1972). *Comm. Math. Phys.* **28**, 251.

Landau, L. Fernando Perez, J. and Wreszinski, W. F. (1981). *J. Stat. Phys.* **26**, 755.

Levin, M. A. and Wen, X.-G. (2005). "String-net condensation: A physical mechanism for topological phases", *Phys. Rev.* **B71**, 045110.

Levin, M. and Wen, X.-G. (2006). arXiv:cond-mat/0510613, *Phys. Rev. Lett.* **96**, 110405.

Majumdar, C. K. and Ghosh, D. J. (1969). *Math. Phys.* 10, 1388.

Nachtergaele B. and Sims, R. (2006). *Comm. Math. Phys.* **265**, 119.

Osborne, T. J. (2007) 'Simulating adiabatic evolution of gapped spin systems', *Phys. Rev. A*, **75**, 032321.

Rowell, E. Stong, R. and Wang, Z. (2009). "On classification of modular tensor categories", *Comm. Math. Phys.* **292**, 343. arXiv: 0712.1377.

Rokhsar, D. and Kivelson, S. (1989). Phys. Rev. Lett. **61**(1988); Read, N. and Chakraborty, B. *Phys Rev.* B **40**, 7133.

Sutherland, B. Phys. Rev. B **37**, 3786 (1988).

Wen, X. G. and Niu, Q. (1990). 'Ground-state degeneracy of the fractional quantum Hall states in the presence of a random potential and on high-genus Riemann surfaces', *Phys. Rev.* **B41**, p. 9377.

Wreszinski, W. F. (1987). *Fortschr. Phys.* **35**, 379.

4

Entropic fluctuations in quantum statistical mechanics—an introduction

V. JAKŠIĆ[1], Y. OGATA[2], Y. PAUTRAT[3] and C.-A. PILLET[4],[*]

[1]Department of Mathematics and Statistics, McGill University, 805 Sherbrooke Street West, Montreal, QC, H3A 2K6, Canada

[2]Department of Mathematical Sciences, University of Tokyo, Komaba,Tokyo,153-8914, Japan

[3]Laboratoire de Mathématiques, Université Paris-Sud, 91405 Orsay Cedex, France

[4]Centre de Physique Théorique[5], FRUMAM, Université du Sud Toulon–Var, B.P. 20132, 83957 La Garde Cedex, France

[5] CNRS – Universités de Provence, de la Méditerranée et du Sud Toulon–Var
[*] Lecture given by C.-A. Pillet.

Chapter Contents

4.1 Introduction

These lecture notes are the second installment in a series of papers dealing with entropic fluctuations in nonequilibrium statistical mechanics. The first installment Jakšić, Pillet, and Rey-Bellet (2011a) concerned classical statistical mechanics. This one deals with the quantum case and is an introduction to the results of Jakšić, Ogata, Pautrat, and Pillet (2011b). Although these lecture notes could be read independently of Jakšić, Pillet, and Rey-Bellet (2011a), a reader who wishes to get a proper grasp of the material is strongly encouraged to consult Jakšić, Pillet, and Rey-Bellet (2011a) for the classical analogues of the results presented here. In fact, to emphasize the link between the mathematical structure of classical and quantum theory of entropic fluctuations, we shall start the lectures with a *classical* example: a thermally driven harmonic chain. This example will serve as a prologue for the rest of the lecture notes.

The mathematical theory of entropic fluctuations developed in (Jakšić, Pillet, and Rey-Bellet, 2011a), (Jakšić, Ogata, Pautrat, and Pillet, 2011b) is axiomatic in nature. Starting with a general classical/quantum dynamical system, the basic objects of the theory—entropy production observable, finite time entropic functionals, finite time fluctuation theorems and relations, finite time linear response theory—are introduced/derived at a great level of generality. The axioms concern the large time limit $t \to \infty$, that is the existence and the regularity properties of the limiting entropic functionals. The introduced axioms are natural and minimal (that is necessary to have a meaningful theory), ergodic in nature, and typically difficult to verify in physically interesting models. Some of the quantum models for which the axioms have been verified (spin-fermion model, electronic black box model) are described in Section 4.7.

However, apart for Section 4.6, we shall not discuss the axiomatic approach of Jakšić, Ogata, Pautrat, and Pillet (2011b) here. The main body of the lecture notes is devoted to a pedagogical self-contained introduction to the finite time entropic functionals and fluctuation relations for *finite* quantum systems. A typical example the reader should have in mind is a quantum spin system or a Fermi gas with finite configuration space $\Lambda \subset \mathbb{Z}^d$. After the theory is developed, one proceeds by taking first the thermodynamic limit ($\Lambda \to \mathbb{Z}^d$), and then the large time limit $t \to \infty$. The thermodynamic limit of the finite time/finite volume theory is typically an easy exercise in the techniques developed in the 1970s (the two volume monograph of Bratteli and Robinson provides a good introduction to this subject). On the other hand, the large time limit, as is to be expected, is typically a very difficult ergodic-type problem. In these notes we shall discuss the thermodynamic and the large time limits only in Section 4.6. This section is intended for more advanced readers who are familiar with our previous works and lectures notes. It may be entirely skipped, although even technically less prepared readers my benefit from Sections 4.6.1 and 4.6.6 up to and including the proof of Theorem 4.50.

Let us comment on our choice of the topic. From a mathematical point of view, there is a complete parallel between classical and quantum theory of entropic fluc-tuations. The quantum theory applied to commutative structures (algebras) reduces to the classical theory, that is the classical theory is a special case of the quantum one. There is, however, a big difference in mathematical tools needed to describe the respective theories. Only basic results of measure theory are needed for the finite time theory in classical statistical mechanics. In the noncommutative setting these familiar tools are replaced by the Tomita–Takesaki modular theory of von Neumann algebras. For example, Connes cocycles and relative modular operators replace Radon–Nikodym derivatives. The quantum transfer operators act on Araki–Masuda noncommutative L^p-spaces which replace the familiar L^p-spaces of measure theory on which Ruelle–Perron–Frobenius (classical) transfer operators act and so on. The remarkably beautiful and powerful modular theory needed to describe quantum theory of entropic fluctuations was developed in the 1970s and 1980s, primarily by Araki, Connes, and Haagerup. Although modular theory has played a key role in the mathematical development of nonequilibrium quantum statistical mechanics over the last decade, the extent of its application to quantum theory of entropic fluctuations is somewhat striking. Practically all fundamental results of modular theory play a role. Some of them, like the Araki-Masuda theory of noncommutative L^p-spaces, have found in this context their first application to quantum statistical mechanics.

The power of modular theory is somewhat shadowed by its technical aspects. Out of necessity, a reader of (Jakšić, Ogata, Pautrat, and Pillet, 2011b) must be familiar with the full machinery of algebraic quantum statistical mechanics and modular theory. Finite quantum systems, that is quantum systems described by finite dimensional Hilbert spaces, are special since all the structures and results of this machinery can be described by elementary tools. The purpose of these lecture notes is to provide a self-contained pedagogical introduction to the algebraic structure of quantum statistical mechanics, finite time entropic functionals, and finite time fluctuation relations for *finite quantum systems*. For the most part, the lecture notes should be easily accessible to an undergraduate student with basic training in linear algebra and analysis. Apart from occasional remarks/exercises and Section 4.6, more advanced tools enter only in the computations of the thermodynamic limit and the large time limit of the examples in Sections 4.2 and 4.7. A student who has taken a course in quantum mechanics and/or operator theory should have no difficulties with those tools either.

Apart from from a few comments in Section 4.2.15 we shall not discuss here the Gallavotti–Cohen fluctuation theorem and the principle of regular entropic fluctuations. These important topics concern nonequilibrium steady states and require a technical machinery not covered in these notes.

The lecture notes are organized as follows. In the Prologue, Section 4.2, we describe the classical theory of entropic fluctuations on the example of a classical harmonic chain. The rest of the notes can be read independently of this section. Section 4.3 is devoted to the algebraic quantum statistical mechanics of finite quantum systems. In Sections 4.4 and 4.5 this algebraic structure is applied to the study of entropic functionals and fluctuation relations of finite quantum systems. In Section 4.7 we illustrate the results of Sections 4.4 and 4.5 on examples of fermionic systems. Large deviation theory and the Gärtner–Ellis theorem play a key role in entropic fluctuation theorems and for this reason we review the Gärtner–Ellis theorem in Appendix 4.8. Another tool, a convergence result based on Vitali's theorem, will be often used in the lecture notes, and we provide its proof in Appendix 4.9.

4.2 Prologue: a thermally driven classical harmonic chain

In this section we will discuss a very simple *classical* example: a finite harmonic chain $\mathcal{C}$ coupled at its left and right ends to two harmonic heat reservoirs $\mathcal{R}_L$, $\mathcal{R}_R$. This model is exactly solvable and allows for a transparent review of the classical theory of entropic fluctuations developed in (Jakšić, Pillet, and Rey-Bellet, 2011a). Needless to say, models of this type have a long history in the physics literature and we refer the reader to (Lebowitz, and Spohn, 1977) for references and additional information. The reader should compare Section 4.5, which deals with the nonequilibrium statistical mechanics of an open quantum systems, with the example of an open classical system described here. The same remark applies to Section 4.7.6, where we study the nonequilibrium statistical mechanics of ideal Fermi gases.

4.2.1 The finite harmonic chain

We start with the description of an isolated harmonic chain on the finite 1D-lattice $\Lambda = [A, B] \subset \mathbb{Z}$ (see Fig. 4.1 below). Its phase space is

$$\Gamma_\Lambda = \{(p, q) = (\{p_x\}_{x \in \Lambda}, \{q_x\}_{x \in \Lambda}) \,|\, p_x, q_x \in \mathbb{R}\} = \mathbb{R}^\Lambda \oplus \mathbb{R}^\Lambda,$$

and its Hamiltonian is given by

$$H_\Lambda(p, q) = \sum_{x \in \mathbb{Z}} \left(\frac{p_x^2}{2} + \frac{q_x^2}{2} + \frac{(q_x - q_{x-1})^2}{2} \right),$$

where we set $p_x = q_x = 0$ for $x \notin \Lambda$.

Thus, w.r.t. the natural Euclidian structure of Γ_Λ, the function $2H_\Lambda(p, q)$ is the quadratic form associated to the symmetric matrix

$$h_\Lambda = \begin{bmatrix} \mathbb{1} & 0 \\ 0 & \mathbb{1} - \Delta_\Lambda \end{bmatrix},$$

where Δ_Λ denotes the discrete Laplacian on $\Lambda = [A, B]$ with Dirichlet boundary conditions

$$(-\Delta_\Lambda u)_x = \begin{cases} 2u_A - u_{A+1} & \text{for } x = A; \\ 2u_x - u_{x-1} - u_{x+1} & \text{for } x \in\,]A, B[; \\ 2u_B - u_{B-1} & \text{for } x = B. \end{cases} \tag{4.1}$$

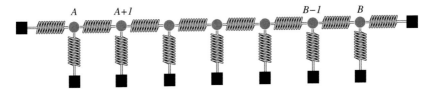

Fig. 4.1 The finite harmonic chain on $\Lambda = [A, B]$.

The equations of motion of the chain,

$$\dot{p} = -(\mathbb{1} - \Delta_\Lambda)q, \qquad \dot{q} = p,$$

define a Hamiltonian flow on Γ_Λ, the one-parameter group $e^{t\mathcal{L}_\Lambda}$ generated by

$$\mathcal{L}_\Lambda = jh_\Lambda, \qquad j = \begin{bmatrix} 0 & -\mathbb{1} \\ \mathbb{1} & 0 \end{bmatrix}.$$

This flow has two important properties:

(i) Energy conservation: $e^{t\mathcal{L}_\Lambda^*} h_\Lambda \, e^{t\mathcal{L}_\Lambda} = h_\Lambda$.

(ii) Liouville's theorem: $\det\left(e^{t\mathcal{L}_\Lambda}\right) = e^{t\,\mathrm{tr}(\mathcal{L}_\Lambda)} = 1$.

An observable of the harmonic chain is a real (or vector) valued function on its phase space Γ_Λ and a state is a probability measure on Γ_Λ. If f is an observable and ω a state, we denote by

$$\omega(f) = \int_{\Gamma_\Lambda} f(p,q)\, d\omega(p,q),$$

the expectation of f w.r.t. ω. Under the flow of the Hamiltonian H_Λ the observables evolve as

$$f_t = f \circ e^{t\mathcal{L}_\Lambda}.$$

In terms of the Poisson bracket

$$\{f, g\} = \nabla_p f \cdot \nabla_q g - \nabla_q f \cdot \nabla_p g,$$

the evolution of an observable f satisfies

$$\partial_t f_t = \{H_\Lambda, f_t\} = \{H_\Lambda, f\}_t.$$

The evolution of a state ω is given by duality

$$\omega_t(f) = \omega(f_t),$$

and satisfies

$$\partial_t \omega_t(f) = \omega_t(\{H_\Lambda, f\}).$$

ω is called steady state or stationary state if it is invariant under this evolution, that is $\omega_t = \omega$ for all t. If ω has a density w.r.t. Liouville's measure on Γ_Λ, that is $d\omega(p,q) = \rho(p,q)\, dpdq$, then Liouville's theorem yields

$$\omega_t(f) = \int_{\Gamma_\Lambda} f \circ e^{t\mathcal{L}_\Lambda}(p,q)\rho(p,q)\,\mathrm{d}p\mathrm{d}q$$

$$= \int_{\Gamma_\Lambda} f(p,q)\rho \circ e^{-t\mathcal{L}_\Lambda}(p,q)\det\left(e^{-t\mathcal{L}_\Lambda}\right)\mathrm{d}p\mathrm{d}q$$

$$= \int_{\Gamma_\Lambda} f(p,q)\rho \circ e^{-t\mathcal{L}_\Lambda}(p,q)\,\mathrm{d}p\mathrm{d}q,$$

and so ω_t also has a density w.r.t. Liouville's measure given by $\rho \circ e^{-t\mathcal{L}_\Lambda}$. If D is a positive definite matrix on Γ_Λ and ω is the centred Gaussian measure with covariance D,

$$\mathrm{d}\omega(p,q) = \det(2\pi D)^{-1/2}\,e^{-D^{-1}[p,q]/2}\,\mathrm{d}p\mathrm{d}q,$$

where $D^{-1}[p,q]$ denotes the quadratic form associated to D^{-1}, then ω_t is the centred Gaussian measure with covariance $D_t = e^{t\mathcal{L}_\Lambda}De^{t\mathcal{L}_\Lambda^*}$.

The thermal equilibrium state of the chain at inverse temperature β is the Gaussian measure with covariance $(\beta h_\Lambda)^{-1}$,

$$\mathrm{d}\omega_{\Lambda\beta}(p,q) = \sqrt{\det\left(\frac{\beta h_\Lambda}{2\pi}\right)}\,e^{-\beta H_\Lambda(p,q)}\mathrm{d}p\mathrm{d}q.$$

Thermal equilibrium states are invariant under the Hamiltonian flow of H_Λ.

4.2.2 Coupling to the reservoirs

As a small system, we consider the harmonic chain $\mathcal{C}$ on $\Lambda = [-N, N]$. The left and right reservoirs are harmonic chains $\mathcal{R}_L$ and $\mathcal{R}_R$ on $\Lambda_L = [-M, -N-1]$ and $\Lambda_R = [N+1, M]$ respectively. In our discussion we shall keep N fixed, but eventually let $M \to \infty$. In any case, the reader should always have in mind that $M \gg N$.

The Hamiltonian of the joint but decoupled system is

$$H_0(p,q) = H_\Lambda(p,q) + H_{\Lambda_L}(p,q) + H_{\Lambda_R}(p,q).$$

The Hamiltonian of the coupled system is

$$H(p,q) = H_{\Lambda_L \cup \Lambda \cup \Lambda_R}(p,q) = H_0(p,q) + V_L(p,q) + V_R(p,q),$$

where $V_L(p,q) = -q_{-N-1}q_{-N}$ and $V_R(p,q) = -q_N q_{N+1}$.

We denote by h_0, h_L, h_R, and h the symmetric matrices associated to the quadratic forms $2H_0$, $2H_L$, $2H_R$, and $2H$ and by $\mathcal{L}_0 = jh_0$ and $\mathcal{L} = jh$ the generators of the corresponding Hamiltonian flows. We also set $v = v_L + v_R = h - h_0$ where v_L and v_R are associated to $2V_L$ and $2V_R$ respectively.

4.2.3 Non-equilibrium reference measure

We shall assume that initially each subsystem is in thermal equilibrium, the reservoirs at temperatures $T_{L/R} = 1/\beta_{L/R}$, and the small system at temperature $T = 1/\beta$. The initial (reference) state is therefore

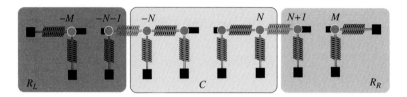

Fig. 4.2 The chain $\mathcal{C}$ coupled at its left and right ends to the reservoirs $\mathcal{R}_L$ and $\mathcal{R}_R$.

$$d\omega_{\Lambda_L \beta_L} \otimes d\omega_{\Lambda\beta} \otimes d\omega_{\Lambda_R \beta_R}(p, q) = Z^{-1} e^{-(\beta_L H_{\Lambda_L}(p,q) + \beta H_\Lambda(p,q) + \beta_R H_{\Lambda_R}(p,q))} dpdq.$$
(4.2)

If the temperatures of the reservoirs are different, the system is initially out of equilibrium. We set $X_L = \beta - \beta_L$, $X_R = \beta - \beta_R$ and $X = (X_L, X_R)$. We call X the thermodynamic force acting on the chain $\mathcal{C}$. X_L and X_R are sometimes called affinities in nonequilibrium thermodynamics (see, e.g., de Groot and Mazur (1969)). When $X = 0$, one has $\beta_L = \beta_R = \beta$ and the joint system is in equilibrium at inverse temperature β for the decoupled dynamics generated by H_0.

In view of the coupled dynamics generated by H, it will be more convenient to use a slightly modified initial state

$$d\omega_X(p, q) = Z_X^{-1} e^{-(\beta_L H_{\Lambda_L}(p,q) + \beta H_\Lambda(p,q) + \beta_R H_{\Lambda_R}(p,q) + \beta V(p,q))} dpdq$$

$$= Z_X^{-1} e^{-(\beta H(p,q) - X_L H_{\Lambda_L}(p,q) - X_R H_{\Lambda_R}(p,q))} dpdq,$$

which, for $X = 0$, reduces to the thermal equilibrium state at inverse temperature β of the joint system under the coupled dynamics. Note that ω_X is the Gaussian measure with covariance

$$D_X = (\beta h - k(X))^{-1}, \qquad k(X) = X_L h_L + X_R h_R,$$

whereas equation (4.2) is Gaussian with covariance $(\beta h_0 - k(X))^{-1}$. Since $h - h_0 = v$ is a rank 4 matrix which is well localized at the boundary of Λ, these two states describe the same thermodynamics.

4.2.4 Comparing states

Under the Hamiltonian flow of H, the state ω_X evolves into $\omega_{X,t}$, the Gaussian measure with covariance

$$D_{X,t} = e^{t\mathcal{L}} D_X e^{t\mathcal{L}^*} = \left(\beta h - e^{-t\mathcal{L}^*} k(X) e^{-t\mathcal{L}}\right)^{-1}.$$

As time goes on, the state $\omega_{X,t}$ diverges from the initial state ω_X. In order to quantify this divergence, we need a way to describe the 'rate of change' of the state, that is a concept of 'distance' between states. Classical information theory provides several candidates for such a distance. In this section, we introduce two of them and explore their physical meaning.

Let ν and ω be two states. Recall that ν is said to be absolutely continuous w.r.t. ω, written $\nu \ll \omega$, if there exists a density, a non-negative function ρ satisfying $\omega(\rho) = 1$, such that $\nu(f) = \omega(\rho f)$ for all observables f. The function ρ is called Radon–Nikodym derivative of ν w.r.t. ω and is denoted $\mathrm{d}\nu/\mathrm{d}\omega$.

The relative entropy of ν w.r.t. ω is defined by

$$S(\nu|\omega) = \begin{cases} \nu\left(-\log \dfrac{\mathrm{d}\nu}{\mathrm{d}\omega}\right) & \text{if } \nu \ll \omega, \\ -\infty & \text{otherwise.} \end{cases} \tag{4.3}$$

Exercise 4.1

1. Show that $\log(x^{-1}) \le x^{-1} - 1$ for $x > 0$, where equality holds iff $x = 1$.
2. Using the previous inequality, show that $S(\nu|\omega) \le 0$ with equality iff $\nu = \omega$. This justifies the use of relative entropy (or rather of $-S(\nu|\omega)$) as a measure of the 'distance' between ν and ω. Note however that $-S(\nu|\omega)$ is not a metric in the usual sense since it is not symmetric and does not satisfy the triangle inequality.

Applying Definition (4.3) to $\omega_{X,t}$ and ω_X, we get

$$-\log\left(\frac{\mathrm{d}\omega_{X,t}}{\mathrm{d}\omega_X}\right) = X_L(H_{\Lambda_L} - H_{\Lambda_L,-t}) + X_R(H_{\Lambda_R} - H_{\Lambda_R,-t}), \tag{4.4}$$

and hence

$$S(\omega_{X,t}|\omega_X) = \omega_{X,t}\left(X_L(H_{\Lambda_L} - H_{\Lambda_L,-t}) + X_R(H_{\Lambda_R} - H_{\Lambda_R,-t})\right)$$
$$= X_L\omega_X\left(H_{\Lambda_L,t} - H_{\Lambda_L}\right) + X_R\omega_X\left(H_{\Lambda_R,t} - H_{\Lambda_R}\right).$$

Since the observable $H_{\Lambda_R,t} - H_{\Lambda_R}$ measures the increase of the energy in the right reservoir during the time interval $[0, t]$ and

$$H_{\Lambda_R,t} - H_{\Lambda_R} = \int_0^t \frac{\mathrm{d}}{\mathrm{d}s} H_{\Lambda_R,s}\, \mathrm{d}s = \int_0^t \{H, H_{\Lambda_R}\}_s\, \mathrm{d}s,$$

we interpret

$$\Phi_R = -\{H, H_{\Lambda_R}\} = \{H_{\Lambda_R}, V_R\} = -p_{N+1}q_N,$$

as the energy flux out of the right reservoir. Similarly,

$$\Phi_L = -\{H, H_{\Lambda_L}\} = \{H_{\Lambda_L}, V_L\} = -p_{-N-1}q_{-N},$$

is the energy flux out of the left reservoir.

Exercise 4.2 Compare the equation of motion of the isolated reservoir $\mathcal{R}_R$ with that of the same reservoir coupled to $\mathcal{C}$. Deduce that the force exerted on the reservoir by the system $\mathcal{C}$ is given by q_N and therefore that $q_N p_{N+1}$ is the power dissipated into the right reservoir.

In terms of fluxes, we have obtained the following *entropy balance relation*

$$S(\omega_{X,t}|\omega_X) = -\int_0^t \omega_X(\sigma_{X,s})\,\mathrm{d}s, \qquad (4.5)$$

where

$$\sigma_X = X_L \Phi_L + X_R \Phi_R.$$

This bilinear expression in the thermodynamic forces and the corresponding fluxes has precisely the form of entropy production as derived in phenomenological nonequilibrium thermodynamics (see, e.g., Section IV.3 of de Groot and Mazur (1969)). For this reason, we shall call σ_X the *entropy production* observable and

$$\Sigma^t = \frac{1}{t}\int_0^t \sigma_{X,s}\,\mathrm{d}s, \qquad (4.6)$$

the *mean entropy production rate*[2] over the time interval $[0,t]$. The important fact is that the mean entropy production rate has non-negative expectation for $t > 0$:

$$\omega_X(\Sigma^t) = \frac{1}{t}\int_0^t \omega_X(\sigma_{X,s})\,\mathrm{d}s = -\frac{1}{t}S(\omega_{X,t}|\omega_X) \geq 0. \qquad (4.7)$$

Another widely used measure of the discrepancy between two states ω and ν is Rényi relative α-entropy, defined for any $\alpha \in \mathbb{R}$ by

$$S_\alpha(\nu|\omega) = \begin{cases} \log \omega\left(\left(\frac{\mathrm{d}\nu}{\mathrm{d}\omega}\right)^\alpha\right) & \text{if } \nu \ll \omega, \\ -\infty & \text{otherwise.} \end{cases}$$

Starting from equation (4.4) one easily derives the formula

$$\log \frac{\mathrm{d}\omega_{X,t}}{\mathrm{d}\omega_X} = \int_0^{-t} \sigma_{X,s}\,\mathrm{d}s = t\Sigma^{-t}, \qquad (4.8)$$

so that

$$e_t(\alpha) = S_\alpha(\omega_{X,t}|\omega_X) = \log \omega_X\left(\left(\frac{\mathrm{d}\omega_{X,t}}{\mathrm{d}\omega_X}\right)^\alpha\right) = \log \omega_X\left(e^{\alpha t\Sigma^{-t}}\right). \qquad (4.9)$$

---

[2] Various other names are commonly used in the literature for the observable σ_X: phase space contraction rate, dissipation function, etc.

Exercise 4.3

1. Assuming $\nu \ll \omega$ and using Hölder's inequality, show that $\alpha \mapsto S_\alpha(\nu|\omega)$ is convex.
2. Show that $S_0(\nu|\omega) = S_1(\nu|\omega) = 0$ and conclude that $S_\alpha(\nu|\omega)$ is nonpositive for $\alpha \in]0, 1[$ and non-negative for $\alpha \notin]0, 1[$.
3. Assuming also $\omega \ll \nu$, show that $S_{1-\alpha}(\nu|\omega) = S_\alpha(\omega|\nu)$.

4.2.5 Time reversal invariance

Our dynamical system is time reversal invariant: the map $\vartheta(p, q) = (-p, q)$ is an anti-symplectic involution, that is $\{f \circ \vartheta, g \circ \vartheta\} = -\{f, g\} \circ \vartheta$ and $\vartheta \circ \vartheta = \text{Id}$. Since $H \circ \vartheta = H$, it satisfies

$$e^{t\mathcal{L}} \circ \vartheta = \vartheta \circ e^{-t\mathcal{L}},$$

and leaves our reference state ω_X invariant,

$$\omega_X(f \circ \vartheta) = \omega_X(f).$$

It follows that $\omega_{X,t}(f \circ \vartheta) = \omega_{X,-t}(f)$, $\Phi_{L/R} \circ \vartheta = -\Phi_{L/R}$ and $\sigma_X \circ \vartheta = -\sigma_X$. Note in particular that $\omega_X(\Phi_{L/R}) = 0$ and $\omega_X(\sigma_X) = 0$. Applying time reversal to Definition (4.6) we further get

$$\Sigma^t \circ \vartheta = \frac{1}{t} \int_0^t \sigma_X \circ e^{s\mathcal{L}} \circ \vartheta \, ds = \frac{1}{t} \int_0^t \sigma_X \circ \vartheta \circ e^{-s\mathcal{L}} \, ds$$

$$= -\frac{1}{t} \int_0^t \sigma_X \circ e^{-s\mathcal{L}} \, ds = \frac{1}{t} \int_0^{-t} \sigma_X \circ e^{s\mathcal{L}} \, ds \qquad (4.10)$$

$$= -\Sigma^{-t},$$

and equation (4.9) becomes

$$e_t(\alpha) = \log \omega_X \left(e^{\alpha t \Sigma^{-t} \circ \vartheta} \right) = \log \omega_X \left(e^{-\alpha t \Sigma^t} \right). \qquad (4.11)$$

Thus, $\alpha \mapsto t^{-1} e_t(\alpha)$ is the cumulant generating function of the observable $-t\Sigma^t$ in the state ω_X, and in particular

$$\frac{d}{d\alpha} t^{-1} e_t(\alpha) \bigg|_{\alpha=0} = -\omega_X \left(\frac{1}{t} \int_0^t \sigma_{X,s} \, ds \right),$$

$$\frac{d^2}{d\alpha^2} t^{-1} e_t(\alpha) \bigg|_{\alpha=0} = \omega_X \left(\left(\frac{1}{\sqrt{t}} \int_0^t (\sigma_{X,s} - \omega_X(\sigma_{X,s})) \, ds \right)^2 \right).$$

4.2.6   A universal symmetry

Let us look more closely at the positivity property (4.7). To this end, we introduce the distribution of the observable Σ^t induced by the state ω_X, that is the probability measure P^t defined by

$$P^t(f) = \omega_X(f(\Sigma^t)).$$

To comply with (4.7), this distribution should be asymmetric and give more weight to positive values than to negative ones. Thus, let us compare P^t with the distribution $\bar{P}^t(f) = \omega_X(f(-\Sigma^t))$ of $-\Sigma^t$. Observing that

$$\Sigma^{-t} = -\frac{1}{t}\int_0^{-t} \sigma_X \circ e^{s\mathcal{L}}\, ds = \frac{1}{t}\int_0^{t} \sigma_X \circ e^{-s\mathcal{L}}\, ds$$

$$= \left(\frac{1}{t}\int_0^{t} \sigma_X \circ e^{(t-s)\mathcal{L}}\, ds\right) \circ e^{-t\mathcal{L}} = \Sigma^t \circ e^{-t\mathcal{L}}, \tag{4.12}$$

we obtain, using (4.8) and (4.10)

$$\bar{P}^t(f) = \omega_X(f(-\Sigma^t)) = \omega_X(f(\Sigma^{-t} \circ \vartheta)) = \omega_X(f(\Sigma^{-t})) = \omega_X(f(\Sigma^t \circ e^{-t\mathcal{L}}))$$

$$= \omega_{X,-t}(f(\Sigma^t)) = \omega_X\left(\frac{d\omega_{X,-t}}{d\omega_X} f(\Sigma^t)\right) = \omega_X\left(e^{-t\Sigma^t} f(\Sigma^t)\right),$$

from which we conclude that $\bar{P}^t \ll P^t$ and

$$\frac{d\bar{P}^t}{dP^t}(s) = e^{-ts}. \tag{4.13}$$

This relation shows that negative values of Σ^t are exponentially suppressed as $t \to \infty$. One easily deduces from (4.13) that

$$-s - \delta \leq \frac{1}{t}\log\frac{\omega_X(\{\Sigma^t \in [-s-\delta, -s+\delta]\})}{\omega_X(\{\Sigma^t \in [s-\delta, s+\delta]\})} \leq -s + \delta,$$

for $t, \delta > 0$ and any $s \in \mathbb{R}$. Such a property was discovered in numerical experiments on shear flows by Evans et al. (Evans, Cohen, and Morriss, 1993). Evans and Searles (Evans, and Searles, 1994) were the first to provide a theoretical analysis of the underlying mechanism. Since then, a large body of theoretical and experimental literature has been devoted to similar 'fluctuation relations' or 'fluctuation theorems'. They have been derived for various types of systems: Hamiltonian and non-Hamiltonian mechanical systems, discrete and continuous time dynamical systems, Markov processes, and so on. We refer the reader to the review by Rondoni and Meíja-

Monasterio (2007) for historical perspective and references and to (Jakšić, Pillet, and Rey-Bellet, 2011a) for a more mathematically oriented presentation.

We can rewrite equation (4.11) in terms of the Laplace transform of the measure P^t,

$$e_t(\alpha) = \log \int e^{-\alpha ts} \, dP^t(s).$$

Relation (4.13) is equivalent to

$$\int e^{-(1-\alpha)ts} \, dP^t(s) = \int e^{\alpha ts} \, d\bar{P}^t(s) = \int e^{-\alpha ts} \, dP^t(s),$$

and therefore can be expressed in the form

$$e_t(1 - \alpha) = e_t(\alpha). \tag{4.14}$$

We shall call the last relation the *finite time Evans–Searles symmetry* of the function $e_t(\alpha)$. The above derivation directly extends to a general time-reversal invariant dynamical system, see (Jakšić, Pillet, and Rey-Bellet, 2011a).

4.2.7 A generalized Evans–Searles symmetry

Relation (4.13) deals with the mean entropy production rate Σ^t. It can be generalized to the mean energy flux, the vector valued observable

$$\boldsymbol{\Phi}^t = \frac{1}{t} \int_0^t \left(\Phi_L \circ e^{s\mathcal{L}}, \Phi_R \circ e^{s\mathcal{L}} \right) \, ds.$$

Exercise 4.4 Denote by Q^t (respectively $\bar{Q}^t$) the distribution of $\boldsymbol{\Phi}^t$ (respectively $-\boldsymbol{\Phi}^t$) induced by the state ω_X, that is $Q^t(f) = \omega_X(f(\boldsymbol{\Phi}^t))$ and $\bar{Q}^t(f) = \omega_X(f(-\boldsymbol{\Phi}^t))$. Using the fact that $X \cdot \boldsymbol{\Phi}^t = \Sigma^t$ and mimicking the proof of (4.13) show that

$$\frac{d\bar{Q}^t}{dQ^t}(\mathbf{s}) = e^{-tX \cdot \mathbf{s}}. \tag{4.15}$$

Again, this derivation can be extended to an arbitrary time-reversal invariant dynamical system, see Jakšić, Pillet, and Rey-Bellet (2011a).

Introducing the cumulant generating function

$$g_t(X, Y) = \log \omega_X \left(e^{-tY \cdot \boldsymbol{\Phi}^t} \right), \tag{4.16}$$

and proceeding as in the previous section, we see that Relation (4.15) is equivalent to

$$\int e^{-t(X-Y) \cdot \mathbf{s}} \, dQ^t(\mathbf{s}) = \int e^{tY \cdot \mathbf{s}} \, d\bar{Q}^t(\mathbf{s}) = \int e^{-tY \cdot \mathbf{s}} \, dQ^t(\mathbf{s}),$$

which leads to the *generalized finite time Evans–Searles symmetry*

$$g_t(X, X - Y) = g_t(X, Y). \tag{4.17}$$

Exercise 4.5 Check that

$$g_t(X, Y) = -\frac{1}{2} \log \det \left(\mathbb{1} - D_X \left(e^{t\mathcal{L}^*} k(Y) e^{t\mathcal{L}} - k(Y) \right) \right), \tag{4.18}$$

where we adopt the convention that $\log x = -\infty$ whenever $x \le 0$. Using this formula verify directly Relation (4.17).

4.2.8 Thermodynamic limit

So far we were dealing with a finite dimensional harmonic system. Its Hamiltonian flow $e^{t\mathcal{L}}$ is quasi-periodic and it is therefore not a surprise that entropy production vanishes in the large time limit,

$$\lim_{t \to \infty} \omega_X(\Sigma^t) = \lim_{t \to \infty} \frac{1}{2t} \mathrm{tr} \left(D_X \left(k(X) - e^{t\mathcal{L}^*} k(X) e^{t\mathcal{L}} \right) \right) = 0,$$

see also Fig. 4.3. To achieve a strictly positive entropy production rate in the asymptotic regime $t \to \infty$, the thermodynamic limit of the reservoirs must be taken prior to the large time limit.

To take $M \to \infty$ while keeping N fixed we observe that the phase space $\Gamma_{[-M,M]}$ is naturally embedded in the real Hilbert space $\Gamma = \ell^2_{\mathbb{R}}(\mathbb{Z}) \oplus \ell^2_{\mathbb{R}}(\mathbb{Z})$ and that h_0, h_L,

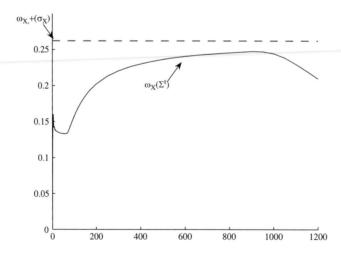

Fig. 4.3 The typical behaviour of the mean entropy production rate $t \mapsto \omega_X(\Sigma^t)$ for a finite system ($N = 20$, $M = 300$). The dashed line represents the steady state value $\omega_{X,+}(\sigma_X) = \lim_{t \to \infty} \omega_X(\Sigma^t)$ for the same finite chain ($N = 20$) coupled to two infinite reservoirs.

h_R and h are uniformly bounded and strongly convergent as operators on this space. For example

$$\mathrm{s} - \lim_{M \to \infty} h_0 = \begin{bmatrix} \mathbb{1} & 0 \\ 0 & \mathbb{1} - \Delta_0 \end{bmatrix},$$

where $\Delta_0 = \Delta_{]-\infty,-N-1]} \oplus \Delta_{[-N,N]} \oplus \Delta_{[N+1,\infty[}$ is the discrete Laplacian on $\mathbb{Z}$ with Dirichlet decoupling at $\pm N$ and

$$\mathrm{s} - \lim_{M \to \infty} h = \begin{bmatrix} \mathbb{1} & 0 \\ 0 & \mathbb{1} - \Delta \end{bmatrix},$$

where $\Delta = \Delta_{\mathbb{Z}}$ is the discrete Laplacian on $\mathbb{Z}$. It follows that $\mathcal{L}_0 = jh_0$ and $\mathcal{L} = jh$ are also strongly convergent. Hence, the Hamiltonian flows $e^{t\mathcal{L}_0}$ and $e^{t\mathcal{L}}$ converge strongly and uniformly on compact time intervals to the uniformly bounded, norm continuous groups on Γ generated by the strong limits of $\mathcal{L}_0$ and $\mathcal{L}$. Finally, since the covariance $D_X = (\beta h - k(X))^{-1}$ of the state ω_X converges strongly, the state ω_X converges weakly to the Gaussian measure with the limiting covariance. In the following, we shall use the same notation for these objects after the limit $M \to \infty$, that is h, h_0, $\mathcal{L}$, $\mathcal{L}_0$, $k(X)$, ω_X, and so on denote the thermodynamic limits of the corresponding finite volume objects.

After the thermodynamic limit, we are left with a linear dynamical system on the L^2-space of the Gaussian measure ω_X. Denoting by $\phi_{L/R}$ the finite rank operators corresponding to the flux observable $2\Phi_{L/R}$ and setting $\phi(Y) = Y_L \phi_L + Y_R \phi_R$, we can write

$$e^{t\mathcal{L}^*} k(Y) e^{t\mathcal{L}} - k(Y) = -\int_0^t e^{s\mathcal{L}^*} \phi(Y) e^{s\mathcal{L}} \, \mathrm{d}s. \tag{4.19}$$

Since the right-hand side of this identity is trace class for every $Y \in \mathbb{R}^2$ and $t \in \mathbb{R}$, we conclude from

$$D_{X,t}^{-1} - D_X^{-1} = e^{-t\mathcal{L}^*} k(X) e^{-t\mathcal{L}} - k(X), \tag{4.20}$$

and the Feldman–Hajek–Shale theorem (see, e.g., Simon (1979)) that the Gaussian measure $\omega_{X,t}$ and ω_X are equivalent and that Relation (4.4) still holds in the following form

$$-\log\left(\frac{\mathrm{d}\omega_{X,t}}{\mathrm{d}\omega_X}\right) = \int_0^{-t} \sigma_{X,s} \, \mathrm{d}s.$$

For the same reason, equation (4.18) for the generalized Evans–Searles functional $g_t(X,Y)$ remains valid in the thermodynamic limit.

4.2.9   Large time limit I: scattering theory

Taking the limit $t \to \infty$ in (4.19), (4.20) we obtain the formal result

$$D_{X,+}^{-1} = \lim_{t \to \infty} D_{X,t}^{-1} = D_X^{-1} + \int_{-\infty}^{0} e^{s\mathcal{L}^*} \phi(X) e^{s\mathcal{L}} \, ds,$$

which we can interpret in the following way: the state $\omega_{X,t}$, Gaussian with covariance $D_{X,t}$, converges as $t \to \infty$ towards a nonequilibrium steady state (NESS) $\omega_{X,+}$, Gaussian with covariance $D_{X,+}$, which formally writes

$$d\omega_{X,+}(p,q) =$$

$$\frac{1}{Z_{X,+}} e^{-\left(\beta H(p,q) - X_L H_{\Lambda_L}(p,q) - X_R H_{\Lambda_R}(p,q) + \int_{-\infty}^{0}(X_L \Phi_{L,s}(p,q) + X_R \Phi_{R,s}(p,q)) \, ds\right)} dpdq.$$

This formal expression is a special case of the McLennan–Zubarev nonequilibrium ensemble (see (McLennan, 1963; Zubarev, 1962; Zubarev, 1974)). In this and the following sections we shall turn this formal argument into a rigorous construction.

The study of the limit $t \to \infty$ in our infinite dimensional harmonic system reduces to an application of trace class scattering theory. We refer to (Reed, and Simon, 1979) for basic facts about scattering theory. We start with a few simple remarks:

(i) We denote by $\mathcal{H} = \ell_{\mathbb{C}}^2(\mathbb{Z}) \oplus \ell_{\mathbb{C}}^2(\mathbb{Z}) \simeq \ell_{\mathbb{C}}^2(\mathbb{Z}) \otimes \mathbb{C}^2$ the complexified phase space and extend all operators on Γ to $\mathcal{H}$ by $\mathbb{C}$-linearity. The inner product on the complex Hilbert space $\mathcal{H}$ is written $\langle \phi | \psi \rangle$.

(ii) $h - h_0 = v$ is finite rank and hence trace class. Since $h_0 \geq \mathbb{1}$ and $h \geq \mathbb{1}$, $h^{1/2} - h_0^{1/2}$ is also trace class.

(iii) $h^{1/2} h_0^{-1/2} - \mathbb{1} = (h^{1/2} - h_0^{1/2}) h_0^{-1/2}$ is trace class. The same is true for $h_0^{1/2} h^{-1/2} - \mathbb{1}$, $h_0^{-1/2} h^{1/2} - \mathbb{1}$ and $h^{-1/2} h_0^{1/2} - \mathbb{1}$.

(iv) $L_0 = i h_0^{1/2} j h_0^{1/2}$ and $L = i h^{1/2} j h^{1/2}$ are self-adjoint, $L - L_0$ is trace class and

$$e^{-itL_0} = h_0^{1/2} e^{t\mathcal{L}_0} h_0^{-1/2}, \qquad e^{-itL} = h^{1/2} e^{t\mathcal{L}} h^{-1/2}.$$

Note that iL_0 (respectively iL) acting on $\mathcal{H}$ is unitarily equivalent to $\mathcal{L}_0$ (respectively $\mathcal{L}$) acting on the 'energy' Hilbert space $\ell_{\mathbb{C}}^2(\mathbb{Z}) \oplus \ell_{\mathbb{C}}^2(\mathbb{Z})$ equipped with the inner product $\langle \phi | \psi \rangle_{h_0} = \langle \phi | h_0 | \psi \rangle$ (respectively $\langle \phi | \psi \rangle_h = \langle \phi | h | \psi \rangle$).

(v) L has purely absolutely continuous spectrum.

(vi) The Hilbert space $\mathcal{H}$ has a direct decomposition into three parts, $\mathcal{H} = \mathcal{H}_L \oplus \mathcal{H}_C \oplus \mathcal{H}_R$, corresponding to the three subsystems $\mathcal{R}_L$, $\mathcal{C}$, and $\mathcal{R}_R$. We denote by P_L, P_C, and P_R the corresponding orthogonal projections.

(vii) This decomposition reduces L_0 so that $L_0 = L_L \oplus L_C \oplus L_R$. The operators L_L and L_R have purely absolutely continuous spectrum and L_C has purely discrete spectrum. In particular, $P_L + P_R$ is the spectral projection of L_0 onto its absolutely continuous part.

By Kato–Birman theory, the wave operators

$$W_\pm = \text{s} - \lim_{t\to\pm\infty} e^{itL} e^{-itL_0} (P_L + P_R)$$

exist and are complete, that is

$$W_\pm^* = \text{s} - \lim_{t\to\pm\infty} e^{itL_0} e^{-itL},$$

also exists and satisfies $W_\pm^* W_\pm = P_L + P_R$, $W_\pm W_\pm^* = \mathbb{1}$. The scattering matrix $S = W_+^* W_-$ is unitary on $\mathcal{H}_L \oplus \mathcal{H}_R$. A few more remarks are needed to actually compute S:

(viii) One has

$$U^* L_0 U = \begin{bmatrix} \Omega_0 & 0 \\ 0 & -\Omega_0 \end{bmatrix}, \qquad U^* L U = \begin{bmatrix} \Omega & 0 \\ 0 & -\Omega \end{bmatrix},$$

where $\Omega = \sqrt{1 - \Delta}$ and $\Omega_0 = \sqrt{1 - \Delta_0}$ are discrete Klein–Gordon operators and U is the unitary

$$U = \frac{1}{\sqrt{2}} \begin{bmatrix} 1 & i \\ i & 1 \end{bmatrix}.$$

(ix) It follows that

$$W_\pm = U \begin{bmatrix} w_\pm & 0 \\ 0 & w_\mp \end{bmatrix} U^*,$$

where

$$w_\pm = \text{s} - \lim_{t\to\pm\infty} e^{it\Omega} e^{-it\Omega_0} (P_L + P_R).$$

In particular, one has

$$S = U \begin{bmatrix} w_+^* w_- & 0 \\ 0 & w_-^* w_+ \end{bmatrix} U^*. \tag{4.21}$$

(x) By the invariance principle for wave operators, we have

$$w_\pm = \text{s} - \lim_{t\to\pm\infty} e^{it\Omega^2} e^{-it\Omega_0^2} P_{\text{ac}}(\Omega_0^2)$$

$$= \text{s} - \lim_{t\to\pm\infty} e^{it(-\Delta)} e^{-it(-\Delta_0)} P_{\text{ac}}(-\Delta_0).$$

We proceed to compute the scattering matrix. A complete set of (properly normalized) generalized eigenfunctions for the absolutely continuous part of $-\Delta_0$ is given by

$$\phi_{\sigma,k}(x) = \sqrt{\frac{2}{\pi}}\, \theta(\sigma x - N) \sin k|\sigma x - N|, \qquad (\sigma, k) \in \{-, +\} \times [0, \pi],$$

where θ denotes the Heaviside step function and $-\Delta_0 \phi_{\sigma,k} = 2(1 - \cos k)\phi_{\sigma,k}$. For the operator $-\Delta$, such a set is given by

$$\chi_{\sigma,k}(x) = \frac{1}{\sqrt{2\pi}} e^{i\sigma kx}, \qquad (\sigma, k) \in \{-, +\} \times [0, \pi].$$

Since

$$w_{\pm}\phi_{\sigma,k} = \mp\sigma i e^{\mp ikN} \chi_{\pm\sigma,k},$$

we deduce that

$$w_{\pm}^* w_{\mp} \phi_{\sigma,k} = e^{\pm 2ikN} \phi_{-\sigma,k}. \tag{4.22}$$

We shall denote by $\mathfrak{h}_{k\pm}$ the two-dimensional generalized eigenspace of L_0 to the 'eigenvalue' $\pm\varepsilon(k) = \pm\sqrt{3 - 2\cos k}$. The space $\mathfrak{h}_{k+}$ is spanned by the two basis vectors

$$\psi_{\sigma,k,+} = U \begin{bmatrix} \phi_{\sigma,k} \\ 0 \end{bmatrix}, \qquad \sigma \in \{-, +\},$$

and $\mathfrak{h}_{k-}$ is the span of

$$\psi_{\sigma,k,-} = U \begin{bmatrix} 0 \\ \phi_{\sigma,k} \end{bmatrix}, \qquad \sigma \in \{-, +\}.$$

In the direct integral representation

$$\mathcal{H}_L \oplus \mathcal{H}_R = \bigoplus_{\mu=\pm} \left(\int_{[0,\pi]}^{\oplus} \mathfrak{h}_{k\mu}\, dk \right),$$

the scattering matrix is given by

$$S = \bigoplus_{\mu=\pm} \left(\int_{[0,\pi]}^{\oplus} S_\mu(k)\, dk \right),$$

where, thanks to (4.21) and (4.22), the on-shell S-matrix $S_\mu(k)$ is given by

$$S_\pm(k) = S|_{\mathfrak{h}_{k\pm}} = e^{\pm 2ikN} \begin{bmatrix} 0 & 1 \\ 1 & 0 \end{bmatrix}. \tag{4.23}$$

4.2.10   Large time limit II: non-equilibrium steady state

We shall now use scattering theory to compute the weak limit, as $t \to \infty$, of the state $\omega_{X,t}$. Setting $\hat{X} = X_L P_L + X_R P_R$ for $X = (X_L, X_R) \in \mathbb{R}^2$, one has

$$k(X) = X_L h_L + X_R h_R = h_0^{1/2} \hat{X} h_0^{1/2}.$$

Energy conservation yields $e^{-t\mathcal{L}_0^*} k(X) e^{-t\mathcal{L}_0} = k(X)$ and

$$e^{t\mathcal{L}^*} k(X) e^{t\mathcal{L}} = e^{t\mathcal{L}^*} e^{-t\mathcal{L}_0^*} h_0^{1/2} \hat{X} h_0^{1/2} e^{-t\mathcal{L}_0} e^{t\mathcal{L}}$$

$$= e^{t\mathcal{L}^*} h_0^{1/2} e^{-itL_0} \hat{X} e^{itL_0} h_0^{1/2} e^{t\mathcal{L}}.$$

$$= \mathrm{e}^{t\mathcal{L}^*} h^{1/2} h^{-1/2} h_0^{1/2} \mathrm{e}^{-\mathrm{i}tL_0} \widehat{X} \mathrm{e}^{\mathrm{i}tL_0} h_0^{1/2} h^{-1/2} h^{1/2} \mathrm{e}^{t\mathcal{L}}$$

$$= h^{1/2} \mathrm{e}^{\mathrm{i}tL} h^{-1/2} h_0^{1/2} \mathrm{e}^{-\mathrm{i}tL_0} \widehat{X} \mathrm{e}^{\mathrm{i}tL_0} h_0^{1/2} h^{-1/2} \mathrm{e}^{-\mathrm{i}tL} h^{1/2}.$$

By Property (ii) of the previous section, one has

$$\mathrm{s} - \lim_{t \to \pm\infty} \mathrm{e}^{\mathrm{i}tL} h^{-1/2} h_0^{1/2} \mathrm{e}^{-\mathrm{i}tL_0} (P_L + P_R) = W_{\pm},$$

$$\mathrm{s} - \lim_{t \to \pm\infty} \mathrm{e}^{\mathrm{i}tL_0} h_0^{1/2} h^{-1/2} \mathrm{e}^{-\mathrm{i}tL} = W_{\pm}^*,$$

and so

$$\mathrm{s} - \lim_{t \to \pm\infty} \mathrm{e}^{t\mathcal{L}^*} k(X) \mathrm{e}^{t\mathcal{L}} = h^{1/2} W_{\pm} \widehat{X} W_{\pm}^* h^{1/2}. \tag{4.24}$$

It follows that

$$\mathrm{s} - \lim_{t \to \infty} D_{X,t} = \mathrm{s} - \lim_{t \to \infty} (\beta h - \mathrm{e}^{-t\mathcal{L}^*} k(X) \mathrm{e}^{-t\mathcal{L}})^{-1}$$

$$= (\beta h - h^{1/2} W_- \widehat{X} W_-^* h^{1/2})^{-1} \tag{4.25}$$

$$= h^{-1/2} W_- (\beta - \widehat{X})^{-1} W_-^* h^{-1/2} = D_{X,+},$$

which implies that the state $\omega_{X,t}$ converges weakly to the Gaussian measure $\omega_{X,+}$ with covariance $D_{X,+}$. The state $\omega_{X,+}$ is invariant under the Hamiltonian flow $\mathrm{e}^{t\mathcal{L}}$ and is called the nonequilibrium steady state (NESS) associated to the reference state ω_X. Note that in the equilibrium case $\beta_L = \beta_R$ the operator $\widehat{X}$ is a multiple of the identity and

$$D_{X,+} = (\beta_L h)^{-1},$$

which means that the stationary state $\omega_{X,+}$ is the thermal equilibrium state of the coupled system at inverse temperature $\beta_L = \beta_R$.

Exercise 4.6 If $X_L \neq X_R$ then $\omega_{X,+}$ is singular w.r.t. ω_X, i.e.,

$$D_{X,+}^{-1} - D_X^{-1} = h_0^{1/2} \widehat{X} h_0^{1/2} - h^{1/2} W_- \widehat{X} W_-^* h^{1/2},$$

is not Hilbert–Schmidt. Prove this fact by deriving explicit formulas for $W_- P_{L/R} W_-^*$.

Exercise 4.7 Compute $\omega_{X,+}(\Phi_{L/R}) = \frac{1}{2}\mathrm{tr}(D_{X,+}\phi_{L/R})$ and show that

$$\omega_{X,+}(\Phi_L) = -\omega_{X,+}(\Phi_R) = \kappa(T_L - T_R),$$

where $T_{L/R} = \beta_{L/R}^{-1}$ is the temperature of the left/right reservoir and

$$\kappa = \frac{\sqrt{5} - 1}{2\pi}.$$

Note in particular that $\omega_{X,+}(\Phi_L) + \omega_{X,+}(\Phi_R) = 0$. What is the physical origin of this fact? Show that, more generally, if ω is a stationary state such that $\omega(p_x^2 + q_x^2) < \infty$ for all $x \in \mathbb{Z}$, then $\omega(\Phi_L) + \omega(\Phi_R) = 0$.

Using the result of Exercise 5.7 we conclude that

$$\omega_{X,+}(\sigma_X) = X_L \omega_{X,+}(\Phi_L) + X_R \omega_{X,+}(\Phi_R)$$
$$= (X_L - X_R)\omega_{X,+}(\Phi_L)$$
$$= \kappa \frac{(T_L - T_R)^2}{T_L T_R} > 0,$$

provided $T_L \neq T_R$. This implies that the mean entropy production rate in the state ω_X is strictly positive in the asymptotic regime,[3]

$$\lim_{t \to \infty} \omega_X(\Sigma^t) = \lim_{t \to \infty} \frac{1}{t} \int_0^t \omega_X(\sigma_{X,s}) \, ds = \lim_{t \to \infty} \omega_{X,t}(\sigma_X) = \omega_{X,+}(\sigma_X) > 0,$$

and that it is constant and strictly positive in the NESS $\omega_{X,+}$,

$$\omega_{X,+}(\Sigma^t) = \frac{1}{t} \int_0^t \omega_{X,+}(\sigma_{X,s}) \, ds = \omega_{X,+}(\sigma_X) > 0.$$

4.2.11   Large time limit III: generating functions

In this section we use scattering theory to study the large time asymptotic of the Evans–Searles functional $e_t(\alpha)$ (Eq. (4.11)) and the generalized Evans–Searles functional $g_t(X, Y)$ (Eq. (4.16)).

Starting from Eq. (4.18), and using (4.19) to write

$$T_t = -D_X \left(e^{t\mathcal{L}^*} k(Y) e^{t\mathcal{L}} - k(Y) \right) = \int_0^t D_X e^{s\mathcal{L}^*} \phi(Y) e^{s\mathcal{L}} \, ds,$$

we get

$$\frac{1}{t} g_t(X, Y) = -\frac{1}{2t} \log \det \left(\mathbb{1} + T_t \right)$$
$$= -\frac{1}{2t} \operatorname{tr} \log \left(\mathbb{1} + T_t \right)$$
$$= -\frac{1}{2t} \int_0^1 \frac{d}{du} \operatorname{tr} \log \left(\mathbb{1} + u T_t \right) \, du.$$

---

[3] Recall that if $\lim_{t \to +\infty} f(t) = a$ exists then it coincide with the Cesáro limit of f at $+\infty$, $\lim_{T \to +\infty} T^{-1} \int_0^T f(t) \, dt = a$, and with its Abel limit, $\lim_{\eta \downarrow 0} \eta \int_0^\infty e^{-\eta t} f(t) \, dt = a$.

Using the result of Exercise 5.8, we further get

$$\frac{1}{t}g_t(X,Y) = -\frac{1}{2t}\int_0^1 \mathrm{tr}\left((\mathbb{1}+uT_t)^{-1}T_t\right)du$$

$$= -\frac{1}{2t}\int_0^1\int_0^t \mathrm{tr}\left((\mathbb{1}+uT_t)^{-1}D_X e^{s\mathcal{L}^*}\phi(Y)e^{s\mathcal{L}}\right)ds\,du$$

$$= -\frac{1}{2t}\int_0^1\int_0^t \mathrm{tr}\left[\left(D_X^{-1} - u\left(e^{t\mathcal{L}^*}k(Y)e^{t\mathcal{L}} - k(Y)\right)\right)^{-1}e^{s\mathcal{L}^*}\phi(Y)e^{s\mathcal{L}}\right]ds\,du$$

$$= -\frac{1}{2}\int_0^1\int_0^1 \mathrm{tr}\left[e^{st\mathcal{L}}\left(D_X^{-1} - u\left(e^{t\mathcal{L}^*}k(Y)e^{t\mathcal{L}} - k(Y)\right)\right)^{-1}e^{st\mathcal{L}^*}\phi(Y)\right]ds\,du.$$

Writing

$$e^{st\mathcal{L}}\left(D_X^{-1} - u\left(e^{t\mathcal{L}^*}k(Y)e^{t\mathcal{L}} - k(Y)\right)\right)^{-1}e^{st\mathcal{L}^*}$$

$$= \left(e^{-st\mathcal{L}^*}D_X^{-1}e^{-st\mathcal{L}} - u\,e^{-st\mathcal{L}^*}\left(e^{t\mathcal{L}^*}k(Y)e^{t\mathcal{L}} - k(Y)\right)e^{-st\mathcal{L}}\right)^{-1}$$

$$= \left(D_{X,st}^{-1} - u\left(e^{(1-s)t\mathcal{L}^*}k(Y)e^{(1-s)t\mathcal{L}} - e^{-st\mathcal{L}^*}k(Y)e^{-st\mathcal{L}}\right)\right)^{-1},$$

and using (4.24) and (4.25), we obtain

$$\mathrm{s}-\lim_{t\to\infty}e^{st\mathcal{L}}\left(D_X^{-1} - u\left(e^{t\mathcal{L}^*}k(Y)e^{t\mathcal{L}} - k(Y)\right)\right)^{-1}e^{st\mathcal{L}^*}$$

$$= \left(D_{X,+}^{-1} - uh^{1/2}\left(W_+\widehat{Y}W_+^* - W_-\widehat{Y}W_-^*\right)h^{1/2}\right)^{-1}$$

$$= \left(h^{1/2}\left(W_-(\beta - \widehat{X} + u\widehat{Y})^{-1}W_-^* - uW_+\widehat{Y}W_+^*\right)h^{1/2}\right)^{-1}$$

$$= h^{-1/2}W_-\left(\beta - \widehat{X} - u(S^*\widehat{Y}S - \widehat{Y})\right)^{-1}W_-^*\,h^{-1/2},$$

for all $s \in]0,1[$. Since $\phi(Y)$ is trace class (actually finite rank), we conclude that

$$g(X,Y) = \lim_{t\to\infty}\frac{1}{t}g_t(X,Y) = -\frac{1}{2}\int_0^1 \mathrm{tr}\left[\left(\beta - \widehat{X} - u(S^*\widehat{Y}S - \widehat{Y})\right)^{-1}\mathcal{T}\right]du,$$

where

$$\mathcal{T} = W_-^* h^{-1/2}\phi(Y)h^{-1/2}W_-.$$

To evaluate the trace, we note that the scattering matrix S and the operators $\widehat{X}$, $\widehat{Y}$ all commute with L_0 while the trace class operator $\mathcal{T}$ acts nontrivially only on the absolutely continuous spectral subspace of L_0. It follows that

$$
\mathrm{tr}\left[\left(\beta - \widehat{X} - u(S^*\widehat{Y}S - \widehat{Y})\right)^{-1}\mathcal{T}\right]
$$

$$
=\mathrm{tr}\left[\left(\mathbb{1} - u(\beta - \widehat{X})^{-1}(S^*\widehat{Y}S - \widehat{Y})\right)^{-1}(\beta - \widehat{X})^{-1}\mathcal{T}\right] \tag{4.26}
$$

$$
= \sum_{\mu=\pm}\int_0^\pi \sum_{\sigma=\pm}\langle\psi_{\sigma,k,\mu}|\left(\mathbb{1} - u(\beta - \widehat{X})^{-1}(S^*\widehat{Y}S - \widehat{Y})\right)^{-1}(\beta - \widehat{X})^{-1}\mathcal{T}|\psi_{\sigma,k,\mu}\rangle\,\mathrm{d}k.
$$

Set

$$
A(\eta) = \int_{-\infty}^\infty e^{-\eta|t|}\langle\psi_{\sigma,k,\pm}|e^{itL_0}\mathcal{T}e^{-itL_0}|\psi_{\sigma',k',\pm}\rangle\frac{\mathrm{d}t}{2\pi},
$$

$$
B(\eta) = \eta\int_0^\infty e^{-\eta t}\langle\psi_{\sigma,k,\pm}|\mathcal{F}|\psi_{\sigma',k',\pm}\rangle\frac{\mathrm{d}t}{2\pi},
$$

where

$$
\mathcal{F} = W_-^*h^{-1/2}\left(e^{-t\mathcal{L}^*}k(Y)e^{-t\mathcal{L}} - e^{t\mathcal{L}^*}k(Y)e^{t\mathcal{L}}\right)h^{-1/2}W_-.
$$

By the intertwining property of the wave operator, we have

$$
e^{itL_0}\mathcal{T}e^{-itL_0} = W_-^*e^{itL}h^{-1/2}\phi(Y)h^{-1/2}e^{-itL}W_-
$$

$$
= W_-^*h^{-1/2}e^{t\mathcal{L}^*}\phi(Y)e^{t\mathcal{L}}h^{-1/2}W_-
$$

$$
= -\frac{\mathrm{d}}{\mathrm{d}t}W_-^*h^{-1/2}e^{t\mathcal{L}^*}k(Y)e^{t\mathcal{L}}h^{-1/2}W_-,
$$

and an integration by parts yields that

$$
A(\eta) = B(\eta), \tag{4.27}
$$

for any $\eta > 0$. Let us now take the limit $\eta \downarrow 0$ in this formula. Since $\psi_{\sigma,k,\pm}$ is a generalized eigenfunction of L_0 to the eigenvalue $\pm\varepsilon(k)$, we get, on the left-hand side of (4.27),

$$
\langle\psi_{\sigma,k,\pm}|\mathcal{T}|\psi_{\sigma',k',\pm}\rangle\int_{-\infty}^\infty e^{-\eta|t|\pm it(\varepsilon(k)-\varepsilon(k'))}\frac{\mathrm{d}t}{2\pi} \to \langle\psi_{\sigma,k,\pm}|\mathcal{T}|\psi_{\sigma',k,\pm}\rangle\delta(\varepsilon(k) - \varepsilon(k')).
$$

Using (4.24), the Abel limit[4] on the right-hand side of (4.27) yields

$$\frac{1}{2\pi}\langle\psi_{\sigma,k,\pm}|W_-^* h^{-1/2}\left(h^{1/2}W_-\widehat{Y}W_-^* h^{1/2} - h^{1/2}W_+\widehat{Y}W_+^* h^{1/2}\right) h^{-1/2}W_-|\psi_{\sigma',k',\pm}\rangle$$

$$= \frac{1}{2\pi}\langle\psi_{\sigma,k,\pm}|\widehat{Y} - S^*\widehat{Y}S|\psi_{\sigma',k',\pm}\rangle$$

$$= \frac{1}{2\pi}\langle\psi_{\sigma,k,\pm}|\widehat{Y} - S_\pm(k)^*\widehat{Y}S_\pm(k)|\psi_{\sigma',k,\pm}\rangle\delta(k-k'),$$

and we conclude that

$$\langle\psi_{\sigma,k,\pm}|T|\psi_{\sigma',k,\pm}\rangle = \frac{1}{2\pi}\langle\psi_{\sigma,k,\pm}|\widehat{Y} - S_\pm(k)^*\widehat{Y}S_\pm(k)|\psi_{\sigma',k,\pm}\rangle\varepsilon'(k). \tag{4.28}$$

Note that the operator $\widehat{Y}$ acts on the fibre $\mathfrak{h}_{k\pm}$ as the matrix

$$\widehat{Y}\Big|_{\mathfrak{h}_{k\pm}} = \begin{bmatrix} Y_L & 0 \\ 0 & Y_R \end{bmatrix}. \tag{4.29}$$

Relation (4.28) allows us to write

$$\sum_{\sigma=\pm}\langle\psi_{\sigma,k,\mu}|\left(I - u(\beta-\widehat{X})^{-1}(S^*\widehat{Y}S - \widehat{Y})\right)^{-1}(\beta-\widehat{X})^{-1}T|\psi_{\sigma,k,\mu}\rangle$$

$$= \mathrm{tr}_{\mathfrak{h}_{k\mu}}\left[\left(\mathbb{1} - u(\beta-\widehat{X})^{-1}(S_\mu(k)^*\widehat{Y}S_\mu(k) - \widehat{Y})\right)^{-1}\right.$$

$$\left. \times (\beta-\widehat{X})^{-1}(\widehat{Y} - S_\mu(k)^*\widehat{Y}S_\mu(k))\right]\frac{\varepsilon'(k)}{2\pi}$$

$$= \frac{\mathrm{d}}{\mathrm{d}u}\mathrm{tr}_{\mathfrak{h}_{k\mu}}\left[\log\left(\mathbb{1} - u(\beta-\widehat{X})^{-1}(S_\mu(k)^*\widehat{Y}S_\mu(k) - \widehat{Y})\right)\right]\frac{\varepsilon'(k)}{2\pi}.$$

Inserting the last identity into (4.26) and integrating over u we derive

$$g(X,Y) = -\sum_{\mu=\pm}\int_0^\pi \mathrm{tr}_{\mathfrak{h}_{k\mu}}\left[\log\left(\mathbb{1} - (\beta-\widehat{X})^{-1}(S_\mu(k)^*\widehat{Y}S_\mu(k) - \widehat{Y})\right)\right]\frac{\mathrm{d}\varepsilon(k)}{4\pi}$$

$$= -\sum_{\mu=\pm}\int_0^\pi \log\det_{\mathfrak{h}_{k\mu}}\left(\mathbb{1} - (\beta-\widehat{X})^{-1}(S_\mu(k)^*\widehat{Y}S_\mu(k) - \widehat{Y})\right)\frac{\mathrm{d}\varepsilon(k)}{4\pi}.$$

Remark The last formula retains its validity in a much broader context. It holds for an arbitrary number of infinite harmonic reservoirs coupled to a finite harmonic system as long as the scattering approach sketched here applies. Furthermore, the formal analogy between our Hilbert space treatment of harmonic dynamics and quantum

---

[4] See footnote 3 on page 232.

mechanics suggests that quasi-free quantum systems could be also studied by a similar scattering approach. That is indeed the case, see Section 4.7.6.

Invoking (4.23) and (4.29) leads to our final result

$$g(X,Y) = -\kappa \log \left(1 + \frac{(Y_R - Y_L)\left[(X_R - X_L) - (Y_R - Y_L)\right]}{(\beta - X_R)(\beta - X_L)} \right). \tag{4.30}$$

Note that $g(X,Y)$ is finite for $-T_R^{-1} < Y_R - Y_L < T_L^{-1}$ and $+\infty$ otherwise. Since $e_t(\alpha) = g_t(X, \alpha X)$, one has

$$e(\alpha) = \lim_{t \to \infty} \frac{1}{t} e_t(\alpha) = -\kappa \log \left(1 + \frac{(T_L - T_R)^2}{T_L T_R} \alpha(1 - \alpha) \right),$$

which is finite provided $2|\alpha - 1/2| < (T_L + T_R)/|T_L - T_R|$ and $+\infty$ otherwise. Note also the explicit symmetries $g(X, X - Y) = g(X,Y)$ and $e(1 - \alpha) = e(\alpha)$ inherited from the finite time Evans–Searles symmetries (4.14) and (4.17).

Exercise 4.8 Let $\mathbb{R} \ni x \mapsto A(x)$ be a differentiable function with values in the trace class operators on a Hilbert space. Show that if $\|A(x_0)\| < 1$ then $x \mapsto \operatorname{tr} \log(1 + A(x))$ is differentiable at x_0 and

$$\frac{d}{dx} \operatorname{tr} \log(1 + A(x)) \Big|_{x=x_0} = \operatorname{tr}((1 + A(x_0))^{-1} A'(x_0)).$$

Hint: use the formula

$$\log(1 + a) = \int_1^\infty \left(\frac{1}{t} - \frac{1}{t+a} \right) dt,$$

valid for $|a| < 1$.

4.2.12 The central limit theorem

As a first application of the generalized Evans–Searles functional $g(X,Y)$, we derive a central limit theorem (CLT) for the current fluctuations. To this end, let us decompose the mean currents into its expected value and a properly normalized fluctuating part, writing

$$\frac{1}{t} \int_0^t \Phi_{j,s} \, ds = \frac{1}{t} \int_0^t \omega_X(\Phi_{j,s}) \, ds + \frac{1}{\sqrt{t}} \delta \Phi_j^t,$$

for $j \in \{L, R\}$. By Definition (4.16), the expected mean current is given by

$$\frac{1}{t} \int_0^t \omega_X(\Phi_{j,s}) \, ds = - \left. \partial_{Y_j} \frac{1}{t} g_t(X,Y) \right|_{Y=0},$$

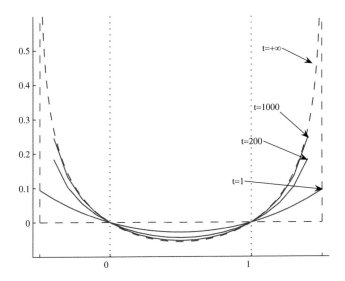

Fig. 4.4 Solid lines: the generating function $\alpha \mapsto t^{-1}e_t(\alpha)$ for various values of t and finite reservoirs ($N = 20$, $M = 300$). The slope at $\alpha = 1$ is $\omega_X(\Sigma^t)$, compare with Fig. 4.3. Dashed line: the limiting function $\alpha \mapsto e(\alpha)$ for infinite reservoirs.

while the fluctuating part is centred, $\omega_X(\delta\Phi_{j,t}) = 0$, with covariance

$$\omega_X(\delta\Phi_j^t \delta\Phi_k^t) = \partial_{Y_k}\partial_{Y_j}\frac{1}{t}g_t(X,Y)\Big|_{Y=0}.$$

For large t, the expected mean current converges to the NESS expectation

$$\lim_{t\to\infty}\frac{1}{t}\int_0^t \omega_X(\Phi_{j,s})\,\mathrm{d}s = \omega_{X,+}(\Phi_j).$$

To study the large time asymptotics of the current fluctuations $\delta\Phi^t = (\delta\Phi_L^t, \delta\Phi_R^t)$ we consider the characteristic function

$$\omega_X\left(e^{\mathrm{i}Y\cdot\delta\Phi^t}\right) = \omega_X\left(e^{\mathrm{i}\sum_j Y_j \frac{1}{\sqrt{t}}\int_0^t(\Phi_{j,s}-\omega_X(\Phi_{j,s}))\,\mathrm{d}s}\right), \tag{4.31}$$

that is the Fourier transform of their distribution. To control the limit $t \to \infty$, we need a technical result which is the object of the following exercise.

Exercise 4.9 Show that for a given $\beta_L > 0$ and $\beta_R > 0$ there exists $\epsilon > 0$ such that the function $Y \mapsto g_t(X,Y)$ is analytic in $D_\epsilon = \{Y = (Y_L, Y_R) \in \mathbb{C}^2 \,|\, |Y_L| < \epsilon, |Y_R| < \epsilon\}$ and satisfies

$$\sup_{\substack{Y\in D_\epsilon \\ t>0}} \left|\frac{1}{t}g_t(X,Y)\right| < \infty. \tag{4.32}$$

Hint: start with (4.18) and use the identity $\log \det(\mathbb{1} - T) = \mathrm{tr}(\log(\mathbb{1} - T))$ and the factorization $\log(\mathbb{1} - z) = -zf(z)$ to obtain the bound $|\log \det(\mathbb{1} - T)| \leq \|T\|_1 f(\|T\|)$ where $\|T\|_1 = \mathrm{tr}(\sqrt{T^*T})$ denotes the trace norm of T.

The convergence result of the preceding section and the uniform bound (4.32) imply that $\lim_{t\to\infty} \frac{1}{t}g_t(X,Y) = g(X,Y)$ uniformly for Y in compact subsets of D_ϵ, that all the derivatives w.r.t. Y of $\frac{1}{t}g_t(X,Y)$ are uniformly bounded on such compact subsets and converge uniformly to the corresponding derivatives of $g(X,Y)$ (see Proposition 4.69 in Appendix 4.9). For $Y \in \mathbb{C}^2$ and $t > 0$ large enough, equation (4.31) can be written as

$$\omega_X\left(e^{iY\cdot\delta\Phi^t}\right) = \exp\left[t\left(\frac{1}{t}g_t\left(X,\frac{Y}{i\sqrt{t}}\right) - \frac{Y}{i\sqrt{t}}\cdot\left(\nabla_Y\frac{1}{t}g_t\right)(X,0)\right)\right],$$

and the Taylor expansion of $\frac{1}{t}g_t(X,Y)$ around $Y = 0$ yields

$$\frac{1}{t}g_t\left(X,\frac{Y}{i\sqrt{t}}\right) - \frac{Y}{i\sqrt{t}}\cdot\left(\nabla_Y\frac{1}{t}g_t\right)(X,0) = -\frac{1}{2t}\sum_{j,k}\left(\partial_{Y_j}\partial_{Y_k}\frac{1}{t}g_t\right)(X,0)Y_jY_k + O(t^{-3/2}),$$

from which we conclude that

$$\lim_{t\to\infty}\omega_X\left(e^{iY\cdot\delta\Phi^t}\right) = e^{-\frac{1}{2}Y\cdot DY}, \tag{4.33}$$

with a covariance matrix $D = [D_{jk}]$ given by

$$D_{jk} = \lim_{t\to\infty}\left(\partial_{Y_j}\partial_{Y_k}\frac{1}{t}g_t\right)(X,0) = \left(\partial_{Y_j}\partial_{Y_k}g\right)(X,0).$$

Evaluating the right-hand side of these identities yields

$$D_{11} = D_{22} = -D_{12} = -D_{21} = \kappa\left(T_L^2 + T_R^2\right).$$

Since the right-hand side of (4.33) is the Fourier transform of the centred Gaussian measure on $\mathbb{R}^2$ with covariance D, the Lévy–Cramér continuity theorem (see e.g., Theorem 7.6 in Billingsley (1968)) implies that the current fluctuations $\delta\Phi^t$ converge in law to this Gaussian, that is that for all bounded continuous functions $f : \mathbb{R}^2 \to \mathbb{R}$

$$\lim_{t\to\infty}\omega_X(f(\delta\Phi^t)) = \int f(\phi,-\phi)e^{-\phi^2/2\mathfrak{d}}\frac{d\phi}{\sqrt{2\pi\mathfrak{d}}}, \tag{4.34}$$

where $\mathfrak{d} = \kappa\left(T_L^2 + T_R^2\right)$. Note in particular that the fluctuations of the left and right mean currents are opposite to each other in this limit.

Exercise 4.10 Use the CLT (4.34) and the results of Exercise 5.7 to show that

$$\frac{1}{t}\int_0^t (\Phi_{L,s} + \Phi_{R,s})\,ds \longrightarrow 0,$$

in probability as $t \to \infty$, that is that for any $\epsilon > 0$ the probability

$$\omega_X \left(\left\{ \left| \frac{1}{t} \int_0^t (\Phi_{L,s} + \Phi_{R,s}) \, \mathrm{d}s \right| \geq \epsilon \right\} \right), \tag{4.35}$$

tends to zero as $t \to \infty$.

It is interesting to compare the equilibrium $(T_L = T_R)$ and the nonequilibrium $(T_L \neq T_R)$ case. In the first case the expected mean currents vanish (recall that in this case $\omega_{X,+}$ is the equilibrium state) while in the second they are nonzero. In both cases the fluctuations of the mean currents have similar qualitative features at the CLT scale $t^{-1/2}$. In particular they are always symmetrically distributed w.r.t. 0.

4.2.13   Linear response theory near equilibrium

The linear response theory for our harmonic chain model follows trivially from the formula for steady heat fluxes derived in Exercise 5.7. Our goal in this section, however, is to present a derivation of the linear response theory based on the functionals $g_t(X,Y)$ and $g(X,Y)$. This derivation, which follows the ideas of Gallavotti (Gallavotti (1996)), is applicable to any time-reversal invariant dynamical system for which the conclusions of Exercise 5.9 hold. For additional information and a general axiomatic approach to derivation of linear response theory based on functionals $g_t(X,Y)$ and $g(X,Y)$ we refer the reader to Jakšić, Pillet, and Rey-Bellet (2011a).

Starting from

$$-\lim_{t\to\infty} \partial_{Y_{L/R}} \frac{1}{t} g_t(X,Y) \Big|_{Y=0} = \omega_{X,+}(\Phi_{L/R}),$$

and using the fact that the derivative and the limit can be interchanged (as we learned in the previous section) one gets

$$-\partial_{Y_{L/R}} g(X,Y)\big|_{Y=0} = \omega_{X,+}(\Phi_{L/R}). \tag{4.36}$$

Remark: The main result of Section 4.2.11 which expresses the Evans–Searles function $g(X,Y)$ in terms of the on-shell scattering matrix immediately implies

$$\omega_{X,+}(\Phi_{L/R})$$

$$= \partial_{Y_{L/R}} \sum_{\mu=\pm} \int_0^\pi \mathrm{tr}_{\mathfrak{h}_{k\mu}} \left[\log \left(\mathbb{1} - (\beta - \widehat{X})^{-1} (S_\mu(k)^* \widehat{Y} S_\mu(k) - \widehat{Y}) \right) \right] \frac{\mathrm{d}\varepsilon(k)}{4\pi} \bigg|_{Y=0}$$

$$= \sum_{\mu=\pm} \int_0^\pi \mathrm{tr}_{\mathfrak{h}_{k\mu}} \left[(\beta - \widehat{X})^{-1} (P_{L/R} - S_\mu(k)^* P_{L/R} S_\mu(k)) \right] \frac{\mathrm{d}\varepsilon(k)}{4\pi},$$

which can be interpreted as a classical version of the Landauer–Büttiker formula (see Exercise 5.59).

The Onsager matrix $\mathbf{L} = [L_{jk}]_{j,k\in\{L,R\}}$ defined by

$$L_{jk} = \partial_{X_k}\omega_{X,+}(\Phi_j)|_{X=0},$$

describes the response of the system to weak thermodynamic forces. Taylor's formula

$$\omega_{X,+}(\Phi_j) = \sum_k L_{jk}X_k + o(X), \qquad (X \to 0),$$

expresses the steady currents to the lowest order in the driving forces. From (4.36), we deduce that

$$L_{jk} = -\partial_{X_k}\partial_{Y_j}g(X,Y)|_{X=Y=0}. \tag{4.37}$$

The ES symmetry $g(X, X - Y) = g(X,Y)$ further leads to

$$\begin{aligned}
\partial_{X_k}\partial_{Y_j}g(X,Y) &= \partial_{X_k}\partial_{Y_j}g(X, X - Y) \\
&= -\partial_{X_k}(\partial_{Y_j}g)(X, X - Y) \\
&= -(\partial_{X_k}\partial_{Y_j}g)(X, X - Y) - (\partial_{Y_k}\partial_{Y_j}g)(X, X - Y),
\end{aligned}$$

so that

$$\partial_{X_k}\partial_{Y_j}g(X,Y)|_{X=Y=0} = -\frac{1}{2}\partial_{Y_k}\partial_{Y_j}g(X,Y)\Big|_{X=Y=0}, \tag{4.38}$$

and hence

$$L_{jk} = \frac{1}{2}\partial_{Y_k}\partial_{Y_j}g(0,Y)\Big|_{Y=0}. \tag{4.39}$$

Since the function $g(0,Y)$ is C^2 at $Y = 0$, we conclude from (4.39) that the Onsager reciprocity relation

$$L_{jk} = L_{kj},$$

holds.

Exercise 4.11 In regard to Onsager relation, open systems with *two* thermal reservoirs are special. Show that the Onsager relation follows from the conservation law

$$\omega_{X,+}(\Phi_L) + \omega_{X,+}(\Phi_R) = 0.$$

Time-reversal invariance plays no role in this argument! What is the physical origin of this derivation? Needless to say, the derivation of the Onsager reciprocity relation described in this section directly extends to open classical systems coupled to more than two thermal reservoirs to which this exercise does not apply.

The positivity of entropy production implies

$$0 \leq \omega_{X,+}(\sigma_X) = \sum_j \omega_{X,+}(\Phi_j) X_j = \sum_{j,k} L_{jk} X_j X_k + o(|X|^2),$$

so that the Onsager matrix is positive semi-definite. In fact, looking back at Section 4.2.12, we observe that the Onsager matrix coincides, up to a constant factor, with the covariance of the current fluctuations at equilibrium,

$$\mathbf{L} = \frac{1}{2} \mathbf{D} \Big|_{X=0}.$$

This is of course the celebrated Einstein relation.

For our harmonic chain model the Green–Kubo formula for the Onsager matrix can be derived by an explicit computation. In the following exercises we outline a derivation that extends to general time-reversal invariant dynamical systems.

Exercise 4.12 Show that the Green–Kubo formula holds in the Cesàro sense

$$L_{jk} = \lim_{t \to \infty} \frac{1}{t} \int_0^t \left[\frac{1}{2} \int_{-s}^s \omega_0(\Phi_j \Phi_{k,\tau}) \, d\tau \right] ds.$$

Hint: using the results of the previous section, rewrite (4.39) as

$$L_{jk} = \lim_{t \to \infty} \partial_{Y_k} \partial_{Y_j} \frac{1}{2t} g_t(0, Y) \Big|_{Y=0},$$

and work out the derivatives.

Exercise 4.13 Using the fact[5] that $\langle \delta_x | e^{\mathrm{i}t\sqrt{I-\Delta}} | \delta_y \rangle = O(t^{-1/2})$ as $t \to \infty$ (δ_x is the Kronecker delta at $x \in \mathbb{Z}$), show that $\omega_0(\Phi_j \Phi_{k,t}) = O(t^{-1})$. Invoke the Hardy–Littlewood Tauberian theorem (see, e.g., Korevaar (2004)) to conclude that the Kubo formula

$$L_{jk} = \lim_{t \to \infty} \frac{1}{2} \int_{-t}^t \omega_0(\Phi_j \Phi_{k,\tau}) \, d\tau,$$

holds.

4.2.14 The Evans–Searles fluctuation theorem

The central limit theorem derived in Section 4.2.12 shows that, for large t, typical fluctuations of the mean current $\mathbf{\Phi}^t$ with respect to its expected value $\omega_X(\mathbf{\Phi}^t)$ are small, of the order $t^{-1/2}$. In the same regime $t \to \infty$, the theory of large deviations provides information on the probability of occurrence of bigger fluctuations, of the

---

[5] This follows from a simple stationary phase estimate.

order 1. More precisely, the existence of the limit,[6]

$$g(X, Y) = \lim_{t \to \infty} \frac{1}{t} \log \int e^{-tY \cdot s} \, dQ^t(s),$$ (4.40)

and the regularity of the function $Y \mapsto g(X, Y)$ allow us to apply the Gärtner–Ellis theorem (see Exercise 5.15 below) to obtain the large deviation principle (LDP)

$$- \inf_{s \in \text{int}(G)} I_X(s) \le \liminf_{t \to \infty} \frac{1}{t} \log Q^t(G) \le \limsup_{t \to \infty} \frac{1}{t} \log Q^t(G) \le - \inf_{s \in \text{cl}(G)} I_X(s),$$

for any Borel set $G \subset \mathbb{R}^2$. Here, $\text{int}(G)$ denotes the interior of G, $\text{cl}(G)$ its closure, and the rate function $I_X : \mathbb{R}^2 \to [-\infty, 0]$ is given by

$$I_X(s) = - \inf_{Y \in \mathbb{R}^2} (Y \cdot s + g(X, Y)).$$ (4.41)

The symmetry $g(X, Y) = g(X, X - Y)$ implies

$$I_X(-s) = X \cdot s + I_X(s).$$ (4.42)

The last relation is sometimes called the Evans–Searles symmetry for the rate function.

Exercise 4.14 Show that

$$I_X(s_L, s_R) = \begin{cases} +\infty \text{ if } s_L + s_R \ne 0, \\ F(\theta) \text{ if } s_L = -s_R = \dfrac{\kappa}{\beta_0} \sinh \theta, \end{cases}$$

where

$$F(\theta) = \kappa \left[2 \sinh^2 \frac{\theta}{2} - \frac{\delta}{\beta_0} \sinh \theta - \log \left(\left(1 - \frac{\delta^2}{\beta_0^2} \right) \cosh^2 \frac{\theta}{2} \right) \right],$$

$\beta_0 = \beta = (X_L + X_R)/2$ and $\delta = (X_L - X_R)/2$. Show that $I_X(s_L, s_R)$ is strictly positive (or $+\infty$) except for $s_L = -s_R = \omega_{X,+}(\Phi_L)$ where it vanishes. Compare with Fig. 4.5.

The LDP provides the most powerful formulation of the Evans–Searles or transient fluctuation theorem. In particular, it gives fairly precise information on the rate at which the measure Q^t concentrates on the diagonal $\{(\phi, -\phi) \mid \phi \in \mathbb{R}\}$ (recall Exercise 5.10): the probability (4.35) decays super-exponentially as $t \to \infty$ for any $\epsilon > 0$. Taking this fact as well as the continuity of the function $F(\theta)$ into account, we observe that for any interval $J \subset \mathbb{R}$ one has

$$\lim_{t \to \infty} \frac{1}{t} \log Q^t(J \times \mathbb{R}) = - \inf_{s \in J} I_X(s, -s).$$

---

[6] The distribution Q^t of the mean current Φ^t was introduced in Exercise 5.4

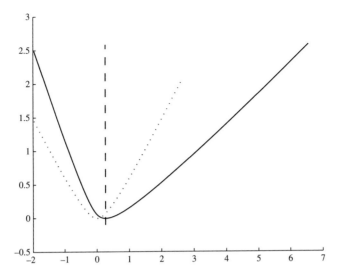

Fig. 4.5 The rate function $I_X(s, -s)$ (solid line). Notice the asymmetry which reflects the fact that $X_L > X_R$. The dashed vertical line marks the position of the mean current $\omega_{X,+}(\Phi_L) > 0$. In contrast, the rate function $I_X(s, -s)$ in the absence of forcing, $X_L = X_R$, (dotted line) is symmetric around zero.

A rough interpretation of this formula

$$\omega_X\left(\left\{\frac{1}{t}\int_0^t \Phi_{L,s}\, \mathrm{d}s = \phi\right\}\right) \sim \mathrm{e}^{-t I_X(\phi, -\phi)},$$

identifies $I_X(-\phi, \phi)$ as the rate of exponential decay of the probability for the mean current to deviate from its expected value $\omega_{X,+}(\Phi_L)$. More precisely, one has

$$\lim_{\delta\downarrow 0} \lim_{t\to\infty} \frac{1}{t} \log Q^t([\phi - \delta, \phi + \delta] \times \mathbb{R}) = -I_X(\phi, -\phi). \qquad (4.43)$$

The symmetry (4.42) implies

$$I_X(-\phi, \phi) = I_X(\phi, -\phi) + (X_L - X_R)\phi \geq (X_L - X_R)\phi,$$

and it follows that

$$\lim_{\delta\downarrow 0} \lim_{t\to\infty} \frac{1}{t} \log \frac{Q^t([-\phi - \delta, -\phi + \delta] \times \mathbb{R})}{Q^t([\phi - \delta, \phi + \delta] \times \mathbb{R})} = -(X_L - X_R)\phi, \qquad (4.44)$$

or, in a more sloppy notation,

$$\frac{\omega_X\left(\left\{\frac{1}{t}\int_0^t \Phi_{L,s}\, \mathrm{d}s = -\phi\right\}\right)}{\omega_X\left(\left\{\frac{1}{t}\int_0^t \Phi_{L,s}\, \mathrm{d}s = \phi\right\}\right)} \sim \mathrm{e}^{-t(X_L - X_R)\phi}.$$

This shows that the mean current is exponentially more likely to flow from the hotter to the colder reservoir than in the opposite direction, i.e., on a large time scale, the probability of violating the second law of thermodynamics becomes exceedingly small. Note also that (4.44) is (essentially) a considerably weaker statement then (4.43). Relation (4.44), after replacing lim with lim sup / lim inf can be derived directly from the finite time symmetry $g_t(X, Y) = g_t(X, X - Y)$ and without invoking the large deviation theory.

Exercise 4.15 Check that the Gärtner–Ellis theorem (Theorem 4.66 in Appendix 4.8.3) applies to (4.40), that is show that the function $Y \mapsto g(X, Y)$ given in equation (4.30) is differentiable on the domain $\mathcal{D} = \{(Y_L, Y_R) \in \mathbb{R}^2 \mid -T_R^{-1} < Y_R - Y_L < T_L^{-1}\}$ where it is finite and that it is steep, *i.e.*,

$$\lim_{\mathcal{D} \ni Y \to Y_0} |\nabla_Y g(X, Y)| = \infty,$$

for $Y_0 \in \partial \mathcal{D}$.

Exercise 4.16 Apply the Gärtner–Ellis theorem to the generating function $e(\alpha)$ to derive a LDP for the mean entropy production rate Σ^t, that is for the probability distribution P^t of Section 4.2.6.

4.2.15 The Gallavotti–Cohen fluctuation theorem

In this section we briefly comment on the Gallavotti–Cohen fluctuation theorem for a thermally driven harmonic chain. Let us consider the cumulant generating function of the currents in the NESS $\omega_{X,+}$,

$$g_{+,t}(X, Y) = \omega_{X,+}\left(e^{-tY \cdot \Phi^t}\right).$$

Evaluating the Gaussian integral yields

$$g_{+,t}(X, Y) = -\frac{1}{2}\log \det \left(\mathbb{1} - D_{X,+}\left(e^{t\mathcal{L}^*} k(Y) e^{t\mathcal{L}} - k(Y)\right)\right).$$

Proceeding as in Section 4.2.11, one shows that

$$g_+(X, Y) = \lim_{t \to \infty} \frac{1}{t} g_{+,t}(X, Y) = \lim_{t \to \infty} \frac{1}{t} g_t(X, Y) = g(X, Y).$$

Hence, $g_+(X, Y)$ and the corresponding rate functions $I_{X,+}(s) = I_X(s)$ inherit the symmetries

$$g_+(X, Y) = g_+(X, X - Y), \qquad I_{X,+}(-s) = X \cdot s + I_{X,+}(s).$$

Via the Gärtner–Ellis theorem, the functional $I_{X,+}(s)$ control the fluctuations of Φ^t as $t \to \infty$ w.r.t. $\omega_{X,+}$ and, after replacing ω_X with $\omega_{X,+}$ (so now $Q^t(f) = \omega_{X,+}(f(\Phi^t))$, etc.) one can repeat the discussion of the previous section line by line. The obtained results are called the Gallavotti–Cohen fluctuation theorem.

Since $\omega_{X,+}$ is singular w.r.t. ω_X in the nonequilibrium case $X_L \neq X_R$, the Gallavotti–Cohen fluctuation theorem refers to configurations (points in the phase space) which are not seen by the Evans–Searles fluctuation theorem (and vice versa, of course). The identity $g_+(X,Y) = g(X,Y)$, which was for the first time observed in (Jakšić, Pillet, and Rey-Bellet, 2011a), may seem surprising on first sight. It turned out, however, that it holds for any *nontrivial* model for which the existence of $g_+(X,Y)$ and $g(X,Y)$ has been established. This point has been raised in (Jakšić, Pillet, and Rey-Bellet, 2011a) to the *principle of regular entropic fluctuations*. Since we will not discuss quantum Gallavotti–Cohen fluctuation theorem in these lecture notes, we refer the reader to (Jakšić, Pillet, and Rey-Bellet, 2011a), (Jakšić, Ogata, Pautrat, and Pillet, 2011b) for additional discussion of these topics.

4.3 Algebraic quantum statistical mechanics of finite systems

We now turn to the main topic of these lecture notes: quantum statistical mechanics. This section is devoted to a detailed exposition of the mathematical structure of algebraic quantum statistical mechanics of finite quantum systems.

4.3.1 Notation and basic facts

Let $\mathcal{K}$ be a finite dimensional complex Hilbert space with inner product $\langle \psi | \phi \rangle$ linear in the second argument.[7] Recall the Schwarz inequality $\langle \psi | \phi \rangle \leq \|\psi\| \, \|\phi\|$, where equality holds iff ψ and ϕ are collinear. In particular $\|\phi\| = \sup_{\|\psi\|=1} \langle \psi | \phi \rangle$. We will use Dirac's notation: for $\psi \in \mathcal{K}$, $\langle \psi |$ denotes the linear functional $\mathcal{K} \ni \phi \mapsto \langle \psi | \phi \rangle \in \mathbb{C}$ and $|\psi\rangle$ its adjoint $\mathbb{C} \ni \alpha \mapsto \alpha \psi \in \mathcal{K}$.

We denote by $\mathcal{O}$ the $*$-algebra[8] of all linear maps $A : \mathcal{K} \to \mathcal{K}$. For $A \in \mathcal{O}$, $\|A\| = \sup_{\|\psi\|=1} \|A\psi\|$ denotes its operator norm and $\mathrm{sp}(A)$ its spectrum, that is the set of all eigenvalues of A. Let us recall some important properties of the operator norm. Since $\|A\psi\| \leq \|A\| \, \|\psi\|$, it follows that $\|AB\| \leq \|A\| \, \|B\|$ for all $A, B \in \mathcal{O}$. Since

$$\|A^*\phi\| = \sup_{\|\psi\|=1} \langle \psi | A^* \phi \rangle \, \sup_{\|\psi\|=1} \langle A\psi | \phi \rangle \leq \|A\| \, \|\phi\|,$$

and $A^{**} = A$, one has $\|A^*\| = \|A\|$ for all $A \in \mathcal{O}$. Finally, from the two inequalities $\|A^*A\| \leq \|A^*\| \, \|A\| = \|A\|^2$ and

$$\|A\|^2 = \sup_{\|\psi\|=1} \|A\psi\|^2 = \sup_{\|\psi\|=1} \langle A\psi | A\psi \rangle = \sup_{\|\psi\|=1} \langle \psi | A^* A \psi \rangle \leq \|A^*A\|,$$

on deduces the C^*-property $\|A^*A\| = \|A\|^2$.

The identity operator is denoted by $\mathbb{1}$ and, whenever the meaning is clear within the context, we shall write α for $\alpha \mathbb{1}$ and $\alpha \in \mathbb{C}$. Occasionally, we shall indicate the dependence on the underlying Hilbert space $\mathcal{K}$ by the subscript $\mathcal{K}$ ($\mathcal{O}_\mathcal{K}$, $\mathbb{1}_\mathcal{K}$, etc.).

---

[7] Many different Hilbert spaces will appear in the lecture notes and in latter parts we will often denote inner product by $(\cdot | \cdot)$

[8] See Exercise 5.17 below.

To any orthonormal basis $\{e_1, \ldots, e_N\}$ of the Hilbert space $\mathcal{K}$ one can associate the basis $\{E_{ij} = |e_i\rangle\langle e_j| \, | \, i, j = 1, \ldots, N\}$ of $\mathcal{O}$ so that, for any $X \in \mathcal{O}$,

$$X = \sum_{i,j=1}^{N} X_{ij} E_{ij},$$

where $X_{ij} = \langle e_i | X e_j \rangle$. Equipped with the inner product

$$(X|Y) = \operatorname{tr}(X^* Y),$$

$\mathcal{O}$ becomes a Hilbert space and $\{E_{ij}\}$ an orthonormal basis of this space.

The self-adjoint and positive parts of $\mathcal{O}$ are the subsets

$$\mathcal{O}_{\mathrm{self}} = \{A \in \mathcal{O} \, | \, A^* = A\},$$

$$\mathcal{O}_{+} = \{A \in \mathcal{O} \, | \, \langle \psi | A \psi \rangle \geq 0 \text{ for all } \psi \in \mathcal{K}\} \subset \mathcal{O}_{\mathrm{self}}.$$

We write $A \geq 0$ if $A \in \mathcal{O}_{+}$ and $A \geq B$ if $A - B \geq 0$. Note that $A \in \mathcal{O}_{+}$ iff $A \in \mathcal{O}_{\mathrm{self}}$ and $\operatorname{sp}(A) \subset [0, \infty[$. If $A \geq 0$ and $\operatorname{Ker} A = \{0\}$ we write $A > 0$.

A linear bijection $\vartheta : \mathcal{O} \to \mathcal{O}$ is called a $*$-automorphism of $\mathcal{O}$ if $\vartheta(AB) = \vartheta(A)\vartheta(B)$ and $\vartheta(A^*) = \vartheta(A)^*$. $\operatorname{Aut}(\mathcal{O})$ denotes the group of all $*$-automorphisms of $\mathcal{O}$ and id denotes its identity. Any $\vartheta \in \operatorname{Aut}(\mathcal{O})$ preserves $\mathcal{O}_{\mathrm{self}}$ and satisfies $\vartheta(\mathbb{1}) = \mathbb{1}$ and $\vartheta(A^{-1}) = \vartheta(A)^{-1}$ for all invertible $A \in \mathcal{O}$. In particular, $\vartheta((z - A)^{-1}) = (z - \vartheta(A))^{-1}$ and $\operatorname{sp}(\vartheta(A)) = \operatorname{sp}(A)$. It follows that ϑ preserves $\mathcal{O}_{+}$ and is isometric, that is $\|\vartheta(A)\| = \|A\|$ for all $A \in \mathcal{O}$.

Let $\mathcal{K}_1$ and $\mathcal{K}_2$ be two complex Hilbert spaces of dimension N_1 and N_2. Let $\{e_1^{(1)}, \ldots, e_{N_1}^{(1)}\}$ and $\{e_1^{(2)}, \ldots, e_{N_2}^{(2)}\}$ be orthonormal basis of $\mathcal{K}_1$ and $\mathcal{K}_2$. The tensor product $\mathcal{K}_1 \otimes \mathcal{K}_2$ is defined, up to isomorphism, as the $N_1 \times N_2$-dimensional complex Hilbert space with orthonormal basis $\{e_{i_1}^{(1)} \otimes e_{i_2}^{(2)} \, | \, i_1 = 1, \ldots, N_1; i_2 = 1, \ldots, N_2\}$, i.e., $\mathcal{K}_1 \otimes \mathcal{K}_2$ consists of all linear combinations

$$\psi = \sum_{i_1=1}^{N_1} \sum_{i_2=1}^{N_2} \psi_{i_1 i_2} \, e_{i_1}^{(1)} \otimes e_{i_2}^{(2)},$$

the inner product being determined by $\langle e_{i_1}^{(1)} \otimes e_{i_2}^{(2)} | e_{j_1}^{(1)} \otimes e_{j_2}^{(2)} \rangle = \delta_{i_1 j_1} \delta_{i_2 j_2}$. The *tensor product* of two vectors $\psi = \sum_{i=1}^{N_1} \psi_i e_i^{(1)} \in \mathcal{K}_1$ and $\phi = \sum_{i=1}^{N_2} \phi_i e_i^{(2)} \in \mathcal{K}_2$ is the vector in $\mathcal{K}_1 \otimes \mathcal{K}_2$ defined by

$$\psi \otimes \phi = \sum_{i_1=1}^{N_1} \sum_{i_2=1}^{N_2} \psi_{i_1} \phi_{i_2} \, e_{i_1}^{(1)} \otimes e_{i_2}^{(2)}.$$

The tensor product extends to a bilinear map from $\mathcal{K}_1 \times \mathcal{K}_2$ to $\mathcal{K}_1 \otimes \mathcal{K}_2$. We recall the characteristic property of the space $\mathcal{K}_1 \otimes \mathcal{K}_2$: for any Hilbert space $\mathcal{K}_3$, any bilinear map $F : \mathcal{K}_1 \times \mathcal{K}_2 \to \mathcal{K}_3$ uniquely extends to a linear map $\widehat{F} : \mathcal{K}_1 \otimes \mathcal{K}_2 \to \mathcal{K}_3$ by setting $\widehat{F}\psi \otimes \phi = F(\psi, \phi)$.

The tensor product of two linear operators $X \in \mathcal{O}_{\mathcal{K}_1}$ and $Y \in \mathcal{O}_{\mathcal{K}_2}$ is the linear operator on $\mathcal{K}_1 \otimes \mathcal{K}_2$ defined by

$$(X \otimes Y)\psi \otimes \phi = X\psi \otimes Y\phi,$$

and $\mathcal{O}_{\mathcal{K}_1} \otimes \mathcal{O}_{\mathcal{K}_2}$ is the $*$-algebra generated by such operators. Denoting by $\{E_{i_1 i_2}^{(1)}\}$ and $\{E_{j_1 j_2}^{(2)}\}$ the basis of $\mathcal{O}_{\mathcal{K}_1}$ and $\mathcal{O}_{\mathcal{K}_2}$ corresponding to the orthonormal basis $\{e_i^{(1)}\}$ and $\{e_j^{(2)}\}$ of $\mathcal{K}_1$ and $\mathcal{K}_2$, the $N_1^2 \times N_2^2$ operators

$$E_{i_1 i_2, j_1 j_2} = E_{i_1 j_1}^{(1)} \otimes E_{i_2 j_2}^{(2)} = |e_{i_1}^{(1)} \otimes e_{i_2}^{(2)}\rangle \langle e_{j_1}^{(1)} \otimes e_{j_2}^{(2)}|,$$

form a basis of $\mathcal{O}_{\mathcal{K}_1 \otimes \mathcal{K}_2}$. This leads to a natural identification of $\mathcal{O}_{\mathcal{K}_1 \otimes \mathcal{K}_2}$ and $\mathcal{O}_{\mathcal{K}_1} \otimes \mathcal{O}_{\mathcal{K}_2}$.

If $\lambda \in \mathrm{sp}(A)$ we denote by P_λ the associated spectral projection. When we wish to indicate its dependence on A we shall write $P_\lambda(A)$. If $A \in \mathcal{O}_{\mathrm{self}}$, we shall denote by $\lambda_j(A)$ the eigenvalues of A listed with *multiplicities* and in *decreasing* order.

If $f(z) = \sum_{n=0}^\infty a_n z^n$ is analytic in the disk $|z| < r$ and $\|A\| < r$, then

$$f(A) = \sum_{n=0}^\infty a_n A^n = \oint_{|w|=r'} f(w)(w-A)^{-1} \frac{dw}{2\pi i}, \qquad (4.45)$$

for any $\|A\| < r' < r$. If $A \in \mathcal{O}_{\mathrm{self}}$ and $f : \mathbb{R} \to \mathbb{C}$, then $f(A)$ is defined by the spectral theorem, that is

$$f(A) = \sum_{\lambda \in \mathrm{sp}(A)} f(\lambda) P_\lambda.$$

In particular, for $A \in \mathcal{O}_+$,

$$\log A = \sum_{\lambda \in \mathrm{sp}(A)} \log(\lambda) P_\lambda,$$

where $\log$ denotes the natural logarithm. We shall always use the following conventions: $\log 0 = -\infty$ and $0 \log 0 = 0$. By the Lie product formula, for any $A, B \in \mathcal{O}$,

$$e^{A+B} = \lim_{n\to\infty} \left(e^{A/n} e^{B/n} \right)^n = \lim_{n\to\infty} \left(e^{A/2n} e^{B/n} e^{A/2n} \right)^n. \qquad (4.46)$$

For any $A \in \mathcal{O}$, $A^* A \geq 0$ and we set $|A| = \sqrt{A^* A} \in \mathcal{O}_+$ and denote by $\mu_j(A)$ the singular values of A (the eigenvalues of $|A|$) listed with multiplicities and in decreasing order. Since $\|A\psi\|^2 = \||A|\psi\|^2$ one has $\mathrm{Ker}\,|A| = \mathrm{Ker}\,A$ and $\mathrm{Ran}\,|A| = \mathrm{Ran}\,A^*$. It follows that the map $U : \mathrm{Ran}\,A^* \ni |A|\psi \mapsto A\psi \in \mathrm{Ran}\,A$ is well defined and isometric. It provides the polar decomposition $A = U|A|$.

Exercise 4.17 A *complex algebra* is a complex vector space $\mathcal{A}$ with a product $\mathcal{A} \times \mathcal{A} \to \mathcal{A}$ satisfying the following axioms: for any $A, B, C \in \mathcal{A}$ and any $\alpha \in \mathbb{C}$,

(1) $A(BC) = (AB)C$.
(2) $A(B+C) = AB + AC$.
(3) $\alpha(AB) = (\alpha A)B = A(\alpha B)$.

The algebra $\mathcal{A}$ is called *abelian* or *commutative* if $AB = BA$ for all $A, B \in \mathcal{A}$ and *unital* if there exists $\mathbb{1} \in \mathcal{A}$ such that $A\mathbb{1} = \mathbb{1}A = A$ for all $A \in \mathcal{A}$.
A *∗-algebra* is a complex algebra with a map $\mathcal{A} \ni A \mapsto A^* \in \mathcal{A}$ such that, for any $A, B \in \mathcal{A}$ and any $\alpha \in \mathbb{C}$,

(4) $A^{**} = A$.
(5) $(AB)^* = B^* A^*$.
(6) $(\alpha A + B)^* = \bar{\alpha} A^* + B^*$.

A norm on a ∗-algebra $\mathcal{A}$ is a norm on the vector space $\mathcal{A}$ satisfying $\|AB\| \le \|A\| \, \|B\|$ and $\|A^*\| = \|A\|$ for all $A, B \in \mathcal{A}$. A finite dimensional normed ∗-algebra $\mathcal{A}$ is a C^*-algebra if $\|A^* A\| = \|A\|^2$ for all $A \in \mathcal{A}$. (If $\mathcal{A}$ is infinite dimensional, one additionally requires $\mathcal{A}$ to be complete w.r.t. the norm topology).
Show that if $\mathcal{K}$ is a finite dimensional Hilbert space then the set $\mathcal{O}$ of all linear maps $A : \mathcal{K} \to \mathcal{K}$ is a unital C^*-algebra.

Exercise 4.18 Prove the Löwner–Heinz inequality: if $A, B \in \mathcal{O}_+$ are such that $A \ge B$ then $A^s \ge B^s$ for any $s \in [0, 1]$.
Hint: show that $(B + t)^{-1} \ge (A + t)^{-1}$ for all $t > 0$ and use the identity

$$A^s - B^s = \frac{\sin \pi s}{\pi} \int_0^\infty t^s \left(\frac{1}{B + t} - \frac{1}{A + t} \right) dt.$$

Exercise 4.19

1. Let $A, B \in \mathcal{O}$. Prove Duhamel's formula

$$e^B - e^A = \int_0^1 e^{sB}(B - A)e^{(1-s)A} \, ds.$$

Hint: integrate the derivative of the function $f(s) = e^{sB} e^{(1-s)A}$.

2. Iterating Duhamel's formula, prove the second order Duhamel expansion

$$e^B - e^A = \int_0^1 e^{sB}(B - A)e^{(1-s)B} \, ds + \int_0^1 \int_0^s e^{uB}(B - A)e^{(s-u)A}(A - B)e^{(1-s)B} \, du ds.$$

3. Let P be a projection and set $Q = \mathbb{1} - P$. Apply the previous formula to the case $B = PAP$ to show that

$$Pe^A P = Pe^{PAP} P + \int_0^1 \int_0^u e^{(u-s)PAP} PAQe^{(1-u)A} QAPe^{sPAP} \, ds du.$$

Exercise 4.20 Let $\vartheta \in \mathrm{Aut}(\mathcal{O})$. Show that there exists a unitary $U_\vartheta \in \mathcal{O}$, unique up to a phase, such that $\vartheta(A) = U_\vartheta A U_\vartheta^{-1}$.
Hint: show first that if $P \in \mathcal{O}$ is an orthogonal projection, then so is $\vartheta(P)$ and $\mathrm{tr}(P) = \mathrm{tr}(\vartheta(P))$. Pick an orthonormal basis $\{e_1, \cdots, e_N\}$ of $\mathcal{K}$ and show that $\vartheta(|e_i\rangle\langle e_j|) = |e_i'\rangle\langle e_j'|$, where $\{e_1', \cdots, e_N'\}$ is also an orthonormal basis of $\mathcal{K}$. Set $U_\vartheta e_i = e_i'$ and complete the proof.

Exercise 4.21 Let $A \in \mathcal{O}_{\mathrm{self}}$. Prove the min-max principle:

$$\lambda_j(A) = \sup_S \inf_{\substack{\psi \perp S \\ \|\psi\| = 1}} \langle \psi | A\psi \rangle,$$

where supremum is taken over all subspaces $S \subset \mathcal{K}$ such that $\dim S = \dim \mathcal{K} - j$. Using the min-max principle, prove that for $A, B \in \mathcal{O}_{\text{self}}$,

$$|\lambda_j(A) - \lambda_j(B)| \leq \|A - B\|.$$

4.3.2 Trace inequalities

Let $\{\psi_j\}$ be an orthonormal basis of $\mathcal{K}$. We recall that the trace of $A \in \mathcal{O}$, denoted $\text{tr}(A)$, is defined by

$$\text{tr}(A) = \sum_j \langle \psi_j | A \psi_j \rangle.$$

For any unitary $U \in \mathcal{O}$, $\text{tr}(A) = \text{tr}(UAU^{-1})$ and $\text{tr}(A)$ is independent of the choice of the basis. In particular, if A is self-adjoint then $\text{tr}(A) = \sum_j \lambda_j(A) \in \mathbb{R}$ and if $A \in \mathcal{O}_+$ then $\text{tr}\,(A) \geq 0$.

For $p \in]0, \infty[$ we set

$$\|A\|_p = (\text{tr}|A|^p)^{1/p} = \left(\sum_j \mu_j(A)^p \right)^{1/p}. \tag{4.47}$$

$\|A\|_\infty = \max_j \mu_j(A)$ is the usual operator norm of A. The function $]0, \infty[\ni p \mapsto \|A\|_p$ is real analytic, monotonically decreasing, and

$$\lim_{p \to \infty} \|A\|_p = \|A\|_\infty. \tag{4.48}$$

For $p \in [1, \infty]$, the map $\mathcal{O} \ni A \mapsto \|A\|_p$ is a unitary invariant norm. Since $\dim \mathcal{K} < \infty$, all these norms are equivalent and induce the same topology on $\mathcal{O}$.

Let $A = U|A|$ be the polar decomposition of A and denote by $\{\psi_j\}$ an orthonormal basis of eigenvectors of $|A|$. Then

$$\text{tr}(BA) = \sum_j \langle \psi_j | BU|A| \psi_j \rangle = \sum_j \mu_j(A) \langle \psi_j | BU \psi_j \rangle,$$

from which we conclude that

$$|\text{tr}(BA)| \leq \sum_j \mu_j(A) |\langle \psi_j | BU \psi_j \rangle| \leq \|B\| \sum_j \mu_j(A) = \|B\| \, \|A\|_1. \tag{4.49}$$

In particular,

$$|\text{tr}(A)| \leq \|A\|_1.$$

The basic trace inequalities are:

Theorem 4.1 *(1) The Peierls-Bogoliubov inequality: for $A, B \in \mathcal{O}_{\text{self}}$,*

$$\log \frac{\text{tr}(e^A e^B)}{\text{tr}(e^B)} \geq \frac{\text{tr}(Ae^B)}{\text{tr}(e^B)}.$$

(2) *The Klein inequality: for $A, B \in \mathcal{O}_+$,*

$$\mathrm{tr}(A \log A - A \log B) \geq \mathrm{tr}(A - B),$$

with equality iff $A = B$.

(3) *The Hölder inequality: for $A, B \in \mathcal{O}$ and $p, q \in [1, \infty]$ satisfying $p^{-1} + q^{-1} = 1$,*

$$\|AB\|_1 \leq \|A\|_p \|B\|_q.$$

(4) *The Minkowski inequality: for $A, B \in \mathcal{O}$ and $p \in [1, \infty]$,*

$$\|A + B\|_p \leq \|A\|_p + \|B\|_p.$$

Proof: (1) For $\lambda \in \mathrm{sp}(A)$ we set

$$p_\lambda = \frac{\mathrm{tr}(P_\lambda(A)\mathrm{e}^B)}{\mathrm{tr}(\mathrm{e}^B)},$$

so that $p_\lambda \in [0, 1]$ and $\sum_\lambda p_\lambda = 1$. The convexity of the exponential function and Jensen's inequality imply

$$\frac{\mathrm{tr}(\mathrm{e}^A \mathrm{e}^B)}{\mathrm{tr}(\mathrm{e}^B)} = \sum_\lambda \mathrm{e}^\lambda p_\lambda \geq \mathrm{e}^{\sum_\lambda \lambda p_\lambda} = \mathrm{e}^{\mathrm{tr}(A\mathrm{e}^B)/\mathrm{tr}(\mathrm{e}^B)}.$$

(2) If $\mathrm{Ker}\, B \not\subset \mathrm{Ker}\, A$, then the left-hand side in (2) is $+\infty$ and the inequality holds trivially. Assuming $\mathrm{Ker}\, B \subset \mathrm{Ker}\, A$, we set

$$p_{\lambda,\mu} = \mathrm{tr}(P_\lambda(A)P_\mu(B)),$$

for $(\lambda, \mu) \in \mathrm{sp}(A) \times \mathrm{sp}(B)$ so that $p_{\lambda,\mu} \in [0, 1]$, $\sum_{\lambda,\mu} p_{\lambda,\mu} = 1$ and $p_{\lambda,0} = \delta_{\lambda,0} p_{0,0}$. Then, we can write

$$\mathrm{tr}(A \log A - A \log B) = \sum_{\substack{\lambda,\mu \\ \mu \neq 0}} \lambda \log(\lambda/\mu) p_{\lambda,\mu}.$$

The inequality $x \log x \geq x - 1$, which holds for $x \geq 0$, implies that for $\lambda \geq 0$ and $\mu > 0$,

$$\lambda \log \frac{\lambda}{\mu} = \mu \frac{\lambda}{\mu} \log \frac{\lambda}{\mu} \geq \mu \left(\frac{\lambda}{\mu} - 1\right) = \lambda - \mu,$$

and so

$$\mathrm{tr}(A \log A - A \log B) \geq \sum_{\substack{\lambda,\mu \\ \mu \neq 0}} (\lambda - \mu) p_{\lambda,\mu} = \sum_{\lambda,\mu} (\lambda - \mu) p_{\lambda,\mu} = \mathrm{tr}(A - B).$$

If the equality holds, then we must have

$$\sum_{\substack{\lambda,\mu \\ \mu \neq 0}} \mu \left[\frac{\lambda}{\mu} \log \frac{\lambda}{\mu} - \left(\frac{\lambda}{\mu} - 1 \right) \right] p_{\lambda,\mu} = 0,$$

where all the terms in the sum are non-negative. Since $x \log x = x - 1$ iff $x = 1$, it follows that $p_{\lambda,\mu} = 0$ for $\lambda \neq \mu \neq 0$. We have already noticed that $p_{\lambda,0} = 0$ for $\lambda \neq 0$, hence we have $p_{\lambda,\mu} = 0$ for $\lambda \neq \mu$ and it follows that

$$P_\lambda(A) P_\mu(B) P_\lambda(A) = 0 = P_\mu(B) P_\lambda(A) P_\mu(B),$$

for $\lambda \neq \mu$. Since

$$P_\lambda(A) = \sum_\mu P_\lambda(A) P_\mu(B) P_\lambda(A) = P_\lambda(A) P_\lambda(B) P_\lambda(A),$$

we must have $P_\lambda(B) \geq P_\lambda(A)$ and $\mathrm{sp}(A) \subset \mathrm{sp}(B)$. By symmetry, the reverse inequalities also hold and hence $B = A$.

(3) Equation (4.48) implies that it suffices to consider the case $1 < p < \infty$. Denote by $AB = U|AB|$, $A = V|A|$ and $B = W|B|$ the polar decompositions of AB, A and B. Then

$$\|AB\|_1 = \mathrm{tr}|AB| = \mathrm{tr}(U^* AB) = \mathrm{tr}(U^* V |A| W |B|) = \lim_{\epsilon \downarrow 0} \mathrm{tr}(U^* V (|A| + \epsilon) W (|B| + \epsilon)).$$

The function

$$F_\epsilon(z) = \mathrm{tr}(U^* V (|A| + \epsilon)^{pz} W (|B| + \epsilon)^{q(1-z)}),$$

is entire analytic and bounded on the strip $0 \leq \mathrm{Re}\, z \leq 1$. For any $y \in \mathbb{R}$, the bound (4.49) yields

$$|F_\epsilon(iy)| \leq \mathrm{tr}((|B| + \epsilon)^q), \qquad |F_\epsilon(1 + iy)| \leq \mathrm{tr}((|A| + \epsilon)^p).$$

Hence, by Hadamard's three lines theorem (see, e.g., Reed and Simon (1975)), for any z in the strip $0 \leq \mathrm{Re}\, z \leq 1$,

$$|F_\epsilon(z)| \leq [\mathrm{tr}((|A| + \epsilon)^p)]^{\mathrm{Re}\, z} \, [\mathrm{tr}((|B| + \epsilon)^q)]^{1 - \mathrm{Re}\, z}.$$

Substituting $z = 1/p$ we get

$$|\mathrm{tr}(U^* V (|A| + \epsilon) W (|B| + \epsilon))| \leq \||A| + \epsilon\|_p \||B| + \epsilon\|_q,$$

and the limit $\epsilon \downarrow 0$ yields the statement.

(4) Again, it suffices to consider the case $1 < p < \infty$. Let q be such that $p^{-1} + q^{-1} = 1$. We first observe that

$$\|A\|_p = \sup_{\|C\|_q = 1} |\mathrm{tr}(AC)|. \tag{4.50}$$

Indeed, the Hölder inequality implies

$$\sup_{\|C\|_q=1} |\text{tr}(AC)| \leq \sup_{\|C\|_q=1} \|AC\|_1 \leq \|A\|_p.$$

On the other hand, if $C = \|A\|_p^{-p/q}|A|^{p/q}U^*$ where $A = U|A|$ denotes the polar decomposition of A, then $\|C\|_q = 1$ and $\text{tr}(AC) = \|A\|_p$, and so (4.50) holds. Finally, (4.50) implies

$$\|A+B\|_p = \sup_{\|C\|_q=1} |\text{tr}((A+B)C)|$$

$$\leq \sup_{\|C\|_q=1} |\text{tr}(AC)| + \sup_{\|C\|_q=1} |\text{tr}(BC)| = \|A\|_p + \|B\|_p.$$

$\square$

We shall also need the following result due to Lieb, and Thirring for $r = 1$ (Lieb and Thirring (1976)) and generalized to $r \geq 1$ by Araki (Araki (1990)).

Theorem 4.2 *The Araki–Lieb–Thirring inequality.* *For $A, B \in \mathcal{O}_+$, $p > 0$ and $r \geq 1$,*

$$\text{tr}\left((A^{1/2}BA^{1/2})^{rp}\right) \leq \text{tr}\left((A^{r/2}B^r A^{r/2})^p\right).$$

Proof: By an obvious limiting argument (replacing A and B with $A + \epsilon$ and $B + \epsilon$) it suffices to prove the theorem in the case $A, B > 0$. We split the proof into four steps.

Step 1. If $A, B > 0$, then for $0 \leq s \leq 1$, $\|A^s B^s\| \leq \|AB\|^s$.
  Proof. Let $\phi, \psi \in \mathcal{K}$ be unit vectors and

$$F(z) = \frac{(\phi|A^z B^z \psi)}{\|AB\|^z}.$$

The function $F(z)$ is entire analytic and bounded on the strip $0 \leq \text{Re}\, z \leq 1$. For $y \in \mathbb{R}$ one has $|F(iy)| \leq 1$, $|F(1 + iy)| \leq 1$, and so by the three lines theorem, $|F(z)| \leq 1$ for $0 \leq \text{Re}\, z \leq 1$. Taking $z = s$, we deduce that

$$|(\phi|A^s B^s \psi)| \leq \|AB\|^s,$$

and

$$\|A^s B^s\| = \sup_{\|\phi\|=\|\psi\|=1} |(\phi|A^s B^s \psi)| \leq \|AB\|^s.$$

Step 2. If $A, B > 0$, then for $s \geq 1$, $\|A^s B^s\| \geq \|AB\|^s$.
  Proof. Let $\tilde{A} = A^s$, $\tilde{B} = B^s$. Then by Step 1, $\|\tilde{A}^{1/s}\tilde{B}^{1/s}\| \leq \|\tilde{A}\tilde{B}\|^{1/s}$, and the result follows.
Step 3. Set $X_r = B^{r/2}A^{r/2}$, $Y_r = X_r^* X_r = A^{r/2}B^r A^{r/2}$. Let $N = \dim \mathcal{K}$ and denote by $\lambda_1(r) \geq \cdots \geq \lambda_N(r)$ the eigenvalues of Y_r listed with multiplicities. Then for $1 \leq n \leq N$,

$$\prod_{j=1}^{n} \lambda_j(r) \geq \prod_{j=1}^{n} \lambda_j(1)^r. \tag{4.51}$$

Proof. Let $\mathcal{H} = \mathcal{K}^{\wedge n}$ be the n-fold anti-symmetric tensor product of $\mathcal{K}$ and $\Gamma_n(Y_q) = Y_q^{\wedge n}$ (the reader not familiar with this concept may consult Section 4.7). Step 2 yields the inequality

$$\|\Gamma_n(Y_r)\| = \|\Gamma_n(X_r)^* \Gamma_n(X_r)\| = \|\Gamma_n(X_r)\|^2 = \|\Gamma_n(B)^{r/2} \Gamma_n(A)^{r/2}\|^2$$

$$\geq \|\Gamma_n(B)^{1/2} \Gamma_n(A)^{1/2}\|^{2r} = \|\Gamma_n(X_1)\|^{2r} = \|\Gamma_n(Y_1)\|^r,$$

Since $\|\Gamma_n(Y_r)\| = \prod_{j=1}^{n} \lambda_j(r)$, (4.51) follows.

Step 4. For $1 \leq n \leq N$,

$$\sum_{j=1}^{n} \lambda_j(r)^p \geq \sum_{j=1}^{n} \lambda_j(1)^{rp}. \tag{4.52}$$

Proof. Set $a_j(r) = \log \lambda_j(r)$. Then, by Step 3, $a_j(r)$ is a decreasing sequence of real numbers satisfying

$$\sum_{j=1}^{n} a_j(r) \geq \sum_{j=1}^{n} r a_j(1),$$

for all n. We have to show that for all n,

$$\sum_{j=1}^{n} e^{p a_j(r)} \geq \sum_{j=1}^{n} e^{p r a_j(1)}. \tag{4.53}$$

Let $y_+ = \max(y, 0)$. We claim that for all $y \in \mathbb{R}$ and all n,

$$\sum_{j=1}^{n} (a_j(r) - y)_+ \geq \sum_{j=1}^{n} (r a_j(1) - y)_+. \tag{4.54}$$

This relation is obvious if $r a_1(1) - y \leq 0$. Otherwise, let $k \leq n$ be such that

$$r a_1(1) - y \geq \cdots \geq r a_k(1) - y \geq 0 \geq r a_{k+1}(1) - y \geq \cdots \geq r a_n(1) - y.$$

Then $\sum_{j=1}^{n} (r a_j(1) - y)_+ = \sum_{j=1}^{k} (r a_j(1) - y)$ and it follows that

$$\sum_{j=1}^{n} (a_j(r) - y)_+ \geq \sum_{j=1}^{k} (a_j(r) - y)_+ \geq \sum_{j=1}^{k} (a_j(r) - y)$$

$$\geq \sum_{j=1}^{k} (r a_j(1) - y) = \sum_{j=1}^{n} (r a_j(1) - y)_+.$$

The relation (4.54) and the identity

$$e^{px} = p^2 \int_{\mathbb{R}} (x - y)_+ e^{py} dy,$$

imply (4.53) and (4.52) follows. In the case $n = N$ the relation (4.52) reduces to the Araki–Lieb–Thirring inequality.

□

Theorem 4.2 and the Lie product formula (4.46) imply:

Corollary 4.3 *For $A, B \in \mathcal{O}_{\text{self}}$ the function*

$$[1, \infty[\ni p \mapsto \|e^{B/p}e^{A/p}\|_p^p = \text{tr}([e^{A/p}e^{2B/p}e^{A/p}]^{p/2})$$

is monotonically decreasing and

$$\lim_{p \to \infty} \|e^{B/p}e^{A/p}\|_p^p = \text{tr}(e^{A+B}).$$

In particular, the Golden–Thompson inequality holds,

$$\text{tr}(e^A e^B) \geq \text{tr}(e^{A+B}).$$

Exercise 4.22

1. Prove the following generalization of Hölder's inequality:

$$\|AB\|_r \leq \|A\|_p \|B\|_q, \tag{4.55}$$

for $p, q, r \in [1, \infty]$ such that $p^{-1} + q^{-1} = r^{-1}$.
Hint: use the polar decomposition $B = U|B|$ to write $|AB|^2 = |B|C^2|B|$ with $C = \sqrt{U^*|A|^2U}$. Invoke the Araki–Lieb–Thirring inequality to show that $\text{tr}(|AB|^r) \leq \text{tr}(|C^r|B|^r|) = \||C^r|B|^r\|_1$. Conclude the proof by applying the Hölder inequality.
2. Using (4.55), show that

$$\|A_1 \cdots A_n\|_r \leq \prod_{j=1}^n \|A_j\|_{p_j},$$

provided $\sum_j p_j^{-1} = r^{-1}$.

Exercise 4.23 Show that for any $A \in \mathcal{O}$ and $p \in [1, \infty]$ one has $\|A^*\|_p = \|A\|_p$. In particular, if $A, B \in \mathcal{O}_{\text{self}}$ then

$$\|AB\|_p = \|BA\|_p. \tag{4.56}$$

Exercise 4.24 Let $A, B \in \mathcal{O}_{\text{self}}$. Prove that the function

$$[1, \infty[\ni p \mapsto \|e^{B/p}e^{A/p}\|_p^p = \text{tr}([e^{A/p}e^{2B/p}e^{A/p}]^{p/2}),$$

is strictly decreasing unless A and B commute (in which case the function is constant). Deduce that the Golden–Thompson inequality is strict unless A and B commute.

Hint: show first that the function is real analytic. Hence, if the function is not strictly decreasing, it must be constant. If the function is constant, then its values at $p = 2$ and $p = 4$ are equal and

$$\text{tr}(e^A e^B) = \text{tr}(e^{A/2} e^{B/2} e^{A/2} e^{B/2}).$$

This identity is equivalent to $\text{tr}([e^{A/2} e^{B/2} - e^{B/2} e^{A/2}][e^{A/2} e^{B/2} - e^{B/2} e^{A/2}]^*) = 0$, and so $e^{A/2} e^{B/2} = e^{B/2} e^{A/2}$.

Corollary 4.4 *For $A, B \in \mathcal{O}_+$ and $p \geq 1$ the function*

$$\mathbb{R} \ni \alpha \mapsto \log \|A^\alpha B^{1-\alpha}\|_p^p,$$

is convex.

Proof: As in the proof of Theorem 4.2 we can assume that A and B are nonsingular. We first note that for any $s \in]0, 1[$ the Araki–Lieb–Thirring inequality implies

$$\|A^s B^s\|_p^p = \text{tr}\left(\left[B^s A^{2s} B^s\right]^{p/2}\right) = \text{tr}\left(\left[\left(B^s A^{2s} B^s\right)^{1/s}\right]^{ps/2}\right)$$

$$\leq \text{tr}\left(\left[B A^2 B\right]^{ps/2}\right) = \|AB\|_{ps}^{ps}.$$

Applying the Hölder inequality (4.55), the identity (4.56) and the previous inequality one gets, for $\alpha, \beta \in \mathbb{R}$ and $\lambda \in]0, 1[$,

$$\|A^{\lambda\alpha + (1-\lambda)\beta} B^{1-(\lambda\alpha+(1-\lambda)\beta)}\|_p^p = \|A^{\lambda\alpha} A^{(1-\lambda)\beta} B^{(1-\lambda)(1-\beta)} B^{\lambda(1-\alpha)}\|_p^p$$

$$= \|B^{\lambda(1-\alpha)} A^{\lambda\alpha} A^{(1-\lambda)\beta} B^{(1-\lambda)(1-\beta)}\|_p^p$$

$$\leq \|B^{\lambda(1-\alpha)} A^{\lambda\alpha}\|_{p/\lambda}^p \|A^{(1-\lambda)\beta} B^{(1-\lambda)(1-\beta)}\|_{p/(1-\lambda)}^p$$

$$= \|A^{\lambda\alpha} B^{\lambda(1-\alpha)}\|_{p/\lambda}^p \|A^{(1-\lambda)\beta} B^{(1-\lambda)(1-\beta)}\|_{p/(1-\lambda)}^p$$

$$\leq \|A^\alpha B^{1-\alpha}\|_p^{\lambda p} \|A^\beta B^{1-\beta}\|_p^{(1-\lambda)p}.$$

Taking the logarithm of both sides yields the result. $\qquad\qquad\square$

4.3.3 Positive and completely positive maps on $\mathcal{O}$

Denoting by $\{e_1, \ldots, e_N\}$ the standard basis of $\mathbb{C}^N$, a vector $\psi \in \mathcal{K} \otimes \mathbb{C}^N$ has a unique representation

$$\psi = \sum_{j=1}^N \psi_j \otimes e_j,$$

where $\psi_j \in \mathcal{K}$ is completely determined by $\langle \phi | \psi_j \rangle = \langle \phi \otimes e_j | \psi \rangle$ for all $\phi \in \mathcal{K}$. Accordingly, an operator $X \in \mathcal{O}_{\mathcal{K} \otimes \mathbb{C}^N}$ can be represented as a $N \times N$ block matrix

$$X = \begin{bmatrix} X_{11} & X_{12} & \cdots & X_{1N} \\ X_{21} & X_{22} & \cdots & X_{2N} \\ \vdots & \vdots & \ddots & \vdots \\ X_{N1} & X_{N2} & \cdots & X_{NN} \end{bmatrix},$$

where $X_{ij} \in \mathcal{O}_{\mathcal{K}}$ is completely determined by $\langle \phi | X_{ij} \psi \rangle = \langle \phi \otimes e_i | X \psi \otimes e_j \rangle$ for all $\phi, \psi \in \mathcal{K}$, so that

$$X\psi = \sum_{i,j=1}^{N} (X_{ij} \psi_j) \otimes e_i.$$

In particular, X is non-negative iff

$$\sum_{i,j} \langle \psi_i | X_{ij} \psi_j \rangle \geq 0,$$

for all $\psi_1, \ldots, \psi_N \in \mathcal{K}$. Note that since $\mathcal{O}_{\mathcal{K}} \otimes \mathcal{O}_{\mathbb{C}^N}$ is isomorphic to $\mathcal{O}_{\mathcal{K} \otimes \mathbb{C}^N}$, the same block matrix representation holds for $X \in \mathcal{O}_{\mathcal{K}} \otimes \mathcal{O}_{\mathbb{C}^N}$.

Let $\Phi : \mathcal{O}_{\mathcal{K}} \to \mathcal{O}_{\mathcal{K}'}$ be a linear map. Φ is called *positive* if $\Phi(\mathcal{O}_{\mathcal{K}+}) \subset \mathcal{O}_{\mathcal{K}'+}$. One easily shows that if Φ is positive, then $\Phi(X^*) = \Phi(X)^*$ for all $X \in \mathcal{O}_{\mathcal{K}}$. Φ is called *N-positive* if the map $\Phi \otimes \mathbb{1}_N : \mathcal{O}_{\mathcal{K}} \otimes \mathcal{O}_{\mathbb{C}^N} \to \mathcal{O}_{\mathcal{K}'} \otimes \mathcal{O}_{\mathbb{C}^N}$ is positive, where $\mathbb{1}_N$ is the identity map on $\mathcal{O}_{\mathbb{C}^N}$. Note that if $X \in \mathcal{O}_{\mathcal{K}} \otimes \mathcal{O}_{\mathbb{C}^N}$ has the block matrix representation $[X_{ij}]$, then $\Phi \otimes \mathbb{1}_N(X) \in \mathcal{O}_{\mathcal{K}'} \otimes \mathcal{O}_{\mathbb{C}^N}$ is represented by the block matrix $[\Phi(X_{ij})]$. If Φ is N-positive for all N, then it is called *completely positive* (CP). Φ is called *unital* if $\Phi(\mathbb{1}_{\mathcal{K}}) = \mathbb{1}_{\mathcal{K}'}$ and *trace preserving* if $\mathrm{tr}\,(\Phi(X)) = \mathrm{tr}\,(X)$ for all $X \in \mathcal{O}_{\mathcal{K}}$.

Example 4.5 Suppose that $\mathcal{K} = \mathcal{K}_1 \otimes \mathcal{K}_2$ and let $\Phi : \mathcal{O}_{\mathcal{K}} \to \mathcal{O}_{\mathcal{K}_1}$ be the unique map satisfying

$$\mathrm{tr}_{\mathcal{K}}((B \otimes \mathbb{1}_{\mathcal{K}_2})A) = \mathrm{tr}_{\mathcal{K}_1}(B\Phi(A)),$$

for all $A \in \mathcal{O}_{\mathcal{K}}$, $B \in \mathcal{O}_{\mathcal{K}_1}$. $\Phi(A)$ is called the partial trace of A over $\mathcal{K}_2$ and we shall denote it by $\mathrm{tr}_{\mathcal{K}_2}(A)$. If $\{\chi_j\}$ is an orthonormal basis of $\mathcal{K}_2$, then the matrix elements of $\mathrm{tr}_{\mathcal{K}_2}(A)$ are

$$\langle \psi | \mathrm{tr}_{\mathcal{K}_2}(A) \varphi \rangle = \sum_k \langle \psi \otimes \chi_k | A \, \varphi \otimes \chi_k \rangle.$$

The map $A \mapsto \mathrm{tr}_{\mathcal{K}_2}(A)$ is obviously linear, positive (in fact $A > 0$ implies that $\mathrm{tr}_{\mathcal{K}_2}(A) > 0$) and trace preserving. To show that it is completely positive, we note that if $[X_{ij}]$ is a positive block matrix then

$$\sum_{i,j} \langle \psi_i | \mathrm{tr}_{\mathcal{K}_2}(X_{ij}) \psi_j \rangle = \sum_k \sum_{i,j} \langle \psi_i \otimes \chi_k | X_{ij} \, \psi_j \otimes \chi_k \rangle \geq 0.$$

Exercise 4.25 Show that the following maps are completely positive:

1. A *-automorphism $\vartheta : \mathcal{O} \to \mathcal{O}$.
2. $\mathcal{O}_{\mathcal{K}} \ni X \mapsto \Phi(X) = X \otimes \mathbb{1}_{\mathcal{K}'} \in \mathcal{O}_{\mathcal{K} \otimes \mathcal{K}'}$.
3. $\mathcal{O}_{\mathcal{K}} \ni X \mapsto \Phi(X) = VXV^* \in \mathcal{O}_{\mathcal{K}}$, where $V \in \mathcal{O}_{\mathcal{K}}$.

The following result, due to Stinespring, gives a characterization of CP maps.

Proposition 4.6 *The linear map* $\Phi : \mathcal{O}_{\mathcal{K}} \to \mathcal{O}_{\mathcal{K}'}$ *is completely positive iff there exists a finite family of operators* $V_\alpha : \mathcal{K} \to \mathcal{K}'$ *such that*

$$\Phi(X) = \sum_\alpha V_\alpha X V_\alpha^*, \tag{4.57}$$

for all $X \in \mathcal{O}_{\mathcal{K}}$. *Moreover,* Φ *is unital iff* $\sum_\alpha V_\alpha V_\alpha^* = \mathbb{1}_{\mathcal{K}'}$ *and trace preserving iff* $\sum_\alpha V_\alpha^* V_\alpha = \mathbb{1}_{\mathcal{K}}$.

Remark. The right hand side of (4.57) is called a Kraus representation of the completely positive map Φ. Such a representation is not unique.

Example 4.7 Let U be a unitary operator on $\mathcal{K}_1 \otimes \mathcal{K}_2$. By Example 4.5 and Exercise 5.25, the map

$$\Phi(X) = \frac{\operatorname{tr}_{\mathcal{K}_2}(U(X \otimes \mathbb{1}_{\mathcal{K}_2})U^*)}{\dim \mathcal{K}_2},$$

is completely positive and unital on $\mathcal{O}_{\mathcal{K}_1}$. A Kraus representation is given by

$$\Phi(X) = \sum_{i,j=1}^{\dim \mathcal{K}_2} V_{i,j} X V_{i,j}^*,$$

where

$$V_{i,j} = \frac{1}{\sqrt{\dim \mathcal{K}_2}} \sum_{k,l=1}^{\dim \mathcal{K}_1} |e_k\rangle\langle e_k \otimes f_i | U e_l \otimes f_j \rangle \langle e_l|,$$

and $\{e_j\}$, $\{f_k\}$ are orthonormal basis of $\mathcal{K}_1$ and $\mathcal{K}_2$.

Proof of Proposition 4.6. The fact that a map Φ defined by equation (4.57) is completely positive follows from Part 2 of Exercise 5.25. To prove the reverse implication, let $\Phi : \mathcal{O}_{\mathcal{K}} \to \mathcal{O}_{\mathcal{K}'}$ be completely positive and denote by $E_{ij} = |\chi_i\rangle\langle\chi_j|$ the basis of $\mathcal{O}_{\mathcal{K}}$ associated to the orthonormal basis $\{\chi_i\}$ of $\mathcal{K}$. Since

$$\sum_{i,j=1}^{\dim \mathcal{K}} \langle \psi_i | E_{ij} \psi_j \rangle = \left| \sum_{i=1}^{\dim \mathcal{K}} \langle \psi_i | \chi_i \rangle \right|^2 \geq 0,$$

the block matrix $[E_{ij}]$ is positive and hence so is the block matrix $M = [\Phi(E_{ij})]$, an operator on $\mathcal{K}' \otimes \mathbb{C}^{\dim \mathcal{K}}$. Let e_i be the standard basis of $\mathbb{C}^{\dim \mathcal{K}}$ and define the operator $Q_i : \mathcal{K}' \otimes \mathbb{C}^{\dim \mathcal{K}} \to \mathcal{K}'$ by $Q_i \sum_j \psi_j \otimes e_j = \psi_i$, so that $\Phi(E_{ij}) = Q_i M Q_j^*$. If

$$M = \sum_{k=1}^{\dim \mathcal{K} \times \dim \mathcal{K}'} \lambda_k |\phi_k\rangle\langle\phi_k|,$$

is a spectral representation of M, then

$$\Phi(E_{ij}) = \sum_{k=1}^{\dim \mathcal{K} \times \dim \mathcal{K}'} \lambda_k Q_i |\phi_k\rangle\langle\phi_k| Q_j^*. \tag{4.58}$$

For each $k = 1, \ldots, \dim \mathcal{K} \times \dim \mathcal{K}'$ define a linear operator $V_k : \mathcal{K} \to \mathcal{K}'$ by $V_k e_i = \sqrt{\lambda_k} Q_i \phi_k$ for $i = 1, \ldots, \dim \mathcal{K}$. Then, we can rewrite (4.58) as

$$\Phi(E_{ij}) = \sum_{k=1}^{\dim \mathcal{K} \times \dim \mathcal{K}'} V_k E_{ij} V_k^*,$$

and since any $X \in \mathcal{O}_{\mathcal{K}}$ can be written as $X = \sum_{i,j} X_{ij} E_{ij}$ we have

$$\Phi(X) = \sum_{k=1}^{\dim \mathcal{K} \times \dim \mathcal{K}'} V_k X V_k^*.$$

The last statement of Proposition 4.6 is obvious. □

Definition 4.8 *A linear map* $\Phi : \mathcal{O}_{\mathcal{K}} \to \mathcal{O}_{\mathcal{K}'}$ *such that, for all* $X \in \mathcal{O}_{\mathcal{K}}$,

$$\Phi(X)^*\Phi(X) \le \Phi(X^*X), \tag{4.59}$$

is called a Schwarz map and (4.59) is called the Schwarz inequality.

Proposition 4.9 *Any 2-positive map* $\Phi : \mathcal{O}_{\mathcal{K}} \to \mathcal{O}_{\mathcal{K}'}$ *is a Schwarz map.*

Proof: For any $X \in \mathcal{O}_{\mathcal{K}}$, the 2×2 block matrix

$$[A_{ij}] = \begin{bmatrix} \mathbb{1} & X \\ X^* & X^*X \end{bmatrix},$$

is non-negative. Indeed, for any $\psi_1, \psi_2 \in \mathcal{K}$ one has

$$\sum_{i,j} \langle \psi_i | A_{ij} \psi_j \rangle = \|\psi_1 + X\psi_2\|^2 \ge 0.$$

If Φ is 2-positive, then the block matrix $[\Phi(A_{ij})]$ is also non-negative and hence

$$\sum_{i,j} \langle \phi_i | A_{ij} \phi_j \rangle = \|\phi_1 + \Phi(X)\phi_2\|^2 + \langle \phi_2 | (\Phi(X^*X) - \Phi(X)^*\Phi(X))\phi_2 \rangle \ge 0,$$

for all $\phi_1, \phi_2 \in \mathcal{K}'$. Setting $\phi_1 = -\Phi(X)\phi_2$ yields the Schwarz inequality.

□

Exercise 4.26 Let $\Phi : \mathcal{O}_{\mathcal{K}} \to \mathcal{O}_{\mathcal{K}'}$ be a linear map and denote by Φ^* its adjoint w.r.t. the inner product $(\,\cdot\,|\,\cdot\,)$, that is

$$(X|\Phi(Y)) = \mathrm{tr}_{\mathcal{K}'}(X^*\Phi(Y)) = \mathrm{tr}_{\mathcal{K}}(\Phi^*(X)^*Y) = (\Phi^*(X)|Y).$$

1. Show that Φ^* is positive iff Φ is positive.
2. Show that Φ^* is N-positive iff Φ is N-positive.
3. Show that Φ^* is trace preserving iff Φ is unital.

4.3.4 States

An element $\rho \in \mathcal{O}_+$ is called a density matrix or a *state* if $\mathrm{tr}(\rho) = 1$. We denote by $\mathfrak{S}$ the collection of all states. We shall identify a state ρ with the linear functional

$$\begin{aligned} \rho : \mathcal{O} &\to \mathbb{C} \\ A &\mapsto \mathrm{tr}(\rho A). \end{aligned}$$

With this identification, $\mathfrak{S}$ can be characterized as the set of all linear functionals $\phi : \mathcal{O} \to \mathbb{C}$ which are positive ($\phi(A) \geq 0$ for all $A \in \mathcal{O}_+$) and normalized ($\phi(\mathbb{1}) = 1$). In models that arise in physics the elements of $\mathcal{O}$ describe observables of the physical system under consideration. The physical states are described by elements of $\mathfrak{S}$. If A is self-adjoint and $A = \sum_{\alpha \in \mathrm{sp}(A)} \alpha P_\alpha$ is its spectral decomposition, then the possible outcomes of a measurement of A are the eigenvalues of A. If the system is in a state ρ, the probability that α is observed is $\mathrm{tr}(\rho P_\alpha) = \mathrm{tr}(P_\alpha \rho P_\alpha) \in [0,1]$. In particular,

$$\rho(A) = \mathrm{tr}(\rho A),$$

is the expectation value of the observable A and its variance is

$$\Delta_\rho(A) = \rho((A - \rho(A))^2) = \rho(A^2) - \rho(A^2).$$

Note that if $A \in \mathcal{O}_+$, then $\rho(A) \geq 0$. For $A, B \in \mathcal{O}_{\mathrm{self}}$ the Heisenberg uncertainty principle takes the form

$$\frac{1}{2}|\rho(\mathrm{i}[A, B])| \leq \sqrt{\Delta_\rho(A)}\sqrt{\Delta_\rho(B)}.$$

If $\Phi : \mathcal{O}_{\mathcal{K}} \to \mathcal{O}_{\mathcal{K}'}$ is a positive, unit preserving map, then its adjoint Φ^* is positive and trace preserving. In particular, it maps states $\rho \in \mathfrak{S}_{\mathcal{K}'}$ into states $\Phi^*(\rho) \in \mathfrak{S}_{\mathcal{K}}$ in such a way that

$$\rho(\Phi(A)) = \Phi^*(\rho)(A),$$

i.e., $\Phi^*(\rho) = \rho \circ \Phi$.

4.3.5   Entropy

Let ρ be a state. The orthogonal projection on the subspace $\operatorname{Ran} \rho = (\operatorname{Ker} \rho)^{\perp}$ is called the support of ρ and is denoted $s(\rho)$. We shall use the notation $\rho \ll \nu$ iff $s(\rho) \leq s(\nu)$, that is, iff $\operatorname{Ran} \rho \subset \operatorname{Ran} \nu$, and $\rho \perp \nu$ iff $s(\rho) \perp s(\nu)$, that is, iff $\operatorname{Ran} \rho \subset \operatorname{Ker} \nu$. Two states ρ and ν are called equivalent if $\rho \ll \nu$ and $\nu \ll \rho$. A state ρ is called faithful if $s(\rho) = \mathbb{1}$, that is if $\rho > 0$. The set $\mathfrak{S}$ and the set of all faithful states $\mathfrak{S}_f$ are convex subsets of $\mathcal{O}_+$. A state ρ is called pure if $\rho = |\psi\rangle\langle\psi|$ for some unit vector ψ. The state

$$\rho_{\mathrm{ch}} = \frac{\mathbb{1}}{\dim \mathcal{K}}, \tag{4.60}$$

is called chaotic. If A is self-adjoint, we denote

$$\rho_A = \frac{e^A}{\operatorname{tr}(e^A)}.$$

The state ρ_A is faithful and $\rho_A = \rho_B$ iff A and B differ by a constant. If $\mathcal{K} = \mathcal{K}_1 \otimes \mathcal{K}_2$ and $\rho \in \mathfrak{S}_{\mathcal{K}}$, then $\rho_{\mathcal{K}_1} = \operatorname{tr}_{\mathcal{K}_2}(\rho) \in \mathfrak{S}_{\mathcal{K}_1}$ and $\rho_{\mathcal{K}_2} = \operatorname{tr}_{\mathcal{K}_1}(\rho) \in \mathfrak{S}_{\mathcal{K}_2}$.

The von Neumann entropy of a state ρ, defined by

$$S(\rho) = -\operatorname{tr}(\rho \log \rho) = - \sum_{\lambda \in \operatorname{sp}(\rho)} \operatorname{tr}(P_\lambda) \lambda \log \lambda.$$

is the noncommutative extension of the Gibbs or Shannon entropy of a probability distribution. It is characterized by the following dual variational principles.

Theorem 4.10 (1) *For any $\rho \in \mathfrak{S}$, one has*

$$S(\rho) = \min_{A \in \mathcal{O}_{\mathrm{self}}} \log \operatorname{tr}(s(\rho) e^A) - \rho(A).$$

(2) *For any $A \in \mathcal{O}_{\mathrm{self}}$, one has*

$$\log \operatorname{tr}(e^A) = \max_{\rho \in \mathfrak{S}} \rho(A) + S(\rho).$$

Remark. Adopting the decomposition $\mathcal{K} = \operatorname{Ran} \rho \oplus \operatorname{Ker} \rho$, the minimum in (1) is achieved at A iff $A = (\log(\rho|_{\operatorname{Ran} \rho}) \oplus B) + c$ where B is an arbitrary self-adjoint operator on $\operatorname{Ker} \rho$ and c an arbitrary real constant. An alternative formulation of (1) is

$$S(\rho) = \inf_{A \in \mathcal{O}_{\mathrm{self}}} \log \operatorname{tr}(e^A) - \rho(A).$$

The maximizer in (2) is unique and given by $\rho = \rho_A$.

Proof: (2) Let $G_A(\rho) = \rho(A) + S(\rho)$. Since $\log \rho_A = A - \log \operatorname{tr}(e^A)$, Klein's inequality implies

$$\log \operatorname{tr}(e^A) - G_A(\rho) = \operatorname{tr}(\rho(\log \rho - \log \rho_A)) \geq \operatorname{tr}(\rho - \rho_A) = 0,$$

for any $\rho \in \mathfrak{S}$, with equality iff $\rho = \rho_A$. Thus,

$$G_A(\rho) \le \log \mathrm{tr}(e^A), \tag{4.61}$$

with equality iff $\rho = \rho_A$.

(1) We decompose $\mathcal{K} = \mathrm{Ran}\,\rho \oplus \mathrm{Ker}\,\rho$ and set $P = s(\rho)$ and $Q = \mathbb{1} - P$. Since $G_A(\rho) = G_{PAP}(\rho)$, we can invoke (4.61) within the subspace $\mathrm{Ran}\,\rho$ to write

$$G_A(\rho) \le \log \mathrm{tr}(Pe^{PAP}P),$$

where equality holds iff $\rho = Pe^{PAP}P/\mathrm{tr}(Pe^{PAP}P)$, that is $PAP = \log(\rho|_{\mathrm{Ran}\,\rho}) + c$ for some real constant c. A second order Duhamel expansion (see Part 3 of Exercise 5.19) further yields

$$\mathrm{tr}(Pe^A P) = \mathrm{tr}(Pe^{PAP}P) + \int_0^1 u\,\mathrm{tr}\left(e^{uPAP/2}PAQe^{(1-u)A}QAPe^{uPAP/2}\right)\,du,$$

so that $\mathrm{tr}(Pe^{PAP}P) \le \mathrm{tr}(Pe^A P)$ with equality iff $QAP = 0$. We conclude that $G_A(\rho) \le \log \mathrm{tr}(Pe^A P)$ and hence $S(\rho) \le \log \mathrm{tr}(Pe^A P) - \rho(A)$ where equality holds iff $A = (\log(\rho|_{\mathrm{Ran}\,\rho}) \oplus B) + c$. $\qquad\square$

An immediate consequence of Theorem 4.10 is

Corollary 4.11 (1) *The function* $\mathfrak{S} \ni \rho \mapsto S(\rho)$ *is concave.*
(2) *The function* $\mathcal{O}_{\mathrm{self}} \ni A \mapsto \log \mathrm{tr}(e^A)$ *is convex.*

Further basic properties of the entropy functional are:

Theorem 4.12 (1) *The map* $\mathfrak{S} \ni \rho \mapsto S(\rho)$ *is continuous.*
(2) $0 \le S(\rho) \le \log \dim \mathcal{K}$. *Moreover,* $S(\rho) = 0$ *iff* ρ *is pure and* $S(\rho) = \log \dim \mathcal{K}$ *iff* ρ *is chaotic.*
(3) *For any unitary* U, $S(U\rho U^{-1}) = S(\rho)$.
(4) $S(\rho_A) = \log \mathrm{tr}(e^A) - \mathrm{tr}(A\rho_A)$.
(5) *If* $\mathcal{K} = \mathcal{K}_1 \otimes \mathcal{K}_2$, *then* $S(\rho) \le S(\rho_{\mathcal{K}_1}) + S(\rho_{\mathcal{K}_2})$ *where the equality holds if and only if* $\rho = \rho_{\mathcal{K}_1} \otimes \rho_{\mathcal{K}_2}$ *(recall that* $\rho_{\mathcal{K}_1} = \mathrm{tr}_{\mathcal{K}_2}(\rho)$*).*

Remark. To (1): the Fannes inequality

$$|S(\rho) - S(\nu)| \le \|\rho - \nu\|_1 \log \frac{\dim \mathcal{K}}{\|\rho - \nu\|_1},$$

holds provided $\|\rho - \nu\|_1 < 1/3$. See, for example Ohya and Petz (2004).

Proof: The proofs of (1)–(4) are easy and left to the reader. To prove (5) we invoke the variational principle to write

$$S(\rho) \le \min_{(A,B) \in \mathcal{O}_1 \times \mathcal{O}_2} \log \mathrm{tr}(s(\rho)e^{A \otimes \mathbb{1} + \mathbb{1} \otimes B}) - \rho(A \otimes \mathbb{1} + \mathbb{1} \otimes B).$$

Setting $\rho_j = \rho_{\mathcal{K}_j}$, the support $s(\rho_1)$ satisfies $1 = \mathrm{tr}_{\mathcal{K}_1}(s(\rho_1)\rho_1) = \mathrm{tr}_{\mathcal{K}}((s(\rho_1) \otimes \mathbb{1})\rho)$ and, by the definition of the support, we must have $s(\rho_1) \otimes \mathbb{1} \ge s(\rho)$ and a similar inequality for $s(\rho_2)$. It follows that $s(\rho_1) \otimes s(\rho_2) \geq s(\rho)$ and therefore

$$\mathrm{tr}(s(\rho)e^{A\otimes \mathbb{1} + \mathbb{1} \otimes B}) \leq \mathrm{tr}(s(\rho_1)e^A \otimes s(\rho_2)e^B).$$

Thus, we can write

$$S(\rho) \leq \min_{(A,B)\in \mathcal{O}_1 \times \mathcal{O}_2} \log \mathrm{tr}(s(\rho_1)e^A) + \log \mathrm{tr}(s(\rho_2)e^B) - \rho_1(A) - \rho_2(B)$$

$$= S(\rho_1) + S(\rho_2).$$

Moreover, equality holds iff the variational principle has a minimizer of the form $A \otimes \mathbb{1} + \mathbb{1} \otimes B$, which, by the remark after Theorem 4.12, is possible only if $\rho = \rho_1 \otimes \rho_2$. □

4.3.6 Relative entropies

The Rényi relative entropy (or α-relative entropy) of two states ρ, ν is defined for $\alpha \in]0, 1[$ by

$$S_\alpha(\rho|\nu) = \log \mathrm{tr}(\rho^\alpha \nu^{1-\alpha}).$$

This quantity will play an important role in these lecture notes. According to our convention $\log(\rho|_{\mathrm{Ker}\,\rho}) = -\infty$, and

$$\rho^\alpha = e^{\alpha \log \rho} = e^{\alpha \log(\rho|_{\mathrm{Ran}\,\rho})} \oplus 0|_{\mathrm{Ker}\,\rho}.$$

Thus, $S_\alpha(\rho|\nu) \in [-\infty, 0]$ and $S_\alpha(\rho|\nu) = -\infty$ iff $\rho \perp \nu$ (that is, if ρ and ν are mutually singular). In terms of the spectral data of ρ and ν,

$$S_\alpha(\rho|\nu) = \log \left[\sum_{\substack{(\lambda,\mu)\in \mathrm{sp}(\rho)\times \mathrm{sp}(\nu) \\ \lambda\neq 0, \mu\neq 0}} \lambda^\alpha \mu^{1-\alpha} \mathrm{tr}(P_\lambda(\rho)P_\mu(\nu)) \right], \quad (4.62)$$

and so if $\rho \not\perp \nu$, then $]0, 1[\ni \alpha \mapsto S_\alpha(\rho|\nu) \in]-\infty, 0]$ extends to a real-analytic function on $\mathbb{R}$. The basic properties of Rényi's relative entropy are:

Proposition 4.13 *Suppose that $\rho \not\perp \nu$. Then:*

(1) $S_0(\rho|\nu) = \log \nu(s(\rho))$ and $S_1(\rho|\nu) = \log \rho(s(\nu))$.
(2) The map $\mathbb{R} \ni \alpha \mapsto S_\alpha(\rho|\nu)$ is convex.
(3) $S_\alpha(U\rho U^{-1}|U\nu U^{-1}) = S_\alpha(\rho|\nu)$ for any unitary U.
(4) Suppose that $s(\rho) = s(\nu)$. Then the map $\mathbb{R} \ni \alpha \mapsto S_\alpha(\rho|\nu)$ is strictly convex iff $\rho \neq \nu$.
(5) $S_\alpha(\rho|\nu) = S_{1-\alpha}(\nu|\rho)$.

Proof: (1), (3), and (5) are obvious. (2) Follows from the following facts, easily derived from (4.62),

$$\partial_\alpha S_\alpha(\rho|\nu) = \sum_{(\lambda,\mu)\in \text{sp}(\rho)\times \text{sp}(\nu)} p_{\lambda,\mu} \log(\lambda/\mu) = \theta_\alpha,$$

$$\partial_\alpha^2 S_\alpha(\rho|\nu) = \sum_{(\lambda,\mu)\in \text{sp}(\rho)\times \text{sp}(\nu)} p_{\lambda,\mu} \left[\log(\lambda/\mu) - \theta_\alpha\right]^2 \geq 0,$$

where

$$p_{\lambda,\mu} = \frac{\lambda^\alpha \mu^{1-\alpha}\text{tr}(P_\lambda(\rho)P_\mu(\nu))}{\displaystyle\sum_{(\lambda,\mu)\in \text{sp}(\rho)\times \text{sp}(\nu)} \lambda^\alpha \mu^{1-\alpha}\text{tr}(P_\lambda(\rho)P_\mu(\nu))} \geq 0,$$

$$\sum_{(\lambda,\mu)\in \text{sp}(\rho)\times \text{sp}(\nu)} p_{\lambda,\mu} = 1.$$

(4) Invoking analyticity, we further deduce that either $\partial_\alpha^2 S_\alpha(\rho|\nu) = 0$ for all $\alpha \in \mathbb{R}$, or $\partial_\alpha^2 S_\alpha(\rho|\nu) > 0$ except possibly on a discrete subset of $\mathbb{R}$. In the former case $S_\alpha(\rho|\nu) = (1-\alpha)S_0(\rho|\nu) + \alpha S_1(\rho|\nu)$ is an affine function of α. In the latter case $S_\alpha(\rho|\nu)$ is strictly convex.

Suppose now that $\text{s}(\rho) = \text{s}(\nu)$. Without loss of generality, we can assume that ρ and ν are faithful. If $\rho = \nu$ then $S_\alpha(\rho|\nu) = 0$ for all $\alpha \in \mathbb{R}$. Reciprocally, if $\partial_\alpha^2 S_\alpha(\rho|\nu)$ vanishes identically then $\theta = \partial_\alpha S_\alpha(\rho|\nu)$ is constant and $\lambda = e^\theta \mu$ whenever $\text{tr}(P_\lambda(\rho)P_\mu(\nu)) \neq 0$. It follows from

$$1 = \text{tr}(\rho) = \sum_{\lambda,\mu}\lambda\,\text{tr}(P_\lambda(\rho)P_\mu(\nu)) = e^\theta \sum_{\lambda,\mu} \mu\,\text{tr}(P_\lambda(\rho)P_\mu(\nu)) = e^\theta\text{tr}(\nu) = e^\theta,$$

that $\theta = 0$. Repeating the argument in the proof of Part (2) of Theorem 4.1 leads to the conclusion that $\rho = \nu$. $\qquad\square$

The following theorem, a variant of the celebrated Kosaki's variational formula (Kosaki 1986; Ohya-Petz 2004), is deeper. The result and its proof were communicated to us by R. Seiringer (unpublished). The proof will be given in Section 4.3.12 as an illustration of the power of the modular structure to be introduced there.

Theorem 4.14 *For $\alpha \in]0, 1[$,*

$$S_\alpha(\rho|\nu) = \inf_{A\in C(\mathbb{R}_+,\mathcal{O})} \log\left[\frac{\sin \pi\alpha}{\pi}\int_0^\infty t^{\alpha-1}\left(\frac{1}{t}\rho(|A(t)^*|^2) + \nu(|\mathbb{1} - A(t)|^2)\right)dt\right],$$

where $C(\mathbb{R}_+,\mathcal{O})$ denotes the set of all continuous functions $\mathbb{R}_+ \ni t \mapsto A(t) \in \mathcal{O}$. Moreover, the infimum is achieved for

$$A(t) = t\int_0^\infty e^{-s\rho}\nu e^{-st\nu}ds,$$

and this is the unique minimizer if either ρ or ν is faithful.

An immediate consequence of Kosaki's variational formula is *Uhlmann's monotonicity theorem*, (Uhlmann 1977):

Theorem 4.15 *If $\Phi : \mathcal{O}_\mathcal{K} \to \mathcal{O}_{\mathcal{K}'}$ is a unital Schwarz map, then*

$$S_\alpha(\rho \circ \Phi | \nu \circ \Phi) \geq S_\alpha(\rho | \nu),$$

for all $\alpha \in [0,1]$ and $\rho, \nu \in \mathfrak{S}_{\mathcal{K}'}$.

Proof: With $\hat{\rho} = \rho \circ \Phi$ and $\hat{\nu} = \nu \circ \Phi$, Kosaki's formula reads

$$S_\alpha(\hat{\rho} | \hat{\nu}) = \inf_{A \in C(\mathbb{R}_+, \mathcal{O}_\mathcal{K})} \log \left[\frac{\sin \pi\alpha}{\pi} \int_0^\infty t^{\alpha-1} \left(\frac{1}{t} \hat{\rho}(|A(t)^*|^2) + \hat{\nu}(|\mathbb{1} - A(t)|^2) \right) dt \right].$$

Since Φ is a unital Schwarz map, for any $A \in \mathcal{O}_\mathcal{K}$ one has

$$\hat{\rho}(|A^*|^2) = \rho(\Phi(AA^*)) \geq \rho(\Phi(A)\Phi(A)^*) = \rho(|\Phi(A)^*|^2),$$

as well as

$$\begin{aligned}
\hat{\nu}(|\mathbb{1} - A|^2) &= \nu(\Phi((\mathbb{1} - A)^*(\mathbb{1} - A))) \\
&\geq \nu(\Phi(\mathbb{1} - A)^* \Phi(\mathbb{1} - A)) \\
&= \nu((\mathbb{1} - \Phi(A))^*(\mathbb{1} - \Phi(A))) = \nu(|\mathbb{1} - \Phi(A)|^2).
\end{aligned}$$

It follows that

$$\begin{aligned}
S_\alpha(\hat{\rho} | \hat{\nu}) &\geq \inf_{A \in C(\mathbb{R}_+, \mathcal{O}_\mathcal{K})} \log \left[\frac{\sin \pi\alpha}{\pi} \int_0^\infty t^{\alpha-1} \left(\frac{1}{t} \rho(|\Phi(A(t))^*|^2) + \nu(|\mathbb{1} - \Phi(A(t))|^2) \right) dt \right] \\
&= \inf_{A \in \Phi(C(\mathbb{R}_+, \mathcal{O}_\mathcal{K}))} \log \left[\frac{\sin \pi\alpha}{\pi} \int_0^\infty t^{\alpha-1} \left(\frac{1}{t} \rho(|A(t)^*|^2) + \nu(|\mathbb{1} - A(t)|^2) \right) dt \right],
\end{aligned}$$

where $\Phi(C(\mathbb{R}_+, \mathcal{O}_\mathcal{K})) = \{\Phi \circ A \mid A \in C(\mathbb{R}_+, \mathcal{O}_\mathcal{K})\}$. Since Φ is continuous, one has $\Phi(C(\mathbb{R}_+, \mathcal{O}_\mathcal{K})) \subset C(\mathbb{R}_+, \mathcal{O}_{\mathcal{K}'})$, and the result follows from Kosaki's formula. $\square$

Another consequence of Theorem 4.14 is:

Theorem 4.16 *For $\alpha \in [0,1]$, the map $\mathfrak{S} \times \mathfrak{S} \ni (\rho, \nu) \mapsto S_\alpha(\rho | \nu)$ is jointly concave, that is*

$$S_\alpha(\lambda \rho + (1 - \lambda)\rho' | \lambda \nu + (1 - \lambda)\nu') \geq \lambda S_\alpha(\rho | \nu) + (1 - \lambda) S_\alpha(\rho' | \nu'),$$

for any $\rho, \rho', \nu, \nu' \in \mathfrak{S}$ and any $\lambda \in [0,1]$.

Proof: The result is obvious for $\lambda = 0$ and for $\lambda = 1$. Hence, we assume $\lambda \in]0,1[$ in the following. For $\alpha = 0$ and for $\alpha = 1$, the result follows from Part (1) of Proposition 4.13, the concavity of the logarithm, and the fact that $s(\lambda \rho + (1 - \lambda)\rho') \geq s(\rho)$. For $\alpha \in]0,1[$ and $A \in C(\mathbb{R}_+, \mathcal{O})$, the map

$$(\rho, \nu) \mapsto F_A(\rho, \nu) = \frac{\sin \pi\alpha}{\pi} \int_0^\infty t^{\alpha-1} \left(\frac{1}{t} \rho(|A(t)^*|^2) + \nu(|\mathbb{1} - A(t)|^2) \right) dt,$$

is affine. The concavity of the logarithm implies that the map $(\rho, \nu) \mapsto \log F_A(\rho, \nu)$ is concave. Therefore, the function $(\rho, \nu) \mapsto S_\alpha(\rho|\nu)$ being the infimum of a family of concave functions, it is itself concave (see the following exercise). □

Exercise 4.27 Let $C \subset \mathbb{R}^n$ be a convex set and $\mathcal{F}$ a nonempty set of real valued functions on C. Set $F(x) = \inf\{f(x) \mid f \in \mathcal{F}\}$.

1. Show that if the elements of $\mathcal{F}$ are concave then F is concave.
2. Show that if the elements of $\mathcal{F}$ are continuous then F is upper semi-continuous, that is

$$\limsup_{x \to x_0} F(x) \le F(x_0),$$

for all $x_0 \in \mathbb{R}^n$.
3. Show that the function $(\rho, \nu) \mapsto S_\alpha(\rho|\nu)$ is upper semi-continuous on $\mathfrak{S} \times \mathfrak{S}$.

The relative entropy of the state ρ w.r.t. the state ν is defined by

$$S(\rho|\nu) = \begin{cases} \mathrm{tr}(\rho(\log \nu - \log \rho)) & \text{if } \rho \ll \nu, \\ -\infty & \text{otherwise.} \end{cases}$$

Equivalently, in terms of the spectral data of ρ and ν, one has

$$S(\rho|\nu) = \sum_{(\lambda, \mu) \in \mathrm{sp}(\rho) \times \mathrm{sp}(\nu)} \lambda(\log \mu - \log \lambda) \mathrm{tr}(P_\lambda(\rho) P_\mu(\nu)).$$

For $\nu \in \mathfrak{S}$ and $A \in \mathcal{O}_{\mathrm{self}}$, we define

$$e^{A + \log \nu} = \lim_{n \to \infty} \left(e^{A/n} \nu^{1/n} \right)^n.$$

It is not difficult to show that, according to the decomposition $\mathcal{K} = \mathrm{Ran}\,\nu \oplus \mathrm{Ker}\,\nu$,

$$e^{A + \log \nu} = e^{s(\nu)As(\nu)|_{\mathrm{Ran}\,\nu} + \log(\nu|_{\mathrm{Ran}\,\nu})} \oplus 0_{\mathrm{Ker}\,\nu}.$$

With this definition, the relative entropy functional has the following variational characterizations:

Theorem 4.17 (1) *For any* $\rho, \nu \in \mathfrak{S}$, *one has*

$$S(\rho|\nu) = \inf_{A \in \mathcal{O}_{\mathrm{self}}} \log \mathrm{tr}(e^{A + \log \nu}) - \rho(A).$$

(2) *For any* $A \in \mathcal{O}_{\mathrm{self}}$ *and* $\nu \in \mathfrak{S}$, *one has*

$$\log \mathrm{tr}(e^{A + \log \nu}) = \max_{\rho \in \mathfrak{S}} S(\rho|\nu) + \rho(A).$$

Remark. If ρ and ν are equivalent, then the infimum in (1) is achieved at A iff $s(\nu)As(\nu)|_{\mathrm{Ran}\,\nu} = \log(\rho|_{\mathrm{Ran}\,\nu}) - \log(\nu|_{\mathrm{Ran}\,\nu}) + c$ where c is an arbitrary real constant. The maximizer in (2) is unique, given by $\rho = e^{A + \log \nu}/\mathrm{tr}(e^{A + \log \nu})$.

Proof: (2) Set $G_{\nu,A}(\rho) = S(\rho|\nu) + \rho(A)$ and $\tilde{\nu} = e^{A+\log\nu}/\mathrm{tr}(e^{A+\log\nu})$. Note that $G_{\nu,A}(\rho) = -\infty$ if $\rho \not\ll \nu$ while $G_{\nu,A}(\nu) = \nu(A) > -\infty$. Thus, it suffices to consider $\rho \ll \nu$ in which case one has

$$\rho(\log\tilde{\nu}) = \rho(A) + \rho(\log\nu) - \log\mathrm{tr}(e^{A+\log\nu}).$$

Klein's inequality yields

$$\log\mathrm{tr}(e^{A+\log\nu}) - G_{\nu,A}(\rho) = \mathrm{tr}\left(\rho(\log\rho - \log\tilde{\nu})\right) \geq \mathrm{tr}(\rho - \tilde{\nu}) = 0,$$

with equality iff $\rho = \tilde{\nu}$.

(1) We first consider the case $\rho \not\ll \nu$. Then, there exists a projection P such that $\mathrm{Ran}\,P \perp \mathrm{Ran}\,\nu$ and $\rho(P) > 0$. Since $e^{\lambda P + \log\nu} = \nu$, it follows that

$$\log\mathrm{tr}(e^{\lambda P + \log\nu}) - \rho(\lambda P) = -\lambda\rho(P) \to -\infty = S(\rho|\nu),$$

as $\lambda \to \infty$. On the other hand, for any $\rho \ll \nu$ and $A \in \mathcal{O}_{\mathrm{self}}$, (2) implies that

$$S(\rho|\nu) \leq \log\mathrm{tr}(e^{A+\log\nu}) - \rho(A),$$

with equality iff $\rho = e^{A+\log\nu}/\mathrm{tr}(e^{A+\log\nu})$. If $\nu \ll \rho$, this means that equality holds iff $s(\nu)As(\nu)|_{\mathrm{Ran}\,\nu} = \log(\rho|_{\mathrm{Ran}\,\nu}) - \log(\nu|_{\mathrm{Ran}\,\nu})$ up to an arbitrary additive constant. If $\nu \not\ll \rho$, that is if $s(\rho) < s(\nu)$, then $A_\lambda = \log(\rho|_{\mathrm{Ran}\,\rho}) \oplus \lambda\mathbb{1}_{\mathrm{Ker}\,\rho} - \log(\nu|_{\mathrm{Ran}\,\nu}) \oplus 0_{\mathrm{Ker}\,\nu}$ is such that, with $d = \dim\mathrm{Ran}\,\nu - \dim\mathrm{Ran}\,\rho$,

$$\log\mathrm{tr}(e^{A_\lambda + \log\nu}) - \rho(A_\lambda) = \log\left(1 + e^\lambda d\right) + S(\rho|\nu) \to S(\rho|\nu),$$

as $\lambda \to -\infty$. □

As an immediate consequence of Theorem 4.17 we note, for later reference

Corollary 4.18 *For any state $\nu \in \mathfrak{S}$ and any self-adjoint observable $A \in \mathcal{O}$ one has the inequality*

$$\mathrm{tr}(e^{\log\nu + A}) \geq e^{\nu(A)}.$$

The basic properties of the relative entropy functional are:

Proposition 4.19 (1) $S(\rho|\nu) \leq 0$ *with equality iff $\rho = \nu$.*
(2) $S(\rho|\rho_{\mathrm{ch}}) = S(\rho) - \log\dim\mathcal{K}$.
(3) $S(U\rho U^{-1}|U\nu U^{-1}) = S(\rho|\nu)$ *for any unitary U.*
(4)

$$S(\rho_A|\rho_B) = \log\frac{\mathrm{tr}(e^A)}{\mathrm{tr}(e^B)} - \mathrm{tr}(\rho_A(A - B)).$$

(5) *For any $\rho, \nu \in \mathfrak{S}$ one has*

$$S(\rho|\nu) = \lim_{\alpha\downarrow 0}\frac{S_\alpha(\nu|\rho)}{\alpha} = \lim_{\alpha\uparrow 1}\frac{S_\alpha(\rho|\nu)}{1 - \alpha}. \tag{4.63}$$

In particular, if $\rho \ll \nu$ then $S_0(\nu|\rho) = S_1(\rho|\nu) = 0$ so that

$$S(\rho|\nu) = \left.\frac{d}{d\alpha}S_\alpha(\nu|\rho)\right|_{\alpha=0} = -\left.\frac{d}{d\alpha}S_\alpha(\rho|\nu)\right|_{\alpha=1}. \qquad (4.64)$$

(6) If $\Phi : \mathcal{O}_\mathcal{K} \to \mathcal{O}_{\mathcal{K}'}$ is a unital Schwarz map then, for any $\rho, \nu \in \mathfrak{S}_\mathcal{K}$,

$$S(\rho \circ \Phi | \nu \circ \Phi) \geq S(\rho|\nu).$$

(7) The map $(\rho, \nu) \mapsto S(\rho|\nu)$ is continuous on $\mathfrak{S} \times \mathfrak{S}_f$ and upper semi-continuous on $\mathfrak{S} \times \mathfrak{S}$.

(8) If $s(\nu) = s(\rho)$, then $S_\alpha(\rho|\nu) \geq \alpha S(\nu|\rho)$.

Proof: Part (1) follows from Klein's inequality. Parts (2), (3), and (4) are obvious. Part (5) is easy and left to the reader. Part (6) follows from (4.63) and Uhlmann's monotonicity theorem (Theorem 4.15). The upper semi-continuity of the map $(\rho, \nu) \mapsto S(\rho|\nu)$ follows from (5) and part 3 of Exercise 4.3.6. A direct proof goes as follows. Let us fix $(\rho_0, \nu_0) \in \mathfrak{S} \times \mathfrak{S}$. Define $\lambda_0 = \min\{\lambda \in sp(\nu_0) \,|\, \lambda > 0\}$, and for $\nu \in \mathfrak{S}$ set

$$Q_\nu = \sum_{\substack{\lambda \in sp(\nu) \\ \lambda > \lambda_0/2}} P_\lambda(\nu).$$

Let $0 < \varepsilon < \lambda_0/2$. We know from perturbation theory that

$$\lim_{\nu \to \nu_0} Q_\nu = s(\nu_0), \qquad \lim_{\nu \to \nu_0} \nu Q_\nu = \nu_0,$$

and that

$$\nu^{(\varepsilon)} = \nu Q_\nu + (\mathbb{1} - Q_\nu)\varepsilon \geq \nu,$$

provided ν is close enough to ν_0. It follows that

$$S(\rho|\nu) = \rho(\log \nu) - S(\rho) \leq \rho(\log \nu^{(\varepsilon)}) - S(\rho).$$

Since $\nu^{(\varepsilon)} \geq \epsilon > 0$ it follows from the analytic functional calculus that

$$\lim_{\nu \to \nu_0} \log \nu^{(\varepsilon)} = \log\left(\lim_{\nu \to \nu_0} \nu^{(\varepsilon)}\right) = \log(\nu_0|_{\mathrm{Ran}\,\nu_0}) \oplus \log \varepsilon|_{\mathrm{Ker}\,\nu_0},$$

and hence, using Theorem 4.12 (1), we deduce

$$\limsup_{(\rho,\nu) \to (\rho_0,\nu_0)} S(\rho|\nu) \leq \lim_{(\rho,\nu) \to (\rho_0,\nu_0)} \rho(\log \nu^{(\varepsilon)}) - S(\rho)$$

$$= \rho_0(\log \nu_0|_{\mathrm{Ran}\,\nu_0} \oplus 0|_{\mathrm{Ker}\,\nu_0}) - S(\rho_0) + (1 - \rho_0(s(\nu_0))) \log \varepsilon.$$

If $\rho_0 \not\ll \nu_0$ then $1 - \rho_0(s(\nu_0)) > 0$ and letting $\varepsilon \downarrow 0$ we conclude that

$$\limsup_{(\rho,\nu) \to (\rho_0,\nu_0)} S(\rho|\nu) \leq -\infty = S(\rho_0|\nu_0).$$

If $\rho_0 \ll \nu_0$ then $1 - \rho_0(s(\nu_0)) = 0$ and $\operatorname{Ker} \nu_0 \subset \operatorname{Ker} \rho_0$ so that

$$\limsup_{(\rho,\nu)\to(\rho_0,\nu_0)} S(\rho|\nu) \le \rho_0(\log \nu_0) - S(\rho_0) = S(\rho_0|\nu_0).$$

Finally, we observe that if $\nu_0 > 0$, then $\nu \ge \lambda_0/2$ for all ν sufficiently close to ν_0 so that $\lim_{\nu\to\nu_0} \log \nu = \log \nu_0$ from which one concludes that

$$\lim_{(\rho,\nu)\to(\rho_0,\nu_0)} S(\rho|\nu) = S(\rho_0|\nu_0).$$

Property (8) is a direct consequence of the convexity of $\alpha \mapsto S_\alpha(\rho|\nu)$ and equation (4.64). $\qquad\square$

Remark. The following example shows that the function $(\rho, \nu) \mapsto S(\rho|\nu)$ is not continuous on $\mathfrak{S} \times \mathfrak{S}$. Setting

$$\rho_n = \begin{bmatrix} 1 - 1/n & 0 \\ 0 & 1/n \end{bmatrix}, \qquad \nu_n = \begin{bmatrix} 1 & 0 \\ 0 & 0 \end{bmatrix},$$

one has $S(\rho_n|\nu_n) = -\infty$ for all $n \in \mathbb{N}^*$, so

$$\lim_{n\to\infty} S(\rho_n|\nu_n) = -\infty \ne S(\lim_{n\to\infty} \rho_n | \lim_{n\to\infty} \nu_n) = 0.$$

The following result, first proved by Lieb and Ruskai (Lieb and Ruskai, 1973), is a direct consequence of Theorem 4.16 and Relation (4.63).

Theorem 4.20 *The map $\mathfrak{S} \times \mathfrak{S} \ni (\rho, \nu) \mapsto S(\rho|\nu)$ is jointly concave, that is, for $\lambda \in [0, 1]$ and $\rho, \rho', \nu, \nu' \in \mathfrak{S}$,*

$$S(\lambda\rho + (1 - \lambda)\rho' | \lambda\nu + (1 - \lambda)\nu') \ge \lambda S(\rho|\nu) + (1 - \lambda)S(\rho'|\nu'). \qquad (4.65)$$

Exercise 4.28 Use Uhlmann's monotonicity theorem to show that

$$S_\alpha(\rho \circ \vartheta | \nu \circ \vartheta) = S_\alpha(\rho|\nu),$$

for all $\rho, \nu \in \mathfrak{S}$ and $\vartheta \in \operatorname{Aut}(\mathcal{O})$.

Remark. Trace inequalities and entropies play a prominent role in quantum statistical mechanics and quantum information theory. In these notes, we only skim over these topics. The interested reader should consult the lecture notes by Carlen (Carlen, 2010) for a more thorough introduction to this vast subject and for further references.

4.3.7 Quantum hypothesis testing

Since the pioneering work of Pearson (Pearson, 1900), hypothesis testing has played an important role in theoretical and applied statistics (see, e.g., (Bera, 2000)). In the last decade, the mathematical structure and basic results of classical hypothesis testing have been extended to the noncommutative setting. A clear exposition of the basic results of quantum hypothesis testing can be found in (Audenaert, Nussbaum, Szkoła, and Verstraete, 2008), (Hiai, Mosonyi, and Ogawa, 2008).

It was recently observed in (Jakšić, Ogata, Pillet, and Seiringer, 2011c) that there is a close relation between recent developments in the field of quantum hypothesis testing and the developments in nonequilibrium statistical mechanics. In this section we describe the setup of quantum hypothesis testing following essentially (Audenaert, Nussbaum, Szkoła, and Verstraete, 2008). We will discuss the relation to nonequilibrium statistical mechanics in Section 4.6.6.

Let ν and ρ be two states and $p \in\,]0, 1[$. Suppose that we know *a priori* that the system is with probability p in the state ρ and with probability $1 - p$ in the state ν. By performing a measurement we wish to decide with minimal error probability what is the true state of the system. The following procedure is known as *quantum hypothesis testing*. A *test* P is an orthogonal projection in $\mathcal{O}$. On the basis of the outcome of the test (that is, a measurement of P) one decides whether the system is in the state ρ or ν. More precisely, if the outcome of the test is 1, one decides that the system is in the state ρ (Hypothesis I) and if the outcome is 0, one decides that the system is in the state ν (Hypothesis II). $\rho(\mathbb{1} - P)$ is the error probability of accepting II if I is true and $\nu(P)$ is the error probability of accepting I if II is true. The average error probability is

$$D_p(\rho, \nu, P) = p\rho(\mathbb{1} - P) + (1 - p)\nu(P),$$

and we are interested in minimizing $D_p(\rho, \nu, P)$ w.r.t. P. Let

$$D_p(\rho, \nu) = \inf\{D_p(\rho, \nu, P) \mid P \in \mathcal{O}_{\text{self}},\ P^2 = P\}.$$

The set of all orthogonal projections is a norm closed subset of $\mathcal{O}$ and so the infimum on the right-hand side is achieved at some projection P. The quantum Bayesian distinguishability problem is to identify the orthogonal projections P such that $D_p(\rho, \nu, P) = D_p(\rho, \nu)$. Let P_{opt} be the orthogonal projection onto the range of

$$((1 - p)\nu - p\rho)_+ ,$$

where $x_+ = (|x| + x)/2$ denotes the positive part of x. The following result was proven in (Audenaert, Nussbaum, Szkoła, and Verstraete, 2008), where the reader can find references to the previous works on the subject.

Theorem 4.21 *(1)*

$$D_p(\rho, \nu) = D_p(\rho, \nu, P_{\text{opt}}) = \frac{1}{2}\left(1 - \|(1 - p)\nu - p\rho\|_1\right).$$

Moreover, P_{opt} is the unique minimizer of the functional $P \mapsto D_p(\rho, \nu, P)$.
(2)

$$D_p(\rho, \nu) = \min\{D_p(\rho, \nu, T) \mid T \in \mathcal{O}_{\text{self}}, 0 \leq T \leq \mathbb{1}\}.$$

(3) For $\alpha \in [0,1]$,

$$D_p(\rho, \nu) \leq p^\alpha (1-p)^{1-\alpha} \text{tr}(\rho^\alpha \nu^{1-\alpha}).$$

Remark. Part (1) is the quantum version of the Neyman–Pearson lemma. Part (3) is the quantum analogue of the Chernoff bound in classical hypothesis testing. In quantum information theory the quantity

$$\zeta_{QCB}(\rho, \nu) = -\log \min_{\alpha \in [0,1]} \text{tr}(\rho^\alpha \nu^{1-\alpha}) = -\min_{\alpha \in [0,1]} S_\alpha(\rho|\nu),$$

is called the Chernoff distance between the states ρ and ν. We shall prove a lower bound on the function $D_p(\rho, \nu)$ in Section 4.3.12.

Proof: (1)–(2) Set $A = p\rho - (1-p)\nu$ so that, for $T \in \mathcal{O}_{\text{self}}$, $0 \leq T \leq 1$, we can write

$$D_p(\rho, \nu, T) = \text{tr}\,(p\rho(1-T) + (1-p)\nu T) = p - \text{tr}(TA) \geq p - \text{tr}(TA_+),$$

where equality holds iff $\text{Ran}\,T \subset \text{Ker}\,A_- = \text{Ran}\,A_+$. It follows that P_{opt} is the unique minimizer and

$$D_p(\rho, \nu, P_{\text{opt}}) = p - \text{tr}(A_+) = p - \frac{1}{2}\text{tr}\,(A + |A|) = \frac{1}{2}(1 - \text{tr}\,(|A|)).$$

(3) (Following S. Ozawa, private communication. The original proof can be found in (Audenaert, Nussbaum, Szkoła, and Verstraete, 2008)). Setting $B = p\rho$ and $C = (1-p)\nu$ and given (1), one has to show that

$$\text{tr}(B - B^\alpha C^{1-\alpha}) \leq \text{tr}(A_+),$$

for all $B, C \in \mathcal{O}_+$ and $\alpha \in [0,1]$ with $A = C - B$. One clearly has

$$C \leq C + A_+, \qquad B \leq C + A_+. \tag{4.66}$$

We shall make repeated use of the Löwner–Heinz inequality (Exercise 4.3.1). From (4.66) and the fact that $B^\alpha \geq 0$ we get

$$\text{tr}(B^\alpha(B^{1-\alpha} - C^{1-\alpha})) \leq \text{tr}(B^\alpha((C + A_+)^{1-\alpha} - C^{1-\alpha})). \tag{4.67}$$

Since

$$(C + A_+)^{1-\alpha} - C^{1-\alpha} \geq 0,$$

we further have

$$\text{tr}(B^\alpha(B^{1-\alpha} - C^{1-\alpha})) \leq \text{tr}((C + A_+)^\alpha((C + A_+)^{1-\alpha} - C^{1-\alpha}))$$
$$= \text{tr}(C + A_+) - \text{tr}((C + A_+)^\alpha C^{1-\alpha}).$$

Using again Inequality (4.67), and the fact that $C \geq 0$, we obtain

$$\text{tr}(B^\alpha(B^{1-\alpha} - C^{1-\alpha})) \leq \text{tr}(C + A_+) - \text{tr}(C^\alpha C^{1-\alpha}) = \text{tr}(A_+).$$

$\square$

4.3.8 Dynamical systems

A *dynamics* on the $*$-algebra $\mathcal{O}$ is a continuous one-parameter subgroup of $\text{Aut}(\mathcal{O})$, that is a map $\mathbb{R} \ni t \mapsto \tau^t \in \text{Aut}(\mathcal{O})$ satisfying $\tau^t \circ \tau^s = \tau^{t+s}$ for all $t, s \in \mathbb{R}$ and $\lim_{t \to 0} \|\tau^t(A) - A\| = 0$ for all $A \in \mathcal{O}$. Such a map automatically satisfies $\tau^0 = \text{id}$ and $(\tau^t)^{-1} = \tau^{-t}$ for all $t \in \mathbb{R}$. Moreover, since τ^t is isometric and $\mathcal{O}$ is a finite dimensional vector space, the continuity is uniform

$$\lim_{\epsilon \to 0} \sup_{\substack{\|A\|=1 \\ t \in \mathbb{R}}} \|\tau^{t+\epsilon}(A) - \tau^t(A)\| = 0,$$

and the map $t \mapsto \tau^t(A)$ is differentiable (in fact entire analytic). In terms of the generator

$$\delta(A) = \left. \frac{\mathrm{d}}{\mathrm{d}t} \tau^t(A) \right|_{t=0},$$

one has $\tau^t(A) = e^{t\delta}(A)$. Clearly, $\delta(\mathbb{1}) = 0$, $\delta(AB) = \delta(A)B + A\delta(B)$ and $\delta(A)^* = \delta(A^*)$ hold for all $A, B \in \mathcal{O}$. We call *dynamical system* a pair $(\mathcal{O}, \tau^t)$, where τ^t is a dynamics on $\mathcal{O}$.

If $H \in \mathcal{O}_{\text{self}}$, then

$$\tau^t(A) = e^{\mathrm{i}tH} A e^{-\mathrm{i}tH}, \qquad (4.68)$$

is a dynamics on $\mathcal{O}$. One of the special features of finite quantum systems is that the converse is true. Given a dynamical system $(\mathcal{O}, \tau^t)$, there exists $H \in \mathcal{O}_{\text{self}}$ such that (4.68) holds for all $t \in \mathbb{R}$. Moreover, H is uniquely determined up to a constant. It can be explicitly constructed as follows. Let δ be the generator of τ^t. Let $\{\psi_j\}$ be an orthonormal basis of $\mathcal{K}$ and $E_{ij} = |\psi_i\rangle\langle\psi_j|$ the corresponding basis of $\mathcal{O}$. Let

$$H = \frac{1}{\mathrm{i}} \sum_j \delta(E_{ji}) E_{ij}.$$

The relation $\sum_j E_{ji} E_{ij} = \sum_j E_{jj} = \mathbb{1}$ implies

$$\sum_j \delta(E_{ji}) E_{ij} + \sum_j E_{ji} \delta(E_{ij}) = \delta(\mathbb{1}) = 0,$$

and

$$\mathrm{i}[H, E_{kl}] = \sum_j \delta(E_{ji}) E_{ij} E_{kl} + E_{kl} E_{ji} \delta(E_{ij})$$

$$= \delta(E_{ki}) E_{il} + E_{ki} \delta(E_{il}) = \delta(E_{ki} E_{il}) = \delta(E_{kl}).$$

Hence $i[H, X] = \delta(X)$ for all $X \in \mathcal{O}$ and (4.68) follows.

Remark. From the above discussion, the reader familiar with the theory of Lie groups will recognize that $\text{Aut}(\mathcal{O})$ is a simply connected Lie group with Lie algebra

$$\text{aut}(\mathcal{O}) = \{d_X = i[X, \cdot] \mid X \in \mathcal{O}_{\text{self}}\},$$

and bracket $[d_X, d_Y] = d_{i[X,Y]}$. Since $d_X = d_Y$ iff $X - Y$ is a real multiple of the identity, the dimension of $\text{Aut}(\mathcal{O})$ is given by $\dim_{\mathbb{R}}(\mathcal{O}_{\text{self}}) - 1 = (\dim \mathcal{K})^2 - 1$.

According to the basic principles of quantum mechanics, if H is the energy observable of the system, that is its Hamiltonian, then the group $\tau^t(A) = e^{itH} A e^{-itH}$ describes its time evolution in the Heisenberg picture. If the system was in the state ρ at time $t = 0$ then the expectation value of the observable A at time t is given by

$$\text{tr}(\rho \tau^t(A)) = \rho(\tau^t(A)) = \rho \circ \tau^t(A).$$

In the Schrödinger picture the state ρ evolves in time as $\tau^{-t}(\rho)$ and in what follows we adopt the shorthands

$$A_t = \tau^t(A), \qquad \rho_t = \tau^{-t}(\rho) = \rho \circ \tau^t.$$

Clearly, $\rho_t(A) = \rho(A_t)$.

4.3.9 Gibbs states, KMS condition, and variational principle

For the dynamical system $(\mathcal{O}, \tau^t)$, with Hamiltonian H, the state of thermal equilibrium at inverse temperature β is described by the Gibbs canonical ensemble

$$\rho_\beta = \frac{e^{-\beta H}}{\text{tr}(e^{-\beta H})}.$$

Note that, for any $A, B \in \mathcal{O}$, one has

$$\rho_\beta(AB) = \frac{\text{tr}(e^{-\beta H} AB)}{\text{tr}(e^{-\beta H})} = \frac{\text{tr}(Be^{-\beta H} A)}{\text{tr}(e^{-\beta H})} = \frac{\text{tr}(e^{-\beta H} \tau^{-i\beta}(B)A)}{\text{tr}(e^{-\beta H})} = \rho_\beta(\tau^{-i\beta}(B)A).$$

We say that a state ρ satisfies the Kubo–Martin–Schwinger (KMS) condition at inverse temperature β, or, for short, that ρ is a β-KMS state if

$$\rho(AB) = \rho(\tau^{-i\beta}(B)A), \tag{4.69}$$

holds for all $A, B \in \mathcal{O}$. The β-KMS condition (4.69) plays a central role in algebraic quantum statistical mechanics. For the finite quantum system considered in this section it is a characterization of the Gibbs state ρ_β.

Proposition 4.22 ρ *is a β-KMS state iff $\rho = \rho_\beta$.*

Proof: It remains to show that if ρ is β-KMS, then $\rho = \rho_\beta$. Setting $X = \rho e^{\beta H}$ and $A = e^{\beta H} C$ in the KMS condition

$$\text{tr}(\rho e^{\beta H} B e^{-\beta H} A) = \text{tr}(\rho AB),$$

yields $\operatorname{tr}(XBC) = \operatorname{tr}(XCB)$ for all $B, C \in \mathcal{O}$. Since this is equivalent to $\operatorname{tr}([X, B]C) = 0$, we conclude that $[X, B] = 0$ for all $B \in \mathcal{O}$ and hence that $X = \alpha \mathbb{1}$ for some constant α. This means that $\rho = \alpha e^{-\beta H}$. The constant α is now determined by the normalization condition $\operatorname{tr}(\rho) = 1$. $\qquad\square$

The Gibbs canonical ensemble can be also characterized by a variational principle. The number $E = \rho_\beta(H)$ is the expectation value of the energy in the state ρ_β. Since

$$\frac{\mathrm{d}}{\mathrm{d}\beta}\rho_\beta(H) = -\rho_\beta((H - E)^2) \leq 0,$$

the function $\beta \mapsto \rho_\beta(H)$ is decreasing and is strictly decreasing unless H is constant. If $E_{\min} = \min \operatorname{sp}(H)$ and $E_{\max} = \max \operatorname{sp}(H)$, then

$$\lim_{\beta \to -\infty} \rho_\beta(H) = E_{\max}, \qquad \lim_{\beta \to \infty} \rho_\beta(H) = E_{\min}.$$

Note also that $\lim_{\beta \to \pm\infty} \rho_\beta = \rho_{\pm\infty}$ where

$$\rho_{+\infty/-\infty} = \frac{P_{\min/\max}}{\operatorname{tr}(P_{\min/\max})},$$

and $P_{\min/\max}$ denotes the spectral projection of H associated to its eigenvalue $E_{\min/\max}$. Hence to any $E \in [E_{\min}, E_{\max}]$ one can associate a unique $\beta \in [-\infty, \infty]$ such that

$$\rho_\beta(H) = E. \tag{4.70}$$

We adopt the shorthands

$$S(\beta) = S(\rho_\beta), \qquad P(\beta) = \log \operatorname{tr}(e^{-\beta H}).$$

The function $P(\beta)$ is called the *pressure* (or *free energy*). Note that

$$S(\beta) = \beta E + P(\beta). \tag{4.71}$$

If $\mathfrak{S}_E = \{\rho \in \mathfrak{S} \mid \rho(H) = E\}$ and $\nu \in \mathfrak{S}_E$, then

$$S(\nu) = S(\nu) - \beta\nu(H) + \beta E \leq \max_{\rho \in \mathfrak{S}}\{S(\rho) - \beta\rho(H)\} + \beta E = \log \operatorname{tr}(e^{-\beta H}) + \beta E,$$

and so

$$S(\nu) \leq S(\beta),$$

where equality holds iff $\nu = \rho_\beta$. Hence, we have proven the Gibbs variational principle:

Theorem 4.23 *Let $E \in [E_{\min}, E_{\max}]$ and let β be given by (4.70). Then*

$$\max_{\rho \in \mathfrak{S}_E} S(\rho) = S(\beta),$$

and the unique maximizer is the Gibbs state ρ_β.

Note that neither the KMS condition nor the Gibbs variational principle require β to be positive. The justification of the physical restriction $\beta > 0$ typically involves some form of the second law of thermodynamics. Recall that $\beta = \beta(E)$ is uniquely specified by (4.70). Considering $S(E) = S(\beta(E))$ as the function of E, the differentiation of equation (4.71) w.r.t. E yields

$$\frac{\mathrm{d}S}{\mathrm{d}E} = \beta,$$

and the second law $\frac{\mathrm{d}S}{\mathrm{d}E} \geq 0$ (the increase of entropy with energy) requires $\beta \geq 0$. An alternative approach goes as follows. Let an external force act on the system during the time interval $[0, T]$ so that its Hamiltonian becomes time dependent, $H(t) = H + V(t)$. We assume that $V(t)$ depends continuously on t and vanishes for $t \notin \,]0, T[$. Let $U(t)$ be the corresponding unitary propagator, that is the solution of the time-dependent Schrödinger equation

$$\mathrm{i}\frac{\mathrm{d}}{\mathrm{d}t}U(t) = H(t)U(t), \qquad U(0) = \mathbb{1}.$$

Suppose that at $t = 0$ the system was in the Gibbs state ρ_β. At the later time $t > 0$, its state is given by $\rho_{\beta,t} = U(t)\rho_\beta U(t)^*$ and the work performed on the system by the external force during the time interval $[0, T]$ is

$$\Delta E = \rho_{\beta,T}(H) - \rho_\beta(H) = \int_0^T \frac{\mathrm{d}}{\mathrm{d}t}\rho_{\beta,t}(H)\,\mathrm{d}t.$$

The change of relative entropy $S(\rho_{\beta,t}|\rho_\beta)$ over the time interval $[0, T]$ equals

$$\Delta S = S(\rho_{\beta,T}|\rho_\beta) - S(\rho_\beta|\rho_\beta) = \int_0^T \frac{\mathrm{d}}{\mathrm{d}t}S(\rho_{\beta,t}|\rho_\beta)\,\mathrm{d}t = -\beta\int_0^T \frac{\mathrm{d}}{\mathrm{d}t}\rho_{\beta,t}(H)\,\mathrm{d}t,$$

and so

$$\Delta S = -\beta\Delta E.$$

If $V(t)$ is nontrivial in the sense that $\rho_{\beta,T} \neq \rho_\beta$, then $\Delta S = S(\rho_{\beta,T}|\rho_\beta) < 0$. The second law of thermodynamics, more precisely the fact that one can not extract work from a system in thermal equilibrium, requires that $\Delta E \geq 0$. Hence, negative values of β are not allowed by thermodynamics.

The above discussion can be generalized as follows. Let $N \in \mathcal{O}_{\text{self}}$ be an observable such that $[H, N] = 0$ (N is colloquially called a *charge*). Let β and μ be real parameters and let

$$\rho_{\beta,\mu} = \frac{\mathrm{e}^{-\beta(H-\mu N)}}{\mathrm{tr}\big(\mathrm{e}^{-\beta(H-\mu N)}\big)},$$

be the β-KMS state for the dynamics generated by $H - \mu N$. Denote $\rho_{\beta,\mu}(H) = E$, $\rho_{\beta,\mu}(N) = \varrho$, $S(\beta,\mu) = S(\rho_{\beta,\mu})$, $P(\beta,\mu) = \log \mathrm{tr}(\mathrm{e}^{-\beta(H-\mu N)})$. Then

$$S(\beta,\mu) = \beta(E - \mu\varrho) + P(\beta,\mu). \tag{4.72}$$

If $\mathfrak{S}_{E,\varrho} = \{\rho \in \mathfrak{S} \mid \rho(H) = E, \rho(N) = \varrho\}$, then

$$\max_{\rho \in \mathfrak{S}_{E,\varrho}} S(\rho) = S(\beta,\mu),$$

with unique maximizer $\rho_{\beta,\mu}$. The parameter μ is interpreted as chemical potential associated to the charge N and the state $\rho_{\beta,\mu}$ describes the system in thermal equilibrium at inverse temperature β and chemical potential μ. Considering $\beta = \beta(E,\varrho)$ and $\mu = \mu(E,\varrho)$ as functions of E and ϱ we see from (4.72) that

$$\frac{\partial S}{\partial E} = \beta, \qquad \frac{\partial S}{\partial \varrho} = -\beta\mu.$$

Although in general $\rho_{\beta,\mu}$ is not a β-KMS state for the dynamics τ^t, if A and B commute with N, then $\tau^t(A) = \mathrm{e}^{\mathrm{i}t(H-\mu N)} A \mathrm{e}^{-\mathrm{i}t(H-\mu N)}$ and the β-KMS condition

$$\rho_{\beta,\mu}(\tau^{-\mathrm{i}\beta}(B)A) = \rho_{\beta,\mu}(AB),$$

is satisfied. In other words, if $\mu \neq 0$, the physical observables must be invariant under the gauge group $\gamma^\theta(A) = \mathrm{e}^{\mathrm{i}\theta N} A \mathrm{e}^{-\mathrm{i}\theta N}$. The generalization of these results to the case of several charges is straightforward.

4.3.10   Perturbation theory

Let $(\mathcal{O}, \tau^t)$ be a dynamical system with Hamiltonian H and let $V \in \mathcal{O}_{\mathrm{self}}$ be a perturbation. In this section we consider the perturbed dynamics τ_V^t generated by the Hamiltonian $H + V$,

$$\tau_V^t(A) = \mathrm{e}^{\mathrm{i}t(H+V)} A \mathrm{e}^{-\mathrm{i}t(H+V)}.$$

If δ denotes the generator of τ^t, then the generator of τ_V^t is given by

$$\delta_V = \mathrm{i}[H + V, \cdot] = \delta + \mathrm{i}[V, \cdot] = \delta + \mathrm{d}_V,$$

and one easily checks that the map $\mathbb{R} \ni t \mapsto \gamma_V^t \in \mathrm{Aut}(\mathcal{O})$ defined by

$$\gamma_V^t = \tau_V^t \circ \tau^{-t} = \mathrm{e}^{t(\delta + \mathrm{d}_V)} \circ \mathrm{e}^{-t\delta},$$

has the following properties:

(1) $\tau_V^t = \gamma_V^t \circ \tau^t$.
(2) $(\gamma_V^t)^{-1} = \tau^t \circ \gamma_V^{-t} \circ \tau^{-t}$.
(3) $\gamma_V^{t+s} = \gamma_V^s \circ \tau^s \circ \gamma_V^t \circ \tau^{-s}$.
(4) $\gamma_V^0 = \mathrm{id}$ and $\partial_t \gamma_V^t = \gamma_V^t \circ \mathrm{d}_{\tau^t(V)}$.

Integration of Relation (4) yields the integral equation

$$\gamma_V^t = \mathrm{id} + \int_0^t \gamma_V^s \circ \mathrm{d}_{\tau^s(V)}\, \mathrm{d}s,$$

which can be iterated to obtain

$$\gamma_V^t = \mathrm{id} + \sum_{n=1}^{N-1} \int_{0 \le s_1 \le \cdots \le s_n \le t} \mathrm{d}_{\tau^{s_n}(V)} \circ \cdots \circ \mathrm{d}_{\tau^{s_1}(V)}\, \mathrm{d}s_1 \cdots \mathrm{d}s_n$$

$$+ \int_{0 \le s_1 \le \cdots \le s_N \le t} \gamma_V^{s_N} \circ \mathrm{d}_{\tau^{s_N}(V)} \circ \cdots \circ \mathrm{d}_{\tau^{s_1}(V)}\, \mathrm{d}s_1 \cdots \mathrm{d}s_N.$$

Since γ_V^t is isometric and $\|\mathrm{d}_{\tau^t(V)}\| = \|\mathrm{i}[\tau^t(V), \cdot]\| \le 2\|V\|$, we can bound the norm of the last term by

$$\int_{0 \le s_1 \le \cdots \le s_N \le t} (2\|V\|)^N\, \mathrm{d}s_1 \cdots \mathrm{d}s_N \le \frac{(2\|V\|t)^N}{N!},$$

and conclude that the Dyson expansion

$$\gamma_V^t = \mathrm{id} + \sum_{n=1}^{\infty} \int_{0 \le s_1 \le \cdots \le s_n \le t} \mathrm{d}_{\tau^{s_n}(V)} \circ \cdots \circ \mathrm{d}_{\tau^{s_1}(V)}\, \mathrm{d}s_1 \cdots \mathrm{d}s_n,$$

converges in norm for all $t \in \mathbb{R}$, and uniformly for t in compact intervals. Using Relation (1), we conclude that

$$\tau_V^t = \tau^t + \sum_{n=1}^{\infty} \int_{0 \le s_1 \le \cdots \le s_n \le t} \mathrm{d}_{\tau^{s_n}(V)} \circ \cdots \circ \mathrm{d}_{\tau^{s_1}(V)} \circ \tau^t\, \mathrm{d}s_1 \cdots \mathrm{d}s_n,$$

which we can rewrite as

$$\tau_V^t(A) = \sum_{n=0}^{\infty} (\mathrm{i}t)^n \int_{0 \le s_1 \le \cdots \le s_n \le 1} [\tau^{ts_n}(V), [\cdots, [\tau^{ts_1}(V), \tau^t(A)] \cdots]]\, \mathrm{d}s_1 \cdots \mathrm{d}s_n.$$

Finally, we note that since $\tau^z(V)$, $\tau^z(A)$ and $\tau_V^z(A)$ are entire analytic functions of z and $\|\tau^z\| \le \mathrm{e}^{2|\mathrm{Im}\, z| \|H\|}$, the above expression provides an expansion of $\tau_V^z(A)$ which converges uniformly for z in compact subsets of $\mathbb{C}$.

Similar conclusions hold for the interaction picture propagator

$$E_V(t) = \mathrm{e}^{\mathrm{i}t(H+V)}\mathrm{e}^{-\mathrm{i}tH}.$$

It satisfies:

(1') $e^{it(H+V)} = E_V(t)e^{itH}$ and $\tau_V^t(A) = E_V(t)\tau^t(A)E_V(t)^{-1}$.
(2') $E_V(t)^{-1} = E_V(t)^* = \tau^t(E_V(-t))$.
(3') $E_V(t+s) = E_V(s)\tau^s(E_V(t))$.
(4') $E_V(0) = \mathbb{1}$ and $\partial_t E_V(t) = iE_V(t)\tau^t(V)$.

Integrating relation (4') yields, after iteration,

$$E_V(t) = \sum_{n=0}^{\infty} (it)^n \int_{0 \le s_1 \le \cdots \le s_n \le 1} \tau^{ts_n}(V) \cdots \tau^{ts_1}(V)\, \mathrm{d}s_1 \cdots \mathrm{d}s_n.$$

This expansion is uniformly convergent for t in compact subsets of $\mathbb{C}$. In particular,

$$E_V(i\beta) = \sum_{n=0}^{\infty} (-\beta)^n \int_{0 \le s_1 \le \cdots \le s_n \le 1} \tau^{i\beta s_n}(V) \cdots \tau^{i\beta s_1}(V)\, \mathrm{d}s_1 \cdots \mathrm{d}s_n. \tag{4.73}$$

Using Relation (1') with $t = i\beta$ we can express the perturbed KMS-state

$$\rho_{\beta V} = \frac{e^{-\beta(H+V)}}{\mathrm{tr}(e^{-\beta(H+V)})},$$

in terms of the unperturbed one $\rho_\beta = e^{-\beta H}/\mathrm{tr}(e^{-\beta H})$ as

$$\rho_{\beta V}(A) = \frac{\rho_\beta(A\, E_V(i\beta))}{\rho_\beta(E_V(i\beta))}. \tag{4.74}$$

Using this last formula one can compute the perturbative expansion of $\rho_{\beta V}(A)$ w.r.t. V. To control this expansion, we need the following estimate.

Proposition 4.24 *The bound*

$$|\rho_\beta(E_{\alpha V}(i\beta)) - 1| \le e^{|\alpha\beta|\|V\|} - 1, \tag{4.75}$$

holds for any $\beta \in \mathbb{R}$, $V \in \mathcal{O}_{\mathrm{self}}$ *and* $\alpha \in \mathbb{C}$.

Proof: Using Duhamel formula

$$\frac{\mathrm{d}}{\mathrm{d}s} e^{-\beta(H+s\alpha V)} = -\alpha \int_0^\beta e^{-(\beta-u)(H+s\alpha V)} V e^{-u(H+s\alpha V)}\, \mathrm{d}u,$$

we can write

$$\rho_\beta(E_{\alpha V}(i\beta) - 1) = \int_0^1 \frac{\mathrm{d}}{\mathrm{d}s} \rho_\beta(E_{s\alpha V}(i\beta))\, \mathrm{d}s = -\alpha\beta \int_0^1 f_\beta(s)\, \mathrm{d}s, \tag{4.76}$$

where

$$f_\beta(s) = \frac{\mathrm{tr}(V e^{-\beta(H+s\alpha V)})}{\mathrm{tr}(e^{-\beta H})}.$$

Starting with the simple bound

$$|f_\beta(s)| \le \|V\| \frac{\|e^{-\beta(H+s\alpha V)}\|_1}{\mathrm{tr}(e^{-\beta H})},$$

and setting $\alpha = a + ib$ with $a, b \in \mathbb{R}$, we estimate the numerator on the right-hand side by the Hölder inequality (Part 2 of Exercise 4.3.2) applied to the Lie product formula,

$$\|e^{-\beta(H+saV+isbV)}\|_1 = \lim_{n\to\infty} \|(e^{-\beta(H+saV)/n} e^{-i\beta sbV/n})^n\|_1$$

$$\le \limsup_{n\to\infty} \|e^{-\beta(H+saV)/n}\|_n^n \, \|e^{-i\beta sbV/n}\|_\infty^n = \mathrm{tr}(e^{-\beta(H+saV)}).$$

For $s \in [0,1]$, the Golden–Thompson inequality further leads to

$$\frac{\mathrm{tr}(e^{-\beta(H+saV)})}{\mathrm{tr}(e^{-\beta H})} \le \frac{\mathrm{tr}(e^{-\beta H} e^{-\beta saV})}{\mathrm{tr}(e^{-\beta H})} = \rho_\beta(e^{-\beta saV}) \le e^{s|\beta\alpha|\,\|V\|},$$

so that, finally,

$$|f_\beta(s)| \le \|V\| \, e^{s|\beta\alpha|\,\|V\|}.$$

Using equation (4.76), we derive

$$|\rho_\beta(\mathrm{E}_V(\mathrm{i}\beta)) - 1| \le |\alpha\beta|\|V\| \int_0^1 e^{s|\alpha\beta|\|V\|} \mathrm{d}s = e^{|\alpha\beta|\|V\|} - 1.$$

$\square$

Replacing V with αV and using the expansion (4.73), we can write

$$\rho_\beta(A\,\mathrm{E}_{\alpha V}(\mathrm{i}\beta)) = \sum_{n=0}^\infty \alpha^n c_n(A),$$

where $c_0(A) = \rho_\beta(A)$ and

$$c_n(A) = (-\beta)^n \int_{0 \le s_1 \le \cdots \le s_n \le 1} \rho_\beta(A\tau^{\mathrm{i}\beta s_n}(V) \cdots \tau^{\mathrm{i}\beta s_1}(V)) \, \mathrm{d}s_1 \cdots \mathrm{d}s_n.$$

It follows from the estimate (4.75) that the entire function $\mathbb{C} \ni \alpha \mapsto \rho_\beta(\mathrm{E}_{\alpha V}(\mathrm{i}\beta))$ has no zero in the disc

$$|\alpha| < \frac{\log 2}{|\beta|\|V\|}.$$

Hence, equation (4.74) shows that the function $\mathbb{C} \ni \alpha \mapsto \rho_{\beta(\alpha V)}(A)$ is analytic on this disc. Writing

$$\rho_{\beta(\alpha V)}(A) = \sum_{n=0}^\infty \alpha^n b_n(A), \qquad (4.77)$$

Relation (4.74) yields

$$\sum_{n=0}^{\infty} \alpha^n c_n(A) = \left(\sum_{n=0}^{\infty} \alpha^n b_n(A)\right) \left(\sum_{n=0}^{\infty} \alpha^n c_n(\mathbb{1})\right),$$

and we conclude that for all n,

$$c_n(A) = \sum_{j=0}^{n} b_j(A)c_{n-j}(\mathbb{1}).$$

Thus, with coefficients $b_n(A)$ given by the recursive formula

$$b_0(A) = c_0(A) = \rho_\beta(A), \qquad b_n(A) = c_n(A) - \sum_{j=0}^{n-1} b_j(A)c_{n-j}(\mathbb{1}),$$

we can write

$$\rho_{\beta V}(A) = \sum_{n=0}^{\infty} b_n(A), \tag{4.78}$$

provided $|\beta| \, \|V\| < \log 2$.

Exercise 4.29 Show that the expression

$$\langle A|B\rangle_\beta = \int_0^1 \rho_\beta(A^* \tau^{i\beta s}(B)) \, ds = \frac{1}{\beta} \int_0^\beta \rho_\beta(A^* \tau^{is}(B)) \, ds, \tag{4.79}$$

defines an inner product on $\mathcal{O}$. It is called Kubo–Mari or Bogoliubov scalar product, Duhamel two point function, or canonical correlation.

Exercise 4.30 Show that the first coefficients $b_1(A)$ and $b_2(A)$ can be written as

$$b_1(A) = -\beta \langle V|\widehat{A}\rangle_\beta,$$

$$b_2(A) = \beta^2 \int_0^1 ds \int_0^s ds' \left[\rho_\beta(\widehat{A}\tau^{i\beta s}(V)\tau^{i\beta s'}(V)) - \rho_\beta(\widehat{A}\tau^{i\beta s}(V))\rho_\beta(V) - \rho_\beta(\widehat{A}\tau^{i\beta s'}(V))\rho_\beta(V)\right],$$

where $\widehat{A} = A - \rho_\beta(A)$.

4.3.11 The standard representations of $\mathcal{O}$

In this and the following sections we introduce the so called modular structure associated with the $*$-algebra $\mathcal{O} = \mathcal{O}_\mathcal{K}$. Historically, the structure was unveiled in the work of Araki and Woods (1964) on the equilibrium states of a free Bose gas and linked to the KMS condition in Haag *et al.* (1967). After the celebrated works of

Tomita (1967) and Takesaki (1970), modular theory became an essential tool in the study of operator algebras.

For us, the main purpose of modular theory is to provide a framework which will allow us to describe a quantum system in a way that is robust enough to survive the thermodynamic limit. While familiar objects like Hamiltonians or density matrices will lose their meaning in this limit, the notions that we are about to introduce: standard representation, modular groups and operators, Connes cocycles, relative Hamiltonians, Liouvilleans, and so on will continue to make sense in the context of extended quantum systems. As a *rule of thumb*, a result that holds for finite quantum systems and can be formulated in terms of robust objects of modular theory will remain valid for extended systems.

Let $\mathcal{H}$ be an auxiliary Hilbert space and denote by $\mathcal{L}(\mathcal{H})$ the $*$-algebra of all linear operators on $\mathcal{H}$. A subset $\mathcal{A} \subset \mathcal{L}(\mathcal{H})$ is called *self-adjoint*, written $\mathcal{A}^* = \mathcal{A}$, if $A^* \in \mathcal{A}$ for all $A \in \mathcal{A}$. A self-adjoint subset $\mathcal{A} \subset \mathcal{L}(\mathcal{H})$ is a $*$-*subalgebra* if it is a vector subspace such that $AB \in \mathcal{A}$ for all $A, B \in \mathcal{A}$. A *representation* of $\mathcal{O}$ in $\mathcal{H}$ is a linear map $\phi : \mathcal{O} \to \mathcal{L}(\mathcal{H})$ such that $\phi(AB) = \phi(A)\phi(B)$ and $\phi(A^*) = \phi(A)^*$ for all $A, B \in \mathcal{O}$. A representation is *faithful* if the map ϕ is injective, that is if $\operatorname{Ker} \phi = \{0\}$. A faithful representation of $\mathcal{O}$ in $\mathcal{H}$ is therefore an isomorphism between $\mathcal{O}$ and the $*$-subalgebra $\phi(\mathcal{O}) \subset \mathcal{L}(\mathcal{H})$. A vector $\psi \in \mathcal{H}$ is called *cyclic* for the representation ϕ if $\mathcal{H} = \phi(\mathcal{O})\psi$. It is called *separating* if $\phi(A)\psi = 0$ implies that $A = 0$. Two representations $\phi_1 : \mathcal{O} \to \mathcal{L}(\mathcal{H}_1)$ and $\phi_2 : \mathcal{O} \to \mathcal{L}(\mathcal{H}_2)$ are called *equivalent* if there exists a unitary $U : \mathcal{H}_1 \to \mathcal{H}_2$ such that $U\phi_1(A) = \phi_2(A)U$ for all $A \in \mathcal{O}$.

Let $\mathcal{A}$ and $\mathcal{B}$ be subsets of $\mathcal{L}(\mathcal{H})$. $\mathcal{A} \vee \mathcal{B}$ denotes the smallest $*$-subalgebra of $\mathcal{L}(\mathcal{H})$ containing $\mathcal{A}$ and $\mathcal{B}$. $\mathcal{A}'$ denotes the *commutant* of $\mathcal{A}$, that is the set of all elements of $\mathcal{L}(\mathcal{H})$ which commute with all elements of $\mathcal{A}$. If $\mathcal{A}$ is self-adjoint, then $\mathcal{A}'$ is a $*$-subalgebra.

A *cone* in the Hilbert space $\mathcal{H}$ is a subset $\mathcal{C} \subset \mathcal{H}$ such that $\lambda\psi \in \mathcal{C}$ for all $\lambda \geq 0$ and all $\psi \in \mathcal{C}$. If $\mathcal{M} \subset \mathcal{H}$, then

$$\widehat{\mathcal{M}} = \{\phi \in \mathcal{H} \mid \langle\psi|\phi\rangle \geq 0 \text{ for all } \psi \in \mathcal{M}\},$$

is a cone. A cone $\mathcal{C} \subset \mathcal{H}$ is called *self-dual* if $\widehat{\mathcal{C}} = \mathcal{C}$.

We have already noticed that $\mathcal{O}$, viewed as a complex vector space, becomes a Hilbert space when equipped with the inner product

$$(\xi|\eta) = \operatorname{tr}(\xi^*\eta).$$

In the sequel, in order to distinguish this Hilbert space from the $*$-algebra $\mathcal{O}$ we shall denote the former by $\mathcal{H}_\mathcal{O}$. Thus, $\mathcal{O}$ and $\mathcal{H}_\mathcal{O}$ are the same set, but carry distinct algebraic structures. We will use lower case greeks $\xi, \eta, \dots$ to denote elements of the Hilbert space $\mathcal{H}_\mathcal{O}$ and upper case romans $A, B, \dots$ to denote elements of the $*$-algebra $\mathcal{O}$.

Remark. Let $\psi \mapsto \overline{\psi}$ denote an arbitrary complex conjugation (i.e., an anti-unitary involution) on the Hilbert space $\mathcal{K}$. One easily checks that the map $|\psi\rangle\langle\varphi| \mapsto \psi \otimes \overline{\varphi}$ extends to a unitary operator from $\mathcal{H}_{\mathcal{O}_\mathcal{K}}$ to $\mathcal{K} \otimes \mathcal{K}$. Thus, the Hilbert space $\mathcal{H}_{\mathcal{O}_\mathcal{K}}$ is isomorphic to $\mathcal{K} \otimes \mathcal{K}$.

To any $A \in \mathcal{O}$ we can associate two elements $L(A)$ and $R(A)$ of $\mathcal{L}(\mathcal{H}_\mathcal{O})$ by

$$L(A) : \xi \mapsto A\xi, \qquad R(A) : \xi \mapsto \xi A^*.$$

The map $\mathcal{O} \ni A \mapsto L(A) \in \mathcal{L}(\mathcal{H}_\mathcal{O})$ is clearly linear and satisfies $L(AB) = L(A)L(B)$. Moreover, for all $\xi, \eta \in \mathcal{H}_\mathcal{O}$ one has

$$(\xi | L(A)\eta) = \mathrm{tr}\,(\xi^* A\eta) = \mathrm{tr}\,((A^*\xi)^*\eta) = (L(A^*)\xi | \eta),$$

so that $L(A^*) = L(A)^*$. In short, L is a representation of the $*$-algebra $\mathcal{O}$ on the Hilbert space $\mathcal{H}_\mathcal{O}$. In the same way one checks that $R : \mathcal{O} \to \mathcal{L}(\mathcal{O})$ is antilinear and satisfies $R(AB) = R(A)R(B)$ as well as $R(A^*) = R(A)^*$.

Proposition 4.25   *(1) The maps L and R are isometric and hence injective.*
(2) $L(\mathcal{O}) = \{L(A) \,|\, A \in \mathcal{O}\}$ and $R(\mathcal{O}) = \{R(A) \,|\, A \in \mathcal{O}\}$ are $*$-subalgebras of $\mathcal{L}(\mathcal{H}_\mathcal{O})$ isomorphic to $\mathcal{O}$.
(3) $L(\mathcal{O}) \cap R(\mathcal{O}) = \mathbb{C}\mathbb{1}$.
(4) $L(\mathcal{O}) \vee R(\mathcal{O}) = \mathcal{L}(\mathcal{H}_\mathcal{O})$.
(5) $L(\mathcal{O})' = R(\mathcal{O})$.
(6) $R(\mathcal{O})' = L(\mathcal{O})$.

Proof: (1)–(2) For $A \in \mathcal{O}$, one has

$$\|L(A)\|^2 = \sup_{\|\xi\|=1} \|L(A)\xi\|^2 = \sup_{\mathrm{tr}\,(\xi^*\xi)=1} \mathrm{tr}((A\xi)^*(A\xi))$$

$$= \sup_{\mathrm{tr}\,(\xi\xi^*)=1} \mathrm{tr}\,((\xi\xi^*)(A^*A)) \leq \|A^*A\| = \|A\|^2.$$

On the other hand, if ψ is a normalized eigenvector of A^*A to its maximal eigenvalue $\|A^*A\|$ and $\xi = |\psi\rangle\langle\psi|$, then $\|\xi\| = 1$ and

$$\|L(A)\xi\| = \|A\xi\| = \langle\psi|A^*A\psi\rangle = \|A^*A\|,$$

so that we can conclude that $\|L(A)\| = \|A\|$. L is a linear map and $\mathrm{Ker}\, L = \{0\}$. Thus, L is injective and is an $*$-isomorphism between $\mathcal{O}$ and its image $L(\mathcal{O})$. The same argument holds for R.
(3) If $T \in L(\mathcal{O}) \cap R(\mathcal{O})$, then there exists $A, B \in \mathcal{O}$ such that $A\xi = \xi B$ for all $\xi \in \mathcal{H}_\mathcal{O}$. Setting $\xi = \mathbb{1}$ we deduce $A = B$. It follows that $[A, \xi] = 0$ for all $\xi \in \mathcal{O}$ and hence A must be a multiple of the identity.
(4) Let $T \in \mathcal{L}(\mathcal{H}_\mathcal{O})$ and denote by $\{E_{ij}\}$ the orthogonal basis of $\mathcal{H}_\mathcal{O}$ associated to some orthogonal basis $\{e_i\}$ of $\mathcal{K}$. Setting $T_{ij,kl} = (E_{ij}|TE_{kl})$, one has

$$TE_{kl} = \sum_{i,j,k,l} T_{ij,kl} E_{ij}.$$

Since $E_{ij} = |e_i\rangle\langle e_j| = |e_i\rangle\langle e_k|e_k\rangle\langle e_l|e_l\rangle\langle e_j| = E_{ik}E_{kl}E_{lj} = L(E_{ik})R(E_{jl})E_{kl}$, we can write

$$T = \sum_{i,j,k,l} T_{ij,kl} L(E_{ik})R(E_{jl}),$$

which shows that the subalgebras $L(\mathcal{O})$ and $R(\mathcal{O})$ generate all of $\mathcal{L}(\mathcal{H}_{\mathcal{O}})$.

(5)–(6) For any $A, B \in \mathcal{O}$ and $\xi \in \mathcal{H}_{\mathcal{O}}$ on has $L(A)R(B)\xi = A\xi B^* = R(B)L(A)\xi$ which shows that $R(\mathcal{O}) \subset L(\mathcal{O})'$ and $L(\mathcal{O}) \subset R(\mathcal{O})'$. Let $T \in L(\mathcal{O})'$ so that $[T, L(A)] = 0$ for all $A \in \mathcal{O}$. Set $B = T\mathbb{1}$, then

$$T\xi = TL(\xi)\mathbb{1} = L(\xi)T\mathbb{1} = L(\xi)B = \xi B = R(B^*)\xi,$$

for all $\xi \in \mathcal{H}_{\mathcal{O}}$. Hence, $T = R(B^*)$ and we conclude that $L(\mathcal{O})' \subset R(\mathcal{O})$. A similar argument shows that $R(\mathcal{O})' \subset L(\mathcal{O})$. $\qquad\square$

Proposition 4.26   *(1) The map $J : \xi \mapsto \xi^*$ is a anti-unitary involution of the Hilbert space $\mathcal{H}_{\mathcal{O}}$, that is J is antilinear and $J = J^* = J^{-1}$.*
(2) $JL(\mathcal{O})J^* = L(\mathcal{O})'$.
(3) $\mathcal{H}_{\mathcal{O}}^+ = \mathcal{O}_+$ is a self-dual cone of the Hilbert space $\mathcal{H}_{\mathcal{O}}$.
(4) $J\xi = \xi$ for all $\xi \in \mathcal{H}_{\mathcal{O}}^+$.
(5) $JXJ = X^*$ for all $X \in L(\mathcal{O}) \cap L(\mathcal{O})'$.
(6) $L(A)JL(A)\mathcal{H}_{\mathcal{O}}^+ \subset \mathcal{H}_{\mathcal{O}}^+$ for all $A \in \mathcal{O}$.

Proof: (1) J is clearly antilinear and involutive. Since

$$(\xi|J\eta) = \mathrm{tr}\,(\xi^*\eta^*) = \overline{\mathrm{tr}\,(\xi\eta)} = \overline{(J\xi|\eta)},$$

J is also anti-unitary.

(2) For all $A \in \mathcal{O}$ and $\xi \in \mathcal{H}_{\mathcal{O}}$ one has $JL(A)J\xi = (A\xi^*)^* = \xi A^* = R(A)\xi$ which implies $JL(A)J = R(A)$.

(3) The fact that $\mathcal{H}_{\mathcal{O}}^+ = \mathcal{O}_+$ is a cone is obvious. It is also clear that if $\xi, \eta \in \mathcal{H}_{\mathcal{O}}^+$ then $(\xi|\eta) \geq 0$ so that $\mathcal{H}_{\mathcal{O}}^+ \subset \widehat{\mathcal{H}_{\mathcal{O}}^+}$. To prove the reverse inclusion, let $\xi \in \widehat{\mathcal{H}_{\mathcal{O}}^+}$. Then $(\eta|\xi) \geq 0$ for all $\eta \in \mathcal{H}_{\mathcal{O}}^+$. In particular, with $\eta = |\psi\rangle\langle\psi|$, we get $(\eta|\xi) = \langle\psi|\xi\psi\rangle \geq 0$ from which we conclude that $\xi \in \mathcal{H}_{\mathcal{O}}^+$.

(4)–(5) are obvious and (6) follows from the fact that

$$L(A)JL(A)\xi = A\xi A^* \geq 0,$$

for all $\xi \geq 0$. $\qquad\square$

The faithful representation $L : \mathcal{O} \to \mathcal{L}(\mathcal{H}_{\mathcal{O}})$ is called *standard representation* of $\mathcal{O}$, J is called the *modular conjugation* and the cone $\mathcal{H}_{\mathcal{O}}^+$ is called the *natural cone*. The map

$$\mathfrak{S} \ni \nu \mapsto \xi_\nu = \nu^{1/2} \in \mathcal{H}_{\mathcal{O}}^+,$$

is clearly a bijection between the set of states and the unit vectors in $\mathcal{H}_{\mathcal{O}}^+$. For all $A \in \mathcal{O}$, one has

$$(\xi_\nu | L(A)\xi_\nu) = \operatorname{tr}(\nu^{1/2} A \nu^{1/2}) = \nu(A).$$

ξ_ν is called the *vector representative* of the state ν in the standard representation. Note that a unit vector $\xi \in \mathcal{H}_{\mathcal{O}}^+$ is cyclic for the standard representation iff $\xi > 0$, *i.e.*, iff the corresponding state is faithful and in this case, for any $\eta \in \mathcal{H}_{\mathcal{O}}$, one has $\eta = L(A)\xi$ with $A = \eta\xi^{-1}$. Since $L(A)\xi = 0$ iff $\operatorname{Ran}\xi \subset \operatorname{Ker}A$, ξ is a separating vector iff $\xi > 0$.

Exercise 4.31 *(The GNS representation)* Let ν be a state and define $\mathcal{H}_\nu$ to be the vector space of all linear maps $\xi : \operatorname{Ran}\nu \to \mathcal{K}$, equipped with the inner product

$$(\xi|\eta)_\nu = \operatorname{tr}_{\operatorname{Ran}\nu}(\nu\xi^*\eta) = \operatorname{tr}_{\mathcal{K}}(\eta\nu\xi^*).$$

1. Show that $\mathcal{H}_\nu$ is a Hilbert space and that $\pi_\nu : \mathcal{O} \to \mathcal{L}(\mathcal{H}_\nu)$ defined by $\pi_\nu(A)\xi = A\xi$ is a representation of $\mathcal{O}$ in $\mathcal{H}_\nu$.
2. Denote by $\eta_\nu : \operatorname{Ran}\nu \hookrightarrow \mathcal{K}$ the canonical injection $\eta_\nu\psi = \psi$. Show that η_ν is a cyclic vector for the representation π_ν and that

$$\nu(A) = (\eta_\nu|\pi_\nu(A)\eta_\nu)_\nu,$$

   for all $A \in \mathcal{O}$.
3. A cyclic representation of $\mathcal{O}$ associated to a state ν is a representation π of $\mathcal{O}$ in a Hilbert space $\mathcal{H}$ such that:
   (i) there exists a vector $\psi \in \mathcal{H}$ which is cyclic for π.
   (ii) $\nu(A) = (\psi|\pi(A)\psi)$ for all $A \in \mathcal{O}$.
   Show that any cyclic representation of $\mathcal{O}$ associated to the state ν is equivalent to the above representation π_ν.
   Hint: show that $\pi(A)\psi \mapsto \pi_\nu(A)\eta_\nu$ defines a unitary map from $\mathcal{H}$ to $\mathcal{H}_\nu$.
   Thus, up to equivalence, there is only one cyclic representation of $\mathcal{O}$ associated to a state ν. This representation is called the Gelfand–Naimark–Segal (GNS) representation of $\mathcal{O}$ induced by ν.
4. Show that the map $U : \mathcal{H}_\nu \ni \xi \mapsto \xi\nu^{1/2} \in \mathcal{H}_{\mathcal{O}}$ is a partial isometry which intertwines the GNS representation and the standard representation

$$U\pi_\nu(A)\xi = L(A)U\xi.$$

   Show that if ν is faithful, then U is unitary so that these two representations are equivalent.
5. Let $\psi \mapsto \bar{\psi}$ be a complex conjugation on $\mathcal{K}$. We have already remarked that the map $U(|\psi\rangle\langle\varphi|) = \psi \otimes \bar{\varphi}$ extends to a unitary operator from $\mathcal{H}_{\mathcal{O}_{\mathcal{K}}}$ to $\mathcal{K} \otimes \mathcal{K}$. Show that under this unitary the standard representation transforms as follows.
   (i) $UR(A)U^{-1} = A \otimes \mathbb{1}$ and $UL(A)U^{-1} = \mathbb{1} \otimes A$.
   (ii) $UJU^{-1}\psi \otimes \phi = \bar{\phi} \otimes \bar{\psi}$.
   (iii) $U\xi_\nu = \sum_j \lambda_j^{1/2} \psi_j \otimes \overline{\psi_j}/\operatorname{tr}(\nu)^{1/2}$, where λ_j's are the eigenvalues of ν listed with multiplicities and ψ_j's are the corresponding eigenfuntions.

Let τ^t be a dynamics on $\mathcal{O}$ generated by the Hamiltonian H. Since

$$L(\tau^t(A)) = L(e^{itH}Ae^{-itH}) = L(e^{itH})L(A)L(e^{-itH}) = e^{itL(H)}L(A)e^{-itL(H)},$$

the self-adjoint operator $L(H)$ seems to play the role of the Hamiltonian in the standard representation. If ν is a state and $\xi_\nu \in \mathcal{H}_\mathcal{O}^+$ its vector representative, then

$$\nu(\tau^t(A)) = (\xi_\nu|L(\tau^t(A))\xi_\nu) = (e^{-itL(H)}\xi_\nu|L(A)e^{-itL(H)}\xi_\nu).$$

The state vector thus evolves according to $e^{-itL(H)}\xi_\nu = e^{-itH}\xi_\nu$. Note that this vector is generally not an element of the natural cone. Indeed, since $\nu_t = e^{-itH}\nu e^{itH}$ its vector representative is given by

$$\xi_{\nu_t} = \nu_t^{1/2} = e^{-itH}\nu^{1/2}e^{itH} = L(e^{-itH})R(e^{-itH})\xi_\nu,$$

which is generally distinct from $e^{-itH}\xi_\nu$. On the other hand, by Part (5) of Proposition 4.25, one has

$$L(e^{itH})R(e^{itH})L(A)R(e^{-itH})L(e^{-itH}) = L(e^{itH})L(A)L(e^{-itH}) = L(\tau^t(A)),$$

so that the unitary group (recall that R is antilinear)

$$L(e^{itH})R(e^{itH}) = e^{itL(H)}e^{-itR(H)} = e^{it(L(H)-R(H))},$$

also implements the dynamics τ^t in the standard representation. We call the self-adjoint generator

$$K = L(H) - R(H) = [H, \cdot], \tag{4.80}$$

the *standard Liouvillean* of the dynamics.

Exercise 4.32

1. Show that if ν is a faithful state on $\mathcal{O}$ then the natural cone of $\mathcal{H}_\mathcal{O}$ can be written as

$$\mathcal{H}_\mathcal{O}^+ = \{L(A)JL(A)\xi_\nu \mid A \in \mathcal{O}\}.$$

Conclude that the unitary group e^{itX} preserves the natural cone iff $JX + XJ = 0$.
2. Show that the standard Liouvillean K is the only self-adjoint operator on $\mathcal{H}_\mathcal{O}$ such that, for all $A \in \mathcal{O}$ and $t \in \mathbb{R}$,

$$e^{itK}L(A)e^{-itK} = L(\tau^t(A)),$$

with the additional property that $e^{-itK}\mathcal{H}_\mathcal{O}^+ \subset \mathcal{H}_\mathcal{O}^+$. (See Proposition 4.32 for a generalization of this result.)
3. Show that the spectrum of K is given by

$$\mathrm{sp}(K) = \{\lambda - \mu \mid \lambda, \mu \in \mathrm{sp}(H)\}.$$

Note in particular that if $\dim \mathcal{K} = n$ then 0 is at least n-fold degenerate eigenvalue of K.

4.3.12 The modular structure of $\mathcal{O}$

Modular group and modular operator

In Section 4.3.9 we showed that, given a dynamics τ^t generated by the Hamiltonian H, $\rho_\beta = e^{-\beta H}/\text{tr}(e^{-\beta H})$ is the unique β-KMS state. Modular theory starts with the reverse point of view. Given a faithful state ρ, the dynamics generated by the Hamiltonian $-\beta^{-1}\log\rho$ is the unique dynamics with respect to which ρ is a β-KMS state. This dynamics might not be in itself physical but it will lead to a remarkable mathematical structure with profound physical implications. For historical reasons the reference value of β is taken to be -1. The dynamics

$$\varsigma_\rho^t(A) = e^{it \log \rho} A e^{-it \log \rho},$$

is called the *modular dynamics* or *modular group* of the state ρ. Its generator is given by

$$\delta_\rho(A) = i[\log\rho, A].$$

The ($\beta = -1$)-KMS condition can be written as

$$\rho(AB) = \rho(\varsigma_\rho^i(B)A).$$

According to the previous section, the standard Liouvillean of the modular dynamics is the self-adjoint operator on $\mathcal{H}_\mathcal{O}$ defined by

$$K_\rho = L(\log\rho) - R(\log\rho),$$

and one has

$$L(\varsigma_\rho^t(A)) = \Delta_\rho^{it} L(A)\Delta_\rho^{-it},$$

where the positive operator $\Delta_\rho = e^{K_\rho}$ is called the *modular operator* of the state ρ. Its action on a vector $\xi \in \mathcal{H}_\mathcal{O}$ is described by

$$\Delta_\rho\xi = e^{L(\log\rho) - R(\log\rho)}\xi = L(\rho)R(\rho^{-1})\xi = \rho\xi\rho^{-1}.$$

More generally, for $z \in \mathbb{C}$,

$$\Delta_\rho^z\xi = e^{z(L(\log\rho) - R(\log\rho))}\xi = L(\rho^z)R(\rho^{-z})\xi = \rho^z\xi\rho^{-z},$$

and in particular

$$J\Delta_\rho^{1/2}A\xi_\rho = (\Delta_\rho^{1/2}A\xi_\rho)^* = (\rho^{1/2}(A\rho^{1/2})\rho^{-1/2})^* = A^*\xi_\rho, \tag{4.81}$$

for any $A \in \mathcal{O}$. The last relation completely characterizes the modular conjugation J and the square root of the modular operator $\Delta_\rho^{1/2}$ as the anti-unitary and positive factors of the (unique) polar decomposition of the antilinear map $A\xi_\rho \mapsto A^*\xi_\rho$.

Generalizing the Kubo-Mari inner product (4.79), we shall call

$$\langle A|B\rangle_\rho = \int_0^1 \rho(A^* \varsigma_\rho^{-iu}(B))du, \tag{4.82}$$

the standard correlation of $A, B \in \mathcal{O}$ w.r.t. ρ

Connes cocycle and relative modular operator

The modular groups of two faithful states ρ and ν are related by their *Connes' cocycle*, the family of unitary elements of $\mathcal{O}$ defined by

$$[D\rho : D\nu]^t = \rho^{it}\nu^{-it} = e^{it\log\rho}e^{-it\log\nu}.$$

Indeed, one has

$$[D\rho : D_\nu]^t\varsigma_\nu^t(A)[D\nu : D\rho]^t = \varsigma_\rho^t(A), \tag{4.83}$$

for all $A \in \mathcal{O}$ and any $t \in \mathbb{R}$. The Connes cocycles have the following immediate properties:

(1) $[D\rho : D\nu]^t[D\nu : D\omega]^t = [D\rho : D\omega]^t$.
(2) $([D\rho : D\nu]^t)^{-1} = [D\nu : D\rho]^t$.
(3) $[D\rho : D\nu]^t\varsigma_\nu^t([D\rho : D_\nu]^s) = [D\rho : D\nu]^{t+s}$.

They are obviously defined for any $t \in \mathbb{C}$ and (4.83) as well as (1)–(3) remain valid. The operator

$$[D\rho : D\nu]^{-i} = \rho\nu^{-1},$$

satisfies

$$\nu(A[D\rho : D\nu]^{-i}) = \rho(A),$$

and is the noncommutative Radon–Nikodym derivative of ρ w.r.t. ν. The Rényi relative entropy can be expressed in terms of the Connes cocycle as

$$S_\alpha(\rho|\nu) = \log\nu([D\rho : D\nu]^{-i\alpha}). \tag{4.84}$$

The *relative modular dynamics* of two faithful states ρ and ν is defined by

$$\varsigma_{\rho|\nu}^t(A) = \rho^{it}A\nu^{-it} = e^{it\log\rho}Ae^{-it\log\nu}.$$

It is related to the modular dynamics of ρ and ν by the Connes cocycles,

$$\varsigma_{\rho|\nu}^t(A) = [D\rho : D\nu]^t\varsigma_\nu^t(A) = \varsigma_\rho^t(A)[D\rho : D\nu]^t.$$

Its standard Liouvillean is given by

$$K_{\rho|\nu} = L(\log\rho) - R(\log\nu),$$

and the corresponding *relative modular operator* $\Delta_{\rho|\nu} = e^{K_{\rho|\nu}}$ is a positive operator acting in $\mathcal{H}_\mathcal{O}$ as

$$\Delta_{\rho|\nu}\xi = L(\rho)R(\nu^{-1})\xi = \rho\xi\nu^{-1}. \tag{4.85}$$

More generally, for $z \in \mathbb{C}$,

$$\Delta_{\rho|\nu}^z\xi = L(\rho^z)R(\nu^{-z})\xi = \rho^z\xi\nu^{-z},$$

and in particular

$$J\Delta_{\rho|\nu}^{1/2}A\xi_\nu = A^*\xi_\rho,$$

for any $A \in \mathcal{O}$. Again, this relation characterizes completely $\Delta_{\rho|\nu}^{1/2}$ as the positive factor of the polar decomposition of the antilinear map $A\xi_\nu \mapsto A^*\xi_\rho$.

In the standard representation of $\mathcal{O}$ the relative modular dynamics is described by

$$L(\varsigma_{\rho|\nu}^t(A)) = \Delta_{\rho|\nu}^{it}L(A)\Delta_{\rho|\nu}^{-it},$$

and the relative entropies of ρ w.r.t. ν are given by

$$S_\alpha(\rho|\nu) = \log(\xi_\nu|\Delta_{\rho|\nu}^\alpha\xi_\nu),$$

$$S(\rho|\nu) = (\xi_\rho|\log\Delta_{\nu|\rho}\xi_\rho).$$

The *relative Hamiltonian* of ρ with respect to ν is the self-adjoint element of $\mathcal{O}$ defined by

$$\ell_{\rho|\nu} = \frac{1}{i}\frac{d}{dt}[D\rho : D\nu]^t\Big|_{t=0} = \log\rho - \log\nu. \tag{4.86}$$

Since $\delta_\rho = \delta_\nu + i[\ell_{\rho|\nu}, \cdot]$, $\ell_{\rho|\nu}$ is the perturbation that links the modular dynamics ς_ν^t and ς_ρ^t, that is with the notation of Section 4.3.10,

$$\varsigma_{\nu\,\ell_{\rho|\nu}}^t = \varsigma_\rho^t.$$

Further immediate properties of the relative Hamiltonian are:

(1) For any $\vartheta \in \mathrm{Aut}(\mathcal{O})$, $\ell_{\rho\circ\vartheta^{-1}|\nu\circ\vartheta^{-1}} = \vartheta(\ell_{\rho|\nu})$.
(2) $S(\rho|\nu) = -\rho(\ell_{\rho|\nu})$.
(3) $\log\Delta_{\rho|\nu} = \log\Delta_\nu + L(\ell_{\rho|\nu})$.
(4) $\log\Delta_\rho = \log\Delta_\nu + L(\ell_{\rho|\nu}) - R(\ell_{\rho|\nu})$.
(5) $\ell_{\rho|\nu} + \ell_{\nu|\omega} = \ell_{\rho|\omega}$.

At this point, the reader could ask about the need for such abstract constructions. To answer these concerns let us make more precise the introductory remarks made at the beginning of Section 4.3.11. After taking the thermodynamic limit, the Hamiltonian H generating the dynamics and the density matrices defining the states will lose their meaning. So will any expression explicitly involving H or density matrices. What will remain is an infinite dimensional algebra $\mathcal{O}$ describing the quantum observables of the system, a group τ^t of $*$-automorphisms of $\mathcal{O}$ describing quantum dynamics and states, positive, normalized linear functionals on $\mathcal{O}$. The modular group ς_ρ will also survive as a group of $*$-automorphisms of $\mathcal{O}$ and the modular operator Δ_ρ will survive as a positive self-adjoint operator on the Hilbert space carrying the standard representation of $\mathcal{O}$. In the same way, relative modular groups and operators will be available after the thermodynamic limit. These objects will become our handles to manipulate states. Modular theory allows us to recover, in the infinite dimensional case, the algebraic structure of the set of states which is clearly visible in the finite dimensional case. For example, the formula

$$[0,1] \ni \alpha \mapsto S_\alpha(\rho|\nu) = \log \operatorname{tr}(\rho^\alpha \nu^{1-\alpha}),$$

obviously makes sense if ρ and ν are density matrices (even in an infinite dimensional Hilbert space—it follows from Hölder's inequality that the product $\rho^\alpha \nu^{1-\alpha}$ is trace class). Thinking of ρ and ν as linear functionals, it is not clear how to make sense of such a product. The alternative formula

$$S_\alpha(\rho|\nu) = \log(\xi_\nu|\Delta_{\rho|\nu}^\alpha \xi_\nu),$$

provides a more general expression which makes sense even if ρ and ν are not associated to density matrices.

From a purely mathematical point of view, modular theory unravels the structures hidden in the traditional presentations of quantum statistical mechanics. These structures often allow for simpler and mathematically more natural proofs of classical results in quantum statistical mechanics with an additional advantage that the proofs typically extend to the general von Neumann algebra setting. We should illustrate this point on three examples at the end of this section.[9]

Exercise 4.33   Let ρ and ν be two faithful states on $\mathcal{O}$.

1. Show that $\Delta_{\rho|\nu}^{-1} = J\Delta_{\nu|\rho}J$.
2. Let τ^t be a dynamics on $\mathcal{O}$. Show that

$$\Delta_{\rho\circ\tau^t|\nu\circ\tau^t} = \mathrm{e}^{-itK}\Delta_{\rho|\nu}\mathrm{e}^{itK}.$$

where K is the standard Liouvillean of τ^t.

Noncommutative L^p-spaces.   For $p \in [1,\infty]$, we denote by $L^p(\mathcal{O})$ the Banach space $\mathcal{O}$ equipped with the p-norm (4.47). It follows from Hölder's inequality (Part (3) of Theorem 4.1) that if $p^{-1} + q^{-1} = 1$ then $L^q(\mathcal{O})$ is the dual Banach space to $L^p(\mathcal{O})$ with respect to the duality $(\xi|\eta) = \operatorname{tr}(\xi^*\eta)$. Note in particular that $L^2(\mathcal{O}) = \mathcal{H}_\mathcal{O}$.

---

[9] A perhaps most famous application of modular theory in mathematics is Alain Connes work on the general classification and structure theorem of type III factors for which he was awarded the Fields medal in 1982.

While the standard representation will provide a natural extension of $L^2(\mathcal{O})$ in the infinite dimensional setting that arises in the thermodynamic limit, there are no such extensions for the Banach spaces $L^p(\mathcal{O})$ for $p \neq 2$. Infinite dimensional extensions of those spaces which depend on a reference state were introduced by Araki and Masuda (1982). We describe here their finite dimensional counterparts and relate them to the spaces $L^p(\mathcal{O})$.

Let ω be a faithful state. For $p \in [2, \infty]$ we set

$$\|\xi\|_{\omega,p} = \max_{\nu \in \mathfrak{S}} \|\Delta_{\nu|\omega}^{\frac{1}{2} - \frac{1}{p}} \xi\|_2.$$

One easily checks that this is a norm on $\mathcal{O}$ and we denote by $L^p(\mathcal{O}, \omega)$ the corresponding Banach space. Note that $\|\xi\|_{\omega,2} = \|\xi\|_2$ so that $L^2(\mathcal{O}, \omega) = L^2(\mathcal{O}) = \mathcal{H}_{\mathcal{O}}$ for any faithful state ω. For $p \in [1, 2]$, we define $L^p(\mathcal{O}, \omega)$ to be the dual Banach space of $L^q(\mathcal{O}, \omega)$ for $p^{-1} + q^{-1} = 1$ w.r.t. the duality $(\xi|\eta) = \text{tr}(\xi^* \eta)$.

Theorem 4.27 *For $p \in [1, \infty]$ one has $\|\xi\|_{\omega,p} = \|\xi \omega^{1/p - 1/2}\|_p$, that is the map*

$$\begin{aligned} L^p(\mathcal{O}) &\to L^p(\mathcal{O}, \omega) \\ \xi &\mapsto \xi \omega^{1/2 - 1/p}, \end{aligned}$$

is a surjective isometry.

Proof: For $p \in [2, \infty]$ one has $r = p/(p-2) \in [1, \infty]$ and if $\nu \in \mathfrak{S}$, then $\nu^{(p-2)/p} \in L^r(\mathcal{O})$ with $\|\nu^{(p-2)/p}\|_r = \|\nu\|_1 = 1$. By definition of the relative modular operator, one further has

$$\|\Delta_{\nu|\omega}^{\frac{p-2}{2p}} \xi\|_2^2 = (\nu^{\frac{p-2}{2p}} \xi \omega^{-\frac{p-2}{2p}} | \nu^{\frac{p-2}{2p}} \xi \omega^{-\frac{p-2}{2p}}) = \text{tr}(\nu^{\frac{p-2}{p}} \xi \omega^{-\frac{p-2}{p}} \xi^*).$$

Noting that $1 - 1/r = 2/p$, we can write

$$\|\xi\|_{\omega,p}^2 = \max_{\|\eta\|_r = 1} \text{tr}(\eta \xi \omega^{-\frac{p-2}{p}} \xi^*) = \|\xi \omega^{-\frac{p-2}{p}} \xi^*\|_{p/2} = \|\omega^{-\frac{p-2}{2p}} \xi^*\|_p^2.$$

We conclude using the fact that $\|\omega^{-\frac{p-2}{2p}} \xi^*\|_p = \|\xi \omega^{-\frac{p-2}{2p}}\|_p$ (recall Exercise 4.3.2). For $p \in [1, 2]$ we have $q \in [2, \infty]$, with $q^{-1} = 1 - p^{-1}$,

$$\|\xi\|_{\omega,p} = \sup_{\eta \neq 0} \frac{|\text{tr}(\xi \eta^*)|}{\|\eta\|_{\omega,q}} = \sup_{\eta \neq 0} \frac{|\text{tr}(\xi \eta^*)|}{\|\eta \omega^{-\frac{q-2}{2q}}\|_q} = \sup_{\nu \neq 0} \frac{|\text{tr}(\xi \omega^{\frac{q-2}{2q}} \nu^*)|}{\|\nu\|_q} = \|\xi \omega^{\frac{q-2}{2q}}\|_p.$$

Since $(q-2)/2q = -(p-2)/2p$, we get

$$\|\xi\|_{\omega,p} = \|\xi \omega^{-\frac{p-2}{2p}}\|_p.$$

$\square$

Exercise 4.34

1. Denote by $L_+^p(\mathcal{O}, \omega)$ the image of the cone $L_+^p(\mathcal{O}) = \{\xi \in L^p(\mathcal{O}) \,|\, \xi \geq 0\}$ by the isometry of Theorem 4.27,

$$L_+^p(\mathcal{O}, \omega) = \{A\omega^{1/2-1/p} \mid A \in \mathcal{O}_+\}.$$

Show that, with $p^{-1} + q^{-1} = 1$, the dual cone to $L_+^p(\mathcal{O}, \omega)$ is $L_+^q(\mathcal{O}, \omega)$, that is that

$$(\eta|\xi) \geq 0$$

for all $\xi \in L_+^p(\mathcal{O}, \omega)$ iff $\eta \in L_+^q(\mathcal{O}, \omega)$. (Note that $L_+^2(\mathcal{O}, \omega) = \mathcal{H}_\mathcal{O}^+$, the natural cone.)

2. Show that

$$L_+^p(\mathcal{O}, \omega) = \{\lambda \Delta_{\rho|\omega}^{1/p} \xi_\omega \mid \rho \in \mathfrak{S}, \lambda > 0\}.$$

We finish this section with several examples of applications of the modular structure. The first one is a proof of Kosaki's variational formula.

Proof of Theorem 4.14 We extend the definition of the relative modular operator to pairs of nonfaithful states. As already noticed (just before Exercise 4.3.11), if $\nu \in \mathfrak{S}$ is not faithful then its vector representative $\xi_\nu \in \mathcal{H}_\mathcal{O}$ is not cyclic for the standard representation. In fact $\mathcal{O}\xi_\nu = \{A\xi_\nu \mid A \in \mathcal{O}\}$ is the proper subspace of $\mathcal{H}_\mathcal{O}$ given by

$$\mathcal{O}\xi_\nu = \{\eta \in \mathcal{H}_\mathcal{O} \mid \text{Ker}\,\eta \supset \text{Ker}\,\nu\} = \{\eta \in \mathcal{H}_\mathcal{O} \mid \eta(\mathbb{1} - s(\nu)) = 0\}.$$

Accordingly, one has the orthogonal decomposition

$$\mathcal{H}_\mathcal{O} = \mathcal{O}\xi_\nu \oplus [\mathcal{O}\xi_\nu]^\perp,$$

where

$$[\mathcal{O}\xi_\nu]^\perp = \{\eta \in \mathcal{H}_\mathcal{O} \mid \text{Ker}\,\eta \supset \text{Ran}\,\nu\} = \{\eta \in \mathcal{H}_\mathcal{O} \mid \eta s(\nu) = 0\}.$$

For $\rho, \nu \in \mathfrak{S}$, we define the linear operator $\Delta_{\rho|\nu}$ on $\mathcal{H}_\mathcal{O}$ by

$$\Delta_{\rho|\nu} : \xi \mapsto \rho\xi[(\nu|_{\text{Ran}\,\nu})^{-1} \oplus 0|_{\text{Ker}\,\nu}].$$

One easily checks that $\Delta_{\rho|\nu}$ is non-negative, with $\text{Ker}\,\Delta_{\rho|\nu} = \{\xi \in \mathcal{H}_\mathcal{O} \mid s(\rho)\xi s(\nu) = 0\}$. We note in particular that

$$J\Delta_{\rho|\nu}^{1/2}(L(A)\xi_\nu \oplus \eta) = s(\nu)L(A)^*\xi_\rho, \tag{4.87}$$

for any $A \in \mathcal{O}$ and $\eta \in [\mathcal{O}\xi_\nu]^\perp$.

Starting from the identity $\text{tr}(\rho^\alpha \nu^{1-\alpha}) = (\xi_\nu|\Delta_{\rho|\nu}^\alpha \xi_\nu)$ and using the integral formula of Exercise 4.3.1 we write, for $\alpha \in]0, 1[$,

$$\text{tr}(\rho^\alpha \nu^{1-\alpha}) = \frac{\sin \pi \alpha}{\pi} \int_0^\infty t^{\alpha-1} \left(\xi_\nu|\Delta_{\rho|\nu}(\Delta_{\rho|\nu} + t)^{-1}\xi_\nu\right) dt.$$

For $A \in \mathcal{O}$ one has,

$$\rho(|A^*|^2) = \|L(A)^*\xi_\rho\|^2 = \|s(\nu)L(A)^*\xi_\rho\|^2 + \|QL(A)^*\xi_\rho\|^2,$$

where $Q = \mathbb{1} - \mathrm{s}(\nu)$ is the orthogonal projection on $\mathrm{Ker}\,\nu$. By equation (4.87), we obtain

$$\rho(|A^*|^2) = \|J\Delta_{\rho|\nu}^{1/2}L(A)\xi_\nu\|^2 + \|QL(A)^*\xi_\rho\|^2$$
$$= (\xi_\nu|L(A^*)\Delta_{\rho|\nu}L(A)\xi_\nu) + \rho(AQA^*),$$

from which we deduce

$$\frac{1}{t}\rho(|A^*|^2) + \nu(|\mathbb{1} - A|^2) = \frac{1}{t}(\xi_\nu|L(A^*)\Delta_{\rho|\nu}L(A)\xi_\nu) + (\xi_\nu|L(|\mathbb{1} - A|^2)\xi_\nu)$$
$$+ \frac{1}{t}\rho(AQA^*).$$

With some elementary algebra, this identity leads to

$$(\xi_\nu|\Delta_{\rho|\nu}(\Delta_{\rho|\nu} + t)^{-1}\xi_\nu) = \frac{1}{t}\rho(|A^*|^2) + \nu(|\mathbb{1} - A|^2) - R_A,$$

where

$$R_A = \frac{1}{t}\rho(AQA^*) + \left\|(\mathbb{1} + \Delta_{\rho|\nu}/t)^{1/2}(L(A) - (\mathbb{1} + \Delta_{\rho|\nu}/t)^{-1})\xi_\nu\right\|^2.$$

Since $R_A \geq 0$, we get

$$\mathrm{tr}(\rho^\alpha \nu^{1-\alpha}) \leq \frac{\sin \pi \alpha}{\pi} \int_0^\infty t^{\alpha-1}\left[\frac{1}{t}\rho(|A(t)^*|^2) + \nu(|\mathbb{1} - A(t)|^2)\right] \mathrm{d}t,$$

for all $A \in C(\mathbb{R}_+, \mathcal{O})$, with equality iff $R_{A(t)} = 0$ for all $t > 0$. Since $\Delta_{\rho|\nu} \geq 0$, this happens iff $(\mathbb{1} + \Delta_{\rho|\nu}/t)L(A(t))\xi_\nu = \xi_\nu$ and $\rho^{1/2}A(t)Q = 0$ for all $t > 0$. The first condition is equivalent to

$$(\mathbb{1} - A(t))\nu = \frac{1}{t}\rho A(t)\mathrm{s}(\nu). \tag{4.88}$$

An integration by parts shows that the function

$$A_{\mathrm{opt}}(t) = t\int_0^\infty \mathrm{e}^{-s\rho}\nu\mathrm{e}^{-st\nu}\mathrm{d}s,$$

satisfies this condition as well as $A_{\mathrm{opt}}(t)Q = 0$ so that $R_{A_{\mathrm{opt}}(t)} = 0$. This proves Kosaki's variational principle.

Suppose that $B(t) \in C(\mathbb{R}_+, \mathcal{O})$ is such that $A(t) = A_{\mathrm{opt}}(t) + B(t)$ is also minimizer. It follows that $B(t)$ satisfies the two conditions

$$tB(t)\nu + \rho B(t)\mathrm{s}(\nu) = 0, \tag{4.89}$$

$$\rho^{1/2}B(t)(\mathbb{1} - \mathrm{s}(\nu)) = 0, \tag{4.90}$$

for all $t > 0$. Let ϕ be an eigenvector of ν to the eigenvalue $p > 0$. Condition (4.89) yields $(\rho + tp)B(t)\phi = 0$ which implies $B(t)\phi = 0$. We conclude that $B(t)\mathrm{s}(\nu) = 0$ and

Condition 4.90 further yields $\rho^{1/2} B(t) = 0$. It follows that if either ν or ρ is faithful then $B(t) = 0$. $\qquad\qquad\square$

As a second application of modular theory, we give an alternative proof of Uhlmann's monotonicity theorem.

Proof of Theorem 4.15. To simplify notation, we shall set $\hat{\nu} = \Phi^*(\nu)$ and $\hat{\rho} = \Phi^*(\rho)$. In terms of the extended modular operator, one has

$$S_\alpha(\rho|\nu) = \log \operatorname{tr}(\rho^\alpha \nu^{1-\alpha}) = \log(\xi_\nu | \Delta^\alpha_{\rho|\nu} \xi_\nu),$$

so that we have to show that

$$(\xi_{\hat{\nu}} | \Delta^\alpha_{\hat{\rho}|\hat{\nu}} \xi_{\hat{\nu}}) \geq (\xi_\nu | \Delta^\alpha_{\rho|\nu} \xi_\nu), \tag{4.91}$$

for all $\alpha \in [0, 1]$.

Consider the orthogonal decomposition $\mathcal{H}_{\mathcal{O}_\mathcal{K}} = \mathcal{O}_\mathcal{K} \xi_{\hat{\nu}} \oplus [\mathcal{O}_\mathcal{K} \xi_{\hat{\nu}}]^\perp$. For $A \in \mathcal{O}_\mathcal{K}$ and $\eta \in [\mathcal{O}_\mathcal{K} \xi_{\hat{\nu}}]^\perp$, the Schwarz inequality (4.59) yields

$$\begin{aligned}
\|\Phi(A)\xi_\nu\|^2 &= (\xi_\nu | \Phi(A)^* \Phi(A) \xi_\nu) \\
&\leq (\xi_\nu | \Phi(A^*A)\xi_\nu) = \nu(\Phi(A^*A)) = \hat{\nu}(A^*A) \\
&= (\xi_{\hat{\nu}} | A^*A\xi_{\hat{\nu}}) = \|A\xi_{\hat{\nu}}\|^2 \\
&\leq \|A\xi_{\hat{\nu}}\|^2 + \|\eta\|^2 = \|A\xi_{\hat{\nu}} \oplus \eta\|^2,
\end{aligned}$$

which shows that the map $A\xi_{\hat{\nu}} \oplus \eta \mapsto \Phi(A)\xi_\nu$ is well defined as a linear contraction $T_\nu : \mathcal{H}_{\mathcal{O}_\mathcal{K}} \to \mathcal{H}_{\mathcal{O}_{\mathcal{K}'}}$. The map T_ρ is defined in a similar way.

For $A \in \mathcal{O}_\mathcal{K}$ and $\eta \in [\mathcal{O}_\mathcal{K} \xi_{\hat{\nu}}]^\perp$, one has

$$\begin{aligned}
J\Delta^{1/2}_{\rho|\nu} T_\nu(A\xi_{\hat{\nu}} \oplus \eta) &= J\Delta^{1/2}_{\rho|\nu} T_\nu(As(\hat{\nu})\xi_{\hat{\nu}} \oplus \eta) \\
&= J\Delta^{1/2}_{\rho|\nu} \Phi(As(\hat{\nu}))\xi_\nu \\
&= s(\nu)\Phi(As(\hat{\nu}))^* \xi_\rho = s(\nu)\Phi(s(\hat{\nu})A^*)\xi_\rho \\
&= s(\nu)T_\rho s(\hat{\nu})A^* \xi_{\hat{\rho}} \\
&= s(\nu)T_\rho J\Delta^{1/2}_{\hat{\rho}|\hat{\nu}}(A\xi_{\hat{\nu}} + \eta),
\end{aligned}$$

from which we conclude that $\Delta^{1/2}_{\rho|\nu} T_\nu = K\Delta^{1/2}_{\hat{\rho}|\hat{\nu}}$ where $K = Js(\nu)T_\rho J$ is a contraction. It follows that for $\varepsilon > 0$

$$\Delta^{1/2}_{\rho|\nu} T_\nu (\Delta^{1/2}_{\hat{\rho}|\hat{\nu}} + \varepsilon)^{-1} = K\Delta^{1/2}_{\hat{\rho}|\hat{\nu}}(\Delta^{1/2}_{\hat{\rho}|\hat{\nu}} + \varepsilon)^{-1},$$

and since $\sup_{x \geq 0} x/(x + \varepsilon) = 1$ one has $\|\Delta^{1/2}_{\rho|\nu} T_\nu(\Delta^{1/2}_{\hat{\rho}|\hat{\nu}} + \varepsilon)^{-1}\| \leq 1$. The entire analytic function

$$F(z) = (\xi | (\Delta^{1/2}_{\hat{\rho}|\hat{\nu}} + \varepsilon)^{-z} T^*_\nu \Delta^z_{\rho|\nu} T_\nu (\Delta^{1/2}_{\hat{\rho}|\hat{\nu}} + \varepsilon)^{-z} \xi),$$

thus satisfies

$$|F(z)| \leq \frac{1}{\varepsilon^2} \|\Delta_{\rho|\nu} + \mathbb{1}\| \|\xi\|^2, \qquad |F(it)| \leq \|\xi\|^2, \qquad |F(1 + it)| \leq \|\xi\|^2,$$

on the strip $0 \leq \operatorname{Re} z \leq 1$. By the three lines theorem $|F(z)| \leq \|\xi\|^2$ on this strip. Setting $z = \alpha \in [0,1]$, we conclude that

$$(T_\nu \xi | \Delta_{\rho|\nu}^\alpha T_\nu \xi) \leq (\xi | (\Delta_{\hat{\rho}|\hat{\nu}}^{1/2} + \varepsilon)^{2\alpha} \xi).$$

Letting $\varepsilon \downarrow 0$ we get

$$(T_\nu \xi | \Delta_{\rho|\nu}^\alpha T_\nu \xi) \leq (\xi | \Delta_{\hat{\rho}|\hat{\nu}}^\alpha \xi),$$

and (4.91) follows from the fact that $T_\nu \xi_{\hat{\nu}} = \Phi(\mathbb{1}) \xi_\nu = \xi_\nu$. $\qquad \square$

As a last illustration of the use of modular theory, we prove a lower bound for quantum hypothesis testing which complements Theorem 4.21. Our proof is an abstract version of similar results proven in Audenaert, Nussbaum, Szkoła, and Verstraete (2008), Hiai, Mosonyi, and Ogawa (2008), where readers can find references for the previous works on the subject. The extension of our proof to the general von Neumann algebra setting can be found in Jakšić, Ogata, Pillet, and Seiringer (2011c).

Let $D_p(\rho, \nu) = D_p(\rho, \nu, P_{\text{opt}})$ be as in Section 1.3.7. Let $\Delta_{\rho|\nu}$ be the modular operator defined in the proof of Theorem 1.14, and let $\mu_{\rho|\nu}$ be the spectral measure for $\Delta_{\rho|\nu}$ and ξ_ν.

Proposition 4.28

$$D_p(\rho, \nu) \geq \frac{1}{2} \min(p, 1 - p) \mu_{\rho|\nu}([1, \infty[).$$

Proof: Let P be an orthogonal projection (a test). By equation (4.87), one has

$$
\begin{aligned}
D_p(\rho, \nu, P) &= p \|(\mathbb{1} - P)\xi_\rho\|^2 + (1 - p)\|P\xi_\nu\|^2 \\
&\geq p \|s(\nu)(\mathbb{1} - P)\xi_\rho\|^2 + (1 - p)\|P\xi_\nu\|^2 \\
&\geq p \|\Delta_{\rho|\nu}^{1/2}(\mathbb{1} - P)\xi_\nu\|^2 + (1 - p)\|P\xi_\nu\|^2 \\
&\geq \min(p, 1 - p) \left(\|\Delta_{\rho|\nu}^{1/2}(\mathbb{1} - P)\xi_\nu\|^2 + \|P\xi_\nu\|^2 \right) \\
&\geq \min(p, 1 - p)(\xi_\nu | ((\mathbb{1} - P)\Delta_{\rho|\nu}(\mathbb{1} - P) + P\mathbb{1}P)\xi_\nu).
\end{aligned}
$$

Let F be the characteristic function of the interval $[1, \infty[$. Since $\mathbb{1} \geq F(\Delta_{\rho|\nu})$ and $\Delta_{\rho|\nu} \geq F(\Delta_{\rho|\nu})$, we further have

$$D_p(\rho, \nu, P) \geq \min(p, 1 - p)(\xi_\nu | ((\mathbb{1} - P)F(\Delta_{\rho|\nu})(\mathbb{1} - P) + PF(\Delta_{\rho|\nu})P)\xi_\nu).$$

From the identity

$$(\mathbb{1} - P)F(\Delta_{\rho|\nu})(\mathbb{1} - P) + PF(\Delta_{\rho|\nu})P - \frac{1}{2}F(\Delta_{\rho|\nu}) = (\mathbb{1} - 2P)F(\Delta_{\rho|\nu})(\mathbb{1} - 2P),$$

we deduce $(\mathbb{1} - P)F(\Delta_{\rho|\nu})(\mathbb{1} - P) + PF(\Delta_{\rho|\nu})P \geq \frac{1}{2}F(\Delta_{\rho|\nu})$ which allows us to conclude

$$D_p(\rho, \nu, P) \geq \frac{1}{2} \min(p, 1 - p)(\xi_\nu | F(\Delta_{\rho|\nu})\xi_\nu),$$

for all orthogonal projections $P \in \mathcal{O}$. Finally we note that

$$D_p(\rho, \nu) = \min_P D_p(\rho, \nu, P) \geq \frac{1}{2} \min(p, 1 - p)(\xi_\nu | F(\Delta_{\rho|\nu})\xi_\nu)$$

$$= \frac{1}{2} \min(p, 1 - p)\mu_{\rho|\nu}([1, \infty[),$$

which concludes the proof. $\qquad\qquad\qquad\qquad\qquad\qquad\qquad\qquad\qquad\square$

Exercise 4.35 Prove the following generalization of Kosaki's variational formula: for any $\rho, \nu \in \mathfrak{S}$, $B \in \mathcal{O}$ and $\alpha \in]0, 1[$ one has

$$\operatorname{tr}\left(B^* \rho^\alpha B \nu^{1-\alpha}\right) = \inf_{A \in C(\mathbb{R}_+, \mathcal{O})} \frac{\sin \pi \alpha}{\pi} \int_0^\infty t^{\alpha-1}\left[\frac{1}{t}\rho(|A(t)^*|^2) + \nu(|B - A(t)|^2)\right] \mathrm{d}t.$$

4.4 Entropic functionals and fluctuation relations of finite quantum systems

4.4.1 Quantum dynamical systems

Our starting point is a quantum dynamical system $(\mathcal{O}, \tau^t, \omega)$ on a finite dimensional Hilbert space $\mathcal{K}$, where $\mathbb{R} \ni t \mapsto \tau^t$ is a continuous group of $*$-automorphisms of $\mathcal{O}$, and ω a faithful state. We denote by δ the generator of τ^t and by H the corresponding Hamiltonian.

As in our discussion of the thermally driven harmonic chain in Section 4.2, time-reversal invariance (TRI) will play an important role in the sequel. An antilinear $*$-automorphism Θ of $\mathcal{O}$ is called time-reversal of $(\mathcal{O}, \tau^t)$ if

$$\Theta \circ \Theta = \mathrm{id}, \qquad \tau^t \circ \Theta = \Theta \circ \tau^{-t}.$$

A state ω is called TRI iff $\omega(\Theta(A)) = \omega(A^*)$. The quantum dynamical system $(\mathcal{O}, \tau^t, \omega)$ is called TRI if there exists a time-reversal Θ of $(\mathcal{O}, \tau^t)$ such that ω is TRI.

Exercise 4.36 Suppose that Θ is a time-reversal of $(\mathcal{O}, \tau^t)$. Show that there exists an anti-unitary $U_\Theta : \mathcal{K} \to \mathcal{K}$, unique up to a phase, such that $\Theta(A) = U_\Theta A U_\Theta^{-1}$ and deduce that $\operatorname{tr}(\Theta(A)) = \operatorname{tr}(A^*)$. Show that $\Theta(H) = H$ and that a state ω is TRI iff $\Theta(\omega) = \omega$.
Hint: recall Exercise 5.20.

4.4.2 Entropy balance

The relative Hamiltonian of ω_t w.r.t. ω, $\ell_{\omega_t|\omega} = \log \omega_t - \log \omega$, is easily seen to satisfy:

Proposition 4.29 (1) *For all $t, s \in \mathbb{R}$ the additive cocycle property*

$$\ell_{\omega_{t+s}|\omega} = \ell_{\omega_t|\omega} + \tau^{-t}(\ell_{\omega_s|\omega}), \tag{4.92}$$

holds.
(2) *If $(\mathcal{O}, \tau, \omega)$ is TRI, then*

$$\Theta(\ell_{\omega_t|\omega}) = -\tau^t(\ell_{\omega_t|\omega}), \tag{4.93}$$

for all $t \in \mathbb{R}$.

Differentiating the cocycle relation (4.92) we obtain

$$\frac{d}{dt}\ell_{\omega_t|\omega} = \tau^{-t}(\sigma),$$

where

$$\sigma = \frac{d}{dt}\ell_{\omega_t|\omega}\Big|_{t=0} = -i[H, \log \omega] = \delta_\omega(H), \tag{4.94}$$

(recall that δ_ω denotes the generator of the modular group of ω). Thus, we can write

$$\ell_{\omega_t|\omega} = \int_0^t \sigma_{-s} \, ds, \tag{4.95}$$

and the relation $S(\omega_t|\omega) = -\omega_t(\ell_{\omega_t|\omega})$ yields the quantum mechanical version of equation (4.5),

$$S(\omega_t|\omega) = -\int_0^t \omega(\sigma_s) \, ds.$$

We shall refer to this identity as the *entropy balance equation* and call σ the *entropy production observable*.

Proposition 4.30 $\omega(\sigma) = 0$ *and if $(\mathcal{O}, \tau^t, \omega)$ is TRI then $\Theta(\sigma) = -\sigma$.*

Proof: $\omega(\sigma) = -i \operatorname{tr}(\omega[H, \log \omega]) = i \operatorname{tr}(H[\omega, \log \omega]) = 0$. Differentiating (4.93) at $t = 0$ one derives the second statement. □

An immediate consequence of the entropy balance equation is that the mean entropy production rate over the time interval $[0, t]$,

$$\Sigma^t = \frac{1}{t}\int_0^t \sigma_s \, ds,$$

has a non-negative expectation

$$\omega(\Sigma^t) = \frac{1}{t} \int_0^t \omega(\sigma_s) \, ds \geq 0. \tag{4.96}$$

Introducing the entropy observable $S = -\log \omega$ (so $S_t = \tau^t(S) = -\log \omega_{-t}$), we see that

$$\Sigma^t = \frac{1}{t}(S_t - S), \qquad \frac{d}{dt} S_t|_{t=0} = \sigma. \tag{4.97}$$

The observable S cannot survive the thermodynamic limit. However, the relative Hamiltonian and all other objects defined in this section do. All relations except (4.97) remain valid after the thermodynamic limit is taken.

Exercise 4.37   Assume that the quantum dynamical system $(\mathcal{O}, \tau^t, \omega)$ is in a steady state, $\omega(\tau^t(A)) = \omega(A)$ for all $A \in \mathcal{O}$ and $t \in \mathbb{R}$. Denote by K the standard Liouvillean of τ^t and by δ_ω the generator of the modular group of ω: $\varsigma_\omega^t = e^{t\delta_\omega}$.

Consider the perturbed dynamical system $(\mathcal{O}, \tau_V^t, \omega)$ associated to $V \in \mathcal{O}_{\text{self}}$ (see Section 4.3.10).

1. Show that its entropy production observable is given by

$$\sigma = \delta_\omega(V).$$

2. Show that its standard Liouvillean is given by

$$K_V \xi = K\xi + V\xi - JVJ\xi.$$

4.4.3   Finite time Evans–Searles (ES) symmetry

At this point, looking back at Section 4.2.6, one may think that, for TRI quantum dynamical systems, the universal ES relation (4.13) holds between the spectral measure P^t of Σ^t associated to ω

$$\omega(f(\Sigma^t)) = P^t(f) = \int f(s) \, dP^t(s),$$

and its reversal $\bar{P}^t(f) = \omega(f(-\Sigma^t))$. To check this point, we first note that, by Proposition 4.30,

$$\Theta(\Sigma^t) = \frac{1}{t} \int_0^t \Theta(\tau^s(\sigma)) \, ds = -\frac{1}{t} \int_0^t \tau^{-s}(\sigma) \, ds = -\tau^{-t}(\Sigma^t),$$

which is the quantum counterpart of Eq.(4.10) and (4.12). Note that this relation implies that $s \in \text{sp}(\Sigma^t)$ iff $-s \in \text{sp}(\Sigma^t)$ and that the eigenvalues $\pm s$ have equal multiplicities. Furthermore,

$$\bar{P}^t(f) = \omega_t \circ \tau^{-t}(f(-\Sigma^t)) = \omega_t \circ \Theta(f(\Sigma^t)) = \omega_{-t}(f(\Sigma^t)) = \omega(f(\Sigma^t)\omega_{-t}\omega^{-1}),$$

which, using (4.95), can be rewritten as

$$\bar{P}^t(f) = \omega\left(f(\Sigma^t)e^{\log \omega - t\Sigma^t}e^{-\log \omega}\right).$$

If ω is not a steady state then $\log \omega$ and Σ^t do not commute and hence we cannot conclude, as in the classical case, that $\bar{P}^t(f)$ equals $\omega\left(f(\Sigma^t)e^{-t\Sigma^t}\right)$. Our naive attempt to generalize the ES relation (4.13) to quantum dynamical systems thus failed because quantum mechanical observables do not commute.

Exercise 4.38 Show that the ES relation

$$\omega\left(e^{-\alpha t\Sigma^t}\right) = \omega\left(e^{-(1-\alpha)t\Sigma^t}\right).$$

holds for all t if and only if $[H, \omega] = 0$.

As noticed in Section 4.2.6, the ES relation (4.13) is equivalent to the ES symmetry (4.14) of the Laplace transform of the measure P^t. We recall also that this Laplace transform is related to the relative entropy through equation (4.9). It is therefore natural to check for the ES symmetry of the function

$$\alpha \mapsto S_\alpha(\omega_t|\omega).$$

Assuming TRI, we have

$$\mathrm{tr}(\omega_t^\alpha \omega^{1-\alpha}) = \mathrm{tr}(\Theta(\omega^{1-\alpha}\omega_t^\alpha)) = \mathrm{tr}(\omega^{1-\alpha}\omega_{-t}^\alpha) = \mathrm{tr}(\omega_t^{1-\alpha}\omega^\alpha),$$

where we used that $\mathrm{tr}(\Theta(A)) = \mathrm{tr}(A^*)$ (Exercise 4.4.1). Thus,

$$S_\alpha(\omega_t|\omega) = \log \mathrm{tr}(\omega_t^\alpha \omega^{1-\alpha}) = \log \mathrm{tr}(\omega^{(1-\alpha)/2}\omega_t^\alpha \omega^{(1-\alpha)/2}),$$

satisfies the ES-symmetry.

In our noncommutative framework one may also define the entropy-like functional

$$\mathbb{R} \ni \alpha \mapsto e_{p,t}(\alpha) = \log \mathrm{tr}\left[\left(\omega^{(1-\alpha)/p}\omega_t^{2\alpha/p}\omega^{(1-\alpha)/p}\right)^{p/2}\right].$$

For reasons that will become clear later, we restrict the real parameter p to $p \geq 1$. Since $\log \omega_t = \log \omega + \ell_{\omega_t|\omega}$, Corollary 4.3 yields

$$e_{\infty,t}(\alpha) = \lim_{p\to\infty} e_{p,t}(\alpha) = \log \mathrm{tr}(e^{(1-\alpha)\log \omega + \alpha \log \omega_t}) = \log \mathrm{tr}(e^{\log \omega + \alpha \ell_{\omega_t|\omega}}).$$

We shall call the $e_{p,t}(\alpha)$ *entropic pressure functionals*. Their basic properties are:

Proposition 4.31   *(1) The function* $[1, \infty] \ni p \mapsto e_{p,t}(\alpha)$ *is continuous and mono-tonically decreasing.*

(2) *The function* $\mathbb{R} \ni \alpha \mapsto e_{p,t}(\alpha)$ *is real-analytic and convex. It satisfies* $e_{p,t}(0) = e_{p,t}(1) = 0$ *and*

$$e_{p,t}(\alpha) \begin{cases} \leq 0 \ \text{for } \alpha \in [0,1], \\ \geq 0 \ \text{otherwise.} \end{cases}$$

(3) $e_{p,t}(\alpha) = e_{p,-t}(1 - \alpha)$.
(4) $\partial_\alpha e_{p,t}(\alpha)|_{\alpha=0} = \omega(\ell_{\omega_t|\omega}) = S(\omega|\omega_t)$ *and* $\partial_\alpha e_{p,t}(\alpha)|_{\alpha=1} = \omega_t(\ell_{\omega_t|\omega}) = -S(\omega_t|\omega)$.
(5) $\partial_\alpha^2 e_{\infty,t}(\alpha)|_{\alpha=0} = \langle \ell_{\omega_t|\omega} | \ell_{\omega_t|\omega} \rangle_\omega - \omega(\ell_{\omega_t|\omega})^2$.
(6) $\partial_\alpha^2 e_{2,t}(\alpha)|_{\alpha=0} = \omega(\ell_{\omega_t|\omega}^2) - \omega(\ell_{\omega_t|\omega})^2$.
(7) *If* $(\mathcal{O}, \tau^t, \omega)$ *is TRI, then the finite time quantum Evans–Searles (ES) symmetry holds,*

$$e_{p,t}(\alpha) = e_{p,t}(1 - \alpha). \tag{4.98}$$

Proof: (1) Continuity is obvious. Writing

$$e_{p,t}(\alpha) = \log \|\omega_t^{\alpha/p} \omega^{(1-\alpha)/p}\|_p^p, \tag{4.99}$$

monotonicity follows from Corollary 4.3.
(2) Analyticity easily follows from the analytic functional calculus and convexity is a consequence of Corollary 4.4. The value taken by $e_{p,t}$ at $\alpha = 0$ and $\alpha = 1$ is evident and the remaining inequalities follow from convexity.
(3) Unitary invariance of the trace norms and Identity (4.56) give

$$\|\omega_t^{\alpha/p} \omega^{(1-\alpha)/p}\|_p = \|e^{-itH} \omega^{\alpha/p} e^{itH} \omega^{(1-\alpha)/p}\|_p$$

$$= \|\omega^{\alpha/p} e^{itH} \omega^{(1-\alpha)/p} e^{-itH}\|_p$$

$$= \|\omega^{\alpha/p} \omega_{-t}^{(1-\alpha)/p}\|_p = \|\omega_{-t}^{(1-\alpha)/p} \omega^{\alpha/p}\|_p.$$

(4) We consider only $p \in [1, \infty[$. The limiting case $p = \infty$ will be treated in the proof of Assertion (5). We set $T(\alpha) = \omega^{(1-\alpha)/p} \omega_t^{2\alpha/p} \omega^{(1-\alpha)/p}$ so that

$$\partial_\alpha e_{p,t}(\alpha) \Big|_{\alpha=0} = \partial_\alpha \mathrm{tr} \left(T(\alpha)^{p/2} \right) \Big|_{\alpha=0}.$$

Let Γ be a closed contour on the right half-plane $\mathrm{Re}\, z > 0$ encircling the strictly positive spectrum of $T(0) = \omega^{2/p}$. Since $\alpha \mapsto T(\alpha)$ is continuous, Γ can be chosen in such a way that it encloses the spectrum of $T(\alpha)$ for α small enough. Hence, with $f(z) = z^{p/2}$, we can write

$$\mathrm{tr}\, T(\alpha)^{p/2} = \oint_\Gamma f(z) \mathrm{tr} \left((z - T(\alpha))^{-1} \right) \frac{dz}{2\pi i},$$

so that

$$\partial_\alpha \mathrm{tr}\,(T(\alpha)^{p/2})\Big|_{\alpha=0} = \oint_\Gamma f(z)\mathrm{tr}\left[(z - T(0))^{-1}T'(0)(z - T(0))^{-1}\right]\frac{dz}{2\pi i}.$$

An elementary calculation gives

$$T'(0) = \frac{2}{p}\omega^{1/p}\ell_{\omega_t|\omega}\omega^{1/p},$$

and the cyclicity of the trace allows us to write

$$\partial_\alpha \mathrm{tr}\,(T(\alpha)^{p/2})\Big|_{\alpha=0} = \frac{2}{p}\oint_\Gamma f(z)\mathrm{tr}\left[(z - \omega^{2/p})^{-2}\omega^{2/p}\ell_{\omega_t|\omega}\right]\frac{dz}{2\pi i}$$

$$= \frac{2}{p}\mathrm{tr}\left[f'(\omega^{2/p})\omega^{2/p}\ell_{\omega_t|\omega}\right] = \mathrm{tr}\,\omega\ell_{\omega_t|\omega} = S(\omega|\omega_t).$$

The second statement also follows by taking (3) into account and observing that $S(\omega|\omega_{-t}) = S(\omega_t|\omega)$.

(5) Setting $T(\alpha) = e^{\log\omega + \alpha\ell_{\omega_t|\omega}}$, we have $T(0) = \omega$ and

$$\partial_\alpha e_{\infty,t}(\alpha)\Big|_{\alpha=0} = \mathrm{tr}\,(T'(0)),$$

$$\partial_\alpha^2 e_{\infty,t}(\alpha)\Big|_{\alpha=0} = \mathrm{tr}\,(T''(0)) - (\mathrm{tr}\,(T'(0)))^2.$$

Iterating Duhamel's formula (recall Exercise 4.3.1), we can write

$$T(\alpha) = \omega + \alpha\int_0^1 \omega^{1-s}\ell_{\omega_t|\omega}\omega^s\,ds + \alpha^2\int_0^1\int_0^u \omega^{1-u}\ell_{\omega_t|\omega}\omega^s\ell_{\omega_t|\omega}\omega^{u-s}\,ds\,du + O(\alpha^3),$$

so that

$$\mathrm{tr}\,(T'(0)) = \int_0^1 \mathrm{tr}\left[\omega^{1-s}\ell_{\omega_t|\omega}\omega^s\right]\,ds = \omega(\ell_{\omega_t|\omega}) = S(\omega|\omega_t),$$

which proves (4) in the special case $p = \infty$, and

$$\mathrm{tr}\,(T''(0)) = 2\int_0^1\int_0^u \mathrm{tr}\left[\omega^{1-s}\ell_{\omega_t|\omega}\omega^s\ell_{\omega_t|\omega}\right]\,ds\,du$$

$$= 2\int_0^1\int_s^1 \mathrm{tr}\left[\omega^{1-s}\ell_{\omega_t|\omega}\omega^s\ell_{\omega_t|\omega}\right]\,du\,ds$$

$$= 2\int_0^1 (1 - s)\mathrm{tr}\left[\omega^{1-s}\ell_{\omega_t|\omega}\omega^s\ell_{\omega_t|\omega}\right]\,ds$$

$$= 2\int_0^1 s\,\mathrm{tr}\left[\omega^s\ell_{\omega_t|\omega}\omega^{1-s}\ell_{\omega_t|\omega}\right]\,ds.$$

Taking the mean of the last two expressions, we get

$$\mathrm{tr}\,(T''(0)) = \int_0^1 \mathrm{tr}\left[\omega^{1-s}\ell_{\omega_t|\omega}\omega^s\ell_{\omega_t|\omega}\right]ds$$

$$= \int_0^1 \omega\left(\varsigma_\omega^{\mathrm{is}}(\ell_{\omega_t|\omega})\ell_{\omega_t|\omega}\right)ds,$$

and hence

$$\partial_\alpha^2 e_{\infty,t}(\alpha)\Big|_{\alpha=0} = \int_0^1 \left[\omega\left(\varsigma_\omega^{\mathrm{is}}(\ell_{\omega_t|\omega})\ell_{\omega_t|\omega}\right) - \omega(\ell_{\omega_t|\omega})^2\right]ds.$$

(6) Follows easily from the fact that $e_{2,t}(\alpha) = S_\alpha(\omega_t|\omega) = \log\mathrm{tr}\,(\omega_t^\alpha\omega^{1-\alpha})$.
(7) Under the TRI assumption one has $\Theta(\omega) = \omega$, $\Theta(\omega_t) = \omega_{-t}$,

$$\Theta\left(\left(\omega^{(1-\alpha)/p}\omega_t^{2\alpha/p}\omega^{(1-\alpha)/p}\right)^{p/2}\right) = \left(\omega^{(1-\alpha)/p}\omega_{-t}^{2\alpha/p}\omega^{(1-\alpha)/p}\right)^{p/2},$$

and hence $e_{p,t}(\alpha) = e_{p,-t}(\alpha)$. The result now follows from Assertion (3).   □

According to our rule of thumb, we reformulate the definition of the functionals $e_{p,t}(\alpha)$ in terms which are susceptible to survive the thermodynamic limit. We first note that

$$e_{2,t}(\alpha) = S_\alpha(\omega_t|\omega) = \log(\xi_\omega|\Delta_{\omega_t|\omega}^\alpha\xi_\omega),$$

while Theorem 4.17 (2) yields the variational principle

$$e_{\infty,t}(\alpha) = \max_{\rho\in\mathfrak{S}} S(\rho|\omega) + \alpha\rho(\ell_{\omega_t|\omega}).$$

Moreover, equation (4.99) and Theorem 4.27 immediately lead to

$$e_{p,t}(\alpha) = \log\|\Delta_{\omega_t|\omega}^{\alpha/p}\xi_\omega\|_{\omega,p}^p,$$

for $p \in [1,\infty[$.

Exercise 4.39 Show that

$$e_{\infty,t}(\alpha) = \log(\xi_\omega|e^{\log\Delta_\omega + \alpha L(\ell_{\omega_t|\omega})}\xi_\omega).$$

Exercise 4.40 Show that the function $[1,\infty] \ni p \mapsto e_{p,t}(\alpha)$ is strictly decreasing unless H and ω commute.
Hint: recall Exercise 5.24.

4.4.4 Quantum transfer operators

For $p \in [1, \infty]$ we define a linear map $U_p(t) : \mathcal{H}_\mathcal{O} \to \mathcal{H}_\mathcal{O}$ by

$$U_p(t)\xi = e^{-itH} \xi \omega^{-\frac{1}{2}+\frac{1}{p}} e^{itH} \omega^{\frac{1}{2}-\frac{1}{p}}.$$

In terms of Connes cocycles and relative modular dynamics, one has

$$U_p(t)\xi = e^{-itH} \xi e^{itH} [D\omega_t : D\omega]^{i(\frac{1}{2}-\frac{1}{p})} = e^{-itH} \xi e^{it\varsigma_\omega^{i(\frac{1}{2}-\frac{1}{p})}(H)}. \tag{4.100}$$

One easily checks that $\mathbb{R} \ni t \mapsto U_p(t)$ is a group of operators on $\mathcal{H}_\mathcal{O}$ which satisfies

$$(\xi|U_p(t)\eta) = (U_q(-t)\xi|\eta), \tag{4.101}$$

for all $\xi, \eta \in \mathcal{H}_\mathcal{O}$ with $p^{-1} + q^{-1} = 1$. The following result elucidates the nature of this group: it is the unique isometric implementation of the dynamics on the Banach space $L^p(\mathcal{O}, \omega)$ which preserves the positive cone $L^p_+(\mathcal{O}, \omega)$.

Proposition 4.32 (1) $t \mapsto U_p(t)$ *is a group of isometries of* $L^p(\mathcal{O}, \omega)$.
(2) $U_p(t)L^p_+(\mathcal{O}, \omega) \subset L^p_+(\mathcal{O}, \omega)$.
(3) $U_p(-t)L(A)U_p(t) = L(\tau^t(A))$ *for any* $A \in \mathcal{O}$.
(4) $U_p(t)$ *is uniquely characterized by Properties (1)–(3).*

The groups U_p are natural noncommutative generalizations of the classical Ruelle transfer operators. We call L^p-*Liouvillean* of the quantum dynamical system $(\mathcal{O}, \tau^t, \omega)$ the generator L_p of U_p,

$$U_p(t) = e^{-itL_p}.$$

From equation (4.100) we immediately get

$$L_p\xi = H\xi - \xi\varsigma_\omega^{i(\frac{1}{2}-\frac{1}{p})}(H).$$

Interpreting (4.101) in terms of the duality between $L^p(\mathcal{O}, \omega)$ and $L^q(\mathcal{O}, \omega)$, we can write

$$L_p^* = L_q.$$

Note that, in the special case $p = 2$, $L_2 = L_2^*$ coincide with the standard Liouvillean K of the dynamics τ^t.

Theorem 4.33 *For any* $p \in [1, \infty]$ *one has*

$$\mathrm{sp}(L_p) = \mathrm{sp}(K) = \{\lambda - \mu \,|\, \lambda, \mu \in \mathrm{sp}(H)\}.$$

Exercise 4.41 This is the continuation of Exercise 4.4.2. Show that the L^p-Liouvillean of the perturbed dynamical system $(\mathcal{O}, \tau_V^t, \omega)$ is given by

$$L_p\xi = K\xi + V\xi - J\varsigma_\omega^{-i(\frac{1}{2}-\frac{1}{p})}(V)J\xi. \tag{4.102}$$

Interestingly enough, one can relate the groups U_p to the entropic pressure functionals introduced in the previous section. The resulting formulas are particularly well suited to investigate the large time limit of these functionals.

Theorem 4.34 *For $\alpha \in [0,1]$,*

$$e_{p,t}(\alpha) = \log \|e^{-itL_{p/\alpha}}\xi_\omega\|_{\omega,p}^p, \tag{4.103}$$

holds provided $p \in [1,\infty[$. In the special case $p = 2$, this reduces to

$$e_{2,t}(\alpha) = \log\left(\xi_\omega|e^{-itL_{1/\alpha}}\xi_\omega\right). \tag{4.104}$$

With the help of Theorem 4.27, the proof of the last theorem reduces to elementary calculations.

Proof of Proposition 4.32 and Theorem 4.33 Let K be the standard Liouvillean of $(\mathcal{O}, \tau^t, \omega)$. Since $e^{-itK}\xi = e^{-itH}\xi e^{itH}$, it is obvious that e^{-itK} is a group of isometries of $L^p(\mathcal{O})$ which preserves the positive cone $L^p_+(\mathcal{O})$. Denote by $V_p : L^p(\mathcal{O}) \to L^p(\mathcal{O}, \omega)$ the isometry defined in Theorem 4.27. Theorem 4.33 and Properties (1) and (2) of Proposition 4.32 follow from the facts that $U_p(t) = V_p e^{-itK}V_p^{-1}$ and $L^p_+(\mathcal{O}, \omega) = V_p L^p_+(\mathcal{O})$. To prove Property (3) we note that $V_p \in R(\mathcal{O}) = L(\mathcal{O})'$, so that

$$U_p(-t)L(A)U_p(t) = V_p e^{itK}V_p^{-1}L(A)V_p e^{-itK}V_p^{-1}$$
$$= V_p e^{itK}L(A)e^{-itK}V_p^{-1}$$
$$= V_p L(\tau^t(A))V_p^{-1} = L(\tau^t(A)).$$

(4) Let $\mathbb{R} \ni t \mapsto U^t$ be a group of linear operators on $\mathcal{H}_\mathcal{O}$ satisfying Properties (1)–(3) and set $V^t = L(e^{itH})U^t$. The group property implies that

$$(V^t)^{-1} = (U^t)^{-1}L(e^{itH})^{-1} = U^{-t}L(e^{-itH}),$$

so that, by Property (3),

$$L(\tau^t(A)) = U^{-t}L(A)U^t = (V^t)^{-1}L(e^{itH})L(A)L(e^{-itH})V^t$$
$$= (V^t)^{-1}L(e^{itH}Ae^{-itH})V^t = (V^t)^{-1}L(\tau^t(A))V^t,$$

for all $A \in \mathcal{O}$. Setting $A = \tau^{-t}(B)$ we conclude that

$$V^t L(B) = L(B)V^t,$$

for all $B \in \mathcal{O}$, i.e., $V^t \in L(\mathcal{O})' = R(\mathcal{O})$. Using the group property of U^t one easily shows that $t \mapsto V^t$ is also a group. It follows that $V^t = R(e^{it\widetilde{H}})$ for some $\widetilde{H} \in \mathcal{O}$. Thus, for any $A \in \mathcal{O}$, one has

$$U^t A\omega^{\frac{1}{2}-\frac{1}{p}} = e^{itH}A\omega^{\frac{1}{2}-\frac{1}{p}}e^{-it\widetilde{H}^*} = e^{itH}Ae^{-itH^\#}\omega^{\frac{1}{2}-\frac{1}{p}},$$

where $H^{\#} = \varsigma_\omega^{-\mathrm{i}(\frac{1}{2}-\frac{1}{p})}(\widetilde{H}^*)$. Exercise 5.34 (i) and Property (2) imply that, for any $A \in \mathcal{O}_+$, $\mathrm{e}^{\mathrm{i}tH}A\mathrm{e}^{-\mathrm{i}tH^{\#}} \in \mathcal{O}_+$. Since any self-adjoint element of $\mathcal{O}$ is a real linear combination of elements of $\mathcal{O}_+$, it follows that

$$\mathrm{e}^{\mathrm{i}tH}A\mathrm{e}^{-\mathrm{i}tH^{\#}} = \left(\mathrm{e}^{\mathrm{i}tH}A\mathrm{e}^{-\mathrm{i}tH^{\#}}\right)^* = \mathrm{e}^{\mathrm{i}tH^{\#*}}A\mathrm{e}^{-\mathrm{i}tH},$$

for any $A \in \mathcal{O}_{\mathrm{self}}$. This identity extends by linearity to arbitrary $A \in \mathcal{O}$. Differentiation at $t = 0$ yields

$$(H - H^{\#*})A = A(H^{\#} - H). \tag{4.105}$$

Setting $A = \mathbb{1}$, we deduce that $H^{\#} + H^{\#*} = 2H$, and hence that $H^{\#} = H + \mathrm{i}T$ with $T \in \mathcal{O}_{\mathrm{self}}$. Relation (4.105) now implies $TA = AT$ for all $A \in \mathcal{O}$ so that $T = \lambda \mathbb{1}$ for some $\lambda \in \mathbb{R}$. It follows that $H^{\#} = H + \mathrm{i}\lambda$ and hence $U^t = \mathrm{e}^{\lambda t}U_p(t)$. Property (1) finally imposes $\lambda = 0$. $\qquad\square$

4.4.5 Full counting statistics

The functional

$$e_{2,t}(\alpha) = S_\alpha(\omega_t|\omega) = \log(\xi_\omega|\Delta_{\omega_t|\omega}^\alpha \xi_\omega) = \log(\xi_\omega|\mathrm{e}^{\alpha \log \Delta_{\omega_t|\omega}}\xi_\omega),$$

can be interpreted in spectral terms. If we denote by Q^t the spectral measure of the self-adjoint operator

$$-\frac{1}{t}\log\Delta_{\omega_t|\omega} = -\frac{1}{t}\log\Delta_\omega - \frac{1}{t}L(\ell_{\omega_t|\omega}) = -\frac{1}{t}\log\Delta_\omega - L(\Sigma^{-t}),$$

for the vector ξ_ω then

$$e_{2,t}(\alpha) = \log\left[\int_{\mathbb{R}} \mathrm{e}^{-\alpha t s}\mathrm{d}Q^t(s)\right]. \tag{4.106}$$

As explained at the end of Section 4.2.6, the ES symmetry (4.98) can be expressed in terms of the measure Q^t in the following familiar form (see Tasaki and Matsui (2003)). Let $\mathfrak{r} : \mathbb{R} \to \mathbb{R}$ be the reflection $\mathfrak{r}(s) = -s$, and let $\bar{Q}^t = Q^t \circ \mathfrak{r}$ be the reflected spectral measure.

Proposition 4.35 *Suppose that $(\mathcal{O}, \tau^t, \omega)$ is TRI. Then the measures Q^t and $\bar{Q}^t$ are mutually absolutely continuous and*

$$\frac{\mathrm{d}\bar{Q}^t}{\mathrm{d}Q^t}(s) = \mathrm{e}^{-ts}. \tag{4.107}$$

The measure Q^t is not the spectral measure of any observable in $\mathcal{O}$ and on the first sight one may question its physical relevance. Its interpretation is somewhat striking and is linked to the concept of *Full Counting Statistics* (FCS) of repeated quantum measurement of the entropy observable $S = -\log \omega$. To our knowledge, this interpretation goes back to Kurchan (2000) (see also Dereziński *et al.* (2008)).

At time $t = 0$, with the system in the state ω, we perform a measurement of S. The possible outcomes of the measurement are eigenvalues of S and $s \in \mathrm{sp}(S)$ is observed with probability $\omega(P_s)$, where P_s is the spectral projection of S onto its eigenvalue s. After the measurement, the state of the system reduces to

$$\frac{\omega P_s}{\omega(P_s)},$$

and this state now evolves according to

$$\frac{e^{-itH} \omega P_s e^{itH}}{\omega(P_s)}.$$

A second measurement of S at time t yields the result $s' \in \mathrm{sp}(S)$ with probability

$$\frac{\mathrm{tr}\left(e^{-itH} \omega P_s e^{itH} P_{s'}\right)}{\omega(P_s)}.$$

Thus, the joint probability distribution of the two measurement is given by

$$\mathrm{tr}\left(e^{-itH} \omega P_s e^{itH} P_{s'}\right), \tag{4.108}$$

and the probability distribution of the mean rate of change of entropy, $\phi = (s' - s)/t$, is given by

$$\mathbb{P}_t(\phi) = \sum_{s'-s=t\phi} \mathrm{tr}\left(e^{-itH} \omega P_s e^{itH} P_{s'}\right).$$

It follows that

$$\mathrm{tr}(\omega_t^{1-\alpha} \omega^{\alpha}) = \sum_{s,s'} e^{-\alpha(s'-s)} \mathrm{tr}\left(e^{-itH} \omega P_s e^{itH} P_{s'}\right) = \sum_{\phi} \mathbb{P}_t(\phi) e^{-t\alpha\phi}.$$

and we conclude that

$$e_{2,-t}(\alpha) = e_{2,t}(1-\alpha) = \log\left[\sum_{\phi} \mathbb{P}_t(\phi) e^{-t\alpha\phi}\right].$$

Comparison with equation (4.106) allows us to conclude that the spectral measure $\bar{Q}^{-t}$ coincide with the distribution $\mathbb{P}_t(\phi)$. Consequently, applying Proposition 4.31, the expectation and variance of ϕ w.r.t. $\mathbb{P}_t$ are given by

$$\mathbb{E}_t(\phi) = -\frac{1}{t}\partial_\alpha e_{2,-t}(\alpha)\Big|_{\alpha=0} = -\frac{1}{t}\omega(\ell_{\omega-t|\omega}) = \omega(\Sigma^t),$$

$$\mathbb{E}_t(\phi^2) - \mathbb{E}_t(\phi)^2 = \frac{1}{t^2}\partial_\alpha^2 e_{2,-t}(\alpha)\Big|_{\alpha=0} = \frac{1}{t^2}\left(\omega(\ell_{\omega-t|\omega}^2) - \omega(\ell_{\omega-t|\omega})^2\right) = \omega(\Sigma^{t^2}) - \omega(\Sigma^t)^2.$$

They coincide with the expectation and variance of Σ^t w.r.t. ω. However, we warn the reader that such a relation does not hold true for higher order cumulants.

Note that time-reversal invariance played no role in the identification of $\bar{Q}^{-t}$ with $\mathbb{P}_t(\phi)$. However if $(\mathcal{O}, \tau, \omega)$ is TRI, then $\bar{Q}^{-t} = Q^t$ and Proposition 4.35 translates into the fluctuation relation

$$\frac{\mathbb{P}_t(-\phi)}{\mathbb{P}_t(\phi)} = e^{-t\phi}, \tag{4.109}$$

where $\phi \in (\mathrm{sp}(S) - \mathrm{sp}(S))/t$.

4.4.6 On the choice of reference state

Starting with entropy production, all the objects that we have introduced so far depend on the choice of the reference state ω. In this subsection we shall indicate by a subscript this dependence on ω (hence, σ_ω is the entropy production of $(\mathcal{O}, \tau^t, \omega)$, etc.).

If ω and ρ are two faithful states on $\mathcal{O}$, then

$$\sigma_\omega - \sigma_\rho = i[\ell_{\omega|\rho}, H] = -\frac{d}{dt}\tau^t(\ell_{\omega|\rho})\Big|_{t=0},$$

and hence

$$\Sigma^t_\omega - \Sigma^t_\rho = \frac{1}{t}\int_0^t \frac{d}{ds}\tau^s(\ell_{\omega|\rho})\,ds = \frac{\tau^t(\ell_{\omega|\rho}) - \ell_{\omega|\rho}}{t}.$$

Consequently,

$$\|\Sigma^t_\omega - \Sigma^t_\rho\| = \|\ell_{\omega|\rho}\|O(t^{-1}).$$

Thus, Σ^t_ω and Σ^t_ρ become indistinguishable for large t. A similar result holds for the properly normalized entropic functionals. For example:

Proposition 4.36 *For all $\alpha \in \mathbb{R}$ and $t \in \mathbb{R}$ one has the estimate*

$$\left|\frac{1}{t}e_{\infty,t,\omega}(\alpha) - \frac{1}{t}e_{\infty,t,\rho}(\alpha)\right| \le (|1 - \alpha| + |\alpha|)\frac{\|\ell_{\omega|\rho}\|}{t}.$$

Proof: We have

$$\mathrm{tr}(e^{\log\omega + \alpha\ell_{\omega_t|\omega}}) = \mathrm{tr}(e^{\log\rho + \alpha\ell_{\rho_t|\rho} + (1-\alpha)\ell_{\omega|\rho} + \alpha\ell_{\omega_t|\rho_t}})$$

$$\le \mathrm{tr}(e^{\log\rho + \alpha\ell_{\rho_t|\rho}}e^{(1-\alpha)\ell_{\omega|\rho} + \alpha\ell_{\omega_t|\rho_t}})$$

$$\le e^{(|1-\alpha|+|\alpha|)\|\ell_{\omega|\rho}\|}\mathrm{tr}(e^{\log\rho + \alpha\ell_{\rho_t|\rho}}),$$

where we have used the Golden–Thompson inequality (Corollary 4.3). Taking logarithms, we get

$$e_{\infty,t,\omega}(\alpha) - e_{\infty,t,\rho}(\alpha) \le (|1 - \alpha| + |\alpha|)\|\ell_{\omega|\rho}\|.$$

Reversing the roles of w and ρ and using that $\|\ell_{w|\rho}\| = \|\ell_{\rho|w}\|$ we deduce the statement.

$\square$

4.4.7   Compound systems

Consider the quantum dynamical system $(\mathcal{O}, \tau^t, w)$ describing a compound system made of n subsystems. The underlying Hilbert space is given by a tensor product

$$\mathcal{K} = \bigotimes_{j=1}^{n} \mathcal{K}_j,$$

and

$$\mathcal{O} = \bigotimes_{j=1}^{n} \mathcal{O}_j, \tag{4.110}$$

where $\mathcal{O}_j = \mathcal{O}_{\mathcal{K}_j}$ is the algebra of observables of the j-th subsystem. We identify $A_j \in \mathcal{O}_j$ with $\mathbb{1}_{\otimes_{i=1}^{j-1}\mathcal{K}_i} \otimes A_j \otimes \mathbb{1}_{\otimes_{i=j+1}^{n}\mathcal{K}_i} \in \mathcal{O}$.

We assume that the reference state w has the product structure

$$w(A_1 \otimes \cdots \otimes A_n) = \prod_{j=1}^{n} w_j(A_j), \tag{4.111}$$

where w_j is a faithful state on $\mathcal{O}_j$. According to the above convention, w_j is identified with the positive operator $\mathbb{1}_{\otimes_{i=1}^{j-1}\mathcal{K}_i} \otimes w_j \otimes \mathbb{1}_{\otimes_{i=j+1}^{n}\mathcal{K}_i}$, so that $\log w_j$ is a self-adjoint element of $\mathcal{O}$ and

$$\log w = \sum_{j=1}^{n} \log w_j.$$

Accordingly, the entropy production observable of the system can be written as

$$\sigma = i[\log w, H] = \sum_{j} \sigma_j,$$

where $\sigma_j = i[\log w_j, H]$. Similarly, the relative Hamiltonian $\ell_{w_t|w}$ decomposes as

$$\ell_{w_t|w} = \sum_{j=1}^{n} \ell_{w_{jt}|w_j},$$

where

$$\ell_{w_{jt}|w_j} = \tau^{-t}(\log w_j) - \log w_j = \int_0^t \tau^{-s}(\sigma_j)\mathrm{d}s.$$

If the system $(\mathcal{O}, \tau^t, w)$ is TRI with time-reversal Θ, we shall always assume that

$$\Theta(w_j) = w_j.$$

This implies

$$\Theta(\sigma_j) = -\sigma_j.$$

For $\boldsymbol{\alpha} = (\alpha_1, \cdots, \alpha_n) \in \mathbb{R}^n$ we denote $\omega^{\boldsymbol{\alpha}} = \omega_1^{\alpha_1} \cdots \omega_n^{\alpha_n}$. Similarly,

$$\omega_t^{\boldsymbol{\alpha}} = \mathrm{e}^{-\mathrm{i}tH} \omega^{\boldsymbol{\alpha}} \mathrm{e}^{\mathrm{i}tH} = \prod_{j=1}^{n} \omega_{jt}^{\alpha_j}.$$

We also denote $\mathbf{1} = (1, \ldots, 1)$ and $\mathbf{0} = (0, \ldots, 0)$. The multi-parameter entropic pressure functionals are defined for $t \in \mathbb{R}$ and $\boldsymbol{\alpha} \in \mathbb{R}^n$ by

$$e_{p,t}(\boldsymbol{\alpha}) = \begin{cases} \log \mathrm{tr}\left[\left(\omega^{\frac{1-\boldsymbol{\alpha}}{p}} \omega_t^{\frac{2\boldsymbol{\alpha}}{p}} \omega^{\frac{1-\boldsymbol{\alpha}}{p}} \right)^{\frac{p}{2}} \right] & \text{for } 1 \le p < \infty, \\[2em] \log \mathrm{tr}\left(\mathrm{e}^{\log \omega + \sum_j \alpha_j \ell_{\omega_{jt}|\omega_j}} \right) & \text{for } p = \infty. \end{cases}$$

These functionals are natural generalizations of the functionals introduced in Section 4.4.3 and have very similar properties:

Proposition 4.37   *(1) The function $[1, \infty] \ni p \mapsto e_{p,t}(\boldsymbol{\alpha})$ is continuous and monotonically increasing.*
 (2) The function $\mathbb{R}^n \ni \boldsymbol{\alpha} \mapsto e_{p,t}(\boldsymbol{\alpha})$ is real-analytic, convex, and $e_{p,t}(\mathbf{0}) = e_{p,t}(\mathbf{1}) = 0$.
 (3) $e_{p,t}(\boldsymbol{\alpha}) = e_{p,-t}(\mathbf{1} - \boldsymbol{\alpha})$.
 (4) $\partial_{\alpha_j} e_{p,t}(\boldsymbol{\alpha})|_{\boldsymbol{\alpha}=0} = \omega(\ell_{\omega_{jt}|\omega_j})$.
 (5)

$$\partial_{\alpha_k} \partial_{\alpha_j} e_{\infty,t}(\boldsymbol{\alpha})|_{\boldsymbol{\alpha}=0} = \langle \ell_{\omega_{kt}|\omega_k} | \ell_{\omega_{jt}|\omega_j} \rangle_\omega - \omega(\ell_{\omega_{kt}|\omega_k}) \omega(\ell_{\omega_{jt}|\omega_j}).$$

 (6)

$$\partial_{\alpha_k} \partial_{\alpha_j} e_{2,t}(\boldsymbol{\alpha})|_{\boldsymbol{\alpha}=0} = \frac{1}{2} \int_0^t \int_0^t \omega\left((\sigma_{ks} - \omega(\sigma_{ks}))(\sigma_{ju} - \omega(\sigma_{ju})) \right) \mathrm{d}s \mathrm{d}u.$$

 (7) If $(\mathcal{O}, \tau^t, \omega)$ is TRI, then the finite time Evans–Searles (ES) symmetry holds:

$$e_{p,t}(\boldsymbol{\alpha}) = e_{p,t}(\mathbf{1} - \boldsymbol{\alpha}).$$

The proof, which is similar to the proof of Proposition 4.31, is left as an exercise.

In order to express the multi-parameter entropic pressure functionals in terms of the modular structure of $(\mathcal{O}, \omega)$, we have to extend the definition of relative modular operator. Let us briefly indicate how to proceed. The main problem is that ω_j is not a state on $\mathcal{O}$ (it is not properly normalized, and cannot be normalized in the thermodynamic limit since the dimensions of the Hilbert spaces $\mathcal{K}_i$ diverge in this limit). However, as a state on $\mathcal{O}_j$, ω_j has a modular group ς_{ω_j} and a modular operator Δ_{ω_j} such that

$$\varsigma_{\omega_j}^s(A) = \Delta_{\omega_j}^{is} A \Delta_{\omega_j}^{-is}.$$

The formula

$$\mathbb{R}^n \ni \mathbf{s} = (s_1, \dots, s_n) \mapsto \varsigma_\omega^s = \bigotimes_{j=1}^n \varsigma_{\omega_j}^{s_j},$$

defines an abelian group of $*$-automorphisms of $\mathcal{O}$. With a slight abuse of language, we shall refer to the multi-parameter group ς_ω^s as the modular group of ω. We denote by

$$\Delta_\omega^{is} = \bigotimes_{j=1}^n \Delta_{\omega_j}^{is_j}.$$

the corresponding abelian unitary group. Setting

$$\varsigma_{\omega_t}^s = \tau^{-t} \circ \varsigma_\omega^s \circ \tau^t,$$

we clearly have $\varsigma_\omega^s(A) = \omega^{is} A \omega^{-is}$ and $\varsigma_{\omega_t}^s(A) = \omega_t^{is} A \omega_t^{-is}$.

The two modular groups ς_ω^s and $\varsigma_{\omega_t}^s$ are related by

$$\varsigma_{\omega_t}^s(A) = [D\omega_t : D\omega]^s \varsigma_\omega^s(A) [D\omega : D\omega_t]^s,$$

where the unitary Connes cocycle

$$[D\omega_t : D\omega]^s = \omega_t^{is} \omega^{-is} = e^{i \sum_j s_j \tau^{-t}(\log \omega_j)} e^{-i \sum_j s_j \log \omega_j} = e^{-itH} e^{its_\omega^s(H)}, \quad (4.112)$$

satisfies the two multiplicative cocycle relations

$$[D\omega_t : D\omega]^s \varsigma_\omega^s([D\omega_t : D\omega]^{s'}) = [D\omega_t : D\omega]^{s+s'},$$

$$\tau^{-t}([D\omega_{t'} : D\omega]^s)[D\omega_t : D\omega]^s = [D\omega_{t+t'} : D\omega]^s. \quad (4.113)$$

Thanks to the first relation,

$$\mathbb{R}^n \ni \mathbf{s} \mapsto \Delta_{\omega_t | \omega}^{is} = L([D\omega_t : D\omega]^s) \Delta_\omega^{is},$$

defines an abelian group of unitaries on $\mathcal{H}_\mathcal{O}$. One easily checks that $\Delta_{\omega_t | \omega}^{is} \xi = \omega_t^{is} \xi \omega^{-is}$. The relative Hamiltonian $\ell_{\omega_{jt} | \omega_j} = \tau^{-t}(\log \omega_j) - \log \omega_j$ is given by

$$\ell_{\omega_{jt} | \omega_j} = \frac{1}{i} \frac{d}{ds_j} [D\omega_t : D\omega]^s \Big|_{s=0}.$$

Using Theorem 4.27 and the fact that $\Delta_\omega^{\alpha/p} \xi_\omega = \xi_\omega$ it is now easy to show that, for $p \in [1, \infty[$,

$$e_{p,t}(\alpha) = \log \|\Delta_{\omega_t | \omega}^{\alpha/p} \xi_\omega\|_{\omega, p}^p = \log \|[D\omega_t : D\omega]^{-i\alpha/p} \xi_\omega\|_{\omega, p}^p, \quad (4.114)$$

while Theorem 4.17 leads to

$$e_{\infty,t}(\boldsymbol{\alpha}) = \max_{\rho \in \mathfrak{S}} \left(S(\rho|\omega) + \sum_{j=1}^{n} \alpha_j \rho(\ell_{\omega_{jt}|\omega_j}) \right).$$

In particular, one has

$$e_{2,t}(\boldsymbol{\alpha}) = \log(\xi_\omega|\Delta^{\alpha}_{\omega_t|\omega}\xi_\omega) = \log \omega([D\omega_t : D\omega]^{-i\alpha}).$$

One can also generalize Theorem 4.34 to the present setup. To this end, let K be the standard Liouvillean of the dynamics τ^t. With $\mathbf{s} \in \mathbb{R}^n$, the second cocycle relation (4.113) allows us to construct the unitary group

$$e^{-itK_{\mathbf{s}}} = R([D\omega_t : D\omega]^{\mathbf{s}})^* e^{-itK},$$

on $\mathcal{H}_\mathcal{O}$. By (4.112), one has

$$e^{-itK_{\mathbf{s}}}\xi = e^{-itH}\xi e^{itH}[D\omega_t : D\omega]^{\mathbf{s}} = e^{-itH}\xi e^{it\varsigma^{\mathbf{s}}_\omega(H)},$$

so that $K_{\mathbf{s}} = L(H) - R(\varsigma^{\mathbf{s}}_\omega(H))$. Analytic continuation of $e^{-itK_{\mathbf{s}}}$ to $\mathbf{s} = i(1/2 - 1/p)\mathbf{1}$ with $p \in [1, \infty]$ yields the group $U_p(t)$ of isometric implementation of the dynamics on the Araki-Masuda space $L^p(\mathcal{O}, \omega)$ introduced in Section 4.4.4.

For $\boldsymbol{\alpha} \in [0,1]^n$ and $p \in [1, \infty]$, let us define

$$L_{\frac{p}{\alpha}} = K_{\mathbf{s}}, \qquad \mathbf{s} = i\left(\frac{1}{2} - \frac{\boldsymbol{\alpha}}{p}\right).$$

From the identity

$$e^{-itL_{\frac{p}{\alpha}}}\xi_\omega = \omega_t^{\alpha/p}\omega^{1/2-\alpha/p} = [D\omega_t : D\omega]^{-i\alpha/p}\xi_\omega,$$

and equation (4.114) we deduce

$$e_{p,t}(\boldsymbol{\alpha}) = \log \|e^{-itL_{\frac{p}{\alpha}}}\xi_\omega\|^p_{\omega,p}.$$

In the special case $p = 2$, this can be rewritten as

$$e_{2,t}(\boldsymbol{\alpha}) = \log(\xi_\omega|e^{-itL_{\frac{1}{\alpha}}}\xi_\omega).$$

Exercise 4.42 Show that the Connes cocycle $\Gamma(\mathbf{s}, t) = [D\omega_t : D\omega]^{\mathbf{s}}$ satisfies the following differential equations,

$$-i\frac{d}{dt}\Gamma(\mathbf{s}, t) = \tau^{-t}(\varsigma^{\mathbf{s}}_\omega(H) - H)\Gamma(\mathbf{s}, t), \qquad \Gamma(\mathbf{s}, 0) = 1,$$

$$-i\frac{d}{ds_j}\Gamma(\mathbf{s}, t) = \Gamma(\mathbf{s}, t)\varsigma^{\mathbf{s}}_\omega(\tau^{-t}(\log \omega_j) - \log \omega_j), \qquad \Gamma(0, t) = 1.$$

Exercise 4.43 Assume that $H = H_0 + V$ with $[H_0, \omega] = 0$, *i.e.*, ω is a steady state for the dynamics τ_0^t generated by H_0. Show that

$$L_{\frac{1}{\alpha}} = K_0 + L(V) - R(\varsigma_\omega^{\mathrm{i}(\alpha - 1/2)}(V)),$$

where K_0 is the standard Liouvillean of τ_0^t.

4.4.8 Multi-parameter full counting statistics

We continue with the framework of the last section and extend to compound systems our discussion of full counting statistics started in Section 4.4.5.

With $\mathbf{1}_j = (0, \ldots, 1, \ldots, 0)$ (a single 1 at the j-th entry) we set

$$\Delta_{\omega_j t | \omega_j} = \Delta_{\omega_t | \omega}^{\mathbf{1}_j}.$$

In terms on the joint spectral measure Q^t of the commuting family of self-adjoint operators

$$-\frac{1}{t} \log \Delta_{\omega_1 t | \omega_1}, \ldots, -\frac{1}{t} \log \Delta_{\omega_n t | \omega_n},$$

associated to the vector ξ_ω one has, for $\boldsymbol{\alpha} \in \mathbb{R}^n$,

$$(\xi_\omega | \Delta_{\omega_t | \omega}^{\boldsymbol{\alpha}} \xi_\omega) = (\xi_\omega | e^{\sum_j \alpha_j \log \Delta_{\omega_j t | \omega_j}} \xi_\omega) = \int e^{-t\boldsymbol{\alpha} \cdot \mathbf{s}} \, dQ^t(\mathbf{s}).$$

Let $\mathfrak{r}$ denote the reflection $\mathfrak{r}(\mathbf{s}) = -\mathbf{s}$ on $\mathbb{R}^n$, and let $\bar{Q}^t = Q^t \circ \mathfrak{r}$ be the reflected spectral measure. The ES symmetry $e_{2,t}(\mathbf{1} - \boldsymbol{\alpha}) = e_{2,t}(\boldsymbol{\alpha})$ translates into

Proposition 4.38 *Suppose that* $(\mathcal{O}, \tau^t, \omega)$ *is TRI. Then the measures* Q^t *and* $\bar{Q}^t$ *are mutually absolutely continuous and*

$$\frac{d\bar{Q}^t}{dQ^t}(\mathbf{s}) = e^{-t\mathbf{1} \cdot \mathbf{s}}. \tag{4.115}$$

To interpret this result, considered the vector observable

$$\mathbf{S} = (-\log \omega_1, \cdots, -\log \omega_n).$$

Since the ω_j's commute, the components of $\mathbf{S}$ can be simultaneously measured. Let $P_\mathbf{s}$ denote the joint spectral projection of $\mathbf{S}$ to the eigenvalue $\mathbf{s} \in \mathrm{sp}(\mathbf{S})$. The joint probability distribution of two measurements is

$$\mathrm{tr}\left(e^{-\mathrm{i}tH} \omega P_\mathbf{s} e^{\mathrm{i}tH} P_{\mathbf{s}'}\right).$$

Denote by $\mathbb{P}_t(\boldsymbol{\phi})$ the induced probability distribution of the vector $\boldsymbol{\phi} = (\mathbf{s}' - \mathbf{s})/t$ which describes the mean rate of change of $\mathbf{S}$ between the two measurements. For $\boldsymbol{\alpha} \in \mathbb{R}^n$ one has, by Proposition 4.37 (3),

$$(\xi_\omega | \Delta^\alpha_{\omega_{-t}|\omega} \xi_\omega) = (\xi_\omega | \Delta^{1-\alpha}_{\omega_t|\omega} \xi_\omega) = \mathrm{tr}(\omega_t^{1-\alpha} \omega^\alpha)$$

$$= \sum_{s,s'} \mathrm{e}^{-\sum_j \alpha_j (s'_j - s_j)} \mathrm{tr}(\mathrm{e}^{-\mathrm{i}tH} \omega P_s \mathrm{e}^{\mathrm{i}tH} P_{s'})$$

$$= \sum_\phi \mathrm{e}^{-\sum_j t\alpha_j \phi_j} \mathbb{P}_t(\phi).$$

As in Section 4.4.5, we can conclude that the spectral measure $\bar{Q}^{-t}$ coincides with the probability distribution $\mathbb{P}_t$. Assertion (4) and (6) of Proposition 4.37 yield the expectation and covariance of ϕ w.r.t. $\mathbb{P}_t$,

$$\mathbb{E}_t(\phi_j) = -\frac{1}{t} \partial_{\alpha_j} e_{2,-t}(\alpha)\Big|_{\alpha=0} = -\frac{1}{t} \omega(\ell_{\omega_{j(-t)}|\omega_j}) = \frac{1}{t} \int_0^t \omega(\sigma_{js}) \mathrm{d}s,$$

$$\mathbb{E}_t(\phi_j \phi_k) - \mathbb{E}_t(\phi_j)\mathbb{E}_t(\phi_k) = \frac{1}{t^2} \partial_{\alpha_j} \partial_{\alpha_k} e_{2,-t}(\alpha)\Big|_{\alpha=0}$$

$$= \frac{1}{2t^2} \int_0^t \int_0^t \omega\left((\sigma_{js} - \omega(\sigma_{js}))(\sigma_{ku} - \omega(\sigma_{ku}))\right) \mathrm{d}s\mathrm{d}u.$$

If the system is TRI then $\bar{Q}^{-t} = Q^t$ and Proposition 4.38 yields the ES fluctuation relation

$$\frac{\mathbb{P}_t(-\phi)}{\mathbb{P}_t(\phi)} = \mathrm{e}^{-t\mathbf{1}\cdot\phi}.$$

Exercise 4.44 The above formula for the covariance of the full counting statistics implies that

$$A_{jk} = \int_0^t \int_0^t \omega\left((\sigma_{js} - \omega(\sigma_{js}))(\sigma_{ku} - \omega(\sigma_{ku}))\right) \mathrm{d}s\mathrm{d}u,$$

is symmetric, $A_{jk} = A_{kj}$. Prove this directly, starting from the definition $\sigma_j = -\mathrm{i}[H, \log \omega_j]$. *Hint*: show that

$$\int_0^t \int_0^t [\sigma_{js}, \sigma_{ku}] \mathrm{d}s\mathrm{d}u = [\log \omega_j, \log \omega_k] + \tau^t([\log \omega_j, \log \omega_k])$$

$$- [\tau^t(\log \omega_j), \log \omega_k] - [\log \omega_j, \tau^t(\log \omega_k)].$$

Exercise 4.45 Check that the tensor product structure (4.110) was never used in the last two sections. More precisely, replacing Assumption (4.111) with

$$\log \omega = \sum_{j=1}^n Q_j,$$

where $(Q_1, \ldots, Q_n)$ is a commuting family of self-adjoint elements of $\mathcal{O}$, and defining $\omega_j = e^{Q_j}$ so that

$$\omega^{\alpha} = e^{\sum_{j=1}^{n} \alpha_j Q_j},$$

show that all the results of the two sections hold without modification.

4.4.9   Control parameters and fluxes

Suppose that our quantum dynamical system $(\mathcal{O}_X, \tau_X, \omega_X)$ depends on some control parameters $X = (X_1, \cdots, X_n) \in \mathbb{R}^n$. One can think of X_j's as mechanical or thermo-dynamical forces acting on the system. We denote by H_X the Hamiltonian generating the dynamics τ_X^t, by σ_X the entropy production observable and so on. We assume that ω_0 is τ_0^t invariant and refer to the value $X = 0$ as *equilibrium*. Note that this implies $\sigma_0 = 0$. We adopt the shorthands $\tau^t = \tau_0^t$, $\omega = \omega_0$.

Definition 4.39 *A vector-valued observable* $\mathbf{\Phi}_X = (\Phi_X^{(1)}, \cdots, \Phi_X^{(n)}) \in \mathcal{O}_{\mathrm{self}}^n$, *is called a flux relation if, for all* X,

$$\sigma_X = \sum_{j=1}^{n} X_j \Phi_X^{(j)}. \tag{4.116}$$

In what follows we will consider a family of quadruples $(\mathcal{O}, \tau_X^t, \omega_X, \mathbf{\Phi}_X)_{X \in \mathbb{R}^n}$, where $\mathbf{\Phi}_X$ is a given flux relation. Somewhat colloquially, we will refer to $\Phi_X^{(j)}$ as the flux (or current) observable associated to the force X_j. In concrete models, physical requirements typically select a unique flux relation $\mathbf{\Phi}_X$.

If $(\mathcal{O}_X, \tau_X^t, \omega_X)_{X \in \mathbb{R}^N}$ are time-reversal invariant (TRI), we shall always assume that

$$\Theta_X(\mathbf{\Phi}_X) = -\mathbf{\Phi}_X. \tag{4.117}$$

This assumption implies that $\omega_X(\mathbf{\Phi}_X) = 0$ for all X.

Notation. For $\nu \in \mathfrak{G}$, $\vartheta \in \mathrm{Aut}(\mathcal{O})$, $\mathbf{A} = (A_1, \ldots, A_n) \in \mathcal{O}^n$, and $Y = (Y_1, \ldots, Y_n) \in \mathbb{C}^n$ we shall use the shorthands

$$\nu(\mathbf{A}) = (\nu(A_1), \cdots, \nu(A_n)) \in \mathbb{C}^n,$$

$$\vartheta(\mathbf{A}) = (\vartheta(A_1), \cdots \vartheta(A_n)) \in \mathcal{O}^n,$$

$$\tau^t(\mathbf{A}) = \mathbf{A}_t = (\tau^t(A_1), \cdots, \tau^t(A_n)) \in \mathcal{O}^n,$$

$$Y \cdot \mathbf{A} = \sum_{j=1}^{n} Y_j A_j \in \mathcal{O}.$$

The relative Hamiltonian of ω_{Xt} w.r.t. ω_X is given by

$$\ell_{\omega_{Xt}|\omega_X} = \int_0^t \tau_X^{-s}(\sigma_X)\,\mathrm{d}s = X \cdot \int_0^t \Phi_{X(-s)}\,\mathrm{d}s = \sum_{j=1}^n X_j \int_0^t \tau_X^{-s}(\Phi_X^{(j)})\,\mathrm{d}s.$$

We generalize the $p = \infty$ entropic pressure functional

$$e_{\infty,t}(\alpha) = \log\mathrm{tr}\left(e^{\log\omega_X + \alpha\ell_{\omega_{Xt}|\omega_X}}\right),$$

by introducing

$$e_t(X,Y) = \log\mathrm{tr}\left(e^{\log\omega_X + Y\cdot\int_0^t \Phi_{X(-s)}\,\mathrm{d}s}\right), \tag{4.118}$$

where $Y \in \mathbb{R}^n$. The basic properties of $e_t(X,Y)$ are summarized in the next proposition.

Proposition 4.40   *(1)*

$$e_t(X,Y) = \sup_{\nu\in\mathfrak{S}}\left[S(\nu|\omega_X) + Y\cdot\int_0^t \nu(\Phi_{X(-s)})\,\mathrm{d}s\right]. \tag{4.119}$$

(2) The function $\mathbb{R}^n \ni Y \mapsto e_t(X,Y)$ is convex and real analytic.
(3) $e_{-t}(X,Y) = e_t(X, X - Y)$.
(4)

$$\partial_{Y_j} e_t(X,Y)\big|_{Y=0} = \int_0^t \omega_X(\Phi_{X(-s)}^{(j)})\,\mathrm{d}s, \tag{4.120}$$

$$\partial_{Y_k}\partial_{Y_j} e_t(X,Y)\big|_{Y=0} = \int_0^t\int_0^t \left(\langle\Phi_{X(-s_1)}^{(k)}|\Phi_{X(-s_2)}^{(j)}\rangle_{\omega_X}\right.$$
$$\left. - \omega_X(\Phi_{X(-s_1)}^{(k)})\omega_X(\Phi_{X(-s_2)}^{(j)})\right)\,\mathrm{d}s_2\mathrm{d}s_1. \tag{4.121}$$

(5) If $(\mathcal{O}_X,\tau_X^t,\omega_X)_{X\in\mathbb{R}^n}$ is TRI, then $e_{-t}(X,Y) = e_t(X,Y)$ and

$$e_t(X,Y) = e_t(X, X - Y). \tag{4.122}$$

We shall refer to Relation (4.122) as the finite time Generalized Evans–Searles (GES) symmetry. Notice that

$$e_t(X,\alpha X) = \log\mathrm{tr}(e^{\log\omega_X + \alpha\ell_{\omega_{Xt}|\omega_X}}) = e_{\infty,t}(\alpha),$$

which shows that the ES symmetry of $e_{\infty,t}(\alpha) = e_{\infty,t}(1-\alpha)$ is a special case of the GES-symmetry.

Proof: (1) follows from Theorem 4.17. (2) Convexity follows from (1) and analyticity is obvious. (3) is a consequence of the following elementary calculation:

$$
\log \omega_X + (X - Y) \cdot \int_0^t \mathbf{\Phi}_{X(-s)} \mathrm{d}s = \log \omega_X + \ell_{\omega_{Xt}|\omega_X} - Y \cdot \int_0^t \mathbf{\Phi}_{X(-s)} \mathrm{d}s
$$

$$
= \log \omega_{Xt} - Y \cdot \int_0^t \mathbf{\Phi}_{X(-s)} \mathrm{d}s
$$

$$
= \mathrm{e}^{-\mathrm{i}tH_X} \left(\log \omega_X - Y \cdot \int_0^t \mathbf{\Phi}_{X(t-s)} \mathrm{d}s \right) \mathrm{e}^{\mathrm{i}tH_X}
$$

$$
= \mathrm{e}^{-\mathrm{i}tH_X} \left(\log \omega_X - Y \cdot \int_0^t \mathbf{\Phi}_{Xs} \mathrm{d}s \right) \mathrm{e}^{\mathrm{i}tH_X}
$$

$$
= \mathrm{e}^{-\mathrm{i}tH_X} \left(\log \omega_X + Y \cdot \int_0^{-t} \mathbf{\Phi}_{X(-s)} \mathrm{d}s \right) \mathrm{e}^{\mathrm{i}tH_X}.
$$

To prove (4) invoke Duhamel formula to differentiate (4.118) (see the proof of Assertion (5) of Proposition 4.31). (5) follows from (2) and Assumption (4.117) which implies that $\Theta_X(\mathbf{\Phi}_{X(-s)}) = -\mathbf{\Phi}_{Xs}$, so that

$$
\Theta_X \left(\log \omega_X + Y \cdot \int_0^t \mathbf{\Phi}_{X(-s)} \mathrm{d}s \right) = \log \omega_X - Y \cdot \int_0^t \mathbf{\Phi}_{Xs} \mathrm{d}s
$$

$$
= \log \omega_X + Y \cdot \int_0^{-t} \mathbf{\Phi}_{X(-s)} \mathrm{d}s.
$$

$\square$

4.4.10   Finite time linear response theory

Finite time linear response theory is concerned with the first order perturbation theory (w.r.t. X) of the expectation values

$$
\langle \mathbf{\Phi}_X \rangle_t = \frac{1}{t} \int_0^t \omega_X(\mathbf{\Phi}_{Xs}) \mathrm{d}s.
$$

In the discussion of linear response theory we shall always assume that functions

$$
X \mapsto H_X, \qquad X \mapsto \omega_X, \qquad X \mapsto \mathbf{\Phi}_X, \tag{4.123}
$$

are continuously differentiable. This implies that the function $X \mapsto \langle \mathbf{\Phi}_X \rangle_t$ is continuously differentiable for all t.

The finite time kinetic transport coefficients are defined by

$$
L_{jkt} = \partial_{X_k} \langle \mathbf{\Phi}_X^{(j)} \rangle_t \big|_{X=0}.
$$

Since

$$\langle \sigma_X \rangle_t = \sum_j X_j \langle \Phi_X^{(j)} \rangle_t = \sum_{j,k} L_{jkt} X_j X_k + o(|X|^2) \geq 0, \tag{4.124}$$

the *real* quadratic form determined by the finite time Onsager matrix $[L_{jkt}]$ is positive definite. This fact does not depend on the TRI assumption and *does not* imply that $L_{jkt} = L_{kjt}$. We shall call the relations

$$L_{jkt} = L_{kjt}$$

the finite time Onsager reciprocity relations (ORR). As a general structural relations, they can hold only for TRI systems.

Another direct consequence of (4.124) is:

Proposition 4.41 *Let* $\Phi_X, \widetilde{\Phi}_X$ *be two flux relations. Then the corresponding finite time transport coefficients satisfy*

$$L_{jkt} + L_{kjt} = \widetilde{L}_{jkt} + \widetilde{L}_{kjt}.$$

If the finite time ORR hold, then $L_{jkt} = \widetilde{L}_{jkt}.$

The next proposition shows that the finite time ORR and Green-Kubo formula follow from the finite time GES symmetry. Recall our notational convention $\tau_{X=0}^t = \tau^t$, $\omega_{X=0} = \omega$, $\Phi_{X=0}^{(j)} = \Phi^{(j)}$, etc.

Proposition 4.42 *If* $(\mathcal{O}_X, \tau_X^t, \omega_X)_{X \in \mathbb{R}^n}$ *is TRI, then*

(1)

$$L_{jkt} = \frac{1}{2} \int_{-t}^{t} \left(1 - \frac{|s|}{t} \right) \langle \Phi^{(k)} | \tau^s(\Phi^{(j)}) \rangle_\omega \, ds.$$

(2) $L_{jkt} = L_{kjt}.$

Proof: By Relation (4.120) and the TRI property one has

$$\langle \Phi_X^{(j)} \rangle_t = -\partial_{Y_j} \frac{1}{t} e_t(X, Y)\big|_{Y=0},$$

so that

$$L_{jkt} = -\partial_{X_k} \partial_{Y_j} \frac{1}{t} e_t(X, Y)\big|_{X=Y=0}.$$

The GES-symmetry implies that

$$-\partial_{X_k} \partial_{Y_j} \frac{1}{t} e_t(X, Y)\big|_{X=Y=0} = \frac{1}{2t} \partial_{Y_k Y_j} e_t(0, Y)\big|_{Y=0},$$

(recall the derivation of (4.38)). Since $\omega(\mathbf{\Phi}) = 0$ and ω is τ^t invariant, Relation (4.121) yields

$$L_{jkt} = \frac{1}{2t} \int_0^t \int_0^t \langle \Phi_{-s_1}^{(k)} | \Phi_{-s_2}^{(j)} \rangle_\omega \, ds_1 ds_2 = \frac{1}{2t} \int_0^t \int_0^t \langle \Phi^{(k)} | \Phi_{s_1-s_2}^{(j)} \rangle_\omega \, ds_1 ds_2.$$

A simple change of integration variable leads to (1). (2) follows from the equality of the mixed partial derivatives $\partial_{Y_k} \partial_{Y_j} e_t(0, Y) = \partial_{Y_j} \partial_{Y_k} e_t(0, Y)$. ☐

4.5   Open quantum systems

4.5.1   Coupling to reservoirs

Let $\mathcal{R}_j$, $j = 1, \cdots, n$, be finite quantum systems with Hilbert spaces $\mathcal{K}_j$. Each $\mathcal{R}_j$ is described by a quantum dynamical system $(\mathcal{O}_j, \tau_j^t, \omega_j)$. Besides the Hamiltonian H_j which generates τ_j, we assume the existence of a 'conserved charge' N_j, a self-adjoint element of $\mathcal{O}_j$ such that $[H_j, N_j] = 0$. It follows that N_j is invariant under the dynamics τ_j^t and that the gauge group $\vartheta_j^t(A) = e^{itN_j} A e^{-itN_j}$ commutes with τ_j^t. We suppose that $\mathcal{R}_j$ is in thermal equilibrium at inverse temperature β_j and chemical potential μ_j, that is that

$$\omega_j = \frac{e^{-\beta_j(H_j - \mu_j N_j)}}{\operatorname{tr}\left(e^{-\beta_j(H_j - \mu_j N_j)}\right)}.$$

The modular group of this state is given by

$$\varsigma_{\omega_j}^t = \tau_j^{-\beta_j t} \circ \vartheta_j^{\beta_j \mu_j t}.$$

Thus, denoting by $\delta_j = i[H_j, \cdot]$ the generator of τ_j^t and by $\xi_j = i[N_j, \cdot]$ the generator of ϑ_j^t, one has

$$\delta_{\omega_j} = -\beta_j(\delta_j - \mu_j \xi_j).$$

Note that in cases where there is no conserved charge, one may simply set $N_j = \mathbb{1}_{\mathcal{K}_j}$ so that the gauge group becomes trivial, $\xi_j = 0$, and the states ω_j independent of the chemical potential μ_j. In such cases, one can simply set $\mu_j = 0$.

The joint system $\mathcal{R} = \mathcal{R}_1 + \cdots + \mathcal{R}_n$ is described by

$$(\mathcal{O}_\mathcal{R}, \tau_\mathcal{R}^t, \omega_\mathcal{R}) = \bigotimes_{j=1}^n (\mathcal{O}_j, \tau_j^t, \omega_j).$$

The generators of the dynamics $\tau_\mathcal{R}^t$, the gauge group $\vartheta_\mathcal{R}^t = \otimes_{j=1}^n \vartheta_j^t$ and the modular group $\varsigma_{\omega_\mathcal{R}}^t = \otimes_{j=1}^n \varsigma_{\omega_j}^t$ are given by

$$\delta_{\mathcal{R}} = \sum_{j=1}^{n} \delta_j = \mathrm{i}[H_{\mathcal{R}}, \cdot], \qquad H_{\mathcal{R}} = \sum_{j=1}^{n} H_j,$$

$$\xi_{\mathcal{R}} = \sum_{j=1}^{n} \xi_j = \mathrm{i}[N_{\mathcal{R}}, \cdot], \qquad N_{\mathcal{R}} = \sum_{j=1}^{n} N_j,$$

$$\delta_{\omega_{\mathcal{R}}} = \sum_{j=1}^{n} \delta_{\omega_j} = \mathrm{i}[\log \omega_{\mathcal{R}}, \cdot], \qquad \log \omega_{\mathcal{R}} = -\sum_{j=1}^{n} \beta_j (H_j - \mu_j N_j),$$

with the notational convention of Section 4.4.7.

Let $\mathcal{S}$ be a finite quantum system described by $(\mathcal{O}_{\mathcal{S}}, \tau_{\mathcal{S}}^t, \omega_{\mathcal{S}})$, the dynamics $\tau_{\mathcal{S}}^t$ being generated by the Hamiltonian $H_{\mathcal{S}}$. We assume the existence of a conserved charge $N_{\mathcal{S}}$ such that $\mathrm{i}[H_{\mathcal{S}}, N_{\mathcal{S}}] = 0$ and denote $\vartheta_{\mathcal{S}}^t$ the corresponding gauge group on $\mathcal{O}_{\mathcal{S}}$.

A gauge invariant coupling of $\mathcal{S}$ to the system of reservoirs $\mathcal{R}$ is a collection of self-adjoint elements $V_j \in \mathcal{O}_{\mathcal{S}} \otimes \mathcal{O}_j$ such that $[N_j + N_{\mathcal{S}}, V_j] = 0$. Denoting $V = \sum_j V_j$, the Hamiltonian

$$H_V = H_{\mathcal{R}} + H_{\mathcal{S}} + V,$$

generates a perturbation τ_V^t of the dynamics $\tau^t = \tau_{\mathcal{S}}^t \otimes \tau_{\mathcal{R}}^t$ on $\mathcal{O} = \mathcal{O}_{\mathcal{S}} \otimes \mathcal{O}_{\mathcal{R}}$. Moreover, τ_V^t preserves the total charge $N = N_{\mathcal{R}} + N_{\mathcal{S}}$ and hence commutes with the gauge group $\vartheta^t = \vartheta_{\mathcal{S}}^t \otimes \vartheta_{\mathcal{R}}^t$.

The quantum dynamical system $(\mathcal{O}, \tau_V^t, \omega)$, where $\omega = \omega_{\mathcal{S}} \otimes \omega_{\mathcal{R}}$, is called *open quantum system*. Open quantum systems are examples of compound systems considered in Sections 4.4.7–4.4.8.

The definition of open quantum system requires some minor modification if the particle statistics (bosons/fermions) are taken into account. These modifications are straightforward (see Section 4.7.6 for an example) and for simplicity of exposition we shall not discuss them in abstract form.

The entropy production observable of $(\mathcal{O}, \tau_V^t, \omega)$ is

$$\sigma = \delta_\omega(H_V).$$

Since

$$\delta_\omega = \delta_{\omega_{\mathcal{R}}} + \delta_{\omega_{\mathcal{S}}} = -\sum_j \beta_j(\delta_j - \mu_j \xi_j) - \mathrm{i}[Q, \cdot],$$

where $Q = -\log \omega_{\mathcal{S}}$, we have

$$\sigma = -\sum_j \beta_j(\Phi_j - \mu_j \mathcal{J}_j) + \sigma_{\mathcal{S}}, \tag{4.125}$$

where

$$\Phi_j = \delta_j(V), \qquad \mathcal{J}_j = \xi_j(V), \qquad \sigma_{\mathcal{S}} = \mathrm{i}[H_V, Q].$$

Observing that

$$\Phi_j = -\mathrm{i}[H_V, H_j], \qquad \mathcal{J}_j = -\mathrm{i}[H_V, N_j], \tag{4.126}$$

we derive

$$H_{jt} - H_j = -\int_0^t \Phi_{js}\mathrm{d}s, \qquad N_{jt} - N_j = -\int_0^t \mathcal{J}_{js}\mathrm{d}s. \tag{4.127}$$

The observables Φ_j and $\mathcal{J}_j$ describe the energy and charge currents out of the j-th reservoir $\mathcal{R}_j$. The observable $\beta_j(\Phi_j - \mu_j\mathcal{J}_j)$ describes entropy flux out of $\mathcal{R}_j$.

The entropy balance equation (more precisely Inequality (4.96)) implies

$$\rho_t(Q) - \rho(Q) \geq \sum_j \beta_j \int_0^t \rho_s(\Phi_j - \mu_j\mathcal{J}_j)\mathrm{d}s$$

$$= \sum_j \beta_j \left[(\rho(H_j) - \rho_t(H_j)) - \mu_j(\rho(N_j) - \rho_t(N_j))\right], \tag{4.128}$$

for any state ρ on $\mathcal{O}$. We note in particular that if ρ is a steady state for the dynamics τ_V^t then both sides of this inequality vanish as long as the joint system remains finite. However, if the reservoirs become infinitely extended while the system $\mathcal{S}$ remains confined then the observable Q remains well defined while H_j and N_j lose their meaning. A very important feature of the proper mathematical formulation of (4.128) in the thermodynamic limit is that the left-hand side still vanishes while the right-hand side is typically nonzero.

Note also that

$$\omega_t = Z^{-1}\mathrm{e}^{-Q-t-\sum_j \beta_j[(H_j - \mu_j N_j) + \int_0^t (\Phi_{j(-s)} - \mu_j \mathcal{J}_{j(-s)})\mathrm{d}s]}, \tag{4.129}$$

where

$$Z = \mathrm{tr}(\mathrm{e}^{-\sum_j \beta_j(H_j - \mu_j N_j)}).$$

The density matrix ω_t expressed in the form (4.129) is known as McLennan–Zubarev dynamical ensemble.

4.5.2   Full counting statistics

We continue with the framework of the previous subsection and adapt our discussion of full counting statistics from Section 4.4.8 to the open quantum system $(\mathcal{O}, \tau_V^t, \omega)$. We note that the reference state ω factorizes into a product of commuting self-adjoint operators

$$\omega = Z^{-1}\mathrm{e}^{-Q - \sum_{j=1}^n \beta_j H_j + \sum_{j=1}^n \beta_j \mu_j N_j} = Z^{-1}\mathrm{e}^{-Q}\left(\prod_{j=1}^n \mathrm{e}^{-\beta_j H_j}\right)\left(\prod_{j=1}^n \mathrm{e}^{\beta_j \mu_j N_j}\right).$$

Defining, according to Exercise 5.45,

$$w^{\alpha} = Z^{-\gamma_0} e^{-\gamma_0 Q} \left(\prod_{j=1}^{n} e^{-\gamma_j \beta_j H_j} \right) \left(\prod_{j=1}^{n} e^{\gamma'_j \beta_j \mu_j N_j} \right),$$

for $\alpha = (\gamma_0, \boldsymbol{\gamma}, \boldsymbol{\gamma}') \in \mathbb{R} \times \mathbb{R}^n \times \mathbb{R}^n$ we have,

$$\mathrm{tr}(w_t^{1-\alpha} w^{\alpha}) = \sum_{q, \boldsymbol{\varepsilon}, \boldsymbol{\nu}} e^{-t(\gamma_0 q + \boldsymbol{\gamma} \cdot \boldsymbol{\varepsilon} + \boldsymbol{\gamma}' \cdot \boldsymbol{\nu})} \mathbb{P}_t(q, \boldsymbol{\varepsilon}, \boldsymbol{\nu}), \tag{4.130}$$

where $\mathbb{P}_t(q, \boldsymbol{\varepsilon}, \boldsymbol{\nu})$ is the joint probability distribution for the mean rates of change of the commuting set of observables

$$\mathbf{S} = (Q, \beta_1 H_1, \ldots, \beta_n H_n, -\beta_1 \mu_1 N_1, \ldots, -\beta_n \mu_n N_n),$$

between two successive joint measurements at time 0 and t. The sum in (4.130) extends over all $(q, \boldsymbol{\varepsilon}, \boldsymbol{\nu}) \in (\mathrm{sp}(\mathbf{S}) - \mathrm{sp}(\mathbf{S}))/t$. As shown in Section 4.4.8, the distribution $\mathbb{P}_t$ coincides with the joint spectral measure of a family of commuting relative modular operators.

Expectation and covariance of $(\boldsymbol{\varepsilon}, \boldsymbol{\nu})$ w.r.t. $\mathbb{P}_t$ are given by

$$\mathbb{E}_t(\varepsilon_j) = -\frac{\beta_j}{t} \int_0^t \omega(\Phi_{js}) \mathrm{d}s,$$

$$\mathbb{E}_t(\nu_j) = \frac{\beta_j \mu_j}{t} \int_0^t \omega(\mathcal{J}_{js}) \mathrm{d}s, \tag{4.131}$$

and,

$$\mathbb{E}_t(\varepsilon_j \varepsilon_k) - \mathbb{E}_t(\varepsilon_j) \mathbb{E}_t(\varepsilon_k) = \frac{\beta_j \beta_k}{t^2} \int_0^t \int_0^t \omega\left((\Phi_{js} - \omega(\Phi_{js}))(\Phi_{ku} - \omega(\Phi_{ku}))\right) \mathrm{d}s \mathrm{d}u,$$

$$\mathbb{E}_t(\nu_j \nu_k) - \mathbb{E}_t(\nu_j) \mathbb{E}_t(\nu_k) = \frac{\beta_j \mu_j \beta_k \mu_k}{t^2} \int_0^t \int_0^t \omega\left((\mathcal{J}_{js} - \omega(\mathcal{J}_{js}))(\mathcal{J}_{ku} - \omega(\mathcal{J}_{ku}))\right) \mathrm{d}s \mathrm{d}u, \tag{4.132}$$

$$\mathbb{E}_t(\varepsilon_j \nu_k) - \mathbb{E}_t(\varepsilon_j) \mathbb{E}_t(\nu_k) = -\frac{\beta_j \beta_k \mu_k}{t^2} \int_0^t \int_0^t \omega\left((\Phi_{js} - \omega(\Phi_{js}))(\mathcal{J}_{ku} - \omega(\mathcal{J}_{ku}))\right) \mathrm{d}s \mathrm{d}u.$$

In terms of Liouvillean, the moment generating function (4.130) reads

$$\mathrm{tr}(w_t^{1-\alpha} w^{\alpha}) = (\xi_\omega | e^{itL_{\frac{1}{\alpha}}} \xi_\omega), \tag{4.133}$$

with (as derived in Exercise 5.43)

$$L_{\frac{1}{\alpha}} = K_0 + L(V) - R(W_\alpha), \tag{4.134}$$

where K_0 denotes the standard Liouvillean of the decoupled dynamics τ^t,

$$W_\alpha = \varsigma_\omega^{\mathrm{i}(\alpha-1/2)}(V) = \sum_{j=1}^n T_j(\alpha)V_jT_j(\alpha)^{-1},$$

and

$$T_j(\alpha) = \mathrm{e}^{-(1/2-\gamma_0)Q-\beta_j[(1/2-\gamma_j)H_j-\mu_j(1/2-\gamma_j')N_j]}.$$

If $(\mathcal{O}, \tau_V^t, \omega)$ is TRI, then the fluctuation relation

$$\frac{\mathbb{P}_t(-q, -\varepsilon, -\boldsymbol{\nu})}{\mathbb{P}_t(q, \varepsilon, \boldsymbol{\nu})} = \mathrm{e}^{-t(q+1\cdot\varepsilon+1\cdot\boldsymbol{\nu})},$$

holds.

4.5.3 Linear response theory

We continue our discussion of open quantum systems. We now adopt the point of view of Section 4.4.9 and describe finite time linear response theory. Let β_{eq} and μ_{eq} be given equilibrium values of the inverse temperature and chemical potential. The thermodynamical forces $X = (X_1, \cdots, X_{2n})$ are

$$X_j = \beta_{\mathrm{eq}} - \beta_j, \quad X_{n+j} = -\beta_{\mathrm{eq}}\mu_{\mathrm{eq}} + \beta_j\mu_j, \quad (j = 1, \ldots, n).$$

The reference state of the system is taken to be

$$\omega_X = Z_X^{-1}\mathrm{e}^{-\beta_{\mathrm{eq}}(H_V-\mu_{\mathrm{eq}}N)+\sum_{j=1}^n(X_jH_j+X_{n+j}N_j)},$$

where $N = N_\mathcal{R} + N_\mathcal{S}$ and $Z_X = \mathrm{tr}(\mathrm{e}^{-\beta_{\mathrm{eq}}(H_V-\mu_{\mathrm{eq}}N)+\sum_{j=1}^n(X_jH_j+X_{n+j}N_j)})$. Clearly,

$$\omega_0 = Z_0^{-1}\mathrm{e}^{-\beta_{\mathrm{eq}}(H_V-\mu_{\mathrm{eq}}N)},$$

is the thermal equilibrium state of $(\mathcal{O}, \tau_V^t)$ at inverse temperature β_{eq} and chemical potential μ_{eq}. Hence, we shall use the notation $\omega_0 = \omega_{\mathrm{eq}}$. The dynamical system $(\mathcal{O}, \tau_V^t, \omega_X)$ fits into the framework of Section 4.4.9 (with $\tau_X^t = \tau_V^t$ independent of X).

Note that the family of states ω_X is distinct from the one used in the previous section: it contains the coupling V. In particular, ω_X is not a product state. This is however in complete parallel with our discussion of linear response theory in classical harmonic chain. If the perturbation V remains local in the thermodynamic limit, the product state ω and the state ω_X describe the same thermodynamics. We shall discuss this issue in more detail in Section 4.6.9.

The entropy production observable of the dynamical system $(\mathcal{O}, \tau_V^t, \omega_X)$ is

$$\sigma_X = \mathrm{i}[\log\omega_X, H_V] = \sum_{j=1}^n X_j\Phi_j + X_{n+j}\mathcal{J}_j, \tag{4.135}$$

where

$$\Phi_j = -\mathrm{i}[H_V, H_j], \qquad \mathcal{J}_j = -\mathrm{i}[H_V, N_j],$$

are the energy and charge flux observables of the j-th reservoir. Clearly, (4.135) is a natural (and X-independent) flux relation. Φ_j is the flux associated to the thermodynamical force $\beta_{\mathrm{eq}} - \beta_j$ and $\mathcal{J}_j$ is the flux associated to the thermodynamical force $-\beta_{\mathrm{eq}}\mu_{\mathrm{eq}} + \beta_j\mu_j$.

The generalized entropic pressure is given by

$$e_t(X, Y) = \log \operatorname{tr}\left(e^{\log \omega_X + \sum_{j=1}^n (Y_j \int_0^t \Phi_{j(-s)}\mathrm{d}s + Y_{n+k}\int_0^t \mathcal{J}_{j(-s)}\mathrm{d}s)}\right).$$

Recall that the equilibrium canonical correlation is

$$\langle A|B\rangle_{\mathrm{eq}} = \int_0^1 \omega_{\mathrm{eq}}(A^*\tau_V^{\mathrm{i}\beta s}(B))\mathrm{d}s.$$

Proposition 4.42 implies the finite Green–Kubo formulas and finite time Onsager reciprocity relations for energy and charge fluxes.

Proposition 4.43 *Suppose that* $(\mathcal{O}, \tau_V^t, \omega_{\mathrm{eq}})$ *is TRI with time reversal* Θ *satisfying* $\Theta(V_j) = V_j$, $\Theta(H_j) = H_j$ *and* $\Theta(N_j) = N_j$ *for all* j. *Then*

$$L_{jkt}^{\mathrm{ee}} = \partial_{X_k}\left.\left(\frac{1}{t}\int_0^t \omega_X(\Phi_{js})\mathrm{d}s\right)\right|_{X=0} = \frac{1}{2}\int_{-t}^t \langle\Phi_k|\Phi_{js}\rangle_{\mathrm{eq}}\left(1 - \frac{|s|}{t}\right)\mathrm{d}s,$$

$$L_{jkt}^{\mathrm{ec}} = \partial_{X_{n+k}}\left.\left(\frac{1}{t}\int_0^t \omega_X(\Phi_{js})\mathrm{d}s\right)\right|_{X=0} = \frac{1}{2}\int_{-t}^t \langle\mathcal{J}_k|\Phi_{js}\rangle_{\mathrm{eq}}\left(1 - \frac{|s|}{t}\right)\mathrm{d}s,$$

$$L_{jkt}^{\mathrm{ce}} = \partial_{X_k}\left.\left(\frac{1}{t}\int_0^t \omega_X(\mathcal{J}_{js})\mathrm{d}s\right)\right|_{X=0} = \frac{1}{2}\int_{-t}^t \langle\Phi_k|\mathcal{J}_{js}\rangle_{\mathrm{eq}}\left(1 - \frac{|s|}{t}\right)\mathrm{d}s,$$

$$L_{jkt}^{\mathrm{cc}} = \partial_{X_{n+k}}\left.\left(\frac{1}{t}\int_0^t \omega_X(\mathcal{J}_{js})\mathrm{d}s\right)\right|_{X=0} = \frac{1}{2}\int_{-t}^t \langle\mathcal{J}_k|\mathcal{J}_{js}\rangle_{\mathrm{eq}}\left(1 - \frac{|s|}{t}\right)\mathrm{d}s,$$

(4.136)

(the indices e/c stand for energy/charge) and

$$L_{jkt}^{\mathrm{ee}} = L_{kjt}^{\mathrm{ee}},$$

$$L_{jkt}^{\mathrm{cc}} = L_{kjt}^{\mathrm{cc}},$$

$$L_{jkt}^{\mathrm{ec}} = L_{kjt}^{\mathrm{ce}}.$$

The special structure of open quantum systems allows for a further insight into linear response theory. Consider the auxiliary Hamiltonian

$$H_X = H_V - \mu_{\mathrm{eq}}N - \frac{1}{\beta_{\mathrm{eq}}}\sum_{j=1}^n (X_j H_j + X_{n+j}N_j),$$

and note that

$$\omega_X = \frac{1}{Z_X} e^{-\beta_{eq} H_X},$$

where $Z_X = \text{tr}(e^{-\beta_{eq} H_X})$. Hence, ω_X is the β_{eq}-KMS state of the dynamics τ_X^t generated by the Hamiltonian H_X. By equation (4.127) one has

$$\omega_{Xt} = e^{-itH_V} \omega_X e^{itH_V} = \frac{1}{Z_X} e^{-\beta_{eq}(H_X + P_t)},$$

where

$$P_t = -\frac{1}{\beta_{eq}} \sum_j \left(X_j \int_0^t \Phi_{j(-s)} ds + X_{n+j} \int_0^t \mathcal{J}_{j(-s)} ds \right).$$

We conclude that ω_{Xt} is the KMS state at inverse temperature β_{eq} of the perturbed dynamics generated by $H_X + P_t$. Moreover, the perturbation satisfies $P_t = O(X)$ as $X \to 0$. Applying the perturbation expansion (4.78) and the formula for the coefficient $b_1(A)$ derived in Exercise 4.3.10, we obtain

$$\omega_{Xt}(A) = \omega_X(A) - \beta_{eq} \int_0^1 \omega_X \left(P_t(\tau_X^{is\beta_{eq}}(A) - \omega_X(A)) \right) ds + O(|X|^2).$$

Since $\omega_X = \omega_{eq} + O(X)$ and $P_t = O(X)$, one has

$$\omega_X \left(P_t(\tau_X^{is\beta_{eq}}(A) - \omega_X(A)) \right) = \omega_{eq} \left(P_t(\tau_X^{is\beta_{eq}}(A) - \omega_{eq}(A)) \right) + O(|X|^2)$$

$$= \omega_{eq} \left(P_t \tau_X^{is\beta_{eq}}(A) \right) - \omega_{eq}(P_t) \omega_{eq}(A) + O(|X|^2).$$

From the fact that $\omega_{eq}(\Phi_{js}) = \omega_{eq}(\Phi_j) = 0$ and $\omega_{eq}(\mathcal{J}_{js}) = \omega_{eq}(\mathcal{J}_j) = 0$ we deduce $\omega_{eq}(P_t) = 0$. Since

$$\tau_X^{is\beta_{eq}}(A) = e^{-s\beta_{eq}(H_V - \mu_{eq} N)} A e^{s\beta_{eq}(H_V - \mu_{eq} N)} + O(X),$$

and $[P_t, N] = 0$, we can further write,

$$\omega_{Xt}(A) = \omega_X(A) - \beta_{eq} \int_0^1 \omega_{eq} \left(P_t \tau_V^{is\beta_{eq}}(A) \right) ds + O(|X|^2). \tag{4.137}$$

By Duhamel's formula one has

$$\partial_{X_k} e^{-\beta_{eq} H_X} \big|_{X=0} = \int_0^{\beta_{eq}} e^{-s(H_V - \mu_{eq} N)} \frac{\partial H_X}{\partial X_k} \bigg|_{X=0} e^{-(\beta_{eq} - s)(H_V - \mu_{eq} N)} ds,$$

from which one easily derives

$$\partial_{X_k} \omega_X(A) \big|_{X=0} = \begin{cases} \langle H_k | A - \omega_{eq}(A) \rangle_{eq} & \text{for } 1 \leq k \leq n, \\ \langle N_k | A - \omega_{eq}(A) \rangle_{eq} & \text{for } n+1 \leq k \leq 2n. \end{cases}$$

Finally, (4.137) yields that for $1 \leq k \leq n$,

$$\partial_{X_k} \omega_X(A_t)|_{X=0} = \langle H_k | A - \omega_{\mathrm{eq}}(A) \rangle_{\mathrm{eq}} + \int_0^t \langle \Phi_k | A_s \rangle_{\mathrm{eq}} \mathrm{d}s,$$

$$\partial_{X_{n+k}} \omega_X(A_t)|_{X=0} = \langle N_k | A - \omega_{\mathrm{eq}}(A) \rangle_{\mathrm{eq}} + \int_0^t \langle \mathcal{J}_k | A_s \rangle_{\mathrm{eq}} \mathrm{d}s.$$

(4.138)

These linear response formulas hold without time reversal assumption and for any observable $A \in \mathcal{O}$. Under the assumptions of Proposition 4.43, ω_X is TRI. If $A = \Phi_j$ or $A = \mathcal{J}_j$ then $\omega_X(A) = 0$. This implies $\partial_{X_k}\omega_X(A)|_{X=0} = 0$ for $k = 1, \ldots, 2n$, and (4.138) reduces to the Green–Kubo formulas (4.136). Using (4.138) it is easy to exhibit examples of open quantum systems for which finite time Onsager reciprocity relations fail in the absence of time reversal.

4.6 The thermodynamic limit and the large time limit

Apart from Subsection 4.6.1 and the first part of Subsection 4.6.6 which should be accessible to all readers, this section is intended for more advanced readers and may be skipped on first reading.

We shall describe, typically without proofs, the thermodynamic limit procedure and how one extends the results of the last two sections to general quantum systems. We shall also discuss the large time limit for infinitely extended quantum system.

4.6.1 Overview

From a mathematical point of view, the dynamics of a finite quantum system $(\mathcal{O}, \tau^t, \omega)$ and that of the finite classical harmonic chain of Section 4.2 are very similar: both are described by a linear quasi-periodic propagator. In particular, the limit

$$\lim_{t\to\infty} \omega(\tau^t(A)),$$

does not exist, except in trivial cases. However, the Cesàro limit

$$\omega_+(A) = \lim_{T\to\infty} \frac{1}{T} \int_0^T \omega(\tau^t(A)) \, \mathrm{d}t,$$

(4.139)

exists for all $A \in \mathcal{O}$ and defines a steady state ω_+ of the system.

Exercise 4.46

1. Show that for a finite quantum system $(\mathcal{O}, \tau^t, \omega)$ with Hamiltonian H, the limit (4.139) exists and that the limiting state ω_+ is described by the density matrix

$$\omega_+ = \sum_{\lambda \in \mathrm{sp}(H)} P_\lambda(H) \omega P_\lambda(H).$$

(4.140)

2. For $A \in \mathcal{O}_{\text{self}}$, set $\Phi_A = \mathrm{i}[H, A]$. Show that

$$\omega_+(\Phi_A) = 0,$$

for any A. Conclude that, in particular, the mean entropy production rate vanishes,

$$\omega_+(\sigma) = \lim_{t \to \infty} \omega(\Sigma^t) = 0.$$

3. Show that the same conclusions hold if the system is infinite (that is the Hilbert space $\mathcal{K}$ is infinite dimensional) but confined in the sense that its Hamiltonian H has a purely discrete spectrum.

Thus, in order to obtain a thermodynamically nontrivial steady state—with non-vanishing currents and strictly positive entropy production rate—we need to perform a thermodynamic (TD) limit before taking the large time limit (4.139). In other words, some parts of the system, for example the reservoirs of an open system, have to be infinitely extended.

There are two difficulties associated with the TD limit: the first is to describe the reference state of the extended system, the second is to define its dynamics. These problems were extensively studied in the 1970s and have led to the algebraic approach to quantum statistical mechanics and quantum field theory. Algebraic quantum statistical mechanics provides a very attractive mathematical framework for the description of infinitely extended quantum systems.

In algebraic quantum statistical mechanics an infinitely extended system is described by a triple $(\mathcal{O}, \tau^t, \omega)$, where $\mathcal{O}$ is a C^*-algebra with identity $\mathbb{1}$ (recall Exercise 5.17), ω is a state (that is a positive normalized linear functional on $\mathcal{O}$) and τ^t is a C^*-dynamics, that is, a norm continuous group of $*$-automorphisms of $\mathcal{O}$. The triple $(\mathcal{O}, \tau^t, \omega)$ is often called the *quantum dynamical system*.[10] The observables are elements of $\mathcal{O}$, ω describes the initial thermodynamical state of our system, and the group τ^t describes its time evolution. The observables evolve in time as $A_t = \tau^t(A)$ and the states as $\omega_t = \omega \circ \tau^t$.

Infinitely extended systems of physical interest arise as the TD limit of finite dimensional systems. There is a number of different ways the TD limit can be realized in practice. In the next section we describe one of them that is suitable for spin systems and quasi-free or locally interacting fermionic systems.

4.6.2   Thermodynamic limit: setup

One starts with a family $\{\mathcal{Q}_M\}_{M \in \mathbb{N}}$ of finite quantum systems described by a sequence of finite dimensional Hilbert spaces $\mathcal{K}_M$, algebras $\mathcal{O}_{\mathcal{K}_M}$, Hamiltonians H_M, and faithful states ω_M. σ_M is the entropy production observable of $\mathcal{Q}_M$. In the presence of control parameters $X \in \mathbb{R}^n$ ($H_{M,X}$ and $\omega_{M,X}$ depend on X), $\Phi_{M,X}$ denotes a chosen flux

---

[10] Such quantum dynamical systems are suitable for the description of spin systems or fermionic systems. In the case of bosonic system, $\mathcal{O}$ is a W^*-algebra, ω is a normal state, and τ^t is weakly continuous. We shall not discuss such systems in these lecture notes (see, e.g., Pillet (2006)).

relation. The number M typically corresponds to the 'size' of $\mathcal{Q}_M$. For example, $\mathcal{Q}_M$ could be a spin system or Fermi gas confined to a box $[-M, M]^d$ of the lattice $\mathbb{Z}^d$.[11] The limiting infinitely extended system is described by a quantum dynamical system $(\mathcal{O}, \tau^t, \omega)$ satisfying the following:

(A1) For all M there is a faithful representation $\pi_M : \mathcal{O}_{\mathcal{K}_M} \to \mathcal{O}$ such that

$$\pi_M(\mathcal{O}_{\mathcal{K}_M}) \subset \pi_{M+1}(\mathcal{O}_{\mathcal{K}_{M+1}}).$$

(A2) $\mathcal{O}_{\mathrm{loc}} = \cup_M \pi_M(\mathcal{O}_{\mathcal{K}_M})$ is dense in $\mathcal{O}$. The elements of $\mathcal{O}_{\mathrm{loc}}$ are sometimes called *local observables* of $\mathcal{O}$.

(A3) For $A \in \mathcal{O}_{\mathrm{loc}}$, $\lim_{M \to \infty} \omega_M \circ \pi_M^{-1}(A) = \omega(A)$ and

$$\lim_{M \to \infty} \pi_M \circ \tau_M^t \circ \pi_M^{-1}(A) = \tau^t(A),$$

where the convergence is uniform for t in compact intervals of $\mathbb{R}$.

(A4) $\lim_{M \to \infty} \pi_M(\sigma_M) = \sigma$, exists in the norm of $\mathcal{O}$. σ is the entropy production observable of $(\mathcal{O}, \tau^t, \omega)$.

(A5) In the presence of control parameters X, $\lim_{M \to \infty} \pi_M(\mathbf{\Phi}_{M,X}) = \mathbf{\Phi}_X$ exists in the norm of $\mathcal{O}$. $\mathbf{\Phi}_X$ is a flux relation of $(\mathcal{O}, \tau_X^t, \omega_X)$,

$$\sigma_X = \sum_{j=1}^{n} X_j \Phi_X^{(j)}.$$

(A6) For $p \in [1, \infty]$ and $\alpha, t \in \mathbb{R}$ the limit

$$e_{t,p}(\alpha) = \lim_{M \to \infty} e_{M,t,p}(\alpha),$$

exists. In the presence of control parameters, the limit

$$e_t(X, Y) = \lim_{M \to \infty} e_{M,t}(X, Y),$$

exists for all $t \in \mathbb{R}$ and $X, Y \in \mathbb{R}^n$.

The verification of (A1)–(A5) in the context of spin systems and Fermi gases is discussed in virtually any mathematically oriented monograph on statistical mechanics (see, e.g., (Bratteli, and Robinson, 1997)). For such systems, the proof of (A6) is typically an easy exercise in the techniques developed in 1970s (see Exercise 5.55 below). In some models $e_{t,p}(\alpha)/e_t(X, Y)$ may be defined/finite only for a restricted range of the parameter $\alpha/(X, Y)$ and in this case the fluctuation theorems need to be suitable modified (this was the case in our introductory example of a thermally driven harmonic chain!).

---

[11] For continuous models one may need to slightly modify this setup. For example, in the case of a free Fermi gas on $\mathbb{R}$, $M = (L, \mathcal{E})$, where L is the spatial cutoff, $\mathcal{E}$ is the energy cutoff, and $M \to \infty$ stands for the ordered limit $\lim_{L \to \infty} \lim_{\mathcal{E} \to \infty}$, see Exercise 5.50. The extension of our axiomatic scheme to this more general setup is straightforward.

In what follows we assume that (A1)–(A6) hold. For reasons of space and notational simplicity we shall assume from the onset that all quantum systems $\mathcal{Q}_M$ are TRI. Also, we shall discuss only the TD/large time limit of the functionals $e_{M,2,t}(\alpha)$ and $e_{M,t}(X,Y)$.

4.6.3 Thermodynamic limit: full counting statistics

The reader should recall the notation and results of Section 4.4.5 where we introduced full counting statistics. We have

$$e_{M,2,t}(\alpha) = e_{M,2,t}(1-\alpha) = \log \int_{\mathbb{R}} e^{-t\alpha\phi} d\mathbb{P}_{M,t}(\phi),$$

where $\mathbb{P}_{M,t}$ is the probability distribution of the mean rate of entropy change associated to the repeated measurement process described in Section 4.4.5.

By (A6),

$$e_{2,t}(\alpha) = \lim_{M\to\infty} e_{M,2,t}(\alpha), \tag{4.141}$$

exists for all t and α. The implications are:

Proposition 4.44 *(1) The sequence of Borel probability measures $\{\mathbb{P}_{M,t}\}$ converges weakly to a Borel probability measure $\mathbb{P}_t$, that is for any bounded continuous function $f : \mathbb{R} \to \mathbb{R}$,*

$$\lim_{M\to\infty} \int_{\mathbb{R}} f d\mathbb{P}_{M,t} = \int_{\mathbb{R}} f d\mathbb{P}_t.$$

(2) For all $\alpha \in \mathbb{R}$,

$$e_{2,t}(\alpha) = \log \int_{\mathbb{R}} e^{-t\alpha\phi} d\mathbb{P}_t(\phi).$$

(3) $e_{2,t}(\alpha)$ is real-analytic and

$$e_{2,t}(\alpha) = e_{2,t}(1-\alpha). \tag{4.142}$$

(4) All the cumulants of $\mathbb{P}_{M,t}$ converge to corresponding cumulants of $\mathbb{P}_t$. In particular,

$$\partial_\alpha e_{2,t}(\alpha)|_{\alpha=0} = -\int_0^t \omega(\sigma_s) ds \le 0.$$

(5) Let $\mathfrak{r} : \mathbb{R} \to \mathbb{R}$ be the reflection $\mathfrak{r}(\phi) = -\phi$ and $\bar{\mathbb{P}}_t = \mathbb{P}_t \circ \mathfrak{r}$ the reflected measure. The measures $\bar{\mathbb{P}}_t$ and $\mathbb{P}_t$ are equivalent and

$$\frac{d\bar{\mathbb{P}}_t(\phi)}{d\mathbb{P}_t(\phi)} = e^{-t\phi}. \tag{4.143}$$

The limiting probability measure $\mathbb{P}_t$ is called full counting statistics of the infinitely extended system $(\mathcal{O}, \tau^t, \omega)$. Relations (4.142) and (4.143) are finite time Evans–Searles symmetries.

Recall that $\mathbb{P}_{M,t}$ is related to the modular structure of $\mathcal{Q}_M$: $\mathbb{P}_{M,t} = Q_M^t$, where Q_M^t is the spectral measure for

$$-\frac{1}{t} \log \Delta_{\omega_{M,t}|\omega_M},$$

and the vector ξ_{ω_M}. Our next goal is to relate $\mathbb{P}_t$ to the modular structure of the infinitely extended systems $(\mathcal{O}, \tau^t, \omega)$. We start with a brief description of this structure assuming that the reader is familiar with the topic.

(1) Let $(\mathcal{H}_\omega, \pi_\omega, \xi_\omega)$ be the GNS-representation of $\mathcal{O}$ associated to ω. $\mathfrak{M}_\omega = \pi_\omega(\mathcal{O})''$ denotes the *enveloping von Neumann algebra*. A vector $\xi \in \mathcal{H}_\omega$ is called cyclic if $\mathfrak{M}_\omega \xi$ is dense in $\mathcal{H}_\omega$ and separating if $A\xi = 0$ for $A \in \mathfrak{M}_\omega$ implies $A = 0$. ξ_ω is automatically cyclic. The state ω is called *modular* if ξ_ω is also separating. We assume ω to be modular.

(2) The antilinear operator $S_\omega : A\xi_\omega \mapsto A^*\xi_\omega$ with domain $\mathfrak{M}_\omega \xi_\omega$ is closable. We denote by the same letter its closure. Let $S_\omega = J\Delta_\omega^{1/2}$ be the polar decomposition of S_ω. J is the *modular conjugation*, an anti-unitary involution on $\mathcal{H}_\omega$, and Δ_ω is the *modular operator* of ω. Δ_ω has a trivial kernel and $\varsigma_\omega^t(A) = \Delta_\omega^{it} A \Delta_\omega^{-it}$ is a group of $*$-automorphism of $\mathfrak{M}_\omega$, the *modular group* of ω.

(3) The set $\mathcal{H}_+ = \{AJA\xi_\omega \mid A \in \mathfrak{M}_\omega\}^{\mathrm{cl}}$ (cl denotes the closure in $\mathcal{H}_\omega$) is the *natural cone*. It is a self-dual cone in $\mathcal{H}_\omega$. A state ν on $\mathcal{O}$ is called normal (or, more precisely, ω-normal) if there exists a density matrix ρ on $\mathcal{H}_\omega$ such that $\nu(A) = \mathrm{tr}(\rho \pi_\omega(A))$. $\mathcal{N}_\omega$ denotes the collection of all ω-normal states. $\mathcal{N}_\omega$ is a norm closed subset of the dual $\mathcal{O}^*$. Any state $\nu \in \mathcal{N}_\omega$ has a unique *vector representative* $\xi_\nu \in \mathcal{H}_+$ such that $\nu(A) = (\xi_\nu | \pi_\omega(A)\xi_\nu)$. ξ_ν is cyclic iff it is separating, that is iff ν is modular.

(4) Let $\nu \in \mathcal{N}_\omega$ be a modular state. The antilinear operator $S_{\nu|\omega} : A\xi_\omega \mapsto A^*\xi_\nu$ is closable on $\mathfrak{M}_\omega \xi_\omega$ and we denote by the same letter its closure. This operator has the polar decomposition $S_{\nu|\omega} = J\Delta_{\nu|\omega}^{1/2}$, where J is the modular conjugation introduced in (2) and $\Delta_{\nu|\omega} > 0$ is the relative modular operator of ν w.r.t. ω.

(5) The Rényi relative entropy of order $\alpha \in \mathbb{R}$ of a state ν w.r.t. ω is defined by

$$S_\alpha(\nu|\omega) = \begin{cases} \log(\xi_\omega | \Delta_{\nu|\omega}^\alpha \xi_\omega) & \text{if } \nu \in \mathcal{N}_\omega, \\ -\infty & \text{otherwise.} \end{cases}$$

Its relative entropy w.r.t. ω is defined by

$$S(\nu|\omega) = \begin{cases} (\xi_\nu | \log \Delta_{\nu|\omega} \xi_\nu) & \text{if } \nu \in \mathcal{N}_\omega, \\ -\infty & \text{otherwise.} \end{cases}$$

To link the modular structure of the finite quantum systems $\mathcal{Q}_M$ to that of $(\mathcal{O}, \tau^t, \omega)$, in addition to (A1)–(A6) we assume:

(A7) Let $\varsigma^t_{\omega_M}$ be the modular group of ω_M. Then for all $A \in \mathcal{O}_{\mathrm{loc}}$,

$$\lim_{M \to \infty} \pi_\omega \circ \pi_M \circ \varsigma^t_{\omega_M} \circ \pi_M^{-1}(A) = \varsigma^t_\omega \circ \pi_\omega(A),$$

and the convergence is uniform for t in compact intervals of $\mathbb{R}$.

Again, the verification of (A7) for spin/fermionic systems is typically an easy exercise. Given (A1)–(A7), we have:

Proposition 4.45 *(1) Let Q^t be the spectral measure for $-\frac{1}{t} \log \Delta_{\omega_t | \omega}$ and ξ_ω. Then $Q^t = \mathbb{P}_t$.*

(2) $\lim_{M \to \infty} S_\alpha(\omega_{M,t} | \omega_M) = S_\alpha(\omega_t | \omega)$ and $\lim_{M \to \infty} S(\omega_{M,t} | \omega_M) = S(\omega_t | \omega)$. In particular,

$$S(\omega_t | \omega) = -\int_0^t \omega(\sigma_s) \mathrm{d}s.$$

The proof of the last proposition is somewhat technical and can be found in (Jakšić, Ogata, Pautrat, and Pillet, 2011b).

Finally, we link $e_{2,t}(\alpha)$ and the full counting statistics $\mathbb{P}_t$ to quantum transfer operators. To avoid introduction of the full machinery of the Araki–Masuda L^p-spaces we shall focus here on the special case described in Exercise 5.43 (this special case covers open quantum systems). Suppose that the finite quantum systems $\mathcal{Q}_M$ have the following additional structure:

(A8) $H_M = H_{M,0} + V_M$, where $[H_{M,0}, \omega_M] = 0$ and

$$\lim_{M \to \infty} \pi_M(V_M) = V,$$

in the norm of $\mathcal{O}$. Moreover, for any $a > 0$,

$$\sup_{|\alpha| < a, M} \| \varsigma^{i\alpha}_{\omega_M}(V_M) \| < \infty. \tag{4.144}$$

(A8) is essentially an assumption on the structure of the model and is easily verifiable in practice. (A3), (A8), and perturbation theory imply that the dynamics $\tau^t_{M,0}$ generated by $H_{M,0}$ converges to the C^*-dynamics τ^t_0, that is that for $A \in \mathcal{O}_{\mathrm{loc}}$ and uniformly for t in compact intervals,

$$\lim_{M \to \infty} \pi_M \circ \tau^t_{M,0} \circ \pi_M^{-1}(A) = \tau^t_0(A).$$

Clearly, $\omega \circ \tau^t_0 = \omega$. The assumption (4.144) and Vitali's theorem ensure that the map

$$\mathbb{R} \ni t \mapsto \varsigma^{it}_\omega(\pi_\omega(V)) \in \mathfrak{M}_\omega,$$

has an analytic continuation to the entire complex plane and that for $z \in \mathbb{C}$,

$$\lim_{M \to \infty} \pi_\omega \circ \pi_M \circ \varsigma^z_{\omega_M}(V_M) = \varsigma^z_\omega \circ \pi_\omega(V).$$

Let K_0 be the standard Liouvillean of $(\mathcal{O}, \tau_0, \omega)$. K_0 is the unique self-adjoint operator on $\mathcal{H}_\omega$ satisfying

$$\pi_\omega(\tau_0^t(A)) = e^{itK_0}\pi_\omega(A)e^{-itK_0}, \qquad e^{itK_0}\mathcal{H}_+ = \mathcal{H}_+,$$

for all $t \in \mathbb{R}$ and $A \in \mathcal{O}$. For $\alpha \in \mathbb{R}$ we set

$$L_{\frac{1}{\alpha}} = K_0 + \pi_\omega(V) - J\varsigma_\omega^{i(\alpha-\frac{1}{2})}(\pi_\omega(V))J.$$

$L_{\frac{1}{\alpha}}$ is a closed operator with the same domain as K_0. Except in trivial cases, $L_{\frac{1}{\alpha}}$ is not self-adjoint unless $\alpha = 1/2$. $L_2 = K$ is the standard Liouvillean of $(\mathcal{O}, \tau^t, \omega)$, that is the unique self-adjoint operator on $\mathcal{H}_\omega$ such that

$$\pi_\omega(\tau^t(A)) = e^{itK}(A)e^{-itK}, \qquad e^{itK}\mathcal{H}_+ = \mathcal{H}_+,$$

for all $t \in \mathbb{R}$ and $A \in \mathcal{O}$.

The following result, which is of considerable conceptual and computational importance, is the extension of Exercise 5.43 to the setting of infinitely extended systems.

Proposition 4.46 *For all t and α,*

$$e_{2,t}(\alpha) = (\xi_\omega | e^{-itL_{\frac{1}{\alpha}}}\xi_\omega).$$

The extension of the results of this section to the multi-parameter/open quantum system full counting statistics is straightforward.

4.6.4 Thermodynamic limit: Control parameters

By (A6), the limit

$$e_t(X, Y) = \lim_{M\to\infty} e_{M,t}(X, Y),$$

exists for all t and $X, Y \in \mathbb{R}^n$. The basic properties of $e_t(X, Y)$ are summarized in:

Proposition 4.47 *(1)*

$$e_t(X, Y) = \sup_{\nu \in \mathcal{N}_{\omega_X}} \left[S(\nu | \omega_X) + Y \cdot \int_0^t \nu(\mathbf{\Phi}_{Xs})\,ds \right].$$

(2) The function $\mathbb{R}^n \ni Y \mapsto e_t(X, Y)$ is convex and real analytic.
(3) $e_t(X, Y) = e_t(X, X - Y)$.
(4)

$$\partial_{Y_j} e_t(X, Y)\big|_{Y=0} = \int_0^t \omega_X(\Phi_{Xs}^{(j)})\,ds, \tag{4.145}$$

$$\partial_{Y_k}\partial_{Y_j} e_t(X, Y)\big|_{Y=0} = \int_0^t \int_0^t \left(\langle \Phi_{Xs_1}^{(k)} | \Phi_{Xs_2}^{(j)} \rangle_{\omega_X} - \omega_X(\Phi_{Xs_1}^{(k)})\omega_X(\Phi_{Xs_2}^{(j)}) \right) ds_2 ds_1.$$

These results are the extension of Proposition 4.40 to the setting of infinitely extended systems. The only difference is that, for simplicity of the exposition, we have exploited the time reversal in the formulation of the results.

The proof of Proposition 4.47 can be found in (Jakšić, Ogata, Pautrat, and Pillet, 2011b) and we restrict ourselves to several comments. Part (3), the generalized finite time Evans–Searles symmetry, is of course an immediate consequences of the same property of the functionals $e_{M,t}(X,Y)$. The convexity of $Y \mapsto e_t(X,Y)$ follows in the same way (note that convexity also follows from (1)). The most natural way to prove the remaining parts is to use Araki's perturbation theory of the KMS/modular structure (this theory is, in part, an extension of the results of Section 4.3.10 to general von Neumann algebras). The Kubo–Mari inner product $\langle \Phi_{Xs_1}^{(k)} | \Phi_{Xs_2}^{(j)} \rangle_{\omega_X}$ in Part (4) is formally similar to its finite-dimensional counterpart. It is a part of the modular structure that for all $A, B \in \mathfrak{M}_{\omega_X}$, the function $t \mapsto (\xi_{\omega_X} | A^* \varsigma_{\omega_X}^t(B) \xi_{\omega_X})$ has an analytic continuation to the strip $-1 < \text{Im } z < 0$ which is bounded and continuous on its closure. Then

$$\langle \Phi_{Xs_1}^{(k)} | \Phi_{Xs_2}^{(j)} \rangle_{\omega_X} = \int_0^1 (\xi_{\omega_X} | \pi_{\omega_X}(\Phi_{Xs_1}^{(k)}) \varsigma_{\omega_X}^{-iu}(\pi_{\omega_X}(\Phi_{Xs_2}^{(j)})) \xi_{\omega_X}) du.$$

The finite time linear response theory for family of infinitely extended systems $(\mathcal{O}, \tau_X^t, \omega_X)$ can be developed along two complementary routes. We shall use the same notational conventions as in Section 4.4.10: $\omega_0 = \omega$, $\tau_0 = \tau$, $\Phi_0 = \Phi$. Since

$$\langle \Phi_X \rangle_t = \frac{1}{t} \int_0^t \omega_X(\Phi_{Xs}) ds = \frac{1}{t} \nabla_Y e_t(X,Y)|_{Y=0},$$

we have the following:

Proposition 4.48 *Suppose that the map* $(X,Y) \mapsto e_t(X,Y)$ *is* C^2 *in an open set containing* $(0,0)$. *Then the finite time kinetic transport coefficients*

$$L_{jkt} = \partial_{X_k} \langle \Phi_X^{(j)} \rangle_t |_{X=0} = \partial_{X_k} \partial_{Y_j} e_t(X,Y)_{X=Y=0},$$

satisfy:

(1)
$$L_{jkt} = \frac{1}{2} \int_{-t}^t \langle \Phi^{(k)} | \Phi_s^{(j)} \rangle_\omega \left(1 - \frac{|s|}{t} \right) ds.$$

(2) $L_{jkt} = L_{kjt}$ *and the quadratic form determined by* $[L_{jkt}]$ *is positive definite.*

Given Proposition 4.47, the proof of Proposition 4.48 is exactly the same as the proof of its finite dimensional counterpart (Proposition 4.42 in Section 4.4.10).

A complementary route is based on the thermodynamical limit of the finite time finite volume linear response theory. This route is both technically and conceptually less satisfactory and we shall not discuss it here.

4.6.5   Large time limit: full counting statistics

To describe fluctuations of $\mathbb{P}_t$ as $t \to \infty$ we need to assume:

(A9) The limit

$$e_{2,+}(\alpha) = \lim_{t \to \infty} \frac{1}{t} e_{2,t}(\alpha),$$

exists for α in some open interval $\mathcal{I}$ containing $[0, 1]$. Moreover, the limiting entropic functional $e_{2,+}(\alpha)$ is differentiable on $\mathcal{I}$.

The verification of (A9) (and (A10) below) is the central step of the programme. Unlike (A1)–(A8), which are typically easily verifiable structural/thermodynamical limit properties of a given model, the verification of (A9) is usually a difficult analytical problem.

The quantum Evans–Searles fluctuation theorem for the full counting statistics follows from (A9) and the Gärtner–Ellis theorem. We describe its conclusions. Without loss of generality we may assume that $\mathcal{I}$ is centred at $\alpha = 1/2$ (recall that we assume the system to be TRI).

Proposition 4.49 *(1) $e_{2,+}(\alpha)$ is convex on $\mathcal{I}$, the Evans–Searles symmetry*

$$e_{2,+}(\alpha) = e_{2,+}(1 - \alpha),$$

holds, and

$$e'_{2,+}(0) = -\lim_{t \to \infty} \mathbb{E}_t(\phi) = -\lim_{t \to \infty} \frac{1}{t} S(\omega_t | \omega) = -\lim_{t \to \infty} \frac{1}{t} \int_0^t \omega(\sigma_s) \mathrm{d}s.$$

The non-negative number $\langle \sigma \rangle_+ = -e'_{2,+}(0)$ is called the entropy production of $(\mathcal{O}, \tau^t, \omega)$. Notice that $\langle \sigma \rangle_+ = 0$ iff the function $e_{2,+}(\alpha) = 0$ for $\alpha \in [0, 1]$.
(2) Let

$$\theta = \sup_{\alpha \in \mathcal{I}} e'_{2,+}(\alpha) = -\inf_{\alpha \in \mathcal{I}} e'_{2,+}(\alpha).$$

The function

$$I(s) = -\inf_{\alpha \in \mathcal{I}} (\alpha s + e_{2,+}(\alpha)),$$

is non-negative, convex, and differentiable on $] - \theta, \theta[$. [12] $I(s) = 0$ iff $s = -\langle \sigma \rangle_+$ and the Evans–Searles symmetry implies

$$I(-s) = s + I(s).$$

The last relation is sometimes called the Evans–Searles symmetry for the rate function.
(3) For any open set $J \subset] - \theta, \theta[$,

$$\lim_{t \to \infty} \frac{1}{t} \log \mathbb{P}_t(J) = -\inf_{s \in J} I(s).$$

---

[12] If $\theta < \infty$, then $I(s)$ is linear on $] - \infty, -\theta]$ and $[\theta, \infty[$.

The interpretation of the quantum ES theorem for the full counting statistics is similar to the classical case. The full counting statistics concerns the operationally defined 'mean entropy flow' across the system. Its expectation value converges, as $t \to \infty$, to the entropy production $\langle \sigma \rangle_+$ of the model. Its fluctuations of order 1 are described by the theory of large deviations. The specific aspect of the ES theorem is that the time reversal invariance implies the universal symmetry of the rate function which in turn implies that the 'mean entropy flow' is exponentially more likely to be positive than negative, that is, the probability of violating the second law of thermodynamics is exceedingly small for large t.

We now describe schematically how Proposition 4.46 can be used to verify the key Assumption (A9).

(i) In typical situations where spectral techniques are applicable the standard Liouvillean K_0 has purely absolutely continuous spectrum filling the real line except for finitely many embedded eigenvalues of finite multiplicity. This is precisely what happens in the study of open quantum systems describing a finite quantum system S coupled to an infinitely extended reservoir $\mathcal{R}$. Typically, $\mathcal{R}$ will consists of several independent sub-reservoirs $\mathcal{R}_j$ which are in thermal equilibrium at inverse temperatures β_j and chemical potentials μ_j, but we do not need at this point to specify further the structure of $\mathcal{R}$. The reservoir system is described by C^*-dynamical system $(\mathcal{O}_\mathcal{R}, \tau_\mathcal{R}^t, w_\mathcal{R})$ where $w_\mathcal{R}$ is stationary for the dynamics $\tau_\mathcal{R}^t$ and assumed to be modular. Let $(\mathcal{H}_\mathcal{R}, \pi_\mathcal{R}, \xi_\mathcal{R})$ be the corresponding GNS representation and let $K_\mathcal{R}$ be the corresponding standard Liouvillean. Since $w_\mathcal{R}$ is steady, $K_\mathcal{R}\xi_\mathcal{R} = 0$. We assume that apart from a simple eigenvalue at 0, $K_\mathcal{R}$ has purely absolutely continuous spectrum filling the entire real line. This assumption ensures that $\mathcal{R}$ has strong ergodic properties and in particular that $(\mathcal{O}_\mathcal{R}, \tau_\mathcal{R}^t, w_\mathcal{R})$ is mixing, that is that

$$\lim_{|t| \to \infty} w_\mathcal{R}(A\tau_\mathcal{R}^t(B)) = w_\mathcal{R}(A)w_\mathcal{R}(B),$$

for $A, B \in \mathcal{O}_\mathcal{R}$. In the simplest nontrivial case, S is a 2-level system, described by the Hilbert space $\mathbb{C}^2$ and the Hamiltonian $\sigma^{(3)}$ (the third Pauli matrix). Then the standard Liouvillean of the joint but decoupled system $S + \mathcal{R}$ acts on the Hilbert space $\mathcal{H} = \mathbb{C}^2 \otimes \mathbb{C}^2 \otimes \mathcal{H}_\mathcal{R}$ and has the form

$$K_0 = (\sigma^{(3)} \otimes \mathbb{1} - \mathbb{1} \otimes \sigma^{(3)}) \otimes \mathbb{1} + \mathbb{1} \otimes K_\mathcal{R}.$$

This will be precisely the case in the spin-fermion model which we will discuss in Section 4.7.5. For simplicity of exposition, we assume in the following that the point spectrum of K_0 is $\{-2, 0, 2\}$, where the eigenvalues ± 2 are simple and 0 is doubly degenerate. The rest of the spectrum of K_0 is purely absolutely continuous and fills the real line, see Fig. 4.6.

(ii) An application of the numerical range theorem yields that the spectrum of $L_{\frac{1}{\alpha}}$ is contained in the strip $\{z \,|\, |\operatorname{Im} z| \leq M_\alpha\}$, where

$$M_\alpha = \|\varsigma_\omega^{i(\alpha - \frac{1}{2})}(\pi_\omega(V))\| + \|\pi_\omega(V)\|.$$

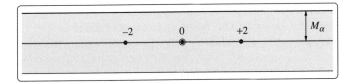

Fig. 4.6 The point spectrum of the uncoupled standard Liouvillean K_0. The spectrum of the transfer operator $L_{\frac{1}{\alpha}}$ is contained in the grey strip.

Thus, the resolvent $(z - L_{\frac{1}{\alpha}})^{-1}$ is an analytic function of z on the half-plane $\mathrm{Im}\, z > M_\alpha$.

(iii) By using complex deformation techniques one proves that for some $\mu > 0$ and all vectors ξ, η in some dense subspace of $\mathcal{H}$ the functions

$$z \mapsto (\xi | (z - L_{\frac{1}{\alpha}})^{-1} \eta),$$

have a meromorphic continuation from the half-plane $\mathrm{Im}\, z > M_\alpha$ to the half-plane $\mathrm{Im}\, z > -\mu$. This extension has four simple poles located at the points $e_\pm(\alpha)$, $e(\alpha)$, $e_1(\alpha)$, where $e(\alpha)$ is the pole closest to the real axis, see Fig. 4.7. For symmetry reasons $e(\alpha)$ is purely imaginary. These poles are resonances of $L_{\frac{1}{\alpha}}$, or in other words, eigenvalues of a complex deformation of $L_{\frac{1}{\alpha}}$. They can be computed by an application of analytic perturbation theory. For this purpose it is convenient to introduce a control parameter $\lambda \in \mathbb{R}$ and replace the interaction term V with λV. The parameter λ controls the strength of the coupling and analytic perturbation theory applies for small values of λ. One proves that given $\alpha_0 > 1/2$ one can find $\Lambda > 0$ such that for $|\alpha - \frac{1}{2}| < \alpha_0$ and $|\lambda| < \Lambda$, μ can be chosen independently of α and λ and that the poles are analytic functions of α. In particular, for α small enough,

$$e(\alpha) = \mathrm{i} \sum_{n=1}^{\infty} E_n(\lambda)\alpha^n,$$

where each coefficient $E_n(\lambda)$ is real-analytic function of λ.

(iv) One now starts with the expression

$$(\xi_\omega | e^{-\mathrm{i}t L_{\frac{1}{\alpha}}} \xi_\omega) = \int_{\mathrm{Re}\, z = a} e^{-\mathrm{i}tz} (\xi_\omega | (z - L_{\frac{1}{\alpha}})^{-1} \xi_\omega) \frac{\mathrm{d}z}{2\pi\mathrm{i}}, \qquad (4.146)$$

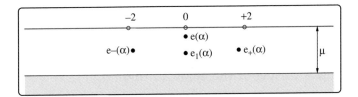

Fig. 4.7 The resonances of the transfer operator $L_{\frac{1}{\alpha}}$.

where $a > M_\alpha$. Moving the line of integration to $\mathrm{Re}\, z = -\mu'$, where $\mu' \in]0, \mu[$ is such that the poles of the integrand are contained in $\{z \mid \mathrm{Im}\, z > -\mu'\}$ for $|\lambda| < \Lambda$ and $|\alpha - \frac{1}{2}| < \alpha_0$, and picking the contribution from theses poles one derives

$$(\xi_\omega | e^{-itL \frac{1}{\alpha}} \xi_\omega) = e^{-ite(\alpha)}(1 + R(t, \alpha)), \tag{4.147}$$

where $R(t, \alpha)$ decays exponentially in t as $t \to \infty$. It then follows that

$$e_{2,+}(\alpha) = \lim_{t\to\infty} \frac{1}{t} e_{2,t}(\alpha) = -ie(\alpha).$$

A proper mathematical justification of (4.146) and (4.147) is typically the technically most demanding part of the argument.

(v) Recall that

$$\partial_\alpha e_{2,+}(\alpha)|_{\alpha=0} = E_1 = -\langle\sigma\rangle_+ = -\lim_{t\to\infty} \mathbb{E}_t(\phi).$$

Given (iv), an application of Vitali's theorem yields

$$\partial_\alpha^2 e_{2,+}(\alpha)_{\alpha=0} = E_2 = \lim_{t\to\infty} \frac{1}{t} \int_0^t \int_0^t (\omega(\sigma_s\sigma_u) - \omega(\sigma_s)\omega(\sigma_u))dsdu$$

$$= \lim_{t\to\infty} t(\mathbb{E}_t(\phi^2) - (\mathbb{E}_t(\phi))^2).$$

(vi) The arguments/estimates in (iv) extend to complex α's satisfying $|\alpha - \frac{1}{2}| < \alpha_0$ and one shows that for α real,

$$\lim_{t\to\infty} \int_\mathbb{R} e^{-i\alpha\sqrt{t}(\phi - \langle\sigma\rangle+)} d\mathbb{P}_t(\phi) = \lim_{t\to\infty} e^{i\alpha\sqrt{t}\langle\sigma\rangle+} (\xi_\omega | e^{-itL \frac{\sqrt{t}}{i\alpha}} \xi_\omega) = e^{-E_2\alpha^2}.$$

Hence, the central limit theorem holds for the full counting statistics $\mathbb{P}_t$, that is, for any interval $[a, b]$,

$$\lim_{t\to\infty} \mathbb{P}_t\left(\langle\sigma\rangle_+ + \frac{1}{\sqrt{tE_2}}[a, b]\right) = \frac{1}{\sqrt{2\pi}} \int_a^b e^{-\frac{x^2}{2}} dx.$$

The above spectral scheme is technically delicate and its implementation requires a number of regularity assumptions on the structure of reservoirs and the interaction V. On the positive side, when applicable the spectral scheme provides a wealth of information and a very satisfactory conceptual picture. In the classical case, the quantum transfer operators reduce to Ruelle–Perron–Frobenius operators and the above spectral scheme is a well-known chapter in the theory of classical dynamical systems, see Section 5.4 in (Jakšić, Pillet, and Rey-Bellet, 2011a) and (Baladi, 2000).

4.6.6 Hypothesis testing of the arrow of time

Theorem 4.21 clearly links the $p = 2$ entropic functional to quantum hypothesis testing. This link, somewhat surprisingly, can be interpreted as quantum hypothesis testing of the second law of thermodynamics and arrow of time: how well can we distinguish the state $\omega_t = \omega \circ \tau^t$, from the same initial state evolved backward in time $\omega_{-t} = \omega \circ \tau^{-t}$? More precisely, we shall investigate the asymptotic behaviour of the minimal error probability for the hypothesis testing associated to the pair (ω_{-t}, ω_t) as $t \to \infty$.

We start with the family of pairs $\{(\omega_{M,-t}, \omega_{M,t}) \mid t > 0\}$. Again, the thermodynamic limit $M \to \infty$ has to be taken prior to the limit $t \to \infty$.

Given their *a priori* probabilities, $1 - p$ and p, the minimal error probability in distinguishing the states $\omega_{M,-t/2}$ and $\omega_{M,t/2}$ is given by Theorem 4.21,

$$D_{M,p}(t) = \frac{1}{2}\left(1 - \operatorname{tr}\left|(1 - p)\,\omega_{M,-t/2} - p\,\omega_{M,t/2}\right|\right).$$

We set

$$\underline{D}_p(t) = \liminf_{M \to \infty} D_{M,p}(t), \qquad \overline{D}_p(t) = \limsup_{M \to \infty} D_{M,p}(t),$$

and define the Chernoff error exponents by

$$\underline{d}_p = \liminf_{t \to \infty} \frac{1}{t} \log \underline{D}_p(t), \qquad \overline{d}_p = \limsup_{t \to \infty} \frac{1}{t} \log \overline{D}_p(t).$$

Theorem 4.50 *For any* $p \in\,]0, 1[$,

$$\underline{d}_p = \overline{d}_p = \inf_{\alpha \in [0,1]} e_{2,+}(\alpha).$$

Moreover, since the system is TRI the infimum is achieved at $\alpha = 1/2$.

Proof: We first notice that

$$D_{M,p}(t) = \frac{1}{2}\left(1 - \operatorname{tr}\left|(1 - p)\,\omega_{M,0} - p\,\omega_{M,t}\right|\right).$$

Theorem 4.21 (3) and the existence of the limiting functional $e_{2,t}(\alpha)$ (for $M \to \infty$) yield the inequality

$$\log \overline{D}_p(t) \le e_{2,2t}(\alpha) + (1 - \alpha) \log(1 - p) + \alpha \log p,$$

for all $\alpha \in [0, 1]$. Dividing by t and letting $t \to \infty$ we obtain the upper bound

$$\overline{d}_p \le \inf_{\alpha \in [0,1]} e_{2,+}(\alpha).$$

For finite M, a lower bound is provided by Proposition 4.28,

$$D_{M,p}(t) \ge \frac{1}{2}\min(p, 1 - p)\,\mathbb{P}_{M,t}(]0, \infty[),$$

where $\mathbb{P}_{M,t}$ is the full counting statistics of $\mathcal{Q}_M$. As we have already discussed, the convergence of $e_{M,2,t}(\alpha)$ to $e_{2,t}(\alpha)$ as $M \to \infty$ implies that $\mathbb{P}_{M,t}$ converges weakly to the full counting statistics $\mathbb{P}_t$ of the extended system. The Portmanteau theorem (Billingsley (1968), Theorem 2.1) implies

$$\liminf_{M \to \infty} \mathbb{P}_{M,t}(]0, \infty[) \geq \mathbb{P}_t(]0, \infty[),$$

and hence

$$\underline{D}_p(t) \geq \frac{1}{2} \min(p, 1-p) \, \mathbb{P}_t(]0, \infty[) \geq \frac{1}{2} \min(p, 1-p) \, \mathbb{P}_t(]0, 1[).$$

Assumption (A9) and the Gärtner–Ellis theorem (or more specifically Proposition 4.64 in Appendix 4.8) imply

$$\liminf_{t \to \infty} \frac{1}{t} \log \mathbb{P}_t(]0, 1[) \geq -\varphi(0),$$

where

$$\varphi(s) = \sup_{\alpha \in \mathbb{R}} (s\alpha - e_{2,+}(\alpha)).$$

Since

$$\varphi(0) = - \inf_{\alpha \in \mathbb{R}} e_{2,+}(\alpha) = - \inf_{\alpha \in [0,1]} e_{2,+}(\alpha),$$

(recall that, by Proposition 4.31, $e_{2,+}(\alpha) \leq 0$ for $\alpha \in [0, 1]$ and $e_{2,+}(\alpha) \geq 0$ otherwise) we have

$$\underline{d}_p \geq \liminf_{t \to \infty} \frac{1}{t} \left(-\log 2 + \min(\log p, \log(1-p)) + \log(\mathbb{P}_t(]0, 1[)) \right) \geq \inf_{\alpha \in [0,1]} e_{2,+}(\alpha).$$

The convexity and the symmetry $e_{2,+}(1 - \alpha) = e_{2,+}(\alpha)$ imply that the infimum is achieved at $\alpha = 1/2$. $\square$

Note that the above result and its proof link the fluctuations of the full counting statistics $\mathbb{P}_t$ as $t \to \infty$ to Chernoff error exponents in quantum hypothesis testing of the arrow of time. The TD limit plays an important role in the discussion of full counting statistics since its physical interpretation in terms of repeated quantum measurement is possible only for finite quantum systems. However, apart from the above mentioned connection with full counting statistics, quantum hypothesis testing can be formulated in the framework of extended quantum systems without reference to the TD limit. In fact, by considering directly an infinitely extended system, one can considerably refine the quantum hypothesis testing of the arrow of time. In the remaining part of this section we indicate how this can be done, referring the reader to (Jakšić, Ogata, Pillet, and Seiringer, 2011c) for proofs and additional information.

(i) We start with an infinitely extended system $\mathcal{Q}$ described by the C^*-dynamical system $(\mathcal{O}, \tau^t, \omega)$. The GNS-representation of $\mathcal{O}$ associated to the state ω is denoted $(\mathcal{H}_\omega, \pi_\omega, \xi_\omega)$, and the enveloping von Neumann algebra is $\mathfrak{M}_\omega = \pi_\omega(\mathcal{O})''$. We assume that ω is modular. The group $\pi_\omega \circ \tau^t$ extends to a weakly continuous group τ_ω^t of $*$-automorphisms of $\mathfrak{M}_\omega$. With a slight abuse of notation we denote the vector state $(\xi_\omega | \cdot \xi_\omega)$ on $\mathfrak{M}_\omega$ again by ω. The triple $(\mathfrak{M}_\omega, \tau_\omega^t, \omega)$ is the W^*-quantum dynamical system induced by $(\mathcal{O}, \tau^t, \omega)$. We denote $\omega_t = \omega \circ \tau_\omega^t$. The quantum hypothesis testing of the arrow of time concerns the family of pairs $\{(\omega_{-t}, \omega_t) \,|\, t > 0\}$.

(ii) Consider the following competing hypothesis:

- Hypothesis I : Q is in the state $\omega_{t/2}$.
- Hypothesis II : Q is in the state $\omega_{-t/2}$.

We know *a priori* that Hypothesis I is realized with probability p and II with probability $1 - p$. A *test* is a self-adjoint projection $P \in \mathfrak{M}_\omega$ and a result of a measurement of the corresponding observable is a number in $\mathrm{sp}(P) = \{0, 1\}$. If the outcome is 1, one accepts I, otherwise one accepts II. The error probability of the test P is

$$D_p(\omega_{t/2}, \omega_{-t/2}, P) = p\,\omega_{t/2}(\mathbb{1} - P) + (1 - p)\,\omega_{-t/2}(P), \tag{4.148}$$

and

$$D_p(\omega_{t/2}, \omega_{-t/2}) = \inf_P D_p(\omega_{t/2}, \omega_{-t/2}, P),$$

is the minimal error probability.

(iii) The quantum Neyman–Pearson lemma holds:

$$D_p(\omega_{t/2}, \omega_{-t/2}) = D_p(\omega_{t/2}, \omega_{-t/2}, P_{\mathrm{opt}}) = \frac{1}{2}(1 - \|(1 - p)\omega_{-t/2} - p\omega_{t/2}\|)$$

$$= \frac{1}{2}(1 - \|(1 - p)\omega - p\omega_t\|),$$

where P_{opt} is the support projection of the linear functional $((1 - p)\omega_{-t/2} - p\omega_{t/2})_+$ (the positive part of $(1 - p)\omega_{-t/2} - p\omega_{t/2}$). Just like in the classical case, the proof of the quantum Neyman–Pearson lemma is straightforward.

(iv) Let $\mu_{\omega_t | \omega}$ be the spectral measure for $\Delta_{\omega_t | \omega}$ and ξ_ω. Then

$$\frac{1}{2}\min(p, 1 - p)\mu_{\omega_t | \omega}([1, \infty[) \leq D_p(\omega_{t/2}, \omega_{-t/2}) \leq p^\alpha (1 - p)^{1-\alpha}(\xi_\omega | \Delta_{\omega_t | \omega}^\alpha \xi_\omega).$$

The proof of the lower bound in exactly the same as in the finite case (recall Proposition 4.28). The proof of the upper bound is based on an extension of Ozawa's argument (see the proof of Part (3) of Theorem 4.21) to the modular setting and is more subtle, see (Ogata, 2010).

(v) Assuming (A9), that is that

$$e_{2,+}(\alpha) = \lim_{t \to \infty} \frac{1}{t} \log(\xi_\omega | \Delta_{\omega_t | \omega}^\alpha \xi_\omega),$$

exist and is differentiable for α is some interval containing $[0,1]$, then a straightforward application of the Gärtner–Ellis theorem yields

$$\lim_{t\to\infty} \frac{1}{t} \log D_p(\omega_t, \omega_{-t}) = \inf_{\alpha \in [0,1]} e_{2,+}(\alpha). \qquad (4.149)$$

Results of this type are often called quantum Chernoff bounds. Our TRI assumption implies that the infimum is achieved for $\alpha = 1/2$.

The Chernoff bound (4.149) quantifies the separation between the past and the future as time $t \uparrow \infty$. Taking $p = 1/2$ and noticing that

$$\frac{1}{2}(2 - \|\omega_{t/2} - \omega_{-t/2}\|) = \omega_{t/2}(\mathrm{s}_-(t/2)) + \omega_{-t/2}(\mathrm{s}_+(t/2)),$$

where $\mathrm{s}_\pm(t)$ is the support projection of the positive linear functional $(\omega_t - \omega_{-t})_\pm$ on $\mathfrak{M}_\omega$, we see that the Chernoff bound implies

$$\limsup_{t\to\infty} \frac{1}{t} \log \omega_t(\mathrm{s}_-(t)) \le 2 \inf_{s \in [0,1]} e_{2,+}(s),$$

$$\limsup_{t\to\infty} \frac{1}{t} \log \omega_{-t}(\mathrm{s}_+(t)) \le 2 \inf_{s \in [0,1]} e_{2,+}(s).$$

Therefore, as $t \uparrow \infty$, the state ω_t concentrates exponentially fast on $\mathrm{s}_+(t)\mathfrak{M}_\omega$ while the state ω_{-t} concentrates exponentially fast on $\mathrm{s}_-(t)\mathfrak{M}_\omega$.

(vi) In the infinite dimensional setting one can introduce other error exponents. For $r \in \mathbb{R}$ the Hoeffding exponents are defined by

$$\overline{B}(r) = \inf_{\{P_t\}} \left\{ \limsup_{t\to\infty} \frac{1}{t} \log \omega_{t/2}(\mathbb{1} - P_t) \;\middle|\; \limsup_{t\to\infty} \frac{1}{t} \log \omega_{-t/2}(P_t) < -r \right\},$$

$$\underline{B}(r) = \inf_{\{P_t\}} \left\{ \liminf_{t\to\infty} \frac{1}{t} \log \omega_{t/2}(\mathbb{1} - P_t) \;\middle|\; \limsup_{t\to\infty} \frac{1}{t} \log \omega_{-t/2}(P_t) < -r \right\},$$

$$B(r) = \inf_{\{P_t\}} \left\{ \lim_{t\to\infty} \frac{1}{t} \log \omega_{t/2}(\mathbb{1} - P_t) \;\middle|\; \limsup_{t\to\infty} \frac{1}{t} \log \omega_{-t/2}(P_t) < -r \right\},$$

where the infimum are taken over families $\{P_t\}_{t>0}$ of orthogonal projections in $\mathfrak{M}_\omega$ subject, in the last case, to the constraint that $\lim_{t\to\infty} t^{-1} \log \omega_{t/2}(\mathbb{1} - P_t)$ exists.

The Hoeffding exponents are increasing functions of r, $\underline{B}(r) \le \overline{B}(r) \le B(r) \le 0$, and $\underline{B}(r) = \overline{B}(r) = B(r) = -\infty$ if $r < 0$. The functions $\underline{B}(r), \overline{B}(r), B(r)$ are left-continuous and upper semi-continuous. If (A9) holds and $\langle \sigma \rangle_+ > 0$, then for all $r \in \mathbb{R}$,

$$\underline{B}(r) = \overline{B}(r) = B(r) = b(r) = -\sup_{0 \le s < 1} \frac{-sr - e_{2,+}(s)}{1 - s},$$

see (Jakšić, Ogata, Pillet, and Seiringer, 2011c). Results of this type are called quantum Hoeffding bounds.

Let $r > 0$ and let P_t be projections in $\mathfrak{M}_\omega$ such that

$$\limsup_{t\to\infty} \frac{1}{t} \log \omega_{-t/2}(P_t) < -r. \tag{4.150}$$

The Hoeffding bound asserts

$$\liminf_{t\to\infty} \frac{1}{t} \log \omega_{t/2}(\mathbb{1} - P_t) \geq b(r).$$

Moreover, one can show that for a suitable choice of P_t,

$$\lim_{t\to\infty} \frac{1}{t} \log \omega_{t/2}(\mathbb{1} - P_t) = b(r).$$

Hence, if $\omega_{-t/2}$ is concentrating exponentially fast on $(\mathbb{1} - P_t)\mathfrak{M}_\omega$ with an exponential rate $< -r$, then $\omega_{t/2}$ is concentrating on $P_t\mathfrak{M}_\omega$ with the optimal exponential rate $b(r)$.

(vii) For $\epsilon \in]0,1[$ set

$$\overline{B}_\epsilon = \inf_{\{P_t\}} \left\{ \limsup_{t\to\infty} \frac{1}{t} \log \omega_{t/2}(\mathbb{1} - P_t) \ \Big| \ \omega_{-t/2}(P_t) \leq \epsilon \right\},$$

$$\underline{B}_\epsilon = \inf_{\{P_t\}} \left\{ \liminf_{t\to\infty} \frac{1}{t} \log \omega_{t/2}(\mathbb{1} - P_t) \ \Big| \ \omega_{-t/2}(P_t) \leq \epsilon \right\}, \tag{4.151}$$

$$B_\epsilon = \inf_{\{P_t\}} \left\{ \lim_{t\to\infty} \frac{1}{t} \log \omega_{t/2}(\mathbb{1} - P_t) \ \Big| \ \omega_{-t/2}(P_t) \leq \epsilon \right\},$$

where the infimum is taken over families of tests $\{P_t\}_{t>0}$ subject, in the last case, to the constraint that $\lim_{t\to\infty} t^{-1} \log \omega_{t/2}(\mathbb{1} - P_t)$ exists. Note that if

$$\beta_t(\epsilon) = \inf_P \{ \omega_{t/2}(\mathbb{1} - P) \,|\, \omega_{-t/2}(P) \leq \epsilon \},$$

then

$$\liminf_{t\to\infty} \frac{1}{t} \log \beta_t(\epsilon) = \underline{B}_\epsilon, \qquad \limsup_{t\to\infty} \frac{1}{t} \log \beta_t(\epsilon) = \overline{B}_\epsilon.$$

We also define

$$\overline{B} = \inf_{\{P_t\}} \left\{ \limsup_{t\to\infty} \frac{1}{t} \log \omega_{t/2}(\mathbb{1} - P_t) \ \Big| \ \lim_{t\to\infty} \omega_{-t/2}(P_t) = 0 \right\},$$

$$\underline{B} = \inf_{\{P_t\}} \left\{ \liminf_{t \to \infty} \frac{1}{t} \log \omega_{t/2}(\mathbb{1} - P_t) \,\Big|\, \lim_{t \to \infty} \omega_{-t/2}(P_t) = 0 \right\}, \tag{4.152}$$

$$B = \inf_{\{P_t\}} \left\{ \lim_{t \to \infty} \frac{1}{t} \log \omega_{t/2}(\mathbb{1} - P_t) \,\Big|\, \lim_{t \to \infty} \omega_{-t/2}(P_t) = 0 \right\},$$

where again in the last case the infimum is taken over all families of tests $\{P_t\}_{t>0}$ for which $\lim_{t \to \infty} t^{-1} \log \omega_t(\mathbb{1} - P_t)$ exists.
We shall call the numbers defined in (4.151) and (4.152) the Stein exponents. Clearly, $\underline{B}_\epsilon \leq \overline{B}_\epsilon \leq B_\epsilon$, $\underline{B} \leq \overline{B} \leq B$, $\underline{B}_\epsilon \leq \underline{B}$, $\overline{B}_\epsilon \leq \overline{B}$, $B_\epsilon \leq B$. If (A9) holds, then for any $\epsilon \in]0, 1[$,

$$\underline{B} = \overline{B} = B = \underline{B}_\epsilon = \overline{B}_\epsilon = B_\epsilon = -\langle \sigma \rangle_+,$$

see (Jakšić, Ogata, Pillet, and Seiringer, 2011c). Results of this type are called quantum Stein Lemma.
Stein's Lemma asserts that for any family of projections P_t such that

$$\sup_{t>0} \omega_{-t}(P_t) < 1, \tag{4.153}$$

one has

$$\liminf_{t \to \infty} \frac{1}{t} \log \omega_t(\mathbb{1} - P_t) \geq -2\langle \sigma \rangle_+,$$

and that for any $\delta > 0$ one can find a sequence of projections $P_t^{(\delta)}$ satisfying (4.153) and

$$\lim_{t \to \infty} \frac{1}{t} \log \omega_t(\mathbb{1} - P_t^{(\delta)}) \leq -2\langle \sigma \rangle_+ + \delta.$$

Hence, if no restrictions are made on P_t w.r.t. ω_{-t} except (4.153) (which is needed to avoid trivial result), the optimal exponential rate of concentration of ω_t as $t \uparrow \infty$ is precisely twice the negative entropy production.

4.6.7 Large time limit: control parameters

We continue with the framework of Section 4.6.4. The infinitely extended systems $(\mathcal{O}, \tau_X, \omega_X)$ are parameterized by control parameters $X \in \mathbb{R}^n$. Recall the shorthands $\omega = \omega_0$, $\tau = \tau_0$, $\Phi = \Phi_0$, etc. We assume

(A10) For all $t > 0$ the functional $(X, Y) \mapsto e_t(X, Y)$ has an analytic continuation to the polydisc $D_{\delta, \epsilon} = \{(X, Y) \in \mathbb{C}^n \times \mathbb{C}^n \mid \max_j |X_j| < \delta, \max_j |Y_j| < \epsilon\}$ satisfying

$$\sup_{\substack{(X,Y) \in D_{\delta, \epsilon} \\ t>0}} \left| \frac{1}{t} e_t(X, Y) \right| < \infty.$$

In addition, the limit

$$e_+(X,Y) = \lim_{t\to\infty} \frac{1}{t} e_t(X,Y),$$

exists for all $(X,Y) \in D_{\delta,\epsilon} \cap (\mathbb{R}^n \times \mathbb{R}^n)$.

As in the case of (A9), establishing (A10) for physically interesting models is typically a very difficult analytical problem. Although (A10) is certainly not a minimal assumption under which the results of this section hold (for the minimal axiomatic scheme see (Jakšić, Ogata, Pautrat, and Pillet, 2011b)), it can be verified in interesting examples and allows for a transparent exposition of the material of this section.

A consequence of the first part of (A10) is that finite time linear response theory holds for $(\mathcal{O}, \tau_X^t, \omega_X)$. By Vitali's theorem, $e_+(X,Y)$ is analytic on $D_{\delta,\epsilon}$ and we have:

Proposition 4.51   *(1) For any $X \in \mathbb{R}^n$ such that $\max_j |X_j| < \delta$,*

$$\langle \mathbf{\Phi}_X \rangle_+ = \lim_{t\to\infty} \frac{1}{t} \int_0^t \omega_X \left(\mathbf{\Phi}_{Xs}\right) ds = \nabla_Y e_+(X,Y)|_{Y=0}.$$

(2) The kinetic transport coefficients defined by

$$L_{jk} = \partial_{X_k} \langle \Phi_X^{(j)} \rangle_+|_{X=0},$$

satisfy

$$L_{jk} = \lim_{t\to\infty} L_{jkt} = \lim_{t\to\infty} \frac{1}{2} \int_{-t}^t \langle \Phi^{(k)} | \Phi_s^{(j)} \rangle_\omega \left(1 - \frac{|s|}{t}\right) ds.$$

(3) The Onsager matrix $[L_{jk}]$ is symmetric and positive semi-definite.
(4) Suppose that ω is a (τ, β)-KMS state for some $\beta > 0$ and that $(\mathcal{O}, \tau, \omega)$ is mixing, that is that $\lim_{t\to\infty} \omega(A\tau^t(B)) = \omega(A)\omega(B)$ for all $A, B \in \mathcal{O}$. Then

$$L_{jk} = \lim_{t\to\infty} \frac{1}{2} \int_{-t}^t \omega(\Phi^{(j)} \Phi_s^{(k)}) ds.$$

Parts (1)–(3) are an immediate consequence of Vitali's theorem (see Proposition 4.69 in Appendix 4.9). Part (4) recovers the familiar form of the Green–Kubo formula under the assumption that for vanishing control parameters the infinitely extended system is in thermal equilibrium (and is strongly ergodic). For the proof of (4) see (Jakšić, Ogata, Pautrat, and Pillet, 2011b) or the proof of Theorem 2.3 in (Jakšić, Ogata, and Pillet, 2007).

4.6.8   Large time limit: nonequilibrium steady states (NESS)

Consider our infinitely extended system $(\mathcal{O}, \tau^t, \omega)$ and suppose

(A11) The limit

$$\lim_{t\to\infty} \omega_t(A) = \omega_+(A),$$

exists for all $A \in \mathcal{O}$. ω_+ is a stationary state called the NESS of $(\mathcal{O}, \tau^t, \omega)$.

Albeit a hard ergodic-type problem, the verification of (A11) is typically easier then the proof of (A9) or (A10). In fact, in all known nontrivial models satisfying (A9)/(A10), the proof of (A11) is a consequence of the proof of (A9)/(A10).

The structural theory of NESS was one of the central topics of the lecture notes (Aschbacher, Jakšić, Pautrat, and Pillet, 2006) and we will not discuss it here. In relation with entropic fluctuations, the NESS plays a central role in the Gallavotti–Cohen fluctuation theorem. We will not enter into this subject in these lecture notes.

4.6.9   Stability with respect to the reference state

In addition to (A11), one expects that under normal conditions any normal state $\nu \in \mathcal{N}_\omega$ is in the basin of attraction of the NESS ω_+, that is that the following holds:

(A12)

$$\lim_{t \to \infty} \nu(\tau^t(A)) = \omega_+(A),$$

for all $\nu \in \mathcal{N}_\omega$ and $A \in \mathcal{O}$.

As for (A11), in all known nontrivial models, (A12) follows from the proofs of (A9)/(A10).

(A12) is a mathematical formulation of the fact that under normal conditions the NESS and more generally the large time thermodynamics do not depend on local perturbations of the initial state ω. More specifically, in the context of open quantum systems, if the coupling V is well localized in the reservoirs, then in the TD limit (the $\mathcal{R}_j$'s becoming infinitely extended and the system S remaining finite), the effect of including V in the reference state becomes negligible for large times. In other words, the product state ω used in Sections 4.5.1–4.5.2 and the state ω_X of Section 4.5.3 become equivalent for large times. More generally, the system loses memory of any localized perturbation of its initial state.

In a similar vein one expects that, under normal conditions, the limiting entropic functionals do not depend on local perturbations of the initial state. To illustrate this point, we consider the functional $e_{\infty,+}(\alpha)$ (and assume that the reader is familiar with Araki's perturbation theory of the KMS structure). ω has a modular group ς_ω^t and if ω_W is the KMS state (at temperature -1) of the perturbed group $\varsigma_{\omega W}^t$ for some $W \in \mathcal{O}_{\text{self}}$ (which, for finite systems, amounts to set $\omega_W = e^{\log \omega + W}/\text{tr}(e^{\log \omega + W})$), then

$$\omega_W(A) = \frac{\omega(AE_W(-i))}{\omega(E_W(-i))},$$

where the cocycle E_W is given by (4.73). The set of states $\{\omega_W \mid W \in \mathcal{O}_{\text{self}}\}$ is norm dense in the (norm closed) set $\mathcal{N}_\omega$ of all normal state on $\mathcal{O}$. Since $\ell_{\omega_W \mid \omega} = W$, one has $\ell_{\omega_{W t} \mid \omega_W} = \ell_{\omega_t \mid \omega} + \tau^{-t}(W) - W$ and hence

$$\omega_+(\sigma_\omega) = \omega_+(\sigma_{\omega W}).$$

Similarly, for $\alpha \in]0, 1[$, Proposition 4.36 holds for infinitely extended systems (this can be proven either via a TD limit argument or by direct application of modular theory), and so

$$\lim_{t \to \infty} \frac{1}{t} \left(e_{\infty,t,\omega}(\alpha) - e_{\infty,t,\omega_W}(\alpha) \right) = 0.$$

Hence,

$$e_{\infty,+,\omega}(\alpha) = \lim_{t \to \infty} \frac{1}{t} e_{\infty,t,\omega}(\alpha),$$

exists iff

$$e_{\infty,+,\omega_W}(\alpha) = \lim_{t \to \infty} \frac{1}{t} e_{\infty,t,\omega_W}(\alpha),$$

exists and the limiting entropic functionals are equal. Similar stability results for other entropic functionals can be established under additional regularity assumptions (Jakšić, Ogata, Pautrat, and Pillet, 2011b).

4.6.10 Full counting statistics and quantum fluxes: a comparison

In this section we shall focus on open quantum systems described in Section 4.5. For simplicity of notation we set the chemical potentials μ_j of the reservoirs $\mathcal{R}_j$ to zero and deal only with energy fluxes Φ_j.

Full counting statistics deals with the mean entropy/energy flow operationally defined by a repeated quantum measurement. It does not refer to the measurement of a single quantum observable. In fact, surprisingly, it gives a physical interpretation to quantities which are considered unobservable from the traditional point of view: the spectral projections of a relative modular operator. Full counting statistics is of purely quantum origin and has no counterpart in classical statistical mechanics. In contrast, the energy flux observables Φ_j introduced in Section 4.5 arise by direct operator quantization of the corresponding classical observables. In this section, we take a closer look at the relation between full counting statistics and energy flux observables.

For open quantum systems, the TD limit concerns only the reservoirs $\mathcal{R}_j$, the finite quantum system $\mathcal{S}$ remaining fixed. As discussed in the previous section, if we are not interested in transient properties then we may assume, without loss of generality, that $\omega_{\mathcal{S}}$ is the chaotic state (4.60). After the TD limit is taken, the infinitely extended reservoir $\mathcal{R}_j$ is described by the quantum dynamical system $(\mathcal{O}_j, \tau_j^t, \omega_j)$, where ω_j is a (τ_j, β_j)-KMS state on $\mathcal{O}_j$. The joint system $\mathcal{R} = \mathcal{R}_1 + \cdots + \mathcal{R}_n$ is described by

$$(\mathcal{O}_{\mathcal{R}}, \tau_{\mathcal{R}}^t, \omega_{\mathcal{R}}) = \bigotimes_{j=1}^{n} (\mathcal{O}_j, \tau_j^t, \omega_j).$$

The joint but decoupled system $\mathcal{S} + \mathcal{R}$ is described by $(\mathcal{O}, \tau^t, \omega)$ where

$$\mathcal{O} = \mathcal{O}_{\mathcal{S}} \otimes \mathcal{O}_{\mathcal{R}}, \qquad \tau^t = \tau_{\mathcal{S}}^t \otimes \tau_{\mathcal{R}}^t, \qquad \omega = \omega_{\mathcal{S}} \otimes \omega_{\mathcal{R}}.$$

The interaction of $\mathcal{S}$ with $\mathcal{R}_j$ is described by a self-adjoint element $V_j \in \mathcal{O}_{\mathcal{S}} \otimes \mathcal{O}_j$. The full interaction $V = \sum_j V_j$ and the corresponding perturbed C^*-dynamics τ_V^t finally yield the quantum dynamical system $(\mathcal{O}, \tau_V^t, \omega)$ which describes the infinitely extended open quantum system. Without further comment, we shall always assume that all relevant quantities are realized as TD limit of the corresponding quantities of a sequence $\{\mathcal{Q}_M\}$ of finite, TRI open quantum systems. In particular, that is so for the energy flux observables

$$\Phi_j = \delta_j(V_j),$$

where δ_j is the generator of τ_j ($\tau_j^t = \mathrm{e}^{t\delta_j}$), and the entropy production observable

$$\sigma = -\sum_j \beta_j \Phi_j,$$

of the infinitely extended open quantum system $(\mathcal{O}, \tau_V^t, \omega)$.

Recall Section 4.5.2. Let $\mathbb{P}_t$ be the full counting statistics of the infinitely extended open systems $(\mathcal{O}, \tau_V^t, \omega)$. The probability measure $\mathbb{P}_t$ arises as the weak limit of the full counting statistics $\mathbb{P}_{M,t}$ of $\mathcal{Q}_M$ (this realization is essential for the physical interpretation of $\mathbb{P}_t$). Thus, it follows from Relations (4.131), (4.132), that

$$\langle \varepsilon_j \rangle_+ = \lim_{t\to\infty} \mathbb{E}_t(\varepsilon_j) = -\beta_j \omega_+(\Phi_j), \tag{4.154}$$

$$D_{\mathrm{fcs},jk} = \lim_{t\to\infty} t\left(\mathbb{E}_t(\varepsilon_j\varepsilon_k) - \mathbb{E}_t(\varepsilon_j)\mathbb{E}_t(\varepsilon_k)\right)$$

$$= \beta_j \beta_k \int_{-\infty}^{\infty} \omega_+\left((\Phi_j - \omega_+(\Phi_j))(\Phi_{kt} - \omega_+(\Phi_k))\right) \mathrm{d}t. \tag{4.155}$$

Here, ω_+ is the NESS of $(\mathcal{O}, \tau_V^t, \omega)$ and we have assumed that the correlation function

$$t \mapsto \omega_+\left((\Phi_j - \omega_+(\Phi_j))(\Phi_{kt} - \omega_+(\Phi_k))\right),$$

is integrable on $\mathbb{R}$.

The fluctuations of $\mathbb{P}_t$ as $t \to \infty$ are described by a central limit theorem and a large deviation principle. The central limit theorem holds if for all $\alpha \in \mathbb{R}^n$,

$$\lim_{t\to\infty} \int_{\mathbb{R}^n} \mathrm{e}^{\mathrm{i}\sqrt{t}\alpha\cdot(\varepsilon - \langle \varepsilon \rangle_+)} \mathrm{d}\mathbb{P}_t(\varepsilon) = \int_{\mathbb{R}^n} \mathrm{e}^{\mathrm{i}\alpha\cdot\varepsilon} \mathrm{d}\mu_{\mathbf{D}_{\mathrm{fcs}}}(\varepsilon),$$

where $\mu_{\mathbf{D}_{\mathrm{fcs}}}$ is the centred Gaussian measure on $\mathbb{R}^n$ with covariance $\mathbf{D}_{\mathrm{fcs}} = [D_{\mathrm{fcs},jk}]$. To discuss the large deviation principle, recall that

$$e_{2,t}(\alpha) = \log \int_{\mathbb{R}^n} \mathrm{e}^{-t\alpha\cdot\varepsilon} \mathrm{d}\mathbb{P}_t(\varepsilon).$$

Suppose that

$$e_{2,+}(\alpha) = \lim_{t\to\infty} \frac{1}{t} e_t(\alpha),$$

exists for $\boldsymbol{\alpha} \in \mathbb{R}^n$ and satisfies the conditions of Gärtner–Ellis theorem (Theorem 4.66 in Appendix 4.8.3). Then for any Borel set $G \subset \mathbb{R}^d$,

$$- \inf_{\mathbf{s} \in \mathrm{int}(G)} I(\mathbf{s}) \leq \liminf_{t \to \infty} \frac{1}{t} \log \mathbb{P}_t(G) \leq \limsup_{t \to \infty} \frac{1}{t} \log \mathbb{P}_t(G) \leq - \inf_{\mathbf{s} \in \mathrm{cl}(G)} I(\mathbf{s}),$$

where

$$I(\mathbf{s}) = - \inf_{\boldsymbol{\alpha} \in \mathbb{R}^n} (\mathbf{s} \cdot \boldsymbol{\alpha} + e_{2,+}(\boldsymbol{\alpha})).$$

Note that $I(\mathbf{s})$ satisfies the Evans–Searles symmetry

$$I(-\mathbf{s}) = \mathbf{1} \cdot \mathbf{s} + I(\mathbf{s}).$$

For some models the central limit theorem and the large deviation principle can be proven following the spectral scheme outlined in Section 4.6.5 (for example, this is the case for the Spin-Fermion model, see Section 4.7.5). For other models, scattering techniques are effective (see Section 4.7.6). In general, however, verifications of central limit theorem and large deviation principle are difficult problems.

Let now

$$X_j = \beta_{\mathrm{eq}} - \beta_j,$$

be the thermodynamic forces. The new reference state ω_X is the TD limit of the states $\omega_{M,X}$ of the finite open quantum systems $\mathcal{Q}_M$. Alternatively, ω_X can be described directly in terms of the modular structure, see (Jakšić, Ogata, and Pillet, 2006). ω_X is modular and normal w.r.t. ω. The entropy production observables of $(\mathcal{O}, \tau_V^t, \omega_X)$ is

$$\sigma_X = \sum_{j=1}^{n} X_j \Phi_j.$$

The NESS ω_{X+} also depends on X and, for $X = 0$, reduces to a $(\tau_V, \beta_{\mathrm{eq}})$-KMS state $\omega_{\beta_{\mathrm{eq}}}$. Let $e_t(X, Y)$ be the entropic functional of the infinitely extended system $(\mathcal{O}, \tau_V^t, \omega_X)$ and suppose that (A10) holds. Then Proposition 4.51 implies that the transport coefficients

$$L_{jk} = \partial_{X_k} \omega_{X+}(\Phi_j)|_{X=0},$$

are defined, satisfy the Onsager reciprocity relations

$$L_{jk} = L_{kj},$$

and the Green–Kubo formulas

$$L_{jk} = \frac{1}{2} \int_{-\infty}^{\infty} \omega_{\beta_{\mathrm{eq}}}(\Phi_j \Phi_{kt}) \mathrm{d}t.$$

Here we have assumed that the quantum dynamical system $(\mathcal{O}, \tau_V^t, \omega_{\beta_{\mathrm{eq}}})$ is mixing and that the correlation function $t \mapsto \omega_{\beta_{\mathrm{eq}}}(\Phi_j \Phi_{kt})$ is integrable.

The linear response theory derived for quantum fluxes Φ_j immediately yields the linear response theory for the full counting statistics. Indeed, it follows from the formulas (4.154) and (4.155) that

$$L_{\text{fcs},kj} = \partial_{X_k} \langle \varepsilon_j \rangle_+ |_{X=0} = -\beta_{\text{eq}} L_{kj} = -\frac{1}{\beta_{\text{eq}}} D_{\text{fcs},kj} |_{X=0}.$$

The last relation also yields the fluctuation–dissipation theorem for the full counting statistics. The Einstein relation takes the form

$$L_{\text{fcs},kj} = -\frac{1}{2\beta_{\text{eq}}} D_{\text{fcs},kj} |_{X=0}.$$

and relates the kinetic transport coefficients of the full counting statistics to its fluctuations in thermal equilibrium. The factor $-\beta_{\text{eq}}^{-1}$ is due to our choice to keep the entropic form of the full counting statistics in the discussion of energy transport. In the energy form of the full counting statistics one considers $\mathbb{E}_t(-\varepsilon_j/\beta_j)$ and then the Einstein relation holds in the usual form $L_{\text{fcs},kj} = \frac{1}{2} D_{\text{fcs},kj} |_{X=0}$. The disadvantage of the energy form is that the Evans–Searles symmetry has to be scaled. The choice between scaling Einstein relations or scaling symmetries is of course of no substance.

At this point let us introduce a 'naive' cumulant generating function

$$e_{\text{naive},t}(\alpha) = \log \omega \left(e^{-\sum_{j=1}^{n} \alpha_j \beta_j \int_0^t \Phi_{js} ds} \right), \tag{4.156}$$

and the 'naive' cumulants

$$\chi_t(k_1, \ldots, k_n) = \partial_{\alpha_1}^{k_1} \cdots \partial_{\alpha_n}^{k_n} e_{\text{naive},t}(\alpha)|_{\alpha=0}.$$

The function $e_{\text{naive},t}(\alpha)$ is the direct quantization of the classical cumulant generating function for the entropy transfer

$$\mathbf{S}^t = (S_1^t, \ldots, S_n^t) = \int_0^t (-\beta_1 \Phi_{1s}, \ldots, -\beta_n \Phi_{ns}) ds,$$

in the state ω. Except in the special case $\alpha = \alpha \mathbf{1}$, $e_{\text{naive},t}(\alpha)$ cannot be described in terms of classical probability, that is $e_{\text{naive},t}(\alpha)$ *is not* the cumulant generating function of a probability measure on $\mathbb{R}^n$. If $\alpha = \alpha \mathbf{1}$, then

$$e_{\text{naive},t}(\alpha \mathbf{1}) = \log \omega \left(e^{\alpha \int_0^t \sigma_s ds} \right) = \log \int_{\mathbb{R}} e^{t\alpha s} d\mu_{\omega,t}(s),$$

where, in the GNS-representation of $\mathcal{O}$ associated to ω, $\mu_{\omega,t}$ is the spectral measure for $t^{-1} \int_0^t \pi_\omega(\sigma_s) ds$ and ξ_ω.

In general the functional $e_{\text{naive},t}(\alpha)$ will not satisfy the Evans–Searles symmetry, that is $e_{\text{naive},t}(\mathbf{1} - \alpha) \neq e_{\text{naive},t}(\alpha)$, and the same remark applies to the limiting functional

$$e_{\text{naive},+}(\alpha) = \lim_{t \to \infty} \frac{1}{t} e_{\text{naive},t}(\alpha),$$

which, we assume, exists and is differentiable on some open set containing $\mathbf{0}$. One easily checks that the first and second order cumulants satisfy

$$\partial_{\alpha_j} e_{\mathrm{naive},t}(\boldsymbol{\alpha})|_{\boldsymbol{\alpha}=0} = \partial_{\alpha_j} e_{2,t}(\boldsymbol{\alpha})|_{\boldsymbol{\alpha}=0},$$

$$\partial_{\alpha_k} \partial_{\alpha_j} e_{\mathrm{naive},t}(\boldsymbol{\alpha})|_{\boldsymbol{\alpha}=0} = \partial_{\alpha_k} \partial_{\alpha_j} e_{2,t}(\boldsymbol{\alpha})|_{\boldsymbol{\alpha}=0},$$

and if the limits and derivatives could be exchanged,

$$\partial_{\alpha_j} e_{\mathrm{naive},+}(\boldsymbol{\alpha})|_{\boldsymbol{\alpha}=0} = \partial_{\alpha_j} e_{2,+}(\boldsymbol{\alpha})|_{\boldsymbol{\alpha}=0},$$

$$\partial_{\alpha_k} \partial_{\alpha_j} e_{\mathrm{naive},+}(\boldsymbol{\alpha})|_{\boldsymbol{\alpha}=0} = \partial_{\alpha_k} \partial_{\alpha_j} e_{2,+}(\boldsymbol{\alpha})|_{\boldsymbol{\alpha}=0}.$$

We summarize our observations:

(i) The first and second order cumulants of the full counting statistics are the same as the corresponding 'naive' quantum energy flux cumulants, that is the direct quantization of the classical energy flux cumulants. In general, the higher order 'naive' cumulants do not coincide with the corresponding cumulants of the full counting statistics.

(ii) The limiting expectation $\langle \boldsymbol{\varepsilon} \rangle_+$ and covariance $\mathbf{D}_{\mathrm{fcs}}$ of the full counting statistics are expressed in terms of the NESS ω_+ and quantized fluxes Φ_j. They are a direct quantization of the corresponding classical expressions. The same remark applies to the central limit theorem, linear response theory and fluctuation–dissipation theorem. If the full counting statistics is restricted to the entropy production observable, then its limiting expectation, covariance, and central limit theorem coincide with those of the spectral measure for $t^{-1} \int_0^t \sigma_s ds$ and ω.

(iii) We emphasize: to detect the difference between full counting statistics and the 'naive' cumulant generating function one needs to consider cumulants of at least third order. In Section 4.7 we shall illustrate this point on some examples of physical interest.

4.7 Fermionic systems

In this section we discuss nonequilibrium statistical mechanics of fermionic systems and describe several physically relevant models to which the structural theory developed in these lecture notes applies.

4.7.1 Second quantization

We start with some notation. Let $\mathcal{Q}$ be a finite set. $\ell^2(\mathcal{Q})$ denotes the Hilbert space of all function $f : \mathcal{Q} \to \mathbb{C}$ equipped with the inner product

$$\langle f|g \rangle = \sum_{q \in \mathcal{Q}} \overline{f(q)} g(q).$$

The functions $\{\delta_q \mid q \in \mathcal{Q}\}$, where $\delta_q(x) = 1$ if $x = q$ and 0 otherwise, form an orthonormal basis for $\ell^2(\mathcal{Q})$. Any Hilbert space of dimension $|\mathcal{Q}|$ is isomorphic to $\ell^2(\mathcal{Q})$.

Let the configuration space of a single particle be the finite set $\mathcal{Q}$. Typically, $\mathcal{Q}$ will be a subset of some lattice, but at this point we do not need to specify its structure further. The Hilbert space of a single particle is $\mathcal{K} = \ell^2(\mathcal{Q})$. If $\psi \in \mathcal{K}$ is a normalized wave function, then $|\psi(q)|^2$ is probability that the particle is located at $q \in \mathcal{Q}$. The configuration space of a system of n distinguishable particles is $\mathcal{Q}^n$ and $\ell^2(\mathcal{Q}^n)$ is its Hilbert space. For $q = (q_1, \ldots, q_n) \in \mathcal{Q}^n$ we set $\delta_q(x_1, \ldots, x_n) = \delta_{q_1}(x_1) \cdots \delta_{q_n}(x_n)$. $\{\delta_q \mid q \in \mathcal{Q}^n\}$ is an orthonormal basis of $\ell^2(\mathcal{Q}^n)$. Let $\mathcal{K}^{\otimes n}$ be the n-fold tensor product of $\mathcal{K}$ with itself. Identifying δ_q with $\delta_{q_1} \otimes \cdots \otimes \delta_{q_n}$ we obtain an isomorphism between $\ell^2(\mathcal{Q}^n)$ and $\mathcal{K}^{\otimes n}$. In the following we shall identify these two spaces.

If $\psi \in \mathcal{K}^{\otimes n}$ is the normalized wave function of the system of n particles and $\psi_1, \ldots, \psi_n \in \mathcal{K}$ are normalized one-particle wave functions, then $|\langle \psi | \psi_1 \otimes \cdots \otimes \psi_n \rangle|^2$ is the probability for the j-th particle to be in the state ψ_j, $j = 1, \ldots, n$. According to Pauli's principle, if the particles are identical fermions, then this probability must vanish if at least two of the ψ_j's are equal. It follows that the multi-linear functional $F(\psi_1, \ldots, \psi_n) = \langle \psi | \psi_1 \otimes \cdots \otimes \psi_n \rangle$ has to vanish if at least two of its arguments coincide. Hence, for $j \neq k$,

$$F(\psi_1, \ldots, \psi_j + \psi_k, \ldots, \psi_k + \psi_j, \ldots, \psi_n) = 0,$$

for any $\psi_1, \ldots, \psi_n \in \mathcal{K}$. By multi-linearity, this is equivalent to

$$0 = F(\psi_1, \ldots, \psi_j, \ldots, \psi_k, \ldots, \psi_n) + F(\psi_1, \ldots, \psi_j, \ldots, \psi_j, \ldots, \psi_n)$$
$$+ F(\psi_1, \ldots, \psi_k, \ldots, \psi_k, \ldots, \psi_n) + F(\psi_1, \ldots, \psi_k, \ldots, \psi_j, \ldots, \psi_n)$$
$$= F(\psi_1, \ldots, \psi_j, \ldots, \psi_k, \ldots, \psi_n) + F(\psi_1, \ldots, \psi_k, \ldots, \psi_j, \ldots, \psi_n),$$

and we conclude that F must be alternating, that is changing sign under transposition of two of its arguments,

$$F(\psi_1, \ldots, \psi_j, \ldots, \psi_k, \ldots, \psi_n) = -F(\psi_1, \ldots, \psi_k, \ldots, \psi_j, \ldots, \psi_n). \tag{4.157}$$

Let S_n be the group of permutations of the set $\{1, \ldots, n\}$. For $\pi \in S_n$ we set

$$\pi \psi_1 \otimes \cdots \otimes \psi_n = \psi_{\pi(1)} \otimes \cdots \otimes \psi_{\pi(n)},$$

and extend this definition to $\mathcal{K}^{\otimes n}$ by linearity. One easily checks that this action of S_n on $\mathcal{K}^{\otimes n}$ is unitary. If $\pi = (jk) = \pi^{-1}$ is the transposition whose only effect is to interchange j and k, then (4.157) is equivalent to

$$\langle \pi \psi | \psi_1 \otimes \cdots \otimes \psi_n \rangle = \langle \psi | \pi \psi_1 \otimes \cdots \otimes \psi_n \rangle = -\langle \psi | \psi_1 \otimes \cdots \otimes \psi_n \rangle,$$

and so $\pi \psi = -\psi$. More generally, if π is the composition of m transpositions, $\pi = (j_1 k_1) \cdots (j_m k_m)$, then we must have $\pi \psi = (-1)^m \psi$. Any permutation $\pi \in S_n$ can be decomposed into a product of transpositions and the corresponding number $(-1)^m$, the signature of π, is denoted by $\text{sign}(\pi)$ (one can show that $\text{sign}(\pi) = (-1)^t$ where t is the number of pairs $(j, k) \in \{1, \ldots n\}$ such that $j < k$ and $\pi(j) > \pi(k)$). We conclude that the wave function ψ of a system of n identical fermions must satisfy

$$\pi\psi = \text{sign}(\pi)\psi,$$

for all $\pi \in S_n$. More explicitly, for $\pi \in S_n$ the wave function ψ satisfies

$$\psi(x_{\pi(1)}, \ldots, x_{\pi(n)}) = \text{sign}(\pi)\psi(x_1, \ldots, x_n). \tag{4.158}$$

Functions satisfying (4.158) are called completely antisymmetric. The set of all completely antisymmetric functions on $\mathcal{Q}^n$ is a subspace of $\ell^2(\mathcal{Q}^n)$ which we denote by $\ell^2_-(\mathcal{Q}^n)$.

Exercise 4.47

1. Show that the orthogonal projection P_- on $\ell^2_-(\mathcal{Q}^n)$ is given by

$$P_-\psi = \frac{1}{n!} \sum_{\pi \in S_n} \text{sign}(\pi)\pi\psi.$$

Hint: use the morphism property of the signature, $\text{sign}(\pi \circ \pi') = \text{sign}(\pi)\text{sign}(\pi')$, to show that $\pi P_- = \text{sign}(\pi)P_-$.

2. Define the wedge product of $\psi_1, \ldots, \psi_n \in \mathcal{K}$ by

$$\psi_1 \wedge \cdots \wedge \psi_n = \sqrt{n!} \, P_-\psi_1 \otimes \cdots \otimes \psi_n,$$

and show that

$$\langle \psi_1 \wedge \cdots \wedge \psi_n | \phi_1 \wedge \cdots \wedge \phi_n \rangle = \det[\langle \psi_i | \phi_j \rangle]_{1 \leq i,j \leq n}. \tag{4.159}$$

Hint: use the Leibnitz formula

$$\det A = \sum_{\pi \in S_n} \text{sign}(\pi) A_{1\pi(1)} \cdots A_{n\pi(n)},$$

for the determinant of the $n \times n$ matrix $A = [A_{jk}]$.

3. Denote by $\mathcal{K}^{\wedge n}$ the linear span of the set $\{\psi_1 \wedge \cdots \wedge \psi_n \mid \psi_1, \ldots, \psi_n \in \mathcal{K}\}$. Suppose that $n \leq d = |\mathcal{Q}| = \dim \mathcal{K}$ and let $\{\phi_1, \ldots, \phi_d\}$ be an orthonormal basis of $\mathcal{K}$. Prove that

$$\{\phi_{j_1} \wedge \cdots \wedge \phi_{j_n} \mid 1 \leq j_1 < \cdots < j_n \leq d\},$$

is an orthonormal basis of $\mathcal{K}^{\wedge n}$ and deduce that

$$\dim \mathcal{K}^{\wedge n} = \binom{\dim \mathcal{K}}{n}.$$

In particular, the vector space $\mathcal{K}^{\wedge \dim \mathcal{K}}$ is one dimensional. For $n > \dim \mathcal{K}$ the vector spaces $\mathcal{K}^{\wedge n}$ are trivial, that is, consist only of the zero vector.

According to our identification of $\mathcal{K}^{\otimes n}$ with $\ell^2(\mathcal{Q}^n)$, the subspaces $\ell_-(\mathcal{Q}^n)$ and $\mathcal{K}^{\wedge n}$ coincide (they are both the range of the projection P_-). We denote by

$$\Gamma_n(\mathcal{K}) = \mathcal{K}^{\wedge n},$$

the Hilbert space of a system of n fermions with the single particle Hilbert space $\mathcal{K}$. By definition, $\Gamma_0(\mathcal{K}) = \mathbb{C}$ is the vacuum sector.

For $A \in \mathcal{O}_\mathcal{K}$ and $n \geq 1$, let $\Gamma_n(A)$ and $d\Gamma_n(A)$ be the elements of $\mathcal{O}_{\Gamma_n(\mathcal{K})}$ defined by

$$\Gamma_n(A)(\psi_1 \wedge \cdots \wedge \psi_n) = A\psi_1 \wedge \cdots \wedge A\psi_n,$$

$$d\Gamma_n(A)(\psi_1 \wedge \cdots \wedge \psi_n) = A\psi_1 \wedge \cdots \wedge \psi_n + \cdots + \psi_1 \wedge \cdots \wedge A\psi_n.$$

For $n = 0$, we define $\Gamma_0(A)$ to be the identity map on $\Gamma_0(\mathcal{K})$ and set $d\Gamma_0(A) = 0$. One easily checks the relations

$$\Gamma_n(A^*) = \Gamma_n(A)^*, \qquad\qquad d\Gamma_n(A^*) = d\Gamma_n(A)^*,$$

$$\Gamma_n(AB) = \Gamma_n(A)\Gamma_n(B), \qquad d\Gamma_n(A + \lambda B) = d\Gamma_n(A) + \lambda d\Gamma_n(B), \qquad (4.160)$$

$$d\Gamma_n(A) = \frac{d}{dt}\Gamma_n(e^{tA})\Big|_{t=0}, \qquad\qquad \Gamma_n(e^A) = e^{d\Gamma_n(A)},$$

for $A, B \in \mathcal{O}_\mathcal{K}$ and $\lambda \in \mathbb{C}$. The fermionic Fock space over $\mathcal{K}$ is defined by

$$\Gamma(\mathcal{K}) = \bigoplus_{n=0}^{\dim \mathcal{K}} \Gamma_n(\mathcal{K}),$$

that is as the set of vectors $\Psi = (\psi_0, \psi_1, \ldots)$ with $\psi_n \in \Gamma_n(\mathcal{K})$ and the inner product

$$(\Psi|\Phi) = \sum_{n=0}^{\dim \mathcal{K}} \langle \psi_n | \phi_n \rangle.$$

Clearly,

$$\dim \Gamma(\mathcal{K}) = \sum_{n=0}^{\dim \mathcal{K}} \dim \Gamma_n(\mathcal{K}) = \sum_{n=0}^{\dim \mathcal{K}} \binom{\dim \mathcal{K}}{n} = 2^{\dim \mathcal{K}}.$$

A normalized vector $\Psi = (\psi_0, \psi_1, \ldots) \in \Gamma(\mathcal{K})$ is interpreted as a state of a gas of identical fermions with one particle Hilbert space $\mathcal{K}$ in the following way. Setting $p_n = \|\psi_n\|^2$, $\phi_n = \psi_n/\|\psi_n\|$ and $\Phi^{(n)} = (0, \ldots, \phi_n, \ldots, 0)$ one can write Ψ as

$$\Psi = \sum_{n=0}^{\dim \mathcal{K}} \sqrt{p_n}\, \Phi^{(n)},$$

a coherent superposition of:

- a state $\Phi^{(0)}$ with no particle. Up to a phase factor, $\Phi^{(0)}$ is the so called vacuum vector $\Omega = (1, 0, \ldots, 0)$;
- a state $\Phi^{(1)}$ with 1 particle in the state $\phi_1 \in \mathcal{K}$;
- a state $\Phi^{(2)}$ with 2 particles in the state $\phi_2 \in \Gamma_2(\mathcal{K})$, etc.

Since the vectors $\Phi^{(n)}$ are mutually orthogonal, p_n is the probability for n particles to be present in the system. Pauli's principle forbids more than $\dim \mathcal{K}$ particles. With a slight abuse of notation, we shall identify the n-particle wave-function $\phi \in \Gamma_n(\mathcal{K})$ with the vector $\Phi = (0, \ldots, \phi, \ldots, 0) \in \Gamma(\mathcal{K})$.

For $A \in \mathcal{O}_\mathcal{K}$ one defines $\Gamma(A)$ and $\mathrm{d}\Gamma(A)$ in $\mathcal{O}_{\Gamma(\mathcal{K})}$ by

$$\Gamma(A) = \bigoplus_{n=0}^{\dim \mathcal{K}} \Gamma_n(A), \qquad \mathrm{d}\Gamma(A) = \bigoplus_{n=0}^{\dim \mathcal{K}} \mathrm{d}\Gamma_n(A).$$

Relations (4.160) yield

$$\Gamma(A^*) = \Gamma(A)^*, \qquad\qquad \mathrm{d}\Gamma(A^*) = \mathrm{d}\Gamma(A)^*,$$

$$\Gamma(AB) = \Gamma(A)\Gamma(B), \qquad\qquad \mathrm{d}\Gamma(A + \lambda B) = \mathrm{d}\Gamma(A) + \lambda \mathrm{d}\Gamma(B),$$

$$\mathrm{d}\Gamma(A) = \left.\frac{\mathrm{d}}{\mathrm{d}t}\Gamma(\mathrm{e}^{tA})\right|_{t=0}, \qquad\qquad \Gamma(\mathrm{e}^A) = \mathrm{e}^{\mathrm{d}\Gamma(A)}.$$

Note that $\Gamma(A)$ is invertible iff A is invertible and in this case $\Gamma(A)^{-1} = \Gamma(A^{-1})$. Moreover, one easily checks that

$$\Gamma(A)\mathrm{d}\Gamma(B)\Gamma(A^{-1}) = \mathrm{d}\Gamma(ABA^{-1}). \tag{4.161}$$

In particular, one has

$$\mathrm{e}^{t\mathrm{d}\Gamma(A)}\mathrm{d}\Gamma(B)\mathrm{e}^{-t\mathrm{d}\Gamma(A)} = \Gamma(\mathrm{e}^{tA})\mathrm{d}\Gamma(B)\Gamma(\mathrm{e}^{-tA}) = \mathrm{d}\Gamma(\mathrm{e}^{tA}B\mathrm{e}^{-tA}).$$

which, upon differentiation at $t = 0$, yields

$$[\mathrm{d}\Gamma(A), \mathrm{d}\Gamma(B)] = \mathrm{d}\Gamma([A, B]). \tag{4.162}$$

The reader familiar with Lie groups will recognize $A \mapsto \Gamma(A)$ as a representation of the linear group $\mathrm{GL}(\mathcal{K})$ in $\Gamma(\mathcal{K})$ and $B \mapsto \mathrm{d}\Gamma(B)$ as the induced representation of its Lie algebra $\mathcal{O}_\mathcal{K}$.

Example 4.52 $N = \mathrm{d}\Gamma(\mathbb{1})$ is called the number operator. Since

$$N|_{\Gamma_n(\mathcal{K})} = n\mathbb{1}_{\Gamma_n(\mathcal{K})},$$

N is the observable describing the number of particles in the system.

We finish this section with a result that will be important in Section 4.7.3.

Lemma 4.53 *For any $A \in \mathcal{O}_{\mathcal{K}}$, one has*

$$\mathrm{tr}(\Gamma(A)) = \det(\mathbb{1} + A).$$

Proof: We first prove the result for self-adjoint A. Let $\{\psi_1, \ldots, \psi_d\}$ be an eigenbasis of A such that $A\psi_j = \lambda_j \psi_j$. Since

$$\det(\mathbb{1} + A) = \prod_{j=1}^{d}(1 + \lambda_j) = \sum_{J \subset \{1,\ldots,d\}} \prod_{k \in J} \lambda_k$$

$$= \sum_{n=0}^{d} \sum_{\substack{J \subset \{1,\ldots,d\} \\ |J|=n}} \prod_{k \in J} \lambda_k = \sum_{n=0}^{d} \sum_{1 \leq j_1 < \cdots < j_n \leq d} \lambda_{j_1} \cdots \lambda_{j_n},$$

and $\lambda_{j_1} \cdots \lambda_{j_n} = \langle \psi_{j_1} \wedge \cdots \wedge \psi_{j_n} | \Gamma_n(A) \psi_{j_1} \wedge \cdots \wedge \psi_{j_n} \rangle$, it follows from Part 3 of Exercise 4.7.1 that

$$\sum_{1 \leq j_1 < \cdots < j_n \leq d} \lambda_{j_1} \cdots \lambda_{j_n} = \mathrm{tr}_{\Gamma_n(\mathcal{K})}(\Gamma_n(A)).$$

Hence,

$$\det(\mathbb{1} + A) = \sum_{n=0}^{d} \mathrm{tr}_{\Gamma_n(\mathcal{K})}(\Gamma_n(A)) = \mathrm{tr}(\Gamma(A)),$$

holds for self-adjoint A. If A is not self-adjoint, we set

$$A(\lambda) = \frac{A + A^*}{2} + \lambda \frac{A - A^*}{2\mathrm{i}}.$$

Clearly, $A(\lambda)$ is self-adjoint for $\lambda \in \mathbb{R}$ and so $\det(\mathbb{1} + A(\lambda)) = \mathrm{tr}(\Gamma(A(\lambda)))$. Since both sides of this identity are analytic functions of λ (in fact, polynomials), the identity extends to the value $\lambda = \mathrm{i}$ for which $A(\mathrm{i}) = A$.   □

4.7.2   The canonical anticommutation relations (CAR)

For $\psi, \psi_1, \ldots, \psi_n \in \mathcal{K}$ we set

$$a^*(\psi)\Omega = \psi,$$

$$a^*(\psi)(\psi_1 \wedge \cdots \wedge \psi_n) = \psi \wedge \psi_1 \wedge \cdots \wedge \psi_n.$$

By linearity, $a^*(\psi)$ extends to an element of $\mathcal{O}_{\Gamma(\mathcal{K})}$ which maps $\Gamma_n(\mathcal{K})$ into $\Gamma_{n+1}(\mathcal{K})$ and in particular $\Gamma_{\dim \mathcal{K}}(\mathcal{K})$ to $\{0\}$. Since $a^*(\psi)$ acts on a state Ψ by adding to it a particle in the state ψ, it is called creation operator. We note that

$$\psi_1 \wedge \cdots \wedge \psi_n = a^*(\psi_1) \cdots a^*(\psi_n)\Omega.$$

Similarly, one defines an element $a(\psi)$ of $\mathcal{O}_{\Gamma(\mathcal{K})}$ by

$$a(\psi)\Omega = 0,$$

$$a(\psi)\psi_1 = \langle \psi | \psi_1 \rangle \Omega,$$

$$a(\psi)(\psi_1 \wedge \cdots \wedge \psi_n) = \sum_{j=1}^{n}(-1)^{1+j}\langle \psi | \psi_j \rangle \, \psi_1 \wedge \cdots \wedge \hat{\psi_j} \wedge \cdots \wedge \psi_n.$$

$a(\psi)$ maps $\Gamma_n(\mathcal{K})$ into $\Gamma_{n-1}(\mathcal{K})$ and in particular $\Gamma_0(\mathcal{K})$ to $\{0\}$. Since it acts on a state Ψ by removing from it a particle in the state ψ, it is called annihilation operator. In the sequel, $a^{\#}(\psi)$ denotes either $a^*(\psi)$ or $a(\psi)$. The basic properties of creation and annihilation operators are summarized in

Proposition 4.54   *(1) The map $\psi \mapsto a^*(\psi)$ is linear and the map $\psi \mapsto a(\psi)$ is antilinear.*

(2) $a(\psi)^* = a^*(\psi)$.

(3) The canonical anticommutation relations (CAR) *hold:*

$$\{a(\psi), a(\phi)\} = \{a^*(\psi), a^*(\phi)\} = 0, \qquad \{a(\psi), a^*(\phi)\} = \langle \psi | \phi \rangle \mathbb{1},$$

where $\{A, B\} = AB + BA$ denotes the anticommutator of A and B.

(4) The family of operators $\mathfrak{A} = \{a^{\#}(\psi) \,|\, \psi \in \mathcal{K}\}$ is irreducible in $\mathcal{O}_{\Gamma(\mathcal{K})}$, that is,

$$\mathfrak{A}' = \{B \in \mathcal{O}_{\Gamma(\mathcal{K})} \,|\, [A, B] = 0 \text{ for all } A \in \mathfrak{A}\} = \mathbb{C}\mathbb{1}_{\Gamma(\mathcal{K})}.$$

(5) $\|a^*(\psi)\| = \|a(\psi)\| = \|\psi\|$.

(6) For any $A \in \mathcal{O}_{\mathcal{K}}$,

$$\Gamma(A)a^*(\psi) = a^*(A\psi)\Gamma(A), \qquad \Gamma(A^*)a(A\psi) = a(\psi)\Gamma(A^*).$$

In particular, if U is unitary,

$$\Gamma(U)a^{\#}(\psi)\Gamma(U^*) = a^{\#}(U\psi).$$

(7) For any $A \in \mathcal{O}_{\mathcal{K}}$,

$$[\mathrm{d}\Gamma(A), a^*(\psi)] = a^*(A\psi), \qquad [\mathrm{d}\Gamma(A), a(\psi)] = -a(A^*\psi).$$

In particular, if A is self-adjoint,

$$\mathrm{i}[\mathrm{d}\Gamma(A), a^{\#}(\psi)] = a^{\#}(\mathrm{i}A\psi).$$

(8) $a^*(\phi)a(\psi) = \mathrm{d}\Gamma(|\phi\rangle\langle\psi|)$.

(9) For any $A \in \mathcal{O}_{\mathcal{K}}$ and any orthonormal basis $\{\psi_1, \ldots, \psi_d\}$ of $\mathcal{K}$ one has

$$\mathrm{d}\Gamma(A) = \sum_{j,k=1}^{d} \langle \psi_j | A\psi_k \rangle a^*(\psi_j)a(\psi_k).$$

Proof: (1) is obvious from the definitions of the creation/annihilation operators. (2) follows from Laplace formula for developing the determinant of a $n \times n$ matrix A along one of its rows,

$$\det A = \sum_{j=1}^{n} (-1)^{i+j} A_{ij} \det A_{(ij)}, \qquad (4.163)$$

where $A_{(ij)}$ denotes the matrix obtained from A by removing its i-th row and j-th column. Indeed, by (4.159)

$$\langle \phi_1 \wedge \cdots \wedge \phi_{n-1} | a^*(\psi)^* \psi_1 \wedge \cdots \wedge \psi_n \rangle = \langle a^*(\psi)\phi_1 \wedge \cdots \wedge \phi_{n-1} | \psi_1 \wedge \cdots \wedge \psi_n \rangle$$
$$= \langle \psi \wedge \phi_1 \wedge \cdots \wedge \phi_{n-1} | \psi_1 \wedge \cdots \wedge \psi_n \rangle$$
$$= \det A,$$

where

$$A = \begin{bmatrix} \langle \psi | \psi_1 \rangle & \langle \psi | \psi_2 \rangle & \cdots & \langle \psi | \psi_n \rangle \\ \langle \phi_1 | \psi_1 \rangle & \langle \phi_1 | \psi_2 \rangle & \cdots & \langle \phi_1 | \psi_n \rangle \\ \vdots & \vdots & \ddots & \vdots \\ \langle \phi_{n-1} | \psi_1 \rangle & \langle \phi_{n-1} | \psi_2 \rangle & \cdots & \langle \phi_{n-1} | \psi_n \rangle \end{bmatrix}.$$

Developing the determinant of A along its first row and using the fact that

$$\det A_{(1j)} = \langle \phi_1 \wedge \cdots \wedge \phi_{n-1} | \psi_1 \wedge \cdots \wedge \widehat{\psi_j} \wedge \cdots \wedge \psi_n \rangle,$$

we obtain

$$\det A = \sum_{j=1}^{n} (-1)^{1+j} \langle \psi | \psi_j \rangle \langle \phi_1 \wedge \cdots \wedge \phi_{n-1} | \psi_1 \wedge \cdots \wedge \widehat{\psi_j} \wedge \cdots \wedge \psi_n \rangle,$$

from which we conclude that

$$a^*(\psi)^* \psi_1 \wedge \cdots \wedge \psi_n = \sum_{j=1}^{n} (-1)^{1+j} \langle \psi | \psi_j \rangle \psi_1 \wedge \cdots \wedge \widehat{\psi_j} \wedge \cdots \wedge \psi_n.$$

Hence, $a^*(\psi)^* = a(\psi)$ and the statement follows.
(3) The relation $\{a^*(\psi), a^*(\phi)\} = 0$ follows from the fact that $\psi \wedge \phi \wedge \psi_1 \cdots \wedge \psi_n$ changes sign when ψ and ϕ are exchanged. The relation $\{a(\psi), a(\phi)\} = 0$ is obtained by conjugating the previous relation. Finally, adding the two formulas

$$a^*(\phi)a(\psi)\psi_1 \wedge \cdots \wedge \psi_n = \sum_{j=1}^{n}(-1)^{j+1}\langle\psi|\psi_j\rangle\phi \wedge \psi_1 \wedge \cdots \wedge \not\psi_j \wedge \cdots \wedge \psi_n,$$

$$a(\psi)a^*(\phi)\psi_1 \wedge \cdots \wedge \psi_n = (-1)^{1+1}\langle\psi|\phi\rangle\psi_1 \wedge \cdots \wedge \psi_n$$

$$+ \sum_{j=1}^{n}(-1)^{j+2}\langle\psi|\psi_j\rangle\phi \wedge \psi_1 \wedge \cdots \wedge \not\psi_j \wedge \cdots \wedge \psi_n,$$

yields the last relation $\{a^*(\phi), a(\psi)\} = \langle\psi|\phi\rangle\mathbb{1}$.

(4) We first notice that if $\Psi \in \Gamma(\mathcal{K})$ is such that $a(\psi)\Psi = 0$ for all $\psi \in \mathcal{K}$, then

$$\langle\psi_n \wedge \cdots \wedge \psi_1|\Psi\rangle = \langle a^*(\psi_n)\psi_{n-1} \wedge \cdots \wedge \psi_1|\Psi\rangle = \langle\psi_{n-1} \wedge \cdots \wedge \psi_1|a(\psi_n)\Psi\rangle = 0,$$

from which we conclude that $\Psi \perp \Gamma_n(\mathcal{K})$ for $n \geq 1$. Hence, $\Psi \in \Gamma_0(\mathcal{K})$, that is $\Psi = \lambda\Omega$ for some $\lambda \in \mathbb{C}$. Let $B \in \mathcal{O}_{\Gamma(\mathcal{K})}$ commute with all creation/annihilation operators. It follows that $a(\psi)B\Omega = Ba(\psi)\Omega = 0$ for all $\psi \in \mathcal{K}$. From the previous remark, we conclude that $B\Omega = \lambda\Omega$ for some $\lambda \in \mathbb{C}$. Then, we can write

$$B\psi_1 \wedge \cdots \wedge \psi_n = Ba^*(\psi_1)\cdots a^*(\psi_n)\Omega$$
$$= a^*(\psi_1)\cdots a^*(\psi_n)B\Omega$$
$$= \lambda a^*(\psi_1)\cdots a^*(\psi_n)\Omega = \lambda\psi_1 \wedge \cdots \wedge \psi_n,$$

which shows that $B|_{\Gamma_n(\mathcal{K})} = \lambda\mathbb{1}_{\Gamma_n(\mathcal{K})}$ and that $B = \lambda\mathbb{1}_{\Gamma(\mathcal{K})}$.

(5) is obvious if $\psi = 0$. The CAR imply

$$(a^*(\psi)a(\psi))^2 = a^*(\psi)(\{a(\psi), a^*(\psi)\} - a^*(\psi)a(\psi))a(\psi)$$
$$= \langle\psi|\psi\rangle a^*(\psi)a(\psi) - a^*(\psi)^2 a(\psi)^2$$
$$= \|\psi\|^2 a^*(\psi)a(\psi),$$

from which we deduce $\|a^*(\psi)a(\psi)\|^2 = \|(a^*(\psi)a(\psi))^2\| = \|\psi\|^2\|a^*(\psi)a(\psi)\|$. If $\psi \neq 0$ then $a(\psi) \neq 0$ and hence $\|a^*(\psi)a(\psi)\| \neq 0$ so that we can conclude

$$\|a(\psi)\|^2 = \|a^*(\psi)\|^2 = \|a^*(\psi)a(\psi)\| = \|\psi\|^2.$$

(6) It follows from the definitions that $\Gamma(A)a^*(\psi)\Omega = \Gamma(A)\psi = A\psi = a^*(A\psi)\Gamma(A)\Omega$ and

$$\Gamma(A)a^*(\psi)\psi_1 \wedge \cdots \wedge \psi_n = \Gamma(A)\psi \wedge \psi_1 \wedge \cdots \wedge \psi_n$$
$$= A\psi \wedge A\psi_1 \wedge \cdots \wedge A\psi_n$$
$$= a^*(A\psi)\Gamma(A)\psi_1 \wedge \cdots \wedge \psi_n.$$

Thus, one has $\Gamma(A)a^*(\psi) = a^*(A\psi)\Gamma(A)$. By conjugation, we also get $\Gamma(A^*)a(A\psi) = a(\psi)\Gamma(A^*)$.

(7) It follows from (6) that

$$e^{td\Gamma(A)}a^*(\psi) = a^*(e^{tA}\psi)e^{td\Gamma(A)}.$$

Differentiation at $t = 0$ yields the first relation in (7). The second is obtained by conjugation.

(8) The CAR imply

$$[a^*(\phi)a(\psi), a^*(\chi)] = a^*(\phi)a(\psi)a^*(\chi) - a^*(\chi)a^*(\phi)a(\psi)$$
$$= a^*(\phi)a(\psi)a^*(\chi) + a^*(\phi)a^*(\chi)a(\psi)$$
$$= a^*(\phi)\{a(\psi), a^*(\chi)\} = \langle\psi|\chi\rangle a^*(\phi).$$

On the other hand, (7) implies that $[d\Gamma(|\phi\rangle\langle\psi|), a^*(\chi)] = \langle\psi|\chi\rangle a^*(\phi)$. Thus, setting $B = a^*(\phi)a(\psi) - d\Gamma(|\phi\rangle\langle\psi|)$ we get $[B, a^*(\chi)] = 0$ for all $\chi \in \mathcal{K}$. Interchanging ϕ and ψ, we obtain in the same way $[B, a(\chi)]^* = -[B^*, a^*(\chi)] = 0$, and so $[B, a(\chi)] = 0$. Hence $B \in \mathfrak{A}'$ and (4) implies that $B = \lambda \mathbb{1}$ for some $\lambda \in \mathbb{C}$. Since $B\Omega = 0$ we conclude that $B = 0$.

(9) Follows from (8) and the representation $A = \sum_{j,k=1}^{d}\langle\psi_j|A\psi_k\rangle|\psi_j\rangle\langle\psi_k|$.   □

Given a Hilbert space $\mathcal{K}$, a representation of the CAR over $\mathcal{K}$ on a Hilbert space $\mathcal{H}$ is a pair of maps

$$\psi \mapsto b(\psi), \qquad \psi \mapsto b^*(\psi), \tag{4.164}$$

from $\mathcal{K}$ to $\mathcal{O}_{\mathcal{H}}$ satisfying Properties (1)–(3) of Proposition 4.54. Such a representation is called irreducible if it also satisfies Property (4) with $\mathcal{O}_{\Gamma(\mathcal{K})}$ replaced by $\mathcal{O}_{\mathcal{H}}$. The particular irreducible representation $\psi \mapsto a^{\#}(\psi)$ on $\Gamma(\mathcal{K})$ is called the Fock representation. We will construct another important representation of the CAR in Section 4.7.4.

Proposition 4.55 *Let $\mathcal{K}$ be a finite dimensional Hilbert space and $\psi \mapsto b^{\#}(\psi)$ an irreducible representation of the CAR over $\mathcal{K}$ on $\mathcal{H}$. Then, there exists a unitary operator $U : \Gamma(\mathcal{K}) \to \mathcal{H}$ such that $Ua^{\#}(\psi)U^* = b^{\#}(\psi)$ for all $\psi \in \mathcal{K}$. Moreover, U is unique up to a phase factor.*

In other words, any two irreducible representations of the CAR over a finite dimensional Hilbert space are unitarily equivalent. A proof of Proposition 4.55 is sketched in the next exercise.

Exercise 4.48 Let $\mathcal{K} \ni \psi \mapsto b(\psi) \in \mathcal{O}_{\mathcal{H}}$ be an irreducible representation of CAR over the d-dimensional Hilbert space $\mathcal{K}$ in the Hilbert space $\mathcal{H}$. Denote by $\{\chi_1, \ldots, \chi_d\}$ an orthonormal basis of $\mathcal{K}$ an set

$$\tilde{N} = \sum_{n=1}^{d} b^*(\chi_n)b(\chi_n).$$

1. Show that $0 \le \tilde{N} \le d\mathbb{1}$ and $\tilde{N}b(\psi) = b(\psi)(\tilde{N} - \mathbb{1})$ for any $\psi \in \mathcal{K}$.

2. Let $\phi \in \mathcal{H}$ be a normalized eigenvector to the smallest eigenvalue of $\widetilde{N}$. Show that $b(\psi)\phi = 0$ for all $\psi \in \mathcal{K}$.
3. Set $\mathcal{H}_0 = \mathbb{C}\phi$ and denote by $\mathcal{H}_n$ the linear span of $\{b^*(\psi_1)\cdots b^*(\psi_n)\phi \mid \psi_1, \ldots, \psi_n \in \mathcal{K}\}$. Show that $\mathcal{H}_n \perp \mathcal{H}_m$ for $n \neq m$ and $\mathcal{H}_n = \{0\}$ for $n > d$.
   Hint: show that $\widetilde{N}|_{\mathcal{H}_n} = n\mathbb{1}_{\mathcal{H}_n}$.
4. Show that

$$\langle b^*(\psi_1)\cdots b^*(\psi_n)\phi | b^*(\psi_1')\cdots b^*(\psi_n')\phi\rangle = \det[\langle\psi_i|\psi_j'\rangle]_{1 \leq i,j \leq n},$$

   and conclude that the map $\psi_1 \wedge \cdots \wedge \psi_n \mapsto b^*(\psi_1)\cdots b^*(\psi_n)\phi$ extends to an isometry $U : \Gamma(\mathcal{K}) \to \mathcal{H}$.
5. Show that $Ua^\#(\psi)U^* = b^\#(\psi)$.
6. Show that $[UU^*, b(\psi)] = 0$ for all $\psi \in \mathcal{K}$ and conclude that U is unitary.

One can hardly overestimate the importance of the CAR. Indeed, as we shall see, they characterize completely the algebra of observables of a Fermi gas with a given finite-dimensional one-particle Hilbert space $\mathcal{K}$.

Proposition 4.56 *A representation $\psi \mapsto b^\#(\psi)$ of the CAR over the finite dimensional Hilbert space $\mathcal{K}$ in $\mathcal{H}$ is irreducible iff the smallest $*$-subalgebra of $\mathcal{O}_\mathcal{H}$ containing the set $\mathfrak{B} = \{b^\#(\psi) \mid \psi \in \mathcal{K}\}$ is $\mathcal{O}_\mathcal{H}$.*

Note that the smallest $*$-subalgebra of $\mathcal{O}_\mathcal{H}$ containing $\mathfrak{B}$ must contain all polynomials in the operators $b^\#(\psi)$, that is all linear combinations of monomials of the form $b^\#(\psi_1)\cdots b^\#(\psi_k)$. But the set of all these polynomials is obviously a $*$-algebra. Hence, a representation $\psi \mapsto b^\#(\psi)$ is irreducible iff any operator on $\mathcal{H}$ can be written as a polynomial in the operators $b^\#$. We can draw important conclusions from this fact:

1. Since the Fock representation $\psi \mapsto a^\#(\psi)$ is irreducible, any operator on the Fock space $\Gamma(\mathcal{K})$ is a polynomial in the creation/annihilation operators $a^\#$.
2. Any representation of the CAR over $\mathcal{K}$ on a Hilbert space $\mathcal{H}$ extends to a representation of the $*$-algebra $\mathcal{O}_{\Gamma(\mathcal{K})}$ on $\mathcal{H}$, i.e., to a $*$-morphism $\pi : \mathcal{O}_{\Gamma(\mathcal{K})} \to \mathcal{O}_\mathcal{H}$.
3. If the representation is irreducible, this morphism is an isomorphism.

To prove Proposition 4.56, we shall need the following result, von Neumann's bicommutant theorem. A subset $\mathfrak{A} \subset \mathcal{O}_\mathcal{K}$ is called self-adjoint if $A \in \mathfrak{A}$ implies $A^* \in \mathfrak{A}$ and unital if $\mathbb{1} \in \mathfrak{A}$.

Theorem 4.57 *Let $\mathcal{K}$ be a finite dimensional Hilbert space and $\mathfrak{A}$ a unital self-adjoint subset of $\mathcal{O}_\mathcal{K}$. Then its bicommutant $\mathfrak{A}''$ is the smallest $*$-subalgebra of $\mathcal{O}_\mathcal{K}$ containing $\mathfrak{A}$.*

Proof: Denote by $\mathcal{A}$ the smallest $*$-subalgebra of $\mathcal{O}_\mathcal{K}$ containing $\mathfrak{A}$, that is the set of polynomials in elements of $\mathfrak{A}$. One clearly has $\mathcal{A}' = \mathfrak{A}'$ and hence $\mathcal{A}'' = \mathfrak{A}''$. Thus, it suffices to show that $\mathcal{A} = \mathcal{A}''$ (a $*$-algebra satisfying this condition is a von Neumann algebra, and we are about to show that any finite dimensional unital $*$-algebra is a von Neumann algebra).

Since any element of $\mathcal{A}$ commutes with all elements of $\mathcal{A}'$ one obviously has $\mathcal{A} \subset \mathcal{A}''$. We must prove the reverse inclusion. Let $\{\psi_1, \ldots, \psi_n\}$ be a basis of $\mathcal{K}$, $\{e_1, \ldots, e_n\}$ a basis of $\mathbb{C}^n$ and set

$$\Psi = \sum_{j=1}^{n} \psi_j \otimes e_j \in \mathcal{H} = \mathcal{K} \otimes \mathbb{C}^n.$$

To any $A \in \mathcal{O}_\mathcal{K}$ we associate the linear operator $\widehat{A} = A \otimes \mathbb{1} \in \mathcal{O}_\mathcal{H}$. It follows that $\widehat{\mathcal{A}} = \{\widehat{A} \mid A \in \mathcal{A}\}$ is a $*$-subalgebra of $\mathcal{O}_\mathcal{H}$ and $\widehat{\mathcal{A}}\Psi = \{\widehat{A}\Psi \mid A \in \mathcal{A}\}$ a subspace of $\mathcal{H}$. Denote by P the orthogonal projection of $\mathcal{H}$ onto this subspace. We claim that $P \in \widehat{\mathcal{A}}'$. Indeed, for any $\widehat{A} \in \widehat{\mathcal{A}}$ and $\Phi \in \mathcal{H}$, one has $\widehat{A}P\Phi \in \widehat{\mathcal{A}}\Psi$, and hence

$$\widehat{A}P\Phi = P\widehat{A}P\Phi.$$

We deduce that $\widehat{A}P = P\widehat{A}P$ for all $\widehat{A} \in \widehat{\mathcal{A}}$ and since $\widehat{\mathcal{A}}$ is self-adjoint, one also has

$$P\widehat{A} = (\widehat{A}^*P)^* = (P\widehat{A}^*P)^* = P\widehat{A}P = \widehat{A}P.$$

Since $\mathcal{A}$ is unital, so is $\widehat{\mathcal{A}}$. It follows that $\Psi \in \widehat{\mathcal{A}}\Psi$ and hence $P\Psi = \Psi$. Recall that $X \in \mathcal{O}_\mathcal{H}$ is described by a $n \times n$ matrix $[X_{jk}]$ of elements of $\mathcal{O}_\mathcal{K}$ (see Section 4.3.3) via the formula

$$X(\psi \otimes e_k) = \sum_{j=1}^{n}(X_{jk}\psi) \otimes e_j.$$

Consequently, one has $\widehat{\mathcal{A}}' = \{X = [X_{jk}] \mid X_{jk} \in \mathcal{A}'\}$. Let $B \in \mathcal{A}''$. By the previous formula, $\widehat{B} \in \widehat{\mathcal{A}}''$, and so $\widehat{B}$ commutes with P. We conclude that

$$\widehat{B}\Psi = \widehat{B}P\Psi = P\widehat{B}\Psi \in \widehat{\mathcal{A}}\Psi,$$

and so there exists $A \in \mathcal{A}$ such that $\widehat{B}\Psi = \widehat{A}\Psi$, i.e.,

$$B\psi_j = A\psi_j,$$

for $j = 1, \ldots, n$. We conclude that $B = A \in \mathcal{A}$. $\qquad\square$

Proof of Proposition 4.56. Note that $\{b^*(\psi), b(\psi)\} = \|\psi\|^2 \mathbb{1}$ so that any $*$-subalgebra of $\mathcal{O}_\mathcal{H}$ containing

$$\mathfrak{B} = \{b^\#(\psi) \mid \psi \in \mathcal{K}\},$$

also contains the unital self-adjoint subset $\widetilde{\mathfrak{B}} = \mathfrak{B} \cup \{\mathbb{1}\}$. It follows that the smallest $*$-subalgebra of $\mathcal{O}_\mathcal{H}$ containing $\mathfrak{B}$ coincide with the smallest $*$-subalgebra of $\mathcal{O}_\mathcal{H}$ containing $\widetilde{\mathfrak{B}}$. Moreover, one clearly has $\widetilde{\mathfrak{B}}' = \mathfrak{B}'$ and hence $\widetilde{\mathfrak{B}}'' = \mathfrak{B}''$. By the von Neumann bicommutant theorem, $\mathfrak{B}''$ is the smallest $*$-subalgebra of $\mathcal{O}_\mathcal{H}$ containing $\mathfrak{B}$. Now the representation $\psi \mapsto b^\#(\psi)$ is irreducible iff $\mathfrak{B}' = \mathbb{C}\mathbb{1}$, that is iff $\mathfrak{B}'' = \mathcal{O}_\mathcal{H}$. $\qquad\square$

Exercise 4.49 Let $\mathcal{K}_1$ and $\mathcal{K}_2$ be two finite dimensional Hilbert spaces. Show that there exists a unitary map $U : \Gamma(\mathcal{K}_1 \oplus \mathcal{K}_2) \to \Gamma(\mathcal{K}_1) \otimes \Gamma(\mathcal{K}_2)$ such that $U\Omega = \Omega \otimes \Omega$ and

$$Ua(\psi \oplus \phi)U^* = a(\psi) \otimes \mathbb{1} + \mathrm{e}^{\mathrm{i}\pi N} \otimes a(\phi).$$

Hint: try to apply Proposition 4.55.

Remark. Apart from a few important exceptions, the material of this and the previous section extends with minor changes to the case where $\mathcal{K}$ is an infinite dimensional Hilbert space. For example:

1. The definition of the Fock space $\Gamma(\mathcal{K})$ has to be complemented with the obvious topological condition that $\Psi = (\psi_0, \psi_1, \ldots) \in \Gamma(\mathcal{K})$ iff $\|\Psi\|^2 = \sum_{n \in \mathbb{N}} \|\psi_n\|^2 < \infty$.

2. The definition of $\Gamma_n(A)$ carries over to bounded operators A on $\mathcal{K}$ and $\|\Gamma_n(A)\| \leq \|A\|^n$. Thus, $\Gamma(A) = \oplus_{n \geq 0} \Gamma_n(A)$ is well defined if:

   - $\|A\| \leq 1$, and then $\|\Gamma(A)\| = \sup_{n \geq 0} \|\Gamma_n(A)\| = 1$. In particular, if U is unitary, so is $\Gamma(U)$.
   - A has finite rank m so that $\Gamma_n(A) = 0$ for $n > m$ and then $\|\Gamma(A)\| = \sup_{n \geq 0} \|\Gamma_n(A)\| \leq \max(1, \|A\|^m)$. In fact, using the polar decomposition $A = U|A|$ together with Lemma 4.53, one sees that $\Gamma(A)$ is trace class with $\|\Gamma(A)\|_1 = \operatorname{tr} \Gamma(|A|) = \det(\mathbb{1} + |A|)$. By a simple approximation argument, one can then show that $\Gamma(A)$ is well defined and trace class provided A is trace class, and Lemma 4.53 carries over.

3. If A generates a strongly continuous contraction semigroup e^{tA} on $\mathcal{K}$, then $\mathrm{d}\Gamma(A)$ is defined as the generator of the strongly continuous contraction semigroup $\Gamma(\mathrm{e}^{tA})$ on $\Gamma(\mathcal{K})$. In particular, if A is self-adjoint, so is $\mathrm{d}\Gamma(A)$. However, some care is required since $\mathrm{d}\Gamma(A)$ is unbounded unless $A = 0$. If A is bounded, the dense subspace $\Gamma_{\mathrm{fin}}(\mathcal{K}) = \cup_{n \geq 0}(\oplus_{k \leq n} \Gamma_k(\mathcal{K}))$ of $\Gamma(\mathcal{K})$ is a core of $\mathrm{d}\Gamma(A)$ and on this subspace, $\mathrm{d}\Gamma(A)$ acts as in the finite dimensional case.

4. The definition of the creation/annihilation operators carries over without change. Parts (1)–(5) of Proposition 4.54 hold with the same proofs while Parts (6)–(8) are easily adapted. Part (9) still holds if A is trace class and it follows that $\|\mathrm{d}\Gamma(A)\| \leq \|A\|_1$.

5. The unitary equivalence described in Exercise 5.49 still holds for infinite dimensional $\mathcal{K}_1$ and $\mathcal{K}_2$ (prove it!).

Proposition 4.55 does not hold for infinite dimensional $\mathcal{K}$. In fact, there are many unitarily inequivalent irreducible representations of the CAR over $\mathcal{K}$. Also Proposition 4.56 and Theorem 4.57 do not hold for infinite dimensional $\mathcal{K}$. In the latter, one has to replace 'smallest $*$-subalgebra of $\mathcal{O}_\mathcal{K}$' by 'smallest weakly closed $*$-subalgebra of $\mathcal{O}_\mathcal{K}$' (see, e.g., Theorem 2.4.11 in Bratteli–Robinson (1987)). Proposition 4.56 has to be modified accordingly: the representation $\psi \mapsto b^\#(\psi)$ in $\mathcal{H}$ is irreducible iff any bounded operator on $\mathcal{H}$ is a weak limit of a net of polynomials in the elements of $\mathfrak{B}$.

4.7.3 Quasi-free states of the CAR algebra

We now turn to states of a free Fermi gas. Let $T \in \mathcal{O}_\mathcal{K}$ be a nonzero operator satisfying $0 \leq T < 1$. In our context, we shall refer to T as *density operator* or just *density*. To such T we associate density matrix on $\Gamma(\mathcal{K})$ by

$$\omega_T = \frac{1}{Z_T} \Gamma\left(\frac{T}{1-T}\right),$$

where

$$Z_T = \mathrm{tr}\left(\Gamma\left(\frac{T}{1-T}\right)\right).$$

As usual, we denote by the same letter the corresponding state on $\mathcal{O}_{\Gamma(\mathcal{K})}$. ω_T is called quasi-free state associated to the density T. Its properties are summarized in

Proposition 4.58   *(1) If $\phi_1, \ldots, \phi_n, \psi_1, \ldots, \psi_m \in K$, then*

$$\omega_T(a^*(\phi_n) \cdots a^*(\phi_1)a(\psi_1) \cdots a(\psi_m)) = \delta_{nm} \det[\langle \psi_i | T\phi_j \rangle].$$

In particular, $\omega_T(a^*(\phi)a(\psi)) = (\psi|T\phi)$.
(2) $\log Z_T = -\log \det(1-T) = -\mathrm{tr}(\log(1-T))$.
(3) $\omega_T(\Gamma(A)) = \det(1 + T(A-1))$.
(4) $\omega_T(\mathrm{d}\Gamma(A)) = \mathrm{tr}(TA)$.
(5) $S(\omega_T) = -\mathrm{tr}(T\log T + (1-T)\log(1-T))$.
(6) $\omega_{T_1} \ll \omega_{T_2}$ iff $\mathrm{Ker}\, T_1 \subset \mathrm{Ker}\, T_2$, and then

$$S(\omega_{T_1}|\omega_{T_2}) = \mathrm{tr}\left(T_1(\log(T_2) - \log(T_1)) + (1 - T_1)(\log(1 - T_2) - \log(1 - T_1))\right).$$

Proof: (1) We set $Q = T(1-T)^{-1}$, $A = a^*(\phi_n) \cdots a^*(\phi_1)a(\psi_1) \cdots a(\psi_m)$ and note that

$$e^{-\mathrm{i}tN} \omega_T\, e^{\mathrm{i}tN} = \frac{1}{Z_T}\Gamma(e^{-\mathrm{i}t})\Gamma(Q)\Gamma(e^{\mathrm{i}t}) = \frac{1}{Z_T}\Gamma(e^{-\mathrm{i}t}Qe^{\mathrm{i}t}) = \frac{1}{Z_T}\Gamma(Q) = \omega_T,$$

so that

$$\omega_T(e^{\mathrm{i}tN} A e^{-\mathrm{i}tN}) = \omega_T(A).$$

By Proposition 4.54 (6), we have

$$e^{\mathrm{i}tN} a^*(\phi_j) e^{-\mathrm{i}tN} = a^*(e^{\mathrm{i}t}\phi_j) = e^{\mathrm{i}t}a^*(\phi_j), \qquad e^{\mathrm{i}tN} a(\psi_k) e^{-\mathrm{i}tN} = a(e^{\mathrm{i}t}\psi_k) = e^{-\mathrm{i}t}a(\psi_k),$$

from which we deduce that $e^{itN} A e^{-itN} = e^{it(n-m)} A$, and hence that $\omega_T(A) = 0$ if $n \neq m$. We shall handle the case $n = m$ by induction on n. For $n = 1$, one has

$$
\begin{aligned}
\omega_T(a^*(\phi)a(\psi)) &= Z_T^{-1} \mathrm{tr}(\Gamma(Q)a^*(\phi)a(\psi)) \\
&= Z_T^{-1} \mathrm{tr}(a^*(Q\phi)\Gamma(Q)a(\psi)) \\
&= Z_T^{-1} \mathrm{tr}(\Gamma(Q)a(\psi)a^*(Q\phi)) \\
&= Z_T^{-1} \mathrm{tr}(\Gamma(Q)(\{a(\psi), a^*(Q\phi)\} - a^*(Q\phi)a(\psi))) \\
&= \langle \psi | Q\phi \rangle - \omega_T(a^*(Q\phi)a(\psi)),
\end{aligned}
$$

from which we deduce that $\omega_T(a^*((\mathbb{1}+Q)\phi)a(\psi)) = \langle \psi | Q\phi \rangle$. Since $(\mathbb{1}+Q) = (\mathbb{1} - T)^{-1}$, we finally get

$$
\omega_T(a^*(\phi)a(\psi)) = \langle \psi | Q(\mathbb{1} - T)\phi \rangle = \langle \psi | T\phi \rangle.
$$

Assuming now that the result holds for $n - 1$, we write

$$
\begin{aligned}
&\omega_T(a^*(\phi_n) \cdots a^*(\phi_1)a(\psi_1) \cdots a(\psi_n)) \\
&= Z_T^{-1} \mathrm{tr}(\Gamma(Q)a^*(\phi_n) \cdots a^*(\phi_1)a(\psi_1) \cdots a(\psi_n)) \\
&= Z_T^{-1} \mathrm{tr}(a^*(Q\phi_n)\Gamma(Q)a^*(\phi_{n-1}) \cdots a^*(\phi_1)a(\psi_1) \cdots a(\psi_n)) \\
&= \omega_T(a^*(\phi_{n-1}) \cdots a^*(\phi_1)a(\psi_1) \cdots a(\psi_n)a^*(Q\phi_n)).
\end{aligned}
$$

Making repeated use of the CAR,

$$
a(\psi_j)a^*(Q\phi_n) = \langle \psi_j | Q\phi_n \rangle - a^*(Q\phi_n)a(\psi_j), \qquad a^*(\phi_j)a^*(Q\phi_n) = -a^*(Q\phi_n)a^*(\phi_j),
$$

we move the last factor $a^*(Q\phi_n)$ back to its original position to get

$$
\omega_T(a^*(\phi_n) \cdots a^*(\phi_1)a(\psi_1) \cdots a(\psi_n)) = -\omega_T(a^*(Q\phi_n) \cdots a^*(\phi_1)a(\psi_1) \cdots a(\psi_n))
$$

$$
+ \sum_{j=1}^n (-1)^{n+j} \langle \psi_j | Q\phi_n \rangle \omega_T(a^*(\phi_{n-1}) \cdots a^*(\phi_1)a(\psi_1) \cdots \widehat{a(\psi_j)} \cdots a(\psi_n)).
$$

By the same argument as in the $n = 1$ case, we deduce

$$
\omega_T(a^*(\phi_n) \cdots a^*(\phi_1)a(\psi_1) \cdots a(\psi_n))
$$

$$
= \sum_{j=1}^n (-1)^{n+j} \langle \psi_j | T\phi_n \rangle \omega_T(a^*(\phi_{n-1}) \cdots a^*(\phi_1)a(\psi_1) \cdots \widehat{a(\psi_j)} \cdots a(\psi_n)),
$$

and the induction step is achieved by Laplace formula (4.163).
(2) and (3) are immediate consequences of Lemma 4.53, (4) follows from (1) and Proposition 4.54 (9).

(5) We again set $Q = T(\mathbb{1} - T)^{-1}$ and notice that

$$\log \Gamma(Q) = d\Gamma(\log Q),$$

so that, by (4),

$$S(\omega_T) = -\omega_T \left(\log \left(Z_T^{-1}\Gamma(Q)\right)\right) = -\omega_T(d\Gamma(\log Q) - \log Z_T) = \log Z_T - \operatorname{tr}\left(T \log Q\right).$$

Using (2), we conclude that

$$S(\omega_T) = -\operatorname{tr}(\log(\mathbb{1} - T)) - \operatorname{tr}(T(\log(T) - \log(\mathbb{1} - T))),$$

from which the desired formula immediately follows.

(6) We set $Q_j = T_j(\mathbb{1} - T_j)^{-1}$ and notice that $\operatorname{Ker} Q_j = \operatorname{Ker} T_j$. It easily follows from $\operatorname{Ker} T_1 \subset \operatorname{Ker} T_2$ that $\operatorname{Ker} \Gamma(Q_1) \subset \operatorname{Ker} \Gamma(Q_2)$ and hence $\omega_{T_1} \ll \omega_{T_2}$. The remaining statement is proved in a similar way as (5). $\qquad\square$

Let $h = h^* \in \mathcal{O}_{\mathcal{K}}$ be the one-particle Hamiltonian—the total energy observable of a single fermion. The Hamiltonian of the free Fermi gas is

$$H = d\Gamma(h).$$

Indeed, if $\{\psi_1, \ldots, \psi_d\}$ denotes an eigenbasis of h such that $h\psi_j = \varepsilon_j \psi_j$, then the state $\Psi = a^*(\psi_{j_1}) \cdots a^*(\psi_{j_n})\Omega$ describes n fermions with energies $\varepsilon_{j_1}, \ldots, \varepsilon_{j_n}$, and one has

$$H\Psi = d\Gamma_n(h)\psi_{j_1} \wedge \cdots \wedge \psi_{j_n} = \left(\sum_{i=1}^{n} \varepsilon_{j_i}\right)\Psi.$$

The thermal equilibrium state at inverse temperature $\beta \in \mathbb{R}$ and chemical potential $\mu \in \mathbb{R}$ is described by the Gibbs grand canonical ensemble

$$\rho_{\beta,\mu} = \frac{e^{-\beta(H-\mu N)}}{\operatorname{tr}(e^{-\beta(H-\mu N)})}.$$

Since

$$e^{-\beta(H-\mu N)} = e^{-d\Gamma(\beta(h-\mu\mathbb{1}))} = \Gamma(e^{-\beta(h-\mu\mathbb{1})}),$$

solving the equation

$$e^{-\beta(h-\mu\mathbb{1})} = \frac{T}{\mathbb{1} - T},$$

for T we see that the density operator of a free Fermi gas in thermal equilibrium at inverse temperature β and chemical potential μ is given by

$$T_{\beta,\mu} = (\mathbb{1} + e^{\beta(h-\mu\mathbb{1})})^{-1}.$$

Fermionic systems   **363**

$T_{\beta,\mu}$ is commonly called the Fermi–Dirac distribution. Following the notation introduced in Section 4.3.9, one has

$$E = \rho_{\beta,\mu}(H) = \mathrm{tr}(hT_{\beta,\mu}),$$

$$\varrho = \rho_{\beta,\mu}(N) = \mathrm{tr}(T_{\beta,\mu}),$$

$$P(\beta,\mu) = \log \mathrm{tr}(e^{-\beta(H-\mu N)}) = \mathrm{tr}\left(\log(\mathbb{1} + e^{-\beta(h-\mu\mathbb{1})})\right),$$

(4.165)

$$S(\beta,\mu) = S(\rho_{\beta,\mu}) = \beta(E - \mu\varrho) + P(\beta,\mu).$$

Exercise 4.50   The purpose of this exercise is to provide a complete discussion of the thermodynamic limit of a 1D free Fermi gas starting from the description of a finite Fermi gas. The target system is the ideal Fermi gas with one-particle Hamiltonian $h = k^2/2$ on the one-particle Hilbert space $\mathcal{K} = L^2(\mathbb{R}, dk/2\pi)$ in the thermal equilibrium state at inverse temperature β and chemical potential μ.

To describe the finite approximation, consider the operator

$$(h_L\psi)(x) = -\frac{1}{2}\psi''(x),$$

on $L^2([-L/2, L/2], dx)$ with periodic boundary conditions $\psi(x + L) = \psi(x)$. h_L is self-adjoint with a purely discrete spectrum consisting of simple eigenvalues $\varepsilon(k) = k^2/2$ with eigenfunctions $\psi_k(x) = L^{-1/2}e^{ikx}$, $k \in \mathcal{Q}_L = \{2\pi j/L \mid j \in \mathbb{Z}\}$. The Fourier transform

$$\hat{\psi}(k) = \langle \psi_k | \psi \rangle = \frac{1}{\sqrt{L}} \int_{-L/2}^{L/2} \psi(x) e^{-ikx} \, dx,$$

provides a unitary map from the position representation $L^2([-L/2, L/2])$ to the 'momentum' representation $\ell^2(\mathcal{Q}_L)$ such that $\widehat{h_L\psi}(k) = \varepsilon(k)\hat{\psi}(k)$. In what follows, we work in the momentum representation and set $\mathcal{K}_L = \ell^2(\mathcal{Q}_L)$ and $(h_L\psi)(k) = \varepsilon(k)\psi(k)$. Let $\mathcal{E} > 0$ be an energy cutoff, set $\mathcal{Q}_{L,\mathcal{E}} = \{k \in \mathcal{Q}_L \mid \varepsilon(k) \leq \mathcal{E}\}$ and consider the free Fermi gas with single particle Hilbert space $\mathcal{K}_{L,\mathcal{E}} = \ell^2(\mathcal{Q}_{L,\mathcal{E}})$, and one particle Hamiltonian $(h_{L,\mathcal{E}}\psi)(k) = \varepsilon(k)\psi(k)$. Let $E_{L,\mathcal{E}}$, $\varrho_{L,\mathcal{E}}$, $P_{L,\mathcal{E}}(\beta,\mu)$ be defined by (4.165).

1. Prove that

$$\lim_{L\to\infty} \lim_{\mathcal{E}\to\infty} \frac{E_{L,\mathcal{E}}}{L} = \int_{-\infty}^{\infty} \frac{\varepsilon(k)}{1 + e^{\beta(\varepsilon(k)-\mu)}} \frac{dk}{2\pi},$$

$$\lim_{L\to\infty} \lim_{\mathcal{E}\to\infty} \frac{\varrho_{L,\mathcal{E}}}{L} = \int_{-\infty}^{\infty} \frac{1}{1 + e^{\beta(\varepsilon(k)-\mu)}} \frac{dk}{2\pi},$$

$$\lim_{L\to\infty} \lim_{\mathcal{E}\to\infty} \frac{P_{L,\mathcal{E}}(\beta,\mu)}{L} = \int_{-\infty}^{\infty} \log(1 + e^{-\beta(\varepsilon(k)-\mu)}) \frac{dk}{2\pi}.$$

2. A wave-function $\psi \in \mathcal{K}_{L,\varepsilon}$ can be isometrically extended to an element of $\mathcal{K}$ by setting

$$\widetilde{\psi}(k) = \sqrt{L} \sum_{\xi \in \mathcal{Q}_{L,\varepsilon}} \psi(\xi) \chi_{[\xi - \pi/L, \xi + \pi/L[}(k),$$

where χ_I denotes the indicator function of the interval I. Thus, we can identify $\mathcal{K}_{L,\varepsilon}$ with a finite dimensional subspace of the Hilbert space $\mathcal{K}$. Denote by $\mathbb{1}_{L,\varepsilon}$ the orthogonal projection on this subspace. Then $\Gamma(\mathbb{1}_{L,\varepsilon})$ is an orthogonal projection in $\Gamma(\mathcal{K})$ whose range can be identified with $\Gamma(\mathcal{K}_{L,\varepsilon})$. Show that we can identify the equilibrium density matrix

$$\rho_{\beta,\mu,L,\varepsilon} = \frac{\Gamma(e^{-\beta(h_{L,\varepsilon} - \mu \mathbb{1})})}{\mathrm{tr}(\Gamma(e^{-\beta(h_{L,\varepsilon} - \mu \mathbb{1})}))};$$

of the finite Fermi gas on $\Gamma(\mathcal{K}_{L,\varepsilon})$ with the density matrix

$$\widetilde{\rho}_{\beta,\mu,L,\varepsilon} = \frac{\Gamma(e^{-\beta(h - \mu \mathbb{1})} \mathbb{1}_{L,\varepsilon})}{\mathrm{tr}(\Gamma(e^{-\beta(h - \mu \mathbb{1})} \mathbb{1}_{L,\varepsilon}))},$$

on $\Gamma(\mathcal{K})$ in the sense that

$$\mathrm{tr}\left(\rho_{\beta,\mu,L,\varepsilon} a^*(\psi_1) \cdots a^*(\psi_n) a(\phi_m) \cdots a(\phi_1)\right) = \mathrm{tr}\left(\widetilde{\rho}_{\beta,\mu,L,\varepsilon} a^*(\widetilde{\psi}_1) \cdots a^*(\widetilde{\psi}_n) a(\widetilde{\phi}_m) \cdots a(\widetilde{\phi}_1)\right),$$

for all $\psi_1, \ldots, \psi_n, \phi_1, \ldots, \phi_m \in \mathcal{K}_{L,\varepsilon}$.

3. Show that, in $\Gamma(\mathcal{K})$, the limit

$$\widetilde{\rho}_{\beta,\mu,L} = \lim_{\varepsilon \to \infty} \widetilde{\rho}_{\beta,\mu,L,\varepsilon},$$

exists in the trace norm and that $\widetilde{\rho}_{\beta,\mu,L}$ is a density matrix that can be identified with

$$\rho_{\beta,\mu,L} = \frac{\Gamma(e^{-\beta(h_L - \mu \mathbb{1})})}{\mathrm{tr}(\Gamma(e^{-\beta(h_L - \mu \mathbb{1})}))},$$

on $\Gamma(\mathcal{K}_L)$. Show that

$$\mathrm{s} - \lim_{L \to \infty} \widetilde{\rho}_{\beta,\mu,L} = 0,$$

that is the equilibrium density matrix disappears in the thermodynamic limit $L \to \infty$.

4. Show that,

$$\mathcal{D} = \bigcup_{L > 0, \varepsilon > 0} \mathcal{K}_{L,\varepsilon},$$

is a dense subspace of $\mathcal{K}$ and that for $\phi, \psi \in \mathcal{D}$ one has

$$\lim_{L \to \infty} \lim_{\varepsilon \to \infty} \mathrm{tr}\left(\widetilde{\rho}_{\beta,\mu,L,\varepsilon} a^*(\phi) a(\psi)\right) = \int_{-\infty}^{\infty} \frac{\overline{\psi(k)} \phi(k)}{1 + e^{\beta(\varepsilon(k) - \mu)}} \frac{dk}{2\pi} = \langle \psi | T \phi \rangle, \tag{4.166}$$

where $T = (\mathbb{1} + e^{\beta(h - \mu)})^{-1}$.

5. Since we have identified $\mathcal{K}_{L,\mathcal{E}}$ with a subspace of $\mathcal{K}$, we can also identify the *-algebra $\mathcal{O}_{\mathcal{K}_{L,\mathcal{E}}}$ with a subalgebra of the *-algebra $\mathcal{O}_{\mathcal{K}}$ of all bounded linear operators on $\mathcal{K}$. This identification is isometric and

$$\mathcal{O}_{\infty} = \bigcup_{L>0, \mathcal{E}>0} \mathcal{O}_{\mathcal{K}_{L,\mathcal{E}}},$$

is the *-algebra of all polynomials in the creation/annihilation operators $a^{\#}(\psi)$, $\psi \in \mathcal{D}$. Show that the limit

$$\rho_{\beta,\mu}(A) = \lim_{L \to \infty} \lim_{\mathcal{E} \to \infty} \mathrm{tr}\,(\tilde{\rho}_{\beta,\mu,L,\mathcal{E}} A),$$

exists for all $A \in \mathcal{O}_{\infty}$.
Hint: show that

$$\lim_{L \to \infty} \lim_{\mathcal{E} \to \infty} \mathrm{tr}\,(\tilde{\rho}_{\beta,\mu,L,\mathcal{E}} a^{*}(\psi_{1}) \cdots a^{*}(\psi_{n}) a(\phi_{m}) \cdots a(\phi_{1})) = \delta_{n,m} \det[\langle \phi_{j} | T \psi_{k} \rangle],$$

for all $\psi_{1}, \ldots, \psi_{n}, \phi_{1}, \ldots, \phi_{m} \in \mathcal{D}$.
6. Denote by $\mathrm{cl}(\mathcal{O}_{\infty})$ the norm closure of $\mathcal{O}_{\infty}$ in $\mathcal{O}_{\mathcal{K}}$ ($\mathrm{cl}(\mathcal{O}_{\infty})$ is the C^{*}-algebra generated by $\mathcal{O}_{\infty}$). Show that for any $A \in \mathrm{cl}(\mathcal{O}_{\infty})$ and any sequence $A_{n} \in \mathcal{O}_{\infty}$ which converges to A the limit

$$\rho_{\beta,\mu}(A) = \lim_{n \to \infty} \rho_{\beta,\mu}(A_{n}),$$

exists and is independent of the approximating sequence A_{n}. The C^{*}-algebra $\mathrm{cl}(\mathcal{O}_{\infty})$ is the algebra of observables of the infinitely extended ideal Fermi gas and $\rho_{\beta,\mu}$ is its thermal equilibrium state.

4.7.4 The Araki–Wyss representation

Araki and Wyss (1964) have discovered a specific cyclic representation of $\mathcal{O}_{\Gamma(\mathcal{K})}$ associated to the quasi-free state ω_{T} which is of considerable conceptual and computational importance. Although any two cyclic representations of $\mathcal{O}_{\Gamma(\mathcal{K})}$ associated to the state ω_{T} are unitarily equivalent, the specific structure inherent to the Araki–Wyss (AW) representation has played a central role in many developments in nonequilibrium quantum statistical mechanics over the last decade.

For the purpose of this section we may assume that $T > 0$ (otherwise, replace $\mathcal{K}$ with $\mathrm{Ran}\,T$). Then the quasi-free state ω_{T} on $\mathcal{O}_{\Gamma(\mathcal{K})}$ is faithful. Set

$$\mathcal{H}_{\mathrm{AW}} = \Gamma(\mathcal{K}) \otimes \Gamma(\mathcal{K}),$$

$$\Omega_{\mathrm{AW}} = \Omega \otimes \Omega,$$

$$b_{\mathrm{AW}}^{*}(\psi) = a^{*}((\mathbb{1} - T)^{1/2}\psi) \otimes \mathbb{1} + e^{i\pi N} \otimes a(\overline{T^{1/2}\psi}),$$

$$b_{\mathrm{AW}}(\psi) = a((\mathbb{1} - T)^{1/2}\psi) \otimes \mathbb{1} + e^{i\pi N} \otimes a^{*}(\overline{T^{1/2}\psi}),$$

where $\overline{\psi}$ denotes the complex conjugate of $\psi \in \mathcal{K} = \ell^{2}(\mathcal{Q})$. For $\Psi \in \Gamma(\mathcal{K})$, $\overline{\Psi}$ denotes the complex conjugate of Ψ (defined in the obvious way). If A is a linear operator, we define the linear operator $\overline{A}$ by $\overline{A}\,\overline{\psi} = \overline{A\psi}$.

Proposition 4.59 (1) *The maps $\psi \mapsto b_{\mathrm{AW}}^{\#}(\psi)$ define a representation of the CAR over $\mathcal{K}$ on the Hilbert space $\mathcal{H}_{\mathrm{AW}}$.*

(2) *Let π_{AW} be the induced representation of $\mathcal{O}_{\Gamma(\mathcal{K})}$ on $\mathcal{H}_{\mathrm{AW}}$. Ω_{AW} is a cyclic vector for this representation and*

$$\omega_T(A) = (\Omega_{\mathrm{AW}} | \pi_{\mathrm{AW}}(A)\Omega_{\mathrm{AW}}), \qquad (4.167)$$

for all $A \in \mathcal{O}_{\Gamma(\mathcal{K})}$. In other words, π_{AW} is a cyclic representation of $\mathcal{O}_{\Gamma(\mathcal{K})}$ associated to the faithful state ω_T.

Proof: The verification of (1) is simple and we leave it as an exercise for the reader. To check that Ω_{AW} is cyclic, we shall show by induction on $n + m$ that each subspace $D_{n,m} = \Gamma_n(\mathcal{K}) \otimes \Gamma_m(\mathcal{K})$ belongs to $\pi_{\mathrm{AW}}(\mathcal{O}_{\Gamma(\mathcal{K})})\Omega_{\mathrm{AW}}$. For $n + m = 1$, we deduce from $\mathrm{Ran}(\mathbb{1} - T)^{1/2} = \mathrm{Ran}\,\overline{T}^{1/2} = \mathcal{K}$ that

$$D_{1,0} = \{b_{\mathrm{AW}}^*(\psi)\Omega_{\mathrm{AW}} \mid \psi \in \mathcal{K}\}, \qquad D_{0,1} = \{b_{\mathrm{AW}}(\psi)\Omega_{\mathrm{AW}} \mid \psi \in \mathcal{K}\}.$$

Assuming $D_{n,m} \subset \pi_{\mathrm{AW}}(\mathcal{O}_{\Gamma(\mathcal{K})})\Omega_{\mathrm{AW}}$ for $n + m \leq k$, we observe that $\Psi \in D_{n+1,m}$ can be written as

$$\Psi = a^*((\mathbb{1} - T)^{1/2}\psi) \otimes \mathbb{1}\Phi,$$

for some $\psi \in \mathcal{K}$ and $\Phi \in D_{n,m}$. Equivalently, we can write

$$\Psi = b_{\mathrm{AW}}^*((\mathbb{1} - T)^{1/2}\psi)\Phi - \Phi'$$

where $\Phi' = (-\mathbb{1})^N \otimes a(\overline{T^{1/2}\psi})\Phi \in D_{n,m-1}$. It follows that $\Psi \in \pi_{\mathrm{AW}}(\mathcal{O}_{\Gamma(\mathcal{K})})\Omega_{\mathrm{AW}}$. A similar argument shows that $D_{n,m+1} \subset \pi_{\mathrm{AW}}(\mathcal{O}_{\Gamma(\mathcal{K})})\Omega_{\mathrm{AW}}$. Hence, the induction property is verified for $n + m \leq k + 1$. Finally, (4.167) follows from an elementary calculation based on equation (4.159). $\qquad \square$

The triple $(\mathcal{H}_{\mathrm{AW}}, \pi_{\mathrm{AW}}, \Omega_{\mathrm{AW}})$ is called the Araki–Wyss representation of the CAR over $\mathcal{K}$ associated to the quasi-free state ω_T. Since ω_T is faithful, it follows from Part (2) of Proposition 4.59 and Part 4 of Exercise 5.31 that this representation is unitarily equivalent to the standard representation and hence carries the entire modular structure. The modular structure in the Araki–Wyss representation takes the following form.

Proposition 4.60 (1) *The modular conjugation is given by*

$$J(\Psi_1 \otimes \Psi_2) = u\overline{\Psi}_2 \otimes u\overline{\Psi}_1,$$

where $u = e^{\mathrm{i}\pi N(N-1)/2}$.

(2) *The modular operator of ω_T is*

$$\Delta_{\omega_T} = \Gamma(e^{kT}) \otimes \Gamma(e^{-\overline{kT}}),$$

where $k_T = \log(T(1-T)^{-1})$. In particular

$$\log \Delta_{\omega_T} = d\Gamma(k_T) \otimes \mathbb{1} - \mathbb{1} \otimes d\Gamma(\overline{k_T}).$$

(3) If ω_{T_1} is the quasi-free state of density $T_1 > 0$, then its relative Hamiltonian w.r.t. ω_T is

$$\ell_{\omega_{T_1}|\omega_T} = \log \det \left((\mathbb{1} - T_1)(\mathbb{1} - T)^{-1}\right) + d\Gamma(k_{T_1} - k_T),$$

and its relative modular operator is determined by

$$\log \Delta_{\omega_{T_1}|\omega_T} = \log \Delta_{\omega_T} + \pi_{\mathrm{AW}}(\ell_{\omega_{T_1}|\omega_T}).$$

(4) Suppose that the self-adjoint operator h commutes with T. Then the quasi-free state ω_T is invariant under the dynamics τ^t generated by $H = d\Gamma(h)$. Moreover, the operator

$$K = d\Gamma(h) \otimes \mathbb{1} - \mathbb{1} \otimes d\Gamma(\overline{h}),$$

is the standard Liouvillean of this dynamics.

Remark. Since the $*$-subalgebra $\mathcal{O}_{\mathrm{AW}} = \pi_{\mathrm{AW}}(\mathcal{O}_{\Gamma(\mathcal{K})})$ is the set of polynomials in the $b_{\mathrm{AW}}^\#$, the dual $*$-subalgebra $\mathcal{O}'_{\mathrm{AW}} = J\mathcal{O}_{\mathrm{AW}}J$ is the set of polynomials in the $b'^\#_{\mathrm{AW}} = Jb_{\mathrm{AW}}^\#J$. By Propositions 4.25 and 4.26, one has

$$\mathcal{O}_{\mathrm{AW}} \cap \mathcal{O}'_{\mathrm{AW}} = \mathbb{C}\mathbb{1},$$
$$\mathcal{O}_{\mathrm{AW}} \vee \mathcal{O}'_{\mathrm{AW}} = \mathcal{O}_{\mathcal{H}_{\mathrm{AW}}}.$$

Proof: We set $\Delta = \Gamma(e^{k_T}) \otimes \Gamma(e^{-\overline{k_T}})$ and $s = e^{\mathrm{i}\pi N}$. Since J is clearly an anti-unitary involution and $\Delta > 0$, we deduce from the observation following equation (4.81) that in order to prove (1) and (2) it suffices to show that $J\Delta^{1/2}A\Omega_{\mathrm{AW}} = A^*\Omega_{\mathrm{AW}}$ for any monomial $A = b_{\mathrm{AW}}^\#(\psi_n) \cdots b_{\mathrm{AW}}^\#(\psi_1)$. We shall do that by induction on the degree n.

We first compute

$$b'_{\mathrm{AW}}(\psi) = Jb_{\mathrm{AW}}(\psi)J = a^*(T^{1/2}\psi)s \otimes s + \mathbb{1} \otimes sa(\overline{(\mathbb{1}-T)^{1/2}\psi}),$$

$$b'^*_{\mathrm{AW}}(\psi) = Jb^*_{\mathrm{AW}}(\psi)J = sa(T^{1/2}\psi) \otimes s + \mathbb{1} \otimes a^*(\overline{(\mathbb{1}-T)^{1/2}\psi})s,$$

and check that $[b'_{\mathrm{AW}}(\psi), b_{\mathrm{AW}}^\#(\phi)] = [b'^*_{\mathrm{AW}}(\psi), b_{\mathrm{AW}}^\#(\phi)] = 0$ for all $\psi, \phi \in \mathcal{K}$. We thus conclude that $b'^\#_{\mathrm{AW}}(\psi) \in \mathcal{O}'_{\mathrm{AW}}$. Next, we observe that

$$\Delta^{1/2}b_{\mathrm{AW}}(\psi)\Delta^{-1/2} = b_{\mathrm{AW}}(e^{-k_T/2}\psi), \qquad \Delta^{1/2}b^*_{\mathrm{AW}}(\psi)\Delta^{-1/2} = b^*_{\mathrm{AW}}(e^{k_T/2}\psi).$$

For $n = 1$, the claim follows from the fact that

$$J\Delta^{1/2}b_{\mathrm{AW}}(\psi)\Omega_{\mathrm{AW}} = J\Delta^{1/2}b_{\mathrm{AW}}(\psi)\Delta^{-1/2}J\Omega_{\mathrm{AW}}$$
$$= b'_{\mathrm{AW}}(\mathrm{e}^{-k_T/2}\psi)\Omega_{\mathrm{AW}}$$
$$= a^*(\mathrm{e}^{-k_T}T^{1/2}\psi) \otimes \mathbb{1}\Omega_{\mathrm{AW}}$$
$$= b^*_{\mathrm{AW}}(\psi)\Omega_{\mathrm{AW}}.$$

To deal with the induction step, let A be a monomial of degree less than n in the $b^{\#}_{\mathrm{AW}}$ and assume that $J\Delta^{1/2}A\Omega_{\mathrm{AW}} = A^*\Omega_{\mathrm{AW}}$ for all such monomials. Then, we can write

$$J\Delta^{1/2}b^{\#}_{\mathrm{AW}}(\psi)A\Omega_{\mathrm{AW}} = (J\Delta^{1/2}b^{\#}_{\mathrm{AW}}(\psi)\Delta^{-1/2}J)J\Delta^{1/2}A\Omega_{\mathrm{AW}}$$
$$= (Jb^{\#}_{\mathrm{AW}}(\mathrm{e}^{\mp k_T/2}\psi)J)A^*\Omega_{\mathrm{AW}}$$
$$= b'^{\#}_{\mathrm{AW}}(\mathrm{e}^{\mp k_T/2}\psi)A^*\Omega_{\mathrm{AW}}$$
$$= A^*b'^{\#}_{\mathrm{AW}}(\mathrm{e}^{\mp k_T/2}\psi)\Omega_{\mathrm{AW}}$$
$$= A^*J\Delta^{1/2}b^{\#}_{\mathrm{AW}}(\psi)\Delta^{-1/2}J\Omega_{\mathrm{AW}}$$
$$= A^*J\Delta^{1/2}b^{\#}_{\mathrm{AW}}(\psi)\Omega_{\mathrm{AW}}$$
$$= A^*b^{\#}_{\mathrm{AW}}(\psi)^*\Omega_{\mathrm{AW}},$$

which shows that the induction property holds for all monomials of degree less than $n + 1$.

(3) The first claim is an immediate consequence of the definition (4.86) of the relative Hamiltonian. Since, by Part (4) of Exercise 5.31, the Araki–Wyss representation is unitarily equivalent to the standard representation on $\mathcal{H}_{\mathcal{O}}$, the second claim follows from Property (3) of the relative Hamiltonian given on page 287.

(4) The fact that ω_T is invariant under the dynamics τ^t is evident. Recall from Exercise 5.32 that the standard Liouvillean is the unique self-adjoint operator K on $\mathcal{H}_{\mathrm{AW}}$ (the Hilbert space carrying the standard representation) such that the unitary group $\mathrm{e}^{\mathrm{i}tK}$ implements the dynamics and preserves the natural cone. These two conditions can be formulated as

$$\mathrm{e}^{\mathrm{i}tK}b_{\mathrm{AW}}(\psi)\mathrm{e}^{-\mathrm{i}tK} = b_{\mathrm{AW}}(\mathrm{e}^{\mathrm{i}th}\psi), \qquad JK + KJ = 0,$$

and are easily verified by $K = \mathrm{d}\Gamma(h) \otimes \mathbb{1} - \mathbb{1} \otimes \mathrm{d}\Gamma(\bar{h})$. □

Remark. The Araki–Wyss representation of the CAR over $\mathcal{K}$ immediately extends to infinite dimensional $\mathcal{K}$ and the proof of Proposition 4.59 carries over without modification. The same is true for Proposition 4.60 provided one assumes, in Part (3), that $\log(T_1) - \log(T)$ and $\log(\mathbb{1} - T_1) - \log(\mathbb{1} - T)$ are both trace class.

4.7.5 Spin-fermion model

The spin-fermion (SF) model describes a two level atom (or a spin $1/2$), denoted $\mathcal{S}$, interacting with $n \geq 2$ independent free Fermi gas reservoirs $\mathcal{R}_j$. The Hilbert space of $\mathcal{S}$ is $\mathcal{H}_\mathcal{S} = \mathbb{C}^2$ and its Hamiltonian is the third Pauli matrix

$$H_\mathcal{S} = \sigma^{(3)} = \begin{bmatrix} 1 & 0 \\ 0 & -1 \end{bmatrix}.$$

Its initial state is $\omega_\mathcal{S} = 1/2$. The reservoir $\mathcal{R}_j$ is described by the single particle Hilbert space $\mathcal{K}_j = \ell^2(\mathcal{Q}_j)$ and single particle Hamiltonian h_j. The Hamiltonian and the number operator of $\mathcal{R}_j$ are denoted by $H_j = \mathrm{d}\Gamma(h_j)$ and N_j. The creation/annihilation operators on the Fock space $\Gamma(\mathcal{K}_j)$ are denoted by $a_j^\#$. We assume that $\mathcal{R}_j$ is in the state

$$\omega_{\beta_j,\mu_j} = \frac{e^{-\beta_j(H_j - \mu_j N_j)}}{\mathrm{tr}\left(e^{-\beta_j(H_j - \mu_j N_j)}\right)},$$

that is, that $\mathcal{R}_j$ is in thermal equilibrium at inverse temperature β_j and chemical potential μ_j. The complete reservoir system $\mathcal{R} = \sum_j \mathcal{R}_j$ is described by the Hilbert space

$$\mathcal{H}_\mathcal{R} = \bigotimes_{j=1}^n \Gamma(\mathcal{K}_j),$$

its Hamiltonian is

$$H_\mathcal{R} = \sum_{j=1}^n H_j,$$

and its initial state is

$$\omega_\mathcal{R} = \otimes_{j=1}^n \omega_{\beta_j,\mu_j} = \frac{1}{Z} e^{-\sum_{j=1}^n \beta_j(H_j - \mu_j N_j)},$$

where $Z = \mathrm{tr}(e^{-\sum_j \beta_j(H_j - \mu_j N_j)})$. The Hilbert space of the joint system $\mathcal{S} + \mathcal{R}$ is

$$\mathcal{H} = \mathcal{H}_\mathcal{S} \otimes \mathcal{H}_\mathcal{R},$$

its initial state is $\omega = \omega_\mathcal{S} \otimes \omega_\mathcal{R}$, and in the absence of interaction its Hamiltonian is

$$H_0 = H_\mathcal{S} + H_\mathcal{R}.$$

The interaction of $\mathcal{S}$ with $\mathcal{R}_j$ is described by

$$V_j = \sigma^{(1)} \otimes P_j,$$

where $\sigma^{(1)}$ is the first Pauli matrix and P_j is a self-adjoint polynomial in the field operators

$$\varphi_j(\psi) = \frac{1}{\sqrt{2}}(a_j(\psi) + a_j^*(\psi)) \in \mathcal{O}_{\Gamma(\mathcal{K}_j)}.$$

For example $P_j = \varphi_j(\psi_j)$ or $P_j = i\varphi_j(\psi_j)\varphi_j(\phi_j)$ with $\psi_j \perp \phi_j$. The complete interaction is $V = \sum_{j=1}^n V_j$ and the full (interacting) Hamiltonian of the SF model is

$$H_\lambda = H_0 + \lambda V,$$

where $\lambda \in \mathbb{R}$ is a coupling constant.

Exercise 4.51 Check that the SF model is an example of open quantum system as defined in Section 4.5.1. Warning: gauge invariance is broken in the SF model.

Exercise 4.52

1. Denote by $\{e_1, e_2\}$ the standard basis of $\mathcal{H}_S = \mathbb{C}^2$. Show that the triple $(\mathcal{H}_S \otimes \mathcal{H}_S, \pi_S, \Omega_S)$, where $\pi_S(A) = A \otimes \mathbb{1}$ and

$$\Omega_S = \frac{1}{\sqrt{2}}(e_1 \otimes e_1 + e_2 \otimes e_2),$$

   is a GNS representation of $\mathcal{O}_{\mathcal{H}_S}$ associated to ρ_S. Since ρ_S is faithful, this representation carries the modular structure of $\mathcal{O}_S$. Show that the modular conjugation and the modular operator are given by $J_S : f \otimes g \mapsto \bar{g} \otimes \bar{f}$ and $\Delta_{\omega_S} = \mathbb{1}$.
   Note that this part of the exercise is the simplest non trivial example of Exercise 5.31 (5).
2. Let $(\mathcal{H}_{\mathrm{AW},j}, \pi_{\mathrm{AW},j}, \Omega_{\mathrm{AW},j})$ be the Araki–Wyss representation of the j-th reservoir associated to the quasi-free state ω_{β_j,μ_j}. Show that $\pi_{\mathrm{SF}} = \pi_S \otimes \pi_{\mathrm{AW},1} \otimes \cdots \otimes \pi_{\mathrm{AW},n}$ is the standard representation of $\mathcal{O}_{\mathcal{H}}$ on the Hilbert space

$$\mathcal{H}_{\mathrm{SF}} = (\mathcal{H}_S \otimes \mathcal{H}_S) \otimes \mathcal{H}_{\mathrm{AW},1} \otimes \cdots \mathcal{H}_{\mathrm{AW},n},$$

   with the cyclic vector

$$\Omega_{\mathrm{SF}} = \Omega_S \otimes \Omega_{\mathrm{AW},1} \otimes \cdots \otimes \Omega_{\mathrm{AW},n}.$$

3. Consider the SF model with interaction $P_j = \varphi_j(\psi_j)$. Show that in the above standard representation the operator $L_{\frac{1}{\alpha}}$, defined by (4.134), takes the form

$$L_{\frac{1}{\alpha}} = (H_S \otimes \mathbb{1}_{\mathcal{H}_S} - \mathbb{1}_{\mathcal{H}_S} \otimes H_S) \otimes \mathbb{1}_{\mathcal{H}_{\mathrm{AW},1}} \otimes \cdots \otimes \mathbb{1}_{\mathcal{H}_{\mathrm{AW},n}} \tag{4.168}$$

$$+ \sum_{j=1}^n (\mathbb{1}_{\mathcal{H}_S} \otimes \mathbb{1}_{\mathcal{H}_S}) \otimes \mathbb{1}_{\mathcal{H}_{\mathrm{AW},1}} \otimes \cdots \otimes (d\Gamma(h_j) \otimes \mathbb{1} - \mathbb{1} \otimes d\Gamma(\bar{h}_j)) \otimes \cdots \otimes \mathbb{1}_{\mathcal{H}_{\mathrm{AW},n}}$$

$$+ \lambda \sum_{j=1}^{n} (\sigma^{(1)} \otimes \mathbb{1}_{\mathcal{H}_S}) \otimes \mathbb{1}_{\mathcal{H}_{\mathrm{AW},1}} \otimes \cdots \otimes \frac{1}{\sqrt{2}} \left(b_{\mathrm{AW},j}(\psi_j) + b_{\mathrm{AW},j}^*(\psi_j) \right) \otimes \cdots \otimes \mathbb{1}_{\mathcal{H}_{\mathrm{AW},n}}$$

$$- \lambda \sum_{j=1}^{M} (\mathbb{1}_{\mathcal{H}_S} \otimes \sigma^{(1)}) \otimes \mathbb{1}_{\mathcal{H}_{\mathrm{AW},1}} \otimes \cdots \otimes \frac{1}{\sqrt{2}} \left(b'_{\mathrm{AW},j}(\psi_j^+) + b'^*_{\mathrm{AW},j}(\psi_j^-) \right) \otimes \cdots \otimes \mathbb{1}_{\mathcal{H}_{\mathrm{AW},n}},$$

where $\boldsymbol{\alpha} = (0, \boldsymbol{\gamma}, \boldsymbol{\gamma}')$ and

$$\psi_j^{\pm} = e^{\pm \beta_j [(1/2 - \gamma_j) h_j - \mu_j (1/2 - \gamma_j')]} \psi_j.$$

Starting with the seminal papers of Davies (1974), Lebowitz–Spohn (1978), and Davies–Spohn (1978), the SF model (together with the closely related spin-boson model) became a paradigm in mathematically rigorous studies of nonequilibrium quantum statistical mechanics. Despite the number of new results obtained in the last decade many basic questions about this model are still open.

The study of the SF model requires sophisticated analytical tools and for reasons of space we shall not make a detailed exposition of specific results in these lecture notes. Instead, we will restrict ourselves to a brief description on the main new conceptual ideas that made the proofs of these results possible. We refer the reader to the original articles for more details.

The key new idea, which goes back to Jakšić–Pillet (2002), is to use modular theory and quantum transfer operators to study large time limits. As we have repeatedly emphasized, before taking the limit $t \to \infty$ one must take reservoirs to their thermodynamic limit. The advantage of the modular structure is that it remains intact in the thermodynamic limit. In other words, the basic objects and relations of the theory remain valid for infinitely extended systems.

In the thermodynamic limit the Hilbert spaces $\mathcal{K}_j$ become infinite dimensional. Under very general conditions the operator $L_{\frac{1}{\alpha}}$ converges to a limiting operator. In the example of Exercise 5.52, this limit has exactly the same form (4.168) on the limiting Hilbert space $\mathcal{H}_{\mathrm{SF}}$ which carries representations $\psi \mapsto b_j^{\#}(\psi)$ of the CAR over the infinite dimensional $\mathcal{K}_j$. Moreover, the limiting moment generating function for the full counting statistics (4.130) is related to this operator as in (4.133). Under suitable technical assumptions on the ψ_j's one then can prove a large deviation principle for full counting statistics by a careful study of the spectral resonances of $L_{\frac{1}{\alpha}}$. It is precisely this last step that is technically most demanding and requires a number of additional assumptions. The existing proofs are based on perturbation arguments that require small λ and, for technical reasons, vanishing chemical potentials μ_j. We remark that the proofs follow line by line the spectral scheme outlined in Section 4.6.5 and we refer the interested reader to (Jakšić, Ogata, Pautrat, and Pillet, 2011b) for details and additional information.

For $\boldsymbol{\alpha} = (0, \mathbf{1/2}, \mathbf{1/2})$, the operator $L_{\frac{1}{\alpha}}$ is the standard Liouvillean K. The spectral analysis of this operator is a key ingredient in the proof of return to equilibrium when all reservoirs are at the same temperature. For related results, see Jakšić–Pillet

(1996), Bach *et al.* (2000), Dereziński–Jakšić (2003), and Fröhlich–Merkli (2004). More generally, the spectrum of K provides information about the normal invariant states of the system, that is the density matrices on the space $\mathcal{H}_{\mathrm{SF}}$ which correspond to steady states. In particular, if K has no point spectrum then the system has no normal invariant state and hence its steady states have to be singular w.r.t. the reference state ρ (see, e.g., Pillet (2006) and Aschbacher *et al.* (2006) for details).

In the case $\boldsymbol{\alpha} = (0, \mathbf{0}, \mathbf{0})$, the operator $L_{\frac{1}{\alpha}}$ reduces to the L^{∞}-Liouvillean (or C-Liouvillean) L_{∞} introduced in Jakšić–Pillet (2002). In this work the relaxation to a non-equilibrium steady state was proven by using the identity

$$\omega_t(A) = \langle \Omega_{\mathrm{SF}} | e^{itL_{\infty}} \pi_{\mathrm{SF}}(A) \Omega_{\mathrm{SF}} \rangle,$$

and by a careful study of resonances and resonance eigenfunctions of the operator L_{∞}. This approach was adapted to the spin-boson model by Merkli *et al.* (2007b).

For a different approach to the large deviation principle for the spin-fermion and the spin-boson model we refer the reader to De Roeck (2009).

4.7.6   Electronic black box model

Model

Let $\mathcal{S}$ be a finite set and $h_{\mathcal{S}}$ a one-particle Hamiltonian on $\mathcal{K}_{\mathcal{S}} = \ell^2(\mathcal{S})$. We think of $\mathcal{S}$ as a 'black box' representing some electronic device (e.g., a quantum dot). To feed this device, we connect it to several, say n, reservoirs $\mathcal{R}_1, \ldots, \mathcal{R}_n$. For simplicity, each reservoir $\mathcal{R}_j$ is a finite lead described, in the tight binding approximation, by a box $\Lambda = [0, M]$ in $\mathbb{Z}$ (see Fig. 4.8). The one-particle Hilbert space of a finite lead is $\mathcal{K}_j = \ell^2(\Lambda)$ and its one-particle Hamiltonian is $h_j = -\frac{1}{2}\Delta_\Lambda$, where Δ_Λ denotes the discrete Laplacian on Λ with Dirichlet boundary conditions (see Section 4.2.1). The electronic black box (EBB) model is a free Fermi gas with single particle Hilbert space

$$\mathcal{K} = \mathcal{K}_{\mathcal{S}} \oplus \left(\oplus_{j=1}^{n} \mathcal{K}_j \right).$$

In the following, we identify $\mathcal{K}_{\mathcal{S}}$ and $\mathcal{K}_j$ with the corresponding subspaces of $\mathcal{K}$ and we denote by $\mathbb{1}_{\mathcal{S}}$ and $\mathbb{1}_j$ the orthogonal projections of $\mathcal{K}$ on these subspaces. In the absence of coupling between $\mathcal{S}$ and the reservoirs, the Hamiltonian of the EBB model is

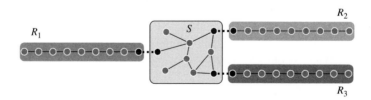

Fig. 4.8   The EBB model with three leads.

$$H_0 = d\Gamma(h_0),$$

where

$$h_0 = h_{\mathcal{S}} \oplus \left(\oplus_{j=1}^n h_j \right).$$

The reference state of the system, denoted ω_0, is the quasi-free state associated to the density

$$T_0 = T_{\mathcal{S}} \oplus \left(\oplus_{j=1}^n T_j \right),$$

where $T_{\mathcal{S}} > 0$ is a density operator on $\mathcal{K}_{\mathcal{S}}$ which commute with $h_{\mathcal{S}}$ and

$$T_j = (\mathbb{1} + e^{\beta_j(h_j - \mu_j \mathbb{1})})^{-1},$$

is the Fermi–Dirac density describing the thermal equilibrium of the j-th reservoir at inverse temperature β_j and chemical potential μ_j.

The coupling of the black box $\mathcal{S}$ to the j-th reservoir is described as follows. Let $\chi_j \in \mathcal{K}_{\mathcal{S}}$ be a unit vector and let $\delta_0^{(j)} \in \mathcal{K}_j$ be the Dirac delta function at site $0 \in \Lambda$, both identified with elements of $\mathcal{K}$. Set $v_j = |\chi_j\rangle\langle\delta_0^{(j)}| + |\delta_0^{(j)}\rangle\langle\chi_j|$. The single particle Hamiltonian of the coupled EBB model is

$$h_\lambda = h_0 + \lambda \sum_{j=1}^n v_j,$$

where $\lambda \in \mathbb{R}$ is a coupling constant. Denoting by $a^\#$ the creation/annihilation operators on $\Gamma(\mathcal{K})$ and using Part (8) of Proposition 4.54 we see that the full Hamiltonian of the coupled EBB model is

$$H_\lambda = d\Gamma(h_\lambda) = H_0 + \lambda \sum_{j=1}^n \left[a^*(\chi_j)a(\delta_0^{(j)}) + a^*(\delta_0^{(j)})a(\chi_j) \right],$$

and that the induced dynamics on the CAR algebra over $\mathcal{K}$ is completely determined by

$$\tau_\lambda^t(a^\#(\psi)) = e^{itH_\lambda} a^\#(\psi) e^{-itH_\lambda} = a^\#(e^{ith_\lambda}\psi).$$

Assume that the black box $\mathcal{S}$ is TRI, that is that there exists an anti-unitary involution $\theta_{\mathcal{S}}$ on $\mathcal{K}_{\mathcal{S}}$ such that $\theta_{\mathcal{S}} h_{\mathcal{S}} \theta_{\mathcal{S}}^* = h_{\mathcal{S}}$ and $\theta_{\mathcal{S}} T_{\mathcal{S}} \theta_{\mathcal{S}}^* = T_{\mathcal{S}}$. If $\theta_{\mathcal{S}} \chi_j = \chi_j$ for $j = 1, \dots, n$, then one easily shows that the EBB model is TRI, with the time reversal

$$\Theta = \Gamma(\theta), \qquad \theta = \theta_{\mathcal{S}} \oplus (\oplus_{j=1}^n \theta_j),$$

where θ_j denotes the complex conjugation on $\mathcal{K}_j = \ell^2(\Lambda)$.

Fluxes

The energy operator of the j-th reservoir is $H_j = d\Gamma(h_j)$. Applying equation (4.126), using Relation (4.162) and Part (8) of Proposition 4.54, we see that the energy flux observables are given by

$$\Phi_j = -i[H_\lambda, H_j] = -d\Gamma(i[h_\lambda, h_j]) = \lambda \, d\Gamma(i[h_j, v_j]) \tag{4.169}$$

$$= i\lambda \left(a^*(h_j \delta_0^{(j)}) a(\chi_j) - a^*(\chi_j) a(h_j \delta_0^{(j)}) \right).$$

The charge operator of $\mathcal{S}$ is $N_{\mathcal{S}} = d\Gamma(\mathbb{1}_{\mathcal{S}})$ and $N_j = d\Gamma(\mathbb{1}_j)$ is the charge operator of $\mathcal{R}_j$. Note that the total charge $N = N_{\mathcal{S}} + \sum_{j=1}^n N_j = d\Gamma(\mathbb{1})$ commutes with H. The charge flux observables are

$$\mathcal{J}_j = -i[H_\lambda, N_j] = -d\Gamma(i[h_\lambda, \mathbb{1}_j]) = \lambda \, d\Gamma(i[\mathbb{1}_j, v_j]) \tag{4.170}$$

$$= i\lambda \left(a^*(\delta_0^{(j)}) a(\chi_j) - a^*(\chi_j) a(\delta_0^{(j)}) \right).$$

It follows from Part (6) of Proposition 4.54 and Part (1) of Proposition 4.58 that the heat and charge fluxes at time t are

$$\omega_0(\tau_\lambda^t(\Phi_j)) = 2\lambda \, \mathrm{Im}\langle e^{ith_\lambda} h_j \delta_0^{(j)} | T_0 e^{ith_\lambda} \chi_j \rangle,$$

$$\omega_0(\tau_\lambda^t(\mathcal{J}_j)) = 2\lambda \, \mathrm{Im}\langle e^{ith_\lambda} \delta_0^{(j)} | T_0 e^{ith_\lambda} \chi_j \rangle.$$

Entropy production

One easily concludes from Part (1) of proposition 4.58 that $\omega_t = \omega_0 \circ \tau_\lambda^t$ is the quasi-free state with density $T_t = e^{-ith_\lambda} T_0 e^{ith_\lambda}$. We set

$$k_0 = \log \left(T_0 (\mathbb{1} - T_0)^{-1} \right) = \log \left(T_{\mathcal{S}} (\mathbb{1}_{\mathcal{S}} - T_{\mathcal{S}})^{-1} \right) \oplus \left(\oplus_{j=1}^n [-\beta_j (h_j - \mu_j \mathbb{1}_j)] \right),$$

so that

$$k_t = \log \left(T_t (\mathbb{1} - T_t)^{-1} \right) = e^{-ith_\lambda} k_0 e^{ith_\lambda}.$$

Proposition 4.60 allows us to write the relative Hamiltonian of ω_t w.r.t. ω_0 as

$$\ell_{\omega_t | \omega_0} = d\Gamma(k_t - k_0).$$

It follows that the entropy production observable is

$$\sigma = \left. \frac{d}{dt} \ell_{\omega_t | \omega_0} \right|_{t=0} = d\Gamma(-i[h_\lambda, k_0]) = -i[H_\lambda, Q_{\mathcal{S}}] - \sum_{j=1}^n \beta_j (\Phi_j - \mu_j \mathcal{J}_j), \tag{4.171}$$

where $Q_{\mathcal{S}} = d\Gamma(\log(T_{\mathcal{S}}(\mathbb{1}_{\mathcal{S}} - T_{\mathcal{S}})^{-1}))$ (compare this expression with equation (4.125)). The entropy balance equation thus reads

$$S(\omega_t | \omega_0) = \omega_0(\tau_\lambda^t(Q_{\mathcal{S}}) - Q_{\mathcal{S}}) + \sum_{j=1}^n \beta_j \int_0^t \omega_s(\Phi_j - \mu_j \mathcal{J}_j) \, ds. \tag{4.172}$$

Entropic pressure functionals

Not surprisingly, these functionals can be expressed in terms of one-particle quantities. For $p \in [1, \infty[$ one has, by Lemma 4.53,

$$
e_{p,t}(\alpha) = \log \operatorname{tr} \left[\left(\omega_0^{(1-\alpha)/p} \omega_t^{2\alpha/p} \omega_0^{(1-\alpha)/p} \right)^{p/2} \right]
$$

$$
= \log \operatorname{tr} \left[\frac{1}{Z_{T_0}} \Gamma \left(\left(e^{k_0(1-\alpha)/p} e^{k_t 2\alpha/p} e^{k_0(1-\alpha)/p} \right)^{p/2} \right) \right]
$$

$$
= - \log Z_{T_0} + \log \det \left(\mathbb{1} + \left(e^{k_0(1-\alpha)/p} e^{k_t 2\alpha/p} e^{k_0(1-\alpha)/p} \right)^{p/2} \right)
$$

$$
= \log \left[\frac{\det \left(\mathbb{1} + \left(e^{k_0(1-\alpha)/p} e^{k_t 2\alpha/p} e^{k_0(1-\alpha)/p} \right)^{p/2} \right)}{\det \left(\mathbb{1} + e^{k_0} \right)} \right]. \tag{4.173}
$$

After some elementary algebra, this can also be expressed as

$$
e_{p,t}(\alpha) = \log \det \left[\mathbb{1} + T_0 \left(e^{-k_0} \left(e^{k_0(1-\alpha)/p} e^{k_t 2\alpha/p} e^{k_0(1-\alpha)/p} \right)^{p/2} - \mathbb{1} \right) \right].
$$

In particular, for $p = 2$, this reduces to

$$
e_{2,t}(\alpha) = \log \det \left(\mathbb{1} + T_0 (e^{-\alpha k_0} e^{\alpha k_t} - \mathbb{1}) \right),
$$

and for $p = \infty$ we obtain

$$
e_{\infty,t}(\alpha) = \lim_{p \to \infty} e_{p,t}(\alpha) = \log \det \left(\mathbb{1} + T_0 (e^{-k_0} e^{(1-\alpha)k_0 + \alpha k_t} - \mathbb{1}) \right).
$$

Exercise 4.53 The multi-parameter formalism of Section 4.4.7 is easily adapted to the EBB model. Indeed, one has

$$
\log \omega_0 = (Q_S - \log Z_{T_0}) - \sum_{j=1}^{n} \beta_j H_j + \sum_{j=1}^{n} \beta_j \mu_j N_j,
$$

and the $n+1$ terms in this sum form a commuting family (the scalar term $-\log Z_T$ plays no role in the following, we can pack it with the Q_S term which will turn out to become irrelevant in the large time limit). Following Exercise 5.45, define

$$
\omega_0^{\boldsymbol{\alpha}} = e^{\alpha_S (Q_S - \log Z_{T_0}) - \sum_{j=1}^{n} \alpha_j \beta_j H_j + \sum_{j=1}^{n} \alpha_{n+j} \beta_j \mu_j N_j}, \qquad \omega_t^{\boldsymbol{\alpha}} = e^{-itH_\lambda} \omega_0^{\boldsymbol{\alpha}} e^{itH_\lambda},
$$

for $\boldsymbol{\alpha} = (\alpha_S, \alpha_1, \ldots, \alpha_{2n}) \in \mathbb{R}^{2n+1}$.

1. Show that the generating functional for multi-parameter full counting statistics is given by

$$
e_{2,t}(\boldsymbol{\alpha}) = \log \operatorname{tr} \left(\omega_0^{1-\boldsymbol{\alpha}} \omega_t^{\boldsymbol{\alpha}} \right) = \log \det \left(\mathbb{1} + T_0 (e^{-k(\boldsymbol{\alpha})} e^{k_t(\boldsymbol{\alpha})} - \mathbb{1}) \right)
$$

where

$$k(\boldsymbol{\alpha}) = \alpha_{\mathcal{S}} \log(T_{\mathcal{S}}(1 - T_{\mathcal{S}})^{-1}) - \sum_{j=1}^{n} \alpha_j \beta_j h_j + \sum_{j=1}^{n} \alpha_{n+j}\beta_j\mu_j \mathbb{1}_j,$$

and $k_t(\boldsymbol{\alpha}) = e^{-\mathrm{i}th_\lambda} k(\boldsymbol{\alpha}) e^{\mathrm{i}th_\lambda}$.
2. Show that the 'naive' generating function (4.156) is given by

$$e_{\text{naive},t}(\boldsymbol{\alpha}) = \log \det \left(\mathbb{1} + T_0(e^{k_{-t}(\boldsymbol{\alpha})-k(\boldsymbol{\alpha})} - \mathbb{1}) \right).$$

Exercise 4.54   Following Section 4.5.3, introduce the control parameters $X_j = \beta_{\text{eq}} - \beta_j$ and $X_{n+j} = \beta_{\text{eq}}\mu_{\text{eq}} - \beta_j\mu_j$, where β_{eq} and μ_{eq} are some equilibrium values of the inverse temperature and chemical potential. Denote by w_X the quasi-free state on the CAR algebra over $\mathcal{K}$ with density

$$T_X = \left(\mathbb{1} + e^{\beta_{\text{eq}}(h_\lambda - \mu_{\text{eq}}\mathbb{1}) - \sum_{j=1}^{n}(X_j h_j + X_{n+j}\mathbb{1}_j)} \right)^{-1},$$

and set $k_X = \log \left(T_X (1 - T_X)^{-1} \right) = -\beta_{\text{eq}}(h_\lambda - \mu_{\text{eq}}\mathbb{1}) + \sum_{j=1}^{n}(X_j h_j + X_{n+j}\mathbb{1}_j)$.

1. Show that

$$\sigma_X = \mathrm{d}\Gamma(-\mathrm{i}[h_\lambda, k_X]) = \sum_{j=1}^{n} X_j \Phi_j + X_{n+j}\mathcal{J}_j,$$

where the individual fluxes are given by (4.169) and (4.170).
2. Show that the generalized entropic pressure is given by

$$e_t(X, Y) = \log \det \left(\mathbb{1} + T_X \left(e^{-k_X} e^{k_X - Y + k_{Y,t} - k_0} - \mathbb{1} \right) \right),$$

where $k_{Y,t} = e^{-\mathrm{i}th_\lambda} k_Y e^{\mathrm{i}th_\lambda}$.
3. Develop the finite time linear response theory of the EBB model.

Thermodynamic limit

The thermodynamic limit of the EBB model is achieved by letting $M \to \infty$, keeping the system $\mathcal{S}$ untouched. We shall not enter into a detailed description of this step which is completely analogous to the thermodynamic limit of the classical harmonic chain discussed in Section 4.2.8 (see Exercise 5.55 below). The one particle Hilbert space of the reservoir $\mathcal{R}_j$ becomes $\mathcal{K}_j = \ell^2(\mathbb{N})$ and its one particle Hamiltonian $h_j = -\frac{1}{2}\Delta$, where Δ is the discrete Laplacian on $\mathbb{N}$ with Dirichlet boundary condition. Using the discrete Fourier transform

$$\widehat{\psi}(\xi) = \sqrt{\frac{2}{\pi}} \sum_{x \in \mathbb{N}} \psi(x) \sin(\xi(x+1)),$$

we can identify $\mathcal{K}_j$ with $L^2([0,\pi], \mathrm{d}\xi)$ and h_j becomes the operator of multiplication by $\varepsilon(\xi) = 1 - \cos \xi$. In particular, the spectrum of h_j is purely absolutely continuous and fills the interval $[0, 2]$ with constant multiplicity one. Thus, the spectrum of the decoupled Hamiltonian h_0 consists of an absolutely continuous part filling the same interval with constant multiplicity n and of a discrete part given by the eigenvalues of $h_\mathcal{S}$. We denote by $\mathbb{1}_\mathcal{R} = \mathbb{1} - \mathbb{1}_\mathcal{S} = \sum_{j=1}^{n} \mathbb{1}_j$ the projection on the absolutely continuous part of h_0. In the momentum representation one has $\mathcal{K}_\mathcal{R} = \operatorname{Ran} \mathbb{1}_\mathcal{R} = L^2([0,\pi]) \otimes \mathbb{C}^n$. Denoting by $\mathbb{1}_j = |e_j\rangle\langle e_j|$ the orthogonal projection of $\mathbb{C}^n$ onto the subspace generated by the j-th vector of its standard basis, we have $\mathbb{1}_j = \mathbb{1} \otimes \mathbb{1}_j$ and $h_j = \varepsilon(\xi) \otimes \mathbb{1}_j$.

Exercise 4.55 Denote by the subscript $(\cdot)_M$ the dependence on the parameter M of the various objects associated to the EBB model, for example $\omega_{M,0}$ is the reference state with density $T_{M,0} = T_\mathcal{S} \oplus (\oplus_{j=1}^{n} T_{M,j})$, etc.

1. Show that

$$\lim_{M \to \infty} T_{M,0}(\mathrm{e}^{-\alpha k_{M,0}} \mathrm{e}^{\alpha k_{M,t}} - \mathbb{1}) = T_0(\mathrm{e}^{-\alpha k_0} \mathrm{e}^{\alpha k_t} - \mathbb{1}),$$

holds in trace norm, where $T_0 = \mathrm{s} - \lim_{M \to \infty} T_{M,0}$, $k_0 = \log(T_0(\mathbb{1} - T_0))$, $k_t = \mathrm{e}^{-\mathrm{i}th_\lambda} k_0 \mathrm{e}^{\mathrm{i}th_\lambda}$ and $h_\lambda = \mathrm{s} - \lim_{M \to \infty} h_{M,\lambda}$.

Hint: write $\mathrm{e}^{-\alpha k_{M,0}} \mathrm{e}^{\alpha k_{M,t}} - \mathbb{1}$ as the integral of its derivative w.r.t. t and observe that $[h_{M,\lambda}, k_{M,0}]$ is a finite rank operator that does not depend on M.

2. Show that, for any $\alpha, t \in \mathbb{R}$,

$$\lim_{M \to \infty} e_{M,2,t}(\alpha) = \log \det \left(\mathbb{1} + T_0(\mathrm{e}^{-\alpha k_0} \mathrm{e}^{\alpha k_t} - \mathbb{1}) \right).$$

Hint: recall that

$$\det(\mathbb{1} + T_{M,0}(\mathrm{e}^{-\alpha k_{M,0}} \mathrm{e}^{\alpha k_{M,t}} - \mathbb{1})) = \omega_{M,0}(\Gamma(\mathrm{e}^{-\alpha k_{M,0}/2} \mathrm{e}^{\alpha k_{M,t}} \mathrm{e}^{-\alpha k_{M,0}/2})) > 0. \quad (4.174)$$

Remark. The implications of this exercise are described in Proposition 4.44.

Exercise 4.56 Let $\mathsf{P}_{M,t}$ denote the spectral measure of $\log(\Delta_{\omega_{M,t}|\omega_{M,0}})$ and $\xi_{\omega_{M,0}}$. Through the following steps, show that the spectral measure P_t of $\log(\Delta_{\omega_t|\omega_0})$ and ξ_{ω_0} is the weak limit of the sequence $\{\mathsf{P}_{M,t}\}$.[13]

1. Show that, for all $\alpha \in \mathbb{R}$, the characteristic function of $\mathsf{P}_{M,t}$,

$$\chi_{M,t}(\alpha) = \int \mathrm{e}^{\mathrm{i}\alpha x} \mathrm{d}\mathsf{P}_{M,t}(x) = (\xi_{\omega_{M,0}} | \Delta_{\omega_{M,t}|\omega_{M,0}}^{\mathrm{i}\alpha} \xi_{\omega_{M,0}})$$

$$= \operatorname{tr} \left(\omega_{M,0}^{1-\mathrm{i}\alpha} \, \omega_{M,t}^{\mathrm{i}\alpha} \right)$$

$$= \det \left(\mathbb{1} + T_{M,0}(\mathrm{e}^{\mathrm{i}\alpha k_{M,t}} \mathrm{e}^{-\mathrm{i}\alpha k_{M,0}} - \mathbb{1}) \right),$$

---

[13] Up to a rescaling, $\mathsf{P}_{M,t}$ is the full counting statistics of the finite EBB model.

converges, as $M \to \infty$, towards

$$\chi_t(\alpha) = \det\left(\mathbb{1} + T_0(e^{i\alpha k_t}e^{-i\alpha k_0} - \mathbb{1})\right) = \omega_0\left(\Gamma(e^{i\alpha k_t}e^{-i\alpha k_0})\right).$$

2. In the Araki–Wyss representation associated to the state ω_0, show that

$$(\xi_{\omega_0}|\Delta^{i\alpha}_{\omega_t|\omega_0}\xi_{\omega_0}) = (\xi_{\omega_0}|\Gamma_t(\alpha)\xi_{\omega_0}),$$

where the cocycle $\Gamma_t(\alpha) = \Delta^{i\alpha}_{\omega_t|\omega_0}\Delta^{-i\alpha}_{\omega_0}$ satisfies the Cauchy problem

$$\frac{d}{d\alpha}\Gamma_t(\alpha) = i\,\Gamma_t(\alpha)\left(\Delta^{i\alpha}_{\omega_0}\pi_{\mathrm{AW}}(\ell_{\omega_t|\omega_0})\Delta^{-i\alpha}_{\omega_0}\right), \qquad \Gamma_t(0) = \mathbb{1}.$$

3. Show that $\Gamma_t(\alpha) = \pi_{\mathrm{AW}}(\gamma_t(\alpha))$ where

$$\frac{d}{d\alpha}\gamma_t(\alpha) = i\,\gamma_t(\alpha)\left(e^{i\alpha d\Gamma(k_0)}\ell_{\omega_t|\omega_0}e^{-i\alpha d\Gamma(k_0)}\right), \qquad \gamma_t(0) = \mathbb{1}.$$

Conclude that $(\xi_{\omega_0}|\Delta^{i\alpha}_{\omega_t|\omega}\xi_{\omega_0}) = \omega_0(\gamma_t(\alpha))$.

4. Show that

$$\gamma_t(\alpha) = [D\omega_t : D\omega_0]^\alpha = e^{i\alpha d\Gamma(k_t)}e^{-i\alpha d\Gamma(k_0)} = \Gamma(e^{i\alpha k_t}e^{-i\alpha k_0}),$$

and conclude that

$$\chi_t(\alpha) = \int e^{i\alpha x}\,dP_t(x).$$

5. Invoke the Lévy–Cramér continuity theorem (Theorem 7.6 in Billingsley (1968)) to conclude that $P_{M,t}$ converges weakly towards P_t.

Large time limit

Let us briefly discuss the limit $t \to \infty$. For simplicity, we shall assume that the one particle Hamiltonian h_λ has purely absolutely continuous spectrum. This is the generic situation for small coupling λ in the fully resonant case where $\mathrm{sp}(h_S) \subset\,]0, 2[$. Since $h_\lambda - h_0 = v = \sum_{j=1}^n v_j$ is finite rank, the wave operators

$$w_\pm = \mathrm{s} - \lim_{t \to \pm\infty} e^{ith_\lambda}e^{-ith_0}\mathbb{1}_{\mathcal{R}},$$

exist and are complete, $w_\pm w_\pm^* = \mathbb{1}$, $w_\pm^* w_\pm = \mathbb{1}_{\mathcal{R}}$. The scattering matrix $s = w_+^* w_-$ is unitary on $\mathcal{K}_{\mathcal{R}}$. It acts as the operator of multiplication by a unitary $n \times n$ matrix $s(\xi) = [s_{jk}(\xi)]$. Since $[h_0, T_0] = 0$, one has

$$\lim_{t \to \infty}\langle\psi|T_t\phi\rangle = \lim_{t \to \infty}\langle e^{ith_\lambda}\psi|T_0 e^{ith_\lambda}\phi\rangle$$

$$= \lim_{t \to \infty} \langle e^{-ith_0} e^{ith_\lambda} \psi | T_0 e^{-ith_0} e^{ith_\lambda} \phi \rangle$$

$$= \langle w_-^* \psi | T_0 w_-^* \phi \rangle = \langle \psi | T_+ \phi \rangle,$$

with $T_+ = w_- T_0 w_-^*$. It follows that for any polynomial A in the creation/annihilation operators on $\Gamma(\mathcal{K})$, one has

$$\lim_{t \to \infty} \omega_0 \circ \tau_\lambda^t(A) = \omega_+(A),$$

where ω_+ is the quasi-free state with density T_+. We conclude that the NESS ω_+ of the EBB model is the quasi-free state with density

$$T_+ = w_- T_0 w_-^*. \tag{4.175}$$

The large time asymptotics of the entropic pressure functionals can be obtained along the same line as in Section 4.2.11. We shall only consider the case $p = 2$ and leave the general case as an exercise.

Starting with (4.173) and using the result of Exercise 5.8, we can write

$$\frac{d}{d\alpha} e_{2,t}(\alpha) = \frac{d}{d\alpha} \operatorname{tr} \log \left(\mathbb{1} + e^{(1-\alpha)k_0} e^{\alpha k_t} \right)$$

$$= \operatorname{tr} \left((\mathbb{1} + e^{(1-\alpha)k_0} e^{\alpha k_t})^{-1} e^{(1-\alpha)k_0} (k_t - k_0) e^{\alpha k_t} \right)$$

$$= \operatorname{tr} \left((\mathbb{1} + e^{-(1-\alpha)k_0} e^{-\alpha k_t})^{-1} (k_t - k_0) \right)$$

$$= -t \int_0^1 \operatorname{tr} \left((\mathbb{1} + e^{-(1-\alpha)k_0} e^{-\alpha k_t})^{-1} e^{-ituh_\lambda} \mathrm{i}[h_\lambda, k_0] e^{ituh_\lambda} \right) du$$

$$= -t \int_0^1 \operatorname{tr} \left(e^{ituh_\lambda} (\mathbb{1} + e^{-(1-\alpha)k_0} e^{-\alpha k_t})^{-1} e^{-ituh_\lambda} \mathrm{i}[h_\lambda, k_0] \right) du$$

$$= -t \int_0^1 \operatorname{tr} \left((\mathbb{1} + e^{-(1-\alpha)k_{-tu}} e^{-\alpha k_{t(1-u)}})^{-1} \mathrm{i}[h_\lambda, k_0] \right) du.$$

The final relation

$$\frac{d}{d\alpha} e_{2,t}(\alpha) = -t \int_0^1 \operatorname{tr} \left((\mathbb{1} + e^{-(1-\alpha)k_{-tu}} e^{-\alpha k_{t(1-u)}})^{-1} \mathrm{i}[h_\lambda, k_0] \right) du,$$

remains valid after the thermodynamic limit is taken. Since k_0 is a bounded operator commuting with h_0, one easily shows that

$$\mathrm{s} - \lim_{t \to \pm \infty} k_t = k_\pm = w_\mp k_0 w_\mp^*,$$

which leads to

$$s - \lim_{t \to \infty}(1 + e^{-(1-\alpha)k_{-ts}}e^{-\alpha k_{t(1-s)}})^{-1} = (1 + e^{-(1-\alpha)k_{-}}e^{-\alpha k_{+}})^{-1}$$

$$= (1 + w_{+}e^{-(1-\alpha)k_0}w_{+}^{*}w_{-}e^{-\alpha k_0}w_{-}^{*})^{-1}$$

$$= (1 + w_{-}s^{*}e^{-(1-\alpha)k_0}se^{-\alpha k_0}w_{-}^{*})^{-1}$$

$$= w_{-}(1 + s^{*}e^{-(1-\alpha)k_0}se^{-\alpha k_0})^{-1}w_{-}^{*}.$$

Since $i[h_\lambda, k_0]$ is finite rank, it follows that

$$\lim_{t \to \infty} \mathrm{tr}_{\mathcal{K}}\left((1 + e^{-(1-\alpha)k_{-tu}}e^{-\alpha k_{t(1-u)}})^{-1}i[h_\lambda, k_0]\right)$$

$$= \mathrm{tr}_{\mathcal{K}_{\mathcal{R}}}\left((1 + s^{*}e^{-(1-\alpha)k_0}se^{-\alpha k_0})^{-1}\mathcal{T}\right),$$

where $\mathcal{T} = w_{-}^{*}i[h_\lambda, k_0]w_{-}$. Since $e_{2,t}(0) = 0$, we can write

$$e_{2,+}(\alpha) = \lim_{t \to \infty} \frac{1}{t}e_{2,t}(\alpha) = \lim_{t \to \infty} \frac{1}{t}\int_0^\alpha \frac{\mathrm{d}}{\mathrm{d}\gamma}e_{2,t}(\gamma)\,\mathrm{d}\gamma,$$

and the dominated convergence theorem yields

$$e_{2,+}(\alpha) = -\int_0^\alpha \int_0^1 \mathrm{tr}\left((1 + s^{*}e^{-(1-\gamma)k_0}se^{-\gamma k_0})^{-1}\mathcal{T}\right)\mathrm{d}u\mathrm{d}\gamma$$

$$= -\int_0^\alpha \mathrm{tr}\left((1 + s^{*}e^{-(1-\gamma)k_0}se^{-\gamma k_0})^{-1}\mathcal{T}\right)\mathrm{d}\gamma.$$

The trace class operator $\mathcal{T}$ on $\mathcal{K}_{\mathcal{R}}$ has an integral kernel $\mathcal{T}(\xi, \xi')$ in the momentum representation. Following the argument leading to (4.28), one shows that its diagonal is given by

$$\mathcal{T}(\xi, \xi) = \frac{\varepsilon'(\xi)}{2\pi}(s^{*}(\xi)k(\xi)s(\xi) - k(\xi)), \tag{4.176}$$

where $k(\xi)$ is the operator on $\mathbb{C}^n$ defined by

$$k(\xi) = -\sum_{j=1}^n \beta_j(\varepsilon(\xi) - \mu_j)1_j.$$

Thus, one has

$$\mathrm{tr}_{\mathcal{K}_{\mathcal{R}}}\left((\mathbb{1} + s^* e^{-(1-\alpha)k_0} s e^{-\alpha k_0})^{-1} \mathcal{T}\right)$$

$$= -\int_0^\pi \mathrm{tr}_{\mathbb{C}^n}\left((\mathbb{1} + s^*(\xi) e^{-(1-\alpha)k(\xi)} s(\xi) e^{-\alpha k(\xi)})^{-1} (k(\xi) - s^*(\xi) k(\xi) s(\xi))\right) \varepsilon'(\xi) \frac{d\xi}{2\pi}$$

$$= -\frac{d}{d\alpha}\int_0^\pi \mathrm{tr}_{\mathbb{C}^n}\left(\log(\mathbb{1} + s^*(\xi) e^{(1-\alpha)k(\xi)} s(\xi) e^{\alpha k(\xi)})\right) \varepsilon'(\xi) \frac{d\xi}{2\pi},$$

and we conclude that

$$e_{2,+}(\alpha) = \int_0^\pi \log\left[\frac{\det\left(\mathbb{1} + e^{(1-\alpha)k(\xi)} s(\xi) e^{\alpha k(\xi)} s^*(\xi)\right)}{\det\left(\mathbb{1} + e^{k(\xi)}\right)}\right] \frac{d\varepsilon(\xi)}{2\pi}.$$

After a simple algebraic manipulation, this can be rewritten as

$$e_{2,+}(\alpha) = \int_0^\pi \log \det\left(\mathbb{1} + T(\xi)(e^{-\alpha k(\xi)} s(\xi) e^{\alpha k(\xi)} s^*(\xi) - \mathbb{1})\right) \frac{d\varepsilon(\xi)}{2\pi}, \qquad (4.177)$$

where $T(\xi) = (\mathbb{1} + e^{-k(\xi)})^{-1}$. In the following exercise, this calculation is extended to various other entropic functionals.

Exercise 4.57

1. Show that for $p \in [1, \infty[$ one has

$$e_{p,+}(\alpha) = \lim_{t \to \infty} \frac{1}{t} e_{p,t}(\alpha)$$

$$= \int_0^\pi \log \det\left[\mathbb{1} + T(\xi)\left(e^{-k(\xi)}\left(e^{k(\xi)(1-\alpha)/p} s(\xi) e^{k(\xi)2\alpha/p} s^*(\xi) e^{k(\xi)(1-\alpha)/p}\right)^{p/2} - \mathbb{1}\right)\right] \frac{d\varepsilon(\xi)}{2\pi}.$$

2. Show that

$$e_{\infty,+}(\alpha) = \lim_{t \to \infty} \frac{1}{t} e_{\infty,t}(\alpha)$$

$$= \int_0^\pi \log \det\left(\mathbb{1} + T(\xi)(e^{-k(\xi)} e^{(1-\alpha)k(\xi) + \alpha s(\xi) k(\xi) s(\xi)^*} - \mathbb{1})\right) \frac{d\varepsilon(\xi)}{2\pi}.$$

3. Compute

$$e_{\mathrm{naive},+}(\alpha) = \lim_{t \to \infty} \frac{1}{t} e_{\mathrm{naive},t}(\alpha).$$

4. Show that the large time asymptotics of the multi-parameter functional of Exercise 5.54 is given by

$$e_{2,+}(\boldsymbol{\alpha}) = \lim_{t\to\infty} \frac{1}{t} e_{2,t}(\boldsymbol{\alpha})$$

$$= \int_0^\pi \log\det\left(\mathbb{1} + T(\xi)(e^{-k(\boldsymbol{\alpha},\xi)}s(\xi)e^{k(\boldsymbol{\alpha},\xi)}s^*(\xi) - \mathbb{1})\right)\frac{d\varepsilon(\xi)}{2\pi},$$

where

$$k(\boldsymbol{\alpha},\xi) = -\sum_{j=1}^n \beta_j \left(\alpha_j \varepsilon(\xi) - \alpha_{n+j}\mu_j\right) 1_j.$$

Note in particular that $e_{2,+}(\boldsymbol{\alpha})$ does not depend on the first component α_S of $\boldsymbol{\alpha}$.
5. Show that the large time asymptotics of the generalized functional of Exercise 5.54 is given by

$$e_+(X,Y) = \lim_{t\to\infty} \frac{1}{t} e_t(X,Y)$$

$$= \int_0^\pi \log\det\left(\mathbb{1} + T_X(\xi)(e^{-k_X(\xi)}e^{k_X-Y(\xi)+s(\xi)k_Y(\xi)s^*(\xi)-k_0(\xi)} - \mathbb{1})\right)\frac{d\varepsilon(\xi)}{2\pi},$$

where $k_X(\xi)$ is the diagonal $n \times n$ matrix with entries $-(\beta_{eq} - X_j)\varepsilon(\xi) + (\beta_{eq}\mu_{eq} + X_{n+j})$ and $T_X(\xi) = (\mathbb{1} + e^{-k_X(\xi)})^{-1}$.
6. Develop the linear response theory of the EBB model.

For $\xi \in [0,\pi]$, denote by ω_ξ the density matrix

$$\omega_\xi = \frac{\Gamma(e^{k(\xi)})}{\operatorname{tr}_{\Gamma(\mathbb{C}^n)}(\Gamma(e^{k(\xi)}))},$$

on $\Gamma(\mathbb{C}^n)$. Clearly, ω_ξ defines a state on $\Gamma(\mathbb{C}^n)$ which is quasi-free with density $T(\xi)$. By Part (3) of Proposition 4.58, the Rényi relative entropy of the state $\Gamma(s(\xi))\omega_\xi\Gamma(s(\xi))^*$ w.r.t. ω_ξ is given by

$$S_\alpha(\Gamma(s(\xi))\omega_\xi\Gamma(s(\xi))^*|\omega_\xi) = \log\operatorname{tr}\left(\omega_\xi^{1-\alpha}\Gamma(s(\xi))\omega_\xi^\alpha\Gamma(s(\xi))^*\right)$$

$$= \log\omega_\xi\left(\Gamma(e^{-\alpha k(\xi)}s(\xi)e^{\alpha k(\xi)}s^*(\xi))\right)$$

$$= \log\det\left(\mathbb{1} + T(\xi)(e^{-\alpha k(\xi)}s(\xi)e^{\alpha k(\xi)}s^*(\xi) - \mathbb{1})\right).$$

Thus, we can rewrite Formula (4.177) as

$$e_{2,+}(\alpha) = \int_0^\pi S_\alpha(\Gamma(s(\xi))\omega_\xi\Gamma(s(\xi))^*|\omega_\xi)\frac{d\varepsilon(\xi)}{2\pi}.$$

Using the second identity in (4.64), we deduce

$$\frac{\mathrm{d}}{\mathrm{d}\alpha} e_{2,+}(\alpha)\Big|_{\alpha=1} = -\int_0^\pi S(\Gamma(s(\xi))\omega_\xi\Gamma(s(\xi))^*|\omega_\xi)\frac{\mathrm{d}\varepsilon(\xi)}{2\pi}.$$

Since $\log\omega_\xi = \mathrm{d}\Gamma(k(\xi)) - \log\det(\mathbb{1}+e^{k(\xi)})$, Relation (4.161) and Part (4) of Proposition 4.58 yield

$$\begin{aligned} S(\Gamma(s(\xi))\omega_\xi\Gamma(s(\xi))^*|\omega_\xi) &= \mathrm{tr}\left[\Gamma(s(\xi))\omega_\xi\Gamma(s(\xi))^* \left(\log\omega_\xi - \Gamma(s(\xi))\log\omega_\xi\Gamma(s(\xi))^*\right)\right] \\ &= \mathrm{tr}\left[\omega_\xi\left(\Gamma(s(\xi))^*\log\omega_\xi\Gamma(s(\xi)) - \log\omega_\xi\right)\right] \\ &= \omega_\xi\left(\mathrm{d}\Gamma(s^*(\xi)k(\xi)s(\xi) - k(\xi))\right) \\ &= \mathrm{tr}\left(T(\xi)(s^*(\xi)k(\xi)s(\xi) - k(\xi))\right). \end{aligned}$$

Hence, it follows from (4.176) and (4.175) that

$$\begin{aligned} \int_0^\pi S(\Gamma(s(\xi))\omega_\xi\Gamma(s(\xi))^*|\omega_\xi)\frac{\mathrm{d}\varepsilon(\xi)}{2\pi} &= -\int_0^\pi \mathrm{tr}\left(T(\xi)\mathcal{T}(\xi,\xi)\right)\mathrm{d}\xi \\ &= -\mathrm{tr}(T_0\mathcal{T}) \\ &= -\mathrm{tr}(T_0 w_-^* \mathrm{i}[h_\lambda, k_0]w_-) \\ &= -\mathrm{tr}(T_+\mathrm{i}[h_\lambda, k_0]). \end{aligned}$$

Finally, (4.171) allows us to write

$$-\mathrm{tr}\left(T_+\mathrm{i}[h_\lambda, k_0]\right) = \omega_+\left(\mathrm{d}\Gamma(-\mathrm{i}[h_\lambda, k_0])\right) = \omega_+(\sigma).$$

Thus, we have shown that

$$\frac{\mathrm{d}}{\mathrm{d}\alpha} e_{2,+}(\alpha)\Big|_{\alpha=1} = \omega_+(\sigma) = -\int_0^\pi S(\Gamma(s(\xi))\omega_\xi\Gamma(s(\xi))^*|\omega_\xi)\frac{\mathrm{d}\varepsilon(\xi)}{2\pi}.$$

Invoking Part (1) of Proposition 4.19 we observe that if $w_+(\sigma) = 0$ then we must have

$$S(\Gamma(s(\xi))\omega_\xi\Gamma(s(\xi))^*|\omega_\xi) = 0, \tag{4.178}$$

for almost all $\xi \in [0, \pi]$ which in turn implies that $\Gamma(s(\xi))\omega_\xi\Gamma(s(\xi))^* = \omega_\xi$, that is that $[k(\xi), s(\xi)] = 0$ for almost all $\xi \in [0, \pi]$. The last condition can be written as

$$[(\beta_k - \beta_j)\varepsilon(\xi) - (\beta_k\mu_k - \beta_j\mu_j)]\,s_{jk}(\xi) = 0,$$

for all $j, k \in \{1, \dots, n\}$, and we conclude that if there exists $j, k \in \{1, \dots, n\}$ and a set $\Omega \subset [0, \pi]$ of positive Lebesgue measure such that $j \neq k$, $s_{jk}(\xi) \neq 0$ for $\xi \in \Omega$ and $(\beta_j, \mu_j) \neq (\beta_k, \mu_k)$, then $w_+(\sigma) > 0$. In more physical terms, if there is an open scattering channel between two leads $\mathcal{R}_j$ and $\mathcal{R}_k$ which are not in mutual thermal equilibrium, then entropy production in the NESS is strictly positive.

Note that since (4.171) implies

$$\omega_+(\sigma) = -\sum_{j=1}^{n} \beta_j \left(\omega_+(\Phi_j) - \mu_j \omega_+(\mathcal{J}_j) \right),$$

the expected currents $\omega_+(\Phi_j)$, $\omega_+(\mathcal{J}_j)$ cannot all vanish if entropy production is strictly positive.

Exercise 4.58 Deduce from Relation (4.172) that

$$- \lim_{t \to \infty} \frac{1}{t} S(\omega_t | \omega_0) = \omega_+(\sigma).$$

Thus, if $\omega_+(\sigma) > 0$ then the entropy of ω_t w.r.t. ω_0 diverges as $t \to \infty$.

Exercise 4.59 Derive the Landauer–Büttiker formulas for the expected energy and charge currents in the steady state ω_+,

$$\omega_+(\Phi_j) = \sum_{k=1}^{n} \int_0^{\pi} t_{jk}(\xi)(\varrho_j(\xi) - \varrho_k(\xi)) \varepsilon(\xi) \frac{d\varepsilon(\xi)}{2\pi},$$

$$\omega_+(\mathcal{J}_j) = \sum_{k=1}^{n} \int_0^{\pi} t_{jk}(\xi)(\varrho_j(\xi) - \varrho_k(\xi)) \frac{d\varepsilon(\xi)}{2\pi},$$

where $\varrho_j(\xi) = (1 + e^{\beta_j(\varepsilon(\xi) - \mu_j)})^{-1}$ is the Fermi–Dirac density of the j-th reservoir and

$$t_{jk}(\xi) = |s_{jk}(\xi) - \delta_{jk}|^2 .$$

Hint: start with $\omega_+(\Phi_j) = -\mathrm{tr}\,(T_+ \mathrm{i}[h_\lambda, h_j]) = -\mathrm{tr}\,(T_0 \mathcal{T}_j)$ where $\mathcal{T}_j = w_-^* \mathrm{i}[h_\lambda, h_j] w_-$, and deduce from (4.176) that the diagonal part of the integral kernel of $\mathcal{T}_j$ is given by

$$\mathcal{T}_j(\xi, \xi) = \frac{\varepsilon'(\xi)}{2\pi} \varepsilon(\xi) \, (s^*(\xi) 1_j s(\xi) - 1_j) .$$

Proceed in a similar way for the charge currents. (For more information on the Landauer-Büttiker formalism, see Datta (1995) and Imry (1997). More general mathematical derivations can be found in Aschbacher *et al.* (2007), Nenciu (2007), and Ben Sâad (2008)).

Exercise 4.60 Starting with the Landauer–Büttiker formulas develop the linear response theory of the EBB model.

Exercise 4.61 Consider the full counting statistics of charge transport in the framework of Section 4.4.8. Let $\mathbb{P}_t^c(\mathbf{q})$, $\mathbf{q} = (q_1, \dots, q_n)$, denote the probability for the results, $\mathbf{n}$ and $\mathbf{n}'$, of two successive joint measurements of $\mathbf{N} = (N_1, \dots, N_n)$, at time 0 and t, to be such that $\mathbf{n}' - \mathbf{n} = t\mathbf{q}$. Loosely speaking, $\mathbb{P}_t^c(q_1, \dots, q_n)$ is the probability for the charge (number of fermions) of the reservoir $\mathcal{R}_j$ to increase by tq_j ($j = 1, \dots, n$) during the time interval $[0, t]$. Denote by

$$\chi_t(\boldsymbol{\nu}) = \sum_{\mathbf{q}} \mathbb{P}_t^c(\mathbf{q}) e^{-t\boldsymbol{\nu} \cdot \mathbf{q}},$$

the Laplace transform of this distribution (that is, the moment generating function of $\mathbb{P}_t^c$).

1. Show that the logarithm of $\chi_t(\boldsymbol{\nu})$ is related to the functional $e_{2,t}(\boldsymbol{\alpha})$ of Exercise 5.53 by

$$\log \chi_t(\boldsymbol{\nu}) = e_{2,t}(1 - \boldsymbol{\alpha}),$$

provided $\boldsymbol{\nu} = (\nu_1, \ldots, \nu_n)$ is related to $\boldsymbol{\alpha} = (\alpha_{\mathcal{S}}, \alpha_1, \ldots, \alpha_{2n})$ according to

$$\alpha_j = \alpha_{\mathcal{S}} = 0, \quad \nu_j = -\alpha_{n+j}\beta_j\mu_j, \quad (j = 1, \ldots, n).$$

2. Show that in the thermodynamic limit

$$\chi_t(\boldsymbol{\nu}) = \det\left(\mathbb{1} + T_0(e^{q_t(\boldsymbol{\nu})}e^{-q(\boldsymbol{\nu})} - \mathbb{1})\right),$$

where

$$q(\boldsymbol{\nu}) = \sum_{j=1}^{n} \nu_j \mathbb{1}_j,$$

and $q_t(\boldsymbol{\nu}) = e^{-ith_\lambda}q(\boldsymbol{\nu})e^{ith_\lambda}$.

Hint: combine Part 1 with the result of Exercise 5.53.

3. Derive the Levitov–Lesovik formula

$$\lim_{t\to\infty} \frac{1}{t}\log \chi_t(\boldsymbol{\nu}) = \int_0^\pi \log\det\left(\mathbb{1} + T(\xi)(s^*(\xi)s^\nu(\xi) - \mathbb{1})\right)\frac{d\varepsilon(\xi)}{2\pi},$$

where the matrix $s^\nu(\xi) = [s_{jk}^\nu(\xi)]$ is defined by

$$s_{jk}^\nu(\xi) = s_{jk}(\xi)e^{\nu_k - \nu_j}.$$

(See Levitov and Lesovik (1993), where the Fourier transform of the probability distribution $\mathbb{P}_t^c$ is considered instead of its Laplace transform. See also Avron *et al.* (2008).)

Exercise 4.62 Consider the EBB model with two reservoirs. Prove that the following statements are equivalent.

1. $e_{p,+}(\alpha)$ does not depend on p.
2. $s_{11}(\xi) = s_{22}(\xi) = 0$ for Lebesgue a.e. $\xi \in [0, \pi]$.
3. The fluctuation relation $e_{\text{naive},+}(\alpha) = e_{\text{naive},+}(1 - \alpha)$ holds.
4. $e_{\text{naive},+}(\alpha) = e_{\infty,+}(\alpha)$.

Exercise 4.63 Consider the following variant of the EBB model. $\mathcal{S}$ is a box $\Lambda = [-l, l]$ in $\mathbb{Z}$ and $h_{\mathcal{S}} = -\frac{1}{2}\Delta_\Lambda$ is the discrete Laplacian on Λ with Dirichlet boundary condition. The box $\mathcal{S}$ is connected to the left and right lead which, before the thermodynamical limit is taken, are described by the boxes $\Lambda_L =]-M, -l-1]$, $\Lambda_R = [l+1, M[$, where $l \ll M$, and after the thermodynamic limit is taken, by the boxes $\Lambda_L =]-\infty, -l-1]$, $\Lambda_R = [l+1, \infty[$. The one particle Hamiltonians are $h_L = -\frac{1}{2}\Delta_{\Lambda_L}$, $h_R = -\frac{1}{2}\Delta_{\Lambda_R}$, where, as usual, Δ_{Λ_L} and Δ_{Λ_R} are the discrete Laplacians on Λ_L and Λ_R with Dirichlet boundary conditions. The corresponding EBB model is a free Fermi gas with single particle Hilbert space

$$\ell^2(\Lambda_L) \oplus \ell^2(\Lambda) \oplus \ell^2(\Lambda_R) = \ell^2(\Lambda_L \cup \Lambda \cup \Lambda_R).$$

In the absence of coupling its Hamiltonian is $H_0 = d\Gamma(h_0)$, where $h_0 = h_L \oplus h_{\mathcal{S}} \oplus h_R$. The Hamiltonian of the joint system is $H = d\Gamma(h)$, where $h = -\frac{1}{2}\Delta_{\Lambda_L \cup \Lambda \cup \Lambda_R}$ and $\Delta_{\Lambda_L \cup \Lambda \cup \Lambda_R}$ is the discrete Laplacian on $\Lambda_L \cup \Lambda \cup \Lambda_R$ with Dirichlet boundary conditions. The reference state of the system is a quasi free state with density

$$T_0 = T_L \oplus T_S \oplus T_R,$$

where $T_S > 0$ is a density operator on $\ell^2(\Lambda)$ that commutes with h_S and

$$T_L = (\mathbb{1} + e^{\beta_L(h_L - \mu_L \mathbb{1})})^{-1}, \qquad T_R = (\mathbb{1} + e^{\beta_R(h_R - \mu_R \mathbb{1})})^{-1},$$

are the Fermi–Dirac densities of the left and right reservoir.

1. Discuss in detail the thermodynamic limit $M \to \infty$ and compare the model with the classical harmonic chain discussed in Section 4.2.
   The remaining parts of this exercise concern the infinitely extended model.
2. Using the discrete Fourier transform

$$\ell^2(\Lambda_L) \oplus \ell^2(\Lambda_R) \ni \psi_L \oplus \psi_R \mapsto \widehat{\psi}_L \oplus \widehat{\psi}_R \in L^2([0,\pi], \mathrm{d}\xi) \oplus L^2([0,\pi], \mathrm{d}\xi),$$

$$\widehat{\psi}_L(\xi) = \sqrt{\frac{2}{\pi}} \sum_{x \in \Lambda_L} \psi_L(x) \sin(\xi(x-1)), \qquad \widehat{\psi}_R(\xi) = \sqrt{\frac{2}{\pi}} \sum_{x \in \Lambda_R} \psi_R(x) \sin(\xi(x+1)),$$

identify $h_L \oplus h_R$ with the operator of multiplication by $(1 - \cos \xi) \oplus (1 - \cos \xi)$ on $L^2([0,\pi], \mathrm{d}\xi) \oplus L^2([0,\pi], \mathrm{d}\xi)$. The wave operators

$$w_\pm = \mathrm{s} - \lim_{t \to \pm\infty} e^{\mathrm{i}th} e^{-\mathrm{i}th_0} \mathbb{1}_{\mathcal{R}},$$

exist and are complete ($\mathbb{1}_{\mathcal{R}}$ is the orthogonal projection onto $\ell^2(\Lambda_L) \oplus \ell^2(\Lambda_R)$). The scattering matrix

$$s = w_+^* w_- : \ell^2(\Lambda_L) \oplus \ell^2(\Lambda_R) \to \ell^2(\Lambda_L) \oplus \ell^2(\Lambda_R),$$

is a unitary operator commuting with $h_L \oplus h_R$. Following computations in Section 4.2.9 verify that in the Fourier representation s acts as the operator of multiplication by the unitary matrix

$$s(\xi) = e^{2\mathrm{i}l\xi} \begin{bmatrix} 0 & 1 \\ 1 & 0 \end{bmatrix}.$$

3. Show that for $p \in [1, \infty]$,

$$e_{p,+}(\alpha) = \frac{1}{2\pi} \int_0^2 \log\left(1 - \frac{\sinh \frac{\alpha(\beta_R(\varepsilon - \mu_R) - \beta_L(\varepsilon - \mu_L))}{2} \sinh \frac{(1-\alpha)(\beta_R(\varepsilon - \mu_R) - \beta_L(\varepsilon - \mu_L))}{2}}{\cosh \frac{\beta_L(\varepsilon - \mu_L)}{2} \cosh \frac{\beta_R(\varepsilon - \mu_R)}{2}}\right) \mathrm{d}\varepsilon. \tag{4.179}$$

Note that, in accordance with Exercise 5.62, $e_{p,+}(\alpha)$ does not depend on p. The function (4.179) can be expressed in terms of Euler dilogarithm, see the end of Section 4.7.7.
4. Verify directly that $e_{\mathrm{naive},+}(\alpha) = e_{p,+}(\alpha)$.

5. (Recall Exercise 5.57). Show that

$$e_{2,+}(\boldsymbol{\alpha}) = \frac{1}{2\pi} \int_0^2 \log\left(1 - \mathcal{D}(\varepsilon)\right) d\varepsilon, \tag{4.180}$$

where

$$\mathcal{D}(\varepsilon) = \frac{\sinh \frac{\beta_R(\alpha_2 \varepsilon - \alpha_4 \mu_R) - \beta_L(\alpha_1 \varepsilon - \alpha_3 \mu_L)}{2} \sinh \frac{\beta_R((1-\alpha_2)\varepsilon - (1-\alpha_4)\mu_R) - \beta_L((1-\alpha_1)\varepsilon - (1-\alpha_3)\mu_L)}{2}}{\cosh \frac{\beta_L(\varepsilon - \mu_L)}{2} \cosh \frac{\beta_R(\varepsilon - \mu_R)}{2}}.$$

6. Using (4.180) show that the steady state charge and heat fluxes out of the left reservoir are

$$\omega_+(\mathcal{J}_L) = \frac{1}{2\pi} \int_0^2 \left[\frac{1}{1 + e^{\beta_L(\varepsilon - \mu_L)}} - \frac{1}{1 + e^{\beta_R(\varepsilon - \mu_R)}} \right] d\varepsilon,$$

$$\omega_+(\Phi_L) = \frac{1}{2\pi} \int_0^2 \varepsilon \left[\frac{1}{1 + e^{\beta_L(\varepsilon - \mu_L)}} - \frac{1}{1 + e^{\beta_R(\varepsilon - \mu_R)}} \right] d\varepsilon,$$

and that $\omega_+(\mathcal{J}_R) = -\omega_+(\mathcal{J}_L)$, $\omega_+(\Phi_R) = -\omega_+(\Phi_L)$.

Exercise 4.64 This exercise is intended for the technically advanced reader. Consider an infinitely extended EBB model with two reservoirs except that now we keep the single particle Hilbert spaces $\mathcal{K}_j$ and Hamiltonians h_j general. The coupling is defined in the same way as previously except that now $\delta_0^{(j)}$ is just a given vector in $\mathcal{K}_j$. We absorb λ in $\delta_0^{(j)}$ and denote by h the single particle Hamiltonian of the joint system. We shall suppose that the spectral measure ν_j for h_j and $\delta_0^{(j)}$ is purely absolutely continuous and denote by $d\nu_j(\varepsilon)/d\varepsilon$ its Radon-Nikodym derivative w.r.t. the Lebesgue measure. We also suppose that h has purely absolutely continuous spectrum. Since h preserves the cyclic subspace spanned by $\{\mathcal{K}_S, \delta_0^{(1)}, \delta_0^{(2)}\}$ and h_0, without loss of generality we may assume that $\mathcal{K}_j = L^2(\mathbb{R}, d\nu_j)$ and that h_j is the operator of multiplication by ε.

1. Show that the scattering matrix is given by

$$s(\varepsilon) = \mathbb{1} + 2i\pi \begin{bmatrix} \langle \chi_1 | (h - \varepsilon + i0)^{-1} \chi_1 \rangle \frac{d\nu_1(\varepsilon)}{d\varepsilon} & \langle \chi_1 | (h - \varepsilon + i0)^{-1} \chi_2 \rangle \sqrt{\frac{d\nu_1(\varepsilon)}{d\varepsilon} \frac{d\nu_2(\varepsilon)}{d\varepsilon}} \\ \langle \chi_2 | (h - \varepsilon + i0)^{-1} \chi_1 \rangle \sqrt{\frac{d\nu_1(\varepsilon)}{d\varepsilon} \frac{d\nu_2(\varepsilon)}{d\varepsilon}} & \langle \chi_2 | (h - \varepsilon + i0)^{-1} \chi_2 \rangle \frac{d\nu_2(\varepsilon)}{d\varepsilon} \end{bmatrix}.$$

2. Compute $e_{p,+}(\alpha)$ for $p \in [1, \infty]$.
3. Verify that Exercise 5.62 applies to this more general model. Classify the examples for which $e_{p,+}(\alpha)$ does not depend on p.
4. Compute $e_{\text{naive},+}(\alpha)$.
5. Compute $e_{2,+}(\boldsymbol{\alpha})$ and derive the formulas for the steady state charge and heat fluxes.
6. Verify the results by comparing them with Exercise 5.63.

Remark. For more information about Exercises 5.62, 5.63, and 5.64 we refer the reader to Bruneau, Jakšić, and Pillet (2011).

Local interactions. One can easily modify the EBB model to allow for interactions between fermions in the device $\mathcal{S}$. For example, let q be a pair interaction on $\mathcal{S}$, that is a self-adjoint operator on $\Gamma_2(\mathcal{K})$ acting like

$$(q\psi)(x_1, x_2) = \begin{cases} q(x_1, x_2)\psi(x_1, x_2) & \text{if } x_1, x_2 \in \mathcal{S}, \\ 0 & \text{otherwise.} \end{cases}$$

Then the operator

$$Q = \frac{1}{2} \sum_{x,y \in \mathcal{S}} q(x, y) a^*(\delta_x) a^*(\delta_y) a(\delta_y) a(\delta_x),$$

is self-adjoint on $\Gamma(\mathcal{K})$ and leaves all the $\Gamma_k(\mathcal{K})$ invariant. It vanishes on $\Gamma_0(\mathcal{K})$ and $\Gamma_1(\mathcal{K})$ and acts like

$$(Q\psi)(x_1, \ldots, x_k) = \left(\frac{1}{2} \sum_{\substack{x,y \in \{x_1,\ldots,x_k\} \cap \mathcal{S} \\ x \neq y}} q(x, y) \right) \psi(x_1, \ldots, x_k),$$

on $\Gamma_k(\mathcal{K})$ for $k \geq 2$. For $\kappa \in \mathbb{R}$, the Hamiltonian

$$H_{\lambda,\kappa} = H_\lambda + \kappa Q,$$

is self-adjoint on $\Gamma(\mathcal{K})$ and defines a dynamics $\tau^t_{\lambda,\kappa}$ on the CAR algebra over $\mathcal{K}$. It is easy to perform the thermodynamic limit of this locally interacting EBB model, the interaction term Q being confined to the finite sample $\mathcal{S}$. The large time limit is a more delicate problem. Hilbert space scattering techniques are no more adapted to this problem and one has to deal with the much harder C^*-scattering theory, for example the existence of the limit

$$\gamma_\pm(A) = \lim_{t \to \pm\infty} \tau_\lambda^{-t} \circ \tau^t_{\lambda,\kappa}(A).$$

Such problems first appeared in the works of Hepp (1970) and Robinson (1973). In the specific context of nonequilibrium statistical mechanics, the scattering approach was advocated by Ruelle (2000) (see also Ruelle (2001) and (2002)). A systematic approach to the scattering problem for local perturbations of free Fermi gases has been developed in Botvich and Malyshev (1983), Aizenstadt and Malyshev (1987), and Malyshev (1988). It relies on the well known Cook argument and a uniform (in t) control of the Dyson expansion

$$\tau^t_{\lambda,\kappa}(A) = \tau^t_\lambda(A)$$
$$+ \sum_{k \geq 1} (\mathrm{i}\kappa)^k \int_{0 \leq s_k \leq \cdots \leq s_1 \leq t} [\tau^{s_k}_\lambda(Q), [\cdots [\tau^{s_1}_\lambda(Q), \tau^t_\lambda(A)] \cdots]] \mathrm{d}s_1 \cdots \mathrm{d}s_k.$$

Optimal bounds for the uniform convergence of such expansions have been obtained by Maassen and Botvich (2009). The interested reader should consult Fröhlich *et al.* (2003), Jakšić *et al.* (2007) and (2009) and the references therein.

4.7.7   The XY-spin chain

In this section, we describe a simple example of extended quantum spin system on a 1D-lattice. We shall follow closely the approach of Section 4.2, starting from the standard quantum mechanical description of a finite sub-lattice.

Finite spin systems

Let Λ be a finite set. A spin $\frac{1}{2}$ system on Λ is a finite quantum system obtained by attaching to each site $x \in \Lambda$ a spin $\frac{1}{2}$. Thus, the Hilbert space of such a spin system is given by

$$\mathcal{H}_\Lambda = \bigotimes_{x \in \Lambda} \mathcal{H}_x,$$

where each $\mathcal{H}_x$ is a copy of $\mathbb{C}^2$. The corresponding $*$-algebra is

$$\mathcal{O}_\Lambda = \bigotimes_{x \in \Lambda} \mathcal{O}_x,$$

where $\mathcal{O}_x = \mathrm{M}_2(\mathbb{C})$ is the algebra of 2×2 complex matrices. Together with the identity $\mathbb{1}_x \in \mathcal{O}_x$, the Pauli matrices

$$\sigma_x^{(1)} = \begin{bmatrix} 0 & 1 \\ 1 & 0 \end{bmatrix}, \quad \sigma_x^{(2)} = \begin{bmatrix} 0 & -\mathrm{i} \\ \mathrm{i} & 0 \end{bmatrix}, \quad \sigma_x^{(3)} = \begin{bmatrix} 1 & 0 \\ 0 & -1 \end{bmatrix},$$

form a basis of $\mathcal{O}_x$ satisfying the well known relations

$$\sigma_x^{(j)}\sigma_x^{(k)} = \delta_{jk}\mathbb{1}_x + \mathrm{i}\varepsilon^{jkl}\sigma_x^{(l)}.$$

For $D \subset \Lambda$ we set $\mathbb{1}_D = \otimes_{x \in D}\mathbb{1}_x$. We shall identify $T_x \in \mathcal{O}_x$ with the element $T_x \otimes \mathbb{1}_{\Lambda \setminus \{x\}}$ of $\mathcal{O}_\Lambda$. With this convention, one has the relations

$$\sigma_x^{(j)}\sigma_x^{(k)} = \delta_{jk}\mathbb{1}_\Lambda + \mathrm{i}\varepsilon^{jkl}\sigma_x^{(l)}, \qquad [\sigma_x^{(j)}, \sigma_y^{(k)}] = 2\mathrm{i}\delta_{xy}\varepsilon^{jkl}\sigma_x^{(l)}. \tag{4.181}$$

Moreover, any element of $\mathcal{O}_\Lambda$ can be written as a finite sum

$$\sum_a \prod_{x \in \Lambda} T_x^{(a)},$$

with $T_x^{(a)} \in \{\mathbb{1}_\Lambda, \sigma_x^{(1)}, \sigma_x^{(2)}, \sigma_x^{(3)}\}$. Since $\mathbb{1}_\Lambda = \sigma_x^{(j)2}$, the smallest $*$-subalgebra of $\mathcal{O}_\Lambda$ containing the set $\mathbf{S}_\Lambda = \{\sigma_x^{(j)} \mid x \in \Lambda, j = 1, 2, 3\}$ is $\mathcal{O}_\Lambda$. By von Neumann's bicommutant theorem (Theorem 4.57), we conclude that $\mathbf{S}_\Lambda'' = \mathcal{O}_\Lambda$ and hence $\mathbf{S}_\Lambda' = \mathbb{C}\mathbb{1}_\Lambda$.

The dynamics of a spin chain is completely determined by its Hamiltonian H_Λ, a self-adjoint element of $\mathcal{O}_\Lambda$. The equilibrium state of the system at inverse temperature β is given by the density matrix

$$\omega_{\beta\Lambda} = \frac{e^{-\beta H_\Lambda}}{\mathrm{tr}\left(e^{-\beta H_\Lambda}\right)}.$$

The particular example we shall consider in the remaining part of this section is the XY-chain on the finite 1D-lattice $\Lambda = [A, B] \subset \mathbb{Z}$. It is defined by the XY-Hamiltonian

$$H_\Lambda = -\frac{1}{4} \sum_{x \in [A,B[} J\left(\sigma_x^{(1)}\sigma_{x+1}^{(1)} + \sigma_x^{(2)}\sigma_{x+1}^{(2)}\right) - \frac{1}{2} \sum_{x \in [A,B]} \lambda \sigma_x^{(3)}, \qquad (4.182)$$

where $J \in \mathbb{R}$ is the nearest-neighbour coupling constant and $\lambda \in \mathbb{R}$ is the strength of an external magnetic field in direction (3)[14]. The case $J > 0$ corresponds to a ferromagnetic coupling while $J < 0$ describes an anti-ferromagnetic system.

The Jordan-Wigner representation

The natural 'spin' interpretation of the $*$-algebra $\mathcal{O}_\Lambda$ described in the previous section is not very convenient for computational purposes. In this section, following Jordan and Wigner (Jordan, and Wigner, 1928), we shall see that $\mathcal{O}_\Lambda$ also carries an irreducible representation of a CAR algebra. Moreover, it turns out that the XY Hamiltonian (4.182) takes a particularly simple form in this representation. In fact, we shall see that the XY-spin chain can be mapped to a free Fermi gas.

Let $\sigma_x^{(\pm)} = (\sigma_x^{(1)} \pm i\sigma_x^{(2)})/2$ denote the spin raising/lowering operators at $x \in \Lambda$. Note that $\sigma_x^{(-)}$ and $\sigma_x^{(+)} = \sigma_x^{(-)*}$ satisfy the anti-commutation relations

$$\{\sigma_x^{(+)}, \sigma_x^{(+)}\} = \{\sigma_x^{(-)}, \sigma_x^{(-)}\} = 0 \qquad \{\sigma_x^{(+)}, \sigma_x^{(-)}\} = \mathbb{1}_\Lambda.$$

Thus, if Λ reduces to the singleton $\{x\}$, then the maps $\alpha \mapsto \alpha \sigma_x^{(+)}$ and $\alpha \mapsto \bar{\alpha}\sigma_x^{(-)}$ define a representation of the CAR over the Hilbert space $\mathbb{C} = \ell^2(\{x\})$ (and one easily checks that this representation is irreducible). This does not directly generalize to larger Λ. Indeed, if Λ contains two distinct sites $x \neq y$ one has

$$[\sigma_x^{(+)}, \sigma_y^{(+)}] = [\sigma_x^{(-)}, \sigma_y^{(-)}] = 0 \qquad [\sigma_x^{(+)}, \sigma_y^{(-)}] = 0,$$

that is operators at distinct sites commute whereas they should anti-commute to define a representation of the CAR over $\ell^2(\Lambda)$.

To transform commutation at distinct sites into anti-commutation, we make the following observation: for $T_x \in \mathcal{O}_x$ and $S_y \in \mathcal{O}_y$ one has

$$\{\sigma_A^{(3)}\cdots\sigma_{x-1}^{(3)}T_x, \sigma_A^{(3)}\cdots\sigma_{y-1}^{(3)}S_y\} = \begin{cases} \{\sigma_x^{(3)}, T_x\}\sigma_{x+1}^{(3)}\cdots\sigma_{y-1}^{(3)}S_y \text{ for } x < y, \\ \{T_x, S_x\} \qquad\qquad\qquad \text{ for } x = y, \\ \{\sigma_y^{(3)}, S_y\}\sigma_{y+1}^{(3)}\cdots\sigma_{x-1}^{(3)}T_x \text{ for } x > y. \end{cases}$$

---

[14] The name XY comes from the coupling between components $(1) = (X)$ and $(2) = (Y)$ of the spins.

Since $\{\sigma_x^{(3)}, \sigma_x^{(\pm)}\} = 0$, it follows that the Jordan–Wigner operators

$$b_x = \sigma_A^{(3)} \cdots \sigma_{x-1}^{(3)} \sigma_x^{(-)}, \qquad b_x^* = \sigma_A^{(3)} \cdots \sigma_{x-1}^{(3)} \sigma_x^{(+)}, \tag{4.183}$$

satisfy

$$\{b_x, b_y\} = \{b_x^*, b_y^*\} = 0, \qquad \{b_x, b_y^*\} = \delta_{xy} \mathbb{1}_\Lambda.$$

Hence, the maps $\ell^2(\Lambda) \ni \alpha \mapsto b^*(\alpha) = \sum_x \alpha_x b_x^*$ and $\ell^2(\Lambda) \ni \alpha \mapsto b(\alpha) = \sum_x \bar{\alpha}_x b_x$ define a representation of the CAR over $\ell^2(\Lambda)$ on the Hilbert space $\mathcal{H}_\Lambda$. We shall call it the Jordan–Wigner representation.

One easily inverts Relations (4.183) to express the spin operators in terms of the Jordan-Wigner operators:

$$\sigma_x^{(1)} = V_x(b_x + b_x^*), \qquad \sigma_x^{(2)} = \mathrm{i} V_x(b_x - b_x^*), \qquad \sigma_x^{(3)} = 2b_x^* b_x - \mathbb{1}_\Lambda, \tag{4.184}$$

where

$$V_x = \begin{cases} \mathbb{1}_\Lambda & \text{if } x = A, \\ \prod_{y \in [A,x[} (2b_y^* b_y - 1) & \text{otherwise.} \end{cases}$$

If follows in particular that $\mathfrak{B}_\Lambda = \{b_x^\# \mid x \in \Lambda\}$ satisfies $\mathfrak{B}_\Lambda' = \mathbf{S}_\Lambda' = \mathbb{C}\mathbb{1}_\Lambda$. Hence, the Jordan–Wigner representation is irreducible. By Proposition 4.55, there exists a unitary operator $U : \Gamma(\ell^2(\Lambda)) \to \mathcal{H}_\Lambda$ such that $b^\#(\alpha) = U a^\#(\alpha) U^*$, where the $a^\#$ are the usual creation/annihilation operators on the fermionic Fock space $\Gamma(\ell^2(\Lambda))$.

A simple calculation shows that

$$\sigma_x^{(1)} \sigma_{x+1}^{(1)} + \sigma_x^{(2)} \sigma_{x+1}^{(2)} = -2(b_{x+1}^* b_x + b_x^* b_{x+1}),$$

so that we can rewrite the XY-Hamiltonian as

$$H_\Lambda = \frac{J}{2} \sum_{x \in [A,B[} (b_{x+1}^* b_x + b_x^* b_{x+1}) - \frac{\lambda}{2} \sum_{x \in [A,B]} (2b_x^* b_x - 1).$$

By Part (8) of Proposition 4.54 we thus have $H_\Lambda = U d\Gamma(h_\Lambda) U^*$, up to an irrelevant additive constant, where the one-particle Hamiltonian h_Λ is the self-adjoint operator on $\ell^2(\Lambda)$ given by

$$h_\Lambda = \frac{J}{2} \sum_{x \in [A,B[} (|\delta_{x+1}\rangle\langle\delta_x| + |\delta_x\rangle\langle\delta_{x+1}|) - \lambda \sum_{x \in [A,B]} |\delta_x\rangle\langle\delta_x| = \frac{J}{2}\Delta_\Lambda + (J - \lambda)\mathbb{1},$$

Δ_Λ being the discrete Laplacian on Λ with Dirichlet boundary conditions (4.1). Thus, the unitary map U provides an equivalence between the XY-chain on Λ and the free Fermi gas with one-particle Hamiltonian h_Λ. In particular, it maps the equilibrium state $\omega_{\beta\Lambda}$ to the quasi-free state on the CAR algebra over $\ell^2(\Lambda)$ with density

$$T_{\beta\Lambda} = (\mathbb{1} + e^{\beta h_\Lambda})^{-1}.$$

Exercise 4.65

1. Use the Jordan–Wigner representation of the XY-chain to show that, for all $x \in \Lambda$,

$$\omega_{\beta\Lambda}(\sigma_x^{(1)}) = \omega_{\beta\Lambda}(\sigma_x^{(2)}) = 0, \qquad \frac{1}{2}\omega_{\beta\Lambda}(\mathbb{1}_\Lambda + \sigma_x^{(3)}) = \frac{2}{|\Lambda|} \sum_{\xi \in \Lambda^*} \frac{\sin^2(\xi(x - A + 1))}{1 + e^{\beta(J \cos \xi - \lambda)}},$$

where $\Lambda^* = \{n\pi/(|\Lambda| + 1) \,|\, n = 1, \ldots, |\Lambda|\}$, $|\Lambda| = B - A + 1$.

2. Show that the mean magnetization per spin is given by

$$m_\Lambda(\beta, J, \lambda) = \frac{1}{|\Lambda|} \sum_{x \in \Lambda} \omega_{\beta\Lambda}(\sigma_x^{(3)}) = \frac{1}{|\Lambda|} \sum_{\xi \in \Lambda^*} \tanh(\beta(\lambda - J \cos \xi)/2).$$

3. Show that, in the thermodynamic limit,

$$\lim_{\Lambda \to \mathbb{Z}} m_\Lambda(\beta, J, \lambda) = \frac{\sinh(\beta\lambda)}{\pi} \int_0^\pi \frac{\mathrm{d}\xi}{\cosh(\beta\lambda) + \cosh(\beta J \cos \xi)}.$$

Hint: use the discrete Fourier transform to diagonalize the Laplacian Δ_Λ.

The open XY-chain

To construct a model of open XY-chain we shall consider the same geometry as in the classical harmonic chain of Section 4.2: a finite system $\mathcal{C}$, consisting of the XY-chain on $\Lambda = [-N, N]$, is coupled at its left and right ends to two reservoirs $\mathcal{R}_L$ and $\mathcal{R}_R$ which are themselves XY-chains on $\Lambda_L = [-M, -N - 1]$ and $\Lambda_R = [N + 1, M]$ (see Fig. 4.9). The size N will be kept fixed and we shall discuss the thermodynamic limit $M \to \infty$.

The Hamiltonian of the decoupled joint system $\mathcal{R}_L + \mathcal{C} + \mathcal{R}_R$ is given by

$$H_0 = H_{\Lambda_L} + H_\Lambda + H_{\Lambda_R}.$$

The coupled Hamiltonian is

$$H = H_{\Lambda_L \cup \Lambda \cup \Lambda_R} = H_0 + V_L + V_R,$$

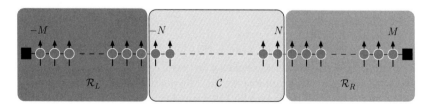

Fig. 4.9 The XY-chain $\mathcal{C}$ coupled at its left and right ends to the reservoirs $\mathcal{R}_L$ and $\mathcal{R}_R$.

with the coupling terms

$$V_L = -\frac{J}{4}\left(\sigma^{(1)}_{-N-1}\sigma^{(1)}_{-N} + \sigma^{(2)}_{-N-1}\sigma^{(2)}_{-N}\right), \qquad V_R = -\frac{J}{4}\left(\sigma^{(1)}_{N}\sigma^{(1)}_{N+1} + \sigma^{(2)}_{N}\sigma^{(2)}_{N+1}\right).$$

We consider the family of initial states

$$\omega_X = \frac{e^{-\beta H + X_L H_{\Lambda_L} + X_R H_{\Lambda_R}}}{\mathrm{tr}(e^{-\beta H + X_L H_{\Lambda_L} + X_R H_{\Lambda_R}})}, \tag{4.185}$$

with control parameter $X = (X_L, X_R) \in \mathbb{R}^2$. The entropy production observable is

$$\sigma_X = X_L \Phi_L + X_R \Phi_R,$$

where the heat fluxes from $\mathcal{R}_{L/R}$ to $\mathcal{C}$ are easily computed using the commutation relations (4.181),

$$\Phi_L = -\mathrm{i}[H, H_{\Lambda_L}]$$
$$= \frac{J^2}{8}\left(\sigma^{(2)}_{-N-2}\sigma^{(1)}_{-N} - \sigma^{(1)}_{-N-2}\sigma^{(2)}_{-N}\right)\sigma^{(3)}_{-N-1} + \frac{\lambda J}{4}\left(\sigma^{(1)}_{-N-1}\sigma^{(2)}_{-N} - \sigma^{(2)}_{-N-1}\sigma^{(1)}_{-N}\right),$$

$$\Phi_R = -\mathrm{i}[H, H_{\Lambda_R}]$$
$$= \frac{J^2}{8}\left(\sigma^{(1)}_{N}\sigma^{(2)}_{N+2} - \sigma^{(2)}_{N}\sigma^{(1)}_{N+2}\right)\sigma^{(3)}_{N+1} + \frac{\lambda J}{4}\left(\sigma^{(1)}_{N+1}\sigma^{(2)}_{N} - \sigma^{(2)}_{N+1}\sigma^{(1)}_{N}\right).$$

In the Jordan–Wigner representation, the decoupled system is a free Fermi gas with one-particle Hilbert space $\ell^2(\Lambda_L \cup \Lambda \cup \Lambda_R) = \ell^2(\Lambda_L) \oplus \ell^2(\Lambda) \oplus \ell^2(\Lambda_R)$ and one particle Hamiltonian

$$h_0 = h_{\Lambda_L} \oplus h_\Lambda \oplus h_{\Lambda_R}.$$

The one-particle Hamiltonian of the coupled system is

$$h = h_{\Lambda_L \cup \Lambda \cup \Lambda_R} = h_0 + v_L + v_R,$$

where the coupling terms

$$v_L = \frac{J}{2}\left(|\delta_{-N-1}\rangle\langle\delta_{-N}| + |\delta_{-N}\rangle\langle\delta_{-N-1}|\right), \qquad v_R = \frac{J}{2}\left(|\delta_N\rangle\langle\delta_{N+1}| + |\delta_{N+1}\rangle\langle\delta_N|\right),$$

are finite rank operators. The initial state ω_X is quasi-free with density

$$T_X = \left(\mathbb{1} + e^{-k_X}\right)^{-1},$$

where

$$k_X = -\beta h + X_L h_{\Lambda_L} + X_R h_{\Lambda_R}$$
$$= -\beta(h_\Lambda + v_L + v_R) - (\beta - X_L)h_{\Lambda_L} - (\beta - X_R)h_{\Lambda_R}.$$

It is now apparent that the results of Section 4.7.6 apply to the open XY-chain. By Part (2) of Exercise 5.54, the generalized entropic pressure is given by

$$e_t(X, Y) = \log \det \left(1 + T_X \left(e^{-k_X} e^{k_X - Y + k_{Y,t} - k_0} - 1 \right) \right),$$

where $k_{Y,t} = e^{-ith} k_Y e^{ith}$. The same formula holds in the thermodynamic limit, provided k_X is replaced by its strong limit. The large time limit follows from Part (5) of Exercise 5.57,

$$e_+(X, Y) = \lim_{t \to \infty} \frac{1}{t} e_t(X, Y)$$

$$= \int_0^{\pi} \log \det \left(1 + T_X(\xi)(e^{-k_X(\xi)} e^{k_X - Y(\xi) + s(\xi) k_Y(\xi) s^*(\xi) - k_0(\xi)} - 1) \right) \frac{d\varepsilon(\xi)}{2\pi},$$

where $\varepsilon(\xi) = 1 - \cos \xi$, $k_X(\xi)$ is the diagonal 2×2 matrix with entries $(\beta - X_j)(\lambda - J \cos \xi)$ and $T_X(\xi) = (1 + e^{-k_X(\xi)})^{-1}$. Using the explicit form

$$s(\xi) = e^{\pm 2iN\xi} \begin{bmatrix} 0 & 1 \\ 1 & 0 \end{bmatrix},$$

of the scattering matrix (see Section 4.2.9, the sign $\pm$ is opposite to the sign of the coupling constant J), we obtain

$$e_+(X, Y) = \frac{1}{J\pi} \int_{u_-}^{u_+} \log \left(1 - \frac{\sinh(u\Delta Y)\sinh(u(\Delta X - \Delta Y))}{\cosh(u(\beta - X_L))\cosh(u(\beta - X_R))} \right) du,$$

where we have set $\Delta X = X_R - X_L$, $\Delta Y = Y_R - Y_L$ and $u_\pm = (\lambda \pm J)/2$. The steady heat current through the chain is given by

$$\langle \Phi_L \rangle_+ = \lim_{t \to \infty} \omega_{X,t}(\Phi_L)$$

$$= -\partial_{Y_L} e_+(X, Y)|_{Y=0} = \frac{1}{J\pi} \int_{u_-}^{u_+} u \left(\tanh(\beta_L u) - \tanh(\beta_R u) \right) du,$$

where $\beta_{L/R} = \beta - X_{L/R}$. It follows that the entropy production

$$\langle \sigma \rangle_+ = \frac{1}{J\pi} \int_{u_-}^{u_+} (\beta_L u - \beta_R u) \left(\tanh(\beta_L u) - \tanh(\beta_R u) \right) du,$$

is strictly positive iff $\beta_L \neq \beta_R$ and $J \neq 0$.

Exercise 4.66 Develop the linear response theory of the open XY-chain.

Exercise 4.67 Instead of (4.185) consider the reference state

$$\omega = \frac{e^{-\beta_L H_{\Lambda_L} - \beta H_\Lambda - \beta_R H_{\Lambda_R}}}{\text{tr}(e^{-\beta_L H_{\Lambda_L} - \beta H_\Lambda - \beta_R H_{\Lambda_R}})}.$$

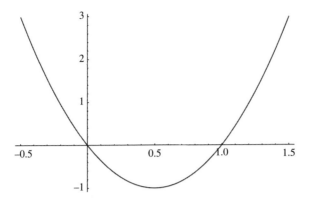

Fig. 4.10 The entropic functional $e_{p,+}(\alpha)$ of the open XY-chain.

In this case, up to irrelevant scaling, the Jordan–Wigner transformation maps the XY-chain to the EBB model considered in Exercise 5.63. Show that for $p \in [1, \infty]$,

$$e_{\text{naive},+}(\alpha) = e_{p,+}(\alpha) = \frac{1}{J\pi} \int_{u_-}^{u_+} \log\left(1 - \frac{\sinh(\alpha u \Delta\beta) \sinh((1-\alpha)u\Delta\beta)}{\cosh(u\beta_L)\cosh(u\beta_R)}\right) du, \qquad (4.186)$$

where $\Delta\beta = \beta_R - \beta_L$ (see Fig. 4.10). Note that $e_{p,+}(\alpha) = e_+(X, \alpha X)$.

The formula (4.186) can be rewritten in terms of Euler's dilogarithm

$$\text{Li}_2(z) = -\int_0^z \frac{\log(1-w)}{w}\, dw,$$

an analytic function on the cut plane $\mathbb{C} \setminus [1, \infty[$ with a branching point at $z = 1$ (see (Lewin, 1981)). More precisely, one has

$$e_{p,+}(\alpha) = G(\bar\beta + (\alpha - 1/2)\Delta\beta) + G(\bar\beta - (\alpha - 1/2)\Delta\beta) - G(\beta_L) - G(\beta_R),$$

where $\bar\beta = (\beta_L + \beta_R)/2$ and

$$G(x) = \frac{\text{Li}_2\left(-e^{2xu_+}\right) - \text{Li}_2\left(-e^{2xu_-}\right)}{\pi x(u_+ - u_-)}.$$

It follows that $e_{p,+}(\alpha)$ is analytic on the strip $|\text{Im}\,\alpha| < \frac{\pi}{(|\lambda|+|J|)|\Delta\beta|}$.

Remark. We were able to compute the TD and large time limits of the entropic functionals of the XY-chain thanks to its Fermi-gas representation. We note however that the operator

$$V_x = (2b^*_{-M}b_{-M} - \mathbb{1}) \cdots (2b^*_{x-1}b_{x-1} - \mathbb{1}),$$

has no limit in the CAR algebra over $\ell^2(\mathbb{Z})$ as $M \to \infty$, and the Jordan–Wigner transformation (4.184) does not survive the TD limit. In fact, to recover the full spin algebra in the TD limit, one needs to enlarge the CAR algebra over $\ell^2(\mathbb{Z})$ with an element V formally equal to

$$\lim_{M \to \infty} (2b^*_{-M}b_{-M} - \mathbb{1}) \cdots (2b^*_{-1}b_{-1} - \mathbb{1}).$$

We refer to Araki (1984) for a complete exposition of this construction. An alternative resolution of the TD limit/Jordan–Wigner transformation conflict goes as follows.

We set $\Lambda_M = [-M, M] \subset \mathbb{Z}$. The operator $W = \sigma^{(3)}_{-M} \cdots \sigma^{(3)}_M \in \mathcal{O}_{\Lambda_M}$ satisfies $W = W^* = W^{-1}$. It implements the rotation by an angle π around the (3)-axis of all the spins of the chain,

$$W\sigma^{(j)}_x W^* = \begin{cases} -\sigma^{(j)}_x & \text{for } j = 1 \text{ or } j = 2, \\ \sigma^{(j)}_x & \text{for } j = 3. \end{cases}$$

Thus, $\theta(A) = WAW^*$ defines an involutive $*$-automorphism of $\mathcal{O}_{\Lambda_M}$. In the fermionic picture, θ is completely characterized by $\theta(b_x) = -b_x$.

Since θ is a linear involution on the vector space $\mathcal{O}_{\Lambda_M}$, it follows that $\mathcal{O}_{\Lambda_M} = \mathcal{O}_{\Lambda_M+} \oplus \mathcal{O}_{\Lambda_M-}$, where

$$\mathcal{O}_{\Lambda_M\pm} = \{A \in \mathcal{O}_{\Lambda_M} \,|\, \theta(A) = \pm A\},$$

are vector subspaces. Note that $\mathcal{O}_{\Lambda_M+}$ is a $*$-subalgebra of $\mathcal{O}_{\Lambda_M}$. Since $H_\Lambda \in \mathcal{O}_{\Lambda_M+}$, the dynamics $\tau^t_\Lambda(A) = e^{itH_\Lambda} A e^{-itH_\Lambda}$ satisfies $\tau^t_\Lambda \circ \theta = \theta \circ \tau^t_\Lambda$ and, in particular, it preserves both subspaces $\mathcal{O}_{\Lambda_M\pm}$. Moreover, our initial state satisfies $\omega_X \circ \theta = \omega_X$ which implies that $\omega_X|_{\mathcal{O}_{\Lambda_M-}} = 0$. Thus, observables with nontrivial expectation belong to the subalgebra $\mathcal{O}_{\Lambda_M+}$ and we may restrict ourselves to such observables.

In the fermionic picture, $\mathcal{O}_{\Lambda_M+}$ is the $*$-algebra of all polynomials in the $b^\#_x$ which contain only monomials of even degree. In the spin picture, it is generated by the operators $\sigma^{(3)}_x$ and $\sigma^{(s)}_x \sigma^{(s')}_{x'}$ with $s, s' \in \{-, +\}$ and $x < x'$, which have a Jordan–Wigner representation surviving the TD limit, for example

$$\sigma^{(-)}_x \sigma^{(+)}_y = b_x (2b^*_{x+1}b_{x+1} - \mathbb{1}) \cdots (2b^*_{y-1}b_{y-1} - \mathbb{1}) b^*_y.$$

Thus, at the price of restricting the dynamical system to the even subalgebra $\mathcal{O}_{\Lambda_M+}$, the XY-chain remains equivalent to a free Fermi gas in the TD limit. This fact is a starting point in the construction of the NESS of the XY-chain. We refer the reader to (Araki and Ho, 2000) and (Aschbacher and Pillet, 2003) for the details of this construction and to (Aschbacher and Barbaroux, 2006), (Aschbacher and Barbaroux, 2007) for additional information about the NESS of the XY-chain.

4.8   Appendix: Large deviations

In this first appendix, we formulate some well known large deviation results that were used in these lecture notes. We provide a proof in the simplest case of scalar random variables.

4.8.1   Fenchel–Legendre transform

In this section, we shall use freely some well known properties of convex real functions of a real variable, see, for example (Roberts and Varberg, 1973).

Let $I = [a, b] \subset \mathbb{R}$ be a closed finite interval, denote by $\text{int}(I) =]a, b[$ its interior, and let $e : I \to \mathbb{R}$ be a continuous convex function. Then e admits finite left and right derivatives

$$D^{\pm}e(s) = \lim_{h \downarrow 0} \frac{e(s \pm h) - e(s)}{\pm h},$$

at every $s \in \text{int}(I)$. $D^+e(a)$ and $D^-e(b)$ exist, although they may be respectively $-\infty$ and $+\infty$. By convention, we set $D^-e(a) = -\infty$ and $D^+e(b) = +\infty$. The functions $D^{\pm}e(s)$ are increasing on I and satisfy $D^-e(s) \leq D^+e(s)$. Moreover, $D^-e(s) = D^+e(s) = e'(s)$ outside a countable set in $\text{int}(I)$. If $e'(s)$ exists for all $s \in \text{int}(I)$, then it is continuous there and

$$\lim_{s \downarrow a} e'(s) = D^+e(a), \qquad \lim_{s \uparrow b} e'(s) = D^-e(b).$$

The subdifferential of e at $s_0 \in I$, denoted $\partial e(s_0)$, is the set of $\theta \in \mathbb{R}$ such that the affine function $\underline{e}(s) = e(s_0) + \theta(s - s_0)$ satisfies $e(s) \geq \underline{e}(s)$ for all $s \in I$, that is the graph of $\underline{e}$ is tangent to the graph of e at the point $(s_0, e(s_0))$. For any $s_0 \in I$, one has $\partial e(s_0) = [D^-e(s_0), D^+e(s_0)] \cap \mathbb{R}$.

It is convenient to extend the function e to $\mathbb{R}$ by setting $e(s) = +\infty$ for $s \notin I$. Then the function $e(s)$ is convex and lower semi-continuous on $\mathbb{R}$, that is

$$e(s_0) = \liminf_{s \to s_0} e(s),$$

holds for all $s_0 \in \mathbb{R}$. The subdifferential of e is naturally extended by setting $\partial e(s) = \emptyset$ for $s \notin I$.

The function

$$\varphi(\theta) = \sup_{s \in I}(\theta s - e(s)) = \sup_{s \in \mathbb{R}}(\theta s - e(s)), \qquad (4.187)$$

is called the Fenchel–Legendre transform of $e(s)$. $\varphi(\theta)$ is finite and convex (hence continuous) on $\mathbb{R}$. Obviously, if $a \geq 0$ then $\varphi(\theta)$ is increasing and if $b \leq 0$ then $\varphi(\theta)$ is decreasing. The subdifferential of φ at $\theta \in \mathbb{R}$ is $\partial\varphi(\theta) = [D^-\varphi(\theta), D^+\varphi(\theta)]$. The basic properties of the pair (e, φ) are summarized in:

Theorem 4.61   *(1) $\theta s \leq e(s) + \varphi(\theta)$ for all $s, \theta \in \mathbb{R}$.*
(2) $\theta s = e(s) + \varphi(\theta) \Leftrightarrow \theta \in \partial e(s)$.
(3) $e(s) = \sup_{\theta \in \mathbb{R}}(\theta s - \varphi(\theta))$.

(4) $\theta \in \partial e(s) \Leftrightarrow s \in \partial\varphi(\theta)$.
(5) If $0 \in]a, b[$, then $\varphi(\theta)$ is decreasing on $] - \infty, D^-e(0)]$, increasing on $[D^+e(0), \infty[$, $\varphi(\theta) = -e(0)$ for $\theta \in \partial e(0)$, and $\varphi(\theta) > -e(0)$ for $\theta \notin \partial e(0)$.

Proof: (1) Follows directly from the definition of φ.
(2) Combining the inequality (1) with the equality $\theta s_0 = e(s_0) + \varphi(\theta)$ we obtain that $e(s) \geq e(s_0) + \theta(s - s_0)$ for all $s \in \mathbb{R}$ which implies $\theta \in \partial e(s_0)$. Reciprocally, if $\theta \in \partial e(s_0)$ then $e(s) \geq e(s_0) + \theta(s - s_0)$ holds for all $s \in \mathbb{R}$ and hence $\theta s_0 \geq e(s_0) + \sup_s(\theta s - e(s)) = e(s_0) + \varphi(\theta)$. Combined with inequality (1), this yields $\theta s_0 = e(s_0) + \varphi(\theta)$.
(3) It follows from Exercise 5.27 that the function $\tilde{e}(s) = \sup_{\theta \in \mathbb{R}}(\theta s - \varphi(\theta))$ is lower semi-continuous on $\mathbb{R}$. (1) implies that $\tilde{e}(s) \leq e(s)$ for any $s \in \mathbb{R}$. $\partial e(s) \neq \emptyset$ for $s \in]a, b[$ we conclude from (2) that $\tilde{e}(s) = e(s)$.

Note that $-e(s) \leq -\min_{u \in I}(-e(u)) = \varphi(0)$. Thus, for $\theta > 0$, we have $\varphi(\theta) = \sup_{s \in [a,b]}(\theta s - e(s)) \leq \theta b + \varphi(0)$ and hence $\theta s - \varphi(\theta) \geq \theta(s - b) - \varphi(0)$. It follows that $\tilde{e}(s) = +\infty = e(s)$ for $s > b$. A similar argument applies to the case $s < a$.

Consider now the case $s = a$. From our previous conclusions, we can write $\tilde{e}(a) = \liminf_{s \to a} \tilde{e}(s) = \lim_{s \downarrow a} \tilde{e}(s) = \lim_{s \downarrow a} e(s) = e(a)$. A similar argument applies to $s = b$.
(4) By (2), $\theta_0 \in \partial e(s)$ is equivalent to the equality $s\theta_0 = e(s) + \varphi(\theta_0)$ which, combined with the inequality (1) yields $\varphi(\theta) \geq \varphi(\theta_0) + s(\theta - \theta_0)$ for all $\theta \in \mathbb{R}$ and hence $s \in \partial\varphi(\theta_0)$. Reciprocally, if $s \in \partial\varphi(\theta_0)$ then $\varphi(\theta) \geq \varphi(\theta_0) + s(\theta - \theta_0)$ for all $\theta \in \mathbb{R}$ and we conclude from (3) that $e(s) \leq \sup_\theta(\theta s - \varphi(\theta_0) - s(\theta - \theta_0)) = -\varphi(\theta_0) + s\theta_0$. Using (1) and (2), we conclude that $\theta_0 \in \partial e(s)$.
(5) It follows from (4) that if $\theta_0 \in \partial e(0) = [D^-e(0), D^+e(0)]$ then $0 \in \partial\varphi(\theta_0)$, that is $\varphi(\theta) \geq \varphi(\theta_0)$ for all $\theta \in \mathbb{R}$. Thus, $\varphi(\theta_0) = \min_\theta \varphi(\theta) = -e(0)$ and since $D^\pm\varphi(\theta)$ are increasing, φ is decreasing for $\theta \leq D^-e(0)$ and increasing for $\theta \geq D^+e(0)$. $\square$

4.8.2 Gärtner–Ellis theorem in dimension $d = 1$

Let $\mathcal{I} \subset \mathbb{R}_+$ be an unbounded index set, $(M_t, \mathcal{F}_t, P_t)$, $t \in \mathcal{I}$, a family of measure spaces, and $X_t : M_t \to \mathbb{R}$ a family of measurable functions. We assume that the measures P_t are finite for all t. For $s \in \mathbb{R}$ let

$$e_t(s) = \log \int_{M_t} e^{sX_t} dP_t.$$

$e_t(s)$ is a convex function taking values in $] - \infty, \infty]$. We make the following assumption:
(LD) For $s \in I = [a, b]$ the limit

$$e(s) = \lim_{t \to \infty} \frac{1}{t} e_t(s),$$

exists and is finite. Moreover, the function $e(s)$ is continuous on I.

Until the end of this section we shall assume that (LD) holds and set $e(s) = \infty$ for $s \notin I$. The function $\varphi(\theta)$ is defined by (4.187).

Proposition 4.62   *(1) Suppose that $0 \in [a, b[$. Then*

$$\limsup_{t \to \infty} \frac{1}{t} \log P_t(\{x \in M_t \mid X_t(x) > t\theta\}) \le \begin{cases} -\varphi(\theta) & \text{if } \theta \ge D^+e(0) \\ e(0) & \text{if } \theta < D^+e(0). \end{cases} \quad (4.188)$$

(2) Suppose that $0 \in]a, b]$. Then

$$\limsup_{t \to \infty} \frac{1}{t} \log P_t(\{x \in M_t \mid X_t(x) < t\theta\}) \le \begin{cases} -\varphi(\theta) & \text{if } \theta \le D^-e(0) \\ e(0) & \text{if } \theta > D^-e(0). \end{cases} \quad (4.189)$$

Proof: We shall prove (1), the proof of (2) follows from (1) applied to $-X_t$ and $-\theta$. For $s \in [0, b]$,

$$P_t(\{x \in M_t \mid X_t(x) > t\theta\}) = P_t(\{x \in M_t \mid e^{sX_t(x)} > e^{st\theta}\}) \le e^{-st\theta} \int_{M_t} e^{sX_t} dP_t,$$

and so

$$\limsup_{t \to \infty} \frac{1}{t} \log P_t(\{x \in M_t \mid X_t(x) > t\theta\}) \le - \sup_{0 \le s \le b} (\theta s - e(s)).$$

For $\theta < D^+e(0)$ and $s \ge 0$, one has $e(s) \ge e(0) + sD^+e(0) \ge e(0) + \theta s$, so that

$$-e(0) \le \sup_{0 \le s \le b} (\theta s - e(s)) \le \sup_{0 \le s \le b} (\theta s - e(0) - \theta s) = -e(0),$$

and hence $\sup_{0 \le s \le b}(\theta s - e(s)) = -e(0)$. One shows in a similar way that, for $\theta \ge D^+e(0)$, $\sup_{a \le s \le 0}(\theta s - e(s)) = -e(0)$. It follows that

$$\varphi(\theta) = \sup_{a \le s \le b} (\theta s - e(s)) = \max\left(-e(0), \sup_{0 \le s \le b} (\theta s - e(s))\right) = \sup_{0 \le s \le b} (\theta s - e(s)).$$

The statement follows.   □

Proposition 4.63 *Suppose that $0 \in]a, b[$, $e(0) \le 0$, and that $e(s)$ is differentiable at $s = 0$. Then for any $\delta > 0$ there is $\gamma > 0$ such that for t large enough,*

$$P_t(\{x \in M_t \mid |t^{-1}X_t(x) - e'(0)| \ge \delta\}) \le e^{-\gamma t}.$$

Proof: Part (2) of Theorem 4.61 implies that $\varphi(e'(0)) = -e(0)$. By Part (5) of the same theorem, one has $\varphi(\theta) > \varphi(e'(0)) \ge 0$ for $\theta \ne e'(0)$. Since

$$P_t(\{x \in M_t \mid |t^{-1}X_t(x) - e'(0)| \ge \delta\}) \le P_t(\{x \in M_t \mid X_t(x) \le t(e'(0) - \delta)\})$$
$$+ P_t(\{x \in M_t \mid X_t(x) \ge t(e'(0) + \delta)\}),$$

Proposition 4.62 implies

$$\limsup_{t \to \infty} \frac{1}{t} \log P_t(\{x \in M_t \mid |t^{-1}X_t(x) - e'(0)| \ge \delta\}) \le - \min\{\varphi(e'(0) + \delta), \varphi(e'(0) - \delta)\},$$

and the statement follows.   □

Proposition 4.64 *Suppose that $0 \in]a, b[$ and $e(s)$ is differentiable on $]a, b[$. Then*

$$\liminf_{t \to \infty} \frac{1}{t} \log P_t(\{x \in M_t \mid X_t(x) > t\theta\}) \geq -\varphi(\theta), \tag{4.190}$$

for any $\theta \in]D^+e(a), D^-e(b)[$.

Proof: Let $\theta \in]D^+e(a), D^-e(b)[$ be given and let α and ϵ be such that

$$\theta < \alpha - \epsilon < \alpha < \alpha + \epsilon < D^-e(b).$$

Let $s_\alpha \in]a, b[$ be such that $e'(s_\alpha) = \alpha$ (so $\varphi(\alpha) = \alpha s_\alpha - e(s_\alpha)$). Let

$$\mathrm{d}\hat{P}_t = \mathrm{e}^{-e_t(s_\alpha)} \mathrm{e}^{s_\alpha X_t} \mathrm{d}P_t.$$

Then $\hat{P}_t$ is a probability measure on $(M_t, \mathcal{F}_t)$ and

$$P_t(\{x \in M_t \mid X_t(x) > t\theta\}) \geq P_t(\{x \in M_t \mid t^{-1} X_t(x) \in [\alpha - \epsilon, \alpha + \epsilon]\})$$

$$= \mathrm{e}^{e_t(s_\alpha)} \int_{\{t^{-1} X_t \in [\alpha - \epsilon, \alpha + \epsilon]\}} \mathrm{e}^{-s_\alpha X_t} \mathrm{d}\hat{P}_t \tag{4.191}$$

$$\geq \mathrm{e}^{e_t(s_\alpha) - s_\alpha t\alpha - |s_\alpha| t\epsilon} \hat{P}_t(\{x \in M_t \mid t^{-1} X_t \in [\alpha - \epsilon, \alpha + \epsilon]\}).$$

Now, if $\hat{e}_t(s) = \log \int_{M_t} \mathrm{e}^{s X_t} \mathrm{d}\hat{P}_t$, then $\hat{e}_t(s) = e_t(s + s_\alpha) - e_t(s_\alpha)$ and so

$$\lim_{t \to \infty} \frac{1}{t} \hat{e}_t(s) = e(s + s_\alpha) - e(s_\alpha),$$

for $s \in [a - s_\alpha, b - s_\alpha]$. Since $\hat{e}(0) = 0$ and $\hat{e}'(0) = e'(s_\alpha) = \alpha$, it follows from Proposition 4.63 that

$$\lim_{t \to \infty} \frac{1}{t} \log \hat{P}_t(\{x \in M_t \mid t^{-1} X_t(x) \in [\alpha - \epsilon, \alpha + \epsilon]\}) = 0,$$

and (4.191) yields

$$\liminf_{t \to \infty} \frac{1}{t} \log P_t(\{x \in M_t \mid X_t(x) > t\theta\}) \geq -s_\alpha \alpha + e(s_\alpha) - |s_\alpha|\epsilon = -\varphi(\alpha) - |s_\alpha|\epsilon.$$

The statement follows by taking first $\epsilon \downarrow 0$ and then $\alpha \downarrow \theta$. $\qquad \square$

The following local version of the Gärtner–Ellis theorem is a consequence of Propositions 4.62 and 4.64.

Theorem 4.65 *If $e(s)$ is differentiable on $]a, b[$ and $0 \in]a, b[$ then, for any open set $\mathbb{J} \subset]D^+e(a), D^-e(b)[$,*

$$\lim_{t \to \infty} \frac{1}{t} \log P_t(\{x \in M_t \mid t^{-1} X_t(x) \in \mathbb{J}\}) = -\inf_{\theta \in \mathbb{J}} \varphi(\theta).$$

Proof: *Lower bound.* For any $\theta \in \mathbb{J}$ and $\delta > 0$ so small that $]\theta - \delta, \theta + \delta[\subset \mathbb{J}$ one has

$$P_t(\{x \in M_t \mid t^{-1}X_t(x) \in \mathbb{J}\}) \geq P_t(\{x \in M_t \mid t^{-1}X_t(x) \in]\theta - \delta, \theta + \delta[\}),$$

and it follows from Proposition 4.64 that

$$\liminf_{t \to \infty} \frac{1}{t} \log P_t(\{x \in M_t \mid t^{-1}X_t(x) \in \mathbb{J}\}) \geq -\varphi(\theta - \delta).$$

Letting $\delta \downarrow 0$ and optimizing over $\theta \in \mathbb{J}$, we obtain

$$\liminf_{t \to \infty} \frac{1}{t} \log P_t(\{x \in M_t \mid t^{-1}X_t(x) \in \mathbb{J}\}) \geq -\inf_{\theta \in \mathbb{J}} \varphi(\theta). \tag{4.192}$$

Upper bound. Note that $e(0) = 0 \in]a, b[$. By Part (5) of Proposition 4.61, we have $\varphi(\theta) = 0$ for $\theta = e'(0)$ and $\varphi(\theta) > 0$ otherwise. Hence, if $e'(0) \in \text{cl}(\mathbb{J})$, then

$$\limsup_{t \to \infty} \frac{1}{t} \log P_t(\{x \in M_t \mid t^{-1}X_t(x) \in \mathbb{J}\}) \leq 0 = -\inf_{\theta \in \mathbb{J}} \varphi(\theta).$$

In the case $e'(0) \notin \text{cl}(\mathbb{J})$, there exist $\alpha, \beta \in \text{cl}(\mathbb{J})$ such that $e'(0) \in]\alpha, \beta[\subset \mathbb{R} \setminus \text{cl}(\mathbb{J})$. It follows that

$$P_t(\{x \in M_t \mid t^{-1}X_t(x) \in \mathbb{J}\})$$
$$\leq P_t(\{x \in M_t \mid t^{-1}X_t(x) < \alpha\}) + P_t(\{x \in M_t \mid t^{-1}X_t(x) > \beta\})$$
$$\leq 2\max\left(P_t(\{x \in M_t \mid t^{-1}X_t(x) < \alpha\}), P_t(\{x \in M_t \mid t^{-1}X_t(x) > \beta\})\right),$$

and Proposition 4.62 yields

$$\limsup_{t \to \infty} \frac{1}{t} \log P_t(\{x \in M_t \mid t^{-1}X_t(x) \in \mathbb{J}\}) \leq -\min(\varphi(\alpha), \varphi(\beta)).$$

Finally, by Part (5) of Proposition 4.61, one has

$$\inf_{\theta \in \mathbb{J}} \varphi(\theta) = \min(\varphi(\alpha), \varphi(\beta)),$$

and therefore

$$\limsup_{t \to \infty} \frac{1}{t} \log P_t(\{x \in M_t \mid t^{-1}X_t(x) \in \mathbb{J}\}) \leq \inf_{\theta \in \mathbb{J}} \varphi(\theta), \tag{4.193}$$

holds for any $\mathbb{J} \subset]D^+e(a), D^-e(b)[$. The result follows from the bounds (4.192) and (4.193). $\qquad\qquad\square$

4.8.3 Gärtner–Ellis theorem in dimension $d > 1$

Let $\mathbf{X}_t : M_t \to \mathbb{R}^d$ be a family of measurable functions w.r.t. the probability spaces $(M_t, \mathcal{F}_t, P_t)$, $t \in \mathcal{I}$. If $G \subset \mathbb{R}^d$ is a Borel set, we denote by $\text{int}(G)$ its interior, by $\text{cl}(G)$ its closure, and by ∂G its boundary. The following result is a multi-dimensional version of the Gärtner–Ellis theorem.

Theorem 4.66 *Assume that the limit*

$$h(\mathbf{Y}) = \lim_{t \to \infty} \frac{1}{t} \log \int_{M_t} e^{\mathbf{Y} \cdot \mathbf{X}_t} \, dP_t,$$

exists in $[-\infty, +\infty]$ for all $\mathbf{Y} \in \mathbb{R}^d$, that the function $h(\mathbf{Y})$ is lower semi-continuous on $\mathbb{R}^d$, differentiable on the interior of the set $\mathcal{D} = \{\mathbf{Y} \in \mathbb{R}^d \mid |h(\mathbf{Y})| < \infty\}$ and satisfies

$$\lim_{\text{int}(\mathcal{D}) \ni \mathbf{Y} \to \mathbf{Y}_0} |\nabla h(\mathbf{Y})| = \infty,$$

for all $\mathbf{Y}_0 \in \partial \mathcal{D}$. Suppose also that $\mathbf{0}$ is an interior point of $\mathcal{D}$. Then, for all Borel sets $G \subset \mathbb{R}^d$ we have

$$-\inf_{\mathbf{Z} \in \text{int}(G)} I(\mathbf{Z}) \leq \liminf_{t \to \infty} \frac{1}{t} \log P_t \left(\{ x \in M_t \mid t^{-1} \mathbf{X}_t(x) \in G \} \right)$$

$$\leq \limsup_{t \to \infty} \frac{1}{t} \log P_t \left(\{ x \in M \mid t^{-1} \mathbf{X}_t(x) \in G \} \right) \leq -\inf_{\mathbf{Z} \in \text{cl}(G)} I(\mathbf{Z}),$$

where

$$I(\mathbf{Z}) = \sup_{\mathbf{Y} \in \mathbb{R}^d} (\mathbf{Y} \cdot \mathbf{Z} - h(\mathbf{Y})).$$

We now describe a local version of Gärtner–Ellis theorem in $d > 1$. Set

$$\bar{h}(\mathbf{Y}) = \limsup_{t \to \infty} \frac{1}{t} \log \int_{M_t} e^{\mathbf{Y} \cdot \mathbf{X}_t} \, dP_t,$$

$$\bar{I}(\mathbf{Z}) = \sup_{\mathbf{Y} \in \mathbb{R}^d} (\mathbf{Y} \cdot \mathbf{Z} - \bar{h}(\mathbf{Y})).$$

Let $\bar{\mathcal{D}} = \{\mathbf{Y} \in \mathbb{R}^d \mid \bar{h}(\mathbf{Y}) < \infty\}$ and let $\mathcal{D}$ be the set of all $\mathbf{Y} \in \mathbb{R}^d$ for which the limit (4.66) exists and is finite. Let $S \subset \mathcal{D}$ be the set of points at which $h(\mathbf{Y})$ is differentiable and let $\mathcal{F} = \{\nabla h(\mathbf{Y}) \mid \mathbf{Y} \in S\}$.

Theorem 4.67 *Suppose that $\mathbf{0} \in \text{int}(\bar{\mathcal{D}})$. Then*

1. *For any Borel set $G \subset \mathbb{R}^d$,*

$$\limsup_{t \to \infty} \frac{1}{t} \log P_t \left(\{ x \in M \mid t^{-1} \mathbf{X}_t(x) \in G \} \right) \leq -\inf_{\mathbf{Z} \in \text{cl}(G)} \bar{I}(\mathbf{Z}).$$

2. *For any Borel set $G \subset \mathcal{F}$,*

$$\liminf_{t \to \infty} \frac{1}{t} \log P_t \left(\{ x \in M \mid t^{-1} \mathbf{X}_t(x) \in G \} \right) \geq -\inf_{\mathbf{Z} \in \text{int}(G)} \bar{I}(\mathbf{Z}).$$

We refer to (Dembo and Zeitouni, 1988) for proofs and various extensions of these fundamental results.

4.8.4 Central limit theorem

Bryc (Bryc (1993)) has observed that under a suitable analyticity assumption the central limit theorem follows from the large deviation principle. In this appendix we state and prove Bryc's result.

The setup is the same as in Appendix 4.8.3. Let

$$h_t(\mathbf{Y}) = \frac{1}{t} \log \int_{M_t} e^{\mathbf{Y} \cdot \mathbf{X}_t} \, \mathrm{d}P_t,$$

and let D_ϵ be the open polydisc of $\mathbb{C}^d$ of radius ϵ centred at $\mathbf{0}$, that is

$$D_\epsilon = \{z = (z_1, \ldots, z_d) \in \mathbb{C}^d \mid \max_j |z_j| < \epsilon\}.$$

The analyticity assumption is:

(A) For some $\epsilon > 0$ and all $t \in \mathcal{I}$ the function $\mathbf{Y} \mapsto h_t(\mathbf{Y})$ has an analytic continuation to the polydisc D_ϵ such that

$$\sup_{\substack{z \in D_\epsilon \\ t \in \mathcal{I}}} |h_t(z)| < \infty.$$

Moreover, for $\mathbf{Y} \in D_\epsilon \cap \mathbb{R}^d$, the limit

$$h(\mathbf{Y}) = \lim_{t \to \infty} h_t(\mathbf{Y}),$$

exists.

This assumption and Vitali's convergence theorem (see Appendix 4.9 below) imply that $h(\mathbf{Y})$ has an analytic extension to D_ϵ and that all derivatives of $h_t(z)$ converge to the corresponding derivatives of $h(z)$ as $t \to \infty$ uniformly on compact subsets of D_ϵ. We denote

$$\mathbf{m}_t = \nabla h_t(\mathbf{Y})|_{\mathbf{Y}=\mathbf{0}}, \qquad \mathbf{m} = \nabla h(\mathbf{Y})|_{\mathbf{Y}=\mathbf{0}}.$$

Clearly, $\mathbf{m}_t$ is the expectation of $\mathbf{X}_t$ w.r.t. P_t and

$$\lim_{t \to \infty} \frac{1}{t} \mathbf{m}_t = \mathbf{m}.$$

Similarly, if $\mathbf{D}_t = [D_{jkt}]$ is the covariance of $\mathbf{X}_t$, then

$$\lim_{t \to \infty} \frac{1}{t} \mathbf{D}_t = \mathbf{D},$$

where $\mathbf{D} = [D_{jk}]$ is given by

$$D_{jk} = \partial^2_{Y_j Y_k} h(\mathbf{Y})|_{\mathbf{Y}=\mathbf{0}}.$$

Theorem 4.68 *Assumption* (A) *implies the central limit theorem: for any Borel set* $G \subset \mathbb{R}^d$,

$$\lim_{t \to \infty} P_t \left(\left\{ x \in M_t \,\middle|\, \frac{\mathbf{X}_t(x) - \mathbf{m}_t}{\sqrt{t}} \in G \right\} \right) = \mu_{\mathbf{D}}(G),$$

where $\mu_{\mathbf{D}}$ *is the centred Gaussian with covariance* $\mathbf{D}$.

Remark 1. In general, the large deviation principle does not imply the central limit theorem. In fact, Assumption (A) cannot be significantly relaxed, see (Bryc, 1993) for a discussion.

Remark 2. Assumption (A) is typically difficult to check in practice. We emphasize, however, that a verification of assumptions of this type has played a central role in the works (Jakšić, Ogata, and Pillet, 2006), (Jakšić, Ogata, and Pillet, 2007), (Jakšić, Ogata, Pautrat, and Pillet, 2011b).

Remark 3. The proof below should be compared with Section 4.2.12.

Proof: By absorbing $\mathbf{m}_t$ into $\mathbf{X}_t$ we may assume that $\mathbf{m}_t = 0$. Let $\mathbf{k} = (k_1, \cdots, k_d)$, $k_j \geq 0$, be a multi-index and

$$\chi_{\mathbf{k}}(t) = \frac{\partial^{k_1 + \cdots + k_d}}{\partial Y_1^{k_1} \cdots \partial Y_d^{k_d}} \log \int_{M_t} e^{\frac{Y \cdot \mathbf{X}_t}{\sqrt{t}}} \, dP_t \Bigg|_{Y=0},$$

the $\mathbf{k}$-th cumulant of $t^{-1/2}\mathbf{X}_t$. Set

$$\Gamma_r = \{z = (z_1, \ldots, z_d) \in \mathbb{C}^d \,|\, |z_j| = r \text{ for all } j\}.$$

The Cauchy integral formula for polydiscs yields

$$\frac{\partial^{k_1 + \cdots + k_d}}{\partial z_1^{k_1} \cdots \partial z_d^{k_d}} h(z) \Bigg|_{z=0} = \frac{k_1! \cdots k_d!}{(2\pi i)^d} \oint_{\Gamma_{\frac{\varepsilon}{2}}} \frac{h(z)}{z_1^{k_1+1} \cdots z_d^{k_d+1}} \, dz_1 \cdots dz_n$$

$$= \lim_{t \to \infty} \frac{k_1! \cdots k_d!}{(2\pi i)^d} \oint_{\Gamma_{\frac{\varepsilon}{2}}} \frac{h_t(z)}{z_1^{k_1+1} \cdots z_d^{k_d+1}} \, dz_1 \cdots dz_n.$$

Note that

$$\oint_{\Gamma_{\frac{\varepsilon}{2}}} \frac{h_t(z)}{z_1^{k_1+1} \cdots z_d^{k_d+1}} \, dz_1 \cdots dz_n = \oint_{\Gamma_{\frac{\varepsilon}{2\sqrt{t}}}} \frac{h_t(z)}{z_1^{k_1+1} \cdots z_d^{k_d+1}} \, dz_1 \cdots dz_n$$

$$= t^{\frac{k_1 + \cdots + k_d}{2}} \oint_{\Gamma_{\frac{\varepsilon}{2}}} \frac{h_t(t^{-1/2}z)}{z_1^{k_1+1} \cdots z_d^{k_d+1}} \, dz_1 \cdots dz_n,$$

and so

$$\frac{\partial^{k_1 + \cdots + k_d}}{\partial z_1^{k_1} \cdots \partial z_d^{k_d}} h(z) \Bigg|_{z=0} = \lim_{t \to \infty} t^{\frac{k_1 + \cdots + k_d}{2}} \frac{k_1! \cdots k_d!}{(2\pi i)^d} \oint_{\Gamma_{\frac{\varepsilon}{2}}} \frac{h_t(t^{-1/2}z)}{z_1^{k_1+1} \cdots z_d^{k_d+1}} \, dz_1 \cdots dz_n.$$

The Cauchy formula implies

$$\chi_{\mathbf{k}}(t) = t \frac{k_1! \cdots k_d!}{(2\pi i)^d} \oint_{\Gamma_{\frac{\epsilon}{2}}} \frac{h_t(t^{-1/2}z)}{z_1^{k_1+1} \cdots z_d^{k_d+1}} \, dz_1 \cdots dz_n,$$

and we see that

$$\left. \frac{\partial^{k_1+\cdots+k_d}}{\partial z_1^{k_1} \cdots \partial z_d^{k_d}} h(z) \right|_{z=0} = \lim_{t\to\infty} t^{\frac{k_1+\cdots+k_d}{2}-1} \chi_{\mathbf{k}}(t).$$

Hence, if $k_1 + \cdots + k_d \geq 3$, then

$$\lim_{t\to\infty} \chi_{\mathbf{k}}(t) = 0.$$

and if $k_1 + \cdots + k_d = 2$ with the pair k_i, k_j strictly positive, then

$$\lim_{t\to\infty} \chi_{\mathbf{k}}(t) = \left. \frac{\partial^2}{\partial z_{k_i} \partial z_{k_j}} h(z) \right|_{z=0}.$$

Since the expectation of $\mathbf{X}_t$ is zero, we see that the cumulants of $t^{-1/2}\mathbf{X}_t$ converge to the cumulants of the centred Gaussian on $\mathbb{R}^d$ with covariance $\mathbf{D}$. This implies that the moments of $t^{-1/2}\mathbf{X}_t$ converge to the moments of the centred Gaussian with covariance $\mathbf{D}$, and the theorem follows (see Section 30 in (Billingsley, 1986)). □

4.9   Vitali convergence theorem

For $\epsilon > 0$ let D_ϵ be the open polydisk in $\mathbb{C}^n$ of radius ϵ centred at $\mathbf{0}$, that is,

$$D_\epsilon = \{z = (z_1, \ldots, z_n) \in \mathbb{C}^n \mid \max_j |z_j| < \epsilon\}.$$

Theorem 4.69 *Let $\mathcal{I} \subset \mathbb{R}_+$ be an unbounded set and let $F_t : D_\epsilon \to \mathbb{C}$, $t \in \mathcal{I}$, be analytic functions such that*

$$\sup_{\substack{z \in D_\epsilon \\ t>0}} |F_t(z)| < \infty.$$

Suppose that the limit

$$\lim_{t\to\infty} F_t(z) = F(z), \tag{4.194}$$

exists for all $z \in D_\epsilon \cap \mathbb{R}^n$. Then the limit (4.194) exists for all $z \in D_\epsilon$ and is an analytic function on D_ϵ. Moreover, as $t \to \infty$, all derivatives of F_t converge uniformly on compact subsets of D_ϵ to the corresponding derivatives of F.

Proof: Set

$$\Gamma_r = \{z = (z_1, \ldots, z_n) \in \mathbb{C}^n \mid |z_j| = r \text{ for all } j\}.$$

For any $0 < r < \epsilon$, the Cauchy integral formula for polydiscs yields

$$\frac{\partial^{k_1 + \cdots + k_n} F_t}{\partial z_1^{k_1} \cdots \partial z_n^{k_n}}(z) = \frac{k_1! \cdots k_n!}{(2\pi i)^n} \oint_{\Gamma_r} \frac{F_t(w)}{(w_1 - z_1)^{k_1+1} \cdots (w_n - z_n)^{k_n+1}} \, dw_1 \cdots dw_n,$$

(4.195)

for all $z \in D_r$. It follows that the family of functions $\{F_t\}_{t \in \mathcal{I}}$ is equicontinuous on $D_{r'}$ for any $0 < r' < r$. By the Arzela–Ascoli theorem, the set $\{F_t\}$ is precompact in the Banach space $C(\bar{D}_{r'})$ of all bounded continuous functions on $\bar{D}_{r'}$ equipped with the sup norm. The Cauchy integral formula (4.195), where now $z \in D_{r'}$ and the integral is over $\Gamma_{r'}$, yields that any limit point of the net $\{F_t\}_{t \in \mathcal{I}}$ (as $t \to \infty$) in $C(\bar{D}_{r'})$ is an analytic function on $D_{r'}$. By the assumption, any two limit functions coincide for z real, and hence they are identical. This yields the first part of the theorem. The convergence of the partial derivatives of $F_t(z)$ is an immediate consequence of the Cauchy integral formula. □

References

Aizenstadt, V.V. and Malyshev, V.A. (1987). Spin interaction with an ideal Fermi gas. *J. Stat. Phys.* **48**: 51–68.

Araki, H. (1984). On the XY-model on two-sided infinite chain. *Publ. RIMS Kyoto Univ.* **20**: 277–296.

Araki, H. (1990). On an inequality of Lieb and Thirring. *Lett. Math. Phys.* **19**: 167–170.

Araki, H. and Ho, T.G. (2000). Asymptotic time evolution of a partitioned infinite two-sided isotropic XY-chain. *Proc. Steklov Inst. Math.* **228**: 91–204.

Araki, H. and Masuda, T. (1982). Positive cones and L^p-spaces for von Neumann algebras. *Publ. RIMS, Kyoto Univ.* **18**: 339–411.

Araki, H. and Woods, E.J. (1963). Representation of the canonical commutation relations describing a non relativistic infinite free Bose gas. *J. Math. Phys.* **4**: 637–662.

Araki, H. and Wyss, W. (1964). Representations of canonical anticommutation relations. *Helv. Phys. Acta* **37**: 139–159.

Aschbacher, W. and Barbaroux, J.-M. (2006). Out of equilibrium correlations in the XY chain. *Lett. Math. Phys.* **77**: 11–20.

Aschbacher, W. and Barbaroux, J.-M. (2007). Exponential spatial decay of spin-spin correlations in translation invariant quasi-free states. *J. Math. Phys.* **48**: 113302 1–14.

Aschbacher, W., Jakšić, V., Pautrat, Y. and Pillet, C.-A. (2006). Topics in non-equilibrium quantum statistical mechanics. In *Open Quantum Systems III. Recent Developments.* S. Attal, A. Joye and C.-A. Pillet editors. *Lecture Notes in Mathematics* **1882**: Springer, Berlin.

Aschbacher, W., Jakšić, V., Pautrat, Y. and Pillet, C.-A. (2007). Transport properties of quasi-free Fermions. *J. Math. Phys.* **48**: 032101-1–28.

Aschbacher, W. and Pillet, C.-A. (2003). Non-equilibrium steady states of the XY Chain. *J. Stat. Phys.* **112**: 1153–1175.

Audenaert, K. M. R., Nussbaum, M., Szkoła, A. and Verstraete, F. (2008). Asymptotic error rates in quantum hypothesis testing. *Commun. Math. Phys.* **279**: 251–283.

Avron, J.E., Bachmann, S., Graf, G.M. and Klich, I. (2008). Fredholm determinants and the statistics of charge transport. *Commun. Math. Phys.* **280**: 807–829.

Bach, V., Fröhlich, J. and Sigal, I.M. (2000). Return to equilibrium. *J. Math. Phys.* **41**: 3985–4060.

Bruneau, L., Jakšić, V. and Pillet, C.-A (2011). Landauer-Büttiker formula and Schrödinger conjecture. Submitted.

Baladi, V. (2000). *Positive Transfer Operators and Decay of Correlations.* Advanced Series in Nonlinear Dynamics **16**. World Scientific, River Edge, NJ.

Ben Sâad, R. (2008). Etude mathématique du transport dans les systèmes ouverts de fermions. PhD thesis (unpublished), Université de la Méditerranée, Marseille.

Bera, A.K. (2000) Hypothesis testing in the 20th century with a special reference to testing with misspecified models. In *Statistics for the 21st Century.* C.R. Rao and G.J. Székely editors. M. Dekker, New York.

Billingsley, P. (1968). *Convergence of Probability Measures.* Wiley, New York.

Billingsley, P. (1986). *Probability and Measure.* Wiley, New York.

Botvich, D.D. and Malyshev, V.A. (1983). Unitary equivalence of temperature dynamics for ideal and locally perturbed Fermi-gas. *Commun. Math. Phys.* **91**: 301–312.

Bratteli, O. and Robinson, D.W. (1987). *Operator Algebras and Quantum Statistical Mechanics I.* Second Edition. Springer, Berlin.

Bratteli, O. and Robinson, D.W. (1997). *Operator Algebras and Quantum Statistical Mechanics II.* Second Edition. Springer, Berlin.

Bryc, W. (1993). A remark on the connection between the large deviation principle and the central limit theorem. *Stat. Prob. Lett.* **18**: 253–256.

Carlen, E.A. (2010). Trace inequalities and quantum entropy: An introductory course. In *Entropy and the Quantum.* R. Sims and D. Ueltschi (eds). Contemporary Mathematics, volume 529. AMS, Providence RI.

Datta, S. (1995). *Electronic Transport in Mesoscopic Systems.* Cambridge University Press, Cambridge.

Davies, E.B. (1974). Markovian master equations. *Commun. Math. Phys.* **39**: 91–110.

Davies, E.B. and Spohn, H. (1978). Open quantum systems with time-dependent Hamiltonians and their linear response. *J. Stat. Phys.* **19**: 511–523.

Dereziński, J., De Roeck, W. and Maes, C. (2008). Fluctuations of quantum currents and unravelings of master equations. *J. Stat. Phys.* **131**: 341–356.

Dereziński, J. and Jakšić, V. (2003). Return to equilibrium for Pauli-Fierz systems. *Ann. Henri Poincaré* **4**: 739–793.

Dembo, A., and Zeitouni, O. (1988) *Large Deviations Techniques and Applications.* Second edition. Springer, New York.

De Roeck, W. (2009). Large deviation generating function for currents in the Pauli-Fierz model. *Rev. Math. Phys.* **21**: 549–585

de Groot, S.R. and Mazur, P. (1969). *Non-Equilibrium Thermodynamics.* North-Holland, Amsterdam.

Evans, D.J., Cohen, E.G.D., and Morriss, G.P. (1993). Probability of second law violation in shearing steady flows. *Phys. Rev. Lett.* **71**: 2401–2404.

Evans, D.J., and Searles, D.J. (1994). Equilibrium microstates which generate second law violating steady states. *Phys Rev. E* **50**: 1645–1648.

Fröhlich, J. and Merkli, M. (2004). Another return of "return to equilibrium". *Commun. Math. Phys.* **251**: 235–262.

Fröhlich, J., Merkli, M. and Sigal, I.M (2004). Ionization of atoms in a thermal field. *J. Stat. Mech.* **116**: 311–359.

Fröhlich, J., Merkli, M. and Ueltschi, D. (2003). Dissipative transport: Thermal contacts and tunneling junctions. *Ann. Henri Poincaré* **4**: 897–945.

Gallavotti, G. (1996). Chaotic hypothesis: Onsager reciprocity and fluctuation-dissipation theorem. *J. Stat. Phys.* **84**: 899–925.

Haag, R., Hugenholtz, N.M. and Winnink, M. (1967). On equilibrium states in quantum statistical mechanics. *Commun. Math. Phys.* **5**: 215–236.

Hepp, K. (1970). Rigorous results on the s-d model of the Kondo effect. *Solid State Communications* **8**: 2087–2090.

Hiai, F., Mosonyi, M. and Ogawa, T. (2008). Error exponents in hypothesis testing for correlated states on a spin chain. *J. Math. Phys.* **49**: 032112-1–22.

Imry, Y. (1997). *Introduction to Mesoscopic Physics.* Oxford University Press, Oxford.

Jakšić, V., Pautrat, Y. and Pillet, C.-A. (2009). Central limit theorem for locally interacting Fermi gas. *Commun. Math. Phys.* **285**: 175–217.

Jakšić, V. and Pillet, C.-A. (1996). On a model for quantum friction III. Ergodic properties of the spin–boson system. *Commun. Math. Phys.* **178**: 627–651.

Jakšić, V. and Pillet, C.-A. (2001). On entropy production in quantum statistical mechanics *Commun. Math. Phys.* **217**: 285–293.

Jakšić, V. and Pillet, C.-A. (2002). Non-equilibrium steady states of finite quantum systems coupled to thermal reservoirs. *Commun. Math. Phys.* **226**: 131–162.

Jakšić, V., Ogata, Y. and Pillet, C.-A (2006). The Green-Kubo formula and the Onsager reciprocity relations in quantum statistical mechanics. *Commun. Math. Phys.* **265**: 721–738.

Jakšić, V., Ogata, Y. and Pillet, C.-A (2007). The Green-Kubo formula for locally interacting fermionic open systems. *Ann. Henri Poincaré* **8**: 1013–1036.

Jakšić, V., Pillet, C.-A. and Rey-Bellet, L. (2011a). Entropic Fluctuations in Statistical Mechanics I. Classical Dynamical Systems. *Nonlinearity* **24**: 699–763.

Jakšić, V., Ogata, Y., Pautrat, Y. and Pillet, C.-A. (2011b). Entropic Fluctuations in Statistical Mechanics II. Quantum Dynamical Systems. In preparation.

Jakšić, V., Ogata, Y., Pillet, C.-A. and Seiringer, R. (2011c). Hypothesis testing and nonequilibrium statistical mechanics. In preparation.

Jordan, P. and Wigner, E. (1928). Über das Paulische Äquivalenzverbot. *Z. Phys.* **47**: 631–651.

Korevaar, J. (2004). *Tauberian Theory. A Century of Developments.* Springer, Berlin.

Kosaki, H. (1986). Relative entropy of states: A variational expression. *J. Operator Theory* **16**: 335–348.

Kurchan, J. (2000). A quantum Fluctuation theorem. arXiv:cond-mat/0007360v2

Lebowitz, J.L. and Spohn, H. (1977). Stationary non-equilibrium states of infinite harmonic systems. *Commun. math. Phys.* **54**: 97–120.

Lebowitz, J.L. and Spohn, H. (1978). Irreversible thermodynamics for quantum systems weakly coupled to thermal reservoirs. *Adv. Chem. Phys.* **38**: 109–142.

Levitov, L.S. and Lesovik, G.B. (1993). Charge distribution in quantum shot noise. *JETP Lett.* **58**: 230–235.

Lewin, L. (1981). *Polylogarithms and Associated Functions.* North-Holland, New York.

Lieb, E. and Ruskai, M. (1973). Proof of the strong subadditivity of quantum-mechanical entropy. With an appendix by B. Simon. *J. Math. Phys.* **14**, 1938–1941.

Lieb, E. and Thirring, W. (1976). Inequalities for the moments of the eigenvalues of the Schrödinger Hamiltonian and their relation to Sobolev inequalities. In *Studies in Mathematical Physics.* E. Lieb, B. Simon and A.S. Wightman (eds). Princeton University Press, Princeton, NJ.

Maassen, H. and Botvich, D. (2009). A Galton-Watson estimate for Dyson series. *Ann. Henri Poincaré* **10**: 1141–1158.

Merkli, M., Mück, M. and Sigal, I.M. (2007-a). Instability of equilibrium states for coupled heat reservoirs at different temperatures *J. Funct. Anal.* **243**: 87–120.

Malyshev, V.A. (1988). Convergence in the linked cluster theorem for many body Fermion systems. *Commun. Math. Phys.* **119**: 501–508.

McLennan, J.A. Jr. (1963). The formal statistical theory of transport processes. In *Advances in Chemical Physics, Volume 5.* I. Prigogine editor. Wiley, Hoboken, NJ.

Merkli, M., Mück, M. and Sigal, I.M. (2007-b). Theory of non-equilibrium stationary states as a theory of resonances. *Ann. Henri Poincaré* **8**: 1539–1593.

Nenciu, G. (2007). Independent electron model for open quantum systems: Landauer-Büttiker formula and strict positivity of the entropy production. *J. Math. Phys.* **48**: 033302-1–8.

Ogata, Y. (2010). A generalization of the inequality of Audenaert *et al.* Preprint, arXiv:1011.1340v1.

Ohya, M. and Petz, D. (2004). *Quantum Entropy and Its Use.* Second edition. Springer, Heidelberg.

Pearson, K. (1900). On the criterion that a given system of deviations from the probable in the case of a correlated system of variables is such that it can be reasonably supposed to have arisen from random sampling. *Phil. Mag. Ser. 5,* **50**: 157–175.

Pillet, C.-A. (2006). Quantum dynamical systems. In *Open Quantum Systems I. The Hamiltonian Approach.* S. Attal, A. Joye, and C.-A. Pillet (eds). Lecture Notes in Mathematics **1880**. Springer, Berlin.

Reed, M. and Simon, B. (1975). *Methods of Modern Mathematical Physics. II: Fourier Analysis, Self-Adjointness.* Academic Press, New York.

Reed, M. and Simon, B. (1979). *Methods of Modern Mathematical Physics. III: Scattering Theory.* Academic Press, New York.

Reed, M. and Simon, B. (1978). *Methods of Modern Mathematical Physics. IV: Analysis of Operators.* Academic Press, New York.

Roberts, A.W. and Varberg, D.E. (1973). *Convex Functions.* Academic Press, New York.

Robinson, D.W. (1973). Return to equilibrium. *Commun. Math. Phys.* **31**: 171–189.

Rondoni, L. and Mejía-Monasterio, C. (2007). Fluctuations in non-equilibrium statistical mechanics: models, mathematical theory, physical mechanisms. *Nonlinearity* **20**: 1–37.

Ruelle, D. (2000). Natural nonequilibrium states in quantum statistical mechanics. *J. Stat. Phys.* **98**: 57–75.

Ruelle, D. (2001). Entropy production in quantum spin systems. *Commun. Math. Phys.* **224**: 3–16.

Ruelle, D. (2002). How should one define entropy production for nonequilibrium quantum spin systems? *Rev. Math. Phys.* **14**: 701–707.

Simon, B. (1979). *Functional Integration and Quantum Physics*. Academic Press, New York.

Tasaki, S. and Matsui,T. (2003). Fluctuation theorem, nonequilibrium steady states and MacLennan-Zubarev ensembles of a class of large quantum systems. *Quantum Prob. White Noise Anal.* **17**: 100–119.

Takesaki M. (1970). *Tomita's Theory of Modular Hilbert Algebras and its Applications*. Lectures Notes in Mathematics **128**. Springer, Berlin.

Tomita, M. (1967). "Quasi-standard von Neumann algebras" and "Standard forms of von Neumann algebras". Unpublished.

Uhlmann, A. (1977). Relative entropy and the Wigner-Yanase-Dyson-Lieb concavity in an interpolation theory. *Commun. Math. Phys.* **54**: 21–32.

Zubarev, D.N. (1962). The statistical operator for nonequilibrium systems. *Sov. Phys. Dokl.* **6**: 776–778.

Zubarev, D.N. (1974). *Nonequilibrium Statistical Thermodynamics*. Consultants, New York.

5

Quantum phase transitions: introduction and some open problems

Achim ROSCH

Institute of Theoretical Physics,
University of Cologne,
50937 Cologne, Germany

Chapter Contents

Preface

These notes give a short overview over five lectures held during the Les Houches summer school 'Quantum theory from small to large scales' in 2010. After an introductory section, field driven magnetic quantum phase transitions of insulators are used as an example to discuss some of the concepts underlying quantum phase transitions. We emphasize that spin precession in the presence of a magnetic field changes the dynamics and implies that the quantum critical theory is in the universality class of a diluted Bose systems. Both experimentally and theoretically, quantum phase transitions in metals are much less understood compared to insulating systems. After a brief review of the standard approach (Hertz theory) to describe those systems, we discuss the importance of multiple time scales and associated multiple critical exponents z using a Pomeranchuk instability as an example. Finally, we investigate the possibility of emergent gauge theories close to critical points. As an example, we discuss why a gauge theory describes the (classical) phase transitions of a nematic, if topological defects are suppressed.

While some introductory sections of the lectures are described in detail, for other parts of the lecture only a short qualitative discussion and an introduction to the relevant literature is presented.

5.1 Quantum criticality and scaling

5.1.1 Classical and quantum phase transitions

A striking observation in solid state physics is that physical properties can sometimes change suddenly upon small changes of external parameters. In the thermodynamic limit, where the number of particles is formally set to infinity, the free energy and its derivative can become a nonanalytic function of parameters like temperature, doping or magnetic field. Such nonanalyticities define phase transitions.

A continuous second order phase transition is characterized by a diverging correlation length

$$\xi \sim (\delta - \delta_c)^{-\nu} \to \infty \quad \text{for } \delta \to \delta_c \tag{5.1}$$

where the 'control parameter' δ (temperature, pressure, chemical composition, magnetic fields, etc.) controls the proximity to the critical point at $\delta = \delta_c$ and the correlation length exponent ν describes how ξ grows close to the phase transition. A classical example is a magnetic material which develops long-range magnetic order below a critical temperature.

Associated to the divergence of correlation length, there is also a divergence of the time scale describing the fluctuation rate of ordered domains of size ξ. This time scale diverges for diverging ξ, a phenomena called 'critical slowing down'

$$\tau \sim \xi^z. \tag{5.2}$$

The laws of quantum mechanics imply that the typical quantum mechanical energy scale

$$\Delta = \frac{\hbar}{\tau} \sim (\delta - \delta_c)^{\nu z} \to 0 \qquad \text{for } \delta \to \delta_c \tag{5.3}$$

vanishes at the critical point. For a phase transition at finite T, this implies that $\Delta \ll k_B T$ sufficiently close to the transition. Close to such a finite-T phase transition, the critical behaviour can therefore be described classically and thermal fluctuations determine the critical properties alone. A famous example for this is the superfluid transition in ^{4}He, probably the most precisely measured phase transition (Lipa *et al.* 1996), whose critical properties are well described by a classical xy model, despite the fact that superfluidity itself is a pure quantum effect. For a phase transition at $T = 0$, in contrast, only quantum fluctuation are responsible for the critical properties.

A quantum phase transition (QPT) (Hertz 1976; Sachdev 1999) is a phase transition at zero temperature, $T = 0$. In the thermodynamic limit the ground state at a critical point, $\delta = \delta_c$, is nonanalytic in δ. We will be mainly interested in second order transition, in this case the quantum phase transition is called a quantum critical point (QCP). While quantum phase transitions occur by definition formally at $T = 0$, they strongly influence physical properties in a wide neighbourhood of the QCP and, especially, in a large temperature range which can cover several decades. Typical phase diagrams are shown in Fig. 5.1.

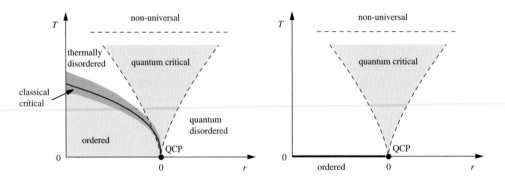

Fig. 5.1 Schematic phase diagram of a system with a quantum critical point. Left panel: Upon increasing the control parameter r an ordered phase is suppressed and vanishes at the QCP. The QCP is the zero-temperature end point of a line of classical phase transitions at $T > 0$. Experimentally, the so-called quantum critical regime above the QCP extends often over a large temperature range. Right panel: A quantum phase transition can also occur in the absence of any classical transition when as a function of the control parameter the ground-state wave function changes in a singular way, while at $T > 0$ only crossovers are observed. Examples include transitions in low-dimensional materials where thermal fluctuations destroy long-range order at $T > 0$, metal-insulator transitions or 'topological transitions', for example a critical point where the topology of Fermi surfaces changes.

Two cases have to be distinguished: In the first case (left panel of Fig. 5.1) the transition temperature, T_c, of a finite-temperature phase transition is suppressed until it reaches $T_c = 0$. In the second case (right panel), there is no phase transition at finite temperature but only at $T = 0$. There are many examples for both situation: in three dimensions magnetic transitions are usually of type 1 while in low dimensions ($d = 1$ for discrete symmetries, $d \leq 2$ for Heisenberg symmetry) thermal fluctuations destroy long range order. Other prominent examples for the second case include metal insulator transitions, transitions between Hall plateaus or between other topologically ordered phases. A simple example is a transition in a metal where as a function of pressure the topology of the Fermi surface changes

While the critical regime of classical transitions (left panel of Fig. 5.1) is usually very small, in quantum phase transitions anomalous behaviour is often experimentally observed (and theoretically predicted) for a wide range of temperatures. Most interesting and mostly studied is usually the T dependence in the so-called 'quantum critical regime' directly above the quantum critical point. Note, however, that in this regime by definition the temperature $k_B T$ is much larger than the intrinsic quantum scale Δ, $\Delta \ll k_B T$. Therefore $k_B T$ is the only remaining energy scale (exceptions are discussed below) and thermal fluctuations control the physics. Nevertheless, the term 'quantum critical regime' is appropriate as the power laws as a function of T observed in this regime are characteristic for the QCP.

Twenty years ago it was believed that quantum critical points were extremely rare phenomena which were difficult to observe. In contrast there are now hundreds of examples of quantum phase transitions (Sachdev 1999; Löhneysen *et al.* 2007; Sachdev and B. 2011; Coleman and Schofield 2005; Stewart 2001) and, whenever a material shows unconventional behaviour over, for example, a decade of temperatures, one suspects the proximity of some quantum phase transition. For an experimental investigation, most promising are systems whose properties can easily be tuned as energy scales are small. This happens for example in heavy fermion materials (Stewart 2001; Löhneysen *et al.* 2007) where the effective masses of electrons can become as large as 1000 times the bare electron mass resulting in small Fermi energies. Also organic compounds are often governed by small scales. Now classical examples of quantum phase transitions are quantum Hall systems where doping can be controlled by gate voltages. Due to the apparently unlimited possibilities to control and manipulate ultracold atoms (Greiner *et al.* 2002; Bloch *et al.* 2008), these systems are also promising candidates to investigate quantum phase transitions and, especially their dynamics, even if inhomogeneities arising from trapping potential make the observation of quantum critical behaviour difficult.

But why are quantum critical points interesting? A few answers are given in the following:

- As for classical phase transitions, the physics close to a QCP is universal and therefore independent of microscopic details at it is governed by the diverging length scale ξ.
- From an experimental point of view, one can ask what a promising route is to find novel phases, new states of matter, or materials which act sensitively to small

changes of parameters. Here a useful strategy is to investigate systems located just at the transition between two known phases (metal and insulator, metal and paramagnet, etc.)

- Indeed, close to quantum critical points not only critical behaviour (power laws) are observed, but often a new phase emerges. For example in many of the quantum critical magnetic metals a superconducting phase appears close to the QCP— obviously superconductivity is driven by quantum critical fluctuations (rather than phonons) in this case.

- While especially for many insulators there is perfect agreement of theory and experiments, there are some carefully investigated materials, most notably $CeCu_{6-x}Au_x$ and $YbRh_2Si_2$, where experiments can apparently not be described by established quantum-critical theories (Löhneysen *et al.* 2007).

- On the theory side, there are a number of exciting proposals which show that close to QCPs completely new quantum effects become important, which may, for example, be described by 'emergent' gauge theories.(Lammert *et al.* 1993; Senthil *et al.* 2004; Levin and Senthil 2004), see Section 5.4.2.

- Compared to classical transitions, the list of possible phase transitions also becomes much larger at $T = 0$. Especially metal insulator transitions and some other transitions with no obvious order parameter (e.g. transitions between states with different topological order) have no classical counterpart.

- A special theoretical challenge is the investigation of transport properties or, more generally, the investigation of dynamical quantities. While for quite a number of QPTs the imaginary time dynamics can be well understood from concepts well known from classical phase transitions, the real time dynamics at finite temperature is a challenging problem.

- This is also related to the rather complex role which temperature has close to QCPs. One way how temperature enters can directly be understood from a functional integral formulation. When writing the partition function as an imaginary time functional integral, $Z = \int \mathcal{D}\Phi\, e^{-S}$, the action S is written as an in integral over space and imaginary time

$$S = \int d^d\mathbf{r} \int_0^\beta d\tau\, L[\Phi] \tag{5.4}$$

While the system size can be considered as infinite, the inverse temperature, $\beta = 1/k_B T$, leads to a finite size in time direction. This is the reason why at a *classical* finite T transition, where the correlation time diverges, the temporal quantum fluctuations are not important for the thermodynamics. The diverging time scales do not influence thermodynamic critical exponents. In the quantum case, the dynamics (including associated Berry phases, Wess Zumino terms, etc.) becomes important and couples back to thermodynamics and statics. Actually the role of temperature is often quite subtle and cannot be reduced to a simple finite size scaling known from classical critical points. Temperature controls the distance to the QCP, the destruction of quantum by thermal fluctuations including dephasing effects and also the presence of thermally excited quasi particles.

5.1.2 Scaling

The universality expected for phase transitions manifests itself in the scaling properties of physical observables. Here one uses that close to quantum critical point one expects that only a single length scale, the correlation length ξ, determines the universal properties and fixes also the relevant time scale, $\tau \sim \xi^z$, as discussed above. This implies that observables change in simple ways when the length scale is changed. $\xi \to b\xi$ implies, for example, $\tau \sim b^z \tau$. For a dynamical susceptibility which depends on space and time, this has the consequence that its universal, quantum critical part takes the form

$$\chi(q, \omega) \approx b^\Phi \tilde{\chi}(qb, \omega b^z, Tb^z, (\delta - \delta_c)b^{1/\nu}) \tag{5.5}$$

for a set of critical exponents Φ, z, ν. b is an arbitrary scale factor which can, for example, be chosen to eliminate one of the variables, for example $b = T^{-1/z}$ gives $\chi \sim T^{-\Phi/z}\tilde{\chi}(q/T^z, \omega/T, (\delta - \delta_c)/T^{1/z\nu})$. In equation (5.5) we have assumed that the frequency and the temperature scale with the same exponent. This so-called ω/T scaling, is expected to hold if there is indeed only one time scale in the problem as the inverse temperature just determines the system size in (imaginary) time direction, see equation (5.4). In general, there can be more than one diverging length and time scale close to a quantum critical point as will be discussed below. In this case, the simple scaling relations have to be modified.

A simple way to decide experimentally whether a system is quantum critical (Zhu *et al.* 2003), is to measure the so-called Grüneisen ratio Γ (or similar related quantities) which is defined as the ratio of specific heat c_p at fixed pressure and the thermal expansion, $\alpha = \frac{1}{V}\frac{\partial V}{\partial T}$

$$\Gamma = \frac{1}{T}\frac{c}{\alpha} = -\frac{1}{T}\frac{\partial S/\partial p}{\partial S/\partial T} \tag{5.6}$$

where p is the pressure, V the volume and S the entropy. For a noncritical system, where $S \approx c(T/T_0(p))^\alpha$ with some constant c for $T \to 0$, one obtains $\Gamma = \frac{1}{T_0}\frac{\partial T_0}{\partial p}$. Γ is finite in this case while it diverges when the typical energy scale T_0 vanishes as expected for a QCP. More precisely, its divergence signals that the pressure p is relevant (Zhu *et al.* 2003), that is the system becomes very sensitive to pressure at the QCP which will always occur when the location of the critical point $\delta_c(p)$ is pressure sensitive. It is a useful exercise to calculate the divergence of Γ assuming that scaling is valid.

5.2 Field driven QPT as Bose–Einstein condensation

We will now discuss a simple example of a quantum critical point, namely a quantum phase transition of a paramagnetic insulator to a magnetic phase driven by magnetic fields. This model is chosen for several reasons: first, it is well understood from the theoretical point of view. Second, a series of very beautiful experiments, for example (Rüegg *et al.* 2003; Lorenz *et al.* 2008; Rüegg *et al.* 2008), exist where the chosen model systems are realized in an ideal way and where theory and experiment show a very good agreement (Lorenz *et al.* 2008; Anfuso *et al.* 2008; Rüegg *et al.* 2008). Finally, the quantum critical degrees of freedom turns out to be diluted interacting bosons and one therefore obtains Bose–Einstein condensation, a topic covered by several other lectures during this Les Houches school.

5.2.1   A model for field-driven quantum criticality

We consider the following one-dimensional spin-1/2 model defined on a 'ladder'

$$H = J_1 \sum_i \mathbf{S}_{1,i}\mathbf{S}_{2,i} + J_2 \sum_i (\mathbf{S}_{1,i}\mathbf{S}_{1,i+1} + \mathbf{S}_{2,i}\mathbf{S}_{2,i+1}) - B \sum_i (S_{1,i}^z + S_{2,i}^z) \quad (5.7)$$

where $S_{\alpha i}$ ($\alpha = 1,2$, $i \in Z$) are spin 1/2 operators. Here pairs of two spins which are antiferromagnetically ($J_i > 0$) coupled by J_1 interact with neighbouring spins via the exchange coupling J_2. B is an external magnetic field measured in units of Bohr's magneton.

We will also be interested in higher dimensional versions of this model, where the pairs of spins $\mathbf{S}_{1i}, \mathbf{S}_{2i}$ are located on some d-dimensional cubic lattice with short-range antiferromagnetic interactions

$$H = J_1 \sum_{i \in Z^d} \mathbf{S}_{1,i}\mathbf{S}_{2,i} + \sum_{ij \in Z^d, \alpha, \beta = 1,2} J_2^{\alpha\beta}(i-j)\mathbf{S}_{\alpha,i}\mathbf{S}_{\eta,j} - B\sum_i (S_{1,i}^z + S_{2,i}^z). \quad (5.8)$$

The one-dimensional spin model is realized with very high precision, for example in the compound $(C_5H_{12}N)_2CuBr_4$ (Lorenz *et al.* 2008; Rüegg *et al.* 2008), an example for a three-dimensional compound is $TlCuCl_3$ (Rüegg *et al.* 2003).

To understand the low-energy properties and phase diagram of (5.7,5.8) it is useful to consider the limit of small J_2. For $J_2 = 0$ and $B = 0$ the two spins in a dimer split in a ground state singlet $|0,0\rangle$ and three triplet states $|1,m\rangle$, $m = -1,0,1$ with energy difference J_1, see inset of Fig. 5.2. For finite B, $J_2 = 0$, the energy of the $|1,1\rangle$ states, $E_{1,1} - E_{0,0} = J_1 - B$, is lowered and it becomes the ground state for $B > J_1$. For $J_2 = 0$ and $T = 0$ one therefore obtains a first order phase transition at $B = J_1$, where the magnetization jumps from 0 to its maximal value 1. The situation is qualitatively different in the presence of a small but finite J_2: the degeneracy of the excited triplet states is lifted and they form a band. For small $J_2 \ll J_1$ and small $B \ll J_1$ the ground state magnetization remains exactly 0 as the system is gapped and the total magnetization commutes with the Hamiltonian. Similarly, for $B \gg J_1$ the magnetization is exactly 1. Therefore there have to be two quantum phase transitions at two critical magnetic fields, B_{c1} and B_{c2} with $M = 0$ for $B < B_{c1}$, $M = 1$ for $B > B_{c2}$ and $0 < M < 1$ for $B_{c1} < B < B_{c2}$ as M grows with increasing B. In the following, we will argue that the two quantum phase transitions are in the same universality class as Bose–Einstein condensates (BEC) where the role of $B - B_c$ is played by the chemical potential. It is therefore justified to talk about a 'condensation of triplets'. In three dimensions, the condensation can occur at finite temperatures and one observes therefore a finite T phase transition for $B_{c1} < B < B_{c2}$ where the critical temperature T_c vanishes at B_{c1} and B_{c2}.

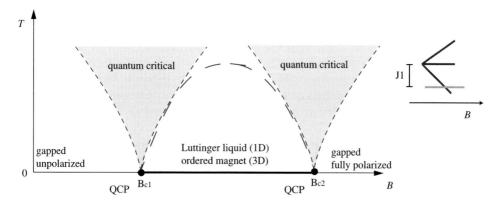

Fig. 5.2 Schematic picture of the phase diagram of a field-driven quantum phase diagram of the model defined by equation 5.7. While in one dimension (1D) a gapless phase is obtain which can be described as a Luttinger liquid, in three dimensions a long range antiferromagnetic order perpendicular to the applied field emerges. Inset: Energy levels for $J_2 = 0$. Upon increasing B the lowest triplet state gets in energy lower than the singlet state.

To obtain some intuition on the physics, one can convince oneself with a short calculation that a linear superposition of a singlet state with a small admixture of the lowest triplet state, $|s\rangle + \alpha|\uparrow\uparrow\rangle$, describes a state with magnetization predominantly in the xy plane perpendicular to the applied field pointing in opposite directions for spin 1 and spin 2. Thereby, $|\alpha|$ controls the amplitude of the perpendicular magnetization while the phase determines the direction of the magnetization. Condensation of triplets therefore describes nothing but (staggered) magnetic order in the direction perpendicular to the applied field.

5.2.2 Effective low-energy theory and scaling analysis

To obtain a description of the physics close to the quantum critical point (in this case there are two quantum critical points), one can follow two different routes. The first—often very successful—route is to 'guess' the relevant effective field theory based on some knowledge of the order parameter and on symmetry considerations. Here the relevant order parameter is the (staggered) magnetization (ϕ_x, ϕ_y) in the xy plane for a magnetic field in the z direction. When deriving the effective action in a gradient expansion, it is important to take into account the rotation symmetry in the xy plane and the fact that time-reversal symmetry is explicitly broken by the external magnetic field. Therefore terms with a *linear* time derivative $\int \phi_x \partial_t \phi_y$ are allowed to occur in the effective action which physically describes the precession of spins around the B-field. Introducing a complex field $\phi = \phi_x + i\phi_y$ the effective action in imaginary time takes the form

$$S = \int d\tau \int d^d r \, c_1 \phi^* \partial_\tau \phi + c_2 |\phi|^2 + c_3 |\nabla \phi|^2 + c_4 |\phi|^4 + \dots \tag{5.9}$$

with some coefficients c_i. This effective action can be identified as a field theory of interacting, nonrelativistic bosons! Upon rescaling fields, one can set $c_1 = 1$. Then c_3 can be identified with $\hbar^2/(2m)$ with the mass m, c_2 with the chemical potential and c_4 with an effective, local interaction. In the absence of a magnetic field, one can also induce magnetic order by pressure (Rüegg *et al.* 2003). In this case, however, the linear time derivative is forbidden by symmetry, there is no spin-precession, and therefore the system belongs to a different universality class unrelated to the condensation of nonrelativistic bosons.

For the present problem it is also easy to derive the effective action starting from the spin Hamiltonian. This is especially simple close to B_{c2}, the upper critical field, as the ground state wave function for $B > B_{c2}$ is known—it is the fully polarized state. Defining this as the vacuum and identifying a single spin flip with a boson creation operator $a^\dagger$, $a^\dagger| \uparrow \rangle = | \downarrow \rangle$, one can conveniently rewrite the Hamiltonian (5.8) in terms of bosons. Technically, one embeds the Hilbert space of spins into the Hilbert space of bosons. Multiple occupations of bosons on a single site are easily suppressed by a onsite repulsion $U \sum_i n_i(n_i - 1)$ for $U \to \infty$. Therefore one can easily rewrite the spin model as a model of interacting bosons.

Finally, one has to obtain the continuum action from the bosonic lattice model in the low-density limit. To determine all prefactors in the effective action (5.9) one just has to solve the one- and two-boson problem both for the lattice model and the continuum version (for a given cutoff scheme which is needed to make the action well-defined). Matching physical observable (e.g. scattering phase shifts) allows us to determine all c_i.

With our analysis, we have been able to map the problem of interacting spins in a magnetic field to the problem of diluted bosons. In the bosonic language, the distance from the critical point, for example $B_{c2} - B$ (similar arguments apply also close to B_{c1}) can directly be identified with the chemical potential μ of the bosons. The identification of Bose–Einstein condensation with field-driven quantum phase transitions goes back to the work of Giamarchi and Tsvelik (1999).

The next step in the analysis of the problem is a simple scaling analysis–the first step of a renormalization group treatment. One observes that under the scaling transformation $r \to r' = r/\lambda$, $\tau \to \tau' = \tau/\lambda^z$ and $\phi \to \phi' = \phi\lambda^{(d+z-2)/2}$ the quadratic part of the action remains invariant at the QCP where $c_2 = 0$. The interaction term, c_4, is, however, rescaled,

$$c_4 \to c_4\lambda^{4-(d+z)} \tag{5.10}$$

This implies that for $d + z > 4$ the effective interactions get less and less important when one goes to larger and larger length scales. For $d = 1$ and $z = 2$, however, the experimentally relevant case for spin ladders (Giamarchi and Tsvelik 1999; Lorenz *et al.* 2008), the importance of interactions grows upon lowering the energy. This observation is easy to interpret for the present problem. When one solves the problem of two bosons which scatter from each other, one realizes that for $d > 2$ scattering becomes less and less effective for low energies while in $d = 1$ one obtains less and less transmission when the energy of the two particles is lowered. In the low energy limit one can therefore set the effective interaction to infinity, consistent with the simple scaling result. Fortunately, one can solve the problem of strongly interaction bosons in $d = 1$ by rewriting them as noninteraction fermions. For a detailed analysis of the quantum critical theory and, especially, a quantitative comparison both of numerics and analytics to experiments, we refer to Anfuso *et al.* (2008).

5.3   Order parameter theories and perturbative renormalization group

In his pioneering paper of 1976, Hertz developed a strategy to derive and analyse quantum critical theories not only for insulators and spin systems, but also for quantum phase transitions in metals. The strategy starts from a microscopic theory (e.g. a Hubbard model) in a functional integral formulation. In a first step, the interaction term of the microscopic theory is rewritten using a simple identity for Gaussian integrals (often called the Hubbard–Stratonovich transformation) and a collective variable, the order parameter field is introduced. In a second step, all degrees of freedom with the exception of the order parameter field are formally integrated out. The remaining action is expanded in powers of the order parameter field. Complicated frequency and momentum dependent vertices are approximated by their $\omega, k \to 0$ limit. Finally, the resulting action is analysed using scaling or more refined perturbative renormalization group approaches. Based on this approach, it is then argued that omitted terms are technically 'irrelevant' implying that it is justified to neglect them.

The approach of Hertz, which follows the strategy associated with the names of Landau, Ginzburg, and Wilson in the context of classical phase transitions, is very well described in the literature and is not repeated here. The insightful original paper (Hertz 1976) contains a clear discussion of the overall strategy. Millis (1993) corrected some technical problems in the original paper with the perturbative renormalization group analysis at finite temperature, therefore one often talks about the Hertz–Millis theory. A detailed discussion can also be found in our review (Löhneysen *et al.* 2007). One of the main results of the analysis of Hertz is that quantum critical points are often above or at the upper critical dimension especially in three dimensional materials as typically $d + z \geq 4$, where according to equation (5.10) the interaction effects become (marginally) irrelevant. More precisely, they are dangerously irrelevant, in the sense that physical properties (e.g. the order parameter susceptibility at finite temperatures) are singular functions of the irrelevant couplings and they can therefore not be set to zero, see, for example (Löhneysen *et al.* 2007) for a discussion of this problem.

An important and not fully resolved question is under what conditions the approach of Hertz is valid and can be applied. The derivation of effective field theories is rather straightforward for simple insulators such as discussed in Section 5.2. While mathematical proofs are missing for almost all situations, there is little doubt that one can obtain valid effective field theories by integrating out massive degrees of freedoms, that is degrees of freedom with gaps. The situation is much less clear in the case of metals. Can these massless low-energy degrees of freedom be integrated out in a reliable and controlled way? A complete analysis of this question and a general answer, for example a systematic and comprehensive renormalization group treatment of both the collective bosonic and, simultaneously, fermionic excitations, is still missing. In Löhneysen *et al.* (2007) these questions are extensively discussed and several examples are given of systems where, by now, it is rather well established that the Hertz approach breaks down, see also the next section. Only very recently, Metlitski and Sachdev (Metlitski and Sachdev, 2010a, Metlitski and Sachdev, 2010b) put forward a detailed analysis of the coupled bosonic and fermionic model, which allowed them to identify a new fixed point not described by the Hertz approach, see also (Mross *et al.* 2010) for a large-N analysis of this problem.

5.4 Beyond the standard models

The 'standard model' for quantum phase transition is based on the idea of deriving theories in terms of simple order parameters, an approach which is connected to the names of Landau and Ginzburg for classical phase transitions. The renormalization group pioneered by Wilson provides the conceptional framework for such theories. As discussed in the previous section, John Hertz (Hertz 1976) generalized this concept to quantum critical points. It was also pointed out that the validity of this Landau–Ginzburg–Wilson–Hertz approach is far from obvious when one considers, for example, quantum phase transitions in metals.

Both theoretical considerations and, especially, experiments (Löhneysen *et al.* 2007) strongly suggest that one has to consider concepts which go beyond the 'standard model' of Hertz. Two aspects of this are briefly discussed in the following.

5.4.1 Multiple critical exponents

In many of the cases (Löhneysen *et al.* 2007), where the Hertz approach fails, one encounters an important problem: the assumption that there is only one diverging time scale and therefore only on dynamical critical exponente z is apparently not valid. Consider, for example, a magnetic quantum critical point in a metal and a magnetic domain of size ξ. There are two relevant time scales: the order parameter, that is the magnetic degrees of freedom, fluctuates on the time scale

$$\tau \sim \xi^{z_{OP}}. \tag{5.11}$$

The fermions, however, traverse the magnetic domain on a completely different time scale

$$\tau \sim \xi^{z_F}. \tag{5.12}$$

and at least within simple approximations the two dynamical critical exponents z_{OP} and z_F are different. At the same time, however, the dynamics of the order parameter arises from the coupling to the fermions and also the dynamics of the fermions is dominated from the singular scattering from the order parameter fluctuations: the two subsystems are intimately coupled. Presently, no fully developed theoretical framework exists which deals with such a situation. Practically, one has to expect that all simple scaling predictions, see equation 5.5, break down in such a case and also the concepts underlying the renormalization group and the notion of fixed points have to be reconsidered.

Three possibilities arise. First, it may occur that the feedback of the two subsystems remains nevertheless trivial. In Rosch (2001), for example, simple scaling arguments are given, why this might be the case in for antiferromagnetic quantum critical point in sufficiently high dimensions. Second, it may happen that at the interacting fixed point the two exponents become equal, $z_{OP} = z_F$. The recent work of Metlitski and Sachdev (Metlitski and Sachdev, 2010a, Metlitski and Sachdev, 2010b) seems to be consistent with such a scenario.

In this chapter, however, a third option is discussed: that the two exponents remain different at the QCP but that they influence each other. To study this question, a simple, purely bosonic model is considered, where it is possible to study this physics in some detail. The model arises from a theory of the so-called Pomeranchuk instability in a two-dimensional metal, where at the QCP the Fermi surface deforms from a perfect circle into an ellipse. Here it turns out that deformations perpendicular and parallel to a small momentum $\mathbf{q}$ obtain different dynamical critical exponents, $z_1 = 2$ and $z_2 = 3$. For $\mathbf{q} = 0$, however, the two modes are degenerate by symmetry and become critical simultaneously. Does one of the two dynamical critical exponents dominate over the other? At zero temperature, the usual scaling argument, see equation (5.10), clearly suggests that the physics is dominated by the *smaller* of the two exponents. For finite T one expects, however, that at the QCP the *larger* of the two exponents dominates as this mode is associated with the shorter length scale, $T^{-1/z}$. For example, the free energy scales as $T/\xi^d \sim T^{1+d/z}$.

A detailed analysis of the problem is not repeated here as it is covered in Zacharias *et al.* (2010). The lecture was based on this paper. A main message is that the interplay of the two modes turns out to be even more interesting than the arguments given above suggest. Especially important are modes on a length scale L which behave 'classical', $L > T^{-1/z}$, for one z but quantum, $L > T^{-1/z}$, for the other z. The interplay of classical and qunatum regimes lead to qualitative new properties (Zacharias *et al.* 2010) and appears to be typical for theories with multiple dynamical exponents.

5.4.2 Emerging gauge theories

One of the most fascinating ideas in the field of quantum criticality is, that close to quantum critical points completely new degrees of freedom might emerge which cannot be identified with order parameter fluctuations. In high energy physics, gauge theories are often considered as the fundamental building block to describe the interaction of elementary particle. In contrast, in solid state physics, gauge theories are used to describe emergent low-energy degrees of freedom and their interaction. As gauge theories arise naturally, when redundant new degrees of freedom are introduced, the decisive question is whether these degrees of freedom are 'physical'. Many gauge theories (the most prominent example being quantum chromodynamics, QCD) are confining, implying that at low energies the degrees of freedom, for which the gauge theory is formulated, are unphysical. Therefore the central question for emergent gauge theories is, whether they are 'deconfined', which implies that the degrees of freedom used for the formulation of the theory are indeed 'physical'.

In a famous theoretical paper of Senthil *et al.* (2004), the concept of 'deconfined quantum criticality' was put forward as a theory for a second order quantum phase transition from an antiferromagnet to a valence bond solid. According to the Ginzburg–Landau paradigm such a direct transition is generically never of second order but Senthil *et al.* (2004) were able to show that second order transitions arise in a different scenario. In their approach, directly at the quantum critical point, an effective gauge theory emerges with deconfined degrees of freedom. From an alternative point of view (Levin and Senthil 2004), one can view this effective gauge theory as a theory of topological defect with nontrivial quantum numbers. Presently it is not clear whether this theory is realized experimentally.

In this chapter, a substantially simpler and probably more transparent example is discussed: an emergent gauge theory arising from the description of topological defects in nematics (Lammert *et al.* 1993; Lammert *et al.* 1995). Historically, this has been the first example where it has been shown that an emergent gauge theory naturally describes the phase diagram and the critical properties of a simple physical system, namely a liquid of long molecules. While in this example only classical phase transitions are considered, all concepts equally apply to quantum theories.

Surprisingly, the work of Lammert, Rokhsar, and Toner from 1993 has largely been ignored and is only rarely cited, despite its dramatic predictions, its direct experimental relevance, and its clear presentation. Our lecture was directly based on this first paper (Lammert *et al.* 1993) and the reader is encouraged to study the original literature. The clear arguments of this paper will not be repeated here. Instead we will restrict ourselves to describe only the basic ideas.

In a phase transition from a liquid to a nematic state (Chaikin and Lubensky 1995), long molecules align parallel to each other when the temperature is lowered. Here the order parameter is the orientation of the molecules, described by a 'director', that is a vector, where opposite orientations are identified with each other. One important results is that this phase transition is generically of first order in a three dimensional system (as a cubic term arises in the effective Ginzburg–Landau free energy). Surprisingly, this is not the only possibility! Instead it is possible to reach the ordered phase via two second order transitions. It turns out that a simple gauge theory provides the most natural language to describe this physics.

The first important step in the analysis is to realize that the presence or absence of topological defects can play an important role both to describe phase transitions and even phases. For our discussion, the so-called π defect, see Fig. 5.3a, plays the central role. It is a line defect characterized by a net rotation of molecules by the angle π (the relevant homotopy group is $\Pi_1(S_2/Z_2) = Z_2$, (Chaikin and Lubensky 1995)). By increasing the length of the molecules, one can suppress such defects.

What happens if π defects are completely suppressed in a phase where there is no long-range nematic order (we assume, however, short range order so that neighbouring directors are approximately parallel)? In such a situation it is naturally to use a different set of variables to describe the system: one can consistently replace directors by vectors as shown in Figs. 5.3c and d. To this end one starts with one director and assigns an arrow in arbitrary direction. Neighbouring directors which are approximately parallel obtain arrows in the same direction. While in the presence

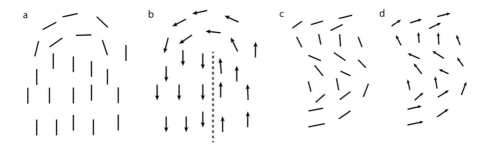

Fig. 5.3 a) π-defect in a nematic. The lines symbolize long molecules (or 'directors'). b) If one tries to decorate the directors by an arrow in such a way that arrows are locally *parallel* to each other, one realizes that this is not possible in the presence of a π defect. Along the dashed line, arrows are antiparallel. For a configuration of molecules without π defects shown in c), one can, in contrast, find in d) a consistent decoration by arrows.

of π defects this necessarily leads to inconsistencies, see Fig. 5.3b, this is not the case if there are no π defects. While this new phase has no long-ranged nematic order, the absence of π defects implies that it is 'topologically ordered'. While no local measurement can distinguish this phase from the conventional disordered phase, its topological properties (here defined by the presence of absence of π defects) are different.

   Probably the most dramatic consequence of the topologically ordered phase is that the ordering transition of a theory of vectors is in the Heisenberg universality class and typically continuous in contrast to the usual transition from a liquid to the nematic which is has to be first order.

   As vectors arise as the natural variables in the problem, a theory of the relevant phases should be formulated in terms of vectors. As the original degree of freedom is a direction, one is forced to formulate the problem as a gauge theory by realizing that a vector multiplied by ± 1 can be identified with a director. Therefore one obtains, as described in detail in (Lammert *et al.* 1993) a so-called Z_2-gauge theory coupled to vectors. Defining vectors S_i on a cubic lattice and Z_2 fields $U_{ij} = \pm 1$ on the link connecting neighbouring lattice sites i and j one obtains a simple model

$$\frac{H}{k_B T} = -J \sum_{\langle i,j \rangle} S_i S_j U_{ij} - K \sum_{\text{plaquettes} \langle ijkl \rangle} U_{ij} U_{jk} U_{kl} U_{ki}. \tag{5.13}$$

The Hamilton function is gauge invariant under the transformation $S_i \rightarrow S_i z_i$, $U_{ij} \rightarrow U_{ij} z_i z_j$ for arbitrary $z_i = \pm 1$. This gauge symmetry reflects that microscopically the order is described by a director and not a vector. J describes that directors favour a parallel alignment and it turns out that, upon increasing K, π-defects are suppressed (Lammert *et al.* 1993). It is not difficult to work out the phase diagram as a function of J and K (Lammert *et al.* 1993; Lammert *et al.* 1995). For small K one obtains the usual first order transition to a nematic phase when J is increased. For $K \rightarrow \infty$ one can, in contrast, choose a gauge where all $U_{ij} = 1$ and one obtains a standard Heisenberg model with a second-order phase transition. For small J one obtains upon increasing K a phase transition from the disordered phase with π defects to a phase without such defects, which can be identified with a confinement–deconfinement transition of the gauge theory (Lammert *et al.* 1993). At this transition, the gauge degrees of freedom become massless. While this transition has the thermodynamic signatures of an Ising transition, there is no local order parameter of any kind. As shown by Lammert *et al.* (1993), it is not difficult to work out the full phase diagram and to relate it to the confinement–deconfinement transition.

This example makes—in an experimentally relevant case—transparent how gauge theories arise naturally as effective low energy degrees of freedom. Gauge theories occur typically when new degrees of freedom with new quantum numbers are introduced. Close to quantum critical points these variables can become the 'natural' degrees of freedom to describe the low-energy physics. A good example of how the emergence of topological defects with non-trivial quantum numbers can lead to new quantum critical points is given by Levin and Senthil (2004).

5.4.3 Concluding remarks

Quantum phase transitions are a very active research area driven both by experiments and theory. Studying systems which are balanced between two ground states is probably the most promising route to discover new states of matter. Also the fact that physical properties are expected to become universal makes this an attractive research area. There are many open problems in the field. One central challenge is to develop methods, for example renormalization group approaches, able to cope with the interplay of multiple degrees of freedom with different dynamics. But even for quantum critical points where thermodynamics properties are well understood, much less is known, for example on transport properties or on the behavior of the system, when parameters are suddenly changed in a quench.

The concepts underlying the theory of quantum phase transitions have also been very useful to describe systems where a connection to quantum critical behaviour is less obvious. For example, it is useful to view graphene (or surface states of a topological insulator) as a quantum critical point with the chemical potential as a tuning parameter. Furthermore, quantum criticality, that is the physics of competing phases, provides a useful language to discuss the properties of complex materials with many competing phases, for example, the high-temperature superconductors. Besides material physics, one can expect that ultracold atoms (Bloch *et al.* 2008) will provide in the future a rich playground for quantum critical physics.

Acknowledgements

I would like to thank especially Markus Garst, Hilbert von Löhneysen, Matthias Vojta and Peter Wölfle for many useful discussion. Financial support by the DFG from the research group FOR 960, *Quantum Phase Transitions*, and the SFB 608 is gratefully acknowledged.

References

Anfuso, F., M., Garst, Rosch, A., Heyer, O., T., Lorenz, Ruegg, C., and K., Kraemer (2008). Spin-spin correlations of the spin-ladder compound $(C_5H_{12}N)_2CuBr_4$ measured by magnetostriction and comparison to quantum monte carlo results. *Phys. Rev. B*, **77**: 235113.

Bloch, I., Dalibard, J., and Zwerger, W. (2008). Many-body physics with ultracold gases. *Rev. Mod. Phys.*, **80**: 885–964.

Chaikin, P. M and Lubensky, T. C. (1995). *Principles of Condensed Matter Physics*. Cambridge University Press, Cambridge.

Coleman, P. and Schofield, A. J. (2005). Quantum criticality. *Nature*, **433**: 226.

Giamarchi, T. and Tsvelik, A. M. (1999). Coupled ladders in a magnetic field. *Phys. Rev. B*, **59**: 11398–11407.

Greiner, M., Mandel, O., Esslinger, T., Hänsch, T.W., and Bloch, I. (2002). Quantum phase transition from a superfluid to a mott insulator in a gas of ultracold atoms. *Nature*, **415**: 39–44.

Hertz, J. A. (1976). Quantum critical phenomena. *Phys. Rev. B*, **14**: 1165–1184.

Lammert, P. E., Rokhsar, D. S., and Toner, J. (1993). Topology and nematic ordering. *Phys. Rev. Lett.*, **70**: 1650–1653.

Lammert, P. E., Rokhsar, D. S., and Toner, J. (1995). Topology and nematic ordering: I. a gauge theory. *Phys. Rev. E*, **52**: 1778–1800.

Levin, M. and Senthil, T. (2004). Deconfined quantum criticality and Neel order via dimer disorder. *Phys. Rev. B*, **70**: 220403.

Lipa, J. A., Swanson, D. R., Nissen, J. A., Chui, T. C. P., and Israelsson, U. E. (1996, Feb). Heat capacity and thermal relaxation of bulk helium very near the lambda point. *Phys. Rev. Lett.*, **76**(6): 944–947.

Löhneysen, H. von, Rosch, A., Vojta, M., and Wölfle, P. (2007). Fermi-liquid instabilities at magnetic quantum phase transitions. *Rev. Mod. Phys.*, **79**: 1015–1075.

Lorenz, T., Heyer, O., Garst, M., Anfuso, F., Rosch, A., Rüegg, Ch., and Kramer, K. (2008). Diverging thermal expansion of the spin-ladder system $(C_5H_{12}N)_2CuBr_4$. *Phys. Rev. Lett.*, **100**: 067208.

Metlitski, M. A. and Sachdev, S. (2010*a*). Quantum phase transitions of metals in two spatial dimensions: I. ising-nematic order. *Phys. Rev. B*, **82**: 075127.

Metlitski, M. A. and Sachdev, S. (2010*b*). Quantum phase transitions of metals in two spatial dimensions: II. spin density wave order. *Phys. Rev. B*, **82**: 075128.

Millis, A. J. (1993). Effect of a nonzero temperature on quantum critical points in itinerant fermion systems. *Phys. Rev. B*, **48**: 7183–7196.

Mross, D. M., McGreevy, J., Liu, H., and Senthil, T. (2010). Controlled expansion for certain non-fermi-liquid metals. *Phys. Rev. B*, **82**: 045121.

Rosch, A. (2001). Some remarks about pseudo gap behavior of nearly antiferromagnetic metals. *Phys. Rev. B*, **64**: 174407.

Rüegg, Ch., Cavadini, N., Furrer, A., Gödel, H.-U., Krämer, K., Mutka, H., Wildes, A., Habicht, K., and Vorderwisch, P. (2003). Bose-einstein condensation of the triplet states in the magnetic insulator $TlCuCl_3$. *Nature*, **423**: 62–65.

Rüegg, Ch., Kiefer, K., Thielemann, B., McMorrow, D.F., Zapf, V.S., Normand, B., Zvonarev, M.B., Bouillot, P., Kollath, C., Giamarchi, T., Capponi, S., Poilblanc, D.,

Biner, D., and Krämer, K.W. (2008). Thermodynamics of the spin luttinger-liquid in a model ladder material. *Phys. Rev. Lett.*, **101**: 247202.

Sachdev, S. (1999). *Quantum Phase Transitions*. Cambridge University Press, Cambridge.

Sachdev, S. and B., Keimer (2011). Quantum criticality. *Physics Today*, **64**: 29.

Senthil, T., Vishwanath, A., L., Balents, Sachdev, S., and Fisher, M. P.A. (2004). Deconfined quantum critical points. *Science*, **303**: 1490–1494.

Stewart, G. R. (2001). Non-fermi-liquid behavior in d- and f-electron metals. *Rev. Mod. Phys.*, **73**: 797.

Zacharias, M., Wölfle, P., and Garst, M. (2010). Multiscale quantum criticality: Pomeranchuk instability in isotropic metals. *Phys. Rev. B*, **80**: 165116.

Zhu, L., Garst, M., Rosch, A., and Q., Si (2003). Universally diverging grüneisen parameter and the magnetocaloric effect close to quantum critical points. *Phys. Rev. Lett.*, **100**: 067208.

6

Cold quantum gases and Bose–Einstein condensation

Robert SEIRINGER

Department of Mathematics and Statistics,
McGill University,
805 Sherbrooke Street West,
Montreal, Quebec H3A 2K6, Canada

Chapter Contents

6.1 Introduction

Bose-Einstein condensation (BEC) in cold atomic gases was first achieved experimentally in 1995 (Anderson *et al.* 1995; Davis *et al.* 1995). After initial failed attempts with spin-polarized atomic hydrogen, the first successful demonstrations of this phenomenon used gases of rubidium and sodium atoms, respectively. Since then there has been a surge of activity in this field, with ingenious experiments putting forth more and more astonishing results about the behaviour of matter at very cold temperatures. BEC has now been achieved by more than a dozen different research groups working with gases of different types of atoms. Literally thousands of scientific articles, concerning both theory and experiment, have been published in recent years.

The theoretical investigation of BEC goes back much further, and even predates the modern formulation of quantum mechanics. It was investigated in two papers by Einstein (Einstein 1925) in 1924 and 1925, respectively, following up on a work by Bose (Bose 1924) on the derivation of Planck's radiation law. Einstein's result, in its modern formulation, can be found in any textbook on quantum statistical mechanics, and was concerned with ideal, that is noninteracting gases.

The understanding of BEC in the presence of interparticle interactions poses a formidable challenge to mathematical physics. Some progress was made in the last ten years or so, and the purpose of these lecture notes is to explain part of what was achieved and how it is related to the actual experiments on cold gases. The content of these notes is naturally strongly biased towards the work of the author and no attempt of completeness is made.

6.2 Quantum many-body systems

6.2.1 The Hamiltonian

For a system of N particles, the Hamiltonian typically takes the form

$$H_N = \sum_{i=1}^{N} \frac{p_i^2}{2m_i} + V(x_1, \ldots, x_N)$$

where m_i is the mass of particle i, $p_i \in \mathbb{R}^3$ and $x_i \in \mathbb{R}^3$ are its momentum and position, respectively, and V denotes the total potential energy. Classically, H is a function on phase space, but in quantum mechanics it becomes a linear operator with the substitution $p_j = -i\hbar\nabla_j$. We shall choose units such that $\hbar = 1$ from now on.

The Hamiltonian H_N is a linear operator on Hilbert space, which is a suitable subspace of $L^2(\mathbb{R}^{3N})$, the square integrable functions of N variables $x_i \in \mathbb{R}^3$. Only a subspace is relevant physically, since for two identical bosons, say i and j, there is the symmetry requirement

$$\psi(x_1, \ldots, x_i, \ldots, x_j, \ldots, x_N) = \psi(x_1, \ldots, x_j, \ldots, x_i, \ldots x_N)$$

For fermions, there is an additional minus sign, that is the wave function is antisymmetric with respect to exchange of coordinates. For simplicity of notation, we ignore here internal degrees of freedom of the particles, like spin, but these could easily be taken into account by adding to the coordinates x_i these additional parameters.

The form of the potential energy depends on the physical system under consideration. Typically, it is a sum of various terms, containing one-particle potentials of the form $\sum_i W(x_i)$, corresponding to an external force, two body interaction potentials $\sum_{i<j} W(x_i, x_j)$ for pairwise interaction, or even some more complicated interactions involving more than two particles at the same time. We will usually assume that

$$V(x_i, \ldots, x_N) = \sum_{i=1}^{N} W(x_i) + \sum_{1 \leq i < j \leq N} v(|x_i - x_j|).$$

6.2.2   Quantities of interest

Given the Hamiltonian H_N of a quantum systems, there are many questions one can try to address. The first one might be concerning its ground state energy, that is the lowest values of the spectrum, which we denote by

$$E_0(N) = \inf \operatorname{spec} H_N$$

If $E_0(N)$ is an eigenvalue, the corresponding ground state wave function ψ_0 is determined by Schrödinger's equation $H_N \psi_0 = E_0(N) \psi_0$.

More generally, if the system is at some positive temperature $T > 0$, one would like to compute the free energy of the system, given by

$$F = -T \ln \operatorname{tr} e^{-H_N/T}$$

We choose units such that Boltzmann's constant equals 1, and small often write $T = 1/\beta$. The trace is over the physical Hilbert space, of course, respecting symmetry constraints arising form the indistinguishability of particles. The equilibrium state at temperature T is the Gibbs state

$$\rho_\beta = e^{-\beta(H-F)}$$

It is normalized to have $\operatorname{tr} \rho_\beta = 1$. For large particle numbers, it is usually hopeless to try to calculate ρ_β directly, but one will try to investigate properties of the reduced n-particle density matrices, obtained by taking the partial trace of ρ_β over $N - n$ variables.

It is often convenient not to fix the particle number N, but rather work in the grand-canonical ensemble, where one takes a certain average over the number of particles in the system. For simplicity, consider a system of just one species of particles. The N-particle Hilbert space, $\mathcal{H}_N$, is then the set of square-integrable functions that are either totally symmetric or antisymmetric under permutations, depending on whether the particles are bosons or fermions.

In the grand-canonical ensemble, one has as Hilbert space the Fock space

$$\mathcal{F} = \bigoplus_{N=0}^{\infty} \mathcal{H}_N$$

Here, $\mathcal{H}_0 = \mathbb{C}$ by definition, and the corresponding vector is called the vacuum vector. As Hamiltonian on Fock space one simply takes

$$H = \bigoplus_{N=0}^{\infty} H_N$$

with H_N the N-particle Hamiltonian. Typically, $H_0 = 0$, that is the vacuum has zero energy.

For $\mu \in \mathbb{R}$, the grand-canonical potential is defined as

$$J = -T \ln \operatorname{tr}_{\mathcal{F}} e^{-\beta(H - \mu N)}$$

where N denotes the number operator, that is

$$N = \bigoplus_{n=0}^{\infty} n$$

Since H is particle number conserving, we can also write this as

$$J = -T \ln \sum_{N \geq 0} z^N \operatorname{tr}_{\mathcal{H}_N} e^{-\beta H_N}$$

where $z = e^{\beta \mu}$ is called the fugacity.

The grand-canonical Gibbs state is

$$\rho_{\beta,\mu} = e^{-\beta(H - \mu N - J)}$$

The chemical potential μ is adjusted to achieve a given average particle number $\langle N \rangle$. The latter equals

$$\langle N \rangle = \operatorname{tr} N \rho_{\beta,\mu} = -\frac{\partial}{\partial \mu} J$$

6.2.3 Creation and annihilation operators on Fock space

On Fock space $\mathcal{F}$, a particularly useful concept are the creation and annihilation operators $a^{\dagger}(f)$ and $a(f)$, with $f \in \mathcal{H}_1$, the one-particle Hilbert space. For any $N \geq 0$, we have

$$a^{\dagger}(f) : \mathcal{H}_N \to \mathcal{H}_{N+1}$$

that is it creates a particle. Likewise, $a(f)$ annihilates a particle, that is,

$$a(f) : \mathcal{H}_N \to \mathcal{H}_{N-1}.$$

Explicitly, they are defined as follows. If ψ_N is an N-particle wave-function in $\mathcal{H}_N$,

$$\left(a^\dagger(f)\psi_N\right)(x_1,\ldots,x_{N+1}) = \frac{1}{\sqrt{N+1}}\sum_{i=1}^{N+1} f(x_i)\psi_N(x_1,\ldots,\cancel{x_i},\ldots,x_{N+1})$$

and

$$\left(a(f)\psi_N\right)(x_1,\ldots,x_{N-1}) = \sqrt{N}\int_{\mathbb{R}^3} \bar{f}(x_N)\psi_N(x_1,\ldots,x_{N-1},x_N)dx_N.$$

This definition works for bosons, for fermions one has to introduce the appropriate (-1) factors to preserve the antisymmetry of the wave functions. One readily checks that these operators satisfy $a(f)^\dagger = a^\dagger(f)$, that is $a^\dagger(f)$ is the adjoint of $a(f)$, as well as the canonical (anti-)commutation relations

$$[a(f), a^\dagger(g)] = \langle f|g\rangle \ , \ [a(f), a(g)] = 0 \ , \ [a^\dagger(f), a^\dagger(g)] = 0$$

Here, $[\cdot, \cdot]$ denotes the usual commutator $[A, B] = AB - BA$ for bosons, while it is the anticommutator $[A, B] = AB + BA$ for fermions.

Consider now a typical many-body Hamiltonian containing one- and two-body terms. For h a one-body operator and W a two body operator, the N-particle Hamiltonian is thus of the form

$$H_N = \sum_{i=1}^{N} h_i + \sum_{1\leq i<j\leq N} W_{ij}$$

where the subscripts indicate what particles the operator acts on. Using creation and annihilation operators, the Fock space Hamiltonian $H = \bigoplus_{N\geq 0} H_N$ can conveniently be written as

$$H = \sum_{i,j}\langle\varphi_i|h|\varphi_j\rangle a_i^\dagger a_j + \frac{1}{2}\sum_{i,j,k,l}\langle\varphi_i \otimes \varphi_j|W|\varphi_k \otimes \varphi_l\rangle a_i^\dagger a_j^\dagger a_k a_l$$

where $\{\varphi_i\}$ is an orthonormal basis of $\mathcal{H}_1$, and $a_i^\dagger = a^\dagger(\varphi_i)$, $a_i = a(\varphi_i)$. A possible choice of the basis $\{\varphi_i\}$ is to diagonalize h, that is

$$\langle\varphi_i|h|\varphi_j\rangle = e_i\delta_{ij}$$

The number operator N is simply $N = \sum_i a_i^\dagger a_i$.

In terms of the creation and annihilation operators, the reduced n-particle density matrices $\gamma^{(n)}$ of a state on Fock space are defined via the expectation values

$$\langle f_1 \otimes \cdots \otimes f_n|\gamma^{(n)}|g_1 \otimes \cdots \otimes g_n\rangle = \langle a^\dagger(g_1)\ldots a^\dagger(g_n)a(f_n)\ldots a(f_1)\rangle.$$

Since product functions span the whole n-particle space, this defines $\gamma^{(n)}$ uniquely. For a state with a fixed particle number, the definition agrees with the previous definition in the canonical ensemble using partial traces (except for an overall normalization factor).

6.2.4   Ideal quantum gases

Consider now an ideal quantum system without interactions. The N-particle Hamiltonian is simply $H_N = \sum_{i=1}^{N} h_i$, where, for example,

$$h = \frac{1}{2m} p^2 = -\frac{1}{2m} \Delta$$

on the cube $[0, L]^3$ with appropriate boundary conditions. In particular, we assume that h has discrete spectrum. Let us denote the eigenvalues of h by e_i,

$$e_0 \le e_1 \le e_2 \le \cdots$$

On Fock space, we then have

$$H = \sum_{i \ge 0} e_i a_i^\dagger a_i$$

and also

$$\beta H - \mu N = \sum_{i \ge 0} \epsilon_i a_i^\dagger a_i$$

with $\epsilon_i = \beta e_i - \mu$.

We wish to calculate

$$\ln \operatorname{tr} e^{-\sum_i \epsilon_i a_i^\dagger a_i}.$$

The spectrum of $\sum_i \epsilon_i a_i^\dagger a_i$ is of the form $\sum_i \epsilon_i n_i$, with $n_i \in \{0, 1, 2, \ldots\}$ for bosons, and $n_i \in \{0, 1\}$ for fermions. Summing over all possible occupation numbers is the same as summing over all eigenstates, hence we have

$$\operatorname{tr} e^{-\sum_i \epsilon_i a_i^\dagger a_i} = \prod_i \sum_n e^{-\epsilon_i n} = \prod_i \begin{cases} (1 - e^{-\epsilon_i})^{-1} & \text{bosons} \\ 1 + e^{-\epsilon_i} & \text{fermions} \end{cases}.$$

For bosons, we have to assume that $\epsilon_i > 0$ for all i for the geometric series to converge. In particular

$$\ln \operatorname{tr} e^{-\sum_i \epsilon_i a_i^\dagger a_i} = \sum_i \mp \ln(1 \mp e^{-\epsilon_i})$$

where $-$ is for bosons and $+$ for fermions.

Consider now an ideal gas in a cubic box of side length L, with periodic boundary conditions. The spectrum of $p^2 = -\Delta$ equals

$$\left(\frac{2\pi}{L} \right)^2 (n_x^2 + n_y^2 + n_z^2)$$

with $(n_x, n_y, n_z) \in \mathbb{Z}^3$. The corresponding eigenstates are the plane waves $e^{ip \cdot x}$, with $p \in \left(\frac{2\pi}{L} \mathbb{Z} \right)^3$. The grand-canonical potential (which equals the negative of the pressure times the volume in this case) thus equals

$$J = \pm T \sum_{p \in (\frac{2\pi}{L}\mathbb{Z})^3} \ln\left(1 \mp e^{-\beta(p^2 - \mu)}\right)$$

where we set the mass of the particles equal to $1/2$ for simplicity.

For bosons, we have to assume that $\mu < 0$. This is not really a restriction, however, as any particle number can be achieved even for negative μ. In fact, the average particle number equals

$$\langle N \rangle = -\frac{\partial}{\partial \mu} J = \sum_{p \in (\frac{2\pi}{L}\mathbb{Z})^3} \underbrace{\frac{1}{e^{\beta(p^2 - \mu)} \mp 1}}_{\langle a_p^\dagger a_p \rangle}$$

Here, the summands are just $\langle a_p^\dagger a_p \rangle$, the average occupation number of momentum p. As μ varies within $(-\infty, 0)$ (for bosons), and $(-\infty, \infty)$ (for fermions), clearly $\langle N \rangle$ varies between 0 and $+\infty$.

We now perform a thermodynamic limit $L \to \infty$. The sum over p can then be interpreted as a Riemann sum for the corresponding integral. In fact,

$$\frac{1}{L^3} \sum_{p \in (\frac{2\pi}{L}\mathbb{Z})^3} \longrightarrow \frac{1}{(2\pi)^3} \int_{\mathbb{R}^3} dp$$

as $L \to \infty$. The thermodynamic pressure of the system is thus

$$P = -\lim_{L \to \infty} \frac{J}{L^3} = \mp \frac{T}{(2\pi)^3} \int_{\mathbb{R}^3} \ln\left(1 \mp e^{-\beta(p^2 - \mu)}\right) dp$$

and the average density equals

$$\rho = \lim_{L \to \infty} \frac{\langle N \rangle}{L^3} = \frac{1}{(2\pi)^3} \int_{\mathbb{R}^3} \frac{1}{e^{\beta(p^2 - \mu)} \mp 1} dp. \tag{6.1}$$

Let us now restrict our attention to the bosonic case, where there is a minus sign in the denominator in (6.1). Notice that the density stays bounded as $\mu \to 0$! That is,

$$\rho_c(\beta) := \lim_{\mu \nearrow 0} \rho = \frac{1}{(2\pi)^3} \int_{\mathbb{R}^3} \frac{1}{e^{\beta p^2} - 1} dp < \infty \tag{6.2}$$

since the integrand behaves like $|p|^{-2}$ for small p, which is integrable in three dimensions.

What is happening here? Recall that μ has to be chosen as to fix the density ρ and, hence, has to depend on L, in general. If $\rho < \rho_c(\beta)$, then $\mu(L) \to \mu < 0$ in the thermodynamic limit. But when $\rho \geq \rho_c(\beta)$, $\mu(L)$ has to tend to zero as $L \to \infty$. In this case, the limits $L \to \infty$ and $\mu \to 0$ must be taken simultaneously and, in particular, do not commute.

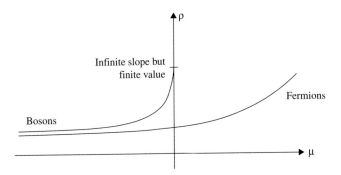

Fig. 6.1 Particle density of an ideal quantum gas at infinite volume, as a function of the chemical potential μ.

In fact, if $\rho > \rho_c(\beta)$, then μ is asymptotically equal to

$$\mu = \left(-\beta L^3(\rho - \rho_c(\beta))\right)^{-1} \quad \text{as } L \to \infty.$$

For this value of μ, we see that

$$\lim_{L \to \infty} \frac{1}{L^3}\langle a_0^\dagger a_0 \rangle = \lim_{L \to \infty} \frac{1}{L^3}\frac{1}{e^{-\beta\mu} - 1} = \rho - \rho_c(\beta).$$

That is, the zero momentum state is occupied by a macroscopic fraction of all the particles. This phenomenon is called **Bose–Einstein condensation** (BEC). It occurs for $\rho > \rho_c(\beta)$, that is, for ρ bigger than the critical density or, equivalently, for

$$T < T_c(\rho) = \frac{4\pi}{\zeta(3/2)^{(2/3)}}\rho^{2/3}$$

since $\rho_c(\beta) = \zeta(3/2)(4\pi)^{-3/2}\beta^{-3/2}$. Here, ζ denotes the Riemann zeta functions

$$\zeta(z) = \sum_{k \geq 1} k^{-z}$$

That is, BEC occurs below a critical temperature $T_c(\rho)$.

We note that only the zero momentum mode is macroscopically occupied, and the other occupations are much smaller. The smallest positive eigenvalue of the Laplacian equals $(2\pi/L)^2$, and

$$\frac{1}{e^{\beta(2\pi/L)^2} - 1} \sim L^2 \ll L^3 \quad \text{for large } L.$$

BEC represents a phase transition in the usual sense, namely that the thermodynamic functions exhibit a nonanalytic behaviour. Consider, for instance, the free energy, which is given in a standard way as the Legendre transform of the pressure.

Specifically, the free energy per unit volume equals

$$f(\beta, \rho) = \mu\rho + \frac{T}{(2\pi)^3} \int_{\mathbb{R}^3} \ln\left(1 - e^{-\beta(p^2 - \mu)}\right) dp \qquad (6.3)$$

where μ is determined by equation (6.1) if $\rho < \rho_c(\beta)$, and $\mu = 0$ if $\rho \geq \rho_c(\beta)$. In the latter case, we see that $f(\beta, \rho)$ does not actually depend on ρ, and is constant for $\rho > \rho_c(\beta)$. In particular, f is not analytic. Intuitively, what is happening as one increases the density beyond $\rho_c(\beta)$ is that all additional particles occupy the zero momentum mode and hence do not contribute to the energy or the entropy, hence also not to the the free energy.

We conclude this section by noting that the grand-canonical ensemble is somewhat unphysical for the ideal Bose gas for $\rho > \rho_c(\beta)$, because of large particle number fluctuations. One readily computes that

$$\langle n_p(n_p - 1)\rangle = 2\langle n_p\rangle^2 \qquad (6.4)$$

for any p, where $n_p = a_p^\dagger a_p$ denotes the number of particles with momentum p. Note that (6.4) is independent of β and μ. It can be easily checked from the explicit form of the Gibbs state, or follows from Wick's rule, for instance.

The factor 2 on the right side of (6.4) is crucial. It means that the variance of the occupation number is of the same order as its value. In particular, if there is a macroscopic occupation, the variance is also macroscopic! By summing over p, we also see that

$$\langle N(N - 1)\rangle = \langle N\rangle^2 + \sum_p \langle n_p\rangle^2.$$

The last term is of the order $\langle N\rangle^2$ if and only if $\rho > \rho_c(\beta)$. That is, there are macroscopically large particle number fluctuations in this case. For real, interacting systems, such fluctuation will typically be suppressed and this problem is not expected to occur.

The macroscopic particle number fluctuations in particular mean that there is not a full equivalence of ensembles for the ideal Bose gas. Nevertheless, working in the canonical ensemble will produce the same free energy (6.3) and the same occupation numbers $\langle n_p\rangle$ in the thermodynamic limit. In particular, the conclusion concerning

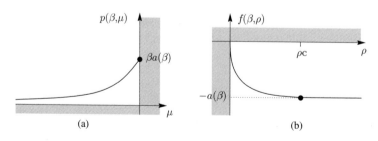

Fig. 6.2   The pressure and the free energy of the ideal Bose gas in three dimensions.

BEC remains the same in the canonical ensemble, although the analysis is much more tedious. (See (Buffet 1983) or also Appendix B of (Ueltschi 2006).)

6.3 BEC for interacting systems

6.3.1 The criterion for BEC

Recall the definition of the one-particle density matrix γ. For $\langle \cdot \rangle$ a state on Fock space

$$\langle g|\gamma|f\rangle = \langle a^\dagger(f)a(g)\rangle.$$

Note that γ is a positive trace class operator on the one-particle Hilbert space $\mathcal{H}_1$, with

$$\operatorname{tr}\gamma = \sum_i \langle \varphi_i|\gamma|\varphi_i\rangle = \left\langle \sum_i a_i^\dagger a_i \right\rangle = \langle N\rangle.$$

This definition applies to any state on Fock space, not only thermal equilibrium states. In particular, one can also consider states of definite particle number, and hence recover the definition for the canonical ensemble.

According to a criterion by Penrose and Onsager (Penrose and Onsager 1956), BEC is said to occur if γ has an eigenvalue of the order of $\langle N\rangle$ of large $\langle N\rangle$. The corresponding eigenfunction is called the **condensate wave-function**.

Since this definition involves large particle numbers $\langle N\rangle$, it refers, strictly speaking, not to a single state but rather a sequence of states for larger and larger system size. When one speaks about the occurrence of BEC one hence always has to specify how various parameters depend on this size.

The standard case where one would like to understand BEC is a translation invariant system at given inverse temperature β and chemical potential μ, in the limit that the system size L tends to infinity. In this case, BEC means that

$$\lim_{L\to\infty} \frac{1}{L^3} \underbrace{\sup_{\|f\|=1} \langle a^\dagger(f)a(f)\rangle_{\beta,\mu}}_{\text{largest eigenvalue of }\gamma} > 0.$$

For translation invariant systems, γ is also translation invariant and hence has an integral kernel of the form

$$\gamma(x-y) = \frac{1}{L^3} \sum_{p\in\left(\frac{2\pi}{L}\mathbb{Z}\right)^3} \gamma_p\, e^{ip(x-y)}$$

where γ_p denote the eigenvalues of γ. Moreover, for a Hamiltonian of the form

$$H = \sum_{i=1}^N p_i^2 + V(x_1,\ldots,x_N)$$

it is also true that γ is positivity improving, meaning that it has a strictly positive integral kernel $\gamma(x-y)$, and hence $\gamma_0 > \gamma_p$ for all $p \neq 0$. Hence the largest eigenvalue is always associated to the constant eigenfunction, and BEC can only occur in the zero momentum mode.

BEC is extremely hard to establish rigorously. In fact, the only known case of an interacting, translation invariant Bose gas where BEC has been proved in the standard thermodynamic limit is the hard-core lattice gas. For completeness, we shall briefly describe it in the following section.

6.3.2   The hard-core lattice gas

For a lattice gas, one replaces the continuous configuration space $\mathbb{R}^3$ of a particle by the cubic lattice $\mathbb{Z}^3$. That is, the one-particle Hilbert space $\mathcal{H}_1$ becomes $\ell^2(\mathbb{Z}^3)$ instead of $L^2(\mathbb{R}^3)$. Other types of lattices are possible, of course, but we restrict our attention to the simple cubic one for simplicity. The appropriate generalization of the Laplacian operator on $\mathbb{R}^3$ is the discrete Laplacian

$$(\Delta\psi)(x) = \sum_e (\psi(x+e) - \psi(x))$$

where the sum is over unit vectors e pointing to the nearest neighbours on the lattice.

We assume that the interaction between particles takes place only on a single site, and that it is sufficiently strong to prevent any two particles from occupying the same site. In this sense, these are hard-core particles. That is the interaction energy is zero if all particles occupy different sites, and $+\infty$ otherwise.

Since there is at most one particle at a site x, we can represent the creation and annihilation operators of a particle at a site x as 2×2 matrices

$$a_x^\dagger = \begin{pmatrix} 0 & 1 \\ 0 & 0 \end{pmatrix}, \qquad a_x = \begin{pmatrix} 0 & 0 \\ 1 & 0 \end{pmatrix}$$

where the vector $\begin{pmatrix} 1 \\ 0 \end{pmatrix}_x$ refers to the state where x is occupied, and $\begin{pmatrix} 0 \\ 1 \end{pmatrix}_x$ to the state where x is empty. Also

$$n_x = a_x^\dagger a_x = \begin{pmatrix} 1 & 0 \\ 0 & 0 \end{pmatrix}.$$

In other words, the appropriate Fock space for this system becomes

$$\mathcal{F} = \bigotimes_{x \in [0,L)^3 \cap \mathbb{Z}^3} \mathbb{C}_x^2$$

and the Hamiltonian (minus the chemical potential times N) equals

$$H = -\sum_{\langle x,y \rangle} a_x^\dagger a_y - \mu \sum a_x^\dagger a_x.$$

Here, $\langle x, y \rangle$ stands for nearest neighbour pairs on the lattice. The diagonal terms in the discrete Laplacian have been dropped for simplicity, as they can be absorbed into the chemical potential μ.

Note that Fock space is finite dimensional! Moreover, the Hamiltonian looks extremely simple, as it is quadratic in the $a_x^\dagger$ and a_x. However, these are not the usual creation and annihilation operators anymore, as they do not satisfy the canonical commutation relations. In fact,

$$[a_x, a_x^\dagger] = 1 - 2n_x$$

and

$$a_x a_x^\dagger + a_x^\dagger a_x = 1.$$

Therefore, at a given site, the system looks like it is fermionic, but for different sites the operators still commute, as appropriate for bosons.

To gain some intuition about the behaviour of the system, it is instructive to rewrite the Hamiltonian H in terms of spin operators. Recall that for a spin $1/2$ particle, the three components of the spin are represented by the $1/2$ times the Pauli matrices, that is, by

$$S^1 = \frac{1}{2} \begin{pmatrix} 0 & 1 \\ 1 & 0 \end{pmatrix}, \qquad S^2 = \frac{1}{2} \begin{pmatrix} 0 & -i \\ i & 0 \end{pmatrix}, \qquad S^3 = \frac{1}{2} \begin{pmatrix} 1 & 0 \\ 0 & -1 \end{pmatrix}$$

If we define, as usual, the spin raising and lowering operators by $S^\pm = S^1 \pm iS^2$, we see that $a^\dagger$ corresponds to S^+, a to S^-, and n to $S^+ S^- = S^3 + 1/2$. Hence, up to an irrelevant constant

$$H = -\sum_{\langle x,y \rangle} S_x^+ S_y^- - \mu \sum_x S_x^3$$

This is known as the (spin $1/2$) XY model. The chemical potential plays the role of an external magnetic field (in the three-direction).

The following theorem establishes the existence of BEC in this system at small enough temperature, for a particular value of the chemical potential.

Theorem 6.1. (Dyson, Lieb, and Simon, 1978) *For $\mu = 0$ and T small enough,*

$$\lim_{L \to \infty} \frac{1}{L^3} \left\langle \underbrace{\left(L^{-3/2} \sum_x S_x^+ \right)}_{a_{p=0}^\dagger} \underbrace{\left(L^{-3/2} \sum_x S_x^- \right)}_{a_{p=0}} \right\rangle > 0 \tag{6.5}$$

We note that $\mu = 0$ corresponds to half-filling, that is, $\langle N \rangle = L^3/2$, and there is a particle-hole symmetry, implying there are half as many particles as there are lattice sites, on average. The result has only been proved in this special case and it is not known how to extend it to $\mu \neq 0$.

Equation (6.5) can be rewritten as

$$\lim_{L\to\infty} \frac{1}{L^6} \sum_{x,y} \left(\langle S_x^+ S_y - \rangle - \underbrace{\langle S_x^+ \rangle \langle S_y^- \rangle}_{=0} \right) > 0$$

since $\langle S_x^+ \rangle = \langle S_y^- \rangle = 0$ by rotation symmetry of H in the 1–2 plane. What this says is that, on average, $\langle S_x^+ S_y - \rangle - \langle S_x^+ \rangle \langle S_y^- \rangle > 0$ even though x and y are macroscopically separated. This property is known as long-range order, and is equivalent to ferromagnetism of the spin system. On average, the value of a spin is zero, but the spins tend to align in the sense that if a spin at some point x points in some directions, all other spins tend to align in the same direction.

The proof of Theorem 6.1 relies crucially on a special property of the system known as reflection positivity. It extends earlier results by Fröhlich, Simon, and Spencer (Fröhlich, Simon, and Spencer, 1976) on classical spin systems, where this property was first used to proof the existence of phase transitions. Reflection positivity holds only in the case of particle-hole symmetry, that is, $\mu = 0$, and hence the proof is restricted to this particular case.

6.4 Dilute Bose gases

6.4.1 The model

In this section, we return to the description of Bose gases in continuous space. For simplicity, let us consider a system of just one species of particles, with pairwise interaction potential. In practice, the gas will consist of atoms, but we can treat them as point particles as long as the temperature and the density are low enough so that excitations of the atoms are rare. The atoms will behave like bosons if the number of neutrons in their nucleus is even, since then they will have an integer total spin.

The Hamiltonian for such a system is

$$H_N = \sum_{i=1}^{N} -\Delta_i + \sum_{1 \leq i < j \leq N} v\left(|x_i - x_j|\right) \tag{6.6}$$

where we again choose units such that $\hbar = 1$ and $m = 1/2$. The particles are confined to a cubic box of side length L, and appropriate boundary conditions have to be chosen to make $-\Delta$ a self-adjoint operator. Usually these are Dirichlet boundary conditions (rigid walls) or periodic boundary conditions (torus).

We assume that the interaction is of short range, by which we mean that

$$\int_{|x| \geq R} v(|x|)dx < \infty$$

for some $R \geq 0$. In other words, v should be integrable at infinity. Locally it can be very strong, however. A typical example is a system of hard spheres of diameter R_0,

where

$$v(|x|) = \begin{cases} +\infty & \text{if } |x| \leq R_0 \\ 0 & \text{if } |x| > R_0 \end{cases} .$$

The interaction has to be sufficiently repulsive to ensure that the system is a gas for low temperatures and densities. In particular, there should be no bound states of any kind. This is certainly the case if $v(|x|) \geq 0$ for all particle separations $|x|$, which we shall assume henceforth.

So far it has not been possible to prove the existence of BEC (in the usual thermodynamic limit) for such a system, even at low density and for weak interaction v.[1] So our goals have to be more modest here. Let us first investigate the ground state energy of the system, that is,

$$E_0(N) = \inf \operatorname{spec} H_N$$

We will be particularly interested in large systems, that is, in the thermodynamic limit

$$\left. \begin{array}{c} L \to \infty \\ N \to \infty \end{array} \right\} \text{ with } \rho = \frac{N}{L^3} \text{ fixed.}$$

At low density, one might expect that the ground state energy is mainly determined by two-particle collisions, and hence

$$E_0(N) \approx \frac{N(N-1)}{2} E_0(2)$$

That is, the energy should approximately equal the energy of just 2 particles in a large box, multiplied by the total number of pairs of particles. We shall compute $E_0(2)$ in the following.

6.4.2 The two-particle case

Consider now two particles in a large cubic box of side length L. Ignoring boundary conditions, the two-particle wave function will be of the form $\psi(x_1, x_2) = \phi(x_1 - x_2)$. Hence

$$\frac{\langle \psi | H_2 | \psi \rangle}{\langle \psi | \psi \rangle} = \frac{\int \left(2|\nabla\phi(x)|^2 + v(|x|)|\phi(x)|^2 \right) dx}{\int |\phi(x)|^2 dx}$$

since the centre-of-mass integration yields L^3 both in the numerator and the denominator. Moreover, since the interaction is short range we can assume that $\phi(x)$ tends to a constant for large $|x|$, and we can take the constant to be 1 without loss of generality. Hence $\int |\phi|^2 = L^3$ to leading order in L.

---

[1] It is possible to prove an upper bound on the critical temperature, however. That is, one can establish the absence of BEC for large enough temperatures, see (Seiringer and Ueltschi 2009).

Definition 6.2 *The scattering length a is defined to be*

$$a = \frac{1}{8\pi} \inf_{\phi} \left\{ \int_{\mathbb{R}^3} \left(2|\nabla\phi(x)|^2 + v(|x|)|\phi(x)|^2 \right) dx \; : \; \lim_{|x|\to\infty} \phi(x) = 1 \right\} \qquad (6.7)$$

Note that integrability of $v(|x|)$ at infinity is equivalent to the scattering length a being finite.

With this definition and the preceding arguments, we see that

$$E_0(2) \approx \frac{8\pi a}{L^3}$$

for large L. Hence we expect that

$$E_0(N) \approx \frac{N(N-1)}{2} E_0(2) \approx 4\pi N a\rho \quad \text{for small } \rho = N/L^3$$

We will investigate the validity of this formula in the next subsection.

We note that the Euler–Lagrange equation for the minimization problem (6.7) is

$$-2\Delta\phi(x) + v(|x|)\phi(x) = 0.$$

This is the zero-energy scattering equation. Asymptotically, as $|x| \to \infty$, the solution is of the form

$$\phi(x) \approx 1 - \frac{a}{|x|}$$

with a the scattering length of v. This is easily seen to be an equivalent definition of a, but we shall find the variational characterization (6.7) to be more useful in the following.

6.4.3 The ground state energy of a dilute gas

Consider the ground state energy per particle, $E_0(N)/N$, of the Hamiltonian (6.6) in the thermodynamic limit

$$e_0(\rho) = \lim_{N\to\infty} \frac{1}{N} E_0(N) \quad \text{with } L^3 = N/\rho$$

Based on the discussion above, we expect that

$$e_0(\rho) \approx 4\pi a\rho$$

for small density ρ. This is in fact true.

Theorem 6.3. (Dyson 1957, Lieb and Yngvason 1998)

$$e_0(\rho) = 4\pi a\rho(1 + o(1))$$

with $o(1)$ going to zero as $\rho \to 0$.

The upper bound was proved by Dyson in 1957 (Dyson 1957) using a variational calculation. He also proved a lower bound, which was 14 times too small, however. The correct lower bound was finally shown in 1998 by Lieb and Yngvason (Lieb and Yngvason 1998).

We remark that the low density limit is very different from the perturbative weak-coupling limit. In fact, at low density the energy is of a particle is very small compared with the strength of v. The interaction potential is hence very strong but short range. First order perturbation theory, in fact, would predict a ground state energy of the form

$$e_0(\rho) = \frac{\rho}{2} \int v(|x|)dx$$

This is strictly bigger than $4\pi a\rho$, as can be seen from the variational principle (6.7); $(8\pi)^{-1} \int v$ is the first order Born approximation to the scattering length a.

The proof of Theorem 6.3 is too lengthy to be given here in full detail, but we shall explain the main ideas. For the upper bound, one can use the variational principle, which says that

$$E_0(N) \leq \frac{\langle \Psi | H_N | \Psi \rangle}{\langle \Psi | \Psi \rangle} \tag{6.8}$$

for any Ψ. As a trial function that captures the right two body physics, one could try a function of the form

$$\Psi(x_1, \ldots, x_N) = \prod_{1 \leq i < j \leq N} \phi(x_i - x_j).$$

The computation of the corresponding energy turns out to be rather tricky, however. One of the reasons for this is that both numerator and denominator on the right side of (6.8) are exponentially small in the particle number N, and hence cancellations have to be taken into account very carefully. Dyson in fact used a slightly different form of the trial function, and his computation of the upper bound fills several pages.

Before explaining the main ideas in the lower bound by Lieb and Yngvason, let us give some intuition as to why this is a hard problem. It is related to the relevant length scales in the system. Since the energy per particle is of the order of $a\rho$, the associated uncertainty principle length ℓ, obtained by setting this energy equal to ℓ^{-2}, equals

$$\ell \sim \frac{1}{\sqrt{a\rho}}$$

At low density ρ, this is

$$\ell \sim \underbrace{\frac{1}{\sqrt{a\rho}}} \quad \gg \quad \underbrace{\rho^{-1/3}}_{\text{mean interparticle separation}} \quad \gg \quad \underbrace{a}_{\text{interaction length}} \quad .$$

Thus, the typical wave functions of a particle are necessarily spread out over a region much bigger than the mean particle distance. The particles hence completely lose their individuality, and behave very quantum (i.e., nonclassical) in this sense. Fermions, on the other hand, behave much more classical, since for them $\ell \sim \rho^{-1/3}$.

The proof of the lower bound on $E_0(N)$ contains two main steps. First, one would like to replace the hard interaction potential v by a soft one, at the expense of kinetic energy. This softer interaction will have a range R, with $a \ll R \ll \rho^{-1/3}$. Specifically, consider $x_2, \ldots, x_N$ to be fixed for the moment, and assume also that $|x_j - x_k| \geq 2R$ for all $j, k \geq 2$. That is, assume that the balls $B_R(x_j)$ of radius R centred at x_j are nonoverlapping. Then

$$
\int \left(|\nabla_1 \psi|^2 + \tfrac{1}{2} \sum_{j \geq 2} v(|x_1 - x_j|)|\psi|^2 \right) dx_1
$$

$$
\geq \sum_{j \geq 2} \int_{B_R(x_j)} \left(|\nabla_1 \psi|^2 + \tfrac{1}{2} v(|x_1 - x_j|)|\psi|^2 \right) dx_1 \tag{6.9}
$$

$$
\geq \sum_{j \geq 2} \int U_R(x_1 - x_j)|\psi|^2 dx_1,
$$

where

$$
U_R(x) = \begin{cases} e(a, R) & |x| \leq R \\ 0 & |x| > R \end{cases}
$$

and $e(a, R)$ is the lowest eigenvalue of $-\Delta + \tfrac{1}{2} v$ on B_R, with Neumann boundary conditions. As we have already argued in Subsection 6.4.2, the latter is easily seen to be equal to

$$
e(a, R) \approx \frac{4\pi a}{|B_R|} \quad \text{for } a \ll R, \tag{6.10}
$$

where $|B_R| = (4\pi/3)R^3$ is the volume of B_R.

This is the desired replacing of v by the soft potential U_R. Repeating the above argument for all other particles, we conclude that

$$
H_N \geq \sum_{i \neq j} U_R(x_i - x_j)\chi \tag{6.11}
$$

where χ is a characteristic function that makes sure that the balls above do not intersect. That is, χ removes three-body collisions. In other words, when three particles come close together, we just drop part of the interaction energy. Since $v \geq 0$, this is legitimate for a lower bound.

The soft potential U_R now predicts the correct energy in first order perturbation theory. In fact, for a constant wave function, the expected value of the right side of (6.11) is approximately equal to $4\pi a N^2/L^3$, with small corrections coming from χ, the region close to the boundary of the box $[0, L]^3$, as well as the fact that (6.10) is only valid approximately.

To make this perturbative argument rigorous, one keeps a bit of the kinetic energy, and uses

$$H_N \geq -\epsilon \sum_{i=1}^{N} \Delta_i + (1 - \epsilon) \sum_{i \neq j} U_R(x_i - x_j)\chi \tag{6.12}$$

(using positivity of v). First order perturbation theory can easily be seen to be correct if the perturbation is small compared to the gap above the ground state energy of the unperturbed operator. The gap in the spectrum of $-\epsilon \sum_{i=1}^{N} \Delta_i$ above zero is of the order ϵ/L^2, which has to be compared with $aN\rho$. That is, if

$$L^3 a\rho^2 \ll \frac{\epsilon}{L^2} , \quad \text{or} \quad L^5 \ll \frac{\epsilon}{a\rho^2} \tag{6.13}$$

then the second term in (6.12) is truly a small perturbation to the first term and first order perturbation theory can be shown to yield the correct result for the ground state energy.[2]

Condition (6.13) is certainly not valid in the thermodynamic limit. To get around this problem, one divides the large cube $[0, L]^3$ into many small cubes of side length ℓ, with ℓ satisfying

$$\ell^5 \ll \frac{\epsilon}{a\rho^2} = \rho^{-5/3} \frac{\epsilon}{a\rho^{1/3}}.$$

For an appropriate choice of ϵ, the last fraction is big, hence ℓ can be chosen much larger than the mean particle spacing $\rho^{-1/3}$.

Dividing up space and distributing particles optimally over the cells gives a lower bound to the energy, due to the introduction of additional Neumann boundary conditions on the boundary of the cells. That is,

$$E_0(N, L) \geq \min_{\{n_i\}} \sum_i E_0(n_i, \ell)$$

where the minimum is over all distribution of the $N = \sum_i n_i$ particles over the small boxes. Since the interaction is repulsive, it is best to distribute the particles uniformly over the boxes. Hence

$$E_0(N, L) \geq \left(\frac{\ell}{L}\right)^3 E_0(\rho\ell^3, \ell).$$

For our choice of ℓ, we have $E_0(\rho\ell^3, \ell) \approx 4\pi a\rho^2\ell^3$, as explained above, and hence

$$E_0(N, L) \approx \left(\frac{\ell}{L}\right)^3 4\pi a\rho^2\ell^3 = 4\pi aN\rho.$$

---

[2] Strictly speaking, it is not the expectation value of the perturbation that is the relevant measure of its smallness, but rather the variance. Hence the condition for validity of perturbation theory is slightly more stringent than what is displayed here.

This concludes the proof, or at least the sketch of the main ideas.

6.4.4   Further rigorous results

Extending the method presented in the previous subsection, further results about the low-density behaviour of quantum gases have been proved. These include

- **Two-dimensional Bose gas.** For a Bose gas in two spatial dimensions, it turns out that (Lieb and Yngvason 2001)

$$e_0(\rho) = \frac{4\pi\rho}{|\ln(a^2\rho)|} \quad \text{for } a^2\rho \ll 1$$

An interesting feature of this formula is that it does not satisfy $E_0(N) \approx \frac{1}{2}N(N-1)$ $E_0(2)$, as it does in three dimensions. The reason for the appearance of the logarithm is the fact that the solution of the zero energy scattering equation

$$-\Delta\phi(x) + \tfrac{1}{2}v(|x|)\phi(x) = 0$$

in two dimensions does not converge to a constant as $|x| \to \infty$, but rather goes like $\ln(|x|/a)$, with a the scattering length.
- **Dilute Fermi gases.** For a (three-dimensional) Fermi gas at low density ρ, one has (Lieb, Seiringer, and Solovej, 2005)

$$e_0(\rho) = \frac{3}{5}\left(\frac{6\pi^2}{q}\right)^{2/3}\rho^{2/3} + 4\pi a\left(1 - \frac{1}{q}\right)\rho + \text{higher order in } \rho$$

Here, q is the number of spin states, that is, the fermions are considered to have spin $\frac{1}{2}(q-1)$. The first term is just the ground state energy of an ideal Fermi gas. The leading order correction due to the interaction is the same as for bosons, except for the presence of the additional factor $(1 - q^{-1})$. Its presence is due to the fact that the interaction between fermions in the same spin states is suppressed, since for them the spatial part of the wave function is antisymmetric and hence vanishes when the particles are at the same location.

A similar result can also be obtained for a two-dimensional Fermi gas (Lieb, Seiringer, and Solovej, 2005).
- **Bose Gas at positive temperature.** For a dilute Bose gas at positive temperature $T = 1/\beta$, the natural quantity to investigate is the free energy. For an ideal Bose gas, the free energy per unit volume is given by (6.3), and we shall denote this expression by $f_0(\beta, \rho)$. For an interaction gas, one has

$$f(\beta, \rho) = \underbrace{f_0(\beta, \rho)}_{\text{ideal gas}} + 4\pi a\left(2\rho^2 - [\rho - \rho_c(\beta)]_+^2\right) + \text{higher order} \tag{6.14}$$

and this formula is valid for $a^3\rho \ll 1$ but $\beta\rho^{2/3} \gtrsim O(1)$. Here, $\rho_c(\beta)$ is the critical density for BEC of the ideal gas, given in (6.2), and $[\cdot]_+$ denotes the positive part. That is, $[\rho - \rho_c(\beta)]_+$ is nothing but the condensate density (of the ideal gas).

Since $\rho_c(\beta) \to 0$ as $\beta \to \infty$, (6.14) reproduces the ground state energy formula $4\pi a \rho^2$ in the zero-temperature limit. Above the critical temperature, that is, for $\rho < \rho_c(\beta)$, the leading order correction is $8\pi a \rho^2$ instead of $4\pi a \rho^2$. The additional factor 2 can be understood as arising from the symmetry requirement on the wavefunctions. Because of symmetrization, the probability that 2 bosons are at the same locations is twice as big than on average. This applies only to bosons in different modes, however, since if they are in the same mode, symmetrization has no effect. Hence the subtraction of the square of the condensate density, which does not contribute to the factor 2.

The lower bound on $f(\beta, \rho)$ of the form (6.14) was proved in (Seiringer 2008), and the corresponding upper bound in (Yin 2010). Both articles are rather lengthy and involved, and there are lots of technicalities to turn the above simple heuristics into rigorous bounds. A corresponding result can also be obtained for fermions, as was shown in (Seiringer 2006).

For further results and more details, we refer the interested reader to (Lieb *et al.* 2005).

6.4.5 The next order term

One of the main open problems concerning the ground state energy of a dilute Bose gas concerns the next order term in an expansion for small ρ. It is predicted to equal

$$e_0(\rho) = 4\pi a \rho \left(1 + \frac{128}{15\sqrt{\pi}} \left(a^3 \rho \right)^{1/2} + \text{higher order} \right) \tag{6.15}$$

This formula was first derived by Lee–Huang–Yang (Lee and Yang 1957, Lee, Huang, and Yang 1957), but is essentially already contained in Bogoliubov's famous 1947 paper (Bogoliubov 1947). The correction term in (6.15) does not have a simple heuristic explanation, but is a truly quantum-mechanical many-body correlation effect.

The way Bogoliubov arrived at this prediction is the following. The starting point is the Hamiltonian on Fock space. We use plane waves as a basis set, and assume periodic boundary conditions. Then

$$H = \sum_p p^2 a_p^\dagger a_p + \frac{1}{2V} \sum_{p,r,s} \widehat{v}(p) a_{s+p}^\dagger a_{r-p}^\dagger a_r a_s$$

where $V = L^3$ is the volume and

$$\widehat{v}(p) = \int_{\mathbb{R}^3} v(|x|) e^{-ip \cdot x} dx$$

denotes the Fourier transform of v. All sums are over $\left(\frac{2\pi}{L} \mathbb{Z} \right)^3$. Bogoliubov introduced two approximations, based on the assumption that in the ground state most particles occupy the zero momentum mode. For this reason, one first neglects all terms in H that are higher than quadratic in $a_p^\dagger$ and a_p for $p \neq 0$. Second, one replaces all $a_0^\dagger$ and a_0 by a number $\sqrt{N_0}$, since these operators are expected to have macroscopic values, while there commutator is only one and hence negligible.

The resulting expression for H is the Bogoliubov Hamiltonian

$$H_B = \frac{N^2 - (N - N_0)^2}{2V} \hat{v}(0)$$

$$+ \sum_{p \neq 0} \left[\left(p^2 + \frac{N_0}{V} \hat{v}(p) \right) a_p^\dagger a_p + \frac{N_0}{2V} \left(a_p^\dagger a_{-p}^\dagger + a_p a_{-p} \right) \right] \qquad (6.16)$$

This Hamiltonian is now quadratic in creation and annihilation operators, and can be diagonalized easily with the help of a Bogoliubov transformation. The resulting expression for the ground state energy per particles in the thermodynamic limit is

$$4\pi\rho(a_0 + a_1) + 4\pi\rho a_0 \frac{128}{15\sqrt{\pi}} \left(a_0^3 \rho \right)^{1/2} + \text{higher order in } \rho \qquad (6.17)$$

where a_0 and a_1 are, respectively, the first and second order Born approximation to the scattering length a. Explicitly,

$$a_0 = \frac{1}{8\pi} \int_{\mathbb{R}^3} v(|x|) dx$$

and

$$a_1 = -\frac{1}{(8\pi)^2} \int_{\mathbb{R}^6} \frac{v(|x|)v(|y|)}{|x-y|} dx\, dy$$

Moreover, in the ground state

$$\left\langle \sum_{p \neq 0} a_p^\dagger a_p \right\rangle \approx N\sqrt{a^3 \rho}$$

hence $(N - N_0)^2 \approx N^2 a^3 \rho$ is negligible to the order we are interested in.

The expression (6.17) looks like an expansion of $e_0(\rho)$ simultaneously in small density and weak coupling. It is hence reasonable to expect the validity of (6.15) without the weak coupling assumption. The proof of this fact is still an open problem, however. For smooth interaction potentials, an upper bound of the correct form was recently proved in (Yau and Yin 2009). There was also some recent progress in (Giuliani and Seiringer 2009) concerning the lower bound, where it was shown that Bogoliubov's approximation is correct, as far as the ground state energy is concerned, if one is allowed to rescale the interaction potential v with ρ is a suitable way.

We remark that the Bogoliubov Hamiltonian (6.16) not only gives a prediction about the ground state energy, but also about the excitation spectrum. Diagonalizing H_B leads to an excitation spectrum of the form

$$\sqrt{p^4 + 2p^2 \rho\hat{v}(p)}$$

which is linear for small momentum p. The nonzero slope at $p = 0$ is in fact extremely important physically and has many interesting consequences, concerning superfluidity, for instance. It is also confirmed experimentally. A rigorous proof that the Bogoliubov approximation indeed predicts the correct low energy excitation spectrum is still lacking, however. With the notable exception of exactly solvable models in one dimension (Girardeau 1960; Lieb and Liniger 1963; Calogero 1969; Sutherland 1971), the only case where Bogoliubov's prediction about the excitation spectrum is rigorously verified is the mean-field or Hartree limit, where the repulsive interaction is very weak and of long range (Seiringer 2011).

6.5 Dilute Bose gases in traps

6.5.1 The Gross–Pitaevskii energy functional

Actual experiments on cold atomic gases concern inhomogeneous systems, since the particles are confined to a trap with soft walls. Let us extend the analysis of the previous sections to see what happens in the inhomogeneous case.

Let $V(x)$ denote the trap potential, and $\rho(x) = |\phi(x)|^2$ the particle density at a point $x \in \mathbb{R}^3$. If V varies slowly, we can use a local density approximation and assume the validity of the formula $4\pi a \rho^2$ for the energy density of a dilute gas even locally. In this way, we arrive at the expression

$$\mathcal{E}^{\mathrm{GP}}(\phi) = \int_{\mathbb{R}^3} |\nabla\phi(x)|^2 dx + \int_{\mathbb{R}^3} V(x)|\phi(x)|^2 dx + 4\pi a \int_{\mathbb{R}^3} |\phi(x)|^4 dx \qquad (6.18)$$

which is known as the *Gross–Pitaevskii (GP) functional*. The last two terms are simply the trap energy and the interaction energy density of a dilute gas in the local density approximation. The first gradient term is added to ensure accuracy even at weak or zero interaction. In fact, for an ideal, $a = 0$ and hence (6.18) is certainly the correct description of the energy of the system in this case.

Minimizing (6.18) under the normalization constraint $\int |\phi(x)|^2 dx = N$ leads to the GP energy

$$E^{\mathrm{GP}}(N, a) = \min\left\{\mathcal{E}^{\mathrm{GP}}(\phi) : \int |\phi(x)|^2 dx = N\right\} \qquad (6.19)$$

Using standard techniques of functional analysis (see Appendix A of (Lieb, Seiringer, and Yngvason, 2000)) one can show that there is a minimizer for this problem, which is moreover unique up to a constant phase factor. This holds under suitable assumptions on the trap potential $V(x)$, for example, if V is locally bounded and tends to infinity as $|x| \to \infty$. The minimizer satisfies the corresponding Euler–Lagrange equation

$$-\Delta\phi + V\phi + 8\pi a|\phi|^2\phi = \mu\phi,$$

which is a nonlinear Schrödinger equation called the GP equation. The chemical potential μ equals $\partial E^{\mathrm{GP}}/\partial N$ and is the appropriate Lagrange parameter to take the normalization condition on ϕ into account.

Based on the discussion above, one would expect that

$$E_0 \approx E^{\mathrm{GP}}$$

and also

$$\rho_0(x) \approx |\phi^{GP}(x)|^2$$

where E_0 and ρ_0 are the ground state energy and corresponding particle density, respectively. This approximation should be valid if V varies slowly and the gas is sufficiently dilute.

Notice that the GP energy $E^{GP}(N, a)$ and the corresponding minimizer $\phi^{GP}_{N,a}$ satisfy the simple scaling relations

$$E^{GP}(N, a) = N E^{GP}(1, Na) \quad \text{and} \quad \phi^{GP}_{N,a}(x) = \sqrt{N}\phi^{GP}_{1,Na}(x)$$

that is, Na is the only relevant parameter for the GP theory. In particular, for the purpose of deriving the GP theory from the many-body problem, it makes sense to take N large while Na is fixed. The latter quantity should really be thought of as Na/L, where L is the length scale of the trap V. Hence $a/L \sim N^{-1}$, that is, V varies indeed much slower that the interaction potential. We shall choose units to make $L = 1$, which simplifies the notation.

Since V is now fixed, we have to rescale the interaction potential v. The appropriate way to do this is to write

$$v_a(|x|) = \frac{1}{a^2}w(|x|/a)$$

for some fixed w. It is then easy to see that v_a has scattering length a if w has scattering length 1. The appropriate many-body Hamiltonian under consideration is thus

$$H_N = \sum_{i=1}^{N}(-\Delta_i + V(x_i)) + \sum_{i<j}v_a(|x_i - x_j|).$$

In this way, a enters as a parameter which can now be varied with N. In particular, the ground state energy $E_0 = \inf \operatorname{spec} H_N$ is now a function of N and a. We shall therefore write $E_0(N, a)$, but suppress the dependence on N and a of the ground state density $\rho_0(x)$ in the notation for simplicity.

Theorem 6.4. (Lieb, Seiringer, and Yngvason, 2000)

$$\lim_{N\to\infty}\frac{E_0(N, g/N)}{N} = E^{GP}(1, g) \quad \text{for any } g \geq 0$$

In the same limit

$$\lim_{N\to\infty}\frac{1}{N}\rho_0(x) = |\phi^{GP}_{1,g}(x)|^2$$

Note that in the limit under consideration $a^3\bar{\rho} \sim N^{-2}$, where $\bar{\rho} \sim N$ denotes the average density. In particular, the gas is very dilute if $g = Na$ is fixed. The result of Theorem 6.4 is actually uniform in g as long $a^3\bar{\rho} \to 0$ as $N \to \infty$. That is, g is allowed to go to ∞ with N at a suitable rate, as long as the gas stays dilute.

The proof of Theorem 6.4 is similar to the homogeneous case, and uses the same ideas in the lower bound. In particular, space is divided up into small boxes and the particles are then distributed optimally over these boxes. In the inhomogeneous case consider here, the distribution will be nonuniform, of course.

6.5.2   BEC of dilute trapped gases

So far the discussion has focused on the ground state energy and the corresponding particle density. But what about BEC? As discussed in Section 6.3, BEC is a property of the reduced one-particle density matrix of the system. Specifically, if Ψ_0 is the ground state of H_N, then the one particle density matrix γ is the operator on $L^2(\mathbb{R}^3)$ with integral kernel

$$\gamma_0(x, x') = N \int \Psi_0(x, x_2, \dots, x_N) \Psi_0(x', x_2, \dots, x_N)^* dx_2 \cdots dx_N$$

Recall that γ is a positive trace-class operator, with $\operatorname{tr} \gamma = N$.

Theorem 6.5. (Lieb and Seiringer, 2002) *In the same limit as in Theorem 6.4*

$$\lim_{N \to \infty} \frac{1}{N} \gamma_0 = |\phi_{1,g}^{\mathrm{GP}}\rangle \langle \phi_{1,g}^{\mathrm{GP}}|$$

What the theorem says is that there is complete BEC, in the sense that the largest eigenvalue of γ_0, divided by N, is not only nonzero but actually equal to one in the dilute limit considered. That is, the one-particle density matrix becomes a rank-one projection in the limit, just like for a noninteracting gas. The condensate wave function $\phi_{1,g}^{\mathrm{GP}}$ still depends on the interacting strength via g, however, and might have very little overlap with the noninteracting state at $g = 0$ if g is large.

Theorem 6.5 represents the only known case of a continuous system with genuine interactions where BEC has been proved. The proof is so far restricted to zero temperature and to the very dilute limit where $a^3 \bar{\rho} \sim N^{-2}$ as $N \to \infty$.

Before discussing the proof of Theorem 6.5, we shall first generalize the setting in a nontrivial way by allowing the system to rotate about a fixed axis. Theorem 6.5 can thus be considered as a special case of a more general result to be discussed next.

6.5.3   Rotating Bose gases

An interesting property of dilute cold Bose gases is their response to rotation. In fact, rotating Bose–Einstein condensates are nowadays routinely created in the lab, by stirring the system much like coffee with a spoon.

Even though the system under consideration is now rotating, we can still think of it as being at equilibrium if we go to the rotating frame of reference. Just like in classical mechanics, the only effect of this transformation on the Hamiltonian is to add a term proportional to the total angular momentum. More precisely,

$$H_N \longrightarrow H_N - \Omega \cdot L$$

where $\Omega \in \mathbb{R}^3$ denotes the angular velocity (having an axis and a magnitude) and $L = \sum_{i=1}^{N} L_i$ denotes the total angular momentum of the system.

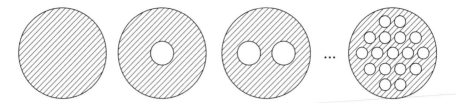

Fig. 6.3 Quantized vortices in a rotating Bose condensate, showing up as holes in the density. More and more vortices appear as the angular velocity is increased. For actual snapshots of experiments, see http://jila.colorado.edu/bec/hi_res_pic_album_macromedia or http://www.bec.nist.gov/gallery.html.

In the experiments on rotating gases, one observes the appearance of quantized vortices, related to the superfluid properties of the system. This is schematically sketched in Fig. 6.3.

The quantized vortices can also be seen by minimizing the appropriate GP functional, which now reads

$$\mathcal{E}^{\mathrm{GP}}(\phi) = \langle \phi | -\Delta + V(x) - \Omega \cdot L | \phi \rangle + 4\pi a \int_{\mathbb{R}^3} |\phi(x)|^4 dx$$

with $L = -ix \wedge \nabla$, as usual. In order for the confining force to overcome the centrifugal force, we have to assume that

$$V(x) - \frac{1}{4}|\Omega \wedge x|^2 \tag{6.20}$$

is bounded below and goes to infinity at infinity. Under this condition, one can still prove the existence of a minimizer of the GP functional. In general it will not be unique anymore, however. This nonuniqueness is related to spontaneous symmetry breaking. In fact, if V is rotation symmetric about the Ω axis, that is $[V, \Omega \cdot L] = 0$, then $\mathcal{E}^{\mathrm{GP}}$ is invariant under rotations about this axis. In general, a minimizer ϕ^{GP} will not have this symmetry, however, due to the appearance of quantized vortices. If there is more than one, these obviously can not be arranged in a symmetric way. That is in general we expect a whole continuum of minimizers in the case the GP functional is rotation symmetric.

The N-body Hamiltonian under consideration now is

$$H_N = \sum_{i=1}^{N}(-\Delta_i + V(x_i) - \Omega \cdot L_i) + \sum_{i<j} v_a(|x_i - x_j|).$$

It ground state energy will be denotes by $E_0(N, a, \Omega)$.

Theorem 6.6. (Lieb and Seiringer, 2006) *For any $g \geq 0$ and $\Omega \in \mathbb{R}^3$ (subject to the constraint that (6.20) in bounded from below and goes to infinity at infinity)*

$$\lim_{N \to \infty} \frac{E_0(N, g/N, \Omega)}{N} = E^{\mathrm{GP}}(1, g, \Omega)$$

Moreover, up to a subsequence, the one-particle density matrix of a ground state (or any approximate ground state, in fact) satisfies

$$\lim_{N \to \infty} \frac{1}{N} \gamma = \int d\mu \, |\phi^{\mathrm{GP}}\rangle \langle \phi^{\mathrm{GP}}| \tag{6.21}$$

where $d\mu$ is a probability measure on the set of minimizers of $\mathcal{E}^{\mathrm{GP}}$.

By an approximate minimizer we mean a state that has the same energy as the ground state energy, to leading order. In other words, a state with energy equal to

$$\lim \frac{1}{N} \langle H_N \rangle = E^{\mathrm{GP}}(1, g, \Omega)$$

in the limit $N \to \infty$, $Na \to g$.

Equation (6.21) is the natural generalization of complete BEC in the case of nonuniqueness of GP minimizers. Because of the linearity of quantum mechanics, the best one can hope for is a convex combination of completely condensed states. In fact, (6.21) can also be seen as establishing the spontaneous symmetry breaking mentioned earlier. Under an infinitesimal perturbation, the GP functional will generically have a unique minimizer, and Theorem 6.6 in this case implies that there is then complete BEC in the usual sense.

We note that the bosonic symmetry requirement on the N-particle wave functions is crucial for Theorem 6.6 to hold. In contrast, for the discussion of the ground state of nonrotating systems, Bose symmetry does not have to be enforced explicitly, it comes out automatically as the ground state of an operator of the form $-\Delta + W(x)$ is always unique and positive and hence can only be permutation symmetric. For rotating systems, this is generally not the case, and Bose symmetry cannot be ignored.

6.5.4   Main ideas in the proof

We split the proof of Theorem 6.6 into three parts.

Step 1. The first step is again to try to replace the hard interaction potential $v_a(|x|)$ by a softer one, $U_R(|x|)$, at the expense of some kinetic energy. We must not use up all the kinetic energy as we did in the homogeneous case, however, since we still need to obtain the gradient term in the GP functional. The key idea is to split the kinetic energy into a high momentum and a low momentum part. Only the high momentum part $|p| \geq p_c$ will be needed to achieve the replacement $v_a \to U_R$, while the low momentum part $|p| \leq p_c$ will kept as it is needed in the GP functional. We will, in fact, choose

$$1 \ll p_c \ll \frac{1}{R} \tag{6.22}$$

The first condition implies that only momenta irrelevant for the GP functional are being used, while the second makes sure that all momenta relevant on the length scale of U_R are actually employed. The crucial Lemma that improves (7.9) is the following. Its proof is in (Lieb, Seiringer, and Solovej, 2005).

Lemma 6.7 *Let* $\chi_{B_R(0)}$ *denote the characteristic function of the ball of radius* R *centered at the origin. For any* $0 < \epsilon < 1$,

$$-\nabla \cdot \xi(p)\chi_{B_R(0)}(x)\xi(p)\nabla + \frac{1}{2}v(|x|) \geq (1-\epsilon)U_R(|x|) - \frac{1}{\epsilon}w_R(x) \tag{6.23}$$

where

$$w_R(x) = \frac{2a}{\pi^2}f_R(x)\int_{\mathbb{R}^3} f_R(y)dy$$

and $f_R(x) = \sup_{|y| \leq R}|h(x-y) - h(x)|$, $\widehat{h}(p) = 1 - \xi(p)$.

The function ξ is chosen to be a smooth characteristic function of the set $\{|p| \geq p_c\}$. Hence the first term in (6.23) is a version of the Laplacian that has been restricted to high momentum and localized to a ball of radius R (centred at the origin). The price one has to pay for the cutoff ξ is the error term w_R, which is supported also outside the ball but can be made to decay very fast by choosing ξ smooth. For our choice of p_c in (6.22), it will be negligible compared to U_R.

Lemma 6.7 implies the operator lower bound

$$H_N \geq \sum_{i=1}^{N}\left(-\Delta_i(1 - \xi(p_i)^2) + V(x) - \Omega \cdot L_i\right)$$

$$+ \sum_{i\neq j}\left((1-\epsilon)U_R(|x_i - x_j|) - \frac{1}{\epsilon}w_R(|x_i - x_j|)\right)\chi \tag{6.24}$$

where χ is again a characteristic function that excludes three and more particle collisions.

Step 2. In order to proceed, we want to get rid of both the w_R term and the characteristic function χ in (6.24). For this purpose, we need some *a priori* bounds that tell us that the expected values of w_R and $1 - \chi$ in the ground state of H_N are not too big. For this purpose, we obtained some rough bounds on the three-particle density of a ground state of H_N, using path integrals. These bounds are of the form

$$\langle f(x_1, x_2, x_3)\rangle \leq \Lambda(\alpha, f)\, e^{\alpha(E_0(N) - E_0(N-3))}$$

where $\langle \cdot \rangle$ denotes expectation in the zero-temperature state, f is an arbitrary positive bounded function, $\alpha > 0$ is arbitrary and $\Lambda(\alpha, f)$ denotes the largest eigenvalue of the operator

$$\sqrt{f}e^{-\alpha(-\Delta_1 - \Delta_2 - \Delta_3 + V(x_1) + V(x_2) + V(x_3))}\sqrt{f}$$

on $L^2(\mathbb{R}^9)$. This bound is certainly not optimal, but suffices to show that the terms in question due not contribute to the ground state energy to the order we are interested. That is, we conclude that

$$\inf \text{spec } H_N \geq \inf \text{spec } \widetilde{H}_N - \delta N$$

where $\delta \to 0$ in the limit considered. Moreover,

$$\tilde{H}_N = \sum_{i=1}^{N} \left(-\Delta_i (1 - \xi(p_i)^2) + V(x) - \Omega \cdot L_i \right) + \sum_{i \neq j} U_R(|x_i - x_j|) \qquad (6.25)$$

That is, we have managed to genuinely replace the hard interaction potential v_a by the soft one U_R, at the expense of the high-momentum part of the kinetic energy, as well as a minor shift in the ground state energy.

Step 3. Let us denote the one-particle part of the Hamiltonian $\tilde{H}_N$ by h for simplicity, that is,

$$h = -\Delta(1 - \xi(p)^2) + V(x) - \Omega \cdot L$$

In second quantized form, using as a basis the eigenstates of h, we have

$$\tilde{H} = \sum_i \langle \varphi_i | h | \varphi_i \rangle a_i^\dagger a_i + \sum_{ijkl} \langle \varphi_i \otimes \varphi_j | U_R | \varphi_k \otimes \varphi_l \rangle a_i^\dagger a_j^\dagger a_k a_l \qquad (6.26)$$

Notice that if we ignore all commutators between the $a_i^\dagger$ and a_i and treat them as numbers, z_i^* and z_i, respectively, (6.26) becomes

$$\langle \Phi | h | \Phi \rangle + \int_{\mathbb{R}^6} |\Phi(x)|^2 U_R(|x - y|) |\Phi(y)|^2 \, dx \, dy$$

with

$$\Phi(x) = \sum_i z_i \varphi_i(x).$$

This is essentially the GP functional, except for the cutoff in the kinetic energy, which is irrelevant for $p_c \gg 1$, and the fact that the interaction is U_R instead of $4\pi a \delta$. Since $R \ll 1$, however, and $\int U_R = 4\pi a$, U_R is an approximate δ function with the correct coefficient.

In other words, the GP functional emerges from the many-body Hamiltonian on Fock space in a classical limit, replacing all the creation and annihilation operators by complex numbers. In this sense, GP theory is a classical field approximation to the quantum field theory defined by $\tilde{H}$. Note that this is only true for the low momentum part, however. It is important that we have already completed Step 1 above to replace the true interaction potential v_a by U_R. If we had not done so, the classical approximation would also look like a GP functional, but with the wrong coefficient $\frac{1}{2} \int v$ instead of $4\pi a$ in front of the quartic term.

What remains to be done is to investigate the validity of the replacement of the creation and annihilation operators by numbers. This can be conveniently done using coherent states. We shall describe what these are in the next subsection, and complete the sketch of the proof of Theorem 6.6 there.

6.5.5   Coherent states

With $|0\rangle$ denoting the Fock space vacuum, and $z \in \mathbb{C}$, consider the state on Fock space

$$|z\rangle = e^{za^\dagger - z^* a}|0\rangle$$

where a and $a^\dagger$ are the annihilation and creation operators for one particular mode. Since the exponent is anti-hermitian, $|z\rangle$ is a vector of length one. Because of $[a, a^\dagger] = 1$, one can rewrite it also as

$$|z\rangle = e^{-|z|^2/2} e^{za^\dagger}|0\rangle.$$

This state is a superposition of all states with different particle number in the mode under consideration. As z varies over the complex plane $\mathbb{C}$, the states $|z\rangle$ span the whole Fock space associated with the mode a. In fact, one can easily check the completeness relation

$$\int_{\mathbb{C}} \frac{dz}{\pi} |z\rangle\langle z| = 1$$

where dz stands for the standard Lebesgue measure $dx\, dy$, $z = x + iy$. States with different value of z are of course not orthogonal. One can also check that

$$a|z\rangle = z|z\rangle$$

that is, $|z\rangle$ is an eigenstate of a with eigenvalue z.

For a general operator given in terms of a and $a^\dagger$ (typically a polynomial), define its *lower symbol* $h_l(z)$ by

$$h_l(z) = \langle z|h|z\rangle.$$

Note that if h is normal ordered, that is, all creation operators appear to the left of all annihilation operators, then $h_l(z)$ is obtained from h simply by replacing all a's by z and all $a^\dagger$'s by z^*. Many operators (in particular, all polynomials) also have *upper symbols*, which are functions $h_u(z)$ such that

$$h = \int_{\mathbb{C}} \frac{dz}{\pi} h_u(z)|z\rangle\langle z|.$$

In fact, $h_u(z)$ is obtained by replacing a by z and $a^\dagger$ by z^* in the anti-normal ordered form of h.

Examples:

h	$h_l(z)$	$h_u(z)$
a	z	z
$a^\dagger$	z^*	z^*
$a^\dagger a$	$|z|^2$	$|z|^2 - 1$

In general, one can show that $h_l(z)$ and $h_u(z)$ are related by

$$h_u(z) = e^{-\partial_{z^*}\partial_z} h_l(z)$$

(as long as the right side exists).

Note that for self-adjoint h

$$\inf_z h_u(z) \leq \inf \operatorname{spec} h \leq \inf_z h_l(z).$$

The same is true for partition functions, namely the Berezin–Lieb inequalities

$$\int_{\mathbb{C}} \frac{dz}{\pi} e^{-h_l(z)} \leq \operatorname{tr} e^{-h} \leq \int_{\mathbb{C}} \frac{dz}{\pi} e^{-h_u(z)}$$

hold. These inequalities are, in fact, the origin of the terminology 'upper' and 'lower' symbols; upper symbols give upper bounds to the partition function, while lower symbols give lower bounds.

Effectively, coherent states replace a quantum problem by a classical problem with phase space $\mathbb{C}$, replacing creation and annihilation operators by numbers. Note that the difference in the upper and lower bounds comes from the difference in the upper and lower symbols, in particular the factor -1 for the quadratic operator $a^\dagger a$ in the example above.

Coherent states can be used for many modes at the same time, simply using tensor products. One can not use them for *all* modes, however. Even for the number operator, the upper and lower symbols differ by a constant which is the number of modes, and we want to avoid infinities.

Let us split the Fock space into two parts,

$$\mathcal{F} = \mathcal{F}_< \otimes \mathcal{F}_>$$

corresponding to the splitting of the one-particle Hilbert space $\mathcal{H}_1$ into $\mathcal{H}_< \oplus \mathcal{H}_>$, where $\mathcal{H}_<$ is a finite dimensional space spanned by the modes $\varphi_1, \ldots, \varphi_J$. Here, $J \geq 0$ is some large finite number to be determined later. On $\mathcal{F}_<$, we can use coherent states for all the modes. In particular, for $\tilde{H}$, our Hamiltonian under consideration, we can write

$$\tilde{H} = \int_{\mathbb{C}^J} \prod_{j=1}^J \frac{dz_j}{\pi} |z_1 \otimes \cdots \otimes z_J\rangle\langle z_1 \otimes \cdots \otimes z_J| \otimes K(z_1, \ldots, z_J)$$

where the upper symbol $K(z_1, \ldots, z_J)$ is now an operator on $\mathcal{F}_>$, the Fock space for the large modes. The key point of this decomposition is that

$$\inf \operatorname{spec} \tilde{H} \geq \inf_{z_1, \ldots, z_J} \inf \operatorname{spec} K(z_1, \ldots, z_J).$$

One can show that, for $J \gg 1$ appropriately chosen

$$K(z_1, \ldots, z_J) = \mathcal{E}^{\mathrm{GP}}(\Phi) + \text{error terms}$$

where

$$\Phi(x) = \sum_{j=1}^{J} z_j \varphi_j(x).$$

The error terms are still operators on $\mathcal{F}_>$, but they are small (at least in expectation) for an appropriate choice of the interaction range R. For these estimates, it is important that $R \gg a \sim N^{-1}$, hence the necessity of Step 1. If fact, the larger R the better the control of the error terms, but we can still get away with some $R \ll N^{-1/3}$, as required.

This completes the sketch of the proof of the lower bound to the ground state energy in Theorem 6.6. For the details, we refer to (Lieb and Seiringer 2006). An appropriate upper bound can be derived using the variational principle (Seiringer 2003).

For the proof of BEC, one can proceed as above, but adding to the one-particle Hamiltonian some perturbation S. The proof goes through essentially without change, since the precise form of h has never been used. The result is the validity of the GP theory for the ground state energy, even with h replaced by $h + S$. One can now use standard convexity theory, differentiating with respect to S. The key point is this: if concave functions $f_n(x)$ converge pointwise to a function f, then the right and left derivatives f'_+ and f'_- (which always exist for concave functions) satisfy

$$f'_+(x) \leq \liminf_{n \to \infty} f'_{n,+}(x) \leq \limsup_{n \to \infty} f'_{n,-}(x) \leq f'_-(x). \tag{6.27}$$

In particular, if f is differentiable at a point x, then there is equality everywhere in (6.27).

The left and right derivatives of

$$\lambda \mapsto \inf_{\phi} \left(\mathcal{E}^{\mathrm{GP}}(\phi) + \lambda \langle \phi | S | \phi \rangle \right)$$

at $\lambda = 0$ are both of the form $\langle \phi^{\mathrm{GP}} | S | \phi^{\mathrm{GP}} \rangle$, with ϕ^{GP} a GP minimizer (in the case $\lambda = 0$). They need not be the same, however. We thus conclude that

$$\min_{\phi^{\mathrm{GP}}} \langle \phi^{\mathrm{GP}} | S | \phi^{\mathrm{GP}} \rangle \leq \lim_{N \to \infty} \frac{1}{N} \, \mathrm{tr}\, S\gamma \leq \max_{\phi^{\mathrm{GP}}} \langle \phi^{\mathrm{GP}} | S | \phi^{\mathrm{GP}} \rangle \tag{6.28}$$

for the one particle density matrix γ of a ground state of H_N, where the maximum and minimum, respectively, is over all minimizers of the GP functional. Since (6.28) is valid for all (hermitian, bounded) S, the statement about BEC follows now quite easily. For simplicity, just consider the case of a unique GP minimizer, in which case there is equality in (6.28). It is easy to see that this implies $\lim_{N \to \infty} N^{-1}\gamma = |\phi^{\mathrm{GP}}\rangle\langle\phi^{\mathrm{GP}}|$. The more general case is discussed in detail in (Lieb and Seiringer 2006).

6.5.6 Rapidly rotating Bose gases

Consider now the special case of a harmonic trap potential

$$V(x) = \frac{1}{4}|x|^2$$

This is of particular relevance for the experimental situation, where the trap potential is typically close to being harmonic. The one-particle part of the Hamiltonian can then be written as

$$h = -\Delta + V(x) - \Omega \cdot L = \left(-i\nabla - \tfrac{1}{2}\Omega \wedge x\right)^2 + \tfrac{1}{4}\left(|x|^2 - |\Omega \wedge x|^2\right).$$

The first part on the right side is the same as the kinetic energy of a particle in a homogeneous magnetic field Ω and, in particular, is translation invariant (up to a gauge transformation). It follows that h is bounded from below only for $|\Omega| \leq 1$. The angular velocity has to be less than one, otherwise the trapping force is not strong enough to compensate the centrifugal force and the system flies apart.

What happens to a dilute Bose gas as $|\Omega|$ approaches 1? For $e = \Omega/|\Omega|$ the rotation axis, let us rewrite h as

$$h = \underbrace{\left(-i\nabla - \tfrac{1}{2}e \wedge x\right)^2 + \tfrac{1}{4}|e \cdot x|^2}_{k} + (e - \Omega) \cdot L.$$

For Ω close to e, the last term can be considered as a perturbation of the rest, which we denote by k. The spectrum of k equals $\{3/2, 5/2, 7/2, \ldots\}$, and each energy level is infinitely degenerate. These energy levels are in fact just the Landau levels for a particle in a homogeneous magnetic field (for the motion perpendicular to Ω), combined with a simple harmonic oscillator in the Ω direction.

For $|e - \Omega| \ll 1$, we can thus restrict the one particle Hilbert space to the lowest Landau level (LLL) when investigating the low energy behaviour of the system. This LLL consists of functions of the form

$$f(z)e^{-|x|^2/4}$$

where we use a complex variable z for the coordinates perpendicular to Ω. In particular, $|x|^2 = |z|^2 + |e \cdot x|^2$. Moreover, the function f has to be analytic, that is, it is an entire function of z. The only freedom lies in the choice of f, in fact, the Gaussian factor is fixed. In particular, the motion in the Ω direction is frozen into the ground state of the harmonic oscillator. Because of that, it is convenient to think of the Hilbert space as the space of analytic functions f only, and absorb the Gaussian into the measure. The resulting space is known as the Bargmann space

$$\mathcal{B} = \left\{ f : \mathbb{C} \to \mathbb{C} \text{ analytic}, \int_{\mathbb{C}} |f(z)|^2 e^{-|z|^2/2} dz < \infty \right\}.$$

On the space $\mathcal{B}$, the angular momentum $e \cdot L$ simply acts as

$$L = z \frac{\partial}{\partial z}.$$

In particular, its eigenstates are z^n, $n = 0, 1, 2, \ldots$. These states form an orthonormal bases of $\mathcal{B}$. In particular, note that $L \geq 0$ on $\mathcal{B}$.

Having identified the relevant one particle Hilbert space for the low energy physics of a rapidly rotating Bose gas, what should the relevant many-body Hamiltonian look like? The only term left from the one-particle part of H_N is $(e - \Omega) \cdot L$. If we assume that the interaction is short range, that is, $a \ll 1$ (where 1 is the relevant 'magnetic length' in our units), in can be approximated by δ-function in the LLL. That is, we introduce as a many-body Hamiltonian on $\mathcal{B}^{\otimes N}$

$$H_N^{\text{LLL}} = \omega \sum_{i=1}^{N} z_i \frac{\partial}{\partial z_i} + 8\pi a \sum_{1 \leq i < j \leq N} \delta_{ij} \tag{6.29}$$

where $\omega > 0$ is short for $1 - |\Omega|$ and δ_{ij} is obtained from projecting $\delta(x_i - x_j)$ onto the LLL level. Explicitly, we have

$$(\delta_{12} f)(z_1, z_2) = (2\pi)^{-3/2} f\left(\tfrac{1}{2}(z_1 + z_2), \tfrac{1}{2}(z_1 + z_2)\right).$$

That is, δ_{12} symmetrizes the arguments z_1 and z_2. In particular, it takes analytic functions into analytic functions. Except for the unimportant prefactor $(2\pi)^{-3/2}$, δ_{12} is, in fact, a projection, projecting onto relative angular momentum zero. The factor $8\pi a$ in front of the interaction term in (6.29) is chosen to reproduce the correct expression for the ground state energy of a homogeneous system.

The introduction of the effective many-body Hamiltonian H_N^{LLL} in the lowest Landau level raises interesting questions. First of all, can one rigorously justify the approximations leading to H_N^{LLL}? In other words, can it be rigorously derived from the full many-body problem, defined by H_N on the entire Hilbert space? This was indeed achieved in (Lewin and Seiringer 2009), where it was shown that for small ω and small a, the low energy spectrum and corresponding eigenstates of H_N are indeed well approximated by the corresponding ones of H_N^{LLL}, and converge to these in the limit $\omega \to 0$, $a \to 0$ with a/ω fixed. Note that H_N^{LLL} *cannot* be obtained by simply projecting H_N onto the LLL, as this would not reproduce the correct prefactor $8\pi a$ in front of the interaction. It is important to first integrate out the high energy degrees of freedom, associated with length scales much smaller than 1, as we have done several times earlier. The projection onto the LLL is only a good approximation for length scales of order one and larger.

Having rigorously derived H_N^{LLL} from the full many-body problem, what have we learned? It still remains to investigate the relevant properties of this effective model, in particular its spectrum and corresponding eigenstates. Relatively little is known about these questions, however, despite the apparent simplicity of the model. Interesting behaviour reminiscent of the fractional quantum Hall effect in fermionic systems is expected.

Note that H_N^{LLL} is the sum of two terms

$$H_N^{\mathrm{LLL}} = \omega \underbrace{\sum_{i=1}^{N} z_i \frac{\partial}{\partial z_i}}_{L_N} + 8\pi a \underbrace{\sum_{1 \le i < j \le N} \delta_{ij}}_{\Delta_N} \qquad (6.30)$$

that commute with each other, that is, $[L_N, \Delta_N] = 0$. It therefore makes sense to look at their joint spectrum. Of particular relevance is the so-called 'yrast curve', which is the lowest energy of Δ_N is a given sector of angular momentum. That is,

$$\Delta_N(L) = \inf \mathrm{spec}\, \Delta_N \upharpoonright_{L_N = L}.$$

It is known explicitly for small and large L. In fact, the known values of $\Delta_N(L)$ are

$$\Delta_N(L) = \frac{1}{2(2\pi)^{3/2}} \times \begin{cases} N(N-1) & L \in \{0,1\} \\ N(N-1-\frac{1}{2}L) & 2 \le L \le N \\ 0 & L \ge N(N-1) \end{cases}.$$

The minimizer for $L = N(N-1)$ is in fact the bosonic analogue of the Laughlin state

$$\prod_{i<j} (z_i - z_j)^2$$

for which obviously the interaction energy is zero. Note the exponent 2, which has to be even since we are dealing with bosons.

A sketch of the joint spectrum of L_N and Δ_N is given in Fig. 6.4. The interesting part concerns angular momenta of order N^2, in which case Δ_N is of order N. For $L \ll N^2$, one can show that the GP approximation becomes exact. That is, for large N and $L \ll N^2$, the convex hull of $\Delta_N(L)$ is given by

$$\inf_{\phi \in B} \left\{ \int_{\mathbb{C}} |\phi(z)|^4 e^{-|z|^2} dz \ : \ \|\phi\|^2 = N, \langle \phi | L | \phi \rangle = L \right\}.$$

This was proved in (Lieb, Seiringer, and Yngvason, 2009) using coherent states. The condition $L \ll N^2$ corresponds to the case where the number of particles is much larger than the number of vortices. Once these two numbers becomes comparable, the GP approximation breaks down and interesting new physics with highly correlated many-body states is expected to occur.

No rigorous bounds on $\Delta_N(L)$ are available for $L \sim N^2$ (but $L < N(N-1)$, of course). In fact, even to prove the existence of the limit

$$\lim_{N \to \infty} \frac{\Delta_N(\ell N^2)}{N}$$

is an open problem. Besides $\Delta_N(L)$, one would also like to understand the existence or nonexistence of spectral gaps above the ground state energy (uniformly in the particle number), and other quantities of this type. A lot remains to be done.

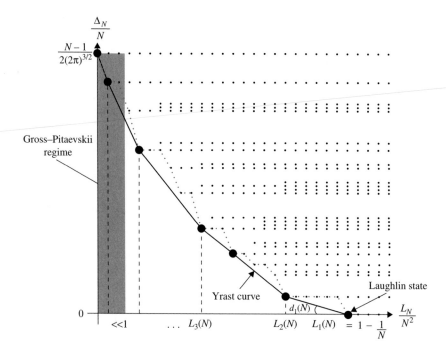

Fig. 6.4 Sketch of the joint spectrum of L_N and Δ_N. The GP approximation is valid in the shaded region on the left. The small dots show the yrast curve, in black is its convex hull. The bold dots on the convex hull are, in fact, the possible ground states of H_N^{LLL} as one varies ω/a.

Acknowledgements

Many thanks to Dana Mendelson, Alex Tomberg and Daniel Ueltschi for allowing me to use their figures in these lecture notes. The support from the ERC starting grant *CoMBoS—Collective Phenomena in Classical and Quantum Many Body Systems* is gratefully acknowledged.

References

Anderson, M.H. Ensher, J.R. Matthews, M.R. Wieman, C.E. Cornell, E.A. (1995). Observation of Bose-Einstein Condensation in a Dilute Atomic Vapor, *Science* **269**: 198–201.

Bogoliubov, N.N. (1947). On the theory of superfluidity, *J. Phys. (U.S.S.R.)* **11**: 23.

Bose, S.N. (1924). Plancks Gesetz und Lichtquantenhypothese, *Z. Phys.* **26**: 178–181.

Buffet, E. Pulè, J.V. (1983) Fluctuation properties of the imperfect Bose gas, *J. Math. Phys.* **24**: 1608–1616.

Calogero, F. (1969). Ground State of a one-dimensional N-body system, *J. Math. Phys.* **10**: 2197–2200 .

Calogero, F. (1971). Solution of the one-dimensional N-body problems with quadratic and/or inversely quadratic pair potentials, *J. Math. Phys.* **12**: 419–436.

Davis, K.B. Mewes, M.O. Andrews, M.R. van Druten, N.J. Durfee, D.S. Kurn, D.M. Ketterle, W. (1995). Bose-Einstein condensation in a gas of sodium atoms, *Phys. Rev. Lett.* **75**: 3969–3973.

Dyson, F.J. (1957). Ground-state energy of a hard-sphere gas, *Phys. Rev.* **106**: 20–26.

Dyson, F.J. Lieb, E.H. Simon, B. (1978). Phase transitions in quantum spin systems with isotropic and nonisotropic interactions, *J. Stat. Phys.* **18**: 335–383.

Einstein, A. (1925). Quantentheorie des einatomigen idealen Gases, *Sitzber. Kgl. Preuss. Akad. Wiss.*, 261–267 (1924), and 3–14.

Fröhlich, J. Simon, B. Spencer, T. (1976). Infrared bounds, phase transitions and continuous symmetry breaking, *Comm. Math. Phys.* **50**: 79–85.

Girardeau, M. (1960). Relationship between systems of impenetrable bosons and fermions in one dimension, *J. Math. Phys.* **1**: 516–523.

Giuliani, A. Seiringer, R. (2009). The ground state energy of the weakly interacting bose gas at high density, *J. Stat. Phys.* **135**: 915–934.

Lee, T.D. Yang, C.N. (1957). Many body problem in quantum mechanics and quantum statistical mechanics, *Phys. Rev.* **105**: 1119–1120.

Lee, T.D. Huang, K. Yang, C.N. (1957). Eigenvalues and Eigenfunctions of a Bose system of hard spheres and its low-temperature properties, *Phys. Rev.* **106**: 1135–1145.

Lewin, M. Seiringer, R. (2009). Strongly correlated phases in rapidly rotating Bose gases, *J. Stat. Phys.* **137**: 1040–1062.

Lieb, E.H. Liniger, W. (1963). Exact analysis of an interacting Bose Gas. I. The general solution and the ground state, *Phys. Rev.* **130**: 1605–1616.

Lieb, E.H. (1963). Exact analysis of an interacting bose gas. II. The excitation spectrum, *Phys. Rev.* **130**: 1616–1624.

Lieb, E.H. Seiringer, R. (2002). Proof of Bose-Einstein condensation for dilute trapped gases, *Phys. Rev. Lett.* **88**: 170409.

Lieb, E.H. Seiringer, R. (2006). Derivation of the Gross-Pitaevskii equation for rotating bose gases, *Comm. Math. Phys.* **264**: 505–537.

Lieb, E.H. Seiringer, R. Solovej, J.P. (2005). Ground-state energy of the low-density fermi gas, *Phys. Rev. A* **71**: 053605.

Lieb, E.H. Seiringer, R. Solovej, J.P. Yngvason, J. (2005). The Mathematics of the bose gas and its condensation, *Oberwolfach Seminars*, Vol. 34, Birkhäuser. Also available at arXiv:cond-mat/0610117.

Lieb, E.H. Seiringer, R. Yngvason, J. (2000). Bosons in a trap: A rigorous derivation of the Gross-Pitaevskii energy functional, *Phys. Rev. A* **61**: 043602.

Lieb, E.H. Seiringer, R. Yngvason, J. (2009). Yrast line of a rapidly rotating Bose gas: Gross-Pitaevskii regime, *Phys. Rev. A* **79**: 063626.

Lieb, E.H. Yngvason, J. (1998). Ground state energy of the low density Bose Gas, *Phys. Rev. Lett.* **80**: 2504–2507.

Lieb, E.H. Yngvason, J. (2001). The ground state energy of a dilute Two-dimensional bose gas, *J. Stat. Phys.* **103**: 509.

Penrose, O. Onsager, L. (1956). Bose-Einstein condensation and liquid helium, *Phys. Rev.* **104**: 576–584.

Seiringer, R. (2003). Ground state asymptotics of a dilute, rotating gas, *J. Phys. A: Math. Gen.* **36**: 9755–9778.

Seiringer, R. (2006). The thermodynamic pressure of a dilute fermi gas, *Comm. Math. Phys.* **261**: 729–758.

Seiringer, R. (2008). Free energy of a dilute Bose gas: lower bound, *Comm. Math. Phys.* **279**: 595–636.

Seiringer, R. (2011). The Excitation spectrum for weakly interacting bosons, *Comm. Math. Phys.* **306**: 565–578.

Seiringer, R. Ueltschi, D. (2009). Rigorous upper bound on the critical temperature of dilute Bose gases, *Phys. Rev. B* **80**: 014502.

Sutherland, B. (1971). Quantum many-body problem in one dimension: ground state, *J. Math. Phys.* **12**, 246–250.

Sutherland, B. (1971). Quantum many-body problem in one dimension: thermodynamics, *J. Math. Phys.* **12**: 251–256.

Ueltschi, D. (2006). Feynman cycles in the bose gas, *J. Math. Phys.* **47**: 123303.

Yau, H.-T. Yin, J. (2009). The Second order upper bound for the ground energy of a Bose gas, *J. Stat. Phys.* **136**: 453–503.

Yin, J. (2010). Free energies of dilute bose Gases, *J. Stat. Phys.* **141**: 683–726.

7

SUSY statistical mechanics and random band matrices

Thomas SPENCER

Institute of Advanced Study,
School of Mathematics,
Einstein Drive, Princeton, NJ 08540, USA

Chapter Contents

7 **SUSY statistical mechanics and random band matrices**   **467**

      Thomas SPENCER

7.1   An overview

The study of large random matrices in physics originated with the work of Eugene Wigner who used them to predict the energy level statistics of a large nucleus. He argued that because of the complex interactions taking place in the nucleus there should be a random matrix model with appropriate symmetries, whose eigenvalues would describe the energy level spacing statistics. Recent developments summarized in (Erdős 2011), give a rather complete description of the universality of eigenvalue spacings for the *mean field* Wigner matrices. Today, random matrices are studied in connection with many aspects of theoretical and mathematical physics. These include Anderson localization, number theory, generating functions for combinatorial enumeration, and low energy models of QCD. See (Brezin *et al.* 2006) for an explanation of how random matrix theory is related to these topics and others.

The goal of these lectures is to present the basic ideas of *supersymmetric* (SUSY) statistical mechanics and its applications to the spectral theory of large Hermitian random matrices—especially *random band matrices*. There are many excellent reviews of SUSY and random matrices in the theoretical physics literature starting with the fundamental work of Efetov (Efetov 1983). See for example (Efetov 2010, 1997; Fyodorov 1994; Guhr 2010; Mirlin 2000; Verbaarschot *et al.* 1985; Zirnbauer 2004). This review will emphasize mathematical aspects of SUSY statistical mechanics and will try to explain ideas in their simplest form. We shall begin by first studying the average Green's function of $N \times N$ GUE matrices—Gaussian matrices whose distribution is invariant under unitary transformations. In this case the SUSY models can be expressed as integrals in just two real variables. The size of the matrix N appears only as a parameter. See Sections 7.1–7.3.

The simple methods for GUE are then generalized to treat the more interesting case of *random band matrices*, RBM, for which much less is rigorously proved. Random band matrices H_{ij} are indexed by vertices i, j of a lattice $\mathbb{Z}^d$. Their matrix elements are random variables which are 0 or small for $|i - j| \geq W$ and hence these matrices reflect the geometry of the lattice. The parameter W will be referred to as the width of the band. As we vary W, random band matrices approximately interpolate between mean field $N \times N$ GUE or Wigner type matrix models where W=N and random Schrödinger matrices, $H = -\Delta + v_j$ on the lattice in which the randomness only appears in the potential v. Following Anderson, random Schrödinger matrices are used to model the dynamics of a quantum particle scattered by random impurities on a crystalline lattice. In Section 7.3 precise definitions of random band and random Schrödinger matrices are given and a qualitative relation between them is explained.

The key to learning about spectral properties of random matrices H lies in the study of averages of its Green's functions, $(E_\epsilon - H)^{-1}(j, k)$ where j, k belong to a lattice. We use the notation $A(j, k)$ to denote the matrix elements of the matrix A. The energy E typically lies inside the spectrum of H and $E_\epsilon = E - i\epsilon$ with $\epsilon > 0$. Efetov (Efetov 1983) showed that averages of products of Green's functions can be exactly expressed in terms of correlations of certain supersymmetric (SUSY) statistical mechanics ensembles. In SUSY statistical mechanics models, the spin or field at each lattice site has both real and Grassmann (anticommuting) components. See (7.20) below for the basic identity. These components appear symmetrically and thus the theory is called supersymmetric. Efetov's formalism builds on the foundational work of Wegner and Schäfer (Schäfer and Wegner 1980; Wegner 1979) which used replicas instead of Grassmann variables. A brief review of Gaussian and Grasssmann integration is given in Appendix 7.10. Although Grassmann variables are a key element of SUSY models, we shall frequently integrate them out so that the statistical mechanics involves only real integrals.

The first step in the SUSY approach to random band matrices is to define a SUSY statistical mechanics model whose correlations equal the averaged Green's functions. This is basically an algebraic step, but some analysis is needed to justify certain analytic continuations and changes of coordinates. This gives us a *dual representation* of the averaged Green's function in terms of SUSY statistical mechanics. In the case of Gaussian randomness, the average of the Green's function can be explicitly computed and produces a quartic action in the fields. The advantage of this dual representation is that many of the concepts of statistical mechanics such as phase transitions, saddle point analysis, cluster expansions, and renormalization group methods, (Salmhofer 1999), can then be used to analyze the behaviour of Green's functions of random band matrices. Section 7.6 will review some results about phase transitions and symmetry breaking for classical statistical mechanics. This may help to give some perspective of how SUSY statistical mechanics is related to its classical counterpart.

The second step is to analyse correlations of the resulting SUSY statistical mechanics model. Typically these models have a formal noncompact symmetry. For large W, the functional integral is expected to be governed by a finite dimensional *saddle manifold*. This manifold is the orbit of a critical point under the symmetry group. The main mathematical challenge is to estimate the fluctuations about the manifold. Most (SUSY) lattice models are difficult to analyse rigorously due to the lack of a spectral gap and the absence of a positive integrand. These lectures will focus on some special cases for which the analysis can be rigorously carried out.

7.1.1 Green's functions

Let H be a random matrix indexed by vertices $j \in \mathbb{Z}^d$. We can think of H as a random Schrödinger matrix or an infinite random band matrix of fixed width acting on $\ell_2(\mathbb{Z}^d)$. To obtain information about time evolution and eigenvectors of H we define the average of the square modulus of the Green's function

$$ < |G(E_\epsilon : 0, j)|^2 > \equiv < |(E - i\epsilon - H)^{-1}(0, j)|^2 > \equiv \Gamma(E_\epsilon, j) \tag{7.1} $$

where $j \in \mathbb{Z}^d$, $\epsilon > 0$, and $< \cdot >$ denotes the average over the random variables of H. We wish to study $\Gamma(E_\epsilon, j)$ for very small $\epsilon > 0$. If for all small $\epsilon > 0$ we have

$$ \Gamma(E_\epsilon, j) \le C\epsilon^{-1} e^{-|j|/\ell} \tag{7.2} $$

then the eigenstates with eigenvalues near E are *localized*. This means the eigenstate ψ decays exponentially fast $|\psi_j|^2 \approx e^{-|j-c|/\ell}$, about some point c of the lattice with probability one. The length $\ell = \ell(E)$ is called the *localization length*.

Quantum diffusion near energy E corresponds to

$$\Gamma(E_\epsilon, j) \approx (-D(E)\Delta + \epsilon)^{-1}(0, j) \quad or \quad \hat{\Gamma}(E_\epsilon, p) \approx (D(E)p^2 + \epsilon)^{-1} \tag{7.3}$$

where $D(E)$ is the diffusion constant and Δ is the discrete lattice Laplacian and $|p|$ is small. Note that the right side is the Laplace transform of the heat kernel. When (7.3) holds with $D(E) \geq \delta > 0$, particles with energy near E are mobile and can conduct. The eigenstates ψ_j at these energies are *extended*. This means that if H is restricted to a large volume Λ, an $\ell_2(\Lambda)$ normalized eigenfunction ψ satisfies $|\psi_j|^2 \approx |\Lambda|^{-1}$ for all $j \in \Lambda$. A brief perturbative discussion of quantum diffusion is presented in Appendix 7.11. In the infinite volume limit, quantum diffusion implies absolutely continuous spectrum near E, whereas localization corresponds to dense point spectrum and the absence of conduction. Although the quantum diffusion in three dimensions is well studied in the physics literature, it remains a major open problem to establish it at a rigorous mathematical level.

The basic identity

$$Im < G(E_\epsilon; 0, 0) > = \epsilon \sum_j < |G(E_\epsilon; 0, j)|^2 > \tag{7.4}$$

is easily proved by applying the resolvent identity to $G - \bar{G}$. It reflects conservation of probability or unitarity and is some times referred to as a Ward identity. In Section 7.5 we shall see that the left side of (7.4) is proportional to the density of states, $\rho(E)$. Thus the identity $\sum_j \Gamma(E_\epsilon, j) \propto \epsilon^{-1}\rho(E)$ always holds and explains the factor ϵ^{-1} in (7.2).

In the language of SUSY statistical mechanics, localization roughly corresponds to high temperature in which distant spins become exponentially decorrelated, whereas quantum diffusion corresponds to an ordered phase in which spins are aligned over long distances. For random band matrices, the inverse temperature is

$$\beta \approx W^2\rho(E)^2 \tag{7.5}$$

where $\rho(E)$ is the density of states, and W is the band width.

For a more intuitive time dependent picture, let $U(j, t)$ be the solution to the Schrödinger equation on the lattice

$$i\partial_t U(j, t) = H\, U(j, t). \tag{7.6}$$

Suppose that at time 0, $U(j, 0)$ is supported near the origin. For any unitary evolution we have conservation of probability: $\sum_j |U(j,t)|^2$ is constant in t. To measure the spread of U we define the average mean-square displacement by

$$R^2(t) = < \sum_j |U(j,t)|^2 |j|^2 > . \tag{7.7}$$

Quantum diffusion corresponds to $R^2 \approx D\, t$, whereas if all states are uniformly localized then the wave packet U does not spread as time gets large, $R^2 \leq Const\, \ell^2$.

The Green's function at energy E, essentially projects the initial wave function U onto a spectral subspace about E. Time is roughly proportional to $t \approx \epsilon^{-1}$.

Note that if $H = -\Delta$ is the lattice Laplacian on $\mathbb{Z}^d$, and there no disorder, the motion is *ballistic*: $R^2 \propto t^2$. The presence of disorder, that is randomness, should change the character of the motion through multiple scatterings. It is expected that in the presence of spatially uniform disorder, the motion is never more than diffusive, $R^2(t) \leq C\, t^{2\alpha}$ with $0 \leq \alpha \leq 1/2$ with possible logarithmic corrections in the presence of spin-orbit coupling. However, in two or more dimensions, this is still a conjecture, even for $\alpha < 1$. It is easy to show that for the lattice random Schrödinger operator we have $R^2(t) \leq Ct^2$ for any potential.

7.1.2   Symmetry and the one-dimensional SUSY sigma model

Symmetries of statistical mechanics models play a key role in the macroscopic behavior of correlations. In Section 7.6 we present a review of phase transitions, symmetry breaking, and Goldstone modes (slow modes) in classical statistical mechanics. The SUSY lattice field action which arises in the study of (7.1) is invariant under a global hyperbolic $SU(1,1|2)$ symmetry. As mentioned earlier, the spin or field has both real and Grassmann components. The symmetry $SU(1,1|2)$ means that for the real components there exists a hyperbolic symmetry U(1, 1) preserving $|z_1|^2 - |z_2|^2$ as discussed in Section 7.7. On the other hand, the Grassmann variables (reviewed in Appendix 7.11) are invariant under a compact SU(2) symmetry. More details about these symmetries are in Section 7.9.

In both physics and mathematics, many statistical mechanics systems are studied in the *sigma model approximation*. In this approximation, spins take values in a symmetric space. The underlying symmetries and spatial structure of the original interaction are respected. The Ising model, the rotator, and the classical Heisenberg model are well known sigma models where the spin s_j lies in $\mathbb{R}^1$, $\mathbb{R}^2$, $\mathbb{R}^3$, respectively with the constraint $|s_j^2| = 1$. Thus they take values in the groups Z_2, S^1 and the symmetric space S^2. One can also consider the case where the spin is matrix valued. In the Efetov sigma models, the fields are 4×4 matrices with both real and Grassmann entries. It is expected that sigma models capture the qualitative physics of more complicated models with the same symmetry, see Section 7.6.

In a one dimensional chain of length L, the SUSY sigma model governing conductance has a simple expression first found in (Efetov 1984). The Grassmann variables can be traced out and the resulting model is a nearest neighbour spin model with *positive* weights given as follows. Let $S_j = (h_j, \sigma_j)$ denote spin vector with h_j and σ_j taking values in a hyperboloid and the sphere S^2 respectively. The Gibbs weight is then proportional to

$$\prod_{j=0}^{L} (h_j \cdot h_{j+1} + \sigma_j \cdot \sigma_{j+1})\, e^{\beta(\sigma_j \cdot \sigma_{j+1} - h_j \cdot h_{j+1})} . \tag{7.8}$$

As in classical statistical mechanics, the parameter $\beta > 0$ is referred to as the inverse temperature. The hyperbolic components $h_j = (x_j, y_j, z_j)$ satisfy the constraint

$z_j^2 - x_j^2 - y_j^2 = 1$. The Heisenberg spins σ are in $\mathbb{R}^3$ with $\sigma \cdot \sigma = 1$ and the dot product is Euclidean. On the other hand the dot product for the h spins is hyperbolic: $h \cdot h' = zz' - xx' - yy'$. It is very convenient to parameterize this hyperboloid with horospherical coordinates $s, t \in \mathbb{R}$:

$$z = \cosh t + s^2 e^t/2, \quad x = \sinh t - s^2 e^t/2, \quad y = se^t. \tag{7.9}$$

The reader can easily check that $z_j^2 - x_j^2 - y_j^2 = 1$ is satisfied for all values of s and t. This parametrization plays an important role in later analysis of hyperbolic sigma models (Disertori, Spencer, and Zirnbauer, 2010; Spencer and Zirnbauer, 2004). The integration measure in σ is the uniform measure over the sphere and the measure over h_j has the density $\prod e^{t_j} ds_j\, dt_j$. At the end points of the chain we have set $s_0 = s_L = t_0 = t_L = 0$. Thus we have nearest neighbor hyperbolic spins (Boson–Boson sector) and Heisenberg spins (Fermion–Fermion sector) coupled via the Fermion–Boson determinant. In general, this coupling quite complicated. However in 1D it is given by $\prod_j (h_j \cdot h_{j+1} + \sigma_j \cdot \sigma_{j+1})$. As in (7.5), the inverse temperature $\beta \approx W^2 \rho(E)^2$ where $\rho(E)$ is the density of states, and W is the band width. When $\beta \gg 1$, (7.8) shows that the spins tend to align over long distances.

By adapting the recent work of (Disertori and Spencer 2010) it can be proved that the model given by equation (7.8) has a localization length proportional to β for all large β. More precisely it is shown that the conductance goes to zero exponentially fast in L, for $L \gg \beta$.

7.1.3 SUSY sigma models in three dimensions

Although the SUSY sigma models are widely used in physics to make detailed predictions about eigenfunctions, energy spacings, and quantum transport, for disordered quantum systems, there is as yet no rigorous analysis of the $SU(1,1|2)$ Efetov models in two or more dimensions. Even in one dimension, where the problem can be reduced to a transfer matrix, rigorous results are restricted to the sigma model mentioned above. A key difficulty arises from the fact that SUSY lattice field models cannot usually be treated using probabilistic methods due to the presence of Grassmann variables and strongly oscillatory factors.

In recent work with Disertori and Zirnbauer (Disertori and Spencer, 2010; Disertori, Spencer, and Zirnbauer, 2010) we have established a phase transition for a *simpler* SUSY hyperbolic sigma model in three dimensions. We shall call this model the $H^{2|2}$ model. The notation refers to the fact that the field takes values in hyperbolic two space augmented with two Grassmann variables to make it supersymmetric. This model, introduced by Zirnbauer (Zirnbauer 1991), is expected to reflect the qualitative behavior, such as localization or diffusion, of random band matrices in any dimension. The great advantage of the $H^{2|2}$ model is that the Grassmann variables can be traced out producing a statistical mechanics model with *positive* but nonlocal weights. (The non locality arises from a determinant produced by integrating out the Grassmann fields.) This means that probabilistic tools can be applied. In fact we shall see that quantum localization and diffusion is closely related to *a random walk in a random environment.* This environment is highly correlated and has strong fluctuations in one and two dimensions.

In Section 7.8 we will describe the $H^{2|2}$ model and sketch a proof of a phase transition as $\beta(E) > 0$ goes from small to large values. Small values of β, high temperature, will correspond to localization—exponential decay of correlations and lack of conductance. In three dimensions, we shall see that large values of β correspond to quantum diffusion and extended states. Standard techniques for proving the existence of phase transitions, such as reflection positivity, do not seem to apply in this case. Instead, the proof of this transition relies crucially on Ward identities arising from SUSY symmetries of the model combined with estimates on random walk in a random environment. The simplest expression of these Ward identities is reflected by the fact that the partition function is identically one for all parameter values which appear in a SUSY fashion. Thus derivatives in these parameters vanish and yield identities. The SUSY $H^{2|2}$ model is nevertheless highly nontrivial because physical observables are not SUSY and produce interesting and complicated correlations. Classical Ward identities will be discussed in Sections 7.6 and 7.7. In Section 7.9 we give a very brief description of Efetov's sigma model.

7.2   Classical models of quantum dynamics

In this section we digress to describe two classical models which have some connection with quantum dynamics. The first is called linearly edge reinforced random walk, ERRW. This history dependent walk favours moving along edges it has visited in the past. Diaconis (Diaconis 1988) introduced this model and discovered that ERRW can be expressed as random walk in a highly correlated random environment. The statistical description of the random environment appears to be closely related but not identical to that of the SUSY, $H^{2|2}$ model described in Section 7.8. In both cases it is important to observe that the generator for the walk is not uniformly elliptic, making it possible to get nondiffusive dynamics. The second model, called the Manhattan model, arises from network models developed to understand the Quantum Hall transition. This model is equivalent to the behavior of a particle undergoing a random unitary evolution. The remarkable feature of this model is that after the randomness is integrated out, the complex phases along paths are cancelled. Thus paths have positive weights and the motion has a probabilistic interpretation.

To define linearly edge reinforced random walk (ERRW), consider a discrete time walk on $\mathbb{Z}^d$ starting at the origin and let $n(e, t)$ denote the number of times the walk has visited the edge e up to time t. Then the probability $P(v, v', t + 1)$ that the walk at vertex v will visit a neighbouring edge $e = (v, v')$ at time $t + 1$ is given by

$$P(v, v', t + 1) = (\beta + n(e, t))/S_\beta(v, t) \qquad (7.10)$$

where S_β is the sum of $\beta + n(e', t)$ over all the edges e' touching v. The parameter β is analogous to β in (7.8) or to the β in the $\mathbb{H}^{2|2}$ model defined later. Note that if β is large, the reinforcement is weak and in the limit $\beta \to \infty$, (7.10) describes simple random walk.

In 1D and 1D strips, ERRW is *localized* for any value of $\beta > 0$. This means that the probability of finding an ERRW, $w(t)$, at a distance r from its initial position is exponentially small in r, thus

$$Prob\left[|w(t) - w(0)| \geq r\right] \leq Ce^{-r/\ell}. \tag{7.11}$$

Merkl and Rolles (Merkl and Rolles 2006) established this result by proving that conductance across an edge goes to zero exponentially fast with the distance of the edge to the origin. Note that in this model, (and in the $H^{(2|2)}$ model with $\epsilon_j = \delta_{j,0}$), the random environment is not translation invariant and the distribution of the local conductance depends on starting point of the walk. Localization is proved, using a variant of the Mermin–Wagner type argument and the localization length ℓ is proportional to $\beta|S|$ where $|S|$ is the width of the strip. Using similar methods, Merkl and Rolles also show that in two dimensions the local conductance goes to zero slowly away from 0. In three dimensions, there are no rigorous theorems for ERRW. Localization probably occurs for strong reinforcement, that is, for β small. It is natural to conjecture that in two dimensions, ERRW is always exponentially localized for all values of reinforcement. On the Bethe lattice, Pemantle (Peamantle 1988) proved that ERRW has a phase transition. For $\beta \gg 1$ the walk is transient whereas for $0 < \beta \ll 1$ the walk is recurrent. See (Merkl and Rolles 2006; Peamantle 1988) for reviews of this subject. It is unknown whether ERRW has a phase transition in three dimensions.

7.2.1 Manhattan model

Another classical model which is closely related to quantum dynamics, is defined on the oriented Manhattan lattice. This model was analysed by Beamond, Owczarek, and Cardy (Beamond *et al.* 2003) following related work of Gruzberg, Ludwig, and Read (Gruzberg, Ludwig, and Read, 1999). In this model disorder occurs by placing obstructions at each vertex, independently with probability $0 < p < 1$. A particle moves straight along the oriented lattice lines according to the orientation of the edges until it meets an obstruction. The particle must turn at an obstruction in the direction of the orientation. Note that orbits can traverse an edge at most once. See Fig. 7.1. This model is closely related to a disordered quantum network model (class C). It can also be expressed as a history dependent walk. If $p > 1/2$ all paths form closed loops with finite expected length. This follows (J. Chalker) by comparing the model to classical bond percolation. For p small (weak disorder) the particle is expected to exhibit diffusion for long time scales. Nevertheless, the renormalization group analysis of (Beamond *et al.* 2003), indicates that for all $p > 0$, every path of this walk is closed with probability one and has a *finite expected diameter*. The average diameter is predicted to be huge $\sim \exp Cp^{-2}$ when p is small. This prediction is consistent with those for the crossover from diffusion to Anderson localization in two dimensions. At a mathematical level, little is known when $0 < p < 1/2$.

Note that the Manhattan model is quite different from the Ruijgrok–Cohen mirror model on the unoriented square lattice. In this case one randomly places mirrors with probability p at the vertices, angled at ± 45 degrees with respect to the x axis. Orbits

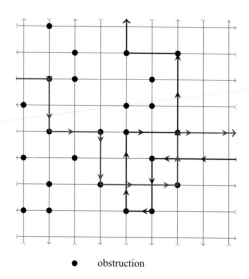

●   obstruction

Fig. 7.1 Manhattan lattice

are obtained by imagining that a light beam light travels along lattice lines until it is reflected by a mirror. If $p = 1$ and the choice of angles is independent with probability $1/2$, the model is equivalent to *critical bond percolation*. Although the loops have finite length with probability one, their average length is *infinite*. For $0 < p < 1$ little is known rigorously but one expects that the average loop length to still be infinite.

Remark: Although the two models described above are quite special, they share one of the essential difficulties of the random Schrödinger models and RBM—they are highly non-Markovian. It is this feature that makes these models challenging to analyse. On the other hand, it is the memory or non-Markovian character which is partly responsible for localization. The problem of proving strictly subballistic dynamics, $R^2(t) \propto t^{2\alpha}$ with $\alpha < 1$, is also unresolved for these classical systems in dimension two or more.

7.3   Introduction to random matrix ensembles and SUSY

In this section we shall define various Gaussian matrix ensembles. The simplest of these is GUE—Gaussian Unitary ensemble. In this case the matrix entries H_{ij} are mean zero, complex, independent random variables for $i \leq j$ and $1 \leq i, j \leq N$. Since H has a Gaussian distribution it suffices to specify its covariance:

$$< H_{ij}\bar{H}_{i'j'} >=< H_{ij}H_{j'i'} >= \delta(ii')\delta(jj')/N. \tag{7.12}$$

The average over the randomness or disorder is denoted by $< \cdot >$ and $\bar{H}$ denotes the complex conjugate of H. The density for this ensemble is given by

$$1/Z_N \ e^{-N \operatorname{tr} H^2/2} \prod dH_{ii} \prod_{i<j} dH_{ij}^{Re} dH_{ij}^{Im}.$$

The factor of $1/Z_N$ ensures that the integral is 1. It is clear that H and U^*HU have identical distributions for any fixed unitary matrix U. This invariance is a crucial feature in the classical analysis of such matrices via orthogonal polynomials. It also holds when $tr\,H^2$ above is replaced by $tr\,V(H)$ for polynomials V, which are bounded from below (Deift *et al.*, 1999; Pastur and Shcherbina, 1997). However, Wigner matrices, and random band matrices do not have unitarily invariant distributions and new methods are needed to obtain the desired spectral information. See (Erdős 2011; Spencer 2010) for recent reviews of Wigner matrices and random band matrices.

To define random band matrices, RBM, with a Gaussian distribution, let i and j range over a periodic box $\Lambda \subset \mathbb{Z}^d$ of volume $|\Lambda|$ and set

$$< H_{ij}\bar{H}_{i'j'} >=< H_{ij}H_{j'i'} >= \delta(ii')\delta(jj')J_{ij}. \tag{7.13}$$

Here J_{ij} is a symmetric function which is *small* for large $|i-j|$. We shall assume that for fixed i, $\sum_j J_{ij} = 1$. With this normalization, it is known that most of the spectrum of H is concentrated in the interval $[-2, 2]$ with high probability. One especially convenient choice of J is given by the lattice Green's function

$$J_{jk} = (-W^2\Delta + 1)^{-1}(j, k) \tag{7.14}$$

where Δ is the lattice Laplacian on Λ with periodic boundary conditions

$$\Delta f(j) = \sum_{|j'-j|=1} (f(j') - f(j)).$$

Note that with this choice of J, $\sum_j J_{ij} = 1$ and the variance of the matrix elements is exponentially small when $|i-j| \gg W$. In fact, in one dimension $J_{ij} \approx e^{-|i-j|/W}/W$. Here $|i-j|$ is the natural distance in the periodic box. Hence W will be referred to as the width of the band. The advantage of (7.14) is that its *inverse*, which appears by duality in certain SUSY formulations, has only *nearest neighbour* matrix elements and W^2 is just a parameter, see (7.31) below.

The density for the Gaussian orthogonal ensemble, GOE, is also proportional to $exp(-N\,tr\,H^2/2)$ but the matrix H is real symmetric. The covariance is given by $< H_{jj}^2 >= 1/N$ and

$$< H_{ij}H_{kl} >= (2N)^{-1}\{\delta(ik)\delta(jl) + \delta(il)\delta(jk)\}, \quad i \neq j.$$

The symmetric RBM are defined by introducing J_{ij} as in (7.13). Although the SUSY formalism can also be applied to the symmetric case, the formulas are a bit more complicated and we shall not discuss them.

7.3.1 Some conjectures about RBM and random Schrödinger

Let us now compare lattice random Schrödinger matrices (RS) on $\mathbb{Z}^d$ given by

$$H_{RS} = -\Delta + \lambda v_j$$

and RBM (7.13) of width W on $\mathbb{Z}^d$. Above, v_j are assumed to be independent identically distributed Gaussian random variables of mean 0 and variance $< v_j^2 >= 1$. The parameter $\lambda > 0$ measures the strength of the disorder. See (Erdős 2012; Spencer 2010) for recent reviews of mathematical aspects of Anderson localization. Although RBM and RS look quite different, they are both local, that is their matrix elements j, k are small (or zero) if $|j - k|$ is large. The models are expected to have the same qualitative properties when $\lambda \approx W^{-1}$. For example, eigenvectors for RS are known to decay exponentially fast in one dimension with a localization length proportional to λ^{-2}. On the other hand for 1D, RBM the localization length is known to be less than W^8 by work of Schenker, (Schenker 2009), and is expected to be approximately W^2. From the point of view of perturbation theory, if we compute the bare diffusion constant D_0, obtained by summing ladder diagrams, then we have $D_0 \propto \lambda^{-2}$, W^2 for the RS and RBM respectively. See Appendix 7.11 for a review of perturbation theory for RS and RBM.

Localization is reasonably well understood for RS in one dimension for any disorder. Localization has also been proved on $\mathbb{Z}^d$ for strong disorder or at energies where the density of states $\rho(E)$ is very small. See (Spencer 2010) for references to the mathematical literature. On the other hand a mathematical proof of quantum diffusion in three dimensions, as defined by (7.3) as $\epsilon \downarrow 0$ for RS or RBM for fixed W, remains a major open problem. In two dimensions it is conjectured that all states are localized with a localization length growing exponentially fast in W^2 or λ^{-2} for real symmetric matrices. Note the similarity of this conjecture with that for the Manhattan model defined in Section 7.2.

7.3.2   Conjectured universality of GUE and GOE models

GUE or GOE matrices are mean field and have no spatial or geometric structure. Nevertheless, the local eigenvalue statistics of these models are expected to match those of certain RBM for large N. For example, in one dimension the RBM of the form (7.13), (7.14) should have the same local energy level statistics as GUE as $N \to \infty$, (modulo a simple rescaling), provided that $W^2 \gg N$ and energies lie in the bulk, $|E| \leq 2 - \delta$. In three dimensions universality is expected to hold for E in the bulk, and for fixed $W \gg 1$ independent of $N = |\Lambda| \to \infty$. A. Sodin (Sodin 2010) recently proved universality of local energy spacing about the spectral edge, $E \approx 2$ provided $W \gg N^{5/6}$. His result is sharp in the sense that for smaller values of W another distribution appears. Universality has been established for Wigner matrices at all energies (Erdős 2011).

In terms of SUSY statistical mechanics, universality of local energy correlations for Hermitian band matrices can be intuitively understood as follows. For appropriate values of E, W, and dimension, energy correlations are essentially governed by the saddle manifold which is the orbit of a single saddle point under the action of $SU(1, 1|2)$. This manifold only depends on the symmetry of the ensemble. See Section 7.3 of (Mirlin 2000) for a more detailed (but not rigorous) explanation of universality and explicit corrections to it arising from finite volume effects. Universality fails when the fluctuations about the saddle manifold are too pronounced. This can happen if

W is not large enough or when $d \leq 2$. In such cases localization may occur and the eigenvalue spacing will be Poisson. Note that in one dimension localization effects should not appear unless $N \gg W^2$.

The local eigenvalue spacing statistics for a random Schrödinger matrix in a large box is expected to match those of GOE after simple rescaling, provided that one looks in a region of energies where diffusion should appear, that is, (7.3) holds. RS corresponds to GOE rather than GUE because it is real symmetric matrix and the corresponding saddle manifold is different from GUE. See (Valko and Virag (to appear)) for some interesting results showing that in some special, limiting cases the local eigenvalue spacings for the random Schrödinger matrix have GOE statistics.

Remark 7.1 *The local eigenvalue statistics of GUE appear to be mysteriously connected to the statistics of the spacings of zeros of the Riemann zeta function. See (Rudnick and Sarnak 1996) and Keating's article in (Brezin et al. 2006) for reviews of this conjecture and others concerning the relation of random matrix theory and number theory.*

7.3.3 Green's functions and Gaussian integrals

For an $N \times N$ Hermitian matrix H, define the inverse matrix:

$$G(E_\epsilon) = (E_\epsilon - H)^{-1} \quad where \quad E_\epsilon = E - i\epsilon . \tag{7.15}$$

This a bounded matrix provided that E is real and $\epsilon > 0$. The Green's matrix is denoted $G(E_\epsilon)$, and $G(E_\epsilon; k, j)$, its matrix elements, will be called the Green's function.

Let $z = (z_1, z_2, ..., z_N)$ with $z_j = x_j + iy_j$ denote an element of $\mathbb{C}^N$ and define the quadratic form

$$[z ; H z] = \sum_{i,j} \bar{z}_k H_{kj} z_j . \tag{7.16}$$

Then we can calculate the following Gaussian integrals:

$$\int e^{-i[z;(E_\epsilon - H)z]} Dz = (-i)^N \det(E_\epsilon - H)^{-1} \quad where \quad Dz \equiv \prod_j^N dx_j \, dy_j / \pi \tag{7.17}$$

and

$$\int e^{-i[z;(E_\epsilon - H)z]} z_k \bar{z}_j \, Dz = (-i)^{N+1} \det(E_\epsilon - H)^{-1} G(E_\epsilon; k, j) . \tag{7.18}$$

It is important to note that the integrals above are *convergent* provided that $\epsilon > 0$. The quadratic form $[z; (E - H)z]$ is real so its contribution only oscillates. The factor of $i = \sqrt{-1}$ in the exponent is needed because the matrix $E - H$ has an indefinite signature when E is in the spectrum of H.

There is a similar identity in which the complex commuting variables z are replaced by anticommuting Grassmann variables ψ_j, $\bar\psi_j$, j = 1, 2 ... N. Let A be an $N \times N$ matrix and as in (7.16) define

$$[\psi; A\,\psi] = \sum \bar\psi_k A_{kj}\,\psi_j .$$

Then we have

$$\int e^{-[\psi;A\psi]} D\psi = \det A \quad where \quad D\psi \equiv \prod_j^N d\bar\psi_j d\psi_j .$$

(7.19)

See Appendix 7.10 for a brief review of Gaussian and Grassmann integration. The Grassmann integral is introduced so that we can cancel the unwanted determinant in (7.18) and so that we can perform the average over the randomness in H. Thus we obtain a SUSY representation for the Green's function:

$$G(E_\epsilon; k, j) = i \int e^{-i[z;(E_\epsilon - H)z]} e^{-i[\psi;(E_\epsilon - H)\psi]} z_k \bar z_j \, Dz \, D\psi .$$

(7.20)

Equation (7.20) is the starting point for all SUSY formulas. Notice that if H has a Gaussian distribution, the expectation of (7.20) can be explicitly performed since H appears linearly. We obtain:

$$< G(E_\epsilon; k, j) >= i \int e^{-iE_\epsilon([z;z]+[\psi;\psi])} e^{-\frac{1}{2}<\{[z;Hz]+[\psi;H\psi]\}^2>} z_k \bar z_j \, Dz \, D\psi.$$

(7.21)

The resulting lattice field model will be quartic in the z and ψ fields. Notice that if the observable $i z_k \bar z_j$ were absent from (7.21), then the determinants would cancel and the integral would be equal to 1 for all parameters. Thus in SUSY systems, the *partition function is identically 1.*

In a similar fashion we can obtain more complicated formulas for

$$\Gamma(E_\epsilon, j) =< G(E_\epsilon; 0, j)\bar G(E_\epsilon; 0, j) > .$$

To do this we must introduce additional variables $w \in \mathbb{C}^N$ and independent Grassmann variables $\chi, \bar\chi$ to obtain the second factor, $\bar G$, which is the complex conjugate of G. See Sections 7.7 and 7.9 below.

7.4   Averaging $Det(E_\epsilon - H)^{-1}$

Before working with Grassmann variables we shall first show how to calculate the average of the inverse determinant over the Gaussian disorder using only the complex variables z_j. Although this average has no physical significance, it will illustrate some basic mathematical ideas.

First consider the simplest case: H is an $N \times N$, GUE matrix. Let us apply (7.17) to calculate

$$(i)^{-N} < \det(E_\epsilon - H)^{-1} >=< \int e^{-i[z;\,(E_\epsilon - H)z]}\,Dz > . \tag{7.22}$$

Since the expectation is Gaussian and the covariance is given by (7.12), the average over H can be expressed as follows:

$$< e^{-i[z;\,Hz]} >_{GUE}= e^{-1/2<[z;\,Hz]^2>} = e^{-\frac{1}{2N}[z;z]^2}. \tag{7.23}$$

The most direct way to estimate the remaining integral over the z variables is to introduce a new coordinate $r = [z, z] = \sum |z_j|^2$. Then we have

$$< \det(E_\epsilon - H)^{-1} >= C_N \int_0^\infty e^{-\frac{1}{2N}r^2 - iE_\epsilon r} r^{N-1}\,dr$$

where C_N is an explicit constant related to the volume of the sphere in 2N dimensions. It is convenient to rescale $r \to Nr$ and obtain an integral of the form

$$\int_0^\infty e^{-N(r^2/2 - \ln r - iE_\epsilon r)}\,dr/r .$$

The method of steepest descent can now be applied. We deform the contour of integration over r so that it passes through the saddle point. The saddle point r_s is obtained by setting the derivative of the exponent to 0: $r_s - 1/r_s - iE_\epsilon = 0$ This is a quadratic equation with a solution $r_s = iE/2 + \sqrt{1 - (E/2)^2}$. The contour must be chosen so that the absolute value of the integrand is dominated by the saddle point.

Exercise 7.2 Stirling's formula:

$$N! = \int_0^\infty e^{-t} t^N dt \approx N^N e^{-N}\sqrt{2N\pi}.$$

To derive this let $t = Ns$ and expand to quadratic order about the saddle point $s = 1$. The square root arises from the identity $N \int e^{-Ns^2/2} ds = \sqrt{2N\pi}$. We shall see many SUSY calculations are related to generalizations of the gamma function.

Remark 7.3 *For other ensembles, radial coordinates do not suffice and one must compute the Jacobian for the new collective variables. See the section on matrix polar coordinates in Appendix 7.10. Collective coordinates can also be set up with the help of the coarea formula.*

An alternative way to compute $< \det(E_\epsilon - H)^{-1} >$ uses the Hubbard–Stratonovich transform. In its simplest form, introduce a real auxiliary variable a to unravel the quartic expression in z as follows:

$$e^{-\frac{1}{2N}[z;z]^2} = \sqrt{N/2\pi} \int e^{-Na^2/2} e^{ia[z;z]}\,da . \tag{7.24}$$

If we apply (7.24) to (7.22) and (7.23), the z variables now appear quadratically and we can integrate over them. Since there are N independent z_j to integrate, we get a factor $(E_\epsilon - a)^{-N}$, hence:

$$< \det(E_\epsilon - H)^{-1} >= i^N \sqrt{N/2\pi} \int e^{-Na^2/2} (E_\epsilon - a)^{-N} da . \qquad (7.25)$$

Let

$$f(a) = N\left[a^2/2 + ln(E_\epsilon - a)\right].$$

The saddle point a_s is obtained by setting $f'(a_s) = 0$. This gives a quadratic equation whose solution is

$$a_s = E_\epsilon/2 + i\sqrt{1 - (E_\epsilon/2)^2}. \qquad (7.26)$$

We shift our contour of integration by $a \to a + a_s$. Note $|a_s| \approx 1$ and that we have chosen the $+$ sign in (7.26) so that the pole of $(E - i\epsilon - a)^{-N}$ has not been crossed. Along this contour one checks that for E satisfying $|E| \leq 1.8$, the maximum modulus of the integrand occurs at the saddle a_s. In particular, this deformation of contour avoids the small denominator $E_\epsilon - a$ occurring when $a \approx E$. For energies such that $1.8 \leq |E| \leq 2$ another contour must be used, (Disertori 2004). Note that the Hessian at the saddle is

$$f''(a_s) = N(1 - a_s^2) = N\{2(1 - (E/2)^2) - iE\sqrt{1 - (E/2)^2}\} \qquad (7.27)$$

has a positive real part for $|E| < 2$. To complete our estimate of (7.25) for large N we calculate the quadratic correction to the saddle point:

$$< \det(E_\epsilon - H)^{-1} >\approx i^N \sqrt{N/2\pi} e^{-Nf(a_s)} \int e^{-N/2f''(a_s)a^2} da .$$

Higher order corrections to the saddle point are suppressed by powers of N^{-1}.

Remark 7.4 *The imaginary part of right-hand side of (7.25), (N even), is proportional to*

$$e^{-E^2/2} H_N(E\sqrt{N/2})$$

where H_N is the Nth Hermite polynomial. This can be seen by integrating by parts $N - 1$ times. Thus such identities point to a connection between the method of orthogonal polynomials and SUSY statistical mechanics. See (Disertori 2004) for more details.

7.4.1   Gaussian band matrices and statistical mechanics

Now let us consider the more general case when H is a finite Gaussian random band matrix indexed by lattice sites in a periodic box $\Lambda \subset \mathbb{Z}^d$ with covariance given by (7.13) and (7.14): $J_{jk} = (-W^2\Delta + 1)^{-1}(j, k)$. Then we have

$$< e^{-i[z;Hz]} >= e^{-1/2<[z;Hz]^2>} = e^{-1/2\sum |z_i|^2 J_{ij}|z_j|^2}. \qquad (7.28)$$

In order to average over the z variables we introduce real Gaussian auxiliary fields a_j with 0 mean and covariance J_{ij}. Let $< \cdot >_J$ be the corresponding expectation so that $< a_i a_j >_J = J_{ij}$ and

$$e^{-1/2 \sum_{ij} |z_i|^2 J_{ij} |z_j|^2} = < e^i \sum a_j |z_j|^2 >_J . \tag{7.29}$$

This is again the Hubbard–Stratonovich trick. We can now average over the z's since they appear quadratically. By combining (7.22), (7.28), and (7.29)

$$< \det(E_\epsilon - H)^{-1} > = < \int e^{-i[z;(E_\epsilon - H)z]} Dz > \tag{7.30}$$

$$= < \int e^{-i[z;(E_\epsilon - a)z]} Dz >_J = < \prod_{j \in \Lambda} (E_\epsilon - a_j)^{-1} >_J .$$

Since by (7.14), the Gaussian a_j fields have covariance $(-W^2 \Delta + 1)^{-1}$, it follows that (7.30) is proportional to:

$$\int e^{-\frac{1}{2} \sum_j (W^2 (\nabla a)_j^2 + a_j^2)} \prod_{j \in \Lambda} (E_\epsilon - a_j)^{-1} da_j . \tag{7.31}$$

The expression $(\nabla a)_j^2$ is the square of finite difference gradient at j. The large parameter W tends to make the a_j fields constant over a large range of lattice sites. The saddle point can be calculated as it was for GUE and we find that it is independent of the lattice site j and is again given by (7.26). We deform the contour of integration $a_j \to a_j + a_s$. We must restrict $|E| \leq 1.8$ as in the case of GUE so that the norm of the integrand along the shifted contour is maximized at the saddle. The Hessian at the saddle is given by

$$H_s'' = -W^2 \Delta + 1 - a_s^2 . \tag{7.32}$$

Since $Re(1 - a_s^2) > 0$, $[H'']^{-1}(j, k)$ decays exponentially fast for separations $|j - k| \gg W$.

Since (7.31) is an integral over many variables a_j, $j \in \Lambda$, its behaviour is not so easy to estimate as (7.25). To control fluctuations away from the saddle one uses the exponential decay of $[H'']^{-1}$ and techniques of *statistical mechanics* such as the cluster or polymer expansion, (Salmhofer 1999), about blocks of side W. This will show that the integral can be approximated by a product of nearly independent integrals over boxes of size W. It enables one to justify the asymptotic expansion for

$$|\Lambda|^{-1} \log < \det(E_\epsilon - H)^{-1} >$$

in powers of W^{-1}. Moreover, one can prove that with respect to the expectation defined using (7.31), a_j, a_k are exponentially independent for $|j - k| \gg W$ uniformly in $\epsilon > 0$. The main reason this works is that the inverse of the Hessian at a_s has exponential decay and the corrections to the Hessian contribution are small when W is large. As in

(7.25), corrections beyond quadratic order are small for large N. See (Constantinescu *et al.*, 1987; Disertori *et al.*, 2002) for the SUSY version of these statements.

Remark 7.5 *In one dimension, (7.31) shows that we can calculate the* $< \det(E_\epsilon - H)^{-1} >$ *using a nearest neighbour transfer matrix. In particular if the lattice sites j belong to a circular chain of length L, there is an integral operator $T = T(a, a')$ such that* $< \det(E_\epsilon - H)^{-1} > = Tr\, T^L$.

Remark 7.6 *For the lattice random Schrödinger matrix, $H = -\Delta + \lambda v$, in a finite box, with a Gaussian potential v of mean 0 and variance 1, it is easy to show that*

$$< \det(E_\epsilon - H)^{-1} > = \int e^{-i\,[z\,;(E_\epsilon+\Delta)z\,] - \frac{\lambda^2}{2}\sum_j |\bar{z}_j z_j|^2}\, Dz\ .$$

The complex variables z_j are indexed by vertices in a finite box contained in $\mathbb{Z}^d$. For small λ this is a highly oscillatory integral which is more difficult than (7.31) to analyze. Rigorous methods for controlling such integrals are still missing when λ is small and $d \geq 2$. Note that for RBM, the oscillations are much less pronounced because the off diagonal terms are also random. In this case the factor

$$\exp -\frac{1}{2}\Big(\sum_j W^2(\nabla a)_j^2 + a_j^2\Big)$$

dominates our integral for large W.

7.5   The density of states for GUE

The *integrated* density of states for an $N \times N$ Hermitian matrix H is denoted $n(E) = \int^E d\rho(E')$ is the fraction of eigenvalues less than E and $\rho(E)$ denotes the density of states. The average of the density of states is given by the expression

$$< \rho_\epsilon(E) > = \frac{1}{N} tr < \delta_\epsilon(H - E) > = \frac{1}{N\pi} tr\, Im\, < G(E_\epsilon) > \qquad (7.33)$$

as $\epsilon \downarrow 0$. Here we are using the well known fact that

$$\delta_\epsilon(x - E) \equiv \frac{1}{\pi}\frac{\epsilon}{(x - E)^2 + \epsilon^2} = \frac{1}{\pi}Im(E_\epsilon - x)^{-1}$$

is an approximate delta function at E as $\epsilon \to 0$.

Remark 7.7 *The Wigner semicircle distribution asserts that the density of states of a GUE matrix is given by $\pi^{-1}\sqrt{1 - (E/2)^2}$. Such results can be proved for many ensembles including RBM by first fixing ϵ in (7.33) and then letting N, and $W \to \infty$. Appendix 7.11 gives a simple formal derivation of this law using self-consistent perturbation theory. Note that the parameter ϵ is the scale at which we can resolve different energies. For a system of size N we would like to understand the number of eigenvalues in an energy window of width $N^{-\alpha}$ with $\alpha \approx 1$. Thus we would like to get estimates for $\epsilon \approx N^{-\alpha}$. Such estimates are difficult to obtain when $\alpha \approx 1$ but* *have recently been established for Wigner ensembles, see (Erdős et al., 2009). Using SUSY (Constantinescu et al., 1987; Disertori et al., 2002) estimates on the density of states for a special class of Gaussian band matrices indexed by $\mathbb{Z}^3$ are proved uniformly in ϵ and the size of the box for fixed $W \geq W_0$. See the theorem in this section.*

Note that the density of states is *not* a universal quantity. At the end of Appendix 7.11 we show that 2 coupled GUE $N \times N$ matrices has a density of states given by a cubic rather than quadratic equation.

We now present an identity for the average Green's function for GUE starting from equation (7.21). Note that by (7.12)

$$-\frac{1}{2} < ([z; Hz] + [\psi; H\psi])^2 >_{GUE}$$

$$= -\frac{1}{2N} \{[z; z]^2 - [\psi; \psi]^2 - 2[\psi; z][z; \psi]\}. \tag{7.34}$$

Let us introduce another real auxiliary variable $b \in \mathbb{R}$ and apply the Hubbard-Stratonovich transform to decouple the Grassmann variables. As in (7.24) we use the identity

$$e^{[\psi;\psi]^2/2N} = \sqrt{N/2\pi} \int db \, e^{-Nb^2/2} \, e^{-b[\psi;\psi]}. \tag{7.35}$$

Now by combining (7.24) and (7.35) we see that the expressions are quadratic in z and ψ. Note that

$$\int e^{-\sum^N \bar{\psi}_j \psi_j (b+iE_\epsilon)} \, D\psi \, Dz = i^N \, (E_\epsilon - i\,b)^N.$$

The exponential series for the cross terms appearing in (7.34), $[\psi; z][z; \psi]$, terminates after one step because $[\psi; z]^2 = 0$. We compute its expectation with respect to the normalized Gaussian measure in ψ, z and get

$$< [\psi; z][z; \psi] >=< \sum_{ij} z_i \bar{\psi}_i \, \bar{z}_j \psi_j >= \sum_j < \bar{\psi}_j \psi_j \bar{z}_j z_j >= -N(E_\epsilon - ib)^{-1}(E_\epsilon - a)^{-1}.$$

Thus after integrating over both the z and ψ fields and obtain the expression:

$$< \frac{1}{N} tr \, G(E_\epsilon) >= N^{-1} \int [z; z] \, e^{-\frac{1}{2N} \{[z;z]^2 - [\psi;\psi]^2 - 2[\psi;z][z;\psi]\}} \, Dz \, D\psi$$

$$= N/2\pi \int dadb \, (E_\epsilon - a)^{-1} \, e^{-N(a^2+b^2)/2} \, (E_\epsilon - ib)^N (E_\epsilon - a)^{-N}$$

$$\times [1 - \frac{N+1}{N}(E_\epsilon - a)^{-1}(E_\epsilon - i\,b)^{-1}] \approx < (E_\epsilon - a)^{-1} >_{SUSY}. \tag{7.36}$$

The first factor of $(E_\epsilon - a)^{-1}$ on the right-hand side roughly corresponds to the trace of the Green's function. The partition function is

$$N/2\pi \int dadb \, e^{-N(a^2+b^2)/2}(E_\epsilon - ib)^N (E_\epsilon - a)^{-N} [1 - (E_\epsilon - a)^{-1}(E_\epsilon - ib)^{-1}] \equiv 1 \tag{7.37}$$

for all values of E, ϵ and N. This is due to the fact that if there is no observable the determinants cancel. The last factor in (7.37) arises from the cross terms and is called the Fermion-Boson **(FB)** contribution. It represents the coupling between the determinants. For band matrices it is useful to introduce auxiliary dual Grassmann variables (analogous to a and b) to treat this contribution. See (Disertori, 2004; Disertori, Pinson, and Spencer, 2002; Mirlin, 2000).

The study of $\rho(E)$ reduces to the analysis about the saddle points of the integrand. As we have explained earlier, there is precisely one saddle point

$$a_s(E) = E_\epsilon/2 + i\sqrt{1 - (E_\epsilon/2)^2} \tag{7.38}$$

in the a field. However, the b field has two saddle points

$$i b_s = a_s, \quad and \quad i b'_s = E_\epsilon - a_s = \bar{a}_s \, .$$

Hence, both saddle points (a_s, b_s) and (a_s, b'_s) contribute to (7.36).

Let us briefly analyse the fluctuations about the saddles as we did in (7.26), (7.27). The first saddle gives the Wigner semicircle law. To see this note that the action at a_s, b_s takes the value 1 since $a_s^2 + b_s^2 = 0$ and $(E_\epsilon - ib_s)/(E_\epsilon - a_s) = 1$. The integral of quadratic fluctuations about the saddle,

$$N \int e^{-N(1-a_s^2)(a^2+b^2)} dadb \tag{7.39}$$

is exactly canceled by the **FB** contribution at (a_s, b_s). Thus to a high level of accuracy we can simply replace the observable in the SUSY expectation (7.36) by its value at the saddle. This gives us Wigner's semicircle law:

$$\rho(E) = \frac{1}{\pi N} Im \, tr < G(E_\epsilon) > \approx \pi^{-1} Im < (E_\epsilon - a)^{-1} >_{SUSY}$$

$$\approx \pi^{-1} Im(E_\epsilon - a_s)^{-1} = \pi^{-1}\sqrt{1 - (E/2)^2}. \tag{7.40}$$

It is easy to check that the second saddle vanishes when inserted into the **FB** factor because $(E - a_s)(E - ib'_s) = \bar{a}_s a_s = 1$. Thus to leading order it does not contribute to the density of states and hence (7.40) holds. However, the second saddle will contribute a highly oscillatory contribution to the action proportional to

$$\frac{1}{N}(\frac{E_\epsilon - a'_s}{E_\epsilon - a_s})^N e^{-N/2(a_s^2 - \bar{a}_s^2)} \, . \tag{7.41}$$

If we take derivatives in E, this contribution is no longer suppressed when $\epsilon \approx 1/N$. This result is not easily seen in perturbation theory. I believe this is a compelling example of the nonperturbative power of the SUSY method.

Remark 7.8 *If $\epsilon = 0$, then (7.41) has modulus $1/N$. We implicitly assume that the energy E is inside the bulk that is $|E| < 2$. Near the edge of the spectrum the Hessian at the saddle point vanishes and a more delicate analysis is called for. The density of states near $E = \pm 2$ then governed by an Airy function. We refer to Disertori's review of GUE, (Disertori 2004), for more details.*

7.5.1 Density of states for RBM

We conclude this section with a brief discussion of the average Green's function for RBM in three dimensions with the covariance J given by (7.14). The SUSY statistical mechanics for RBM is expressed in terms of a_j, b_j and Grassmann fields $\bar{\eta}_j, \eta_j$ with $j \in \Lambda \subset Z^d$ and it has a local (nearest neighbour) weight

$$\exp[-\frac{1}{2}\sum_j \{W^2(\nabla a_j)^2 + W^2(\nabla b_j)^2 + a_j^2 + b_j^2\}] \prod_j \frac{E_\epsilon - ib_j}{E_\epsilon - a_j}$$

$$\times \exp - \sum_j \{W^2 \nabla\bar{\eta}_j \nabla\eta_j + \bar{\eta}_j\eta_j (1 - (E_\epsilon - a_j)^{-1}(E_\epsilon - ib_j)^{-1})\}.$$

After the Grassmann variables have been integrated out we get:

$$\exp[-\frac{1}{2}\sum_j \{W^2(\nabla a_j)^2 + W^2(\nabla b_j)^2 + a_j^2 + b_j^2\}] \prod_j \frac{E_\epsilon - ib_j}{E_\epsilon - a_j}$$

$$\times \det\{-W^2\Delta + 1 - \delta_{ij}(E_\epsilon - a_j)^{-1}(E_\epsilon - ib_j)^{-1}\} \, DaDb. \tag{7.42}$$

Note the similarity of the above weight with (7.31) and (7.37). The determinant is called the Fermion-Boson contribution. In fact, in one dimension, the DOS can be reduced to the analysis of a nearest neighbor transfer matrix. Large W keeps the fields a_j, b_j nearly constant. This helps to control fluctuations about saddle point.

Theorem 7.9. (Disertori, Pinson, and Spencer, 2002) *Let d=3, J be given by (7.14) and $|E| \leq 1.8$. For $W \geq W_0$ the average $< G(E_\epsilon, j, j) >$ for RBM is uniformly bounded in ϵ and Λ. It is approximately given by the Wigner distribution with corrections of order $1/W^2$. Moreover, we have*

$$|\langle G(E_\epsilon; 0, x) \, G(E_\epsilon; x, 0)\rangle| \leq e^{-m|x|} \tag{7.43}$$

for $m \propto W^{-1}$.

The proof of this theorem follows the discussion after (7.31). We deform our contour of integration and prove that for $|E| \leq 1.8$, the dominant contribution comes from the saddle point a_s, b_s which is independent of j. However, as in GUE one must take into account the second saddle which complicates the analysis. Note that the Hessian at the saddle is still given by (7.32). Its inverse governs the decay in (7.43).

Remark 7.10 *The first rigorous results of this type appear in (Constantinescu et al., 1987) for the Wegner N orbital model. Their SUSY analysis is similar to that presented in (Disertori et al., 2002).*

Remark 7.11 *One can derive a formula analogous to (7.42) for the random Schrodinger matrix in 2D or 3D, but we do not have good control over the SUSY statistical mechanics for small l. See (7.42) and Remark 7.6. A formal perturbative analysis of the average Green's function for the random Schrödinger matrix with small λ is given in Appendix 7.11. Note that for large λ, it is relatively easy to control the SUSY integral since in this case the SUSY measure is dominated by the product measure $\prod_j exp[-\lambda^2(\bar{z}_j z_j + \bar{\psi}_j \psi_j)^2]$. As a final remark note that in the product of Green's functions appearing in (7.43), energies lie on the same side of the real axis. The expectation $< |G(E_\epsilon; 0, j)|^2 >$ is quite different and for such cases hyperbolic symmetry emerges, see Section 7.7 and gapless modes may appear.*

7.6   Statistical mechanics, sigma models, and Goldstone modes

In this section we shall review some classical sigma models and their lattice field theory counterparts. This may help to see how classical statistical mechanics is related to the SUSY sigma models which will be described later. The symmetry groups described in this section are compact whereas those of related the Efetov models are noncompact as we shall see in Sections 7.7–7.9.

Consider a lattice model with a field $\phi(j) = (\phi^1(j), ..., \phi^{m+1}(j))$ taking values in $\mathbb{R}^{m+1}$ and whose quartic action in given by

$$A_\Lambda(\phi) = \sum_{j \in \Lambda} \{|\nabla\phi(j)|^2 + \lambda(|\phi(j)|^2 - a)^2 + \epsilon \cdot \phi^1\}. \tag{7.44}$$

As before ∇ denotes the nearest neighbour finite difference gradient and Λ is a large periodic box contained in $\mathbb{Z}^d$. The partition function takes the form

$$Z_\Lambda = \int e^{-\beta A_\Lambda(\phi)} \prod_{j \in \Lambda} d\phi_j .$$

To define its sigma model approximation, consider spins, $S_j \in S^m$ where $j \in \Lambda$. More precisely $S_j = (S_j^{(1)}, ... S_j^{(m+1)})$ with $S_j{}^2 = S_j \cdot S_j = 1$. For m=0, the spin takes values ± 1 and this is the Ising spin. The energy or action of a configuration $\{S_j\}$ is given by

$$A_{\bar{\beta},\Lambda}(S) = \bar{\beta} \sum_{j \sim j' \in \Lambda} (S_j - S_{j'})^2 - \epsilon \sum_j S_j^{(1)}. \tag{7.45}$$

The Gibbs weight is proportional to $e^{-A(S)}$, the parameters $\bar{\beta}$, $\epsilon \geq 0$ are proportional to the inverse temperature and the applied field respectively, and $j \sim j'$ denote adjacent lattice sites. If $\epsilon > 0$ then the minimum energy configuration is unique and the all the spins point in the direction $(1, 0, ...0)$.

If $\epsilon = 0$, A is invariant under a global rotation of the spins. The energy is minimized when all the spins are pointing in the same direction. This is the ground state of the system. When m $= 1$ we have $O(2)$ symmetry and for m $= 2$ the symmetry is $O(3)$ corresponding to the classical rotator (or X-Y) model and classical Heisenberg model respectively. When $\epsilon > 0$ the full symmetry is broken but in the case of $O(3)$ is broken to $O(2)$. The parameter ϵ in (7.45) also appears in SUSY models and is related to the imaginary part of E_ϵ in the Green's function.

Consider the classical scalar $|\phi|^4$ model, (7.44), with $\phi_j \in \mathbb{C}$ with a $U(1)$ symmetry. The idea behind the sigma approximation is that the complex field ϕ_j can be expressed in polar coordinates $r_j S_j$ where $|S_j| = 1$. The radial variables r are 'fast' variables fluctuating about some effective radius $r*$. If the quartic interaction has the form $\lambda(|\phi_j|^2 - 1)^2$ and $\lambda \gg 1$, $r^* \approx 1$. If we freeze $r_j = r*$ we obtain (7.45) with $\bar{\beta} \approx (r*)^2\beta$. The $|\phi|^4$ model is expected to have the same qualitative properties as the $O(2)$ sigma models. Moreover, at their respective phase transitions, the long distance asymptotics of correlations should be identical. This means that the two models have the same critical exponents although their critical temperatures will typically be different. This is a reflection of the principle universality for second order transitions. Its proof is a major mathematical challenge.

Remark 7.12 *The role of large λ above will roughly be played by W in the SUSY models for random band matrices.*

Now let us discuss properties of the sigma models with Gibbs expectation:

$$< \cdot >_\Lambda (\bar{\beta}, \epsilon) = Z_\Lambda(\bar{\beta}, \epsilon)^{-1} \int \cdot \, e^{-A_{\bar{\beta}\Lambda}(S)} \prod_j d\mu(S_j) \, . \tag{7.46}$$

The measure $d\mu(S_j)$ is the uniform measure on the sphere. The partition function Z is defined so that this defines a probability measure. Basic correlations are the spin-spin correlations $< S_0 \cdot S_x >$ and the magnetization $< S_0^1 >$. Let us assume that we have taken the limit in which $\Lambda \uparrow Z^d$. First consider the case in which here is no applied field $\epsilon = 0$. In one dimension there is no phase transition and the spin-spin correlations decay exponentially fast for all nonzero temperatures, (β finite),

$$0 \leq< S_0 \cdot S_x >\leq Ce^{-|x|/\ell} \tag{7.47}$$

where ℓ is the correlation length. For the models with continuous symmetry, $m = 1, 2$, ℓ is proportional to β. However, for the Ising model ℓ is exponentially large in β. At high temperature, β small, it is easy to show that in any dimension the spins are independent at long distances and (7.47) holds.

In dimension $d \geq 2$, the low temperature phase is interesting and more difficult to analyse. In two dimensions we know that the Ising model has long range order and a *spontaneous magnetization* M for $\beta > \beta_c$. This means that for large x

$$< S_0 \cdot S_x > (\beta, \epsilon = 0) \rightarrow M^2(\beta) > 0 \ \ and \ \ \lim_{\epsilon \downarrow 0} < S_0^{(1)} > (\beta, \epsilon) = M \ . \qquad (7.48)$$

Thus at low temperature spins align even at long distances. Note that the order of limits here is crucial. In (7.48) we have taken the infinite volume limit first and then sent $\epsilon \rightarrow 0$. For any finite box it is easy to show that second the limit is 0 because the symmetry of the spins is present. For this reason (7.48) is called symmetry breaking.

In two dimensions, the Mermin–Wagner Theorem states that a continuous symmetry cannot be broken. Hence, $M = 0$ for the $O(2)$ and $O(3)$ models. In classical XY model with $O(2)$ symmetry there is a Kosterlitz–Thouless transition (Kosterlitz and Thouless 1973; Fröhlich, and Spencer 1981): the magnetization, M, vanishes but the spin-spin correlation with $\epsilon = 0$ goes from an exponential decay for small β to a slow power law decay $\approx |x|^{-c/\beta}$ for large β.

A major conjecture of mathematical physics, first formulated by Polyakov, states that for the classical Heisenberg model in two dimensions, the spin-spin correlation always decays exponentially fast with correlation length $\ell \approx e^{\beta}$. This conjecture is related to Anderson localization expected in two dimensions. It is a remote cousin of confinement for non-Abelian gauge theory. The conjecture can partly be understood by a renormalization group argument showing that the effective temperature increases under as we scale up distances. The positive curvature of sphere play a key role in this analysis.

In three dimensions it is known through the methods of reflection positivity, (Fröhlich *et al.* 1976), that there is long range order and continuous symmetry breaking at low temperatures. These methods are rather special and do not apply to the SUSY models described here.

Continuous symmetry breaking is accompanied by Goldstone Bosons. This means that there is a very slow decay of correlations. This is most easily seen in the Fourier transform of the pair correlation. Consider a classical XY or Heisenberg model on $\mathbb{Z}^d$. The following theorem is a special case of the Goldstone–Nambu–Mermin–Wagner Theorem. Let M be defined as in (7.48).

Theorem 7.13 *Let correlations be defined as above and suppose that the spontaneous magnetization $M > 0$. Then*

$$\sum_x e^{-ip \cdot x} < S_0^{(2)} \ S_x^{(2)} > (\beta, \epsilon) \geq CM^2(\beta p^2 + \epsilon M)^{-1} \ . \qquad (7.49)$$

Thus for $p = 0$ the sum diverges as $\epsilon \downarrow 0$. This theorem applies to the case $m = 1, 2$ but not the Ising model. It is also established for lattice field theories with continuous symmetry.

Remark 7.14 *In two dimensions, if the spontaneous magnetization were positive, then the integral over p of the right side of (7.49) diverges for small ϵ. On the other hand, the integral over p of the left side equals 1 since $S_0 \cdot S_0 = 1$. Thus M must vanish in two dimensions. In higher dimensions there is no contradiction since the integral of $|p|^{-2}, |p| \leq \pi$ is finite.*

We shall derive a simple Ward identity for the $O(2)$ sigma model in d dimensions. In angular coordinates we have

$$A(\theta) = -\sum_{j \sim j'} \beta \cos(\theta_j - \theta_{j'}) - \epsilon \sum_j \cos(\theta_j) . \tag{7.50}$$

Consider the integral in a finite volume: $\int sin(\theta_0)e^{-A(\theta)} \prod_{j \in \Lambda} d\theta_j$. Make a simple change of variables $\theta_j \to \theta_j + \alpha$. The integral is of course independent of α. If we take the derivative in α and then set $\alpha = 0$ the resulting expression must vanish. This yields a simple Ward identity

$$M = <cos(\theta_0)> = \epsilon \sum_j < sin(\theta_0) \sin(\theta_j) > . \tag{7.51}$$

After dividing by ϵ we obtain the Theorem for $p = 0$. Ward identities are often just infinitesimal reflections of symmetries in our system. In principle we can apply this procedure to any one parameter change of variables.

Remark 7.15 *Note the similarity of (7.51) with the Ward identity (7.4). The density of states $\rho(E)$ plays the role of the spontaneaous magnetization M. Moreover, the right side of (7.49) is analogous to (7.3). However, there is an important distinction: $\rho(E)$ does not vanish in the high temperature or localized regions. Unlike the magnetization in classical models, $\rho(E)$ is not an order parameter. Indeed, $\rho(E)$ is not expected to reflect the transition between localized and extended states. So in this respect classical and SUSY models differ. For SUSY or RBM, the localized phase is reflected in a vanishing of the diffusion constant in (7.3), $D(E, \epsilon) \propto \epsilon$. Thus in regions of localization $< |G(E_\epsilon, 0, 0)|^2 >$ diverges as $\epsilon \to 0$ as we see in (7.1).*

Theorem 7.13: Let $|\Lambda|$ denote the volume of the periodic box and set $D = |\Lambda|^{-1/2} \sum_j e^{-ip \cdot j} \frac{\partial}{\partial \theta_j}$ and $\hat{S}(p) = |\Lambda|^{-1/2} \sum_j e^{+ip \cdot j} \sin(\theta_j)$. By translation invariance, integration by parts, and the Schwarz inequality we have

$$M = <\cos(\theta_0)>_A = <D\hat{S}(p)>_A = <\hat{S}(p)D(A)>_A \tag{7.52}$$

$$\leq < |\hat{S}(p)|^2 >_A^{1/2} < \overline{D(A)}D(A) >_A^{1/2} .$$

Since $< |\hat{S}(p)|^2 >$ equals the left side of (7.49) the theorem follows by integrating by parts once again,

$$< \overline{D(A)}D(A) >_A = < D\overline{D(A)} >_A$$

$$= |\Lambda|^{-1} \sum_{j \sim j'} < 2\beta cos(\theta_j - \theta_{j'})(1 - \cos(j - j')p) + \epsilon cos(\theta_j) >_A$$

$$\leq C(\beta p^2 + \epsilon < \cos(\theta_0) >_A) \tag{7.53}$$

which holds for small p. Here we have followed the exposition in (Fröhlich and Spencer 1981).

7.7 Hyperbolic symmetry

Let us analyse the average of $|\det(E_\epsilon - H)|^{-2}$ with H a GUE matrix. We study this average because it is the simplest way to illustrate the emergence of hyperbolic symmetry. This section analyses the so called Boson–Boson sector in the sigma model approximation. More complicated expressions involving Grassmann variables appear when analyzing the average of $|G(E_\epsilon; j, k)|^2$. This is touched upon in Section 7.9.

To begin, let us contrast $< G >$ and $< |G|^2 >$ for the trivial scalar case N=1. In the first case we see that $\int e^{-H^2}(E_\epsilon - H)^{-1}dH$ is finite as $\epsilon \to 0$ by shifting the contour of integration off the real axis $H \to H + i\delta$ with $\delta > 0$ so that the pole is not crossed. On the other hand, if one considers the average of $|(E_\epsilon - H)|^{-2}$, we cannot deform the contour integral near E and the integral will diverge like ϵ^{-1}. This divergence will be reflected in the hyperbolic symmetry. Later in this section we shall see that in three dimensions this symmetry will be associated with gapless Goldstone modes. These modes were absent in Sections 7.3–7.5.

Let $z, w \in C^N$. As in (7.17) we can write:

$$|\det(E_\epsilon - H)|^{-2} = \det(E_\epsilon - H) \times \det(E_{-\epsilon} - H)$$

$$= \int e^{-i\,[z;(E_\epsilon-H)z]}\,Dz \times \int e^{i\,[w;(E_{-\epsilon}-H)w]}\,Dw \,. \qquad (7.54)$$

Note that the two factors are complex conjugates of each other. The factor of i has been reversed in the w integral to guarantee its convergence. This sign change is responsible for the hyperbolic symmetry. The Gaussian average over H is

$$< e^{-i([zHz]-[w,Hw])} >= e^{-1/2<([z,Hz]-[w,Hw])^2>}. \qquad (7.55)$$

Note that

$$< ([z, Hz] - [w, Hw])^2 >=< [\sum H_{kj}(\bar{z}_k z_j - \bar{w}_k z_j)]^2 > . \qquad (7.56)$$

For GUE the right side is computed using (7.12)

$$< ([z, Hz] - [w, Hw])^2 >= 1/N([z, z]^2 + [w, w]^2 - 2[z, w][w, z]) \,. \qquad (7.57)$$

Note that the hyperbolic signature is already evident in (7.55). Following Fyodorov, (Fyodorov 2002), introduce the 2×2 non-negative matrix:

$$M(z, w) = \begin{pmatrix} [z, z] & [z, w] \\ [w, z] & [w, w] \end{pmatrix} \qquad (7.58)$$

and let

$$L = diag(1, -1) \,. \qquad (7.59)$$

Then we see that

$$< |\det(E_\epsilon - H)|^{-2} >= \int e^{-\frac{1}{2N} tr(ML)^2 - iEtr(ML) + \epsilon tr M}\,Dz Dw \,. \qquad (7.60)$$

For a positive 2×2 matrix P, consider the delta function $\delta(P - M(z,w))$ and integrate over z and w. It can be shown (Fyodorov 2002; Spencer and Zirnbauer 2004), also see (7.116) in Appendix 7.10, that

$$\int \delta(P - M(z,w)) Dz Dw = (det P)^{N-2}. \tag{7.61}$$

Assuming this holds we can now write the right side in terms of the new collective coordinate P

$$< |\det(E_\epsilon - H)|^{-2} > = C_N \int_{P>0} e^{-\frac{1}{2N} tr(PL)^2} e^{-iEtr(PL) - \epsilon tr P} det P^{N-2} dP. \tag{7.62}$$

After rescaling $P \to NP$ we have

$$< |\det(E_\epsilon - H)|^{-2} > = C'_N \int_{P>0} e^{-N\{tr(PL)^2/2 + iEtr(PL) + \epsilon tr P\}} det P^{N-2} dP. \tag{7.63}$$

In order to compute the integral we shall again change variables and integrate over PL. First note that for $P > 0$, PL has two real eigenvalues of opposite sign. This is because PL has the same eigenvalues as $P^{1/2} L P^{1/2}$ which is self-adjoint with a negative determinant. Moreover, it can be shown that

$$PL = TDT^{-1} \tag{7.64}$$

where $D = diag(p_1, -p_2)$ with p_1, p_2 positive and T belongs to $U(1,1)$. By definition $T \in U(1,1)$ when

$$T^* LT = L \to T \in SU(1,1). \tag{7.65}$$

The proof is similar to that for Hermitian matrices. We shall regard PL as our new integration variable.

Note that (7.63) can be written in terms of p_1, p_2 except for $\epsilon\, tr\, P$ which will involve the integral over $SU(1,1)$. The p_1, p_2 are analogous to the radial variable r introduced in Section 7.6. Converting to the new coordinate system our measure becomes

$$(p_1 + p_2)^2 dp_1 dp_2 d\mu(T) \tag{7.66}$$

where $d\mu(T)$ the Haar measure on U(1,1). For large N, the p variables are approximately given by the complex saddle point

$$p_1 = -iE/2 + \rho(E), \quad -p_2 = -iE/2 - \rho(E), \quad \rho(E) \equiv \sqrt{1 - (E/2)^2}. \tag{7.67}$$

The p variables fluctuate only slightly about (7.67) while the T matrix ranges over the symmetric space $SU(1,1)/U(1)$ and produces a singularity for small ϵ. With the p_1, p_2 fixed the only remaining integral is over $SU(1,1)$. Thus from (7.64) we have:

$$Q \equiv PL \approx \rho(E)\, TLT^{-1} + iE/2 .$$

If we set $\epsilon = \bar{\epsilon}/N$ and take the limit of large N we see that (7.63) is given by

$$\int e^{-\bar{\epsilon}\rho(E)\,tr(LS)}\,d\mu(T), \quad where \quad S \equiv TLT^{-1}. \tag{7.68}$$

This is the basis for the sigma model. The second term iE above is independent of T so it is dropped in (7.68).

7.7.1 Random band case

The band version or Wegner's N orbital (Schäfer and Wegner 1980) version of such hyperbolic sigma models was studied in (Spencer and Zirnbauer 2004). The physical justification of this sigma model and also the Efetov SUSY sigma model comes from a granular picture of matter. The idea is that a metal consists of grains (of size $N \gg 1$). Within the grains there is mean field type interaction and there is a weak coupling across the neighbouring grains. As in (7.68) if the grain interactions are scaled properly and $N \to \infty$ we will obtain a sigma model.

For each lattice site $j \in \Lambda \subset Z^d$ we define a new spin variable given by

$$S_j = T_j^{-1}LT_j \quad and \quad P_jL \approx \rho(E)\,S_j\,.$$

Note that $S_j^2 = 1$ and S_j naturally belongs to $SU(1,1)/U(1)$. This symmetric space is isomorphic to the hyperbolic upper half plane. In the last equation we have used the form of the p_i given as above. The imaginary part of the p_1 and $-p_2$ are equal at the saddle so that T and T^{-1} cancel producing only a trivial contribution. The explicit dependence on E only appears through $\rho(E)$.

There is a similar picture for the square of the average determinant using Grassmann integration. We can integrate out the Grassmann fields and in the sigma model approximation obtain an integral over the symmetric space $SU(2)/U(1) = S^2$—this gives the classical Heisenberg model.

The action of the hyperbolic sigma model, see (Spencer and Zirnbauer 2004), on the lattice is

$$A(S) = \beta \sum_{j \sim j'} tr S_j S_{j'} + \epsilon \sum_j tr L S_j. \tag{7.69}$$

The notation $j \sim j'$ denotes nearest neighbour vertices on the lattice. The Gibbs weight is proportional to $e^{-A(S)}d\mu(T)$. The integration over $SU(1,1)$ is divergent unless $\epsilon > 0$. The last term above is a symmetry breaking term analogous to a magnetic field . For RBM, $\beta \approx W^2\rho(E)^2$.

To parametrize the integral over SU(1,1) we use *horospherical* coordinates $(s_j, t_j) \in \mathbb{R}^2$ given by (7.9). In this coordinate system, the action takes the form:

$$A(s,t) = \beta \sum_{j \sim j'} [\cosh(t_j - t_{j'}) + \frac{1}{2}(s_j - s_{j'})^2 e^{(t_j + t_{j'})}] + \epsilon \sum_{j \in \Lambda} [\cosh(t_j) + \frac{1}{2}s^2 e^{t_j}].$$

$$\tag{7.70}$$

Equivalently, if h_j are the hyperbolic spins appearing in (7.8) note that we have

$$h_j \cdot h_{j'} = z_j z_{j'} - x_j x_{j'} - y_j y_{j'} = cosh(t_j - t_{j'}) + \frac{1}{2}(s_j - s_{j'})^2 e^{(t_j + t_{j'})} .$$

The symmetry breaking term is just $\epsilon \sum z_j$. There is a symmetry $t_j \to t_j + \gamma$ and $s_j \to s_j e^{-\gamma}$ which leaves the action invariant when $\epsilon = 0$. Note that $A(s,t)$ is quadratic in s_j. Let us define the quadratic form associated to the s variables in (7.70):

$$[v \, ; D_{\beta,\varepsilon}(t) \, v]_\Lambda = \beta \sum_{(i \sim j)} e^{t_i + t_j} (v_i - v_j)^2 + \varepsilon \sum_{k \in \Lambda} e^{t_k} v_k^2 \qquad (7.71)$$

where v is a real vector, $v_j, j \in \Lambda$. Integration over the s fields produces $\det^{-1/2}(D_{\beta,\varepsilon}(t))$. The determinant is positive but nonlocal in t. Thus the total *effective action* is given by

$$A(t) = \sum_{j \sim j'} \{\beta cosh(t_j - t_{j'})\} + \frac{1}{2} \ln \det(D_{\beta,\epsilon}(t)) + \epsilon \sum_j cosh(t_j) . \qquad (7.72)$$

The quadratic form $D_{\beta,\varepsilon}(t)$ will also appear in the SUSY sigma model defined later. Note that if $t = 0$, $D_{\beta,\varepsilon}(t) = -\beta\Delta + \epsilon$. For $t \neq 0$, $D_{\beta,\varepsilon}(t)$ is a finite difference elliptic operator which is the generator of a *random walk in a random environment*. The factor $e^{(t_j + t_{j'})}$ is the conductance across the edge (j, j'). Note that since t_j are not bounded, $D_{\beta,\varepsilon}(t)$ is *not* uniformly elliptic.

Lemma 7.16 $Det(D_{\beta,\varepsilon}(t))$ *is a log convex function of the t variables.*

Following D. Brydges, this a consequence of the matrix tree theorem which expresses the determinant of as a sum:

$$\sum_{\mathcal{F},\mathcal{R}} \prod_{j,j' \in \mathcal{F}} \beta \, e^{t_j + t_{j'}} \prod_{j \in \mathcal{R}} \epsilon \, e^{t_j}$$

where the sum above ranges over spanning rooted forests on Λ. See (Abdesselam 2004) for a proof of such identities.

Thus the effective action A (7.72) is convex and in fact its Hessian $A''(t) \geq -\beta\Delta + \epsilon$. The sigma model can now be analysed using Brascamp–Lieb inequalities and a Ward identity. Its convexity will imply that this model does not have a phase transition in three dimensions.

Theorem 7.17 (Brascamp and Lieb, 1976) *Let $A(t)$ be a real convex function of $t_j, j \in \Lambda$ and v_j be a real vector. If the Hessian of the action A is convex $A''(t) \geq K > 0$ then*

$$< e^{([v \, ; t] - <[v \, ; t]>)} >_A \leq e^{\frac{1}{2}[v \, ; K^{-1} v]} . \qquad (7.73)$$

Here K is a positive matrix independent of t and $<>_A$ denotes the expectation with respect to the normalized measure $Z_\Lambda^{-1} e^{-A(t)} Dt$. Note if A is quadratic in t, (7.73) is an identity.

Note that in three dimensions the Brascamp-Lieb inequality only gives us estimates on functions of $([v\,;\,t]- < [v\,;\,t] >)$. Below we obtain estimates on $< [v\,;\,t] >$ by applying Ward identities and translation invariance. Then Theorem 7.17 will yield a bound on moments of the local conductance $e^{(t_j+t_j')}$. These bounds on the conductance imply that in three dimensions we have 'diffusion' with respect to the measure defined via (7.70). See the discussion in Remark 7.21 and the last appendix.

Theorem 7.18 (Spencer and Zirnbauer 2004) *In the three dimensions hyperbolic sigma model described by (7.72), all moments of the form $< \cosh^p(t_0) >$ are bounded for all β. The estimates are uniform in ϵ provided we first take the limit $\Lambda \to \mathbb{Z}^3$.*

Proof First note that if we let $v_j = p\delta_0(j)$ then in three dimensions since $K = -\beta\Delta + \epsilon$ we have $[v\,;\,K^{-1}v] \leq Const$ uniformly as $\epsilon \to 0$. To estimate the average $< t_0 >$ we apply the Ward identity:

$$2 < \sinh(t_0) >=< s_0^2 e^{t_0} >=< D_{\beta,\varepsilon}^{-1}(0,0)e^{t_0} > . \tag{7.74}$$

To obtain this identity we apply $t_j \to t_j + \gamma$ and $s_j \to s_j\, e^{-\gamma}$ to (7.70) then differentiate in γ and set $\gamma = 0$. We have also assumed translation invariance. Since the right side of (7.74) is positive, if we multiply the above identity by $e^{-<t_0>}$ we have

$$< e^{(t_0-<t_0>)} >=< e^{-t_0-<t_0>} > +e^{-<t_0>} < s_0^2 e^{t_0} >\geq e^{-2<t_0>}$$

where the last bound follows from Jensen's inequality. The left side is bounded by Brascamp–Lieb by taking $v_j = \delta_0(j)$. Thus we get an upper bound on $- < t_0 >$ and on $< e^{-t_0} >$. To get the upper bound on $< t_0 >$ we use the inequality

$$e^{t_0}D_{\beta,\varepsilon}^{-1}(0,0) \leq e^{t_0}\, \beta \sum_{j\sim j'} e^{-t_j-t_{j'}}\,(G_0(0,j) - G_0(0,j'))^2 + O(\epsilon)$$

where $G_0 = (-\beta\Delta + \epsilon)^{-1}$. See (7.121) in Appendix 7.12 for a precise statement and proof of this inequality. The sum is convergent in three dimensions since the gradient of $G_0(0,j)$ decays like $|j|^{-2}$. Multiplying (7.74) by $e^{<t_0>}$ we get

$$< e^{(t_0+<t_0>)} > \leq \beta \sum_{j\sim j'} < e^{-t_j-t_{j'}}\, e^{(t_0+<t_0>)} > (G_0(0,j) - G_0(0,j'))^2$$

$$+ < e^{-(t_0-<t_0>)} > +O(\epsilon).$$

By translation invariance $< t_j >=< t_0 >$, the exponent on the right side has the form $([v\,;\,t]- < [v\,;\,t] >)$ and thus the Brascamp–Lieb inequality gives us a bound on the right side and a bound on $< t_0 >$ by Jensen's inequality. Above we have ignored the $O(\epsilon)$ term. If we include it then we obtain an inequality of the form $e^{2<t_0>} \leq C + \epsilon e^{<t_0>}$ and the bound for $e^{<t_0>}$ still follows. Since we now have estimates on $< |t_j| >$, the desired estimate on $< \cosh^p(t_j) >$ follows from the Brascamp–Lieb inequality.

Remark 7.19 *The effective action for the SUSY hyperbolic sigma model, $H^{2|2}$, described in Section 7.8, looks very much like the formula above except that the coefficient of $\ln \det$ is replaced by $-1/2$. For this case the action is not convex and* *so phase transitions are not excluded. In fact in three dimensions for small β there is localization and $< e^{-t_0} >$ will diverge as $\epsilon \to 0$. Note that the above results and those described in Section 7.8 rely heavily on the use of horospherical coordinates.*

7.8   Phase transition for A SUSY hyperbolic sigma model

In this section we study a simpler version of the Efetov sigma models introduced by Zirnbauer (Zirnbauer 1991). This model is the $\mathbb{H}^{2|2}$ model mentioned in the introduction. This model is expected to qualitatively reflect the phenomenology of Anderson's tight binding model. The great advantage of this model is that the Grassmann degrees of freedom can be explicitly integrated out to produce a real effective action in bosonic variables. Thus probabilistic methods can be applied. In three dimensions we shall sketch the proof (Disertori and Spencer, 2010; Disertori, Spencer, and Zirnbauer, 2010) that this model has the analog of the *Anderson transition*. The analysis of the phase transition relies heavily on Ward identities and on the study of a random walk in a strongly correlated random environment.

In order to define the $\mathbb{H}^{2|2}$ sigma model, let u_j be a vector at each lattice point $j \in \Lambda \subset \mathbb{Z}^d$ with three bosonic components and two fermionic components

$$u_j = (z_j, x_j, y_j, \xi_j, \eta_j),$$

where ξ, η are odd elements and z, x, y are even elements of a real Grassmann algebra. The scalar product is defined by

$$(u, u') = -zz' + xx' + yy' + \xi\eta' - \eta\xi', \qquad (u, u) = -z^2 + x^2 + y^2 + 2\xi\eta$$

and the action is obtained by summing over nearest neighbours j, j'

$$\mathcal{A}[u] = \frac{1}{2} \sum_{(j,j') \in \Lambda} \beta(u_j - u_{j'}, u_j - u_{j'}) + \sum_{j \in \Lambda} \varepsilon_j(z_j - 1). \tag{7.75}$$

The sigma model constraint, $(u_j, u_j) = -1$, is imposed so that the field lies on a SUSY hyperboloid, $\mathbb{H}^{2|2}$.

We choose the branch of the hyperboloid so that $z_j \geq 1$ for each j. It is very useful to parametrize this manifold in horospherical coordinates:

$$x = \sinh t - e^t \left(\tfrac{1}{2}s^2 + \bar{\psi}\psi\right), \quad y = se^t, \quad \xi = \bar{\psi}e^t, \quad \eta = \psi e^t,$$

and

$$z = \cosh t + e^t \left(\tfrac{1}{2}s^2 + \bar{\psi}\psi\right)$$

where t and s are even elements and $\bar{\psi}, \psi$ are odd elements of a real Grassmann algebra.

In these coordinates, the sigma model action is given by

$$\mathcal{A}[t, s, \psi, \bar{\psi}] = \sum_{(ij)\in\Lambda} \beta(\cosh(t_i - t_j) - 1)$$

$$+ \frac{1}{2}[s; D_{\beta,\varepsilon}s] + [\bar{\psi}; D_{\beta,\varepsilon}\psi] + \sum_{j\in\Lambda} \varepsilon_j(\cosh t_j - 1). \qquad (7.76)$$

We define the corresponding expectation by $< \cdot > = < \cdot >_{\Lambda,\beta,\epsilon}$.

Note that the action is quadratic in the Grassmann and s variables. Here $D_{\beta,\varepsilon} = D_{\beta,\varepsilon}(t)$ is the generator of a *random walk in random environment*, given by the quadratic form

$$[v; D_{\beta,\varepsilon}(t) v]_\Lambda \equiv \beta \sum_{(j\sim j')} e^{t_j + t_{j'}}(v_j - v_{j'})^2 + \sum_{k\in\Lambda} \varepsilon_k\, e^{t_k} v_k^2 \qquad (7.77)$$

as it is in (7.71). The weights, $e^{t_j + t_{j'}}$, are the *local conductances* across a nearest neighbour edge j, j'. The $\varepsilon_j\, e^{t_j}$ term is a killing rate for the walk at j. For the random walk starting at 0 without killing, we take $\epsilon_0 = 1$ and $\epsilon_j = 0$ otherwise. For the random band matrices if we set $\epsilon_j = \epsilon$ then ϵ may be thought of as the imaginary part of the energy.

After integrating over the Grassmann variables $\psi, \bar{\psi}$ and the variables $s_j \in \mathbb{R}$ in (7.76) we get the effective bosonic field theory with action $\mathcal{E}_{\beta,\varepsilon}(t)$ and partition function

$$Z_\Lambda(\beta, \epsilon) = \int e^{-\mathcal{E}_{\beta,\epsilon}(t)} \prod e^{-t_j} dt_j \equiv \int e^{-\beta\mathcal{L}(t)} \cdot [\det D_{\beta,\varepsilon}(t)]^{1/2} \prod_j e^{-t_j} \frac{dt_j}{\sqrt{2\pi}}. \qquad (7.78)$$

where

$$\mathcal{L}(t) = \sum_{j\sim j'}[\cosh(t_j - t_{j'}) - 1] + \sum_j \frac{\varepsilon_j}{\beta}[(\cosh(t_j) - 1]. \qquad (7.79)$$

Note that the determinant is a positive but nonlocal functional of the t_j hence the effective action, $\mathcal{E} = \mathcal{L} - 1/2 \ln Det D_{\beta,\epsilon}$, is also nonlocal. The additional factor of e^{-t_j} in (7.78) arises from a Jacobian. Because of the internal supersymmetry, we know that for all values of β, ε the partition function

$$Z(\beta, \varepsilon) \equiv 1. \qquad (7.80)$$

This identity holds even if β is edge dependent.

The analog of the Green's function $< |G(E_\epsilon; 0, x)|^2 >$ of the Anderson model is the average of the Green's function of $D_{\beta,\varepsilon}$,

$$< s_0 e^{t_0} s_x e^{t_x} > (\beta, \varepsilon) = < e^{(t_0 + t_x)} D_{\beta,\varepsilon}(t)^{-1}(0, x) > (\beta, \varepsilon) \equiv \mathcal{G}_{\beta,\varepsilon}(0, x) \qquad (7.81)$$

where the expectation is with respect to the SUSY statistical mechanics weight defined above. The parameter $\beta = \beta(E)$ is roughly the bare conductance across an edge and we shall usually set $\varepsilon = \varepsilon_j$ for all j. In addition to the identity $Z(\beta, \varepsilon) \equiv 1$, there are additional Ward identities

$$< e^{t_j} >\equiv 1, \qquad \varepsilon \sum_x \mathcal{G}_{\beta,\varepsilon}(0,x) = 1 \qquad (7.82)$$

which hold for all values of β and ε. The second identity above corresponds to the Ward identity (7.4).

Note that if $|t_j| \leq Const$, then the conductances $e^{t_j+t_{j'}}$ are uniformly bounded from above and below and

$$D_{\beta,\varepsilon}(t)^{-1}(0,x) \approx (-\beta\Delta + \varepsilon)^{-1}(0,x)$$

is the diffusion propagator. Thus localization can only occur due to large deviations of the t field.

An alternative Schrödinger like representation of (7.81) is given by

$$\mathcal{G}_{\beta,\varepsilon}(0,x) =< \tilde{D}_{\beta,\varepsilon}^{-1}(t)(0,x) > \qquad (7.83)$$

where

$$e^{-t}D_{\beta,\varepsilon}(t)e^{-t} \equiv \tilde{D}_{\beta,\varepsilon}(t) = -\beta\Delta + \beta V(t) + \varepsilon\, e^{-t} , \qquad (7.84)$$

and $V(t)$ is a diagonal matrix (or 'potential') given by

$$V_{jj}(t) = \sum_{i:|i-j|=1} (e^{t_i-t_j} - 1).$$

In this representation, the potential is highly correlated and $\tilde{D} \geq 0$ as a quadratic form.

Some insight into the transition for the $\mathbb{H}^{2|2}$ model can be obtained by finding the configuration $t_j = t^*$ which minimizes the effective action $\mathcal{E}_{\beta,\varepsilon}(t)$ appearing in (7.78). It is shown in (Disertori, Spencer, and Zirnbauer, 2010) that this configuration is unique and does not depend on j. For large β, it is given by

$$\text{1D:} \quad \varepsilon\, e^{-t^*} \simeq \beta^{-1}, \qquad \text{2D:} \quad \varepsilon\, e^{-t^*} \simeq e^{-\beta} , \qquad \text{3D:} \quad t^* \simeq 0 . \qquad (7.85)$$

Note that in one and two dimensions, t^* depends sensitively on ϵ and that negative values of t_j are favoured as $\varepsilon \to 0$. This means that at t^* a mass εe^{-t^*} in (7.84) appears even as $\varepsilon \to 0$. Another interpretation is that the classical conductance $e^{t_j+t_{j'}}$ should be small in *some sense*. This is a somewhat subtle point. Due to large deviations of the t field in one dimension and two dimensions, $< e^{t_j+t_{j'}} >$ is expected to diverge, whereas $< e^{t_j/2} >$ should become small as $\varepsilon \to 0$. One way to adapt the saddle approximation so that it is sensitive to different observables is to include the observable when computing the saddle point. For example, when taking the expectation of $e^{p\, t_0}$, the saddle is only slightly changed when $p = 1/2$ but for $p = 2$ it will give a divergent contribution when there is localization.

When β is small, $\varepsilon e^{-t^*} \simeq 1$ in any dimension. Thus the saddle point t^* suggests localization occurs in both one dimension and two dimensions for all β and in three dimensions for small β. In two dimensions, this agrees with the predictions of localization by (Abrahams, Anderson, Licciardello and Ramakrishnan, 1979) at any nonzero disorder. Although the saddle point analysis has some appeal, it is not easy to estimate large deviations away from t^* in one and two dimensions. In three dimensions, large deviations away from $t^* = 0$ are controlled for large β. See the discussion below.

The main theorem established in (Disertori, Spencer, and Zirnbauer, 2010) states that in three dimensions, fluctuations around $t^* = 0$ are rare. Let $G_0 = (-\beta\Delta + \epsilon)^{-1}$ be the Green's function for the Laplacian.

Theorem 7.20 *If $d \geq 3$, and the volume $\Lambda \to \mathbb{Z}^d$, there is a $\bar\beta \geq 0$ such that for $\beta \geq \bar\beta$ then for all j,*

$$< \cosh^8(t_j) >\leq Const \tag{7.86}$$

where the constant is uniform in ϵ. This implies diffusion in a quadratic form sense: Let $\mathcal{G}$ be given by (7.81) or (7.83). There is a constant C so that we have the quadratic form bound

$$\frac{1}{C}[f;G_0f] \leq \sum_{x,y}\mathcal{G}_{\beta,\varepsilon}(x,y)\,f(x)f(y) \leq C[f;G_0f]\,, \tag{7.87}$$

where $f(x)$ is non-negative function.

Remark 7.21 *A weaker version of the lower bound in (7.87) appears in (Disertori, Spencer, and Zirnbauer, 2010). Yves Capdeboscq (private communication) showed (7.87) follows directly from the (7.86). This proof is explained in Appendix 7.12. The power 8 can be increased by making β larger. One expects pointwise diffusive bounds on $\mathcal{G}_{\beta,\varepsilon}(x,y)$ to hold. However, in order to prove this one needs to show that the set $(j : |t_j| \leq M)$ percolates in a strong sense for some large M. This is expected to be true but has not yet been mathematically established partly because of the strong correlations in the t field.*

The next theorem establishes localization for small β in any dimension. See (Disertori and Spencer 2010).

Theorem 7.22 *Let $\varepsilon_x > 0$, $\varepsilon_y > 0$ and $\sum_{j\in\Lambda}\varepsilon_j \leq 1$. Then for all $0 < \beta < \beta_c$ (β_c defined below) the correlation function $\mathcal{G}_{\beta,\varepsilon}(x,y)$, (7.83), decays exponentially with the distance $|x-y|$. More precisely:*

$$\mathcal{G}_{\beta,\varepsilon}(x,y) = \langle \tilde{D}_{\beta,\varepsilon}^{-1}(t)(x,y)\rangle \leq\ C_0\left(\varepsilon_x^{-1}+\varepsilon_y^{-1}\right)\left[I_\beta\,e^{\beta(c_d-1)}\,c_d\right]^{|x-y|} \tag{7.88}$$

where $c_d = 2d - 1$, C_0 is a constant and

$$I_\beta = \sqrt{\beta}\int_{-\infty}^{\infty}\frac{dt}{\sqrt{2\pi}}e^{-\beta(\cosh t-1)}.$$

Finally β_c is defined so that:

$$\left[I_\beta\,e^{\beta(c_d-1)}\,c_d\right] < \left[I_{\beta_c}e^{\beta_c(c_d-1)}c_d\right] = 1 \quad \forall\beta < \beta_c.$$

These estimates hold uniformly in the volume. Compare (7.88) with (7.2).

Remark 7.23 *The first proof of localization for the $\mathbb{H}^{2|2}$ model in one dimension was given by Zirnbauer in (Zirnbauer 1991). Note that in one dimension, $c_d - 1 = 0$ and exponential decay holds for all $\beta < \infty$ and the localization length is proportional to β when β is large. One expects that for 1D strips of width $|S|$ and β large, the localization length is bounded by $\beta |S|$. However, this has not yet been proved. The divergence in ε^{-1} is compatible with the Ward identity (7.4) and is a signal of localization.*

7.8.1 Role of Ward identities in the Proof

The proofs of Theorems 7.20 and 7.22 above rely heavily on Ward identities arising from internal SUSY. These are described below. For Theorem 7.20 we use Ward identities to bound fluctuations of the t field by getting bounds in three dimensions on $< \cosh^m(t_i - t_j) >$. This is done by induction on the distance $|i - j|$. Once these bounds are established we use ϵ to get bounds for $< \cosh^p t >$. For Theorem 7.22 we use the fact that for any region Λ, the partition function $Z_\Lambda = 1$.

If an integrable function S of the variables $x, y, z, \psi, \bar{\psi}$ is supersymmetric, that is it is invariant under transformations preserving

$$x_i x_j + y_i y_j + \bar{\psi}_i \psi_j - \psi_i \bar{\psi}_j$$

then $\int S = S(0)$. In horospherical coordinates the function S_{ij} given by

$$S_{ij} = B_{ij} + e^{t_i + t_j}(\bar{\psi}_i - \bar{\psi}_j)(\psi_i - \psi_j) \tag{7.89}$$

where

$$B_{ij} = \cosh(t_i - t_j) + \frac{1}{2} e^{t_i + t_j}(s_i - s_j)^2 \tag{7.90}$$

is supersymmetric. If i and j are nearest neighbours, $S_{ij} - 1$ is a term in the action $\mathcal{A}$ given in (7.76) and it follows that the partition function $Z_\Lambda(\beta, \epsilon) \equiv 1$. More generally for each m we have

$$< S_{ij}^m >_{\beta,\epsilon} = < B_{ij}^m [1 - m B_{ij}^{-1} e^{t_i + t_j}(\bar{\psi}_i - \bar{\psi}_j)(\psi_i - \psi_j)] >_{\beta,\epsilon} \equiv 1. \tag{7.91}$$

Here we have used that $S_{ij}^m e^{-\mathcal{A}_{\beta,\epsilon}}$ is integrable for $\epsilon > 0$. The integration over the Grassmann variables in (7.91) is explicitly given by

$$G_{ij} = \frac{e^{t_i + t_j}}{B_{ij}} \left[(\delta_i - \delta_j); D_{\beta,\varepsilon}(t)^{-1}(\delta_i - \delta_j) \right]_\Lambda \tag{7.92}$$

since the action is quadratic in $\bar{\psi}, \psi$. Thus we have the identity

$$< B_{ij}^m (1 - m G_{ij}) > \equiv 1. \tag{7.93}$$

Note that $0 \leq \cosh^m(t_i - t_j) \leq B_{ij}^m$. From the definition of $D_{\beta,\varepsilon}$ given in (7.77) we see that for large β, G in (7.92) is typically proportional to $1/\beta$ in three dimensions. However, there are rare configurations of $t_k \ll -1$ with k on a surface separating i and j for which G_{ij} can diverge as $\epsilon \to 0$. In 2D, G_{ij} grows logarithmically in $|i - j|$ as $\epsilon \to 0$ even in the ideal case $t \equiv 0$.

Our induction starts with the fact that if i, j are nearest neighbours then it is easy to show that G_{ij} is less than β^{-1} for all t configurations. This is because of the factor $e^{t_i + t_j}$ in (7.92).

If $|i - j| > 1$ and $m\, G_{ij} \leq 1/2$, then (7.93) implies that

$$0 \leq\ < cosh^m(t_i - t_j) > \leq < B_{ij}^m > \leq 2.$$

It is not difficult to show that one can get bounds on G_{ij} depending only on the t fields in a double cone with vertices at i and j. In fact for k far way from i, j, the dependence of G_{ij} on t_k is mild. Nevertheless, there is still no uniform bound on G_{ij} due of t fluctuations. We must use induction on $|i - j|$ and SUSY characteristic functions of the form $\chi\{S_{ij} \leq r\}$, to prove that configurations with $1/2 \leq m\, G_{ij}$, with $\beta \gg m \gg 1$, are rare for large β in three dimensions. We combine these estimates with the elementary inequality $B_{ik}B_{kj} \geq 2B_{ij}$ for $i, j, k \in \mathbb{Z}^d$ to go to the next scale. See (Disertori *et al.* 2010) for details.

The proof of the localized phase is technically simpler than the proof of Theorem 7.20. Nevertheless, it is of some interest because it shows that $\mathbb{H}^{2|2}$ sigma model reflects the localized as well as the extended states phase in three dimensions. The main idea relies on the following lemma. Let M be an invertible matrix indexed by sites of Λ and let γ denote a self avoiding path starting at i and ending at j. Let M_{ij}^{-1} be matrix elements of the inverse and let $M_{\gamma}\ c$ be the matrix obtained from M by striking out all rows and columns indexed by the vertices covered by γ.

Lemma 7.24 *Let M and $M_{\gamma}\ c$ be as above, then*

$$M_{ij}^{-1} \mathrm{det} M = \sum_{\gamma\ i,j} [(-M_{ij_1})(-M_{j_1 j_2}) \cdots (-M_{j_m j})]\ \mathrm{det} M_{\gamma}\ c$$

where the sum ranges over all self-avoiding paths γ connecting i and j, $\gamma_{i,j} = (i, j_1, \ldots j_m, j)$, with $m \geq 0$.

Apply this lemma to

$$M = e^{-t} D_{\beta,\varepsilon}(t) e^{-t} \equiv \tilde{D}_{\beta,\varepsilon}(t) = -\beta\Delta + V(t) + \varepsilon\, e^{-t} \tag{7.94}$$

and notice that for this M, for all nonzero contributions, γ are nearest neighbour self-avoiding paths and that each step contributes a factor of β. To prove the lemma, write the determinant of M as a sum over permutations of $j \in \Lambda$ indexing M. Each permutation can be expressed as a product of disjoint cycles covering Λ. Let E_{ij} denote the elementary matrix which is 0 except at ij place where it is 1. The derivative in s of $\det(M + sE_{ji})$ at $s = 0$ equals $M_{ij}^{-1} \mathrm{det} M$ and selects the self-avoiding path in the cycle containing j and i. The other loops contribute to $\det M_{\gamma}\ c$. By (7.83) and (7.94) we have

$$\mathcal{G}_{\beta,\varepsilon}(x, y) =< M_{xy}^{-1} >= \int e^{-\beta\mathcal{L}(t)} M_{xy}^{-1} [\det M]^{1/2} \prod_j \frac{dt_j}{\sqrt{2\pi}}. \tag{7.95}$$

Note the factors of e^{-t_j} appearing in (7.78) have been absorbed into the determinant. Now write

$$M_{xy}^{-1}[\det M]^{1/2} = \sqrt{M_{xy}^{-1}}\sqrt{M_{xy}^{-1}\det M}\ .$$

The first factor on the right-hand side is bounded by $\epsilon_x^{-1/2}e^{t_x/2} + \epsilon_y^{-1/2}e^{t_y/2}$. For the second factor, we use the lemma. Let $\mathcal{L} = \mathcal{L}_\gamma + \mathcal{L}_{\gamma^c} + \mathcal{L}_{\gamma,\gamma^c}$ where $\mathcal{L}_\gamma$ denotes the restriction of $\mathcal{L}$ to γ. Then by supersymmetry

$$\int e^{-\beta \mathcal{L}_{\gamma^c}}[\det M_{\gamma^c}]^{1/2}\prod_j \frac{dt_j}{\sqrt{2\pi}} \equiv 1$$

we can bound

$$0 \le \mathcal{G}_{\beta,\varepsilon}(x,y) \le \sum_{\gamma_{xy}}\sqrt{\beta}^{|\gamma_{xy}|}\int e^{-\beta \mathcal{L}_\gamma + \beta \mathcal{L}_{\gamma,\gamma^c}}\,[\epsilon_x^{-1/2}e^{t_x/2} + \epsilon_y^{-1/2}e^{t_y/2}]\prod_j \frac{dt_j}{\sqrt{2\pi}}$$

where $|\gamma_{xy}|$ is the length of the self-avoiding path from x to y. The proof of Theorem 7.22 follows from the fact that the integral along γ is one dimensional and can be estimated as a product. See (Disertori and Spencer 2010) for further details.

7.9 Efetov's sigma model

In this final section we present a very brief description of the Efetov sigma model for Hermitian matrices, following (Efetov 1983, 1997; Mirlin 2000; Zirnbauer 2004). This sigma model is the basis for most random band matrix calculations in theoretical physics. Unfortunately, the mathematical analysis of this model is still limited to one dimension. Even in this case our analysis is far from complete. The ideas of this section are quite similar to those in Section 7.7 leading to (7.68) and (7.69) except that we now include Grassmann variables.

In order to obtain information about averages of the form (7.1) we introduce a field with four components

$$\Phi_j = (z_j, w_j, \psi_j, \chi_j) \tag{7.96}$$

where z, w are complex fields and ψ, χ are Grassmann fields. Let $L = diag(1, -1, 1, 1)$, and $\Lambda = diag(1, -1, 1, -1)$. For a Hermitian matrix H, define the action

$$A(E, \epsilon) = \Phi^* \cdot L\{i(H - E) + \epsilon\Lambda\}\Phi. \tag{7.97}$$

Note that the signature of L is chosen so that the z and w variables appear as complex conjugates of each other as they do in (7.1). Then we have the identity:

$$|(E - i\epsilon - H)^{-1}(0, j)|^2 = \int z_0 \bar{z}_j w_0 \bar{w}_j\, e^{-A(E,\epsilon)}\, D\Phi \tag{7.98}$$

where

$$D\Phi \equiv Dz\, Dw\, D\psi\, D\chi \,.$$

Without the observable $z_0\bar{z}_j w_0 \bar{w}_j$, $\int e^{-A} = 1$. The integral over Gaussian H can be calculated as in (7.55) and will produce a quartic interaction in Φ. However, now the Hubbard–Stratonovich transformation, which is usually used in the Bosonic sector, involves a subtle analytic continuation first worked out in (Schäfer and Wegner 1980), see also (Fyodorov 1994; Zirnbauer 2004).

Now let us define a matrix of the form

$$M = \begin{pmatrix} [BB] & [BF] \\ [FB] & [FF] \end{pmatrix} \tag{7.99}$$

where each block is a 2×2 matrix. M will be called a *supermatrix* if the diagonal blocks, BB and FF are made of commuting (even elements) variables while the off diagonal blocks FB and BF consist of odd elements in the Grassmann algebra. Define the supertrace

$$Str(M) \equiv Tr([BB] - [FF]) \,.$$

Note that for supermatrices A and B we have $Str(AB) = Str(BA)$. We define the adjoint of a supermatrix M by

$$M^\dagger = \begin{pmatrix} [BB]^* & [FB]^* \\ -[BF]^* & [FF]^* \end{pmatrix} \,.$$

The symbol * denotes the usual transpose followed by conjugation. For Grassmann variables we have $\overline{\psi_a \psi_b} = \bar\psi_a \bar\psi_b$. But $\bar{\bar\psi} = -\psi$ so that $\dagger$ is an involution and

$$\Phi_1^\dagger(M\Phi_2) = (M^\dagger \Phi_1)^\dagger \Phi_2.$$

For $\epsilon = 0$ the action is invariant under the action of matrices T in $SU(1;1|2)$ which satisfy:

$$T^\dagger L T = L. \tag{7.100}$$

As in (7.68) and (7.69) the spin or field of the sigma model is given by matrices

$$S_j = T_j^{-1} \Lambda T_j \tag{7.101}$$

as T ranges over $SU(1;1|2)$. It is the orbit of a critical point, which is proportional to Λ under the action of $SU(1,1|2)$. Thus the matrix S ranges over a supersymmetric space $U(1,1|2)/(U(1|1) \times U(1|1))$. A general discussion of more general supersymmetric spaces appears in (Zirnbauer 1996). The SUSY sigma model has a Gibbs density defined by

$$\exp\{-\beta Str \sum_{j \sim j'} (S_j - S_{j'})^2 - \epsilon\, Str \sum_j \Lambda S_j\}. \tag{7.102}$$

In a one dimensional chain of length L with $\epsilon = 0$ except at the end points, the Grassmann variables can be explicitly integrated over producing the formula (7.8). In higher dimensions the action (7.102) is harder to analyse because of the Fermion–Boson coupling of the BB sector (hyperbolic sigma model) and the FF sector (Heisenberg model). An explicit parametrization of the 4×4 matrices S_j and integration measure is given in (Efetov 1983, 1997; Mirlin 2000). Fluctuations about the saddle should produce massless modes—Goldstone Bosons in three dimensions.

7.10 Appendix: Gaussian and Grassmann integration

Let $z = x + iy$ with $x, y \in \mathbb{R}$. Let $dz = dxdy/\pi$ and suppose $Re\, a > 0$, $a \in \mathbb{C}$. Then

$$\int e^{-a\bar{z}z} dz = \pi^{-1} \iint e^{-ar^2} rdrd\theta = a^{-1} . \tag{7.103}$$

Also

$$\frac{1}{\sqrt{2\pi}} \int e^{-ax^2/2} dx = a^{-1/2}.$$

In the multi-dimensional case let $z = (z_1, z_2, ...z_n)$, $z^* = \bar{z}^t$. For an $n \times n$ matrix A with $Re\, A > 0$ as a quadratic form

$$\int e^{-[z;\, Az]} Dz = (\det A)^{-1} \quad where \quad Dz = \prod_1^n dx_i dy_i/\pi . \tag{7.104}$$

Recall the notation $[z; Az] \equiv \sum \bar{z}_j A_{ij} z_j = z^* Az$. To prove the formula, first note that it holds in the Hermitian case by diagonalization. Since both sides of the equation are analytic in the matrix entries, and agree for the Hermitian case, the result follows by uniqueness of analytic continuation. The expectation with respect to A is defined by

$$< \cdot >_A \equiv \det(A) \int e^{-z^* Az} \cdot Dz$$

and from integration by parts, the pair correlation is given by

$$< z_j \bar{z}_k >_A = A_{jk}^{-1}. \tag{7.105}$$

Note that $< z_j z_k >_A = < \bar{z}_j \bar{z}_k >_A = 0$. This is because the integral is invariant under the global transform $z \to e^{i\phi} z$, $\bar{z} \to e^{-i\phi} \bar{z}$. The generating function is given by

$$< e^{z^* v + w^* z} >_A = e^{w^* A^{-1} v} = e^{[w; A^{-1} v]} .$$

This identity follows by changing variables: $z \to z - A^{-1} v$ and $\bar{z} \to \bar{z} - (A^t)^{-1} \bar{w}$.

For real variables $x = (x_1, ...x_n)$ if A is symmetric and positive

$$\int e^{-[x;\, Ax]/2} Dx = (\det A)^{-1/2} \quad where \quad Dx = \prod_i^n dx_i/\sqrt{2\pi} . \tag{7.106}$$

Its generating function $< e^{[x;y]} >_A = e^{[y;A^{-1}y]/2}$, is obtained by completing the square. There are similar formulas for integration over $N \times N$ matrices:

$$\int e^{-NTrH^2/2} e^{iTrMH} \, DH = e^{-TrM^2/2N} \int e^{-NTrH^2/2} DH \ . \tag{7.107}$$

For the case of band matrices the generating function is

$$< e^{iTrHM} >= e^{-<(trHM)^2>/2>} = e^{-1/2 \sum J_{ij} M_{ji} M_{ij}} \ . \tag{7.108}$$

7.10.1   Grassmann integration

Grassmann variables $\psi_i, \bar{\psi}_j$ are anti-commuting variables $1 \leq i, j \leq N$ satisfying $\psi_j^2 = \bar{\psi}_j^2 = 0$, *and* $\bar{\psi}_j \psi_i = -\psi_i \bar{\psi}_j$. Also

$$\psi_j \psi_i = -\psi_i \psi_j, \qquad \bar{\psi}_j \bar{\psi}_i = -\bar{\psi}_i \bar{\psi}_j \ . \tag{7.109}$$

The $\bar{\psi}_j$ is simply convenient notation for another independent family of Grassmann variables which anti-commute with ψ. Even monomials in the Grassmann variables and complex numbers commute with Grassmann variables. The polynomials in these variables form a Z_2 graded algebra, with the even and odd monomials belonging to the even and odd gradings respectively. One way to think about the Grassmann variables is to let $\psi_j = dx_j$ and $\bar{\psi}_j = dy_j$ and consider the product as the wedge product in the theory of differential forms.

The Grassmann integral, defined below, plays an important role in many parts of physics. It is an extremely efficient and useful notation for the analysis of interacting Fermi systems, Ising models (free fermions), and SUSY. Although most of the time we shall eliminate the Grassmann variables by integrating them out, they are nevertheless an essential tool for obtaining the identities we shall analyse. See (Abdesselam, 2004; Berezin, 1987; Feldman, Knörrer, and Trubowitz, 2002; Mirlin, 2000; Salmhofer, 1999) for more details about Grassmann integration.

We define integration with respect to

$$D\psi \equiv \prod_j^N d\bar{\psi}_j d\psi_j \tag{7.110}$$

as follows. For N=1

$$\int (a\psi_1 \bar{\psi}_1 + b\psi_1 + c\bar{\psi}_1 + d) \, D\psi = a \ .$$

The general rule is that the integral of a polynomial in 2N variables with respect to $D\psi$ equals the coefficient of the top monomial of degree 2N ordered as $\prod^N \psi_j \bar{\psi}_j$. Note that since the factors in the product are even, their order does not matter. Any element of the Grassmann algebra can be expressed as a polynomial and the top monomial can always be rearranged using the anti-commutation rules so that it coincides with $\prod^N \psi_j \bar{\psi}_j$.

To differentiate a Grassmann monomial, use the rule $\frac{\partial}{\partial \psi_j} \psi_k = \delta_{jk}$. The derivative anti-commutes with other Grassmann variables. We have

$$\frac{\partial}{\partial \psi_j} \psi_j \prod_{k \neq j} \psi_k = \prod_{k \neq j} \psi_k.$$

To differentiate a general monomial in ψ_j, use linearity and the anti-commutation relations so that it is of the above form. If ψ_j is not a factor then the derivative is 0.

For any $N \times N$ matrix A we have the following analogue of Gaussian integration

$$\int e^{-[\psi; A\psi]} D\psi = det A \quad where \quad [\psi; A\psi] = \sum \bar{\psi}_i A_{ij} \psi_j . \tag{7.111}$$

Moreover,

$$< \psi_i \bar{\psi}_j > \equiv det A^{-1} \int \psi_i \bar{\psi}_j \, e^{-[\psi; A\psi]} D\psi = A_{ij}^{-1} . \tag{7.112}$$

More generally for a polynomial F in $\psi, \bar{\psi}$ we can integrate by parts to obtain

$$\int \psi_i \, F e^{-[\psi; A\psi]} D\psi = \sum_j A_{ij}^{-1} \int \{ \frac{\partial}{\partial \bar{\psi}_j} F \} e^{-[\psi; A\psi]} D\psi . \tag{7.113}$$

To prove this formula we use

$$\sum_j A_{ij}^{-1} \int \frac{\partial}{\partial \bar{\psi}_j} (e^{-[\psi; A\psi]} F) D\psi = 0 .$$

Let us establish (7.111) in the simplest case:

$$\int e^{-a\bar{\psi}_1 \psi_1} D\psi = \int (1 - a\bar{\psi}_1 \psi_1) D\psi = a \int \psi_1 \bar{\psi}_1 D\psi = a .$$

Exercise 7.25. Show that if A is a 2×2 matrix, (7.111) holds.

To prove the general case note that the exponential can be written as a product $\prod_i (1 - \sum_j A_{ij} \bar{\psi}_i \psi_j)$ and we look at the terms:

$$\sum A_{1j_1} A_{2j_2} \ldots A_{N, j_N} \int \bar{\psi}_1 \psi_{j_1} \, \bar{\psi}_2 \psi_{j_2} \ldots \bar{\psi}_N \psi_{j_N} \, D\psi .$$

The j_i are distinct and hence are a permutation of $1 \ldots N$. The integral then is the sign of the permutation and thus we obtain the determinant. The generating function is given by

$$< e^{\bar{\rho}^t \psi + \bar{\psi}^t \rho} > = det A^{-1} \int e^{-[\psi; A\psi]} e^{\bar{\rho}^t \psi + \bar{\psi}^t \rho} D\psi = e^{\bar{\rho}^t A^{-1} \rho} \tag{7.114}$$

where $\rho, \bar{\rho}$ are independent Grassmann variables.

7.10.2   Polar coordinates for Grassmann and bosonic matrices

Grassmann integrals can also be expressed as integrals over unitary matrices. Let $d\mu(U)$ denote the Haar measure on $U(m)$. Given a family of Grassman variables $\bar{\psi}_j^\alpha$, ψ_j^α with $1 \le \alpha \le N$ and $1 \le j \le m$, define the matrix $M_{ij} = \psi_i \cdot \bar{\psi}_j$, where the dot product is the sum over α. Then for a smooth function F of M we have

$$\int F(M)\, D\psi = C_{N,m} \int F(U) \det(U)^{-N}\, d\mu(U) \qquad (7.115)$$

where $U \in U(m)$. For example for $m = 1$

$$\int e^{a\psi \cdot \bar{\psi}}\, D\psi = a^N = \frac{N!}{2\pi} \int exp(ae^{i\theta})e^{-iN\theta} d\theta \ .$$

It suffices to check (7.115) when $F = e^{trQ\,M}$ for a general matrix Q since any function can be approximated by linear combinations of such exponentials. From (7.111) the left side is equal to $\det(Q)^N$. These expressions agree since

$$\int e^{trQ\,U} \det(U)^{-N}\, d\mu(U) \propto \det(Q)^N.$$

To prove this statement first suppose that $Q = e^{itH}$ is a unitary matrix. Our identity now follows by the invariance of Haar measure. Since both sides are analytic in the matrix elements of H and agree in the Hermitian case, the result follows for a general matrix by analytic continuation.

For bosonic fields the analogue of the above relation is given by the generalized gamma function or Ingham–Siegel identity, see (Fyodorov 2002):

$$\int_{P>0} e^{-trQ\,P} \det(P)^{N-m}\, dP \propto \det(Q)^{-N} \ .$$

Here P and Q denote positive Hermitian $m \times m$ matrices and dP denotes the flat measure restricted positive matrices. We must assume that $N \ge m$. Hence, if we set $M_{ij} = \bar{z}_i, \cdot z_j$ we have

$$\int F(M)Dz = C'_{N,m} \int_{P>0} F(P) \det(P)^{N-m}\, dP \ . \qquad (7.116)$$

Note that $\det(P)^{-m}dP$ is proportional to the measure $d\mu(P)$ which is invariant under the transformation $P \to g^*Pg$, for $g \in GL_m(\mathbb{C})$. For example dt/t is invariant under $t \to at$ for $a > 0$.

We may think of U and P as the polar or collective variable for the Grassmann and Bosonic variables respectively and their eigenvalues could be referred to as the radial coordinates.

Let $\Phi = (z, \psi)$ denote a column vector of with n bosonic components, z_j, and n Grassmann components, ψ_j. Define Φ^* to be the transpose conjugate of Φ. If M denotes a supermatrix with $n \times n$ blocks of the form (7.99), then the standard SUSY integration formula

$$\int e^{-\Phi^* M \Phi} \, DzD\psi = SDet^{-1}(M)$$

holds, where the superdeterminant is given by

$$SDet(M) = det^{-1}([FF]) \, det([BB] - [BF][FF]^{-1}[FB]) \, .$$

This formula may be derived by first integrating out the ψ variables using (7.114) and then integrating the z variables. An equivalent formula given by

$$SDet(M) = det([BB]) \, det^{-1}([FF] - [FB][BB]^{-1}[BF])$$

is obtained by reversing the order of integration. If A and B are supermatrices then $SDet(AB) = SDet(A) \, SDet(B)$ and $\ln SDet A = Str A$.

Now consider a supermatrix with both Grassmann and bosonic components. For example, in the simplest case, (m =1):

$$M = \begin{pmatrix} z \cdot \bar{z} & z \cdot \bar{\psi} \\ \psi \cdot \bar{z} & \psi \cdot \bar{\psi} \end{pmatrix}.$$

then for suitably regular F, the SUSY generalization of the above formulas is given by

$$\int F(M) \, DzD\psi \propto \int F(Q) \, SDet(Q)^N \, DQ \, .$$

Here Q is a 2×2 supermatrix and m=1. For $m \geq 1$ of Q has the form

$$Q = \begin{pmatrix} P & \bar{\chi} \\ \chi & U \end{pmatrix}$$

with blocks of size $m \times m$ and

$$DQ \equiv dP \, d\mu(U) \, det^m(U) \, D\bar{\chi}D\chi$$

where $d\mu$ Haar measure on $U(m)$, and dP is the flat measure on positive Hermitian matrices. As a concrete example let us evaluate the integral using the above formula for m=1 in the special case

$$\int e^{-a\bar{z} \cdot z + b\psi\bar{\psi}} \, Dz \, D\psi = b^N a^{-N} = \int e^{-ap + be^{i\theta}} \, SDet^N(Q) \, dp \, d\theta e^{i\theta} d\bar{\chi} \, d\chi$$

where

$$Sdet^N(Q) = p^N e^{-iN\theta} (1 - Np^{-1}\bar{\chi}e^{-i\theta}\chi) \, .$$

Note that in the contribution to the integral, only the second term of $SDet$ above contributes. It is of top degree equal to $N \, p^{N-1} e^{-i(N+1)\theta} \chi\bar{\chi}$.

Remark 7.26 *The above integral identity is a simple example of superbozonization, See (Littelmann, Sommers and Zirnbauer, 2008) for a more precise formulation and proof as well as references.*

7.11   Appendix: Formal perturbation theory

In this appendix we will explain some formal perturbation calculations for the average Green's function of the random Schrödinger matrix when the strength of the disorder λ is small. For a rigorous treatment of perturbation theory, see for example (Erdős 2012). At present it is not known how to rigorously justify this perturbation theory when the imaginary part of the energy, ϵ is smaller than λ^4. In fact the best rigorous estimates require $\epsilon \geq \lambda^p$, $p \leq 2 + \delta$, $\delta > 0$. We shall explain some of the challenges in justifying the perturbation theory. Similar perturbative calculations can be done for random band matrices and in special cases these can be justified using SUSY as explained in Section 7.5.

Let us write the Green's function for the random Schrödinger matrix as

$$G(E_\epsilon; j, k) = [E_\epsilon + \Delta - \lambda^2 \sigma(E_\epsilon) - \lambda v + \lambda^2 \sigma(E_\epsilon)]^{-1}(j, k).$$

We assume that the potential v_j are independent Gaussian random variables of 0 mean and variance 1. Note that we have added and subtracted a constant $\lambda^2 \sigma(E_\epsilon)$ and we shall now expand around

$$G_0(E_\epsilon; j, k) \equiv [E_\epsilon + \Delta - \lambda^2 \sigma(E_\epsilon)]^{-1}.$$

The first order term in λ vanishes since v has mean 0. The basic idea is to choose the constant $\sigma(E_\epsilon)$ so that after averaging, the second order term in λ vanishes:

$$\lambda^2 < G_0 v G_0 v G_0 - G_0 \sigma G_0 >= 0 \quad hence \quad G_0(E_\epsilon; j, j) = \sigma(E_\epsilon) \qquad (7.117)$$

holds for all j. This gives us an equation for $\sigma(E_\epsilon)$ which is the leading contribution to the self-energy. In field theory this is just self-consistent Wick ordering. When E lies in the spectrum of $-\Delta$, the imaginary part of $\sigma(E_\epsilon)$ does not vanish even as $\epsilon \to 0$. In fact, to leading order $Im\sigma(E)$ is proportional to the density of states for $-\Delta$. See (7.33). Thus to second order in perturbation theory we have $< G(E_\epsilon; j, k) > \approx G_0(E_\epsilon; j, k)$. Note that since the imaginary part of $\sigma(E_\epsilon) > 0$, $G_0(E_\epsilon; j, k)$ will decay exponentially fast in $\lambda^2 |j - k|$ uniformly for $\epsilon > 0$.

Exponential decay is also believed to hold for $< G(E_\epsilon; j, k) >$ in all dimensions but has only been proved in one dimension. This decay is *not* related to localization and should also hold at energies where extended states are expected to occur. If λ is large and if v has a Gaussian distribution then it is easy to show that in any dimension the average Green's function decays exponentially fast. One way to see this is to make a shift in the contour of integration $v_j \to v_j + i\delta$. Since λ is large it will produce a large diagonal term. We get a convergent random walk expansion for $< G >$ by expanding in the off diagonal matrix elements.

Remark 7.27 *Note that by the Ward identity (7.4) and (7.117) we have*

$$\sum_j |G_0(E_\epsilon; 0, j)|^2 (\lambda^2 Im\sigma(E_\epsilon) + \epsilon) = Im\sigma(E_\epsilon). \tag{7.118}$$

We now explain a problem in justifying perturbation theory. First note that one could proceed by defining higher order corrections to the self-energy, $\lambda^4 \sigma_4(E_\epsilon, p)$, where p is the Fourier dual of the variable j so that the fourth order perturbation contribution vanishes. In this case σ acts as a convolution. However, this perturbative scheme is not convergent and must terminate at some order. We wish to estimate the remainder. This is where the real difficulty appears to lie. Consider the remainder term written in operator form:

$$< [G_0(E_\epsilon) \lambda v]^n G(E_\epsilon) >.$$

For brevity of notation we have omitted the contributions of the self-energy $\lambda^2 \sigma$ which should also appear in this expression. Suppose we use the Schwarz inequality to separate $[G_0(E_\epsilon) \lambda v]^n$ from G. Then we have to estimate the expression:

$$< [G_0(E_\epsilon) \lambda v]^n \cdot [\bar{G}_0(E_\epsilon) \lambda v]^n >.$$

Here $\bar{G}_0$ denotes the complex conjugate of G_0. The high powers of λ may suggest that this is a small term. However, this is not so. If we write the above in matrix form we have

$$\lambda^{2n} < \sum_{j,k} [G_0(E_\epsilon; 0, j_1) v_{j_1} G_0(E_\epsilon; j_1, j_2) v_{j_2} ... v_{j_{n-1}} G_0(E_\epsilon; j_{n-1}, j_n) v_{j_n}$$

$$\times \bar{G}_0(E_\epsilon; 0, k_1) v_{k_1} \bar{G}_0(E_\epsilon; k_1, k_2) v_{k_2} ... v_{k_{n-1}} \bar{G}_0(E_\epsilon; k_{n-1}, j_n) v_{k_n}] >. \tag{7.119}$$

When computing the average over v the indices $\{j, k\}$ must be paired otherwise the expectation vanishes since $< v_j >= 0$ for each j. There are $n!$ such pairings, each of which gives rise to a graph. The pairing of any adjacent j_i, j_{i+1} or k_i, k_{i+1} is canceled by the self-energy contributions which we have omitted. We shall next discuss some other simple pairings.

After cancelling the self-energy, the leading contribution to (7.119) should be given by *ladder graphs* of order n. These are a special class of graphs obtained by setting $j_i = k_i$ and summing over the vertices. Assuming G_0 is translation invariant, this sum is approximately given by:

$$[\lambda^2 \sum_j |G_0(E_\epsilon; 0, j)|^2]^n = [\frac{\lambda^2 Im\sigma(E_\epsilon)}{\lambda^2 Im\sigma(E_\epsilon) + \epsilon}]^n = [1 + \frac{\epsilon}{\lambda^2 Im\sigma(E_\epsilon)}]^{-n}.$$

The right-hand side of this equation is obtained using (7.118). Note that the right-hand side of this equation is not small if $\epsilon \leq \lambda^p$ unless $n \gg \lambda^{-(p-2)}$. Although contributions obtained from other pairings give smaller contributions, there are about $n!$ terms of size λ^{2n}. Hence we must require $n \ll \lambda^{-2}$ so that the number of graphs is offset by their size $n!\lambda^{2n} \ll 1$. Thus, $p < 4$ and this naive method of estimating $G(E_\epsilon)$ can only work for $\epsilon \gg \lambda^4$.

7.11.1   Leading contribution to diffusion

In three dimensions, to leading order $< |G(E_\epsilon; 0, x)|^2 >$ is expected to given by the sum of ladder graphs described above - denoted $L(E_\epsilon, x)$. We shall calculate this sum and show that to leading order we get quantum diffusion. Let

$$K(E_\epsilon; p) = \sum_j e^{ij \cdot p}|G_0(E_\epsilon; 0, j)|^2.$$

Note that $K(p)$ is analytic for small values of p^2 since $Im\sigma > 0$. Then the sum of the ladders is given for $p \approx 0$ by

$$\hat{L}(p) = K(p)(1 - \lambda^2 K(p))^{-1} \approx \frac{K(0)}{1 - \lambda^2 K(0) - \frac{1}{2}\lambda^2 K''(0)p^2}$$

$$\approx \frac{Im\sigma}{\lambda^4 Im\sigma K''(0) p^2/2 + \epsilon}.$$

In the last approximation we have used the Ward identity (7.118) to get

$$1 - \lambda^2 K(0) = \frac{\epsilon}{\lambda^2 Im\sigma + \epsilon} \quad so \quad K(0) \approx \lambda^{-2}.$$

Thus the sum of ladder graphs produces quantum diffusion as defined by (7.3) with $D_0 = \lambda^4 Im\sigma \hat{K}''(0)/2$. Since $K''(0) \propto \lambda^{-6}$, we see that D_0 is proportional to λ^{-2} for small λ. We refer to the work of Erdős, Salmhofer, and Yau (Erdős *et al.* 2008; Erdős 2012), where such estimates were proved in three dimensions for $\epsilon \approx \lambda^{2+\delta}$, $\delta > 0$. At higher orders of perturbation theory, graphs in three dimensions must be grouped together to exhibit cancellations using Ward identities so that the bare diffusion propagator $1/p^2$ does not produce bogus divergences of the form $\int (1/p^2)^m$ for $m \geq 2$. These higher powers of are offset by a vanishing at p=0 of the vertex joining such propagators. In both one dimension and two dimensions this perturbation theory breaks down at higher orders. There is a divergent sum of crossed ladder graphs obtained by setting $j_i = k_{n-i}$ above. The sum of such graphs produces an integral of the form

$$\int_{|p| \leq 1} (p^2 + \epsilon)^{-1} d^2p$$

which diverges logarithmically in two dimensions as $\epsilon \to 0$. This divergence is closely related to localization which is expected for small disorder in two dimensions, (Abrahams, Anderson, Licciardello and Ramakrishnan, 1979).

7.11.2   Perturbation theory for random band matrices

To conclude this appendix, we briefly show how to adapt the perturbative methods described above to the case of random band matrices. Let H be a Gaussian random band matrix with covariance as in (7.14). Let

$$G(E_\epsilon) = (E_\epsilon - \lambda^2 \sigma - \lambda H + \lambda^2 \sigma)^{-1}.$$

Here the parameter λ is a bookkeeping device and it will be set equal to 1. Define $G_0(E_\epsilon) = [E_\epsilon - \lambda^2 \sigma(E_\epsilon)]^{-1}$ and perturb $G(E_\epsilon)$ about G_0. Note that in this case G_0 is a scalar multiple of the identity. Proceeding as above we require that to second order perturbation theory in λ vanishes. From (7.13) and (7.14) we have

$$\sum_k < H_{jk} H_{kj} > = \sum_k J_{jk} = 1$$

and we obtain the equation

$$G_0 = [E_\epsilon - \lambda^2 \sigma(E_\epsilon)]^{-1} = \sigma(E_\epsilon).$$

This is a quadratic equation for σ and when we set $\lambda = 1$ and take the imaginary part we recover Wigner's density of states. For a band matrix of width W the corrections at λ^4 to the Wigner semicircle law is formally of order $1/W^2$. If we apply the Schwarz inequality as above we will encounter similar difficulties. In fact it is known that $\epsilon \gg W^{-2/5}$ is needed to ensure convergence of perturbation theory. However, by using different methods, Erdős, Yau, and Yin (Erdős *et al.* 2012) and Sodin (Sodin 2011) establish control of the DOS to scale $\epsilon \approx W^{-1}$. For three dimensions Gaussian band matrices with covariance given by (7.14) these difficulties are avoided with nonperturbative SUSY methods (Constantinescu *et al.*, 1987; Disertori *et al.*, 2002) and ϵ may be sent to zero for fixed large W. The calculation of the bare diffusion constant explained above for random Schrödinger matrices can be applied to RBM. In this case $D_0 \approx W^2$.

We now illustrate the perturbation theory for the two site N-orbital model:

$$M = \begin{pmatrix} \lambda A & I \\ I & \lambda B \end{pmatrix}$$

where A and B are independent $N \times N$ GUE matrices and I is the $N \times N$ identity matrix. Define

$$G_0^{-1} = \begin{pmatrix} E_\epsilon - \lambda^2 \sigma & I \\ I & E_\epsilon - \lambda^2 \sigma \end{pmatrix}.$$

By requiring second order perturbation theory to vanish as above we obtain a cubic equation for the self-energy $\sigma = (E_\epsilon - \lambda^2 \sigma)[(E_\epsilon - \lambda^2 \sigma)^2 - 1]^{-1}$. This equation is analogous to (7.117). The imaginary part of $\sigma(E)$ is proportional to the density of states for M for large N. To calculate finer properties of the density of states one can apply the SUSY methods of Sections 7.3–7.5 and it can be shown that the main saddle point will coincide the self-energy σ. Note that the density of states for M no longer looks like the semicircle law, in fact it is peaked near ± 1 when λ is small. Nevertheless, the local eigenvalue spacing distribution is expected to be equal to that of GUE after a local rescaling.

Remark 7.28 *One of the major challenges in the analysis of random band or random Schrödinger matrices is to find a variant of perturbation theory which allows one to analyse smaller values of ϵ as a function of W or λ. For certain Gaussian band matrices SUSY gives much better estimates but the class of models to which it applies is relatively narrow. Perhaps there is a method which combines perturbative and SUSY ideas and emulates both of them.*

7.12 Appendix: Bounds on Green's functions of divergence form

In this appendix we will show how to obtain upper and lower bounds on (7.87)

$$< [fe^t; [D_{\beta,\varepsilon}(t)]^{-1}fe^t] >= \sum_{x,y} \mathcal{G}_{\beta,\varepsilon}(x,y)\, f(x) f(y)$$

in terms of $[f; G_0 f] = [f; (\beta \nabla^* \nabla + \epsilon)^{-1} f]$ for $f(j) \geq 0$. Recall $[\ ;\ ]$ is the scalar product and that $D_{\beta,\varepsilon}(t)$ is given by (7.71) or (7.77). We shall suppose that $d \geq 3$ and that $< \cosh^8(t_j) >$ is bounded as in Theorem 7.20 of Section 7.8. For brevity let

$$G_t(i,j) = [D_{\beta,\varepsilon}(t)]^{-1}(i,j).$$

We first prove the lower bound. For any two real vectors X and Y we have the inequality

$$X \cdot X \geq 2X \cdot Y - Y \cdot Y.$$

Let

$$X_1(j,\alpha) = \sum_k e^{(t_j + t_j + e_\alpha)/2} \nabla_\alpha G_t(j,k) f(k) e^{t_k}$$

where e_α is the unit vector in the α direction. Define

$$X_2(j) = \sqrt{\epsilon}\, e^{t_j/2} \sum_k G_t(j,k) f(k) e^{t_k}\ .$$

If we set $X = (X_1, X_2)$ we see that $< X \cdot X >= [f; \mathcal{G}_{\beta,\varepsilon} f]$. We shall define Y to be proportional to X with G_t replaced by G_0,

$$Y_1(j,\alpha) = a\, e^{-2t_0} \sum_k e^{(t_j + t_j + e_\alpha)/2} \nabla_\alpha G_0(j,k) f(k) e^{t_k}$$

and

$$Y_2(j) = a\, e^{-2t_0} \sqrt{\epsilon}\, e^{t_j/2} \sum_k G_0(j,k) f(k) e^{t_k}\ .$$

The constant a is chosen so that the error term $Y \cdot Y$ is small. By integrating by parts we have

$$X \cdot Y = a \, [fe^t; \, G_0 fe^t] e^{-2t_0}.$$

Since $< e^{t_k + t_j - 2t_0} >\geq 1$ by Jensen's inequality and translation invariance, we get the desired lower bound on $< X \cdot Y >$ in terms of $a \, [f; G_0 f]$. The error term

$$< Y_1 \cdot Y_1 >= a^2 < [\nabla G_0 fe^t; \, e^{t_j + t_{j'}} \nabla G_0 fe^t] \, e^{-4t_0} >\leq C \, a^2 [f; G_0 f] \,.$$

The last inequality follows from estimates on $< \cosh^4(t_0 - t_j) >$ which can be bounded by $< \cosh^8 t_j >$. We must also use the fact that $f \geq 0$ and

$$\sum_{ijk} f(i) |\nabla G_0(i,j)| \, |\nabla G_0(j,k)| f(k) \leq Const \, [f; G_0 f] \tag{7.120}$$

in three dimensions. The final error term is

$$< Y_2 \cdot Y_2 >= a^2 \epsilon < [e^t G_0 e^t f; \, G_0 e^t f] e^{-4t_0} >\leq C \, a^2 \, [f; G_0 f] \,.$$

In the last inequality we bound $< e^{-t_0} >$ as well as $< \cosh^4(t_j - t_k) >$. The parameter a is chosen small so that $< X \cdot Y >$ is the dominant term.

The upper bound is a standard estimate obtained as follows. Let $L_0 = \beta \nabla^* \nabla + \varepsilon$ and $G_0 = L_0^{-1}$. Then

$$[fe^t; \, G_t fe^t] = [L_0 G_0 \, fe^t; \, G_t fe^t] \,.$$

Integrating by parts we see that the right side equals

$$[\sqrt{\beta} e^{-(t_j + t_{j'})/2} \nabla G_0 \, fe^t; \, \sqrt{\beta} e^{(t_j + t_{j'})/2} \nabla G_t fe^t] + [\sqrt{\varepsilon} e^{-t/2} G_0 fe^t; \, \sqrt{\varepsilon} e^{t/2} G_t fe^t] \,.$$

By the Schwarz inequality we get

$$[fe^t; \, G_t fe^t] \leq [fe^t; \, G_t fe^t]^{1/2}$$

$$\times \left(\beta \sum_j |\nabla_j (G_0 fe^t)|^2 e^{-(t_j + t_{j'})} + \varepsilon \sum_j |(G_0 fe^t)(j)|^2 e^{-t_j} \right)^{1/2} \,.$$

Therefore

$$[fe^t; G_t fe^t] \leq \beta \sum_j |\nabla_j (G_0 e^t f)|^2 e^{-(t_j + t_{j'})} + \varepsilon \sum_j |(G_0 e^t f)(j)|^2 e^{-t_j} \,. \tag{7.121}$$

The desired upper bound on the expectation (7.121) in terms of G_0 now follows from bounds on $\langle \cosh^8(t_j) \rangle$ and (7.120). Note that if the factor of e^t times f were not present, we would not need to require $f \geq 0$ and use (7.120).

The proof of the upper bound now follows using (7.120). Note that in both inequalities we need to assume that $f \geq 0$.

Acknowledgements

I wish to thank my collaborators Margherita Disertori and Martin Zirnbauer for making this review on supersymmetry possible. Thanks also to Margherita Disertori and Sasha Sodin for numerous helpful comments and corrections on early versions of this review and to Yves Capdeboscq for improving the lower bound in (7.87). I am most grateful to the organizers of the Les Houches summer school, Jürg Fröhlich, Vieri Mastropietro, Wojciech De Roeck, and Manfred Salmhofer for inviting me to give these lectures and for creating a stimulating mathematical environment.

References

Abdesselam, A. (2004). Grassmann-Berezin Calculus and Theorems of the Matrix-Tree Type, *Advances in Applied Mathematics* **33**: 51–70, http://arxiv.org/abs/math/0306396.

Abrahams, E., Anderson, P. W., Licciardello, D. C. and Ramakrishnan, T. V. (1979). Scaling theory of localization: Absence of quantum diffusion in two dimensions, *Phys. Rev. Lett.* **42**: 673.

Beamond, E. J., Owczarek, A. L., Cardy, J., Quantum and classical localisation and the Manhattan lattice, *J. Phys. A, Math. Gen.* **36**: 10251, arXiv:cond-mat/0210359.

Berezin, F. A. (1987). *Introduction to Superanalysis.* Reidel, Dordrecht, The Netherlands.

Brascamp, H. and Lie, E. (1976). On extensions of the Brunn-Minkowski and Prekopa-Leindler theorems, including inequalities for log concave functions, and with an application to the diffusion equation, *J. Func. Anal.* **22**: 366–389.

Brezin, E. Kazakov, V. Serban, D. Wiegman, P. and Zabrodin, A. (2006). Applications of Random matrices to physics, *Nato Science Series* **221**, Springer.

Constantinescu, F. Felder, G. Gawedzki, K. Kupiainen, A. (1987). Analyticity of density of states in a gauge-invariant model for disordered electronic systems, *Jour. Stat. Phys.* **48**, 365–391.

Deift, P. *Orthogonal Polynomials, and Random Matrices: A Riemann-Hilbert Approach*, (CIMS, New York University, New York, 1999)

Deift, P. Kriecherbauer, T. McLaughlin, K. T.- K. Venakides, S. and Zhou, X. (1999). Uniform asymptotics for polynomials orthogonal with respect to varying exponential weights and applications to universality questions in random matrix theory, *Comm. Pure Appl. Math.* **52**, 1335–1425.

Diaconis, P. Recent progress on de Finetti's notions of exchangeability. In: *Bayesian Statistics* 3, (Oxford University Press, New York, 1988), 111–125.

Disertori, M. (2004). Density of states for GUE through supersymmetric approach, *Rev. Math. Phys.* **16**, 1191–1225.

Disertori, M. Spencer, T. (2010). Anderson localization for a SUSY sigma model, *Comm. Math. Phys.* 300, 659-671 (2010), *Comm. Math. Phys.* to appear, http://arxiv.org/abs/0910.3325.

Disertori, M. Pinson, H. Spencer, T. (2002). Density of states for random band matrices, *Comm. Math. Phys.* **232**, 83–124 (2202) http://arxiv.org/abs/math-ph/0111047.

Disertori, M. Spencer, T. Zirnbauer, M. R. (2010). Quasi-diffusion in a 3D Supersymmetric Hyperbolic Sigma Model *Comm. Math Phys.* **300**, 435–486 http://arxiv.org/abs/0901.1652.

Efetov, K. B. Anderson localization and supersymmetry, In: 50 Years of Anderson Localization, E. Abrahams (ed.), World Scientific Publishing Company 2010, http://arxiv.org/abs/1002.2632.

Efetov, K. B. (1984). Minimum Metalic conductivity in the theory of localization, *JETP Lett*, **40**, 738.

Efetov, K.B. (1983). Supersymmetry and theory of disordered metals, *Adv. Phys.* **32**, 874.

Efetov, K. B. *Supersymmetry in Disorder and Chaos*, (Cambridge, UK, Cambridge University Press, 1997).

Erdős, L. (2011). The Published reference: Universality of Wigner Random Matrices: A Survey of Recent Results, Russian Math. Surveys 66:3 507–626 (2011), http://arxiv.org/abs/1004.0861.

Erdős, L. Schlein, B. and Yau, H-T. (2009). Local Semicircle Law and Complete Delocalization for Wigner Random Matrices, *Comm. Math. Phys.* **287**, 641–655.

Erdős, L. Salmhofer, M. and Yau, H-T. (2008). Quantum diffusion of the random Schrödinger evolution in the scaling limit, *Acta Math.* **200**, 211.

Erdős, L. Lecture Notes on Quantum Brownian Motion, These Les Houches Lectures, http://arxiv.org/abs/1009.0843.

Erdős, L. Yau, H. T. and Yin, J. (to appear). Bulk universality for generalized Wigner matrices, to appear in Prob. Th. and Rel.Fields, arxiv:1001.3453.

Feldman, J. Knörrer, H. and Trubowitz, E. Fermionic Functional Integrals and the Renormalization Group, CRM monograph series (AMS, Providence, Rhode Island, 2002)

Fröhlich, J. and Spencer, T. (1981). The Kosterlitz Thouless transition, *Comm. Math. Phys.* **81**, 527.

Fröhlich, J. and Spencer, T. (1981). On the statistical mechanics of classical Coulomb and dipole gases, *Jour. Stat. Phys.* **24**, 617–701.

Fröhlich, J. Simon, B. Spencer, T. (1976). Infrared bounds, phase transitions and continuous symmetry breaking, *Comm. Math. Phys.* **50**, 79.

Y. V. Fyodorov, Basic Features of Efetov's SUSY. In: Mesoscopic Quantum Physics E. Akkermans et al. (eds), (Les Houches, France, 1994).

Fyodorov, Y.V. (2002). Negative moments of characteristic polynomials of random matrices, *Nucl. Phys. B* **621**, 643–674, http://arXiv.org/abs/math-ph/0106006.

Gruzberg, I. A. Ludwig, A. W. W. and Read, N. (1999). Exact exponents for the spin Quantun Hall transition, *Phys. Rev. Lett.* **82**, 4524–4527.

Guhr, T. Supersymmetry in random matrix theory. In: The Oxford Handbook of Random Matrix Theory, (Oxford University Press, 2010), http://arXiv:1005.0979v1.

Kosterlitz, J. M. and Thouless, D.J. (1973). Ordering, metastability and phase transitions in two-dimensional systems, *J. Phys. C* **6**, 1181.

Littelmann, P. Sommers, H. -J. and Zirnbauer, M. R. (2008). Superbosonization of Invariant Random Matrix Ensembles, *Comm. Math. Phys.* **283**, 343–395.

Merkl, F. and Rolles, S. Linearly edge-reinforced random walks, Lecture Notes-Monograph Series vol. 48, Dynamics and Stochastics (2006), pp. 66–77.

Merkl, F. and Rolles, S. (2009). Edge-reinforced random walk on one-dimensional periodic graphs, *Probab. Theory Relat. Fields*, **145**, 323.

Mirlin, A. D. Statistics of energy levels. In: New Directions in Quantum Chaos, (Proceedings of the International School of Physics "Enrico Fermi", Course CXLIII), by G.Casati, I.Guarneri, and U. Smilansky (eds) (IOS Press, Amsterdam, 2000), pp. 223–298, http://arXiv.org/abs/cond-mat/0006421.

Pastur, L. and Shcherbina, M. (1997). Universality of the Local eigenvalue statistics for a class of unitary invariant matrix ensembles, *J. Stat. Phys.*, **86**, 109–147.

Peamantle, R. (1988). Phase transition in reinforced random walk and RWRE on trees, *Ann. Probab.* **16**, 1229–1241.

Rudnick, Z. and Sarnak, P. (1996). Zeros of principal L-functions and random-matrix theory, *Duke Math. J.* **81**, 269–322.

Salmhofer, M. Renormalization: An Introduction, (Springer, 1999)

Schäfer, L. and Wegner, F. (1980). Disordered system with n orbitals per site: Lagrange formulation, hyperbolic symmetry, and Goldstone modes *Z. Phys. B* **38**, 113–126.

Schenker, J. (2009). Eigenvector localization for random band matrices with power law band width, *Comm. Math. Phys.* **290**, 1065–1097.

Sodin, A. (2010). The spectral edge of some random band matrices, *Annals of Mathematics*, **172**, 2223.

Sodin, A. (2011) An Estimate on the Average Spectral Measure for Random Band Matrices, J. Stat. Phys. **144** 46–59, http://arxiv.org/1101.4413.

Spencer, T. Mathematical aspects of Anderson Localization. In: 50 years of Anderson Localization, E. Abrahams (ed) (World Scientific Publishing Company, 2010)

Spencer, T. Random band and sparse matrices, in The Oxford Handbook of Random Matrix Theory (Oxford University Press, 2010)

Spencer, T. and Zirnbauer, M. R. (2004). Spontaneous symmetry breaking of a hyperbolic sigma model in three dimensions, *Comm. Math. Phys.* **252**, 167–187, http://arXiv.org/abs/math-ph/0410032.

Valko, B. and Virag, B. (to appear). Random Schrödinger operators on long boxes, noise explosion and the GOE, http://arXiv.org/abs/math-ph/09120097 not yet published.

Verbaarschot, J., J. M. Weidenmüller, H.A. and Zirnbauer, M.R. (1985). *Phys. Reports*, **129**, (367–438).

Wegner, F. (1979). The mobility edge problem: Continuous symmetry and a Conjecture, *Z. Phys. B* **35**, 207–210.

Zirnbauer, M. R. (2004). The Supersymmetry method of random matrix theory, http://arXiv.org/abs/math-ph/0404057.

Zirnbauer, M. R. (1991). Fourier analysis on a hyperbolic supermanifold with constant curvature, *Comm. Math. Phys.* **141**, 503–522.

Zirnbauer, M. R. (1996). Riemannian symmetric superspaces and their origin in random-matrix theory, *J. Math. Phys.* **37**, 4986–5018.

Part II

Short lectures

8

Mass renormalization in nonrelativistic quantum electrodynamics

Volker BACH

Institut für Analysis und Algebra,
TU Braunschweig,
Pockelsstraße 14
v.bach@tu-bs.de

Chapter Contents

8.1 Mass renormalization in NR QED—overview of results

8.1.1 Quantized photon field

The photon Hilbert space is the *Fock space*

$$\mathcal{F} = \mathcal{F}_b[\mathfrak{h}] = \bigoplus_{n=0}^{\infty} \mathcal{F}^{(n)} \tag{8.1}$$

where the *one-photon Hilbert space* is defined as $\mathfrak{h} := L^2(\mathbb{R}^3 \times \mathbb{Z}_2)$. The *n-photon sectors* $\mathcal{F}^{(n)}$ in (8.1) are defined as follows. If $n = 0$ then no photons are present which is the reason that $\mathcal{F}^{(0)} := \mathbb{C} \cdot \Omega$ is called the *vacuum sector* and the normalized vector Ω spanning $\mathcal{F}^{(0)}$ is called the *vacuum vector*. For $n \geq 1$,

$$\mathcal{F}^{(n)} := \left\{ \psi_n \in \otimes^n \mathfrak{h} \,\middle|\, \forall \pi \in \mathfrak{S}_n : \psi_n(k_1, \ldots, k_n) - \psi_n(k_{\pi(1)}, \ldots, k_{\pi(n)}) \right\} \tag{8.2}$$

is the subspace of totally symmetric vectors in $\otimes^n \mathfrak{h}$. It is convenient to rewrite $\mathcal{F}^{(n)}$ by means of creation operators $a^*(f)$ and annihilation operators $a(f)$ on $\mathcal{F}$, which constitute the Fock representation of the canonical commutation relations (CCR),

$$\left[a(f), a^*(g)\right] = \langle f|g \rangle \cdot \mathbf{1}, \quad \left[a(f), a(g)\right] = \left[a^*(f), a^*(g)\right] = 0, \quad a(f)\Omega = 0, \tag{8.3}$$

for all $f, g \in \mathfrak{h}$. Note that $f \mapsto a^*(f)$ is linear, while $f \mapsto a(f)$ is antilinear. Using the creation and annihilation operators, the n-photon sector can be rewritten as

$$\mathcal{F}^{(n)} = \overline{\mathrm{span}\left\{ a^*(f_1)\, a^*(f_2) \cdots a^*(f_n)\Omega \,\middle|\, f_1, f_2, \ldots, f_n \in \mathfrak{h} \right\}}. \tag{8.4}$$

It is furthermore convenient to introduce pointwise creation and annihilation operators a_k^* and a_k by

$$\int_{\mathbb{R}^3 \times \mathbb{Z}_2} f(k)\, a_k^*\, dk := a^*(f), \quad \int_{\mathbb{R}^3 \times \mathbb{Z}_2} \overline{f(k)}\, a_k\, dk := a(f), \tag{8.5}$$

to hold for all $f \in \mathfrak{h}$. Then (8.4) is equivalent to

$$[a_k, a_{k'}^*] = \delta(k - k') \cdot \mathbf{1}, \quad [a_k, a_{k'}] = [a_k^*, a_{k'}^*] = 0, \quad a_k\Omega = 0, \tag{8.6}$$

for all $k, k' \in \mathbb{R}^3 \times \mathbb{Z}_2$. In view of (8.5) it is clear that (8.6) is meant to hold in the sense of operator-valued distributions, that is, it is understood that (8.3) is integrated against a sufficiently regular test function $\mathbb{R}^3 \times \mathbb{Z}_2 \to \mathbb{C}$.

Using creation and annihilation operators, the Hamiltonian of the free photon field is given by

$$H_f = \int \omega(k)\, a_k^* a_k\, dk, \tag{8.7}$$

where, for $k = (\vec{k}, \lambda) \in \mathbb{R}^3 \times \mathbb{Z}_2$,

$$\omega(k) := |k| := |\vec{k}| \tag{8.8}$$

is the photon dispersion, reflecting the fact that photons are massless particles, $|\vec{k}| = \sqrt{k^2 + m_{\mathrm{rad}}^2}\big|_{m_{\mathrm{rad}}=0}$. In terms of the representation (8.2), the photon Hamiltonian acts as a Fourier multiplier. That is, $H_f = \bigoplus_{n=0}^{\infty} H_f^{(n)}$ leaves the n-photon sectors invariant, $H_f \Omega := 0$, and

$$\left[H_f^{(n)} \psi_n\right](k_1, \ldots, k_n) = \sum_{j=1}^{n} \omega(k_j) \psi_n(k_1, \ldots, k_n), \qquad (8.9)$$

from which we can immediately read off its spectral properties,

- $\sigma[H_f] = \mathbb{R}_0^+$,
- $\sigma_{\mathrm{ac}}[H_f] = \mathbb{R}^+$,
- 0 is a simple (the only) eigenvalue: $H_f \Omega = 0$.

8.1.2   A single, free nonrelativistic particle in the photon field

The Hilbert space carrying a single (spinless) particle and the quantized photon field, as described in the previous section, is the tensor product space,

$$\mathcal{H} = \mathcal{H}_{\mathrm{el}} \otimes \mathcal{F} = L^2(\mathbb{R}^3) \otimes \mathcal{F} \sim L^2\left(\mathbb{R}_x^3; \mathcal{F}\right) =: \int_{\mathbb{R}^3}^{\oplus} \mathcal{F} \, d^3x, \qquad (8.10)$$

which we represent as a *direct integral*, that is, as the space of Fock space-valued square-integrable functions on configuration space $\mathbb{R}_x^3$.

The Hamiltonian generating the dynamics of a system consisting of a free nonrelativistic particle and a photon field, but not in interaction with each other, is given by

$$H_0 := \frac{-\Delta_x}{2m} + H_f. \qquad (8.11)$$

The interaction is realized by minimal coupling, i.e., by replacing the particle *momentum operator* $-i\vec{\nabla}_x$ by the *velocity operator* $-i\vec{\nabla}_x - g\vec{A}(\vec{x})$, where

$$\vec{A}(\vec{x}) := \int_{\mathbb{R}^3 \times \mathbb{Z}_2} \frac{\kappa(|k|/\Lambda)}{2\pi^{3/2} \, \omega(k)^{1/2}} \left\{ \vec{\varepsilon}_k \, e^{-i\vec{k}\cdot\vec{x}} \, a_k^* + \vec{\varepsilon}_k^* \, e^{i\vec{k}\cdot\vec{x}} \, a_k \right\} dk \qquad (8.12)$$

is the *vector potential of the magnetic field*, with an *ultraviolet cutoff* $\kappa \in C^\infty(\mathbb{R}_0^+; [0,1])$, $\kappa \equiv 1$ on $[0, 1/2]$, $\kappa \equiv 0$ on $[1, \infty)$, and $\kappa' \leq 0$. Furthermore, the coupling constant $g = \sqrt{\alpha} \approx 1/\sqrt{137}$ is the square root of *Sommerfeld's fine structure constant*. The interacting Hamiltonian is thus given by

$$H_g := \frac{1}{2m}\left(-i\vec{\nabla}_x - g\vec{A}(\vec{x})\right)^2 + H_f. \qquad (8.13)$$

8.1.3   Total momentum

The *total momentum operator* is the sum of the particle and the field momentum operator, that is, the triple

$$\vec{P}_{\text{tot}} := -i\vec{\nabla}_x + \vec{P}_f = -i\vec{\nabla}_x + \int \vec{k}\, a_k^* a_k\, dk \tag{8.14}$$

of mutually commuting self-adjoint operators $P_{\text{tot},1}$, $P_{\text{tot},2}$, and $P_{\text{tot},3}$. The importance of the total momentum lies in the fact that it commutes with the Hamiltonian, too,

$$\left[H_g, \vec{P}_{\text{tot}}\right] = \vec{O}. \tag{8.15}$$

Thus, the total momentum defines *good quantum numbers*. To make this explicit, we introduce a unitary

$$U : L^2\big(\mathbb{R}_x^3; \mathcal{F}\big) \to L^2\big(\mathbb{R}_p^3; \mathcal{F}\big), \tag{8.16}$$

$$[U\phi]\,(\vec{p}) := \int \exp\big[-i\vec{x}\cdot(\vec{p}-\vec{P}_f)\big]\,\phi(x)\,d^3x. \tag{8.17}$$

One easily checks that, for all $\psi \in L^2(\mathbb{R}_p^3; \mathcal{F})$ and all $\vec{p} \in \mathbb{R}^3$,

$$[U H_g U^* \psi]\,(\vec{p}) = H_g(\vec{p})\,\psi(\vec{p}), \tag{8.18}$$

where

$$H_g(\vec{p}) := \frac{1}{2m}\big(\vec{p} - \vec{P}_f - g\vec{A}(\vec{O})\big)^2 + H_f \tag{8.19}$$

acts only on $\mathcal{F}$, pointwise in $\vec{p}$. We write this 'block diagonalization' of H_g by U as

$$U H_g U^* = \int_{\mathbb{R}^3}^{\oplus} H_g(\vec{p})\,d^3p. \tag{8.20}$$

Consequently,

$$\sigma(H_g) = \bigcup_{\vec{p}\in\mathbb{R}^3} \sigma\big(H_g(\vec{p})\big), \tag{8.21}$$

but we now have a finer resolution of the spectrum and may introduce the *ground state energy* $E_g(\vec{p})$ for fixed total momentum $\vec{p}$ by

$$E_g(\vec{p}) = \inf \sigma\big(H_g(\vec{p})\big). \tag{8.22}$$

It is instructive to first study the noninteracting system, that is, to set $g := 0$. Then

$$H_0(\vec{p}) := \frac{1}{2m}\big(\vec{p} - \vec{P}_f\big)^2 + H_f, \tag{8.23}$$

which acts on $\psi(\vec{p}\,) = \bigoplus_{n=0}^{\infty} \psi_n(\vec{p}\,)$ as

$$\left[H_0^{(n)}(\vec{p}\,)\psi_n(\vec{p}\,)\right](k_1,\ldots,k_n) = \left\{\frac{1}{2m}\left(\vec{p} - \sum_{j=1}^{n}\vec{k}_j\right)^2 + \sum_{j=1}^{n}|k_j|\right\}\psi_n(\vec{p}; k_1,\ldots,k_n). \quad (8.24)$$

Using

$$\left|\sum_{j=1}^{n}\vec{k}_j\right| \leq \sum_{j=1}^{n}\omega(k_j), \quad (8.25)$$

we find that

$$E_0(\vec{p}\,) = \inf\left\{\frac{1}{2m}(\vec{p} - \vec{k})^2 + \omega \,\middle|\, \vec{k} \in \mathbb{R}^3, \ \omega \geq 0, \ |\vec{k}| \leq \omega\right\}$$

$$= \begin{cases} \frac{1}{2m}\vec{p}^{\,2}, & \text{for } |\vec{p}\,| \leq m, \\ |\vec{p}\,| - \frac{m}{2}, & \text{for } |\vec{p}\,| \geq m. \end{cases} \quad (8.26)$$

Note that $E_0(\vec{p}\,)$ is an eigenvalue of $H_0(\vec{p}\,)$ for $|\vec{p}\,| \leq m$ but not for $|\vec{p}\,| > m$. Moreover,

$$\frac{1}{m} = \left.\frac{\partial^2 E_0(\vec{p}\,)}{\partial|\vec{p}\,|^2}\right|_{\vec{p}=\vec{0}}. \quad (8.27)$$

The following results have been obtained for this or related models:

(a) J. Fröhlich (Fröhlich, 1973; Fröhlich, 1974) studied the Nelson model with massless scalar fields.

(b) T. Chen proved (Chen, 2001) for $0 < g \ll 1$ and $0 < |\vec{p}\,|^2 \ll m$ that $E_g(\vec{0})$ is an eigenvalue but $E_g(\vec{p}\,)$ is not.

(c) From J. Fröhlich's results (Fröhlich 1973; Fröhlich 1974) follows, however, that a ground state of $H_g(\vec{p}\,)$, for $0 < g \ll 1$ and $0 < |\vec{p}\,|^2 \ll m$, does exist after a suitable change of the representation of the CCR for each $\vec{p}$ by an improper $(f_{\vec{p}} \notin \mathfrak{h})$ Bogolubov transformation (in fact, Weyl operator) of the form $W_{\vec{p}} = \exp[a^*(f_{\vec{p}}) + a(f_{\vec{p}})]$.

(d) T. Chen, J. Fröhlich, and A. Pizzo construct (Chen, Fröehlich, and Pizzo, 2010) asymptotic scattering states, that is, solutions of $\exp[-itH_g]\Psi$, for $t \gg 1$, where

$$\Psi = \int^{\oplus} h(\vec{p}\,)\,\psi_g(\vec{p}, \sigma)\,d^3p, \quad (8.28)$$

where $\psi_g(\vec{p}, \sigma)$ is the ground state of an infrared regularized Hamiltonian ($\sigma > 0$ cutting out momenta $|\vec{k}| \leq \sigma$) and $h \in C_0^{\infty}[B(0,r)]$, with $r \ll 1$.

(e) In a follow-up paper (which appeared, however, earlier), T. Chen, J. Fröhlich, and A. Pizzo proved (Chen, Fröehlich, and Pizzo, 2009) that $\vec{p} \mapsto E_g(\vec{p}\,)$ is $C^{1,\gamma}$, for any $\gamma < 1$. It is believed that $\vec{p} \mapsto E_g(\vec{p}\,)$ is actually two times continuously differentiable at $\vec{p} = \vec{0}$, which would be important to give a meaning to the definition of the renormalized particle mass

$$\frac{1}{m_g} := \left.\frac{\partial^2 E_g(\vec{p})}{\partial |\vec{p}|^2}\right|_{\vec{p}=\vec{0}}. \tag{8.29}$$

(f) T. Chen, J. Faupin, J. Fröhlich, and I.M. Sigal established (Chen, Faupin, Fröhlich, and Sigal, 2009) a *limiting absorption principle* above $E_g(\vec{p})$.

In these notes we are concerned with the value m_g of the renormalized mass. To obtain a definition which is accessible to analysis, we introduce an infrared regularization $\sigma > 0$, replacing $vA(\vec{0}) = a^*(\vec{G}_0) + a(\vec{G}_0)$ by $vA_\sigma(\vec{0}) = a^*(\vec{G}_\sigma) + a(\vec{G}_\sigma)$, where

$$\vec{G}_\sigma(k) := \frac{\kappa(|k|/\Lambda)}{2\pi^{3/2}\,|k|^{\frac{1}{2}-\sigma}}\,\vec{\varepsilon}_k. \tag{8.30}$$

Similarly,

$$H_g(\vec{p}, \sigma) := \frac{1}{2m}\left(\vec{p} - \vec{P}_f - g\vec{A}_\sigma(\vec{0})\right)^2 + H_f, \tag{8.31}$$

$$E_g(\vec{p}, \sigma) := \inf \sigma H_g(\vec{p}, \sigma), \tag{8.32}$$

$$\frac{1}{m_g(\sigma)} = \left.\frac{\partial^2 E_g(\vec{p}, \sigma)}{\partial |\vec{p}|^2}\right|_{\vec{p}=\vec{0}}. \tag{8.33}$$

Note that

$$m_g = \lim_{\sigma \to 0}\left\{m_g(\sigma)\right\}, \tag{8.34}$$

due to (Chen, Fröehlich, and Pizzo, 2009). The following further results have been obtained on $E_g(\vec{p}, \sigma)$ and $m_g(\sigma)$.

(a) E. Lieb and M. Loss (Lieb and Loss, 2002) studied the behaviour of $E_g(\vec{0}, 0)$ for fixed $g > 0$ and large ultraviolet cutoff $\Lambda \gg 1$ and derived the estimate

$$C_1\, g\, \Lambda^{3/2} \leq E_g(\vec{0}, 0) \leq C_2\, g^{4/7}\, \Lambda^{12/7}, \tag{8.35}$$

for suitable universal constants $0 < C_1 < C_2 < \infty$.

(b) I. Catto and C. Hainzl (Catto and Hainzl, 2001) obtained the following bound for fixed $\Lambda \geq 1$ and small coupling $0 < g \ll 1$,

$$E_g(\vec{0}, 0) = 8\pi\, g\left[\Lambda - \ln(1 + \Lambda)\right] + \mathcal{O}(g^2\Lambda^2). \tag{8.36}$$

(c) C. Hainzl and R. Seiringer (Hainzl and Seiringer, 2002) and E. Lieb and M. Loss (Lieb and Loss, 2002) gave an alternative definition of the renormalized mass for fixed $\Lambda \geq 1$ and small coupling $0 < g \ll 1$ based on the difference of the ground state energy and the ionization threshold of a hydrogen atom (coupled to the quantized radiation field).

(d) F. Hiroshima and H. Spohn (Hiroshima and Spohn, 2005) fixed $\Lambda \geq 1$, chose $\kappa(r) := \mathbf{1}[r \leq 1]$, considered small coupling $0 < g \ll 1$, and derived an estimate on the renormalized mass:

$$\lim_{\sigma \to 0} \left\{ \frac{m_g(\sigma)}{m} \right\} = 1 + \frac{16\pi}{3} g^2 \ln(1 + \Lambda/2) + o(g^2). \tag{8.37}$$

The purpose of this chapter is to sketch some ideas of the following result of V. Bach, T. Chen, J. Fröhlich, and I.M. Sigal (Bach, Chen, Fröhlich, and Sigal, 2007).

Theorem 8.1 *There exists a constant $\sigma_0 > 0$ such that, for all $|\vec{p}| \le m/3$ and $0 < \sigma < \sigma_0$ there are constants $g_\sigma, p_\sigma > 0$ such that, for all $0 < g < g_\sigma$,*

(i) $E_g(\vec{p}, \sigma)$ is a simple eigenvalue of $E_g(\vec{p}, \sigma)$;
(ii) The map $\vec{p} \mapsto E_g(\vec{p}, \sigma)$ is C^2 for $|\vec{p}| < p_\sigma$;
(iii) For $\vec{p} \mapsto E_g(\vec{p}, \sigma)$ a convergent series expansion is derived;
(iv) The following estimates hold true,

$$0 < E_g(\vec{0}, \sigma) < \mathcal{O}(g^2), \tag{8.38}$$

$$\left. \frac{\partial E_g(\vec{p}, \sigma)}{\partial |\vec{p}|} \right|_{\vec{p} = \vec{0}} = 0, \tag{8.39}$$

$$\lim_{\sigma \to 0} \left\{ \frac{m_g(\sigma)}{m} \right\} = 1 + \tfrac{8\pi}{3} g^2 \int_0^\infty \frac{|\kappa(s/\Lambda)|^2 \, ds}{2 + s} + \mathcal{O}(g^{7/3}). \tag{8.40}$$

These results are in agreement with formal perturbation theory to low orders in g.

8.2   Renormalization group based on the smooth Feshbach–Schur map

We now turn to the study of $H_g(\vec{p}, \sigma)$ and its spectral properties. For the sake of simplicity, we fix the total momentum $\vec{p} = \vec{0}$ to zero and also the infrared cutoff parameter $\sigma > 0$. We then omit to display the dependence of objects on σ and $\vec{p} = \vec{0}$ henceforth and write $\vec{G} \equiv \vec{G}_\sigma$ and $\vec{A}_\sigma \equiv \vec{A}_\sigma(\vec{0}) = a^*(\vec{G}) + a(\vec{G})$, further assuming that

$$\vec{G}(k) := \frac{\kappa(|k|/\Lambda)}{|k|^{\frac{1}{2} - \sigma}} \vec{\varepsilon}_k. \tag{8.41}$$

That is, we study

$$H_g := H_f + \frac{1}{2m} \vec{P}_f + \frac{g}{m} \vec{P}_f \cdot \vec{A}_\sigma + \frac{g^2}{2m} \vec{A}_\sigma^2 \tag{8.42}$$

$$= H_f + \frac{1}{2m} \vec{P}_f + W_{1,0} + W_{0,1} + W_{1,1} + W_{2,0} + W_{0,2} + \frac{g^2}{2m} \|\vec{G}\|^2,$$

where

$$W_{1,0} := \tfrac{g}{m} a^*(\vec{G}) \cdot \vec{P}_f, \qquad W_{0,1} := \tfrac{g}{m} \vec{P}_f \cdot a(\vec{G}), \tag{8.43}$$

$$W_{2,0} := \tfrac{g^2}{2m} a^*(\vec{G}) \cdot a^*(\vec{G}), \qquad W_{0,2} := \tfrac{g^2}{2m} a(\vec{G}) \cdot a(\vec{G}). \tag{8.44}$$

$$W_{1,1} := \tfrac{g^2}{m} a^*(\vec{G}) \cdot a(\vec{G}), \tag{8.45}$$

8.2.1   The smooth Feshbach–Schur map and isospectrality

We first present the smooth Feshbach–Schur map in abstract form as introduced in (Bach, Chen, Fröhlich, and Sigal, 2003).

- We assume $\mathcal{H}$ to be a (separable, complex) Hilbert space.
- On $\mathcal{H}$ we are given a positive operator $0 \leq \chi \leq \mathbf{1}$, and we denote $\overline{\chi} := \sqrt{1 - \chi^2}$. We denote by P and $\overline{P}$ the orthogonal projection onto $\mathrm{Ran}\,\chi$ and $\mathrm{Ran}\,\overline{\chi}$, respectively. Note that $P\vec{P} \neq 0$ unless χ and $\overline{\chi}$ are orthogonal projections themselves.
- Our goal is to study the spectral properties of a closed and densely defined operator H which is given in a perturbative form, $H = T + W$. Here, T is a closed operator commuting with χ and $\overline{\chi}$ the 'unperturbed' operator,

$$[T, \chi] = [T, \overline{\chi}] = 0, \tag{8.46}$$

  and W is a relatively T-form bounded 'perturbation', that is,

$$\left\| (|T| + 1)^{1/2}\, W\, (|T| + 1)^{1/2} \right\|_{\mathrm{op}} < \infty. \tag{8.47}$$

- Finally, $z \in \mathbb{C}$ is a complex number which plays the role of the spectral parameter.

Theorem 8.2 *Assume that $T + \overline{\chi}W\overline{\chi}$ is invertible on $\mathrm{Ran}\,\overline{\chi} = \overline{P}\mathcal{H}$ and set*

$$F_\chi \equiv F_\chi(H - z, T - z) := T - z + \chi W \chi - \chi W \overline{\chi}(T + \overline{\chi}W\overline{\chi})^{-1}\overline{\chi}W\chi, \tag{8.48}$$

$$Q_\chi \equiv Q_\chi(H - z, T - z) := \chi - \overline{\chi}(T + \overline{\chi}W\overline{\chi})^{-1}\overline{\chi}W\chi. \tag{8.49}$$

Then

$$\left(H - z \text{ is invertible on } \mathcal{H} \right) \Leftrightarrow \left(F_\chi \text{ is invertible on } P\mathcal{H} \right), \tag{8.50}$$

$$\left((H - z)\psi = 0, \ \psi \neq 0 \right) \Rightarrow \left(F_\chi \chi\psi = 0, \ \chi\psi \neq 0 \right), \tag{8.51}$$

$$\left(F_\chi \varphi = 0, \ \varphi = P\varphi \neq 0 \right) \Rightarrow \left((H - z)Q_\chi\varphi = 0, \ Q_\chi\varphi \neq 0 \right), \tag{8.52}$$

$$\dim \mathrm{Ker}(H - z) = \dim \mathrm{Ker}\, F_\chi(H - z, T - z). \tag{8.53}$$

We remark that F_χ is called the *Feshbach–Schur map* and that, for orthogonal projections $\chi = \chi^* = \chi^2$, Theorem 8.2 is known as the *Feshbach Projection method* in quantum mechanics. Actually, the Feshbach projection method is known in various fields of mathematics, but under different names, such as, *Ljapunov–Schmidt reduction*, *Schur complement method*, or *Grushin problem*. Properties (8.50)–(8.53) are called *isospectrality of F_χ*.

We apply Theorem 8.2 to H_g with

$$\chi := \chi_\rho := \chi_\rho(H_f) := \chi_1\!\left(\frac{H_f}{\rho}\right), \tag{8.54}$$

where $0 < \rho < \frac{1}{2}$ is fixed and $\chi \in C^\infty\left(\mathbb{R}_0^+; [0,1]\right)$, with $\chi \equiv 1$ on $[0, \frac{1}{2}]$, $\chi \equiv 0$ on $[1, \infty)$, and $\chi' \leq 0$. Furthermore,

$$T \equiv T\big[H_f; \vec{P}_f\big] \; := \; H_f + \frac{1}{2m}\vec{P}_f^2, \tag{8.55}$$

$$W = W_{1,0} + W_{0,1} + W_{1,1} + W_{2,0} + W_{0,2}, \tag{8.56}$$

$$z \in \left[-\tfrac{1}{16}\rho, \, \tfrac{1}{16}\rho\right]. \tag{8.57}$$

To this end, we observe that

$$\left|\sum_{\nu=1}^{n} \vec{k}_\nu\right| \leq \sum_{\nu=1}^{n} |\vec{k}_\nu| \quad \Rightarrow \quad |\vec{P}_f| \leq H_f, \tag{8.58}$$

$$T = H_f + \tfrac{1}{2m}\vec{P}_f^2 \geq H_f \geq \tfrac{1}{2}\rho, \qquad \text{on } \operatorname{Ran}\overline{\chi}. \tag{8.59}$$

Furthermore, for all $\psi \in \mathcal{F}$, we have that

$$\begin{aligned}
\left|\langle\psi|W_{0,1}\psi\rangle\right| &= \tfrac{g}{m}\left|\langle\vec{P}_f\psi| \cdot a(\vec{G})\psi\rangle\right| \; \leq \; \tfrac{g}{m}\|\vec{P}_f\psi\|\,\|a(\vec{G})\psi\| \\
&\leq \tfrac{g}{2m}\|\vec{P}_f\psi\|^2 + \tfrac{g}{2m}\|a(\vec{G})\psi\|^2
\end{aligned} \tag{8.60}$$

and

$$\begin{aligned}
\|a(\vec{G})\psi\|^2 = \left\|\int \vec{G}(k)\, a_k\psi \, dk\right\| &\leq \left(\int \frac{|\vec{G}(k)|^2}{|k|}\, dk\right)\left(\int |k|\,\|a_k\psi\|^2\, dk\right) \\
&= C_1\left(\int_0^\infty k^{2\sigma}\,\kappa(k/\Lambda)\, dk\right)\langle\psi|H_f\psi\rangle \; = \; C_2\,\langle\psi|H_f\psi\rangle,
\end{aligned} \tag{8.61}$$

which, put together, implies the existence of a constant $C_3 \equiv C_3(m,\Lambda)$ such that

$$\left|\langle\psi|W_{0,1}\psi\rangle\right| \; \leq \; C_3\, g\,\left|\left\langle\psi\left|\left(H_f + \tfrac{1}{2m}\vec{P}_f^2\right)\psi\right\rangle\right| \; = \; C_3\, g\,\langle\psi|T\psi\rangle. \tag{8.62}$$

Similarly, but additionally assuming that $\psi \in \operatorname{Ran}\overline{\chi}$ and using $a(f)a^*(f) = a^*(f)a(f) + \|f\|^2$ and (8.57), we find that

$$\left|\langle\psi|W_{0,1}\psi\rangle\right|, \; \left|\langle\psi|W_{1,0}\psi\rangle\right|, \; \left|\langle\psi|W_{1,1}\psi\rangle\right| \leq C_4\, g\,\langle\psi|T\psi\rangle, \tag{8.63}$$

$$\left|\langle\psi|W_{0,2}\psi\rangle\right|, \; \left|\langle\psi|W_{2,0}\psi\rangle\right|, \; \leq C_4\, g^2\, \rho^{-1/2}\,\langle\psi|T\psi\rangle, \tag{8.64}$$

and hence

$$\forall\psi \in \operatorname{Ran}\overline{\chi}: \quad \langle\psi|W\psi\rangle \; \leq \; C_6\, g\,\langle\psi|T\psi\rangle, \tag{8.65}$$

for suitable constants $C_6, C_4 \in [C_3, \infty)$. So, if $g > 0$ is sufficiently small such that $\frac{3}{16}\rho \geq C_6 g$, we obtain the estimate

$$\langle \psi | (T - z + \overline{\chi} W \overline{\chi}) \psi \rangle \geq \left\langle \psi \left| \left(T - \tfrac{1}{16} \rho - C_6 \, g \, \overline{\chi} T \overline{\chi} \right) \psi \right. \right\rangle \tag{8.66}$$

$$\geq \left(\tfrac{1}{2} \rho - \tfrac{1}{16} \rho - C_6 \, g \right) \|\psi\|^2 \geq \tfrac{1}{4} \rho \|\psi\|^2,$$

for all $\psi \in \overline{P} \mathcal{H} = \operatorname{Ran} \overline{\chi}$. For the derivation of the second inequality in (8.66), we use that $\overline{\chi} T \overline{\chi} = T^{1/2} \overline{\chi}^2 T^{1/2} \leq T$. Eq. (8.66) implies the strict positivity and, hence in particular, the invertibility of $T - z + \overline{\chi} W \overline{\chi}$ on $\overline{P} \mathcal{H}$. Thus, Theorem 8.2 is applicable and $H_g - z$ is isospectral to

$$F_\chi \equiv F_{\chi_\rho}(H_g - z, T - z)$$

$$= T - z + \chi_\rho W \chi_\rho - \chi_\rho W \overline{\chi}_\rho (T - z + \overline{\chi}_\rho W \overline{\chi}_\rho)^{-1} \overline{\chi}_\rho W \chi_\rho$$

$$= T - z + \sum_{\ell=0}^{\infty} (-1)^\ell \chi_\rho W \left\{ \left(\frac{\overline{\chi}_\rho^2}{T - z} \right) W \right\}^\ell \chi_\rho, \tag{8.67}$$

which is a norm-convergent Neumann series expansion, by estimates similar to (8.65) and (8.66).

8.2.2 Normal-ordered form of operators

We write $F_{\chi_\rho}(H_g - z, T - z) =: \widetilde{H}^{(0)}[z]$ as an operator of the form

$$\widetilde{H}^{(0)}[z] = \widetilde{E}^{(0)}[z] + \widetilde{T}^{(0)}[H_f, \vec{P}_f, z] + \widetilde{W}^{(0)}[z], \tag{8.68}$$

where $\widetilde{E}^{(0)}[z] \in \mathbb{C}$ (in fact, $\widetilde{E}^{(0)}[z] \in \mathbb{R}$ because $z \in \mathbb{R}$ here), $\widetilde{T}^{(0)}[H_f, \vec{P}_f, z]$ is defined by the functional calculus, that is, for a given function $T : \mathbb{R}_0^+ \times \mathbb{R}^3 \times \mathbb{C} \to \mathbb{C}$, the operator $T[H_f, \vec{P}_f, z] = \bigoplus_{n=0}^{\infty} T_n[H_f, \vec{P}_f, z]$ is defined by

$$\left(T[H_f, \vec{P}_f, z] \psi_n \right)(k_1, \ldots, k_n) := T \left[\sum_{\nu=1}^{n} |\vec{k}_\nu|, \sum_{\nu=1}^{n} \vec{k}_\nu, \, z \right] \psi_n(k_1, \ldots, k_n), \tag{8.69}$$

for all $\psi = \bigoplus_{n=0}^{\infty} \psi_n \in \mathcal{F}$. The function $\widetilde{T}^{(0)}[H_f, \vec{P}_f, z]$ turns out to equal $T = H_f + \frac{1}{2m} \vec{P}_f^2$ plus terms of higher order in g and, especially, to be Lipschitz-continuous in H_f (and hence also in $\vec{P}_f$) and to vanish at $H_f = 0$ (which implies that also $\vec{P}_f = 0$). Moreover,

$$\widetilde{W}^{(0)}[z] = \sum_{m+n \geq 1} \widetilde{W}_{m,n}^{(0)}[z], \tag{8.70}$$

where

$$\widetilde{W}_{m,n}^{(0)}[z] \equiv \mathbb{Q}_\rho \big[\widetilde{w}_{m,n}^{(0)} \big] \tag{8.71}$$

$$= \int_{B_\rho^{m+n}} dK^{(m,n)} \, a^*(k^{(m)}) \, \widetilde{w}_{m,n}^{(0)} \big[H_f, \vec{P}_f, z; K^{(m,n)} \big] \, a(\tilde{k}^{(n)}),$$

where B_ρ denotes $\mathbb{Z}_2$ times the unit ball in $\mathbb{R}^3$, $\widetilde{w}_{m,n}^{(0)}\big[H_f, \vec{P}_f, z; K^{(m,n)}\big]$ is an operator defined by the functional calculus, and here and henceforth we use the notation

$$k^{(m)} := (k_1, \ldots, k_m) \in (\mathbb{R}^3 \times \mathbb{Z}_2)^m, \tag{8.72}$$

$$\tilde{k}^{(n)} := (\tilde{k}_1, \ldots, \tilde{k}_n) \in (\mathbb{R}^3 \times \mathbb{Z}_2)^n, \tag{8.73}$$

$$K^{(m,n)} := (k^{(m)}, \tilde{k}^{(n)}), \qquad dK^{(m,n)} := \prod_{i=1}^m dk_i \prod_{i=1}^n d\tilde{k}_i, \tag{8.74}$$

$$a^*(k^{(m)}) := \prod_{i=1}^m a_{k_i}^*, \qquad a(\tilde{k}^{(n)}) := \prod_{j=1}^n a_{\tilde{k}_j}, \tag{8.75}$$

$$|K^{(m,n)}| := |k^{(m)}| \cdot |\tilde{k}^{(n)}|, \qquad |k^{(m)}| := |k_1| \cdots |k_m|, \tag{8.76}$$

$$\Sigma[k^{(m)}] := |k_1| + \ldots + |k_m|. \tag{8.77}$$

The use of $\mathbb{Q}_\rho$ in (8.71) indicates that $\widetilde{W}_{m,n}^{(0)}[z]$ may be viewed as a *quantization* of the integral kernel $\widetilde{w}_{m,n}^{(0)}$. These integral kernels result from a normal-ordering procedure moving all creation operators to the left and all annihilation operators to the right in any product using the *Pull-Through Formulae*

$$a_k\, F\big[H_f,\ \vec{P}_f\big] = F\big[H_f + |k|,\ \vec{P}_f + \vec{k}\big]\, a_k, \tag{8.78}$$

$$F\big[H_f,\ \vec{P}_f\big]\, a_k^* = a_k^*\, F\big[H_f + |k|,\ \vec{P}_f + \vec{k}\big], \tag{8.79}$$

which hold true for sufficiently regular functions F in the sense of operator-valued distributions. We carry out a sample computation for $\ell = 1$ in equation (8.67), that is, for

$$\chi_\rho W\left(\frac{\overline{\chi}_\rho^2}{T-z}\right) W \chi_\rho = \chi_\rho \big(W_{1,0} + W_{0,1} + W_{1,1} + W_{2,0} + W_{0,2}\big) \tag{8.80}$$

$$\left(\frac{\overline{\chi}_\rho^2}{T-z}\right)\big(W_{1,0} + W_{0,1} + W_{1,1} + W_{2,0} + W_{0,2}\big)\chi_\rho.$$

Out of the 25 terms obtained from expanding the product on the right side of (8.80), we have a closer look at two, namely,

$$\chi_\rho W_{2,0}\left(\frac{\overline{\chi}_\rho^2}{T-z}\right) W_{1,1}\chi_\rho \quad \text{and} \quad \chi_\rho W_{0,1}\left(\frac{\overline{\chi}_\rho^2}{T-z}\right) W_{1,0}\chi_\rho. \tag{8.81}$$

Writing out the integrals and using the pull-through formulae, we obtain

$$\chi_\rho W_{2,0}\left(\frac{\overline{\chi}_\rho^2}{T-z}\right) W_{1,1}\chi_\rho \tag{8.82}$$

$$= \int_{(\mathbb{R}^3)^4} dk_1\, dk_2\, dk_3\, d\tilde{k}_1\, \frac{g^4}{2m^2}\, \big(\vec{G}(k_1) \cdot \vec{G}(k_2)\big)\big(\vec{G}(k_3) \cdot \vec{G}(\tilde{k}_1)^*\big)$$

$$\chi_\rho[H_f]\, a_{k_1}^*\, a_{k_2}^* \left(\frac{\overline{\chi}_\rho[H_f]^2}{T[H_f, \vec{P}_f] - z}\right) a_{k_3}^*\, a_{\tilde{k}_1}\, \chi_\rho[H_f]$$

$$= \int_{B_\rho^4} dk_1 \, dk_2 \, dk_3 \, d\tilde{k}_1 \, \frac{g^4}{2m^2} \left(\vec{G}(k_1) \cdot \vec{G}(k_2) \right) \left(\vec{G}(k_3) \cdot \vec{G}(\tilde{k}_1)^* \right)$$

$$a_{k_1}^* a_{k_2}^* a_{k_3}^* \chi_\rho \left[H_f + |k_1| + |k_2| + |k_3| \right] \left(\frac{\overline{\chi}_\rho \left[H_f + |k_3| \right]^2}{T[H_f, \vec{P}_f] - z} \right) \chi_\rho \left[H_f + |\tilde{k}_1| \right] a_{\tilde{k}_1}.$$

Note that the restriction of k_1, k_2, k_3, and $\tilde{k}_1$ to B_ρ is implied by the support properties of χ_ρ and the positivity of H_f. Recalling (8.71), we can write

$$\chi_\rho W_{2,0} \left(\frac{\overline{\chi}_\rho^2}{T - z} \right) W_{1,1} \chi_\rho \;=\; \mathbb{Q}_\rho \big[\widetilde{w}_{(2,0)\circ(1,1)\to(3,1)}^{(0)} \big], \tag{8.83}$$

where

$$\widetilde{w}_{(2,0)\circ(1,1)\to(3,1)}^{(0)} [\omega, \vec{p}, K^{(3,1)}; z] \;=\; \frac{g^4}{2m^2} \left(\vec{G}(k_1) \cdot \vec{G}(k_2) \right) \left(\vec{G}(k_3) \cdot \vec{G}(\tilde{k}_1)^* \right)$$

$$\times \, \chi_\rho \left[\omega + \Sigma(k^{(3)}) \right] \left(\frac{\overline{\chi}_\rho \left[\omega + |k_3| \right]^2}{T[\omega, \vec{p}] - z} \right) \chi_\rho \left[\omega + |\tilde{k}_1| \right] a_{\tilde{k}_1}. \tag{8.84}$$

The term $\mathbb{Q}_\rho \big[\widetilde{w}_{(2,0)\circ(1,1)\to(3,1)}^{(0)} \big]$ is then one contribution to $\widetilde{W}_{3,1}^{(0)}[z]$.

Similarly, we obtain

$$\chi_\rho W_{0,1} \left(\frac{\overline{\chi}_\rho^2}{T - z} \right) W_{1,0} \chi_\rho \tag{8.85}$$

$$= \int_{(\mathbb{R}^3)^2} dk_1 \, d\tilde{k}_1 \, \frac{g^2}{m^2} \, \chi_\rho[H_f] \left(\vec{P}_f \cdot \vec{G}(\tilde{k}_1)^* \right) a_{\tilde{k}_1}$$

$$\times \left(\frac{\overline{\chi}_\rho[H_f]^2}{T[H_f, \vec{P}_f] - z} \right) a_{k_1}^* \left(\vec{P}_f \cdot \vec{G}(k_1) \right) \chi_\rho[H_f]$$

$$= \int_{B_\rho} dk \, \frac{g^2}{m^2} \, \chi_\rho[H_f] \left(\frac{|\vec{P}_f \cdot \vec{G}(k)|^2 \, \overline{\chi}_\rho \left[H_f + |k| \right]^2}{T \left[H_f + |k|, \, \vec{P}_f + \vec{k} \right] - z} \right) \chi_\rho[H_f]$$

$$+ \int_{B_\rho^2} dk_1 \, d\tilde{k}_1 \, a_{k_1}^* \, \frac{g^2}{m^2} \, \chi_\rho \left[H_f + |k_1| \right] \left((\vec{P}_f + \vec{k}_1) \cdot \vec{G}(\tilde{k}_1)^* \right)$$

$$\times \left(\frac{\overline{\chi}_\rho \left[H_f + |k_1| + |\tilde{k}_1| \right]^2 \left((\vec{P}_f + \vec{k}_1) \cdot \vec{G}(k_1) \right)}{T \left[H_f + |k_1| + |\tilde{k}_1|, \, \vec{P}_f + \vec{k}_1 + \vec{\tilde{k}}_1 \right] - z} \right) \chi_\rho \left[H_f + |\tilde{k}_1| \right] a_{\tilde{k}_1},$$

and so we can rewrite

$$\chi_\rho W_{0,1} \left(\frac{\overline{\chi}_\rho^2}{T - z} \right) W_{1,0} \chi_\rho \;=\; \widetilde{W}_{(0,1)\circ(1,0)\to(0,0)}^{(0)} [H_f, \vec{P}_f; z] \,+\, \mathbb{Q}_\rho \big[\widetilde{w}_{(0,1)\circ(1,0)\to(1,1)}^{(0)} \big], \tag{8.86}$$

where

$$\widetilde{w}^{(0)}_{(0,1)\circ(1,0)\to(1,1)}[\omega,\vec{p},K^{(1,1)};z] = \frac{g^2}{m^2}\chi_\rho[\omega+|k_1|]\,\chi_\rho[\omega+|\tilde{k}_1|]\,\overline{\chi}_\rho[\omega+|k_1|+|\tilde{k}_1|]^2$$

$$\times \frac{\left((\vec{p}+\vec{k}_1)\cdot\vec{G}(\tilde{k}_1)^*\right)\left((\vec{p}+\vec{k}_1)\cdot\vec{G}(k_1)\right)}{T[\omega+|k_1|+|\tilde{k}_1|,\ \vec{p}+\vec{k}_1+\vec{k}_1]-z} \tag{8.87}$$

and $\mathbb{Q}_{1,1}\big[\widetilde{w}^{(0)}_{(0,1)\circ(1,0)\to(1,1)}\big]$ is one of the terms which contribute to $\widetilde{W}^{(0)}_{1,1}[z]$.
Moreover,

$$\widetilde{W}^{(0)}_{(0,1)\circ(1,0)\to(0,0)}[H_f,\vec{P}_f;z] = \frac{g^2}{m^2}\int_{B_\rho}\frac{|\vec{P}_f\cdot\vec{G}(k)|^2\,\chi_\rho[H_f]^2\,\overline{\chi}_\rho[H_f+|k|]^2}{T[H_f+|k|,\ \vec{P}_f+\vec{k}]-z}\,dk, \tag{8.88}$$

and

$$\widetilde{W}^{(0)}_{(0,1)\circ(1,0)\to(0,0)}[H_f,\vec{P}_f;z] - \widetilde{W}^{(0)}_{(0,1)\circ(1,0)\to(0,0)}[0,\vec{0};z] \tag{8.89}$$

goes additively into $\widetilde{T}^{(0)}[H_f,\vec{P}_f;z]$, while

$$\widetilde{W}^{(0)}_{(0,1)\circ(1,0)\to(0,0)}[0,\vec{0};z] \tag{8.90}$$

is a pure number, that is, a multiple of the identity operator and contributes to $\widetilde{E}^{(0)}[z]$.
(Actually, $\widetilde{W}^{(0)}_{(0,1)\circ(1,0)\to(0,0)}[0,\vec{0};z]=0$ by accident.)

Carrying out this procedure for all 25 terms resulting from the expansion of the product for $\ell=1$ and, in general, for all $5^{\ell+1}$ terms resulting from the expansion of the corresponding product in case $\ell\geq1$ and summing up the contributions, we obtain the normal-ordered form

$$\widetilde{H}^{(0)}[z] = \widetilde{E}^{(0)}[z]+\widetilde{T}^{(0)}[H_f,\vec{P}_f,z]+\widetilde{W}^{(0)}[z] \tag{8.91}$$

of $F_{\chi_\rho}(H_g-z,T-z)$, as asserted in (8.68). Details of this computation can be found in (Bach, Chen, Fröhlich, and Sigal, 2003). They are involved, but straightforward.

8.2.3 Normal-ordered form of operators

After applying the normal-ordering procedure described in the previous section, we rescale $\widetilde{H}^{(0)}[z]$ to act on

$$\mathcal{H}_{\mathrm{red}} = \mathrm{Ran}\,\chi_1[H_f] \subset \mathcal{F}, \tag{8.92}$$

rather than $\mathrm{Ran}\,\chi_\rho[H_f]$. To this end, we introduce a scaling transformation

$$S_\rho[A] := \frac{1}{\rho}U_\rho A U_\rho^*, \tag{8.93}$$

where U_ρ is the unitary dilation determined by

$$U_\rho \Omega := \Omega, \qquad U_\rho a_k^* U_\rho^* := \rho^{-3/2} a_{k/\rho}^*, \tag{8.94}$$

where $k/\rho = (\vec{k}, \lambda)/\rho := (\vec{k}/\rho, \lambda)$. It is easy to check that

$$U_\rho H_f U_\rho^* = \rho H_f, \qquad U_\rho \vec{P}_f U_\rho^* = \rho \vec{P}_f, \tag{8.95}$$

$$U_\rho \chi_\rho[H_f] U_\rho^* = U_\rho \chi_1[H_f/\rho] U_\rho^* = \chi_1[H_f], \tag{8.96}$$

$$U_\rho \, \text{Ran} \, \chi_\rho[H_f] = \text{Ran} \, \chi_1[H_f] = \mathcal{H}_{\text{red}}. \tag{8.97}$$

As immediate consequences, we have

$$S_\rho\big[\widetilde{E}^{(0)}[z] - z\big] = \frac{1}{\rho}\big(\widetilde{E}^{(0)}[z] - z\big), \tag{8.98}$$

$$S_\rho\big[\widetilde{T}^{(0)}[H_f, \vec{P}_f; z] - z\big] = \frac{1}{\rho}\widetilde{T}^{(0)}\big[\, \rho h f, \rho \vec{P}_f; z\big], \tag{8.99}$$

$$S_\rho\big[\mathbb{Q}_\rho[\widetilde{w}_{m,n}^{(0)}]\big] = \mathbb{Q}_1\big[s_\rho(\widetilde{w}_{m,n}^{(0)})\big], \tag{8.100}$$

where

$$s_\rho(w_{m,n})\big[H_f, \vec{P}_f, z; K^{(m,n)}\big] := \rho^{\frac{3}{2}(m+n)-1} w_{m,n}\big[\rho H_f, \rho\vec{P}_f, z; \rho K^{(m,n)}\big]. \tag{8.101}$$

Note that, to leading order in g (or W),

$$\widetilde{H}^{(0)}[z] - z = -z + H_f + \frac{1}{2m}\vec{P}_f^2 \tag{8.102}$$
$$+ \frac{g}{m}\chi_\rho\big(\vec{P}_f \cdot a(\vec{G}) + a^*(\vec{G}) \cdot \vec{P}_f\big)\chi_\rho + \mathcal{O}(g^2),$$

$$S_\rho\big[\widetilde{H}^{(0)}[z] - z\big] = -\frac{1}{\rho}z + H_f + \frac{\rho}{2m}\vec{P}_f^2 \tag{8.103}$$
$$+ \rho^{1+\sigma}\frac{g}{m}\chi_1\big(\vec{P}_f \cdot a(\vec{G}) + a^*(\vec{G}) \cdot \vec{P}_f\big)\chi_1 + \mathcal{O}(\rho^{1+\sigma}g^2),$$

as operators on $\text{Ran} \, \chi_\rho$ and $\mathcal{H}_{\text{red}} = \text{Ran} \, \chi_1$, respectively. We observe that the only term in $\widetilde{H}^{(0)}[z] - z$ which is expanding under S_ρ is $z \mapsto \rho^{-1}z$.

We digress and temporarily consider $\vec{p}_{\text{tot}} \neq 0$. In this case,

$$\widetilde{H}^{(0)}[z] - z = -z + H_f + \frac{1}{2m}(\vec{p}_{\text{tot}} - \vec{P}_f)^2 + \mathcal{O}(g^2), \tag{8.104}$$
$$+ \frac{g}{m}\chi_\rho\big((\vec{P}_f - \vec{p}_{\text{tot}}) \cdot a(\vec{G}) + a^*(\vec{G}) \cdot (\vec{P}_f - \vec{p}_{\text{tot}})\big)\chi_\rho,$$

$$S_\rho\big[\widetilde{H}^{(0)}[z] - z\big] = -\frac{1}{\rho}z + H_f + \frac{1}{2m\rho}(\vec{p}_{\text{tot}} - \rho\vec{P}_f)^2 + \mathcal{O}(\rho^\sigma g^2) \tag{8.105}$$
$$+ \rho^\sigma \frac{g}{m}\chi_1\big((\rho\vec{P}_f - \vec{p}_{\text{tot}}) \cdot a(\vec{G}) + a^*(\vec{G}) \cdot (\rho\vec{P}_f - \vec{p}_{\text{tot}})\big)\chi_1.$$

The main reason for concentrating on $\vec{p}_{\text{tot}} = 0$ in this exposition is that if $\vec{p}_{\text{tot}} \neq 0$ then there is the additional expanding term $\frac{1}{2m}\vec{p}_{\text{tot}}^2 \mapsto \frac{1}{2m\rho}\vec{p}_{\text{tot}}^2$ which complicates matters considerably.

Going back to setting $\vec{p}_{\text{tot}} = 0$, we introduce another rescaling transformation

$$Z_\rho(z) := \frac{1}{\rho}\left(z - \widetilde{E}^{(0)}[z]\right) \tag{8.106}$$

exactly for the elimination of the expanding term $-\rho^{-1}z$. Namely, it turns out that Z_ρ is an invertible (in fact, biholomorphic) map, and for $\zeta \in \text{Ran}\, Z_\rho$ we define

$$H^{(0)}[\zeta] - \zeta \equiv E^{(0)}[\zeta] - \zeta + T^{(0)}[H_f, \vec{P}_f, \zeta] + W^{(0)}[\zeta]$$

$$:= S_\rho\big[\widetilde{H}^{(0)}[Z_\rho^{-1}(\zeta)] - Z_\rho^{-1}(\zeta)\big]$$

$$:= S_\rho\big[F_{\chi_\rho}(H_g - z,\ T - z)\big]\Big|_{z = Z_\rho^{-1}(\zeta)}. \tag{8.107}$$

8.2.4   Renormalization group map

We generalize the definition (8.107) as follows: Given

$$H[z] = E[z] + T[H_f, \vec{P}_f, z] + W[z], \tag{8.108}$$

on $\mathcal{H}_{\text{red}} = \text{Ran}\,\chi_1[H_f]$, where $E[z] \in \mathbb{C}$ is a multiple of the identity operator, $T[H_f, \vec{P}_f, z]$ is defined by the functional calculus and obeys $T[0, \vec{O}, z] = 0$, and

$$W[z] = \sum_{m+n \geq 1} \mathcal{Q}_1\big[w_{m,n}[z]\big], \tag{8.109}$$

where

$$w_{m,n}[z] : [0, 1] \times B_{\mathbb{R}^3}(\vec{O}, 1) \times B_1^{m+n} \to \mathbb{C}, \tag{8.110}$$

we define

$$Z_\rho(z) := \frac{1}{\rho}\left(z - E[z]\right) \tag{8.111}$$

and then

$$\mathcal{R}(H)[\zeta] := S_\rho \circ F_{\chi_\rho}\big(H[z] - z,\ T[z] - z\big)\Big|_{z = Z_\rho^{-1}(\zeta)} + \zeta. \tag{8.112}$$

Note that the map $\mathcal{R} : H[z] \mapsto \mathcal{R}(H)[Z_\rho^{-1}(\zeta)]$ is meaningful only if Theorem 8.2 is applicable, that is, if $T[z] - z$ is invertible on $\text{Ran}\,\overline{\chi}_\rho[H_f]$. If this is the case then $H[z] - z$ is isospectral to $\mathcal{R}(H)[Z_\rho^{-1}(\zeta)]$.

Using the normal-ordering procedure described in Section 8.2.3, we can write again

$$\mathcal{R}(H)[\zeta] \;=\; \widehat{E}[\zeta] + \widehat{T}[H_f, \vec{P}_f, \zeta] + \widehat{W}[\zeta], \quad \widehat{W}[z] \;=\; \sum_{m+n\geq 1} \mathbb{Q}_1\big[\widehat{w}_{m,n}[z]\big], \quad (8.113)$$

on $\mathcal{H}_{\mathrm{red}}$, with $T[0, \vec{0}, \zeta] = 0$. We can iterate $\mathcal{R}$ again, provided the Feshbach–Schur map, that is, Theorem 8.2 is applicable, and so on.

8.2.5 Contractivity of $\mathcal{R}$ on a Banach space of operators

To topologize these notions and to prove convergence, we introduce a Banach space

$$\mathcal{W}_\mu^\# \;:=\; \mathbb{C} \oplus \mathcal{T} \oplus \Big(\bigoplus_{m+n\geq 1} \mathcal{W}_{m,n}^{\#,\mu} \Big), \quad (8.114)$$

where

$$\mathcal{T} :=$$

$$\Big\{ T : [0,1] \times B_{\mathbb{R}^3}(\vec{0}, 1) \to \mathbb{C} \;\Big|\; T[0, \vec{0}] = 0, \; \|\partial_\omega T\|_\infty < \infty, \; \|\vec{\nabla}_{\vec{p}} T\|_\infty < \infty \Big\} \quad (8.115)$$

and

$$\mathcal{W}_{m,n}^{\#,\mu} \;:=\; \Big\{ w_{m,n} : [0,1] \times B_{\mathbb{R}^3}(\vec{0}, 1) \times B_1^{m+n} \to \mathbb{C} \;\Big|\; \|w_{m,n}\|_\mu < \infty \Big\}, \quad (8.116)$$

with

$$\|w_{m,n}\|_\mu^\# :=$$

$$\sup \left\{ \frac{|w_{m,n}(\omega, \vec{p}, K^{(m,n)})|}{|K^{(m,n)}|^{-\frac{1}{2}+\mu}} \;\middle|\; (\omega, \vec{p}, K^{(m,n)}) \in [0,1] \times B_{\mathbb{R}^3}(\vec{0}, 1) \times B_1^{m+n} \right\}. \quad (8.117)$$

Given an element $(E, T, \underline{w}) \in \mathcal{W}_\mu^\#$ with $\underline{w} = \bigoplus_{m+n\geq 1} w_{m,n}$, we define its norm by

$$\big\|(E, T, \underline{w})\big\|_\mu^\# \;:=\; \max\Big\{ |E|, \; \|\partial_\omega T\|_\infty, \; \|\vec{\nabla}_{\vec{p}} T\|_\infty, \; \|\underline{w}\|_\mu^\# \Big\} \quad (8.118)$$

where

$$\|\underline{w}\|_\mu^\# \;:=\; \sum_{m+n\geq 1} \Big(\|w_{m,n}\|_\mu^\# + \|\partial_\omega w_{m,n}\|_\mu^\# + \|\vec{\nabla}_{\vec{p}} w_{m,n}\|_\mu^\# \Big). \quad (8.119)$$

We imbed $\mathcal{I} : \mathcal{W}_\mu^\# \to \mathcal{B}[\mathcal{H}_{\mathrm{red}}]$ in the Banach space of bounded operators on $\mathcal{H}_{\mathrm{red}}$ by setting

$$\mathcal{I}\big[(E, T, \underline{w})\big] \;:=\; E \cdot \mathbf{1} + T[H_f, \vec{P}_f] + \mathbb{Q}_1[\underline{w}], \quad (8.120)$$

$$\mathbb{Q}_1[\underline{w}] \;:=\; \sum_{m+n\geq 1} \mathbb{Q}_1[w_{m,n}]. \quad (8.121)$$

An important property this imbedding possesses is that

$$\left\| E \cdot \mathbf{1} + T[H_f, \vec{P}_f] + \mathbb{Q}_1[\underline{w}] \right\|_{\mathrm{op}} \leq \left\| (E, T, \underline{w}) \right\|_{\mu}^{\#}; \tag{8.122}$$

in other words, the $\| \cdot \|_{\mu}^{\#}$-norm controls the operator norm. For the formulation of the contractivity of $\mathcal{R}$ we introduce a polydisc

$$\mathcal{D}_{\mu}^{\#}(\delta, \varepsilon) :=$$

$$\left\{ (E, T, \underline{w}) \in \mathcal{W}_{\mu}^{\#} \;\middle|\; |E| < \varepsilon, \; \|\vec{\nabla}_{\vec{p}} T\|_{\infty} < \varepsilon, \; \|\underline{w}\|_{\mu}^{\#} < \varepsilon, \; \|\partial_{\omega} T - 1\|_{\infty} < \delta \right\}. \tag{8.123}$$

Theorem 8.3. (Bach, Chen, Fröhlich, and Sigal, 2007) *There are $\delta_0, \varepsilon_0 >$ such that, for all $0 \leq \delta \leq \delta_0$ and $0 \leq \varepsilon \leq \varepsilon_0$, the renormalization map is defined on $\mathcal{D}_{\mu}^{\#}(\delta, \varepsilon)$ and possesses the following codimension-one contraction property,*

$$\mathcal{R} : \mathcal{D}_{\mu}^{\#}(\delta, \varepsilon) \;\to\; \mathcal{D}_{\mu}^{\#}(\delta + \rho^{\mu}\varepsilon, \; \rho^{\mu}\varepsilon). \tag{8.124}$$

In particular, if $\varepsilon < \min\left\{ (1 - \rho^{\mu})\delta_0, \; \varepsilon_0 \right\}$ then

$$\mathcal{R}^n : \mathcal{D}_{\mu}^{\#}(\varepsilon, \varepsilon) \;\to\; \mathcal{D}_{\mu}^{\#}\left((1 - \rho^{\mu})^{-1}\varepsilon, \; \rho^{\mu n}\varepsilon \right), \tag{8.125}$$

for all $n \in \mathbb{N}$.

Theorem 8.3 and especially (8.125) enables us to iterate $\mathcal{R}$ arbitrarily often on $\mathcal{D}_{\mu}^{\#}(\varepsilon, \varepsilon)$, provided $\varepsilon > 0$ is chosen sufficiently small. By (8.123) and (8.122), this means that

$$\mathcal{R}^n(H_g - z_n) \;=\; T^{(n)}[H_f, \vec{P}_f, 0] + \mathcal{O}(\varepsilon \rho^{\mu n}), \tag{8.126}$$

where $\|\mathcal{O}(\varepsilon \rho^{\mu n})\|_{\mathrm{op}} \leq C\varepsilon \rho^{\mu n}$ and

$$(1 - C\varepsilon)H_f \;\leq\; T^{(n)}[H_f, \vec{P}_f, z] \;\leq\; (1 + C\varepsilon)H_f, \tag{8.127}$$

for some constant $C < \infty$, provided the spectral parameter z is tuned carefully, that is, $z := z_n$ is chosen to be

$$z_n \;=\; \left(Z_{\rho}^{(1)} \right)^{-1} \circ \left(Z_{\rho}^{(2)} \right)^{-1} \circ \cdots \circ \left(Z_{\rho}^{(n)} \right)^{-1}(0). \tag{8.128}$$

Equations (8.126) and (8.127) imply that the ground state energy of $\mathcal{R}^n(H_g - z_n)$ equals $0 + \mathcal{O}(\varepsilon \rho^{\mu n})$ and that the ground state vector of $\mathcal{R}^n(H_g - z_n)$ is $\Omega + \mathcal{O}(\varepsilon \rho^{\mu n})$.

Applying the isospectrality of Theorem 8.2 n times, we infer that the nth-order approximations $E_{\mathrm{gs}}(n)$ and $\psi_{\mathrm{gs}}(n)$ to the ground state energy and the ground state eigenvector of H_g modulo an error $C\varepsilon \rho^{\mu n}$ are given by

$$E_{\mathrm{gs}}(n) = \left(Z_{\rho}^{(1)} \right)^{-1} \circ \left(Z_{\rho}^{(2)} \right)^{-1} \circ \cdots \circ \left(Z_{\rho}^{(n)} \right)^{-1}(0), \tag{8.129}$$

$$\psi_{\mathrm{gs}}(n) = Q_{\chi_{\rho}}^{(1)} \circ U_{\rho}^{-1} \circ Q_{\chi_{\rho}}^{(2)} \circ U_{\rho}^{-1} \circ \ldots \circ Q_{\chi_{\rho}}^{(n)} \Omega, \tag{8.130}$$

respectively. That is, there exists a universal constant $C < \infty$ such that, for all $n \in \mathbb{N}$,

$$\left\| \left(H_g - E_{\mathrm{gs}}(n) \right) \psi_{\mathrm{gs}}(n) \right\| \;\leq\; C \varepsilon \, \rho^{\mu n} \, \| \psi_{\mathrm{gs}}(n) \|. \tag{8.131}$$

It is then not difficult to prove that $E_{\mathrm{gs}}(n) \to E_{\mathrm{gs}}$ converges to the ground state energy of $H_g \equiv H_g(\vec{p}\,)$ and $\|\psi_{\mathrm{gs}}(n)\|^{-1} \psi_{\mathrm{gs}}(n) \to \psi_{\mathrm{gs}}$ converges to the normalized ground state of $H_g \equiv H_g(\vec{p}\,)$, respectively, as $n \to \infty$.

8.2.6 Mass renormalization

We conclude with a few remarks concerning the derivation of the renormalized mass. For $\sigma > 0$ we choose $\mu = \sigma$ and obtain the renormalized mass from a specific matrix element, namely,

$$\frac{1}{m_g(\sigma)} \;=\; \lim_{n \to \infty} \left\{ \frac{\langle \Omega \, | (\partial^2_{|\vec{p}\,|} T^{(n)}) [0, \vec{0}, 0] \, \Omega \rangle}{\rho^n \, \langle \Omega \, | (\partial_\omega T^{(n)}) [0, \vec{0}, 0] \, \Omega \rangle} \right\}, \tag{8.132}$$

uniformly in $\sigma > 0$. To treat the case $\sigma = 0$ directly, we use that

$$\widetilde{w}^{(0)}_{0,1} [H_f, \vec{P}_f, k; z] \;=\; \tfrac{g}{m} \vec{G}(k) \cdot \vec{P}_f, \tag{8.133}$$

which, together with a similar identity for $\widetilde{w}^{(0)}_{1,0}$, implies that

$$\widetilde{w}^{(0)}_{1,0} [H_f, \vec{P}_f, k; z] \big|_{\vec{P}_f = \vec{0}} \;=\; \widetilde{w}^{(0)}_{0,1} [H_f, \vec{P}_f, k; z] \big|_{\vec{P}_f = \vec{0}} \;=\; 0. \tag{8.134}$$

The preservation of this property under application of $\mathcal{R}$ is shown by means of *Ward–Takahashi* identities. More specifically, we (formally) use that

$$\left(g \vec{e}_k \cdot \vec{\nabla}_{\vec{P}_f} \right) H_g \;=\; \lim_{|k| \to 0} \left\{ |k|^{\frac{1}{2} - \sigma} [a_k, H_g] \right\}, \tag{8.135}$$

leading to

$$g \vec{e}_k \cdot \vec{\nabla}_{\vec{p}} w^{(\ell)}_{m,n} [H_f, \vec{P}_f, K^{(m,n)}; z]$$

$$= (m+1) \lim_{|k| \to 0} \left\{ |k|^{\frac{1}{2} - \sigma} w^{(\ell)}_{m+1,n} [H_f, \vec{P}_f, k, K^{(m,n)}; z] \right\}, \tag{8.136}$$

$$= (n+1) \lim_{|\tilde{k}| \to 0} \left\{ |\tilde{k}|^{\frac{1}{2} - \sigma} w^{(\ell)}_{m,n+1} [H_f, \vec{P}_f, K^{(m,n)}, \tilde{k}; z] \right\}.$$

A consequence of these identities is that the most singular coefficients in $\underline{w}^{(\ell)}[z] = \bigoplus_{n+m \geq 1} w^{(\ell)}_{n,m}[z]$, namely, $w^{(\ell)}_{1,0}[z]$ and $w^{(\ell)}_{0,1}[z]$, are actually more regular than what mere 'power counting', that is, the scaling behaviour under s_ρ in (8.101) seems to suggest. A careful use of (8.136) then shows that the contractivity asserted in Theorem 8.3 holds true with 1 replacing μ, in spite of the fact that $\mu = 0$ for this model.

Acknowledgements

My contributions to this field all result from a fruitful and enjoyable collaboration with J. Fröhlich and I.M. Sigal, and in part with T. Chen, F. Klopp, A. Pizzo, A. Soffer, and H. Zenk—I am indebted to all these individuals. Furthermore, I greatly benefited from discussions with M. Griesemer, M. Salmhofer, and H. Spohn.

References

Bach, V. Chen, T. Fröhlich, J. and Sigal, I. M. (2003). Smooth Feshbach map and operator-theoretic renormalization group methods. *J. Funct. Anal.*, **203**(1): 44–92.

Bach, V. Chen, T. Fröhlich, J. and Sigal, I. M. (2007). The renormalized electron mass in non-relativistic quantum electrodynamics. *J. Func. Anal.*, **243**(2): 426–535.

Catto, I. and Hainzl, C. (2004). Self-energy of one electron in non-relativistic QED. *J. Funct. Analysis*, **207**: 68–110.

Chen, T. (2001). Operator-theoretic infrared renormalization and construction of dressed 1-particle states. *preprint mp-arc 01-310*.

Chen, T. Faupin, J. Fröehlich, J. and Sigal, I.M. (2001). Local decay in non-relativistic QED. *Comm. Math. Phys.* **309**: 543–583 (2012).

Chen, T. Fröehlich, J. and Pizzo, A. (2009). Infraparticle scattering states in non-relativistic QED: II. mass shell properties. *J. Math. Phys.*, 50: 012103.

Chen, T. Fröehlich, J. and Pizzo, A. (2010). Infraparticle scattering states in non-relativistic QED: I. the Block-Nordsieck paradigm. *Comm. Math. Phys.*, 294(3): 761–825, DOI: 10.1007/s00220-009-0950-x.

Fröhlich, J. (1973). On the infrared problem in a model of scalar electrons and massless scalar bosons. *Ann. Inst. H. Poincaré*, 19: 1–103.

Fröhlich, J. (1974). Existence of dressed one-electron states in a class of persistent models. *Fortschr. Phys.*, 22: 159–198.

Hainzl, Ch. and Seiringer, R. (2002). General decomposition of radial functions on r^n and applications to n-body quantum systems. *Lett. Math. Phys.*, 61: 75–84.

Hiroshima, F. and Spohn, H. (2005). Mass renormalization in nonrelativistic QED. *J. Math. Phys.* 46: 042302.

Lieb, E. and Loss, M. (2002). A bound on binding energies and mass renormalization in models of quantum electrodynamics. *J. Stat. Phys.*, 108(5/6): 1057–1069.

9

Long-time diffusion for a quantum particle

Wojciech DE ROECK

Institut für Theoretische Physik,
Universität Heidelberg,
Philosophenweg 16,
D69120 Heidelberg, Germany

Chapter Contents

9.1   Introduction

The study of irreversible phenomena is a beautiful theme in mathematical physics. Whereas phenomena as diffusion and thermalization are universal, we do not understand them well, even less so on a rigorous level. For the sake of brevity, let us narrow the perspective to the derivation of diffusion for systems described by Hamiltonian equations of motion, classical or quantum, and to models that are very similar in spirit (e.g. billiards). By 'deriving' diffusion, we mean showing that a position-like variable X_t, satisfies a central limit theorem, that is that $t^{-1/2}X_t$ becomes Gaussian in the limit of large times t (below, we refer to such statements as long-time results to distinguish them from scaling limits). There are only a few models where this has been achieved. On the side of classical mechanics, we mention the finite-horizon billiards in (Bunimovich and Sinai 1981) (see also the related model (Knauf 1987) and the result about coupled maps in (Bricmont and Kupiainen). On the side of quantum mechanics, we have the models described in this contribution (see also (De Roeck and Fröhlich 2011)), and the recent work of (Disertori, Spencer and Zirnbauer 2010) on toy versions of the Anderson model, also discussed in the contribution of T. Spencer to these notes.

Up to now, most of the rigorous results on diffusion are formulated in a *scaling limit*. This means that one does not fix a single dynamical system and study its behaviour in the long-time limit, but, rather, one compares a family of dynamical systems at different times, as a certain parameter goes to 0. The precise definition of the scaling limit differs from model to model, but, in general, one scales time, space, and the coupling strength (and possibly also the initial state) such that a certain Markovian approximation to the dynamics becomes exact. There are quite some results on scaling limits in the literature, for example (Erdös and Yau 2000; Erdös, Salmhofer, and Yau 2008; Lukkarinen and Spohn 2011; Komorowski and Ryzhik 2006; Dürr, Goldstein and Lebowitz 1980), and we do not attempt an overview here.

In Section 9.2, we introduce our model(s). We discuss the associated scaling limits in Section 9.3 and results concerning long time behaviour in Section 9.4.

9.2   The models

Let us describe here in a quantitative way the quantum models that we study. The feature that accounts for the irreversible behaviour is that one part of the system is large and hence effectively acts as a heat bath or a thermostat to the rest of the composite system. We will in general call the part that acts as a thermostat the 'Environment' and denote it by the subscript E, whereas the rest of the total system is called 'Subsystem' and is denoted by the subscript S. The joint Hilbert space of subsystem and environment is then

$$\mathcal{H} = \mathcal{H}_{\mathrm{S}} \otimes \mathcal{H}_{\mathrm{E}} \qquad (9.1)$$

The environment is usually modelled by a bosonic quantum field, described by a Fock space

$$\mathcal{H}_E = \Gamma(l^2(\mathbb{Z}^d)) = \mathbb{C} \oplus l^2(\mathbb{Z}^d) \oplus l^2(\mathbb{Z}^d) \underset{\text{symm}}{\otimes} l^2(\mathbb{Z}^d) \oplus \dots \qquad (9.2)$$

where $\underset{\text{symm}}{\otimes}$ denotes the symmetrized tensor product, and the spaces on the right are respectively the vacuum, one particle space, two particle space, and so on and Γ implements second quantization. We assume the field to be free, or, in other words, we asssume its Hamiltonian to be quadratic in the creation-annihilation operators (satisfying the canonical commutation relations)

$$\mathcal{H}_E = \int_{\mathbb{T}^d} dq\, \omega(q) a_q^* a_q, \qquad \text{with } [a_q, a_{q'}^*] = \delta(q - q') \qquad (9.3)$$

where $q \in \mathbb{T}^d$ is the (quasi-)momentum of one particle and $\omega(q)$ is its dispersion relation (energy). The 'particles' should be thought of as lattice vibrations, or phonons, with dispersion relation either $\omega(q) = |q|$ for small q (accoustical branch), or $\omega(q) = \sqrt{m_{\text{ph}}^2 + |q|^2}$ for small q (optical branch). The initial state of the environment is conveniently assumed to be the Gibbs state at a given inverse temperature β.

$$\rho_E^\beta \approx \frac{1}{Z(\beta)} e^{-\beta H_E}, \qquad Z(\beta) = \operatorname{Tr} e^{-\beta H_E} \qquad (9.4)$$

where we have put $\approx$ to indicate that in infinite volume, these concepts need to be interpreted carefully, (in particular $Z(\beta) = \infty$ if $\beta < \infty$), either by setting up the problem in finite volume, and taking the infinite-volume limit at the end of the construction, or by using the algebraic theory of infinite-volume states. In fact, if one uses the second option, than the Hilbert space $\mathcal{H}_E$ has to be abandoned as it cannot accomodate a finite density of excitations. If one uses finite-volume approximations, no such complication arises and we assume this henceforth to be the case. Since ρ_E^β is Gaussian, it is completely characterized by its covariance or two-point function

$$\operatorname{Tr}(\rho_E^\beta a_q^* a_{q'}) = \delta(q - q') \frac{1}{e^{\beta\omega(q)} - 1}, \qquad \text{(Bose–Einstein distribution)} \qquad (9.5)$$

We now turn to the subsystem. The archetypical example is of course a free particle, but for reasons to become clear soon, we prefer to give it more structure by endowing it with an internal degree of freedom that is called 'spin', for the sake of simplicity. Hence we have

$$\mathcal{H}_S = l^2(\mathbb{Z}^d) \otimes \mathcal{H}_{\text{spin}}, \qquad \mathcal{H}_{\text{spin}} = \mathbb{C}^N, \qquad N < \infty \qquad (9.6)$$

and the Hamiltonian of the particle is

$$H_S = -\frac{1}{M}\Delta \otimes 1 + 1 \otimes H_{\text{spin}} \qquad (9.7)$$

where Δ is the lattice Laplacian, M is thought of as the mass of the particle, and H_{spin} is some Hermitian $N \times N$ matrix.

The interaction between subsystem and environment is then for simplicity assumed to be linear in creation and annihilation operators

$$H_{\text{S}-\text{E}} = \int_{\mathbb{T}^d} \text{d}q \, \phi(q) \text{e}^{-\text{i}qX} \otimes W \otimes a_q + \text{h.c.} \tag{9.8}$$

where h.c. stands for 'hermitian conjugate', X is the position operator on $l^2(\mathbb{Z}^d)$, $\phi \in L^2(\mathbb{T}^d)$ is a formfactor that is inserted to smear out the interaction between particle and field quanta and W is a hermitian $N \times N$ matrix that mediates the interaction between the spin degrees of freedom and field quanta. The most important quantity that characterizes the interaction is the 'free correlation function', which arises naturally if one expands the evolution in powers of $H_{\text{S}-\text{E}}$ and which has the following form

$$\zeta(x - x', t - t') = \rho_{\text{E}}[\Phi(x, t)\Phi(x', t')], \qquad \Phi(x, t) = \int \text{d}q \, \phi(q) \text{e}^{\text{i}(xq + \omega(q)t)} a_q + \text{h.c.} \tag{9.9}$$

If ϕ is chosen sufficiently smooth and vanishing outside a neighbourhood of $q = 0$, then $\sup_x |\zeta(x, t)| \le C(1 + |t|)^{-\frac{d-1}{2}}$ for accoustical phonons, where $\omega(q) \propto |q|$ for small q, and $\sup_x |\zeta(x, t)| \le C(1 + |t|)^{-\frac{d}{2}}$ for optical phonons, where $\omega(q) - \omega(0) \propto |q|^2$ for small q. One sees that correlations never decay exponentially, which is a general feature in momentum conserving models, as observed in the seminal work of Adler–Wainwright (Alder and Wainwright 1967). In dimension $d < 3$, nonintegrabilitiy of the correlation function can lead to anomalous diffusion (superdiffusion) and anomalous heat conduction.

The full Hamiltonian of our model is

$$H = H_{\text{S}} + H_{\text{E}} + \lambda H_{\text{S}-\text{E}}, \tag{9.10}$$

where we inserted a coupling constant $\lambda \in \mathbb{R}$. We choose the initial state of the particle to be described by some density matrix $\rho_{\text{S},0}$ and the full initial state is the product $\rho_{\text{S},0} \otimes \rho_{\text{E}}^\beta$. The time-evolved state is then

$$\rho_{\text{SE},t} := \text{e}^{-\text{i}tH} \left(\rho_{\text{S},0} \otimes \rho_{\text{E}}^\beta \right) \text{e}^{\text{i}tH} \tag{9.11}$$

Since we are mainly interested in properties of the subsystem, it is natural to 'trace out' the degrees of freedom of the environment; that is we define the time-evolved *reduced* density matrix

$$\rho_{\text{S},t} = \text{Tr}_{\text{E}} \rho_{\text{SE},t} \tag{9.12}$$

The question that concerns us here is about the long-time behaviour of the particle. What happens to $\rho_{\text{S},t}$ for large times t? We discuss three cases, depending on the value of the particle mass M.

- Setting $M = \infty$, we effectively get a confined model with only the spin degree of freedom. The main phenomenon that occurs is thermalization of the spin and this has been largely understood in the past years.
- The choice $M \sim \lambda^{-2}$ is the main focus of our discussion. In this regime one can prove diffusion and thermalization for long times.
- $M \sim 1$ is the most ambitious choice. Results on diffusion and thermalization are available for intermediate times, but not for long times.

9.3   Weak coupling limit

Let us introduce first a simplification of the problem, which turns out to be an important tool in almost any approach: the weak coupling limit. The idea of the weak coupling limit is to take the coupling constant $\lambda \to 0$, but to scale up time, in order to attain a regime where only the lowest orders in pertrubation theory contribute. Since we need second order in perturbation theory to see dissipative effects, the right scaling is $t \sim \lambda^{-2}$. This means that we aim to understand the scaling limit

$$\lim_{\substack{\lambda \searrow 0, \, t \nearrow \infty \\ \lambda^2 t = \tau \text{ fixed}}} \rho_{S,t}^{\lambda} \tag{9.13}$$

where we indicated explicitly the dependence of the density matrix on the coupling constant λ. Let us discuss this limit in the three cases that we consider.

9.3.1   Weak coupling limit for $M = \infty$

In the case of infinite mass, we can actually forget about the translational degrees of freedom of the particle (if started at the origin, it stays there forever) and we are only concerned with the spin degrees of freedom. We view $\rho_{S,t}$ as a $N \times N$ matrix and a convenient choice of basis is to take the eigenvectors of H_{spin} (let us assume that its spectrum $\sigma(H_{\text{spin}})$ is nondegenerate so that this is unambiguous) which we label by the symbols e, e'; we write $\rho_{S,t}(e, e')$ for matrix elements of $\rho_{S,t}$. In fact, the most interesting features can be observed on the diagonal elements of $\rho_{S,t}$ in the e-basis and we restrict our discussion to them. Hence we define

$$\mu_t^{\lambda}(e) = \rho_{S,t}^{\lambda}(e, e), \qquad \mu_t^{\lambda}(e) \geq 0, \qquad \sum_e \mu_t^{\lambda}(e) = \text{Tr}\rho_{S,t}^{\lambda} = 1 \tag{9.14}$$

In this case, the weak coupling limit is a classic theme in mathematical physics, it was first treated in (Davies 1974). If the correlation function is integrable in time, that is $\int dt |\zeta(0, t)| < \infty$ (note that we set $x = 0$ since the particle cannot move), then for any $\tau < \infty$,

$$\lim_{\substack{\lambda \searrow 0, \, t \nearrow \infty \\ \lambda^2 t = \tau \text{ fixed}}} \mu_t^{\lambda} = \mu_\tau = e^{\tau L}\mu_0 \tag{9.15}$$

where $e^{\tau L} = e^{\tau L}(e, e')$ is the transition kernel of a Markov jump process on the spin states $e \in \sigma(H_{\text{spin}})$, with generator L of the form

$$(L\mu)(e) = \sum_{e' \in \sigma(H_{\text{spin}}), e \neq e'} (c(e, e')\mu(e') - c(e', e)\mu(e)) \tag{9.16}$$

where the numbers $c(e, e')$ are jump rates (from e' to e) that can be calculated from the model parameters by the well-known Fermi golden rule. They satisfy the detailed balance conditions $c(e, e')/c(e', e) = e^{\beta(e'-e)}$ and as a straightforward consequence, the distribution tends to the Gibbs state as $\tau \to \infty$:

$$\lim_{\tau \nearrow \infty} \mu_\tau(e) = \frac{1}{Z_S(\beta)} e^{-\beta e}, \qquad Z_S(\beta) = \sum_e e^{-\beta e} \tag{9.17}$$

It is important to stress that this does not imply anything about the long time limit of μ_t^λ because there is an interchange of limit involved! In other words, it is not yet clear (that is: not from the above result) whether

$$\lim_{\substack{\tau \nearrow \infty \\ \lambda^2 t = \tau \text{ fixed}}} \lim_{\lambda \searrow 0, t \nearrow \infty} \mu_t^\lambda = \lim_{\lambda \searrow 0} \lim_{t \nearrow \infty} \mu_t^\lambda \tag{9.18}$$

9.3.2 Weak coupling limit for $M = \lambda^{-2} m$

As far as we know, this regime has not been considered in the mathematical physics literature, but it is very interesting. Since the translational degrees of freedom **do** play a role in this regime, we now consider density matrices that depend both on the position x and spin e, and we take the Fourier transform in $x \in \mathbb{Z}^d \leftrightarrow k \in \mathbb{T}^d$, such that we have

$$\hat{\rho}_{S,t}(k, e; k', e') = \sum_{x, x' \in \mathbb{Z}^d} e^{i(kx - k'x')} \rho_{S,t}(x, e; x', e') \tag{9.19}$$

Just as above, we first consider diagonal elements, this time in e, k-basis, that is we define, at least for sufficiently smooth density matrices,

$$\mu_t(k, e) = \hat{\rho}_{S,t}(k, e; k, e) \tag{9.20}$$

Then, the result (9.15) holds as before if the mass M is additionally scaled as $M = \lambda^{-2} m$, with m fixed. If $\int dt \sup_x |\zeta(x, t)| < \infty$, then the proof of (9.15) can be taken over almost literally, but we think that this condition could be relaxed (the integrable decay of the correlation function is probably not essential for $|x|$ so large that the propagation of the particle over a distance $|x|$ in a time t is strongly suppressed.)

Here, L is the generator of a jump process with state space $\mathbb{T}^d \times \sigma(H_{\text{spin}})$, of the form

$$L\mu(k, e) = \sum_{e' \in \sigma(H_{\text{spin}})} \int dk' \, (c(k, e; k', e')\mu(k', e') - c(k', e'; k, e)\mu(e, k)) \tag{9.21}$$

where now the jump rates $c(k, e; k', e')$ have a signature of the conservation of energy and momentum in the original picture. Indeed, a transition from momentum (k, e) to (k', e') can only occur if the particle emits a phonon of momentum $k - k'$ (or absorbs a phonon of momentum $k' - k$). Such a phonon has energy $\omega(k - k')$ and this constrains the possible transitions to satisfy

$$|(e + E_{\text{kin}}(k)) - (e' + E_{\text{kin}}(k'))| = \omega(k - k') \tag{9.22}$$

where $E_{\text{kin}} \sim \frac{1}{M} \sum_j (2 - 2 \cos k_j)$ is the kinetic energy of the particle at momentum k. However, since $M \nearrow \infty$, the kinetic energies $E_{\text{kin}}(k), E'_{\text{kin}}(k)$ drop out of the energy balance equation (9.22) yielding

$$|e' - e| = \omega(k - k') \tag{9.23}$$

which restricts the support of the jump rates. This is the compelling reason why we introduced the spin degrees of freedom! Without them, the energy balance equation (9.23) would not have any solutions with $k \neq k'$, since $\omega(\cdot)$ vanishes only at the origin (if at all), and hence no dissipative phenomena would take place on this time scale. Thanks to the spin degrees of freedom, the jump process is however nontrivial. If the jump rates do not vanish on a too large set (this is essentially determined by the form factor ϕ), then (9.21) tends to have a unique invariant state, seperated from the rest of the spectrum by a gap (this property will be referred to as exponential ergodicity). The invariant state is $\mu_{\text{inv}}(k, e) \propto e^{-\beta e}$, that is a Gibbs state at inverse temperature β, but without any dependence on k, corresponding to taking $M = \infty$. One could be under the impression that the translational degrees of freedom do not enter the game at all, but this is not true! Since a particle momentum k corresponds to a velocity $\frac{2}{M} \sin k = O(\lambda^2)$ and momentum jumps occur with rate of $O(\lambda^2)$, we expect that the particle travels a distance of $O(1)$ between two jumps. This suggests that we should be able to exhibit this motion in the weak coupling limit. Let us us define the probability of finding the particle at lattice site x after time t,

$$\nu_t(x) := \sum_{e \in \sigma(H_{\text{spin}})} \rho_{\text{S},t}(x, e; x, e), \tag{9.24}$$

and we write $\nu_t(x) = \nu_t^\lambda(x)$. Then, the weak coupling limit yields a limiting density[1]

$$\lim_{\substack{\lambda \searrow 0, t \nearrow \infty \\ \lambda^2 t = \tau \text{ fixed}}} \nu_t^\lambda(x) =: \nu_\tau(x) \tag{9.25}$$

By the reasoning expressed above, we might conjecture that the behaviour of the density $\nu_\tau(x)$ is that of a classical particle that undergoes collisions and moves between those collisions with a velocity $v_j = \frac{2}{m} \sin k_j$. This picture is not quite correct, since one can not neglect quantum effects in general (note that it is correct for the case

---

[1] In fact, the scaling limit can be formulated on the level of density matrices and the dynamics in that limit is a so-called 'quantum Markov semigroup', generated by a Lindblad operator.

$M \sim 1$, briefly discussed in Section (9.3.3)), but the picture does predict the right diffusion constant D_{kin} via the Green–Kubo formula, that is

$$\lim_{\tau \to \infty} \sum_x e^{i \frac{k}{\sqrt{\tau}} x} \nu_\tau(x) = e^{-D_{\text{kin}} k^2} \tag{9.26}$$

with

$$D_{\text{kin}} \delta_{i,j} = \frac{1}{2} \int_{-\infty}^{\infty} dt \langle v_i(t) v_j(0) \rangle \tag{9.27}$$

where $\langle \cdot \rangle$ is the expectation w.r.t. to the stationary Markov process generated by L. As in Section 9.3.1, we stress that this result as such does not imply anything about the $t \to \infty$ limit at fixed $\lambda > 0$.

9.3.3 Weak coupling limit for $M \sim 1$

In this case, the spin Hamiltonian does not play a prominent role (although it does show up in the weak coupling limit) and therefore, we omit it from this description, which amounts to taking $H_{\text{spin}} = 0$. Hence we now take the momentum distribution $\mu_t(k)$ as the natural object of interest. Again, the result (9.15) holds (we omit a discussion of the type of convergence and the allowed class of initial states, issues that are rather delicate in this regime) and L is now of the form

$$L\mu(k) = \int dk' \left(c(k, k') \mu(k') - c(k', k) \mu(k) \right) \tag{9.28}$$

where the rates $c(k, k')$ again satisfy the energy balance equation (9.22), this time with $e = e' = 0$ but with $E_{\text{kin}}(k), E_{\text{kin}}(k')$ nonvanishing (this is the reason why in this regime it would be unnatural to include the spin degrees of freedom). A particle momentum now corresponds to a velocity $\frac{2}{M} \sin k = O(1)$ and momentum jumps occur again with rate $O(\lambda^2)$. The particle travels a distance λ^{-2} between two jumps and we expect a diffusion constant of order $O(\lambda^{-2})$. If one includes the spatial degree of freedom, the weak coupling limit yields a linear Boltzmann equation. We do not go further into that since it is described in the contribution by L. Erdős to these notes. Let us however stress that unlike in the cases above, even the weak coupling limit is highly nontrivial in this case. It was derived in (Erdős and Yau 2000; Erdős, Salmhofer, and Yau 2008).

9.4 Long-time behaviour

We now turn to the results concerning the long-time behaviour. That is, we look for results at a fixed (but possibly small) coupling strength λ.

9.4.1 Long-time behaviour for $M = \infty$

For long times, one expects that 'spin' degrees of freedom thermalize at the temperature of the enivronment. This is called 'return to equilibrium'.

So one expects that, independently of the initial condition (as long as the environment still has a well-defined temperature)

$$\rho_{S,t} \quad \underset{t \to \infty}{\to} \quad \rho_S^\beta \tag{9.29}$$

where ρ_S^β is the restriction of the coupled Gibbs state (see Section 9.4.2) to the system S. Hence, qualitatively, the prediction of the weak coupling limit is correct, but the asymptotic state is different (although the difference vanishes as $\lambda \searrow 0$ and in particular (9.18) is correct). Versions of this result have been proven by many authors, see for example (Bach, Fröhlich, and Sigal 2000; Jakšić and Pillet 1996; Dereziński and Jakšić 2003) and we refer to the contribution of C.-A. Pillet to these notes for a discussion of related topics and sufficient conditions for (9.29).

9.4.2   Long-time behaviour for $M = \lambda^{-2}m$

As explained in Section 9.3.1, we expect the particle to diffuse with a diffusion constant $D \sim \lambda^2$. Recently, we proved (De Roeck and Kupiainen):

Theorem 9.1. (Diffusion) *Assume that*

1. *There is a $\alpha > 1/4$ such that*

$$\int_0^\infty dt\,(1+t)^\alpha \sup_x |\zeta(x,t)| < \infty$$

2. *The Markov process generated by L is exponentially ergodic and the diffusion constant is positive: $D_{kin} > 0$ in (9.27).*

Then, for sufficiently small λ,

$$\sum_x e^{ik\frac{x}{\sqrt{t}}} \nu_t(x) \quad \underset{t \to \infty}{\to} \quad e^{-Dk^2}, \qquad k \in \mathbb{R}^d \tag{9.30}$$

where $D = D_\lambda = \lambda^2 D_{kin} + o(\lambda^2) > 0$.

The form of the diffusion constant shows that weak coupling theory is qualitatively correct but there are subleading corrections in λ. Moreover, we can also prove thermalization. Let

$$\rho_{SE}^\beta = \left[\mathrm{Tr}(e^{-\beta H})\right]^{-1} e^{-\beta H}, \qquad \rho_S^\beta = \mathrm{Tr}_E \rho_{SE}^\beta \tag{9.31}$$

where both density matrices are only well-defined in finite volume, but the reduction to the diagonal

$$\mu^\beta(k,e) = \hat{\rho}_S^\beta(k,e;k,e) \tag{9.32}$$

remains meaningful even in the thermodynamic limit. Then

Theorem 9.2. (Thermalization) *Under the same assumptions as in Theorem 9.1, we have*

$$\mu_t(k, e) \quad \underset{t \to \infty}{\to} \quad \mu^\beta(k, e). \tag{9.33}$$

As one sees, our results are restricted to models where $\sup_x |\zeta(x, t)|$ decays for large $|t|$ at least as fast as $|t|^{-(1/4+\delta)}$, for some $\delta > 0$. As follows from the discussion in Section 9.2, this allows us to treat some optical phonon branches in $d \geq 3$, but accoustical phonons only in $d \geq 4$. It is conceivable (though far from obvious) that the method of proof might be extended to include $\sup_x |\zeta(x, t)| \sim |t|^{-1}$, which would allow for accoustical phonons in $d = 3$. The method of proof is a combination of a renormalization group scheme, developed in (Bricmont and Kupiainen 1991; Ajanki, De Roeck, and Kupiainen), and a perturbation around the weak coupling limit, as done in (De Roeck and Fröhlich 2011).

9.4.3 Long-time behaviour for $M \sim 1$

This regime is widely open. We cannot imagine that our method of proof for $M = \lambda^{-2}m$ extends to this case. In our opinion, this is a truly challenging and beautiful problem of mathematical physics.

References

Ajanki, O., De Roeck, W. and Kupiainen, A. Random walks in dynamic environment with integrable correlations. *in preparation*.

Alder, B. J. and Wainwright, T. E. (1967). Velocity autocorrelations for hard spheres. *Phys. Rev. Lett.*, **18**: 988–990.

Bach, V., Fröhlich, J., and Sigal, I. (2000). Return to equilibrium. *J. Math. Phys.*, **41**: 3985.

Bricmont, J. and Kupiainen, A. Diffusion in energy conserving coupled maps. arXiv:1102.3831.

Bricmont, J. and Kupiainen, A. (1991). Random walks in asymmetric random environments. *Comm. Math. Phys.*, **142**: 345–420.

Bunimovich, L.A. and Sinai, Ya. G. (1981). Statistical properties of Lorentz gas with periodic configuration of scatterers. *Comm. Math. Phys.*, **78**: 479–497.

Davies, E.B. (1974). Markovian master equations. *Comm. Math. Phys.*, **39**: 91–110.

Dereziński, J. and Jakšić, V. (2003). Return to equilibrium for Pauli-Fierz systems. *Ann. H. Poincaré*, **4**: 739–793.

De Roeck, W. and Fröhlich, J. (2011). Diffusion of a massive quantum particle coupled to a quasi-free thermal medium. *Comm. Math. Phys.*, **303**: 613–707.

De Roeck, W. and Kupiainen, A. Diffusion for a quantum particle coupled to phonons. *in preparation*. arXiv:1107.4832.

Disertori, M., Spencer, T., and Zirnbauer, M. (2010). Quasi-diffusion in a 3d supersymmetric hyperbolic sigma model. *Comm. Math. Phys.*, **300**(2): 435–486.

Dürr, D., Goldstein, S., and Lebowitz, J. L. (1980). A mechanical model of Brownian motion. *Comm. Math. Phys.*, **78**: 507–530.

Erdös, L., Salmhofer, M., and Yau, H.-T. (2008). Quantum diffusion of the random Schrödinger evolution in the scaling limit i. the non-recollision diagrams. *Acta Mathematica*, **200**: 211–277.

Erdös, L. and Yau, H.-T. (2000). Linear Boltzmann equation as the weak coupling limit of a random Schrödinger equation. *Comm. Pure Appl. Math.*, **53**(6): 667–735.

Jakšić, V. and Pillet, C.-A. (1996). On a model for quantum friction. iii: Ergodic properties of the spin-boson system. **178**: 627–651.

Knauf, A. (1987). Ergodic and topological properties of Coulombic periodic potentials. *Comm. Math. Phys.*, **110**(1): 89–112.

Komorowski, T. and Ryzhik, L. (2006). Diffusion in a weakly random hamiltonian flow. *Comm. Math. Phys*, **263**: 277–323.

Lukkarinen, J. and Spohn, H. (2011). Weakly nonlinear Weakly nonlinear schrödinger equation with random initial data. *Invent. Math.*, **183**: 79–188.

10

The ground state construction of the two-dimensional Hubbard model on the honeycomb lattice

Alessandro GIULIANI

Dipartimento di Matematica, Università di Roma Tre
L.go S. L. Murialdo 1, 00146 Roma, Italy

Chapter Contents

10.1 Introduction

There are very few quantum interacting systems whose ground state properties (thermodynamic functions, reduced density matrices, etc.) can be computed without approximations. Among these, the Luttinger and the Thirring model (Tomonaga 1950; Luttinger 1963; Mattis and Lieb 1965; Thirring 1958; Johnson 1961) (one-dimensional spinless relativistic fermions), the one-dimensional Hubbard model (Lieb and Wu, 1968; Essler *et al.*, 2005) (nonrelativistic lattice fermions with spin), the Lieb–Liniger model (Lieb and Liniger 1963; Lieb 1963) (one-dimensional bosons with repulsive delta interactions) and the BCS model (Bardeen, Cooper, and Schrieffer, 1957; Richardson and Sherman, 1964; Dukelsky, Pittel, and Sierra, 2004) (d-dimensional spinning fermions with mean field interactions, $d \geq 1$); the construction of the ground states of these systems is based on some remarkably exact solutions, which make use of bosonization techniques and Bethe ansatz. Unfortunately, in most cases these exact solutions do not allow us to compute the long-distance decay of the n-particles correlation functions in a closed form (with some notable exceptions, namely the Luttinger and Thirring models, see (Klaiber 1968; Mastropietro 1994)). Moreover, even the computation of the thermodynamic functions crucially relies on a very special choice of the particle–particle interaction; as soon as the interparticle potential is slightly modified the exact solvability of these models is destroyed and no conclusion on the new 'perturbed' system can be drawn from their solution. This is, of course, very annoying and unsatisfactory from a physical point of view.

There is another rigorous powerful method that allows us in a few cases to fully construct the ground state of a system of interacting particles, known as renormalization group (RG). This method, whenever it works, has the advantage that it provides full information on the zero or low temperatures state of the system, including correlations and that, typically, it is robust under small modifications of the interparticle potential; even better, it usually gives a very precise meaning to what 'small modifications' means: it allows one to classify perturbations into 'relevant' and 'irrelevant' and to show that the addition of a small irrelevant perturbation does not change the asymptotic behaviour of correlations. Unfortunately, it only works in the weak coupling regime and it has only been successfully applied to a limited number of interacting quantum systems. Most of the available results on the ground (or thermal) states of interacting quantum systems obtained via RG concern one-dimensional weakly interacting fermions, for example ultraviolet $O(N)$ models with $N \geq 2$ (Gawedzki and Kupiainen, 1985; Feldman, Magnen, Rivasseau and Sénéor, 1986), ultraviolet QED-like models (Lesniewski 1987) and nonrelativistic lattice systems (Benfatto, Gallavotti, Procacci, and Scoppola, 1994; Gentile and Mastropietro, 2001; Benfatto and Mastropietro, 2005). In more than one dimension, most of the results derived by rigorous RG techniques concern the finite temperature properties of two-dimensional fermionic systems above the BCS critical temperature (Benfatto and Gallavotti, 1990; Feldman, Magnen, Rivasseau, and Trubowitz, 1992; Disertori and Rivasseau, 2000; Benfatto, Giuliani and Mastropietro, 2003; Benfatto, Giuliani, and Mastropietro, 2006). Two remarkable exceptions are the Fermi liquid construction by Feldman, Knörrer and Trubowitz (Feldman, Knörrer, and Trubowitz, 2004), which concerns zero temperature properties of a system of two-dimensional interacting fermions with highly asymmetric Fermi surface, and the ground state construction of the short range half-filled two-dimensional Hubbard model on the honeycomb lattice by Giuliani and Mastropietro (Giuliani and Mastropietro, 2009; Giuliani and Mastropietro, 2010), which will be reviewed here.

The two-dimensional Hubbard model on the honeycomb lattice is a basic model for describing *graphene*, a newly discovered material consisting of a one-atom thick layer of graphite (Novoselov *et al.*, 2004), see (Castro Neto, Guinea, Peres, Novoselov, and Geim, 2009) for a comprehensive and up-to-date review of its low temperature properties. One of the most remarkable features of graphene is that at half-filling the Fermi surface is highly degenerate and it consists of just two isolated points. This makes the infrared properties of the system very peculiar: in the absence of interactions, it behaves in the same way as a system of noninteracting $(2+1)$-dimensional Dirac particles (Wallace 1947). Therefore, the interacting system is a sort of $(2+1)$-dimensional QED, with some peculiar differences that make its study new and nontrivial (Semenoff 1984; Haldane 1988).

The goal of these lectures is to give a self-contained proof of the analyticity of the ground state energy of the Hubbard model on the two-dimensional honeycomb lattice at half filling and weak coupling via constructive RG methods. A simple extension of the proof of convergence of the series for the specific ground state energy presented below allows one to construct the whole set of reduced density matrices at weak coupling (see (Giuliani and Mastropietro 2010)): it turns out that the off-diagonal elements of these matrices decay to zero at infinity, with the same decay exponents as the noninteracting system; in this sense, the construction presented below rigorously exclude the presence of long range order in the ground state, and the absence of anomalous critical exponents (in other words, the interacting system is in the same universality class as the noninteracting one). Let me also mention that more sophisticated extensions of the methods exposed here also allow us to: (i) give an order by order construction of the ground state correlations of the same model in the presence of electromagnetic interactions (Giuliani, Mastropietro and Porta, 2010a; Giuliani, Mastropietro and Porta, 2010b; Giuliani, Mastropietro and Porta, 2012a); (ii) predict a possible mechanism for the spontaneous generation of the Peierls-Kekulé instability (Giuliani, Mastropietro and Porta, 2010b; Giuliani, Mastropietro and Porta, 2012a); (iii) prove the universality of the optical conductivity (Giuliani, Mastropietro and Porta, 2011; Giuliani, Mastropietro and Porta, 2012b).

The plan of these lectures is the following:

- In Section 10.2, I introduce the model and state the main result.
- In Section 10.3, I review the noninteracting case.
- In Section 10.4, I describe the formal series expansion for the ground state energy, I explain how to conveniently re-express it in terms of Grassmann functional integrals and estimate by naive power-counting the generic N-th order in perturbation theory, so identifying two main issues in the convergence of the series: a combinatorial problem, related to the large number of Feynman graphs contributing at a generic perturbative order, and a divergence problem, related to the (very mild) ultraviolet singularity of the propagator and to its (more serious) infrared singularity.
- In Section 10.5, I describe a way to reorganize and estimate the perturbation theory, via the so-called *determinant expansion*, that allows one to prove convergence of the series, for any fixed choice of the ultraviolet and infrared cutoffs.

- In Section 10.6, I describe how to resum the determinant expansion in order to cure the mild ultraviolet divergences appearing in perturbation theory, via a multi-scale expansion, whose result is conveniently expressed in terms of *Gallavotti–Nicolò* trees (Gallavotti and Nicolò 1985; Gallavotti 1985).
- In Section 10.7, I describe how to resum and cure the infrared divergences and conclude the proof of the main result.
- Finally, in Section 10.8, I draw the conclusions. A few technical aspects of the proof are described in the Appendices

The material presented in this lectures is mostly taken from (Giuliani and Mastropietro 2010). Some technical proofs concerning the determinant and the tree expansions are taken from the review (Gentile and Mastropietro 2001). Other reviews of the constructive RG methods discussed here, which the reader may find useful to consult, are (Benfatto and Gallavotti 1995; Gallavotti 1985; Mastropietro 2008; Salmhofer 1999).

10.2 The model and the main results

The grand-canonical Hamiltonian of the two-dimensional Hubbard model on the honeycomb lattice at half filling in second quantized form is given by:

$$H_\Lambda = -t \sum_{\substack{\vec{x}\in\Lambda_A \\ i=1,2,3}} \sum_{\sigma=\uparrow\downarrow} \left(a^+_{\vec{x},\sigma} b^-_{\vec{x}+\vec{\delta}_i,\sigma} + b^+_{\vec{x}+\vec{\delta}_i,\sigma} a^-_{\vec{x},\sigma} \right) + \tag{10.1}$$

$$+ U \sum_{\vec{x}\in\Lambda_A} \left(a^+_{\vec{x},\uparrow} a^-_{\vec{x},\uparrow} - \frac{1}{2} \right) \left(a^+_{\vec{x},\downarrow} a^-_{\vec{x},\downarrow} - \frac{1}{2} \right) + U \sum_{\vec{x}\in\Lambda_B} \left(b^+_{\vec{x},\uparrow} b^-_{\vec{x},\uparrow} - \frac{1}{2} \right) \left(b^+_{\vec{x},\downarrow} b^-_{\vec{x},\downarrow} - \frac{1}{2} \right)$$

where:

1. $\Lambda_A = \Lambda$ is a periodic triangular lattice, defined as $\Lambda = \mathbb{B}/L\mathbb{B}$, where $L \in \mathbb{N}$ and $\mathbb{B}$ is the infinite triangular lattice with basis $\vec{l}_1 = \frac{1}{2}(3, \sqrt{3})$, $\vec{l}_2 = \frac{1}{2}(3, -\sqrt{3})$. $\Lambda_B = \Lambda_A + \vec{\delta}_i$ is obtained by translating Λ_A by a nearest neighbour vector $\vec{\delta}_i$, $i = 1, 2, 3$, where

$$\vec{\delta}_1 = (1,0), \quad \vec{\delta}_2 = \frac{1}{2}(-1, \sqrt{3}), \quad \vec{\delta}_3 = \frac{1}{2}(-1, -\sqrt{3}). \tag{10.2}$$

   The honeycomb lattice we are interested in is the union of the two triangular sublattices Λ_A and Λ_B, see Fig. 10.1
2. $a^\pm_{\vec{x},\sigma}$ are creation or annihilation fermionic operators with spin index $\sigma = \uparrow\downarrow$ and site index $\vec{x} \in \Lambda_A$, satisfying periodic boundary conditions in $\vec{x}$. Similarly, $b^\pm_{\vec{x},\sigma}$ are creation or annihilation fermionic operators with spin index $\sigma = \uparrow\downarrow$ and site index $\vec{x} \in \Lambda_B$, satisfying periodic boundary conditions in $\vec{x}$.
3. U is the strength of the on–site density–density interaction; it can be either positive or negative.

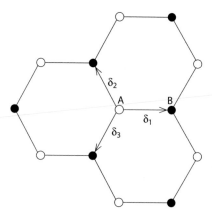

Fig. 10.1   A portion of the honeycomb lattice Λ. The white and black dots correspond to the sites of the Λ_A and Λ_B triangular sublattices, respectively. These two sublattices are one the translate of the other. They are connected by nearest neighbour vectors $\vec{\delta}_1, \vec{\delta}_2, \vec{\delta}_3$ that, in our units, are of unit length.

Note that the terms in the second line of equation (10.1) can be rewritten as the sum of a truly quartic term in the creation/annihilation operators (the density-density interaction), plus a quadratic term (a chemical potential term, of the form $-\mu N$, with $\mu = U/2$ the chemical potential and N the total particles number operator), plus a constant (which plays no role in the thermodynamic properties of the system). The Hamiltonian (10.1) is hole-particle symmetric, that is, it is invariant under the exchange $a^{\pm}_{\vec{x},\sigma} \longleftrightarrow a^{\mp}_{\vec{x},\sigma}$, $b^{\pm}_{\vec{x}+\vec{\delta}_1,\sigma} \longleftrightarrow -b^{\mp}_{\vec{x}+\vec{\delta}_1,\sigma}$. This invariance implies in particular that, if we define the average density of the system to be $\rho = (2|\Lambda|)^{-1} \langle N \rangle_{\beta,\Lambda}$, with $N = \sum_{\vec{x},\sigma}(a^{+}_{\vec{x},\sigma} a^{-}_{\vec{x},\sigma} + b^{+}_{\vec{x}+\vec{\delta}_1,\sigma} b^{-}_{\vec{x}+\vec{\delta}_1,\sigma})$ the total particle number operator and $\langle \cdot \rangle_{\beta,\Lambda} = \text{Tr}\{e^{-\beta H_{\Lambda}}\cdot\}/\text{Tr}\{e^{-\beta H_{\Lambda}}\}$ the average with respect to the (grand canonical) Gibbs measure at inverse temperature β, one has $\rho \equiv 1$, for any $|\Lambda|$ and any β; in other words, the grand canonical Hamiltonian equation (10.1) describes the system at half-filling, for all U, β, Λ. Let

$$f_\beta(U) = -\frac{1}{\beta} \lim_{|\Lambda|\to\infty} |\Lambda|^{-1} \log \text{Tr}\{e^{-\beta H_\Lambda}\}. \tag{10.3}$$

be the specific free energy of the system and $e(U) = \lim_{\beta\to\infty} f_\beta(U)$ the specific ground state energy. We will prove the following Theorem.

Theorem 10.1   *There exists a constant $U_0 > 0$ such that, if $|U| \le U_0$, the specific free energy $f_\beta(U)$ of the two-dimensional Hubbard model on the honeycomb lattice at half-filling is an analytic function of U, uniformly in β as $\beta \to \infty$, and so is the specific ground state energy $e(U)$.*

The proof is based on RG methods, which will be reviewed below. A straightforward extension of the proof of Theorem 10.1 allows one to prove that the correlation functions (i.e., the off-diagonal elements of the reduced density matrices of the system) are analytic functions of U and they decay to zero at infinity with the same decay exponents as in the noninteracting ($U = 0$) case, see (Giuliani and Mastropietro 2010). This rigorously excludes the presence of LRO in the ground state and proves that the interacting system is in the same universality class as the noninteracting system.

10.3 The noninteracting system

Let us begin by reviewing the construction of the finite and zero temperature states for the noninteracting ($U = 0$) case. In this case the Hamiltonian of interest reduces to

$$H_\Lambda^0 = -t \sum_{\substack{\vec{x}\in\Lambda \\ i=1,2,3}} \sum_{\sigma=\uparrow\downarrow} \left(a_{\vec{x},\sigma}^+ b_{\vec{x}+\vec{\delta}_i,\sigma}^- + b_{\vec{x}+\vec{\delta}_i,\sigma}^+ a_{\vec{x},\sigma}^- \right), \tag{10.4}$$

with Λ, $a_{\vec{x},\sigma}^\pm$, $b_{\vec{x}+\vec{\delta}_i,\sigma}^\pm$ defined as in items (1)–(4) after (10.1). We aim at computing the spectrum of H_Λ^0 by diagonalizing the right-hand side (r.h.s.) of (10.4). To this purpose, we pass to Fourier space. We identify the periodic triangular lattice Λ with the set of vectors of the infinite triangular lattice within a 'box' of size L, that is,

$$\Lambda = \{n_1\vec{l}_1 + n_2\vec{l}_2 \ : \ 0 \le n_1, n_2 \le L-1\}, \tag{10.5}$$

with $\vec{l}_1 = \frac{1}{2}(3, \sqrt{3})$ and $\vec{l}_2 = \frac{1}{2}(3, -\sqrt{3})$. The reciprocal lattice Λ^* is the set of vectors $\vec{K}$ such that $e^{i\vec{K}\vec{x}} = 1$, if $\vec{x} \in \Lambda$. A basis $\vec{G}_1, \vec{G}_2$ for Λ^* can be obtained by the inversion formula:

$$\begin{pmatrix} G_{11} & G_{12} \\ G_{21} & G_{22} \end{pmatrix} = 2\pi \begin{pmatrix} l_{11} & l_{21} \\ l_{12} & l_{22} \end{pmatrix}^{-1}, \tag{10.6}$$

which gives

$$\vec{G}_1 = \frac{2\pi}{3}(1, \sqrt{3}), \qquad \vec{G}_2 = \frac{2\pi}{3}(1, -\sqrt{3}). \tag{10.7}$$

We call $\mathcal{B}_L$ the set of quasi-momenta $\vec{k}$ of the form

$$\vec{k} = \frac{m_1}{L}\vec{G}_1 + \frac{m_2}{L}\vec{G}_2, \qquad m_1, m_2 \in \mathbb{Z}, \tag{10.8}$$

identified modulo Λ^*; this means that $\mathcal{B}_L$ can be identified with the vectors $\vec{k}$ of the form (10.2) and restricted to the *first Brillouin zone*:

$$\mathcal{B}_L = \{\vec{k} = \frac{m_1}{L}\vec{G}_1 + \frac{m_2}{L}\vec{G}_2 \ : \ 0 \le m_1, m_2 \le L-1\}. \tag{10.9}$$

Given a periodic function $f : \Lambda \to \mathbb{R}$, its Fourier transform is defined as

$$f(\vec{x}) = \frac{1}{|\Lambda|} \sum_{\vec{k}\in\mathcal{B}_L} e^{i\vec{k}\vec{x}} \hat{f}(\vec{k}), \tag{10.10}$$

which can be inverted into

$$\hat{f}(\vec{k}) = \sum_{\vec{x} \in \Lambda} e^{-i\vec{k}\vec{x}} f(\vec{x}), \qquad \vec{k} \in \mathcal{B}_L, \tag{10.11}$$

where we used the identity

$$\sum_{\vec{x} \in \Lambda} e^{i\vec{k}\vec{x}} = |\Lambda|\delta_{\vec{k},\vec{0}} \tag{10.12}$$

and δ is the periodic Kronecker delta function over Λ^*.

We now associate to the set of creation/annihilation operators $a^{\pm}_{\vec{x},\sigma}$, $b^{\pm}_{\vec{x}+\vec{\delta}_i,\sigma}$ the corresponding set of operators in momentum space:

$$a^{\pm}_{\vec{x},\sigma} = \frac{1}{|\Lambda|} \sum_{\vec{k} \in \mathcal{B}_L} e^{\pm i\vec{k}\vec{x}} \hat{a}^{\pm}_{\vec{k},\sigma}, \qquad b^{\pm}_{\vec{x}+\vec{\delta}_1,\sigma} = \frac{1}{|\Lambda|} \sum_{\vec{k} \in \mathcal{B}_L} e^{\pm i\vec{k}\vec{x}} \hat{b}^{\pm}_{\vec{k},\sigma}. \tag{10.13}$$

Using (10.10)–(10.12), we find that

$$\hat{a}^{\pm}_{\vec{k},\sigma} = \sum_{\vec{x} \in \Lambda} e^{\mp i\vec{k}\vec{x}} a^{\pm}_{\vec{x},\sigma}, \qquad \hat{b}^{\pm}_{\vec{k},\sigma} = \sum_{\vec{x} \in \Lambda} e^{\mp i\vec{k}\vec{x}} b^{\pm}_{\vec{x}+\vec{\delta}_1,\sigma} \tag{10.14}$$

are fermionic creation/annihilation operators, *periodic over* Λ^*, satisfying

$$\{a^{\varepsilon}_{\vec{k},\sigma}, a^{\varepsilon'}_{\vec{k}',\sigma'}\} = |\Lambda|\delta_{\vec{k},\vec{k}'}\delta_{\varepsilon,-\varepsilon'}\delta_{\sigma,\sigma'}, \qquad \{b^{\varepsilon}_{\vec{k},\sigma}, b^{\varepsilon'}_{\vec{k}',\sigma'}\} = |\Lambda|\delta_{\vec{k},\vec{k}'}\delta_{\varepsilon,-\varepsilon'}\delta_{\sigma,\sigma'} \tag{10.15}$$

and $\{a^{\varepsilon}_{\vec{k},\sigma}, b^{\varepsilon'}_{\vec{k}',\sigma'}\} = 0$[1]. With the previous definitions, we can rewrite

$$H^0_\Lambda = -t \sum_{\substack{\vec{x} \in \Lambda \\ i=1,2,3}} \sum_{\sigma=\uparrow\downarrow} (a^+_{\vec{x},\sigma} b^-_{\vec{x}+\vec{\delta}_i,\sigma} + b^+_{\vec{x}+\vec{\delta}_i,\sigma} a^-_{\vec{x},\sigma}) = \tag{10.16}$$

$$= -\frac{v_0}{|\Lambda|} \sum_{\substack{\vec{k} \in \mathcal{B}_L \\ \sigma=\uparrow\downarrow}} \left(\hat{a}^+_{\vec{k},\sigma}, \hat{b}^+_{\vec{k},\sigma}\right) \begin{pmatrix} 0 & \Omega^*(\vec{k}) \\ \Omega(\vec{k}) & 0 \end{pmatrix} \begin{pmatrix} \hat{a}^-_{\vec{k},\sigma} \\ \hat{b}^-_{\vec{k},\sigma} \end{pmatrix}$$

with $v_0 = \frac{3}{2}t$ the unperturbed Fermi velocity (if the hopping strength t is chosen to be the one measured in real graphene, v_0 turns out to be approximately 300 times smaller than the speed of light) and

---

[1] Of course, there is some arbitrariness in the denition of $\hat{a}^{\pm}_{\vec{k},\sigma}, \hat{b}^{\pm}_{\vec{k},\sigma}$: we could change the two sets of operators by multiplying them by two $\vec{k}$-dependent phase factors, $e^{\mp i\vec{k}\vec{A}}, e^{\mp i\vec{k}\vec{B}}$ with $\vec{A}, \vec{B}$ two arbitrary constant vectors, and get a completely equivalent theory in momentum space. The freedom in the choice of these phase factors corresponds to the freedom in the choice of the origins of the two sublattices Λ_A, Λ_B; this symmetry is sometimes referred to as Berry-gauge invariance and it can be thought as a local gauge symmetry in momentum space. Note that, by changing the Berry phase, the boundary conditions on $\hat{a}^{\pm}_{\vec{k},\sigma}, \hat{b}^{\pm}_{\vec{k},\sigma}$ at the boundaries of the first Brillouin zone change; our explicit choice of the Berry phase has the 'advantage' of making $\hat{a}^{\pm}_{\vec{k},\sigma}, \hat{b}^{\pm}_{\vec{k},\sigma}$ periodic over Λ^* rather than quasi-periodic.

$$\Omega(\vec{k}) = \frac{2}{3} \sum_{i=1}^{3} e^{i(\vec{\delta}_i - \vec{\delta}_1)\vec{k}} = \frac{2}{3} \left[1 + 2e^{-i\frac{3}{2}k_1} \cos(\frac{\sqrt{3}}{2}k_2) \right] \qquad (10.17)$$

the complex *dispersion relation*. By looking at the second line of equation (10.16), one realizes that the natural creation/annihilation operator is a two-dimensional spinor of components a and b:

$$\hat{\Psi}^+_{\vec{k},\sigma} = (\hat{a}^+_{\vec{k},\sigma}, \hat{b}^+_{\vec{k},\sigma}), \qquad \hat{\Psi}^-_{\vec{k},\sigma} = \begin{pmatrix} \hat{a}^-_{\vec{k},\sigma} \\ \hat{b}^-_{\vec{k},\sigma} \end{pmatrix}, \qquad (10.18)$$

whose real space counterparts read

$$\Psi^+_{\vec{x},\sigma} = (a^+_{\vec{x},\sigma}, b^+_{\vec{x}+\vec{\delta}_1,\sigma}), \qquad \Psi^-_{\vec{x},\sigma} = \begin{pmatrix} \hat{a}^-_{\vec{x},\sigma} \\ \hat{b}^-_{\vec{x}+\vec{\delta}_1,\sigma} \end{pmatrix}. \qquad (10.19)$$

In order to fully diagonalize the theory, one needs to perform the diagonalization of the 2×2 quadratic form in the second line of equation (10.16), which can be realized by the $\vec{k}$-dependent unitary transformation

$$U_{\vec{k}} = \begin{pmatrix} \frac{1}{\sqrt{2}} & \frac{1}{\sqrt{2}} \frac{\Omega^*(\vec{k})}{|\Omega(\vec{k})|} \\ -\frac{1}{\sqrt{2}} \frac{\Omega(\vec{k})}{|\Omega(\vec{k})|} & \frac{1}{\sqrt{2}} \end{pmatrix}, \qquad (10.20)$$

in terms of which, defining

$$\hat{\Phi}^-_{\vec{k},\sigma} = \begin{pmatrix} \hat{\alpha}^-_{\vec{k},\sigma} \\ \hat{\beta}^-_{\vec{k},\sigma} \end{pmatrix} := U \hat{\Psi}^-_{\vec{k},\sigma} = \frac{1}{\sqrt{2}} \begin{pmatrix} \hat{a}^-_{\vec{k},\sigma} + \frac{\Omega^*(\vec{k})}{|\Omega(\vec{k})|} \hat{b}^-_{\vec{k},\sigma} \\ -\frac{\Omega(\vec{k})}{|\Omega(\vec{k})|} \hat{a}^-_{\vec{k},\sigma} + \hat{b}^-_{\vec{k},\sigma} \end{pmatrix}, \qquad (10.21)$$

we can rewrite

$$H^0_\Lambda = -\frac{v_0}{|\Lambda|} \sum_{\substack{\vec{k} \in \mathcal{B}_L \\ \sigma = \uparrow\downarrow}} \hat{\Psi}^+_{\vec{k},\sigma} U^\dagger_{\vec{k}} U_{\vec{k}} \begin{pmatrix} 0 & \Omega^*(\vec{k}) \\ \Omega(\vec{k}) & 0 \end{pmatrix} U^\dagger_{\vec{k}} U_{\vec{k}} \hat{\Psi}^-_{\vec{k},\sigma} =$$

$$= -\frac{v_0}{|\Lambda|} \sum_{\substack{\vec{k} \in \mathcal{B}_L \\ \sigma = \uparrow\downarrow}} \hat{\Phi}^+_{\vec{k},\sigma} \begin{pmatrix} |\Omega(\vec{k})| & 0 \\ 0 & -|\Omega(\vec{k})| \end{pmatrix} \hat{\Phi}^-_{\vec{k},\sigma} = \qquad (10.22)$$

$$= -\frac{v_0}{|\Lambda|} \sum_{\substack{\vec{k} \in \mathcal{B}_L \\ \sigma = \uparrow\downarrow}} \left(|\Omega(\vec{k})| \hat{\alpha}^+_{\vec{k},\sigma} \hat{\alpha}^-_{\vec{k},\sigma} - |\Omega(\vec{k})| \hat{\beta}^+_{\vec{k},\sigma} \hat{\beta}^-_{\vec{k},\sigma} \right).$$

The two energy bands $\pm v_0 |\Omega(\vec{k})|$ are plotted in Fig. 10.2. They cross the Fermi energy $E_F = 0$ at the *Fermi points* $\vec{k} = \vec{p}^\omega_F$, $\omega = \pm$, with

$$\vec{p}^\omega_F = (\frac{2\pi}{3}, \omega \frac{2\pi}{3\sqrt{3}}), \qquad (10.23)$$

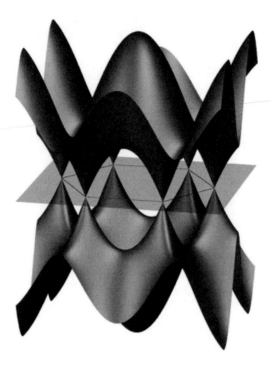

Fig. 10.2   A sketch of the energy bands of the free electron gas with nearest neighbour hopping on the honeycomb lattice. The plotted plane corresponds to the Fermi energy at half-filling. It cuts the bands at a discrete set of points, known as the Fermi points or Dirac points. From the picture, it seems that there are six distinct Fermi points. However, after identification of the points modulo vectors of the reciprocal lattice, it turns out that only two of them are independent.

close to which the complex dispersion relation vanishes linearly:

$$\Omega(\vec{p}_F^{\,\omega} + \vec{k}') = ik_1' + \omega k_2' + O(|\vec{k}'|^2), \qquad (10.24)$$

resembling in this sense the relativistic dispersion relation of $(2 + 1)$-dimensional Dirac fermions. From Eq.(10.22) it is apparent that the ground state of the system consists of a Fermi sea such that all the negative energy states (the 'α-states') are filled and all the positive energy states (the 'β-states') are empty. The specific ground state energy $e_{0,\Lambda}$ is

$$e_{0,\Lambda} = -\frac{2v_0}{|\Lambda|} \sum_{\vec{k}\in\mathcal{B}_L} |\Omega(\vec{k})|, \qquad (10.25)$$

from which we find that the specific ground state energy in the thermodynamic limit is

$$e(0) = -2v_0 \int_{\mathcal{B}} \frac{d\vec{k}}{|\mathcal{B}|} |\Omega(\vec{k})|, \qquad (10.26)$$

where $\mathcal{B} := \{\vec{k} = t_1\vec{G}_1 + t_2\vec{G}_2 \ : \ t_i \in [0,1)\}$ and $|\mathcal{B}| = 8\pi^2/(3\sqrt{3})$. Similarly, the finite volume specific free energy $f_{0,\Lambda}^\beta := -(\beta|\Lambda|)^{-1}\log \mathrm{Tr}\{e^{-\beta H_\Lambda^0}\}$ is

$$f_{0,\Lambda}^\beta = -\frac{2}{\beta|\Lambda|}\sum_{\vec{k}\in\mathcal{B}_L}\log\left[\left(1+e^{\beta v_0|\Omega(\vec{k})|}\right)\left(1+e^{-\beta v_0|\Omega(\vec{k})|}\right)\right], \tag{10.27}$$

from which

$$f_\beta(0) = -\frac{2}{\beta}\int_\mathcal{B}\frac{d\vec{k}}{|\mathcal{B}|}\log\left[\left(1+e^{\beta v_0|\Omega(\vec{k})|}\right)\left(1+e^{-\beta v_0|\Omega(\vec{k})|}\right)\right]. \tag{10.28}$$

Besides these thermodynamic functions, it is also useful to compute the *Schwinger functions* of the free gas, in terms of which, in the next section, we will write down the perturbation theory for the interacting system. These are defined as follows. Let $x_0 \in [0,\beta)$ be an *imaginary time*, let $\mathbf{x} := (x_0, \vec{x}) \in [0,\beta) \times \Lambda$ and let us consider the time-evolved operator $\Psi_{\mathbf{x},\sigma}^\pm = e^{Hx_0}\Psi_{\vec{x},\sigma}^\pm e^{-Hx_0}$, where $\Psi_{\vec{x},\sigma}^\pm$ is the two-components spinor defined in equation (10.19); in the following we shall denote its components by $\Psi_{\mathbf{x},\sigma,\rho}^\pm$, with $\rho \in \{1,2\}$, $\Psi_{\mathbf{x},\sigma,1}^\pm = a_{\mathbf{x},\sigma}^\pm$ and $\Psi_{\mathbf{x},\sigma,2}^\pm = b_{\mathbf{x}+\delta_1,\sigma}^\pm$ (here $\delta_1 = (0,\vec{\delta}_1)$). We define the n-points Schwinger functions at finite volume and finite temperature as:

$$S_n^{\beta,\Lambda}(\mathbf{x}_1,\varepsilon_1,\sigma_1,\rho_1;\ldots;\mathbf{x}_n,\varepsilon_n,\sigma_n,\rho_n) = \left\langle\mathbf{T}\{\Psi_{\mathbf{x}_1,\sigma_1,\rho_1}^{\varepsilon_1}\cdots\Psi_{\mathbf{x}_n,\sigma_n,\rho_n}^{\varepsilon_n}\}\right\rangle_{\beta,\Lambda} \tag{10.29}$$

where: $\mathbf{x}_i \in [0,\beta] \times \Lambda$, $\sigma_i =\uparrow\downarrow$, $\varepsilon_i = \pm$, $\rho_i = 1,2$ and $\mathbf{T}$ is the operator of fermionic time ordering, acting on a product of fermionic fields as:

$$\mathbf{T}(\Psi_{\mathbf{x}_1,\sigma_1,\rho_1}^{\varepsilon_1}\cdots\Psi_{\mathbf{x}_n,\sigma_n,\rho_n}^{\varepsilon_n}) = (-1)^\pi\Psi_{\mathbf{x}_{\pi(1)},\sigma_{\pi(1)},\rho_{\pi(1)}}^{\varepsilon_{\pi(1)}}\cdots\Psi_{\mathbf{x}_{\pi(n)},\sigma_{\pi(n)},\rho_{\pi(n)}}^{\varepsilon_{\pi(n)}} \tag{10.30}$$

where π is a permutation of $\{1,\ldots,n\}$, chosen in such a way that $x_{\pi(1)0} \geq \cdots \geq x_{\pi(n)0}$, and $(-1)^\pi$ is its sign. (If some of the time coordinates are equal each other, the arbitrariness of the definition is solved by ordering each set of operators with the same time coordinate so that creation operators precede the annihilation operators.) Taking the limit $\Lambda \to \infty$ in (10.29) we get the finite temperature n-point Schwinger functions, denoted by $S_n^\beta(\mathbf{x}_1,\varepsilon_1,\sigma_1,\rho_1;\ldots;\mathbf{x}_n,\varepsilon_n,\sigma_n,\rho_n)$, which describe the properties of the infinite volume system at finite temperature. Taking the $\beta \to \infty$ limit of the finite temperature Schwinger functions, we get the zero temperature Schwinger functions, simply denoted by $S_n(\mathbf{x}_1,\varepsilon_1,\sigma_1,\rho_1;\ldots;\mathbf{x}_n,\varepsilon_n,\sigma_n,\rho_n)$, which by definition characterize the properties of the *thermal ground state* of (10.1) in the thermodynamic limit.

In the noninteracting case, that is, if $H_\Lambda = H_\Lambda^0$, the Hamiltonian is quadratic in the creation/annihilation operators. Therefore, the $2n$-point Schwinger functions satisfy the Wick rule, that is,

$$\left\langle\mathbf{T}\{\Psi_{\mathbf{x}_1,\sigma_1,\rho_1}^-\Psi_{\mathbf{y}_1,\sigma_1',\rho_1'}^+\cdots\Psi_{\mathbf{x}_n,\sigma_n,\rho_n}^-\Psi_{\mathbf{y}_n,\sigma_n',\rho_n'}^+\}\right\rangle_{\beta,\Lambda} = \det G,$$

$$G_{ij} = \delta_{\sigma_i\sigma_j'}\left\langle\mathbf{T}\{\Psi_{\mathbf{x}_i,\sigma_i,\rho_i}^-\Psi_{\mathbf{y}_j,\sigma_j',\rho_j'}^+\}\right\rangle_{\beta,\Lambda}. \tag{10.31}$$

Moreover, every n–point Schwinger function $S_n^{\beta,\Lambda}(\mathbf{x}_1,\varepsilon_1,\sigma_1,\rho_1;\ldots;\mathbf{x}_n,\varepsilon_n,\sigma_n,\rho_n)$ with $\sum_{i=1}^n \varepsilon_i \neq 0$ is identically zero. Therefore, in order to construct the whole set of Schwinger functions of H_Λ^0, it is enough to compute the two-point function $S_0^{\beta,\Lambda}(\mathbf{x}-\mathbf{y}) = \left\langle \mathbf{T}\{\Psi_{\mathbf{x},\sigma,\rho}^- \Psi_{\mathbf{y},\sigma,\rho'}^+\}\right\rangle_{\beta,\Lambda}$. This can be easily reconstructed from the two-point function of the α-fields and β-fields, see equation (10.21).

Let $\vec{x} \in \Lambda$, $\alpha_{\vec{x},\sigma}^\pm = |\Lambda|^{-1}\sum_{\vec{k}\in\mathcal{B}_L} e^{\pm i\vec{k}\vec{x}}\hat{\alpha}_{\vec{k},\sigma}$ and $\beta_{\vec{x},\sigma}^\pm = |\Lambda|^{-1}\sum_{\vec{k}\in\mathcal{B}_L} e^{\pm i\vec{k}\vec{x}}\hat{\alpha}_{\vec{k},\sigma}$; if $\mathbf{x} = (x_0,\vec{x})$, let $\alpha_{\mathbf{x},\sigma}^\pm = e^{H_\Lambda^0 x_0}\alpha_{\vec{x},\sigma}^\pm e^{-H_\Lambda^0 x_0}$ and $\beta_{\mathbf{x},\sigma}^\pm = e^{H_\Lambda^0 x_0}\beta_{\vec{x},\sigma}^\pm e^{-H_\Lambda^0 x_0}$. A straightforward computation shows that, if $-\beta < x_0 - y_0 \leq \beta$,

$$\left\langle \mathbf{T}\{\alpha_{\mathbf{x},\sigma}^- \alpha_{\mathbf{y},\sigma'}^+\}\right\rangle_{\beta,\Lambda} = \frac{\delta_{\sigma,\sigma'}}{|\Lambda|}\sum_{\vec{k}\in\mathcal{B}_L} e^{-i\vec{k}(\vec{x}-\vec{y})} \tag{10.32}$$

$$\cdot\left[\mathbb{1}(x_0-y_0>0)\frac{e^{v_0(x_0-y_0)|\Omega(\vec{k})|}}{1+e^{v_0\beta|\Omega(\vec{k})|}} - \mathbb{1}(x_0-y_0\leq 0)\frac{e^{v_0(x_0-y_0+\beta)|\Omega(\vec{k})|}}{1+e^{v_0\beta|\Omega(\vec{k})|}}\right],$$

$$\left\langle \mathbf{T}\{\beta_{\mathbf{x},\sigma}^- \beta_{\mathbf{y},\sigma'}^+\}\right\rangle_{\beta,\Lambda} = \frac{\delta_{\sigma,\sigma'}}{|\Lambda|}\sum_{\vec{k}\in\mathcal{B}_L} e^{-i\vec{k}(\vec{x}-\vec{y})} \tag{10.33}$$

$$\cdot\left[\mathbb{1}(x_0-y_0>0)\frac{e^{-v_0(x_0-y_0)|\Omega(\vec{k})|}}{1+e^{-v_0\beta|\Omega(\vec{k})|}} - \mathbb{1}(x_0-y_0\leq 0)\frac{e^{-v_0(x_0-y_0+\beta)|\Omega(\vec{k})|}}{1+e^{-v_0\beta|\Omega(\vec{k})|}}\right]$$

and $\left\langle \mathbf{T}\{\alpha_{\mathbf{x},\sigma}^- \beta_{\mathbf{y},\sigma'}^+\}\right\rangle_{\beta,\Lambda} = \left\langle \mathbf{T}\{\beta_{\mathbf{x},\sigma}^- \alpha_{\mathbf{y},\sigma'}^+\}\right\rangle_{\beta,\Lambda} = 0$. A priori, equations (10.33) and (10.34) are defined only for $-\beta < x_0 - y_0 \leq \beta$, but we can extend them periodically over the whole real axis; the periodic extension of the propagator is continuous in the time variable for $x_0 - y_0 \notin \beta\mathbb{Z}$, and it has jump discontinuities at the points $x_0 - y_0 \in \beta\mathbb{Z}$. Note that at $x_0 - y_0 = \beta n$, the difference between the right and left limits is equal to $(-1)^n \delta_{\vec{x},\vec{y}}$, so that the propagator is discontinuous only at $\mathbf{x}-\mathbf{y} = \beta\mathbb{Z} \times \vec{0}$. If we define $\mathcal{B}_{\beta,L} := \mathcal{B}_\beta \times \mathcal{B}_L$, with $\mathcal{B}_\beta = \{k_0 = \frac{2\pi}{\beta}(n_0+\frac{1}{2}) : n_0 \in \mathbb{Z}\}$, then for $\mathbf{x}-\mathbf{y} \notin \beta\mathbb{Z} \times \vec{0}$ we can write

$$\left\langle \mathbf{T}\{\alpha_{\mathbf{x},\sigma}^- \alpha_{\mathbf{y},\sigma'}^+\}\right\rangle_{\beta,\Lambda} = \frac{\delta_{\sigma,\sigma'}}{\beta|\Lambda|}\sum_{k\in\mathcal{B}_{\beta,L}} e^{-ik(\mathbf{x}-\mathbf{y})}\frac{1}{-ik_0-v_0|\Omega(\vec{k})|}, \tag{10.34}$$

$$\left\langle \mathbf{T}\{\beta_{\mathbf{x},\sigma}^- \beta_{\mathbf{y},\sigma'}^+\}\right\rangle_{\beta,\Lambda} = \frac{\delta_{\sigma,\sigma'}}{\beta|\Lambda|}\sum_{k\in\mathcal{B}_{\beta,L}} e^{-ik(\mathbf{x}-\mathbf{y})}\frac{1}{-ik_0+v_0|\Omega(\vec{k})|}. \tag{10.35}$$

If we now re-express $\alpha_{\mathbf{x},\sigma}^\pm$ and $\beta_{\mathbf{x},\sigma}^\pm$ in terms of $a_{\mathbf{x},\sigma}^\pm$ and $b_{\mathbf{x}+\delta_1,\sigma}^\pm$, using equation (10.21), we find that, for $\mathbf{x}-\mathbf{y} \notin \beta\mathbb{Z} \times \vec{0}$:

$$S_0^{\beta,\Lambda}(\mathbf{x}-\mathbf{y})_{\rho,\rho'} := S_2^{\beta,\Lambda}(\mathbf{x},\sigma,-,\rho;\mathbf{y},\sigma,+,\rho')\Big|_{U=0} =$$

$$= \frac{1}{\beta|\Lambda|}\sum_{k\in\mathcal{B}_{\beta,L}} \frac{e^{-ik\cdot(\mathbf{x}-\mathbf{y})}}{k_0^2+v_0^2|\Omega(\vec{k})|^2}\begin{pmatrix} ik_0 & -v_0\Omega^*(\vec{k}) \\ -v_0\Omega(\vec{k}) & ik_0 \end{pmatrix}_{\rho,\rho'} \tag{10.36}$$

Finally, if $\mathbf{x} - \mathbf{y} = (0^-, \vec{0})$:

$$S_0^{\beta,\Lambda}(0^-, \vec{0}) = -\frac{1}{2} + \frac{1}{\beta|\Lambda|} \sum_{k \in \mathcal{B}_{\beta,L}} \frac{1}{k_0^2 + v_0^2 |\Omega(\vec{k})|^2} \begin{pmatrix} 0 & -v_0 \Omega^*(\vec{k}) \\ -v_0 \Omega(\vec{k}) & 0 \end{pmatrix}. \qquad (10.37)$$

10.4 Perturbation theory and Grassmann integration

Let us now turn to the interacting case. The first step is to derive a formal perturbation theory for the specific free energy and ground state energy. In other words, we want to find rules to compute the generic perturbative order in U of $f_{\beta,\Lambda} := -(\beta|\Lambda|)^{-1} \log \mathrm{Tr}\{e^{-\beta H_\Lambda}\}$. We write $H_\Lambda = H_\Lambda^0 + V_\Lambda$, with V_Λ the operator in the second line of Eq.(10.1) and we use Trotter's product formula

$$e^{-\beta H_\Lambda} = \lim_{n \to \infty} \left[e^{-\beta H_\Lambda^0/n} (1 - \frac{\beta}{n} V_\Lambda) \right]^n \qquad (10.38)$$

so that, defining $V_\Lambda(t) := e^{t H_\Lambda^0} V_\Lambda e^{-t H_\Lambda^0}$,

$$\frac{\mathrm{Tr}\{e^{-\beta H_\Lambda}\}}{\mathrm{Tr}\{e^{-\beta H_\Lambda^0}\}} = \qquad (10.39)$$

$$= 1 + \sum_{N \geq 1} (-1)^N \int_0^\beta dt_1 \int_0^{t_1} dt_2 \cdots \int_0^{t_{N-1}} dt_N \frac{\mathrm{Tr}\{e^{-\beta H_\Lambda^0} V_\Lambda(t_1) \cdots V_\Lambda(t_N)\}}{\mathrm{Tr}\{e^{-\beta H_\Lambda^0}\}}.$$

Using the fermionic time-ordering operator defined in equation (10.30), we can rewrite equation (10.39) as

$$\frac{\mathrm{Tr}\{e^{-\beta H_\Lambda}\}}{\mathrm{Tr}\{e^{-\beta H_\Lambda^0}\}} = 1 + \sum_{N \geq 1} \frac{(-1)^N}{N!} \left\langle \mathbf{T}\{(V_{\beta,\Lambda}(\Psi))^N\} \right\rangle_{\beta,\Lambda}^0, \qquad (10.40)$$

where $\langle \cdot \rangle_{\beta,\Lambda}^0 = \mathrm{Tr}\{e^{-\beta H_\Lambda^0} \cdot\}/\mathrm{Tr}\{e^{-\beta H_\Lambda^0}\}$,

$$V_{\beta,\Lambda}(\Psi) := U \sum_{\rho=1,2} \int_{(\beta,\Lambda)} d\mathbf{x} \left(\Psi_{\mathbf{x},\uparrow,\rho}^+ \Psi_{\mathbf{x},\uparrow,\rho}^- - \frac{1}{2} \right) \left(\Psi_{\mathbf{x},\downarrow,\rho}^+ \Psi_{\mathbf{x},\downarrow,\rho}^- - \frac{1}{2} \right), \qquad (10.41)$$

and $\int_{(\beta,\Lambda)} d\mathbf{x}$ must be interpreted as $\int_{(\beta,\Lambda)} d\mathbf{x} = \int_{-\beta/2}^{\beta/2} dx_0 \sum_{\vec{x} \in \Lambda}$. Note that the N-th term in the sum in the r.h.s. of equation (10.40) can be computed by using the Wick rule (10.31) and the explicit expression for the 2-point function equations (10.36)-(10.37). It is straightforward to check that the 'Feynman rules' needed to compute $\left\langle \mathbf{T}\{(V_{\beta,\Lambda}(\Psi))^N\} \right\rangle_{\beta,\Lambda}^0$ are the following: (i) draw N graph elements consisting of 4-legged vertices, with the vertex associated to two labels $\mathbf{x}_i$ and ρ_i, $i = 1, \ldots, N$, and the four legs associated to two exiting fields (with labels $(\mathbf{x}_i, \uparrow, \rho_i)$ and $(\mathbf{x}_i, \downarrow, \rho_i)$) and two entering fields (with labels $(\mathbf{x}_i, \uparrow, \rho_i)$ and $(\mathbf{x}_i, \downarrow, \rho_i)$), respectively; (ii) pair the fields in all possible ways, in such a way that every pair consists of one entering

Fig. 10.3 The four-legged graph element (left); an example of a Feynman diagram of order 3 (right).

and one exiting field, with the same spin index, see Fig. 10.3; (iii) associate to every pairing a sign, corresponding to the sign of the permutation needed to bring every pair of contracted fields next to each other; (iv) associate to every paired pair of fields $[\Psi^-_{x_i,\sigma_i,\rho_i}, \Psi^+_{x_j,\sigma_j,\rho_j}]$ an oriented line connecting the i-th with the j-th vertex, with orientation from j to i; (v) associate to every oriented line $[j \to i]$ a value equal to

$$
g_{\rho_i,\rho_j}(\mathbf{x}_i - \mathbf{x}_j) := \frac{1}{\beta|\Lambda|} \sum_{k \in \mathcal{B}^{(M)}_{\beta,L}} e^{-ik\cdot(\mathbf{x}_i - \mathbf{x}_j)} \frac{\chi_0(2^{-M}|k_0|)}{k_0^2 + v_0^2|\Omega(\vec{k})|^2} \begin{pmatrix} ik_0 & -v_0\Omega^*(\vec{k}) \\ -v_0\Omega(\vec{k}) & ik_0 \end{pmatrix}_{\rho_i,\rho_j}
$$

$$(10.42)$$

where $\mathcal{B}^{(M)}_{\beta,L} = \mathcal{B}^{(M)}_{\beta} \times \mathcal{B}_L$, $\mathcal{B}^{(M)}_{\beta} = \mathcal{B}_{\beta} \cap \{k_0 : \chi_0(2^{-M}|k_0|) > 0\}$ and $\chi_0(t)$ is a smooth compact support function that is equal to 1 for (say) $|t| \le 1/3$ and equal to 0 for $|t| \ge 2/3$; (vi) associate to every pairing (i.e. to every Feynman graph) a value, equal to the product of the sign of the pairing times U^N times the product of the values of all the oriented lines; (vii) integrate over $\mathbf{x}_i$ and sum over ρ_i the value of each pairing, then sum over all pairings; (viii) finally, take the $M \to \infty$ limit: the result is equal to $\langle \mathbf{T}\{(V_{\beta,\Lambda}(\Psi))^N\}\rangle^0_{\beta,\Lambda}$. Note that the $M \to \infty$ limit of the *propagator* $g(\mathbf{x})$ is equal to $S^{\beta,\Lambda}_0(\mathbf{x})$ if $\mathbf{x} \ne \mathbf{0}$, while $\lim_{M\to\infty} g(\mathbf{0}) = S^{\beta,\Lambda}_0(0^-, \vec{0}) + \frac{1}{2}$, see equation (10.37): the difference between $\lim_{M\to\infty} g(\mathbf{x})$ and $S^{\beta,\Lambda}_0(\mathbf{x})$ takes into account the $-\frac{1}{2}$ terms in the definition of $V_{\beta,\Lambda}(\Psi)$.

An algebraically convenient way to re-express equation (10.40) is in terms of *Grassmann integrals*. Consider the set $\mathcal{A}_{M,\beta,L} = \{\hat{\psi}^\pm_{k,\sigma,\rho}\}^{\sigma=\uparrow\downarrow, \rho=1,2}_{k\in\mathcal{B}^{(M)}_{\beta,L}}$, where the *Grassmann variables* $\hat{\psi}^\pm_{k,\sigma,\rho}$ satisfy by the definition the anti-commutation rules $\{\hat{\psi}^\varepsilon_{k,\sigma,\rho}, \hat{\psi}^{\varepsilon'}_{k',\sigma',\rho'}\} = 0$. In particular, the square of a Grassmann variable is zero and the only nontrivial Grassmann monomials are at most linear in each variable. Let the Grassmann algebra generated by $\mathcal{A}_{M,\beta,L}$ be the set of all polynomials obtained by linear combinations of such nontrivial monomials. Let us also define the Grassmann integration $\int \left[\prod_{k\in\mathcal{B}^{(M)}_{\beta,L}} \prod^{\rho=1,2}_{\sigma=\uparrow\downarrow} d\hat{\psi}^+_{k,\sigma,\rho} d\hat{\psi}^-_{k,\sigma,\rho} \right]$ as the linear operator on the Grassmann algebra such that, given a monomial $Q(\hat{\psi}^-, \hat{\psi}^+)$ in the variables $\hat{\psi}^\pm_{k,\sigma,\rho}$, its action on $Q(\hat{\psi}^-, \hat{\psi}^+)$ is 0 except in the case $Q(\hat{\psi}^-, \hat{\psi}^+) = \prod_{k\in\mathcal{B}^{(M)}_{\beta,L}} \prod^{\rho=1,2}_{\sigma=\uparrow\downarrow} \hat{\psi}^-_{k,\sigma,\rho} \hat{\psi}^+_{k,\sigma,\rho}$, up to a permutation of the variables. In this case the value of the integral is determined, by using the anticommutation properties of the variables, by the condition

$$\int \left[\prod_{\mathbf{k}\in\mathcal{B}_{\beta,L}^{(M)}} \prod_{\sigma=\uparrow\downarrow}^{\rho=1,2} d\hat{\psi}_{\mathbf{k},\sigma,\rho}^{+} d\hat{\psi}_{\mathbf{k},\sigma,\rho}^{-} \right] \prod_{\mathbf{k}\in\mathcal{B}_{\beta,L}^{(M)}} \prod_{\sigma=\uparrow\downarrow}^{\rho=1,2} \hat{\psi}_{\mathbf{k},\sigma,\rho}^{-} \hat{\psi}_{\mathbf{k},\sigma,\rho}^{+} = 1 \qquad (10.43)$$

Defining the free propagator matrix $\hat{g}_{\mathbf{k}}$ as

$$\hat{g}_{\mathbf{k}} = \chi_0(2^{-M}|k_0|) \begin{pmatrix} -ik_0 & -v_0\Omega^*(\vec{k}) \\ -v_0\Omega(\vec{k}) & -ik_0 \end{pmatrix}^{-1} \qquad (10.44)$$

and the 'Gaussian integration' $P_M(d\psi)$ as

$$P_M(d\psi) = \left[\prod_{\mathbf{k}\in\mathcal{B}_{\beta,L}^{(M)}}^{\sigma=\uparrow\downarrow} \frac{-\beta^2|\Lambda|^2 \left[\chi_0(2^{-M}|k_0|)\right]^2}{k_0^2 + v_0^2|\Omega(\vec{k})|^2} d\hat{\psi}_{\mathbf{k},\sigma,1}^{+} d\hat{\psi}_{\mathbf{k},\sigma,1}^{-} d\hat{\psi}_{\mathbf{k},\sigma,2}^{+} d\hat{\psi}_{\mathbf{k},\sigma,2}^{-} \right] \cdot$$

$$\cdot \exp\left\{ -(\beta|\Lambda|)^{-1} \sum_{\mathbf{k}\in\mathcal{B}_{\beta,L}^{(M)}}^{\sigma=\uparrow\downarrow} \hat{\psi}_{\mathbf{k},\sigma}^{+} \hat{g}_{\mathbf{k}}^{-1} \hat{\psi}_{\mathbf{k},\sigma}^{-} \right\}, \quad (10.45)$$

it turns out that

$$\int P(d\psi)\hat{\psi}_{\mathbf{k}_1,\sigma_1}^{-} \hat{\psi}_{\mathbf{k}_2,\sigma_2}^{+} = \beta|\Lambda|\delta_{\sigma_1,\sigma_2}\delta_{\mathbf{k}_1,\mathbf{k}_2}\hat{g}_{\mathbf{k}_1}, \qquad (10.46)$$

while the average of an arbitrary monomial in the Grassmann variables with respect to $P_M(d\psi)$ is given by the fermionic Wick rule with propagator equal to the r.h.s. of equation (10.46). Using these definitions and the Feynman rules described above, we can rewrite equation (10.40) as

$$\frac{\text{Tr}\{e^{-\beta H_\Lambda}\}}{\text{Tr}\{e^{-\beta H_\Lambda^0}\}} = \lim_{M\to\infty} \int P_M(d\psi)e^{-\mathcal{V}(\psi)}, \qquad (10.47)$$

where

$$\mathcal{V}(\psi) = U \sum_{\rho=1,2} \int_{(\beta,\Lambda)} d\mathbf{x}\, \psi_{\mathbf{x},\uparrow,\rho}^{+} \psi_{\mathbf{x},\uparrow,\rho}^{-} \psi_{\mathbf{x},\downarrow,\rho}^{+} \psi_{\mathbf{x},\downarrow,\rho}^{-}, \qquad (10.48)$$

$$\psi_{\mathbf{x},\sigma,\rho}^{\pm} = \frac{1}{\beta|\Lambda|} \sum_{\mathbf{k}\in\mathcal{B}_{\beta,L}^{(M)}} e^{\pm i\mathbf{k}\mathbf{x}} \hat{\psi}_{\mathbf{k},\sigma,\rho}^{\pm}, \qquad \mathbf{x} \in (-\beta/2,\beta/2] \times \Lambda \qquad (10.49)$$

and the exponential $e^{-\mathcal{V}(\psi)}$ in the r.h.s. of equation (10.47) must be identified with its Taylor series in U (which is finite for every finite M, due to the anti-commutation rules of the Grassmann variables and the fact that the Grassmann algebra is finite for every finite M). *A priori*, equation (10.47) must be understood as an equality between formal power series in U. However, it can be given a nonperturbative meaning, provided that we can prove the convergence of the Grassmann functional integral in the r.h.s., as shown by the following Proposition.

Proposition 10.2 *Let*

$$F_{\beta,\Lambda}^{(M)} := -\frac{1}{\beta|\Lambda|} \log \int P_M(d\psi)\left(e^{-\mathcal{V}(\psi)}\right) \tag{10.50}$$

and let β and $|\Lambda|$ be sufficiently large. Assume that there exists $U_0 > 0$ such that $F_{\beta,\Lambda}^{(M)}$ is analytic in the complex domain $|U| \leq U_0$ and is uniformly convergent as $M \to \infty$. Then, if $|U| \leq U_0$,

$$f_{\beta,\Lambda} = -\frac{2}{\beta|\Lambda|} \sum_{\vec{k} \in \mathcal{B}_L} \log\left(2 + 2\cosh(\beta v_0|\Omega(\vec{k})|)\right) + \lim_{M \to \infty} F_{\beta,\Lambda}^{(M)}. \tag{10.51}$$

Proof: We need to prove that

$$\frac{\mathrm{Tr}\{e^{-\beta H_\Lambda}\}}{\mathrm{Tr}\{e^{-\beta H_{0,\Lambda}}\}} = \exp\left\{ -\beta|\Lambda| \lim_{M \to \infty} F_{\beta,\Lambda}^{(M)}\right\} \tag{10.52}$$

under the given analyticity assumptions on $F_{\beta,\Lambda}^{(M)}$. The first key remark is that, if β, Λ are finite, the left-hand side of equation (10.52) is *a priori* well defined and analytic on the whole complex plane. In fact, by the Pauli principle, the Fock space generated by the fermion operators $a_{\vec{x},\sigma}^{\pm}, b_{\vec{x}+\vec{\delta}_1,\sigma}^{\pm}$, with $\vec{x} \in \Lambda, \sigma = \uparrow\downarrow$, is finite dimensional. Therefore, writing $H_\Lambda = H_\Lambda^0 + V_\Lambda$, with H_Λ^0 and V_Λ two bounded operators, we see that $\mathrm{Tr}\{e^{-\beta H_\Lambda}\}$ is an entire function of U, simply because $e^{-\beta H_\Lambda}$ converges in norm over the whole complex plane:

$$||e^{-\beta H_\Lambda}|| \leq \sum_{n=0}^{\infty} \frac{\beta^n}{n!} \left(||H_\Lambda^0|| + ||V_\Lambda||\right)^n = e^{\beta||H_\Lambda^0||+\beta||V_\Lambda||} \tag{10.53}$$

where the norm $|| \cdot ||$ is, for example the Hilbert–Schmidt norm $||A|| = \sqrt{\mathrm{Tr}(A^\dagger A)}$.

On the other hand, by assumption, $F_{\beta,\Lambda}^{(M)}$ is analytic in $|U| \leq U_0$, with U_0 independent of β, Λ, M, and uniformly convergent as $M \to \infty$. Hence, by the Weierstrass theorem, the limit $F_{\beta,\Lambda} = \lim_{M\to\infty} F_{\beta,\Lambda}^{(M)}$ is analytic in $|U| \leq U_0$ and its Taylor coefficients coincide with the limits as $M \to \infty$ of the Taylor coefficients of $F_{\beta,\Lambda}^{(M)}$. Moreover, $\lim_{M\to\infty} e^{-\beta|\Lambda|F_{\beta,\Lambda}^{(M)}} = e^{-\beta|\Lambda|F_{\beta,\Lambda}}$, again by the Weierstrass theorem.

As discussed above, the Taylor coefficients of $e^{-\beta|\Lambda|F_{\beta,\Lambda}}$ coincide with the Taylor coefficients of $\mathrm{Tr}\{e^{-\beta H_\Lambda}\}/\mathrm{Tr}\{e^{-\beta H_{0,\Lambda}}\}$: therefore, $\mathrm{Tr}\{e^{-\beta H_\Lambda}\}/\mathrm{Tr}\{e^{-\beta H_{0,\Lambda}}\} = e^{-\beta|\Lambda|F_{\beta,\Lambda}}$ in the complex region $|U| \leq U_0$, simply because the l.h.s. is entire in U, the r.h.s. is analytic in $|U| \leq U_0$ and the Taylor coefficients at the origin of the two sides are the same. Taking logarithms at both sides proves equation (10.51). $\quad\square$

By Proposition 10.2, the Grassmann integral equation (10.50) can be used to compute the free energy of the original Hubbard model, provided that the r.h.s.

of equation (10.50) is analytic in a domain that is uniform in M, β, Λ and that it converges to a well defined analytic function uniformly as $M \to \infty$. The rest of these notes are devoted to the proof of this fact. We start from equation (10.50), which can be rewritten as

$$F_{\beta,\Lambda}^{(M)} := -\frac{1}{\beta|\Lambda|} \sum_{N \geq 1} \frac{(-1)^N}{N!} \mathcal{E}^T(\mathcal{V}; N), \tag{10.54}$$

where the *truncated expectation* $\mathcal{E}^T$ is defined as

$$\mathcal{E}^T(\mathcal{V}; N) := \frac{\partial^N}{\partial \lambda^N} \log \int P_M(d\psi) e^{\lambda V(\psi)}\Big|_{\lambda=0}. \tag{10.55}$$

More in general,

$$\mathcal{E}^T(\mathcal{V}_1, \dots, \mathcal{V}_N) := \frac{\partial^N}{\partial \lambda_1 \cdots \partial \lambda_N} \log \int P_M(d\psi) e^{\lambda_1 V_1(\psi) + \cdots + \lambda_N V_N(\psi)}\Big|_{\lambda_i=0} \tag{10.56}$$

and $\mathcal{E}^T(\mathcal{V}_1, \dots, \mathcal{V}_N)\big|_{\mathcal{V}_i=\mathcal{V}} = \mathcal{E}^T(\mathcal{V}; N)$. It can be checked by induction that the truncated expectation is related to the simple expectation $\mathcal{E}(X(\psi)) = \int P_M(d\psi)X(\psi)$ by

$$\mathcal{E}(\mathcal{V}_1 \cdots \mathcal{V}_N) = \sum_{m=1}^{N} \sum_{\{Y^1, \dots, Y^m\}} \mathcal{E}^T(\mathcal{V}_{j_1^1}, \dots, \mathcal{V}_{j_{|Y^1|}^1}) \cdots \mathcal{E}^T(\mathcal{V}_{j_1^m}, \dots, \mathcal{V}_{j_{|Y^m|}^m}), \tag{10.57}$$

where the second sum in the r.h.s. runs over partitions of $\{1, \dots, N\}$ of multiplicity m, that is, over unordered m-ples of disjoint sets such that $\cup_{i=1}^m Y^i = \{1 \dots N\}$, with $Y^i = \{j_1^i, \dots, j_{|Y_i|}^i\}$. Note that $\mathcal{E}(\mathcal{V}^N) = \mathcal{E}(\mathcal{V}_1 \dots \mathcal{V}_N)\big|_{\mathcal{V}_i=\mathcal{V}}$ can be computed as a sum of Feynman diagrams whose values are determined by the same Feynman rules described after equation (10.41) (with the exception of rule (viii): of course, since $\mathcal{E}(X) = \int P_M(d\psi)X$, M should be temporarily kept fixed in the computation); we shall write

$$\mathcal{E}(\mathcal{V}^N) = \sum_{\mathcal{G} \in \Gamma_N} \widehat{\mathrm{Val}}(\mathcal{G}), \tag{10.58}$$

where Γ_N is the set of all Feynman diagrams with N vertices, constructed with the rules described above; $\widehat{\mathrm{Val}}(\mathcal{G})$ includes the integration over the space–time labels $\mathbf{x}_i$ and the sum over the component labels ρ_i: if $\mathcal{G} \in \Gamma_N^T$, we shall symbolically write

$$\widehat{\mathrm{Val}}(\mathcal{G}) = \sigma_{\mathcal{G}} U^N \sum_{\rho_1, \dots, \rho_N} \int d\mathbf{x}_1 \cdots d\mathbf{x}_N \prod_{\ell \in \mathcal{G}} \delta_{\sigma(\ell), \sigma'(\ell)} g_{\rho(\ell), \rho'(\ell)}(\mathbf{x}(\ell) - \mathbf{x}'(\ell)), \tag{10.59}$$

where $\sigma_{\mathcal{G}}$ is the sign of the permutation associated to the graph $\mathcal{G}$ and we denoted by $(\mathbf{x}(\ell), \sigma(\ell), \rho(\ell))$ and $(\mathbf{x}'(\ell), \sigma'(\ell), \rho'(\ell))$ the labels of the two vertices, which the line ℓ enters in and exits from, respectively. Using equations (10.57)–(10.58), it can be proved by induction that

$$\mathcal{E}^T(\mathcal{V}; N) = \sum_{\mathcal{G} \in \Gamma_N^T} \widehat{\mathrm{Val}}(\mathcal{G}), \qquad (10.60)$$

where $\Gamma_N^T \subset \Gamma_N$ is the set of *connected* Feynman diagrams with N vertices. Combining equation (10.54) with equation (10.60) we finally have a formal power series expansion for the specific free energy of our model (more precisely, of its ultraviolet regularization associated to the imaginary-time ultraviolet cutoff $\chi_0(2^{-M}|k_0|)$). The Feynman rules for computing $\widehat{\mathrm{Val}}(\mathcal{G})$ allow us to derive a first *very naive* upper bound on the N-th order contribution to $F_{\beta,\Lambda}^{(M)}$, that is to

$$F_{\beta,\Lambda}^{(M;N)} := -\frac{1}{\beta|\Lambda|} \frac{(-1)^N}{N!} \mathcal{E}^T(\mathcal{V}; N). \qquad (10.61)$$

We have:

$$|F_{\beta,\Lambda}^{(M;N)}| \le \frac{1}{\beta|\Lambda|} \frac{1}{N!} \sum_{\mathcal{G} \in \Gamma_N^T} |\widehat{\mathrm{Val}}(\mathcal{G})| \le$$

$$\le \frac{|\Gamma_N^T|}{N!} 2^N |U|^N \|g\|_\infty^{N+1} \|g\|_1^{N-1}, \qquad (10.62)$$

where $|\Gamma_N^T|$ is the number of connected Feynman diagrams of order N and $\beta|\Lambda|(2|U|)^N \|g\|_\infty^{N+1} \|g\|_1^{N-1}$ is a uniform bound on the value of a generic connected Feynman diagram of order N. The bound is obtained as follows: given $\mathcal{G} \in \Gamma_N^T$, select an arbitrary 'spanning tree' in $\mathcal{G}$, that is a loopless subset of $\mathcal{G}$ that connects all the N vertices; now: the integrals over the space–time coordinates of the product of the propagators on the spanning tree can be bounded by $\beta|\Lambda| \|g\|_1^{N-1}$; the product of the remaining propagators can be bounded by $\|g\|_\infty^{N+1}$; finally, the sum over the ρ_i labels is bounded by 2^N. Using equation (10.62) and the facts that, for a suitable constant $C > 0$: (i) $|\Gamma_N^T| \le C^N(N!)^2$ (see, e.g., (Gentile and Mastropietro 2001, Appendix A.1.3) for a proof of this fact), (ii) $\|g\|_\infty \le CM$, (iii) $\|g\|_1 \le C\beta$, we find:

$$|F_{\beta,\Lambda}^{(M;N)}| \le (2C^3|U|)^N N! M^{N+1} \beta^{N-1}. \qquad (10.63)$$

Remark 10.3 *While the bound $\|g\|_1 \le C\beta$ (see Appendix 10.9 for a proof) is dimensionally optimal, the estimate $\|g\|_\infty \le CM$ could be improved to $\|g\|_\infty \le (\text{const.})$, at the cost of a more detailed analysis of the definition of $g(\mathbf{x})$, which shows that the apparent ultraviolet logarithmic divergence associated to the sum over k_0 in equation (10.42) is in fact related to the jump singularity of $g(x_0, \vec{0})$ at $x_0 = \beta\mathbb{Z}$ mentioned at the end of Section 10.3. This can be proved along the lines of (Benfatto 2008, Section 2). However, to the purpose of the present discussion, the rough (and easier) bound $\|g\|_\infty \le CM$ is enough; see Appendix 10.9 for a proof.*

The pessimistic bound equation (10.63) has two main problems: (i) a combinatorial problem, associated to the $N!$, which makes the r.h.s. of Eq.(10.63) not summable over N, not even for finite M and β; (ii) a divergence problem, associated to the factor $M^{N+1}\beta^{N-1}$, which diverges as $M \to \infty$ (i.e., as the ultraviolet regularization is removed) and as $\beta \to \infty$ (i.e., as the temperature is sent to 0). The combinatorial problem is solved by a smart reorganization of the perturbation theory, in the form of a determinant expansion, together with a systematic use of the Gram–Hadamard bound. The divergence problem is solved by systematic resummations of the series: we will first identify the class of contributions that produce ultraviolet or infrared divergences and then we show how to inductively resum them into a redefinition of the coupling constants of the theory; the inductive resummations are based on a multi-scale integration of the theory: at the end of the construction, they will allow us to express the specific free energy in terms of modified Feynman diagrams, whose values are not affected anymore by ultraviolet or infrared divergences.

10.5   The determinant expansion

Let us now show how to attack the first of the two problems that arose at the end of previous section. In other words, let us show how to solve the combinatorial problem by reorganizing the perturbative expansion discussed above into a more compact and more convenient form. In the previous section we discussed a Feynman diagram representation of the truncated expectation, see equation (10.60). A slightly more general version of equation (10.60) is the following. For a given set of indices $P = \{f_1, \ldots, f_{|P|}\}$, with $f_i = (\mathbf{x}_i, \sigma_i, \rho_i, \varepsilon_i)$, $\varepsilon_i \in \{+, -\}$, let

$$\psi_P := \prod_{f \in P} \psi^{\varepsilon(f)}_{\mathbf{x}(f), \sigma(f), \rho(f)}. \tag{10.64}$$

Each field $\psi^{\varepsilon(f)}_{\mathbf{x}(f), \sigma(f), \rho(f)}$ can be represented as an oriented half-line, emerging from the point $\mathbf{x}(f)$ and carrying an arrow, pointing in the direction entering or exiting the point, depending on whether $\varepsilon(f)$ is equal to $-$ or $+$, respectively; moreover, the half-line carries two labels, $\sigma(f) \in \{\uparrow, \downarrow\}$ and $\rho(f) \in \{1, 2\}$. Now, given s set of indices $P_1, \ldots, P_s$, we can enclose the points $\mathbf{x}(f)$ belonging to the set P_j, for some $j = 1, \ldots, s$, in a box: in this way, assuming that all the points $\mathbf{x}(f)$, $f \in \cup_i P_i$, are distinct, we obtain s disjoint boxes. Given $\mathcal{P} := \{P_1, \ldots, P_s\}$, we can associate to it the set $\Gamma^T(\mathcal{P})$ of connected Feynman diagrams, obtained by pairing the half-lines with consistent orientations, in such a way that the two half-lines of any connected pairs carry the same spin index, and in such a way that all the boxes are connected. Using a notation similar to equation (10.59), we have:

$$\mathcal{E}^T(\psi_{P_1}, \ldots, \psi_{P_s}) = \sum_{\mathcal{G} \in \Gamma^T(\mathcal{P})} \mathrm{Val}(\mathcal{G}),$$

$$\mathrm{Val}(\mathcal{G}) = \sigma_{\mathcal{G}} \prod_{\ell \in \mathcal{G}} \delta_{\sigma(\ell), \sigma'(\ell)} g_{\rho(\ell), \rho'(\ell)}(\mathbf{x}(\ell) - \mathbf{x}'(\ell)), \tag{10.65}$$

A different and more compact representation for the truncated expectation, alternative to equation (10.65), is the following:

$$\mathcal{E}^T(\psi_{P_1}, \ldots, \psi_{P_s}) = \sum_{T \in \mathbf{T}(\mathcal{P})} \alpha_T \prod_{\ell \in T} g_\ell \int dP_T(\mathbf{t}) \det G^T(\mathbf{t}), \qquad (10.66)$$

where:

- any element T of the set $\mathbf{T}(\mathcal{P})$ is a set of lines forming an *anchored tree* between the boxes $P_1, \ldots, P_s$, that is, T is a set of lines that becomes a tree if one identifies all the points in the same clusters;
- α_T is a sign (irrelevant for the subsequent bounds);
- g_ℓ is a shorthand for $\delta_{\sigma(\ell),\sigma'(\ell)} g_{\rho(\ell),\rho'(\ell)}(\mathbf{x}(\ell) - \mathbf{x}'(\ell))$;
- if $\mathbf{t} = \{t_{i,i'} \in [0,1], 1 \le i, i' \le s\}$, then $dP_T(\mathbf{t})$ is a probability measure with support on a set of $\mathbf{t}$ such that $t_{i,i'} = \mathbf{u}_i \cdot \mathbf{u}_{i'}$ for some family of vectors $\mathbf{u}_i \in \mathbb{R}^s$ of unit norm;
- if $2n = \sum_{i=1}^s |P_i|$, then $G^T(\mathbf{t})$ is a $(n-s+1) \times (n-s+1)$ matrix, whose elements are given by $G^T_{f,f'} = t_{i(f),i(f')} g_{\ell(f,f')}$, where: $f, f' \notin \cup_{\ell \in T}\{f_\ell^1, f_\ell^2\}$ and f_ℓ^1, f_ℓ^2 are the two field labels associated to the two (entering and exiting) half-lines contracted into ℓ; $i(f) \in \{1, \ldots, s\}$ is s.t. $f \in P_{i(f)}$; $g_{\ell(f,f')}$ is the propagator associated to the line obtained by contracting the two half-lines with indices f and f'.

If $s = 1$ the sum over T is empty, but we can still use the equation (10.66) by interpreting the r.h.s. as equal to $\det G^T(1)$.

The proof of the determinant representation is described in Appendix 10.10; this representation is due to a fermionic reinterpretation of the interpolation formulas by Battle, Brydges, and Federbush (Battle and Federbush 1984; Brydges 1986; Brydges and Federbush 1978), used originally by Gawedski–Kupianen (Gawedzki and Kupiainen 1985) and by Lesniewski (Lesniewski 1987), among others, to study certain $(1+1)$-dimensional fermionic quantum field theories. Using equation (10.66) we get an alternative representation for the N-th order contribution to the specific free energy:

$$F_{\beta,\Lambda}^{(M;N)} = -\frac{1}{\beta|\Lambda|} \frac{(-1)^N}{N!} \mathcal{E}^T(V; N) = -\frac{1}{\beta|\Lambda|} \frac{(-1)^N}{N!} U^N \sum_{\rho_1,\ldots,\rho_N} \sum_{T \in \mathbf{T}_N} \alpha_T \int dx_1 \cdots dx_N$$

$$\cdot \prod_{\ell \in T} \delta_{\sigma(\ell),\sigma'(\ell)} g_{\rho(\ell),\rho'(\ell)}(\mathbf{x}(\ell) - \mathbf{x}'(\ell)) \int dP_T(\mathbf{t}) \det G^T(\mathbf{t}). \qquad (10.67)$$

Using the fact that the number of anchored trees in $\mathbf{T}_N$ is bounded by $C^N N!$ for a suitable constant C (see, e.g., (Gentile and Mastropietro 2001, Appendix A3.3) for a proof of this fact), from equation (10.67) we get:

$$|F_{\beta,\Lambda}^{(M;N)}| \le (\text{const.})^N |U|^N \, \|g\|_1^{N-1} \| \det G^T(\cdot)\|_\infty. \qquad (10.68)$$

In order to bound $\det G^T$, we use the *Gram–Hadamard inequality*, stating that, if M is a square matrix with elements M_{ij} of the form $M_{ij} = \langle A_i, B_j \rangle$, where A_i, B_j are vectors in a Hilbert space with scalar product $\langle \cdot, \cdot \rangle$, then

$$|\det M| \le \prod_i \|A_i\| \cdot \|B_i\|. \qquad (10.69)$$

where $|| \cdot ||$ is the norm induced by the scalar product. See (Gentile and Mastropietro 2001, Theorem A.1) for a proof of equation (10.69).

Let $\mathcal{H} = \mathbb{R}^n \otimes \mathcal{H}_0$, where $\mathcal{H}_0$ is the Hilbert space of the functions $\mathbf{F} : [-\beta/2, \beta/2] \times \Lambda \to \mathbb{C}^2$, with scalar product $\langle \mathbf{F}, \mathbf{G} \rangle = \sum_{\rho=1,2} \int d\mathbf{z} \, F_\rho^*(\mathbf{z}) G_\rho(\mathbf{z})$, where $F_\rho = [\mathbf{F}]_\rho$, $G_\rho = [\mathbf{G}]_\rho$, $\rho = 1, 2$, are the components of the vectors $\mathbf{F}$ and $\mathbf{G}$. It is easy to verify that

$$G^T_{f,f'} = t_{i(f),i(f')} \, \delta_{\sigma(f),\sigma(f')} \, g_{\rho(f),\rho(f')}(\mathbf{x}(f) - \mathbf{x}(f')) =$$
$$= \left\langle \mathbf{u}_{i(f)} \otimes \mathbf{e}_{\sigma(f)} \otimes \mathbf{A}_{\mathbf{x}(f),\rho(f)}, \mathbf{u}_{i(f')} \otimes \mathbf{e}_{\sigma(f')} \otimes \mathbf{B}_{\mathbf{x}(f'),\rho(f')} \right\rangle, \quad (10.70)$$

where: $\mathbf{u}_i \in \mathbb{R}^n$, $i = 1, \ldots, n$, are vectors such that $t_{i,i'} = \mathbf{u}_i \cdot \mathbf{u}_{i'}$; $\mathbf{e}_\uparrow = (1,0)$, $\mathbf{e}_\downarrow = (0,1)$; $\mathbf{A}_{\mathbf{x},\rho}$ and $\mathbf{B}_{\mathbf{x},\rho}$ have components:

$$[\mathbf{A}_{\mathbf{x},\rho}(\mathbf{z})]_i = \frac{1}{\beta|\Lambda|} \sum_{\mathbf{k} \in \mathcal{B}^{(M)}_{\beta,L}} \frac{\sqrt{\chi_0(2^{-M}|k_0|)} \, e^{-ik(\mathbf{z}-\mathbf{x})}}{\left[k_0^2 + v_0^2|\Omega(\vec{k})|^2\right]^{1/4}} \delta_{\rho,i}, \quad (10.71)$$

$$[\mathbf{B}_{\mathbf{x},\rho}(\mathbf{z})]_i = \frac{1}{\beta|\Lambda|} \sum_{\mathbf{k} \in \mathcal{B}^{(M)}_{\beta,L}} \frac{\sqrt{\chi_0(2^{-M}|k_0|)} \, e^{-ik(\mathbf{z}-\mathbf{x})}}{\left[k_0^2 + v_0^2|\Omega(\vec{k})|^2\right]^{3/4}} \begin{pmatrix} ik_0 & -v_0\Omega^*(\vec{k}) \\ -v_0\Omega(\vec{k}) & ik_0 \end{pmatrix}_{i,\rho},$$

so that

$$||\mathbf{A}_{\mathbf{x},\rho}||^2 = ||\mathbf{B}_{\mathbf{x},\rho}||^2 = \frac{1}{\beta|\Lambda|} \sum_{\mathbf{k} \in \mathcal{B}^{(M)}_{\beta,L}} \frac{\chi_0(2^{-M}|k_0|)}{\left[k_0^2 + v_0^2|\Omega(\vec{k})|^2\right]^{1/2}} \leq CM, \quad (10.72)$$

for a suitable constant C. Using the Gram–Hadamard inequality, we find $|| \det G^T ||_\infty \leq (\text{const.})^N M^{N+1}$; substituting this result into Eq.(10.68), we finally get:

$$|F^{(M;N)}_{\beta,\Lambda}| \leq (\text{const.})^N |U|^N M^{N+1} \beta^{N-1}, \quad (10.73)$$

which is similar to equation (10.63), but for the fact that there is no $N!$ in the r.h.s.! In other words, using the determinant expansion, we recovered the same dimensional estimate as the one obtained by the Feynman diagram expansion and we combinatorially gained a $1/N!$. The r.h.s. of equation (10.73) is now summable over N for $|U|$ sufficiently small, even though non uniformly in M and β. In the next section we will discuss how to systematically improve the dimensional bound by an iterative resummation method.

10.6   The multi-scale integration: the ultraviolet regime

In this section we begin to illustrate the multi-scale integration of the fermionic functional integral of interest. This method will later allow us to perform iterative resummations and to re-express the specific free energy in terms of a modified expansion, whose N-th order term is summable in N and uniformly convergent as $M \to \infty$ and $\beta \to \infty$, as desired.

The first step in the computation of the partition function

$$\Xi_{M,\beta,\Lambda} := \int P_M(d\psi) e^{-\mathcal{V}(\psi)} \tag{10.74}$$

and of its logarithm is the integration of the ultraviolet degrees of freedom corresponding to the large values of k_0. We proceed in the following way. We decompose the free propagator $\hat{g}_{\mathbf{k}}$ into a sum of two propagators supported in the regions of k_0 'large' and 'small', respectively. The regions of k_0 large and small are defined in terms of the smooth support function $\chi_0(t)$ introduced after equation (10.42); note that, by the very definition of χ_0, the supports of $\chi_0\left(\sqrt{k_0^2 + |\vec{k} - \vec{p}_F^+|^2}\right)$ and $\chi_0\left(\sqrt{k_0^2 + |\vec{k} - \vec{p}_F^-|^2}\right)$ are disjoint (here $|\cdot|$ is the euclidean norm over $\mathbb{R}^2/\Lambda^*$). We define

$$f_{u.v.}(\mathbf{k}) = 1 - \chi_0\left(\sqrt{k_0^2 + |\vec{k} - \vec{p}_F^+|^2}\right) - \chi_0\left(\sqrt{k_0^2 + |\vec{k} - \vec{p}_F^-|^2}\right) \tag{10.75}$$

and $f_{i.r.}(\mathbf{k}) = 1 - f_{u.v.}(\mathbf{k})$, so that we can rewrite $\hat{g}_{\mathbf{k}}$ as:

$$\hat{g}_{\mathbf{k}} = f_{u.v.}(\mathbf{k})\hat{g}_{\mathbf{k}} + f_{i.r.}(\mathbf{k})\hat{g}_{\mathbf{k}} \stackrel{def}{=} \hat{g}^{(u.v.)}(\mathbf{k}) + \hat{g}^{(i.r.)}(\mathbf{k}). \tag{10.76}$$

We now introduce two independent set of Grassmann fields $\{\psi_{\mathbf{k},\sigma,\rho}^{(u.v.)\pm}\}$ and $\{\psi_{\mathbf{k},\sigma,\rho}^{(i.r.)\pm}\}$, with $\mathbf{k} \in \mathcal{B}_{\beta,L}^{(M)}$, $\sigma = \uparrow\downarrow$, $\rho = 1, 2$, and the Gaussian integrations $P(d\psi^{(u.v.)})$ and $P(d\psi^{(i.r.)})$ defined by

$$\int P(d\psi^{(u.v.)})\hat{\psi}_{\mathbf{k}_1,\sigma_1}^{(u.v.)-}\hat{\psi}_{\mathbf{k}_2,\sigma_2}^{(u.v.)+} = \beta|\Lambda|\delta_{\sigma_1,\sigma_2}\delta_{\mathbf{k}_1,\mathbf{k}_2}\hat{g}^{(u.v.)}(\mathbf{k}_1),$$

$$\int P(d\psi^{(i.r.)})\hat{\psi}_{\mathbf{k}_1,\sigma_1}^{(u.v.)-}\hat{\psi}_{\mathbf{k}_2,\sigma_2}^{(i.r.)+} = \beta|\Lambda|\delta_{\sigma_1,\sigma_2}\delta_{\mathbf{k}_1,\mathbf{k}_2}\hat{g}^{(i.r.)}(\mathbf{k}_1). \tag{10.77}$$

Similarly to $P_M(d\psi)$, the Gaussian integrations $P(d\psi^{(u.v.)})$, $P(d\psi^{(i.r.)})$ also admit an explicit representation analogous to (10.45), with $\hat{g}_{\mathbf{k}}$ replaced by $\hat{g}^{(u.v.)}(\mathbf{k})$ or $\hat{g}^{(i.r.)}(\mathbf{k})$ and the sum over $\mathbf{k}$ restricted to the values in the support of $f_{u.v.}(\mathbf{k})$ or $f_{i.r.}(\mathbf{k})$, respectively. The definition of Grassmann integration implies the following identity ('addition principle'):

$$\int P(d\psi) e^{-\mathcal{V}(\psi)} = \int P(d\psi^{(i.r.)}) \int P(d\psi^{(u.v.)}) e^{-\mathcal{V}(\psi^{(i.r.)}+\psi^{(u.v.)})} \tag{10.78}$$

so that we can rewrite the partition function as

$$\Xi_{M,\beta,\Lambda} = e^{-\beta|\Lambda|F_{\beta,\Lambda}^{(M)}} = \int P(d\psi^{(i.r.)}) \exp\left\{ \sum_{n\geq 1} \frac{1}{n!} \mathcal{E}_{u.v.}^T(-\mathcal{V}(\psi^{(i.r.)}+\cdot); n)\right\} :=$$

$$:= e^{-\beta|\Lambda|F_{0,M}} \int P(d\psi^{(i.r.)}) e^{-\mathcal{V}_0(\psi^{(i.r.)})}, \tag{10.79}$$

where the *truncated expectation* $\mathcal{E}^T_{u.v.}$ is defined, given any polynomial $V_1(\psi^{(u.v.)})$ with coefficients depending on $\psi^{(i.r.)}$, as

$$\mathcal{E}^T_{u.v.}(V_1(\cdot);n) = \frac{\partial^n}{\partial\lambda^n}\log\int P(d\psi^{(u.v.)})e^{\lambda V_1(\psi^{(u.v.)})}\Big|_{\lambda=0} \tag{10.80}$$

and $\mathcal{V}_0$ is fixed by the condition $\mathcal{V}_0(0) = 0$. We will prove below that $\mathcal{V}_0$ can be written as

$$\mathcal{V}_0(\psi^{(i.r.)}) = \sum_{n=1}^{\infty}(\beta|\Lambda|)^{-2n}\sum_{\sigma_1,\dots,\sigma_n=\uparrow\downarrow}\sum_{\rho_1,\dots,\rho_{2n}=1,2}\sum_{\mathbf{k}_1,\dots,\mathbf{k}_{2n}}\left[\prod_{j=1}^{n}\hat{\psi}^{(i.r.)+}_{\mathbf{k}_{2j-1},\sigma_j,\rho_{2j-1}}\hat{\psi}^{(i.r.)-}_{\mathbf{k}_{2j},\sigma_j,\rho_{2j}}\right].$$

$$\cdot\hat{W}_{M,2n,\underline{\rho}}(\mathbf{k}_1,\dots,\mathbf{k}_{2n-1})\,\delta(\sum_{j=1}^{n}(\mathbf{k}_{2j-1}-\mathbf{k}_{2j})), \tag{10.81}$$

where $\underline{\rho} = (\rho_1,\dots,\rho_{2n})$ and we used the notation

$$\delta(\mathbf{k}) = \delta(\vec{k})\delta(k_0), \qquad \delta(\vec{k}) = |\Lambda|\sum_{n_1,n_2\in\mathbb{Z}}\delta_{\vec{k},n_1\vec{G}_1+n_2\vec{G}_2}, \qquad \delta(k_0) = \beta\delta_{k_0,0}, \tag{10.82}$$

with $\vec{G}_1, \vec{G}_2$ a basis of Λ^*. The possibility of representing $\mathcal{V}_0$ in the form (10.81), with the *kernels* $\hat{W}_{M,2n,\underline{\rho}}$ independent of the spin indices σ_i, follows from a number of remarkable symmetries, discussed in Appendix 10.11, see in particular symmetries (1)–(3) in Lemma 10.5. The regularity properties of the kernels are summarized in the following Lemma, which will be proved below.

Lemma 10.4 The constant $F_{0,M}$ in (10.79) and the kernels $\hat{W}_{M,2n,\underline{\rho}}$ in (10.81) are given by power series in U, convergent in the complex disc $|U| \leq U_0$, for U_0 small enough and independent of M, β, Λ; after Fourier transform, the **x**-space counterparts of the kernels $\hat{W}_{M,2n,\underline{\rho}}$ satisfy the following bounds:

$$\int d\mathbf{x}_1\cdots d\mathbf{x}_{2n}|W_{M,2n,\underline{\rho}}(\mathbf{x}_1,\dots,\mathbf{x}_{2n})| \leq \beta|\Lambda|C^n|U|^{\max\{1,n-1\}} \tag{10.83}$$

for some constant $C > 0$. Moreover, the limits $F_0 = \lim_{M\to\infty}F_{0,M}$ and $W_{2n,\underline{\rho}}(\mathbf{x}_1,\dots,\mathbf{x}_{2n}) = \lim_{M\to\infty}W_{M,2n,\underline{\rho}}(\mathbf{x}_1,\dots,\mathbf{x}_{2n})$ exist and are reached uniformly in M, so that, in particular, the limiting functions are analytic in the same domain $|U| \leq U_0$ and so are their $\beta, |\Lambda| \to \infty$ limits (that, with some abuse of notation, we shall denote by the same symbols).

Remark 10.5 *Once that the ultraviolet degrees of freedom have been integrated out, the remaining infrared problem (i.e., the computation of the Grassmann integral in the second line of equation (10.79)) is essentially independent of M, given the fact that the limit $W_{2n,\underline{\rho}}$ of the kernels $W_{M,2n,\underline{\rho}}$ is reached uniformly and that the limiting kernels are analytic and satisfy the same bounds as equation (10.83). For this reason, in the infrared integration described in the next two sections, M will not play any essential role and, whenever possible, we will simplify the notation by dropping the label M.*

Before we present the proof of Lemma 10.6.1, let us note that the kernels $W_{M,2n,\rho}$ satisfy a number of nontrivial invariance properties. We will be particularly interested in the invariance properties of the quadratic part $\hat{W}_{M,2,(\rho_1,\rho_2)}(\mathbf{k}_1,\mathbf{k}_2)$, which will be used below to show that the structure of the quadratic part of the new effective interaction has the same symmetries as the free integration. The crucial properties that we will need are summarized in the following Lemma, which is proved in Appendix 10.11.

Lemma 10.6 The kernel $\hat{W}_2(\mathbf{k}) := \hat{W}_{M,2}(\mathbf{k},\mathbf{k})$, thought as a 2×2 matrix with components $\hat{W}_{M,2,(i,j)}(\mathbf{k},\mathbf{k})$, satisfies the following symmetry properties:

$$\hat{W}_2(\mathbf{k}) = e^{-i\vec{k}(\vec{\delta}_1-\vec{\delta}_2)n_-} \hat{W}_2\big((k_0, e^{i\frac{2\pi}{3}\sigma_2}\vec{k})\big)e^{i\vec{k}(\vec{\delta}_1-\vec{\delta}_2)n_-} = \hat{W}_2^*(-\mathbf{k}) = \hat{W}_2\big((k_0,k_1,-k_2)\big)$$

$$= \sigma_1 \hat{W}_2\big((k_0,-k_1,k_2)\big)\sigma_1 = \hat{W}_2^T\big((k_0,-\vec{k})\big) = -\sigma_3\hat{W}_2\big((-k_0,\vec{k})\big)\sigma_3, \qquad (10.84)$$

where $\sigma_1,\sigma_2,\sigma_3$ are the standard Pauli matrices and $n_- = (1-\sigma_3)/2$. In particular, if $\hat{W}(\mathbf{k}) = \lim_{\beta,|\Lambda|\to\infty} \hat{W}_2(\mathbf{k})$, in the vicinity of the Fermi point $\mathbf{p}_F^\omega = (0,\vec{p}_F^\omega)$, with $\omega = \pm$,

$$\hat{W}(\mathbf{k}) = -\begin{pmatrix} iz_0 k_0 & \delta_0\Omega^*(\vec{k}) \\ \delta_0\Omega(\vec{k}) & iz_0 k_0 \end{pmatrix} + O(|\mathbf{k}-\mathbf{p}_F^\omega|^2), \qquad (10.85)$$

for some real constants z_0, δ_0.

Note that equation (10.85) can be read by saying that, in the zero temperature and thermodynamic limits, the two-legged kernel has the same structure as the inverse of the free covariance, $\hat{S}_0^{-1}(\mathbf{k})$, modulo higher order terms in $\mathbf{k}-\mathbf{p}_F^\omega$. This fact will be used in the next section to define a dressed infrared propagator $[\hat{S}_0^{-1}(\mathbf{k}) + \hat{W}(\mathbf{k})]^{-1}$, with the same infrared singularity structure as the free one. We will come back to this point in more detail. For the moment, let us turn to the proof of Lemma 10.6, which illustrates the main RG strategy that will be also used below, in the more difficult infrared integration.

Proof of Lemma 14.4 Let us rewrite the Fourier transform of $\hat{g}^{(u.v.)}(\mathbf{k})$ as

$$g^{(u.v.)}(\mathbf{x}) = \sum_{h=1}^M g^{(h)}(\mathbf{x}), \qquad (10.86)$$

where

$$g^{(h)}(\mathbf{x}) = \frac{1}{\beta|\Lambda|}\sum_{k\in\mathcal{B}_{\beta,L}^{(M)}} e^{-i\mathbf{k}\mathbf{x}}\frac{f_{u.v.}(\mathbf{k})H_h(k_0)}{k_0^2 + v_0^2|\Omega(\vec{k})|^2}\begin{pmatrix} ik_0 & -v_0\Omega^*(\vec{k}) \\ -v_0\Omega(\vec{k}) & ik_0 \end{pmatrix}, \qquad (10.87)$$

with $H_1(k_0) = \chi_0(2^{-1}|k_0|)$ and, if $h \geq 2$, $H_h(k_0) = \chi_0(2^{-h}|k_0|) - \chi_0(2^{-h+1}|k_0|)$. Note that $[g^{(h)}(\mathbf{0})]_{\rho\rho} = 0$, $\rho = 1, 2$, and, for $0 \leq K \leq 6$, $g^{(h)}(\mathbf{x})$ satisfies the bound

$$||g^{(h)}(\mathbf{x})|| \leq \frac{C_K}{1 + (2^h |x_0|_\beta + |\vec{x}|_\Lambda)^K}, \tag{10.88}$$

where $|x_0|_\beta = \min_{n \in \mathbb{Z}} |x_0 + n\beta|$ is the distance over the one-dimensional torus of length β and $|\vec{x}|_\Lambda = \min_{\vec{\ell} \in \mathbb{B}} |\vec{x} + L\vec{\ell}|$ is the distance over the periodic lattice Λ (here $\mathbb{B}$ is the triangular lattice defined after Eq.(10.1)); see Appendix 10.9 for a proof. Moreover, $g^{(h)}(\mathbf{x})$ admits a Gram representation that, in notation analogous to equation (10.70), reads

$$g^{(h)}_{\rho,\rho'}(\mathbf{x} - \mathbf{y}) = \left\langle \mathbf{A}^{(h)}_{\mathbf{x},\rho}, \mathbf{B}^{(h)}_{\mathbf{y},\rho'} \right\rangle, \tag{10.89}$$

with

$$[\mathbf{A}^{(h)}_{\mathbf{x},\rho}(\mathbf{z})]_i = \frac{1}{\beta|\Lambda|} \sum_{\mathbf{k} \in \mathcal{B}^{(M)}_{\beta,L}} \sqrt{f_{u.v.}(\mathbf{k}) H_h(k_0)} \frac{e^{-i k(\mathbf{z}-\mathbf{x})}}{[k_0^2 + v_0^2|\Omega(\vec{k})|^2]^{1/4}} \delta_{\rho,i}, \tag{10.90}$$

$$[\mathbf{B}^{(h)}_{\mathbf{x},\rho}(\mathbf{z})]_i = \frac{1}{\beta|\Lambda|} \sum_{\mathbf{k} \in \mathcal{B}^{(M)}_{\beta,L}} \frac{\sqrt{f_{u.v.}(\mathbf{k}) H_h(k_0)}\, e^{-i k(\mathbf{z}-\mathbf{x})}}{[k_0^2 + v_0^2|\Omega(\vec{k})|^2]^{3/4}} \begin{pmatrix} ik_0 & -v_0\Omega^*(\vec{k}) \\ -v_0\Omega(\vec{k}) & ik_0 \end{pmatrix}_{i,\rho},$$

and

$$||\mathbf{A}^{(h)}_{\mathbf{x},\rho}||^2 = ||\mathbf{B}^{(h)}_{\mathbf{x},\rho}||^2 = \frac{1}{\beta|\Lambda|} \sum_{\mathbf{k} \in \mathcal{B}^{(M)}_{\beta,L}} \frac{f_{u.v.}(\mathbf{k}) H_h(k_0)}{[k_0^2 + v_0^2|\Omega(\vec{k})|^2]^{1/2}} \leq C, \tag{10.91}$$

for a suitable constant C. Our goal is to compute

$$e^{-\beta|\Lambda| F_{0,M} - \mathcal{V}_0(\psi^{(i.r.)})} = \int P(d\psi^{[1,M]}) e^{-\mathcal{V}(\psi^{(i.r.)} + \psi^{[1,M]})}$$

$$= \exp\left\{ \log \int P(d\psi^{[1,M]}) e^{-\mathcal{V}(\psi^{(i.r.)} + \psi^{[1,M]})} \right\}, \tag{10.92}$$

where $P(d\psi^{[1,M]})$ is the fermionic 'Gaussian integration' associated with the propagator $\sum_{h=1}^{M} \hat{g}^{(h)}(\mathbf{k})$ (i.e., it is the same as $P(d\psi^{(u.v.)})$); moreover, we want to prove that $F_{0,M}$ and $\mathcal{V}_0(\psi^{(i.r.)})$ are uniformly convergent as $M \to \infty$. We perform the integration of (10.92) in an iterative fashion: in fact, we will inductively prove that for $0 \leq h \leq M$,

$$e^{-\beta|\Lambda| F_{0,M} - \mathcal{V}_0(\psi^{(i.r.)})} = e^{-\beta|\Lambda| F_h} \int P(d\psi^{[1,h]}) e^{-\mathcal{V}^{(h)}(\psi^{(i.r.)} + \psi^{[1,h]})} \tag{10.93}$$

where $P(d\psi^{[1,h]})$ is the fermionic 'Gaussian integration' associated with the propagator $\sum_{k=1}^{h} \hat{g}^{(k)}(\mathbf{k})$ and $\mathcal{V}^{(M)} = \mathcal{V}$; for $0 \le h \le M$,

$$\mathcal{V}^{(h)}(\psi) = \tag{10.94}$$

$$= \sum_{n=1}^{\infty} \sum_{\rho,\underline{\sigma}} \int d\mathbf{x}_1 \cdots d\mathbf{x}_{2n} \left[\prod_{j=1}^{n} \psi^+_{\mathbf{x}_{2j-1},\sigma_j,\rho_{2j-1}} \psi^-_{\mathbf{x}_{2j},\sigma_j,\rho_{2j}} \right] W^{(h)}_{M,2n,\underline{\rho}}(\mathbf{x}_1,\dots,\mathbf{x}_{2n}),$$

where $\mathcal{V}^{(0)}(\psi)$ should be identied with $\mathcal{V}_0(\psi)$. In order to inductively prove (10.93)–(10.94) we use the addition principle to rewrite

$$\int P(d\psi^{[1,h]}) e^{-\mathcal{V}^{(h)}(\psi^{(i.r.)}+\psi^{[1,h]})} = \int P(d\psi^{[1,h-1]}) \int P(d\psi^{(h)}) e^{-\mathcal{V}^{(h)}(\psi^{(i.r.)}+\psi^{[1,h-1]}+\psi^{(h)})}$$

$$\tag{10.95}$$

where $P(d\psi^{(h)})$ is the fermionic Gaussian integration with propagator $\hat{g}^{(h)}(\mathbf{k})$. After the integration of $\psi^{(h)}$ we define

$$e^{-\mathcal{V}^{(h-1)}(\psi^{(i.r.)}+\psi^{[1,h-1]})-\beta|\Lambda|\bar{e}_h} = \int P(d\psi^{(h)}) e^{-\mathcal{V}^{(h)}(\psi^{(i.r.)}+\psi^{[1,h-1]}+\psi^{(h)})}, \tag{10.96}$$

which proves (10.93) with $F_M = 0$ and, for $0 \le h < M$,

$$F_M = 0, \quad F_h = \sum_{k=h+1}^{M} \bar{e}_k, \tag{10.97}$$

where F_0 should be identified with $F_{0,M}$. Let $\mathcal{E}^T_h$ be the truncated expectation associated to $P(d\psi^{(h)})$: then we have

$$\beta|\Lambda|\bar{e}_h + \mathcal{V}^{(h-1)}(\psi) = \sum_{n\ge 1} \frac{1}{n!}(-1)^{n+1} \mathcal{E}^T_h (\mathcal{V}^{(h)}(\psi+\psi^{(h)});n). \tag{10.98}$$

Equation (10.98) can be graphically represented as in Fig. 10.4. The tree on the l.h.s., consisting of a single horizontal branch, connecting the left node (called the *root* and associated to the *scale label* $h - 1$) with a big black dot on scale h, represents $\mathcal{V}^{(h-1)}(\psi)$. In the r.h.s., the term with n final points represents the corresponding term in the r.h.s. of Eq.(10.98): a scale label $h - 1$ is attached to the leftmost node (the root); a scale label h is attached to the central node (corresponding to the action of $\mathcal{E}^T_h$); a scale

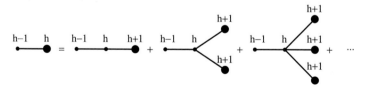

Fig. 10.4 The graphical representation of $\mathcal{V}^{(h-1)}$.

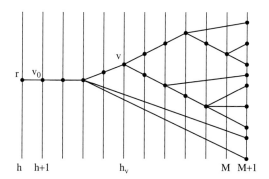

Fig. 10.5 A tree $\tau \in \widetilde{\mathcal{T}}_{M;h,n}$ with $n = 9$: the root is on scale h and the endpoints are on scale $M + 1$.

label $h + 1$ is attached to the n rightmost nodes with the big black dots (representing $\mathcal{V}^{(h)}$). Iterating the graphical equation in Fig. 10.4 up to scale M, and representing the endpoints on scale $M + 1$ as simple dots (rather than big black dots), we end up with a graphical representation of $\mathcal{V}^{(h)}$ in terms of *Gallavotti–Nicolò* trees, Gallavotti and Nicolò (1985) see Fig. 10.5, defined in terms of the following features.

1. Let us consider the family of all trees which can be constructed by joining a point r, the *root*, with an ordered set of $n \geq 1$ points, the *endpoints* of the *unlabelled tree*, so that r is not a branching point. n will be called the *order* of the unlabelled tree and the branching points will be called the *non trivial vertices*. The unlabelled trees are partially ordered from the root to the endpoints in the natural way; we shall use the symbol $<$ to denote the partial order. Two unlabeled trees are identified if they can be superposed by a suitable continuous deformation. It is then easy to see that the number of unlabelled trees with n end-points is bounded by 4^n (see, e.g., (Gentile and Mastropietro 2001, Appendix A.1.2) for a proof of this fact). We shall also consider the *labelled trees* (to be called simply trees in the following); they are defined by associating some labels with the unlabelled trees, as explained in the following items.

2. We associate a label $0 \leq h \leq M - 1$ with the root and we denote by $\widetilde{\mathcal{T}}_{M;h,n}$ the corresponding set of labeled trees with n endpoints. Moreover, we introduce a family of vertical lines, labeled by an integer taking values in $[h, M + 1]$, and we represent any tree $\tau \in \widetilde{\mathcal{T}}_{M;h,n}$ so that, if v is an endpoint, it is contained in the vertical line with index $h_v = M + 1$, while if it is a nontrivial vertex, it is contained in a vertical line with index $h < h_v \leq M$, to be called the *scale* of v; the root r is on the line with index h. In general, the tree will intersect the vertical lines also at points different from the root, the endpoints and the branching points; these points will be called *trivial vertices*. The set of the *vertices* will be the union of the endpoints, of the trivial vertices and of the non trivial vertices; note that the root is not a vertex. Every vertex v of a tree will be associated to its scale label h_v, defined, as above, as the label of the vertical line whom v belongs to. Note that, if v_1 and v_2 are two vertices and $v_1 < v_2$, then $h_{v_1} < h_{v_2}$.

3. There is only one vertex immediately following the root, called v_0 and with scale label equal to $h + 1$.

4. Given a vertex v of $\tau \in \widetilde{\mathcal{T}}_{M;h,n}$ that is not an endpoint, we can consider the subtrees of τ with root v, which correspond to the connected components of the restriction of τ to the vertices $w \geq v$. If a subtree with root v contains only v and one endpoint on scale $h_v + 1$, it will be called a *trivial subtree* (note that by construction a trivial subtree necessarily has root on scale M).

5. With each endpoint v we associate a factor $\mathcal{V}(\psi^{(i.r.)} + \psi^{[1,M]})$. We introduce a *field label f* to distinguish the field variables appearing in the factors $\mathcal{V}(\psi^{(i.r.)} + \psi^{[1,M]})$ associated with the endpoints; the set of field labels associated with the endpoint v will be called I_v; note that if v is an endpoint $|I_v| = 4$. Analogously, if v is not an endpoint, we shall call I_v the set of field labels associated with the endpoints following the vertex v; $\mathbf{x}(f)$, $\varepsilon(f)$, $\sigma(f)$ and $\rho(f)$ will denote the space–time point, the ε index, the σ index and the ρ index, respectively, of the Grassmann field variable with label f.

In terms of trees, the effective potential $\mathcal{V}^{(h)}$, $0 \leq h \leq M$ can be written as

$$\mathcal{V}^{(h)}\left(\psi^{(i.r.)} + \psi^{[1,h]}\right) + \beta|\Lambda|\bar{e}_{h+1} = \sum_{n=1}^{\infty} \sum_{\tau \in \widetilde{\mathcal{T}}_{M;h,n}} \widetilde{\mathcal{V}}^{(h)}\left(\tau, \psi^{(i.r.)} + \psi^{[1,h]}\right), \qquad (10.99)$$

where, if v_0 is the first vertex of τ and $\tau_1, \ldots, \tau_s$ $(s = s_{v_0})$ are the subtrees of τ with root v_0, $\widetilde{\mathcal{V}}^{(h)}\left(\tau, \psi^{(i.r.)} + \psi^{[1,h]}\right)$ is defined inductively as:

$$\widetilde{\mathcal{V}}^{(h)}\left(\tau, \psi^{[0,h]}\right) = \frac{(-1)^{s+1}}{s!}\mathcal{E}_{h+1}^{T}\left[\widetilde{\mathcal{V}}^{(h+1)}\left(\tau_1, \psi^{[0,h+1]}\right); \ldots; \widetilde{\mathcal{V}}^{(h+1)}\left(\tau_s, \psi^{[0,h+1]}\right)\right].$$
$$(10.100)$$

where $\psi^{[0,h]} := \psi^{(i.r.)} + \psi^{[1,h]}$ and, if τ is a trivial subtree with root on scale M, then $\widetilde{\mathcal{V}}^{(M)}\left(\tau, \psi^{[0,M]}\right) = \mathcal{V}(\psi^{[0,M]})$.

For what follows, it is important to specify the action of the truncated expectations on the branches connecting any endpoint v to the closest *non–trivial* vertex v' preceding it. In fact, if τ has only one end–point, it is convenient to rewrite $\widetilde{\mathcal{V}}^{(h)}\left(\tau, \psi^{[0,h]}\right) = \mathcal{E}_{h+1}^{T}\mathcal{E}_{h+2}^{T} \cdots \mathcal{E}_{M}^{T}(\mathcal{V}(\psi^{[0,M]}))$ as:

$$\widetilde{\mathcal{V}}^{(h)}\left(\tau, \psi^{[0,h]}\right) = \mathcal{V}(\psi^{[0,h]}) + \mathcal{E}_{h+1}^{T} \cdots \mathcal{E}_{M}^{T}\left(\mathcal{V}(\psi^{[0,M]}) - \mathcal{V}(\psi^{[0,h]})\right). \qquad (10.101)$$

Now, the key observation is that, since $\mathcal{V}(\psi)$ is defined as in equation (10.48) and since our explicit choice of the ultraviolet cutoff makes the tadpoles equal to zero (i.e., $[g^{(h)}(\mathbf{0})]_{\rho\rho} = 0$), then the second term in the r.h.s. of equation (10.101) is identically zero:

$$\mathcal{E}_{h+1}^{T} \cdots \mathcal{E}_{M}^{T}\left(\mathcal{V}(\psi^{[0,M]}) - \mathcal{V}(\psi^{[0,h]})\right) = 0, \qquad (10.102)$$

for all $0 \leq h < M$. Therefore, it is natural to shrink all the branches of $\tau \in \widetilde{\mathcal{T}}_{M;h,n}$ consisting of a subtree $\tau' \subseteq \tau$, having root r' on scale $h' \in [h, M]$ and only one endpoint on scale $M + 1$, into a trivial subtree, rooted in r' and associated to the

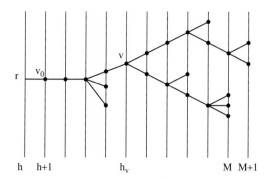

Fig. 10.6 A tree $\tau \in \mathcal{T}_{M;h,n}$ with $n = 9$: the root is on scale h and the endpoints are on scales $\leq M + 1$.

factor $\mathcal{V}(\psi^{[0,h']})$. By doing so, we end up with an alternative representation of the effective potentials, which is based on a slightly modified tree expansion. The set of modified trees with n endpoints contributing to $\mathcal{V}^{(h)}$ will be denoted by $\mathcal{T}_{M;h,n}$; every $\tau \in \mathcal{T}_{M;h,n}$ is characterized in the same way as the elements of $\widetilde{\mathcal{T}}_{M;h,n}$, but for two features: (i) the endpoints of $\tau \in \mathcal{T}_{M;h,n}$ are not necessarily on scale $M + 1$; (ii) every endpoint v of τ is attached to a non-trivial vertex on scale $h_v - 1$ and is associated to the factor $\mathcal{V}(\psi^{[0,h_v-1]})$. See Fig. 10.6.

In terms of these modified trees, we have:

$$\mathcal{V}^{(h)}(\psi^{[0,h]}) + \beta|\Lambda|\bar{e}_{h+1} = \sum_{n=1}^{\infty} \sum_{\tau \in \mathcal{T}_{M;h,n}} \mathcal{V}^{(h)}(\tau, \psi^{[0,h]}), \tag{10.103}$$

where

$$\mathcal{V}^{(h)}(\tau, \psi^{[0,h]}) = \frac{(-1)^{s+1}}{s!} \mathcal{E}_{h+1}^{T} \big[\mathcal{V}^{(h+1)}(\tau_1, \psi^{[0,h+1]}); \dots; \mathcal{V}^{(h+1)}(\tau_s, \psi^{[0,h+1]}) \big] \tag{10.104}$$

and, if τ is a trivial subtree with root on scale $k \in [h, M]$, then $\mathcal{V}^{(k)}(\tau, \psi^{[0,k]}) = \mathcal{V}(\psi^{[0,k]})$.

Using its inductive definition equation (10.104), the right-hand side of equation (10.103) can be further expanded (it is a sum of several contributions, differing for the choices of the field labels contracted under the action of the truncated expectations $\mathcal{E}_{h_v}^{T}$ associated with the vertices v that are not endpoints), and in order to describe the resulting expansion we need some more definitions (allowing us to distinguish the fields that are contracted or not 'inside the vertex v').

We associate with any vertex v of the tree a subset P_v of I_v, the *external fields* of v. These subsets must satisfy various constraints. First of all, if v is not an endpoint and $v_1, \dots, v_{s_v}$ are the $s_v \geq 1$ vertices immediately following it, then $P_v \subseteq \cup_i P_{v_i}$; if v is an endpoint, $P_v = I_v$. If v is not an endpoint, we shall denote by Q_{v_i} the intersection of P_v and P_{v_i}; this definition implies that $P_v = \cup_i Q_{v_i}$. The union $\mathcal{I}_v$ of the subsets

$P_{v_i} \setminus Q_{v_i}$ is, by definition, the set of the *internal fields* of v, and is nonempty if $s_v > 1$. Given $\tau \in \mathcal{T}_{M;h,n}$, there are many possible choices of the subsets P_v, $v \in \tau$, compatible with all the constraints. We shall denote by $\mathcal{P}_\tau$ the family of all these choices and by **P** the elements of $\mathcal{P}_\tau$. With these definitions, we can rewrite $\mathcal{V}^{(h)}(\tau, \psi^{[0,h]})$ as:

$$\mathcal{V}^{(h)}(\tau, \psi^{[0,h]}) = \sum_{\mathbf{P} \in \mathcal{P}_\tau} \mathcal{V}^{(h)}(\tau, \mathbf{P}),$$

$$\mathcal{V}^{(h)}(\tau, \mathbf{P}) = \sum_{\rho_{v_0}, \sigma_{v_0}, \varepsilon_{v_0}} \int d\mathbf{x}_{v_0} \, \psi_{P_{v_0}}^{[0,h]} K_{\tau,\mathbf{P}}^{(h+1)}(\mathbf{x}_{v_0}), \qquad (10.105)$$

where $\mathbf{x}_v = \cup_{f \in I_v} \{\mathbf{x}(f)\}$, $\rho_v = \cup_{f \in I_v} \{\rho(f)\}$, $\sigma_v = \cup_{f \in I_v} \{\sigma(f)\}$, $\varepsilon_v = \cup_{f \in I_v} \{\varepsilon(f)\}$,

$$\psi_{P_v}^{[0,h]} = \prod_{f \in P_v} \psi_{\mathbf{x}(f),\sigma(f),\rho(f)}^{[0,h]\,\varepsilon(f)} \qquad (10.106)$$

and $K_{\tau,\mathbf{P}}^{(h+1)}(\mathbf{x}_{v_0})$ is defined inductively by the equation, valid for any $v \in \tau$ that is not an endpoint,

$$K_{\tau,\mathbf{P}}^{(h_v)}(\mathbf{x}_v) = \frac{(-1)^{s_v+1}}{s_v!} \prod_{i=1}^{s_v} [K_{v_i}^{(h_v+1)}(\mathbf{x}_{v_i})] \, \mathcal{E}_{h_v}^T [\psi_{P_{v_1} \setminus Q_{v_1}}^{(h_v)}, \dots, \psi_{P_{v_{s_v}} \setminus Q_{v_{s_v}}}^{(h_v)}], \qquad (10.107)$$

where $\psi_{P_{v_i} \setminus Q_{v_i}}^{(h_v)}$ has a definition similar to Eq.(10.106). Moreover, if v_i is an endpoint $K_{v_i}^{(h_v+1)}(\mathbf{x}_{v_i}) = U$; if v_i is not an endpoint, $K_{v_i}^{(h_v+1)} = K_{\tau_i,\mathbf{P}_i}^{(h_v+1)}$, where $\mathbf{P}_i = \{P_w, w \in \tau_i\}$. Using in the r.h.s. of equation (10.107) the determinant representation of the truncated expectation discussed in the previous section, we finally get:

$$\mathcal{V}^{(h)}(\tau, \mathbf{P}) = \sum_{T \in \mathbf{T}} \sum_{\rho_{v_0}, \sigma_{v_0}, \varepsilon_{v_0}} \int d\mathbf{x}_{v_0} \, \psi_{P_{v_0}}^{[0,h]} W_{\tau,\mathbf{P},T}^{(h)}(\mathbf{x}_{v_0}) := \sum_{T \in \mathbf{T}} \mathcal{V}^{(h)}(\tau, \mathbf{P}, T), \qquad (10.108)$$

where, for a suitable sign $\alpha_T = \pm$,

$$W_{\tau,\mathbf{P},T}^{(h)}(\mathbf{x}_{v_0}) = \alpha_T U^n \cdot \qquad (10.109)$$

$$\cdot \left\{ \prod_{\substack{v \text{ not} \\ \text{e.p.}}} \frac{(-1)^{s_v+1}}{s_v!} \int dP_{T_v}(\mathbf{t}_v) \det G^{h_v,T_v}(\mathbf{t}_v) \left[\prod_{\ell \in T_v} \delta_{\sigma(\ell),\sigma'(\ell)} \, g_{\rho(\ell),\rho'(\ell)}^{(h_v)}(\mathbf{x}(\ell) - \mathbf{x}'(\ell)) \right] \right\}$$

and $G^{h_v,T_v}(\mathbf{t}_v)$ is a matrix analogous to the one defined in previous section, with g replaced by $g^{(h)}$. Note that $W_{\tau,\mathbf{P},T}$ and, therefore, $\mathcal{V}^{(h)}(\tau, \mathbf{P})$ do not depend on M: therefore, the effective potential $\mathcal{V}^{(h)}(\psi)$ depends on M only through the choice of the scale labels (i.e., the dependence on M is all encoded in $\mathcal{T}_{M;h,n}$). Using equations (10.108)–(10.109), we finally get the bound:

$$\frac{1}{\beta|\Lambda|} \int dx_1 \cdots dx_{2l} |W_{M,2l,\rho}^{(h)}(\mathbf{x}_1, \ldots, \mathbf{x}_{2l})| \le \sum_{n \ge \max\{1, l-1\}} |U|^n \sum_{\tau \in \mathcal{T}_{M;h,n}} \sum_{\substack{\mathbf{P} \in \mathcal{P}_\tau \\ |P_{v_0}| = 2l}} \sum_{T \in \mathbf{T}} \cdot$$

$$\cdot \int \prod_{\ell \in T} d(\mathbf{x}(\ell) - \mathbf{x}'(\ell)) \left[\prod_{\substack{v \text{ not} \\ \text{e.p.}}} \frac{1}{s_v!} \max_{\mathbf{t}_v} |\det G^{h_v, T_v}(\mathbf{t}_v)| \prod_{\ell \in T_v} ||g^{(h_v)}(\mathbf{x}(\ell) - \mathbf{x}'(\ell))|| \right].$$

(10.110)

Now, an application of the Gram–Hadamard inequality (10.69), combined with the representation (10.89) and the dimensional bounds (10.91), implies that

$$|\det G^{h_v, T_v}(\mathbf{t}_v)| \le (\text{const.})^{\sum_{i=1}^{s_v} |P_{v_i}| - |P_v| - 2(s_v - 1)}.$$

(10.111)

By the decay properties of $g^{(h)}(\mathbf{x})$ given by equation (10.88), it also follows that

$$\prod_{\substack{v \text{ not} \\ \text{e.p.}}} \frac{1}{s_v!} \int \prod_{\ell \in T_v} d(\mathbf{x}(\ell) - \mathbf{x}'(\ell)) \, ||g^{(h_v)}(\mathbf{x}(\ell) - \mathbf{x}'(\ell))|| \le (\text{const.})^n \prod_{\substack{v \text{ not} \\ \text{e.p.}}} \frac{1}{s_v!} 2^{-h_v(s_v - 1)}.$$

(10.112)

Plugging equations (10.111)–(10.112) into equation (10.110), we find that the l.h.s. of equation (10.110) can be bounded from above by

$$\sum_{n \ge \max\{1, l-1\}} \sum_{\tau \in \mathcal{T}_{M;h,n}} \sum_{\substack{\mathbf{P} \in \mathcal{P}_\tau \\ |P_{v_0}| = 2l}} \sum_{T \in \mathbf{T}} (\text{const.})^n |U|^n \left[\prod_{\substack{v \text{ not} \\ \text{e.p.}}} \frac{1}{s_v!} 2^{-h_v(s_v - 1)} \right].$$

(10.113)

Using the following relation, which can be easily proved by induction,

$$\sum_{\substack{v \text{ not} \\ \text{e.p.}}} h_v(s_v - 1) = h(n - 1) + \sum_{\substack{v \text{ not} \\ \text{e.p.}}} (h_v - h_{v'})(n(v) - 1),$$

(10.114)

where v' is the vertex immediately preceding v on τ (or the root if v' does not exist) and $n(v)$ the number of endpoints following v on τ, we find that equation (10.113) can be rewritten as

$$\sum_{n \ge \max\{1, l-1\}} \sum_{\tau \in \mathcal{T}_{M;h,n}} \sum_{\substack{\mathbf{P} \in \mathcal{P}_\tau \\ |P_{v_0}| = 2l}} \sum_{T \in \mathbf{T}} (\text{const.})^n |U|^n 2^{-h(n-1)} \left[\prod_{\substack{v \text{ not} \\ \text{e.p.}}} \frac{1}{s_v!} 2^{-(h_v - h_{v'})(n(v) - 1)} \right],$$

(10.115)

where, by construction, we have that $n(v) > 1$ for any vertex v of $\tau \in \mathcal{T}_{M;h,n}$ that is not an endpoint (simply because every endpoint v of τ is attached to a non-trivial vertex on scale $h_v - 1$, see the discussion after equation (10.102)). Now, the number of terms in $\sum_{T \in \mathbf{T}}$ can be bounded by $(\text{const.})^n \prod_{v \text{ not e.p.}} s_v!$ (see (Gentile and Mastropietro

2001, Appendix A3.3); moreover, $|P_v| \le 4n(v)$ and $n(v) - 1 \ge \max\{1, \frac{n(v)}{2}\}$, so that $n(v) - 1 \ge \frac{1}{2} + \frac{|P_v|}{16}$. Therefore,

$$\frac{1}{\beta|\Lambda|} \int d\mathbf{x}_1 \cdots d\mathbf{x}_{2l} |W_{M,2l,\rho}^{(h)}(\mathbf{x}_1, \ldots, \mathbf{x}_{2l})| \le \sum_{n \ge \max\{1, l-1\}} (\text{const.})^n |U|^n 2^{-h(n-1)} .$$

$$\cdot \sum_{\tau \in \mathcal{T}_{M;h,n}} \left(\prod_{v \text{ not e.p.}} 2^{-\frac{1}{2}(h_v - h_{v'})} \right) \sum_{\substack{P \in \mathcal{P}_\tau \\ |P_{v_0}| = 2l}} \left(\prod_{\substack{v \text{ not} \\ \text{e.p.}}} 2^{-|P_v|/16} \right). \tag{10.116}$$

Now, the sum over $\mathbf{P}$ can be bounded as follows: defining $p_v := |P_v|$ (so that $p_v \le p_{v_1} + \cdots + p_{v_{s_v}}$), then

$$\sum_{\substack{P \in \mathcal{P}_\tau \\ |P_{v_0}| = 2l}} \left(\prod_{\substack{v \text{ not} \\ \text{e.p.}}} 2^{-|P_v|/16} \right) \le \prod_{\substack{v \text{ not} \\ \text{e.p.}}} \sum_{p_v} 2^{-p_v/16} \binom{p_{v_1} + \cdots + p_{v_{s_v}}}{p_v}. \tag{10.117}$$

The sum in the r.h.s. can be computed inductively, starting from the root and moving towards the endpoints; in fact, at the first step

$$\sum_{p_{v_0}} 2^{-p_{v_0}/16} \binom{p_{v_1} + \cdots + p_{v_{s_{v_0}}}}{p_{v_0}} = \prod_{j=1}^{s_{v_0}} (1 + 2^{-1/16})^{p_{v_j}} ; \tag{10.118}$$

at the second step,

$$\prod_{j=1}^{s_{v_0}} (1 + 2^{-1/16})^{p_{v_j}} \sum_{p_{v_j}} 2^{-p_{v_j}/16} \binom{p_{v_{j,1}} + \cdots + p_{v_{j,s_{v_j}}}}{p_{v_j}} = \prod_{j=1}^{s_{v_0}} \prod_{j'=1}^{s_{v_j}} (1 + 2^{-\frac{1}{16}} + 2^{-\frac{2}{16}})^{p_{v_{j,j'}}}$$

$$\tag{10.119}$$

so that, iterating,

$$\prod_{\substack{v \text{ not} \\ \text{e.p.}}} \sum_{p_v} 2^{-p_v/16} \binom{p_{v_1} + \cdots + p_{v_{s_v}}}{p_v} = \prod_{v \text{ e.p.}} \left[\sum_{k=0}^{L_v} 2^{-\frac{k}{16}} \right]^4 \tag{10.120}$$

where L_v is the distance from v to the root and we used the fact that $p_v = 4$ for all endpoints v. Plugging equation (10.120) into equation (10.117), we get

$$\sum_{\substack{P \in \mathcal{P}_\tau \\ |P_{v_0}| = 2l}} \left(\prod_{\substack{v \text{ not} \\ \text{e.p.}}} 2^{-|P_v|/16} \right) \le \left(\frac{1}{2^{1/16} - 1} \right)^n . \tag{10.121}$$

Similarly, one can prove that

$$\sum_{\tau \in \mathcal{T}_{M;h,n}} \left(\prod_{\substack{v \text{ not} \\ \text{e.p.}}} 2^{-\frac{1}{2}(h_v - h_{v'})} \right) \le \left(\frac{4}{2^{1/2} - 1} \right)^n , \tag{10.122}$$

uniformly in M as $M \to \infty$, see (Gentile and Mastropietro 2001, Lemma A3). Collecting all the previous bounds, we obtain

$$\frac{1}{\beta|\Lambda|} \int d\mathbf{x}_1 \cdots d\mathbf{x}_{2l} |W^{(h)}_{M,2l,\underline{\rho}}(\mathbf{x}_1, \ldots, \mathbf{x}_{2l})| \leq \sum_{n \geq \max\{1, l-1\}} (\text{const.})^n |U|^n 2^{-h(n-1)},$$
(10.123)

which implies Eq.(10.83). The constant $\bar{e}_h$ can be bounded by the r.h.s. of equation (10.110) with $l = 0$ and $n \geq 2$ (because the contributions to $\bar{e}_h$ with $n = 1$ are zero, by the condition that the tadpoles vanish), which implies

$$\bar{e}_h \leq \sum_{n \geq 2} (\text{const.})^n |U|^n 2^{-h(n-1)} \leq (\text{const.}) |U|^2 2^{-h}.$$
(10.124)

Therefore, $F_{0,M} = \sum_{k=1}^{M} \bar{e}_k$ is given by an absolutely convergent power series in U, as desired. A critical analysis of the proof shows that all the bounds are uniform in M, β, Λ and all the expressions involved admit well-defined limits as $M, \beta, |\Lambda| \to \infty$. In particular, this implies that $F_0 = \lim_{M \to \infty} F_{0,M}$ is analytic in U (and so is its $\beta, |\Lambda| \to \infty$ limit, which is reached uniformly) in the complex domain $|U| \leq U_0$, for U_0 small enough. See (Giuliani and Mastropietro 2010, Appendices C and D) for details on these technical aspects. This concludes the proof of Lemma 10.7. $\qquad\square$

10.7 The multi-scale integration: the infrared regime

We are now left with computing

$$\Xi_{M,\beta,\Lambda} = e^{-\beta|\Lambda|F_0} \int P(d\psi^{(i.r.)}) e^{-\mathcal{V}_0(\psi^{(i.r.)})}.$$
(10.125)

We proceed in an iterative fashion, similar to the one described in the previous section for the integration of the large values of k_0. As a starting point, it is convenient to decompose the infrared propagator as:

$$g^{(i.r.)}(\mathbf{x} - \mathbf{y}) = \sum_{\omega = \pm} e^{-i\vec{p}_F^{\omega}(\vec{x} - \vec{y})} g_{\omega}^{(\leq 0)}(\mathbf{x} - \mathbf{y}),$$
(10.126)

where, if $\mathbf{k}' = (k_0, \vec{k}')$,

$$g_{\omega}^{(\leq 0)}(\mathbf{x} - \mathbf{y}) = \frac{1}{\beta|\Lambda|} \sum_{\mathbf{k}' \in \mathcal{B}_{\beta,L}^{\omega}} \chi_0(|\mathbf{k}'|) e^{-i\mathbf{k}'(\mathbf{x}-\mathbf{y})} \begin{pmatrix} -ik_0 & -v_0\Omega^*(\vec{k}' + \vec{p}_F^{\omega}) \\ -v_0\Omega(\vec{k}' + \vec{p}_F^{\omega}) & -ik_0 \end{pmatrix}^{-1}$$
(10.127)

and $\mathcal{B}_{\beta,L}^{\omega} = \mathcal{B}_{\beta}^{(M)} \times \mathcal{B}_L^{\omega}$, with $\mathcal{B}_L^{\omega} = \{\frac{n_1}{L}\vec{G}_1 + \frac{n_2}{L}\vec{G}_2 - \vec{p}_F^{\omega}, \ 0 \leq n_1, n_2 \leq L - 1\}$. Correspondingly, we rewrite $\psi^{(i.r.)}$ as a sum of two independent Grassmann fields:

$$\psi_{\mathbf{x},\sigma}^{(i.r.)\pm} = \sum_{\omega = \pm} e^{i\vec{p}_F^{\omega}\vec{x}} \psi_{\mathbf{x},\sigma,\omega}^{(\leq 0)\pm}$$
(10.128)

and we rewrite Eq.(10.79) in the form:

$$\Xi_{M,\beta,\Lambda} = e^{-\beta|\Lambda|F_0} \int P_{\chi_0,A_0}(d\psi^{(\leq 0)}) e^{-\mathcal{V}^{(0)}(\psi^{(\leq 0)})}, \tag{10.129}$$

where $\mathcal{V}^{(0)}(\psi^{(\leq 0)})$ is equal to $\mathcal{V}_0(\psi^{(i.r.)})$, once $\psi^{(i.r.)}$ is rewritten as in (10.128), i.e.,

$$\mathcal{V}^{(0)}(\psi^{(\leq 0)}) = \tag{10.130}$$

$$= \sum_{n=1}^{\infty} (\beta|\Lambda|)^{-2n} \sum_{\sigma_1,\ldots,\sigma_n=\uparrow\downarrow} \sum_{\substack{\omega_1,\ldots,\omega_{2n}=\pm \\ \rho_1,\ldots,\rho_{2n}=1,2}} \sum_{\mathbf{k}'_1,\ldots,\mathbf{k}'_{2n}} \left[\prod_{j=1}^{n} \hat{\psi}^{(\leq 0)+}_{\mathbf{k}'_{2j-1},\sigma_j,\rho_{2j-1},\omega_{2j-1}} \hat{\psi}^{(\leq 0)-}_{\mathbf{k}'_{2j},\sigma_j,\rho_{2j},\omega_{2j}} \right] \cdot$$

$$\cdot \hat{W}^{(0)}_{2n,\rho,\omega}(\mathbf{k}'_1,\ldots,\mathbf{k}'_{2n-1}) \, \delta\left(\sum_{j=1}^{2n} (-1)^j (\mathbf{p}_F^{\omega_j} + \mathbf{k}'_j) \right) =$$

$$= \sum_{n=1}^{\infty} \sum_{\underline{\sigma},\underline{\rho},\underline{\omega}} \int d\mathbf{x}_1 \cdots d\mathbf{x}_{2n} \left[\prod_{j=1}^{n} \psi^{(\leq 0)+}_{\mathbf{x}_{2j-1},\sigma_j,\rho_{2j-1},\omega_{2j-1}} \psi^{(\leq 0)-}_{\mathbf{x}_{2j},\sigma_j,\rho_{2j},\omega_{2j}} \right] W^{(0)}_{2n,\rho,\omega}(\mathbf{x}_1,\ldots,\mathbf{x}_{2n})$$

with:

1) $\underline{\omega} = (\omega_1,\ldots,\omega_{2n})$, $\underline{\sigma} = (\sigma_1,\ldots,\sigma_n)$ and $\mathbf{p}_F^{\omega} = (0,\vec{p}_F^{\omega})$;

2) $\hat{W}^{(0)}_{2n,\rho,\omega}(\mathbf{k}'_1,\ldots,\mathbf{k}'_{2n-1}) = \hat{W}_{M,2n,\rho}(\mathbf{k}'_1 + \mathbf{p}_F^{\omega_j},\ldots,\mathbf{k}'_{2n-1} + \mathbf{p}_F^{\omega_{2n-1}})$, see (10.81);

3) the kernels $W^{(0)}_{2n,\rho,\omega}(\mathbf{x}_1,\ldots,\mathbf{x}_{2n})$ are defined as:

$$W^{(0)}_{2n,\rho,\omega}(\mathbf{x}_1,\ldots,\mathbf{x}_{2n}) = \tag{10.131}$$

$$= (\beta|\Lambda|)^{-2n} \sum_{\mathbf{k}'_1,\ldots,\mathbf{k}'_{2n}} e^{i \sum_{j=1}^{2n} (-1)^j \mathbf{k}'_j \mathbf{x}_j} \hat{W}^{(0)}_{2n,\rho,\omega}(\mathbf{k}'_1,\ldots,\mathbf{k}'_{2n-1}) \, \delta\left(\sum_{j=1}^{2n} (-1)^j (\mathbf{p}_F^{\omega_j} + \mathbf{k}'_j) \right).$$

Moreover, $P_{\chi_0,A_0}(d\psi^{(\leq 0)})$ is defined as

$$P_{\chi_0,A_0}(d\psi^{(\leq 0)}) = \mathcal{N}_0^{-1} \left[\prod_{\sigma,\omega,\rho} \prod_{\substack{\mathbf{k}'\in\mathcal{B}^{\omega}_{\beta,L} \\ \chi_0(|\mathbf{k}'|)>0}} d\hat{\psi}^{(\leq 0)+}_{\mathbf{k}',\sigma,\rho,\omega} d\hat{\psi}^{(\leq 0)-}_{\mathbf{k}',\sigma,\rho,\omega} \right] \cdot \tag{10.132}$$

$$\cdot \exp\left\{ -(\beta|\Lambda|)^{-1} \sum_{\omega=\pm} \sum_{\substack{\mathbf{k}'\in\mathcal{B}^{\omega}_{\beta,L} \\ \sigma=\uparrow\downarrow, \, \chi_0(|\mathbf{k}'|)>0}} \chi_0^{-1}(|\mathbf{k}'|) \hat{\psi}^{(\leq 0)+}_{\mathbf{k}',\sigma,\cdot,\omega} A_{0,\omega}(\mathbf{k}') \hat{\psi}^{(\leq 0)-}_{\mathbf{k}',\sigma,\cdot,\omega} \right\},$$

where:

$$A_{0,\omega}(\mathbf{k}') = \begin{pmatrix} -ik_0 & -v_0\Omega^*(\vec{k}' + \vec{p}_F^\omega) \\ -v_0\Omega(\vec{k}' + \vec{p}_F^\omega) & -ik_0 \end{pmatrix} =$$

$$= \begin{pmatrix} -i\zeta_0 k_0 + s_0(\mathbf{k}') & v_0(ik_1' - \omega k_2') + t_{0,\omega}(\mathbf{k}') \\ v_0(-ik_1' - \omega k_2') + t_{0,\omega}^*(\mathbf{k}') & -i\zeta_0 k_0 + s_0(\mathbf{k}') \end{pmatrix},$$

$\mathcal{N}_0$ is chosen in such a way that $\int P_{\chi_0, A_0}(d\psi^{(\leq 0)}) = 1$, $\zeta_0 = 1$, $s_0 = 0$ and $|t_{0,\omega}(\mathbf{k}')| \leq C|\mathbf{k}'|^2$.

It is apparent that the $\psi^{(\leq 0)}$ field has zero mass (i.e. its propagator decays polynomially at large distances in $\mathbf{x}$-space). Therefore, its integration requires an infrared multiscale analysis. As in the analysis of the ultraviolet problem, we define a sequence of geometrically decreasing momentum scales 2^h, with $h = 0, -1, -2, \ldots$ Correspondingly we introduce compact support functions $f_h(\mathbf{k}') = \chi_0(2^{-h}|\mathbf{k}'|) - \chi_0(2^{-h+1}|\mathbf{k}'|)$ and we rewrite

$$\chi_0(|\mathbf{k}'|) = \sum_{h=h_\beta}^{0} f_h(\mathbf{k}'), \tag{10.133}$$

with $h_\beta := \lfloor \log_2\left(\frac{3\pi}{4\beta}\right) \rfloor$ (the reason why the sum in equation (10.133) runs over $h \geq h_\beta$ is that, if $\mathbf{k}' \in \mathcal{B}_{\beta,L}^\omega$ and $h < h_\beta$, then $f_h(\mathbf{k}')$ is zero, simply because $|\mathbf{k}'| \geq \pi/\beta$). The purpose is to perform the integration of (10.129) in an iterative way. We step by step decompose the propagator into a sum of two propagators, the first supported on momenta $\sim 2^h$, $h \leq 0$, the second supported on momenta smaller than 2^h. Correspondingly we rewrite the Grassmann field as a sum of two independent fields: $\psi^{(\leq h)} = \psi^{(h)} + \psi^{(\leq h-1)}$ and we integrate the field $\psi^{(h)}$. In this way we inductively prove that, for any $h \leq 0$, equation (10.129) can be rewritten as

$$\Xi_{M,\beta,\Lambda} = e^{-\beta|\Lambda|F_h} \int P_{\chi_h, A_h}(d\psi^{(\leq h)})e^{-\mathcal{V}^{(h)}(\psi^{(\leq h)})}, \tag{10.134}$$

where $F_h, A_h, \mathcal{V}^{(h)}$ will be defined recursively, $\chi_h(|\mathbf{k}'|) = \sum_{k=h_\beta}^{h} f_k(\mathbf{k}')$ and $P_{\chi_h, A_h}(d\psi^{(\leq h)})$ is defined in the same way as $P_{\chi_0, A_0}(d\psi^{(\leq 0)})$ with $\psi^{(\leq 0)}, \chi_0, A_{0,\omega}, \zeta_0, v_0, s_0, t_{0,\omega}$ replaced by $\psi^{(\leq h)}, \chi_h, A_{h,\omega}, \zeta_h, v_h, s_h, t_{h,\omega}$, respectively. Moreover $\mathcal{V}^{(h)}(0) = 0$ and

$$\mathcal{V}^{(h)}(\psi) = \sum_{n=1}^{\infty} (\beta|\Lambda|)^{-2n} \sum_{\underline{\sigma},\underline{\rho},\underline{\omega}} \sum_{\mathbf{k}_1',\ldots,\mathbf{k}_{2n}'} \left[\prod_{j=1}^{n} \hat{\psi}_{\mathbf{k}_{2j-1}',\sigma_j,\rho_{2j-1},\omega_{2j-1}}^{(\leq h)+} \hat{\psi}_{\mathbf{k}_{2j}',\sigma_j,\rho_{2j},\omega_{2j}}^{(\leq h)-} \right] \cdot$$

$$\cdot \hat{W}_{2n,\underline{\rho},\underline{\omega}}^{(h)}(\mathbf{k}_1',\ldots,\mathbf{k}_{2n-1}') \, \delta(\sum_{j=1}^{2n}(-1)^j(\mathbf{p}_F^{\omega_j} + \mathbf{k}_j')) = \tag{10.135}$$

$$= \sum_{\substack{n \geq 1 \\ \underline{\sigma},\underline{\rho},\underline{\omega}}} \int d\mathbf{x}_1 \cdots d\mathbf{x}_{2n} \left[\prod_{j=1}^{n} \psi_{\mathbf{x}_{2j-1},\sigma_j,\rho_{2j-1},\omega_{2j-1}}^{(\leq h)+} \psi_{\mathbf{x}_{2j},\sigma_j,\rho_{2j},\omega_{2j}}^{(\leq h)-} \right] W_{2n,\underline{\rho},\underline{\omega}}^{(h)}(\mathbf{x}_1,\ldots,\mathbf{x}_{2n}).$$

Note that the field $\psi_{\mathbf{k}',\sigma,\omega}^{(\leq h)}$, whose propagator is given by $\chi_h(|\mathbf{k}'|)[A_\omega^{(h)}(\mathbf{k}')]^{-1}$, has the same support as χ_h, that is on a neighbourhood of size 2^h around the singularity $\mathbf{k}' = \mathbf{0}$ (that, in the original variables, corresponds to the Fermi point $\mathbf{k} = \mathbf{p}_F^\omega$). It is important for the following to think $\hat{W}_{2n,\rho,\underline{\omega}}^{(h)}$, $h \leq 0$, as functions of the variables $\{\zeta_k, v_k\}_{h<k\leq 0}$. The iterative construction below will inductively imply that the dependence on these variables is well defined. The iteration continues up to the scale h_β and the result of the last iteration is $\Xi_{M,\beta,\Lambda} = e^{-\beta|\Lambda|F_{\beta,\Lambda}^{(M)}}$.

Localization and renormalization. In order to inductively prove equation (10.134) we write

$$\mathcal{V}^{(h)} = \mathcal{L}\mathcal{V}^{(h)} + \mathcal{R}\mathcal{V}^{(h)} \tag{10.136}$$

where

$$\mathcal{L}\mathcal{V}^{(h)} = \frac{1}{\beta|\Lambda|} \sum_{\sigma=\uparrow\downarrow}^{\omega=\pm} \sum_{\mathbf{k}'}^{\chi_h(|\mathbf{k}'|)>0} \hat{\psi}_{\mathbf{k}',\sigma,\omega}^{(\leq h)+} \hat{W}_{2,(\omega,\omega)}^{(h)}(\mathbf{k}') \hat{\psi}_{\mathbf{k}',\sigma,\omega}^{(\leq h)-}, \tag{10.137}$$

and $\mathcal{R}\mathcal{V}^{(h)}$ is given by equation (10.135) with $\sum_{n=1}^\infty$ replaced by $\sum_{n=2}^\infty$, that is it contains only the monomials with four fermionic fields or more. The symmetries of the action, which are described in Appendix 10.11 and are preserved by the iterative integration procedure, imply that, in the zero temperature and thermodynamic limit,

$$\hat{W}_{2,(\omega,\omega)}^{(h)}(\mathbf{k}') = \begin{pmatrix} -iz_h k_0 & \delta_h(ik_1' - \omega k_2') \\ \delta_h(-ik_1' - \omega k_2') & -iz_h k_0 \end{pmatrix} + O(|\mathbf{k}'|^2), \tag{10.138}$$

for suitable real constants z_h, δ_h. Equation (10.138) is the analogue of Equation (10.85); its proof is completely analogous to the proof of Lemma 10.6.2 and will not be belaboured here.

Once that the above definitions are given, we can describe our iterative integration procedure for $h \leq 0$. We start from equation (10.134) and we rewrite it as

$$\int P_{\chi_h,A_h}(d\psi^{(\leq h)}) e^{-\mathcal{L}\mathcal{V}^{(h)}(\psi^{(\leq h)}) - \mathcal{R}\mathcal{V}^{(h)}(\psi^{(\leq h)}) - \beta|\Lambda|F_h}, \tag{10.139}$$

with

$$\mathcal{L}\mathcal{V}^{(h)}(\psi^{(\leq h)}) = (\beta|\Lambda|)^{-1} \sum_{\omega,\sigma} \sum_{\mathbf{k}'}^{\chi_h(|\mathbf{k}'|)>0} \cdot \tag{10.140}$$

$$\cdot \hat{\psi}_{\mathbf{k}',\sigma,\omega}^{(\leq h)+} \begin{pmatrix} -iz_h k_0 + \sigma_h(\mathbf{k}') & \delta_h(ik_1' - \omega k_2') + \tau_{h,\omega}(\mathbf{k}') \\ \delta_h(-ik_1' - \omega k_2) + \tau_{h,\omega}^*(\mathbf{k}') & -iz_h k_0 + \sigma_h(\mathbf{k}') \end{pmatrix} \hat{\psi}_{\mathbf{k}',\sigma,\omega}^{(\leq h)-},$$

where $\sigma_h(\mathbf{K}')$, $\tau_{h,\omega}(\mathbf{K}')$ are two functions vanishing at $\mathbf{K}' = 0$ faster than linearly in $\mathbf{K}'$. Then we include $\mathcal{L}\mathcal{V}^{(h)}$ in the fermionic integration, so obtaining

$$\int P_{\chi_h,\bar{A}_{h-1}}(d\psi^{(\leq h)}) e^{-\mathcal{R}\mathcal{V}^{(h)}(\psi^{(\leq h)}) - \beta|\Lambda|(F_h + e_h)}, \tag{10.141}$$

where

$$e_h = \frac{1}{\beta|\Lambda|} \sum_{\omega,\sigma} \sum_{\mathbf{k}'} \sum_{n\geq 1} \frac{(-1)^n}{n} \mathrm{Tr}\left\{ \left[\chi_h(|\mathbf{k}'|)A_{h,\omega}^{-1}(\mathbf{k}')W_{2,(\omega,\omega)}^{(h)}(\mathbf{k}')\right]^n \right\} \qquad (10.142)$$

is a constant taking into account the change in the normalization factor of the measure and

$$\overline{A}_{h-1,\omega}(\mathbf{k}') = \begin{pmatrix} -i\overline{\zeta}_{h-1}k_0 + \overline{s}_{h-1}(\mathbf{k}') & \overline{v}_{h-1}(ik_1' - \omega k_2') + \overline{t}_{h-1,\omega}(\mathbf{k}') \\ \overline{v}_{h-1}(-ik_1' - \omega k_2') + \overline{t}_{h-1,\omega}^*(\mathbf{k}') & -i\overline{\zeta}_{h-1}k_0 + \overline{s}_{h-1}(\mathbf{k}') \end{pmatrix} \qquad (10.143)$$

with:

$$\overline{\zeta}_{h-1}(\mathbf{k}') = \zeta_h + z_h\chi_h(|\mathbf{k}'|), \qquad \overline{v}_{h-1}(\mathbf{k}') = v_h + \delta_h\chi_h(|\mathbf{k}'|), \qquad (10.144)$$
$$\overline{s}_{h-1}(\mathbf{k}') = s_h(\mathbf{k}') + \sigma_h(\mathbf{k}')\chi_h(|\mathbf{k}'|), \qquad \overline{t}_{h-1,\omega}(\mathbf{k}') = t_{h,\omega}(\mathbf{k}') + \tau_{h,\omega}(\mathbf{k}')\chi_h(|\mathbf{k}'|).$$

Now we can perform the integration of the $\psi^{(h)}$ field. We rewrite the Grassmann field $\psi^{(\leq h)}$ as a sum of two independent Grassmann fields $\psi^{(\leq h-1)} + \psi^{(h)}$ and correspondingly we rewrite Eq.(10.141) as

$$e^{-\beta|\Lambda|(F_h+e_h)} \int P_{\chi_{h-1},A_{h-1}}(d\psi^{(\leq h-1)}) \int P_{f_h,\overline{A}_{h-1}}(d\psi^{(h)}) e^{-\mathcal{R}\mathcal{V}^{(h)}(\psi^{(\leq h-1)}+\psi^{(h)})}, \qquad (10.145)$$

where

$$A_{h-1,\omega}(\mathbf{k}') = \begin{pmatrix} -i\zeta_{h-1}k_0 + s_{h-1}(\mathbf{k}') & v_{h-1}(ik_1' - \omega k_2') + t_{h-1,\omega}(\mathbf{k}') \\ v_{h-1}(-ik_1' - \omega k_2') + t_{h-1,\omega}^*(\mathbf{k}') & -i\zeta_{h-1}k_0 + s_{h-1}(\mathbf{k}') \end{pmatrix} \qquad (10.146)$$

with:

$$\zeta_{h-1} = \zeta_h + z_h, \qquad v_{h-1} = v_h + \delta_h,$$
$$s_{h-1}(\mathbf{k}') = s_h(\mathbf{k}') + \sigma_h(\mathbf{k}'), \qquad t_{h-1,\omega}(\mathbf{k}') = t_{h,\omega}(\mathbf{k}') + \tau_{h,\omega}(\mathbf{k}'). \qquad (10.147)$$

The single scale propagator is

$$\int P_{f_h,\overline{A}_{h-1}}(d\psi^{(h)})\psi_{\mathbf{x}_1,\sigma_1,\omega_1}^{(h)-}\psi_{\mathbf{x}_2,\sigma_2,\omega_2}^{(h)+} = \delta_{\sigma_1,\sigma_2}\delta_{\omega_1,\omega_2}g_\omega^{(h)}(\mathbf{x}_1 - \mathbf{x}_2), \qquad (10.148)$$

where

$$g_\omega^{(h)}(\mathbf{x}_1 - \mathbf{x}_2) = \frac{1}{\beta|\Lambda|} \sum_{\mathbf{k}'\in\mathcal{D}_{\beta,L}^\omega} e^{-i\mathbf{k}'(\mathbf{x}_1-\mathbf{x}_2)} f_h(\mathbf{k}') \left[\overline{A}_{h-1,\omega}(\mathbf{k}')\right]^{-1}. \qquad (10.149)$$

After the integration of the field on scale h we are left with an integral involving the fields $\psi^{(\leq h-1)}$ and the new effective interaction $\mathcal{V}^{(h-1)}$, defined as

$$e^{-\mathcal{V}^{(h-1)}(\psi^{(\leq h-1)})-\overline{e}_h\beta|\Lambda|} = \int P_{f_h,\overline{A}_{h-1}}(d\psi^{(h)}) e^{-\mathcal{R}\mathcal{V}^{(h)}(\psi^{(\leq h-1)}+\psi^{(h)})}, \qquad (10.150)$$

with $\mathcal{V}^{(h-1)}(0) = 0$. It is easy to see that $\mathcal{V}^{(h-1)}$ is of the form equation (10.135) and that $F_{h-1} = F_h + e_h + \bar{e}_h$. It is sufficient to use the identity

$$\beta|\Lambda|\bar{e}_h + \mathcal{V}^{(h-1)}(\psi^{(\leq h-1)}) = \sum_{n\geq 1} \frac{1}{n!}(-1)^{n+1}\mathcal{E}_h^T(\mathcal{R}\mathcal{V}^{(h)}(\psi^{(\leq h-1)} + \psi^{(h)}); n), \quad (10.151)$$

where $\mathcal{E}_h^T(X(\psi^{(h)}); n)$ is the truncated expectation of order n w.r.t. the propagator $g_\omega^{(h)}$, which is the analogue of equation (10.80) with $\psi^{(u.v.)}$ replaced by $\psi^{(h)}$ and with $P(d\psi^{(u.v.)})$ replaced by $P_{f_h,\bar{A}_{h-1}}(d\psi^{(h)})$.

Note that the above procedure allows us to write the *effective constants* (ζ_h, v_h), $h \leq 0$, in terms of (ζ_k, v_k), $h < k \leq 0$, namely $\zeta_{h-1} = \beta_h^\zeta((\zeta_h, v_h), \dots, (\zeta_0, v_0))$ and $v_{h-1} = \beta_h^v((\zeta_h, v_h), \dots, (\zeta_0, v_0))$, where $\beta_h^\#$ is the so-called *Beta function*.

An iterative implementation of equation (10.151) leads to a representation of $\mathcal{V}^{(h)}$, $h < 0$, in terms of a new tree expansion. The set of trees of order n contributing to $\mathcal{V}^{(h)}(\psi^{(\leq h)})$ is denoted by $\mathcal{T}_{h,n}$. The trees in $\mathcal{T}_{h,n}$ are defined in a way very similar to those in $\mathcal{T}_{M;h,n}$, but for the following differences: (i) all the endpoints are on scale 1; (ii) the scale labels of the vertices of $\tau \in \mathcal{T}_{h,n}$ are between $h + 1$ and 1; (iii) with each endpoint v we associate one of the monomials with four or more Grassmann fields contributing to $\mathcal{R}\mathcal{V}^{(0)}(\psi^{(\leq 0)})$, corresponding to the terms with $n \geq 2$ in the r.h.s. of equation (10.130) (here $\mathcal{R}$ is the linear operator acting as the identity on Grassmann monomials of order 4 or more, and acting as 0 on Grassmann monomials of order 0 or 2). In terms of these trees, the effective potential $\mathcal{V}^{(h)}$, $h \leq -1$, can be written as

$$\mathcal{V}^{(h)}(\psi^{(\leq h)}) + \beta|\Lambda|\bar{e}_{h+1} = \sum_{n=1}^{\infty} \sum_{\tau \in \mathcal{T}_{h,n}} \mathcal{V}^{(h)}(\tau, \psi^{(\leq h)}), \quad (10.152)$$

where, if v_0 is the first vertex of τ and $\tau_1, \dots, \tau_s$ ($s = s_{v_0}$) are the subtrees of τ with root v_0,

$$\mathcal{V}^{(h)}(\tau, \psi^{(\leq h)}) = \frac{(-1)^{s+1}}{s!}\mathcal{E}_{h+1}^T[\mathcal{R}\mathcal{V}^{(h+1)}(\tau_1, \psi^{(\leq h+1)}); \dots; \mathcal{R}\mathcal{V}^{(h+1)}(\tau_s, \psi^{(\leq h+1)})],$$
$$(10.153)$$

where, if τ_i is trivial, then $h + 1 = 0$ and $\mathcal{R}\mathcal{V}^{(0)}(\tau_i, \psi^{(\leq 0)}) = \mathcal{R}\mathcal{V}^{(0)}(\psi^{(\leq 0)})$. Repeating step by step the discussion leading to equations (10.105),(10.108), and (10.109), and using analogous definitions, we find that

$$\mathcal{V}^{(h)}(\tau, \psi^{(\leq h)}) = \sum_{\mathbf{P} \in \mathcal{P}_\tau} \sum_{T \in \mathbf{T}} \sum_{\rho_{v_0}, \sigma_{v_0}, \varepsilon_{v_0}} \int d\mathbf{x}_{v_0} \psi_{P_{v_0}}^{(\leq h)} W_{\tau, \mathbf{P}, T}^{(h)}(\mathbf{x}_{v_0}), \quad (10.154)$$

with

$$W_{\tau, \mathbf{P}, T}(\mathbf{x}_{v_0}) = \alpha_T \left[\prod_{i=1}^{n} K_{v_i^*}^{(h_i)}(\mathbf{x}_{v_i^*})\right] \left\{\prod_{\substack{v \text{ not} \\ \text{e.p.}}} \frac{(-1)^{s_v+1}}{s_v!}\int dP_{T_v}(\mathbf{t}_v)\det G^{h_v, T_v}(\mathbf{t}_v) \cdot \right.$$

$$\left. \cdot \left[\prod_{\ell \in T_v} \delta_{\omega(\ell),\omega'(\ell)}\delta_{\sigma(\ell),\sigma'(\ell)}[g_{\omega(\ell)}^{(h_v)}(\mathbf{x}(\ell) - \mathbf{x}'(\ell))]_{\rho(\ell),\rho'(\ell)}\right]\right\}. \quad (10.155)$$

and: v_i^*, $i = 1, \ldots, n$, are the endpoints of τ; $K_{v_i^*}^{(h_i)}(\mathbf{x}_{v_i^*})$ is the kernel of one of the monomials contributing to $\mathcal{R}\mathcal{V}^{(0)}(\psi^{(\leq h_v)})$; $G^{h,T}$ is a matrix with elements

$$G_{f,f'}^{h,T} = t_{i(f),i(f')}\delta_{\omega(f),\omega(f')}\delta_{\sigma(f),\sigma(f')}\left[g_{\omega(f)}^{(h)}(\mathbf{x}(f) - \mathbf{x}(f'))\right]_{\rho(f),\rho(f')}, \qquad (10.156)$$

Once again, it is important to note that $G^{h,T}$ is a Gram matrix, that is., defining $\mathbf{e}_+ = \mathbf{e}_\uparrow = (1,0)$ and $\mathbf{e}_- = \mathbf{e}_\downarrow = (0,1)$, the matrix elements in equation (10.156), using a notation similar to equation (10.70), can be written in terms of scalar products:

$$G_{f,f'}^{h,T} = \qquad\qquad (10.157)$$

$$= \left\langle \mathbf{u}_{i(f)} \otimes \mathbf{e}_{\omega(f)} \otimes \mathbf{e}_{\sigma(f)} \otimes \mathbf{A}_{\mathbf{x}(f),\rho(f),\omega(f)}, \mathbf{u}_{i(f')} \otimes \mathbf{e}_{\omega(f')} \otimes \mathbf{e}_{\sigma(f')} \otimes \mathbf{B}_{\mathbf{x}(f'),\rho(f'),\omega(f')} \right\rangle,$$

where

$$[\mathbf{A}_{\mathbf{x},\rho,\omega}(\mathbf{z})]_i = \frac{1}{\beta|\Lambda|} \sum_{\mathbf{k'}\in\mathcal{B}_{\beta,L}^\omega} \frac{e^{-i\mathbf{k'}(\mathbf{z}-\mathbf{x})}\sqrt{f_h(\mathbf{k'})}}{|\det \overline{A}_{h-1,\omega}(\mathbf{k'})|^{1/4}} \delta_{\rho,i},$$

$$[\mathbf{B}_{\mathbf{x},\rho,\omega}(\mathbf{z})]_i = \frac{1}{\beta|\Lambda|} \sum_{\mathbf{k'}\in\mathcal{B}_{\beta,L}^\omega} \frac{e^{-i\mathbf{k'}(\mathbf{z}-\mathbf{x})}\sqrt{f_h(\mathbf{k'})}}{|\det \overline{A}_{h-1,\omega}(\mathbf{k'})|^{3/4}} \cdot \qquad (10.158)$$

$$\cdot \begin{pmatrix} i\overline{\zeta}_{h-1}k_0 - \overline{s}_{h-1}(\mathbf{k'}) & \overline{v}_{h-1}(ik_1' - \omega k_2') + \overline{t}_{h-1,\omega}(\mathbf{k'}) \\ \overline{v}_{h-1}(-ik_1' - \omega k_2') + \overline{t}_{h-1,\omega}^*(\mathbf{k'}) & i\overline{\zeta}_{h-1}k_0 - \overline{s}_{h-1}(\mathbf{k'}) \end{pmatrix}_{i,\rho}.$$

Using the representation equations (10.152),(10.153), and (10.155) and proceeding as in the proof of Lemma 10.6.1, we can get a bound on the kernels of the effective potentials, which is the key ingredient for the proof of Theorem 10.1 and is summarized in the following theorem.

Theorem 10.3 *There exists a constants $U_0 > 0$, independent of M, β and Λ, such that the kernels $W_{2l,\rho,\omega}^{(h)}(\mathbf{x}_1, \ldots, \mathbf{x}_{2l})$ in equation (10.135), $h \leq -1$, are analytic functions of U in the complex domain $|U| \leq U_0$, satisfying, for any $0 \leq \theta < 1$ and a suitable constant $C_\theta > 0$ (independent of M, β, Λ), the following estimates:*

$$\frac{1}{\beta|\Lambda|}\int d\mathbf{x}_1 \cdots d\mathbf{x}_{2l}\left[\prod_{1\leq i<j\leq 2l} ||\mathbf{x}_i - \mathbf{x}_j||^{m_{ij}}\right]|W_{2l,\rho,\omega}^{(h)}(\mathbf{x}_1,\ldots,\mathbf{x}_{2l})| \leq$$

$$2^{h(3-2l-m+\theta)}\left(C_\theta|U|\right)^{max(1,l-1)}. \qquad (10.159)$$

where $m_{ij} \in \{0,1,2\}$ and $m := \sum_{i<j} m_{ij} \leq 2$. Moreover, the constants e_h and $\overline{e}_h$ defined by equation (10.142) and equation (10.151) are analytic functions of U in the same domain $|U| \leq U_0$, uniformly bounded as $|e_h| + |\overline{e}_h| \leq C_\theta|U|2^{h(3+\theta)}$. Both the kernels $W_{2l,\rho,\omega}^{(h)}(\mathbf{x}_1,\ldots,\mathbf{x}_{2l})$ and the constants $e_h, \overline{e}_h$ admit well defined limits as $M, \beta, |\Lambda| \to \infty$, which are reached uniformly (and, with some abuse of notation, will be denoted by the same symbols).

Before we present its proof, let us show how Theorem 10.3 implies, as a corollary, Theorem 10.1. It is enough to observe that, by Proposition 10.2 and by the multiscale integration procedure described above, $f_\beta(U) = F_0 + \sum_{h=h_\beta}^{0} (e_h + \bar{e}_h)$, with $F_0, e_h, \bar{e}_h$ analytic functions of U for U small enough (see the discussion at the end of Section 10.6 and the statement of Theorem 10.3). Since $|e_h| + |\bar{e}_h| \leq C_\theta |U| 2^{h(3+\theta)}$, the sum $\sum_{h=h_\beta}^{0} (e_h + \bar{e}_h)$ is absolutely convergent, uniformly in h_β; therefore, both $f_\beta(U)$ and its $\beta \to \infty$ limit are analytic functions of U for U small enough. This concludes the proof of Theorem 10.1.

Proof of Theorem 10.3. Let us preliminarily assume that, for all $h < h' \leq 0$, the vectors $\mathbf{A}_{\mathbf{x},\rho,\omega}^{(h')}$ and $\mathbf{B}_{\mathbf{x},\rho,\omega}^{(h')}$ in equation (10.158) and the single scale propagator $g_\omega^{(h')}$ in equation (10.149) satisfy the following bounds:

$$||\mathbf{A}_{\mathbf{x},\rho,\omega}^{(h')}||^2 = ||\mathbf{B}_{\mathbf{x},\rho,\omega}^{(h')}||^2 = \frac{1}{\beta|\Lambda|} \sum_{\mathbf{k}' \in \mathcal{B}_{\beta,L}^\omega} \frac{f_{h'}(\mathbf{k}')}{|\det \bar{A}_{h'-1,\omega}(\mathbf{k}')|^{1/2}} \leq (\text{const.}) 2^{2h'} \quad (10.160)$$

$$\int d\mathbf{x} \, ||\mathbf{x}||^m ||g_\omega^{(h')}(\mathbf{x})|| \leq (\text{const.}) 2^{-(1+m)h'}, \quad 0 \leq m \leq 2, \quad (10.161)$$

which will be inductively proved below. Let us first prove equation (10.159) in the case that $m_{ij} \equiv 0$. Using the tree expansion described above and, in particular, equations (10.152),(10.154),(10.155), we find that the l.h.s. of equation (10.159) can be bounded from above by

$$\sum_{n \geq 1} \sum_{\tau \in \mathcal{T}_{h,n}} \sum_{\substack{P \in \mathcal{P}_\tau \\ |P_{v_0}|=2l}} \sum_{T \in \mathbf{T}} \int \prod_{\ell \in T^*} d(\mathbf{x}(\ell) - \mathbf{x}'(\ell)) \left[\prod_{i=1}^{n} |K_{v_i^*}^{(h_i)}(\mathbf{x}_{v_i^*})| \right] \cdot \qquad (10.162)$$

$$\cdot \left[\prod_{\substack{v \text{ not} \\ \text{e.p.}}} \frac{1}{s_v!} \max_{\mathbf{t}_v} |\det G^{h_v, T_v}(\mathbf{t}_v)| \prod_{\ell \in T_v} ||g_{\omega(\ell)}^{(h_v)}(\mathbf{x}(\ell) - \mathbf{x}'(\ell))|| \right]$$

where T^* is a tree graph obtained from $T = \cup_v T_v$, by adding in a suitable (obvious) way, for each endpoint v_i^*, $i = 1, \ldots, n$, one or more lines connecting the space-time points belonging to $\mathbf{x}_{v_i^*}$.

An application of the Gram–Hadamard inequality equation (10.69), combined with the representation equation (10.157) and the dimensional bound equation (10.160), implies that

$$|\det G^{h_v, T_v}(\mathbf{t}_v)| \leq (\text{const.})^{\sum_{i=1}^{s_v} |P_{v_i}| - |P_v| - 2(s_v - 1)} \cdot 2^{h_v \left(\sum_{i=1}^{s_v} |P_{v_i}| - |P_v| - 2(s_v - 1) \right)}. \qquad (10.163)$$

By using equation (14.161) with m = 0, it also follows that

$$\prod_{\substack{v \text{ not} \\ \text{e.p.}}} \frac{1}{s_v!} \int \prod_{\ell \in T_v} d(\mathbf{x}(\ell) - \mathbf{x}'(\ell)) \, ||g_{\omega(\ell)}^{(h_v)}(\mathbf{x}(\ell) - \mathbf{x}'(\ell))|| \leq c^n \prod_{\substack{v \text{ not} \\ \text{e.p.}}} \frac{1}{s_v!} 2^{-h_v(s_v - 1)}. \qquad (10.164)$$

The bound equation (10.83) on the kernels produced by the ultraviolet integration implies that

$$
\int \prod_{\ell \in T^* \setminus \cup_v T_v} d(\mathbf{x}(\ell) - \mathbf{x}'(\ell)) \prod_{i=1}^n |K_{v_i^*}^{(h_i)}(\mathbf{x}_{v_i^*})| \leq \prod_{i=1}^n C^{p_i} |U|^{\frac{p_i}{2}-1},
\tag{10.165}
$$

where $p_i = |P_{v_i^*}|$. Combining the previous bounds, we find that equation (10.162) can be bounded from above by

$$
\sum_{n \geq 1} \sum_{\tau \in \mathcal{T}_{h,n}} \sum_{\substack{P \in \mathcal{P}_\tau \\ |P_{v_0}|=2l}} \sum_{T \in \mathbf{T}} C^n \left[\prod_{\substack{v \text{ not} \\ \text{e.p.}}} \frac{1}{s_v!} 2^{h_v \left(\sum_{i=1}^{s_v} |P_{v_i}| - |P_v| - 3(s_v-1) \right)} \right] \left[\prod_{i=1}^n C^{p_i} |U|^{\frac{p_i}{2}-1} \right]
\tag{10.166}
$$

Let us recall that $n(v) = \sum_{i:v_i^*>v} 1$ is the number of endpoints following v on τ, that v' is the vertex immediately preceding v on τ and that $|I_v|$ is the number of field labels associated to the endpoints following v on τ (note that $|I_v| \geq 4n(v)$). Using the following relations, which can be easily proved by induction,

$$
\sum_{v \text{ not e.p.}} h_v \left[\left(\sum_{i=1}^{s_v} |P_{v_i}| \right) - |P_v| \right] = h(|I_{v_0}| - |P_{v_0}|) + \sum_{v \text{ not e.p.}} (h_v - h_{v'})(|I_v| - |P_v|),
$$

$$
\sum_{v \text{ not e.p.}} h_v (s_v - 1) = h(n-1) + \sum_{v \text{ not e.p.}} (h_v - h_{v'})(n(v) - 1),
\tag{10.167}
$$

we find that equation (10.166) can be bounded above by

$$
\sum_{n \geq 1} (\text{const.})^n \sum_{\tau \in \mathcal{T}_{h,n}} \sum_{\substack{P \in \mathcal{P}_\tau \\ |P_{v_0}|=2l}} \sum_{T \in \mathbf{T}} C^n 2^{h(3-|P_{v_0}|+|I_{v_0}|-3n)} \cdot
$$

$$
\cdot \left[\prod_{\substack{v \text{ not} \\ \text{e.p.}}} \frac{1}{s_v!} 2^{(h_v - h_{v'})(3-|P_v|+|I_v|-3n(v))} \right] \left[\prod_{i=1}^n C^{p_i} |U|^{\frac{p_i}{2}-1} \right].
\tag{10.168}
$$

Using the following identities

$$
2^{hn} \prod_{\substack{v \text{ not} \\ \text{e.p.}}} 2^{(h_v - h_{v'})n(v)} = \prod_{v \text{ e.p.}} 2^{h_{v'}},
\tag{10.169}
$$

$$
2^{h|I_{v_0}|} \prod_{\substack{v \text{ not} \\ \text{e.p.}}} 2^{(h_v - h_{v'})|I_v|} = \prod_{v \text{ e.p.}} 2^{h_{v'}|I_v|},
\tag{10.170}
$$

combined with the remark that all the endpoints are on scale 1 (so that the right-hand sides of equations (10.169)–(10.170) are equal to 1), we obtain

$$\frac{1}{\beta|\Lambda|}\int d\mathbf{x}_1\cdots d\mathbf{x}_{2l}|W^{(h)}_{2l,\underline{\rho},\underline{\omega}}(\mathbf{x}_1,\ldots,\mathbf{x}_{2l})| \leq \sum_{n\geq 1}\sum_{\tau\in\mathcal{T}_{h,n}}\sum_{\substack{\mathbf{P}\in\mathcal{P}_\tau \\ |P_{v_0}|=2l}}\sum_{T\in\mathbf{T}}C^n 2^{h(3-|P_{v_0}|)}.$$

$$\cdot\left[\prod_{\substack{v\text{ not}\\ \text{e.p.}}}\frac{1}{s_v!}2^{-(h_v-h_{v'})(|P_v|-3)}\right]\left[\prod_{i=1}^{n}C^{p_i}|U|^{\frac{p_i}{2}-1}\right]. \tag{10.171}$$

Note that, if v is not an endpoint, $|P_v|-3\geq 1$ by the definition of $\mathcal{R}$. Now, note that the number of terms in $\sum_{T\in\mathbf{T}}$ can be bounded by $C^n\prod_{v\text{ not e.p.}}s_v!$. Using also that $|P_v|-3\geq 1$ and $|P_v|-3\geq|P_v|/4$, we find that the l.h.s. of equation (10.171) can be bounded as

$$\frac{1}{\beta|\Lambda|}\int d\mathbf{x}_1\cdots d\mathbf{x}_{2l}|W^{(h)}_{2l,\underline{\rho},\underline{\omega}}(\mathbf{x}_1,\ldots,\mathbf{x}_{2l})| \leq 2^{h(3-|P_{v_0}|)}\sum_{n\geq 1}C^n\sum_{\tau\in\mathcal{T}_{h,n}}\cdot \tag{10.172}$$

$$\cdot\left(\prod_{\substack{v\text{ not}\\ \text{e.p.}}}2^{-\theta(h_v-h_{v'})}2^{-(1-\theta)(h_v-h_{v'})/2}\right)\sum_{\substack{\mathbf{P}\in\mathcal{P}_\tau \\ |P_{v_0}|=2l}}\left(\prod_{\substack{v\text{ not}\\ \text{e.p.}}}2^{-(1-\theta)|P_v|/8}\right)\prod_{i=1}^{n}C^{p_i}|U|^{\frac{p_i}{2}-1}.$$

Proceeding as in the previous section, we get the analogue of equation (10.121):

$$\sum_{\substack{\mathbf{P}\in\mathcal{P}_\tau \\ |P_{v_0}|=2l}}\left(\prod_{\substack{v\text{ not}\\ \text{e.p.}}}2^{-(1-\theta)|P_v|/8}\right)\prod_{i=1}^{n}C^{p_i}|U|^{\frac{p_i}{2}-1} \leq C^n_\theta|U|^n.$$

Finally, using that $\prod_{v\text{ not e.p.}}2^{-\theta(h_v-h_{v'})}\leq 2^{\theta h}$ and that, for $0<\theta<1$ (in analogy with equation (10.122)),

$$\sum_{\tau\in\mathcal{T}_{h,n}}\prod_{v\text{ not e.p.}}2^{-(1-\theta)(h_v-h_{v'})/2} \leq C^n;$$

and collecting all the previous bounds, we obtain

$$\frac{1}{\beta|\Lambda|}\int d\mathbf{x}_1\cdots d\mathbf{x}_{2l}|W^{(h)}_{2l,\underline{\rho},\underline{\omega}}(\mathbf{x}_1,\ldots,\mathbf{x}_{2l})| \leq 2^{h(3-|P_{v_0}|+\theta)}\sum_{n\geq 1}C^n|U|^n, \tag{10.173}$$

which is the desired result. The proof of equation (10.159) in the presence of the factors $\prod_{i<j}||\mathbf{x}_i-\mathbf{x}_j||^{m_{ij}}$, with $m\neq 0$, goes exactly along the same lines: it is enough to remark that, for each given tree in the expansion for the effective potential, we can 'decompose the zeros along the spanning tree', as follows: $\prod_{i<j}||\mathbf{x}_i-\mathbf{x}_j||^{m_{ij}}\leq C^n n^m\sum_v\sum_{\ell\in T_v}||\mathbf{x}(\ell)-\mathbf{y}(\ell)||^m$, with n the order of the considered tree; then one should use Eq.(10.161) with $m\neq 0$ and the desired result follows.

We now need to prove the assumptions equations (10.160)–(10.161). We proceed as follows. Let us inductively assume that equation (10.159) is valid for all $\bar{h} \leq h \leq 0$. Equation (10.159) is certainly true for $\bar{h} = h = 0$, thanks to the bounds on the ultraviolet integration of Section 10.6; to be fair, in Section 10.6 the analogue of equation (10.159) is spelled out for $m_{ij} \equiv 0$, but a similar result is valid for $\sum_{i<j} m_{ij} = m \leq 2$, as one can prove again by 'decomposing the zeros along the spanning tree' and by using the decay bounds equation (10.88) (note that here is where we use the condition that K can be equal to 6 in equation (10.88)). If now, assuming the validity of equation (10.159) for all $\bar{h} \leq h \leq 0$, we manage to prove equations (10.160)–(10.161) for $h' = \bar{h}$, then we are done, because we already showed how to prove equation (10.159) with h replaced by $\bar{h} - 1$ under the assumption that equations (10.160)–(10.161) for $\bar{h} \leq h \leq 0$.

Let us then focus on proving equations (10.160)–(10.161) for $h' = \bar{h}$. The key remark is that the matrix $\overline{A}_{\bar{h}-1}$ entering the definitions of $g_\omega^{(\bar{h})}$, $\mathbf{A}_{\mathbf{x},\rho,\omega}^{(\bar{h})}$ and $\mathbf{B}_{\mathbf{x},\rho,\omega}^{(\bar{h})}$ is given by:

$$\overline{A}_{\bar{h}-1,\omega}(\mathbf{k}') = -\begin{pmatrix} ik_0 & v_0\Omega^*(\vec{k}'+\vec{p}_F^\omega) \\ v_0\Omega(\vec{k}'+\vec{p}_F^\omega) & ik_0 \end{pmatrix} - \sum_{h=\bar{h}}^{0} \hat{W}_{2,(\omega,\omega)}^{(h)}(\mathbf{k}')\chi_{h'}(|\mathbf{k}'|) =$$

$$=: -\begin{pmatrix} ik_0\overline{\zeta}_{\bar{h}-1} & \overline{v}_{\bar{h}-1}(-ik_1'+\omega k_2') \\ \overline{v}_{\bar{h}-1}(ik_1'+\omega k_2') & ik_0\overline{\zeta}_{\bar{h}-1} \end{pmatrix} - R_2^{(\bar{h})}(\mathbf{k}'), \qquad (10.174)$$

$$=: A_{\bar{h}-1,\omega}^{rel}(\mathbf{k}') - R_2^{(\bar{h})}(\mathbf{k}'), \qquad (10.175)$$

where, if $\overline{W}_2^{(\bar{h})}(\mathbf{k}') := \sum_{h=\bar{h}}^{0} \hat{W}_{2,(\omega,\omega)}^{(h)}(\mathbf{k}')\chi_{h'}(|\mathbf{k}'|)$ and using the symmetry properties of the kernels $\hat{W}_{2,(\omega,\omega)}^{(h)}(\mathbf{k}')$, we defined:

$$R_2^{(\bar{h})}(\mathbf{k}') = v_0\begin{pmatrix} 0 & \Omega^*(\vec{k}'+\vec{p}_F^\omega)+ik_1'-\omega k_2' \\ \Omega(\vec{k}'+\vec{p}_F^\omega)-ik_1'-\omega k_2' & 0 \end{pmatrix} +$$

$$+ \overline{W}_2^{(\bar{h})}(\mathbf{k}') - \overline{W}_2^{(\bar{h})}(\mathbf{0}) - \mathbf{k}'\partial_{\mathbf{k}'}\overline{W}_2^{(\bar{h})}(\mathbf{0}). \qquad (10.176)$$

Recall the recursive definitions of $\overline{\zeta}_k$ and $\overline{v}_k$, i.e., $\overline{\zeta}_{k-1}(\mathbf{k}') = \zeta_k + z_k\chi_k(|\mathbf{k}'|)$, $\overline{v}_{k-1}(\mathbf{k}') = v_k + \delta_k\chi_h(|\mathbf{k}'|)$, with

$$\max\{|z_k|,|\delta_k|\} \leq \frac{1}{\beta|\Lambda|}\int d\mathbf{x}d\mathbf{y} \, ||\mathbf{x}-\mathbf{y}|| \, ||W_{2,(\omega,\omega)}^{(k)}(\mathbf{x},\mathbf{y})|| \leq (\text{const.})|U|2^{\theta k}, \quad (10.177)$$

thanks to the inductive hypothesis. Therefore, $\max\{|\overline{\zeta}_{\bar{h}-1} - 1|, |\overline{v}_{\bar{h}-1} - v_0|\} \leq$ (const.)$|U|$ and, proceeding as in Appendix 10.9, one can easily prove that equation (10.161) is valid with $g_\omega^{(\bar{h})}$ replaced by

$$\hat{g}_{\omega,rel}^{(\bar{h})}(\mathbf{x}) := \frac{1}{\beta|\Lambda|}\sum_{\mathbf{k}'} e^{-i\mathbf{k}'\mathbf{x}} f_{\bar{h}}(\mathbf{k}') \left[A_{\bar{h}-1,\omega}^{rel}(\mathbf{k}')\right]^{-1}.$$

Finally, using the fact that

$$
\left[\overline{A}_{\bar{h}-1,\omega}(\mathbf{k}')\right]^{-1} = \left[A_{\bar{h}-1,\omega}^{rel}(\mathbf{k}')\right]^{-1} \sum_{n \geq 0} \left\{ R_2^{(\bar{h})}(\mathbf{k}') \left[A_{\bar{h}-1,\omega}^{rel}(\mathbf{k}')\right]^{-1} \right\}^n =
$$

$$
=: \left[A_{\bar{h}-1,\omega}^{rel}(\mathbf{k}')\right]^{-1} \left(1 + r_{\bar{h},\omega}(\mathbf{k}')\right), \tag{10.178}
$$

we rewrite $g_\omega^{(\bar{h})}(\mathbf{x}) = g_{\omega,rel}^{(\bar{h})}(\mathbf{x}) + \left[g_{\omega,rel}^{(\bar{h})} * r_{\bar{h},\omega}\right](\mathbf{x})$. Plugging this into equation (10.161), using the definition of $r_{\bar{h},\omega}$ and the inductive bounds on the kernels of the effective potentials, we get the desired estimate for the propagator on scale $\bar{h}$.

It remains to prove the estimates on $e_h, \bar{e}_h$. The bound on $\bar{e}_h$ is an immediate corollary of the discussion above, simply because $\bar{e}_h$ can be bounded by equation (10.162) with $l = 0$. Finally, remember that e_h is given by equation (10.142): an explicit computation of $A_{h,\omega}^{-1}(\mathbf{k}')W_{2,(\omega,\omega)}^{(h)}(\mathbf{k}')$ and by using once again the inductive bounds on $A_{h,\omega}$ and on the kernels of the effective potential, we find $||A_{h,\omega}^{-1}(\mathbf{k}')W_{2,(\omega,\omega)}^{(h)}(\mathbf{k}')|| \leq C|U|2^{\theta h}$, from which: $|e_h| \leq C'2^{3h} \sum_{n \geq 1} (C|U|2^{\theta h})^n$, as desired. This concludes the proof of Theorem 10.3 and, therefore, as discussed after the statement of Theorem 10.3, it also concludes the proof of analyticity of the specific free energy and ground state energy. $\qquad\square$

10.8 Conclusions

In conclusion, I presented a self-contained proof of the analyticity of the specific free energy and ground state energy of the two-dimensional Hubbard model on the honeycomb lattice, at half-filling and weak coupling. The proof is based on rigorous fermionic RG methods and can be extended to the construction of the interacting correlations, that is, the offdiagonal elements of the reduced density matrices of the system (Giuliani and Mastropietro 2010), and to the computation of the universal optical conductivity (Giuliani, Mastropietro and Porta, 2011). Such construction shows that the interacting correlations decay to zero at infinity with the same decay exponents as those of the non-interacting case. The 'only' effect of the interactions is to change by a finite amount the quasi-particle weight Z^{-1} at the Fermi surface and the Fermi velocity v.

The example presented here is the only known example of a realistic two-dimesional interacting Fermi system for which the ground state (including the correlations) can be constructed. The main difference with respect to other more standard two-dimesional Fermi systems is the fact that here, at half-filling, the Fermi surface reduces to a set of two isolated points. This fact dramatically improves the infrared scaling properties of the theory: the four-fermions interaction, rather than being marginal, as in many other similar cases, is irrelevant; this is the technical point that makes the construction of the ground state possible and 'relatively easy'.

It is natural to ask how the system behaves in the presence of electromagnetic interactions among the electrons, which is the case of interest for applications to clean graphene samples. In this case, the system has many analogies with $(2+1)$-dimensional QED. The four-fermions interaction, rather than being irrelevant, is marginal, and the fixed point of the theory is expected to be non-trivial. The long distance decay of correlations is expected to be described in terms of anomalous critical exponents and the effective Fermi velocity is expected to grow up to the maximal possible value, that is, the speed of light. The specific values of the critical exponents suggest that local distortions of the lattice (in the form of the so-called Kekulé pattern (Hou, Chamon, and Mudry, 2007)) are amplified by electromagnetic interactions: this led us to propose a possible mechanism for the spontaneous formation of the Kekulé pattern, via a mechanism analogous to the one-dimensional Peierls' instability (Giuliani, Mastropietro and Porta, 2010b). All these claims have been proved so far only order by order in renormalized perturbation theory (Giuliani, Mastropietro and Porta, 2010a,b); proving them in a non-perturbative fashion is an important open problem, whose solution would represent a cornerstone in the mathematical theory of quantum Coulomb systems.

10.9 Appendix: Dimensional estimates of the propagators

Proof: In this Appendix we prove the dimensional bounds on the propagators used in Sections 10.4–10.7 and, in particular, the bounds $||g||_\infty \leq CM$ and $||g||_1 \leq C\beta$, stated right before equation (10.63), and the estimates equations (10.88). The basic idea is to decompose the propagator in equation (10.42),

$$g(\mathbf{x}) = \frac{1}{\beta|\Lambda|} \sum_{\mathbf{k} \in \mathcal{B}_{\beta,L}^{(M)}} e^{-i\mathbf{k}\mathbf{x}} \frac{\chi_0(2^{-M}|k_0|)}{k_0^2 + v_0^2|\Omega(\vec{k})|^2} \begin{pmatrix} ik_0 & -v_0\Omega^*(\vec{k}) \\ -v_0\Omega(\vec{k}) & ik_0 \end{pmatrix}, \tag{10.179}$$

as a sum of single scale propagators:

$$g(\mathbf{x}) = \sum_{h=1}^{M} g^{(h)}(\mathbf{x}) + \sum_{\omega=\pm} \sum_{h=h_\beta}^{0} \bar{g}_\omega^{(h)}(\mathbf{x}), \tag{10.180}$$

where: (i) if $1 \leq h \leq M$, then $g^{(h)}(\mathbf{x})$ is defined as in equation (10.87); (ii) if $h_\beta := \lfloor \log_2 \left(\frac{3\pi}{4\beta} \right) \rfloor$ and $h_\beta \leq h \leq 0$, then

$$\bar{g}_\omega^{(h)}(\mathbf{x}) = \frac{1}{\beta|\Lambda|} \sum_{\mathbf{k} \in \mathcal{B}_{\beta,L}^{(M)}} e^{-i\mathbf{k}\mathbf{x}} \frac{f_h(k_0, \vec{k} - \vec{p}_F^\omega)}{k_0^2 + v_0^2|\Omega(\vec{k})|^2} \begin{pmatrix} ik_0 & -v_0\Omega^*(\vec{k}) \\ -v_0\Omega(\vec{k}) & ik_0 \end{pmatrix}, \tag{10.181}$$

with $f_h(\mathbf{k}') = \chi_0(2^{-h}|\mathbf{k}'|) - \chi_0(2^{-h+1}|\mathbf{k}'|)$. The main issue is to prove equation (10.88), that is,

$$||g^{(h)}(\mathbf{x})|| \leq \frac{C_K}{1 + (2^h|x_0|_\beta + |\vec{x}|_\Lambda)^K}, \tag{10.182}$$

and

$$||\bar{g}_\omega^{(h)}(\mathbf{x})|| \leq \frac{C_K 2^{2h}}{1 + (2^h||\mathbf{x}||)^K}, \tag{10.183}$$

where $||\mathbf{x}||^2 = |x_0|_\beta^2 + |\vec{x}|_\Lambda^2$, for all $0 \leq K \leq 6$. Note that $||g||_\infty \leq CM$ and $||g||_1 \leq C\beta$ are immediate corollaries of equations (10.182)–(10.183). In fact, plugging these bounds into equation (10.180), we find:

$$||g||_\infty \leq \sum_{h=1}^{M} C_0 + \sum_{\omega=\pm} \sum_{h=h_\beta}^{0} C_0 2^{2h} \leq C_0(M + \frac{8}{3}), \tag{10.184}$$

$$||g||_1 \leq C_4 \sum_{h=1}^{M} \int_{(\beta,\Lambda)} d\mathbf{x} \frac{1}{1 + (2^h|x_0|_\beta + |\vec{x}|_\Lambda)^4} + 2C_4 \sum_{h=h_\beta}^{0} \int_{(\beta,\Lambda)} d\mathbf{x} \frac{2^{2h}}{1 + (2^h||\mathbf{x}||)^4} \leq$$

$$\leq C_4' \sum_{h=h_\beta}^{M} 2^{-h} \leq C_4' 2^{-h_\beta + 1}, \tag{10.185}$$

which are the desired bounds.

So, let us start by proving equation (10.182). We denote by $\partial_{\mathbf{k}}$ the discrete derivative, and by $\tilde{\partial}_\mu = \tilde{e}_\mu \cdot \partial_{\mathbf{k}}$, with $\mu \in \{0, 1, 2\}$, the discrete derivative in the direction $\tilde{e}_\mu$ (here $\tilde{e}_0 = (1, 0, 0)$, $\tilde{e}_1 = (0, \frac{1}{2}, \frac{\sqrt{3}}{2})$ and $\tilde{e}_2 = (0, \frac{1}{2}, -\frac{\sqrt{3}}{2})$), defined as follows: given a compact support function $\hat{F}(\mathbf{k})$ with $\mathbf{k} \in \mathcal{B}_{\beta,L}^{(\infty)}$, we let

$$\tilde{\partial}_\mu \hat{F}(\mathbf{k}) = \tilde{e}_\mu \cdot \partial_{\mathbf{k}} \hat{F}(\mathbf{k}) = \frac{\hat{F}(\mathbf{k} + \tilde{e}_\mu \Delta k_\mu) - \hat{F}(\mathbf{k})}{\Delta k_\mu}, \tag{10.186}$$

with $\Delta k_0 = \frac{2\pi}{\beta}$ and $\Delta k_1 = \Delta k_2 = \frac{4\pi}{3L}$. Note that, if

$$F(\mathbf{x}) = \sum_{\mathbf{k} \in \mathcal{B}_{\beta,L}^{(\infty)}} e^{-i\mathbf{k} \cdot \mathbf{x}} \hat{F}(\mathbf{k}), \tag{10.187}$$

then, defining $\tilde{x}_\mu = \mathbf{x} \cdot \tilde{e}_\mu$,

$$\sum_{\mathbf{k} \in \mathcal{B}_{\beta,L}^{(\infty)}} e^{-i\mathbf{k} \cdot \mathbf{x}} \tilde{\partial}_\mu \hat{F}(\mathbf{k}) = \left(\frac{e^{i\Delta k_\mu \tilde{x}_\mu} - 1}{\Delta k_\mu} \right) \sum_{\mathbf{k} \in \mathcal{B}_{\beta,L}^{(\infty)}} e^{-i\mathbf{k} \cdot \mathbf{x}} \hat{F}(\mathbf{k}), \tag{10.188}$$

so that, using the fact that $|x_0|_\beta \leq \frac{\pi}{2} d_\beta(x_0)$ and $|\vec{x}|_\Lambda \leq \frac{\pi}{\sqrt{2}} (d_L^2(\tilde{x}_1) + d_L^2(\tilde{x}_2))^{1/2}$ (here $d_\beta(x_0) = \frac{\sin(\pi x_0/\beta)}{\pi/\beta}$ and $d_L(\tilde{x}_i) = \frac{\sin(2\pi \tilde{x}_i/3L)}{2\pi/3L}$), we find

$$|x_0|_\beta^2 |F(\mathbf{x})| \leq \frac{\pi^2}{4} d_\beta^2(x_0)|F(\mathbf{x})| = \frac{\pi^2}{4} \left| \frac{e^{i\Delta k_0 x_0} - 1}{\Delta k_0} \right|^2 \cdot \left| \sum_{\mathbf{k} \in \mathcal{B}_{\beta,L}^{(\infty)}} e^{-i\mathbf{k}\cdot\mathbf{x}} \hat{F}(\mathbf{k}) \right| =$$

$$= \frac{\pi^2}{4} \left| \sum_{\mathbf{k} \in \mathcal{B}_{\beta,L}^{(\infty)}} e^{-i\mathbf{k}\cdot\mathbf{x}} \partial_{k_0}^2 \hat{F}(\mathbf{k}) \right| \leq \frac{\pi^2}{4} \sum_{\mathbf{k} \in \mathcal{B}_{\beta,L}^{(\infty)}} |\partial_{k_0}^2 \hat{F}(\mathbf{k})|, \qquad (10.189)$$

and, similarly,

$$|\vec{x}|_\Lambda^2 |F(\mathbf{x})| \leq \frac{\pi^2}{2} \left(d_L^2(\tilde{x}_1) + d_L^2(\tilde{x}_2) \right) |F(\mathbf{x})| =$$

$$= \frac{\pi^2}{2} \sum_{\mu=1}^{2} \left| \frac{e^{i\Delta k_\mu \tilde{x}_\mu} - 1}{\Delta k_\mu} \right|^2 \cdot \left| \sum_{\mathbf{k} \in \mathcal{B}_{\beta,L}^{(\infty)}} e^{-i\mathbf{k}\cdot\mathbf{x}} \hat{F}(\mathbf{k}) \right| = \qquad (10.190)$$

$$= \frac{\pi^2}{2} \sum_{\mu=1}^{2} \left| \sum_{\mathbf{k} \in \mathcal{B}_{\beta,L}^{(\infty)}} e^{-i\mathbf{k}\cdot\mathbf{x}} \tilde{\partial}_\mu^2 \hat{F}(\mathbf{k}) \right| \leq$$

$$\leq \frac{3\pi^2}{8} \sum_{\mathbf{k} \in \mathcal{B}_{\beta,L}^{(\infty)}} \left(|\partial_{k_1}^2 \hat{F}(\mathbf{k})| + 2|\partial_{k_1} \partial_{k_2} \hat{F}(\mathbf{k})| + |\partial_{k_2}^2 \hat{F}(\mathbf{k})| \right).$$

Coming back to equation (10.182), recalling that

$$g^{(h)}(\mathbf{x}) = \frac{1}{\beta|\Lambda|} \sum_{\mathbf{k} \in \mathcal{B}_{\beta,L}^{(M)}} e^{-i\mathbf{k}\mathbf{x}} \frac{f_{u.v.}(\mathbf{k}) H_h(k_0)}{k_0^2 + v_0^2 |\Omega(\vec{k})|^2} \begin{pmatrix} ik_0 & -v_0\Omega^*(\vec{k}) \\ -v_0\Omega(\vec{k}) & ik_0 \end{pmatrix}, \qquad h \geq 2.$$

$$(10.191)$$

Therefore,

$$\|g^{(h)}(\mathbf{x})\| \leq \frac{1}{\beta|\Lambda|} \sum_{\mathbf{k} \in \mathcal{B}_{\beta,L}^{(M)}} \frac{f_{u.v.}(\mathbf{k}) H_h(k_0)}{[k_0^2 + v_0^2 |\Omega(\vec{k})|^2]^{1/2}} \leq (\text{const.}) 2^h \, 2^{-h}, \qquad (10.192)$$

where in the last inequality we used the fact that, on the support of $f_{u.v.}(\mathbf{k}) H_h(k_0)$, $k_0^2 + v_0^2 |\Omega(\vec{k})|^2 \sim 2^{2h}$ (here $\sim$ means that the ratio of the two sides is bounded above and below by universal constants) and that, moreover, the measure of the support of $f_{u.v.}(\mathbf{k}) H_h(k_0)$ is itself of order 2^h, that is $(\beta|\Lambda|)^{-1} \sum_{\mathbf{k}} f_{u.v.}(\mathbf{k}) H_h(k_0) \sim 2^h$.

Moreover, using equation (10.189)–(10.190), we have that for all $N \geq 0$ and a suitable constant C,

$$|x_0|_\beta^{2N}||g^{(h)}(\mathbf{x})|| \leq \frac{C^N}{\beta|\Lambda|} \sum_{\mathbf{k}\in\mathcal{B}_{\beta,L}^{(M)}} \left\|\partial_{k_0}^{2N}\left[\frac{f_{u.v.}(\mathbf{k})H_h(k_0)}{k_0^2 + v_0^2|\Omega(\vec{k})|^2}\begin{pmatrix} ik_0 & -v_0\Omega^*(\vec{k}) \\ -v_0\Omega(\vec{k}) & ik_0 \end{pmatrix}\right]\right\|,$$

$$|\vec{x}|_\Lambda^{2N}||g^{(h)}(\mathbf{x})|| \leq \frac{C^N}{\beta|\Lambda|} \sum_{\mathbf{k}\in\mathcal{B}_{\beta,L}^{(M)}} \sum_{j=0}^{N} \binom{N}{j} .$$

$$\cdot\left\|\partial_{k_1}^{j}\partial_{k_2}^{N-j}\left[\frac{f_{u.v.}(\mathbf{k})H_h(k_0)}{k_0^2 + v_0^2|\Omega(\vec{k})|^2}\begin{pmatrix} ik_0 & -v_0\Omega^*(\vec{k}) \\ -v_0\Omega(\vec{k}) & ik_0 \end{pmatrix}\right]\right\|$$

Now, in the first line, every derivative with respect to k_0 can act either on the denominator $k_0^2 + v_0^2|\Omega(\vec{k})|^2$, or on $f_{u.v.}(\mathbf{k})H_h(k_0)$, or on the diagonal elements of the matrix, ik_0; in all these cases, recalling that, on the support of $f_{u.v.}H_h$, $|k_0| \sim 2^h$, every derivative ∂_{k_0} can be estimated dimensionally by a factor proportional to 2^{-h}. This leads to the bound

$$|x_0|_\beta^{2N}||g^{(h)}(\mathbf{x})|| \leq (\text{const.})^N 2^h\, 2^{-2Nh}\, 2^{-h}, \qquad (10.193)$$

which differs from equation (10.192) precisely by the factor 2^{-2Nh}. Similarly, in the second line, the derivative with respect to $\vec{k}$ can act either on the denominator $k_0^2 + v_0^2|\Omega(\vec{k})|^2$, or on $f_{u.v.}(\mathbf{k})H_h(k_0)$, or on the off-diagonal elements of the matrix, $-v_0\Omega(\vec{k})$ or $-v_0\Omega^*(\vec{k})$; in the first case, $\partial_{\vec{k}}$ can be estimated dimensionally by a factor proportional to 2^{-h} while, in the other cases, by an order 1 factor. This leads to the bound

$$|\vec{x}|_\Lambda^{2N}||g^{(h)}(\mathbf{x})|| \leq (\text{const.})^N 2^h\, 2^{-h}, \qquad (10.194)$$

which, if combined with equations (10.192)–(10.193), finally implies equation (10.182).

The proof of equation (10.183) is very similar. In fact, for all $N \geq 0$,

$$||\mathbf{x}||^{2N}||\bar{g}_\omega^{(h)}(\mathbf{x})|| \leq \frac{C^N}{\beta|\Lambda|} \sum_{\mathbf{k}\in\mathcal{B}_{\beta,L}^{(M)}} \left\|\partial_{\mathbf{k}}^{2N}\left[\frac{f_h(k_0,\vec{k}-\vec{p}_F^\omega)}{k_0^2 + v_0^2|\Omega(\vec{k})|^2}\begin{pmatrix} ik_0 & -v_0\Omega^*(\vec{k}) \\ -v_0\Omega(\vec{k}) & ik_0 \end{pmatrix}\right]\right\|$$

$$(10.195)$$

Now, using the fact that, on the support of $f_h(k_0,\vec{k}-\vec{p}_F^\omega)$, $|k_0|, |\vec{k}-\vec{p}_F^\omega|, |\Omega(\vec{k})| \sim 2^h$, we see that every derivative $\partial_{\mathbf{k}}$, when acting either on f_h, or on the denominator $k_0^2 + v_0^2|\Omega(\vec{k})|^2$, or on the elements of the matrix, can be estimated dimensionally by a factor proportional to 2^{-h}; moreover, the measure of the support of f_h is itself of order 2^{3h}, that is $(\beta|\Lambda|)^{-1}\sum_{\mathbf{k}} f_h(k_0,\vec{k}-\vec{p}_F^\omega) \sim 2^{3h}$. This leads to the bound

$$||\mathbf{x}||^{2N}||\bar{g}_\omega^{(h)}(\mathbf{x})|| \leq (\text{const.})^N 2^{3h}\, 2^{-2Nh}\, 2^{-h}, \qquad (10.196)$$

which finally implies equation (10.183). $\qquad\qquad\qquad\qquad\qquad\qquad\qquad \square$

10.10   Appendix: Truncated expectations and determinants

Proof: In this Appendix we prove equation (10.66), following (Gentile and Mastropietro 2001, Appendix A.3.2). Given s set of indices $P_1, \ldots, P_s$, consider the quantity $\mathcal{E}^T(\psi_{P_1}, \ldots, \psi_{P_s})$. Define

$$P_j^{\pm} = \{f \in P_j \; : \; \varepsilon(f) = \pm\} \tag{10.197}$$

and set $f = (j, i)$ for $f \in P_j^{\pm}$, with $i = 1, \ldots, |P_j^{\pm}|$. Note that $\sum_{j=1}^{s} |P_j^+| = \sum_{j=1}^{s} |P_j^-|$, otherwise the considered truncated expectation is vanishing. Define

$$\mathcal{D}\psi = \prod_{j=1}^{s} \left[\prod_{f \in P_j^+} d\,\psi^+_{\mathbf{x}(f), \sigma(f), \rho(f)} \right] \left[\prod_{f \in P_j^-} d\,\psi^-_{\mathbf{x}(f), \sigma(f), \rho(f)} \right], \tag{10.198}$$

$$(\psi^+, G\psi^-) = \sum_{j,j'=1}^{s} \sum_{i=1}^{|P_j^-|} \sum_{i'=1}^{|P_{j'}^+|} \psi^+_{(j',i')} G_{(j,i),(j',i')} \psi^-_{(j,i)}, \tag{10.199}$$

where $\psi^{\pm}_{(j,i)} := \psi^{\pm}_{\mathbf{x}(j,i), \sigma(j,i), \rho(j,i)}$ and, if $n = \sum_{j=1}^{s} |P_j^+| = \sum_{j=1}^{s} |P_j^-|$, then G is the $n \times n$ matrix with entries

$$G_{(j,i),(j',i')} := \delta_{\sigma(j,i), \sigma(j',i')} g_{\rho(j,i), \rho(j',i')}(\mathbf{x}(j,i) - \mathbf{x}(j',i')), \tag{10.200}$$

so that

$$\mathcal{E}\left(\prod_{j=1}^{s} \psi_{P_j}\right) = \det G = \int \mathcal{D}\psi \, \exp\left[-(\psi^+, G\psi^-)\right]. \tag{10.201}$$

Setting $X := \{1, \ldots, s\}$ and

$$\overline{V}_{jj'} = \sum_{i=1}^{|P_j^-|} \sum_{i'=1}^{|P_{j'}^+|} \psi^+_{(j',i')} G_{(j,i),(j',i')} \psi^-_{(j,i)}, \tag{10.202}$$

we write

$$V(X) = \sum_{j,j' \in X} \overline{V}_{jj'} = \sum_{j \leq j'} V_{jj'}, \tag{10.203}$$

where

$$V_{jj'} = \begin{cases} \overline{V}_{jj'}, & \text{if } j = j', \\ \overline{V}_{jj'} + \overline{V}_{j'j}, & \text{if } j < j'. \end{cases} \tag{10.204}$$

In terms of these definitions, equation (10.201) can be rewritten as

$$\mathcal{E}\left(\prod_{j=1}^{s}\psi_{P_j}\right) = \det G = \int \mathcal{D}\psi \, e^{-V(X)}. \tag{10.205}$$

We now want to express the last expression in terms of the functions W_X, defined as follows:

$$W_X(X_1,\ldots,X_r;t_1,\ldots,t_r) = \sum_{\ell \in L(X)} \prod_{k=1}^{r} t_k(\ell) \, V_\ell, \tag{10.206}$$

where:

1. X_k are subsets of X with $|X_k| = k$ and such that

$$\begin{cases} X_1 = \{1\}, \\ X_{k+1} \supset X_k; \end{cases} \tag{10.207}$$

2. $L(X)$ is the set of unordered pairs in X, that is, the set of pairs (j,j') with $j,j' \in X$, such that (j,j') is identified with (j',j) and, possibly, $j = j'$; we shall say that $\ell = (j,j')$ is *nontrivial* if $j \neq j'$, and *trivial* otherwise;
3. the functions $t_k(\ell)$ are defined as follows:

$$t_k(\ell) = \begin{cases} t_k, & \text{if} \quad \ell \sim \partial X_k, \\ 1, & \text{otherwise}, \end{cases} \tag{10.208}$$

where $\ell \sim X_k$ means that $\ell = (j,j')$ 'intersects the boundary' of X_k, that is it connects $j \in X_k$ with $j' \notin X_k$, or vice versa. See Fig. 10.7.

Let us show how to re-express $e^{-V(X)}$ in terms of the W_X's. The basic step is the following: using the definition equation (10.206), we rewrite

$$W_X(X_1;t_1) = t_1 V(X) + (1 - t_1)\left[V(X_1) + V(X \backslash X_1)\right]; \tag{10.209}$$

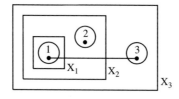

Fig. 10.7 Graphical representation of the sets X_k, $k = 1, 2, 3$. In the example $X_1 = \{1\}$, $X_2 = \{1, 2\}$, $X_3 = \{1, 2, 3\}$ and $\ell = (1, 3)$. The line ℓ intersects both the boundary of X_1 and of X_2, that is $\ell \sim \partial X_1$ and $\ell \sim \partial X_2$.

that is, we recognize that $W_X(X_1; t_1)$ interpolates between the full $V(X)$ and two of its 'proper subsets', $V(X_1)$, $V(X \backslash X_1)$. In this way,

$$e^{-V(X)} = e^{-W_X(X_1;0)} + \int_0^1 d t_1 \left[\frac{\partial}{\partial t_1} e^{-W_X(X_1;t_1)} \right]$$

$$= e^{-W_X(X_1;0)} - \sum_{\ell_1 \sim \partial X_1} V_{\ell_1} \int_0^1 d t_1 \, e^{-W_X(X_1;t_1)}. \qquad (10.210)$$

Let us now iterate this construction: let us consider one of the terms in the summation in the r.h.s. of equation (10.210); if $\ell_1 = (1, j^*)$, we let $X_2 := X_1 \cup \{j^*\}$ and we note that, by definition,

$$W_X(X_1, X_2; t_1, t_2) = t_2 W_X(X_1; t_1) + (1 - t_2) \left[W_{X_2}(X_1; t_1) + V(X \backslash X_2) \right], \qquad (10.211)$$

so that

$$e^{-W_X(X_1;t_1)} = e^{-W_X(X_1,X_2;t_1,0)} + \int_0^1 d t_2 \left[\frac{\partial}{\partial t_2} e^{-W_X(X_1,X_2;t_1,t_2)} \right] \qquad (10.212)$$

$$= e^{-W_X(X_1,X_2;t_1,0)} - \sum_{\ell_2 \sim \partial X_2} V_{\ell_2} \int_0^1 d t_2 \, t_1(\ell_2) \, e^{-W_X(X_1,X_2;t_1,t_2)}.$$

Substituting equation (10.212) into equation (10.210) we get:

$$e^{-V(X)} = e^{-W_X(X_1;0)} + \sum_{\ell_1 \sim \partial X_1} \int_0^1 d t_1 \, (-1) V_{\ell_1} \, e^{-W_X(X_1,X_2;t_1,0)} + \qquad (10.213)$$

$$+ \sum_{\ell_1 \sim \partial X_1} \sum_{\ell_2 \sim \partial X_2} \int_0^1 d t_1 \int_0^1 d t_2 \, (-1)^2 \, V_{\ell_1} V_{\ell_2} \, t_1(\ell_2) \, e^{-W_X(X_1,X_2;t_1,t_2)}.$$

Using the fact that

$$W_X(X_1, \ldots, X_{p+1}; t_1, \ldots, t_{p+1}) = t_{p+1} W_X(X_1, \ldots, X_p; t_1, \ldots, t_p) +$$
$$(1 - t_{p+1}) \left[W_{X_{p+1}}(X_1, \ldots, X_p; t_1, \ldots, t_p) + V(X \backslash X_{p+1}) \right] \qquad (10.214)$$

and iterating we find

$$e^{-V(X)} = \sum_{r=0}^{s-1} \sum_{\ell_1 \sim \partial X_1} \cdots \sum_{\ell_r \sim \partial X_r} \int_0^1 d t_1 \ldots \int_0^1 d t_r \, (-1)^r \, V_{\ell_1} \ldots V_{\ell_r}$$

$$\left(\prod_{k=1}^{r-1} t_1(\ell_{k+1}) \ldots t_k(\ell_{k+1}) \right) e^{-W_X(X_1,\ldots,X_{r+1};t_1,\ldots,t_r,0)}, \qquad (10.215)$$

where, if $r = 0$, the summand should be interpreted as equal to $e^{-W_X(X_1;0)}$. Moreover, by the definition of the W_X's, if $r > 1$,

$$W_X(X_1,\ldots,X_r;t_1,\ldots,t_{r-1},0) = W_{X_r}(X_1,\ldots,X_{r-1};t_1,\ldots,t_{r-1}) + V(X \setminus X_r).$$

$$(10.216)$$

Using this remark and the notion of *anchored tree*, which will be defined in a moment, equation (10.215) can be rewritten in a more suggestive and convenient way. Let $\mathcal{T}(X)$ be the set of tree graphs on X, that is the set of $(|X| - 1)$-ples of lines in $L(X)$ connecting (in a minimal way) all the elements of X. Given a sequence of subsets $X_1 \subset \cdots \subset X_r$ as above, we shall say that $T \in \mathcal{T}(X_r)$ is an *anchored tree* on $(X_1,\ldots,X_r)$ if its lines can be ordered in such a way that $\ell_1 \sim \partial X_1, \ell_2 \sim \partial X_2, \ldots, \ell_{r-1} \sim \partial X_{r-1}$. Moreover, given a sequence $X_1 \subset \cdots \subset X_s$ as above and a nontrivial line $\ell \in L(X)$, we let

$$n(\ell) = \max\{k : \ell \sim \partial X_k\}, \qquad n'(\ell) = \min\{k : \ell \sim \partial X_k\}; \qquad (10.217)$$

if ℓ is trivial, we let $n(\ell) = n'(\ell) = 0$. Using these definitions, we rewrite equation (10.215) as:

$$e^{-V(X)} = \sum_{r=1}^{s} \sum_{X_r \subset X} \sum_{X_2,\ldots,X_{r-1}} \sum_{T \text{ on}(X_1,\ldots,X_r)} (-1)^{r-1} \prod_{\ell \in T} V_\ell \cdot \qquad (10.218)$$

$$\cdot \int_0^1 dt_1 \ldots \int_0^1 dt_{r-1} \left(\prod_{\ell \in T} \frac{\prod_{k=1}^{r-1} t_k(\ell)}{t_{n(\ell)}} \right) e^{-W_{X_r}(X_1,\ldots,X_{r-1};t_1,\ldots,t_{r-1})} e^{-V(X \setminus X_r)},$$

where "T on$(X_1,\ldots,X_r)$" means that T is an anchored tree on $(X_1,\ldots,X_r)$. Defining

$$K(X_r) = \sum_{X_2,\ldots,X_{r-1}} \sum_{T \text{ on}(X_1,\ldots,X_r)} \prod_{\ell \in T} V_\ell \qquad (10.219)$$

$$\int_0^1 dt_1 \ldots \int_0^1 dt_{r-1} \left(\prod_{\ell \in T} \frac{\prod_{k=1}^{r-1} t_k(\ell)}{t_{n(\ell)}} \right) e^{-W_{X_r}(X_1,\ldots,X_{r-1};t_1,\ldots,t_{r-1})},$$

equation (10.218) can be further rewritten as

$$e^{-V(X)} = \sum_{\substack{Y \subset X \\ Y \ni \{1\}}} (-1)^{|Y|-1} K(Y) e^{-V(X \setminus Y)}, \qquad (10.220)$$

and, iterating,

$$e^{-V(X)} = \sum_{m=1}^{s} \sum_{(Y^1,\ldots,Y^m)} (-1)^s (-1)^m \prod_{i=1}^{m} K(Y^i), \qquad (10.221)$$

where the second summation runs over partitions of X of multiplicity m, i.e., over m-ples of disjoint sets $Y^1,\ldots,Y^m$ such that $\cup_{i=1}^{m} Y^i = X$. Plugging equation (10.221) into equation (10.205) gives

$$\mathcal{E}\left(\prod_{j=1}^{s}\psi_{P_j}\right) = \sum_{m=1}^{s}\sum_{(Y^1,\ldots,Y^m)}(-1)^{s+m}(-1)^{\sigma}\int \mathcal{D}\psi_{Y^i}\prod_{i=1}^{m}K(Y^i), \qquad (10.222)$$

where $\mathcal{D}\psi_{Y^i} = \prod_{j\in Y^i}\left[\prod_{f\in P_j^+}\mathrm{d}\,\psi^+_{\mathrm{x}(f),\sigma(f),\rho(f)}\right]\left[\prod_{f\in P_j^-}\mathrm{d}\,\psi^-_{\mathrm{x}(f),\sigma(f),\rho(f)}\right]$ and $(-1)^{\sigma}$ is the sign of the permutation leading from the ordering of the fields in $\mathcal{D}\psi$ to the ones in $\prod_i \mathcal{D}\psi_{Y^i}$. In equation (10.222), each factor $K(Y^i)$, after small manipulations of its definition, equation (10.219), can be rewritten as

$$K(Y^i) = \sum_{T\in\mathcal{T}(Y^i)}\sum_{\substack{Y_2^i,\ldots,Y_{|Y^i|-1}^i\\ \text{compatible with } T}}\prod_{\ell\in T}V_{\ell}\int_0^1 dt_1\cdots\int_0^1 dt_{|Y^i|-1}\cdot$$

$$\cdot\prod_{\ell\in T}(t_{n'(\ell)}\cdots t_{n(\ell)-1})e^{-\sum_{\ell\in L(Y^i)}t_{n'(\ell)}\cdots t_{n(\ell)}V_{\ell}}, \qquad (10.223)$$

where: (i) $Y_1^i := \{\min\{j : j \in Y^i\}\}$; (ii) the second summations is over sequences of subsets Y_q^i of Y^i with $|Y_q^i| = q$ and such that $Y_1^i \subset Y_2^i \subset \cdots \subset Y_{|Y^i|-1}^i \subset Y_{|Y^i|}^i \equiv Y^i$ and T is anchored on $(Y_1^i,\ldots,Y_{|Y^i|}^i)$; (iii) in the second line, if $n(\ell) = n'(\ell)$, the factor $t_{n'(\ell)}\cdots t_{n(\ell)-1}$ should be interpreted as equal to 1; (iv) in the exponent, if ℓ is trivial (and, therefore, $n'(\ell) = n(\ell) = 0$), t_0 should be interpreted as equal to 1. Now, using the analogue of equation (10.57) in a case, like the one at hand, where the monomials ψ do not necessarily all commute among each other, we have that (denoting the elements of Y^i as $Y^i = \{j_1^i,\ldots,j_{|Y_i|}^i\}$)

$$\mathcal{E}\left(\prod_{j=1}^{s}\psi_{P_j}\right) = \sum_{m=1}^{s}\sum_{(Y^1,\ldots,Y^m)}(-1)^{\sigma}\mathcal{E}^T(\psi_{P_{j_1^1}},\ldots,\psi_{P_{j_{|Y^1|}^1}})\cdots\mathcal{E}^T(\psi_{P_{j_1^m}},\ldots,\psi_{P_{j_{|Y^m|}^m}})$$

$$(10.224)$$

where $(-1)^{\sigma}$ is the parity of the permutation leading to the ordering on the r.h.s. from the one on the l.h.s.; note that σ is the same as in equation (10.222). Comparing equation (10.224) with equation (10.222) and using equation (10.223), we get:

$$\mathcal{E}^T(\psi_{P_1},\ldots,\psi_{P_s}) = (-1)^{s+1}\sum_{T\in\mathcal{T}(X)}\int \mathcal{D}\psi\prod_{\ell\in T}V_{\ell}\int dP_T(\mathbf{t})e^{-V(\mathbf{t})}, \qquad (10.225)$$

where we defined:

$$dP_T(\mathbf{t}) = \sum_{\substack{X_2,\ldots,X_{s-1}\\ \text{compatible with } T}}\prod_{\ell\in T}(t_{n'(\ell)}\cdots t_{n(\ell)-1})\prod_{q=1}^{s-1}\mathrm{d}\,t_q \qquad (10.226)$$

and

$$V(\mathbf{t}) \equiv \sum_{\ell\in L(X)}t_{n'(\ell)}\cdots t_{n(\ell)}V_{\ell}. \qquad (10.227)$$

Finally, if we use the definition equations (10.202) and (10.204) and explicitly integrate the Grassmann variables along the lines of the anchored tree, we end up with

$$\mathcal{E}^T\left(\psi_{P_1},\ldots,\psi_{P_s}\right) = \sum_{T\in\mathbf{T}(\mathcal{P})} \alpha_T \prod_{\ell\in T} G_{f^1_\ell,f^2_\ell} \int \mathcal{D}^*(\mathrm{d}\,\psi) \int \mathrm{d}\,P_T(\mathbf{t})\, e^{-V^*(\mathbf{t})}, \qquad (10.228)$$

where $\mathcal{P} = \{P_1,\ldots,P_s\}$ and $\mathbf{T}(\mathcal{P})$ is the set of anchored trees between the 'boxes' $P_1,\ldots,P_s$, that is the graphs that become trees if one identifies all the points in the same 'clusters' P_i (note that now the lines of an anchored tree $T \in \mathbf{T}(\mathcal{P})$ are pairs of field variables (f,f'), rather than pairs of indices (j,j')). Moreover, α_T is a sign (irrelevant to the purpose of the bounds performed in this chapter), f^1_ℓ, f^2_ℓ are the two field labels associated to the two (entering and exiting) half-lines contracted into ℓ, and

$$\mathcal{D}^*(\mathrm{d}\,\psi) = \prod_{\substack{f\in P \\ f\notin T}} \mathrm{d}\psi_{\mathrm{x}(f),\sigma(f),\rho(f)}^{\varepsilon(f)}, \qquad V^*(\mathbf{t}) = \sum_{\ell\in L(X)} t_{n'(\ell)}\cdots t_{n(\ell)}\, V_\ell^T \qquad (10.229)$$

where

$$V_{jj'}^T = \begin{cases} \overline{V}_{jj'}^T, & \text{if } j = j', \\ \overline{V}_{jj'}^T + \overline{V}_{j'j}^T, & \text{if } j < j'. \end{cases} \qquad (10.230)$$

and, if $((j,i),(j',i'))$ is the line obtained by contracting $\psi_{(j,i)}^-$ with $\psi_{(j',i')}^+$,

$$\overline{V}_{jj'}^T = \sum_{i=1}^{|P_j^-|}\sum_{i'=1}^{|P_{j'}^+|} \psi_{(j',i')}^+ G_{(j,i),(j',i')}\, \psi_{(j,i)}^- \times \Big(((j,i),(j',i')) \notin T\Big). \qquad (10.231)$$

The term $\int \mathcal{D}^*(\mathrm{d}\,\psi)\, e^{-V^*(\mathbf{t})}$ in equation (10.228) is the determinant of the $(n - s + 1) \times (n - s + 1)$ matrix $G^T(\mathbf{t})$ (here $2n = \sum_{i=1}^s |P_i|$), with elements $G_{f,f'}^T(\mathbf{t}) = t_{j(f),j(f')} G_{f,f'}$, where $f, f' \notin T$, $j(f) \in \{1,\ldots,s\}$ and $t_{j,j'} = t_{n'((j,j'))}\cdots t_{n((j,j'))}$:

$$\int \mathcal{D}^*(\mathrm{d}\,\psi)\, e^{-V^*(\mathbf{t})} = \det G^T(\mathbf{t}). \qquad (10.232)$$

Plugging this into equation (10.228) finally gives equation (10.66). In order to complete the proof of the claims following equation (10.66) we are left with proving the following Lemma. □

Lemma 10.4 *$dP_T(\mathbf{t})$ is a normalized, positive and σ-additive measure on the natural σ-algebra of $[0,1]^{s-1}$. Moreover there exists a set of unit vectors $\mathbf{u}_j \in \mathbb{R}^s$, $j = 1,\ldots,s$, such that $t_{j,j'} = \mathbf{u}_j \cdot \mathbf{u}_{j'}$.*

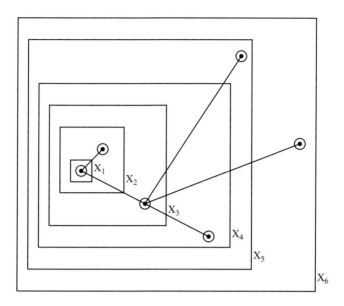

Fig. 10.8 The sets $X_1, \ldots, X_6$, the anchored tree T and the lines $\ell_1, \ldots, \ell_5$ belonging to T. In the example, the coefficients $b_1, \ldots, b_5$ are respectively equal to: $2, 1, 3, 2, 1$.

Proof: Let us denote by b_k the number of lines $\ell \in T$ exiting from the points $x(j, i)$, $j \in X_k$, such that $\ell \sim X_k$. Let us consider the integral

$$\sum_{\substack{X_2, \ldots, X_{s-1} \\ \text{compatible with } T}} \int_0^1 d t_1 \ldots \int_0^1 d t_{s-1} \prod_{\ell \in T} \left(t_{n'(\ell)} \ldots t_{n(\ell)-1}\right) = 1, \qquad (10.233)$$

and note that, by construction, the parameter t_k inside the integral in the l.h.s. appears at the power $b_k - 1$. In fact any line intersecting ∂X_k contributes by a factor t_k, except for the line connecting X_k with the point in $X_{k+1} \backslash X_k$. See Fig. 10.8.
Then

$$\prod_{\ell \in T} \left(t_{n'(\ell)} \ldots t_{n(\ell)-1}\right) = \prod_{k=1}^{s-1} t_k^{b_k - 1}, \qquad (10.234)$$

and in equation (10.233) the $s - 1$ integrations are independent. One has:

$$\int_0^1 d t_1 \ldots \int_0^1 d t_{s-1} \prod_{\ell \in T} \left(t_{n'(\ell)} \ldots t_{n(\ell)-1}\right) = \prod_{k=1}^{s-1} \left(\int_0^1 d t_k \, t_k^{b_k - 1}\right) = \prod_{k=1}^{s-1} \frac{1}{b_k}, \qquad (10.235)$$

which is well defined, since $b_k \geq 1$. Moreover we can write:

$$\underset{\substack{X_2,\dots,X_{s-1} \\ \text{compatible with } T}}{\sum} = \overset{*}{\underset{\substack{X_2 \\ X_1 \text{ fixed}}}{\sum}} \quad \overset{*}{\underset{\substack{X_3 \\ X_1,X_2 \text{ fixed}}}{\sum}} \quad \cdots \quad \overset{*}{\underset{\substack{X_{s-1} \\ X_1,\dots,X_{s-2} \text{ fixed}}}{\sum}} \quad , \tag{10.236}$$

where the $*$'s over the sums remind that all the summations are subject to the constraint that the subsets $X_1, X_2, \dots, X_s$ must be compatible with the structure of T. Now, the number of terms in the sum over X_k, once that T and the sets $X_1, \dots, X_{k-1}$ are fixed, is exactly b_{k-1}, so that

$$\underset{\substack{X_2,\dots,X_{s-1} \\ \text{compatible with } T}}{\sum} 1 = b_1 \dots b_{s-2}, \tag{10.237}$$

and, recalling that $b_{s-1} = 1$,

$$\underset{\substack{X_2,\dots,X_{s-1} \\ \text{compatible with } T}}{\sum} \int_0^1 d\,t_1 \dots \int_0^1 d\,t_{s-1} \prod_{\ell \in T} \left(t_{n'(\ell)} \dots t_{n(\ell)-1} \right) = \prod_{k=1}^{s-2} \frac{b_k}{b_k}, \tag{10.238}$$

yielding to $\int dP_T(\mathbf{t}) = 1$. The positivity and σ–additivity of $dP_T(\mathbf{t})$ is obvious by definition.

We are left with proving that, for any given sequence of subsets $X_1 \subset X_2 \subset \cdots \subset X_s$ compatible with T, we can find unit vectors $\mathbf{u}_j \in \mathbb{R}^s$ such that $t_{j,j'} = \mathbf{u}_j \cdot \mathbf{u}_{j'}$. With no loss of generality, we can assume that $X_1 = \{1\}$, $X_2 = \{1,2\}$, $\dots$, $X_{s-1} = \{1,\dots,s-1\}$. We introduce a family of unit vectors in $\mathbb{R}^s$ defined as follows:

$$\begin{cases} \mathbf{u}_1 = \mathbf{v}_1, \\ \mathbf{u}_j = t_{j-1}\mathbf{u}_{j-1} + \mathbf{v}_j\sqrt{1 - t_{j-1}^2}, \quad j = 2,\dots,s, \end{cases} \tag{10.239}$$

where $\{\mathbf{v}_i\}_{i=1}^s$ is an orthonormal basis. From this equation, as desired:

$$\mathbf{u}_j \cdot \mathbf{u}_{j'} = t_j \dots t_{j'-1}. \tag{10.240}$$

This concludes the proof of the determinant formula. □

10.11   Appendix: Symmetry properties

In this Appendix we prove Lemma 10.6.2. The key remark underlying the nice properties stated in the Lemma is that both the Gaussian integration $P_M(d\psi)$ and the interaction $\mathcal{V}(\psi)$ are invariant under the action of a number of remarkable symmetry transformations, which are also preserved by the multiscale integration. In the following, we denote by $\sigma_1, \sigma_2, \sigma_3$ the standard Pauli matrices:

$$\sigma_1 = \begin{pmatrix} 0 & 1 \\ 1 & 0 \end{pmatrix}, \qquad \sigma_2 = \begin{pmatrix} 0 & -i \\ i & 0 \end{pmatrix}, \qquad \sigma_3 = \begin{pmatrix} 1 & 0 \\ 0 & -1 \end{pmatrix}. \tag{10.241}$$

and $n_\pm = (1 - \sigma_3)/2$. Moreover, in order to avoid confusion with the Pauli matrices, we shall use the symbol τ as spin index rather than σ. As a preliminary result, let us start by listing all the transformation properties under which our theory is invariant.

Lemma 10.5 *For any choice of M, β, L, the fermionic Gaussian integration $P(d\psi)$ and the interaction $\mathcal{V}(\psi)$ are separately invariant under the following transformations:*

(1) *Spin flip:* $\hat{\psi}^\varepsilon_{\mathbf{k},\tau} \leftrightarrow \hat{\psi}^\varepsilon_{\mathbf{k},-\tau}$;

(2) *Global $U(1)$:* $\hat{\psi}^\varepsilon_{\mathbf{k},\tau} \to e^{i\varepsilon\alpha_\tau} \hat{\psi}^\varepsilon_{\mathbf{k},\tau}$, *with $\alpha_\tau \in \mathbb{R}/2\pi\mathbb{Z}$ independent of $\mathbf{k}$;*

(3) *Spin $SO(2)$:* $\begin{pmatrix} \hat{\psi}^\varepsilon_{\mathbf{k},\uparrow,\cdot} \\ \hat{\psi}^\varepsilon_{\mathbf{k},\downarrow,\cdot} \end{pmatrix} \to e^{-i\theta\sigma_2} \begin{pmatrix} \hat{\psi}^\varepsilon_{\mathbf{k},\uparrow,\cdot} \\ \hat{\psi}^\varepsilon_{\mathbf{k},\downarrow,\cdot} \end{pmatrix}$, *with $\theta \in \mathbb{R}/2\pi\mathbb{Z}$ independent of $\mathbf{k}$;*

(4) *Discrete rotations:* $\hat{\psi}^-_{\mathbf{k},\tau} \to e^{i\vec{k}(\vec{\delta}_3 - \vec{\delta}_1)n_-} \hat{\psi}^-_{T\mathbf{k},\tau}$ *and* $\hat{\psi}^+_{\mathbf{k},\tau} \to \hat{\psi}^+_{T\mathbf{k},\tau} e^{-i\vec{k}(\vec{\delta}_3 - \vec{\delta}_1)n_-}$, *with $T\mathbf{k} = (k_0, e^{-i\frac{2\pi}{3}\sigma_2} \vec{k})$;*

(5) *Complex conjugation:* $\hat{\psi}^\varepsilon_{\mathbf{k},\tau} \to \hat{\psi}^\varepsilon_{-\mathbf{k},\tau}$ *and $c \to c^*$, where c is a generic constant appearing in $P(d\Psi)$ or in $\mathcal{V}(\Psi)$;*

(6.a) *Horizontal reflections:* $\hat{\psi}^-_{\mathbf{k},\tau} \to \sigma_1 \hat{\psi}^-_{R_h\mathbf{k},\tau}$ *and* $\hat{\psi}^+_{\mathbf{k},\tau} \to \hat{\psi}^+_{R_h\mathbf{k},\tau} \sigma_1$, *with $R_h\mathbf{k} = (k_0, -k_1, k_2)$;*

(6.b) *Vertical reflections:* $\hat{\psi}^\varepsilon_{\mathbf{k},\tau} \to \hat{\psi}^\varepsilon_{R_v\mathbf{k},\tau}$, *with $R_v\mathbf{k} = (k_0, k_1, -k_2)$;*

(7) *Particle-hole:* $\hat{\psi}^-_{\mathbf{k},\tau} \to i\hat{\psi}^{+,T}_{P\mathbf{k},\tau}$, $\hat{\psi}^+_{\mathbf{k},\tau} \to i\hat{\psi}^{-,T}_{P\mathbf{k},\tau}$, *with $P\mathbf{k} = (k_0, -k_1, -k_2)$;*

(8) *Inversion:* $\hat{\psi}^-_{\mathbf{k},\tau} \to -i\sigma_3 \hat{\psi}^-_{I\mathbf{k},\tau}$, $\hat{\psi}^+_{\mathbf{k},\tau} \to -i\hat{\psi}^+_{I\mathbf{k},\tau} \sigma_3$, *with $I\mathbf{k} = (-k_0, k_1, k_2)$.*

Proof of Lemma 10.5 Let us first recall the definitions of $P_M(d\psi)$ and $\mathcal{V}(\psi)$:

$$P_M(d\psi) = \frac{1}{\mathcal{N}} \prod_{\mathbf{k} \in \mathcal{B}^{(M)}_{\beta,L}}^{\tau=\uparrow\downarrow} d\hat{\psi}^+_{\mathbf{k},\tau,1} d\hat{\psi}^-_{\mathbf{k},\tau,1} d\hat{\psi}^+_{\mathbf{k},\tau,2} d\hat{\psi}^-_{\mathbf{k},\tau,2} \, e^{-\frac{1}{\beta|\Lambda|} \hat{\psi}^+_{\mathbf{k},\tau} \hat{g}^{-1}_{\mathbf{k}} \hat{\psi}^-_{\mathbf{k},\tau}},$$

$$\mathcal{V}(\psi) = \frac{U}{(\beta|\Lambda|)^3} \sum_{\mathbf{k},\mathbf{k}',\mathbf{p}}^{\alpha=\pm} (\hat{\psi}^+_{\mathbf{k}+\mathbf{p},\uparrow} n_\alpha \hat{\psi}^-_{\mathbf{k},\uparrow}) (\hat{\psi}^+_{\mathbf{k}'-\mathbf{p},\downarrow} n_\alpha \hat{\psi}^-_{\mathbf{k}',\downarrow}),$$

where $\mathcal{N}$ is the normalization constant in equation (10.45) and $n_\pm = (1 \pm \sigma_3)/2$. Given these definitions and recalling the definition of $\hat{g}^{-1}_{\mathbf{k}}$, we see that the invariance of $P_M(d\psi)$ and $\mathcal{V}(\psi)$ are equivalent to the invariance of the following combinations:

$$(*) := \sum_{\mathbf{k},\tau} \hat{\psi}^+_{\mathbf{k},\tau} \begin{pmatrix} ik_0 & v_0\Omega^*(\vec{k}) \\ v_0\Omega(\vec{k}) & ik_0 \end{pmatrix} \hat{\psi}^-_{\mathbf{k},\tau}, \tag{10.242}$$

$$(**) := \sum_{\mathbf{k},\mathbf{k}',\mathbf{p}}^{\alpha=\pm} (\hat{\psi}^+_{\mathbf{k}+\mathbf{p},\uparrow} n_\alpha \hat{\psi}^-_{\mathbf{k},\uparrow}) (\hat{\psi}^+_{\mathbf{k}'-\mathbf{p},\downarrow} n_\alpha \hat{\psi}^-_{\mathbf{k}',\downarrow}). \tag{10.243}$$

Now, the invariance of $(*)$ and $(**)$ under symmetries (1) and (2) is completely apparent. Let us check one by one the invariance under the other symmetries.

Symmetry (3). Note that (∗) and (∗∗) can be rewritten as

$$(*) = \sum_{k,\rho,\rho'} (\hat{g}_k^{-1})_{\rho\rho'} \sum_{\tau} \hat{\psi}_{k,\tau,\rho}^+ \hat{\psi}_{k',\tau,\rho'}^-, \tag{10.244}$$

$$(**) = \frac{1}{2} \sum_{\substack{k,k',p \\ \alpha,\rho}} (n_\alpha)_{\rho\rho} \sum_{\tau,\tau'} (\hat{\psi}_{k+p,\tau,\rho}^+ \hat{\psi}_{k,\tau,\rho}^-)(\hat{\psi}_{k'-p,\tau',\rho}^+ \hat{\psi}_{k',\tau',\rho}^-) \tag{10.245}$$

Invariance of these expression under symmetry (3) follows from the invariance of the combination $\sum_\tau \hat{\psi}_{k,\tau,\rho}^+ \hat{\psi}_{k',\tau,\rho'}^-$: in fact, denoting by R^θ the matrix $e^{-i\theta\sigma_2}$, we see that under (3)

$$\sum_{\tau} \hat{\psi}_{k,\tau,\rho}^+ \hat{\psi}_{k',\tau,\rho'}^- \to \sum_{\tau,\tau_1,\tau_2} \hat{\psi}_{k,\tau_1,\rho}^+ R_{\tau_1,\tau}^{\theta,T} R_{\tau,\tau_2}^\theta \hat{\psi}_{k',\tau_2,\rho'}^- . \tag{10.246}$$

which is invariant, simply by the fact that R^θ is orthogonal.

Symmetry (4). Under the action of (4),

$$(*) \to \sum_{k,\tau} \hat{\psi}_{Tk,\tau}^+ e^{-i\vec{k}(\vec{\delta}_3-\vec{\delta}_1)n_-} \begin{pmatrix} ik_0 & v_0\Omega^*(\vec{k}) \\ v_0\Omega(\vec{k}) & ik_0 \end{pmatrix} e^{i\vec{k}(\vec{\delta}_3-\vec{\delta}_1)n_-} \hat{\psi}_{Tk,\tau}^-, \tag{10.247}$$

with $Tk = (k_0, R^{\frac{2\pi}{3}}\vec{k})$. The r.h.s. of equation (10.247) can be rewritten as

$$\sum_{k,\tau} \hat{\psi}_{Tk,\tau}^+ \begin{pmatrix} ik_0 & v_0\Omega^*(\vec{k})e^{i\vec{k}(\vec{\delta}_3-\vec{\delta}_1)} \\ v_0\Omega(\vec{k})e^{-i\vec{k}(\vec{\delta}_3-\vec{\delta}_1)} & ik_0 \end{pmatrix} \hat{\psi}_{Tk,\tau}^-, \tag{10.248}$$

which is the same as (∗), as it follows by the remark that $\Omega(\vec{k})e^{-i\vec{k}(\vec{\delta}_3-\vec{\delta}_1)} = \Omega(R^{\frac{2\pi}{3}}\vec{k})$. Regarding the interaction term, note that

$$\left[e^{-i(\vec{k}+\vec{p})(\vec{\delta}_3-\vec{\delta}_1)n_-} n_\alpha e^{i\vec{k}(\vec{\delta}_3-\vec{\delta}_1)n_-} \right] = e^{-i\vec{p}(\vec{\delta}_3-\vec{\delta}_1)\delta_{\alpha,-1}} n_\alpha, \tag{10.249}$$

which immediately shows that (∗∗) is invariant under (4).

Symmetry (5). The term (∗) is changed under (5) as:

$$(*) \to \sum_{k,\tau} \hat{\psi}_{-k,\tau}^+ \begin{pmatrix} -ik_0 & v_0\Omega(\vec{k}) \\ v_0\Omega^*(\vec{k}) & -ik_0 \end{pmatrix} \hat{\psi}_{-k,\tau}^-, \tag{10.250}$$

which is the same as (∗) (simply because $\Omega^*(\vec{k}) = \Omega(-\vec{k})$). Similarly, the term (∗∗) is changed under (5) as:

$$(**) \to \sum_{k,k',p}^{\alpha=\pm} (\hat{\psi}_{-k-p,\uparrow}^+ n_\alpha \hat{\psi}_{-k,\uparrow}^-)(\hat{\psi}_{-k'+p,\downarrow}^+ n_\alpha \hat{\psi}_{-k',\downarrow}^-), \tag{10.251}$$

which is the same as (∗∗).

Symmetry (6.a). The term (∗) is changed under (6.a) as:

$$(*) \rightarrow \sum_{k,\tau} \hat{\psi}^+_{R_h k, \tau} \sigma_1 \begin{pmatrix} ik_0 & v_0 \Omega^*(\vec{k}) \\ v_0 \Omega(\vec{k}) & ik_0 \end{pmatrix} \sigma_1 \hat{\psi}^-_{R_h k, \tau} =$$

$$= \sum_{k,\tau} \hat{\psi}^+_{R_h k, \tau} \begin{pmatrix} ik_0 & v_0 \Omega(\vec{k}) \\ v_0 \Omega^*(\vec{k}) & ik_0 \end{pmatrix} \hat{\psi}^-_{R_h k, \tau}. \tag{10.252}$$

Using the fact that $\Omega^*(\vec{k}) = \Omega((-k_1, k_2))$, we see that this term is invariant under (6.a). Regarding the interaction term, if we note that $\sigma_1 n_\alpha \sigma_1 = n_{-\alpha}$, we immediately see that it is invariant under (6.a).

Symmetry (6.b). Invariance of the term (∗) follows from the remark that $\Omega^*(\vec{k}) = \Omega((k_1, -k_2))$; invariance of (∗∗) is trivial.

Symmetry (7). The term (∗) is changed under (7) as:

$$(*) \rightarrow \sum_{k,\tau} \hat{\psi}^+_{Pk,\tau} \begin{pmatrix} ik_0 & v_0 \Omega^*(\vec{k}) \\ v_0 \Omega(\vec{k}) & ik_0 \end{pmatrix}^T \hat{\psi}^-_{Pk,\tau}, \tag{10.253}$$

which is the same as (∗) (simply because $\Omega^*(\vec{k}) = \Omega(-\vec{k})$). Similarly, using the fact that $n^T_\alpha = n_\alpha$, the term (∗∗) is changed under (7) as:

$$(**) \rightarrow \sum_{k,k',p}^{\alpha=\pm} \left(\hat{\psi}^+_{Pk,\uparrow} n_\alpha \hat{\psi}^-_{P(k+p),\uparrow}\right) \left(\hat{\psi}^+_{Pk',\downarrow} n_\alpha \hat{\psi}^-_{P(k'-p),\downarrow}\right), \tag{10.254}$$

which is the same as (∗∗).

Symmetry (8). The term (∗) is changed under (8) as:

$$(*) \rightarrow - \sum_{k,\tau} \hat{\psi}^+_{Ik,\tau} \sigma_3 \begin{pmatrix} ik_0 & v_0 \Omega^*(\vec{k}) \\ v_0 \Omega(\vec{k}) & ik_0 \end{pmatrix} \sigma_3 \hat{\psi}^-_{Ik,\tau} =$$

$$= \sum_{k,\tau} \hat{\psi}^+_{Ik,\tau} \begin{pmatrix} -ik_0 & v_0 \Omega^*(\vec{k}) \\ v_0 \Omega(\vec{k}) & -ik_0 \end{pmatrix} \hat{\psi}^-_{Ik,\tau}, \tag{10.255}$$

which is the same as (∗). Moreover, note that $\sigma_3 n_\alpha \sigma_3 = n_\alpha$, so that the term (∗∗) is obviously invariant under (8). This concludes the proof of Lemma 10.5. □

Now we are ready to prove Lemma 10.6.2.

Proof of Lemma 10.6.2. The key remark is that, in addition to the fact that $P_M(d\psi)$ is invariant under the transformations (1)–(8), the ultraviolet and infrared integrations $P(d\psi^{(u.v.)})$ and $P(d\psi^{(i.r.)})$ are separately invariant under the analogous transformations of the fields $\psi^{(u.v.)}$ and $\psi^{(i.r.)}$. Therefore, the effective potential

$\mathcal{V}_0(\psi^{(i.r.)})$ is also invariant under the same transformations. Let us restrict our attention to the quadratic contribution to $\mathcal{V}_0$, that is

$$\frac{1}{\beta|\Lambda|}\sum_{\mathbf{k},\tau}\hat{\psi}_{\mathbf{k},\tau}^{(i.r.)+}\hat{W}_2(\mathbf{k})\hat{\psi}_{\mathbf{k},\tau}^{(i.r.)-}, \tag{10.256}$$

which, as we said, must be invariant under the symmetries (1)–(8) listed in Lemma 10.5. The very fact that we can write the quadratic term in the form equation (10.256), with $\hat{W}_2(\mathbf{k})$ independent of the spin index τ, is a consequence of symmetries (1)–(3). It is straightforward to check that equation (10.84) follows from the invariance of equation (10.256) under (4)–(8).

So, we are left with proving equation (10.85). Let us start by showing that $\hat{W}(\mathbf{p}_F^\omega) = 0$. Writing $\hat{W}(\mathbf{p}_F^\omega) = a_0^\omega I + a_1^\omega \sigma_1 + a_2^\omega \sigma_2 + a_3^\omega \sigma_3$ and using the fact that, by equation (10.84), $\hat{W}(\mathbf{p}_F^\omega) = \sigma_1 \hat{W}(\mathbf{p}_F^\omega)\sigma_1 = -\sigma_3 \hat{W}(\mathbf{p}_F^\omega)\sigma_3$, we immediately see that $a_0^\omega = a_2^\omega = a_3^\omega = 0$. Moreover, using the fact that (still by equation (10.84)) $\hat{W}(\mathbf{p}_F^\omega) = e^{-i\omega\frac{2\pi}{3}n_-}\hat{W}(\mathbf{p}_F^\omega)e^{i\omega\frac{2\pi}{3}n_-}$, we get $a_1^\omega\sigma_1 = a_1^\omega(\cos(2\pi/3)\sigma_1 - \omega\sin(2\pi/3)\sigma_2)$, which implies $a_1^\omega = 0$.

Let us now look at $A_\mu^\omega := \partial_{k_\mu}\hat{W}(\mathbf{p}_F^\omega)$. Writing $A_0^\omega = b_0^\omega I + b_1^\omega\sigma_1 + b_2^\omega\sigma_2 + b_3^\omega\sigma_3$ and using that, by equation (10.84), $A_0^\omega = -(A_0^{-\omega})^* = A_0^{-\omega} = \sigma_1 A_0^\omega \sigma_1 = \sigma_3 A_0^\omega \sigma_3$, we see that $A_0^\omega = -iz_0 I$, with z_0 real and independent of ω. In a similar way, writing $A_1^\omega = c_0^\omega I + c_1^\omega\sigma_1 + c_2^\omega\sigma_2 + c_3^\omega\sigma_3$ and $A_2^\omega = d_0^\omega I + d_1^\omega\sigma_1 + d_2^\omega\sigma_2 + d_3^\omega\sigma_3$, using the fact that $A_1^\omega = -(A_1^{-\omega})^* = A_1^{-\omega} = -\sigma_1 A_1^\omega \sigma_1 = -(A_1^{-\omega})^T = -\sigma_3 A_1^\omega \sigma_3$ and $A_2^\omega = -(A_2^{-\omega})^* = -A_2^{-\omega} = \sigma_1 A_2^\omega \sigma_1 = -(A_2^{-\omega})^T = -\sigma_3 A_2^\omega \sigma_3$, we see that $A_1^\omega = c_2\sigma_2$ and $A_2^\omega = \omega d_1\sigma_1$, with c_2 and d_1 real and independent of ω. Finally, using the fact that, again by equation (10.84), $A_i^\omega = \sum_{j=1}^2 e^{i\omega\frac{2\pi}{3}\sigma_3}(R^{\frac{2\pi}{3}})_{ij}A_j^\omega$, we get $c_2 = d_1 =: -\delta_0$ that, if combined with our previous findings, implies equation (10.85). This concludes the proof of Lemma 10.6.2. $\qquad\square$

Acknowledgements

A particular thank goes to Vieri Mastropietro: we started the research project on graphene together and the material reviewed in these lecture notes is based on joint work with him. Many thanks also to Rafael Greenblatt and especially to Ian Jauslin for a careful reading of the manuscript and several technical comments. I gratefully acknowledge financial support from the ERC Starting Grant CoMBoS-239694.

References

Bardeen, T. Cooper, L. N. and Schrieffer, J. R. (1957). Theory of Superconductivity, *Phys. Rev.* **108**: 1175–1204.

Battle, G. A. and Federbush, P. (1984). A note on cluster expansions, tree graph identities, extra 1/N! factors!!!, *Lett. Math. Phys.* **8**: 55–57.

Benfatto, G. (2008). On the Ultraviolet Problem for the 2D Weakly Interacting Fermi Gas, *Annales Henri Poincaré* **10**: 1–17.

Benfatto, G. and Gallavotti, G. (1990). Perturbation theory of the Fermi surface in a quantum liquid. A general quasiparticle formalism and one dimensional systems, *Jour. Stat. Phys.* **59**: 541–664.

Benfatto, G. and Gallavotti, G. (1995). *Renormalization Group*, Physics Notes **1**, Princeton University Press, Princeton.

Benfatto, G. Gallavotti, G. Procacci, A. Scoppola, B. (1994). Beta functions and Schwinger functions for a many fermions system in one dimension, *Comm. Math. Phys.* **160**: 93–171.

Benfatto, G. Giuliani, A. and Mastropietro, V. (2003). Low Temperature Analysis of Two-Dimensional Fermi Systems with Symmetric Fermi Surface, *Ann. Henri Poincaré* **4**: 137–193.

Benfatto, G. Giuliani, A. and Mastropietro, V. (2006). Fermi liquid behavior in the 2D Hubbard model at low temperatures, *Ann. Henri Poincaré* **7**: 809–898.

Benfatto, G. and Mastropietro, V. (2005). Ward identities and chiral anomaly in the Luttinger liquid, *Comm. Math. Phys.* **258**: 609–655.

Brydges, D. C. *A short course on cluster expansions*, Phénomènes critiques, systèmes aléatoires, théories de jauge, Part I, II (Les Houches, 1984), 129–183, North-Holland, Amsterdam, 1986.

Brydges, D. C. and Federbush, P. (1978). A new form of the Meyer expansion in classical statistical mechanics, *Jour. Math. Phys.* **19**: 2064–2067.

Castro Neto, A. H. Guinea, F. Peres, N. M. R. Novoselov, K. S. and Geim, A. K. (2009). The electronic properties of graphene, *Rev. Mod. Phys.* **81**: 109–162.

Disertori, M. and Rivasseau, V. (2000). Interacting Fermi liquid in two dimensions at finite temperature, Part I - Convergent attributions and Part II - Renormalization, *Comm. Math. Phys.* **215**: 251–290 (2000) and 291–341.

Disertori, M. and Rivasseau, V. (2000). Rigorous Proof of Fermi Liquid Behavior for Jellium two-dimensional interacting fermions, *Phys. Rev. Lett.* **85**: 361–364.

Dukelsky, J., Pittel, S. and Sierra, G. (2004). Colloquium: Exactly solvable Richardson-Gaudin models for many-body quantum systems, *Rev. Mod. Phys.* **76**: 643–662.

Essler, F. H. L., Frahm, H., Göhmann, F., Klümper, A. and Korepin, V. E. (2005). *The One-Dimensional Hubbard Model*, Cambridge University Press.

Feldman, J. Magnen, J. Rivasseau, V. and Sénéor, R. (1986). A renormalizable field theory: the massive Gross-Neveu model in two dimensions, *Comm. Math. Phys.* **103**: 67–103.

Feldman, J. Magnen, J. Rivasseau, V. and Trubowitz, E. (1992). An infinite volume expansion for many fermions Freen functions, *Helv. Phys. Acta* **65**: 679–721.

Feldman, J. Knörrer, H. and Trubowitz, E. (2004). A two dimensional fermi liquid, *Comm. Math. Phys* **247**: 1–319.

Gallavotti, G. (1985). Renormalization group and ultraviolet stability for scalar fields via renormalization group methods, *Rev. Mod. Phys.* **57**: 471–562.

Gallavotti, G. and Nicolò, F. (1985). Renormalization theory for four dimensional scalar fields. Part I and II, *Comm. Math. Phys.* **100**: 545–590 (1985) and **101**: 471–562.

Gawedzki, K. and Kupiainen, A. (1985). Gross-Neveu model through convergent perturbation expansions, *Comm. Math. Phys.* **102**: 1–30.

Gentile, G. and Mastropietro, V. (2001). Renormalization group for one-dimensional fermions. A review on mathematical results. In: Renormalization group theory in the new millennium, III, *Phys. Rep.* **352**: 273–437.

Giuliani, A. and Mastropietro, V. (2009). Rigorous construction of ground state correlations in graphene: renormalization of the velocities and Ward Identities, *Phys. Rev. B* **79**: 201403(R) (2009); Erratum, ibid **82**: 199901.

Giuliani, A. and Mastropietro, V. (2010). The two-dimensional Hubbard model on the honeycomb lattice, *Comm. Math. Phys.* **293**: 301–346.

Giuliani, A. and Mastropietro, V. (2012). Exact RG computation of the optical conductivity of graphene, *Phys. Rev. B* **85**: 045420.

Giuliani, A. Mastropietro, V. and Porta M. (2010a). Anomalous behavior in an effective model of graphene with Coulomb interactions, *Ann. Henri Poincaré* **11**: 1409–1452.

Giuliani, A. Mastropietro, V. and Porta, M. (2010b). Lattice gauge theory model for graphene, *Phys. Rev. B* **82**: 121418(R).

Giuliani, A. Mastropietro, V. and Porta, M. (2011). Absence of interaction corrections in the optical conductivity of graphene, *Phys. Rev. B* **83**: 195401.

Giuliani, A. Mastropietro, V. and Porta, M. (2012a). Lattice quantum electro-dynamics for graphene, Annals of Physics **327**: 461–511.

Giuliani, A. Mastropietro, V. and Porta, M. (2012b). Universality of conductivity in interacting graphene, *Comm. Math. Phys.* **311**: 317–355.

Haldane, F. D. M. (1988). Model for a Quantum Hall Effect without Landau Levels: Condensed-Matter Realization of the 'Parity Anomaly', *Phys. Rev. Lett.* **61**: 2015–2018.

Hou, C.-Y. Chamon, C. and Mudry, C. (2007). Electron fractionalization in two-dimensional graphenelike structures, *Phys. Rev. Lett.* **98**: 186809.

Johnson, K. (1961). Solution of the equations for the Green's functions of a two-dimensional relativistic field theory, *Nuovo Cimento* **20**: 773–790.

Klaiber, B. The Thirring model. In: A. O. Barut and W. E. Britting (eds), *lectures in theoretical physics*, **10**A: 141–176 (New York, Gordon and Breach, 1968).

Lesniewski, A. (1987). Effective action for the Yukawa$_2$ quantum field theory, *Comm. Math. Phys.* **108**: 437–467.

Lieb, E. H. (1963). Exact analysis of an interacting Bose gas. II. The excitation spectrum, *Phys. Rev.* **130**: 1616–1624.

Lieb, E. H. and Liniger, W. (1963). Exact analysis of an interacting Bose gas. I. The general solution and the ground state, *Phys. Rev.* **130**: 1605–1616.

Lieb, E. H. and Wu, F. Y. (1968). Absence of mott transition in an exact solution of the short-range, one-band Model in one dimension, *Phys. Rev. Lett.* **20**: 1445–1448.

Luttinger, J. M. (1963). An exactly soluble model of a many fermions system, *J. Math. Phys.* **4**: 1154–1162.

Mastropietro, V. (1994). Interacting soluble Fermi systems in one dimension, *Nuovo Cim.* **109**B: 304–312.

Mastropietro, V. (2008). Non-Perturbative Renormalization, *World Scientific*.

Mattis, D. C. and Lieb, E. H. (1965). Exact solution of a many fermion system and its associated boson field, *J. Math. Phys.* **6**: 304–312.

Novoselov, K. S. Geim, A. K. Morozov, S. V. Jiang, D. Zhang, Y. Dubonos, S. V. Grigorieva, I. V. and Firsov, A. A. (2004). Electric field effect in atomically thin carbon films, Science **306**: 666.

Richardson, R. W. and Sherman, N. (1964). Exact eigenstates of the pairing-force Hamiltonian, *Nucl. Phys.* **52**: 221–238.

Salmhofer M. (1999). Renormalization: An Introduction, Springer.

Semenoff, G. W. (1984). Condensed-matter simulation of a three-dimensional anomaly, *Phys. Rev. Lett.* **53**: 2449–2452.

Thirring, W. (1958). A soluble relativistic field theory, *Ann. Phys.* **3**, 91–112.

Tomonaga, S. (1950). Remarks on Bloch's methods of sound waves applied to many fermion systems, *Progr. Theo. Phys.* **5**, 544–569.

Wallace, P. R. (1947). The Band Theory of Graphite, *Phys. Rev.* **71**: 622–634.

11
On transport in quantum devices

Gian Michele GRAF

Theoretische Physik, ETH-Zürich
8093 Zürich, Switzerland

Chapter Contents

Preface

These notes are mainly intended as a guide to the literature complementing the lectures presented by the author at the School. Two main topics were discussed: quantum pumps and transport statistics. For detailed statements the reader is asked to consult the original works.

11.1   Quantum pumps

An adiabatic *quantum pump* is a time-dependent scatterer connected to several leads; each lead may have several channels, which account for spin or degrees of freedom transverse to the leads (modes). The idealized setting is as follows: a pump, whose internal configuration X varies slowly in time in a prescribed manner, is connected to n channels, along each of which independent electrons of charge $e = 1$ can enter or leave the pump. The incoming electron distribution, at zero temperature, is a Fermi sea with Fermi energy μ common to all channels. The same distribution would apply to the outgoing states if X were constant in time. As that is not the case, the energies of the electrons may have shifted while they scattered at the pump. Because of this imbalance a net current is flowing in the channels. The expected charge transport is expressed by the formula of Büttiker et al. (Büttiker, Prêtre, and Thomas 1993; Büttiker, Thomas, and Prêtre 1994; Brouwer 1998)

$$dQ_j = \frac{i}{2\pi}((dS)S^*)_{jj}\,. \tag{11.1}$$

Here $S = (S_{ji})$ is the $n \times n$ scattering matrix at energy μ computed as if the pump were *frozen* into its instantaneous configuration X. The entry S_{ji} is the transition amplitude between an incoming state in channel i and an outgoing one in channel j. We recall that the definition of the S-matrix involves the comparison of the dynamics associated to X with that of a reference X_0, for example one with channels disconnected from the pump. A change of the configuration is accompanied by a change $S \to S + dS$ of the scattering matrix and by a net charge dQ_j leaving the pump through channel j. The notation d indicates that dQ_j is not an exact differential on the manifold of configurations X. In fact, the charge delivered to a channel during a cycle does not vanish as a rule, $\oint dQ_j \neq 0$, in line with the purpose of a pump. However,

$$\sum_{j=1}^{n} dQ_j = -d\xi\,,$$

where the Krein spectral shift $\xi(\mu)$ counts the excess of electrons in the Fermi sea, relatively to that of X_0, and may thus be interpreted as the number of electrons contained in the pump proper of configuration X. If the Hamiltonian (and not just the scattering matrix) is periodic with the cycle, then $\oint \sum_{j=1}^{n} dQ_j = 0$.

Besides the original work (Büttiker, Prêtre, and Thomas 1993) (see also (Brouwer 1998)), we mention a heuristic derivation (Avron, Elgart, Graf, and Sadun 2004), as well as a mathematical analysis (Avron, Elgart, Graf, Sadun, and Schnee 2004).

A heuristic derivation. The discussion is to be placed in a classical context, where the energy $E \in [0, \infty)$ and the time of passage $t \in \mathbb{R}$ of the electron at the pump are canonical coordinates for the electron in a fixed channel; or, more precisely, in a semiclassical context: an incoming particle of well-defined energy does not have a well-defined time of passage. This spread in time is a consequence of the uncertainty principle, and is not related to the dwell time of the particle in the scatterer. The S-matrix seen by the tail of the wave packet will differ slightly from that seen by the head of the wave packet. This differential scattering produces interference between the parts of the wave, causing the outgoing occupation density to differ from the incoming one. The following discussion quantifies the idea.

Two quantities are natural to the discussion: The first is the Wigner time delay (Wigner 1955)

$$\mathcal{T}(E, t) = -i \frac{\partial S}{\partial E}(E, t) S^*(E, t),$$

where $S(E, t) = S(E, X(t))$ is the scattering matrix. The diagonal element $\mathcal{T}_{jj}(E, t)$ has the meaning of the average time delay of a particle exiting channel j, in comparison to the dynamics associated to X_0. Similarly, the Martin–Sassoli energy shift (Martin and Sassoli de Bianchi 1995)

$$\mathcal{E}(E, t) = i \frac{\partial S}{\partial t}(E, t) S^*(E, t).$$

describes the change in energy accompanying the scattering process. The net charge leaving the pump through channel j in the time interval $[0, T]$ is the difference between the outgoing and incoming contributions:

$$Q_j = \frac{1}{2\pi} \int_0^\infty dE' \int_0^T dt' \, \rho(E) - \frac{1}{2\pi} \int_0^\infty dE \int_0^T dt \, \rho(E),$$

where ρ is the occupation of incoming states, e.g. $\rho(E) = \theta(\mu - E)$ in the zero temperature situation considered above. In the first integral E is expressed through the map

$$(E', t') \mapsto (E, t) = \left(E' - \mathcal{E}_{jj}(E', t'), t' - \mathcal{T}_{jj}(E', t') \right),$$

which describes the effect of the pump on the energy and the time of passage of an electron in terms of the outgoing data (E', t'). The denominator 2π is the size of the phase space cell of a quantum state. By linearizing in the small energy shift $\mathcal{E}_{jj}$ we obtain

$$\dot{Q}_j(t) = -\frac{1}{2\pi} \int_0^\infty dE \rho'(E) \mathcal{E}_{jj}(E, t),$$

($\dot{} = d/dt$, $' = d/dE$), which reduces to

$$\dot{Q}_j(t) = \frac{1}{2\pi} \mathcal{E}_{jj}(\mu, t),$$

that is to equation (11.1), at zero temperature.

We stress that the time dependence of $S(E, t)$ is merely parametric, in that its definition is based on the frozen configuration $X = X(t)$. More radically, the discussion ought to be based on the nonautonomous dynamics of an electron facing a changing configuration $X(t)$. The scattering operator then results from a comparison between two dynamics, generated by the time-dependent Hamiltonian $H(X(t))$ and by the reference $H_0 = H(X_0)$. Let $U(t_2, t_1)$ and $U_0(t_2, t_1) = \mathrm{e}^{-\mathrm{i}(t_2 - t_1)H_0}$ be the respective propagators for the evolution from times t_1 to t_2. The scattering operator is defined as

$$S(t) = \lim_{t_\pm \to \pm\infty} U_0(t, t_+)U(t_+, t_-)U_0(t_-, t),$$

assuming the limit exists. The relation to the family of matrices $S(E, t)$ emerges in the adiabatic limit, which here parallels the semiclassical one, and can be clarified by means of coherent states (Avron, Elgart, Graf, and Sadun 2002) or pseudo-differential operators (Avron, Elgart, Graf, and Sadun 2004). For instance, $S(E, t)$ is the principal symbol of $S(0)$, and $\mathcal{E}(E, t)$ that of $\mathrm{i}(\dot{S}(t)S^*(t))|_{t=0}$.

A mathematical analysis. The basic equation (11.1) may be linked more closely to the Schrödinger equation and to the basic rule $\mathrm{Tr}(\rho I)$ for computing expectations of an observable I in a state ρ. We refer to (Avron *et al.*, 2004) for the details of the setup. Since the pump configuration is supposed to change slowly in time t, we consider the evolution of the electrons in an adiabatic limit, where $s = \varepsilon t$ is kept fixed as $\varepsilon > 0$ tends to 0. The Hamiltonian $H(s)$ applies to states representing a particle in the pump or in the semi-infinite channels. However the time derivative $\dot{H}(s)$ is of finite spatial extent, since changes are limited to the pump proper. In terms of the rescaled time coordinate s the propagator $U_\varepsilon(s, s')$ satisfies the Schrödinger equation

$$\mathrm{i}\partial_s U_\varepsilon(s, s') = \varepsilon^{-1}H(s)U_\varepsilon(s, s'), \qquad U_\varepsilon(s', s') = 1.$$

The system is started from equilibrium at $s = s_0$. Accordingly, the initial state has one-particle density matrix of the form $\rho(H(s_0))$. Here ρ is a function of bounded variation with $\mathrm{supp}\, d\rho \subset (0, \infty)$. In particular, an admissible choice is the Fermi sea, where $\rho(E) = \theta(\mu - E)$. The time evolution then acts as

$$\rho(H(s_0)) \mapsto P_\varepsilon(s) = U_\varepsilon(s, s_0)\rho(H(s_0))U_\varepsilon(s_0, s). \tag{11.2}$$

It should be noted that $P_\varepsilon(s)$ is not an equilibrium state for $H(s)$. Finally, a one-particle operator $Q_j(a)$ is introduced representing (roughly) the charge in channel j beyond a distance a from the pump. In line with the infinite extent of the channels, it does not have a finite expectation in $P_\varepsilon(s)$, but its time derivative, the current $I_j(a) = \mathrm{i}[H(s), Q_j(a)]$, does.

The fundamental result equation (11.1) may thus be given the following reformulation (Avron *et al.*, 2004).

Theorem 11.1 *Under the above assumptions we have for $s > s_0$*

$$\lim_{a \to \infty} \lim_{\varepsilon \to 0} \varepsilon^{-1} \operatorname{Tr}(P_\varepsilon(s) I_j(a)) = -\frac{i}{2\pi} \int_0^\infty d\rho(E) \left(\frac{dS}{ds} S^* \right)_{jj}, \qquad (11.3)$$

where $S(E, s)$ is scattering $n \times n$ matrix (fibre) of the scattering operator for the pair $(H(s), H_0)$ with H_0 being any fixed reference Hamiltonian. The double limit is uniform in $s \in I$, I being a compact interval, whence it carries over to the transferred charge

$$\int_0^{s/\varepsilon} dt' \operatorname{Tr}(P_\varepsilon(\varepsilon t') I_j(a)) = \varepsilon^{-1} \int_0^s ds' \operatorname{Tr}(P_\varepsilon(s') I_j(a)).$$

A few remarks are appropriate.

1) The limit $a \to \infty$ is necessary in order to obtain a scattering description. A positive energy) particle has a finite dwell time in the pump and takes a finite time to complete a distance a to or from it. Their durations remain bounded in ε small. Hence the events become simultaneous in terms of scaled time if the adiabatic limit $\varepsilon \to 0$ is performed first, and the result (11.3) is local in s.

2) The above result is a statement about the adiabatic behaviour of certain open, gapless systems, as explained shortly. Typically, adiabatic theorems are concerned with the evolution $U_\varepsilon(s, s_0) P U_\varepsilon(s, s_0)^*$, where P is the spectral projection of $H(s_0)$ onto a separated part of its spectrum (Avron, Seiler, and Yaffe 1993; Nenciu 1993) or on an embedded eigenvalue (Bornemann 1998; Avron and Elgart 1999), and possibly with its deviation from the corresponding projection of $H(s)$. The same is true here, see equation (11.2) with $\rho(E) = \theta(\mu - E)$, except that $P = \rho(H(s_0))$ corresponds to a gapless part of the continuous spectrum.

Further transport quantities. Let $\dot{E}_j$ be the expectation value of energy current in the j-th channel. Part of the energy is forever lost as the electrons are dumped into a reservoir. The part that can be recovered from the reservoir, by reclaiming the transported charge, is $\mu \dot{Q}_j$. We therefore define the dissipation in a quantum channel as the difference of the two. For an adiabatic quantum pump at zero temperature we have (Avron, Elgart, Graf, and Sadun 2001)

$$\dot{E}_j - \mu \dot{Q}_j = \frac{1}{4\pi} (\mathcal{E}^2)_{jj} \geq 0,$$

where, again, $\mathcal{E}(t) \equiv \mathcal{E}(\mu, t)$. More generally, the combination $\dot{E}_j - \mu \dot{Q}_j$ is of interest, because for any one-dimensional quantum channel connecting a source to a reservoir at Fermi energy μ one has the following lower bound on the dissipation (Avron, Elgart, Graf, and Sadun 2001):

$$\dot{E} - \mu \dot{Q} \geq \frac{R_k}{2} \dot{Q}^2 \equiv \pi \dot{Q}^2, \qquad (11.4)$$

with $R_k = h/e^2$ the (von Klitzing) unit of resistance. Note that $R_k/2$ is the quantum limit of a contact resistance. The bound does not depend on the nature of the particle source, as long as it is stationary, and is saturated if the population distribution that is supplied by the source is filled up to some energy and empty thereafter. In particular, the bound applies if the source is an adiabatic quantum pump, in which case it reduces to

$$(\mathcal{E}^2)_{jj} \geq (\mathcal{E}_{jj})^2 . \tag{11.5}$$

We will soon discuss the condition for saturation. For the moment we observe that the inequality is independently true by $(\mathcal{E}^2)_{jj} = \sum_k |\mathcal{E}_{jk}|^2$.

Both the dissipation and the current admit transport formulae that are local in time. Namely, the response is determined by the energy shift at the same time. This is remarkable, for at zero temperature quantum correlations decay slowly in time and one may worry that transport will retain memory about the scatterer at earlier times. This is indeed the case for the variance (noise) of the transported charge (Ivanov, Lee, and Levitov 1997) (see also (Avron, Elgart, Graf, and Sadun 2004))

$$\langle\langle Q_j^2 \rangle\rangle = \frac{1}{(2\pi)^2} \int_0^T \int_{-\infty}^\infty \frac{1 - |(S(t)S^*(t'))_{jj}|^2}{(t-t')^2} dt \, dt' ,$$

whereas, for comparison, equation (11.1) can be stated as a single integral,

$$\langle Q_j \rangle = \frac{i}{2\pi} \int_0^T (\dot{S}(t)S^*(t))_{jj} dt . \tag{11.6}$$

Noise and entropy production at positive temperature are discussed in (Avron, Elgart, Graf, and Sadun 2004).

Quantization of charge transport. We first describe a (well-known) geometric construction, which in a second step will allow for an interpretation of the above quantities.

A normalized vector $|\psi\rangle \in \mathbb{C}^n$ can be viewed as a point on the sphere $S^{2n-1} \subset \mathbb{C}^n$. The Hopf map $\pi : S^{2n-1} \to \mathbb{C}P^{n-1}$, which 'forgets' the phase of $|\psi\rangle$, turns S^{2n-1} into a fibre (circle) bundle with base space $\mathbb{C}P^{n-1}$. The fibre through $|\psi\rangle$ is the curve $e^{i\theta}|\psi\rangle$, $(0 \leq \theta < 2\pi)$ with tangent vector $i|\psi\rangle$. Any other curve $|\psi(t)\rangle$ through that point has vertical component $-i\langle\psi|\dot\psi\rangle$, which is just the connection one-form associated to the Berry phase.

For a given channel j, let $\langle\psi(t)|$ be the j-th row of $S(t)$. It represents the linear combination of incoming channels feeding the outgoing channel j at time t and energy μ. Then, $\mathcal{E}_{jj} = -i\langle\psi|\dot\psi\rangle$, $(\mathcal{E}^2)_{jj} = \langle\dot\psi|\dot\psi\rangle$, and $(S(t)S^*(t'))_{jj} = \langle\psi(t)|\psi(t')\rangle$. In the language just introduced the charge transport (11.6) is the Berry phase of the curve associated to the pump process. Moreover, for cyclic processes the following properties are equivalent:

- the dissipation inequality (11.4) (or (11.5)) is saturated (minimal dissipation);
- no noise: $\langle\langle Q_j^2 \rangle\rangle = 0$;
- the charge transported in a cycle is quantized, i.e., given by a non-random integer:

$$Q_j = \langle Q_j \rangle \in \mathbb{Z}.$$

In fact, each property is equivalent to a curve $|\psi(t)\rangle$ winding along the fibre, that is $|\psi(t)\rangle = e^{i\theta(t)} |\psi\rangle$, in which case $Q_j = (2\pi)^{-1} \oint \dot{\theta} dt$ is the winding number.

In the case of two channels a pump process is optimal (in the above sense) for one channel if it is for the other. Explicitly, the charges Q_j, $(j = 1, 2)$ transported in a cycle are quantized iff the scattering matrix $S(t)$ is of the form (Andreev and Kamenev 2000)

$$S(t) = \begin{pmatrix} e^{i\varphi_1(t)} & 0 \\ 0 & e^{i\varphi_2(t)} \end{pmatrix} S_0 \,,$$

with arbitrary unitary S_0 independent of t. Then Q_j is the winding number of $\varphi_j(t)$, $(j = 1, 2)$.

The result can be extended to two leads consisting of n_1 and n_2 channels, respectively. The charges delivered to the leads (but not necessarily to the constituent channels) are quantized iff the scattering matrix $S(t)$ is of the block form

$$S(t) = \begin{pmatrix} U_1(t) & 0 \\ 0 & U_2(t) \end{pmatrix} S_0 \tag{11.7}$$

with $U_j(t)$ unitary $n_j \times n_j$-matrices $(j = 1, 2)$. The charges are the winding numbers of $\det U_j(t)$.

It should be stressed that, at this level of generality, quantization is not generic and, as a result, not intrinsically stable against perturbations if it occurs. Instances of (stable) quantized transport, like in the integer quantum Hall effect or electron wave guides, can however be formulated in the above scattering formalism (Avron, Elgart, Graf, and Sadun 2004). The next section provides a further example.

The topological approach. So far all transport properties have been tied to the scattering matrix at Fermi energy. That energy is of course part of the spectrum: the scatterer, being compact, does not affect the continuous spectrum contributed by the leads.

Another, and in fact earlier, approach to pumping by Thouless (Thouless 1983) (see also (Niu and Thouless 1984)) accounts for transport in a rather different, if not opposing, way. The model is that of a noninteracting Fermi gas ranging over an infinitely extended device, with the Fermi energy lying in a spectral gap. In particular, there is no longer a scattering matrix at that energy, and transport is attributed to energies way below it. The resulting transport is quantized over a cycle.

We first review the result of (Thouless 1983), where the device is modelled by means of a periodic potential in one dimension and the quantized charge occurs as as a Chern number. After extending the result (and reformulating the Chern number) we will relate it to the scattering approach (Bräunlich, Graf, and Ortelli 2010; Graf and Ortelli 2008).

To begin, consider the Schrödinger Hamiltonian on the line $\mathbb{R}_x$

$$H(s) = -\frac{d^2}{dx^2} + V(x, s) \,, \tag{11.8}$$

where V is periodic in space and time: $V(x + L, s) = V(x, s)$ and $V(x, s + T) = V(x, s)$. Let $\psi_{nks}(x)$ be the Bloch solution (unique up to phase) of the time-independent Schrödinger equation, normalized with respect to the inner product

$$\langle \phi, \psi \rangle = \frac{1}{L} \int_0^L dx \, \overline{\phi}(x) \psi(x) \,.$$

The index $n = 1, 2, ..$ labels the bands. The Fermi energy is assumed to lie in a spectral gap of $H(s)$ for all s. Let then $s = \varepsilon t$ change slowly with t and let the Fermi sea evolve by $H(\varepsilon t)$, as in equation (11.2). Finally, let $\varepsilon \to 0$.

Theorem 11.2 (***Thouless 1983***) *Under the above assumptions, the charge transported during a cycle across a fiducial point, for example $x = 0$, is*

$$\langle Q \rangle = \sum_n^* \frac{i}{2\pi} \int_0^T ds \int_0^{2\pi/L} dk \Big(\langle \frac{\partial \psi_{nks}}{\partial s}, \frac{\partial \psi_{nks}}{\partial k} \rangle - \langle \frac{\partial \psi_{nks}}{\partial k}, \frac{\partial \psi_{nks}}{\partial s} \rangle \Big),$$

where the star indicates that the sum extends over filled bands n only.

Each term is an integer defining the Chern number of the $U(1)$ fibre bundle ψ_{nks} over the torus $\mathbb{T} = \mathbb{R}^2/(T, 2\pi/L)$. That number reflects the obstruction to choosing the phase of ψ_{nks} in a continuous way on all of $\mathbb{T}$. Underlying the result is the adiabatic theorem for Hamiltonians with a spectral gap.

We next discuss a generalization which forgoes the spatial periodicity and, at the same time, allows for $n \geq 1$ channels. The Hamiltonian, formally still given by (11.8), is now acting on $L^2(\mathbb{R}_x, \mathbb{C}^n)$; the potential $V = V(x, s)$ takes values in the $n \times n$ matrices, $M_n(\mathbb{C})$, is Hermitian, $V = V^*$, and assumed periodic in time, $V(x, s + T) = V(x, s)$. Then, for any $z \in \rho(H(s))$ in the resolvent set, there are matrix-valued solutions $\psi_+(x) \in M_n(\mathbb{C})$ of the Schrödinger equation

$$-\psi_+''(x) + V(x, s)\psi_+(x) = z\psi_+(x) \,. \tag{11.9}$$

We consider those which decay at $x = +\infty$ and are regular, in the sense that for any $x \in \mathbb{R}$

$$\psi_+(x)a = 0, \; \psi_+'(x)a = 0 \;\Rightarrow\; a = 0, \qquad (a \in \mathbb{C}^n) \,. \tag{11.10}$$

These solutions carry a transitive right action of $GL(n) \ni T$, namely $\psi_+(x) \mapsto \psi_+(x)T$. Similarly, consider solutions $\psi(x) \in M_n(\mathbb{C})$ of the adjoint equation

$$-\psi_-''(x) + \psi_-(x)V(x, s) = z\psi_-(x) \,, \tag{11.11}$$

which decay at $x = -\infty$ and are regular (with $a\psi_-(x)$ replacing $\psi_+(x)a$ in equation (11.10)). They carry a left action, $\psi_-(x) \mapsto T\psi_-(x)$.

For any two differentiable functions $\psi_+, \psi_- : \mathbb{R} \to M_n(\mathbb{C})$ we define the Wronskian

$$W(\psi_-, \psi_+; x) = \psi_-(x)\psi_+'(x) - \psi_-'(x)\psi_+(x) \in M_n(\mathbb{C}) \,.$$

It is independent of x if ψ_+ and ψ_- are solutions of (11.9), resp. of (11.11), in which case it is simply denoted as $W(\psi_-, \psi_+)$. If $\psi_\pm$ also satisfy the above decay and regularity properties, then $W(\psi_-, \psi_+)$ is a regular matrix. By the group actions there is a bijective relation between solutions ψ_+ and ψ_- such that $W(\psi_-, \psi_+) = 1$.

We may now state the announced generalization.

Theorem 11.3 *Under the above assumptions, the charge transported during a cycle across the fiducial point $x = x_0$ is*

$$\langle Q \rangle = \frac{i}{2\pi} \oint_\gamma dz \int_0^T ds \ \mathrm{tr}\Big(W\big(\frac{\partial\psi_-}{\partial z}, \frac{\partial\psi_+}{\partial s}; x_0\big) - W\big(\frac{\partial\psi_-}{\partial s}, \frac{\partial\psi_+}{\partial z}; x_0\big)\Big), \tag{11.12}$$

where $\gamma \ni z$ is a complex contour encircling the part of the spectrum of $H(s)$ lying below μ; tr denotes the matrix trace; and the solutions ψ_+, ψ_- are locally smooth in (z, s) and satisfy the stated normalization. Except for these conditions, the trace is independent of ψ_+, ψ_-, as the integral is of x_0. Moreover, the r.h.s. is the first Chern number of the bundle with the torus $\gamma \times [0, T]$ as base and fibres consisting of all solutions ψ_+.

Finally, we explain the link to the scattering approach. To this end we truncate the potential to a large but finite interval, $V(x, s)\chi_{[0,L]}(x)$, whence the gap closes, and denote its scattering matrix at Fermi energy using block notation as

$$S = \begin{pmatrix} R & T' \\ T & R' \end{pmatrix}.$$

In the limit $L \to \infty$ the original physical situation is recovered and the two approaches agree, as stated in the following result.

Theorem 11.4 *Assume that the infinitely extended system is gapped at the Fermi energy. Then, for large L, the scattering matrix $S_L(s)$ is of the form*

$$S_L(s) = \begin{pmatrix} R_\infty(s) & 0 \\ 0 & R'_L(s) \end{pmatrix} + o(1), \qquad (L \to \infty).$$

In particular, the condition (11.7) for quantization is attained in the limit. Moreover, the winding number of $\det R_\infty(s)$ equals the Chern number on the r.h.s of equation (11.12).

In physical terms, the scattering and the topological approach yield the same (quantized) charge transport.

11.2   Transport statistics

There are several reviews on counting statistics, among them several contained in (Nazarov 2003), and (Bachmann and Graf 2008). We refer to those for motivation. Here we concentrate on a more specific aspect, namely the relation between current correlators and the statistics of the transferred charge.

We consider a junction and the charge ΔQ transferred across it during a time t. The generating function of counting statistics (Levitov and Lesovik 1992) is

$$\chi(\lambda, t) = \langle e^{i\lambda\Delta Q}\rangle = \sum_{k=0}^{\infty} \frac{(i\lambda)^k}{k!} \langle(\Delta Q)^k\rangle,$$

where $\langle(\Delta Q)^k\rangle$ is the k-th moment of the charge transported during time t. Similarly, $\log\chi$ generates the cumulants $\langle\langle(\Delta Q)^k\rangle\rangle$. Several quantum mechanical expression for χ have been proposed, based in part on different measurement protocols. The first proposal (Levitov and Lesovik 1992) is

$$\chi(\lambda, t) = \langle\exp\left(i\lambda \int_0^t I(t')dt'\right)\rangle \tag{11.13}$$

where $I(t)$ is the current operator at the junction and at time t, and $\langle\cdot\rangle$ is understood as the expectation in the initial quantum state of the system. The second (Levitov and Lesovik 1993) reads, as recast by (Muzykanskii and Adamov 2003),

$$\chi(\lambda, t) = \langle e^{i\lambda Q(t)}e^{-i\lambda Q}\rangle = \langle e^{iHt}e^{i\lambda Q}e^{-iHt}e^{-i\lambda Q}\rangle, \tag{11.14}$$

where $Q(t) = \exp(iHt)Q\exp(-iHt)$ is the charge operator on the right of the junction and at time t, and H is the Hamiltonian. The definition is appropriate to the situation where the initial state is an eigenstate of Q and the charge is measured after time t. In fact, if the initial eigenvalue is q, then $\chi(\lambda, t) = \langle e^{i\lambda Q(t)}\rangle e^{-i\lambda q}$ just describes the statistics of $q' - q$, where q' is the (random) outcome of the measurement of $Q(t)$. The quantity $q' - q$ is identified with the transported charge ΔQ. A further expression, believed to be equivalent to equation (11.14) in situations where it actually is not (see below), is

$$\chi(\lambda, t) = \langle\mathrm{T}\exp\left(i\lambda \int_0^t I(t')dt'\right)\rangle, \tag{11.15}$$

where T denotes the time-ordering of a product. We mention in passing that there is a generalization (Shelankov and Rammer 2003) of this definition to the case where the assumption on the initial state does not apply and q is the outcome of the measurement of $Q(0) = Q$.

Further proposals (Levitov and Lesovik 1994; Levitov, Lee, and Lesovik 1996) are based on observing a spin coupled to current. They yield expressions paralleling equations (11.14) and (11.15), which happened to differ in some applications, as it is the case in the very simple situation of independent fermions passing across a scatterer of transparency T that is biased by V.

Following (Bachmann, Graf, and Lesovik 2010) we report on the solution of the discrepancy, or actually of that between equations (11.14) and (11.15). To begin we should explain the domain of their equivalence, which ultimately rests on $I(t) = i[H, Q(t)]$. Rearranging Eq. (11.14) we have

$$\chi(\lambda, t) = \langle e^{iHt}\left(e^{i\lambda Q}e^{-iHt}e^{-i\lambda Q}\right)\rangle = \langle e^{iHt}e^{-iH(\lambda)t}\rangle,$$

where $H(\lambda) = e^{i\lambda Q} H e^{-i\lambda Q}$. If $[Q, I] = 0$ the commutator expansion of $H(\lambda)$ terminates,

$$H(\lambda) = H - i\lambda[H, Q] = H - \lambda I \,,$$

and yields equation (11.15) by the Dyson expansion. In order to gain an impression of the commutator $[Q, I]$, we ignore second quantization and consider operators pertaining to a single particle moving on a line. The operator representing charge on the right half-line is $Q = \theta(x)$, meaning multiplication by the Heaviside step function. For a Hamiltonian like $H = p^2$ we have $I = i[H, Q] = p\delta(x) + \delta(x)p$, whence $[Q, I] \neq 0$, even if the step were smoothed out. By contrast, linearizing the dispersion relation, as it is often done, yields a Hamiltonian describing a right (or left) moving particle, $H = \pm p$; we then have $I = \pm\delta(x)$ and $[Q, I] = 0$. After combining both copies of the model we may describe scattering between right and left movers by including a potential, for example

$$H = p\sigma_z + V(x) \qquad \text{on } L^2(\mathbb{R}_x; \mathbb{C}^2) \,, \tag{11.16}$$

where $V = V^*$ and σ_z is a Pauli matrix. The commutator continues to vanish, suggesting equivalence between equations (11.14) and (11.15).

However, in the situation mentioned above the two equations, evaluated on a linear dispersion relation and on an energy independent scattering matrix for the junction, produces conflicting values for the third cumulant ((Levitov, Lee, and Lesovik 1996; Lesovik and Chtchelkatchev 2003)):

$$\langle\!\langle\!\langle(\Delta Q)^3\rangle\!\rangle\!\rangle = T(1 - T)(1 - 2T)\frac{Vt}{2\pi} \,, \qquad \langle\!\langle\!\langle(\Delta Q)^3\rangle\!\rangle\!\rangle = -2T^2(1 - T)\frac{Vt}{2\pi} \,. \tag{11.17}$$

The first result, which is consistent with the binomial distribution, was also found in (Salo, Hekking, and Pekola 2006) and in (Levitov and Lesovik 1993). The second surprisingly agrees with the prediction based on (11.13), which lacks time ordering.

The resolution of the puzzle consists of two steps, the first general and the second applying to a specific model. Regardless of whether $[Q, I]$ vanishes, we have

$$\langle(\Delta Q)^k\rangle = \int_0^t d^k t \, \langle \mathrm{T}^*\big(I(t_1)\cdots I(t_k)\big)\rangle \,,$$

where $d^k t = dt_1 \cdots dt_k$ and T^* means that the derivative has to be taken after the time ordering:

$$\mathrm{T}^*\big(I(t_1)\cdots I(t_n)\big) := \frac{\partial}{\partial t_n} \cdots \frac{\partial}{\partial t_1} \mathrm{T}\big(Q(t_1)\cdots Q(t_n)\big) \,. \tag{11.18}$$

Hence, equation (11.14) may be summarized as $\chi(\lambda, t) = \langle \mathrm{T}^* \exp\big(i\lambda \int_0^t I(t')dt'\big)\rangle$; by (11.18), this amounts to keeping the time ordering outside of the integrals of the exponential series. It should be emphasized that $\mathrm{T}^*\big(I(t_1)\cdots I(t_n)\big)$ is not, strictly speaking, a prescription of ordering currents, since it is defined in terms of charges. It is moreover known (Matthews 1949) that, as a rule, (11.18) differs from

$T(I(t_1)\cdots I(t_n)) = T((dQ(t_1)/dt_1)\cdots (dQ(t_n)/dt_n))$ by contact terms supported at coinciding times, but agrees with it if $[Q, I] = 0$. For instance, for the third cumulant we have

$$\langle\!\langle(\Delta Q)^3\rangle\!\rangle = \int_0^t d^3t\,\langle\!\langle\mathrm{T}\,I_1 I_2 I_3\rangle\!\rangle + 3\int_0^t d^2t\,\langle\!\langle\mathrm{T}\,I_1[Q_2, I_2]\rangle\!\rangle + \int_0^t dt_1\,\langle\!\langle[Q_1, [Q_1, I_1]]\rangle\!\rangle,$$

$$(11.19)$$

where $Q_i = Q(t_i)$, $I_i = I(t_i)$.

In the derivation of equation (11.17) no model having both linear dispersion relation and an energy independent scattering matrix S was explicitly spelled out in terms of a Hamiltonian. It was however in an unrelated context (Falkensteiner and Grosse 1987; Albeverio and Kurasov 1997): Formally the (first quantized) Hilbert space is $L^2(\mathbb{R}_x; \mathbb{C}^2)$ with the two components standing for the left and right channel (as opposed to left and right movers in (11.16)), whence

$$Q = \begin{pmatrix} 0 & 0 \\ 0 & 1 \end{pmatrix}.$$

The Hamiltonian H is the momentum $p = -id/dx$ with boundary conditions $\psi_+(0) = S\psi_-(0)$ at $x = \pm 0$, where

$$S = \begin{pmatrix} r & t' \\ t & r' \end{pmatrix};$$

in particular, the transparency is $T = |t|^2$. Alternatively, but more casually, the Hamiltonian can be written as

$$H = p + 2i\frac{S-1}{S+1}\cdot\delta.$$

Despite the linear dispersion relation one finds

$$[Q(t), I(t)] = ([Q, S^*QS]\theta(-x) + [SQS^*, Q]\theta(x))\delta(x+t) \neq 0.$$

After passing to second quantization, equation (11.19) takes the appearance of

$$\langle\!\langle\Delta Q^3\rangle\!\rangle = \frac{Vt}{2\pi}(-2T^2(1-T) + 0 + T(1-T)) = \frac{Vt}{2\pi}T(1-T)(1-2T),$$

thereby resolving the discrepancy. It should be noted the last term arises as a Schwinger term (Schwinger 1959), as we momentarily explain. Let us denote by $A \mapsto \hat{A}$ the second quantization of a single-particle operator A, meaning its sum over all particles. It satisfies $\langle\hat{A}\rangle = 0$ in the reference state. In the Fock space construction one simply has $[\hat{A}, \hat{B}] = \widehat{[A, B]}$; in the more general case of a fermionic quasi-free state with covariance $\rho = \rho^2$, the operator $\hat{A}$ is defined in the GNS-construction if $[\rho, A]$ is Hilbert–Schmidt, and one has

$$[\hat{A}, \hat{B}] = \widehat{[A, B]} + s(A, B)1$$

with Schwinger term

$$s(A, B) = \text{Tr}([\rho, A]\rho'[B, \rho]) - \text{Tr}([\rho, B]\rho'[A, \rho])$$
$$= \text{Tr}(\rho A\rho' B\rho) - \text{Tr}(\rho' A\rho B\rho'), \qquad (\rho' = 1 - \rho),$$

and in particular $\langle[\widehat{A}, \widehat{B}]\rangle = s(A, B)$. In our case the last integrand in equation (11.19) is a Schwinger term, as announced:

$$\langle[\widehat{Q}(t_1), [\widehat{Q}(t_1), \widehat{I}(t_1)]]\rangle = \langle[\widehat{Q(t_1)}, [\widehat{Q(t_1)}, \widehat{I(t_1)}]]\rangle = s(Q(t_1), [Q(t_1), I(t_1)]).$$

For completeness it should be added that if the current counter, that is the point of separation between right and left channel, is placed at any small distance away from the (point-like) scatterer, then $[Q, I] = 0$: While the cumulant is unchanged, it is now all accounted for by the first term in (11.19).

References

Albeverio, S. and Kurasov, P. (1997). Rank one perturbations of not semibounded operators, *Integr. Equ. Oper. Th.* **27**: 379–400.

Andreev, A. and Kamenev, A. (2000). *Phys. Rev. Lett.* **85**, 1294.

Avron, J.E., Bachmann, S., Graf, G.M. and Klich, I. (2008). Fredholm determinants and the statistics of charge transport, *Comm. Math. Phys.* **280**: 807–829.

Avron, J.E. and Elgart, A. (1999). Adiabatic theorem without a gap condition, *Comm. Math. Phys.* **203**: 445.

Avron, J.E., Elgart, A., Graf, G.M. and Sadun, L. (2001). Optimal quantum pumps, *Phys. Rev. Lett.* **87**: 236601.

Avron, J.E., Elgart, A., Graf, G.M. and Sadun, L. (2002). Time-energy coherent states and adiabatic scattering, *J. Math. Phys.* **43**: 3415.

Avron, J.E., Elgart, A., Graf, G.M. and Sadun, L. (2004). Transport and dissipation in quantum pumps, *J. Stat. Phys.* **116**: 425–473.

Avron, J.E., Elgart, A., Graf, G.M., Sadun, L. and Schnee, K. (2004). Adiabatic charge pumping in open quantum systems, *Comm. Pure Appl. Math.* **57**: 528–561.

Avron, J.E., Seiler, R., and Yaffe, L.G., (1993). Adiabatic theorems and applications to the quantum Hall effect, *Comm. Math. Phys.* **110**: 33–49 (1987); Erratum: "Adiabatic theorems and applications to the quantum Hall effect", *Comm. Math. Phys.* **156**: 649–650.

Bachmann, S., Graf, G.M. and Lesovik, G.B. (2010). Time ordering and counting statistics, *J. Stat. Phys.* **138**: 333–350.

Bachmann, S. and Graf, G.M. (2008). Charge transport and determinants, in *Mathematical Results in Quantum Mechanics: Proceedings of the Qmath10 Conference, Moeciu*, World Scientific, 2008.

Brouwer, P.W. (1998). Scattering approach to parametric pumping, *Phys. Rev. B* **58**: R10135.

Bornemann, F. (1998). *Homogenization in Time of Singularly Perturbed Mechanical Systems*, Lecture Notes in Mathematics, **1687**, Springer.

Bräunlich, G., Graf, G.M. and Ortelli, G. (2010). Equivalence of topological and scattering approaches to quantum pumping, *Comm. Math. Phys.* **295**: 243.

Büttiker, M., Prêtre, A. and Thomas, H. (1993). Dynamic conductance and the scattering matrix of small conductors, *Phys. Rev. Lett.* **70**: 4114.

Büttiker, M., Thomas, H. and Prêtre, A. (1994). Current partition in multiprobe conductors in the presence of slowly oscillating external potentials, *Z. Phys. B* **94**: 133–137.

Falkensteiner, P. and Grosse, H. (1987). Quantization of fermions interacting with point-like external fields, *Lett. Math. Phys.* **14**: 139–148.

Graf, G.M. and Ortelli, G. (2008). Comparison of quantization of charge transport in periodic and open pumps, *Phys. Rev. B* **77**, 033304.

Ivanov, D.A., Lee, H.W. and Levitov, L.S. (1997). Coherent states of alternating current, *Phys. Rev. B* **56**: 6839–6850.

Lesovik, G.B. and Chtchelkatchev, N.M. (2003). Quantum and classical binomial distributions for the charge transmitted through coherent conductor, *JETP Lett.* **77**(7): 393–396.

Levitov, L.S., Lee, H.W. and Lesovik, G.B. (1996). Electron counting statistics and coherent states of electric current, *J. Math. Phys.* **37**(10): 4845–4866.

Levitov, L.S. and Lesovik, G.B. (1993). Charge distribution in quantum shot noise, *JETP Lett.*, **58**(3): 230–235.

Levitov, L.S. and Lesovik, G.B. (1992). Charge-transport statistics in quantum conductors, *JETP Lett.* **55**(9): 555–559.

Levitov, L.S. and Lesovik, G.B. (1994). Quantum measurement in electric circuit. arXiv:cond-mat/9401004.

Martin, P.A. and Sassoli de Bianchi, M. (1995). On the low- and high-frequency limit of quantum scattering by time-dependent potentials, *J. Phys. A* **28**: 2403.

Matthews, P.T. (1949). The application of Dyson's method to meson interactions, *Phys. Rev.* **76**(5): 684–685.

Muzykanskii, B.A. and Adamov, Y. (2003). Scattering approach to counting statistics in quantum pumps, *Phys. Rev. B,* **68**(15): 155304–155313.

Nazarov, Y.V. (ed.), *Quantum noise in Mesoscopic Physics*, NATO Science Series II: Mathematics, Physics and Chemistry, Kluwer Academic Publishers, 2003.

Nenciu, G. (1993). Linear adiabatic theory. Exponential estimates, *Comm. Math. Phys.* **152**: 479–496.

Niu, Q. and Thouless, D.J. (1984). Quantised adiabatic charge transport in the presence of substrate disorder and many-body interaction, *J. Phys. A* **17**: 2453–2462.

Salo, J.F., Hekking, W.J. and Pekola, J.P. (2006). Frequency-dependent current correlation functions from scattering theory, *Phys. Rev. B* **74**: 125427.

Shelankov, A. and Rammer, J. (2003). Charge transfer counting statistics revisited, *Europhys. Lett.* **63**: 485–491.

Schwinger, J. (1959). Field Theory Commutators, *Phys. Rev. Lett.* **3**: 296–297.

Thouless, D.J. (1983). Quantisation of particle transport, *Phys. Rev. B* **27**: 6083–6087.

Wigner, E.P. (1955). Lower limit for the energy derivative of the scattering phase shift, *Phys. Rev.* **98**: 145.

12
Renormalization group and problem of radiation

Israel Michael SIGAL[1]

Department of Mathematics,
University of Toronto,
Toronto, Canada

[1] Supported by NSERC Grant No. NA7901

Chapter Contents

12.1 Overview

We will describe some key results in theory of radiation for the standard model of non-relativistic electrodynamics (QED). The non-relativistic QED was proposed in the early days of Quantum Mechanics[2] and it describes quantum-mechanical particle systems coupled to quantized electromagnetic field. It arises from a standard quantization of the corresponding classical systems (with possible addition of internal—spin—degrees of freedom)[3] and it gives a complete and consistent account of electrons and nuclei interacting with electromagnetic radiation at low energies. It accounts for all physical phenomena in QED, apart from vacuum polarization. A sample of the issues it addresses are:

- stability;
- radiation;
- renormalization of mass;
- anomalous magnetic moment;
- one-particle states;
- scattering theory.

There has been remarkable progress in the last 10 or so years in rigorous understanding of the corresponding phenomena. In this brief review we will deal with results concerning the first two items. We translate them into mathematical terms:

- Stability $\Longleftrightarrow$ Existence of the *ground state*.
- Radiation $\Longleftrightarrow$ Formation of *resonances* out of the excited states of particle systems, scattering theory.

One of the key notions here is that of the *resonance*. It gives a clear-cut mathematical description of processes of emission and absorption of the electromagnetic radiation.

The key and unifying technique we will concentrate on is the spectral *renormalization group*. It is easily combined with other techniques, for example complex deformations (for resonances), the Mourre estimate (for dynamics), analyticity, fibre integral decompositions, and Ward identities (used so far for translationally invariant systems). It was also extended to analysis of existence and stability of thermal states.

12.2 Nonrelativistic QED

12.2.1 Schrödinger equation

We consider a system consisting of n charged particles interacting between themselves and with external fields, which are coupled to a quantized electromagnetic field. The starting point of the nonrelativistic QED is the state Hilbert space $\mathcal{H} = \mathcal{H}_p \otimes \mathcal{H}_f$, which is the tensor product of the state spaces of the particles, $\mathcal{H}_p$, say, $\mathcal{H}_p = L^2(\mathbb{R}^{3n})$,

---

[2] It was used, as already known, by Fermi (Fermi 1932) in 1932 in his review of the theory of radiation.

[3] In fact, it is the most general quantum theory obtained as a quantization of a classical system.

and of bosonic Fock space $\mathcal{H}_f$ of the quantized electromagnetic field, and the standard quantum Hamiltonian $H \equiv H_{g\chi}$ on $\mathcal{H} = \mathcal{H}_p \otimes \mathcal{H}_f$, given (in the units in which the Planck constant divided by 2π and the speed of light are equal to 1: $\hbar = 1$ and $c = 1$) by

$$H = \sum_{j=1}^{n} \frac{1}{2m_j} (i\nabla_{x_j} - gA_\chi(x_j))^2 + V(x) + H_f \tag{12.1}$$

(see (Fermi, 1932) and (Pauli and Fierz, 1938)). Here, m_j and x_j, $j = 1, \ldots, n$, are the ('bare') particle masses and the particle positions, $x = (x_1, \ldots, x_n)$, $V(x)$ is the total potential affecting particles and $g > 0$ is a coupling constant related to the particle charges, $A_\chi := \check{\chi} * A$, where $A(y)$ is the *quantized vector potential*, in the Coulomb gauge $(\mathrm{div} A(y) = 0)$, describing the quantized electromagnetic field, and χ is an *ultraviolet cutoff*,

$$A_\chi(y) = \int (e^{iky} a(k) + e^{-iky} a^*(k)) \chi(k) \frac{d^3 k}{\sqrt{|k|}} \tag{12.2}$$

($a(k)$ and $a^*(k)$ are annihilation and creation operators acting on the Fock space $\mathcal{H}_f \equiv \mathcal{F}$, see Supplement 12.18 for the definitions), and H_f is the quantum Hamiltonian of the quantized electromagnetic field, describing the dynamics of the latter, it is given by

$$H_f = \int d^3 k \, \omega(k) a^*(k) \cdot a(k), \tag{12.3}$$

where $\omega(k) = |k|$ is the dispersion law connecting the energy of the field quantum with its wave vector k. For simplicity we omitted the interaction of the spin with magnetic field. (For a discussion of this Hamiltonian including units, the removal of the center-of-mass motion of the particle system and taking into account the spin of the particles, see Appendix 12.13. Note that our units are not dimensionless. We use these units since we want to keep track of the particle masses. To pass to the dimensionless units we would have to set $m_{\mathrm{el}} = 1$ also.) The Hamiltonian H determines the dynamics via the time-dependent Schrödinger equation

$$i\partial_t \psi = H\psi,$$

where ψ is a differentiable path in $\mathcal{H} = \mathcal{H}_p \otimes \mathcal{H}_f$.

The ultraviolet cutoff, χ, satisfies $\chi(k) = 1$ in a neighbourhood of $k = 0$ and is decaying at infinity on the scale κ and sufficiently fast. We assume that $V(x)$ is a generalized n-body potential, that is it satisfies the assumptions:

(V) $V(x) = \sum_i W_i(\pi_i x)$, where π_i are a linear maps from $\mathbb{R}^{3n}$ to $\mathbb{R}^{m_i}$, $m_i \leq 3n$ and W_i are Kato–Rellich potentials (i.e. $W_i(\pi_i x) \in L^{p_i}(\mathbb{R}^{m_i}) + (L^\infty(\mathbb{R}^{3n}))_\varepsilon$ with $p_i = 2$ for $m_i \leq 3$, $p_i > 2$ for $m_i = 4$ and $p_i \geq m_i/2$ for $m_i > 4$).

Under the assumption (V), the operator H is self-adjoint and bounded below.

We assume for simplicity that our matter consists of electrons and the nuclei and that the nuclei are infinitely heavy and therefore are manifested through the interactions only (put differently, the molecules are treated in the Born–Oppenheimer approximation). In this case, the coupling constant g is related to the electron charge $-e$ as $g := \alpha^{3/2}$, where $\alpha = \frac{e^2}{4\pi\hbar c} \approx \frac{1}{137}$, the fine-structure constant, and $m_j = m$. It is shown (see Section 12.12 and the contribution (Bach 2012) of Volker Bach in this volume for references and discussion) that the physical electron mass, $m_{\rm el}$, is not the same as the parameter $m \equiv m_j$ (the 'bare' electron mass) entering (12.1), but depends on m and κ. Inverting this relation, we can think of m as a function of $m_{\rm el}$ and κ. If we fix the particle potential $V(x)$ (e.g. taking it to be the total Coulomb potential), and $m_{\rm el}$ and e, then the Hamiltonian (12.1) depends on one free parameter, the bare electron mass m (or the ultraviolet cutoff scale, κ).

12.2.2 Stability and radiation

We begin with considering the matter system alone. As was mentioned above, its state space, $\mathcal{H}_p$, is either $L^2(\mathbb{R}^{3n})$ or a subspace of this space determined by a symmetry group of the particle system, and its Hamiltonian operator, H_p, acting on $\mathcal{H}_p$, is given by

$$H_p := \sum_{j=1}^{n} \frac{-1}{2m_j}\Delta_{x_j} + V(x), \tag{12.4}$$

where Δ_{x_j} is the Laplacian in the variable x_j and, recall, $V(x)$ is the total potential of the particle system. Under the conditions on the potentials $V(x)$, given above, the operator H_p is self-adjoint and bounded below. Typically, according to the HVZ theorem, its spectrum consists of isolated eigenvalues, $\epsilon_0^{(p)} < \epsilon_1^{(p)} < \ldots < \Sigma^{(p)}$, and continuum $[\Sigma^{(p)}, \infty)$, starting at the ionization threshold $\Sigma^{(p)}$, as shown in the Fig. 8.1.

The eigenfunctions corresponding to the isolated eigenvalues are exponentially localized. Thus left on its own the particle system, either in its ground state or in one of the excited states, is stable and well localized in space. We expect that this picture changes dramatically when the total system (the universe) also includes the electromagnetic field, which at this level must be considered to be quantum. As was already indicated above what we expect is the following

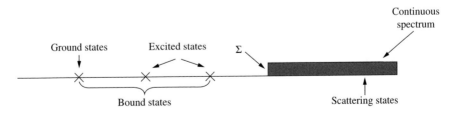

Fig. 12.1 The spectrum of H_p.

- The stability of the system under consideration is equivalent to the statement of existence of the ground state of H, that is an eigenfunction with the smallest possible energy.
- The physical phenomenon of radiation is expressed mathematically as emergence of resonances out of excited states of a particle system due to coupling of this system to the quantum electromagnetic field.

Our goal is to develop the spectral theory of the Hamiltonian H and relate it to the properties of the relevant evolution. Namely, we would like to show that

1) The *ground state* of the particle system is *stable* when the coupling is turned on, while
2) The excited states, generically, are not. They turn into *resonances*.

12.2.3 Ultraviolet cutoff

We reintroduce the Planck constant, $\hbar$, speed of light, c, and electron mass, m_{el}, for a moment. Assuming the ultraviolet cutoff $\chi(k)$ decays on the scale κ, in order to correctly describe the phenomena of interest, such as emission and absorption of electromagnetic radiation, that is for optical and rf modes, we have to assume that the cutoff energy,

$$\hbar c \kappa \gg \alpha^2 m_{el} c^2, \text{ ionization energy, characteristic energy of the particle motion.}$$

On the other hand, we should exclude the energies where the relativistic effects, such as electron-positron pair creation, vacuum polarization, and relativistic recoil, take place, and therefore we assume

$$\hbar c \kappa \ll m_{el} c^2, \text{ the rest energy of the the electron.}$$

Combining the last two conditions we arrive at $\alpha^2 m_{el} c/\hbar \ll \kappa \ll m_{el} c/\hbar$, or in our units,

$$\alpha^2 m_{el} \ll \kappa \ll m_{el}.$$

The Hamiltonian (12.1) is obtained by the rescaling $x \to \alpha^{-1} x$ and $k \to \alpha^2 k$ of the original QED Hamiltonian (see Appendix 12.13). After this rescaling, the new cutoff momentum scale, $\kappa' = \alpha^{-2} \kappa$, satisfies

$$m_{el} \ll \kappa' \ll \alpha^{-2} m_{el},$$

which is easily accommodated by our estimates (e.g. we can have $\kappa' = O(\alpha^{-1/3} m_{el})$).

12.3 Resonances

As was mentioned above, the mathematical language which describes the physical phenomenon of radiation is that of *quantum resonances*. We expect that the latter emerge out of excited states of a particle system due to coupling of this system to the quantum electromagnetic field.

Quantum resonances manifest themselves in three different ways:

1) eigenvalues of complexly deformed Hamiltonians;
2) poles of the meromorphic continuation of the resolvent across the continuous spectrum;
3) metastable states.

12.3.1 Complex deformation

To define resonances we use complex deformation method. In order to be able to apply this method we choose the ultraviolet cutoff, $\chi(k)$, so that

> The function $\theta \to \chi(e^{-\theta}k)$ has an analytic continuation from the real axis, $\mathbb{R}$, to the strip $\{\theta \in \mathbb{C} || \operatorname{Im} \theta| < \pi/4\}$ as a $L^2 \cap L^\infty(\mathbb{R}^3)$ function,

for example $\chi(k) = e^{-|k|^2/\kappa^2}$. For the same purpose, we assume that the potential, $V(x)$, satisfies the condition:

(DA) The the particle potential $V(x)$ is dilation analytic in the sense that the operator-function $\theta \to V(e^\theta x)(-\Delta + 1)^{-1}$ has an analytic continuation from the real axis, $\mathbb{R}$, to the strip $\{\theta \in \mathbb{C} || \operatorname{Im} \theta| < \theta_0\}$ for some $\theta_0 > 0$.

To define the resonances for the Hamiltonian H we pass to the one-parameter (deformation) family

$$H_\theta := U_\theta H U_\theta^{-1}, \tag{12.5}$$

where θ is a real parameter and U_θ, on the total Hilbert space $\mathcal{H} := \mathcal{H}_p \otimes \mathcal{F}$, is the one-parameter group of unitary operators, whose action is rescaling particle positions and of photon momenta:

$$x_j \to e^\theta x_j \text{ and } k \to e^{-\theta}k.$$

One can show show that:

1) Under a certain analyticity condition on coupling functions, the family H_θ has an analytic continuation in θ to the disc $D(0, \theta_0)$, as a type A family in the sense of Kato.
2) The real eigenvalues of H_θ, $\operatorname{Im} \theta > 0$, coincide with eigenvalues of H and that complex eigenvalues of H_θ, $\operatorname{Im} \theta > 0$, lie in the complex half-plane $\mathbb{C}^-$.
3) The complex eigenvalues of H_θ, $\operatorname{Im} \theta > 0$, are locally independent of θ. The typical spectrum of $[H_\theta \equiv H_\theta^{SM}]|_{g=0}$, $\operatorname{Im} \theta > 0$ (here the superindex SM stands for the standard model) is shown in Fig. 8.2 below.

We call complex eigenvalues of H_θ, $\operatorname{Im} \theta > 0$ the *resonances* of H.

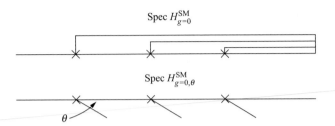

Fig. 12.2 The spectra of $H_{g=0}$ and $H_{g=0,\theta}$, $\mathrm{Im}\,\theta > 0$.

As an example of the above procedure we consider the complex deformation of the hydrogen atom and photon Hamiltonians $H_{hydr} := -\frac{1}{2m}\Delta - \frac{\alpha}{|x|}$ and H_f:

$$H_{hydr\theta} = e^{-2\theta}\Big(-\frac{1}{2m}\Delta\Big) - e^{-\theta}\frac{\alpha}{|x|}, \quad H_{f\theta} = e^{-\theta}H_f.$$

Let e_j^{hydr} be the eigenvalues of the hydrogen atom. Then the spectra of these deformations are

$$\sigma(H_{hydr\theta}) = \{e_j^{hydr}\} \cup e^{-2\,\mathrm{Im}\,\theta}[0,\infty), \quad \sigma(H_{f\theta}) = \{0\} \cup e^{-\,\mathrm{Im}\,\theta}[0,\infty).$$

12.3.2 Resonances as poles

Similarly to eigenvalues, we would like to characterize the resonances in terms of poles of matrix elements of the resolvent $(H - z)^{-1}$ of the Hamiltonian H. To this end we have to go beyond the spectral analysis of H. Let $\Psi_\theta = U_\theta\Psi$, and so on, for $\theta \in \mathbb{R}$ and $z \in \mathbb{C}^+$. Use the unitarity of U_θ for real θ, to obtain (the Combes argument)

$$\langle \Psi,\, (H - z)^{-1}\Phi \rangle = \langle \Psi_{\bar{\theta}},\, (H_\theta - z)^{-1}\Phi_\theta \rangle. \tag{12.6}$$

Assume now that for a dense set of Ψ's and Φ's (say, $\mathcal{D}$, defined below), Ψ_θ and Φ_θ have analytic continuations into a complex neighbourhood of $\theta = 0$ and continue the r.h.s of (12.6) analytically first in θ into the upper half-plane and then in z across the continuous spectrum. This meromorphic continuation has the following properties:

- The real eigenvalues of H_θ give real poles of the r.h.s. of (12.6) and therefore they are the eigenvalues of H.
- The complex eigenvalues of H_θ are poles of the meromorphic continuation of the l.h.s. of (12.6) across the spectrum of H onto the second Riemann sheet.

The poles manifest themselves physically as bumps in the scattering cross-section or poles in the scattering matrix.

The r.h.s. of (12.6) has an analytic continuation into a complex neighbourhood of $\theta = 0$, if $\Psi, \Phi \in \mathcal{D}$, where

$$\mathcal{D} := \bigcup_{n>0, a>0} \mathrm{Ran}\left(\chi_{N \leq n}\chi_{|T| \leq a}\right). \tag{12.7}$$

Here $N = \int d^3k a^*(k)a(k)$ is the photon number operator and T is the self-adjoint generator of the one-parameter group U_θ, $\theta \in \mathbb{R}$. (It is dense, since N and T commute.)

12.3.3 Resonance states as metastable states

While bound states are stationary solutions, one expects that resonances to lead to almost stationary, long-living solutions. Let z_*, $\mathrm{Im}\, z_* \leq 0$, be the ground state or resonance eigenvalue. One expects that for an initial condition, ψ_0, localized in a small energy interval around the ground state or resonance energy, $\mathrm{Re}\, z_*$, the solution, $\psi = e^{-iHt}\psi_0$, of the time-dependent Schrödinger equation, $i\partial_t \psi = H\psi$, is of the form

$$\psi = e^{-iz_* t}\phi_* + O_{\mathrm{loc}}(t^{-\alpha}) + O_{\mathrm{res}}(g^\beta), \tag{12.8}$$

for some α, $\beta > 0$ (depending on ψ_0), where

- ϕ_* is either the ground state or an excited state of the unperturbed system, depending on whether z_* is the ground state energy or a resonance eigenvalue;
- the error term $O_{\mathrm{loc}}(t^{-\alpha})$ satisfies $\|(1 + |T|)^{-\nu} O_{\mathrm{loc}}(t^{-\alpha})\| \leq Ct^{-\alpha}$, where T is the generator of the group U_θ, with an appropriate $\nu > 0$.

For the *ground state*, equation (12.8), without the error term $O_{\mathrm{res}}(g^\beta)$, is called the local decay property (see Section 12.10). One way to prove it is to use the formula connecting the propagator and the resolvent:

$$e^{-iHt}f(H) = \frac{1}{\pi}\int_{-\infty}^{\infty} d\lambda f(\lambda)e^{-i\lambda t}\,\mathrm{Im}(H - \lambda - i0)^{-1}. \tag{12.9}$$

Then one controls the boundary values of the resolvent on the spectrum (the corresponding result is called the limiting absorption principle, see Appendix 12.16.2) and uses properties of the Fourier transform.

For the *resonances*, (12.8) implies that $-\,\mathrm{Im}\, z_*$ has the meaning of the decay probability per unit time, and $(-\,\mathrm{Im}\, z_*)^{-1}$, as the lifetime, of the resonance. To prove it, one uses (12.6) and the analyticity of its r.h.s. in z and performs in (12.9) a suitable deformation of the contour of integration to the second Riemann sheet to pick up the contribution of poles there. This works when the resonances are isolated. In the present case, they are not. This is a consequence of the infrared problem. Hence, determining the long-time behaviour of $e^{-iHt}\psi_0$ is a subtle problem in this case.

To prove the equation (12.8), that is to determine the asymptotic behaviour of solutions $e^{-iHt}\psi_0$, we use the formula connecting the propagator and the resolvent:

$$e^{-iHt}f(H) = \frac{1}{\pi}\int_{-\infty}^{\infty} d\lambda f(\lambda)e^{-i\lambda t}\,\mathrm{Im}(H - \lambda - i0)^{-1}.$$

For the *ground state* the absolute continuity of the spectrum outside the ground state energy, or a stronger property of the limiting absorption principle, suffices to establish the result above.

For the *resonances*, one performs a suitable deformation of the contour of integration to the second Riemann sheet to pick up the contribution of poles there. This works when the resonances are isolated. In the present case, they are not. This is a consequence of the infrared problem. Hence, determining the long-time behaviour of $e^{-iHt}\psi_0$ is a subtle problem.

12.3.4 Comparison with quantum mechanics

This situation is quite different from the one in quantum mechanics (e.g. Stark effect or tunnelling decay) where the resonances are isolated eigenvalues of complexly deformed Hamiltonians. This makes the proof of their existence and establishing their properties, for example independence of θ (and, in fact, of the transformation group U_θ), relatively easy. In the non-relativistic QED (and other massless theories), giving meaning of the resonance poles and proving independence of their location of θ is a rather involved matter.

12.4 Existence of the ground and resonance states

12.4.1 Bifurcation of eigenvalues and resonances

Stated informally what we show is

- The ground state of $H|_{g=0} \Rightarrow$ the ground state H ($\epsilon_0 = \epsilon_0^{(p)} + O(g^2)$ and $\epsilon_0 < \epsilon_0^{(p)}$).
- The excited states of $H|_{g=0} \Rightarrow$ (generically) the resonances of H ($\epsilon_{j,k} = \epsilon_j^{(p)} + O(g^2)$);
- There is $\Sigma > \inf \sigma(H)$ (the ionization threshold, $\Sigma + \Sigma^{(p)} + O(g^2)$) s.t. for energies $< \Sigma$ that particles are exponentially localized around the common centre of mass.

For energies $> \Sigma$ the system either sheds off locally the excess of energy and descends into a localized state or breaks apart with some of the particles flying off to infinity.
   To formulate this result more precisely, denote

$$\epsilon_{gap}^{(p)}(\nu) := \min\{|\epsilon_i^{(p)} - \epsilon_j^{(p)}| \mid i \neq j, \ \epsilon_i^{(p)}, \epsilon_j^{(p)} \leq \nu\}.$$

Theorem 12.1. (*Fate of particle bound states*). *Fix* $\epsilon_0^{(p)} < \nu < \inf \sigma_{ess}(H_p)$ *and let* $g \ll \epsilon_{gap}^{(p)}(\nu)$. *Then for* $g \neq 0$,

- *H has a ground state, originating from a ground state of $H|_{g=0}$ ($\epsilon_0 = \epsilon_0^{(p)} + O(g^2)$, $\epsilon_0 < \epsilon_0^{(p)}$).*
- *Generically, H has no other bound state (besides the ground state).*
- *Eigenvalues, $\epsilon_j^{(p)} < \nu$, $j \neq 0$, of $H|_{g=0} \Longrightarrow$ resonance eigenvalues, $\epsilon_{j,k}$, of H.*
- *$\epsilon_{j,k} = \epsilon_j^{(p)} + O(g^2)$ and the total multiplicity of $\epsilon_{j,k}$ equals the multiplicity of $\epsilon_j^{(p)}$.*
- *The ground and resonance states are exponentially localized in the physical space:*

$$\|e^{\delta|x|}\psi\| < \infty, \ \forall \psi \in \operatorname{Ran} E_\Delta(H), \ \delta < \Sigma^{(p)} - \sup \Delta.$$

Remark 12.2 *The relation $\epsilon_0 < \epsilon_0^{(p)}$ is due to the fact that the electron surrounded by clouds of photons becomes heavier.*

• **Binding and Instability.** H has ground state, but no excited states.

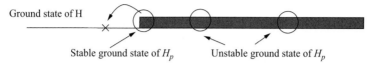

• **Resonances.** The excited states $H_p \Rightarrow$ resonances H.

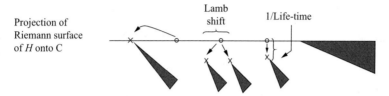

Fig. 12.3 The spectra of H and H_θ, $\operatorname{Im}\theta > 0$, and migration of the isolated eigenvalues of H_p.

12.4.2 Meromorphic continuation across spectrum

Theorem 12.3 *(Meromorphic continuation of the matrix elements of the resolvent) Assume $g \ll \epsilon_{gap}^{(p)}(\nu)$ and let $\epsilon_0 := \inf \sigma(H)$ be the ground state energy of H. Then*

• *For a dense set (defined in (12.7) below) of vectors Ψ and Φ, the matrix elements*

$$F(z, \Psi, \Phi) := \langle \Psi, (H - z)^{-1}\Phi \rangle$$

have meromorphic continuations from $\mathbb{C}^+$ across the interval $(\epsilon_0, \nu) \subset \sigma_{ess}(H)$ into

$$\{z \in \mathbb{C}^- \mid \epsilon_0 < \operatorname{Re} z < \nu\} / \bigcup_{0 \leq j \leq j(\nu)} S_{j,k},$$

where $S_{j,k}$ are the wedges starting at the resonances

$$S_{j,k} := \{z \in \mathbb{C} \mid \tfrac{1}{2}\operatorname{Re}(e^\theta(z - \epsilon_{j,k})) \geq |\operatorname{Im}(e^\theta(z - \epsilon_{j,k}))|\}; \qquad (12.10)$$

• *This continuation has poles at $\epsilon_{j,k}$: $\lim_{z \to \epsilon_{j,k}} (\epsilon_{j,k} - z)F(z, \Psi, \Phi)$ is finite and $\neq 0$.*

12.4.3 Discussion

• Generically, excited states turn into the resonances, not bound states.
• The second theorem implies the absolute continuity of the spectrum and its proof gives also the limiting absorption principle for H (see Appendix 12.16.2 for the definitions).
• The proof of the first theorem gives fast convergent expressions in the coupling constant g for the ground state energy and resonances.

- One can show analyticity of $\epsilon_{j,k}$ in the coupling constant g (see Appendix 12.10 for a result on the ground state energies).
- The meromorphic continuation in question is constructed in terms of matrix elements of the resolvent of a complex deformation, H_θ, $\operatorname{Im}\theta > 0$, of the Hamiltonian H.
- A description of resonance poles is given in Section 12.10.

12.4.4   Approach

The main steps in our analysis of the spectral structure of the quantum Hamiltonian H are:

- Perform a new canonical transformation (a generalized Pauli–Fierz transform)

$$H \to H^{PF} := e^{-igF} H e^{igF},$$

  in order to bring H to a more convenient for our analysis form.
- Apply the spectral renormalization group (RG) on new–momentum anisotropic–Banach spaces.

  The main ideas of the spectral RG are as follows:

- Pass from a single operator H_θ^{PF} to a Banach space $\mathcal{B}$ of Hamiltonian-type operators.
- Construct a map, $\mathcal{R}_\rho$, (RG transformation) on $\mathcal{B}$, with the following properties:

  (a) $\mathcal{R}_\rho$ is 'isospectral';
  (b) $\mathcal{R}_\rho$ removes the photon degrees of freedom related to energies $\geq \rho$.

- Relate the dynamics of semi-flow, $\mathcal{R}_\rho^n, n \geq 1$, (called renormalization group) to spectral properties of individual operators in $\mathcal{B}$.

12.5   Generalized Pauli–Fierz transformation

We perform a canonical transformation (generalized Pauli–Fierz transform) of H in order to bring it to a form which is accessible to spectral renormalization group (it removes the marginal operators). For simplicity, consider one particle of mass 1. We define the generalized Pauli–Fierz transformation as:

$$H^{PF} := e^{-igF} H e^{igF}, \tag{12.11}$$

where $F(x)$ is the self-adjoint operator given by

$$F(x) = \sum_\lambda \int (\bar{f}_{x,\lambda}(k) a_\lambda(k) + f_{x,\lambda}(k) a_\lambda^*(k)) \frac{\chi(k) d^3 k}{\sqrt{|k|}}, \tag{12.12}$$

with the coupling function $f_{x,\lambda}(k)$ chosen as

$$f_{x,\lambda}(k) := e^{-ikx}\frac{\varphi(|k|^{\frac{1}{2}}e_\lambda(k)\cdot x)}{\sqrt{|k|}}, \qquad (12.13)$$

with $\varphi \in C^2$ bounded, with bounded derivatives and satisfying $\varphi'(0) = 1$. For the *standard* Pauli–Fierz transformation, we have $\varphi(s) = s$.

The Hamiltonian H^{PF} is of the same form as H. Indeed, using the commutator expansion
$e^{-igF(x)}H_f e^{igF(x)} = -ig[F,H_f] - g^2[F,[F,H_f]]$, we compute

$$H^{PF} = \frac{1}{2m}(p + gA_{\chi\varphi}(x))^2 + V_g(x) + H_f + gG(x), \qquad (12.14)$$

where

$$A_{\chi\varphi}(x) = \sum_\lambda \int (\bar{\varphi}_{x,\lambda}(k)a_\lambda(k) + \varphi_{x,\lambda}(k)a_\lambda^*(k))\frac{\chi(k)d^3k}{\sqrt{|k|}},$$

with the new coupling function $\varphi_{\lambda,x}(k) := e_\lambda(k)e^{-ikx} - \nabla_x f_{x,\lambda}(k)$ and

$$V_g(x) := V(x) + 2g^2\sum_\lambda \int |k||f_{x,\lambda}(k)|^2d^3k,$$

$$G(x) := -i\sum_\lambda \int |k|(\bar{f}_{x,\lambda}(k)a_\lambda(k) - f_{x,\lambda}(k)a_\lambda^*(k))\frac{\chi(k)d^3k}{\sqrt{|k|}}.$$

The potential $V_g(x)$ is a small perturbation of $V(x)$ and the operator $G(x)$ is easy to control. The new coupling function has better infrared behaviour for bounded $|x|$:

$$|\varphi_{\lambda,x}(k)| \leq \text{const}\min(1, \sqrt{|k|}\langle x \rangle). \qquad (12.15)$$

To prove the results above we first establish the spectral properties of the generalized Pauli–Fierz Hamiltonian H^{PF} and then transfer the obtained information to the original Hamiltonian H.

12.6 Renormalization group map

The *renormalization map* is defined on Hamiltonians acting on $\mathcal{H}_f$ as follows

$$\mathcal{R}_\rho = \rho^{-1}S_\rho \circ F_\rho, \qquad (12.16)$$

where $\rho > 0$, $S_\rho : \mathcal{B}[\mathcal{H}] \to \mathcal{B}[\mathcal{H}]$ is the *scaling transformation*:

$$S_\rho(1) := 1, \qquad S_\rho(a^\#(k)) := \rho^{-3/2}a^\#(\rho^{-1}k), \qquad (12.17)$$

and F_ρ is the *Feshbach–Schur map*, or decimation, map,

$$F_\rho(H) := \chi_\rho(H - H\overline{\chi}_\rho(\overline{\chi}_\rho H\overline{\chi}_\rho)^{-1}\overline{\chi}_\rho H)\chi_\rho, \qquad (12.18)$$

where χ_ρ and $\overline{\chi}_\rho$ are a pair of orthogonal projections, defined as

$$\chi_\rho = \chi_{H_{p\theta}=e_j} \otimes \chi_{H_f \le \rho} \quad \text{and} \quad \overline{\chi}_\rho := 1 - \chi_\rho.$$

Remark 12.4 *For simplicity we defined the decimation map as the Feshbach–Schur map. Technically it is more convenient to use the smooth Feshbach–Schur map, which we defined and discussed in Appendix 12.16. The smooth Feshbach–Schur map uses 'smooth' projections which form a partition of unity $\chi_\rho^2 + \overline{\chi}_\rho^2 = 1$, instead of true projections as defined above.*

The map F_ρ is *isospectral* in the sense of the following theorem:

Theorem 12.5

  (i) $\lambda \in \rho(H) \Leftrightarrow 0 \in \rho(F_\rho(H - \lambda))$;
 (ii) $H\psi = \lambda\psi \Longleftrightarrow F_\rho(H - \lambda)\varphi = 0$;
(iii) $\dim \mathrm{Null}(H - \lambda) = \dim \mathrm{Null}\, F_\rho(H - \lambda)$;
(iv) $(H - \lambda)^{-1}$ exists $\Longleftrightarrow F_\rho(H - \lambda)^{-1}$ exists.

For the proof of this theorem as well as for the relation between ψ and φ in (ii) and between $(H - \lambda)^{-1}$ and $F_\rho(H - \lambda)^{-1}$ in (iv) see Appendix 12.15.

12.7   A Banach space of Hamiltonians

We will study operators on the subspace $\mathrm{Ran}\,\chi_1$ of the Fock space $\mathcal{F}$. Such operators are said to be in the *generalized normal form* if they can be written as:

$$H = \sum_{m+n \ge 0} W_{m+n}, \qquad (12.19)$$

$$W_{m+n} = \int_{B_1^{m+n}} \prod_{i=1}^{m+n} d^3 k_i \prod_{i=1}^{m} a^*(k_i)\, w_{m,n}\left(H_f; k_{(m+n)}\right) \prod_{i=m+1}^{m+n} a(k_i),$$

where B_1^r denotes the Cartesian product of r unit balls in $\mathbb{R}^3$, $k_{(m)} := (k_1, \ldots, k_m)$ and $w_{m,n} : I \times B_1^{m+n} \to \mathbb{C}$, $I := [0, 1]$. We sometimes we display the dependence of H and $W_{m,n}$ on the coupling functions $\underline{w} := (w_{m,n}, m+n \ge 0)$ by writing $H[\underline{w}]$ and $W_{m,n}[\underline{w}]$.

We assume that the functions $w_{m,n}(r, , k_{(m+n)})$ are continuously differentiable in $r \in I$, symmetric w. r. t. the variables $(k_1, \ldots, k_m)$ and $(k_{m+1}, \ldots, k_{m+n})$ and obey $\|w_{m,n}\|_{\mu,1} := \sum_{n=0}^{1} \|\partial_r^n w_{m,n}\|_\mu < \infty$, where $\mu \ge 0$ and

$$\|w_{m,n}\|_\mu := \max_j \sup_{r \in I, k_{(m+n)} \in B_1^{m+n}} \left| |k_j|^{-\mu} \prod_{i=1}^{m+n} |k_i|^{1/2} w_{m,n}(r; k_{(m+n)}) \right|. \qquad (12.20)$$

Here $k_j \in \mathbb{R}^3$ is the j−th 3−vector in $k_{(m,n)}$ over which we take the supremum. Note that these norms are anisotropic in the total momentum space.

For $\mu \geq 0$ and $0 < \xi < 1$ we define the Banach space

$$\mathcal{B}^{\mu\xi} := \{H \ : \ \|H\|_{\mu,\xi} := \sum_{m+n\geq 0} \xi^{-(m+n)} \|w_{m,n}\|_{\mu,1} < \infty\}. \tag{12.21}$$

We mention some properties of these spaces

- For any $\mu \geq 0$ and $0 < \xi < 1$, the map $H : \underline{w} \to H[\underline{w}]$, given in (12.19), is one-to-one.
- If H is self-adjoint, then so are $W_{0,0}$ and $\sum_{m+n\geq 1} \chi_1 W_{m,n} \chi_1$ (see (12.19)).

Remarks. 1) Unlike the Banach spaces defined in (Bach, Fröhlich, and Sigal 1998; Bach, Fröhlich, and Sigal 1998; Bach, Chen, Fröhlich, and Sigal 2003), the Banach spaces are anisotropic in the momentum space. This is needed to overcome the problem of marginal operators which arise in the renormalization group approach.

2) The self-adjointness statement follows from (Bach, Chen, Fröhlich, and Sigal 2003), Eq. (3.33).

12.7.1 Basic bound

The following bound shows that our Banach space norm controls the operator norm and the terms with higher numbers of creation and annihilation operators make progressively smaller contributions:

Theorem 12.6 Let $\chi_\rho \equiv \chi_{H_f \leq \rho}$. Then for all $\rho > 0$ and $m + n \geq 1$

$$\|\chi_\rho H_{m,n} \chi_\rho\| \leq \frac{\rho^{m+n+\mu}}{\sqrt{m!\,n!}} \|w_{m,n}\|_\mu. \tag{12.22}$$

Sketch of Proof For simplicity we prove this inequality for $m = n = 1$. Let $\phi \in \mathcal{F}$ and $\Phi_k = a(k)\chi_\rho\phi$. We have

$$\langle \chi_\rho\phi, H_{1,1} \chi_\rho\phi \rangle = \int_{B_1^2} \prod_{i=1}^{2} d^3k_i \, \langle \Phi_{k_1}, w_{1,1}(H_f; k_1, k_2) \Phi_{k_2} \rangle.$$

Now we write $\chi_\rho = \chi_\rho\chi_{2\rho}$ and pull $\chi_{2\rho}$ toward $w_{1,1}(H_f; k_1, k_2)$ using the pull-through formulae

$$a(k)\,f(H_f) = f(H_f + |k|)\,a(k), \ \ f(H_f)\,a^*(k) = a^*(k)\,f(H_f + |k|)$$

(see Appendix 12.17). This gives

$$|\langle \phi, \chi_\rho H_{1,1} \chi_\rho\phi \rangle|$$

$$= |\int_{|k_i| \leq 2\rho, i=1,2} \prod_{i=1}^{2} d^3k_i \, \langle \Phi_{k_1}, \chi_{2\rho-|k_1|} w_{1,1}(H_f; k_1, k_2)\chi_{2\rho-|k_2|}\Phi_{k_2} \rangle|$$

$$\leq \int_{|k_i|\leq 2\rho, i=1,2} \prod_{i=1}^{2} d^3 k_i \, \|\Phi_{k_1}\| \, \|w_{1,1}(H_f; k_1, k_2)\| \, \|\Phi_{k_2}\|$$

$$\leq \left(\int_{|k_i|\leq 2\rho, i=1,2} \prod_{i=1}^{2} d^3 k_i \frac{\|w_{1,1}(H_f; k_1, k_2)\|^2}{|k_1||k_2|} \right)^{1/2} \int d^3 k \|\sqrt{|k|}\Phi_k\|^2.$$

Now, using $\|w_{1,1}(H_f; k_1, k_2)\| \leq \|w_{1,1}\|_\mu \frac{|k_1|^\mu + |k_2|^\mu}{|k_1|^{\frac{1}{2}} |k_2|^{\frac{1}{2}}}$ and $\int d^3 k \|\sqrt{|k|}\Phi_k\|^2 = \|\sqrt{H_f}\chi_\rho\phi\|^2$, we find

$$|\langle \phi, \chi_\rho H_{1,1} \chi_\rho \phi \rangle| \lesssim \rho^{2+\mu} \|w_{1,1}\|_\mu. \qquad \square$$

12.7.2 Unstable, neutral, and stable components

We decompose $H \in \mathcal{B}^{\mu\xi}$ into the components $E := \langle \Omega, H\Omega \rangle$, $T := H_{0,0} - \langle \Omega, H\Omega \rangle$, $W := \sum_{m+n \geq 1} W_{m,n}$, so that

$$H = E\mathbf{1} + T + W. \qquad (12.23)$$

If we assume $\sup_{r\in[0,\infty)} |T'(r) - 1| \ll 1$, then we have $T \sim H_f$. These Hamiltonian components scale as follows

- $\rho^{-1} S_\rho(H_f) = H_f$ (H_f is a *fixed point* of $\rho^{-1} S_\rho$);
- $\rho^{-1} S_\rho(E \cdot \mathbf{1}) = \rho^{-1} E \cdot \mathbf{1}$ ($E \cdot \mathbf{1}$ *expand* under $\rho^{-1} S_\rho$ at a rate ρ^{-1});
- $\|S_\rho(W_{m,n})\|_\mu \leq \rho^\alpha \|w_{m,n}\|_\mu$, $\alpha := m + n - 1 + \mu\delta_{m+n=1}$ (W_{mn} *contract* under $\rho^{-1} S_\rho$, if $\mu > 0$).

Thus for $\mu > 0$, E, T, W behave, in the terminology of the renormalization group approach, as relevant, marginal, and irrelevant operators, respectively. For $\mu = 0$, the operators W_{mn}, $m + n = 1$, become marginal.

12.8 Action of renormalization map

To control the components E, T, W of H we introduce, for $\alpha, \beta, \gamma > 0$, the following polydisc:

$$\mathcal{D}^\mu(\alpha, \beta, \gamma) := \Big\{ H = E + T + W \in \mathcal{B}^{\mu\xi} \mid |E| \leq \alpha,$$

$$\sup_{r\in[0,\infty)} |T'(r) - 1| \leq \beta, \ \|W\|_{\mu,\xi} \leq \gamma \Big\}.$$

(Strictly speaking we should write $\chi_{H_{p\theta}=e_j} \otimes \mathcal{D}^\mu(\alpha, \beta, \gamma)$ instead of $\mathcal{D}^\mu(\alpha, \beta, \gamma)$ (for various sets of parameters α, β, γ) in the statement below.)

Theorem 12.7 Let $0 < \rho < 1/2$, α, β, $\gamma \leq \rho/8$ and $\mu_0 = 1/2$. Then there is $c > 0$, s.t.

- $\mathcal{R}_\rho(H_\theta^{PF}) \in \mathcal{D}^{\mu_0}(\alpha_0, \beta_0, \gamma_0)$, $\alpha_0 = cg^2\rho^{\mu_0-2}$, $\beta_0 = cg^2\rho^{\mu_0-1}$, $\gamma_0 = cg\rho_0^\mu$, provided $g \ll 1$;

- $\mathcal{D}^\mu(\alpha, \beta, \gamma) \subset D(\mathcal{R}_\rho)$, provided $\mu > 0$;
- $\mathcal{R}_\rho : \mathcal{D}^\mu(\alpha, \beta, \gamma) \to \mathcal{D}^\mu(\alpha', \beta', \gamma')$, continuously, with
$\alpha' = \rho^{-1}\alpha + c\left(\gamma^2/2\rho\right), \beta' = \beta + c\left(\gamma^2/2\rho\right), \gamma' = c\rho^\mu\gamma$.

Sketch of Proof of the second and third properties   1) $\mathcal{D}^\mu(\rho/8, 1/8, \rho/8) \subset D(\mathcal{R}_\rho)$. Since $W := H - E - T$ defines a bounded operator on $\mathcal{F}$, we only need to check the invertibility of $H_{T\overline{\chi}_\rho}$ on Ran $\overline{\chi}_\rho$. The operator $E + T$ is invertible on Ran $\overline{\chi}_\rho$: for all $r \in [3\rho/4, \infty)$

$$\text{Re } T(r) + \text{Re } E \geq r - |T(r) - r| - |E|$$

$$\geq r\left(1 - \sup_r |T'(r) - 1|\right) - |E|$$

$$\geq \frac{3\rho}{4}(1 - 1/8) - \frac{\rho}{8} \geq \frac{\rho}{2}$$

$$\Rightarrow E + T \text{ is invertible and } \|(E + T)^{-1}\| \leq 2/\rho.$$

Now, by the basic estimate, $\|W\| \leq \rho/8$ and therefore,

$$\|\overline{\chi}_\rho W \overline{\chi}_\rho (E + T)^{-1}\| \leq 1/4$$

$$\Rightarrow E + T + \overline{\chi}_\rho W \overline{\chi}_\rho \text{ is invertible on Ran } \overline{\chi}_\rho$$

$$\Rightarrow \mathcal{D}^\mu(\rho/8, 1/8, \rho/8) \subset D(F_\rho) = D(\mathcal{R}_\rho). \qquad \square$$

2) $\mathcal{R}_\rho : \mathcal{D}^\mu(\alpha, \beta, \gamma) \to \mathcal{D}^\mu(\alpha', \beta', \gamma')$ (normal form of $\mathcal{R}_\rho(H)$). Recall that $\chi_\rho \equiv \chi_{H_f \leq \rho}$ and $\overline{\chi}_\rho := 1 - \chi_\rho$. Let $H_0 := E + T$, so that $H = H_0 + W$. We have shown above

$$\|H_0^{-1}\overline{\chi}_\rho\| \leq \frac{2}{\rho} \quad \text{and} \quad \|W\| \leq \frac{\rho}{8}.$$

In the Feshbach–Schur map, F_ρ,

$$F_\rho(H) = \chi_\rho\left(H_0 + W - W\,\overline{\chi}_\rho\left(\overline{\chi}_\rho(H_0 + W)\overline{\chi}_\rho\right)^{-1}\overline{\chi}_\rho\, W\right)\chi_\rho,$$

we expand the resolvent $\left(\overline{\chi}_\rho(H_0 + W)\overline{\chi}_\rho\right)^{-1}$ in the norm convergent Neumann series

$$F_\rho(H) = \chi_\rho\left[H_0 + \sum_{s=0}^{\infty}(-1)^s\, W\left(H_0^{-1}\overline{\chi}_\rho^2\, W\right)^s\right]\chi_\rho.$$

Next, we transform the right side to the generalized normal form using generalized Wick's theorem.

Generalized Wick's theorem. To write the product $W\left(H_0^{-1}\overline{\chi}_\rho^2\, W\right)^s$ in the generalized normal form we pull the annihilation operators, a, to the right and the creation operators, a^*, to the left, apart from those which enter H_f. We use the rules (see Appendix 12.17):

$$a(k)a^*(k') = a^*(k')a(k) + \delta(k - k'),$$

$$a(k)\,f(H_f) = f(H_f + |k|)\,a(k), \quad f(H_f)\,a^*(k) = a^*(k)\,f(H_f + |k|).$$

Some of the creation and annihilation operators reach the extreme left and right positions, while the remaining ones contract (see Fig. 8.4 below). The terms with m creation operators on the left and n annihilation operators on the right contribute to the $(m,n)-$ formfactor, $w_{m,n}^{(s)}$, of the operator $W\left(H_0^{-1}\overline{\chi}_\rho^2\,W\right)^s$. As a result we obtain the generalized normal form of $F_\rho(H)$:

$$F_\rho(H) = \sum_{m+n\geq 0} W'_{m,n}.$$

The term $W'_{0,0} = \langle W_{0,0}^{(s)}\rangle\Omega + (W_{0,0}^{(s)} - \langle W_{0,0}^{(s)}\rangle\Omega)$ contributes the corrections to $E + T$.

This is the standard way for proving the Wick theorem, taking into account the presence of H_f- dependent factors. (See (Bach, Fröhlich, and Sigal 1998) for a different, more formal proof.)

Estimating formfactors. The problem here is that the number of terms generated by various contractions is $O(s!)$. Therefore a simple majoration of the series for the $(m,n)-$ formfactor, $w_{m,n}^{(s)}$, of the operator $W\left(H_0^{-1}\overline{\chi}_\rho^2\,W\right)^s$ will diverge badly.

To overcome this we re-sum the series by, roughly, representing the sum over all contractions, for a given m and n, as

$$w_{m,n}^{(s)} \sim \langle\Omega, [W\left(H_0^{-1}\overline{\chi}_\rho^2\,W\right)^s]_{m,n}\Omega\rangle,$$

where $[W\left(H_0^{-1}\overline{\chi}_\rho^2\,W\right)^s]_{m,n}$ is $W\left(H_0^{-1}\overline{\chi}_\rho^2\,W\right)^s$, with m escaping creation operators and n escaping annihilation operators deleted. Now the estimate of $w_{m,n}^{(s)}$ is straightforward and can be written (symbolically) as

$$\|w_{m,n}^{(s)}\| \lesssim \|\chi_{\rho'}\,W'\chi_{\rho'}\|^{s+1},$$

for the operator norm, and similarly for $\mathcal{B}^{\mu,\xi}-$norm.

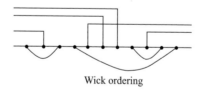

Wick ordering

Given *external* lines corresponding to $x_1...x_m, x_{m+1}...x_{m+n}$,

$w_{mn}^{(s)}$ is obtained by summing over all possible contraction schemes producing internal lines

Fig. 12.4 Wick ordering and contraction scheme.

12.9 Renormalization group

To analyse spectral properties of individual operators in $\mathcal{B}^{\mu\xi}$ we use the discrete semi-flow, $\mathcal{R}_\rho^n, n \geq 1$ (called the renormalization group), generated by the renormalization transformation, $\mathcal{R}_\rho$. By Theorem 12.7, in order to iterate $\mathcal{R}_\rho$ we have to control the expanding direction: $\mathcal{R}_\rho(\zeta\mathbf{1}) = \rho^{-1}\zeta\mathbf{1}$. To control this direction, we adjust, inductively, at each step the constant component $\langle H \rangle_\Omega := \langle \Omega, H\Omega \rangle$ of the initial Hamiltonian, H:

$$|\langle H \rangle_\Omega - e_{n-1}| \leq \frac{1}{12}\rho^{n+1},$$

e_{n-1} is the unique zero of the function $\lambda \to \left\langle \mathcal{R}_\rho^{n-1}(H - \langle H \rangle_\Omega + \lambda) \right\rangle_\Omega$,

so that

$$H \in \quad \text{the domain of} \quad \mathcal{R}_\rho^n.$$

This way one adjusts the initial conditions closer and closer to the stable manifold $\mathcal{M}_s = \cap_n D(\mathcal{R}_\rho^n)$.

The procedure above leads to the following results:

- $\mathcal{R}_\rho^n$ has the fixed-point manifold $\mathcal{M}_{fp} := \mathbb{C}H_f$;
- $\mathcal{R}_\rho^n$ has an unstable manifold $\mathcal{M}_u := \mathbb{C}\mathbf{1}$;
- $\mathcal{R}_\rho^n$ has a (complex) co-dimension 1 stable manifold $\mathcal{M}_s$ for $\mathcal{M}_{fp}$;
- $\mathcal{M}_s$ is foliated by (complex) co-dimension 2 stable manifolds for each fixed point.

Define the polydisc $\mathcal{D}_s := \mathcal{D}^\mu(0, \beta_0, \gamma_0)$, with $\mu > 0$, $\beta_0 = cg^2\rho^{\mu_0-1}$ and $\gamma_0 = cg\rho_0^\mu$, the same as in Theorem 8.5. Using the above results, we draw the following conclusions about the spectrum of an operator $H \in \mathcal{D}_s$. Find $\lambda \in$ a neighbourhood of 0,

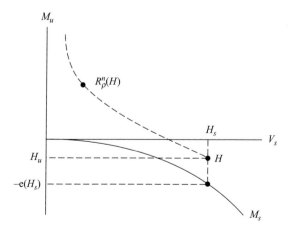

Fig. 12.5 Orbit of H under the renormalization group.

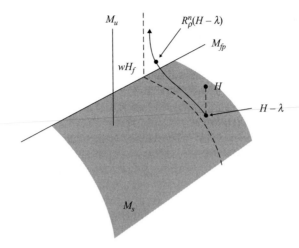

Fig. 12.6 Unstable, stable and fixed point manifolds of the renormalization group.

s.t. $H(\lambda) := H - \lambda \in \mathcal{D}(\mathcal{R}^n_\rho)$ and define $H^{(n)}(\lambda) := \mathcal{R}^n_\rho(H(\lambda))$. We illustrate this as putting $H(\lambda)$ through the RG (black) box:

$$H(\lambda) \Longrightarrow \textbf{RG BOX} \Longrightarrow H^{(n)}(\lambda).$$

Use that for n sufficiently large, $H^{(n)}(\lambda) \approx \zeta H_f$, for some $\zeta \in \mathbb{C}$, Re $\zeta > 0$, to find wanted spectral information about $H^{(n)}(\lambda)$ in a neighbourhood of 0. Then, use the isospectrality of $\mathcal{R}_\rho$ to pass this information up the chain. These steps are described schematically as follows:

Spectral information about $H^{(n)}(\lambda) \Longrightarrow$ (by 'isospectrality' of $\mathcal{R}_\rho$)
Spectral information about $H^{(n-1)}(\lambda)$

$\dots$

$\Longrightarrow$ Spectral information about $H(\lambda)$.
We sum up this as

$$\textbf{Spec info } (H) \Longleftarrow \textbf{RG BOX} \Longleftarrow \textbf{Spec info } (H^{(n)}(\lambda)).$$

If we start with the Hamiltonian H_θ, that we are interested in, we derive in this way the required spectral information.

12.10   Related results

Existence of the ionization threshold. There is $\Sigma = \Sigma^{(p)} + O(g^2) > \inf \sigma(H)$ (the ionization threshold) s.t. for any energy interval in $\Delta \subset (\inf \sigma(H), \Sigma)$,

$$\|e^{\delta|x|}\psi\| < \infty, \ \forall \psi \in \text{Ran} \, E_\Delta(H), \ \delta < \Sigma - \sup \Delta.$$

Analyticity

If the ground state energy ϵ_g of H is non-degenerate, then it is analytic in g and has the following expansion

$$\epsilon_g = \sum_{j=1}^{\infty} \epsilon_\alpha^{2j} \alpha^{3j}, \tag{12.24}$$

where ϵ_α^{2j} are smooth function of α (recall that $g := \alpha^{3/2}$).

The proof relies on the renormalization group analysis together with the analyticity and the formfactor rotation symmetry transfer by the Feshbach–Schur map (see Appendix 12.16.3) and on treating two sources of the dependence of H on the coupling constant g - in the prefactor of $A_\chi hi(y)$ and in its argument (see the line after (12.30)), differently, that is by considering the following family of Hamiltonians

$$H := \sum_{j=1}^{n} \frac{1}{2m_j} (i\nabla_{x_j} - g A_\chi(\alpha x_j))^2 + V(x) + H_f, \tag{12.25}$$

and consider g and α as independent variables. The powers of g in (12.24) come from g in (12.25).

Analyticity of all parts of H

Suppose that $\lambda \mapsto H_\lambda \equiv H(w^\lambda)$ is of the form (12.19) and is analytic in $\lambda \in S \subset \mathbb{C}$ and that $H(w^\lambda)$ belongs to some polydisc $\mathcal{D}(\alpha, \beta, \gamma)$ for all $\lambda \in S$. Then $\lambda \mapsto E_\lambda := w_{0,0}^\lambda(0)$, $T_\lambda := w_{0,0}^\lambda(H_f) - w_{0,0}^\lambda(0)$, $W_\lambda := H_\lambda - E_\lambda - T_\lambda$ are analytic in $\lambda \in S$.

Resonance poles

Can we make sense of the resonance poles in the present context? Let

$$Q := \{z \in \mathbb{C}^- \mid \epsilon_0 < \operatorname{Re} z < \nu\} / \bigcup_{j \le j(\nu), k} S_{j,k}.$$

Theorem 12.8 For each Ψ and Φ from a dense set of vectors (say, (12.7)), the meromorphic continuation, $F(z, \Psi, \Phi)$, of the matrix element $\langle \Psi, (H-z)^{-1}\Phi \rangle$ is of the following form near the resonance ϵ_j of H:

$$F(z, \Psi, \Phi) = (\epsilon_{j,k} - z)^{-1} p(\Psi, \Phi) + r(z, \Psi, \Phi). \tag{12.26}$$

Here p and $r(z)$ are sesquilinear forms in Ψ and Φ, s.t.

- $r(z)$ is analytic in Q and bounded on the intersection of a neighbourhood of $\epsilon_{j,k}$ with Q as

$$|r(z, \Psi, \Phi)| \le C_{\Psi, \Phi} |\epsilon_{j,k} - z|^{-\gamma} \text{ for some } \gamma < 1;$$

- $p \neq 0$ at least for one pair of vectors Ψ and Φ and $p = 0$ for a dense set of vectors Ψ and Φ in a finite co-dimension subspace.

Local decay

For any compactly supported function $f(\lambda)$ with $\operatorname{supp} f \subseteq (\inf H, \infty)/(\text{a neighbour-hood of } \Sigma)$, and for $\theta > \frac{1}{2}$, $\nu < \theta - \frac{1}{2}$, we have that

$$\|\langle Y \rangle^{-\theta} e^{-iHt} f(H) \langle Y \rangle^{-\theta}\| \leq Ct^{-\nu}. \tag{12.27}$$

Here $\langle Y \rangle := (1 + Y^2)^{1/2}$, and Y denotes the self-adjoint operator on Fock space $\mathcal{F}$ of photon coordinate,

$$Y = \int d^3k \, a^*(k) \, i\nabla_k \, a(k), \tag{12.28}$$

extended to the Hilbert space $\mathcal{H} = \mathcal{H}_p \otimes \mathcal{F}$. (A self-adjoint operator H obeying (12.27) is said to have the (Δ, ν, Y, θ)—*local decay* (LD) property.) Equation (12.27) shows that for well-prepared initial conditions ψ_0, the probability of finding photons within a ball of an arbitrary radius $R < \infty$, centred say at the centre-of-mass of the particle system, tends to 0, as time t tends to ∞.

Note that one can also show that as $t \to \infty$, the photon coordinate and wave vectors in the support of the solution $e^{-iHt}\psi_0$ of the Schrödinger equation become more and more parallel. This follows from the local decay for the self-adjoint generator of dilatations on Fock space $\mathcal{F}$,

$$B = \frac{i}{2} \int d^3k \, a^*(k) \left\{ k \cdot \nabla_k + \nabla_k \cdot k \right\} a(k). \tag{12.29}$$

(In fact, one first proves the local decay property for B and then transfers it to the photon coordinate operator Y.)

12.11   Conclusion

Apart from the vacuum polarization, the nonrelativistic QED provides a good qualitative description of the physical phenomena related to the interaction of quantized electrons and nuclei and the electromagnetic field. (Though construction of the scattering theory is not yet completed and the correction to the gyromagnetic ratio is not established, it is fair to conjecture that while both are difficult problems, they should go through without a hitch.)

The quantitative results though are still missing. Does the free parameter, m (or κ), suffice to give a good fit with the experimental data say on the radiative corrections?

12.12   Comments on literature

These lectures follow the papers (Sigal 2009; Fröhlich, Griesemer and Sigal 2009; Abou Salem *et al.* 2007), which in turn extend (Bach, Fröhlich, and Sigal 1998; Bach,

Fröhlich, and Sigal 1998, Bach, Chen, Fröhlich, and Sigal 2003). The papers (Sigal 2009; Fröhlich, Griesemer and Sigal 2009; Abou Salem, Faupin, Fröhlich, and Sigal 2007) use the smooth Feshbah–Schur map ((Bach, Chen, Fröhlich, and Sigal 2003, Griesemer and Hasler 2008)), which is much more powerful (see Appendix 12.16), while in these lectures we use, for simplicity, the the original, Feshbach–Schur map (see Bach, Fröhlich, and Sigal 1998, Bach, Fröhlich, and Sigal 1998), which is simpler to formulate.

The self-adjointness of H is not difficult and was proven in (Bach, Fröhlich, and Sigal 1999) for sufficiently small coupling constant (g) and in (Hiroshima 2002) (see also (Hiroshima 2000)), for an arbitrary one.

Theorems 4.1 and 4.2 were proven in (Bach, Fröhlich, and Sigal 1998; Bach, Fröhlich, and Sigal 1998; Bach, Fröhlich, and Sigal 1999) for 'confined particles' (the exact conditions are somewhat technical) and in the present form in (Sigal Sigal).

The results of (Bach, Fröhlich, and Sigal 1999) on existence of the ground state were considerably improved in (Hiroshima 1999; Hiroshima 2000; Hiroshima 2001; Hiroshima and Spohn 2001; Arai and Hirokawa 2000; Loss, Miyao and Spohn 2008) (by compactness techniques) and (Bach, Fröhlich, and Pizzo 2006) (by multi-scale techniques), with the sharpest result given in (Griesemer, Lieb and Loss 2001; Lieb and Loss 2003). The papers (Bach, Fröhlich, and Sigal 1999; Hiroshima 1999; Griesemer, Lieb and Loss 2001; Lieb and Loss 1999; Lieb and Loss 2002; Lieb and Loss 2003; Hiroshima and Spohn 2001; Loss, Miyao and Spohn 2008) include the interaction of the spin with magnetic field in the Hamiltonian.

Related results:

- The asymptotic stability of the ground state (local decay, see Section 12.10): (Bach, Fröhlich, Sigal, and Soffer 1999; Fröhlich, Griesemer and Sigal 2008; Fröhlich, Griesemer and Sigal 2011).
- The survival probabilities of excited states (see Section (12.8)): (Bach, Fröhlich, and Sigal 1999; Hasler, Herbst and Huber 2008; Abou Salem, Faupin, Fröhlich, and Sigal 2007).
- Atoms with dynamic nuclei: (Faupin 2008; Amour, Grébert, and Guillot 2006; Loss, Miyao and Spohn 2007).
- Analyticity of the ground state eigenvalues in parameters and asymptotic expansions (see Section 12.10): (Bach, Fröhlich, and Sigal 1999; Bach, Fröhlich, and Pizzo 2006; Bach, Fröhlich, and Pizzo 2009; Griesemer and Hasler 2009; Hasler and Herbst 2008; Hasler and Herbst 2010; Hasler and Herbst 2010). Chen
- Existence of the ionization threshold (see Section 12.10): (Griesemer 2004).
- Resonance poles (see Section 12.10): (Abou Salem, Faupin, Fröhlich, and Sigal 2007) (see also (Bach, Fröhlich, and Sigal 1999).
- Self-energy and binding energy: (Hiroshima and Spohn 2001; Lieb and Loss 1999; Lieb and Loss 2002; Lieb and Loss 2003; Hainzl 2002; Barbaroux, Chen, and Vugalter 2003; Chen, Vougalter and Vugalter 2003; Hainzl, Vougalter, and Vugalter 2003; Hainzl 2003; Hainzl and Seiringer 2003; Hainzl, Hirokawa, and Spohn 2005; Catto, Exner, and Hainzl 2004; Barbaroux and Vugalter 2010; Barbaroux, Chen, Vougalter and Vugalter 2010).

- Electron mass renormalization: (Hainzl and Seiringer 2003; Hiroshima and Spohn 2005; Bach, Chen, Fröhlich, and Sigal 2007; Chen 2008; Fröhlich and Pizzo 2010).
- One particle states: (Fröhlich 1973; Fröhlich 1974, Chen, Fröhlich, and Pizzo 2010, Chen, Fröhlich, and Pizzo 2009, Fröhlich and Pizzo 2010).
- Scattering amplitudes: (Bach, Fröhlich, and Pizzo 2007).
- Semirelativistic Hamiltonians: (Barbaroux, Dimassi, and Guillot 2004; Miyao, and Spohn 2008; Matte, and Stockmeyer 2010; Könenberg, Matte, and Stockmeyer 2011; Stockmeyer 2009; Hiroshima and Sasaki 2010).
- Other aspects of nonrelativistic QED:

    i Photo-electric effect: (Bach, Klopp, and Zenk 2001; Griesemer and Zenk 2010).
    ii Scattering theory: (Fröhlich, Griesemer and Schlein 2001; Fröhlich, Griesemer and Schlein 2002; Fröhlich, Griesemer and Schlein 2004).
    iii Stability of matter: (Bugliaro, Fröhlich and Graf 1996; Fefferman 1997; Lieb and Loss 1999).

There is an extensive literature on related models, which we do not mention here: Nelson models describing a particle linearly coupled to a free massless scalar field (phonons), semirelativistic models, based on the Dirac equation, and quantum statistics (open systems, positive temperature) models. (In the latter case, one deals with Liouvillians, rather than Hamiltonians, on positive temperature Hilbert spaces. The main results above were proved simultaneously for the QED and Nelson models and extended, at least partially, to positive temperatures.)

12.13 Appendix: Hamiltonian of the standard model

In this appendix we demonstrate the origin the quantum Hamiltonian H_g given in (12.1). To be specific consider an atom or molecule with n electrons interacting with radiation field. In this case the Hamiltonian of the system in our units is given by

$$H(\alpha) = \sum_{j=1}^{n} \frac{1}{2m}(i\nabla_{x_j} - \sqrt{\alpha}A_{\chi'}(x_j))^2 + \alpha V(x) + H_f, \qquad (12.30)$$

where $\alpha V(x)$ is the total Coulomb potential of the particle system, m is the electron bare mass, $\alpha = \frac{e^2}{4\pi\hbar c} \approx \frac{1}{137}$ (the fine-structure constant), and $A_{\chi'}(y)$ is the original vector potential with the ultraviolet cutoff χ'. Rescaling $x \to \alpha^{-1}x$ and $k \to \alpha^2 k$, we arrive at the Hamiltonian (12.1), where $g := \alpha^{3/2}$ and $A(y) = A_\chi(\alpha y)$, with $\chi(k) := \chi'(\alpha^2 k)$. After that we relax the restriction on $V(x)$ by allowing it to be a standard generalized n-body potential (see Subsection 12.2.1). Note that though this is not displayed, $A(x)$ does depend on g. This however does not effect the analysis of the Hamiltonian H. (If anything, this makes certain parts of it simpler, as derivatives of $A(x)$ bring down g.)

$$* * * *$$

In order not to deal with the problem of centre-of-mass motion which is not essential in the present context, we assume that either some of the particles (nuclei) are infinitely heavy (a molecule in the Born–Oppenheimer approximation), or the system is placed in a binding, external potential field. (In the case of the Born–Oppenheimer molecule, the resulting $V(x)$ also depends on the rescaled coordinates of the nuclei, but this does not effect our analysis except by making the complex deformation of the particle system more complicated (see (Hunziker and Sigal 2000)).) This means that the operator H_p has isolated eigenvalues below its essential spectrum. The general case is considered below.

In order to take into account the particle spin we change the state space of the particle system to $\mathcal{H}_p = \otimes_1^n L^2(\mathbb{R}^3, \mathbb{C}^2)$ (or the antisymmetric in identical particles subspace thereof), and the standard quantum Hamiltonian on $\mathcal{H} = \mathcal{H}_p \otimes \mathcal{H}_f$, is taken to be (see e.g. (Cohen-Tannoudji, Dupont-Roc, and Grynberg 1991; Cohen-Tannoudji, Dupont-Roc, and Grynberg 1992, Sakurai 1987)

$$H_{spin} = \sum_{j=1}^n \frac{1}{2m}[\sigma_j \cdot (i\nabla_{x_j} - gA_\chi(x_j))]^2 + V(x) + H_f, \tag{12.31}$$

where $\sigma_j := (\sigma_{j1}, \sigma_{j2}, \sigma_{j3})$, σ_{ji} are the Pauli matrices of the j–th particle and the identity operator on $\mathbb{C}^{2n}$ is omitted in the last two terms. It is easy to show that

$$[\sigma \cdot (i\nabla_x - gA_\chi(x))]^2 = (i\nabla_x - gA_\chi(x))^2 + g\sigma \cdot B(x). \tag{12.32}$$

where $B(x) := \mathrm{curl} A_\chi(x)$ is the magnetic field. As a result the operator (12.31) can be rewritten as

$$H_{spin} = H \otimes \mathbf{1} + \mathbf{1} \otimes g \sum_{j=1}^n \frac{1}{2m} \sigma_j \cdot B(x_j). \tag{12.33}$$

For the semi-relativistic Hamiltonian, the nonrelativistic kinetic energy $\frac{1}{2m}|p|^2$ is replaced by the relativistic one, $\sqrt{|p|^2 + m^2}$ or $\sqrt{(\sigma \cdot p)^2 + m^2}$.

12.14 Translationally invariant Hamiltonians

If we do not assume that the nuclei are infinitely and there are no external forces acting on the system, then the Hamiltonian (12.1) is translationally symmetric. This leads to conservation of the total momentum (a quantum version of the classical Noether theorem). Indeed, the system of particles interacting with the quantized electromagnetic fields is invariant under translations of the particle coordinates, $\underline{x} \to \underline{x} + \underline{y}$, where $\underline{y} = (y, \ldots, y)$ ($n-$ tuple) and the fields, $A(x) \to A(x - y)$, that is H commutes with the translations $T_y : \Psi(\underline{x}) \to e^{iy \cdot P_f} \Psi(\underline{x} + \underline{y})$, where P_f is the momentum operator associated to the quantized radiation field,

$$P_f = \sum_\lambda \int dk\, k\, a_\lambda^*(k) a_\lambda(k).$$

It is straightforward to show that T_y are unitary operators and that they satisfy the relations $T_{x+y} = T_x T_y$, and therefore $y \to T_y$ is a unitary Abelian representation of $\mathbb{R}^3$. Finally, we observe that the group T_y is generated by the total momentum operator of the electrons and the photon field: $T_y = e^{iy \cdot A_{tot}}$. Here

$$A_{tot} := \sum_i p_i \otimes 1_f + 1_{el} \otimes P_f \tag{12.34}$$

where, as above, $p_j := -i\nabla_{x_j}$, the momentum of the j-th electron and P_f is the field momentum given above. (A_{tot} is the self-adjoint operator on $\mathcal{H}$.) Hence $[H, A_{tot}] = 0$.

Let $\mathcal{H}$ be the direct integral $\mathcal{H} = \int_{\mathbb{R}^3}^{\oplus} \mathcal{H}_P dP$, with the fibres $\mathcal{H}_P := L^2(X) \otimes \mathcal{F}$, where $X := \{x \in \mathbb{R}^{3n} \mid \sum_i m_i x_i = 0\} \simeq \mathbb{R}^{3(n-1)}$, (this means that $\mathcal{H} = L^2(\mathbb{R}^3, dP; L^2(X) \otimes \mathcal{F}))$ and define $U : \mathcal{H}_{el} \otimes \mathcal{H}_f \to \mathcal{H}$ on smooth functions with compact domain by the formula

$$(U\Psi)(\underline{x}', P) = \int_{\mathbb{R}^3} e^{i(P - P_f) \cdot x_{cm}} \Psi(\underline{x}' + \underline{x}_{cm}) \, d\, y, \tag{12.35}$$

where $\underline{x}'$ are the coordinates of the N particles in the centre-of-mass frame and $\underline{x}_{cm} = (x_{cm}, \dots, x_{cm})$ ($n-$ tuple), with $x_{cm} = \frac{1}{\sum_i m_i} \sum_i m_i x_i$, the centre-of-mass coordinate, so that $\underline{x} = \underline{x}' + \underline{x}_{cm}$. Then U extends uniquely to a unitary operator (see below). Its converse is written, for $\Phi(\underline{x}', P) \in L^2(X) \otimes \mathcal{F}$, as

$$(U^{-1}\Phi)(\underline{x}) = \int_{\mathbb{R}^3} e^{-i x_{cm} \cdot (P - P_f)} \Phi(\underline{x}', P) \, d\, P. \tag{12.36}$$

The functions $\Phi(\underline{x}', P) = (U\Psi)(\underline{x}', P)$ are called fibres of Ψ. One can easily prove the following:

Lemma 12.9 The operations (12.35) and (12.36) define unitary maps $L^2(\mathbb{R}^{3n}) \otimes \mathcal{F} \to \mathcal{H}$ and $\mathcal{H} \to L^2(\mathbb{R}^{3n}) \otimes \mathcal{F}$, and are mutual inverses.

Since H commutes with A_{tot}, it follows that it admits the fibre decomposition

$$UHU^{-1} = \int_{\mathbb{R}^3}^{\oplus} H(P) \, d\, P, \tag{12.37}$$

where the fibre operators $H(P)$, $P \in \mathbb{R}^3$, are self-adjoint operators on $\mathcal{F}$. Using $a(k)e^{-iy \cdot P_f} = e^{-iy \cdot (P_f+k)} a(k)$ and $a^*(k)e^{-iy \cdot P_f} = e^{-iy \cdot (P_f-k)} a^*(k)$, we find $\nabla_y e^{iy \cdot (P-P_f)} A_\chi(x' + y) e^{iy \cdot (P-P_f)} = 0$ and therefore

$$A_\chi(x) e^{iy \cdot (P - P_f)} = e^{iy \cdot (P - P_f)} A_\chi(x - y). \tag{12.38}$$

Using this and (12.36), we compute $H(U^{-1}\Phi)(x) = \int_{\mathbb{R}^3} e^{ix \cdot (P - P_f)} H(P) \Phi(P) dP$, where $H(P)$ are Hamiltonians on the space fibres $\mathcal{H}_P := \mathcal{F}$ given explicitly by

$$H(P) = \sum_j \frac{1}{2m_i} \left(P - P_f - i\nabla_{x_j'} - e_i A_\chi(x_j')\right)^2 + V_{coul}(\underline{x}') + H_f \tag{12.39}$$

where $x_i' = x_i - x_{cm}$ is the coordinate of the i-th particle in the centre-of-mass frame. Now, this Hamiltonian can be investigated similarly to the one in Section (12.1).

12.15 Proof of Theorem 12.5

In this appendix we *omit the subindex* ρ at χ_ρ and $\overline{\chi}_\rho$, and *replace the subindex* ρ in other operators by the subindex χ. Moreover, we replace $H - \lambda$ by H. Though χ and $\overline{\chi}$ we deal with are projections, we often keep the powers χ^2 and $\overline{\chi}^2$, which occur often below, having in mind showing possible generalization to χ and $\overline{\chi}$ which are 'almost (or smooth) projections' satisfying $\chi^2 + \overline{\chi}^2 = 1$ (see Appendix 12.16).

First we note that the relation between ψ and φ in Theorem 12.5 (ii) is $\varphi = \chi\psi$, $\psi = Q_\chi(H)\varphi$, and between H^{-1} and $F_\chi(H)^{-1}$ in (iv) is

$$H^{-1} = Q_\chi(H)\, F_\chi H^{-1}\, Q_\chi(H)^\# + \overline{\chi}\, H_{\overline{\chi}}^{-1}\overline{\chi}, \tag{12.40}$$

where $H_{\overline{\chi}} := \overline{\chi}_\rho H \overline{\chi}_\chi$ and $Q_\chi(H)$ and $Q_\chi(H)^\#$ are the operators, given by

$$Q_\chi(H) := \chi - \overline{\chi}\, H_{\overline{\chi}}^{-1}\overline{\chi} H\chi,$$
$$Q_\chi^\#(H) := \chi - \chi H \overline{\chi}\, H_{\overline{\chi}}^{-1}\overline{\chi}.$$

Proof of Theorem 12.5

Throughout the proof we use the notation $F := F_{\chi(H)}$, $Q := Q_{\chi(H)}$, and $Q^\# := Q_\chi^\#(H)$. Note that (i) $(0 \in \rho(H) \Leftrightarrow 0 \in \rho(F_\chi(H)))$ follows from (iv) $(H^{-1}$ exists $\Leftrightarrow F_\chi(H)^{-1}$ exists) and (iv) follows from (12.40), so we start with the latter.

Proof of equation (12.40) The next two identities,

$$HQ = \chi F \quad \text{and} \quad Q^\# H = F\chi, \tag{12.41}$$

are of key importance in the proof. They both derive from a simple computation, which we give only for the first equality in (12.41). We observe the relations

$$H\chi = \chi H_\chi + \overline{\chi}^2 H\chi, \quad \text{and} \quad H\overline{\chi} = \overline{\chi} H_{\overline{\chi}} + \chi^2 H\overline{\chi}, \tag{12.42}$$

which follow from $\chi^2 + \overline{\chi}^2 = 1$. Now, using the definition of the operator Q and the relations (12.42), we obtain

$$HQ = \chi H_\chi + \overline{\chi}^2 H\chi - \left(\overline{\chi} H_{\overline{\chi}} + \chi^2 H\overline{\chi}\right) H_{\overline{\chi}}^{-1}\overline{\chi} H\chi. \tag{12.43}$$

Cancelling the second term on the r.h.s. with the first term in the parentheses in the third term, we see that the r.h.s is equal χF, which gives the first equality in (12.41).

Now, suppose first that the operator F has bounded invertible and define

$$R := Q F^{-1} Q^\# + \overline{\chi}\, H_{\overline{\chi}}^{-1}\overline{\chi}. \tag{12.44}$$

Using (12.41) and (12.42), we obtain

$$H R = H Q F^{-1} Q^{\#} + \left(\overline{\chi} H_{\overline{\chi}} + \chi^2 H \overline{\chi} \right) H_{\overline{\chi}}^{-1} \overline{\chi} \tag{12.45}$$

$$= \chi Q^{\#} + \overline{\chi}^2 + \chi^2 H \overline{\chi} H_{\overline{\chi}}^{-1} \overline{\chi}$$

$$= \chi^2 + \overline{\chi}^2 = \mathbf{1},$$

and, similarly, $RH = \mathbf{1}$. Thus $R = H^{-1}$, and (12.40) holds true.

Conversely, suppose that H is bounded invertible. Then, using the definition of F and the relation $\chi^2 + \overline{\chi}^2 = \mathbf{1}$, we obtain

$$F \chi H^{-1} \chi = \chi H \chi^2 H^{-1} \chi - \chi H \overline{\chi} H_{\overline{\chi}}^{-1} \overline{\chi} H \chi^2 H^{-1} \chi \tag{12.46}$$

$$= \chi H \chi^2 H^{-1} \chi - \chi H \overline{\chi} H_{\overline{\chi}}^{-1} \overline{\chi} H H^{-1} \chi + \chi H \overline{\chi} H_{\overline{\chi}}^{-1} \overline{\chi} H \overline{\chi}^2 H^{-1} \chi$$

$$= \chi H \chi^2 H^{-1} \chi + \chi H \overline{\chi}^2 H^{-1} \chi = \chi^2.$$

Similarly, one checks that $\chi H^{-1} \chi F = \mathbf{1}$. Thus F is invertible on $\mathrm{Ran}\,\chi$ with inverse $F^{-1} = \chi H^{-1} \chi$. □

Proof of (ii) $(H\psi = \lambda\psi \iff F_\rho(H - \lambda)\varphi = 0)$. If $\psi \in \mathcal{H} \setminus \{0\}$ solves $H\psi = 0$ then equation (12.41) implies that

$$F \chi \psi = Q^{\#} H \psi = 0. \tag{12.47}$$

Furthermore, by (12.42), $0 = \overline{\chi} H \psi = H_{\overline{\chi}} \overline{\chi}\psi + \overline{\chi} H \chi^2 \psi$, and hence

$$Q \chi \psi = \chi^2 \psi - \overline{\chi} H_{\overline{\chi}}^{-1} \overline{\chi} H \chi^2 \psi = \chi^2 \psi + \overline{\chi}^2 \psi = \psi. \tag{12.48}$$

Therefore, $\psi \neq 0$ implies $\chi\psi \neq 0$.

If $\varphi \in \mathrm{Ran}\,\chi \setminus \{0\}$ solves $F\varphi = 0$ then the definition of Q implies that

$$\chi Q \varphi = \chi \varphi = \varphi, \tag{12.49}$$

which implies that $Q\varphi \neq 0$ provided $\varphi \neq 0$. □

Proof of (iii) $(\dim \mathrm{Null}(H - \lambda) = \dim \mathrm{Null}\, F_\rho(H - \lambda))$ By (i), $\dim \mathrm{Null}\, H = 0$ is equivalent to $\dim \mathrm{Null}\, F = 0$, assuming that $H \in D(F)$. We may therefore assume that $\mathrm{Null}\, H \neq 0$ and $\mathrm{Null}\, F \neq 0$ are both nontrivial. Equation (12.48) shows that $\chi : \mathrm{Null}\, H \to \mathrm{Null}\, F$ is injective, hence $\dim \mathrm{Null}\, H \leq \dim \mathrm{Null}\, F$, and equation (12.49) shows that $Q : \mathrm{Null}\, F \to \mathrm{Null}\, H$ is injective, hence $\dim \mathrm{Null}\, H \geq \dim \mathrm{Null}\, F$. This establishes (iv) and moreover that $\chi : \mathrm{Null}\, H \to \mathrm{Null}\, F$ and $Q : \mathrm{Null}\, F \to \mathrm{Null}\, H$ are actually bijections. □

12.16   Smooth Feshbach–Schur Map

12.16.1   Definition and isospectrality

We define the smooth Feshbach–Schur map and formulate its important isospectral property. Let χ, $\overline{\chi}$ be a partition of unity on a separable Hilbert space $\mathcal{H}$, that is χ and $\overline{\chi}$ are positive operators on $\mathcal{H}$ whose norms are bounded by one, $0 \leq \chi, \overline{\chi} \leq \mathbf{1}$, and $\chi^2 + \overline{\chi}^2 = \mathbf{1}$. We assume that χ and $\overline{\chi}$ are nonzero. Let τ be a (linear) projection acting on closed operators on $\mathcal{H}$ with the property that operators in its image commute with χ and $\overline{\chi}$. We also assume that $\tau(\mathbf{1}) = \mathbf{1}$. Let $\overline{\tau} := \mathbf{1} - \tau$ and define

$$H_{\tau,\chi^\#} \; := \tau(H) \; + \; \chi^\# \overline{\tau}(H) \chi^\#, \tag{12.50}$$

where $\chi^\#$ stands for either χ or $\overline{\chi}$.

Given χ and τ as above, we denote by $D_{\tau,\chi}$ the space of closed operators, H, on $\mathcal{H}$ which belong to the domain of τ and satisfy the following three conditions:

(i) τ and χ (and therefore also $\overline{\tau}$ and $\overline{\chi}$) leave the domain $D(H)$ of H invariant:

$$D(\tau(H)) = D(H) \text{ and } \chi D(H) \subset D(H), \tag{12.51}$$

(ii)

$$H_{\tau,\overline{\chi}} \text{ is (bounded) invertible on } \operatorname{Ran} \overline{\chi}, \tag{12.52}$$

(iii)

$$\overline{\tau}(H)\chi \text{ and } \chi\overline{\tau}(H) \text{ extend to bounded operators on } \mathcal{H}. \tag{12.53}$$

(For more general conditions see (Bach, Chen, Fröhlich, and Sigal 2003; Griesemer and Hasler 2008.)

The *smooth Feshbach–Schur map (SFM)* maps operators from $D_{\tau,\chi}$ into operators on $\mathcal{H}$ by

$$F_{\tau,\chi}(H) \; := \; H_0 + \chi W \chi - \chi W \overline{\chi} H_{\tau,\overline{\chi}}^{-1} \overline{\chi} W \chi, \tag{12.54}$$

where $H_0 := \tau(H)$ and $W := \overline{\tau}(H)$. Note that H_0 and W are closed operators on $\mathcal{H}$ with coinciding domains, $D(H_0) = D(W) = D(H)$, and $H = H_0 + W$. We remark that the domains of $\chi W \chi$, $\overline{\chi} W \overline{\chi}$, $H_{\tau,\chi}$, and $H_{\tau,\overline{\chi}}$ all contain $D(H)$.

Define operators $Q_{\tau,\chi}(H) := \chi - \overline{\chi} H_{\tau,\overline{\chi}}^{-1} \overline{\chi} W \chi$ and $Q_{\tau,\chi}^\#(H) := \chi - \chi W \overline{\chi} H_{\tau,\overline{\chi}}^{-1} \overline{\chi}$. The following result (Bach, Chen, Fröhlich, and Sigal 2003) generalizes Theorem 12.5 above; its proof is similar to the one of that theorem:

Theorem 12.10. (Isospectrality of SFM) Let $0 \leq \chi \leq 1$ and $H \in D_{\tau,\chi}$ be an operator on a separable Hilbert space $\mathcal{H}$. Then we have the following results:

(i) H is bounded invertible on $\mathcal{H}$ if and only if $F_{\tau,\chi}(H)$ is bounded invertible on Ran χ. In this case

$$H^{-1} = Q_{\tau,\chi}(H)\, F_{\tau,\chi}(H)^{-1}\, Q_{\tau,\chi}(H)^{\#} + \overline{\chi}\, H_{\overline{\chi}}^{-1}\overline{\chi}, \qquad (12.55)$$

$$F_{\tau,\chi}(H)^{-1} = \chi\, H^{-1}\chi + \overline{\chi}\tau(H)^{-1}\overline{\chi}. \qquad (12.56)$$

(ii) If $\psi \in \mathcal{H} \setminus \{0\}$ solves $H\psi = 0$ then $\varphi := \chi\psi \in$ Ran $\chi \setminus \{0\}$ solves $F_{\tau,\chi}(H)\varphi = 0$.
(iii) If $\varphi \in$ Ran $\chi \setminus \{0\}$ solves $F_{\tau,\chi}(H)\varphi = 0$ then $\psi := Q_{\tau,\chi}(H)\varphi \in \mathcal{H} \setminus \{0\}$ solves $H\psi = 0$.
(iv) The multiplicity of the spectral value $\{0\}$ is conserved in the sense that dim Null $H =$ dim Null $F_{\tau,\chi}(H)$.

We also mention the following useful property of $F_{\tau,\chi}$:

$$H \text{ is self-adjoint} \quad \Rightarrow \quad F_{\tau,\chi}(H) \text{ is self-adjoint.} \qquad (12.57)$$

12.16.2   Transfer of local decay

We have shown above that the smooth Feshbach–Schur map is isospectral. In fact, under certain additional conditions it preserves (or transfers) much stronger spectral property—the limiting absorption principle (LAP) (Fröhlich, Griesemer and Sigal 2011), which is defined as follows. Let $\Delta \subset \mathbb{R}$ be an interval, $\nu > 0$ and B, a self-adjoint operator. We say that a C^1 family of self-adjoint operators, s.t. $H(\lambda) \in D_{\tau,\chi}$ has the (Δ, ν, B, θ) *limiting absorption principle* (LAP) property iff

$$\lim_{\varepsilon \to 0+} \langle B\rangle^{-\theta}\big(H(\lambda) - i\varepsilon\big)^{-1}\langle B\rangle^{-\theta} \text{ exists and } \in C^{\nu}(\Delta). \qquad (12.58)$$

Usually LAP holds for $\nu < \theta - \frac{1}{2}$. One can show that the LAP implies the local decay property (see, e.g. (Reed and Simon 1978), vol III; recall that the definition of the local decay property is given in Section 12.10).

Theorem 12.11 Let $\Delta \subset \mathbb{R}$ and , $\forall \lambda \in \Delta$, $H(\lambda)$ be a C^1 family of self-adjoint operators, s.t. $H(\lambda) \in D_{\tau,\chi}$. Assume that there is a self-adjoint operator B s.t.

$$\mathrm{ad}_B^j(A) \text{ is bounded and differentiable in } \lambda, \ \forall j \leq 2, \qquad (12.59)$$

where A is one of the operators χ, $\overline{\chi}$, $\chi\overline{\tau}(H(\lambda))$, $\overline{\tau}(H(\lambda))\chi$, $\partial_\lambda^k(\overline{\chi}H_{\tau,\overline{\chi}}(\lambda)^{-1}\overline{\chi})$, $k = 0, 1$. Then, for any $0 \leq \nu \leq 1$ and $0 < \theta \leq 1$ and in the operator norm, we have

$$\lim_{\varepsilon \to 0+} \langle B\rangle^{-\theta}\, (F_{\tau,\chi}(H(\lambda)) - i\varepsilon)^{-1}\langle B\rangle^{-\theta} \text{ exists and } \in C^{\nu}(\Delta) \qquad (12.60)$$

$$\Rightarrow \lim_{\varepsilon \to 0+} \langle B\rangle^{-\theta}\big(H(\lambda) - i\varepsilon\big)^{-1}\langle B\rangle^{-\theta} \text{ exists and } \in C^{\nu}(\Delta). \qquad (12.61)$$

This allows one to reduce the proof of the LAP for the original operator, $H - \lambda$, to the proof of this property for a much simpler one, $\mathcal{R}_\rho^n(H - \lambda)$.

12.16.3   Transfer of analyticity

Theorem 12.12 Let Λ be an open set in $\mathbb{C}$ and $H(\lambda)$, $\lambda \in \Lambda$, a family of operators with a fixed domain, which belong to the domain of $F_{\tau\chi}$. Assume $H(\lambda)$ and $\tau(H(\lambda))$, with the same domain, are analytic in the sense of Kato (see e.g. (Reed and Simon 1978), vol IV). Then we have that

- $F_{\tau,\chi}(H(\lambda))$ is an analytic in $\lambda \in \Lambda$ family of operators.

Proof: Since $H(\lambda)$, $H_0(\lambda) := \tau(H(\lambda))$ and $W(\lambda)\chi$ and $\chi W(\lambda)$, where $W(\lambda) := \overline{\tau}(H(\lambda))$, are analytic in $\lambda \in \Lambda$, we see from the definition of the smooth Feshbach–Schur map $F_{\tau\chi}$ in (12.54) that $F_{\tau\chi}(H(\lambda))$ is analytic in $\lambda \in \Lambda$, provided $\overline{\chi}_\rho H(\lambda)^{-1}_{\tau,\overline{\chi}_\rho} \overline{\chi}_\rho$ is analytic in Λ. The analyticity of the latter family follows by the Neumann series argument. $\qquad\square$

One can generalize the above result to Λ's which are open sets in a complex Banach space. Recall that a complex vector-function f in an open set Λ in a complex Banach space $\mathcal{W}$ is said to be *analytic* iff it is locally bounded and Gâteaux-differentiable. One can show that f is analytic iff $\forall \xi \in \mathcal{W}$, $f(H + \tau\xi)$ is analytic in the complex variable τ for $|\tau|$ sufficiently small (see (Berger 1977; Hille and Phillips 1957)). Furthermore if f is analytic in Λ and g is an analytic vector-function from an open set Ω in $\mathbb{C}$ into Λ, then the composite function $f \circ g$ is analytic on Ω.

12.17   Pull-through formulae

In this appendix we prove the very useful 'pull-through' formulae (see (Bach, Fröhlich, and Sigal 1998))

$$a(k)f(H_f) = f(H_f + \omega(k))a(k) \qquad (12.62)$$

and

$$f(H_f)a^*(k) = a^*(k)f(H_f + \omega(k)), \qquad (12.63)$$

valid for any piecewise continuous, bounded function, f, on $\mathbb{R}$. First, using the commutation relations for $a(k)$, $a^*(k)$, one proves relations (12.62)–(12.63) for $f(H) = (H_f - z)^{-1}$, $z \in \mathbb{C}/\mathbb{R}^+$. Then using the Stone–Weierstrass theorem, one can extend (12.62)–(12.63) from functions of the form $f(\lambda) = (\lambda - z)^{-1}$, $z \in \mathbb{C}\backslash\overline{\mathbb{R}}^+$, to the class of functions mentioned above.

12.18   Supplement: creation and annihilation operators

Let $\mathfrak{h}$ be either $L^2(\mathbb{R}^3, \mathbb{C}, d^3k)$ or $L^2(\mathbb{R}^3, \mathbb{C}^2, d^3k)$. In the first case we consider $\mathfrak{h}$ to be the Hilbert space of one-particle states of a scalar boson or phonon, and in the second case, of a photon. The variable $k \in \mathbb{R}^3$ is the wave vector or momentum of the particle. (Recall that throughout these lectures, the propagation speed c, of photon or photons and Planck's constant, $\hbar$, are set equal to 1.) The Bosonic Fock space, $\mathcal{F}$, over $\mathfrak{h}$ is defined by

$$\mathcal{F} := \bigoplus_{n=0}^{\infty} \mathcal{S}_n \, \mathfrak{h}^{\otimes n}, \tag{12.64}$$

where $\mathcal{S}_n$ is the orthogonal projection onto the subspace of totally symmetric n-particle wave functions contained in the n-fold tensor product $\mathfrak{h}^{\otimes n}$ of $\mathfrak{h}$; and $\mathcal{S}_0 \mathfrak{h}^{\otimes 0} := \mathbb{C}$. The vector $\Omega := (1, 0, \dots)$ is called the *vacuum vector* in $\mathcal{F}$. Vectors $\Psi \in \mathcal{F}$ can be identified with sequences $(\psi_n)_{n=0}^{\infty}$ of n-particle wave functions, which are totally symmetric in their n arguments, and $\psi_0 \in \mathbb{C}$. In the first case these functions are of the form, $\psi_n(k_1, \dots, k_n)$, while in the second case, of the form $\psi_n(k_1, \lambda_1, \dots, k_n, \lambda_n)$, where $\lambda_j \in \{-1, 1\}$ are the polarization variables.

In what follows we present some key definitions in the first case, limiting ourselves to remarks at the end of this appendix on how these definitions have to be modified for the second case. The scalar product of two vectors Ψ and Φ is given by

$$\langle \Psi, \Phi \rangle := \sum_{n=0}^{\infty} \int \prod_{j=1}^{n} d^3 k_j \, \overline{\psi_n(k_1, \dots, k_n)} \, \varphi_n(k_1, \dots, k_n). \tag{12.65}$$

Given a one particle dispersion relation $\omega(k)$, the energy of a configuration of n *non-interacting* field particles with wave vectors $k_1, \dots, k_n$ is given by $\sum_{j=1}^{n} \omega(k_j)$. We define the *free-field Hamiltonian*, H_f, giving the field dynamics, by

$$(H_f \Psi)_n(k_1, \dots, k_n) = \left(\sum_{j=1}^{n} \omega(k_j) \right) \psi_n(k_1, \dots, k_n), \tag{12.66}$$

for $n \geq 1$ and $(H_f \Psi)_n = 0$ for $n = 0$. Here $\Psi = (\psi_n)_{n=0}^{\infty}$ (to be sure that the r.h.s. makes sense we can assume that $\psi_n = 0$, except for finitely many n, for which $\psi_n(k_1, \dots, k_n)$ decrease rapidly at infinity). Clearly that the operator H_f has the single eigenvalue 0 with the eigenvector Ω and the rest of the spectrum absolutely continuous.

With each function $\varphi \in L^2(\mathbb{R}^3, \mathbb{C}, d^3 k)$ one associates an *annihilation operator* $a(\varphi)$ defined as follows. For $\Psi = (\psi_n)_{n=0}^{\infty} \in \mathcal{F}$ with the property that $\psi_n = 0$, for all but finitely many n, the vector $a(\varphi)\Psi$ is defined by

$$(a(\varphi)\Psi)_n(k_1, \dots, k_n) := \sqrt{n+1} \int d^3 k \, \overline{\varphi(k)} \, \psi_{n+1}(k, k_1, \dots, k_n). \tag{12.67}$$

This equation defines a closable operator $a(\varphi)$ whose closure is also denoted by $a(\varphi)$ and which satisfies the relation

$$a(\varphi)\Omega = 0, \tag{12.68}$$

The creation operator $a^*(\varphi)$ is defined to be the adjoint of $a(\varphi)$ with respect to the scalar product defined in equation (12.65). Since $a(\varphi)$ is antilinear, and $a^*(\varphi)$ is linear in φ, we write formally

$$a(\varphi) = \int d^3k \, \overline{\varphi(k)} \, a(k), \qquad a^*(\varphi) = \int d^3k \, \varphi(k) \, a^*(k), \qquad (12.69)$$

where $a(k)$ and $a^*(k)$ are unbounded, operator-valued distributions. The latter are well-known to obey the *canonical commutation relations* (CCR):

$$\left[a^\#(k), \, a^\#(k') \right] = 0, \qquad \left[a(k), \, a^*(k') \right] = \delta^3(k - k'), \qquad (12.70)$$

where $a^\# = a$ or a^*.

Now, using this one can rewrite the quantum Hamiltonian H_f in terms of the creation and annihilation operators, a and a^*, as

$$H_f = \int d^3k \, a^*(k) \, \omega(k) \, a(k), \qquad (12.71)$$

acting on the Fock space $\mathcal{F}$.

More generally, for any operator, t, on the one-particle space $L^2(\mathbb{R}^3, \mathbb{C}, d^3k)$ we define the operator T on the Fock space $\mathcal{F}$ by the following formal expression $T := \int a^*(k) t a(k) dk$, where the operator t acts on the $k-$variable (T is the second quantization of t). The precise meaning of the latter expression can obtained by using a basis $\{\phi_j\}$ in the space $L^2(\mathbb{R}^3, \mathbb{C}, d^3k)$ to rewrite it as $T := \sum_j \int a^*(\phi_j) a(t^* \phi_j) dk$.

To modify the above definitions to the case of photons, one replaces the variable k by the pair (k, λ) and adds to the integrals in k sums over λ. In particular, the creation and annihilation operators have now two variables: $a_\lambda^\#(k) \equiv a^\#(k, \lambda)$; they satisfy the commutation relations

$$\left[a_\lambda^\#(k), \, a_{\lambda'}^\#(k') \right] = 0, \qquad \left[a_\lambda(k), \, a_{\lambda'}^*(k') \right] = \delta_{\lambda,\lambda'} \delta^3(k - k'). \qquad (12.72)$$

One can introduce the operator-valued transverse vector fields by

$$a^\#(k) := \sum_{\lambda \in \{-1,1\}} e_\lambda(k) a_\lambda^\#(k),$$

where $e_\lambda(k) \equiv e(k, \lambda)$ are polarization vectors, that is, orthonormal vectors in $\mathbb{R}^3$ satisfying $k \cdot e_\lambda(k) = 0$. Then, in order to reinterpret the expressions in this paper for photons, one either adds the variable λ, as was mentioned above, or replaces, in appropriate places, the usual product of scalar functions or scalar functions and scalar operators by the dot product of vector-functions or vector-functions and operator-valued vector-functions.

Acknowledgements

The author is grateful to Walid Abou Salem, Thomas Chen, Jérémy Faupin, Marcel Griesemer, and especially Volker Bach and Jürg Fröhlich for fruitful collaboration and for all they have taught him in the course of joint work.

References

Abou Salem, W. Faupin, J. Fröhlich, J. Sigal, I. M. (2007). On theory of resonances in non-relativisitc QED, *Adv. Appl. Math.*, 43, no. 3, 201–230.

Amour, L. Grébert, B. Guillot, J.-C. (2006). The dressed mobile atoms and ions, *J. Math. Pures Appl.* (9) **86**(3): 177–200.

Arai, A. (1999). *Mathematical Analysis of a Model in Relativistic Quantum Electrodynamics. Applications of Renormalization Group Methods in Mathematical Sciences* (in Japanese) Kyoto.

Arai, A. Hirokawa, M. (2000). Ground states of a general class of quantum field Hamiltonians, *Rev. Math. Phys.* **12**, (8): 1085–1135.

Arai, A. Hirokawa, M. and Hiroshima, F. (2007). Regularities of ground states of quantum field models, Kyushu *J. Math.* 61, no. 2, 321–372.

Bach, V. (2012). Mass renormalization in nonrelativisitic quantum electrodynamics. This volume.

Bach, V. Chen, T. Fröhlich, J. and Sigal, I. M. (2003). Smooth Feshbach map and operator-theoretic renormalization group methods, *Journal of Functional Analysis*, **203**: 44–92.

Bach, V. Chen, T. Fröhlich, J. and Sigal, I. M. (2007). The renormalized electron mass in non-relativistic quantum electrodynamics, *Journal of Functional Analysis*, 243, 426–535.

Bach, V. Fröhlich, J. and Pizzo, A. (2006). Infrared-Finite Algorithms in QED: The Groundstate of an atom interacting with the quantized radiation field, *Communications in Mathematical Physics* 264, Issue: 1, 145–165.

Bach, V. Fröhlich, J. and Pizzo, A. (2007). An infrared-finite algorithm for Rayleigh scattering amplitudes, and Bohr's frequency condition, *Comm. Math. Phys.* 274, no. 2, 457–486.

Bach, V. Fröhlich, J. and Pizzo, A. (2009). Infrared-finite algorithms in QED II. The expansion of the groundstate of an atom interacting with the quantized radiation field, *Adv. in Math.*, 220(4), 1023–1074.

Bach, V. Fröhlich, J. and Sigal, I. M. (1995). Mathematical theory of non-relativistic matter and radiation, *Letters in Math. Physics*, 34: 183–201.

Bach, V. Fröhlich, J. and Sigal, I. M. (1998). Quantum electrodynamics of confined non-relativistic particles, *Adv. in Math.*, 137: 299–395.

Bach, V. Fröhlich, J. and Sigal, I. M. (1998). Renormalization group analysis of spectral problems in quantum field theory, *Adv. in Math.*, 137: 205–298.

Bach, V. Fröhlich, J. and Sigal, I. M. (1999). Spectral analysis for systems of atoms and molecules coupled to the quantized radiation field, *Commun. Math. Phys.*, 207(2): 249–290.

Bach, V. Fröhlich, J. and Sigal, I. M. (2000). Return to equilibrium, *J. Math. Phys.*, 41(6): 3985–4060.

Bach, V. Fröhlich, J. Sigal, I. M. and Soffer, A. (1999). Positive commutators and spectrum of Pauli-Fierz Hamiltonian of atoms and molecules, *Commun. Math. Phys.*, 207(3): 557–587.

Bach, V. Klopp, F. and Zenk, H. (2001). Mathematical analysis of the photoelectric effect. *Adv. Theor. Math. Phys.*, 5(6): 969–999.

Barbaroux, J.-M. Chen, T. and Vugalter, S. A. (2003). Binding conditions for atomic N-electron systems in non-relativistic QED, *Ann. H. Poinc.*, 4 (6), 1101–1136.

Barbaroux, J.-M. Chen, T. Vougalter, V. and Vugalter, S. A. (2010). Quantitative estimates on the hydrogen ground state energy in non-relativistic QED, *Ann. H. Poincare*, 11(8), 1487–1544.

Barbaroux, J.-M. Dimassi, M. and Guillot, J. C. (2004). Quantum electrodynamics of relativistic bound states with cutoffs, *J. Hyperbolic Differ. Equ.* **1**, 271–314.

Barbaroux, J.-M. and Vugalter, S. A. (2010). Non analyticity of the ground state energy of the Hamiltonian for Hydrogen atom in nonrelativistic QED (2010), *J. Phys. A* 43no. 47, 474004.

Berger, M. (1977). *Nonlinearity and Functional Analysis. Lectures on Nonlinear Problems in Mathematical Analysis.* Pure and Applied Mathematics. Academic Press, New York-London.

Bugliaro, L. Fröhlich, J. and Graf, G. M. (1996). Stability of quantum electrodynamics with nonrelativistic matter, *Phys.Rev. Lett.* 77, 3494–3497.

Catto, I. Exner, P. Hainzl, Ch. (2004). Enhanced binding revisited for a spinless particle in nonrelativistic QED, *J. Math. Phys.* 45, no. 11, 4174–4185.

Catto, I. Hainzl, Ch. (2004). Self-energy of one electron in non-relativistic QED, *J. Funct. Anal.* 207, no. 1, 68–110.

Chen, T. (2008). Infrared renormalization in non-relativistic QED and scaling criticality, *J. Funct. Anal.* 254, no. 10, 2555–2647.

Chen, T. Faupin, J. Frohlich, J. and Sigal I.M. (2012). Local decay in non-relativistic QED, *Commun. Math. Phys.*, 309, no. 2, 543–582.

Chen, T. Fröhlich, J. (2007). Coherent infrared representations in non-relativistic QED, *Spectral Theory and Mathematical Physics: A Festschrift in Honor of Barry Simon's 60th Birthday, Proc. Symp. Pure Math.*, AMS.

Chen, T. Fröhlich, J. and Pizzo, A. (2010). Infraparticle scattering states in non-relativistic QED - I. The Bloch-Nordsieck paradigm, *Commun. Math. Phys.*, **294** (3), 761–825.

Chen, T. Fröhlich, J. and Pizzo, A. (2009). Infraparticle scattering states in non-relativistic QED - II. Mass shell properties, *J. Math. Phys.*, **50**(1), 012103. arXiv: 0709.2812

Chen, T. Vougalter, V. and Vugalter, S. A. (2003). The increase of binding energy and enhanced binding in non-relativistic QED, *J. Math. Phys.*, 44(5), 1961–1970.

Cohen-Tannoudji, C. Dupont-Roc, J. and Grynberg, G. (1991). *Photons and Atoms – Introduction to Quantum Electrodynamics.* John Wiley, New York.

Cohen-Tannoudji, C. Dupont-Roc, J. and Grynberg, G. (1992). *Atom-Photon Interactions – Basic Processes and Applications.* John Wiley, New York.

Faupin, J. (2008). Resonances of the confined hydrogen atom and the Lamb-Dicke effect in non-relativisitc quantum electrodynamics, *Ann. Henri Poincaré* 9, no. 4, 743–773.

Fefferman, C. Fröhlich, J. and Graf, G. M. (1997). Stability of ultraviolet cutoff quantum electrodynamics with non-relativistic matter,*Commun. Math. Phys.* 190, 309–330.

Fermi, E. (1932). Quantum theory of radiation, *Rev. Mod. Phys.*, 4: 87–132.

Feshbach, H. (1958). Unified theory of nuclear reactions, *Ann. Phys.*, 5: 357–390.

Fröhlich, J. (1973). On the infrared problem in a model of scalar electrons and massless, scalar bosons, *Ann. Inst. Henri Poincaré*, Section Physique Théorique, **19** (1), 1–103.

Fröhlich, J. (1974). Existence of dressed one electron states in a class of persistent models, *Fortschritte der Physik* **22**, 159–198.

Fröhlich, J. Griesemer, M. and Schlein, B. (2001). Asymptotic electromagnetic fields in models of quantum-mechanical matter interacting with the quantized radiation field, *Advances in Mathematics* 164, Issue: 2, 349–398.

Fröhlich, J. Griesemer, M. and Schlein, B. (2002). Asymptotic completeness for Rayleigh scattering, *Ann. Henri Poincar* 3, no. 1, 107–170.

Fröhlich, J. Griesemer, M. and Schlein, B. (2004). Asymptotic completeness for Compton scattering, *Comm. Math. Phys.* 252, no. 1–3, 415–476.

Fröhlich, J. Griesemer, M. and Sigal, I. M. (2008). Spectral theory for the standard model of non-relativisitc QED, *Comm. Math. Phys.*, 283, no 3, 613–646.

Fröhlich, J. Griesemer, M. and Sigal, I. M. (2009). Spectral renormalization group analysis, *Rev. Math. Phys.* 21(4), 511–548, e-print, ArXiv, 2008.

Fröhlich, J., Griesemer, M. and Sigal, I. M. (2011). Spectral renormalization group and limiting asorption principle for the standard model of non-relativisitc QED, *Rev. Math. Phys.* 23(2), e-print, ArXiv, 2010.

Fröhlich, J. and Pizzo, A. The renormalized electron mass in nonrelativistic QED, *Commun. Math. Phys.*, 294, no.2, 439–470

Gérard, C. Hiroshima, F. Panati, A. and Suzuki, A. (2011). Infrared problem for the Nelson model on static space-times, *Commun. Math. Phys.*, 308, no. 2, Pages 543–566.

Gergescu, V. Gérard, C. and Møller, J. S. (2004). Spectral Theory of massless Pauli-Fierz models, *Commun. Math. Phys.*, 249: 29–78.

Griesemer, M. (2004). Exponential decay and ionization thresholds in non-relativistic quantum electrodynamics, *J. Funct. Anal.*, 210(2): 321–340.

Griesemer, M. and Hasler, D. (2008). On the smooth Feshbach-Schur map, *J. Funct. Anal.*, 254(9): 2329–2335.

Griesemer, M. and Hasler, D.G. (2009). Analytic perturbation theory and renormalization analysis of matter coupled to quantized radiation, Ann. Henri Poincaré 10, no. 3, 577–621.

Griesemer, M. Lieb, E. H. and Loss, M. (2001). Ground states in non-relativistic quantum electrodynamics, *Invent. Math.* 145, no. 3, 557–595.

Griesemer, M. and Zenk, H. (2010). On the atomic photoeffect in non-relativistic QED, *Comm. Math. Phys.* 300, no. 3, 615–639.

Gustafson, S. and Sigal, I. M. (2006). *Mathematical Concepts of Quantum Mechanics.* 2nd edition. Springer.

Hainzl, Ch. (2002). Enhanced binding through coupling to a photon field. Mathematical results in quantum mechanics, Contemp, Math., 307, Amer. Math. Soc., Providence, RI, 149-154.

Hainzl, Ch. (2003). One non-relativistic particle coupled to a photon field, *Ann. Henri Poincaré* 4, no. 2, 217–237.

Hainzl, Ch. Hirokawa, M. and Spohn, H. (2005). Binding energy for hydrogen-like atoms in the Nelson model without cutoffs, *J. Funct. Anal.* 220, no. 2, 424459.

Hainzl, Ch. and Seiringer, R. (2003). Mass renormalization and energy level shift in non-relativistic QED, *Adv. Theor. Math. Phys.* 6 (5) 847–871.

Hainzl, Ch. Vougalter, V. and Vugalter, S. (2003). Enhanced binding in non-relativistic QED, *Commun. Math. Phys.* 233, no. 1, 13–26.

Hasler, D. and Herbst, I. (2008). Absence of ground states for a class of translation invariant models of non-relativistic QED, *Commun. Math. Phys.* 279, no. 3, 769–787.

Hasler, D. Herbst, I. and Huber, M. (2008). On the lifetime of quasi-stationary states in non-relativistic QED, *Ann. Henri Poincaré* 9, no. 5, 1005–1028, ArXiv:0709.3856.

Hasler, D.; Herbst, I. (2011). Convergent expansions in non-relativistic QED: analyticity of the ground state. *J. Funct. Anal.* 261, no. 11, 3119–3154

Hasler, D. and Herbst, I. (2011). Smoothness and analyticity of perturbation expansions in qed, *Adv. Math.* 228, no. 6, 3249–3299.

Hasler, D. and Herbst, I. Convergent expansions in non-relativistic QED, 2010, ArXiv:1005.3522v1.

Hille, E. and Phillips, R. S. *Functional Analalysis and Semi-groups*. AMS 1957.

Hirokawa, M. (2003). Recent developments in mathematical methods for models in non-relativistic quantum electrodynamics, *A Garden of Quanta*, 209–242, World Sci. Publishing, River Edge, NJ.

Hiroshima, F. (1999). Ground states of a model in nonrelativistic quantum electrodynamics. I, *J. Math. Phys.* 40, no. 12, 6209–6222.

Hiroshima, F. (2000). Ground states of a model in nonrelativistic quantum electrodynamics. II, J. Math. Phys. 41, no. 2, 661–674.

Hiroshima, F. (2000). Essential self-adjointness of translation-invariant quantum field models for arbitrary coupling constants, Commun. Math. Phys. 211, 585–613.

Hiroshima, F. (2001). Ground states and spectrum of quantum electrodynamics of non-relativistic particles, *Trans. Amer. Math.* Soc. 353, no. 11, 4497–4528 (electronic).

Hiroshima, F. (2002). Self-adjointness of the Pauli-Fierz Hamiltonian for arbitrary values of coupling constants, *Ann. Henri Poincaré* 3, no. 1, 171–201.

Hiroshima, F. (2003). Nonrelativistic QED at large momentum of photons, *A Garden of Quanta*, 167–196, World Sci. Publishing, River Edge, NJ.

Hiroshima, F. (2003). Localization of the number of photons of ground states in nonrelativistic QED, *Rev. Math. Phys.* 15, no. 3, 271–312.

Hiroshima, F. (2004). Analysis of ground states of atoms interacting with a quantized radiation field, *Topics in the Theory of Schrödinger Operators*, World Sci. Publishing, River Edge, NJ, 145–272.

Hiroshima, F. and Sasaki, I. (2010). On the ionization energy of the semi-relativistic Pauli-Fierz model for a single particle, Preprint, arXiv:1003.1661v4.

Hiroshima, F. and Spohn, H. (2001). Ground state degeneracy of the Pauli-Fierz Hamiltonian with spin. *Adv. Theor. Math. Phys.* 5, no. 6, 1091–1104.

Hiroshima, F. and Spohn, H. (2001). Enhanced binding through coupling to a quantum field, *Ann. Inst. Henri Poincaré* 2(6), no. 6, 1159–1187.

Hiroshima, F. and Spohn, H. (2005). Mass renormalization in nonrelativistic quantum electrodynamics, *J. Math. Phys.* 46 (4).

Hübner, M. and Spohn, H. (1995). Radiative decay: nonperturbative approaches, *Rev. Math. Phys.* 7, no. 3, 363–387.

Hunziker, W. (1990). Resonances, metastable states and exponential decay laws in perturbation theory, *Comm. Math. Phys.* 132, no. 1, 177–188.

Hunziker, W. and Sigal, I.M. (2000). The quantum N-body problem, *J. Math. Phys.* 41, no. 6, 3448–3510.

King, Ch. (1994). Resonant decay of a two state atom interacting with a massless non-relativistic quantized scalar field, Comm. Math. Phys., 165(3), 569–594.

Könenberg, M. Matte, O. and Stockmeyer, E. (2011). Existence of ground states of hydrogen-like atoms in relativistic QED I: The semi-relativistic Pauli-Fierz operator, *Rev. Math. Phys.* 23, no. 4, 375–407.

Lieb, E. H. and Loss, M. (2000). Self-energy of electrons in non-perturbative QED, in: Conference Moshe Flato 1999, vol. I, Dijon, in: *Math. Phys. Stud.* vol. 21, Kluwer Acad. Publ., Dordrecht, pp. 327–344.

Lieb, E. H. and Loss, M. (2002). A bound on binding energies and mass renormalization in models of quantum electrodynamics, *J. Statist. Phys.* 108, no. 5–6, 1057–1069.

Lieb, E. H. and Loss, M. (2003). Existence of atoms and molecules in non-relativistic quantum electrodynamics, *Adv. Theor. Math. Phys.* **7**, 667–710. arXiv math-ph/0307046.

Loss, M. Miyao, T. and Spohn, H. (2007). Lowest energy states in nonrelativistic QED: Atoms and ions in motion, *J. Funct. Anal.* 243, Issue 2, 353–393.

Loss, M. Miyao, T. and Spohn, H. (2008). Kramers degeneracy theorem in nonrelativistic QED, arXiv:0809.4471.

Matte, O. and Stockmeyer, E. (2010). Exponential localization for a hydrogen-like atom in relativistic quantum electrodynamics, Comm. Math. Phys. **295**, 551–583.

Merkli, M. Mück, M. and Sigal, I. M. (2007). Instability of Equilibrium States for Coupled Heat Reservoirs at Different Temperatures, *J. Funct. Anal.*, 243, no. 1, 87–120.

Merkli, M. Mück, M. and Sigal, I. M. (2008). Nonequilibrium stationary states and their stability, *Ann. Inst. H. Poincaré*, 243, no. 1, 87–120.

Miyao, T. and Spohn, H. (2008). Spectral analysis of the semi-relativistic Pauli-Fierz Hamiltonian, *J. Funct. Anal.*, 256, 2123–2156 (2009), arxiv.

Mück, M. *Thermal Relaxation for Particle Systems in Interaction with Several Bosonic Heat Reservoirs.* Ph.D. Dissertation, Department of Mathematics, Johannes-Gutenberg University, Mainz, July 2004, ISBN 3-8334-1866-4.

Mück, M. (2004). Construction of metastable states in quantum electrodynamics, *Rev. Math. Phys.* 16, no. 1, 1–28.

Pauli, W. and Fierz, M. (1938). Zur Theorie der Emission langwelliger Lichtquanten, *Il Nuovo Cimento* 15, 167–188.

Pizzo, A. (2003). One-particle (improper) States in Nelson's massless model, *Annales Henri Poincaré* 4, Issue: 3, 439–486, June.

Pizzo, A. (2005). Scattering of an infraparticle: The one particle sector in Nelson's massless model, *Annales Henri Poincaré* 6, Issue: 3, 553–606.

Reed, M. and Simon, B. (1978). *Methods of Modern Mathematical Physics, volumes III, IV*. Academic Press.

Sakurai, J. J. (1987). *Advanced Quantum Mechanics*, Addison-Wesley.

Sasaki, I. (2006). Ground state of a model in relativistic quantum electrodynamics with a fixed total momentum, Preprint, arXiv:math-ph/0606029v4.

Schur, J. (1917). Über Potenzreihen die im Inneren des Einheitskreises beschränkt sind, *J. reine u. angewandte Mathematik*, 205–232.

Sigal, I. M. (2009). Ground state and resonances in the standard model of the non-relativistic quantum electrodynamics, *J. Stat. Physics*, 134, no. 5–6, 899–939. arXiv 0806.3297.

Spohn, H. (2004). *Dynamics of Charged Particles and their Radiation Field*, Cambridge University Press, Cambridge.

Stockmeyer, E. (2009). On the non-relativistic limit of a model in quantum electrodynamics, Preprint, arXiv:0905.1006v1.

13
Bell inequalities

Michael Marc WOLF

Zentrum Mathematik, M5, Technische Universität München,
Boltzmannstrasse 3, 85748 Garching, Germany

14

Universality of generalized Wigner matrices

Horng-Tzer YAU[1]

Department of Mathematics, Harvard University,
Cambridge, MA 02138, USA
htyau@math.harvard.edu

[1] Partially supported by NSF grants DMS-0757425, 0804279.

Chapter Contents

14.1 Introduction

Random matrices were introduced in mathematical statistics by J. Wishart in the early twentieth century. The subject was reinvented by E. Wigner (Wigner 1955) to model the excitation spectrum of large nuclei. Wigner's central thesis is that the eigenvalue gap distributions for large complicated systems are universal in the sense that they depend only on the symmetry classes of the physical systems but not other detailed structures. While this is an extremely bold hypothesis, later developments, mostly via numerical simulations, have strongly vindicated it. As a special case of this general belief, the eigenvalue gap distributions of random matrices should be independent of the probability distributions of the ensembles and thus are given by the classical Gaussian ensembles. Besides the eigenvalue gap distributions, similar predictions hold for short distance correlation functions of the eigenvalues. Since the gap distributions can be expressed in terms of correlation functions, mathematical analysis is usually performed on correlation functions.

The universality question can be roughly divided into the bulk universality in the interior of the spectrum and the edge universality near the spectral edges. Over the past two decades, spectacular progress on bulk and edge universality was made for invariant ensembles, see, (Bleher 1999; Deift *et al.* 1999; Deift Pastur and Shcherbina 2008) and (Anderson, Guionnet, and Zeitouni, 2009; Deift, 1999; Deift, and Gioev, 2009) for a review. For noninvariant ensembles with i.i.d. matrix elements (*Standard Wigner ensembles*) edge universality can be proved via the moment method and its various generalizations, see, for example, (Sinai and Soshnikov 1998; Soshnikov 1999; Sodin 2010). In a striking contrast, the only rigorous results for the bulk universality of non-invariant Wigner ensembles are the work by Johansson (Johansson 2001) and subsequent improvements (Ben Arous and Péché 2005; Johansson 2011) on *Gaussian divisible Hermitian ensembles*, that is, Hermitian ensembles of the form

$$H_s = H_0 + sV, \tag{14.1}$$

where H_0 is a Wigner matrix, V is an independent standard GUE matrix, and s is a fixed positive constant independent of N. The Hermitian assumption is essential since the key formula used in (Johansson 2001) and the earlier work (Brézin and Hikami 1996/97) is valid only for *Hermitian ensembles*.

In this chapter, we will report the recent progress on the universality developed in (Erdős, Schlein, and Yau, 2009; Erdős, Schlein, and Yau, 2009; Erdős, Schlein, and Yau, 2010; Erdős, Schlein, and Yau, 2010; Erdős, Schlein, Yau, and Yin, 2010; Erdős, Yau, and Yin, 2011a; Erdős, Yau, and Yin, 2011b; Erdős, Yau, and Yin, 2011c). This approach eliminates explicit computations and is valid for Hermitian, symmetric and symplectic ensembles as well as sample covariance ensembles. The basic strategy of this approach can be divided into the following three steps:

Step 1: Local Semicircle Law The first step was to derive a *local semicircle law*, a precise estimate of the local eigenvalue density, down to energy scales containing around N^ε eigenvalues.

Step 2: Universality of Gaussian divisible ensembles The second step is a general approach for the universality of Gaussian divisible ensembles by embedding the matrix (14.1) into a stochastic flow of matrices and using that the eigenvalues evolve according to a distinguished coupled system of stochastic differential equations, the Dyson Brownian motion (Dyson 1962). The central idea is to estimate the time to local equilibrium for the Dyson Brownian motion with the introduction of a new stochastic flow, the *local relaxation flow*, which locally behaves like a Dyson Brownian motion but has a faster decay to global equilibrium.

Step 3: Approximation by Gaussian divisible ensembles Once the universality for the Gaussian divisible ensemble is established, the last step is to approximate all matrix ensembles by Gaussian divisible ones. This step can be done via a *reverse heat flow* argument (Erdős et al., 2010; Erdős, Schlein, Yau, and Yin, 2010) for ensembles with smooth probability distributions or more generally via the *Green function comparison theorem* (Erdős, Yau, and Yin, 2011a) which compares the distributions of eigenvalues of two ensembles around a fixed energy.

A different approach was developed in (Tao and Vu, 2010a; Tao and Vu, 2010b; Tao and Vu, 2010c; Tao and Vu, 2010d) which mainly consists of an eigenvalue comparison theorem with a four moment matching condition and using the results of (Johansson 2001; Ben Arous and Péché 2005) to provide large classes of comparison matrices. This approach also relies on the local semicircle law to provide basic estimates on eigenvalues and eigenfunctions. But it requires much more detailed analysis of eigenvalues which we will not go into. Notice that the results of (Johansson 2001; Ben Arous and Péché 2005) were restricted to complex Wigner matrices, thus there are substantial restrictions in this approach for ensembles other than the Hermitian ones. Recently, there is an extension of the Green function comparison theorem (Knowles and Yin 2011) which in particular implies the eigenvalue comparison theorem of (Tao and Vu 2010a), see Section 4 for some discussion. In general, we will not make detailed comparison between these approaches in this chapter and we refer the reader to the original papers or the review (Erdős 2010) for some history and comparisons.

We first introduce the random matrix models considered in this lecture. Let $H = (h_{ij})_{i,j=1}^N$ be an $N \times N$ Hermitian or symmetric matrix whose upper-triangular matrix elements $h_{ij} = \bar{h}_{ji}$, $i \leqslant j$, are independent random variables with law ν_{ij} having mean zero and variance σ_{ij}^2:

$$\mathbb{E}\, h_{ij} = 0, \qquad \sigma_{ij}^2 := \mathbb{E}|h_{ij}|^2. \tag{14.2}$$

The distribution ν_{ij} and its variance σ_{ij}^2 may depend on N, but we omit this fact in the notation. Denote by $B := (\sigma_{ij}^2)_{i,j=1}^N$ the matrix of variances. The ensemble is called a *generalized Wigner ensemble* if the following two assumptions hold:

(A) For any fixed j we have

$$\sum_{i=1}^N \sigma_{ij}^2 = 1. \tag{14.3}$$

Thus B is symmetric and doubly stochastic and, in particular, satisfies $-1 \leqslant B \leqslant 1$.

(B) Define

$$C_{\text{inf}}(N) := \inf_{i,j}\{N\sigma_{ij}^2\} \leqslant \sup_{i,j}\{N\sigma_{ij}^2\} =: C_{\text{sup}}(N). \qquad (14.4)$$

We assume that there exists two positive constants, C_- and C_+, independent of N, such that

$$0 < C_- \leqslant C_{\text{inf}}(N) \leqslant C_{\text{sup}}(N) \leqslant C_+ < \infty, \qquad (14.5)$$

The special case $C_{\text{inf}} = C_{\text{sup}} = 1$ reduces to the standard Wigner matrices.

In this lecture, we assume that distributions of the matrix elements satisfy the uniform subexponential decay condition

$$\mathbb{P}(|h_{ij}| \geqslant x\sigma_{ij}) \leqslant \vartheta^{-1}\exp(-x^\vartheta) \qquad (14.6)$$

for some $\vartheta > 0$. The main result of the bulk universality is the following theorem.

Theorem 14.1. (universality for generalized Wigner matrices) *Consider a generalized Hermitian Wigner ensemble that the distributions ν_{ij} of the matrix elements have a uniformly subexponential decay in the sense of (14.6). Then for any $k \geqslant 1$ and for any compactly supported continuous test function $O : \mathbb{R}^k \to \mathbb{R}$ we have for any $\varepsilon > 0$ that*

$$\lim_{N\to\infty} \int_{E-b}^{E+b} \frac{\mathrm{d}E'}{2b} \int_{\mathbb{R}^n} \mathrm{d}\alpha_1\ldots\mathrm{d}\alpha_n\, O(\alpha_1,\ldots,\alpha_n)\frac{1}{\varrho(E)^n}$$
$$\times \left(p_N^{(k)} - p_{GUE,N}^{(k)}\right)\left(E' + \frac{\alpha_1}{N\varrho(E)},\ldots,E' + \frac{\alpha_n}{N\varrho(E)}\right) = 0, \quad b = N^{-1+\varepsilon}$$
$$(14.7)$$

where $p_{GUE,N}^{(k)}$ is the k-point correlation function for the GUE ensemble. The same statement holds for symmetric matrices, with GOE replacing the GUE ensemble.

Theorem 14.1 in this form was proved in (Erdős, Yau, and Yin, 2011c). Partial results and extensions to sample covariance matrices were proved in our earlier work (Erdős *et al.*, 2010; Erdős, Schlein, Yau, and Yin, 2010; Erdős, Schlein, and Yau, 2010; Erdős, Yau, and Yin, 2011a; Erdős *et al.*, 2011b). For Wigner matrices and sample covariance matrices, partial results were also obtained in (Tao and Vu, 2010a,d). We refer the reader to the lecture (Erdős 2010) for a review of recent history.

Denote by λ_j the j-th eigenvalue of the random matrix H, labelled in increasing order, $\lambda_1 \leqslant \lambda_2 \leqslant \ldots \leqslant \lambda_N$. The probability distribution functions of the largest eigenvalue λ_N for the classical Gaussian ensembles are identified by Tracy and Widom (Tracy and Widom 1994, 1996) to be

$$\lim_{N\to\infty} \mathbb{P}(N^{2/3}(\lambda_N - 2) \leqslant s) = F_\beta(s), \qquad (14.8)$$

where the function $F_\beta(s)$ can be computed in terms of Painlevé equations and $\beta = 1, 2, 4$ corresponds to the standard classical ensembles. The distribution of λ_N is believed to be universal and independent of the Gaussian structure. The precise form is given in the following theorem (Erdős, Yau, and Yin, 2011c).

Theorem 14.2. (Universality of extreme eigenvalues) *Suppose that we have two $N \times N$ matrices, $H^{(v)}$ and $H^{(w)}$, with matrix elements h_{ij} given by the random variables $N^{-1/2}v_{ij}$ and $N^{-1/2}w_{ij}$, respectively, with v_{ij} and w_{ij} satisfying the uniform subexponential decay condition (14.6). Let $\mathbb{P}^v$ and $\mathbb{P}^w$ denote the probability and $\mathbb{E}v$ and $\mathbb{E}bw$ the expectation with respect to these collections of random variables. Suppose the first two moments of v_{ij} and w_{ij} are the same, that is*

$$\mathbb{E}v \bar{v}_{ij}^l v_{ij}^u = \mathbb{E}w \bar{w}_{ij}^l w_{ij}^u, \qquad 0 \leqslant l + u \leqslant 2, \tag{14.9}$$

then there is an $\varepsilon > 0$ and $\delta > 0$ depending on ϑ in (14.6) such that for any real parameter s (may depend on N) we have

$$\mathbb{P}^v(N^{2/3}(\lambda_N - 2) \leqslant s - N^{-\varepsilon}) - N^{-\delta} \leqslant \mathbb{P}^w(N^{2/3}(\lambda_N - 2) \leqslant s) \leqslant$$
$$\mathbb{P}^v(N^{2/3}(\lambda_N - 2) \leqslant s + N^{-\varepsilon}) + N^{-\delta} \tag{14.10}$$

for $N \geqslant N_0$ sufficiently large, where N_0 is independent of s. An analogous result holds for the smallest eigenvalue λ_1.

Theorem 14.2 can be extended to finite correlation functions of extreme eigenvalues. For example, we have the following extension to (14.10):

$$\mathbb{P}^v\left(N^{2/3}(\lambda_N - 2) \leqslant s_1 - N^{-\varepsilon}, \ldots, N^{2/3}(\lambda_{N-k} - 2) \leqslant s_{k+1} - N^{-\varepsilon}\right) - N^{-\delta}$$

$$\leqslant \mathbb{P}^w\left(N^{2/3}(\lambda_N - 2) \leqslant s_1, \ldots, N^{2/3}(\lambda_{N-k} - 2) \leqslant s_{k+1}\right) \tag{14.11}$$

$$\leqslant \mathbb{P}^v\left(N^{2/3}(\lambda_N - 2) \leqslant s_1 + N^{-\varepsilon}, \ldots, N^{2/3}(\lambda_{N-k} - 2) \leqslant s_{k+1} + N^{-\varepsilon}\right) + N^{-\delta}$$

for all k fixed and N sufficiently large.

The edge universality for Wigner matrices was traditionally approached via the moment method (Soshnikov 1999) (see also the earlier work (Sinai and Soshnikov 1998)) for Hermitian and orthogonal ensembles with symmetric distributions to ensure that all odd moments vanish. The symmetry assumption was partially removed in (Péché and Soshnikov, 2008; Péché and Soshnikov, 2007) and in (Tao and Vu 2010b) which assumes only that the first three moments of two Wigner ensembles are identical. For a special class of ensembles, the Gaussian divisible Hermitian ensembles, edge universality was proved (Johansson 2011) under the sole condition that the second moment is finite. In comparison with these results, Theorem 14.2 assumes no symmetry nor vanishing third moment condition and is valid for both Hermitian and symmetric matrices. Recently, Theorem 14.2 was extended to eigenvectors (Knowles and Yin 2011) as well. The bulk and edge universalities are supposed to be valid for band and sparse matrices. But so far only partial results are available, see, for example,

(Erdős, Yau, and Yin, 2011a; Sodin, 2010; Sodin, 2011a) for some partial results and comments. Due to the recent fast progress in this direction, we will not be able to give detailed account of all partial results and important steps made in various papers. We refer the reader to the review (Erdős 2010) or the recent paper (Erdős, Yau, and Yin, 2011c) for more detailed citations.

14.2 Local semicircle law

Define the Green function of H by

$$G_{ij}(z) = \left(\frac{1}{H - z} \right)_{ij}, \qquad z = E + i\eta, \qquad E \in \mathbb{R}, \quad \eta > 0. \tag{14.12}$$

The Stieltjes transform of the empirical eigenvalue distribution of H is given by

$$m(z) = m_N(z) := \frac{1}{N} \sum_j G_{jj}(z) = \frac{1}{N} \operatorname{Tr} \frac{1}{H - z}. \tag{14.13}$$

Define $m_{sc}(z)$ as the unique solution of

$$m_{sc}(z) + \frac{1}{z + m_{sc}(z)} = 0, \tag{14.14}$$

with positive imaginary part for all z with $\operatorname{In} z > 0$, that is

$$m_{sc}(z) = \frac{-z + \sqrt{z^2 - 4}}{2}, \tag{14.15}$$

where the square root function is chosen with a branch cut in the segment $[-2, 2]$ so that asymptotically $\sqrt{z^2 - 4} \sim z$ at infinity. This guarantees that the imaginary part of m_{sc} is non-negative for $\eta = \operatorname{In} z > 0$ and in the $\eta \to 0$ limit it is the Wigner semicircle distribution

$$\varrho_{sc}(E) := \lim_{\eta \to 0+0} \frac{1}{\pi} \operatorname{In} m_{sc}(E + i\eta) = \frac{1}{2\pi} \sqrt{(4 - E^2)_+}. \tag{14.16}$$

The Wigner semicircle law (Wigner 1955) states that $m_N(z) \to m_{sc}(z)$ for any fixed z, that is provided that $\eta = \operatorname{In} z > 0$ is independent of N. Let $z = E + i\eta$ $(\eta > 0)$ and denote $\kappa := ||E| - 2|$ the distance of E to the spectral edges ± 2.

 The local semicircle law is a local version of the Wigner law, meaning that we will take the parameter η down to $1/N$ which is the smallest scale we can hope that the semicircle law is valid without taking expectation of the random variable $m(z)$. We will see that such a strong result on the semicircle law is the key input to the universality of random matrices. Due to the fundamental role of the local semicircle law, its error estimates were improved many times since its first proof in (Erdős, Schlein, and Yau, 2009).

Let

$$v_k := G_{kk} - m_{sc}, \qquad m := \frac{1}{N}\sum_{k=1}^{N} G_{kk}, \qquad [v] := \frac{1}{N}\sum_{k=1}^{N} v_k = m - m_{sc}.$$

Our goal is to estimate the following quantities

$$\Lambda_d := \max_k |v_k| = \max_k |G_{kk} - m_{sc}|, \qquad \Lambda_o := \max_{k\neq\ell} |G_{k\ell}|, \qquad \Lambda := |m - m_{sc}|,$$

$$\tag{14.17}$$

where the subscripts refer to 'diagonal' and 'off-diagonal' matrix elements. All these quantities depend on the spectral parameter z and on N but for simplicity we often omit this fact from the notation. We now state the uniform semicircle law obtained in (Erdős, Yau, and Yin, 2011c).

Definition 14.3 *Abbreviate $L \equiv L_N := A_0 \log\log N$ for some fixed A_0 as well as $\varphi \equiv \varphi_N := (\log N)^{\log\log N}$.*

Theorem 14.4. (uniform local semicircle law) *Let $H = (h_{ij})$ be a generalized Hermitian or symmetric $N \times N$ random matrix. Suppose that the distributions of the matrix elements have a uniformly subexponential decay (14.6). Then there exist positive constants $A_0 > 1$ (see Definition 14.3), C, c and $\phi < 1$ such that the following estimates hold for any sufficiently large $N \geqslant N_0(\vartheta, C_\pm)$:*

1. *The Stieltjes transform of the empirical eigenvalue distribution of H satisfies*

$$\mathbb{P}\left(\bigcup_{z\in S_L} \left\{ \Lambda(z) \geqslant \frac{(\log N)^{4L}}{N\eta} \right\} \right) \leqslant \exp\left[-c(\log N)^{\phi L} \right], \tag{14.18}$$

   where

$$S_L := \left\{ z = E + i\eta : |E| \leqslant 5, \quad N^{-1}(\log N)^{10L} < \eta \leqslant 10 \right\}. \tag{14.19}$$

2. *The individual matrix elements of the Green function satisfy that*

$$\mathbb{P}\left(\bigcup_{z\in S_L} \left\{ \Lambda_d(z) + \Lambda_o(z) \geqslant (\log N)^{4L} \sqrt{\frac{\operatorname{Im} m_{sc}(z)}{N\eta}} + \frac{(\log N)^{4L}}{N\eta} \right\} \right) \leqslant$$

$$\exp\left[-c(\log N)^{\phi L} \right]. \tag{14.20}$$

3. *The largest eigenvalue of H is bounded by $2 + N^{-2/3}(\log N)^{9L}$ in the sense that*

$$\mathbb{P}\left(\max_{j=1,\ldots,N} |\lambda_j| \geqslant 2 + N^{-2/3}(\log N)^{9L} \right) \leqslant \exp[-c(\log N)^{\phi L}]. \tag{14.21}$$

The local semicircle estimates imply that the empirical counting function of the eigenvalues is close to the semicircle counting function and that the locations of the eigenvalues are close to their classical locations in mean square deviation sense. Recall that $\lambda_1 \leqslant \lambda_2 \leqslant \ldots \leqslant \lambda_N$ are the ordered eigenvalues of H. We define the *normalized empirical counting function* by

$$\mathfrak{n}(E) := \frac{1}{N} \#\{\lambda_j \leqslant E\}. \tag{14.22}$$

Denote by γ_j the *classical location* of the j-th eigenvalue, that is γ_j is defined by

$$N \int_{-\infty}^{\gamma_j} \varrho_{sc}(x)\,\mathrm{d}x = j, \qquad 1 \leqslant j \leqslant N, \tag{14.23}$$

where $\varrho_{sc}(x)$ is the semicircle law.

Theorem 14.5. (rigidity of eigenvalues) *Suppose that Assumptions* **(A)** *and* **(B)** *and the condition* (14.6) *hold. Then there exist positive constants* $A_0 > 1$ *(see Definition 14.3),* $C, c,$ *and* $\phi < 1$, *depending only on* ϑ *and on* $C_\pm$ *from Assumption* **(B)**, *such that*

$$\mathbb{P}\left\{\exists j : |\lambda_j - \gamma_j| \geqslant (\log N)^L \left[\min\left(j, N - j + 1\right)\right]^{-1/3} N^{-2/3}\right\} \leqslant \exp\left[- c(\log N)^{\phi L}\right] \tag{14.24}$$

and

$$\mathbb{P}\left\{\sup_{|E|\leqslant 5} \left|\mathfrak{n}(E) - n_{sc}(E)\right| \geq \frac{(\log N)^L}{N}\right\} \leqslant \exp\left[- c(\log N)^{\phi L}\right] \tag{14.25}$$

for any sufficiently large $N \geqslant N_0(\vartheta, C_\pm)$.

It is well-known that the gaps between extremal eigenvalues and their fluctuations are of order $N^{-2/3}$. Thus the edge deterioration factor in (14.24) is the natural interpolation between N^{-1} in the bulk and $N^{-2/3}$ on the edges. The surprising feature of the rigidity estimate is that even if one eigenvalue is at a slightly wrong location, the probability is already extremely small. Our interest in the rigidity estimate was due to its implication concerning the ergodicity of Dyson's Brownian motion which we will review in the next section. For this reason, partial results were obtained in (Erdős, Ramirez, Schlein, and Yau, 2010; Erdős, Schlein, and Yau, 2010; Erdős, Yau, and Yin, 2011a) and later work. For Wigner matrices, partial results were also obtained in (Tao and Vu 2010d) and we refer to the reader to the original papers for details.

We remark that both local semicircle law and rigidity of eigenvalues were not completely resolved for band matrices. In (Erdős, Yau, and Yin, 2011a), the local semicircle law was obtained only up to the inverse of the band (a similar result in the average sense was recently proved in (Sodin 2011b)). We note that a certain three dimensional version of Gaussian band matrices was considered by Disertori, Pinson, and Spencer (Disertori, Pinson, and Spencer, 2002) using the supersymmetric method. They proved that the expectation of the density of eigenvalues is smooth and it coincides with the Wigner semicircle law.

14.3 Universality of Gaussian divisible ensembles via Dyson Brownian motion

We consider a flow of random matrices H_t satisfying the following matrix valued stochastic differential equation

$$\mathrm{d}\, H_t = \frac{1}{\sqrt{N}}\,\mathrm{d}\,\beta_t - \frac{1}{2}H_t\,\mathrm{d}\,t,\qquad(14.26)$$

where β_t is a Hermitian matrix valued process whose diagonal matrix elements are standard real Brownian motions and the off-diagonal elements are independent standard complex Brownian motions; with all Brownian motions being independent. The initial condition H_0 is the original Hermitian Wigner matrix. For any fixed $t \geq 0$, the distribution of H_t coincides with that of

$$e^{-t/2}H_0 + (1 - e^{-t})^{1/2}\,V,\qquad(14.27)$$

where V is an independent GUE matrix whose matrix elements are centred Gaussian random variables with variance $1/N$. For the symmetric case, the matrix elements of β_t in (14.26) are real Brownian motions and V in (14.27) is a GOE matrix. It is well-known that the eigenvalues of H_t follow a process that is also called the Dyson Brownian motion (in our case with a drift but we will still call it Dyson Brownian motion).

More precisely, let

$$\mu = \mu_N(\mathrm{d}\mathbf{x}) = \frac{e^{-\mathcal{H}(\mathbf{x})}}{Z_\beta}\,\mathrm{d}\mathbf{x},\qquad \mathcal{H}(\mathbf{x}) = N\left[\beta\sum_{i=1}^{N}\frac{x_i^2}{4} - \frac{\beta}{N}\sum_{i<j}\log|x_j - x_i|\right]\qquad(14.28)$$

be the probability measure of the eigenvalues of the general β ensemble, with $\beta \geq 1$ ($\beta = 2$ for GUE, $\beta = 1$ for GOE). Here Z_β is the normalization factor so that μ is probability measure. In this section, we often use the notation x_j instead of λ_j for the eigenvalues to follow the notations of (Erdős, Schlein, Yau, and Yin, 2010). Denote the distribution of the eigenvalues at time t by $f_t(\mathbf{x})\mu(\mathrm{d}\mathbf{x})$. Then f_t satisfies

$$\partial_t f_t = \mathcal{L}f_t.\qquad(14.29)$$

where

$$\mathcal{L} = \sum_{i=1}^{N}\frac{1}{2N}\partial_i^2 + \sum_{i=1}^{N}\left(-\frac{\beta}{4}x_i + \frac{\beta}{2N}\sum_{j\neq i}\frac{1}{x_i - x_j}\right)\partial_i.\qquad(14.30)$$

For any $n \geq 1$ we define the n-point correlation functions (marginals) of the probability measure $f_t\,\mathrm{d}\mu$ by

$$p_{t,N}^{(n)}(x_1, x_2, \ldots, x_n) = \int_{\mathbb{R}^{N-n}} f_t(\mathbf{x})\mu(\mathbf{x})\,\mathrm{d}\,x_{n+1}\ldots\mathrm{d}\,x_N.\qquad(14.31)$$

With a slight abuse of notations, we will sometimes also use μ to denote the density of the measure μ with respect to the Lebesgue measure. The correlation functions of the equilibrium measure are denoted by

$$p_{\mu,N}^{(n)}(x_1, x_2, \ldots, x_n) = \int_{\mathbb{R}^{N-n}} \mu(\mathbf{x}) \, d\, x_{n+1} \ldots d\, x_N. \tag{14.32}$$

The main result in (Erdős, Schlein, Yau, and Yin, 2010) concerning Dyson Brownian motion, Theorem 2.1, states that the local ergodicity of Dyson Brownian motion holds for $t \geqslant N^{-2\mathfrak{a}+\delta}$ for any $\delta > 0$ provided that the following Assumption III (14.33) holds.

Assumption III. There exists an $\mathfrak{a} > 0$ such that

$$\sup_{t \geqslant N^{-2\mathfrak{a}}} \frac{1}{N} \mathbb{E}t \sum_{j=1}^{N} (\lambda_j - \gamma_j)^2 \leqslant CN^{-1-2\mathfrak{a}} \tag{14.33}$$

with a constant C uniformly in N. Here $\mathbb{E}t$ is the expectation w.r.t. Dyson Brownian motion at the time t.

Strictly speaking, there are four assumptions in (Erdős, Schlein, and Yau, 2010; Erdős, Schlein, Yau, and Yin, 2010), but Assumption III is the main assumption. Thus for a given $0 < \varepsilon' < 1$, choosing $\mathfrak{a} = 1/2 - \varepsilon'/2$ we have the following theorem (Erdős, Yau, and Yin, 2011c).

Theorem 14.6. (Local ergodicity of Dyson Brownian motion) *Let H be a generalized Hermitian or symmetric $N \times N$ random matrix with the subexponential property (14.6). Let $E \in [2 - c, 2 + c]$ with some $c > 0$. Then for any $\varepsilon' > 0$, $\delta > 0$, $0 < b < c/2$, any integer $n \geqslant 1$ and for any compactly supported continuous test function $O : \mathbb{R}^n \to \mathbb{R}$ we have*

$$\sup_{t \geqslant N^{-1+\delta+\varepsilon'}} \left| \int_{E-b}^{E+b} \frac{d\, E'}{2b} \int_{\mathbb{R}^n} d\, \alpha_1 \ldots d\, \alpha_n \, O(\alpha_1, \ldots, \alpha_n) \frac{1}{\varrho(E)^n} \right.$$

$$\times \left. \left(p_{t,N}^{(n)} - p_{\mu,N}^{(n)} \right) \left(E' + \frac{\alpha_1}{N\varrho(E)}, \ldots, E' + \frac{\alpha_n}{N\varrho(E)} \right) \right|$$

$$\leqslant C_n N^{2\varepsilon'} \left[b^{-1} N^{-1+\varepsilon'} + b^{-1/2} N^{-\delta/2} \right], \tag{14.34}$$

where $p_{t,N}^{(n)}$ and $p_{\mu,N}^{(n)}$, (14.31)–(14.32), are the correlation functions of the eigenvalues of the Dyson Brownian motion flow (14.27) and those of the equilibrium measure, respectively.

A weaker version of Theorem 14.6 was proved in (Erdős, Yau, and Yin, 2011b), and a similar result, with no error estimate, was obtained in (Erdős *et al.*, 2010) for the Hermitian case by using an explicit formula (Brézin and Hikami 1996/97; Johansson 2001). Theorem 14.6, however, contains explicit estimates and is valid for a time range much larger than the previous results. In particular, we mention the following three special cases:

- If we choose $\delta = 1 - 2\varepsilon'$ and thus $t = N^{-\varepsilon'}$, then we can choose $b \sim N^{-1}$ and the universality is valid with essentially no averaging in E.

- If we choose the energy window of size $b \sim 1$ and the time $t = N^{-\varepsilon'}$, then the error estimate is of order $\sim N^{-1/2}$.

- If we choose $b \sim 1$, then the smallest time scale for which we can prove the universality is $t = N^{-1+\varepsilon'}$. This scale, up to the arbitrary small exponent ε', is optimal in accordance with the time scale to local equilibrium conjectured by Dyson (Dyson 1962).

14.4   Green function comparison theorem

Once the universality for the Gaussian divisible ensemble is established, the last step is to approximate all matrix ensembles by Gaussian divisible ones. This step can be done via a *reverse heat flow* argument (Erdős et al., 2010; Erdős, Schlein, Yau, and Yin, 2010) for ensembles with smooth probability distributions or more generally via the *Green function comparison theorem* (Erdős, Yau, and Yin, 2011a) which compares the distributions of eigenvalues of two ensembles around a fixed energy. The key input for the latter approach was to prove *a priori* estimates on the matrix elements of the Green function which were the content of the local semicircle proved in Theorem 14.4. In this section, we explain the idea behind the Green function comparison theorem by sketching the idea to prove the edge universality, Theorem 14.2. The basic idea is similar to Linderberg's proof of central limit theorem. The application of Linderberg's idea in random matrices appeared recently in a proof of the Wigner semicircle law in (Chatterjee 2006). Linderberg's idea was also used in (Tao and Vu 2010a) to compare eigenvalues of two Wigner matrices. As individual eigenvalue locations are sensitive to the change of matrix elements and resonances of eigenvalues are difficult to rule out, the work (Tao and Vu 2010a) involved very complicated estimates for which the local semicircle law, Theorem 14.4 (or various weaker forms in earlier work (Erdős, Schlein, and Yau, 2009)), was an important ingredient. We note that the result of (Tao and Vu 2010a) gives a direct comparison of individual eigenvalues, while the Green function comparison theorem compares Green functions at a fixed energy and is valid for a more general class of matrices, that is the generalized Wigner matrices. Notice that the difficulty with resonances of eigenvalues (Tao and Vu 2010a) disappears in the Green function approach due to the natural regularization of the Green function by the imaginary part of the spectral parameter. The Green function comparison theorem is thus a simple direct consequence of the local semicircle law. Recently, an extension of the Green function comparison theorem to include the comparison of individual eigenvalues was obtained in (Knowles and Yin 2011). This work shows that a general form of the Green function comparison theorem yields the comparison of eigenvalues both at a fixed energy and for a fixed label. In the following paragraphs, we explain the basic ideas and the power counting of our approach.

We first notice that the total number of eigenvalues in an interval is given by

$$N \int_{E_1}^{E_2} \mathrm{d}y \, \mathrm{Im}\, m(y + i\eta) = \eta \int_{E_1}^{E_2} \mathrm{d}y \, \mathrm{Tr}\, G(z)\overline{G}(z), \quad z = y + i\eta. \tag{14.35}$$

Since we are interested in the eigenvalues near the edges, we only have to consider $|E_2 - E_1| \sim N^{-2/3} \sim \eta$. Instead of the expression in (14.35), we will explain our idea using the functional

$$N\eta^2 \ln m(z) = \eta \ln \operatorname{Tr} G = \eta^2 \sum_{ij} G_{ij}\overline{G_{ij}},$$

The modification to the expression (14.35) is obvious.

We now set up notations to replace the matrix elements one by one. Fix a bijective ordering map on the index set of the independent matrix elements,

$$\phi : \{(i,j) : 1 \leqslant i \leqslant j \leqslant N\} \to \{1, \ldots, \gamma(N)\}, \qquad \gamma(N) := \frac{N(N+1)}{2},$$

and denote by H_γ the generalized Wigner matrix whose matrix elements h_{ij} follow the v-distribution if $\phi(i,j) \leqslant \gamma$ and they follow the w-distribution otherwise; in particular $H_0 = H^{(v)}$ and $H_{\gamma(N)} = H^{(w)}$.

We set $z = E + i\eta$ where $|E - 2| \leqslant CN^{-2/3+\varepsilon}$ and $\eta = N^{-2/3-\varepsilon}$. The size of $\operatorname{In} m_{sc}$ near the edge is given by

$$\operatorname{In} m_{sc}(E + i\eta) \leqslant \sqrt{|E - 2| + \eta} \leqslant CN^{-1/3+\varepsilon/2} \tag{14.36}$$

for $|E - 2| \leqslant CN^{-2/3+\varepsilon}$. The local semicircle law, Theorem 14.4, near the edge asserts that, with $\eta = N^{-2/3-\varepsilon}$, for any constant $\xi > 0$,

$$\mathbb{P}\left(\max_{0 \leqslant \gamma \leqslant \gamma(N)} \max_{1 \leqslant k,l \leqslant N} \max_E \left| \left(\frac{1}{H_\gamma - E - i\eta}\right)_{kl} - \delta_{kl}m_{sc}(E + i\eta) \right| \leqslant N^{-1/3+2\varepsilon}\right)$$

$$\geqslant 1 - C\exp[-c(\log N)^\xi] \tag{14.37}$$

with some constants C, c and large enough $N \geqslant N_0$ (may depend on ξ). From equations (14.37), (14.36) we have that

$$\left| \eta^2 \sum_{ij} G_{ij}\overline{G_{ij}} \right| = |N\eta \operatorname{In} m(z)| \leqslant CN^{2\varepsilon} \tag{14.38}$$

Thus our basic quantity is of order one and this is an important step.

Consider the telescopic sum of differences of expectations

$$\mathbb{E}\left(\eta^2 \sum_{i \neq j} \left(\frac{1}{H^{(v)} - z}\right)_{ij} \overline{\left(\frac{1}{H^{(v)} - z}\right)_{ji}}\right) - \mathbb{E}\left(H^{(v)} \to H^{(w)}\right) \tag{14.39}$$

$$= \sum_{\gamma=1}^{\gamma(N)} \left[\mathbb{E}\left(H^{(v)} \to H_\gamma\right) - \mathbb{E}\left(H^{(v)} \to H_{\gamma-1}\right)\right].$$

Let $E^{(ij)}$ denote the matrix whose matrix elements are zero everywhere except at the (i,j) position, where it is 1, that is, $E_{k\ell}^{(ij)} = \delta_{ik}\delta_{j\ell}$. Fix a $\gamma \geqslant 1$ and let (a, b)

be determined by $\phi(a,b) = \gamma$. For simplicity to introduce the notation, we assume that $a \neq b$. The $a = b$ case can be treated similarly. We note the total number of the diagonal terms is N and the one of the off-diagonal terms is $O(N^2)$. We will compare $H_{\gamma-1}$ with H_γ for each γ and then sum up the differences according to (14.39).

Note that these two matrices differ only in the (a,b) and (b,a) matrix elements and they can be written as

$$H_{\gamma-1} = Q + \frac{1}{\sqrt{N}} V, \qquad V := v_{ab} E^{(ab)} + v_{ba} E^{(ba)} \tag{14.40}$$

$$H_\gamma = Q + \frac{1}{\sqrt{N}} W, \qquad W := w_{ab} E^{(ab)} + w_{ba} E^{(ba)},$$

with a matrix Q that has zero matrix element at the (a,b) and (b,a) positions and where we set $v_{ji} := \bar{v}_{ij}$ for $i < j$ and similarly for w. Define the Green functions

$$R := \frac{1}{Q-z}, \qquad S := \frac{1}{H_{\gamma-1}-z}, \qquad T := \frac{1}{H_\gamma-z}. \tag{14.41}$$

We can check that the estimate (14.37) holds for the Green function R as well. After having introduced these notations, we are in a position to give a heuristic power counting argument that is the core of the proof. In particular, we can explain the origin of the second moment matching condition. By the resolvent expansion, we have

$$\mathbb{E}\eta \sum_i \mathrm{Im}\, S_{ii} = \eta\, \mathbb{E}\, \mathrm{Im} \sum_i \left[R_{ii} - N^{-1/2}(RVR)_{ii} + N^{-1}((RV)^2 R)_{ii} + \dots \right] \tag{14.42}$$

which is an expansion in the order of $N^{-1/2}$ since the matrix V contains only a few nonzero elements of size $N^{-1/2}$. Notice that $\eta \sum_i \mathrm{Im}\, S_{ii}$ estimates the number of eigenvalues near E in a window of size η. For the two ensembles to have the same local eigenvalue distribution on scale η, we need the error term to be less than order one even after performing the telescopic sum. In the bulk, η has to be chosen as $\eta \sim N^{-1}$ and we can view $\eta \sum_i$ as order one in the power counting. Since in the telescopic expansion we will have N^2 terms to sum up, we need that the error term of the expansion is $o(N^{-2})$ for each replacement step, that is, for each fixed label (a,b). This explains the usual condition of four moments to be identical for the Green function comparison theorem in the bulk (Erdős, Yau, and Yin, 2011a) since the first four terms in (14.42) have to be equal. Near the edges, that is, at energies E with $|E - 2| \lesssim N^{-2/3}$, the correct local scale is $\eta \sim N^{-2/3}$ and the strong local semicircle law (14.20) implies that the off-diagonal Green functions are of order $N^{-1/3}$ and the diagonal Green functions are bounded. Hence the size of the third order term $\eta\, \mathbb{E} \sum_i N^{-3/2}((RV)^3 R)_{ii}$ is of order

$$\eta N N^{-3/2} N^{-2/3} = N^{-2+1/6}$$

where we used that, for a generic label (a,b), there are at least two off-diagonal resolvent terms in $((RV)^3 R)_{ii}$. Notice that the error term is still larger than N^{-2}, required for summing over a,b (this argument would be sufficient if we had a matching of three moments and only the fourth order term in (14.42) needed to be estimated).

The key observation is that the leading term, which gives this order $N^{-2+1/6}$, has actually almost zero expectation which improves the error to be less than $o(N^{-2})$. This is due to the fact that with the help of (14.37) we are able to follow the main term in the diagonal elements of the Green functions and thus compute the expectation fairly precisely. The computation of this second order term can be found in the original paper (Erdős, Yau, and Yin, 2011c).

References

Anderson, G., Guionnet, A., Zeitouni, O. (2009). *An Introduction to Random Matrices*. Studies in advanced mathematics, **118**, Cambridge University Press.

Bai, Z. D., Miao, B., and Tsay, J. (2002). Convergence rates of the spectral distributions of large Wigner matrices. *Int. Math. J.* **1** (1): 65–90.

Ben Arous, G. and Péché, S. (2005). Universality of local eigenvalue statistics for some sample covariance matrices. *Comm. Pure Appl. Math.* **LVIII.** 1–42.

Bleher, P., and Its, A. (1999). Semiclassical asymptotics of orthogonal polynomials, Riemann-Hilbert problem, and universality in the matrix model. *Ann. of Math.* **150**: 185–266.

Brézin, E. and Hikami, S. (1996). Correlations of nearby levels induced by a random potential. *Nucl. Phys. B* **479**: 697–706, and (1997) Spectral form factor in a random matrix theory. *Phys. Rev. E* **55**: 4067–4083.

Chatterjee, S. (2006). A generalization of the Lindeberg principle, *Ann Prob.*, **34**, 2061–2076.

Deift, P. (1999). Orthogonal polynomials and random matrices: a Riemann-Hilbert approach. *Courant Lecture Notes in Mathematics* **3**, American Mathematical Society, Providence, RI.

Deift, P. and Gioev, D. (2009). Random Matrix Theory: Invariant ensembles and universality. *Courant Lecture Notes in Mathematics* **18**, American Mathematical Society, Providence, RI.

Deift, P., Kriecherbauer, T., McLaughlin, K.T-R, Venakides, S. and Zhou, X. (1999). Uniform asymptotics for polynomials orthogonal with respect to varying exponential weights and applications to universality questions in random matrix theory. *Comm. Pure Appl. Math.* **52**: 1335–1425.

Deift, P., Kriecherbauer, T., McLaughlin, K.T-R, Venakides, S., and Zhou, X. (1999). Strong asymptotics of orthogonal polynomials with respect to exponential weights. *Comm. Pure Appl. Math.* **52**: 1491–1552.

Disertori, M., Pinson, H., and Spencer, T. (2002). Density of states for random band matrices. *Comm. Math. Phys.* **232**, 83–124.

Dyson, F.J. (1962). A Brownian-motion model for the eigenvalues of a random matrix. *J. Math. Phys.* **3**, 1191–1198.

Erdős, L. (2010). Universality of Wigner random matrices: a Survey of Recent Results, arXiv:1004.0861.

Erdős, L., Schlein, B., and Yau, H.-T. (2009). Semicircle law on short scales and delocalization of eigenvectors for Wigner random matrices. *Ann. Probab.* **37**, No. 3, 815–852.

Erdős, L., Schlein, B., and Yau, H.-T. (2009). Local semicircle law and complete delocalization for Wigner random matrices. *Comm. Math. Phys.* **287**, 641–655.

Erdős, L., Schlein, B., and Yau, H.-T. (2010). Wegner estimate and level repulsion for Wigner random matrices. *Int. Math. Res. Notices.* **2010**, No. 3, 436–479.

Erdős, L., Schlein, B., and Yau, H.-T. (2010). Universality of random matrices and local relaxation flow. Preprint. arxiv.org/abs/0907.5605.

Erdős, L., Ramirez, J., Schlein, B., and Yau, H.-T. (2010). Universality of sine-kernel for Wigner matrices with a small Gaussian perturbation. *Electr. J. Prob.* **15**, Paper 18, 526–604.

Erdős, L., Péché, G., Ramírez, J., Schlein, B., and Yau, H.-T. (2010). Bulk universality for Wigner matrices. *Comm. Pure Appl. Math.* **63**, No. 7, 895–925.

Erdős, L., Schlein, B., Yau, H.-T., and Yin, J. (2010). The local relaxation flow approach to universality of the local statistics for random matrices. Preprint arXiv:0911.3687.

Erdős, L., Yau, H.-T., and Yin, J. (2011a). Bulk universality for generalized Wigner matrices. Preprint arXiv:1001.3453.

Erdős, L., Yau, H.-T., and Yin, J. (2011b). Universality for generalized Wigner matrices with Bernoulli distribution. Preprint arXiv:1003.3813.

Erdős, L., Yau, H.-T., Yin, J. (2011c). Rigidity of eigenvalues of generalized Wigner matrices. Preprint arXiv:1007.4652.

Johansson, K. (2001). Universality of the local spacing distribution in certain ensembles of Hermitian Wigner matrices. *Comm. Math. Phys.* **215**, no.3. 683–705.

Johansson, K. (2011). Universality for certain hermitian Wigner matrices under weak moment conditions. Preprint arXiv:0910.4467.

Knowles, A. and Yin, J. (2011). Eigenvector Distribution of Wigner Matrices. Preprint arXiv:1102.0057.

Pastur, L. and Shcherbina M. (2008). Bulk universality and related properties of Hermitian matrix models. *J. Stat. Phys.* **130**, no.2., 205–250.

Péché, S. and Soshnikov, A. (2008). On the lower bound of the spectral norm of symmetric random matrices with independent entries. *Electron. Comm. Probab.* **13**, 280–290.

Péché, S. and Soshnikov, A. (2007). Wigner random matrices with non-symmetrically distributed entries. *J. Stat. Phys.* **129**, no. 5–6, 857–884.

Sinai, Y. and Soshnikov, A. (1998). A refinement of Wigner's semicircle law in a neighborhood of the spectrum edge. *Functional Anal. and Appl.* **32**, no. 2, 114–131.

Sodin, S. (2010). The spectral edge of some random band matrices. Preprint arXiv:0906.4047.

Sodin, S. (2011a). The Tracy–Widom law for some sparse random matrices. Preprint arXiv:0903.4295.

Sodin, S. (2011b). An estimate for the average spectral measure of random band matrices. Preprint arXiv:1101.4413.

Soshnikov, A. (1999). Universality at the edge of the spectrum in Wigner random matrices. *Comm. Math. Phys.* **207**, no.3. 697–733.

A. Soshnikov, (2004). Poisson statistics for the largest eigenvalues of Wigner matrices with heavy tails, *Elect. Comm. in Probab.*, **9**, 82–91.

Spencer, T. Review article on random band matrices. Draft in preparation.

Tao, T. and Vu, V. (2010a). Random matrices: Universality of the local eigenvalue statistics, to appear in *Acta Math*. Preprint arXiv:0906.0510.

Tao, T. and Vu, V. (2010b). Random matrices: Universality of local eigenvalue statistics up to the edge. Preprint arXiv:0908.1982.

Tao, T. and Vu, V. (2010c). Random covariance matrices: Universality of local statistics of eigenvalues. Preprint arXiv:0912.0966.

Tao, T. and Vu, V. (2010d). Random matrices: localization of the eigenvalues and the necessity of four moments. Preprint arXiv:1005.2901.

Tracy, C. and Widom, H. (1994). Level-spacing distributions and the airy kernel, *Comm. Math. Phys.* **159**, 151–174.

Tracy, C. and Widom, H. (1996). On orthogonal and symplectic matrix ensembles, *Comm. Math. Phys.* **177**, no. 3, 727–754.

Wigner, E. (1955). Characteristic vectors of bordered matrices with infinite dimensions. *Ann. of Math.* **62**, 548–564.